THIRD EDITION

The Invertebrates: a synthesis

R.S.K. Barnes

Department of Zoology,
University of Cambridge, UK

P. Calow

Department of Animal and Plant Sciences,
University of Sheffield, UK

P.J.W. Olive

Department of Marine Sciences and Coastal Management,
University of Newcastle, UK

D.W. Golding

Department of Marine Sciences and Coastal Management,
University of Newcastle, UK

J.I. Spicer

Department of Biological Sciences,
University of Plymouth, UK

Blackwell Publishing

BLACKWELL PUBLISHING
350 Main Street, Malden, MA 02148-5020, USA
9600 Garsington Road, Oxford OX4 2DQ, UK
550 Swanston Street, Carlton, Victoria 3053, Australia

First edition published 1988
Second edition 1993
Third edition 2001

3 2006

Library of Congress Cataloging-in-Publication Data

The invertebrates : a synthesis / R.S.K. Barnes ... [*et al.*].—3rd ed.
p. cm.
Rev. ed. of: The invertebrates / R.S.K. Barnes, P. Calow, P.J.W. Olive.
Includes bibliographical references.
ISBN 0-632-04761-5
1. Invertebrates. I. Barnes, R.S.K. (Richard Stephen Kent). II. Barnes, R.S.K. (Richard Stephen Kent). Invertebrates.

QL362 .B26 2001
592—dc21 00-044516

ISBN-13: 978-0-632-04761-1

A catalogue record for this title is available from the British Library.

Set by Graphicraft Ltd, Hong Kong

For further information on
Blackwell Publishing, visit our website:
www.blackwellpublishing.com

The Invertebrates:

a synthesis

Contents

Part 3: Invertebrate Functional Biology

Preface

To the Third Edition

The period since the second edition of *The Invertebrates* went to press has seen a great increase in our knowledge in several areas, most spectacularly in the effect of molecular sequence data on our understanding of invertebrate relationships. New types of animals continue to be discovered and described. We have therefore taken the opportunity provided by this third edition to incorporate such new knowledge and completely to rewrite several of the chapters in its light, whilst maintaining the same overall structure, style and approach of earlier editions which we feel do continue usefully to fill a niche. RSKB and PJWO are most grateful to DWG for continuing to contribute his much appreciated Chapter 16 and to JIS for revising the contribution to earlier editions of PC for this new one.

To the Second Edition

We have been most gratified by the response to the first edition and feel that we have good reason to believe that it filled a useful niche. Experience with that edition, however, has led us to a new design for the format of this edition. We have also taken the opportunity to update its factual content where appropriate and to add an enlarged section on the various groups of protists formerly regarded as constituting the 'Protozoa'.

To the First Edition

There are several available textbooks devoted to 'The Invertebrates' and hence the production of yet another one requires some words of explanation and justification. Books already on the market tend to fall into one or other of two categories: they are either systematic treatments covering each group of animals phylum by phylum (e.g. R.D. Barnes's *Invertebrate Zoology*, Saunders, 1987), or are functional approaches reviewing the various invertebrate anatomical and physiological 'systems' (respiration, movement, co-ordination, etc.) mainly with reference to the better known groups (e.g. E.J.W. Barrington's *Invertebrate Structure and Function*, Nelson, 1979). Invertebrate courses therefore require one of each category as associated texts.

In general, however, the last 25 years have seen a great reduction in the teaching time devoted specifically to the various groups, in part to make room in courses of fixed length for new and expanded subject areas and in part because systematic reviews of the range and diversity of organisms have declined in popularity since the days of classical zoology. The end result has been that any pair of existing texts, and indeed many individual works, contain far more information than is required by shortened courses and, deluged by detail, students fail to see the wood for the trees.

We have therefore endeavoured to provide within one pair of covers the basic information on both the range and diversity of invertebrates and on their different functional systems that we feel most university courses actually require. Our main problem has thus been what to leave out rather than what to include, and here we have attempted critically to assess the essential features of each group or system and to bias our accounts towards these. Further, we believe strongly that the process of evolution is central to an understanding of all aspects of biology, and that too few existing texts present animals as other than static, mechanistic entities. Wherever possible, we have therefore adopted an evolutionary approach which aims to portray invertebrate diversity and function against a background of selective pressures and selective advantages now and in the past. This has also influenced our selection and treatment of material. The book is not therefore a summary of existing texts but is, we trust, a new, critical look at the essential features of invertebrate biology.

Since, as pointed out above, zoology courses contain less and less coverage of individual types of animals with each passing decade, it is perhaps not inappropriate here briefly to defend the place for a broad knowledge of the invertebrates in zoological education. Much of our present understanding of biological processes in general has been derived from research on invertebrates; we need only mention the fruit fly in respect of genetics and the squid with regard to neurophysiology to make the point. As yet, however, the number of animal phyla, let alone species, which have been studied in detail is extremely small, and certainly is no true reflection of the diversity of pattern and process that is the invertebrates. We believe that many future generalizations will emerge from studies of these so-far neglected groups, and that without an appreciation of the diversity, as well as the unity, of life, it is impossible to obtain a valid perspective both of biology in general and of the extent to which today's knowledge is based on a minute and biased sample.

With the exception of Chapter 16, this book reflects a collaborative effort of its three authors. Although in practice first drafts of the various chapters, or parts of chapters, were prepared each by a single author, all were then recast in the light of communal discussion and criticism: all three of us therefore accept joint responsibility for Chapters 1–15 inclusive and Chapter 17. No book is a product solely of the authors, however, and we are most grateful to the many people who have helped or tolerated us during its preparation. In particular, we would like to record our gratitude to David Golding for taking on the task of preparing Chapter 16, and our appreciation of the labours of Helen Creighton, together with Bob Foster-Smith and Peter Kingston, who prepared all the final text figures. Several of our colleagues were so kind as to read drafts of material: Henry Bennet-Clark read the whole work, and Brian Bayne, Jack Cohen, Simon Conway Morris, Peter Croghan, Mustafa Djamgoz, David George, Peter Gibbs, Roger Hughes, Peter Miller, Todd Newberry, David Nichols, John Ryland, Ray Seed, Seth Tyler and Pat Willmer read various parts. Many others provided individual pieces of information and opinion. Their efforts have saved us from factual errors and textual infelicities. It is too much to hope that no unorthodoxies and inexactitudes remain, not least because we were sometimes intransigent in the face of just criticism. Robert Campbell and Simon Rallison of Blackwell Scientific Publications provided copious aid, advice, cajolery and administrative assistance: we owe a great deal to their iron fists and velvet gloves. The debt which we owe our families can only be appreciated by those who have also devoted most of their available 'spare time' to works of this kind.

Most of our illustrations are based on ones that have already appeared in the scientific literature, although all such have been redrawn. Citations to the original sources are given in the relevant figure legends and a list of these sources, other than those listed in the further reading sections, is provided on pp. 478–81.

R.S.K.B.
P.C.
P.J.W.O.

PART 1

Evolutionary Introduction

The main undercurrent which permeates our survey of invertebrate diversity (Part 2) and functional biology (Part 3) is that of the evolutionary pressures and advantages which have influenced these animals in the past and which continue to mould invertebrate biology today. In this introductory section, we describe briefly this all-pervasive evolutionary ethos.

The word 'evolution' simply means 'change', and change can be analysed by two different approaches, which generally are related to each other as is cause to effect or mechanism to manifestation: (a) there are the *processes* ultimately responsible for such changes as are observed; and (b) there is the overall *pattern* or sequence of the changes which have occurred through time. In fact, although Charles Darwin is popularly credited with having demonstrated the fact of evolution, what he did was to propose a viable mechanism – natural selection – which could account for the evolutionary changes which others, before him, had already suggested had taken place. As indicated above, an evolutionary (or 'phylogenetic') tree of the invertebrate phyla and the process of natural selection are related to each other, but in practice a wide gulf and a large measure of controversy exist between, on the one hand, population geneticists studying processes of selection in living organisms, and, on the other, taxonomists classifying phylogenetic patterns and seeking to account for the origin of new taxa above the level of species.

Here we treat these two subject areas largely separately, in that Chapter 1, besides serving as an introduction to the book as a whole, considers selection as a mechanism of change (this aspect is sometimes termed the 'Special Theory of Evolution'), whilst Chapter 2 deals with the phylogenetic interrelationships of the invertebrate groups (or the 'General Theory of Evolution') and the patterns of diversity and diversification through time. Nevertheless, within each of these chapters we have found it appropriate to introduce elements of the subject matter of the other; for example to comment on such controversial matters as the manner of origin of invertebrate classes and phyla.

CHAPTER 1

Introduction: Basic Approach and Principles

1.1 Why invertebrates?

This book is about the invertebrates – animals *without* backbones. A definition such as this, based on the *absence* rather than the *presence* of a specific characteristic, is unusual and implies a deviation from a standard type that has the characteristic. If there were no standard or norm then such a definition would hardly make sense.

Thus, when Aristotle classified animals into sanguineous and non-sanguineous, the implication was that the presence of blood was a norm for animals. What he believed was that in its evolution, life had been directed towards a perfect animal form that involved having blood. He incorporated this idea into a hierarchical classification of living things called the 'scale of life' (*Scala naturae*) in which there was progression from the non-sanguineous state to the sanguineous *goal* (Table 1.1).

Similarly, when Lamarck (of acquired characters fame) first separated invertebrate from vertebrate animals (in his *Système des Animaux sans Vertèbres*, Paris, 1801) there was an implication that the latter were the norm. And again this probably followed from Lamarck's peculiar theory of evolution in which he supposed that acquired characters were incorporated for hereditary transmission according to principles that not only involved survivorship criteria but progress towards some higher form, of which the vertebrates and humans were closest representatives.

Modern Zoology has forsaken concepts of goal-directed evolution (teleology) and yet the distinction between vertebrates and invertebrates has persisted and has influenced numerous generations of students. This is remarkable since the distinction is hardly natural or even very sharp; that is, it separates a group containing many phyla (the invertebrates) from one containing part of a phylum (some members of the phylum Chordata do not have true vertebral columns).

However, there are two other main reasons for a continuing distinction being made between invertebrate and vertebrate zoology. First, a historical one – Lamarck created a precedent that, once established as a method of approaching zoology, was difficult to escape. Second, and probably more influential, there is still a feeling that because we ourselves have a backbone, vertebrate animals deserve more attention than their taxonomic status might merit.

Table 1.1 Aristotle's 'scale of life' or *Scala naturae*

		Sanguineous
VIVIPAROUS		1 Man
		2 Hairy quadrupeds (land mammals)
		3 Cetacea (sea mammals)
OVIPAROUS	With perfect egg	4 Birds
		5 Scaly quadrupeds and apoda (reptiles and amphibia)
	With imperfect egg	6 Fishes
		Non-sanguineous
		7 Malacia (cephalopods)
		8 Malacostraca (crustaceans)
VERMIPAROUS		9 Insects
Produced by generative slime, budding or spontaneous generation		10 Ostracoderma (molluscs other than cephalopods)
Produced by spontaneous generation		11 Zoophytes

By concentrating on the biology of invertebrate animals, we here perpetuate this distinction, but not because we have any philosophical commitment to goal-directed evolution or to the view that there are fundamental biological distinctions between invertebrates and vertebrates. Rather our position is pragmatic. We wish to demonstrate that:

1 All living things share in common a number of basic features of structure and function.

2 Major variations occur in these themes, and groups of taxa sharing these in common are referred to as phyla.

3 These variations have evolved and hence should be related by common descent.

4 Within the constraints of each major theme, animals have become adapted to the ecological circumstances in which they occur by natural selection. (The extent to which these micro-evolutionary processes can account for the macroevolutionary changes noted in points **2** and **3** is a matter of some debate and we shall return to this later.)

In examining these issues it is expedient to circumscribe the material in some way. We do this on the basis of historical precedent. Moreover, the invertebrates provide us with maximum diversity for examining the issues raised under points **2–4**. Before doing this, however, we need to have an appreciation of the basic features common to all living things (point **1**), how they differ from those of non-living things and how they originated. This is the aim of the next few sections.

1.2 Properties of living things

1.2.1 Introduction

At a basic chemical level, of the 92 naturally occurring elements on Earth, less than a third are found in living things (Table 1.2). Only 11 elements are found in more than trace amounts. With the exception of oxygen (O_2) the most common elements in living things are not those that are most abundant in the Earth's crust. Around 75% of most animals is water and 50% of the remaining dry weight is carbon with little, if any, silicon. The Earth's crust, in contrast, consists of 27.7% silicon, and about 0.03% carbon.

Table 1.2 The elements found in living things

Element	Symbol	Approximate atomic weight (daltons)	Approximate proportion of Earth's crust (% weight)
Most abundant (>90%) elements in living things			
Hydrogen	H	1	0.14
Carbon	C	12	0.03
Nitrogen	N	14	<0.01
Oxygen	O	16	46.6
Next most abundant elements in living things			
Sodium	Na	23	2.8
Magnesium	Mg	24	2.1
Phosphorus	P	31	0.07
Sulphur	S	32	0.03
Chlorine	Cl	35	0.01
Potassium	K	39	2.6
Calcium	Ca	40	3.6
Elements present but normally in trace amounts (in total <0.01%)			
Iron	Fe	56	5.0
Fluorine	F	19	0.07
Silicon	Si	28	27.7
Vanadium	V	51	0.01
Chromium	Cr	52	0.01
Manganese	Mn	55	0.1
Cobalt	Co	59	<0.01
Nickel	Ni	59	<0.01
Copper	Cu	64	0.01
Arsenic	As	75	<0.01
Zinc	Zn	65	<0.01
Selenium	Se	79	<0.01
Molybdenum	Mo	96	<0.01
Tin	Sn	119	<0.01
Iodine	I	127	<0.01

Despite the restricted number of chemical elements found in living organisms, the molecules they contain are structurally and functionally very diverse. This is because carbon is outstanding amongst all elements, with only silicon coming a close second, in being able to form diverse molecular strands and rings. These carbon-based, molecular building blocks of organisms are: sugars, amino acids, fatty acids and nucleotides and these associate as macromolecules to form, in turn, polysaccharides, proteins, lipids and nucleic acids. Free organic chemicals, of this complexity, form only rarely in non-living systems (p. 7).

An even more profound distinction, though, between the living and the non-living is the way that the organic chemicals are organized. In living systems the macromolecules make up membranes that bound further, non-random collections of macromolecules that react together in ordered metabolism. These packages are the cells. Many cells make a multicellular organism, and in this context are collected together as a highly ordered and organized, structural and functional unit. The very existence and persistence of this order, organization and complexity were considered for a long time to be *special*, even *mysterious*, features of organisms, created and maintained by mysterious *vital forces*; for the rule of the non-living world, summarized in physics by the Second Law of Thermodynamics, is that order and organization are unstable. Entropy, or disorder, should progressively increase in *all* reactions and processes.

However, we now recognize that the order and organization of biological systems arise from two non-mysterious features common to them all. *These are crucial in understanding the basic principles of biology and we shall continually allude to them in what follows:*

1 Organisms are *programmed*.

2 These *programmes* specify working systems and subsystems that are *open* to the input of material and energy.

1.2.2 The programme

At the most fundamental level, the genetic programme controls the properties of proteins by specifying the types and sequences of amino acids from which they are constructed. Only 20 different amino acids occur frequently in animals but with a chain of only 100 (which is short for a protein) there could in principle be 20^{100} possible configurations! This explains the great diversity of proteins. Some are enzymes that control all metabolic processes within organisms and others comprise the fabric of the cell and organism.

The programme itself is coded as nucleotide sequences in DNA. There are four different kinds of nucleotides (adenine, thymine, guanine and cytosine) but 20 amino acids so that there could not be one-to-one specification of the latter by the former. Only combinations of three (or more) nucleotides could give sufficient alternative combinations ($4^3 = 64$) and this triplet code has been found to be universal, with the excess number of alternatives (*c.* 40) being accounted for by redundancy (more than one triplet coding for a particular amino acid) and punctuation. However, there is not a direct translation of DNA to proteins. Instead the information is first transcribed to RNA (like DNA but uracil replaces thymine) in a process termed transcription. This RNA acts as a messenger (referred to as messenger or mRNA). It takes the coded information across the nuclear membrane to a site, the ribosome (composed of ribosomal or rRNA) where it can be translated into an amino acid sequence to form a protein. Yet another set of RNAs (the transfer or tRNAs) are responsible for transporting amino acids to the ribosome and positioning them at the correct location in the new protein. The process that takes place on the ribosomes is termed translation. The complexities of constructing protein molecules from information encoded in DNA is depicted schematically in Fig. 1.1.

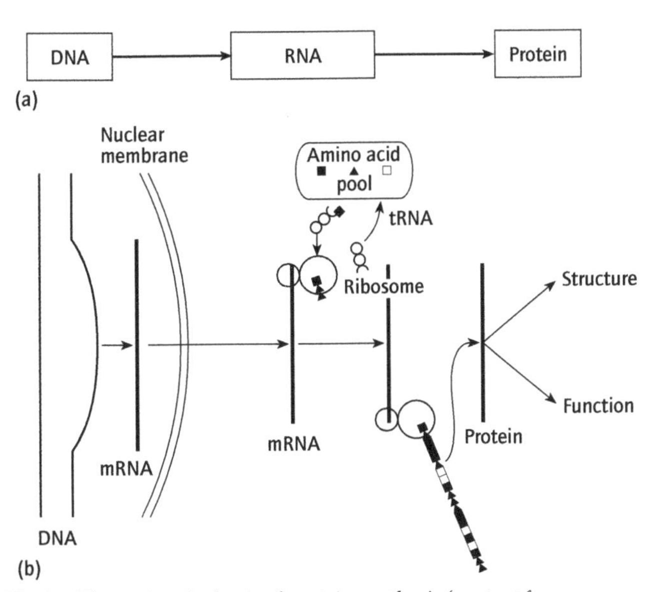

Fig. 1.1 The molecular basis of protein synthesis (see text for explanation).

1.2.3 Openness

The ordered systems of organisms, right down to their macromolecules, are continually subjected to 'entropic insults', but disordered systems can be replaced according to specifications embodied in the genetic programmes. However, this is only possible if there is a continuous import of ordered raw materials and an export of disordered material (excreta) and energy (largely heat). Hence, organisms and the cells that they contain have to be open systems. Even in non-growing animals there should be a continuous turnover of cells and/or molecules – what Schoenheimer described as the *dynamic steady-state* of the body in his book, *The Dynamic State of Body Constituents* (1946).

1.2.4 Evolution by natural selection is an inevitable consequence of systems that persist by replication

Whole organisms can also be replaced – *reproduced* – by replication of the genetic programme. This involves separation of multicellular or unicellular propagules that carry a whole or part of the genetic programme (genome) of their parents. Multicellular propagules invariably contain more or less complete replicas of the parent genome and processes of reproduction involving them are referred to as asexual or vegetative. Unicellular propagules (gametes) most often contain a part-replica of the parent genome and have to fuse with other gametes to reinstate the complete genome (fertilization) before development can proceed. This is sexual reproduction.

We have, then, systems that *replicate* a genome to *reproduce* an organism. However, the genome is not always replicated faithfully; even in asexual reproduction mutations introduce variation, and additionally in sexual reproduction the 'shuffling' processes associated with meiosis and the mixing associated with fertilization introduce considerable differences between parents and offspring. Variations in the genetic programme lead to variation in the form and function of the phenotype and this, in turn, influences the way that it interacts with its environment and hence its chances of survival and its rate of reproduction. It follows that those programmes that best promote survivorship and fecundity in the environment in which they occur, i.e. are best *adapted* to it, will become most common. Moreover, given that the world is finite and that the resources needed for the life processes are limiting, these programmes will tend to replace other less well-adjusted ones. This summarizes, in simple form, the process of evolution by *natural selection* that was first made explicit by Charles Darwin in his *Origin of Species* (1859). Borrowing from Herbert Spencer, he used the catch-phrase 'survival of the fittest' to describe this process. However, the above description makes it clear that fitness is the *ability of one gene-determined trait to spread in a population in comparison with others* which involves both *survivorship* and *fecundity*.

1.3 Origins of life

The most fundamental feature of living systems is that they are able to persist in ordered and organized state by a process of programmed replication and reproduction. *Evolution by natural selection follows as an automatic consequence of this organization.* But how did such a system originate? Discovering how the organic molecules that make up organisms (p. 4) themselves originated is only a partial answer to this question. We need to imagine plausible ways whereby they become *organized* into self-replicating systems.

The carbon-based molecules that comprise living organisms were once thought to be so unique and special that they could only be synthesized by living things. Hence a distinction was made between organic (= from life) and inorganic chemicals. The first breach in this demarcation was made when Wöhler synthesized a very simple organic molecule (urea) from ammonium cyanate (an inorganic molecule) simply by applying heat (in 1832). And this initiated the rational and scientific treatment of the chemistry of life, that formed the foundation of modern biochemistry and molecular biology. Yet the controlled synthesis of urea is a far cry from the spontaneous origin of polysaccharides, lipids, proteins and nucleic acids that would be needed for the origin of living systems.

1.3.1 Prebiotic synthesis of organic polymers

Little is known with certainty about the early atmosphere of the earth, but it was probably formed by 'degassing' of the planet so that it is likely to have borne a close resemblance to the gas mixtures that escape from volcanoes. On this basis it was almost certainly devoid of O_2 (see Chapter 11). Experiments have now shown that with these conditions almost any energy source, lightning, shock waves, ultraviolet radiation (because there was no O_2 there was no ozone which filters these wavelengths out of the insolation) or hot volcanic ash, could have led to the prebiotic synthesis of a variety of 'organic' monomers: sugars, amino acids, fatty acids and even nucleotides. Given appropriate conditions, for example high concentrations of inorganic polyphosphates, it is possible for these to join in long chains to form, for example, polynucleotides and polypeptides. It is widely held that all these substances concentrated in the early ocean to form the famous 'primordial soup'.

There must have been a kind of selection going on in this prebiotic world since molecules that could polymerize fastest, and/or were most stable would be most common. But the tempo of these changes was not very rapid and could not be very 'adventurous', since the formation of each polymer was an independent event; there was no building on a genetic memory. Once certain polymers have formed these are able to influence the formation of other polymers. Polynucleotides in particular have the ability to specify the sequence of nucleotides by acting as a template for polymerization. If one polynucleotide acts as a template for its complement, which then acts as a template for the original, we have lineages linked by a kind of genetic memory. Those polynucleotides that do this most effectively increase in abundance relative to others, i.e. are selectively favoured.

The template systems would have been error-prone and new polynucleotides would have been formed by 'mutation' and would compete with others for possibly limited building blocks. Since deoxyribonucleosides (precursors of nucleotides) are more difficult to synthesize than ribonucleosides, and since RNA plays a central part in modern protein synthesis, a widely (but not universally) favoured hypothesis is that these early, self-replicating polymers were small RNAs. It has been suggested that these early RNA molecules would have acted as both genes and enzyme-like catalysts. However, those who hold this 'RNA world' scenario for the early evolution of life still have some key questions to address, e.g. how were the first RNA molecules formed and exactly how would they have acted as catalysts in even the simplest of metabolic systems?

In recent years there has been a shift away from the theory that biomolecules arose abiotically and that they, interacting with each other in the primordial soup, formed 'proto-cells'. For example, there are some who think that these proto-cells evolved from simple biochemical complexes originally attached to the surface of minerals (e.g. pyrites).

1.3.2 Origin and evolution of cells

The next steps towards the system summarized in Fig. 1.1 are more difficult to imagine. At normal temperatures the spontaneous replication described above would have been slow and the error rate high. Association with a replicase, a protein capable of catalysing replication, would have speeded up the process. Just how this might have originated is not clear, but once it was present it would have been favoured. Moreover, there would have been some advantage in enclosing the template and replicase because then the advantages derived from this liaison could not be of benefit to other slightly different but competing templates. The cell was thus born and we begin to see a distinction between genotype and phenotype. Selection would operate on these primitive cells such that those in which the co-operation between genotype and phenotype enhanced replication rate and fidelity would spread more rapidly than others. Though difficult to specify precisely, it was in this context of 'co-operation' and selection that the complex system involving DNA, as well as various forms of RNA, originated and was refined.

The original cells were small and had a simple internal structure, something like modern bacteria, the so-called prokaryotes. In some cells a further membrane originated to enclose the genetic information and this is likely to have been favoured because it probably gave more protection against genetic damage. These cells, the so-called protoeukaryotes, also, probably later, acquired cytoplasmic organelles and prominent amongst these are the mitochondria. The latter show many similarities with free-living prokaryotes – resembling them in size and shape, containing their own DNA, and reproducing by dividing in two – and are now thought to have arisen through symbiotic associations between small prokaryotes similar to the surviving *Paracoccus* and larger nucleate protoeukaryotes. By breaking up eukaryotic cells it is possible to show that all the machinery for aerobic metabolism is located within mitochondria, so that this supposed symbiosis evolved in association with the accumulation of O_2 in the earth's atmosphere from the photosynthetic activity of early cyanobacteria.

1.3.3 Why doesn't spontaneous generation happen all the time?

If large biomolecules and indeed cells have originated once it is reasonable to consider why this does not happen continuously. The answer is probably that the living things themselves, once originated, created conditions which were unsuitable for this. For example, O_2, a product of life, once formed would destroy the organic building blocks that form living things. In the presence of O_2, solitary organic polymers are broken down into simple inorganic constituents by oxidation. Hence, once free O_2 became plentiful, the 'primordial soup' of organic molecules could no longer be sustained. Moreover, the complex organic molecules in the 'soup' were probably an excellent nutritional source for the early organisms, and were probably eaten or degraded more rapidly than they formed.

1.4 Levels of organismic organization

It is unlikely that the totality of physiological processes, when crammed together within a single cell (as they are in protists; Section 3.1), could be as effective as when they are divided between many cells in multicellular organisms. The multicellular condition provides more space for the reactions, and the division of labour between cells, by compartmentalization of function, means that at least within compartments, physiological conflicts can be minimized. Hence, there would have been considerable selection pressure for the evolution of multicellularity, e.g. for the origin of the Animalia.

The next chapter summarizes the major features of the invertebrate phyla and speculates upon their relatedness and evolution. It illustrates various possible trends in organization that have progressively opened up physiological potentialities of multicellular animals. For example:

1 The evolution of cellular differentiation.
2 The spatial localization of cells of the same type in tissues and then the organization of these into organs (collection of cells contributing to a common function).
3 The evolution of a through-gut which allowed more specialization of different regions.
4 The development of fluid-filled body cavities that allowed the gut and other organs (e.g. heart) to operate independently from body-wall musculature, facilitated diffusional distribution of nutrients and, by providing a hydrostatic skeleton (Section 10.4), more effective locomotion.
5 The evolution of specific systems for the distribution of nutrients and respiratory gases between tissues. This allowed escape from size constraints imposed by the diffusional distribution of these products (see Chapter 11).
6 The evolution of limbs, which opened up considerable potential for locomotion, particularly on land and in the air.

It is quite easy to see how natural selection might have improved function within levels of organization. But was it responsible for the major shifts between levels? Did these shifts occur *gradually* by a continuous sequence of small changes that improved physiological function and enhanced fitness, or by 'quantal' leaps between levels that were more to do with chance and developmental opportunities than natural selection? These alternatives are referred to in turn as the *gradualist* and *punctuated equilibrium* hypotheses. There is certainly some evidence for a punctuated process in the evolution of the invertebrates (Chapter 2). But since the punctuations are usually measured in geological time, i.e. represent several million years, they could still alter under the influence of natural selection. It is likely that selection pressures change considerably from time to time and cause significant differences in the tempo of evolution. Hence, a punctuated pattern of evolution does not exclude a Darwinian mechanism. This has been a hotly debated area of evolutionary biology (see Ridley, 1996 or Futuyma, 1998 for details of the debate and for an excellent introduction to evolution generally) and we shall return to it later in this text.

1.5 Prospects

This chapter has briefly outlined what we consider to be the basic features of living systems:

They are organized systems that depend upon programmes, replication and openness for persistence.

The features of animal systems that follow, almost logically, from them are:

They acquire resources as food from their environment and use them in ways that promote survivorship and fecundity.

Various levels of organization of animal life have evolved and natural selection has operated within these phyletic constraints to cause adaptation in the acquisition and utilization processes. After a preliminary outline of these levels of organization in the next chapter, we describe them in more detail in Part 2. This sets the scene for a more in-depth consideration of the behaviour patterns and physiologies of invertebrates in Part 3, where we concentrate on individual aspects of their functioning. Hence, Part 2 adopts a phylum-by-phylum approach and Part 3 a cross-phylum, functional approach to the invertebrates. The reader might, therefore, choose to concentrate on the phyletic approach in Part 2 and use Part 3 as a source of further information, or alternatively to concentrate on the functional biology of the invertebrates in Part 3 and to use Part 2 as an 'index' to the taxa that are referred to within it. The two Parts are, nevertheless, integrated and aim to give a complete, holistic appreciation of invertebrate organisms.

1.6 Further reading

Cox, T. 1990. Origin of the chemical elements. *New Sci.*, Feb 3, 1–4.

Des Marais, D.J. & Walter, M.R. 1999. Astrobiology: Exploring the origins, evolution and distribution of life in the universe. *Annu. Rev. Ecol. Syst.*, **30**, 397–420.

Edwards, M.R. 1998. From a soup or a seed? Pyritic metabolic complexes in the origin of life. *Trends Ecol. Evol.*, **13**, 178–181.

Garland, T. & Carter, P.A. 1994. Evolutionary physiology. *Annu. Rev. Physiol.*, **56**, 579–621.

Gibbs, A.G. 1999. Laboratory selection for the comparative physiologist. *J. exp. Biol.*, **202**, 2709–2718.

Lewin, B. 1998. *Genes VI*. Oxford University Press, Oxford.

Futuyma, D.J. 1998. *Evolutionary Biology*, 3rd edn. Sinauer Associates Inc., Sunderland Massachusetts.

Kirchner, M. & Gerhart, J. 1998. Evolvability. *Proc. Natl. Acad. Sci. USA*, **95**, 8420–8427.

Maynard Smith, J. 1986. *The Problems of Biology*. Oxford University Press, Oxford.

Morris, S.C. 1998. Early metazoan evolution. Reconciling paleontology and molecular biology. *Am. Zool.*, **38**, 867–877.

Pigliucci, M. 1996. How organisms respond to environmental changes: from phenotypes to molecules (and vice versa). *Trends Ecol. Evol.*, **11**, 168–173.

Ridley, M. 1996. *Evolution*, 2nd edn. Blackwell Science, Massachusetts.

Schmitt, J. 1999. Introduction: Experimental approaches to testing adaptation. *Am. Nat.* (suppl.) 154: S1–S3.

Schopf, J.W. (Ed.) 1992. *Major Events in the History of Life*. Jones & Bartlett, Boston.

Schopf, J.W. 1994. The early evolution of life – solution to Darwin's dilemma. *Trends Ecol. Evol.*, **9**, 375–377.

Sibley, R.M. & Calow, P. 1986. *Physiological Ecology of Animals: An Evolutionary Approach*. Blackwell Science, Oxford.

Smith, D.C. & Douglas, A.E. 1987. *The Biology of Symbiosis*. Edward Arnold, London.

CHAPTER 2

The Evolutionary History and Phylogeny of the Invertebrates

Living animals are the products of their evolutionary pasts, and it is not possible fully to understand modern biology without some appreciation of that past and of the constraints that it has placed upon animal structure, ecology and life styles. In this chapter we describe the major features of the evolutionary history and diversification of the animal kingdom, including those of its origin.

Our knowledge of the relationships of the major groups of animals has undergone a revolution in the last 10–15 years as molecular or gene sequence studies and cladistic analyses have probed a wider range of species. There are certainly still gaps and uncertainties but a broad picture is emerging. We set this against a background of the selective pressures that must have operated in the past and the likely responses of organisms then alive to those pressures. We also stress what the fossil record has to tell us about the nature of diversification and extinction.

Readers should note that unavoidably this chapter draws on some of the anatomical features that are described in Part 2 of the book. We feel that it is more appropriate to present an overview before a detailed consideration of the individual groups, even though this may mean that some structures and concepts may require reference to material to be covered later. Such has been kept to the minimum, however. We should also stress that this is not a textbook of evolution, of cladistics or of molecular sequence analysis. We have presented a synthesis based on the conclusions of those studies that seem to us to be the most comprehensive and informative, and must leave the reader to seek detailed information on procedures and rationale from the works cited in the 'Further reading' section.

2.1 Introduction

It is certain that the multicellular animals, like the two other multicellular kingdoms, the Fungi and Plantae, are the descendents of lines of unicellular eukaryote protists, and it is appearing more and more likely that the fungi and the animals are close to each other in origin. But what the first animal was, or first animals were, is much less clear. Most of the animal groups that are represented in the fossil record first appear, 'fully formed' and identifiable as to their phylum, in the Cambrian, some 550 million years ago. These include such anatomically complex and distinctive types as trilobites, echinoderms, brachiopods, molluscs and chordates. Earlier, Precambrian, fossil animals are not numerous, but it is possible that cnidarians and segmented worms are represented, although the resemblence of several Ediacaran fossils (Fig. 2.1) to living animal groups may be only

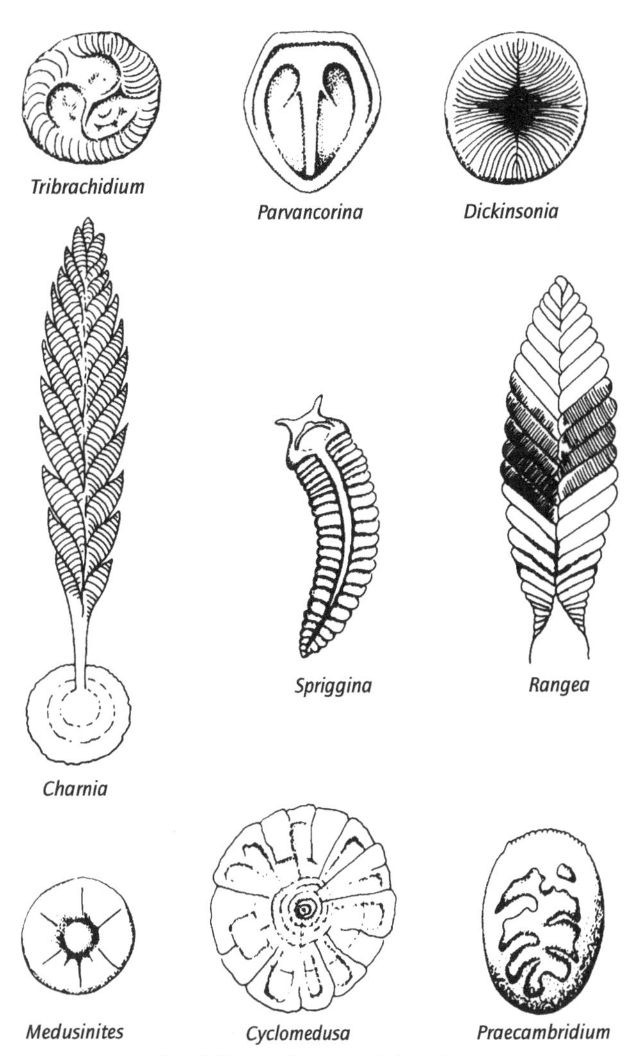

Fig. 2.1 Precambrian (Ediacaran) animals (after Glaessner & Wade, 1966).

superficial. Some to all of these Precambrian forms, it has been argued, are not even animals as that term would be generally understood. The fossil record is therefore of no help with respect to understanding the origin and early diversification of the various animal phyla, except in so far as to indicate that these ancestral events must have occurred in the Precambrian, probably at least some 1000 million years before the present day – the likely date of origin is currently hotly debated. Life itself dates back 3500 million years.

The first multicellular animals would presumably have been (a) small, (b) composed of relatively few cells of a very limited variety of differentiated types, and (c) without any hard parts. Since the fossil record is overwhelmingly one of organisms with hard tests, shells, plates or skeletons, it would be unrealistic to expect that this record will ever be able to contribute to unravelling the ancestry of animals: such ancestral forms are most unlikely ever to have been preserved. Zoologists have therefore been forced to argue solely on the basis of features shown by living members of the different animal and protist phyla. Here it should be remembered that the living representatives of *all* groups of organisms are separated by some 1000 million years, and in some cases considerably more, from the origin of their group, with all the possible changes in biochemistry, physiology, embryology, anatomy, etc. that may have occurred during this interval.

Some surviving animals are clearly extremely simple in their structure: they may have very few cells in total and/or those cells may be of very few types (just like the presumed ancestral multicellular forms); and they may lack many of the organ systems found in most animals (e.g. blood systems and/or body cavities). These animals might therefore be regarded as the unchanged descendents of ancestral groups that have maintained their body plan unmodified over hundreds of millions of years. Some surviving species – often termed 'living fossils' – do seem to all intents and purposes to be identical to ones that lived up to 500 million years ago, at least at the gross level, and hence some may have remained unchanged for even longer. If that was the only possible interpretation the search for the ancestral condition might be more simple, but there is another alternative view of such groups. During their individual development, animals naturally pass from the relatively simple to the more complex, and juveniles and/or larval stages often possess a simple body plan. In the normal course of events, sexual maturity is not reached until the (relatively complex) adult stage, but the onset of sexual maturity whilst still a juvenile in bodily structure (paedomorphosis) is a well known, and under various ecological circumstances a selectively advantageous, phenomenon. It can be seen occurring today in that a number of marine larvae have been captured in plankton nets that have developing, and in some cases functional, gonads (see Section 2.5). Several groups of animals do bear a remarkable resemblance to the larval stages of other types of animals and are thought to have arisen paedomorphically. Bodily simplicity can therefore be primary, but it can also be secondary. Simple animals can evolve from more complex ones. Reading the polarity of ancestor to descendent relationships is thus far from straightforward and has proved the subject of strenuous debate in individual cases.

Classically, animal interrelations were postulated on the basis of shared anatomical features. The process of classification operated by placing all organisms with a very similar structure in the same group (taxon) and by having one such taxon for each such assemblage of essentially similar species; organisms with different morphologies were thus sited in different taxa. Taxa were then nested in an increasingly more inclusive set of groupings, again on the basis of anatomical similarity, and this was assumed to reflect evolutionary affinity since fundamental similarity was likely to be proportional to relatedness. Two problems were inherent in this approach. First, it is clear that the magnitude of morphological change over evolutionary time is not proportional to the time taken to evolve that change (Section 2.5). Two groups may have diverged from each other a very long time ago but may have since maintained their ancestral anatomies almost unaltered, as in the living fossils above. On the other hand, a group that separated from another relatively recently may have rapidly acquired a very different adult body structure, for example through the agency of paedomorphosis. Secondly, anatomical similarity can be deceptive. It could indeed reflect the fact that their structure was inherited from a recent common ancestor. Or it could have been a consequent of convergence: the evolution of similar anatomical responses to common selective pressures in unrelated organisms, especially where there are a limited number of possible adaptive responses to the circumstances in question. Or it could have arisen in parallel where the organisms inherited the genetic ability to respond in the same manner in similar circumstances without inheriting the actual structure from their common ancestor: animals as diverse as flatworms, insects and vertebrates have been shown to possess very similar blocks of genes specifying the linear arrangement of structure along their anterior-posterior axis, for example. Similarity is therefore not necessarily proportional to relatedness either.

Two modern techniques have been widely applied to circumvent these problems. One is the methodology of cladistics which seeks to identify the sequence of evolutionary branching in lineages of organisms, irrespective of the extent of their overall anatomical similarity and dissimilarity. In essence, this it does by trying to establish the ancestral and derived character states of given anatomical (or other) features, and by identifying the derived character states that are shared only by certain groups of organisms. A collection of such shared derived character states can then be nested to portray the pattern of evolutionary branching (allowing of course for reversals of state). The other is the construction of molecular phylogenies based on the degree of similarity in the sequence of the component units, e.g. nucleotides, along certain molecules such as in subunits of ribosomal RNA, on the assumption that mutations at any one point occur at a relatively constant rate and therefore that a relationship exists between divergence in precise molecular structure and length of time since evolutionary branching. Neither method is foolproof. Multiple acquisition of the same feature by parallel evolution can lead to erroneous cladistic analyses, for example, and the 'molecular clock' (the notion that the rate of accumulation of changes along a molecule in question is constant within and across groups) may not be as constant as some molecular phylogenies might imply. Further, the precise methodology of many cladistic analyses is not above criticism (Jenner & Schram, 1999), and sequence data from different molecules can give divergent results. Nevertheless, both techniques have proved invaluable in distinguishing between grades of organization (equivalent levels of body structure) and clades (individual phylogenetic lines), and in identifying features that have been convergently evolved rather than inherited from common ancestors. Revolutions brought about by such analyses include appreciation that the segmented worms and the segmented arthropods are not in fact closely related by virtue of this shared segmentation, and the true affinities of arthropods and those animals bearing lophophores. The remainder of this chapter relies heavily on these approaches, although it should be noted that cladistic analyses and molecular comparisons do not always agree and there remain several areas of uncertainty; not all animal groups have yet been subjected to molecular analysis either.

2.2 The simplest animals

2.2.1 The evolution of multicellularity

There are theoretically three ways in which a multicellular animal or other organism could evolve from a protist. First, different types of protist could together symbiotically form a composite multicellular organism, similarly to the manner thought likely to be the origin of the eukaryote cell from different prokaryotes, and of lichens by algal and fungal partners (Fig. 2.2a). Secondly, the asexual division products of a single individual protist could remain together after fission, and multicellularity arise via an intermediate colonial stage (Fig. 2.2b). Here each protist individual would be the equivalent of one cell of a multicellular organism and the protists could truly be regarded as being 'unicellular'. Thirdly, a multinucleate protist could evolve internal membrane partitions around each of its nuclei, confining its sphere of operation to a certain region of the body, and thereby become internally compartmented (Fig. 2.2c). If multicellularity arose via this route, any derived multicellular organism would be better regarded as being 'cellular' (i.e. divided into cells) and the founding protist as being 'acellular' (not divided into cells), rather than unicellular.

The first of these three potential mechanisms (Fig. 2.2a) presents serious genetic problems. How do the genetically distinct founding protists integrate into one single, reproducing, multicellular organism? Even the two or three distinct symbionts forming the composite lichens have to reproduce separately and then re-associate to form new colonies. There is evidence that a few eukaryote protists have arisen by union of two entities that

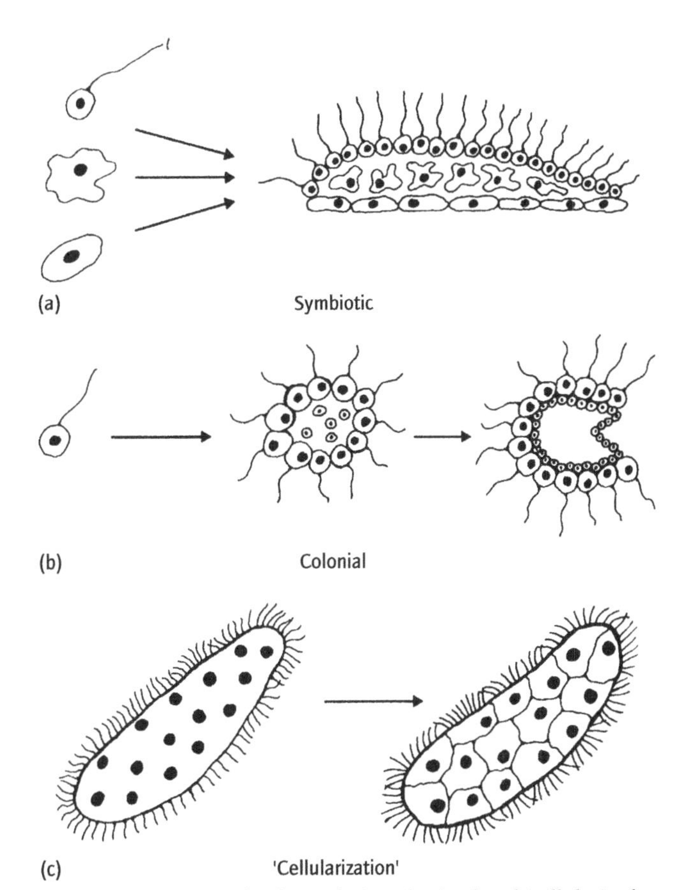

Fig. 2.2 Possible routes for the evolution of animal multicellularity from within the Protista (see text for explanation).

were themselves eukaryote, but this seems to have happened very rarely. In respect of the third potential route (Fig. 2.2c), there are no indications of internal compartmentalization amongst living protists and so there is no comparative evidence to suggest that it might have occurred in the past. It must be said, however, that if such a protist had undergone partial or complete compartmentalization it would probably be regarded as a multicellular organism and not as a protist, so that comparative evidence would be lacking by definition. Many protists, however, are known to form colonies by asexual fission (Fig. 2.3), as indeed do many prokaryotic bacteria, and in some colonies there is differentiation into distinct cell types. The second mechanism (Fig. 2.2b) is, therefore, that favoured by most biologists, and there is a wealth of indications of how unicellular protists might have formed multicellular organisms via coloniality. The multicellular state is after all formed by repeated mitotic (asexual) division of the founding zygote and of its fission products.

It is in fact rather difficult to distinguish between a colonial protist and a multicellular organism. Not all organisms classically regarded as being multicellular exhibit much co-ordination of their component cells and, as we have just seen, cellular differentiation is not confined to the multicellular. Often it seems to be based to a considerable degree on tradition and convenience. Of the 27 phyla of protists recognised in one classification, 16

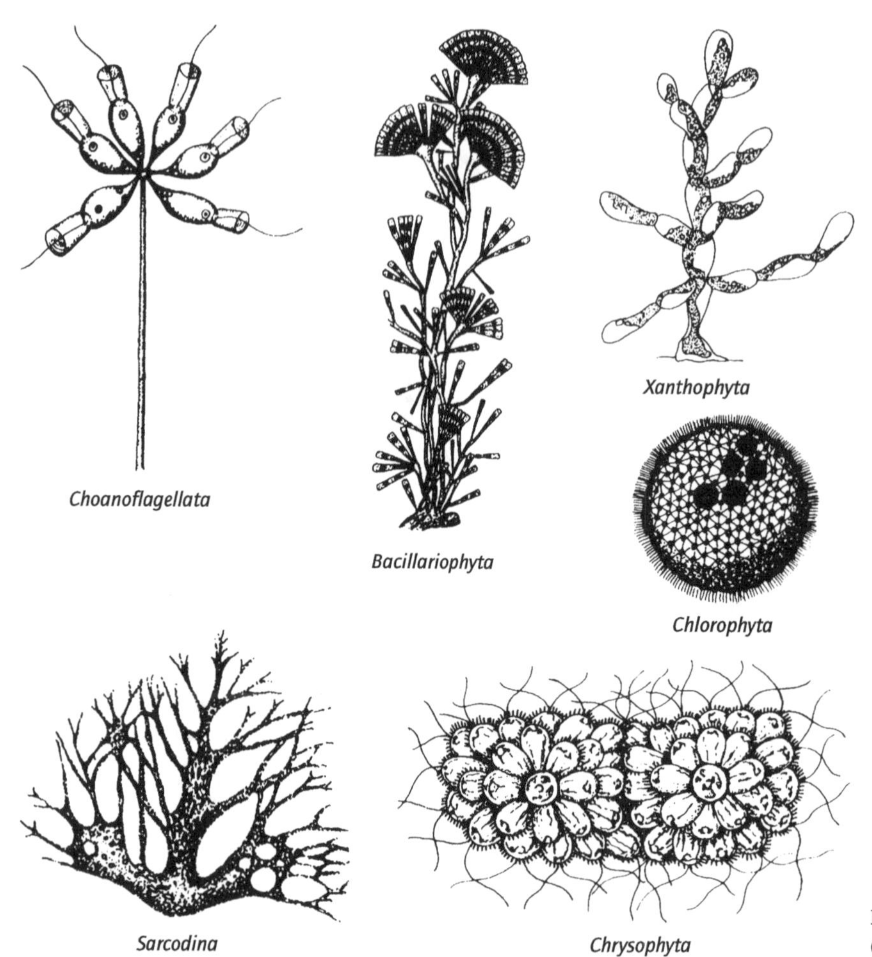

Fig. 2.3 Colonial protists from a variety of phyla (after various sources).

include colony-forming species and in three the level of organization of some, or all, species is regarded as having crossed the multicellular threshold. Excluding the animals, eukaryotic multicellularity may have arisen more than 15 times from a protist base, and indeed there is no necessary reason to assume that all multicellular animals arose from an organism that was itself multicellular.

Nevertheless, all animals do share a number of cytological and biochemical features (e.g. the ability to synthesize collagen and certain extracellular-matrix and cell-adhesion molecules; the presence of septate or 'tight' cell junctions) and all develop via a blastula embryological stage from a diploid zygote (except in certain secondary mechanisms of gametogenic asexual reproduction – see Chapter 14). Further, all share features with one group of solitary or colonial protists, the choanoflagellates (see Box. 3.1), and many therefore consider it highly probable that the ancestry of the animals lies within these heterotrophic aquatic protists. Whether animal multicellularity arose more than once from such colonial choanoflagellates, however, is a different and largely unresolved matter. On present knowledge, it seems possible that they did. So as to make allowance for this possibility whilst arranging for all animals to descend from an organism that was itself an animal (i.e. be a monophyletic group), several zoologists include the choanoflagellates within the kingdom Animalia, notwithstanding that this has the effect of rendering the animals no longer solely a group of multicellular organisms. An alternative reading of this general scenario, however, places the choanoflagellates as secondarily derived from the sponges (Section 2.2.2), and whilst this may account for the animal-like features of these protists, if true it would throw the nature of the ancestral protist group wide open (again).

Several groups of animals separate at or very near the base of the animal phylogenetic tree, and to that extent are potential candidates for the independent evolution of their multicellular state from different lines of colonial choanoflagellates, especially as they display radically different fundamental patterns of bodily architecture. We will consider these in turn in the remainder of this section.

2.2.2 Sponges

Sponges come closest of all animals to being regarded as a colony of protists rather than as a multicellular animal. Indeed the most characteristic sponge cell type, the choanocyte, is virtually identical to the free-living ancestral choanoflagellates and feeds in essentially the same fashion (Fig. 2.4). That the

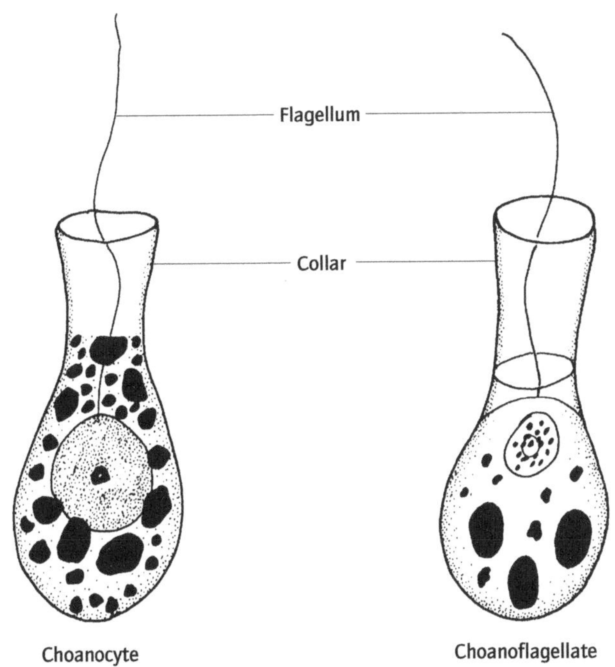

Fig. 2.4 The morphological similarity between a sponge choanocyte and a choanoflagellate protist.

individual sponge cell is the only essential subunit of their bodies is shown by the ability of some sponges to reconstruct themselves after complete disaggregation. They have been described as animals at the cellular grade of organization, in contrast to the tissue grade displayed by the radiates (see Section 2.2.3) and the organ-system grade of the bilateral groups (Sections 2.2.4 and 2.3). They do, however, develop from a blastula and possess the animal-type cellular biochemistry.

The peculiarities of sponge construction are:

- there is no fundamental system of symmetry
- their bodies are formed by, in essence, a monolayer of cells surrounding a secreted matrix; the whole in effect comprising a tube with differentiation of the cells lining the inner and outer walls of that tube
- there is no nervous system
- the cells do not together form well-defined tissues (let alone organs).

This structure is so individual that it is impossible to derive any other living animal group from their body plan. This has led some people to regard sponges as an early, unsuccessful 'attempt' at multicellularity and, with denigratory overtones, a dead-end group (except in respect of the extinct Archaeocyatha). This does not do them justice. The sponges are an extremely successful group of marine animals, with more living species than the echinoderms, for example, and almost as many as the annelids, and they have been a prominent part of the marine fauna ever since the Cambrian. Their undoubted simplicity need not be viewed as the result of some supposed inability to evolve the symmetry, co-ordination and bodily complexity of other animals, but as an adaptation to their admittedly unanimal-like life style.

They are attached, sessile and, except at the exhalant orifices of their tubular plumbing system, completely immobile suspension feeders. Indeed, the function of the skeletal system located within the secreted matrix is the very antithesis of that in all other animals: it serves to prevent movement and to provide rigidity to the body. Environmental water is induced to flow through the tubing that is the body mass by the (unorganized) beating of the choanocyte flagella; were the tubes not rigid, local reduction in water pressure could cause them to constrict rather then serve to draw more water through. Granted that the body is incapable of movement, a nervous system, for example, would be functionless and certainly not selectively advantageous. Protection from predators is brought about not by detection and escape but (as in plants) by distasteful chemicals or by the spicular or fibrous nature of the skeleton. Seen in the same light, it is difficult to imagine how any of the other systems possessed by more organized animals could in any way increase sponge efficiency or survival. Neither is it necessarily the case that sponges were comparatively early animals. Their appearance, and certainly their abundant appearance, in the fossil record post-dates the occurrence of, for example, segmented animals, and it is possible that they evolved from the colonial choanoflagellates that they so closely resemble (if the evolutionary relationship is read in that direction) after several other animal groups.

The sponge is therefore an alternative animal, not an evolutionary reject.

2.2.3 The radiates

Like the sponges, the radiates or coelenterates as they are often termed (the Cnidaria and Ctenophora) are generally, though not universally, regarded as a highly individual and dead-end group. The supposition that they did not provide the ancestry of the other phyla is – again similarly to the sponges – only another way of stating that their general body plan is so successful that no basic change to it would be likely to lead to greater success. This general body plan shows:

- fundamental radial symmetry
- bodies formed by, in essence, a monolayer of cells surrounding a secreted matrix; the whole in effect comprising a tube with differentiation of the cells lining the inner and outer walls of that tube
- a nervous system of naked nerve cells organized into a nerve net or nets completely surrounding the lumen of the tube
- individual partly-muscular cells each containing contractile fibres and arranged so as to function collectively as circular or longitudinal muscles
- the cells together form well-defined tissues, but not organs.

In addition, the Cnidaria are characterized by their shared derived feature of the occurrence of cnidocyte cells each containing an intracellular organelle, the nematocyst (Fig. 2.5), and the somewhat similar colloblast cells in the Ctenophora. Such cells and organelles are otherwise known only in a few groups of

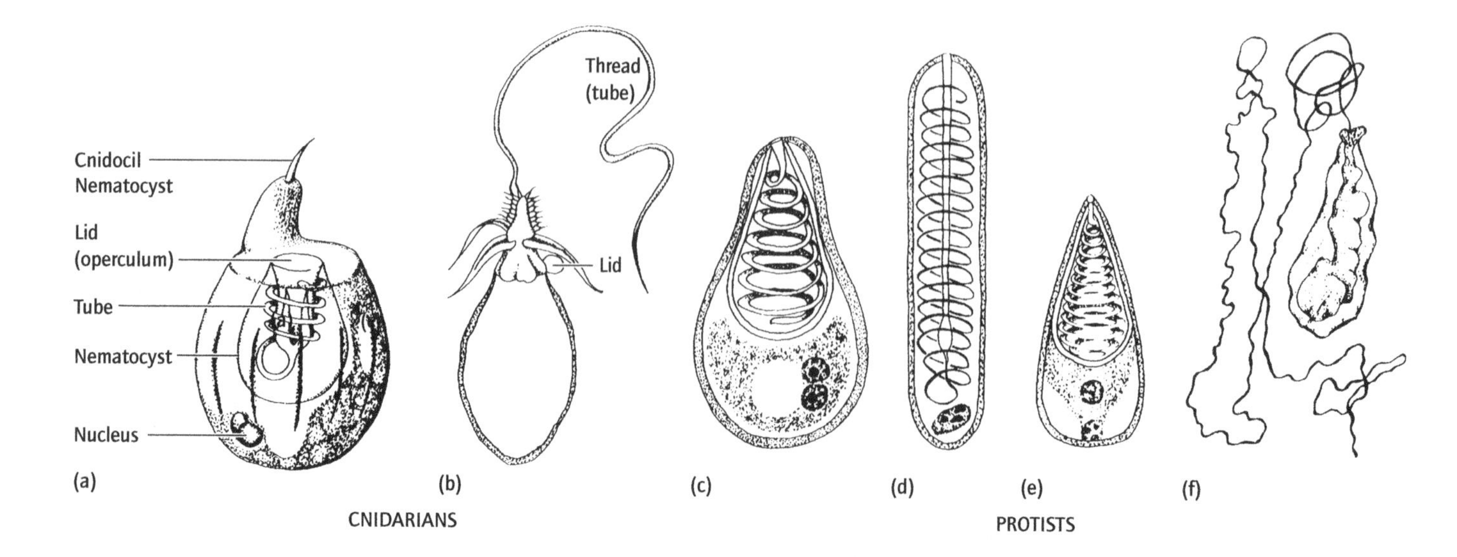

Fig. 2.5 Nematocyst-like organelles in cnidarian radiates and in myxosporan and microsporan protists. Cnidarian nematocysts: (a) cnidocyte containing an undischarged nematocyst; (b) discharged nematocyst. (c) Spore of a myxosporan protist with an undischarged organelle. (d, e) Spores of microsporan protists with undischarged and (f) discharged organelles. (After Hyman, 1940; Calkins, 1926; Wenyon, 1926 and others.)

protists (Fig. 2.5), and one of those – the Myxospora (see Box 3.1) – may be parasitically degenerate cnidarians in origin.

The body form, though not the symmetry, of the radiates is clearly basically similar to that of the sponges, i.e. a simple cup or tube, sometimes flattened or elongated, enclosing a simple cavity communicating with the external environment through a single aperture and containing environmental water, the tube being a virtual monolayer of cells completely surrounding a secreted matrix, with differentiation of the outer and the lining cells. In marked contrast to the sponges, however, the radiates are mobile and they feed on large particles – small animals – which they capture in the water column using, in the cnidarians, the cnidocytes located on tubular extensions of their body wall that surround the opening to the central cavity, their characteristic tentacles.

The cup- or tube-shaped body and their two-cell layered state (the outer and inner cell layers) were seized on by early phylogenetists, who pointed out that this 'diploblastic' condition is the same as that seen in the gastrula stage of animal embryology. Since in the embryological development of most animals, the diploblastic gastrula becomes triploblastic by the formation of a third germ layer, the mesoderm, the radiates were viewed as the relict, permanently diploblastic, ancestral forms that gave rise to all the bilaterally symmetrical, triploblastic, animal phyla. The three stages of animal evolution were envisaged as: (i) a hollow ball of cells (equivalent to the blastula embryological stage), a form in which some colonial flagellates occur; (ii) these for one reason or another became a double-walled cup (= the gastrula = the radiates); and (iii) when mesoderm evolved, the triploblastic animals were formed. This argument by embryological analogy may have been ingenious, and partly lives on as the blastaea – gastraea – trochaea theory (see Fig. 2.18), but there is no evidence to support the notion that embryological germ layers can in any way be equated with the cells of superficially similar living animals. Embryological terms such as 'diplo-' and 'triploblastic' should not be applied to adult morphology, nor – as is often done – should the inner and outer cell layers of radiates be termed 'ectoderm' and 'endoderm'.

Cnidarians may occur in either or both of two forms. The bodily tube may be elongate and attached to the substratum such that the single opening and the surrounding tentacles project up into the water column (the polyp), or it may be flattened to saucer-shape, free-living, and oriented so that the aperture and tentacles are on the underside of the saucer (the medusa). Several cnidarians are modularly colonial. Asexually produced polyps and sometimes medusae are budded off but remain in tissue contact with the other modules. Since all such modules are genetically identical, polymorphism of the modules is possible: some can be specialized for feeding and/or defence; others for reproduction; and so on. Via specialization of individual polyps and retained medusae within the one colony, some cnidarians have evolved the equivalent of the organ systems possessed by most other animals, although via a different route.

In most modern phylogenies, the cnidarians branch off at the base of the animal evolutionary tree and, except at the choanoflagellate level, they may not be directly related to either the sponges or the bilaterally symmetrical animals. Nevertheless, some zoologists do regard the planula dispersal stage of several cnidarians – a free-swimming, non-feeding, solid blastula that whilst being radially symmetrical is elongated along its longitudinal axis – as being the ancestor of flatworms or flatworm-like animals. Others consider that relatively large worms, with large coelomic body cavities, evolved from anthozoan cnidarians and that the flatworms and potentially related groups were derived paedomorphically from such coelomate animals (see Section 2.4). It is notable in this respect that the surviving cnidarians are almost all carnivores and none can digest algae, although some are partially to wholly dependent on symbiotic photosynthesizers for their nutrition. The rise to prominence of the cnidarians in late Precambrian seas was probably consequent on the evolution of a zooplankton, and again their success if not their origin is likely to have post-dated the appearance of other groups of animals.

The ctenophores are more obscure, and they may or may not be related to the cnidarians. In most recent phylogenies, they also branch off at the very base, but affinities with a wide range of other groups have been argued.

2.2.4 Flatworms

Although the surviving phylum Platyhelminthes contains many relatively complex species, there are some free-living marine forms that are the simplest bilaterally symmetrical animals known. The basic flatworm body plan shows:

- bilateral symmetry with a distinct head and tail end, and dorsal and ventral surfaces
- solid bodies formed by more (and often many more) than two cell layers
- a nervous system with sheathed longitudinal nerve cords
- the presence of tissues and organs.

This is the same basic plan as almost all other animals (Section 2.3). In comparison with the groups to be considered in Section 2.3, however, all flatworms lack a through gut, a circulatory system, and any body cavity except for that of the gut. The possession by most types of flatworm of the shared derived feature of biflagellate sperm containing 9 + 1 axonemes places them off the main line of bilateral evolution, but many acoelomorphs (Fig. 2.6) possess monoflagellate sperm with standard animal 9 + 2 axonemes. Acoels are also considerably less complex than most other flatworms, lacking a permanent mouth and gut, for example, and possessing poorly differentiated tissues and organ systems. Their structure is as near to a possible ancestral bilateral animal as can be imagined. As noted above, however, some see this simplicity as a secondary consequence of paedomorphosis and regard all flatworms as descended from relatively complex worms.

Individual sponges and individual or colonial radiates can achieve very large size. This is because regardless of body size each cell is in direct contact with the external environment – those on the outer layer with the water surrounding the animal and those of the inner layer with the water contained within the lumen of the tube – and hence diffusion of excretory products, of respiratory gases, and of food is uninfluenced by overall size. Such is not the case in solid-bodied animals like the flatworms. Lacking a circulatory system, flatworms must maintain at least one bodily dimension small to permit diffusion to continue to service the needs of the body mass. The smaller species also creep across surfaces under the power of their epidermal cilia, and neither will ciliary propulsion be effective in animals above a certain (small) size. The evolution of additional organ systems will permit larger size and alternative methods of locomotion to be adopted, however, and if an animal approximating an acoel was the ancestor of the Bilateria, the selective advantages of larger body size may have provided the pressure to evolve the organ systems that flatworms lack by definition, but other potential flatworm descendents possess.

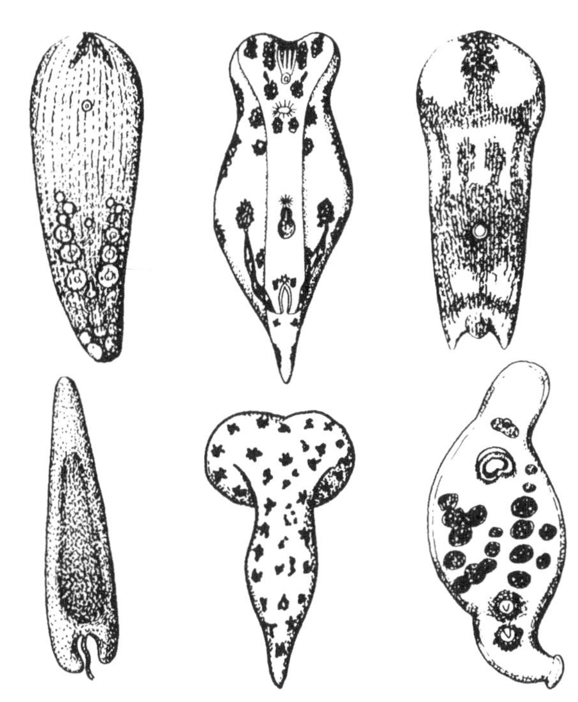

Fig. 2.6 The body form of acoel turbellarians (after Hyman, 1951).

Like the radiates, the larger flatworms are almost exclusively carnivorous, whilst the smaller species are consumers of bacteria and heterotrophic protists. This may appear most paradoxical: and two prime candidates for the surviving animal that most closely resembles the ancestral form are themselves typically consumers of other animals or more generally of non-photosynthetic organisms (except via intracellular symbioses)! In fact the paradox is an illusion. In part it derives from the dated notion that all organisms are either plant or animal, so that being the primary producers plants have to precede animals in appearance, whereas it is now considered that bacteria, protists and fungi are neither animals nor plants. And in part it derives from the notion that the basic form of ecological food chains is what we apparently see around us, running plant → herbivore → carnivore. Although over most of the planet photosynthesis is clearly the ultimate source of most fixed energy, few terrestrial invertebrates and even fewer vertebrates can actually digest plant material. Most can benefit from it only after it has been processed by bacteria, heterotrophic protists or fungi, either via the decomposer food chain or by virtue of the possession of a culture of micro-organisms in their gut (see Chapter 9). Bacteria and protists were presumably the ancestral animal diet, and the larger descendents either maintained this ancestral diet or turned to feeding on larger individual food items, other animals. Digestion of plants unaided by bacteria or protists remains a rare animal achievement, and is largely based on structures such as fruits produced by the plants for animal consumption.

2.2.5 Placozoans

The only known species of placozoan looks and behaves like a large (up to 3-mm diameter), flat, flagellated amoeba. Its body is:

- without symmetry
- formed by a monolayer of cells surrounding a secreted matrix; the whole in effect comprising a plate with differentiation of the cells on the upper and lower surfaces
- without a nervous system
- without tissues or organs.

Placozoans can move in any direction, can change shape (partly as a result of the presence of fibre cells in the matrix), and glide across surfaces feeding on protists. They contain less DNA than any other animal.

The most likely origin of these enigmatic and little known animals is that they are paedomorphic sponge or cnidarian larvae, but little is known of them and an independent choanoflagellate ancestry is possible.

2.2.6 Rhombozoans

Rhombozoans are minute, vermiform endoparasites of the excretory organs of cephalopod molluscs. Their bodies are:

- helically symmetrical
- formed by a monolayer of cells surrounding a single elongate axial cell
- without a nervous system
- without tissues or organs.

The outer layer of ciliated cells that surrounds the single axial reproductive cell numbers only up to 30 – so that rhombozoans contain fewer cells than any other animal. On the basis of their ribosomal RNA, they are the most primitive of all animals. Their ciliated cells and worm-like body shape have suggested to some that they originated from within the flatworms and that their extreme bodily simplicity is a consequence of their parasitic or commensal nature. Many platyhelminths are indeed parasitic and all such are, except for their reproductive systems and life cycles, secondarily simplified in anatomy – lacking a sense organ bearing head, for example. But none is anywhere near as simplified as are the rhombozoans, which really only share with the flatworms their cilia and vermiformity, features common to many animals. Their symmetry, peculiar cellular construction – including the development of reproductive propagules intracellularly within the axial cell – and highly individual life cycle are without parallel in flatworms, or other animals, and a flatworm ancestry seems no more likely than any other possible scenario. If they are not related to other worms, then an origin must presumably be sought within the cilia-bearing protists, but little relevant information is yet available on this small and enigmatic group. Hox gene data, however, does suggest platyhelminthe affinity.

Clearly the fact that they are confined to cephalopod molluscs strongly suggests that they arose relatively late in animal evolution, and no other types of animal are likely to have descended from them.

2.2.7 Conclusions

The five groups of animals considered above are all simple in construction, but are organized around fundamentally different symmetries and forms of basic architecture. All could claim an ancestry within the protists, although secondary simplification is possible for three of them (flatworms, placozoans and rhombozoans). Be that as it may, only the flatworms share the same form of body architecture as the other animals and if a flatworm-like animal was not the ancestor the flatworms are nevertheless related to other bilaterally symmetrical groups. On present evidence it seems likely that something like an acoel flatworm was the ancestor, but a cnidarian origin cannot be ruled out. It is relevant to later discussion to note here a few features of flatworm development. The platyhelminths are hermaphrodite and most show spiral cleavage of the cells that will form the blastula (the cleavage plane being oblique to the blastula's polar axis), development is determinate (the fate of the cells is fixed at a very early blastula stage, usually when only four cells have been produced by asexual division of the zygote), and mesoderm is formed from the 4d blastula cell (see Chapter 15). The acoels, on the other hand, exhibit biradial cleavage (see p. 471), indeterminate development (the fate of the cells is fixed relatively late in development), and mesoderm is formed from the embryological endoderm.

2.3 The bilaterally-symmetrical animals

Some seven major evolutionary lines of bilaterally-symmetrical animals are generally recognised to exist, together with one phylum that is unplaceable on current knowledge. These can be considered to represent 'superphyla'. From what was said in Section 2.1, it follows that these seven major groupings may be difficult to define in terms of their anatomy. Their identity may have been disclosed by molecular sequence data or by their possession of one or a few unique, shared derived features. In Part 2 of this book, we will adopt a classical, morphological approach and treat all animals that have a similar structure together; in this section, however, we will introduce the same animals in their phylogenetic context, and then discuss in Section 2.4 how the seven groupings may be related.

Six of the seven groupings contain animals that could be described as worms, where a worm is any legless and soft-bodied animal with a length greater than some two to three times its breadth. The worms do not therefore form a natural group, as might be expected from a group largely defined on a negative basis. They are the animals without legs, without an exoskeleton, without a shell, etc. If the ancestral animal was worm-like, as many phylogenetic schemes postulate, it would generally follow that a variety of worms might evolve from that vermiform ancestor, and that ultimately some of these worms might give rise to non-worms. But, as has been pointed out above, there are those who would read the sequence in the opposite direction and derive at least some worms from non-worms.

2.3.1 The acoelomate worms

The Gnathostomula and arguably the Gastrotricha are basically similar in their body form to the flatworms, but have evolved specializations that debar them from being included in the Platyhelminthes. The gnathostomulans share the derived features of monociliated epidermal cells (a feature also seen in some gastrotrichs, and in the lophophorates and deuterostomes) and the presence of jaws in the pharynx. The gastrotrichs have evolved a through gut and a peculiar two-layered non-moulted cuticle, the outer layer of which covers each cilium in a thin sleeve. The affinities of the gastrotrichs are contentious and there is little available information on the gnathostomulans.

2.3.2 The trochatans or syndermatans

The Rotifera and Acanthocephala share the derived feature of an intracellular cuticle located within the epidermis. It is fairly clear that the parasitic and secondarily gutless acanthocephalans are likely to have derived from the rotifers in that they and the bdelloid rotifers also share the presence of infoldings of the epidermis termed lemnisci and a proboscis of individual type. Alone amongst animals, rotifers do not seem to have the ability to synthesize collagen.

Both also have fluid-filled spaces within the body, a feature possessed by all the remaining bilateral animals. Such cavities are of many different forms and embryological origins, and they have probably been evolved independently by several separate lineages of vermiform animals. Broadly, however, three general developmental types can be distinguished (Fig. 2.7): (i) 'pseudocoels', often formed from a persistent blastocoel (the cavity within the blastula that is usually almost obliterated during gastrulation) and sometimes forming the blood system in which cases dilated regions are termed 'haemocoels'; (ii) 'schizocoels', formed within blocks of mesodermal cells by cavitation; and (iii) 'enterocoels', formed as outpocketings of the archenteron or embryonic gut. Such spaces within the body may be large and form hydrostatic skeletons, or they may be small and restricted to the cavities within certain organs or to the cavities within which certain organs, like the heart, are located. In some, they only appear fleetingly during development. Schizocoely and enterocoely are alternative methods by which a 'coelom' can be formed, where a coelom is a cavity within mesodermal tissues that is characteristically bounded by a mesodermal membrane, the peritoneum, although this lining is absent in several lines that are otherwise coelomate (e.g. many annelids, lophophorates and echinoderms). In the trochatans, the body cavity is a pseudocoelom.

Although they have an extracellular cuticle and no intracellular one, the gastrotrichs may provide a link between the acoelomate worms and these trochatans, although other systems of affinity have been proposed.

2.3.3 The nemathelminth or aschelminth worms

The phyla Priapula, Kinorhyncha, Loricifera, Nematoda and Nematomorpha generally share a large number of features: (i) the occurrence of a moulted cuticle and the corresponding absence of epidermal cilia; (ii) the possession of a single, pseudocoelomic body cavity, which in some may be reduced to a system of interstitial spaces; (iii) distinctive types of pharyngeal and body wall construction; (iv) a through gut with a terminal or subterminal anus; (v) the absence of asexual multiplication and of any marked ability to regenerate lost bodily regions; (vi) the absence from the life history of a larval stage (although the juvenile may differ from the adult in morphology); (vii) the lack of a blood system (although the body cavity may carry out this function); (viii) determinate development, with very early differentiation of the future germ cells and the formation of mesoderm from the rim of the blastopore; (ix) separate sexes; (x) an asymmetrical form of cleavage that is neither radial nor spiral; (xi) internal fertilization via copulation; (xii) the presence of a protonephridial osmoregulatory/excretory system; (xiii) projecting sense organs that terminate in small papillae ('flosculae'); (xiv) small size (except in some parasitic forms); (xv) a collar-shaped circumpharyngeal brain; and (xvi) at some stage of their life history and in at least some species, an anterior region of the body specialized to be forcibly everted, by increase in pseudocoelomic pressure, and retracted by longitudinal muscles. On eversion, this 'introvert' bears radially-symmetrical whorls of cuticular and epidermal processes ('scalids') taking a variety of shapes – spines, clubs, hooks, scales or even feathers – that have sensory, penetrant or food-capturing functions. Some also share (xvi) a terminal mouth cone with stylets.

Opinions differ as to the basal members of this group, there being two possible candidates. The gastrotrichs do share some similarities with nematodes and they may provide a link to those worms without moulted cuticles or body cavities. On the other hand, the priapulans are unusual amongst nemathelminths in being relatively large and in displaying external fertilization; their body cavities may even be coelomic like those of larger worms. Accordingly, since external fertilization is generally held to be the ancestral animal condition, the priapulans are seen by several as being close to the ancestral state and they would derive the other nemathelminths from larger, coelomate worms via paedomorphosis.

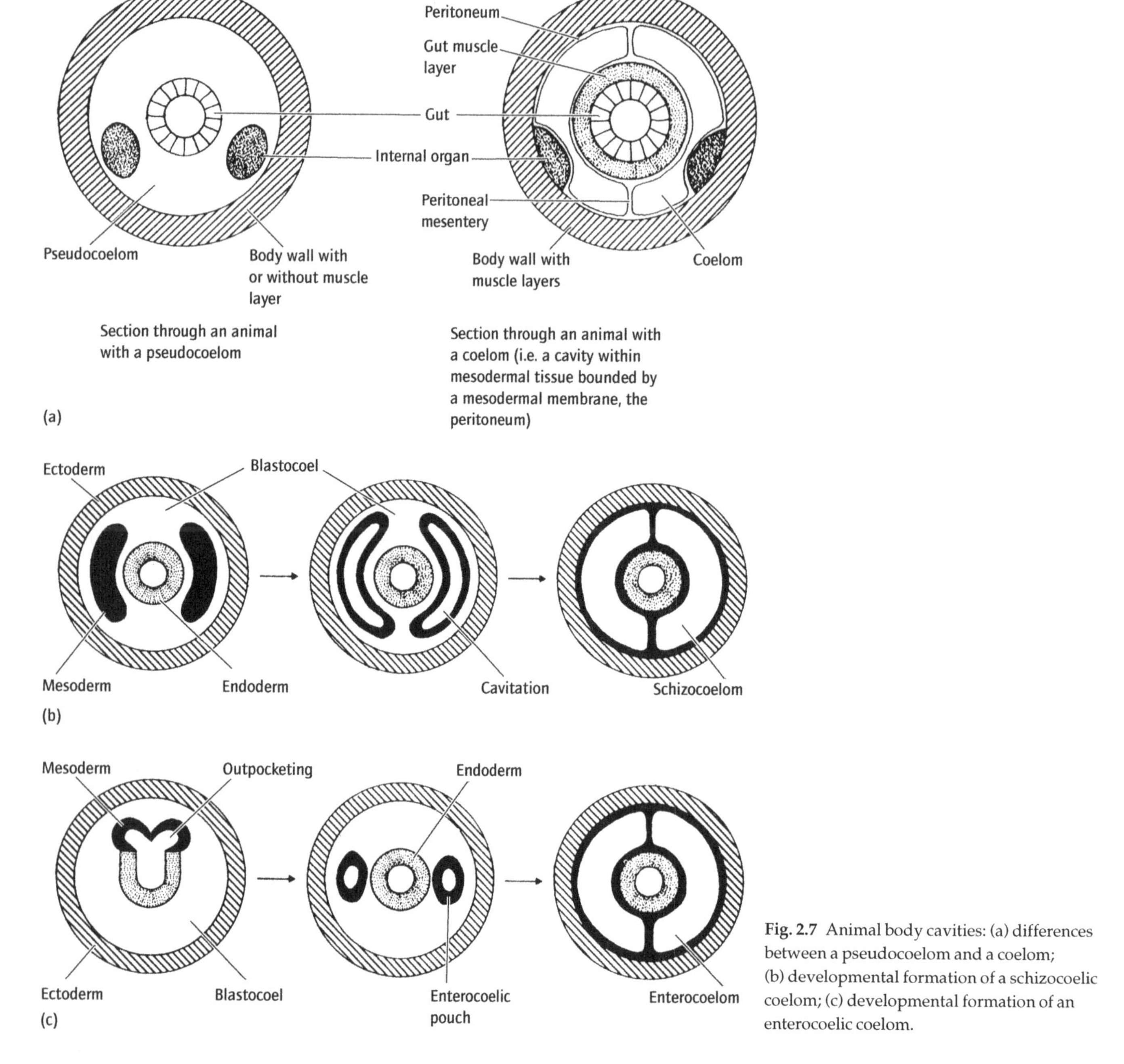

Fig. 2.7 Animal body cavities: (a) differences between a pseudocoelom and a coelom; (b) developmental formation of a schizocoelic coelom; (c) developmental formation of an enterocoelic coelom.

Many of the nemathelminths are small species inhabiting interstitial spaces in aquatic sediments and in the bodies of other organisms. In company with the equally small trochatans, these species often show eutely – the state in which cell number does not increase after development so that it achieves a finite and fixed species-specific level. Further growth is then only by increase in cell size. Indeed, several of the nemathelminth features are those typifying small, interstitial species and have been viewed more as convergent or parallel adaptations to that life style than as being indicative of affinity. The pseudocoelomate condition itself is one such. The introvert, however, is a very characteristic shared derived feature.

Lacking the ancestral cilia and being covered by a cuticle, movement is achieved in novel ways. Species with introverts may evert this anterior region of the body, the scalids on which then anchor it in everted form, and contraction of the longitudinal muscles will pull the rest of the body over its introvert. Almost all nematodes and all adult nematomorphs lack an introvert and they also lack circular muscles (which are mainly associated with generating the introvert-everting pressure). Movement is then effected by coupling contraction of longitudinal muscles with a body that cannot be shortened – effectively the same system as in those chordates without legs. This results in a series of C- or S-shaped thrashings in the dorsoventral plane (in chordates it is in the lateral plane). The kinorhynchs are notable in that their entire body wall, including the cuticle, is segmented, there being 13 or 14 segments, although their locomotion appears to rely solely on use of the introvert.

2.3.4 The eutrochozoan worms and molluscs

The unity of the disparate eutrochozoan phyla (the Nemertea, Mollusca, Sipuncula, Echiura, Annelida, Pogonophora, Entoprocta and Cycliophora) rests largely on molecular sequence data. They do, however, share four developmental features, two of which are also seen in most (i.e. non-acoel) flatworms: the dividing cells that will form the blastula cleave in a spiral manner; and the fate of these cells is fixed by the 16-cell stage, with the future mesoderm deriving from the 4d cell. In addition, their larval stage is a trochophore, and major or minor body cavities – forming the spaces within which certain organs are housed through to large hydrostatic skeletal systems – are with a few exceptions produced schizocoelically. Further, blood systems are typically present, as are through guts, metanephridial excretory organs, tracts of cilia over at least some regions of the body surface, and the sexes are separate although secondary hermaphroditism is widespread. The similarity of the non-acoelomorph Platyhelminthes (with the possible exception of the catenulids) to the Eutrochozoa has been confirmed by molecular sequence data.

Although, therefore, a number of features are held in common, adult morphology varies very widely. Some (the molluscs and nemertines) are essentially acoelomate and have fluid-filled schizocoelomic cavities only in association with a single organ, that of the nemertines housing a long, harpoon-like 'proboscis' that opens near the mouth and that, like the nemathelminth introvert, can be shot out hydraulically and retracted by a longitudinal muscle. Amongst worms, this shared, derived, prey-capture organ is unusual in not being part of the gut or a specialized anterior part of the body. Others (echiurans, sipunculans, annelids and pogonophorans) possess large schizocoelomic hydrostatic skeletons. The body cavity of the entoprocts is a pseudocoel. Most (including the nemertines, molluscs, entoprocts, echiurans and sipunculans) are monomeric and without trace of segmentation, whilst others (e.g. the annelids and pogonophorans) are metamerically segmented. The entoprocts are probably on the periphery of this assemblage, and the cycliophorans are only included because their only arguable affinities are with the entoprocts.

Most of the body of the pogonophorans comprises only three metameric segments, although the small posterior holdfast region contains up to 30. That of the annelids, on the other hand, forms a linear chain of more or less equi-sized segments between the pre-segmental prostomium and the post-segmental pygidium, from in front of which they are budded off. Each segment forms a separate functional unit, being primitively at least isolated from that in front and that behind by a septum, and containing a separate pair of body cavities together with a full complement of body organs including the excretory metanephridia (see Fig. 2.8). This arrangement permits great localization of changes in body shape, and the adoption of vigorous burrowing. The contraction of longitudinal muscles in some segments (and the relaxation of the circulars) causes those segments to increase in diameter, anchoring the worm against the side of the burrow, whilst elsewhere along the body contraction of the circulars (and relaxation of the longitudinals) can extend the affected region, so as for example to move it forwards. Anchorage is aided by the possession of chitinous setae projecting from the body wall, features also present in the pogonophorans and the unsegmented echiurans, suggesting that these three phyla are particularly closely related. Indeed, molecular evidence strongly suggests that the pogonophorans are really highly modified polychaete annelids.

All but one group of the eutrochozoans have remained worms or worm-like. The molluscs, however, are characterized not only by their shared derived feature of a chitinous, toothed, tongue-like ribbon, the radula, in their buccal cavity (permitting algal and other foods to be rasped from hard surfaces) but by the deposition of calcium carbonate over much of their body surface. Many animals can deposit such hard protective material, but with the exception of the molluscs all have coupled such protection with a sessile or sedentary life style: they have encased themselves within a tube or box. The molluscs, on the other hand, have achieved both protection and mobility at the same time: they are the crawling (or in some the swimming) armoured. They can be envisaged as flatworm-like animals that have covered their dorsal surface – the only surface exposed to predators – in calcium carbonate spicules or plates whilst retaining the broad, flat, locomotory ventral foot. That dorsal surface, however, was also the only one exposed to the ambient water and therefore the site of, for example, gaseous exchanges with the environment. Hence hand-in-hand with the development of the dorsal protective covering must have occurred the elaboration of some region of the integument into a system of greatly increased surface area, together with a circulatory system to distribute and collect the respiratory gases to and from the tissues. The paired gills of the molluscs are of an individual and distinctive type ('ctenidia'), and they are housed beneath the protection of the shell in a mantle cavity. Various extinct groups, the Hyolitha, Wiwaxiida and Halkieriida (see Fig. 2.21), appear to have reacted to the selective pressures favouring a protective body covering in a similar manner to the molluscs (see Fig. 2.21). The combination of a rasping radula, a shell (which can be used to resist dehydration), and a mantle cavity (the wall of which can be vascularized to form a lung) pre-adapted the molluscs to life on land, and they are the most successful eutrochozoan group. The annelids also colonized the continents, but their hydrostatic skeletons and permeable integuments have restricted them to moist if not downright wet habitats.

The sister group of the molluscs seems to be the sipunculan worms, and two groups of shell-less molluscs (the chaetodermomorphs and neomeniomorphs) are elongate and very worm-like, with well-developed body wall musculature and a protective covering of individual spicules embedded in a

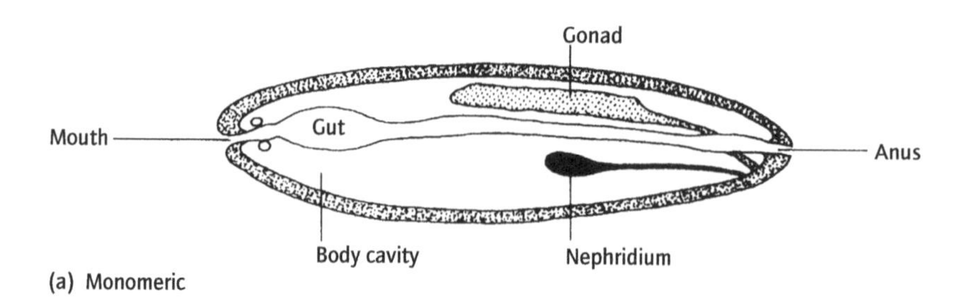

(a) Monomeric

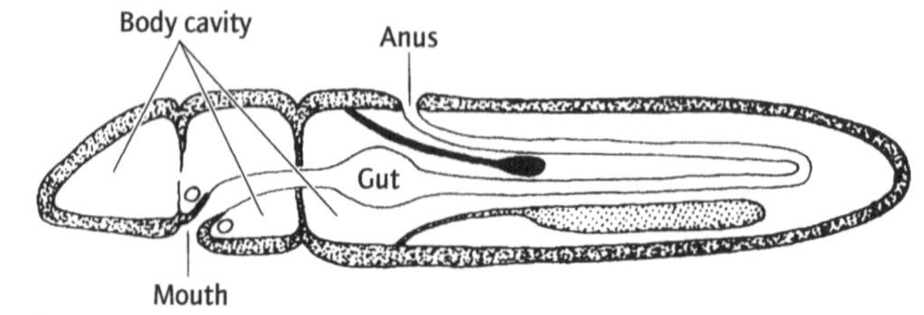

(b) Oligomeric

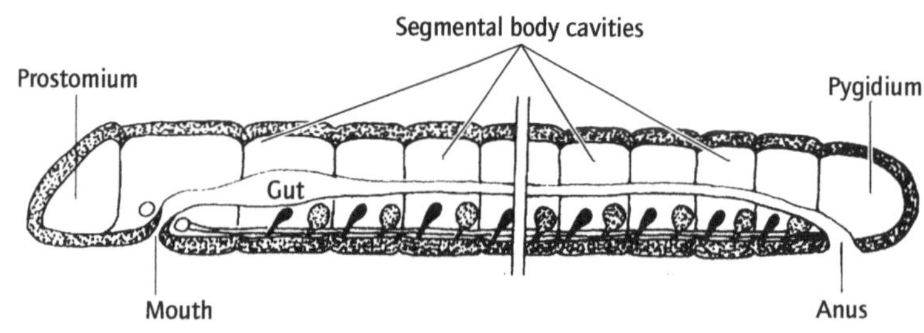

(c) Metameric (as in annelids)

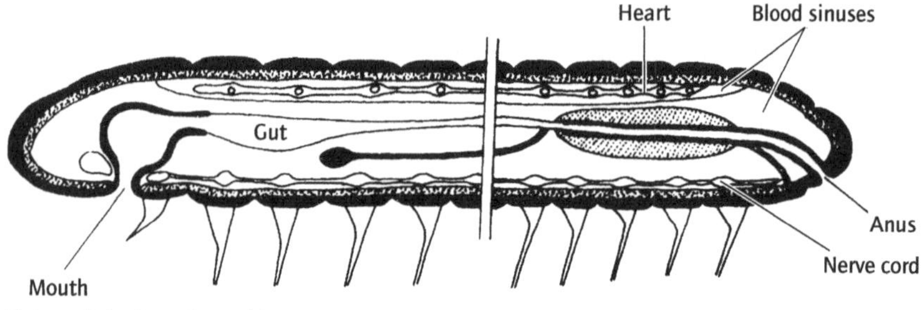

(d) Metameric (as in arthropods)

Fig. 2.8 Fundamentally different vermiform body plans in diagrammatic longitudinal section.

chitinous cuticle. These have been regarded as being close to the vermiform ancestor of the molluscs (as well – almost needless to say – as secondarily simplified, paedomorphic derivatives of shelled and non-worm-like forms).

2.3.5 The panarthropods: arthropods and lobopods

The arthropods and the lobopods, which can be united as the 'panarthropods', are the invertebrates with legs, and they comprise some 75% of all described animal species. All also share the presence of a moulted cuticle and a largely haemocoelic body cavity. The essential differences between a monomeric haemocoelomic worm and an arthropod are that the latter has serially repeated legs and is covered by a hard, jointed exoskeleton. These differences are not as major as might appear at first sight. Of the two it is likely that legs evolved before the exoskeleton. Some living animals, i.e. the lobopod Onychophora and Tardigrada, are soft-bodied, rather worm-like animals with ventral or near ventral, unjointed legs and a pseudocoelomic/haemocoelic hydrostatic skeleton. The onychophorans are covered by a flexible chitinous cuticle, whilst the tardigrades have external cuticular plates. It is therefore relatively easy to visualize how toughened regions of the pre-existing cuticle could extend to cover the body completely, necessarily with articulatory joints between the various sets of serially repeated plates (as seen, for example, in the nemathelminth kinorhynchs) and between the different articles of the limbs, and then be utilized as a hard skeleton in place of the original hydrostatic one. Extension of the limbs of arthropods may still be achieved hydraulically and only leg flexure be by means of the exoskeleton/muscular system.

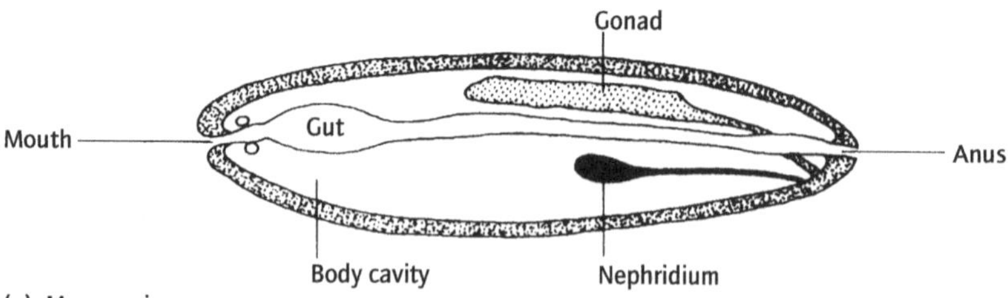

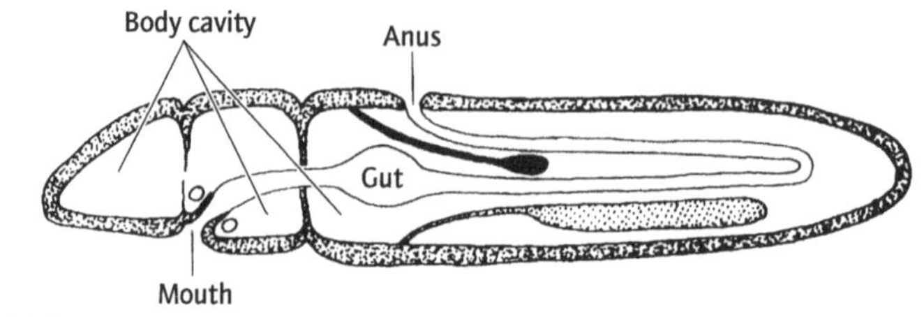

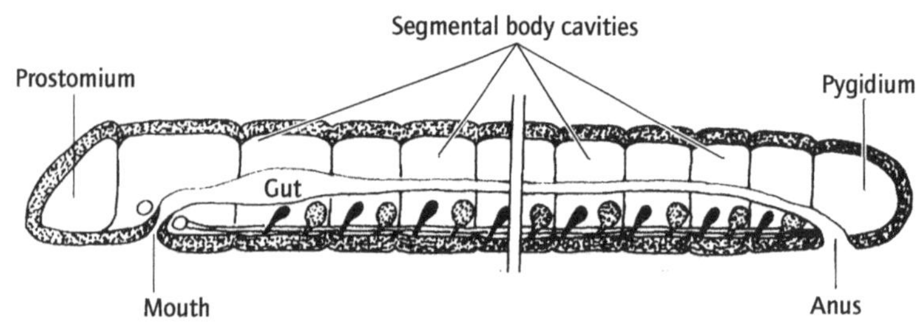

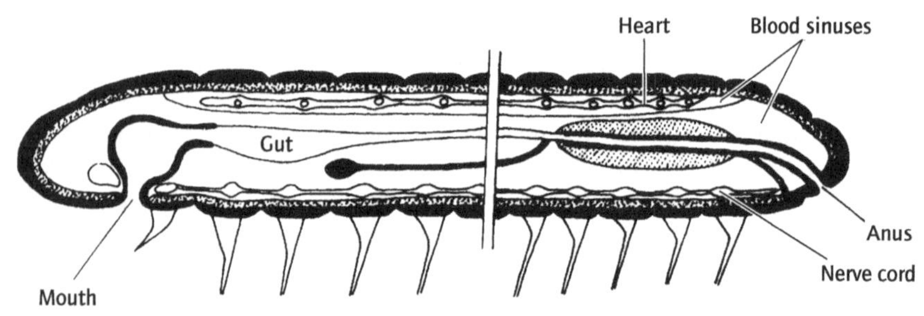

The nature of lobopod and arthropod segmentation is thus essentially different from that of the annelids (see above), but of the same form as that of the kinorhynchs and vertebrates. It is based on the serial repetition of pairs of limbs and of limb-associated skeletal, muscular, nervous and vascular elements along a monomeric, non-compartmented body (see Fig. 2.8) (in respect of the vertebrates, for 'limbs' read the 'muscle blocks powering swimming'). In both the annelids and the arthropods, however, these segments form a linear chain between a pre-segmental anterior region (the acron in arthropods) and a post-segmental posterior portion (telson), in front of which they are produced during development or throughout life. The limbs of newly produced segments take some time to develop to full size, so that the rear end of the body may have only partially developed limbs or limb buds.

Lobopods are now known to have been both abundant and diverse in the Cambrian (see for example, Fig. 2.21 g & i), and the arthropods themselves seem rapidly to have radiated (Fig. 2.9) into numerous types (Fig. 2.10), many of them extremely bizarre. Few of the many Cambrian arthropods outlived that period, however, and only five groups survived to beyond the Devonian. One of these, the trilobites, having survived for 300 million years at least, became extinct at the end of the Permian, leaving only four groups alive today: the Chelicerata, Crustacea and Uniramian myriapods and hexapods. Although all four groups are clearly arthropods in that they have a cuticular exoskeleton and jointed limbs, they show a number of differences particularly in the nature and form of their appendages (presence or absence of antennae, jaws, etc.), in their respiratory and excretory organs, and in their embryology. Thus the development of tracheal systems would appear to have proceeded convergently, and so it has been argued have the evolution of compound eyes and Malpighian tubules. Nevertheless, many people would unite them in a single phylum Arthropoda and there is no doubt that at some level they are a related assemblage. Whether they are all descended from an animal that was itself an arthropod is another question, however. It is a matter not dissimilar to the possible relationship between the choanoflagellates and the ancestral animals (Section 2.1.1). The arthropod phyla presumably arose from the lobopods that cluster with them in molecular analyses, but not necessarily from the same lobopod. Even the panarthropods may not be a natural group, in that sequence data indicate that included within this group are also some nemathelminth worms (see, for example, Fig. 2.13), and so inclusion of the lobopods within the Arthropoda would not be an admissible procedure to create a monophyletic assemblage either. The picture has been further complicated recently by molecular and gene-sequence data that show that the two major components of the hitherto unitary Uniramia, the Myriapoda and Hexapoda, may each be more closely related to other arthropod groups than they are to each other, notwithstanding their great morphological similarity. On these bases, the myriapods are related to the chelicerates, whilst the hexapod insects are to the crustaceans (although on some

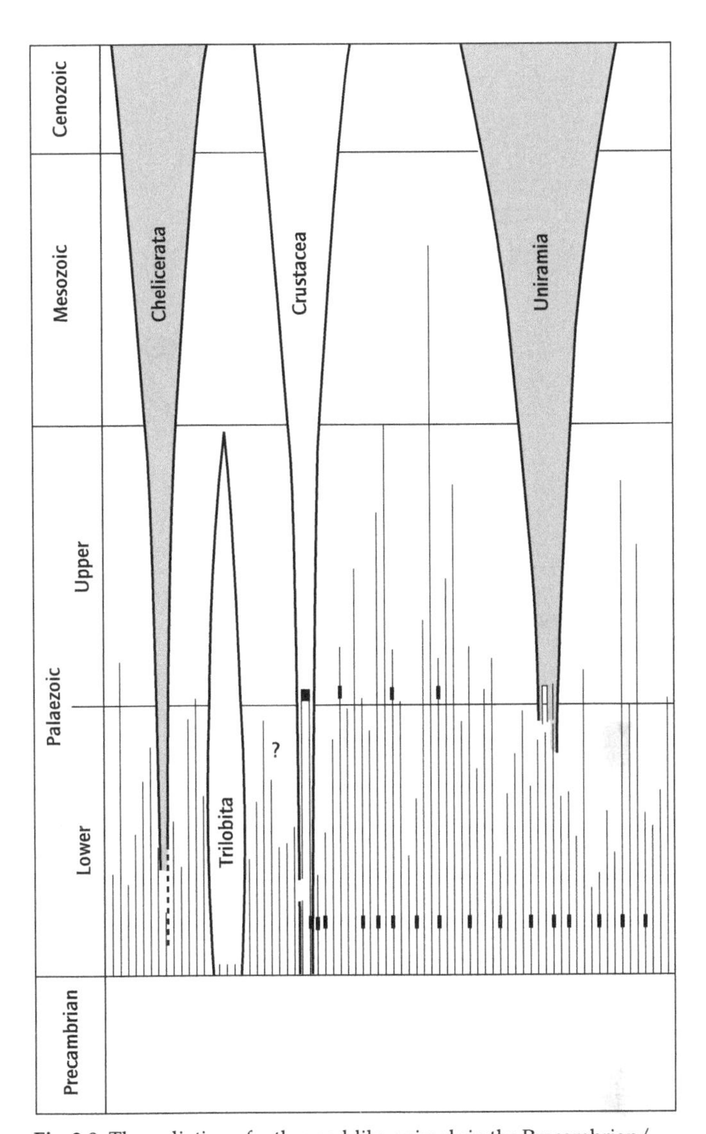

Fig. 2.9 The radiation of arthropod-like animals in the Precambrian/Cambrian (after Whittington, 1979).

analyses no more closely than they are to nematodes), and the myriapods + chelicerates are relatively distantly related to other arthropods. These molecular results appear to clash with cladistic analyses. Cladistically, the arthropods – including the diverse Cambrian forms – are a monophyletic group. Unfortunately all too often this conclusion is an automatic consequence of the data used, in that either only arthropods are included in the data set together with an inappropriate outgroup (e.g. the annelids) or else only a few lobopod groups and no nemathelminths are also included. The Pentastoma – worm-like parasites – have long been allied with the lobopod phyla, and this does adequately reflect their anatomy, but it now appears most likely that this is secondary and that they are really degenerate branchiuran crustaceans.

Whatever their origin, however, the ancestral forms of all the surviving lines of arthropods were probably elongate, bottom-dwelling animals with many segments, each with one pair of

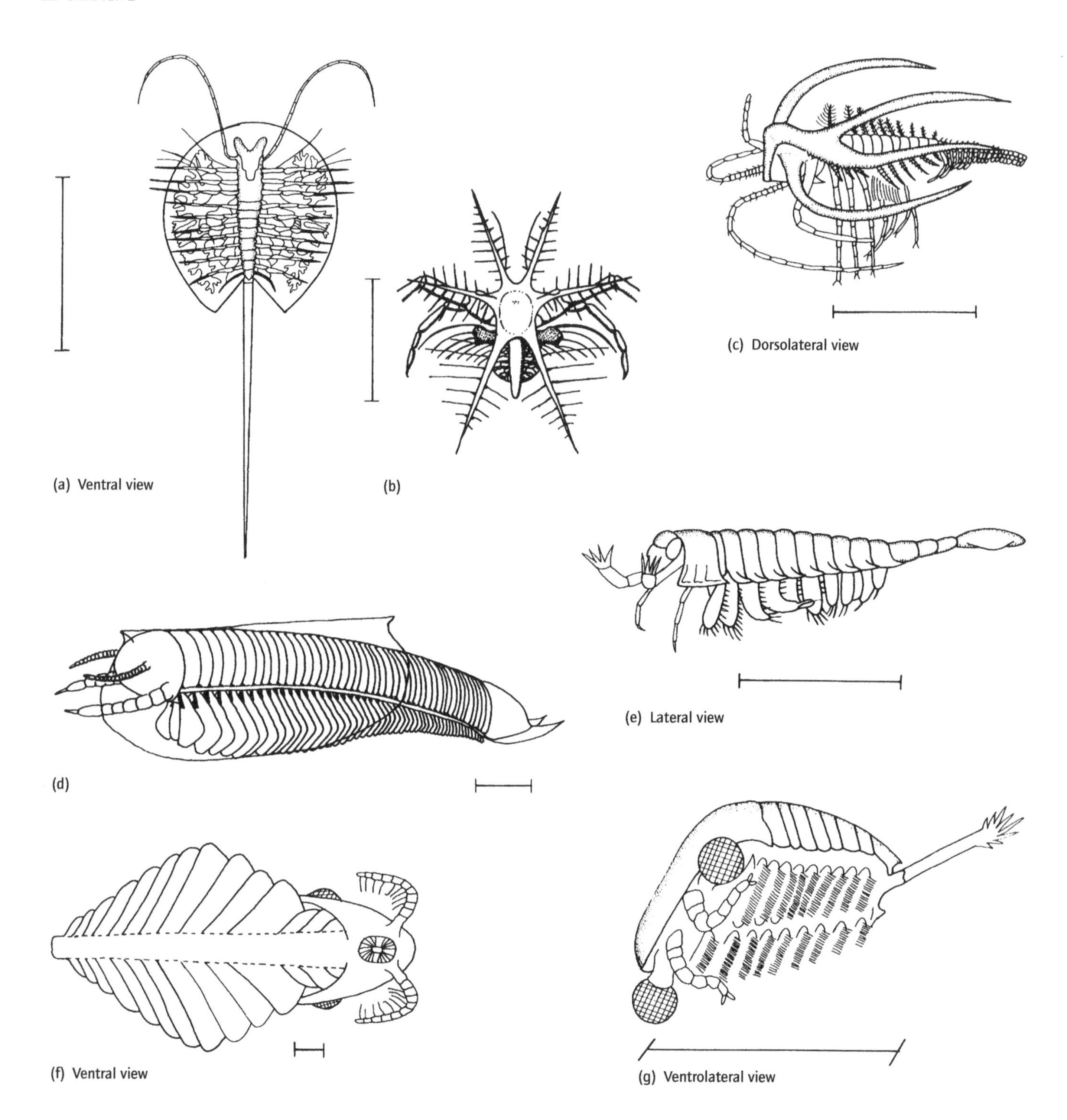

Fig. 2.10 Arthropod-like animals belonging to phylogenetic lines other than those of the Trilobita, Chelicerata, Crustacea or Uniramia: (a) *Burgessia* (Cambrian); (b) *Mimetaster* (Devonian); (c) *Marrella* (Cambrian); (d) *Branchiocaris* (Cambrian); (e) *Yohoia* (Cambrian); (f) *Anomalocaris* (Cambrian); (g) *Sarotrocercus* (Cambrian). (After Manton & Anderson, 1979 and Whittington, 1985.) Scale lines: 1 cm.

appendages, those of the chelicerates and crustaceans being marine in habitat and that of the insects, if not the myriapods as well, being terrestrial. Many members of all arthropod clades have retained this elongate, ancestral shape with various degrees of differentiation of regions of the body and their associated appendages to form a head and trunk, a head, thorax and abdomen, a prosoma and opisthosoma, or whatever. Many descendent lines, however, have greatly reduced the number of segments, by loss or fusion, and to an even greater extent the number of limb pairs (the terminal segments of juvenile arthropods possessing incompletely developed limbs – see above). The origin of these descendent clades, e.g. those of the marine and freshwater planktonic and interstitial maxillopod crustaceans, and of the uniramian insects, has been suggested to have been via paedomorphosis, such that the copepods, for example, are the paedomorphic larvae of large, elongate, bottom-dwelling crustaceans in ancestry.

2.3.6 The lophophorates

Traditionally, the three lophophorate phyla (the Phorona, Brachiopoda and Bryozoa) have not only been regarded as being closely related – so much so that some would unite them into the single phylum Lophophorata – but have been considered to be a link between the groups discussed above in Sections 2.3.1–2.3.5 and the deuterostomes to be treated next in Section 2.3.7. The Phorona and Brachiopoda are still regarded as being closely related, but the affinities of the Bryozoa, which have always been contentious, as well as the relationship with the deuterostomes have been called into question by molecular sequence data and, to a lesser extent, by cladistic analysis.

All the animal groups considered thus far have been protostome, that is the embryonic blastopore forms their mouth, and so it does in the lophophorates, and also like the protostomes chitin is common in the form of secreted tubes or bristles. These same protostome groups also tend to show the following suite of embryological features that are held in common with non-acoel flatworms: spiral cleavage of the blastula cells (or something that could be considered a highly modified version of spiral cleavage); determinate development; and development of a coelomic cavity, if one occurs, by schizocoely. The phoronans and brachiopods, however, like the deuterostomes, exhibit radial cleavage, indeterminate development and develop coelomic cavities by enterocoely (see Fig. 2.7). Further, like the deuterostome pterobranchs, they have oligomeric tripartite bodies (Fig. 2.8) comprising a small anterior prosome (absent in the brachiopods), a larger mesosome, the enterocoelic cavities of which support a lophophore hydrostatically, and a large posterior metasome that contains virtually all the body organs, all three compartments being separated by septa. The characteristic lophophore is a series of hollow, ciliated, tentacle-like outgrowths of the body wall encircling the mouth that is used in suspension feeding. The enterocoels of the three (or two) body segments interconnect via pores through the septa. Other common features include: larvae with an upstream food-collection system of tracts of monociliated cells around the mouth which, on metamorphosis, are transformed into the equally monociliated cilia of the lophophoral arms; a mesosomal brain; and mesoderm derived from the embryonic endoderm. If these lophophorates and the deuterostomes are in fact unrelated, this is an impressive list of convergent features, to say the least.

Part of the reason why the position of the bryozoans has been contentious is that there is no continuity between their embryonic structures and those of the adult. Larval tissues break down during metamorphosis and the adult body cavities, for example, are formed *de novo*. The bryozoans, however, have a bi- or tripartite body with a lophophore supported by the body cavity of the mesosomal compartment, just like the phoronans and brachiopods.

The phoronans are small, tube-dwelling worms with an effectively terminal lophophore, whilst the brachiopods are covered by a large, bivalved, calcareous or phosphatic shell, equivalent to that of the bivalve molluscs although in this group the two shells are dorsal and ventral, not lateral. On the basis of a general evolutionary progression from worms to non-worms, it was generally considered that the phoronans had given rise to the brachiopods, possibly polyphyletically (Fig. 2.11). The contrary scenario that the phoronans are paedomorphic brachiopods has more recently been argued. Some living phoronans can multiply asexually by budding and this is the norm in the bryozoans. In the latter, colonies sometimes containing thousands of interconnected modules ('zooids') are formed by repeated asexual budding of the founding ancestrula and of its descendents. These zooids are frequently polymorphic. It is normally the case that individual colonial modules are very much smaller than the unitary individual animals in groups to which they are most closely related (cf. coral polyps and sea-anemones or colonial sea-squirt zooids and solitary sea-squirts), and such is the position in the lophophorate groups too. Bryozoan zooids, which are each encased in a calcareous box or tube or gelatinous matrix, apart from an aperture through which the lophophore can be extended, are much smaller than individual phoronans or brachiopods (if indeed the groups are related).

2.3.7 The deuterostomes

It was noted above that in contrast to the protostome phyla, the deuterostomes show radial cleavage, indeterminate development, and enterocoelic formation of coelomic body cavities. And, of course, they are deuterostomatous, i.e. the embryonic blastopore does not form the mouth, which is a secondary orifice, although it may form the anus. In addition, they differ from most protostomes in having ciliary photoreceptors, use of creatine phosphate as the phosphate store, and in the absence of chitin. Groups that can be considered to display the basal type of anatomy also have a tripartite, oligomeric body plan with a pair of enterocoels in each of the three body compartments, gonads formed by temporary aggregations of cells lining the body cavities, a diffuse, subepidermal nervous system, a monociliated epidermis, and the ability to multiply asexually by budding or fission.

The three contained phyla (the Hemichordata, Echinodermata and Chordata) therefore form a closely knit group, sharing a large number of features, but paradoxically relatively little is known of their interrelationships. When some or all of the lophophorates were generally regarded as being allied to the deuterostomes, the origin of the group as a whole seemed clear

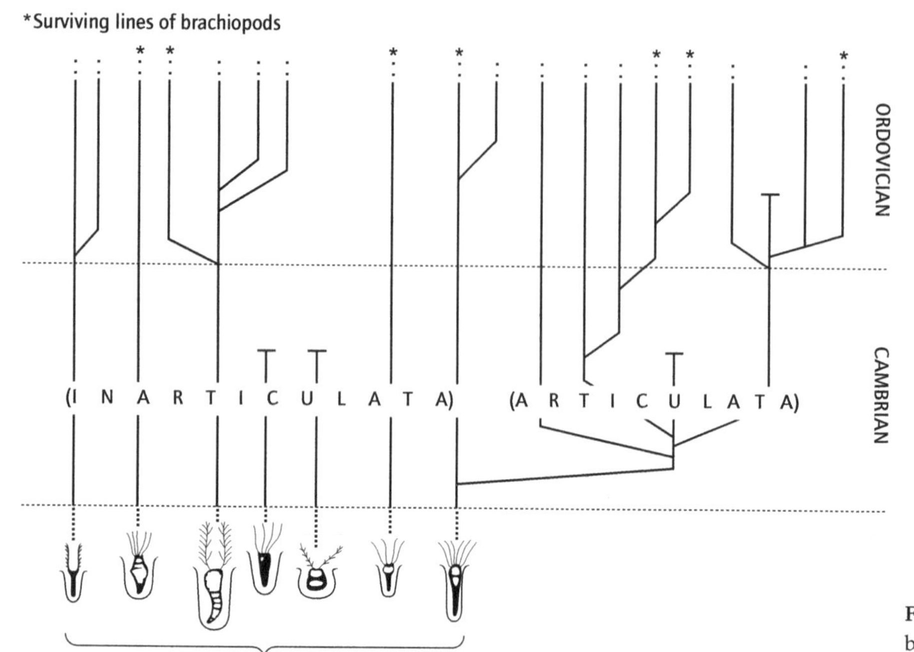

Fig. 2.11 Possible polyphyletic origin of the brachiopods as a wave of 'brachiopodization' of various phoronan-like worms in the Precambrian/Cambrian (after Wright, 1979).

in that the lophophorate pterobranch hemichordates are very similar in basic structure to the phoronans. The only major differences between the two are that the lophophore of the pterobranchs flares away from the mouth, instead of encircling it, their embryonic blastopore does not form the mouth, and their prosomal body compartment is large and used in locomotion. Even if they are not as closely related to the lophophorate phyla as was once thought, the pterobranchs may still provide a model of what the basal deuterostome stock looked like. Some pterobranchs also show a feature that has played an important role in several deuterostome lines: two pores through the body wall connect the lumen of their pharynx to the external environment. The water current entering the mouth with food captured by the lophophoral tentacles can then leave through these pores whilst the food particles continue down the gut, thereby creating an efficient unidirectional flow. This pore system has been elaborated in the lophophore-less, worm-like, enteropneust hemichordates, in the invertebrate chordates and in the vertebrates as an internal filter for suspension feeding or to house internal gills for gaseous exchange; they may even have been present in some echinoderms (Fig. 2.12). Suspension-feeding appears to be the hallmark of all deuterostome clades. (Some gastrotrichs have convergently evolved equivalent pharyngeal pores.)

Although appearing markedly dissimilar, largely as a result of their early adoption of a spherical body with a near radial form of symmetry (in association with their sessile life style), the echinoderms retain in their body cavities clear indications of the basic tripartite body organization of the hemichordates. One of the prosomal enterocoels forms the axial sinus; the other, together with one of the mesosomal cavities, forms the water vascular system; and the two metasomal enterocoels form the main body cavity. The echinoderms are one of the many animal phyla first to appear, clearly recognisable, in the Cambrian, although the early forms did not uniformly display the fivefold (pentaradial) symmetry that characterizes the surviving groups. The ancestral body plan is purely a matter of conjecture. One of the more plausible scenarios derives them from a sedentary pterobranch-like suspension feeder in which the lophophoral arms and their supporting hydraulic mesosomal cavities developed into the radial canal system and feeding podia (converted into the tube-foot system in later free-living clades).

Besides the common features displayed by all deuterostomes, echinoderms share with the chordates a hard, mesodermal, protective system of calcium-based plates in the dermis. Not surprisingly, therefore, some attempts to derive the chordates from another deuterostome group have implicated the echinoderms as being near to the point of origin (Fig. 2.12). But as with the hemichordate/echinoderm transition, all such suggested schemes have come up against a lack of potential intermediate stages; no living species can serve as appropriate analogies; and the fossil record is of no help since the chordates first appear at more or less the same time as the echinoderms. In spite of their clear affinities with the hemichordates and echinoderms, the chordates are a very distinctive and isolated group and their ancestry is unknown. Molecular sequence data fairly uniformly suggest that the two hemichordate groups are related to each other (despite several cladistic views to the contrary) and that the living hemichordates are the sister group of the echinoderms. The two phyla are then together the sister group of the chordates.

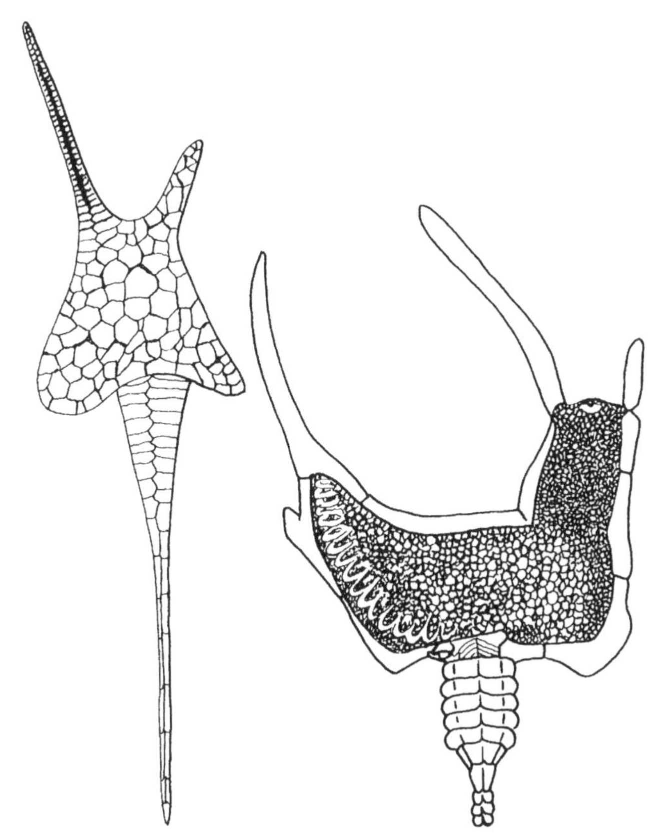

Fig. 2.12 Homalozoan echinoderms (after Clarkson, 1986): the ancestors of the chordates?

2.3.8 The chaetognaths

The chaetognaths are an enigma. The early development of these transparent, soft-bodied, planktonic worms appears typically deuterostome, and they possess the standard tripartite body plan in the form of a head, trunk and post-anal locomotory tail, each separated by a septum. Some of their body cavities originate enterocoelically, although the two tail cavities are secondary derivatives of the paired trunk pouches, and in the adult the cavities lack a mesodermal lining. Further, the body is covered by a cuticle, the spines around the mouth contain chitin, and the germ cells are differentiated very early in development – all non-deuterostome features. They have traditionally been accommodated in the deuterostomes, for want of anywhere better to place them. This is not supported by molecular or cladistic evidence, but to date that same evidence does not indicate affinity to any other larger grouping except possibly to the nemathelminths.

2.4 Interrelationships of the superphyla

If – which not all would accept – each of the acoelomates, trochatans, nemathelminths, eutrochozoans, panarthropods, lophophorates and deuterostomes represents a related group of bilaterally-symmetrical animals, with the chaetognaths as *incertae sedis*, the question of how they are likely to be related to each other remains. Traditionally, the trochatans and nemathelminths have been regarded as being related on the basis of their shared pseudocoelomic condition, eutely, etc. as have the eutrochozoans and panarthropods because of the occurrence in both groupings of segmentation and schizocoelomic cavities, and the argument for the relatedness of the lophophorates and deuterostomes was described above.

Recent molecular and cladistic studies and those concerned with hox genes, whilst not always arriving at identical schemes, have tended towards four conclusions that contrast with these traditional views. First, that the deuterostomes are a very isolated group, distinct from all protostomes, and that their radial cleavage, indeterminate development and formation of mesoderm from the endoderm may well be the ancestral bilateral condition, inherited directly from acoel-like ancestors. It was noted above that the acoel flatworms also share something approaching the latter three states. Second, that the two groups that possess a cuticle that is moulted at least once in their lives, the nemathelminth worms and the panarthropods, are related to each other, and can be united in the same clade, the Ecdysozoa, so that the annelids and arthropods, for example, are only distantly related. Third, that the eutrochozoans and the lophophorates are similarly related to each other and can be united as the Lophotrochozoa clade. Fourth, that the flatworms comprise three very different clades, the catenulids, the acoels and the rest (the Rhabditophora), of which the acoels may well be the sister group of all the other bilateral clades, and the catenulids may be the sister group of the protostomes. The rhabditophore flatworms, in contrast, cluster as an integral part of the Eutrochozoa. The trochatans have rarely been included in molecular studies, but some have shown the rotifers to be nested within the lophotrochozoans. Four recent cladistic and molecular phylogenies are portrayed in Figs 2.13–2.16, and a synthesis is presented in Fig. 2.17.

These phylogenies share the feature that the sponges, cnidarians, ctenophores, placozoans and (?) mesozoans are only distantly related to the bilateral groups, but debate on the nature and habitat of the basal animal stock remains. Three broad hypotheses survive. One, that basically presented here, is that the ancestral bilateral animal was a small flatworm-like species living on the sea bed; from such a benthic ancestor then derived the other protostomes and the deuterostomes. The second, the 'blastea–gastraea–trochaea theory', dating back to Haeckel (1874) and of which a modern version forms Fig. 2.18, is that the ancestral animal was a radially symmetrical, planktonic choanoflagellate colony, and that the various bilaterally symmetrical groups originated relatively late in evolution all from planktonic ancestors. The third, which stresses paedomorphosis, has an equally long pedigree, in this case back to Lankester (1875) and Sedgwick (1884). In this 'archicoelomate theory' (Fig. 2.19), the ancestral form was similar to the modern, colonial, benthic, anthozoan cnidarians. From such an animal arose directly colonial, benthic, oligomeric and enterocoelic groups, somewhat equivalent to the surviving pterobranch hemichordates, that

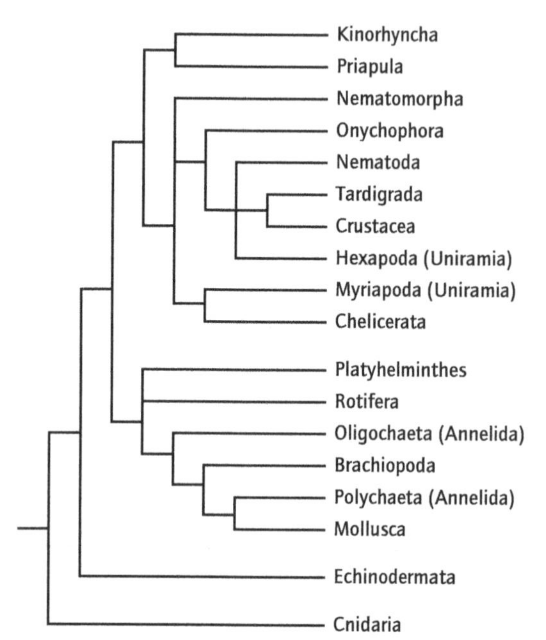

Fig. 2.13 Some phylogenetic relationships suggested by Aguinaldo *et al.* (1997) on the basis of sequence data from 18S ribosomal DNA. Note (a) that within the Ecdysozoa some nemathelminth worms appear within the panarthropod clade, (b) that the myriapod and chelicerate arthropods are less closely related to the hexapod and crustacean arthropods than the latter two are to the nematodes and tardigrades, and that the crustaceans are themselves more closely related to the lobopod tardigrades than they are to the chelicerates, (c) that the trochatan rotifers cluster with the Lophotrochozoa, and (d) that the polychaete annelids are more closely related to the molluscs and brachiopods than they are to the oligochaete annelids. (After Aguinaldo *et al.*, 1997, Figs 2 & 3.)

had acoelomate or pseudocoelomate larvae. The ancestral bilateral animal was thus relatively complex, segmented and with coelomic body cavities, and from this groups such as the flatworms, nemathelminths and molluscs arose via paedomorphosis. All three scenarios can be accommodated with minor modification within the broad phylogenetic picture that is emerging and the preferred scheme is still largely a matter of personal preference.

In spite of this broad measure of agreement, several areas of uncertainty remain. Are the rotifers (and through them the acanthocephalans) paedomorphic larvae of some group of eutrochozoans? Is the similarity of the myriapod to the hexapod uniramians entirely due to convergence? Similarly, why do the pterobranch deuterostomes so closely resemble the lophophorate phyla at both the gross- and fine-structural levels? What are the nearest relatives of the chaetognaths, and where do the affinities of the rhombozoans lie? Have the arthropod phyla descended from different groups of lobopods? And what is the origin of the lophotrochozoans? It has been advocated, for example (see Conway Morris, 1998), that the halkieriids (see Fig. 2.21h) are close to the ancestors of the annelids, molluscs and brachiopods, which do indeed seem to be a closely related group of phyla, but the halkieriid to brachiopod transition seems to us at present to rely on some rather improbable intermediate forms: could the lophophore really be a derivative of a crawling foot or was it a purely larval adaptation?

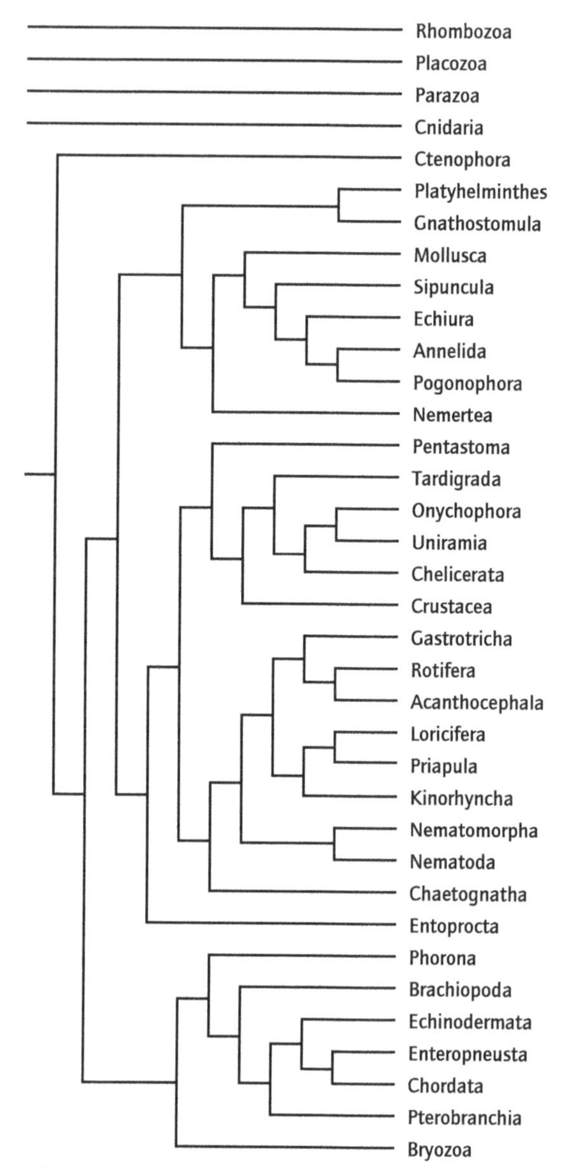

Fig. 2.14 The phylogeny of animals according to Eernisse *et al.*'s (1992) cladistic analysis of a data set of 141 independent morphological and embryological characters (after Eernisse *et al.*, 1992, Figs 2b & 4). Note that the ecdysozoan clade appears in this analysis too. The enteropneusts and pterobranchs are groups of the Hemichordata.

2.5 The origin, radiation and extinction of animal groups

Living animals can be placed in some 30–40 phyla; a phylum being defined either pragmatically as a group of organisms that appear to be related one to each other, but whose relationships with other such groups is debatable, or as a major clade of

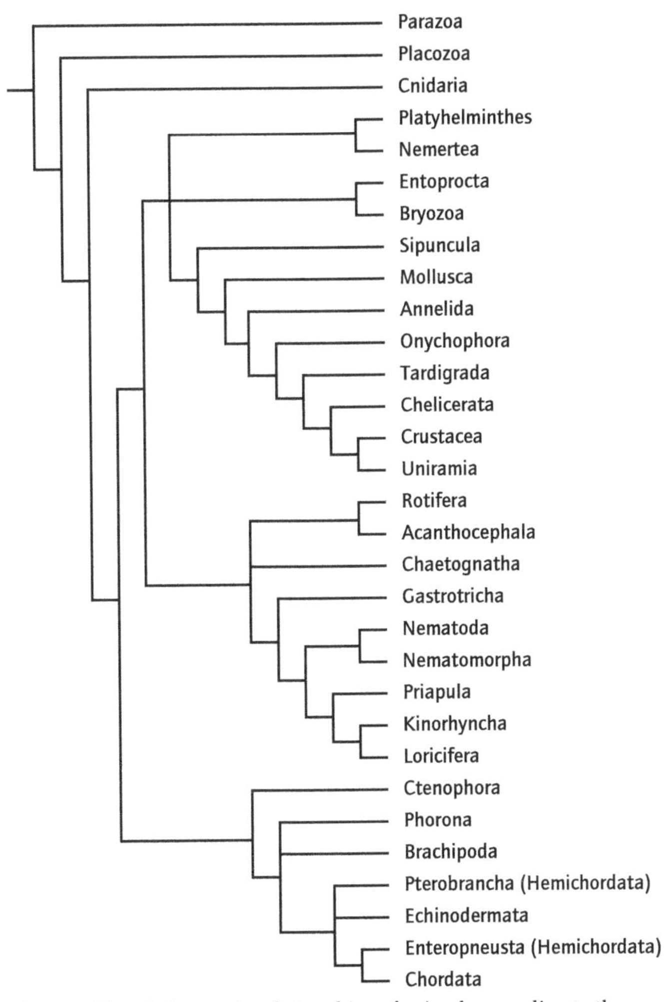

Fig. 2.15 The phylogenetic relationships of animals according to the cladistic analyses of Nielsen (1995). In this scheme, the pogonophorans, echiurans and gnathostomulans are included within the Annelida, and the pentastomans are included within the Crustacea.

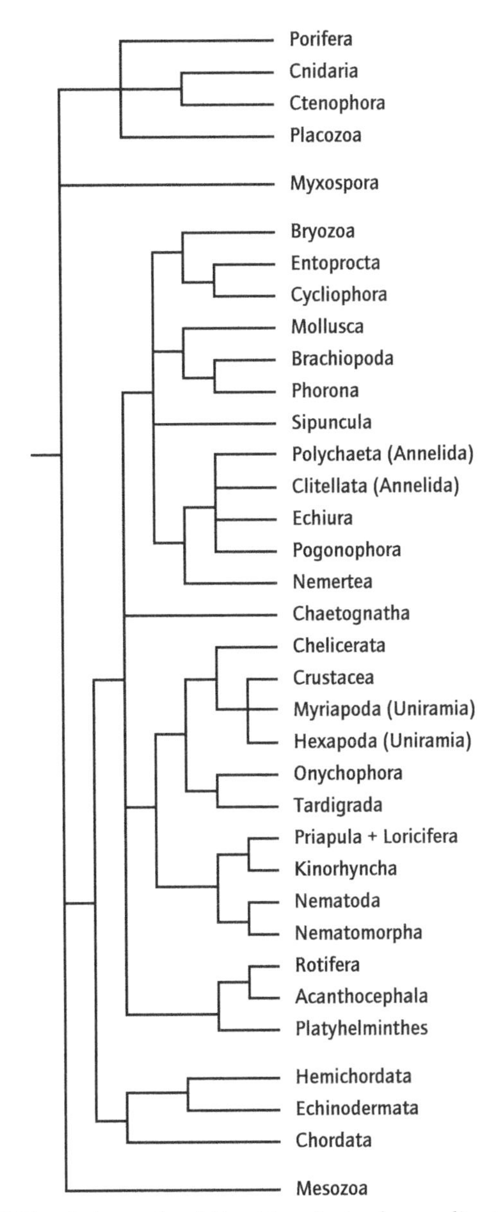

Fig. 2.16 The phylogenetic relationships of animals according to the synthesis of Cavalier-Smith (1998).

uncertain origin except within the framework of general affinity as disclosed by cladistic or molecular analysis. In essence, therefore, phylum status is an admission of our ignorance, although the foregoing sections have traced the pattern of general affinity alluded to above. (In a few cases, e.g. the Pentastoma and Pogonophora, a phylum can be used to house species of very individual and distinctive morphology notwithstanding that their ancestry is reasonably certain.) The three dozen or so known phyla are not the end result of a long, slow process of evolutionary divergence which has culminated in the diversity that we see today. As we noted in Section 2.1, it seems probable that with very few exceptions all the surviving animal phyla were already in existence in the Cambrian, and that the radiation of animal clades took place in the Precambrian. This presents something of a conundrum.

The fossil record usually documents that, over time, animals change slowly. There are several sequences in which the forms present at the beginning and at the end of a given sequence have changed in anatomy to the extent that they are categorized as belonging to different species (termed different 'chronospecies'). The average time required for change of this magnitude across a broad range of invertebrates is of the order of 10 million years. Let us now assume that different genera are, say, ten times more dissimilar to each other than would be two species within the same genus, that animals in different families are some ten times more dissimilar to each other than would be two genera within the same family, and so on up through the taxonomic heirarchy of order, class and phylum. Then on the basis of the average rate of morphological change required to convert an ancestral chronospecies into a descendent one, multiplied up, we can calculate that it would require in excess of

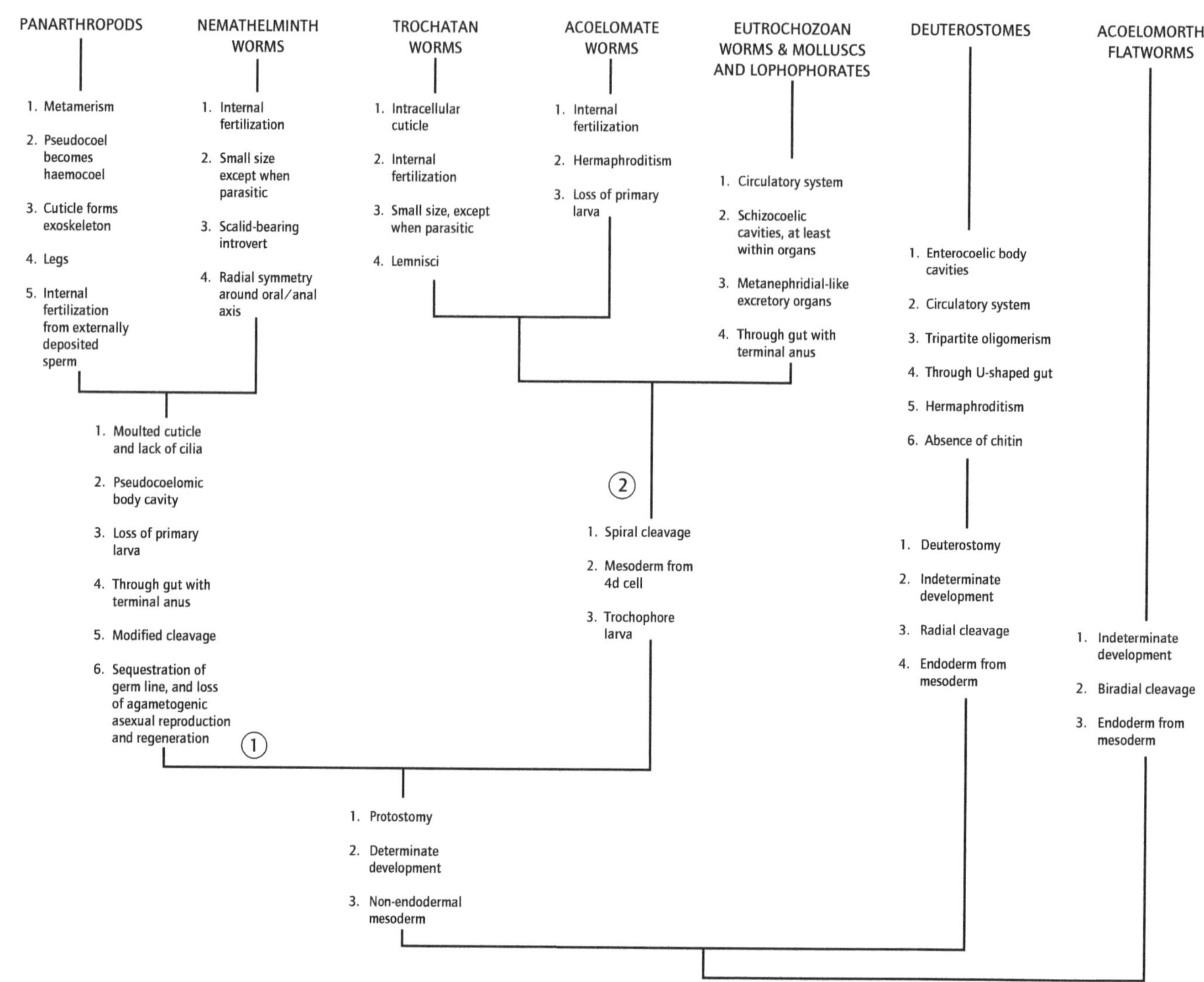

Fig. 2.17 Summary general pattern of relationships within bilaterally symmetrical animals as presented here, showing the major secondary specializations characteristic of the various clades (at least primitively). The enigmatic chaetognaths have been omitted. The position of some of the acoelomate worms, especially the catenulid flatworms and the gastrotrichs, is currently uncertain; these may split off from the base of the protostome and lophotrochozoan clades, respectively. Node 1 is the Ecdysozoa, and node 2 the Lophotrochozoa. The ancestral forms in this scheme are benthic, flatworm-like, solid-bodied animals, with a simple blind-ending gut and separate sexes, fertilization occurring externally. Compare with Figs 2.18 and 2.19.

10^{12} years for two original sibling organisms to diverge from each other to the extent that we would place them in different phyla. If, as is thought likely, the first animal(s) evolved from its/their choanoflagellate ancestors some 1000 or so million years ago, then on the above basis they should have diverged into different phyla only billions of years into the future. By today, their descendents should be as different from each other as are, for example, the various families of beetles. Tinkering with the admittedly arbitrary numbers used in the calculation does not help any. We could reduce both the time taken to evolve into a new species and the 'difference factors' by half and still calculate that new phyla should not appear until thousands of millions of years time. Yet this was in fact accomplished in a maximum time of the order of 500 million years, and probably much less, and at the end of this period the phyla were already as distinct from each other as they are today.

Something is clearly very wrong with the scenario presented above, and that something has to be that not all evolutionary change is as slow as that recorded in the fossil record within individual invertebrate chronospecies. Periods of rapid change in morphology have occurred, and they have taken place so quickly that the intermediate stages escaped fossilization. It may have been that some of these events took place in relatively small populations (the rate of evolutionary change is roughly inversely proportional to population size), further decreasing the likelihood of preservation. Paedomorphosis (see above) is

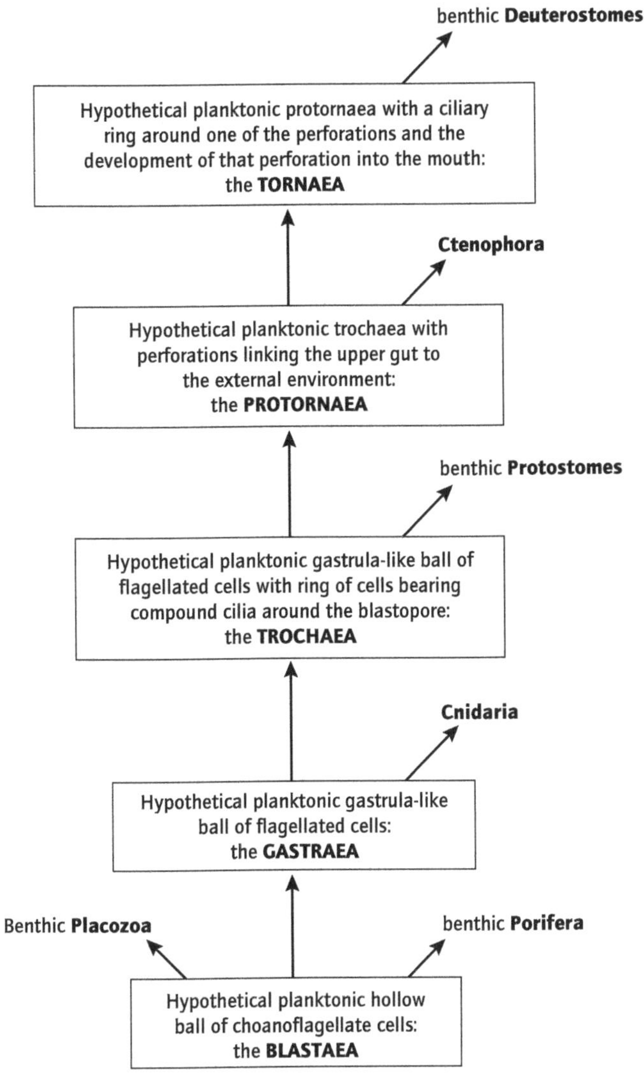

Fig. 2.18 A blastea–gastraea–trochaea hypothesis, as argued, for example, by Nielsen (1985, 1995).

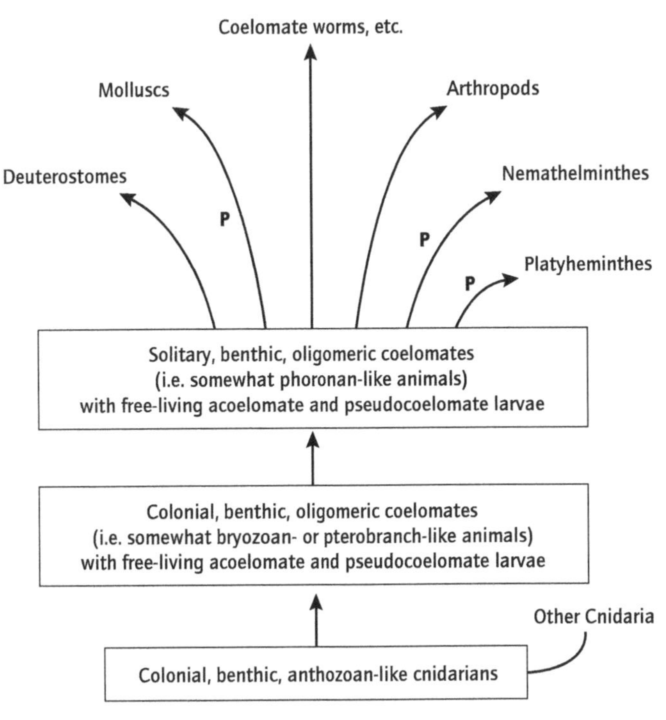

Fig. 2.19 An archicoelomate hypothesis, as argued, for example, by Rieger (1986).

another potentially rapid generator of change in adult body form, and it could theoretically be accomplished between one generation and the next, producing in only a few years a change of such magnitude as to result in the permanently larvalized descendent being placed in a different order or class to those members of the parent stock that had continued the ancestral life history unchanged. In other words, the taxonomic categories still in everyday zoological use are based on morphological distinctiveness not on recency of common ancestry. This is a situation that cladistics (see Section 2.1) seeks to rectify, but it is important to note that one cannot have things both ways. Either a taxonomic system is predicated on putting like, and only like, organisms in each taxon, so that structure can be deduced from taxonomic placement, or it is predicated on putting only monophyletic groups in each classificatory unit, in which case like animals may appear in different taxa and unlike ones in the same group. A cladistic re-analysis would lead to few of the existing phyla, classes, etc. in common use remaining valid (see Fig. 2.20).

These bursts of rapid change have often been associated with the generation of much diversity at the species, genus, family or even order level from a single founding type. Such 'adaptive radiations' also seem to occur over short intervals of geological time, and the manner in which several new types of animal can evolve relatively rapidly from a single stock has given rise (a) to notions that the processes controlling change within a single population (microevolution) are qualitatively different to those resulting in the formation of taxonomic novelty and variety (macroevolution) and (b) to theories that the rates at which micro- and macroevolutionary events occur are quantitatively different (see the 'gradualist' and 'punctuated equilibrium' models, Chapter 17).

Not all clades that were produced in the extensive Precambrian and Cambrian radiations are represented amongst living invertebrates. Some are known only from the fossil record; those of the archaeocyathans and trilobites, for example, and the twenty or so groups comprising the arthropod radiation depicted in Fig. 2.10. There are surprisingly few phyla only known as fossils, however, and most of these are characterized by the possession of hard parts – features that are likely to have appeared relatively late in several lineages. The fossil record may therefore give a false impression of past animal diversity. This is dramatically emphasized by the abundance and diversity of novel types of soft-bodied animals present in those very few

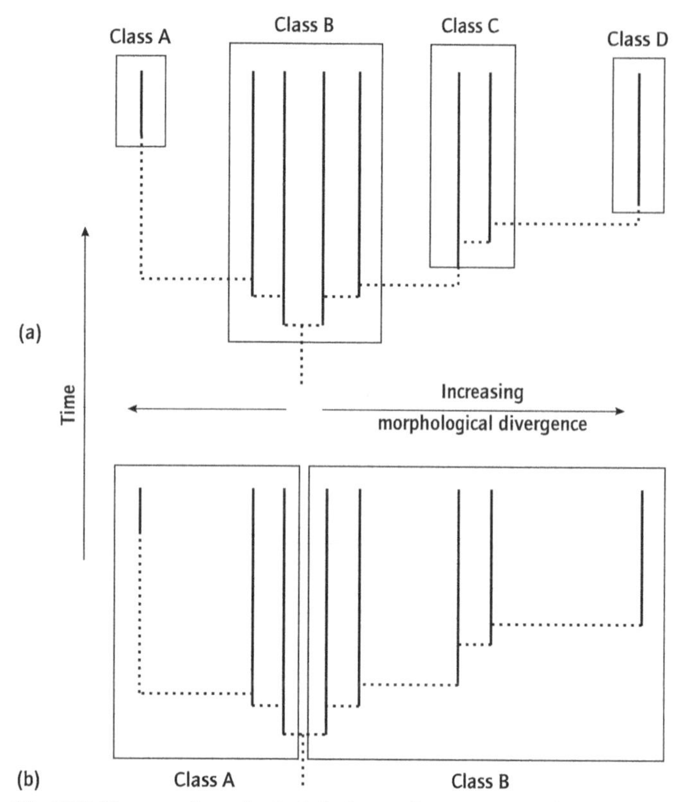

Fig. 2.20 The same hypothetical phylogenetic tree presented in two different forms: (a) classified along traditional lines biased towards morphological similarity and difference; (b) classified according to cladistic principles.

known deposits in which this category of organisms has been preserved, e.g. the Precambrian Ediacaran (Fig. 2.1) and Cambrian Burgess Shale beds and others of the same age in Greenland and China (Fig. 2.21).

Some twenty phyla with living representatives, often termed the 'minor phyla', contain few species (less than 500 each) and, with one exception that 'proves the rule', being soft bodied have left little or no fossil record. The exception is the Brachiopoda with only some 350 living species but over 25 000 fossil forms known from their hard external shells. Attempts to account for the fact that these small clades have managed to persist for so long whilst containing so few species have suggested that statistically this is likely only if once many more similar (i.e. soft-bodied) clades were in existence but have since become extinct. The known extinct phyla may therefore represent a small, and by virtue of their possession of fossilizable structures atypical, sample. What we know of the Precambrian and Cambrian fauna is perhaps only the tip of the iceberg of extinct clades.

The picture then is one of a large-scale radiation of soft-bodied and probably vermiform animals before the mid Cambrian that produced maybe hundreds of separate clades (although see Conway Morris, 1998, for a contrary view). Most of these possessed few component species and the majority probably became extinct without trace and without leaving living descendents (but possibly including the ancestors of the chaetognaths; see Section 2.3.8). Some lineages evolved hard protective or skeletal parts and of these we have considerable knowledge: it has been estimated that some 12% of such species have now been found and described. A few clades radiated to give rise to large numbers of species and much within-phylum diversity. Today, ten phyla are each represented by more than 10 000 living species, whilst in addition the brachiopods and echinoderms once achieved this level of species richness although they have declined to less than this total today. These form the so-called 'major phyla'.

Even though some clades have maintained high levels of species richness over long periods of geological time, it has not usually been achieved by the same groups of species within any one phylum. Different component orders or classes have dominated succeeding periods, and several once dominant subgroups are now extinct, having been replaced by others. Various subgroups have radiated in turn. This can be illustrated by the now extinct ammonite molluscs (Fig. 2.22), but the same picture is seen in the brachiopods, echinoderms, non-cephalopod molluscs, vertebrates and indeed any group for which there is a good fossil record. Rather counterintuitively, with the possible exception of interactions between the bivalve molluscs and the brachiopods, replacement of one group of animals by another does not seem to have occurred by competitive exclusion. Instead, one type of animal has first become extinct and only later, presumably under conditions of an ecological vacuum, has its replacement radiated to fill the ecological niches created by the demise of the hitherto dominant group or subgroup (examine Fig. 2.22). It appears as if the rapid radiation into numerous species is then followed by a much longer period of evolutionary stasis, until that group in turn suffers drastic reduction in numbers of species or extinction, eventually to be replaced by another.

Perhaps equivalently, there is little evidence from modern ecological studies that anatomically complex animals are ousting morphologically simpler ones from shared habitats: segmented arthropods and annelids, for example, still coexist in the same patches of marine sediment with 'primitive' unsegmented sipunculans and priapulans without any signs of competitive exclusion. Although, therefore, the popular concept of evolution is as, for example, given voice in Peter Høeg's 1994 novel *Borderliners* – 'The ascent from simple primitive organisms to the complex and highly developed' – this temptation to regard simple animals, like sponges, jellyfish and flatworms, as in some way inferior to structurally more complicated insects or mammals is to be avoided. The large numbers of animal clades produced from the Precambrian onwards are all more or less equally ancient, alternative body plans; they do not form a series of ever increasing adaptedness or fitness (with, of course, ourselves placed on the pinnacle). To put the matter another way, all eukaryotes have originated from within the prokaryotic bacteria and some of the descendents of these Precambrian prokaryotes are now oak trees, humans and giant squid, but these have in no

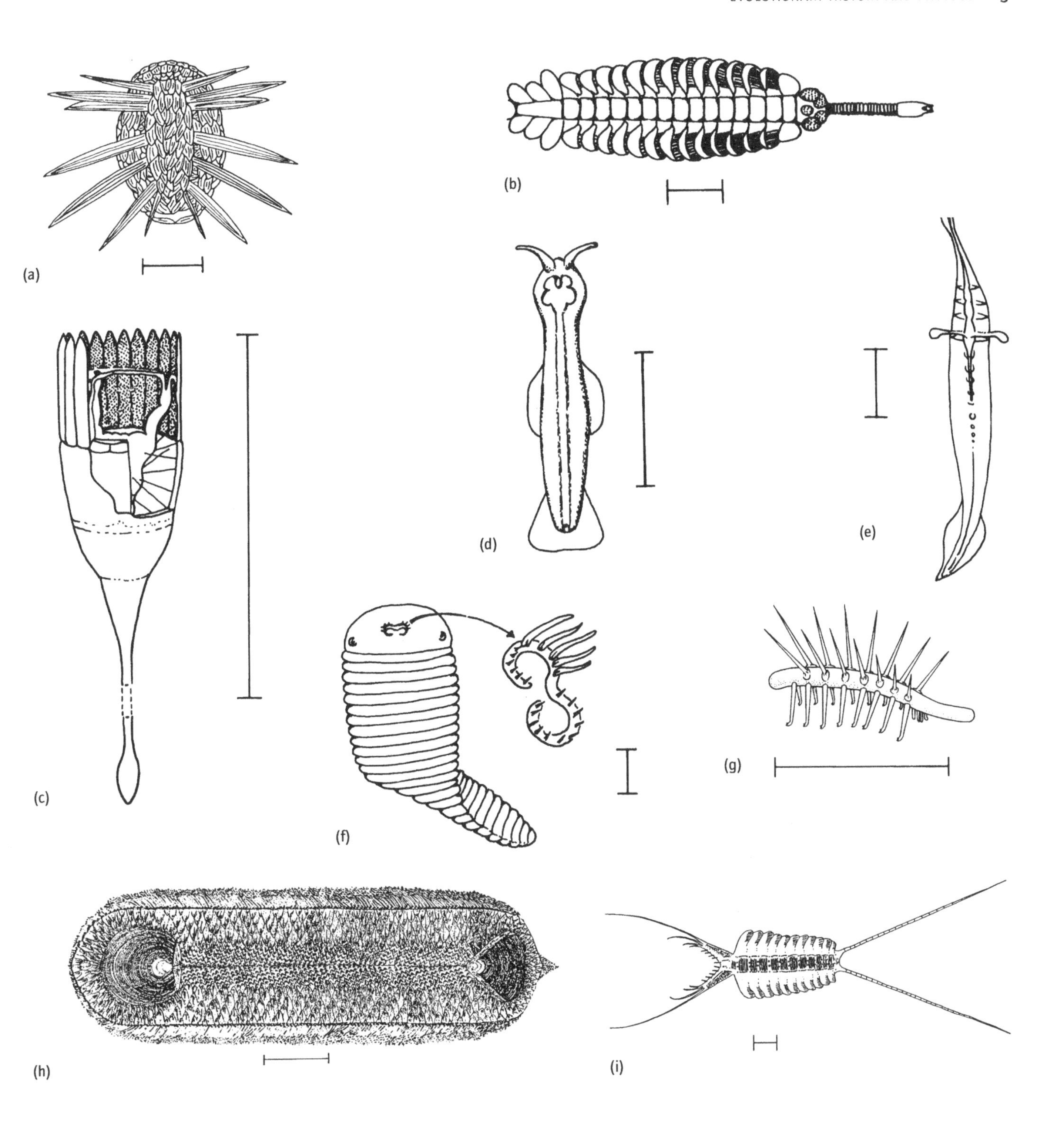

Fig. 2.21 Extinct and mostly soft-bodied animals, several of which can only be accommodated in known phyla with difficulty. *Hallucigenia* (g) and *Kerygmachela* (i), however, are lobopods: (a) *Wiwaxia* (Cambrian); (b) *Opabinia* (Cambrian); (c) *Dinomischus* (Cambrian); (d) *Amiskwia* (Cambrian); (e) *Tullimonstrum* (Carboniferous); (f) *Odontogriphus* (Cambrian); (g) *Hallucigenia* (Cambrian); (h) *Halkieria* (Cambrian); (i) *Kerygmachela* (Cambrian). (After several sources.) Scale lines: 1 cm.

way superseded the prokaryotes: bacteria still successfully dominate most habitat types, and we need them far more than they need us!

The extent to which different types of animals characterizing a wide range of niches and habitat types have become extinct at more or less the same time is somewhat contentious, although it appears that mass extinctions have occurred at various times in the past, especially towards the end of the Ordovician,

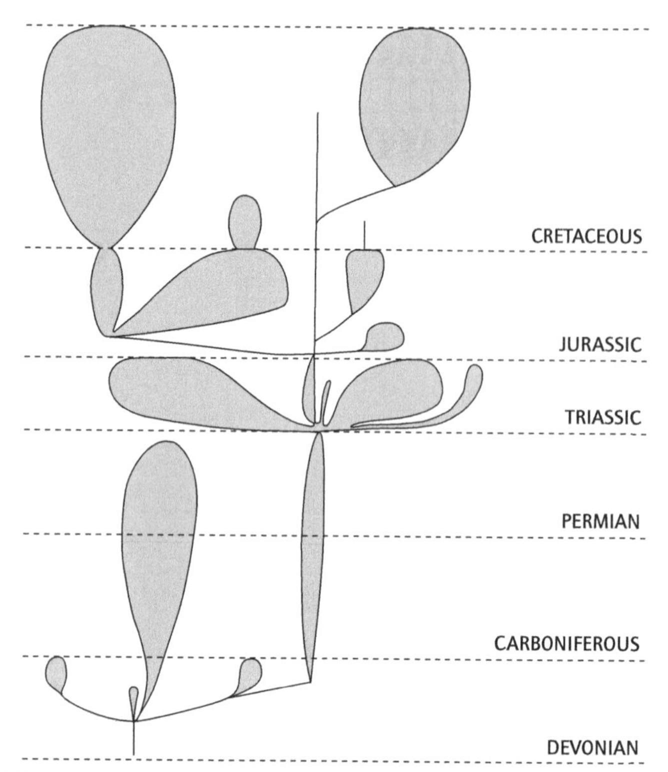

Fig. 2.22 Phylogenetic tree of the ammonites, showing radiation, extinction and replacement of one subgroup by another from the Devonian to the end of the Cretaceous, when the whole group became extinct (from data in Moore, 1957).

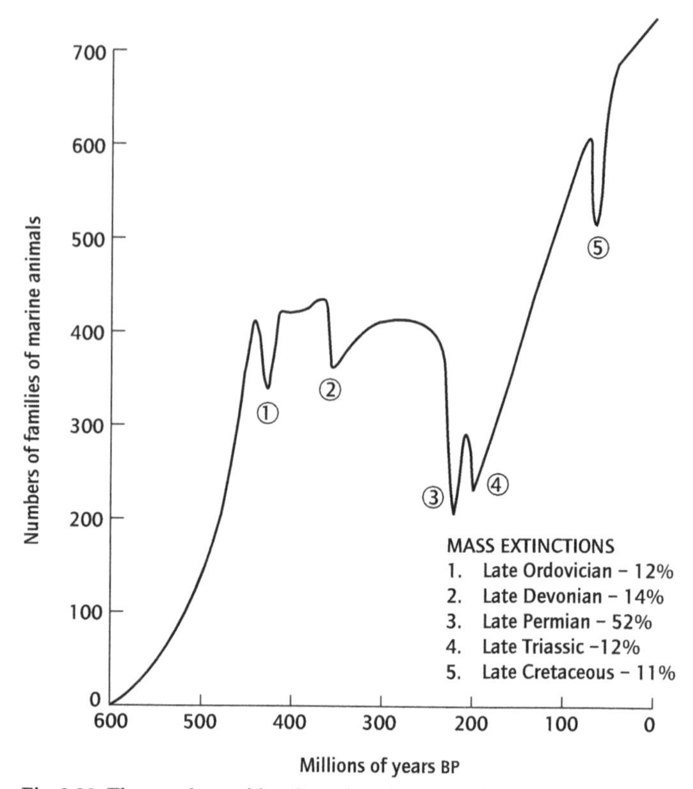

Fig. 2.23 The numbers of families of marine animals at different periods in the past, showing a series of mass extinctions (after Valentine & Moores, 1974 and others).

Devonian, Permian, Triassic and Cretaceous periods (Fig. 2.23), and, it has been claimed, every 26 million years or so. The event that took place some 225 million years ago at the end of the Permian, for example, involved the extinction of over half the then existing families of marine animals and perhaps 80–95% of the then species. Hitherto dominant clades, including the eurypterid chelicerates and the trilobites, wholly disappeared. The end result of the Permian and earlier Ordovician extinctions has been to bring about three contrasting faunas of the world ocean over time. The Cambrian–Ordovician one was dominated by trilobites and other extinct arthropod clades, by inarticulate brachiopods and by monoplacophoran molluscs; this was succeeded until the end of the Permian by one in which shelled cephalopods, crinoid echinoderms, articulate brachiopods and stenolaeme bryozoans were the most abundant fossilizable forms. From the start of the Mesozoic, however, the seas have had an essentially modern look, characterized by the relative importance of gastropod, bivalve and coleoid-cephalopod molluscs, malacostracan crustaceans, echinoid echinoderms, gymnolaeme bryozoans and the vertebrates.

The causes of these mass extinctions are unknown, although they correlate fairly well with major changes in sea level, themselves probably brought about by glacial phases or continental movements, and/or with episodes of intense volcanic activity that would have released dust and potentially harmful gases into the atmosphere. There have, however, been numerous speculative attempts, both popular and scientific, to implicate various cosmic forces and events. On currently available evidence, it seems as if each of the three, great, replacing marine faunas first become established in shallow shelf waters after the sea had transgressed over the continental shelves, only to succumb – except in relatively deep water – during a later regression off the shelves. The extinction of major clades appears therefore to have had a large stochastic element, as indeed did the identity of the replacing groups. Clades did not die out because they were in some way inferior body plans but because of drastic change in their environment. Members of those few groups that happened to survive a major extinction event would initially have experienced undersaturated habitats in which some selective pressures, at least, are likely to have been reduced and many niches were available to be filled. It is under these circumstances that adaptive radiations from a single basic stock have occurred. These can be envisaged as proceeding rapidly until the habitat was once again approaching saturation. Selective pressures would then be imposed ever more strongly, and selection would limit the amount of the diversity formed during each radiation that could outlive the expansion phase. It was then these selected surviving lines within each clade that dominated the time interval until the next environmentally induced phase of extinction initiated the whole cycle again to a greater or lesser degree.

The history of animal life seems therefore to have been distinctly episodic; each episode, including the one in which we are now living, being typified by a mixture of expanding and declining clades. As a result of chance, some of these groups will have managed to persist for a long time, albeit with very few species. Others, although perhaps insignificant now, will have the opportunity to increase and diversify after the next such phase of extinction, human-induced or otherwise. Chance therefore determines the play enacted, provides the initial potential cast and brings down the curtain, whilst selection auditions for occupancy of the vacant parts and oversees the continued adjustments made to the script and characters.

2.6 Biodiversity

With the passage of time, there has been an impoverishment in the number of clades originally established in the Precambrian/Cambrian as some – and possibly many – have become extinct, although the number of derivative lineages of surviving clades and especially of animal species alive today is probably greater than it has ever been. The seas of the world are currently relatively fragmented, shallow shelf-seas are abundant both geographically and areally, and the ocean is oxygenated right down to its abyssal bed – circumstances that have not always prevailed; in the Cretaceous, for example, large volumes of the sea below some 100 m may have been anoxic. There is no reason to believe, therefore, that marine species are not as diverse now as at any time in the past, and indeed most estimates would rank the current marine fauna as the richest ever.

This notwithstanding, the enormous number of living species is almost entirely a reflection of the conquest of the land that began in the Silurian–Devonian and was consolidated in the Carboniferous. Somewhat surprisingly, the route that this colonization took appears largely to have been up the intertidal shore and thence on to the land, and not the more obvious one for an aquatic animal of sea → estuary → river (→ lake) → swamp → land, in that most terrestrial invertebrates have high fluid concentrations and considerable dessication resistance, not vice versa. (The vertebrates and some molluscs are the main groups to have used the aquatic route.) Invasion of bodies of freshwater was then largely accomplished by a suite of animals already adapted to terrestrial conditions. The small interstitial animals (nematodes, tardigrades, rotifers, etc.) have never left the water films surrounding soil or organic particles, and they are terrestrial only in the sense that their aquatic habitat also occurs on land. The impermanence of any individual water film has necessitated an ability to enter suspended animation during periodic evaporation of their environment.

There are many more barriers to dispersal on the land than there are in the sea, and this has resulted in large-scale speciation within the relatively few clades to have become successfully terrestrial (and for the same reason there are many more benthic than pelagic species in the sea). Many terrestrial species are geographically replacing versions of a given animal type. The land

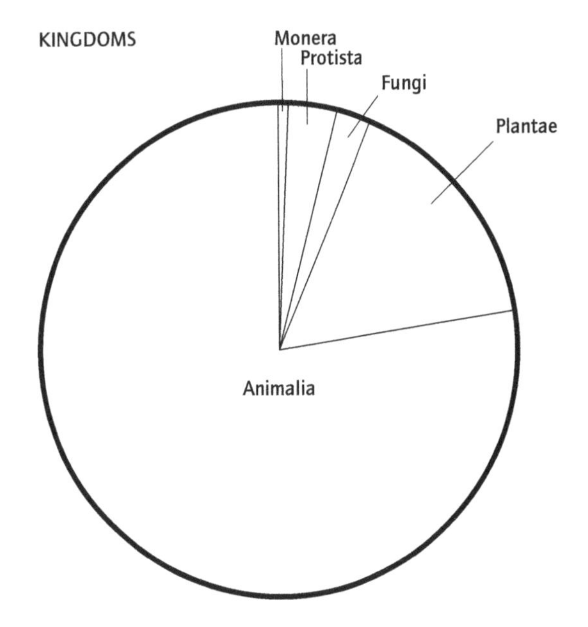

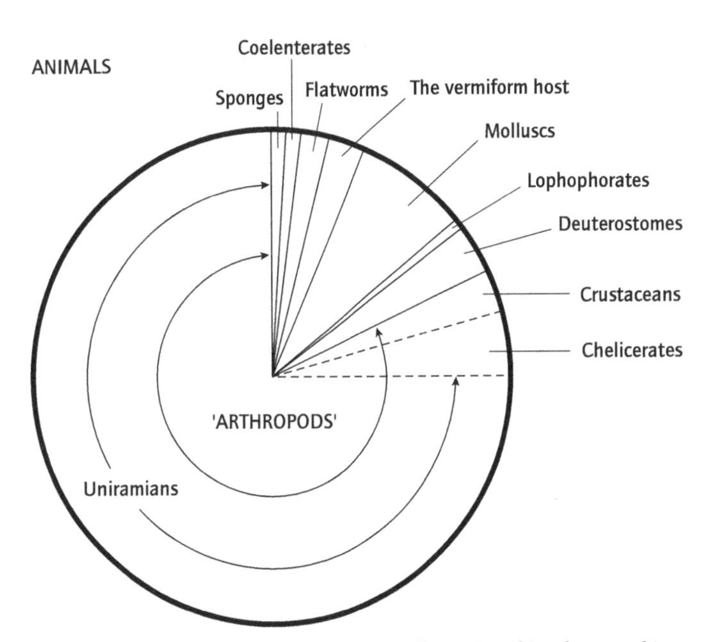

Fig. 2.24 The numbers of living species in the various kingdoms and in the animal phyla (from data in Barnes, 1998).

chelicerates and uniramians, especially the hexapod insects, are now the dominant invertebrates on the planet in terms of numbers of species (Fig. 2.24). In the last 350 million years, the focus of organismal diversity at the generic and specific levels has moved away from the ancestral sea out onto the land.

Diagrams such as Fig. 2.24 reflect current knowledge, but we are probably very ignorant of the true diversity of living animals even – or perhaps especially – on the land. The number of described animal species is of the order of one million, but this may represent a very small proportion of the total. Based on (i) the number of beetle species in the canopy of one type of rain forest tree in one area, (ii) the mean host-specificity of beetles, (iii) the number of species of rain forest trees, (iv) the proportion of the total canopy-living arthropods comprised by beetles, and (v) the proportion of the total rainforest arthropods that live in the canopy, one calculation estimates the total number of arthropod species in tropical rain forests at 30 million. Admittedly this is an extrapolation from 1200 species of beetles on only 19 sampled *Luehea* trees, but other estimates of terrestrial species richness also suggest totals of up to 50 million. It is devastating to reflect that the current rate of rainforest destruction means that many more unknown species of animals have been rendered extinct by human activity in the last few decades than we have identified to date. And many 'known species' are in fact known only from some original description based on a few preserved specimens. We are also having to revise our estimates of marine species richness, particularly of that of interstitial species in the ocean sediments. Similar extrapolations to that above, but here based on nematode species richness in a few samples from limited areas of the sea bed, suggest that the number of nematode species may be between one and one hundred million. We certainly have reasonable knowledge of the biology of considerably less than 0.001% of animals that were alive a century ago, and a good understanding of less than that. At the very least, this should teach us to treat generalizations concerning invertebrate biology cautiously.

2.7 Further reading

Aguinaldo, A.M.A., Turbeville, J.M., Linford, L.S., Rivera, M.C., Garey, J.R., Raff, R.A. & Lake, J.A. 1997. Evidence for a clade of nematodes, arthropods and other moulting animals. *Nature (Lond)*, **387**, 489–493.

Barnes, R.S.K. (Ed.) 1998. *The Diversity of Living Organisms*. Blackwell Science, Oxford.

Bergström, J. 1989. The origin of animal phyla and the new phylum Procoelomata. *Lethaia*, **22**, 259–269.

Bryce, D. 1986. *Evolution and the New Phylogeny*. Llanerch, Dyfed.

Cavalier-Smith, T. 1998. A revised six-kingdom system of life. *Biol. Rev.*, **73**, 203–266.

Clarkson, E.N.K. 1986. *Invertebrate Palaeontology and Evolution*, 2nd edn. Allen & Unwin, London.

Cohen, J. & Massey, B.D. 1983. Larvae and the origins of major phyla. *Biol. J. Linn. Soc., Lond.*, **19**, 321–328.

Conway Morris, S. 1998. *The Crucible of Creation. The Burgess Shale and the Rise of Animals*. Oxford University Press, Oxford.

Conway Morris, S., George, J.D., Gibson, R. & Platt, H.M. (Eds) 1985. *The Origins and Relationships of Lower Invertebrates*. Clarendon Press, Oxford.

Eernisse, D.J., Albert, J.S. & Anderson, F.E. 1992. Annelida and Arthropoda are not sister taxa: a phylogenetic analysis of Spiralian Metazoan morphology. *System. Biol.*, **41**, 305–330.

Glaessner, M.F. 1984. *The Dawn of Animal Life*. Cambridge University Press, Cambridge.

Goldsmith, D. 1985. *Nemesis. The Death Star and Other Theories of Mass Extinction*. Walker, New York.

Haeckel, E. 1874. The gastraea theory, the phylogenetic classification of the Animal Kingdom and the homology of the germ-lamellae. *Quart. J. Microsc. Sci.*, **14**, 142–165; 223–247.

Halanych, K.M., Bacheller, J.D., Aguinaldo, A.M.A., Liva, S.M., Hillis, D.M. & Lake, J.A. 1995. Evidence from 18S ribosomal DNA that the lophophorates are protostome animals. *Science, New York*, **267**, 1641–1643.

Hanson, E.D. 1977. *The Origin and Early Evolution of Animals*. Wesley University Press, Middleton, Connecticut.

House, M.R. (Ed.) 1979. *The Origin of Major Invertebrate Groups*. Academic Press, London.

Jenner, R.A. & Schram, F.R. 1999. The grand game of metazoan phylogeny: rules and strategies. *Biol. Rev.*, **74**, 121–142.

Lankester, E.R. 1875. On the invaginate planula, a diploblastic phase of *Paludina vivipara*. *Quart. J. Microsc. Sci.*, **15**, 159–166.

Manton, S.M. & Anderson, D.T. 1979. Polyphyly and the evolution of arthropods. In: M.R. House (Ed.) *The Origin of Major Invertebrate Groups*, pp. 269–321. Academic Press, London.

McKerrow, W.S. (Ed.) 1978. *The Ecology of Fossils*. Duckworth, London.

Moore, J. & Willmer, P. 1997. Convergent evolution in invertebrates. *Biol. Rev.*, **72**, 1–60.

Nielsen, C. 1985. Animal phylogeny in the light of the trochaea theory. *Biol. J. Linn. Soc., Lond.*, **25**, 243–299.

Nielsen, C. 1995. *Animal Evolution. Interrelationships of the Living Phyla*. Oxford University Press, Oxford.

Raff, R.A. 1996. *The Shape of Life. Genes, Development, and the Evolution of Animal Form*. University of Chicago Press, Chicago & London.

Rieger, R.M. 1986. Über den Ursprung der Bilateria: die Bedeutung der Ultrastrukturforschung für ein neues Verstehen der Metazoenevolution. *Verh. Deutsch. Zool. Ges.*, **79**, 31–50.

Rosa, R.de, Grenier, J.K., Andreeva, T., Cook, C.E., Adoutte, A., Akam, M., Carroll, S.B. & Balavoine, G. 1999. *Hox* genes in brachiopods and priapulids and protostome evolution. *Nature (Lond)*, **399**, 772–776.

Ruiz-Trillo, I., Riutort, M., Littlewood, D.T.J., Herniou, E.A. & Baguna, J. 1999. Acoel flatworms: earliest extant bilateral metazoans, not members of the Platyhelminthes. *Science, New York*, **283**, 1919–1923.

Salvini-Plawen, L. von 1988. Annelida and Mollusca – a prospectus. *Microfauna Marina*, **4**, 383–396.

Scientific American 1982. *The Fossil Record and Evolution*. Freeman, San Francisco.

Sedgwick, A. 1884. On the nature of metameric segmentation and some other morphological questions. *Quart. J. Microsc. Sci.*, **24**, 43–82.

Sleigh, M.A. 1979. Radiation of the eukaryote Protista. In: M.R. House (Ed.) *The Origin of Major Invertebrate Groups*, pp. 23–53. Academic Press, London.

Thomson, K.S. 1988. *Morphogenesis and Evolution*. Oxford University Press, New York.
Trueman, E.R. & Clarke, M.R. (Ed.) 1985. *The Mollusca* (Vol. 10). *Evolution*. Academic Press, Orlando.
Valentine, J.W. (Ed.) 1985. *Phanerozoic Diversity Patterns*. Princeton University Press, Princeton, New Jersey.
Valentine, J.W. 1989. Bilaterians of the Precambrian–Cambrian transition and the annelid–arthropod relationship. *Proc. Nat. Acad. Sci., U.S.A.*, **86**, 2272–2275.
Whittington, H.B. 1985. *The Burgess Shale*. Yale University Press, New Haven.
Willmer, P. 1990. *Invertebrate Relationships*. Cambridge University Press, Cambridge.

PART 2

The Invertebrate Phyla

In the chapters of this Part we illustrate and describe briefly all the known phyla of invertebrates with living representatives, together with their component classes.

You will find that our systematic survey is much less detailed than in many textbooks of invertebrate zoology, since, rather than describe all the various anatomical features of the different types of animals, we have endeavoured to distil only those essential characteristics of each group with which the student should be familiar, i.e. those responsible for the singularity of the phylum or class and for its evolutionary success or ecological significance. We have also provided lists of diagnostic features to permit comparison of the individual phyla.

Further, in this Part our accent is on the diversity of invertebrate body plans, as reflected in the major animal taxa; we have therefore treated all groups equally by granting the same degree of coverage to all the classes of invertebrates, regardless of the number of species which they may contain. The systems-based chapters of Part 3, however, will redress this bias by drawing their material largely from the bigger, better-known and, arguably, more important groups.

For each phylum we have provided an outline classification diagram down to the level of order, and each phylum (or, where appropriate, class) is illustrated by a figure showing the range of diversity of body form of representatives of all or most of these orders.

The phyla, classes, etc., recognized in this section are basically those of Barnes (1998) *The Diversity of Living Organisms*, Blackwell Science, Oxford. You may care to note that new types of animals are still being discovered – in the last 20 years, two new phyla (the Loricifera and Cycliophora) and two new classes in supposedly well-known phyla (Remipedia, phylum Crustacea, and Concentricycloidea, phylum Echinodermata) have been described. Whilst, therefore, our coverage of animal classes is complete as at 1999, there is no reason to believe that all the major types of animals surviving on the planet have now been discovered; it is likely that as little-known habitats such as the deep sea and interstitial zones are investigated in greater detail, so will new, and sometimes radically different animals be seen for the first time. The last word on the range and diversity of organisms will not be said for a long time yet! [Indeed, whilst the proofs of this book were being read, a new group of animals – the micrognathozoans – was being described.]

CHAPTER 3

Parallel Approaches to Animal Multicellularity

'The Protozoa'
Parazoa: Porifera and Symplasma
Phagocytellozoa: Placozoa
Radiata: Cnidaria and Ctenophora
Mesozoa: Rhombozoa
Bilateria: Platyhelminthes

The groups of animals considered in this chapter are mainly those which it was suggested in Chapter 2 could lay claim to have been derived directly from within the kingdom Protista, in that they possess fundamentally different forms of bodily organization, construction and symmetry. They can be viewed, therefore, as a series of parallel evolutionary lines that may have achieved the multicellular animal condition independently of each other, albeit probably from the same type of ancestral protist, the choanoflagellates. The five such lines are here termed different superphyla, and it is only to one of these, represented by the bilaterally symmetrical flatworms, that the animals described in Chapters 4–8 are related.

3.1 'The Protozoa'

Until the early 1970s it was customary to allocate all living organisms to one or other of two kingdoms. If an organism was photosynthetic, or grew in the ground, it was 'a plant'; if it did not photosynthesize and was freely mobile, it was 'an animal'. Whether an organism was prokaryotic or eukaryotic, or whether uni- or multicellular, were not considerations relevant to this fundamental division of life. Within this classification, the multicellular plants then formed one group, the Metaphyta, whilst the unicellular photosynthesizers were termed the Protophyta or Algae (a category which included various prokaryotes). Comparably, the multicellular animals comprised the Metazoa, and the single-celled animals the Protozoa.

Such a two-kingdom system clearly suffers from a number of drawbacks, of which two are major. First, it does not reflect any real phylogenetic dichotomy: there is no evidence at all to suggest that the cyanobacteria, diatoms, red seaweeds, toadstools and conifers, for example, all belong to one natural group, in contrast to the amoebae, gut flagellates, sponges and echinoderms which all belong to a second. Such artificiality need not be a significant problem, however, if the aim of a classification is purely pragmatic – the two kingdoms might still provide an unambiguous, convenient but arbitrary pair of mutually exclusive divisions of living organisms. But the second drawback is that the two categories of plant and animal are not really distinguishable. Within a group of apparently closely related organisms (the euglenoid flagellates, for example), some species within a given genus are apparently plants, others can be considered animals, and yet further species can be both at the same time. Some dinoflagellates also obtain some 5% of their requirements from photosynthesis and 95% from heterotrophic digestion of consumed materials. Moreover, a number of flagellates could even change the kingdom to which they were assigned by remaining in the dark for 24 hours! The two-kingdom system is thus neither a reflection of phylogenetic relationships nor is it a practical working scheme, in large measure because although many organisms are 'animal' and many are 'plant', many are neither.

During the past 30 years or so, an alternative basic classification, which seeks both to reflect phylogeny more accurately and to be unambiguous, has gradually been gaining adherents and is now that in most widespread use. This scheme also postulates a basic dichotomy of living organisms, although in this case between the prokaryote and the eukaryote superkingdoms, the eukaryotes being then subdivided into four kingdoms: the unicellular groups (the Protista), and the multicellular plant, fungal and animal kingdoms (Fig. 3.1). The Protista therefore includes the former algae and protozoans, neither of which any longer has any formal biological status except as a general term equivalent to 'shellfish' or 'worms'; in a nutshell, the 'Protozoa' has for some time ceased to be regarded as a valid or useful grouping of organisms. Modern classifications of the Protista recognize between 27 and 45 phyla, together with an additional one only described in 1988 and some three to six groups of uncertain relationship that may well prove also to be of phylum level (Corliss, 1984; Sleigh, 1989; etc.). A relatively uncontentious scheme is set

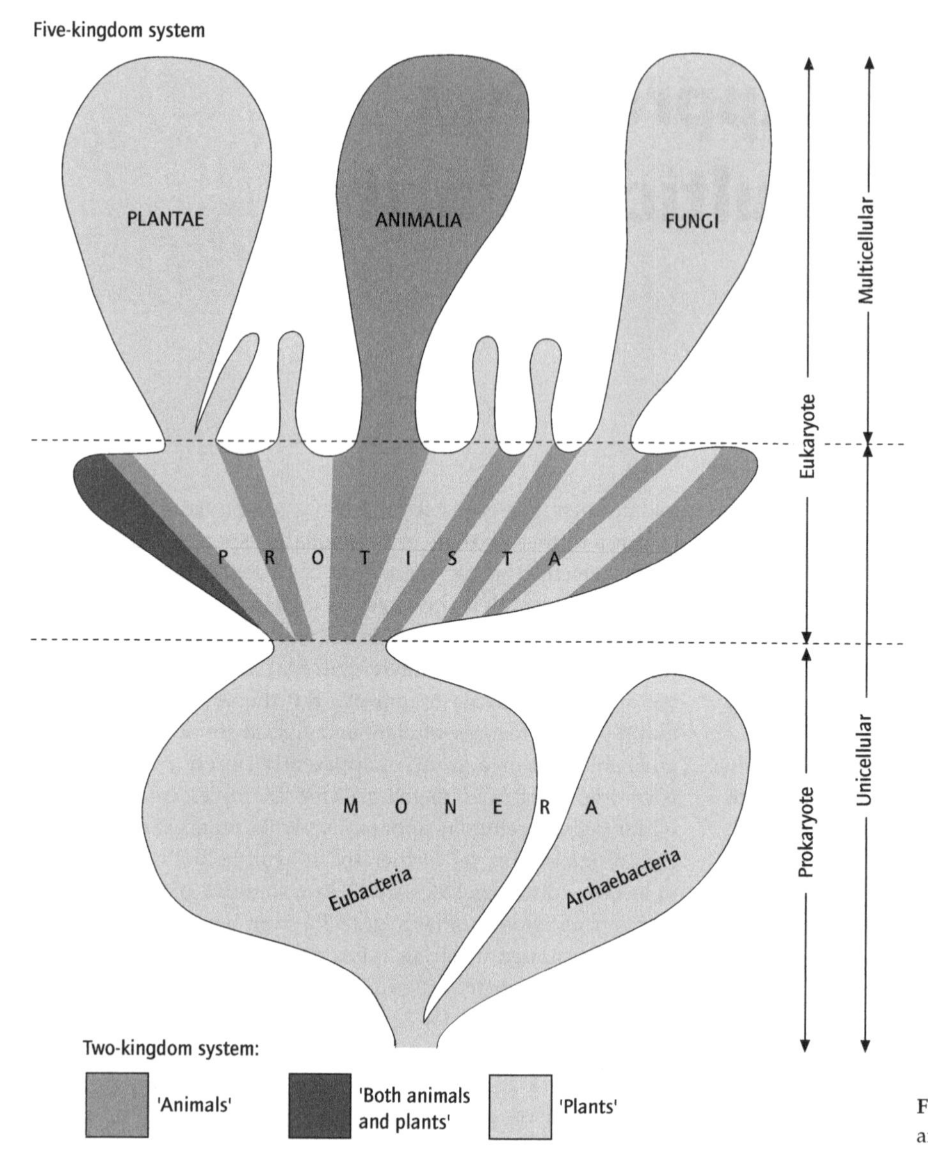

Fig. 3.1 The five-kingdom system of classification and its relations to the plant/animal dichotomy.

out in Table 3.1. More than half of the 42 groups listed there are obligately non-photosynthetic and 17 of them were once regarded as being groups of animals. A further 14 of the listed phyla were claimed by both botanists and zoologists, were given different names by plant and animal biologists (Table 3.2), and figured *twice* in some classifications of organisms (e.g. in Parker, 1982)! Box 3.1 provides further details of those protist groups once included in the 'Protozoa'.

The majority of these 'protozoan protists' are phylogenetically remote from the invertebrates and they share no characteristics uniquely with the multicellular animals. Shared negative features, such as lack of plastids, are also held in common with the 'fungus-like protists' and with the fungi themselves, and shared positive characters are solely those common to all the eukaryotes, including all the photosynthetic groups. By definition, of course, *the* difference between the protists, including the non-photosynthetic forms, and the other eukaryote kingdoms, including the animals, is that the latter are multicellular whilst the protists are solitary or colonial unicells. The individual protist cell is therefore capable of performing all the functions necessary for survival and multiplication – at one and the same time it may be the agent effecting locomotion, digestion, osmoregulation, reproduction, etc. – whilst in the cellularly differentiated animals, these, and the other functions, are carried out by different, specialized groups of cells usually arranged in discrete tissues or organs. This distinction, however, is purely phenotypic. All the many component cells of a multicellular animal are descended from a single, normally diploid cell and hence are genetically as totipotent as a protist individual. Some individual cells, e.g. the amoebocytes of sponges, may even remain undifferentiated and persist as a reservoir from which specialized cells can be differentiated as required during life.

Not surprisingly, the protist body is generally much smaller than that of an animal, although this is by no means always so. A

Table 3.1 The phyla of living protists.

Karyoblastea † (= Pelobiontea), 1 sp.
Amoebozoa † (= Rhizopoda), 5000 spp.
Heterolobosa, † 40 spp.
Eumycetozoa, † 600 spp.
Granuloreticulosa (= Foraminifera), 6000 spp.
Xenophyophora, † 40 spp.
Plasmodiophora, † 40 spp.
Oomycota, 800 spp.
Hyphochytridiomycota, 25 spp.
Chytridiomycota, 900 spp.
Chlorophyta, 3000 spp.
Prasinophyta, 250 spp.
Charophyta, 100 spp.
Conjugatophyta (= Gamophyta, = Zygnematophyta), 5000 spp.
Glaucophyta, 10 app.
Euglenophyta, † 1000 spp.
Kinetoplasta, † 600 spp.
Stephanopogonomorpha † (= Pseudociliata), 5 spp.
Rhodophyta, 4250 spp.
Hemimastigophora, † 2 spp.
Cryptophyta, 200 spp.
Choanoflagellata, † 150 spp.
Chrysophyta, 650 spp.
Haptophyta (= Prymnesiophyta), 450 spp.
Bacillariophyta, 10 000 spp.
Xanthophyta, 650 spp.
Eustigmatophyta, 12 spp.
Phaeophyta, 1600 spp.
Proteromonada, † 50 spp.
Bicosoecidea, † 40 spp.
Raphidiophyta, 30 spp.
Labyrinthomorpha, † 35 spp.
Metamonada, † 200 spp.
Parabasalia, † 1750 spp.
Opalinata, † 300 spp.
Actinopoda, † 4200 spp.
Dinophyta, † 2000 spp.
Ciliophora, † 8000 spp.
Sporozoa † (= Apicomplexa), 5000 spp.
Microspora, † 800 spp.
Ascetospora † (= Haplospora), 25 spp.
Myxospora † (= Myxozoa), 1200 spp.

† see Box 3.1

Table 3.2 Contrasting botanical and zoological nomenclature for the same protist groups.

Botanical	Zoological
Myxomycetes	Mycetozoans
Volvocales	Phytomonads
Cryptophytes	Cryptomonads
Chrysophytes	Chrysomonads
Dictyochales	Silicoflagellates
Haptophytes	Haptomonads
Coccosphaerales	Coccolithophorids
Xanthophytes	Heterochlorids
Raphidiophytes	Chloromonads
Dinophytes	Dinoflagellates
Prasinophytes	Prasinomonads
Craspedophytes	Chaonoflagellates

Box 3.1 The more animal-like phyla of protists

The various protist phyla are often diagnosed by complex cytological features beyond the scope of this book. The descriptions below should therefore be regarded as being general sketches applicable to most members of the groups concerned, not as formal diagnoses.

Karyoblastea Giant (up to 5 mm long) multinucleate amoebae that lack cellular organelles such as mitochondria, contractile vacuoles, endoplasmic reticula and Golgi bodies, and which appear to divide amitotically – no centrioles or chromosomes occur. Endosymbiotic bacteria, including methanogens, are present in their cytoplasm. They live in freshwater pond sediments, under conditions of low oxygen tension, where they feed on bacteria and photosynthetic protists. A free-living mitochondrion-less, Golgi-less amoeba with a single flagellum may also belong to this group.

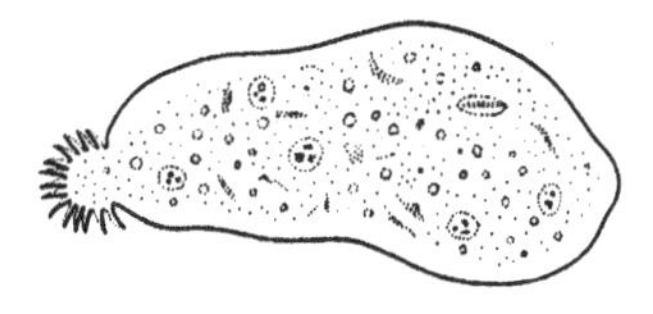

Amoebozoa A large group of asexual single-celled amoebae that lack flagella and associated organelles and move via active pseudopodia that vary in shape from broad (lobose) to fine (filose): pseudopodia may branch but they do not anastomose, nor do they contain microtubular skeletal systems. Several species construct tests of chitin or from particulate material available in their habitat. They occur in most habitats – the sea, freshwaters, soil and, symbiotically and parasitically, in other organisms.

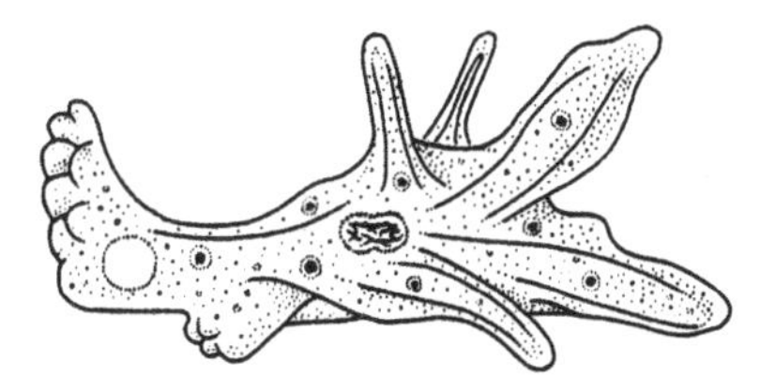

Heterolobosa A small group of asexual, uninucleate amoebae with lobose pseudopodia that occur in freshwaters and damp soil, on dung and decomposing vegetation. The individuals of one component subgroup, the 'acrasid cellular slime moulds', can aggregate in small groups to form a cluster from which issue fruiting bodies, on stalks of live cells, containing encysted individual amoebae. These emerge and resume independent life under appropriate conditions. With the acrasids are associated a group of somewhat similar, solitary, uninucleate amoeboflagellates or 'mastigamoebae', that reversibly can develop a pair of flagella. At least one species can parasitize man, causing amoebic meningoencephalitis.

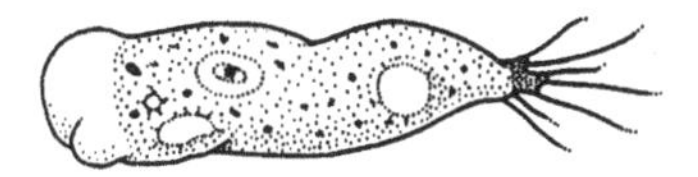

Eumycetozoa These, the 'true slime moulds', are amoeboid protists with a complex life cycle in which the solitary amoebae either (a) may

Continued p. 42

Box 3.1 *(cont'd)*

develop (and lose) a pair of flagella and two such amoebae may fuse and then mitotically form a large, multinucleate, sessile, plasmodial mass, up to several metres long, or alternatively (b) may aggregate and stream together as a pseudoplasmodial cellular 'slug'. From the plasmodium or pseudoplasmodium issue fruiting bodies, containing many uninucleated haploid spores, on largely acellular cellulose stalks or directly from the plasmodial surface. These spores germinate into the solitary amoeboid feeding stage with its filose pseudopodia or flagella. The true slime moulds prey on bacteria, other protists (including each other) and fungi, and occur in a wide range of habitats, but especially in the soil and on dead plant materials.

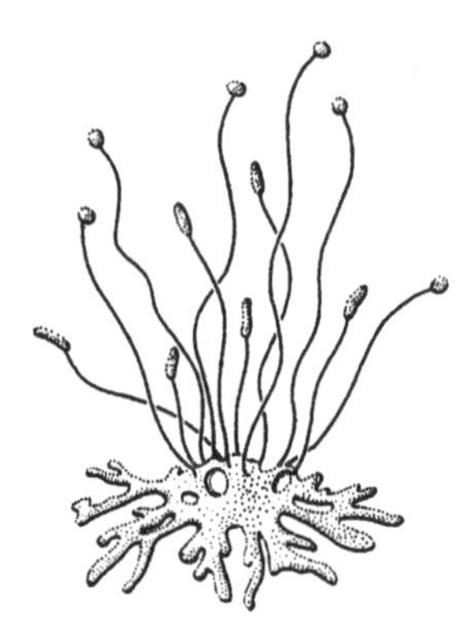

Plasmodiophora Obligate intracellular parasites of fungi and plants that occur in the form of a feeding plasmodium from which issue flagellated spores that fuse, in pairs, and reinfect the host species. Cysts may form and survive dormant for long periods. They cause, amongst others, the club-root disease of cabbages and 'powdery scab' in potatoes.

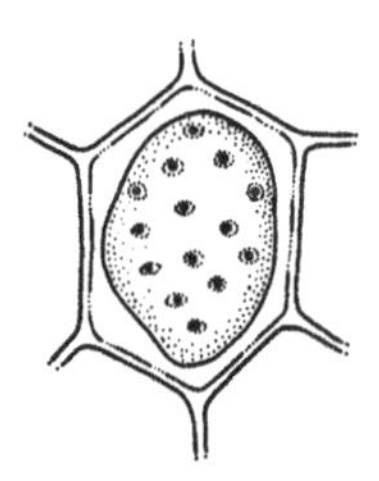

Granuloreticulosa Amoeboid protists that secrete a one- to many-chambered organic test in which calcium carbonate may be deposited or environmental materials such as sand grains may be fixed. Through pores in this test extend elongate, fine, feeding and locomotory pseudopodia that anastomose to form complex nets. 'Foram' tests may measure >1 cm in diameter and their pseudopodial nets may extend many centimetres. Several species contain photosynthetic symbionts; others are entirely carnivorous, catching organisms ranging up to nematodes and small crustaceans in size. In many, an asexually produced, uninucleate, haploid generation alternates with a sexually produced, multinucleate, diploid one. Almost all are marine, both planktonically and benthically.

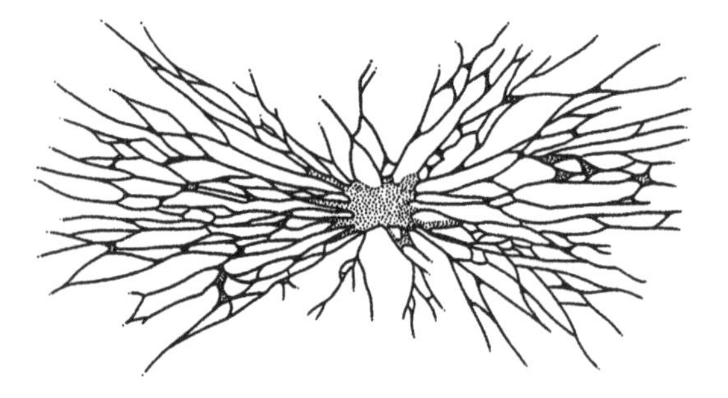

Xenophyophora Giant, benthic marine amoebae, known – in so far as they are known, which is poorly – mostly from the deep sea. Their body is in the form of a multinucleate plasmodium containing barium sulphate crystals and enclosed within a branched organic test impregnated with foreign particles (sponge spicules, foram tests, their own faecal pellets, etc.). Although mostly <10 cm in greatest dimension, some achieve 0.25 m. Feeding is thought to be mostly by means of pseudopodia.

Euglenophyta Asexual, solitary, free-living protists that may or may not contain green chloroplasts (of an unusual type, with three surrounding membranes). They move by means of a large anteriorly directed flagellum; a minute second flagellum also occurs: both are inserted in an anterior depression. In some, the two flagella are of equal length; in others more than two are present. The cell surface is covered by a tough pellicle, but there is no cellulose cell wall. Mostly freshwater in habitat, but some are marine and a few parasitic; a few form colonies and some are attached to the substratum.

Kinetoplasta Small, asexual, solitary protists, with one or two flagella issuing, as in the euglenophytes, from an anterior depression, and with a naked cell surface. A single, very long mitochondrion extends through most of the cell as a loop or network of branched threads. Although several species are free-living in aquatic habitats, many are parasitic in plants and/or animals, and are responsible, for example, for kala azar, Chagas' disease and sleeping sickness in man; a few species are colonial and some live attached to the substratum.

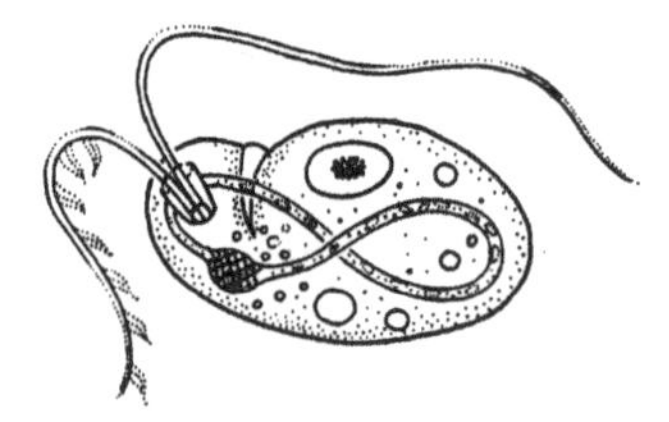

Continued

Box 3.1 *(cont'd)*

Stephanopogonomorpha Asexual, interstitial marine, free-living protists with numerous but rather sparsely distributed short flagella disposed in about 12 rows over the body surface. They resemble ciliates, but most obviously differ in that their two to 16 nuclei are all similar. They appear to feed on photosynthetic protists.

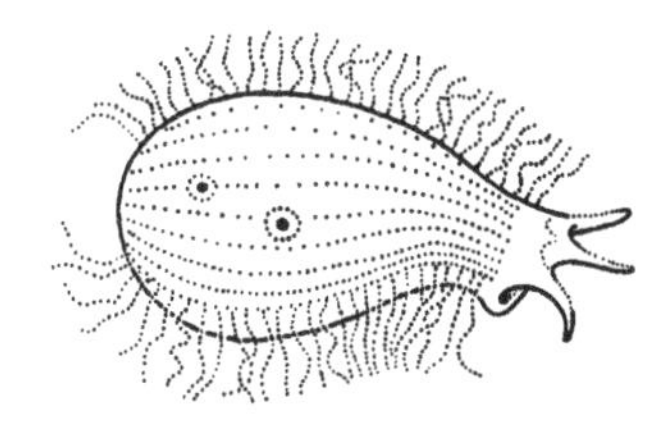

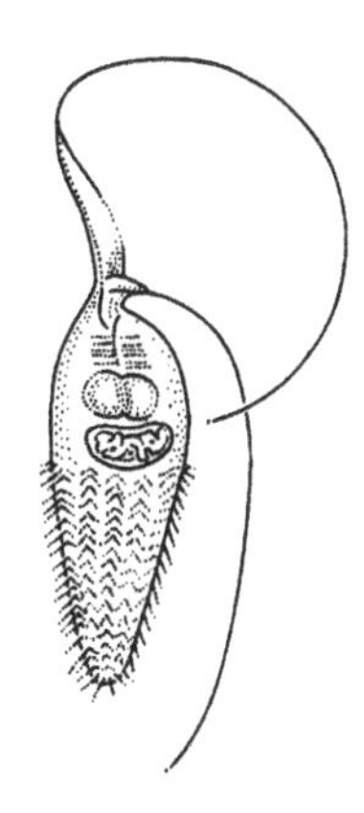

Hemimastigophora A little-known group only described in 1988, mainly on the basis of a single species discovered in soil in Australia and Chile. It is a small, uninucleate flagellate with two opposite and slightly spiralling rows each of about 12 cilia-like flagella.

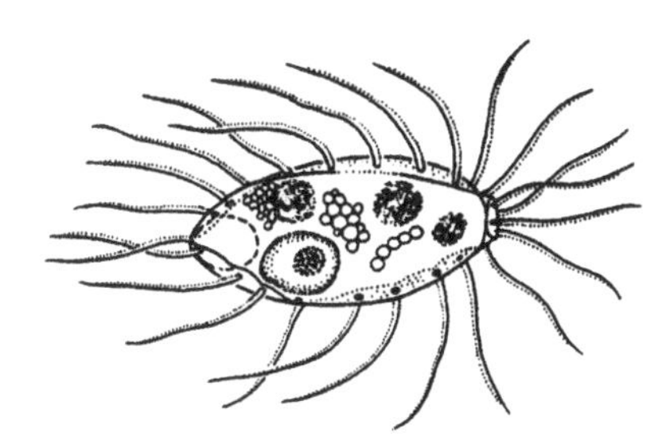

Bicosoecidea Small, free-living, aquatic flagellates, with two flagella arising from an anterior depression. They secrete a cup-shaped lorica within which they attach themselves by means of the posterior of the flagella whilst the anterior flagellum collects food materials. Most species are solitary although a few are colonial.

Choanoflagellata Free-swimming or attached, solitary or colonial flagellates with a strong resemblance to the choanocytes of sponges (see Sections 2.2.2 and 3.2.2); i.e. they bear a single, retractable flagellum surrounded by a circlet of minute filopodia or microvilli that form a bacteria-capturing filter-feeding system. Choanoflagellates usually live within a secreted membranous or gelatinous sheath or a lorica of siliceous ribs. They occur in all aquatic habitats and have been described as being 'so common in the plankton that they may be the most numerous phagotrophs on Earth'.

Labyrinthomorpha Colonial, spindle-shaped (or in one group, oval), non-amoeboid protists that move within an anastomosing slime track secreted by characteristic organelles, sagenetosomes, and containing extracorporeal cytoplasm. The latter includes actin fibres which may partly be responsible for the gliding motion of the cells. A flagellated haploid spore stage is included in the life cycle. 'Slime nets' are mostly marine saprotrophs, growing on eel grasses and seaweeds.

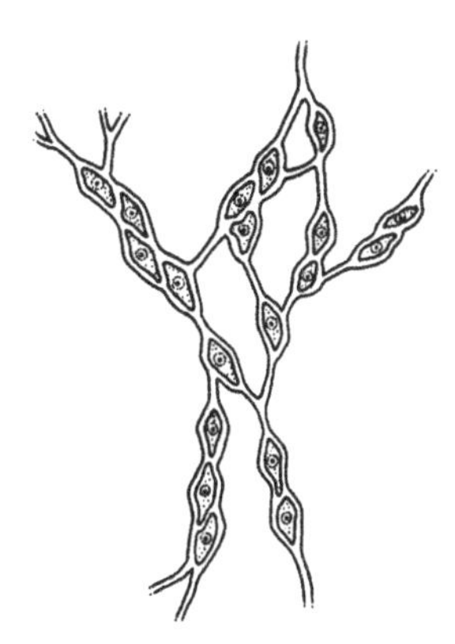

Proteromonada Small and poorly understood flagellates with one or two pairs of flagella arising directly from the cell surface. Whilst some species are free-living in aquatic habitats, sometimes colonially, others parasitize the guts of vertebrates, taking in food materials by pinocytosis.

Metamonada Phagotrophic, solitary commensals, symbionts or parasites in the guts of animals, especially insects, that lack mitochondria and Golgi bodies. They possess one or several 'karyomastigonts' – complex assemblages of fibres, one associated

Continued p. 44

Box 3.1 (*cont'd*)

nucleus, and one or two pairs of flagella – which may be associated with a curved sheet of microtubules, also enclosing the associated nucleus, and extending to the end of the cell, the 'axostyle'. A few species are free-living (but still lack mitochondria).

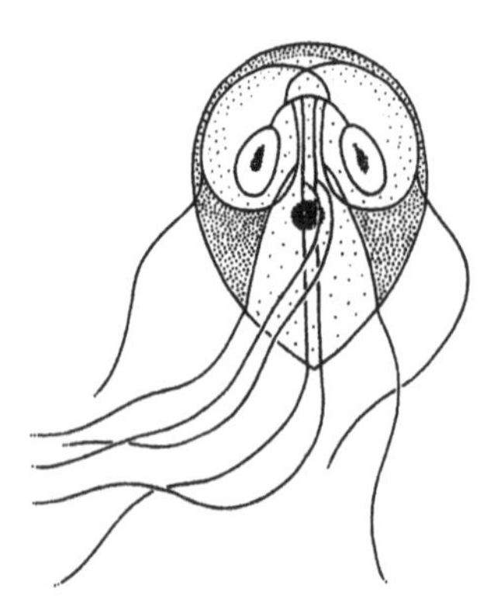

Parabasalia A somewhat similar group to the metamonads in that they possess up to several hundred karyomastigonts and lack mitochondria; the parabasalians are also symbiotic or parasitic in animal guts, especially in those of wood-consuming insects. They differ, however, in the form of their mitosis and mitotic spindles, and in their possession of Golgi apparatus that, here, are associated with the basal bodies of some of their flagella to form the 'parabasal bodies' that give these flagellates their name. From four to thousands of flagella (two to 16 per karyomastigont) occur in an anterior region. Many species contain symbiotic spirochaete bacteria.

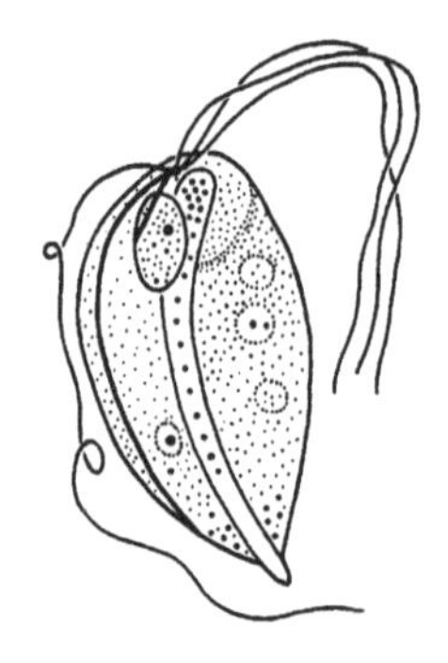

Opalinata Superficially ciliate-like, flattened, solitary flagellates that inhabit the guts of aquatic ectothermal vertebrates, especially amphibians, and absorb food materials from their hosts through their pellicles. They possess two to hundreds of similar nuclei, and many diagonal rows of numerous short flagella across the body, new rows being added during growth from a special proliferation site. Sexual reproduction is known, the gametes being uninucleate and multiflagellate.

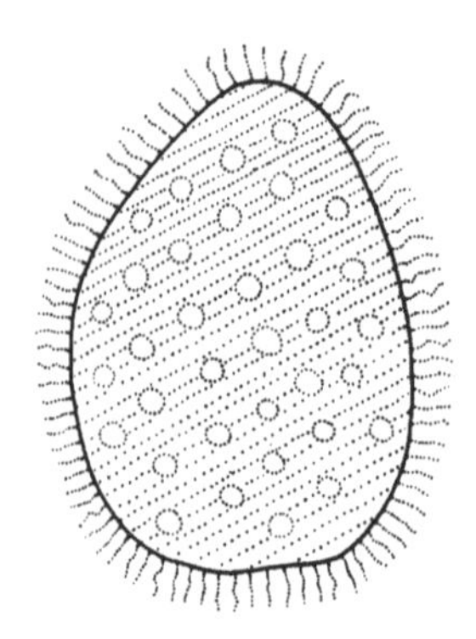

Actinopoda A large, almost certainly polyphyletic group of spherically symmetrical amoebae characterized by numerous, fine, sticky pseudopodia, each supported by an internal skeletal microtubular 'axoneme', radiating in geometrical patterns from the central cell mass. Many are large and possess an inorganic endoskeletal system: radiating spicules of strontium sulphate; a central, perforated, ball-like shell of silica with or without radiating elements; etc. These 'radiolarians', 'heliozoans' and 'acantharians' are mainly planktonic predators, although many also contain photosynthetic symbionts; some live attached to the substratum.

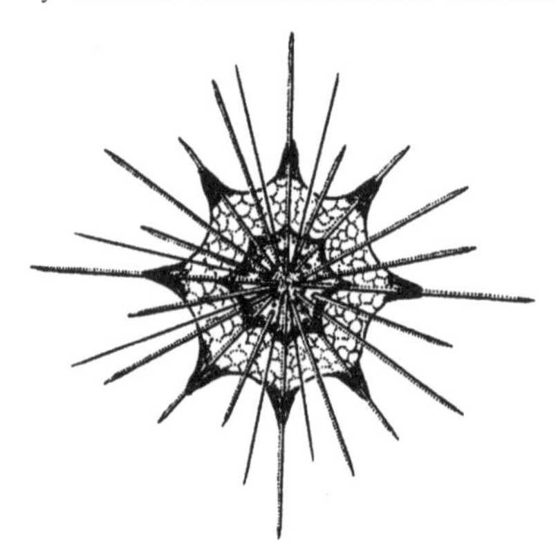

Dinophyta Motile or non-motile, thecate or athecate, mostly solitary and photosynthetic protists that bear permanently condensed chromosomes and two flagella, one commonly in a groove around the cell body, the other free. Several species, however, lack the characteristic brown chloroplasts, and take in preformed food materials saprotrophically or phagotrophically; some are parasitic (in other protists or animals), one species forms 'zooxanthellae' (see Section 9.2.7). A few contain nematocyst-like organelles (see Section 2.2.3). Most species are haploid, marine and planktonic; some live attached to the substratum, a few are colonial.

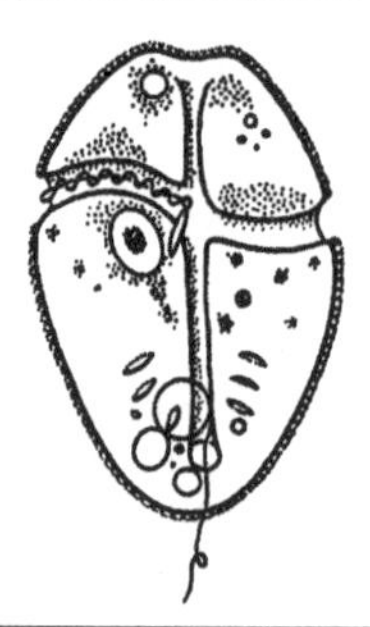

Continued

Box 3.1 *(cont'd)*

Ciliophora The 'ciliates' move by means of the cilia that occur in highly organized rows along their bodies (the number of such rows remaining constant during growth). Feeding is often also accomplished by means of cilia that may be aggregated into compound tentacle- or membrane-like organelles. Each cell contains nuclei of two types: a diploid 'micronucleus' capable of meiosis; and a polyploid, somatic 'macronucleus' not so capable and usually dividing amitotically. Sexual reproduction involves reciprocal exchange of meiotically produced haploid nuclei during conjugation. A large and diverse group containing free-living and attached, solitary and colonial, predatory, parasitic, filter-feeding and symbiont-feeding forms.

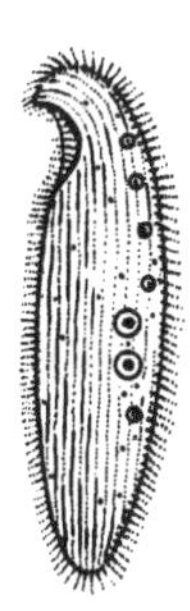

Sporozoa Parasitic protists with a complex life cycle, in which haploid and diploid generations alternate and in which at least one stage possesses an apical complex of organelles used in attachment to, or penetration of, the host cell, and in which at least one stage also possesses specialized micropores in the cell wall through which food materials are taken up. Episodes of multiple fission occur in the cycle. Mitochondria may be lacking; flagellae do not occur, except in the male gamete of some species. Sporozoans parasitize other protists and many groups of animals from all types of habitat: they are responsible for diseases such as malaria and various forms of dysentery in man.

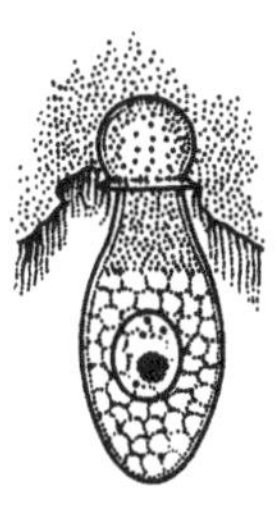

Microspora Intracellular parasites lacking mitochondria and flagella, that produce minute, resistant, chitinous, unicellular spores, each containing a single eversible filament like the cnidarian nematocyst (see Sections 2.2.3 and 3.4.2). When the filament everts, the spore plasm emerges through the tube and is injected into the host cell. It then usually develops into a plasmodium in which, in turn, large numbers of such spores may be formed within vacuoles. Microsporans mainly infect other protists, invertebrates and fish.

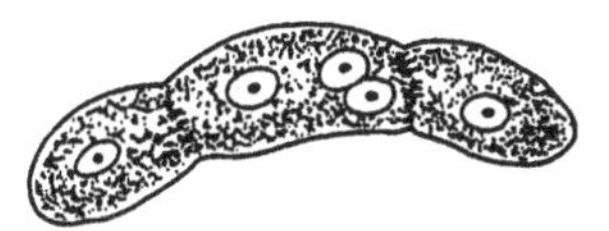

Ascetospora A small group of parasites mainly of marine invertebrates, particularly annelids and molluscs, that lack flagella and produce complex unicellular or multicellular spores that do not contain eversible, coiled, nematocyst-like filaments. The uninucleate ascetosporan spore plasm escapes from its resistant spore through an apical pore, otherwise closed by a diaphragm or operculum. It then develops into an extracellular plasmodium.

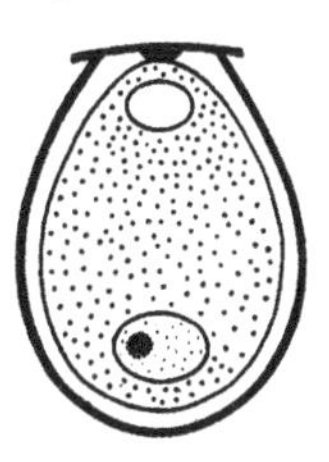

Myxospora A large group of plasmodial parasites, generally regarded as being diploid, that lack flagella and produce multicellular spores containing one to several capsules, each with an eversible, coiled filament. This filament serves to anchor the capsule to the host tissue, whilst the spore plasm escapes from the capsule between its valves, not through the filament. Spores are formed within the plasmodium by portions becoming isolated by membranes and then differentiating; some cells form within other cells by internal division (see also Section 3.5). Because the spore cells are differentiated into several types, and because of the strong resemblance of the spore capsules to cnidarian nematocysts (see Sections 2.2.3 and 3.4.2), including in the manner of their development, some believe that the myxosporans have evolved from the (multicellular) cnidarians (Section 3.4).

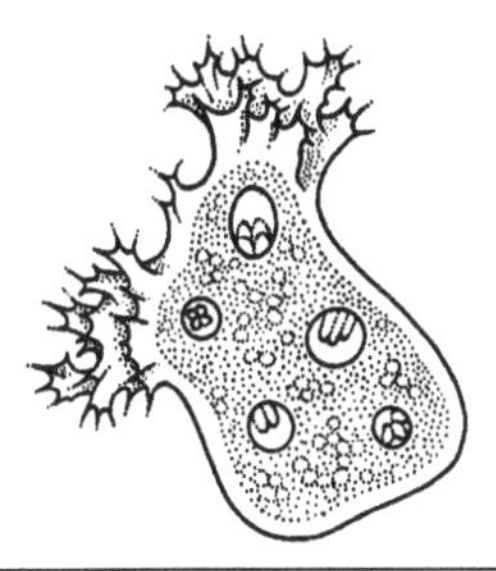

considerable number of animals, composed of thousands of cells and with all the bodily complexity of lophophorates, arthropods or chordates, measure less than 1 mm in largest dimension; most members of the pseudocoelomate groups are smaller than 2 mm. On the other hand, some ciliated protists (phylum **Ciliophora**) attain 5 cm in cell length, and the pseudopodia of some xenophyophoreans (phylum **Xenophyophorea**) may span diameters of 25 cm. The only means by which a non-colonial protist can respond to any selective pressure favouring large size (see e.g. Section 9.1) is by increase in cell size itself, often

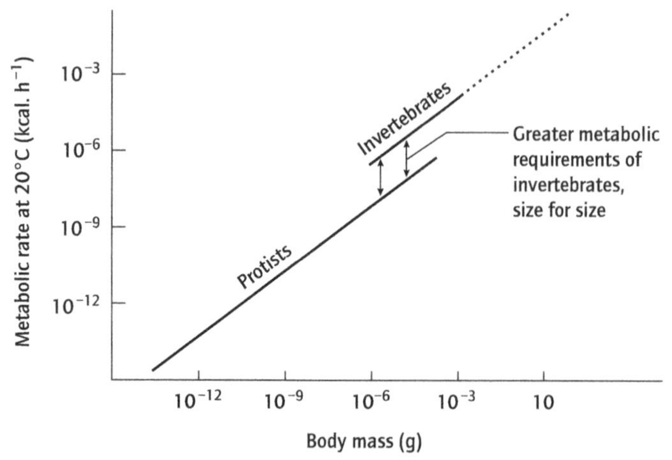

Fig. 3.2 The relationship between resting metabolic rate and body size in protists and invertebrates, showing the greater requirements of multicellular organisms, size for size (after Schmidt-Nielsen, 1984) (NB 1 kcal = 4.2 kJ).

accompanied by polyploidy or multiple nuclei, cytoplasmic streaming then forming an intracellular circulatory system. In a sense, large size is relatively easily achieved in a protist since, weight for weight, the basal metabolic requirements of a single-celled organism are less than those of a multicellular one (Fig. 3.2).

The organisms formerly regarded as constituting the phylum or subkingdom Protozoa are therefore a diverse assortment of protists, each in some respects individually equivalent to one of the component cells of an animal and in other respects to a whole animal. The use of the word 'equivalent' in the above sentence is crucial, however; the surviving protists have had an even longer separate evolutionary history than have the various multicellular organisms and only the phylum **Choanoflagellata** has retained elements of structure that indicate that it might be considered directly related to any specific group of animals (Chapter 2).

3.2 Superphylum PARAZOA (sponges)

3.2.1 Introduction

Sponges are the most inanimate of animals, differing from typical forms in that their bodies (i) show no system of symmetry, so that there are no dorsal or ventral surfaces and no anterior/posterior or oral/aboral polarity, (ii) lack any nervous and muscular cells (although see below), and (iii) are not composed of tissues and organs but are constructed of individual cells, in one group organized into syncytial sheets. They can be – and have been – regarded as being animals at the cellular level of organization.

In essence, the sponge body is in the shape of a cylinder or bag closed at one end and open at the other (Fig. 3.3). This cylinder is supported by dispersed fibrillar collagen together, in most species, with a skeletal system of often diffusely organized, calcium carbonate or silicon dioxide spicules or tough fibres of collagenous spongin, or combinations of these three, located in an internal, gelatinous and proteinaceous matrix, the mesohyl. Around the internal and to a varying extent the outer surface of the cylinder is the majority of the living material in the form of a sheet of cells. Some cells, especially totipotent amoebocytes ('archaeocytes'), are also found in the mesohyl. Water is induced to flow through the wall of this cylinder by Bernoulli's principle and/or by the beating of the flagellum of each of a series of specialized cells, the choanocytes or equivalent syncitial elements, that comprise the inner cell layer, and it then passes into the enclosed central cavity (the 'spongocoel' or atrium) and leaves the body through the open end of the cylinder, the osculum. Surrounding the choanocyte flagellum is a collar of microvilli (see Fig. 2.4) and this collar serves to filter food particles from the stream of water. Sponges, with one recently discovered exception (see below), are suspension feeders.

Describing sponges as being at the cellular level of construction can be taken as a dismissive statement; carrying the overtones that they never managed to evolve the organs, nerves and symmetry of 'proper animals'. As emphasized in Chapter 2, however, their bodily simplicity is not indicative of some failure to evolve true animal multicellularity but is directly adapted to their highly individual life style. The overriding feature of sponge architecture is that, except at the level of the individual cell, they cannot move, although they can close their oscula. Some of the cells – myocytes – surrounding each osculum have actin and myosin and can contract, thereby decreasing the aperture of the osculum to prevent, for example, the entry of unwanted materials. Sponges are thus the biological equivalent of a tubular plastic or metal plumbing system in which lowered pressure in one region serves to draw water through the tubes. It is a choanocyte-based filter feeding tube. In a sponge, the environment is moved relative to the static animal rather than the more usual animal converse. The skeletal spicules or fibres that make this system possible are secreted by sclerocytes and spongocytes, respectively, that differentiate from the basic sponge cell type, the archaeocyte.

The amount of food that can be filtered from suspension is dependent on the surface area of the inner layer of choanocytes, and the obvious way to increase this surface area is to fold it. The simplest sponges have remained simple test-tube-like cylinders (the asconoid form in Fig. 3.4), but many species possess walls folded in complex manners so that the choanocytic layer occurs along canals or within pouches (the syconoid and leuconoid forms in Fig. 3.4) instead of lining the atrium. Correlated with increased folding of the cylinder walls, the volume of the atrium is reduced so that in leuconoid forms it is represented only as a series of canals and chambers in an otherwise solid body. As size increases, so the number of oscula may also increase.

Many sponges have no definite body form. Instead size and shape are at the mercy of the external water currents, the presence of other sessile organisms, etc. In this they resemble several other modularly colonial animals and although their modularity is not visually obvious, they are in effect colonies of daughter

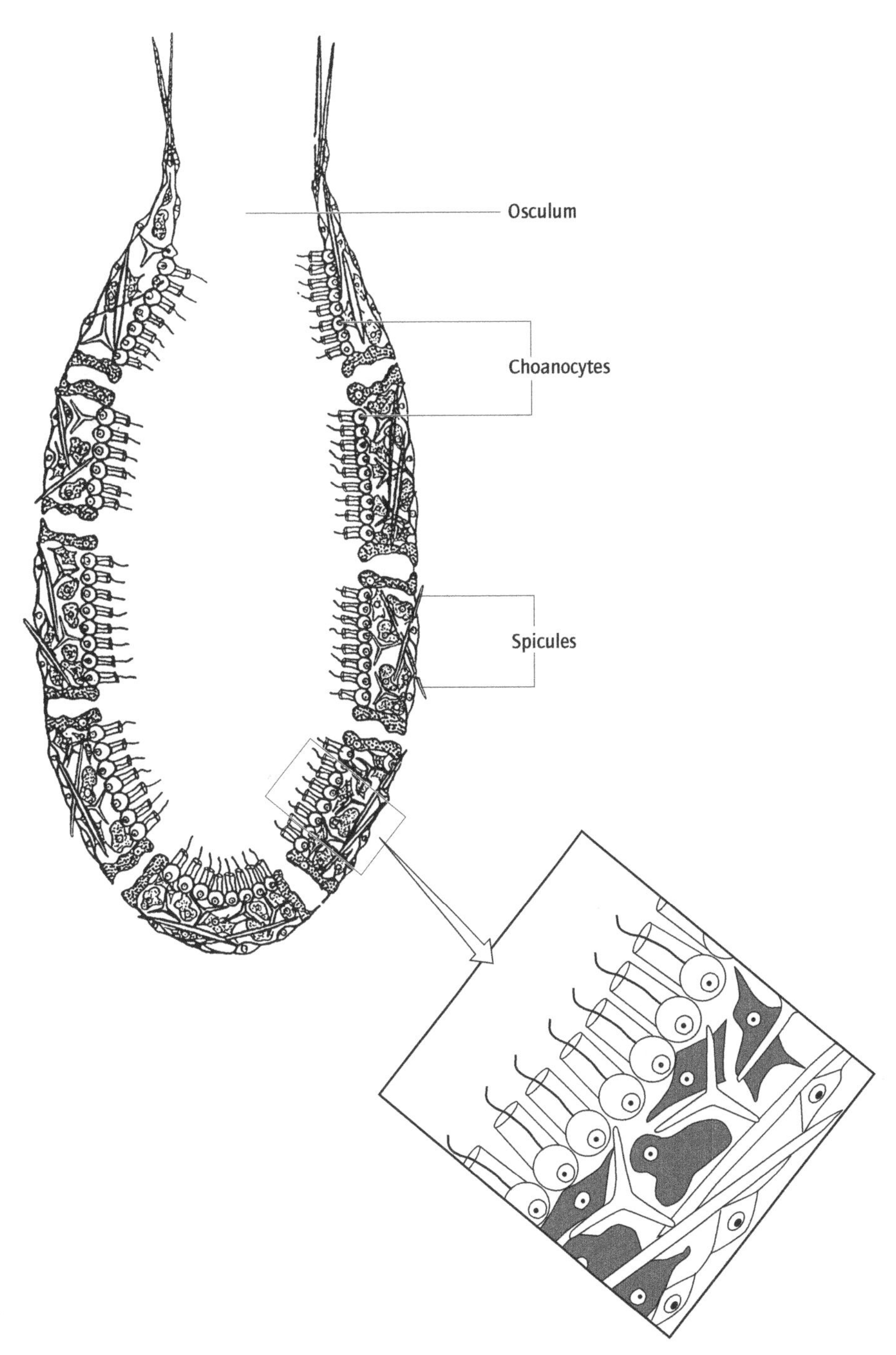

Fig. 3.3 Diagrammatic longitudinal section down the tube of a simple sponge (after Hyman, 1940).

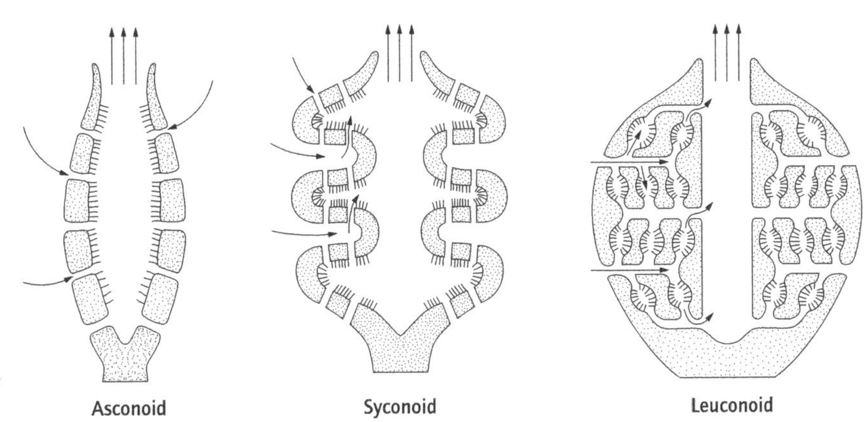

Fig. 3.4 The three types of bodily complexity in sponges, essentially created by folding of the wall of the tube. In the asconoid form, the choanocytes line the internal wall of the atrium; in the two other forms, they line cavities set within the wall, canals in the case of syconoids and pockets in the leuconoids.

cells of a very limited range of types. Cell type is notably plastic, individual cells dedifferentiating to amoeboid form and redifferentiating into other cell types: choanocytes, for example, may redifferentiate into eggs and sperm. Using this ability, some sponges can regenerate whole colonies from a few individual cells and (in freshwater) can produce small overwintering resting bodies.

Being completely sessile, sponges may appear sitting targets for predators. They are relatively immune from casual attack, however, being – like many terrestrial plants – protected by distasteful or poisonous chemicals and usually by having the majority of their biomass in the form of their rubbery or sharp, spicular skeleton that discourages consumption, not only because the proportion of digestible material per mouthful ingested is low (and therefore the costs of obtaining it may exceed the gains) but also because the actual process of ingestion may be likened to trying to eat glass fibre or a rubber ball.

Recent cytological investigations of sponges have disclosed that two rather radically different types occur, the cellular sponges and the syncitial forms which are here treated as different phyla, the Porifera and Symplasma respectively.

3.2.2 Phylum PORIFERA (cellular sponges)

3.2.2.1 *Etymology*

Latin: *porus*, pore; *ferre*, to bear.

3.2.2.2 *Diagnostic and special features*

1 Body without any symmetry, although 'colonies' often have characteristic shapes.

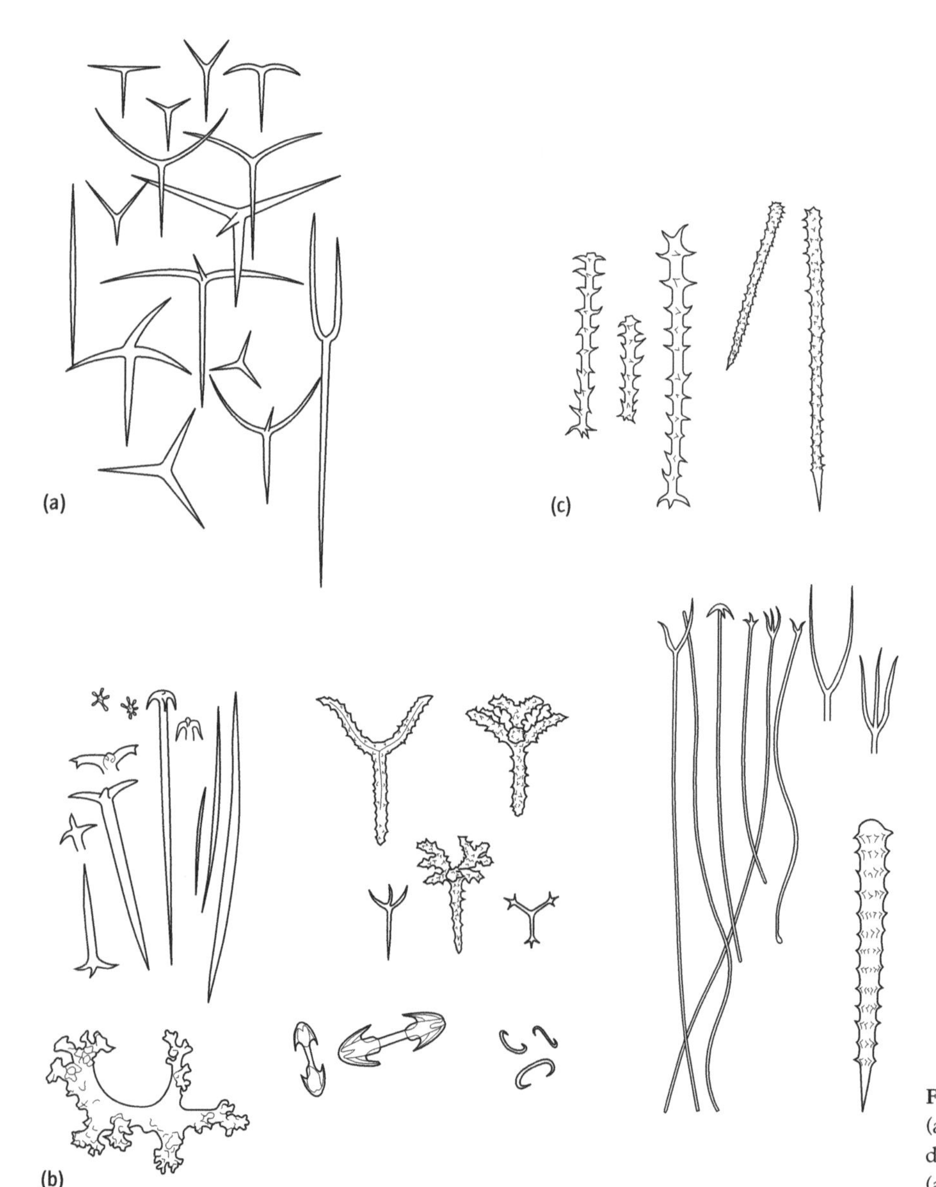

Fig. 3.5 A variety of poriferan skeletal spicules: (a) from calcareous sponges, (b) from the siliceous demosponges, and (c) from a sclerosponge (after various sources).

2 Body essentially with two layers of cells, one on either side of a gelatinous and proteinaceous mesohyl that contains free cells including totipotent amoebocytes; without organs or distinct tissues.
3 Without a gut, but inner cell layer (of choanocytes) capture and ingest food particles.
4 Body in form of solid mass enclosing one central atrium or a system of canals and chambers into which water is drawn through a series of fine pores, 'ostia', within specialized porocyte cells, and out of which it is expelled through one or a few larger apertures, 'oscula', that can be closed by the contraction of myocytes.
5 Choanocytes with flagella, the uncoordinated beating of which provides power to drive water through system, and microvillar collar which collects food particles.
6 Freshwater species with protist-like contractile vacuoles.
7 Without muscular or nervous cells.
8 Mostly hermaphrodite; sperm taken up in similar manner to food; with larval stage; asexual multiplication frequent.
9 With few cell types.
10 Sessile, immobile animals with internal fibrillar collagen skeleton, usually reinforced by inorganic spicules or organic spongin fibres, which serves to prevent movement and deter consumption.
11 Benthic filter feeders in aquatic habitats, mainly the sea.
The Porifera contains those sponges in which the individual cells are not organized into syncytia, and in which the outer margin of the sponge mass is covered by a layer of cells, the pinacocytes. Besides the dispersed fibrillar collagen, their skeletal system is usually constructed of calcareous or siliceous spicules (Fig. 3.5) and/or spongin fibres. Body form varies from small 'individual' tube-shaped species to large irregular masses with many internal canals, chambers and oscula. In all forms, water enters the sponge mass through small pores in the wall formed within characteristic porocyte cells.

Class	Order
Calcarea	Clathrinida
	Leucettida
	Leucosoleniida
	Sycettida
	Inozoida
	Sphinctozoida
Demospongiae	Homoscleropherida
	Choristida
	Spirophorida
	Lithistida
	Hadromerida
	Axinellida
	Agelasida
	Halichondrida
	Poecilosclerida
	Petrosiida
	Haplosclerida
	Verongiida
	Dictyoceratida
	Dendroceratida
Sclerospongiae	Ceratoporellida
	Tabulospongida

3.2.2.3 Classification
The 10 000 species are distributed between two or three classes and 20–22 orders.

3.2.2.3.1 Class Calcarea The two-to-four pointed skeletal spicules of the entirely marine Calcarea (see Fig. 3.5a) are composed of calcite or aragonite: they are the calcareous sponges. In one order (the Sphinctozoida), with only a single surviving species, the body form is that of a linear chain of chambers, the oldest being filled with a solid deposit of aragonite; massive calcareous plates or scales may also occur in other groups. The other orders, however, are of typical poriferan form, are relatively small (<10 cm high) and occur mainly in shallow waters (see Fig. 3.6). Asconoid, syconoid and leuconoid forms all occur.

3.2.2.3.2 Class Demospongiae This class contains some 95% of living sponge species (see Fig. 3.7), characterized by a non-calcareous but otherwise diverse skeletal system of dispersed collagen fibrils usually reinforced by horny protein (spongin) fibres, and sometimes by siliceous spicules (Fig. 3.5b) wholly or partly imbedded within the fibres (the siliceous spicules, if present, differ from those of the Symplasma, Section 3.2.3, in not being six-rayed). All but one have a complex system of canals and chambers, and some achieve large size (>1 m high and/or wide). They are mainly marine, although some 150 species occur in freshwater, the only sponges to do so. One recently (1995) described demosponge lacks choanocytes and the characteristic sponge canal system, looks superficially like a hydroid cnidarian (Section 3.4.2.3.1), and traps small crustaceans on filamentous extensions of the body that are adhesive by virtue of hook-shaped spicules. Captured animals are enveloped by migratory cells and digested.

3.2.2.3.3 Class Sclerospongiae The leuconoid sclerosponges possess colonial masses in two distinct regions (Fig. 3.8): the thin, upper, living layer has a structure equivalent to that of a demosponge, but the basal region, which is separated from the living tissue, is in the form of a massive deposition of calcium carbonate, calcitic in one order (Tabulospongida) and aragonitic in the other (Ceratoporellida). They occur in association with coral reefs, in caves and underwater tunnels, and can easily be mistaken for corals. Several authors regard the sclerosponges as being polyphyletic members of the Demospongiae, and recent molecular evidence suggests that the tabulospongidans should indeed be subsumed within the demosponge order

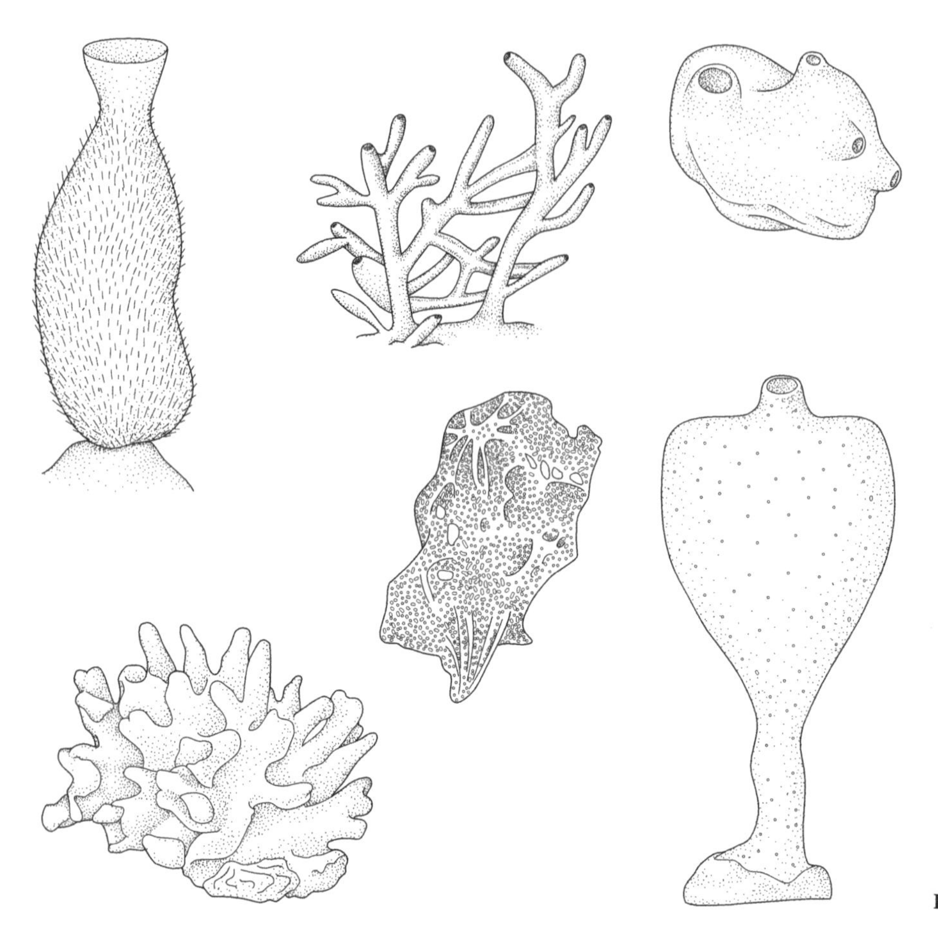

Fig. 3.6 Calcarean body form (various sources).

Hadromerida and that some at least of the ceratoporellidans should likewise be placed in the demosponge order Agelasida.

3.2.3 Phylum SYMPLASMA (syncytial sponges)

3.2.3.1 Etymology

Greek: *syn*, with; *plasma*, form.

3.2.3.2 Diagnostic and special features

1 Body without any symmetry, although 'colonies' usually have characteristic, superficially radially symmetrical shapes.
2 Body essentially with two syncytial cell layers, one on either side of a thin gelatinous and proteinaceous mesohyl that contains free cells including totipotent amoebocytes, although outer layer in the form only of strands; without organs or distinct tissues.
3 Without a gut, but inner choanosyncytial layer captures and ingests food.
4 Syconoid or leuconoid body in form of single layer of thimble-shaped chambers surrounding a central atrium into which water drawn through series of irregular spaces in outer syncytial layer and out of which it is expelled through a single osculum.
5 Choanosyncytial elements without nuclei; with flagella, the coordinated beating of which provides power to drive water through the system, and microvillar collars which collect food particles.
6 Without muscular or nervous cells.
7 Mostly hermaphrodite; sperm taken up in similar manner to food; with larval stage; asexual multiplication frequent.
8 With few cell types.
9 Sessile, immobile animals with internal skeletal matrix of fibrillar collagen reinforced by silicon dioxide spicules which serves to prevent movement and deter consumption.
10 Benthic filter feeders in the sea.

The glass or siliceous sponges differ in several major respects from the cellular sponges treated above, most notably in that their cells take the form of syncytial sheets (with even the amoebocytes in the thin mesohyl being closely associated with the syncytia), their choanosyncytial collars are without nuclei, beating of their choanosyncytial flagella is coordinated, and the outer body surface is covered only by web-like strands of syncytium. The majority have regular, and usually vase-like, shapes. The skeletal system is formed largely by (mostly) six-pointed spicules of silicon dioxide, often fused together into a rigid skeletal mass (Fig. 3.9); the single osculum may also be roofed over by a fused meshwork of such spicules, and the spicules near the base of some species form a tuft-like series of rootlets that anchors the

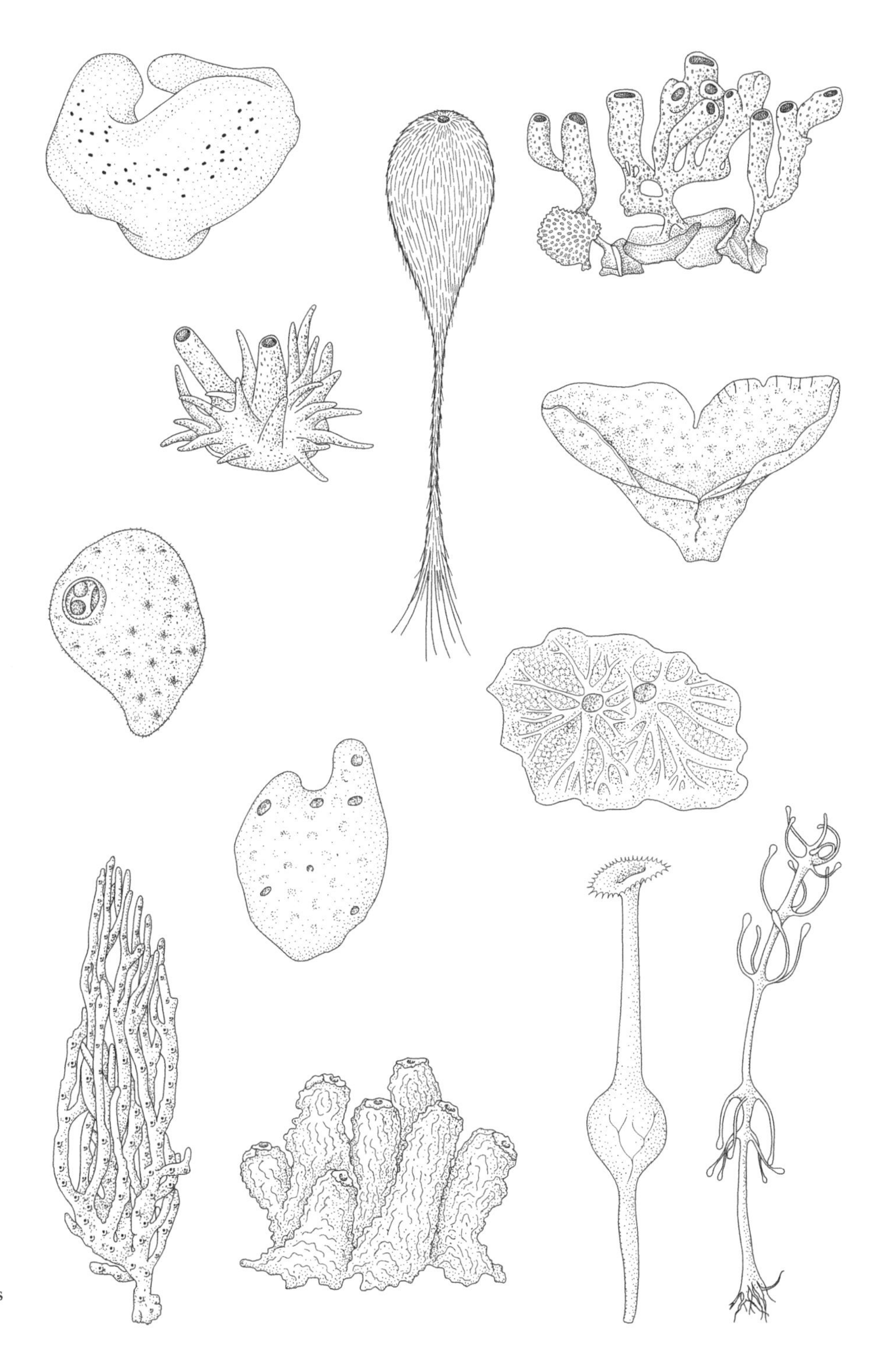

Fig. 3.7 Demosponge body form (after various sources).

sponge in soft sediments. Water enters not through porocytes, as in the Porifera (Section 3.2.2), but through irregular spaces in the meshwork of the outer syncytium. It then enters a single layer of thimble-shaped chambers that contains the choanosyncytia. All species are marine and most occur only in deep water (>500 m).

3.2.3.3 Classification

The 500 or so known species are included in a single class, the Hexactinellida, with four orders (Amphidiscosida, Hexactinosida, Lychniscosida and Lyssacinosida).

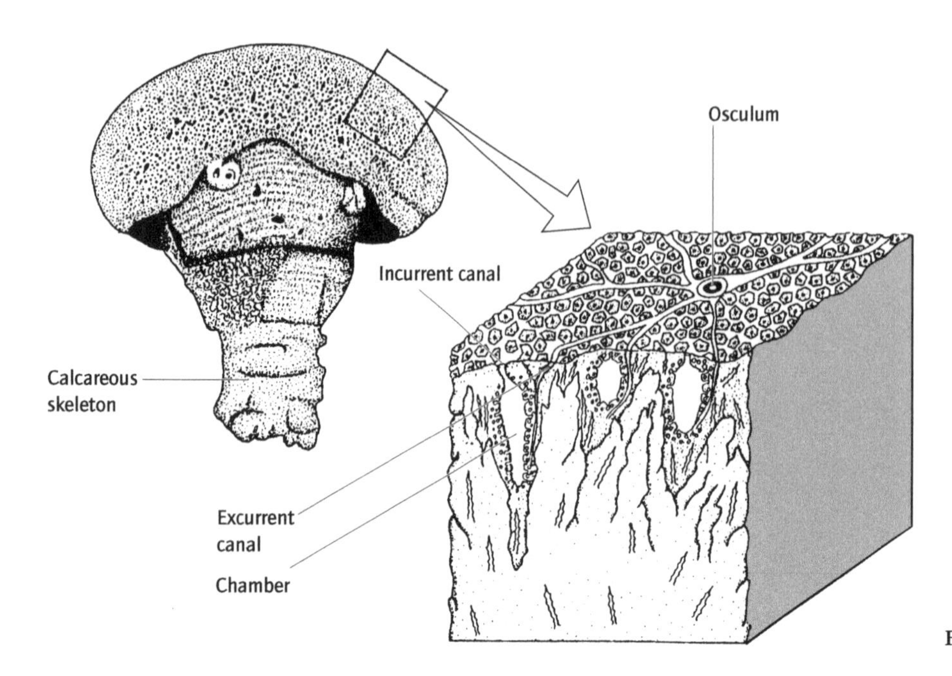

Fig. 3.8 A sclerosponge (after Bergquist, 1978).

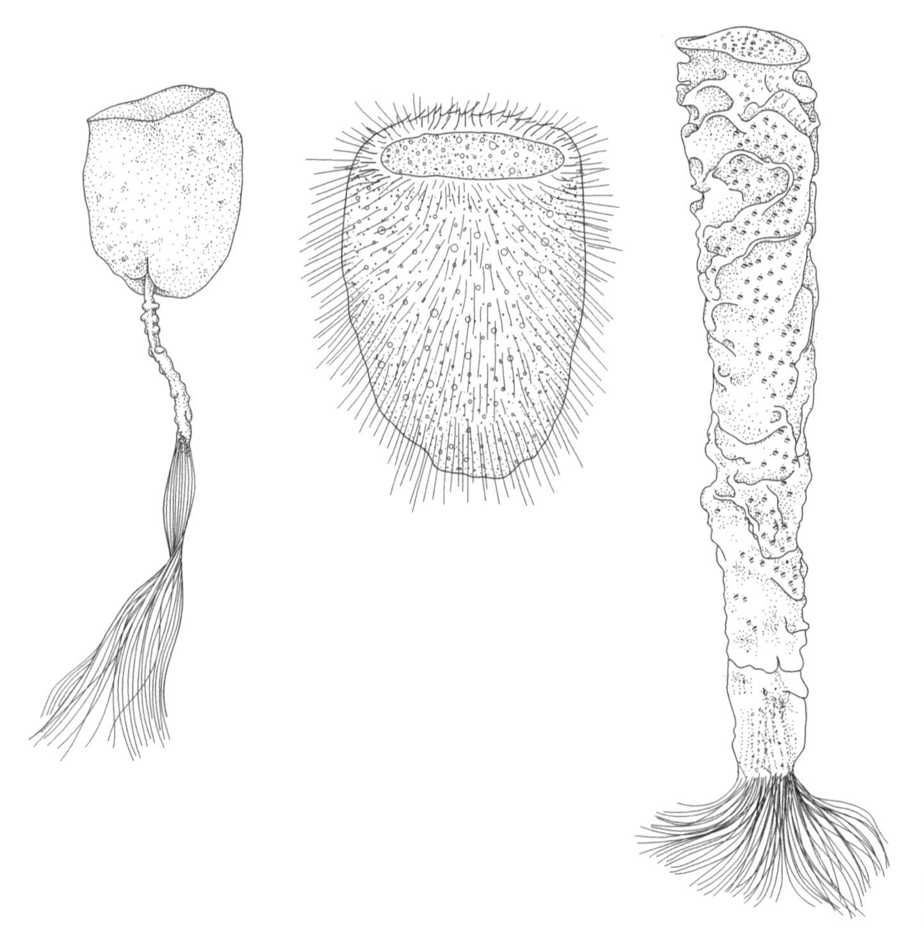

Fig. 3.9 Hexactinellid body form (after Barnes, 1998).

3.3 Superphylum PHAGOCYTELLOZOA

3.3.1 Introduction

The Phagocytellozoa contains but one known phylum and one known species. Their bodily construction is essentially similar to that of the sponges (and to that of the radiates) in that it is simply two layers of cells, one on either side of a gelatinous matrix. Also like the sponges, they lack any symmetry, tissues, organs, and nerve and muscle cells. But in this group the body is not a sessile tube but is flat and solid, taking the form of a giant amoeba and, as in amoebae, it can vary its outline by the production of pseudopodia-like extensions and move in any direction.

3.3.2 Phylum PLACOZOA

3.3.2.1 Etymology
Greek: *plakos*, flat; *zoon*, animal.

3.3.2.2 Diagnostic and special features (Fig. 3.10)
1 Without any system of symmetry, and capable of changing shape in an amoeboid manner.
2 Without distinct tissues or organs.
3 Without any body cavity or digestive cavity.
4 Without any system of nervous co-ordination.
5 Body in the form of a flat plate, which can move in any direction in the plane of the body.
6 A single outer layer of flagellated cells enclosing a fluid-filled mesohyl containing a network of stellate fibre cells.
7 Marine.

The upper surface of the flat bodily disc is formed of a thin layer of squamous cells, many of them bearing a flagellum, whilst the thicker lower layer is composed of two cell types, flagellated columnar cells, interspersed with non-flagellated glandular cells (Fig. 3.11). These gland cells secrete enzymes on to their prey (various protists), which are therefore partly digested outside the body, and the digestion products are then absorbed by the same secretory cells. The ventral flagellated cells are responsible for a gliding form of locomotion, whereas the changes of shape which cause placozoans to resemble large amoebae are brought about by the co-ordinated contraction and relaxation of the fibre cells in the mesohyl.

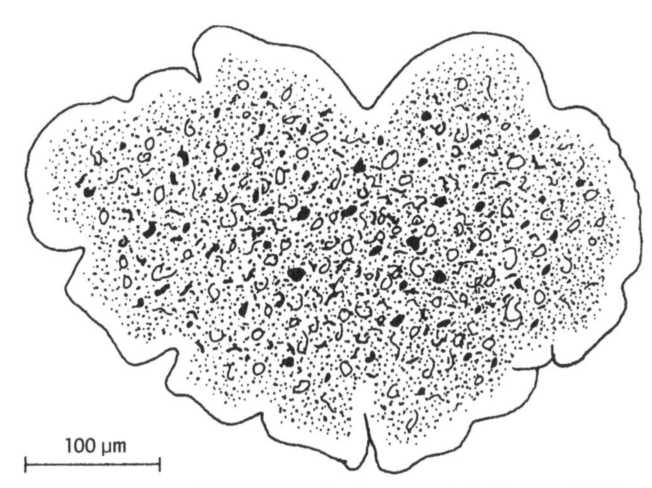

Fig. 3.10 The general appearance of a placozoan (after Barnes, 1980).

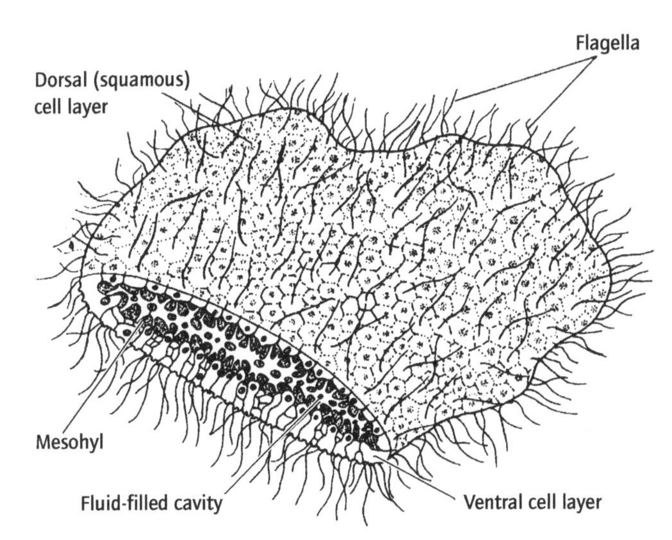

Fig. 3.11 A diagrammatic section through a placozoan (after Margulis & Schwartz, 1982).

Asexual multiplication, by both fission and budding, commonly occurs, and although sexual reproduction does take place, both it and the subsequent development of the embryo are very poorly known. Eggs, probably deriving from the lower cell layer, may be present in the mesohyl, and sperm have been reported from the water in which placozoans were being maintained, but sperm production and fertilization have not yet been observed.

The body contains a few thousand cells and can attain a diameter of 3 mm, although placozoans contain only four cell types and therefore are one of the simplest known multicellular organisms. Their component cells also contain less DNA than any other animal (of the same order of magnitude as in many bacteria) and their chromosomes are minute, less than 1 µm in length. For almost a century they were regarded as the planuloid larva of some type of sponge or cnidarian, until in the late 1960s it was discovered that they could achieve sexual maturity and a new phylum was created for them in 1971.

Placozoans have most often been recorded from marine aquaria and they are known in the wild from the intertidal zone. They are probably quite widespread in the sea, though easily overlooked.

3.3.2.3 Classification
Only one (or possibly two) species has been described. The phylum is sometimes placed in a separate subkingdom, the Phagocytellozoa (an unfortunate name since uptake of food is not usually by phagocytosis).

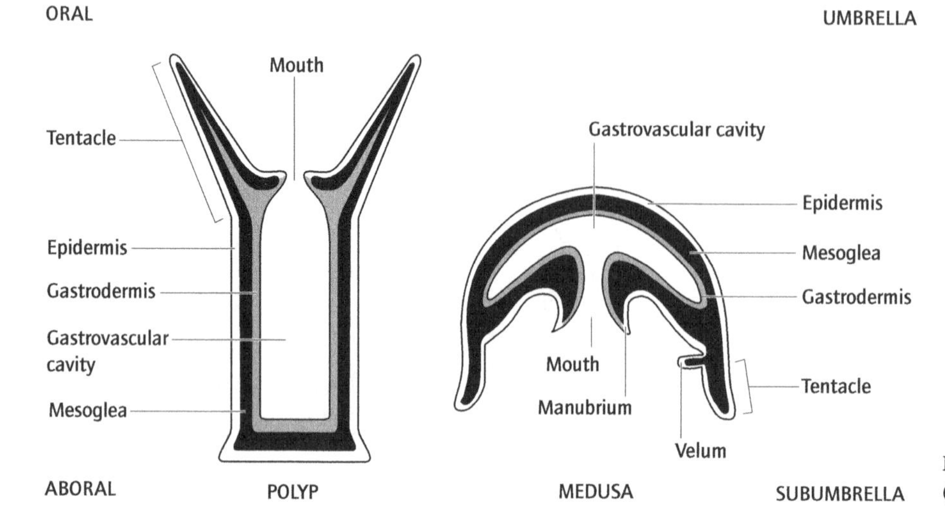

Fig. 3.12 The two body forms of cnidarians (after Fingerman, 1976).

3.4 Superphylum RADIATA

3.4.1 Introduction

As in the Parazoa and Phagocytellozoa above, the radiates (or coelenterates as they are often known) have a bodily organization of just two layers of cells, one on either side of a gelatinous matrix, in this superphylum termed the 'mesogloea'. Like the sponges, this sandwich construction also forms the wall of a cylinder that is closed or almost closed at one end. But unlike the sponges and placozoans, and indeed other animals, (i) the body is radially (perhaps originally tetraradially) symmetrical, (ii) true tissues, though not organ systems, occur, as do individual muscle cells, and (iii) individual, naked, unsheathed, nerve cells are present in the form of a network or networks all around the tube, although no nerve cords are found. They have been called animals at the tissue level of organization. The radiates are organisms capable of movement that characteristically catch and consume animal prey, the lumen of their tubular bodies (the coelenteron) serving as a gut for this purpose, the discharge of specialized cellular organelles being the agents of prey capture as well as defence.

Two phyla are often, as here, included in the superphylum, one, the Cnidaria, uncontentiously so. Inclusion of the Ctenophora is questionable, however, and their treatment under this heading is largely for convenience in that if they are not radiates where they should be placed is equally contentious – affinity with both the acoelomate worms and the deuterostomes has been argued.

3.4.2 Phylum CNIDARIA (hydroids, jellyfish, anemones, corals)

3.4.2.1 Etymology
Greek: *knide*, nettle.

3.4.2.2 Diagnostic and special features (Fig. 3.12)
1 Body radially symmetrical.
2 Body essentially with two layers of cells, one on either side of a gelatinous mesogloea which may or may not contain cells.
3 Body in form of elongate to flattened tube, open at one end and closed at other, enclosing central cavity (coelenteron or gastrovascular cavity); open end of tube drawn out into series of tentacles surrounding single mouth/anus; with tissues but without any distinct organs.
4 With individual partly-muscular cells, and network of naked nerve cells at the base of each cell layer.
5 Body in two forms – elongate 'polyp' attached aborally to substratum and with mouth and tentacles held uppermost, and flattened swimming disc- or bell-shaped 'medusa' with mouth positioned in middle of lower surface. In many, the two forms alternate during life cycle, the polyp multiplying asexually and the medusa sexually, in others, one or other of the forms reduced or suppressed.
6 Polyps often interconnected and forming modular colonies; polyps monomorphic or polymorphic within single colony.
7 With characteristic intracellular nematocyst or nematocyst-like organelles within cnidocyte cells, each nematocyst a coiled and often barbed thread capable of forcible eversion to serve for offence and defence or more rarely in locomotion or tube-building.
8 Often with secreted external or internal support or protection of chitin or calcium carbonate.
9 Gonochoristic or hermaphrodite, usually with external fertilization and development via a planula larva.
10 Medusae swim by pulsations, muscle cells acting against elastic mesogloea; polyps usually sessile, but can use water in coelenteron as a hydrostatic skeleton during limited movement.
11 Almost exclusively carnivorous, although some partially or totally dependent on intracellular symbiotic photosynthesizers.
12 Aquatic, mostly marine, pelagic and benthic.

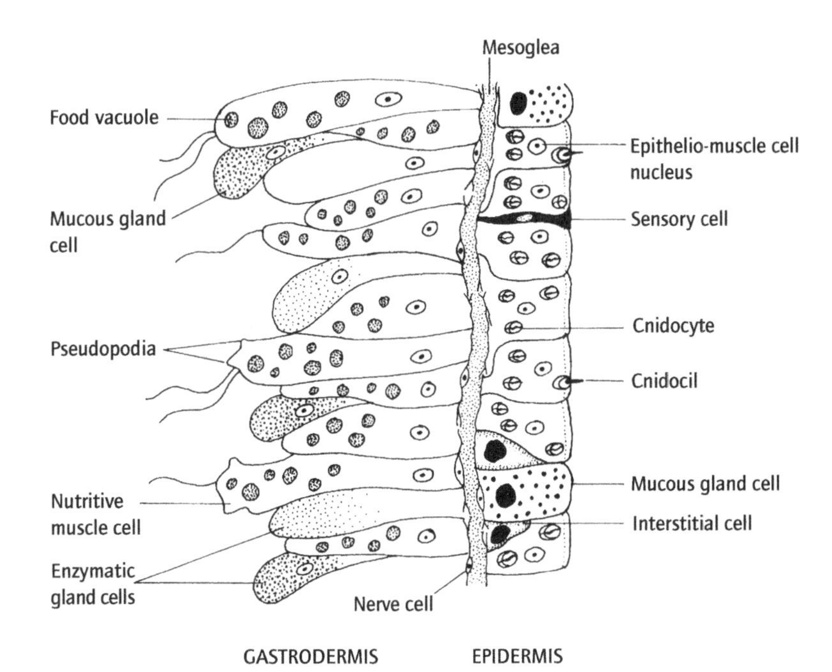

Fig. 3.13 Longitudinal section through the body wall of a hydrozoan; see Box 3.2 (after Barnes, 1980).

Cnidarians are a diverse group of radiates that can achieve a high level of complexity in spite of their very simple body form. This results from repeated asexual production of interconnected modular units, sometimes coupled with considerable modular polymorphism. An individual colony formed in this manner may contain feeding, reproductive, propulsive, defensive and other specialized modular units: these in a sense are the equivalent of the organs of most other animals. Colonies, and sometimes individual units, may secrete a calcareous, chitinous or equivalent supportive and protective external casing, the calyx; those with a calcareous calyx, generally referred to as corals, have created reefs of great ecological and geological importance. The true skeletal system of the cnidarians, however, is the mesogloea and/or water contained within the coelenteron.

Their basic, tissue-level structure of just two layers of cells, an outer epidermis and an inner gastrodermis, one on either side of the gelatinous mesogloea (Fig. 3.13 and Box 3.2), is not radically different from that of a sponge. And as in that group, the three layers together form the wall of a tube closed at one end. But in the cnidarians muscle cells, co-ordinated by nerve nets, occur, and these coupled with the form of their skeletal system permit movement. This makes possible a radically different life style. Since the body wall in the region around the single opening to the tube is drawn out into one or more whorls of mobile, solid tentacles, these can be used to catch prey which is then conveyed into the tube's lumen for digestion. Individual prey items are subdued by the action of phylum-specific cells, the cnidocytes, each containing an intracellular organelle, most commonly a nematocyst, in the form of a coiled thread (Fig. 3.14). On being triggered, nematocysts forcibly evert (as a result of the pressure increase associated with an inrush of water into the high osmotic pressure cnidocyte) and usually inject a toxin into the prey, although some cnidocyte organelles discharge to release an adhesive substance and yet others entangle potential food items. Once its contained organelle has discharged, the cnidocyte is resorbed and a new one is differentiated.

Cnidarians occur in either or both of two body forms that typically alternate during the life history and are often associated with an alternation in the means of multiplication. Sexual reproduction gives rise to a small, ciliated, solid-bodied planula larva which eventually develops into the benthic polyp phase. This is tubular, attached to the substratum aborally, and bears the tentacles and mouth directed upwards into the overlying water. It multiplies asexually both to produce other polyps, which may lead an independent life or form the units of a modular colony, as well as to the bell- to saucer-shaped, planktonic medusoid stage. Medusae swim by rhythmic contractions of the bell or saucer and most live with their mouth and tentacles directed downwards. The medusoid stage in turn develops gonads and reproduces sexually (Fig. 3.15), completing the cycle. One or other of the two phases may be lost from the life history, and if the medusa stage is absent the polyp develops gonads as well as multiplying asexually. Variations on this basic life history, however, are numerous.

3.4.2.3 Classification

Two subphyla are usually recognised: the Medusozoa, in which a medusa occurs in the life history (and is often the main body form), and the Anthozoa in which a medusa is lacking and in which the polyps may be large (up to 1 m in diameter) and relatively complex. In fact although the anthozoans appear to be polyps, it is possible that they are really sessile medusae in origin. There are five cnidarian classes, 28 orders and some 10 000 species.

Box 3.2 *Hydra*, the common but atypical hydroid

Because of its wide availability both in terms of frequency of occurrence and local abundance, the small polyp 'hydra' (*Hydra*, *Chlorohydra* or *Pelmatohydra*) is often used for teaching purposes as an example of a hydroid cnidarian. In some ways this is unfortunate because hydra is unusual in many respects: it is freshwater rather than marine; it is solitary not colonial; it does not possess any secreted external calyx; the medusa stage is completely lacking from its life cycle; the polyp stage is not polymorphic; and there is no planula larval stage.

Because of its familiarity and because at least its structure at the cellular level is typical of hydrozoans, hydra will serve as a convenient example of the nature of basic cnidarian cell types (see Fig. 3.13).

Epidermis

The Epidermis contains:

1 Epithelio-muscular cells. These are columnar, or occasionally squamous, 'cover' cells that also each have two or more basal extensions each containing a contractile myofibril that is oriented along the animal's oral–aboral axis. They therefore function as longitudinal muscles as well as an external epithelium.

2 Interstitial cells. These are recognisable by their large nuclei and small amount of cytoplasm: they can differentiate into other cell types as required, the cell turnover rate being in any event high.

3 Cnidocytes (see Fig. 3.14 and text). In hydra as in hydrozoans generally, as well as scyphozoans, these bear a stiff cilium-like structure – the cnidocil – that serves as the triggering device. Discharge is effected by a very rapid uptake of water resulting from a change in membrane permeability, forcibly everting the nematocyst filament out of the organelle and cell. All nematocysts of hydra are of the toxin-injecting type and are covered by a lid-like operculum. Some anthozoans, however, possess additional types of cnidocyte organelles, and anthozoan cnidocytes characteristically lack opercula and cnidocils although functionally equivalent tripartite flaps and ciliary cones may replace them, respectively. Such additional types are the zoantharian 'spirocysts', which discharge to form an adhesive net used for prey capture and/or attachment to the substratum, and the sticky ceriantharian 'ptychocysts', which are used in tube construction.

4 Mucous gland cells. The mucus secreted by these cells aids in adhesion, prey capture and protection.

5 Nerve cells. These are equivalent to the multipolar neurones described in Chapter 16, although they are usually directionally non-polar. They synapse with other such cells to form an irregular network around the polyp at the base of the epidermis, and also with receptor cells that are elongated at right angles to the epidermal surface and project through it as fine 'hairs'. In some hydrozoans, there are two epidermal nerve nets; in most cnidarians there are both epidermal and gastrodermal nets – when two nets are present, one is usually a fast-conduction pathway.

6 Eggs and sperm. In the breeding season, some of the epidermal interstitial cells differentiate into eggs or sperm. The epidermal mass of sperm – the 'testis' – opens to the outside via a nipple through which the sperm escape, whereas each 'ovary' contains a single egg together with associated nutritive cells. The egg is fertilized *in situ* and in some specimens the resultant embryo may been seen enclosed within a chitinous capsule. This system is most unusual amongst cnidarians, although the epidermal 'gonads' are a typical hydrozoan feature. External fertilization is the cnidarian norm; most hydrozoan polyps only multiply asexually; and the 'gonads' of anthozoans and scyphozoans are located in the gastrodermis.

Mesogloea

In hydra, the mesogloea is very thin and without cells, but in various other cnidarians (especially in the medusoid form and in anthozoans) it is thicker and more fibrous, and in the anthozoans it contains amoebocytes.

Gastrodermis

In hydra, the gastrodermis forms a relatively simple, tubular lining to the gastrovascular cavity, although in non-hydrozoan cnidarians it extends into the cavity in a series of folds in the oral–aboral plane, each fold being supported by mesogloea; these folds are termed 'septa' or 'mesenteries'. The following cell types occur:

1 Nutritive muscle cells. These engulf food particles and carry out intracellular digestion, and their flagella move food through the gastrovascular lumen and mix its contents. They also have basal extensions containing contractile myofibrils that are oriented at right angles to the animal's oral–aboral axis, and therefore function as circular muscles.

2 Enzymatic gland cells. These secrete proteolytic enzymes that effect some extracellular digestion.

3 Mucous gland cells. These mainly occur near the mouth, and are similar to those of the epidermis.

4 Nerve cells. The gastrodermal nerve cells are similar to those of the epidermis and are similarly located adjacent to the mesogloea. Although they occur somewhat infrequently in hydra they are more abundant in other cnidarians.

5 Cnidocytes. Gastrodermal cnidocytes occur in many cnidarians (see above), though not in hydra.

6 Symbiotic algae. *Chlorohydra*, many other cnidarian polyps and a few medusae, bear unicellular, photosynthetic symbionts within their gastrodermal cells: chlorophyte 'zoochlorellae' in the case of the freshwater hydras and dinoflagellate 'zooxanthellae' in the marine cnidarians (see Chapter 9). In some species, they occur in the epidermal cells as well, and even extracellularly in the mesogloea.

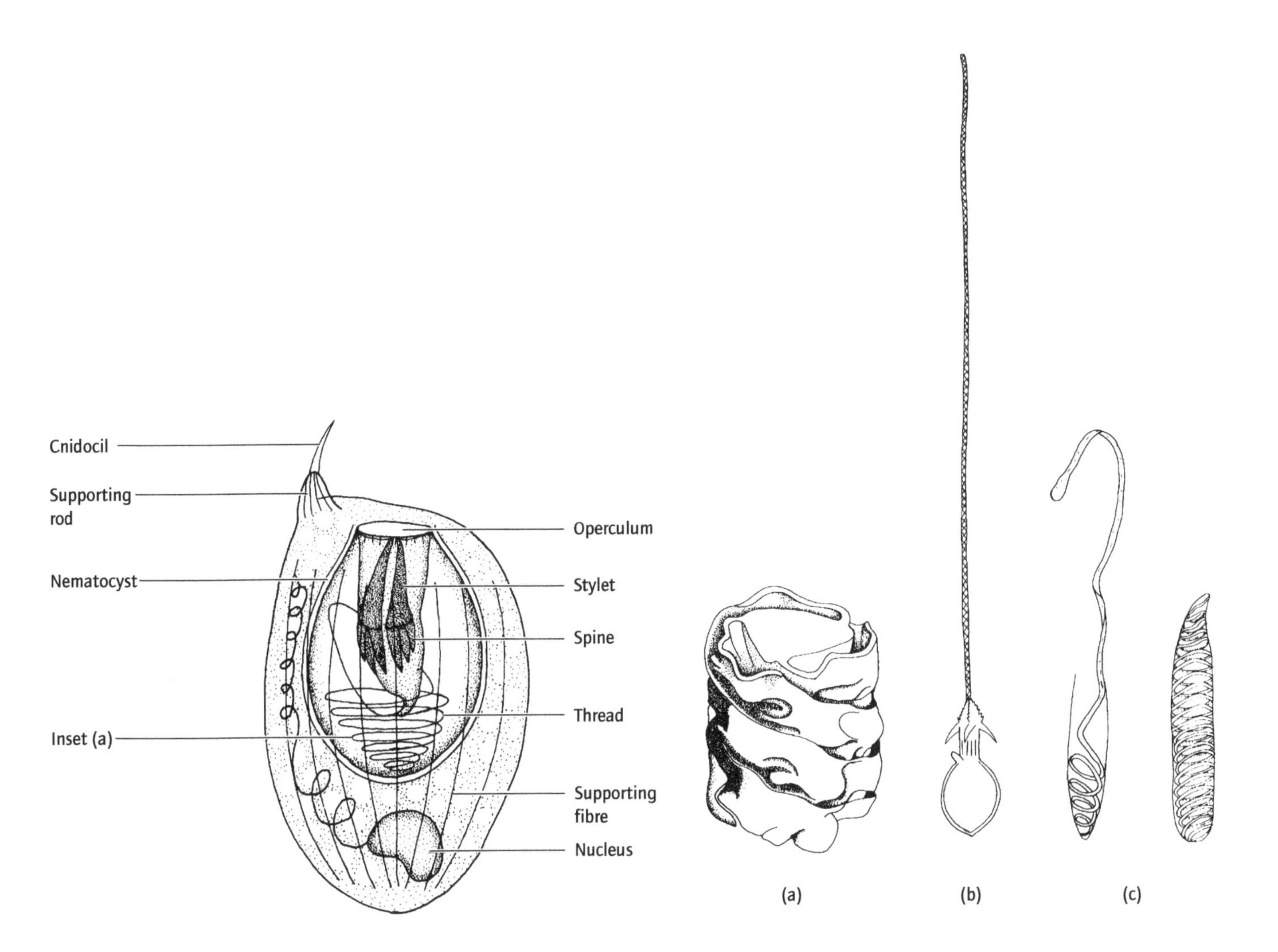

Fig. 3.14 Cnidocyte and nematocyst form; (a) the undischarged thread is pleated; (b) the discharged thread; (c) the spirocyst of an anemone. (After various sources.) (See also Fig. 2.5.)

3.4.2.3.1 Class Hydrozoa The marine and freshwater hydrozoans comprise the hydroids and (hydro)medusae (Fig. 3.16); both stages characteristically alternate through the life history although either may be reduced or absent. Hydrozoans are characterized by having epidermal gonads, a mesogloea without cells, and by the presence of cnidocytes only in the epidermis. The hydroid polyp is relatively simple, with a coelenteron unpartitioned by inpushings of the gastrodermis, and they are usually modularly colonial and polymorphic (the familiar hydras being unusual in this respect; see Box 3.2). In many forms the whole colony is encased in a chitinous calyx that may be reinforced with calcium carbonate. Medusae often remain attached to the polyp stage, and there they may give rise not to planula larvae, but to a later developmental stage, the actinula larva (Fig. 3.17), which although polyp-like is planktonic. The 'medusae' of some orders and the 'polyps' of the order Actinulida appear in origin to be free-living and attached actinula larvae, respectively. The true medusae are usually small and have a shelf of tissue extending inwards from the margin of the bell, the velum. The pelagic siphonophore colonies are the most polymorphic of all hydrozoans, with several types of both polyps and medusae within a single entity that could lay claim to be an 'individual' organism rather than a modular colony (see Fig. 3.18).

3.4.2.3.2 Class Scyphozoa In the 200 species of scyphozoans or 'jellyfish', all of them marine, the medusa is the dominant phase in the life cycle (Fig. 3.19a), the polyp phase or 'scyphistoma' being small, short-lived or absent. When present, the polyp has its coelenteron partitioned by four septa. It divides transversely along the oral-aboral axis ('strobilates') to bud off juvenile medusae, termed 'ephyrae' (Fig. 3.19b), so that a single polyp gives rise to many medusae. The adult medusa, which can be very large (>2 m in diameter and with tentacles up to 70 m long), differs from the equivalent stage in the hydrozoans by having a well-developed, cellular mesogloea (the jelly of the jellyfish), gonads in its gastrodermis, a mouth often drawn out into lobes, and in lacking a velum. Cnidocytes also occur in both of the cell

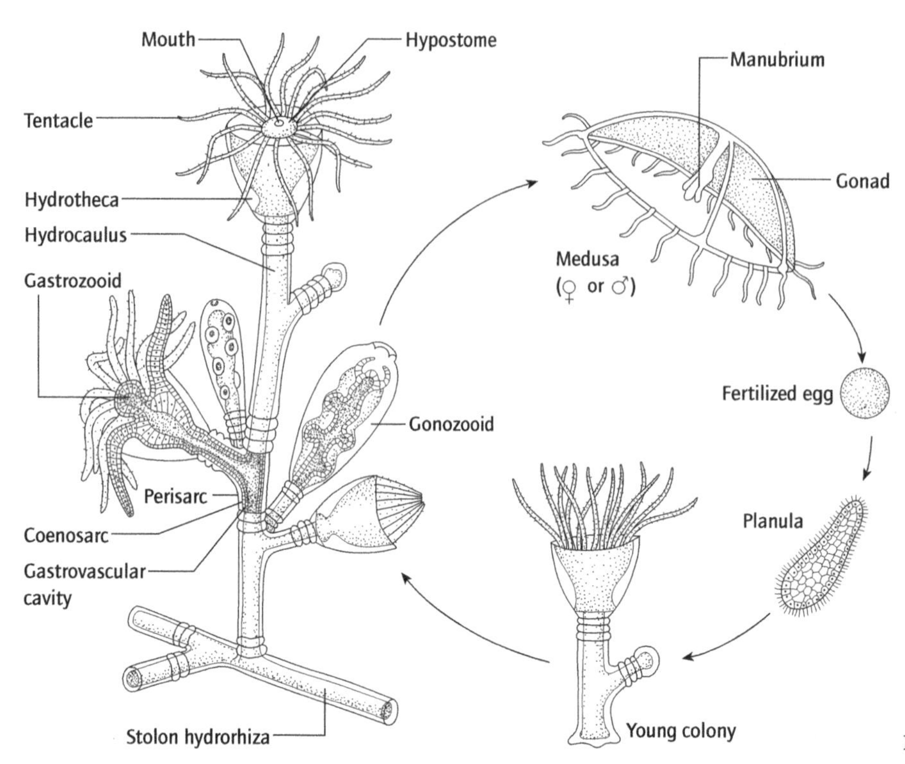

Fig. 3.15 Life cycle of *Obelia* (after Barnes, 1980).

Subphylum	Class	Order
Medusozoa	Hydrozoa	Limnomedusae
		Laingiomedusae
		Narcomedusae
		Trachymedusae
		Actinulida
		Anthoathecata
		Leptothecata
		Siphonophora
	Scyphozoa	Stauromedusae
		Coronatae
		Semaeostomeae
		Rhizostomeae
	Cubozoa	Cubomedusae
Anthozoa	Alcyonaria	Protoalcyonaria
		Stolonifera
		Telestacea
		Gastraxonacea
		Gorgonacea
		Coenothecalia
		Alcyonacea
		Pennatulacea
	Zoantharia	Actiniaria
		Zoanthinaria
		Scleractinia
		Corallimorpharia
		Ptychodactiaria
		Antipatharia
		Ceriantharia

layers. Although most jellyfish are major planktonic predators, some filter feed or are dependent on photosynthetic symbionts for their nutrition. Variations include medusae attached to the substratum aborally (the Stauromedusae), thereby behaving like polyps, and even species in which the medusa is absent from the life history and the scyphistoma reproduces sexually.

3.4.2.3.3. Class Cubozoa The 15 species of largely tropical, marine cubozoans are also jellyfish but their medusa is somewhat intermediate between those of the hydrozoans and scyphozoans in that it possesses a velum (like the hydrozoans) and hydrozoan-like nematocysts, but has a thick cellular mesogloea and gastrodermal gonads (like the scyphozoans). The polyp phase is similar to the actinula larva of some hydrozoans and, insofar as is known, transforms directly into a small medusa (i.e. does not strobilate). This stage is box-like (square in section and flat sided) with a tentacle or tentacles at each of the lower corners of the box (Fig. 3.20), and hence their common name of 'box jellies'. Besides the tentacles, the margin of the medusa may also bear up to 24 quite well-developed eyes, complete with cornea, lens and multilayered retina. Some species, at least, appear in effect to copulate. They are notorious because their nematocysts cause excruciating pain and death to humans brushing against them whilst swimming.

3.4.2.3.4 Class Alcyonaria The alcyonarians are modularly colonial but usually non-polymorphic, marine anthozoans that possess an internal supporting system of horny material or of fused or separate spicules of calcium carbonate secreted

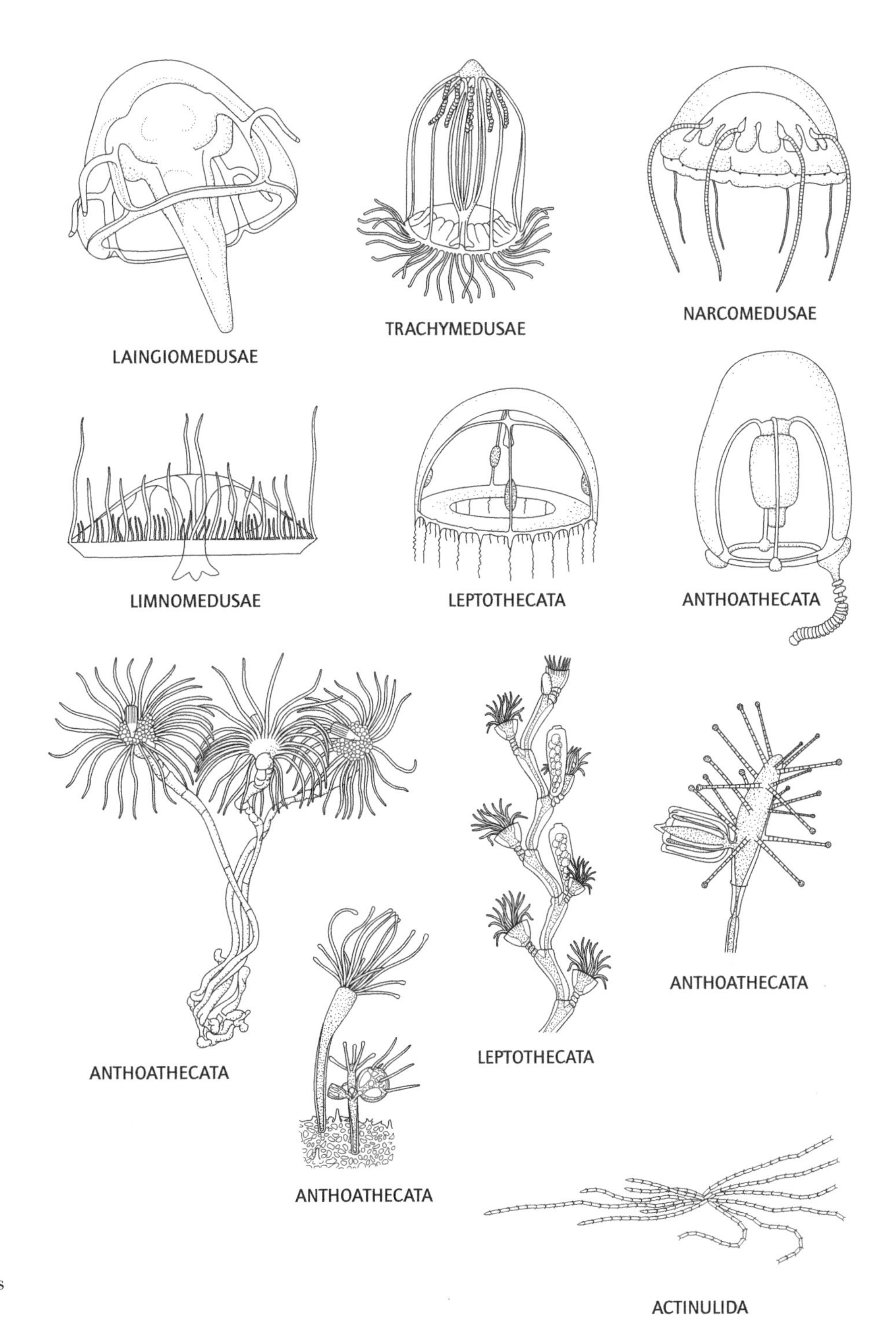

Fig. 3.16 Medusae and polyps of hydrozoans (after various sources).

within the mesogloea by amoebocytes. The polyps, which are all attached to each other and intercommunicate, bear eight pinnate tentacles and possess a coelenteron divided by eight longitudinal septa (hence the alternative name of 'octocorals'), each septum being an inward extension of the inner body wall (gastrodermis + mesogloea). Like the other anthozoans, their mesogloea is thick and cellular, cnidocytes occur in both cell layers, the gonads are gastrodermal, and a sleeve-like pharynx extends from the mouth more than half way down the coelenteron. In alcyonarians, this sleeve bears a single ciliated groove, the siphonoglyph, that creates a water current into or through the coelenteron. Organ-pipe coral, blue coral, precious red coral, sea pens, sea fans and soft corals are all alcyonarians (Fig. 3.21).

Although seemingly anatomically adapted to be the standard cnidarian predator, many shallow-water alcyonarians have

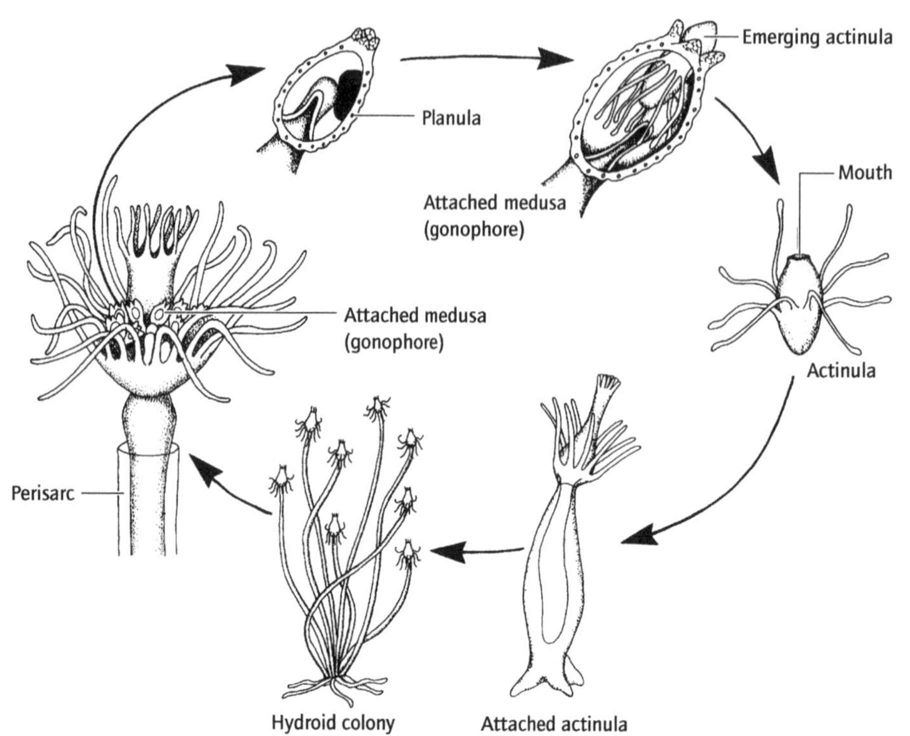

Fig. 3.17 Life cycle of *Tubularia* (after Barnes, 1980).

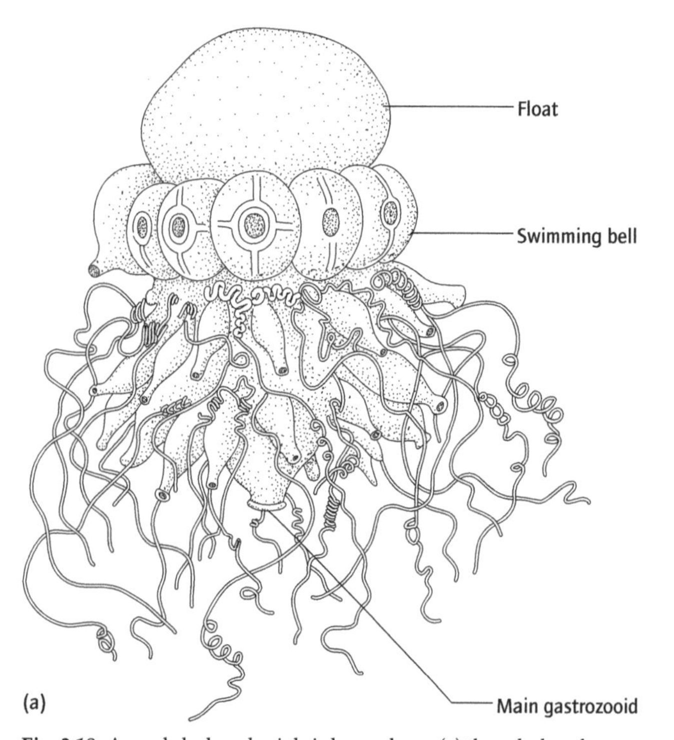

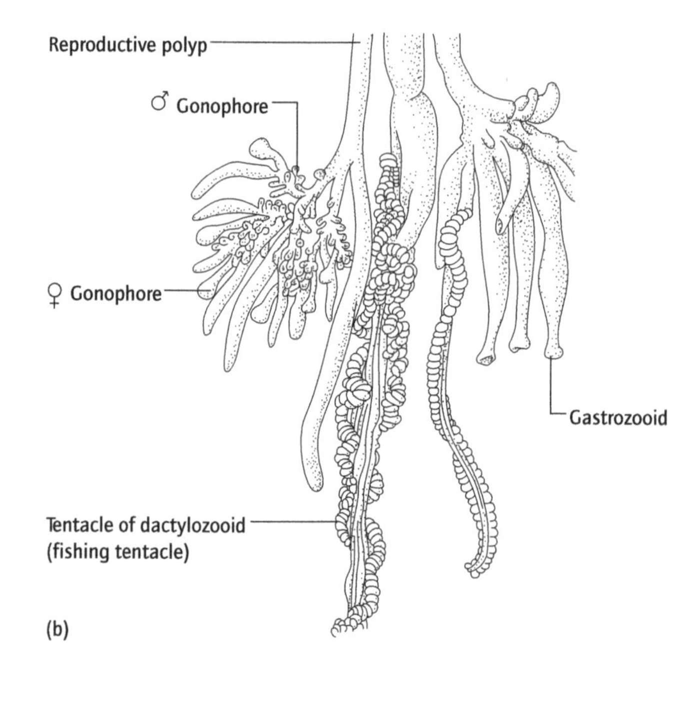

Fig. 3.18 A modularly colonial siphonophore: (a) the whole colony; (b) detail of some of the polymorphic modules (from Barnes, 1998).

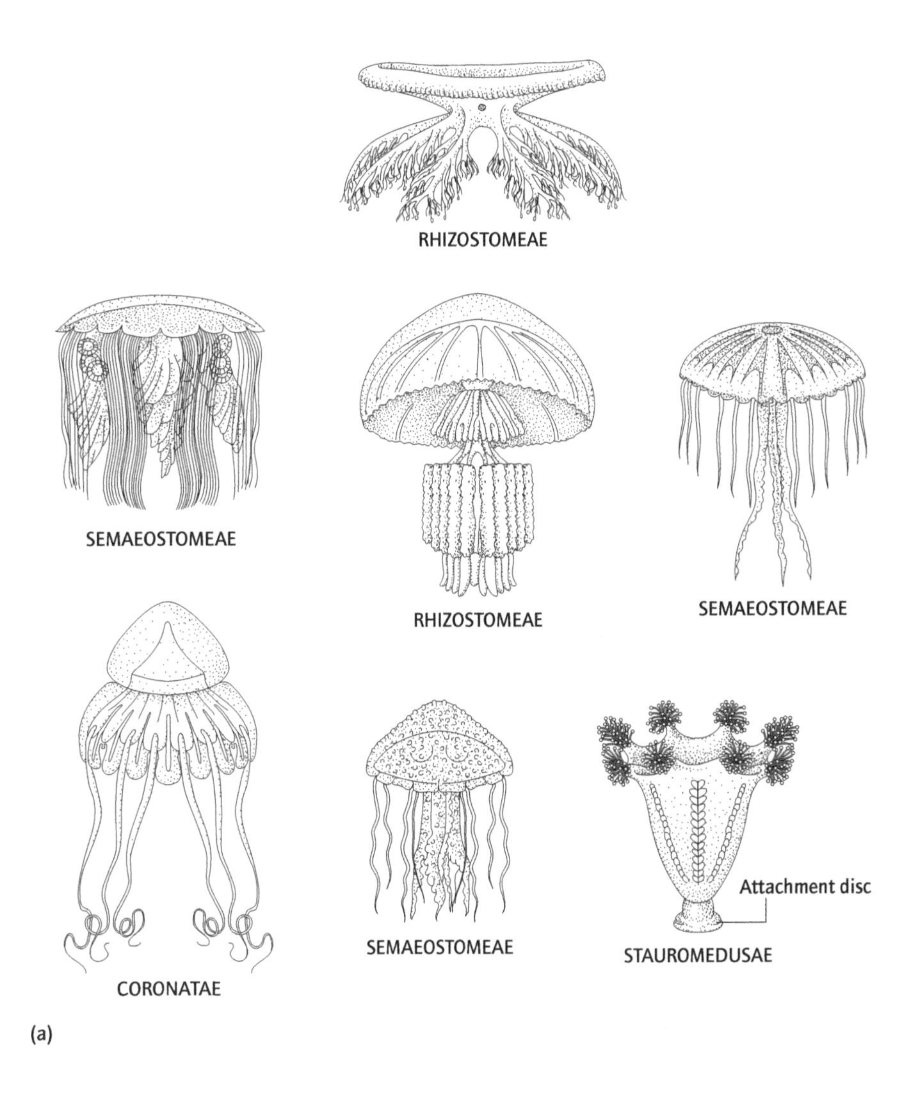

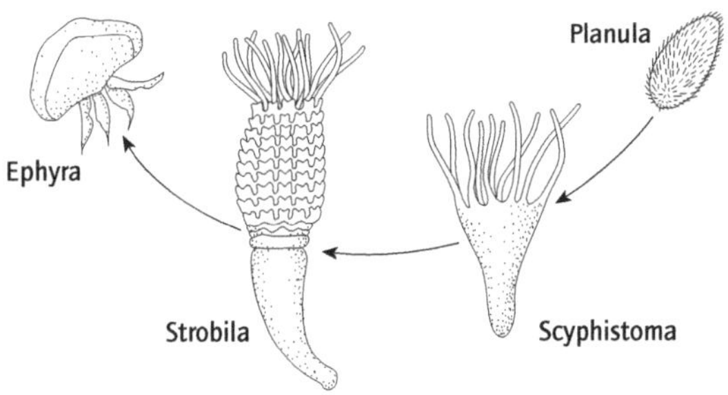

Fig. 3.19 (*right*) The Scyphozoa: (a) Adult scyphozoan body form; (b) typical scyphozoan life cycle (after several sources).

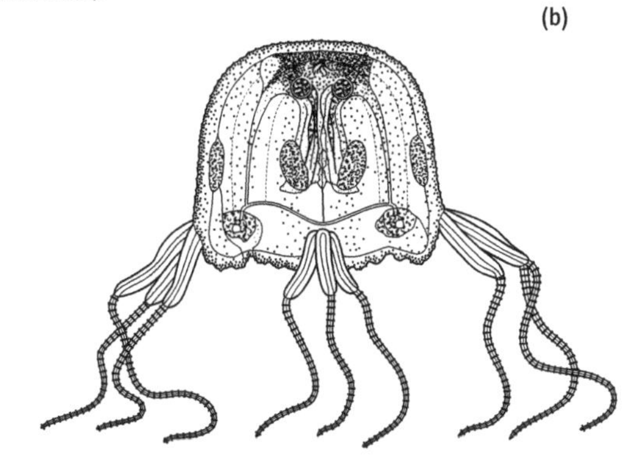

Fig. 3.20 A Cubozoan (after Barnes, 1998).

relatively poorly developed nematocysts and rely in large measure (some entirely) on 'milking' intracellular populations of the symbiotic, photosynthetic dinoflagellate *Symbiodinium* ('zooxanthellae'). They can also feed on dissolved organic matter and planktonic bacteria.

3.4.2.3.5 Class Zoantharia The zoantharians (Fig. 3.22) are solitary or modularly colonial, non-polymorphic, marine anthozoans that may (as in the stony corals for example) or may not (as

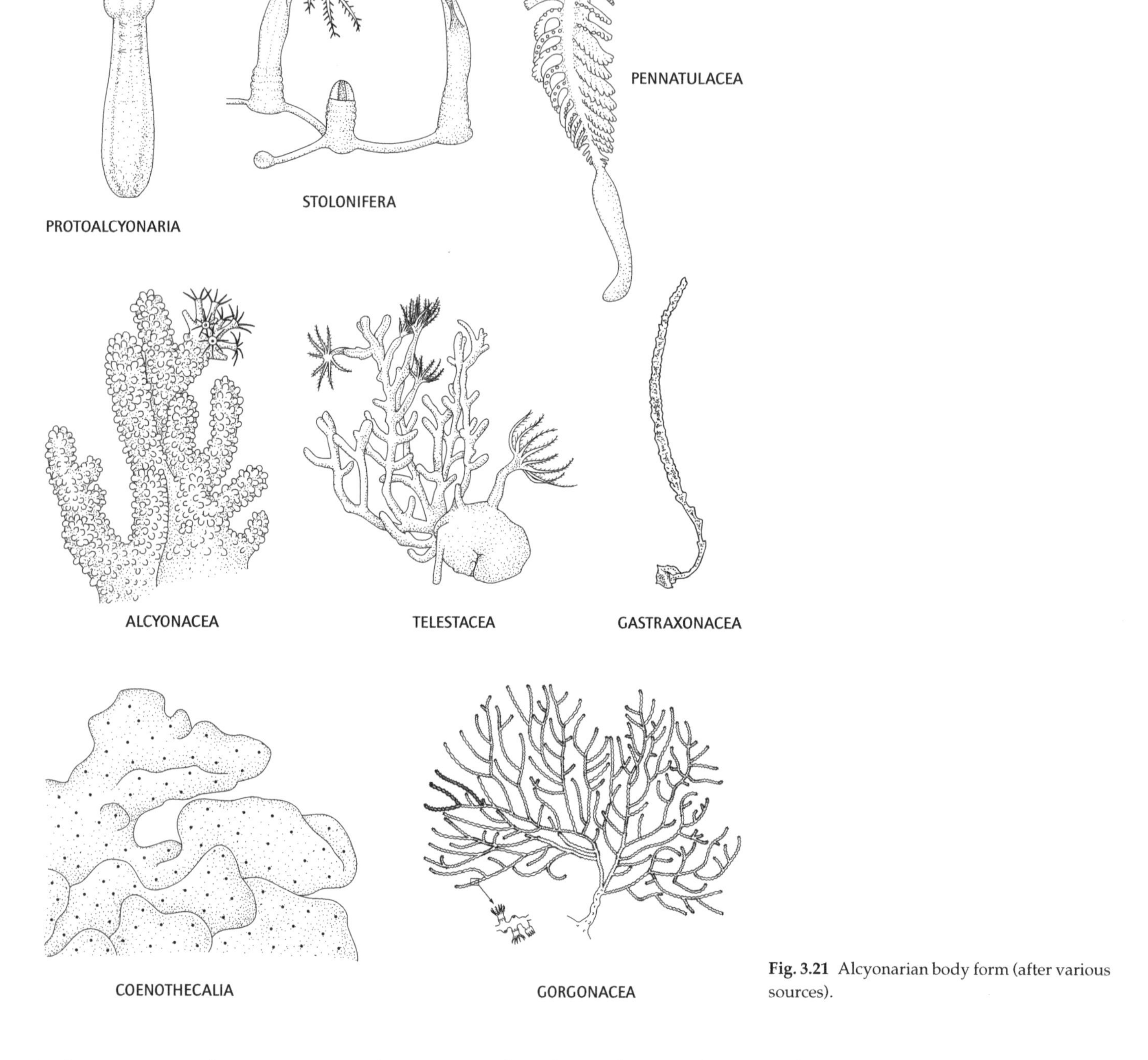

Fig. 3.21 Alcyonarian body form (after various sources).

in the sea anemones) secrete an external supporting calyx of calcium carbonate in which they are embedded (Fig. 3.23). The solitary forms are nevertheless often clonal in that asexual multiplication is frequent. In contrast to the alcyonarians, their often numerous, simple (i.e. non-pinnate) tentacles are in multiples of six and they have a coelenteron divided by paired longitudinal septa also in multiples of six (hence the alternative name of 'hexacorals'), each septum being an inward extension of the inner body wall (gastrodermis + mesogloea). Like the other anthozoans, however, their mesogloea is thick and cellular, cnidocytes occur in both cell layers, the gonads are gastrodermal, and a sleeve-like pharynx extends from the mouth more than half way down the coelenteron. In zoantharians, this sleeve bears up to two ciliated grooves, the siphonoglyphs, that creates a water current into or through the coelenteron. The two siphonoglyphs that are the norm, the slit-like sleeve and the arrangement of the septa impart a considerable measure of bilateral symmetry to the group.

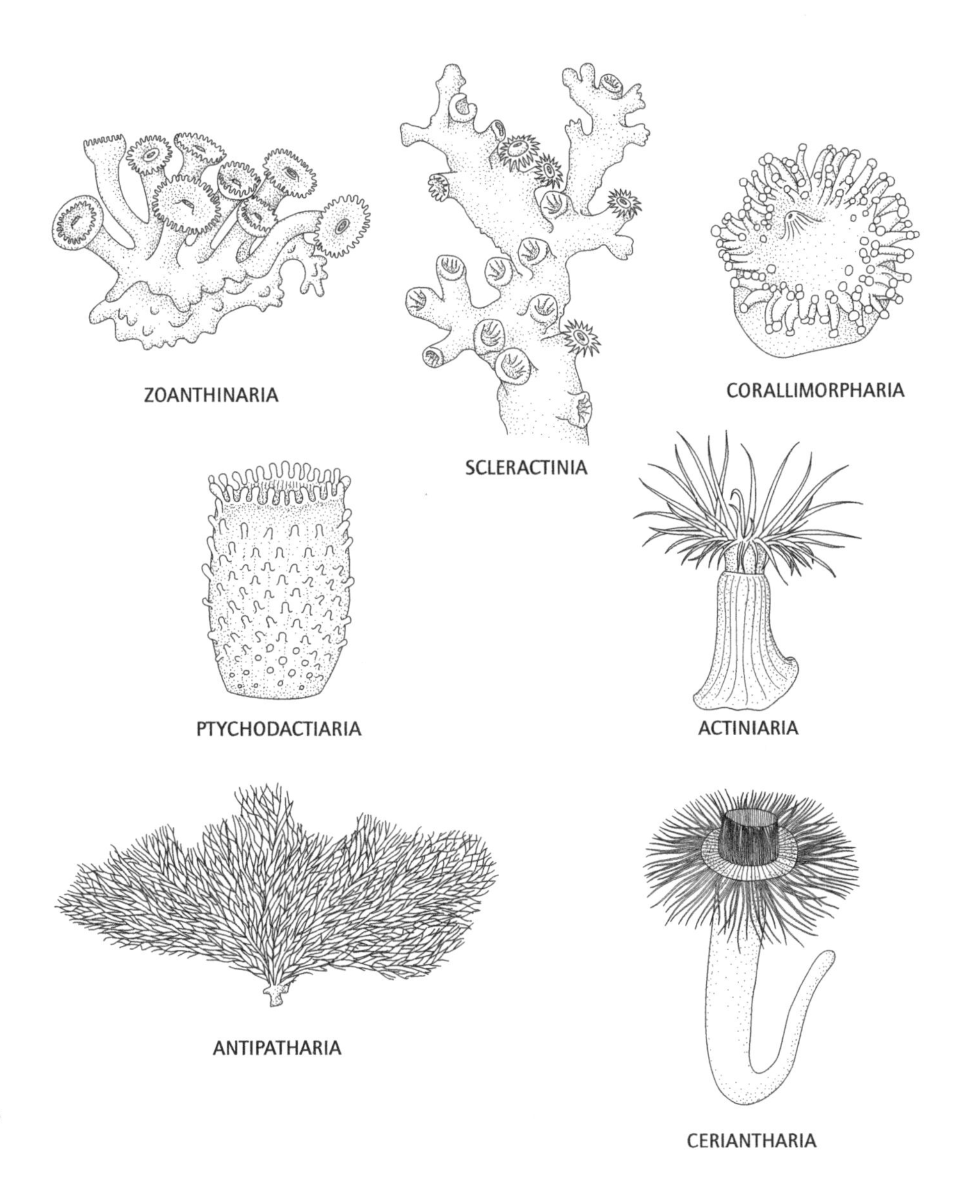

Fig. 3.22 Zoantharian body form (after various sources).

One tentacleless, deep-water coral has a coelenteron perforated through to the outside by numerous 1–2 μm wide pores; by virtue of these it possesses a through gut which it uses for filter feeding in a manner exactly analogous to the sponges (although here water enters through the larger aperture and leaves via the many small ones). Otherwise, and in a similar manner to the alcyonarians (see above and Chapter 9), many shallow-water zoantharians feed in large measure by 'milking' intracellular populations of zooxanthellae. They too can also consume dissolved organic matter (DOM) and planktonic bacteria, although predation on the zooplankton is relatively important; these modes of nutrition are the norm in deeper living species, as well as in some shallow-water types, e.g. several sea anemones. Reef corals, however, may obtain on average some 70% of their requirements from their zooxanthellae, leaving 20% to be gained through predation and 10% from DOM and bacteria.

Two orders are sometimes separated into a class of their own, the Ceriantipatharia, distinguished by a relatively simple arrangement of muscles, septa and tentacles: the solitary, anemone-like ceriantharians which secrete a tube of the discharged threads of ptychocyst organelles (see Box 3.2) and mucus; and the colonial antipatharians (the black or thorny corals) that have an axial chitinous skeleton somewhat like that of some alcyonarians although secreted by the epidermis as in all zoantharians. Unlike the other zoantharians, they do not contain zooxanthellae and are solely predatory.

3.4.3 Phylum CTENOPHORA (sea-gooseberries or comb jellies)

3.4.3.1 Etymology

Greek: *ktenos*, comb; *phoros*, bearing.

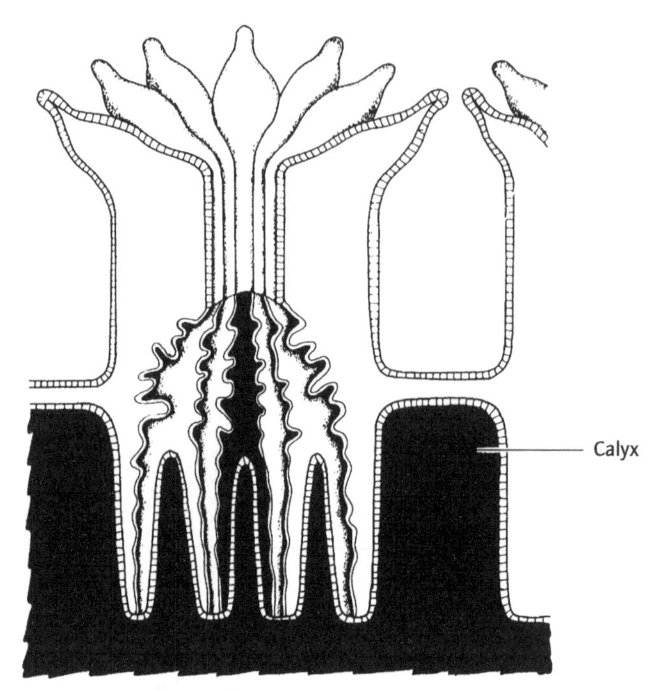

Fig. 3.23 Diagrammatic section through a scleractinian coral set in its external calyx (after Hyman, 1940).

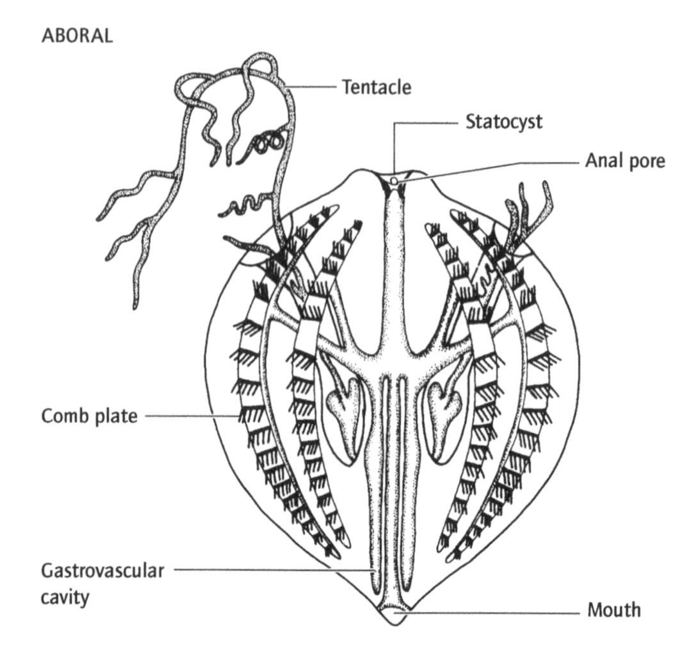

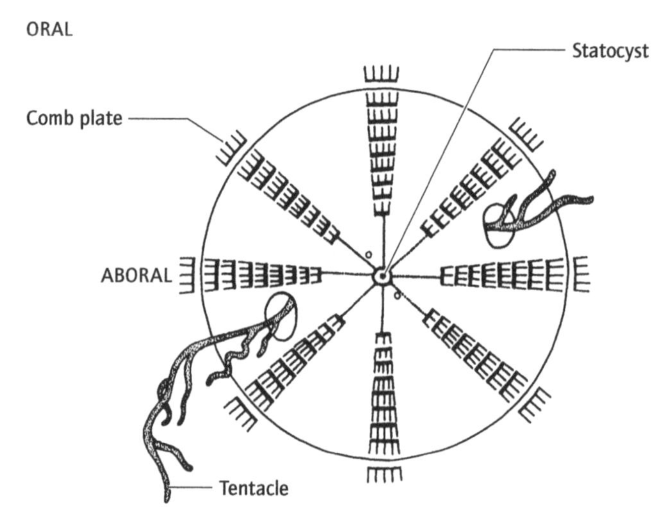

Fig. 3.24 A typical ctenophore, from the side and aboral surfaces (after Buchsbaum, 1951).

3.4.3.2 *Diagnostic and special features* (Fig. 3.24)

1 Body radially or biradially symmetrical, sometimes markedly flattened.

2 Body with two layers of cells, one on either side of thick gelatinous and cellular mesogloea.

3 Body usually spherical or oval, sometimes flattened in variety of planes including being ribbon-shaped; body wall enclosing central cavity (gastrovascular cavity) communicating with exterior by mouth and several 'anal' pores; with tissues but without distinct organs.

4 With muscular cells and network of naked nerve cells.

5 Body medusa-like, often with two highly extensile tentacles housed in sheaths, not associated with mouth, and eight oral-aboral rows of plates of fused cilia (comb plates); in life, mouth positioned uppermost.

6 Tentacles, if present, with colloblast cells discharging sticky threads when fired; without nematocysts.

7 Hermaphrodite with external fertilization; development determinate after biradial cleavage and via larval stage; asexual multiplication only in one order (Platyctida).

8 Movement usually by means of co-ordinated beating of comb plates.

9 Predatory; pelagic, sometimes benthic, in the sea.

Most ctenophores are transparent, bioluminescent, fragile and gelatinous members of the marine plankton; a few are benthic and one somewhat leathery species is sessile. Superficially at least, they appear to be equivalent to the cnidarian medusae (there is no ctenophore equivalent to the polyp stage). Most are oval, but a wide range of shapes are represented; several are leaf to ribbon shaped (see Fig. 3.26), flattening occurring in any one of three planes (oral/aboral; lateral, in that of the tentacles; and at right angles to the tentacular plane). Such ribbon-shaped species may be up to 2 m long.

Their most obvious and characteristic feature, in which they differ from all cnidarians, is the presence of the eight rows of short, transverse plates of long, fused cilia ('comb plates') extending around the body from oral to aboral poles that form the main means of propulsion. The plates beat in synchronous metachronal waves starting at the aboral pole and movement is usually mouth first, with co-ordination of the comb plates being achieved by local concentrations of the nerve net and an apical sense organ.

The body shows the standard radiate construction of two layers of cells, one on either side of a gelatinous mesogloea, and the presence of a diffuse, net-like nervous system, but ctenophores show several advanced features in comparison to the Cnidaria. The branched gut also communicates with the external environment via a series of 'anal' pores instead of being blind ending, for example (although pores connecting the gut to the external environment are found in at least one cnidarian, see Section 3.4.2.3.5). In spite of their name, these pores do not serve an anal function, however; they probably only act as exit points for gastrovascular water when food is ingested. In further contrast to the Cnidaria, the thick mesogloea contains muscle fibres and mesenchyme cells, and a subepidermal layer of muscle cells may be present. It has been argued that these muscle cells and cellular mesogloea are really mesodermal in origin and hence the attempts to ally the group with various of the bilateral phyla referred to above. Also unlike the cnidarians, there is no cnidocyte/nematocyst system (one species does possess them but they are derived from its cnidarian prey).

3.4.3.3 Classification

Ctenophores are easily fragmented on attempted collection, and so knowledge of the group is undoubtedly incomplete. Currently there are some 100 known species, placed in two classes and seven orders.

Class	Order
Tentaculata	Cydippida
	Platyctida
	Lobata
	Ganeshida
	Thalassocalycida
	Cestida
Nuda	Beroida

3.4.3.3.1 Class Tentaculata This class contains the large majority of ctenophores and is distinguished by the possession of a pair of large extensile and contractile, cylindrical, pinnate tentacles, housed in deep, ciliated, tentacular sheaths, that render the symmetry biradial. These tentacles, which on extension are tens of times the body size, bear specialized cells, the lasso cells or colloblasts (Fig. 3.25), that trap the prey in a sticky secretion when discharged. Prey may also be captured by mucus-covered or muscular flaps of the body. Sometimes, however, the tentacles are secondarily reduced even to the point of effective absence. Body form is highly variable (Fig. 3.26) and this, together with the form and development of the tentacles, is used as the basis of the ordinal classification. Benthic forms also occur, and these creep across the sea bed using the cilia of a permanently or temporarily everted oral region. Ribbon-shaped species may swim by means of bodily undulations.

3.4.3.3.2 Class Nuda Members of the class Nuda lack tentacles at all stages of their life history (Fig. 3.27) and capture their food – mainly other ctenophores – by engulfing them with their wide, flexible mouth region and voluminous gut. Just within the mouth are located 'macrocilia' (each comprising thousands of axonemes contained within a single membrane) that function as teeth.

3.5 Superphylum MESOZOA

3.5.1 Introduction

The mesozoans are a most peculiar and enigmatic group of <8-mm long, commensal or parasitic worms, with a bizarre structure and life cycles without parallel in other animals. They possess fewer cells than any other animal, their solid bodies comprising just an outer layer of some 20–30 somatic cells surrounding a single, elongate, cylindrical, axial cell. The somatic

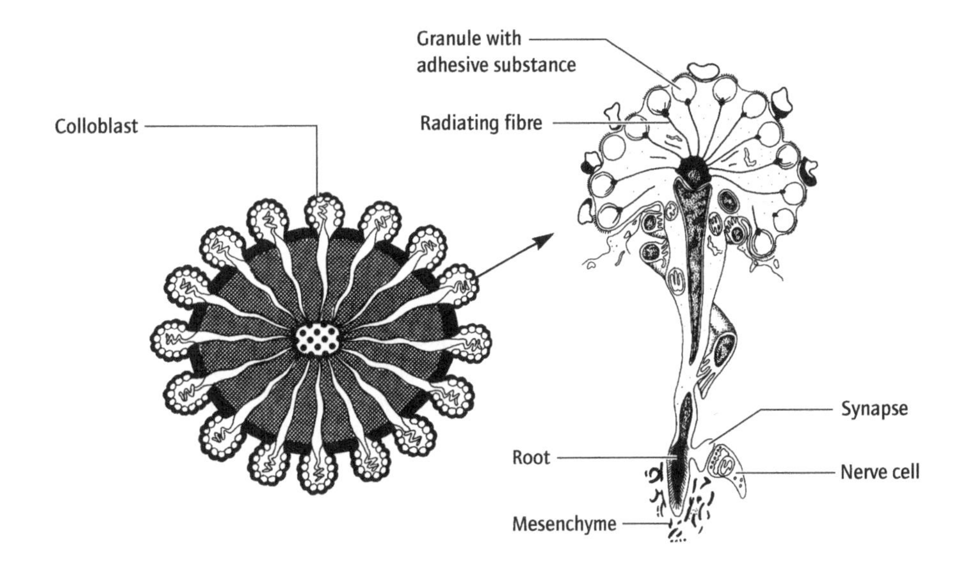

Fig. 3.25 Section through a tentacle of a ctenophore, with details of a colloblast (after various sources).

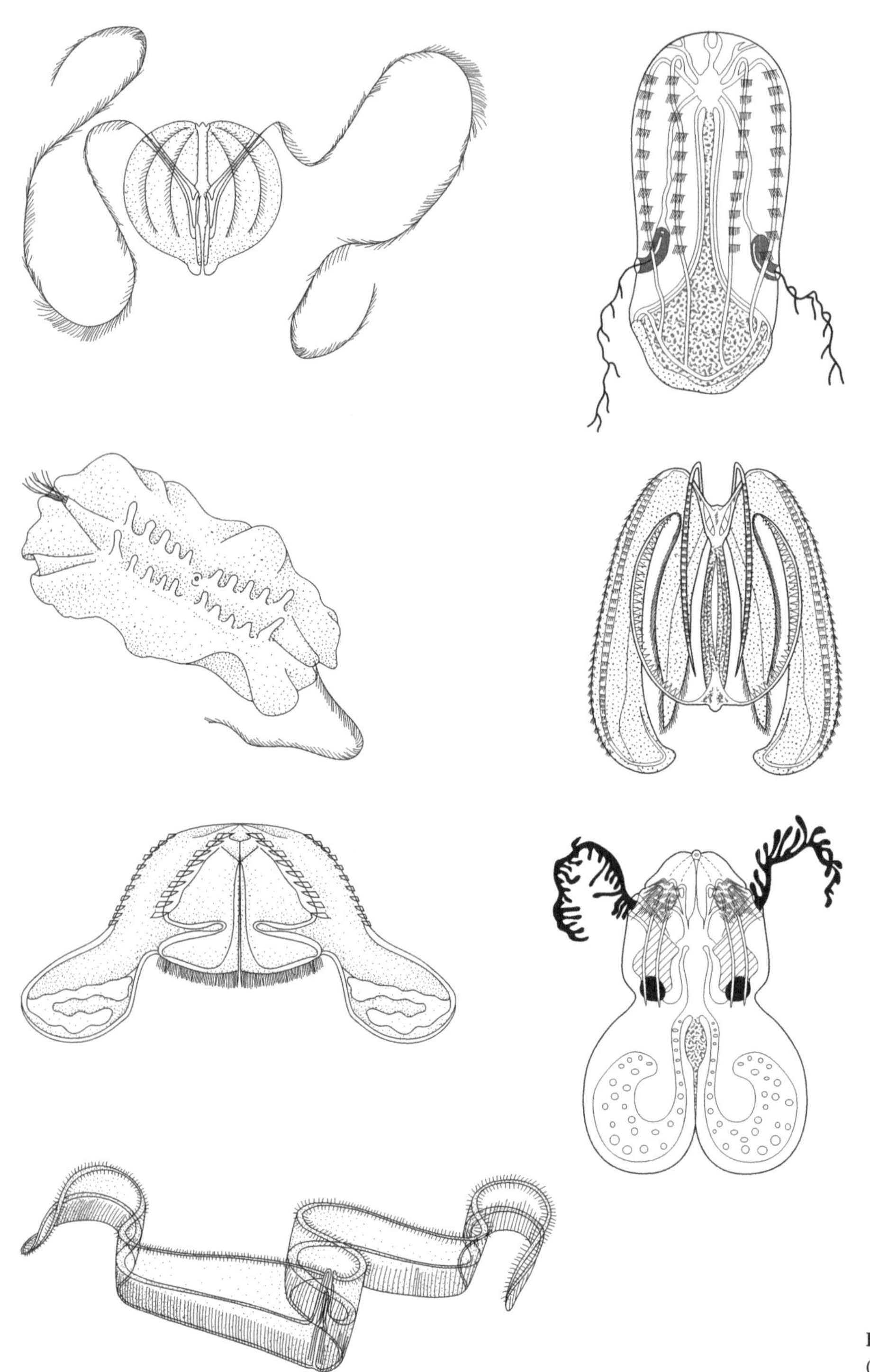

Fig. 3.26 Body form of tentaculatan ctenophores (after various sources).

cells, if separate, display a helical symmetry and the axial cell contains up to 100 intracellular reproductive 'axoblast cells'. There are no tissues, skeletal material or organs and no nerve or muscle cells. As with all endoparasites there is doubt as to whether this extreme simplicity is a result of parasitic degeneration from some more complicated ancestor or whether it indicates that perhaps they represent a line of protists that have only achieved a small degree of multicellularity. If they are a case of parasitic degeneration, however, no trace whatsoever of the ancestral complexity remains.

The life cycle is as complex as their anatomy is simple (Fig. 3.28). Typically, the axoblast cells develop intracellularily (i.e. still within the axial cell) into asexually multiplying 'nematogens', via 'vermiform larvae'. At certain times these

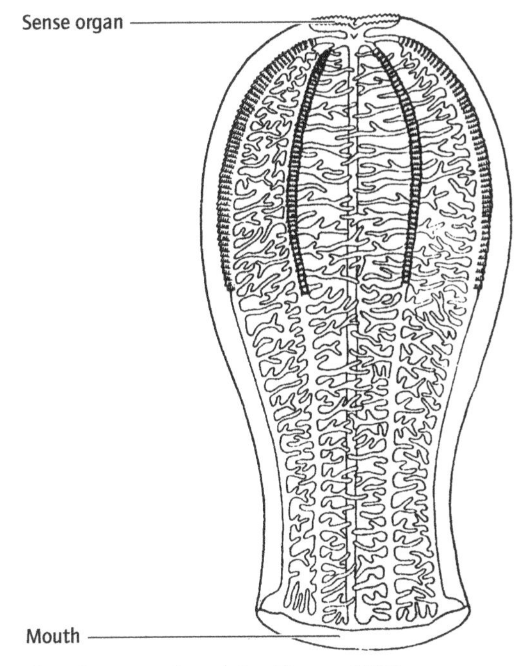

Fig. 3.27 A nudan ctenophore (after Hyman, 1940).

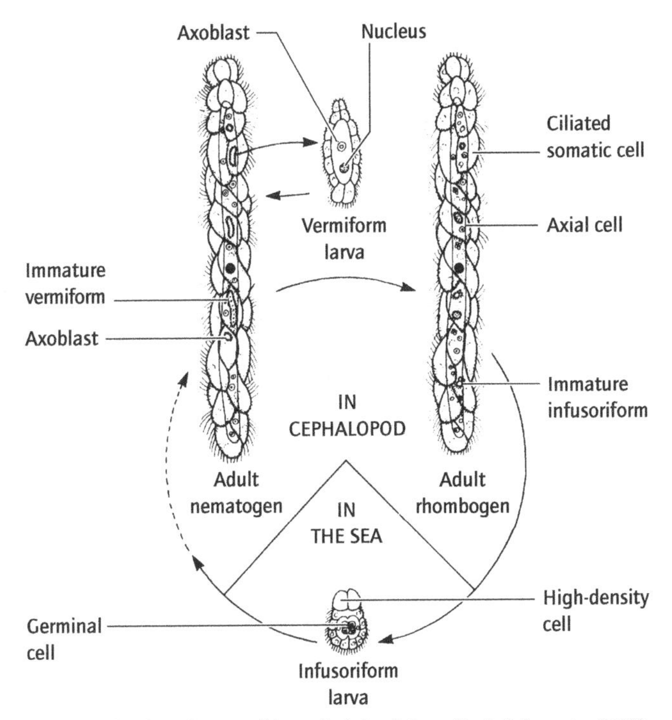

Fig. 3.28 The rhombozoan life cycle (after Margulis & Schwartz, 1982).

nematogens then produce 'rhombogens' that in turn produce gametes. Fertilization occurs within the rhombogen's axial cell and an 'infusoriform larva' results which leaves the rhombogen and the host for a free-living existence. In some way, the infusoriform larvae find new hosts and there develop into the adult stage complete with axoblasts in the axial cell.

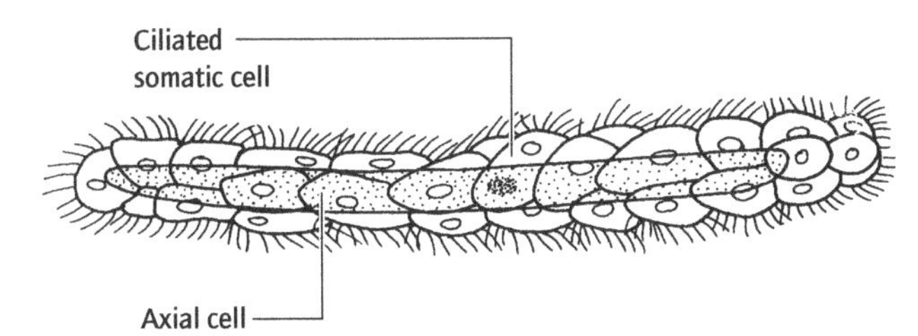

Fig. 3.29 The body plan of a rhombozoan (after Lapan & Morowitz, 1972).

A single phylum, the Rhombozoa, is included; the orthonectidans (Section 4.15.4) that were once grouped with them are now regarded as being unrelated.

3.5.2 Phylum RHOMBOZOA

3.5.2.1 *Etymology*

Greek: *rhombos*, that which revolves; *zoon*, animal.

3.5.2.2 *Diagnostic and special features* (Fig. 3.29)

1 Helically symmetrical worms with anterior/posterior polarity; 0.5–7 mm long.
2 Body with very few cells (30 or less): a single inner axial cell surrounded by a monolayer of (usually) ciliated somatic cells.
3 Without any tissues or organs.
4 Without nerve or muscle cells.
5 Without gut, body cavity or skeletal system.
6 Reproductive cells and their products develop within the axial cell.
7 With complex life history involving both sexual and asexual reproduction and larval stages.
8 Endoparasitic in cephalopod molluscs.
9 Marine.

Rhombozoans occur in the ducts of the excretory organs of certain cephalopod molluscs, subsisting by absorbing materials from the urine. They appear to cause little harm except in very high densities. Most individual octopuses and cuttlefish support numerous individuals.

Their general anatomy and life history is as described above.

3.5.2.3 *Classification*

Two classes, each with a single order, are recognised, containing a total of some 75 species.

3.5.2.3.1 *Class Dicyemida* The somatic cells of dicyemid rhombozoans are ciliated and separate, each species having a characteristic (and constant) number of them. The anteriormost eight or nine cells form a polar cap, followed by two parapolar cells, 10–15 trunk cells, and two terminal uropolar cells (Fig. 3.30). The polar cap is used for attachment to the lining epithelia of the host's kidney ducts. High population densities can build up within the kidney and this appears to trigger the production of the sexual rhombogens.

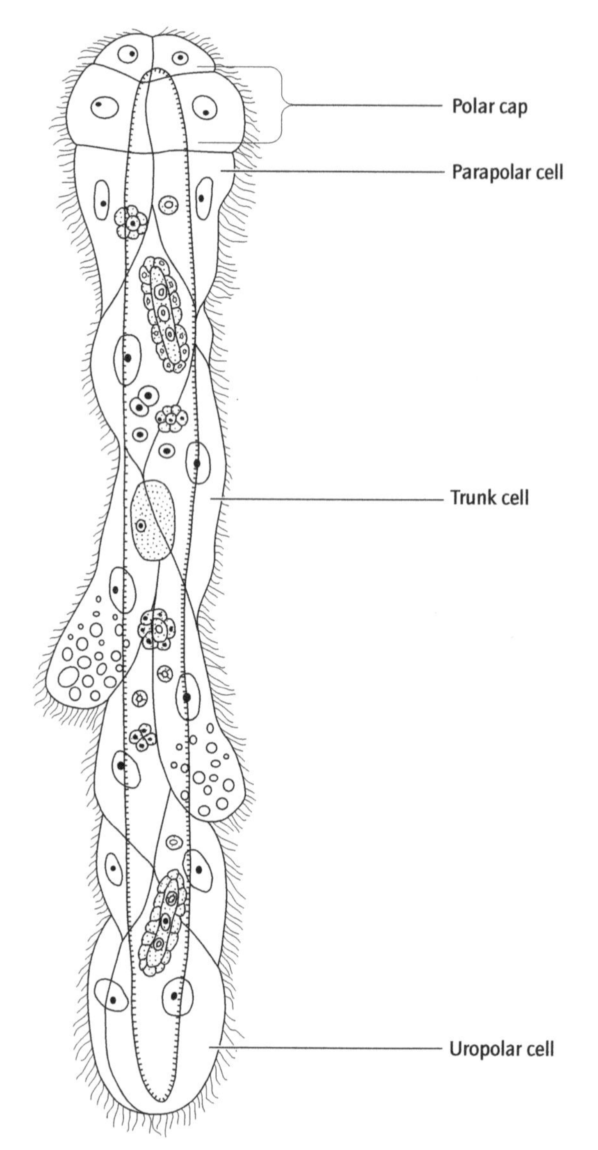

Fig. 3.30 A dicyemid (after Grassé, 1961).

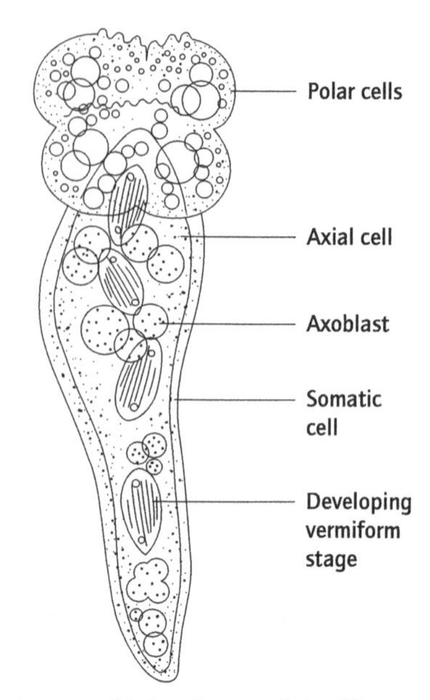

Fig. 3.31 A heterocyemid rhombozoan (after Hyman, 1940).

3.5.2.3.2 Class Heterocyemida The somatic cells of the only two known heterocyemid rhombozoans lack cilia (although the larvae are ciliated) and form a very thin outer layer. In one of the two species, a polar cap of four greatly enlarged cells is present whilst a few flat cells cover the rest of the body (Fig. 3.31); in the other species, the somatic cells are syncytial and a polar cap is absent. Heterocyemids are very poorly known.

3.6 Superphylum BILATERIA

3.6.1 Introduction

All other animals, i.e. besides those covered in Sections 3.2–3.5 above, share a further suite of characteristics at variance with those of the parazoans, phagocytellozoans, radiates and mesozoans, most obviously in their pattern of symmetry and in not having a basically two-cell layered construction. They (i) are bilaterally symmetrical, with dorsal and ventral surfaces and an anterior and posterior end and (ii) possess bodies constructed from more than two (and usually very many more than two) cell layers, the majority of their tissues developing from an embryological germ layer, the mesoderm, found only in this superphylum. Further, in these species mesoderm gives rise to (iii) organ systems, a level of complexity absent from the basically two cell layered animals. All also possess (iv) a nervous system organized into a brain and a number of longitudinal sheathed nerve cords, although nerve nets may be present as well. These four features distinguish the superphylum Bilateria, of which the simplest and arguably the most primitive members are the flatworms.

3.6.2 Phylum PLATYHELMINTHES (flatworms, flukes, tape-worms)

3.6.2.1 Etymology
Greek: *platy*, flat; *helminthes*, worms.

3.6.2.2 Diagnostic and special features (Fig. 3.32)
1 Bilaterally symmetrical, dorsoventrally flattened worms, <1 mm to >5-m long.
2 Body more than two cell layers thick, with tissues and organs.
3 With sac-like or sometimes branched, blind-ending gut, lacking in some parasitic species; mouth on ventral surface, often mid-ventrally, in free-living species, and, if present, terminal in parasitic ones.
4 Body monomeric, although one group can bud off segment-like units from an anterior proliferation zone; free-living species with differentiated head and usually with eyes.
5 Without a body cavity; skeleton parenchymatous tissue.

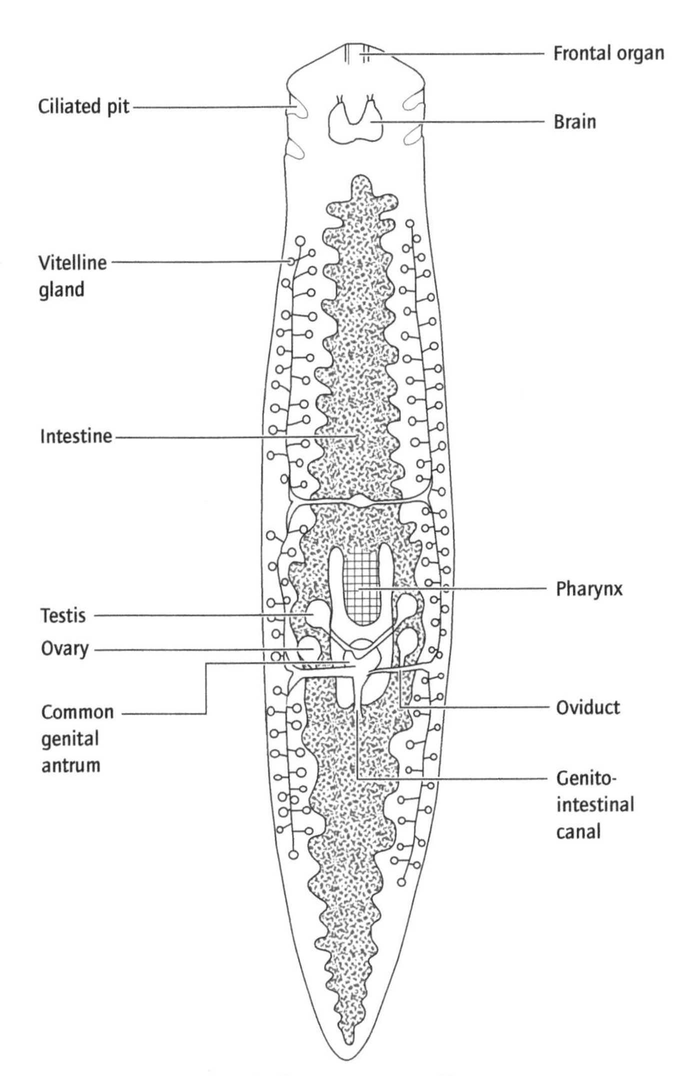

Fig. 3.32 The anatomy of a flatworm, seen as if transparent (after Meglitsch, 1972).

6 Body wall with layers of circular, longitudinal and other muscles overlain, in parasitic species by resistant cuticle and non-ciliated, syncytial epidermis, in free-living species by ciliated epidermis.
7 Without circulatory system.
8 Excretory/osmoregulatory organs protonephridia.
9 Nervous system simple but usually organized into brain, from one to several pairs of distinct longitudinal cords linked by transverse connectives, and a subepidermal network.
10 Mostly hermaphrodite, with internal fertilization and complex reproductive systems; sperm usually biflagellate; development usually direct except in parasitic species in which numerous secondary larval stages may occur, some multiplying asexually; asexual multiplication widespread; cleavage spiral.
11 Move by ciliary gliding or by waves of muscular contraction passing along ventral surface or whole body.
12 Exclusively feeders on animal tissue, whether as predators, scavengers or parasites.
13 Free-living species benthic, mainly marine but also freshwater and terrestrial; most species ecto- or endo-parasites.

Of the animals considered in this chapter, the parazoans and radiates possess 'guts' in the sense that their tube-shaped bodies surround an equally tube-shaped central cavity, whilst the solid bodied placozoans and mesozoans lack any gut. Flatworms, in contrast, are the bilaterally symmetrical, solid-bodied animals with true guts, albeit that this is only a blind ending system with the mouth also serving as the anus. They are then mainly distinguished from members of the many other bilateral animal phyla by what they lack rather than by virtue of any special features, which is perhaps appropriate in a group that can claim to be closest to the ancestral body form. That is they are the Bilateria that lack a body cavity, lack a blood system, lack a through gut, and lack any special distinguishing features. Lacking a fluid system capable of circulating the respiratory gases has required flatworms, or at least the larger forms, to adopt one positive feature: their flatness. Thickness of the body must be kept small in at least one plane (the dorsoventral one) in order to permit diffusion to suffice to meet the oxygen requirements of all the cells in their solid bodies and to remove the carbon dioxide. Relatively large size must therefore be accompanied by leaf or ribbon shape. By the same token the products of digestion must also diffuse from the gut to all cells of the body, and hence relatively large size must equally be accompanied by a much branched gut with diverticulae extending throughout most of the body volume (Fig. 3.33). Most flatworms do possess 'excretory organs' in the form of protonephridia to eliminate waste products although even here diffusion is probably the main process whereby these leave the body, and the main function of the protonephridia is osmoregulatory: they are especially well developed in freshwater species. Granted these adaptations to macroscopic body size, there are still limits to the ability of diffusion to service increasing body mass, and most flatworms have remained microscopic. (In most two cell layered animals, both cell layers are in contact with the external environment, the outer layer directly so and the inner layer with that in the (usually sea) water-filled coelenteron of radiates or the canal system of sponges. Diffusion distances are always very small and large individual size poses no problems. Flatworms are very much smaller, on average, than sponges, jellyfish and solitary – let alone colonial – anthozoans.)

Although the majority of the Bilateria possess one or more body cavities of various forms and origin, the flatworms are solid bodied (acoelomate) (Fig. 3.34). This is usually (as here) interpreted as being a primitive character although some zoologists argue that it is secondary and that platyhelminths are descended from animals that had coelomic body spaces. Be that as it may, there is no trace of a coelomic cavity in any surviving flatworm and instead the space between the body wall and the various organs is filled by irregularly shaped cells forming the 'parenchyma'. Amongst these cells are small 'neoblasts' which are important in regeneration and asexual multiplication, both of which are highly developed in flatworms. In the free-living species, other gland cells within the parenchyma, together with

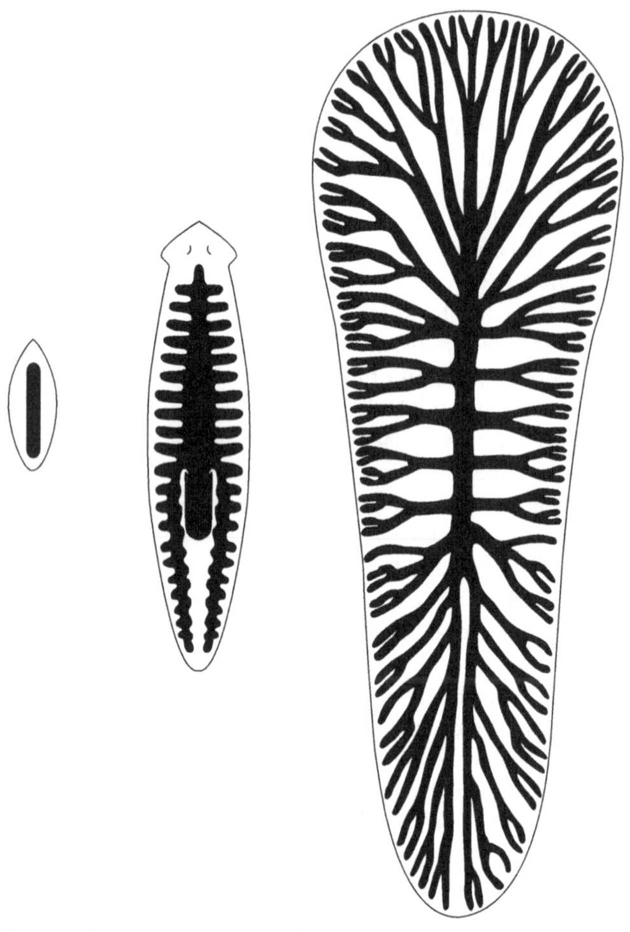

Fig. 3.33 The increase in gut ramification with increase in body size in flatworms.

cells in the epidermis, secrete rod-like structures termed rhabdoids. These rather mysterious structures are discharged to the external environment (those produced in the parenchyma passing through spaces in the epidermis to do so), where they release mucus and possibly defensive chemicals. They do not occur in the parasitic groups, presumably correlated with their specialized external covering. Somewhat unusually amongst animals, chitin is apparently unknown in platyhelminths.

3.6.2.3 Classification

In terms of general body form and life style there are four major types of platyhelminth: the free-living turbellarians; the monogenean flukes that are usually ectoparasitic on fish; the endoparasitic trematode flukes; and the gutless endoparasitic tapeworms. These form the four traditional platyhelminth classes, and they are the standard divisions recognised by parasitologists. Molecular and cladistic analyses, however, have revealed that the major lines of flatworm evolution are not these four but instead are some five groups all dominated by turbellarians. The three classes of parasitic flatworms, which include the large majority of living species in the phylum (see Box. 3.3), form just part of only one of these five lines. One effect of this is that in the past the free-living species have been 'underclassified' whilst the parasitic ones have been 'overclassified'. These phylogenetic data are reflected in the classification given below, in which the 25 000 species and 18 different primary groups of platyhelminth are divided between the five cladistic 'classes'. Such is the importance of the parasitic flatworms, however, that in the text we describe flatworm diversity along the more traditional lines.

3.6.2.3.1 The Turbellarian flatworms Turbellarians are mostly free-living flatworms (Fig. 3.35) that creep across surfaces by means of epidermal cilia or in the larger species by waves of muscular contraction passing along their flat ventral surface: the body wall usually contains both circular and longitudinal muscles, as well as oblique ones that span the body. They prey on organisms ranging in size from bacteria to small animals which they can locate with their sense organ-bearing head; a few are commensal and/or parasitic, mostly in other invertebrates. The majority are marine although many are freshwater and some are terrestrial in humid or moist microhabitats. Most species are 0.5–5 mm long, but a few of the terrestrial species can attain 50 cm, and a freshwater species in Lake Baikal can exceed 60 cm

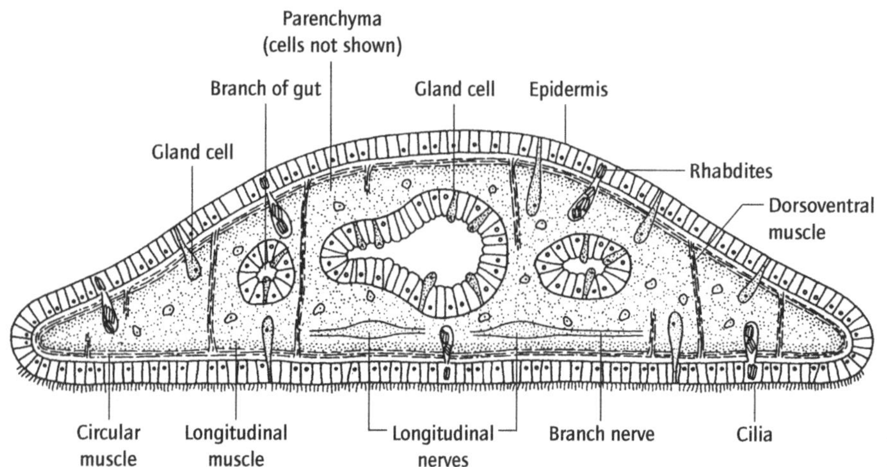

Fig. 3.34 A cross-section through a flatworm (after Kozloff, 1990).

Box 3.3 Parasitism and Platyhelminthes

As pointed out in the text, the majority of living platyhelminths are parasites of other animals, when adult being especially associated with vertebrates as endoparasites, and they have departed quite markedly in anatomy and in their life cycles from the ancestral neoophoran turbellarians. In this Box, we discuss briefly the biology of parasitism with particular regard to the endoparasitic flukes and tapeworms that often impinge directly on human biology (the fluke *Schistosoma*, for example, infects some 200 million people, and the tapeworm *Taenia*, some 70 million).

Everyone knows what a 'parasite' is, yet it is a surprisingly difficult concept to define and to separate from other related concepts such as predator and symbiote. The everyday notion of a parasite is something that utilizes the resources gained by something else at little cost to itself but at greater cost to the resource provider. Some animals often labelled as parasites, however, are really standard predators (finders and consumers of other animals' living tissues, etc.) that just happen to be much smaller than their prey and incapable of killing them. A leech feeding on, say, a large mammal is a parasite whilst the same animal feeding on a snail is a predator; a mosquito drinking the body fluid of a mammal is a parasite whereas a notonectid bug doing the same thing to a mosquito larva is a predator. The above examples involve one free-living animal attacking another from the outside, e.g. by penetrating its skin with piercing and sucking mouthparts, but the same applies to animals feeding on others from within. A small animal feeding on a larger one whilst living inside it is conventionally a parasite (as, for example, is the case with the trematode flukes), yet it is behaving no differently in essence from any other carnivore. It is merely relatively so small as to be unable to eat the whole of its prey. Some insect larvae which are (eventually) so large as to achieve this are often termed 'parasitoids'. Animals, however, which inhabit the guts of a host and there feed on ('steal') part of the food gathered and digested by that host, such as tapeworms, really are behaving parasitically in the normal use of that word. Nature provides continua and not discrete categories at almost every level, and there is little point here in arguing the merits of any particular definition. Our consideration of the parasitic platyhelminths in this Box will simply follow general usage of regarding the trematodes and tapeworms as equally endoparasitic, notwithstanding that one is a micropredator and the other a stealer of the food obtained by another. It is the requirements and consequences of living within another animal that are here of interest, not the precise terms to be applied to such species (see also Section 9.2.3).

The internal environment of other animals is in large measure a relatively uniform habitat with a precise series of problems to be overcome by an invading animal. All parasitic species are therefore faced with the same evolutionary pressures and not surprisingly many have responded with a common set of solutions. The difficulties of transmission from one host to another have also to be faced by all parasites, and again solutions cross taxonomic boundaries.

Anatomical changes

Turbellarians are characterized by a sense organ-bearing head and a ciliated epidermis. These have been lost by the parasitic groups, either totally or such occurs after the first larval stage. The loss of the head is for largely self-evident reasons: they are surrounded by their food and external threats come not from directional sources within the host; energy has then been channelled away from the selectively no longer advantageous head into other activities. The loss of the ciliated epidermis is consequent on the development of a resistant covering, the tegument, shared by all species (see Figs 3.38 and 3.43). (This feature is one of the reasons why flukes and tapeworms are considered to be closely related and its possession gives rise to the name, Neodermata, of the group to which all three parasitic 'classes' belong.) Somewhat unusually, the parasitic platyhelminths do not possess a cuticle (although older works describe the tegument as such); instead they are covered by a living cytoplasmic syncytium, formed from cells located within the parenchyma. The tegument provides protection against host antibodies, and enzymes in gut-inhabiting species, and renders possible the uptake of soluble organics from the host across the body wall. This process is particularly developed in the gutless tapeworms that absorb all their requirements directly from the host.

Like many parasites, all parasitic flatworms possess organs of attachment to their host in the form of hooks and a variety of types of sucker, separately or in combination; they also have either well developed guts (i.e. the flukes) to process the constantly available food supply, or, if they themselves inhabit the alimentary canal of their host, they have lost their gut and absorb the predigested food across their body surface. This is the case in the tapeworms, in which the mitochondrion-rich tegument has a greatly increased surface area to facilitate uptake, via the presence of the microvillus-like microtriches (Fig. 3.43).

Life cycles and asexual multiplication

That transmission from one definitive host (in which the fluke or tapeworm is adult) to another is hazardous and associated with large losses can be inferred from the large number of young produced, a process made possible by the superabundant food surrounding the parasite, the minimal other demands on the energy gained, and the very well developed and usually protandrously hermaphroditic, reproductive systems of parasitic flatworms. Most tapeworms produce hundreds of thousands of eggs each day, in a conveyor-belt system. A linear series of segment-like proglottids are budded off from an anterior proliferation zone just behind the organ of attachment (the scolex), each proglottid being little more than a set of reproductive organs. After fertilization (cross-fertilization is the norm but self-fertilization is possible), proglottids become full of developing eggs and finally fall off the end of the chain and pass out with the faeces. In the human tapeworm, *Taenia saginata*, 3–10 proglottids each containing 100 000 eggs pass out per day. Flukes, with their unitary bodies, can only produce some hundreds per day or at most thousands but each egg may potentially give rise to up to a million infective larvae as a result of asexual multiplication. This takes place in one or more intermediate hosts, which may also serve as the vehicles via which the parasite finally gains the definitive host. Some tapeworms can also produce many adult worms from a single egg in a comparable manner. Such asexual multiplication by polyembryony or budding is unusual amongst invertebrate parasites. The nematodes, acanthocephalans and pentastomes, many or all of which are also endoparasites, can only reproduce sexually.

In a typical fluke, eggs pass out of the definitive host in its faeces, urine or mucus. On being wetted, each egg then develops into a

Continued p. 72

Box 3.3 *(cont'd)*

short-lived *miracidium* larva. This is free swimming, by virtue of a ciliated epidermis, and contains a number of 'germinative cells' from which the subsequent generations of asexually produced larvae will be derived. Should a mollusc (usually a gastropod) be encountered, gland cells enable the miracidium to bore through its skin, or alternatively it may be eaten. In either event, once inside the snail, the miracidium loses its ciliated epidermis, develops the tegument and forms a somewhat irregular *sporocyst*. This is essentially a bag containing germinative cells that divide mitotically to produce either several further generations of sporocysts or further generations of a similar larval stage – the *redia* – which differ in their possession of a gut. Eventually, after extensive asexual multiplication, the germinative cells of sporocysts and/or rediae develop into the *cercaria* larval stage. In effect, cercariae are small flukes with a propulsive tail that leave the intermediate host and then either encyst to form the *metacercaria* stage on vegetation or on animal skin, etc., there to await consumption by the definitive host, or else they burrow directly through that host's skin (or into a second intermediate host). On emerging from the cyst, a young fluke moves to its final destination in the host.

Within the snail intermediate host, the various fluke larval stages consume the digestive diverticulum and, eventually, the gonad, castrating the mollusc. Frequently, parasitism induces changes in snail behaviour that render it more likely to be found and consumed by the next host required by the fluke.

Common variations of this life cycle are illustrated below:

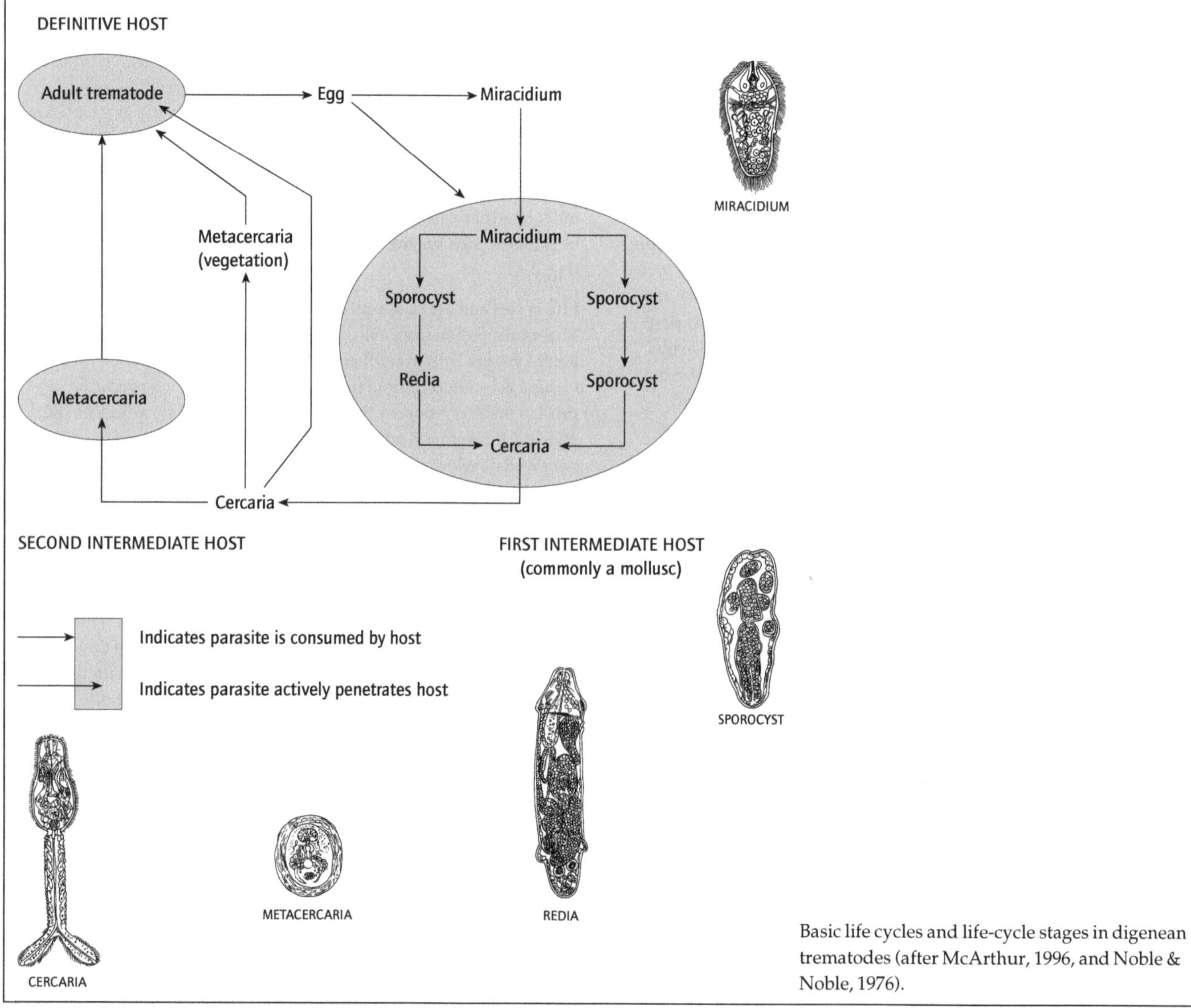

Basic life cycles and life-cycle stages in digenean trematodes (after McArthur, 1996, and Noble & Noble, 1976).

Continued

Box 3.3 *(cont'd)*

The typical tapeworm life cycle differs from that of flukes in that the infective *oncosphere* larva has to be ingested by the host, even when it is covered by a ciliated membrane and free swimming (then termed a *coracidium* larva). Although most tapeworm larvae are not capable of asexual multiplication, such is possible in species in which the *hydatid cyst* stage occurs: the intermediate hosts are therefore mostly vehicles to ensure transmission from one definitive host to another. Variations in the life cycle are shown below:

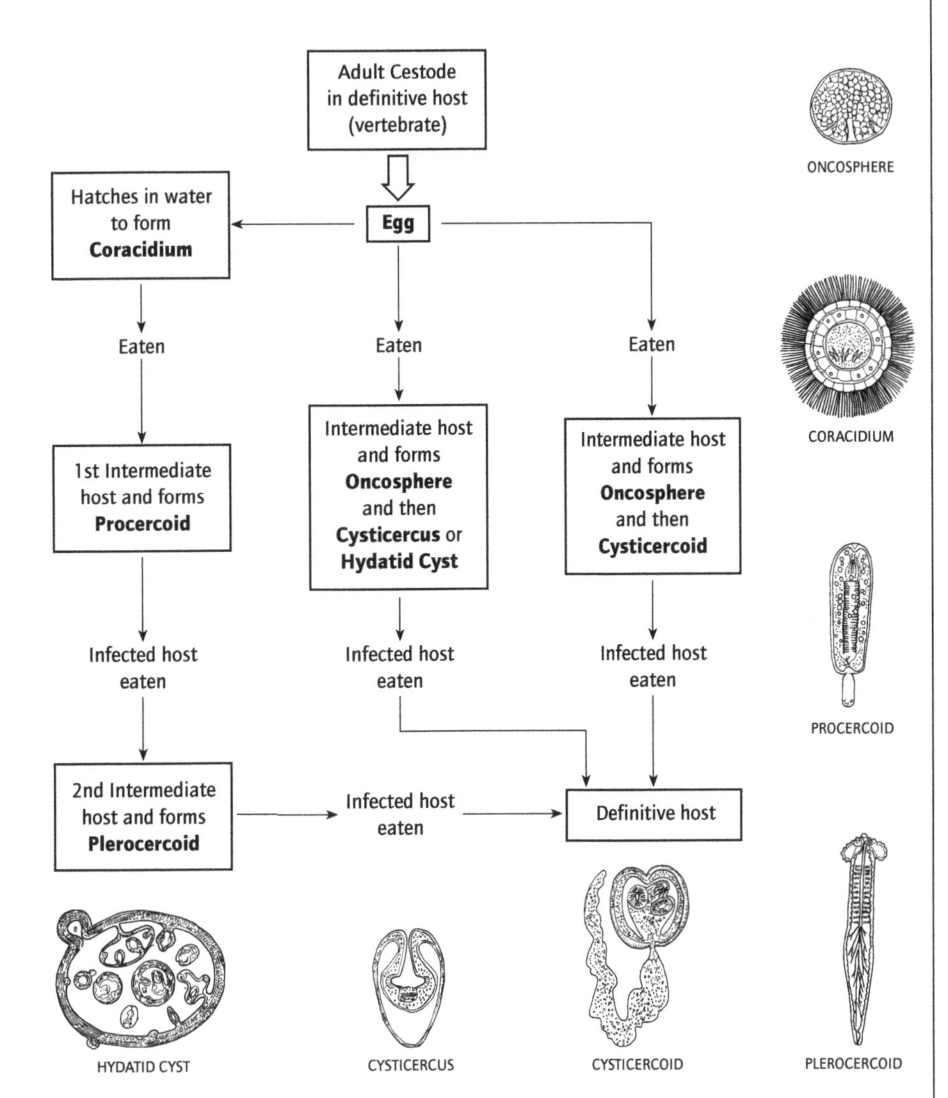

Basic life cycles and life-cycle stages in cestodes (after Crompton & Joyner, 1980, and others).

The *procercoid* larva is small and spindle shaped, with a solid body and retained oncosphere hooks posteriorly; the *plerocercoid* into which it develops lacks the oncosphere hooks but has developed a scolex. The *cysticercoid* larva also retains the oncosphere hooks posteriorly, but has a body in which the middle section grows up around the developing anterior scolex, enclosing it within a cavity whilst the posterior end remains free. A *cysticercus* is a bladder enclosing a single invaginated scolex. Other named types of larvae are also known, but these are relatively minor variations of these four; some cysticercus-like bladders, for example, can be large and contain a number of scolices as a result of asexual budding – these are termed the *hydatid* cysts referred to above. On gaining the gut of the definitive host, the scolex of a plerocercoid, cysticercoid or cysticercus larva attaches to the gut wall, and proliferation of proglottids begins. Paedomorphosis is quite common in tapeworms and paedomorphic species include some that are plerocercoid (or in one genus procercoid) as adults.

The host is not necessarily a passive partner in the relationship. In insects, for example, haemocytes may stick to the parasite's surface and encapsulate it, depriving it of exchanges with its external environment. Equivalently, in vertebrates it may be encased in connective tissue and/or calcium carbonate. This appears more common, however, if the parasitic flatworm invades an unusual host, and within a host species to which it has become adapted most larvae or adults seem to escape encapsulation. Some, at least, also evade vertebrate immune systems by incorporating host antigens onto their body surface so as to disguise themselves and avoid detection and attack.

Class	Order
Acoelomorpha	Nemertodermatida Acoela
Catenulidea	Catenulida
Macrostomomorpha	Macrostomida Haplopharyngida
Polycladidea	Polycladida
Neoophora	Lecithoepitheliata Prolecithophora Proseriata Tricladida Kalyptorhynchida Typhloplanida Temnocephalida Dalyelliida Aspidogastrea Digenea Monogenea Cestoda

length. The mouth, which occurs on the ventral surface, is often located at the end of an eversible pharynx and leads into a sac-like, lobed or much branched intestine, dependent on the flatworm's size. In most cases development is direct, but in some a free-living ciliated larva occurs.

The various lines of turbellarian evolution are distinguished mainly by somewhat esoteric cytological details of their cilia, spermatozoa, and egg cells. Particularly notable features are that whereas most flatworms have biflagellated sperm (with, unusually, a 9+1 axoneme or, in some acoels, 9+0), those of the macrostomomorphs and catenulids lack a flagellum, and only nemertodermatid sperm and that of various other acoels possess the standard single flagellum (with 9+2 axoneme); and whereas all groups except the Neoophora deposit yolk within the egg cytoplasm, the reproductive system of the latter group contains separate 'vitellaria' that add many yolk-rich nurse cells to the fertilised ovum to give rise to a complex egg (Section 14.4). There is also a trend in increasing bodily complexity from the catenulids and acoelomorphs to the neoophorans. Thus the acoelomorphs may lack a permanent mouth, pharynx and/or gut, a temporary gut forming around ingested bacteria and protists;

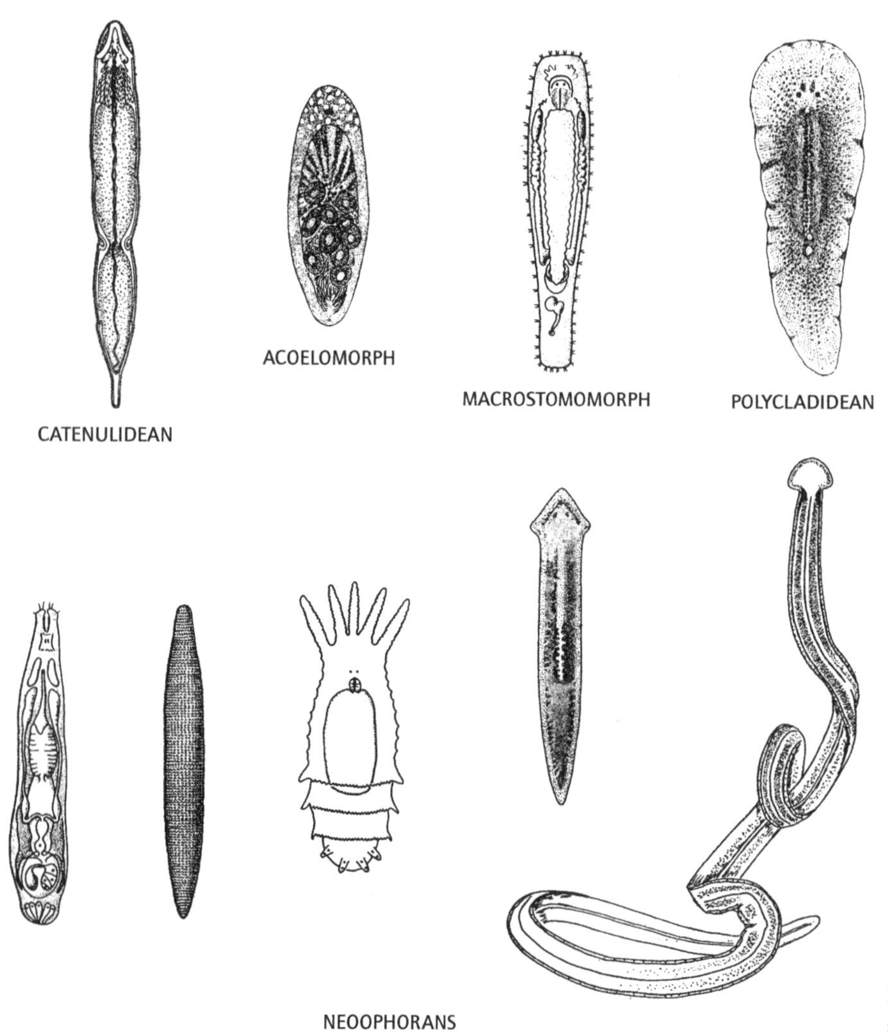

Fig. 3.35 Body form of turbellarian flatworms (after several sources).

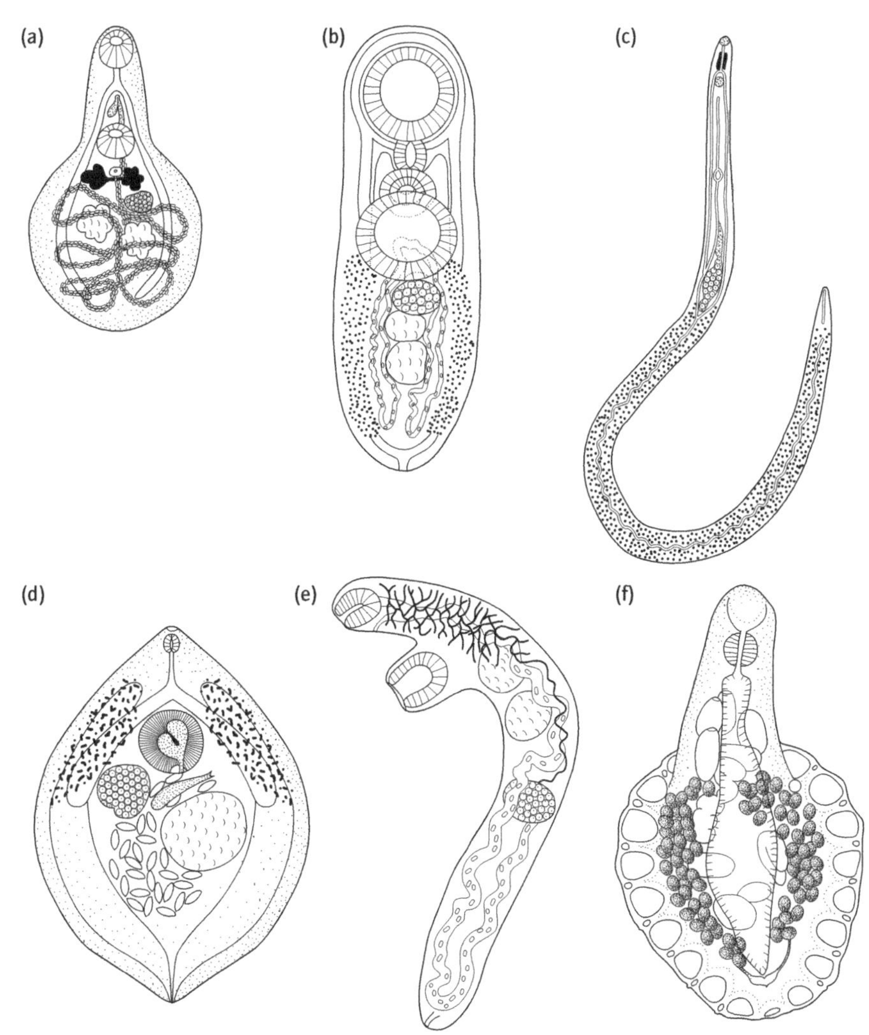

Fig. 3.36 Trematode body forms (after several sources). (a)–(e) are digeneans; (f) an aspidogastrean.

the catenulids have very poorly differentiated parenchymatous tissue and are effectively pseudocoelomic (see Chapter 4); and members of both groups have some of their 'tissues' lacking basement membranes. The tissue and organ grade of organization of these forms is therefore the least of any of the Bilateria. The acoels further differ from other platyhelminths in having biradial (rather than spiral) cleavage, in having indeterminate (rather than determinate) development, and in not forming their mesoderm from the 4d cell.

3.6.2.3.2 The Trematode flukes The trematodes (the digeneans and aspidogastreans) are cylindrical or leaf-shaped, usually 0.5–10 mm (but exceptionally up to 6 m) long, endoparasitic flukes (Fig. 3.36), occurring especially in vertebrates (and particularly fish), that consume the host's tissue and hence have retained a well-developed gut (Fig. 3.37). Adaptations to their parasitic life (see Box 3.3) include a syncytial, non-ciliated, protective 'tegument' in place of the ciliated turbellarian epidermis (Fig. 3.38); organs of attachment to the host in the form of oral and/or mid ventral suckers; and usually a complicated life cycle involving two or more (up to four) hosts of which the first is normally a mollusc (Fig. 3.39). Various secondary larval stages, including miracidia, redia and cercariae, serve the functions of dispersal, asexual multiplication and invasion of new hosts. Unusually amongst platyhelminths, the sexes are separate in a few flukes. The digeneans form the majority of the trematodes; the aspidogastreans are a related group that have the whole of their ventral surface converted into a sucker or longitudinal row of suckers and a relatively simple life cycle involving a single (mollusc) or at most two hosts.

3.6.2.3.3 The Monogenean flukes Like most digeneans, the monogeneans are small (up to 3 cm long), leaf-shaped or elongate flukes with a well-developed gut (Fig. 3.40), but in contrast to that class they are mainly ectoparasites, mostly on fish, and they are considered to be more closely related to tapeworms than to other flukes. A tegument similar to that of the digeneans is present (see Fig. 3.38), as are organs of attachment, here

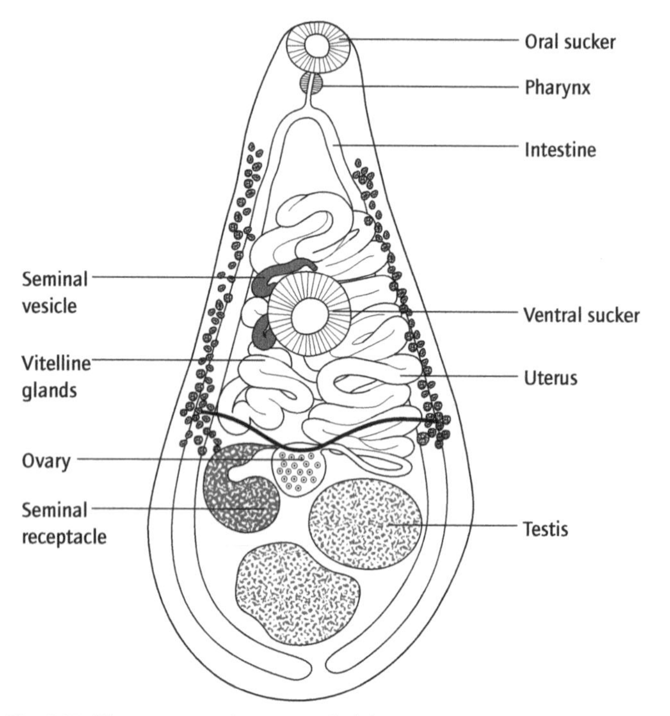

Fig. 3.37 The anatomy of a trematode (after Baer & Joyeux, 1961).

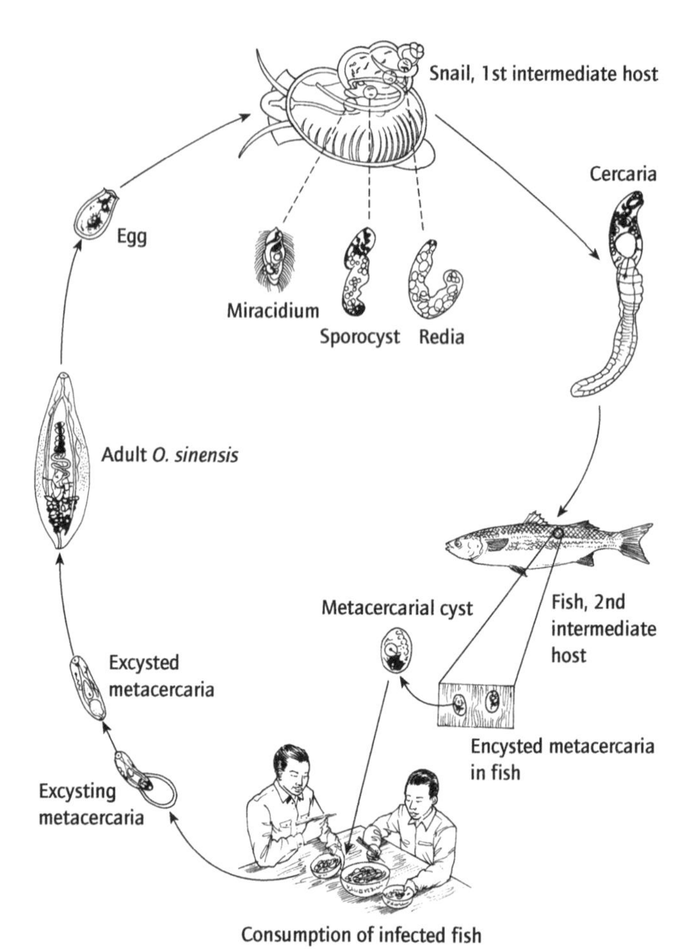

Fig. 3.39 The life cycle of the Chinese liver fluke (after Ruppert & Barnes, 1994).

including multiple suckers, hooks and clamps in 'haptors' located both anteriorly and posteriorly on the body (Fig. 3.41), the posterior, disc-shaped or lobed 'opisthaptor' often being the larger. The life cycle is relatively simple and without phases of asexual multiplication. Eggs hatch into onchomiracidia larvae which find and attach to their fish (or other vertebrate or cephalopod mollusc) host and gradually transform into the adult fluke.

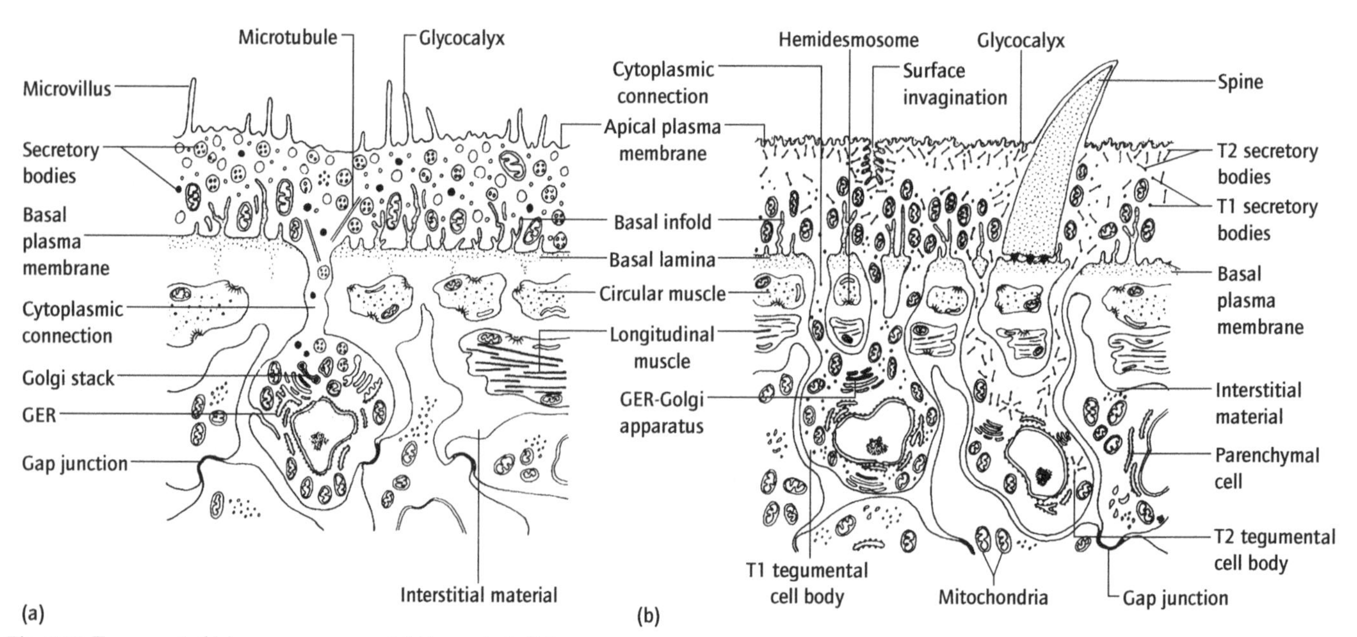

Fig. 3.38 Tegument of (a) monogenean and (b) trematode flukes.

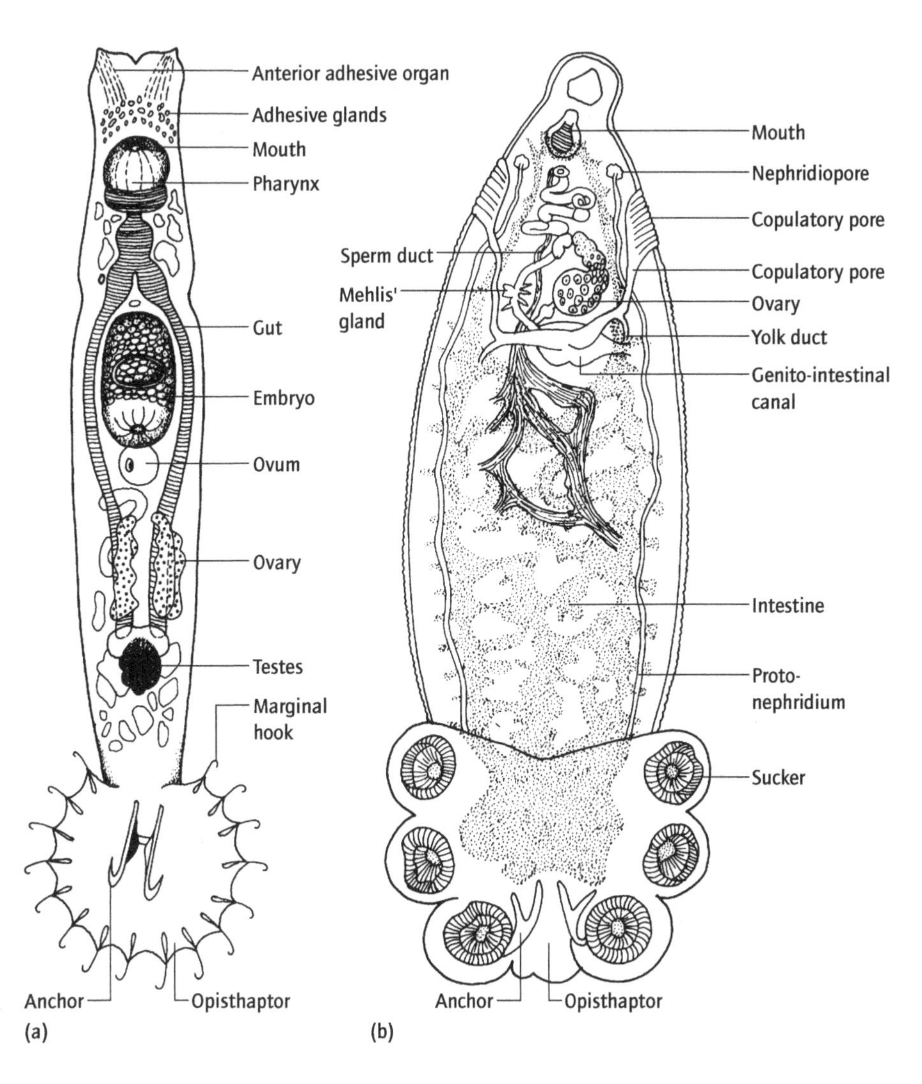

Fig. 3.40 Monogenean anatomy: (a) a species with a single, hook-bearing opisthaptor; (b) one with a multiple, sucker-bearing opisthaptor (after Barnes, Calow & Olive, 1993).

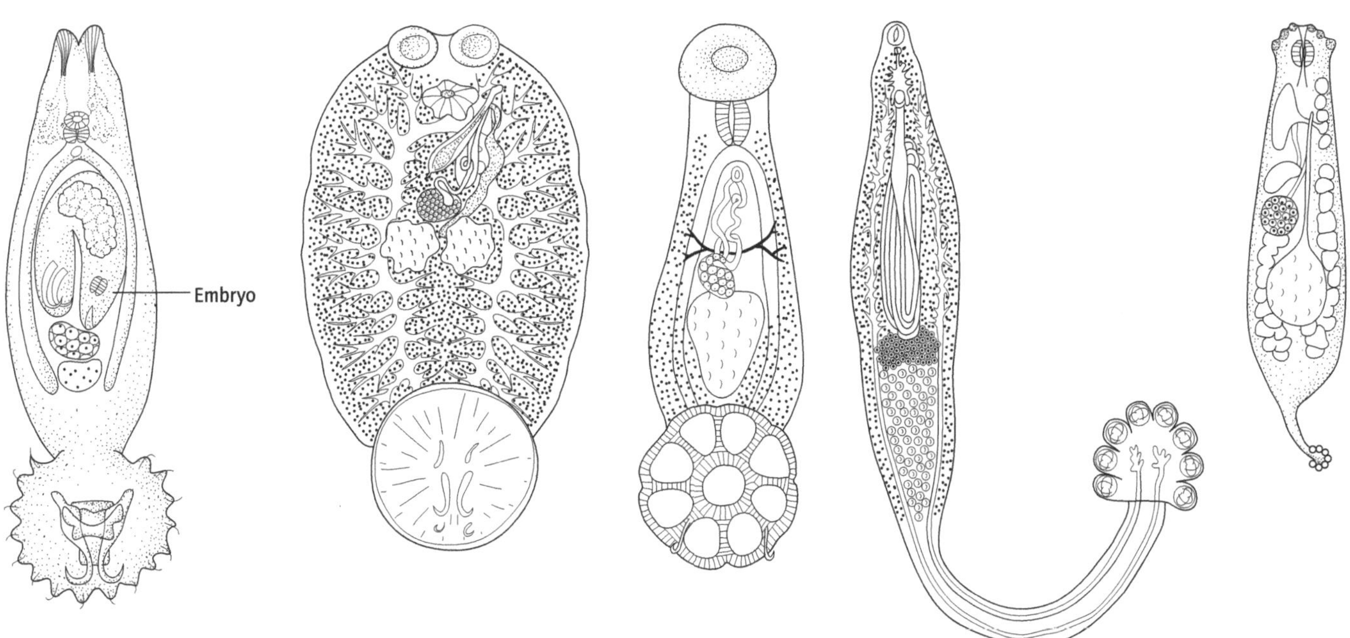

Fig. 3.41 Monogenean body forms (after Barnes, 1998).

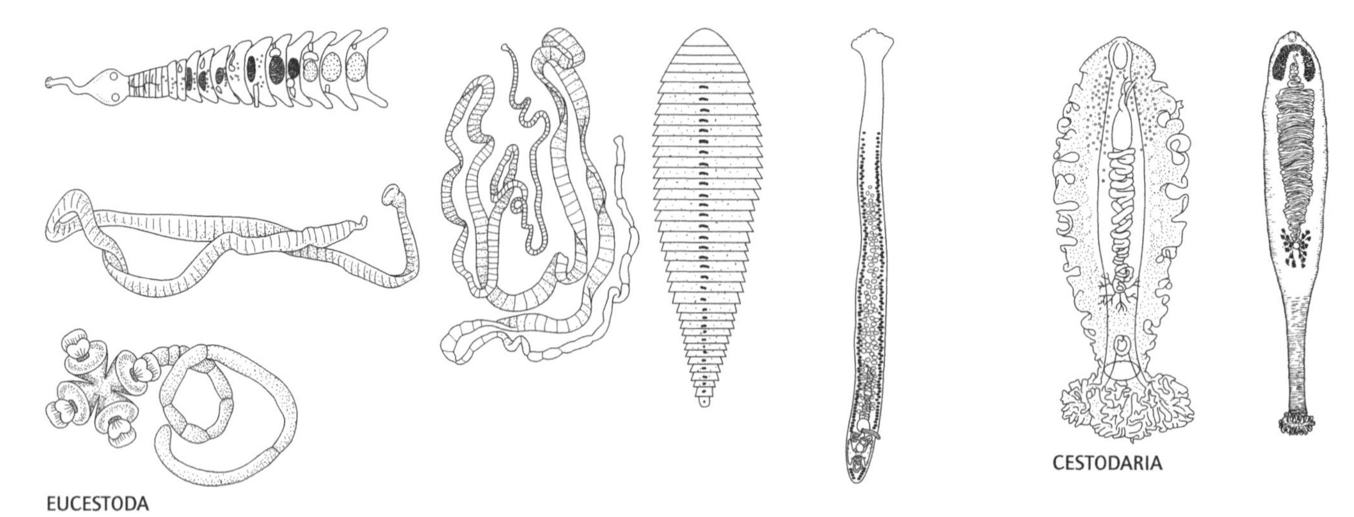

Fig. 3.42 The body form of tapeworms (after several sources).

3.6.2.3.4 The Tapeworms Cestodes or tapeworms (Fig. 3.42) are, when adult, gutless endoparasites of the alimentary canals (or more rarely body cavities) of vertebrates, especially cartilaginous fish, which, like the other parasitic platyhelminths, are covered by a resistant, syncytial, and in the case of the tapeworms absorptive, tegument bearing microvilli-like surface projections (microtriches) (Fig. 3.43). Most tapeworms possess an anterior attachment organ, the scolex, that bears suckers, hooks or equivalent structures, and have the body or 'strobila' divided into a linear chain of segment-like proglottids that are budded off asexually from an anterior proliferation zone (Fig. 3.44). As each proglottid ages, its single or multiple set of reproductive organs matures, until eventually the proglottid becomes full of eggs, drops off the end of the body and passes out of the host with the faeces. Large tapeworms may attain lengths of 5 m and bear 4500 proglottids; lengths of over 30 m have been recorded. The eggs hatch into an oncosphere larva usually with six hooks ('hexacanth larva'), and development proceeds via further procercoid, plerocercoid, cysticercoid or cysticercus larvae in a variety of intermediate hosts (Fig. 3.45 and Box 3.3). Some are paedomorphic and in these the body may comprise just a single proglottid with a single set of gonads. Attachment to the host is not permanent, the scolex can be disengaged and the worms move by muscular contractions through the host's gut; at least some do so in a regular rhythm. Two of the 14 major subgroups of cestodes, separated as the Cestodaria, are particularly fluke-like in shape and differ from the true tapeworms (the Eucestoda) in lacking a scolex, attachment to the host being by means of a simple anterior, monogenean-like sucker, as well as in not being divided into a series of proglottids, and in having an oncosphere (or 'decacanth') larva with 10 hooks. The life cycles of the cestodarians are poorly known, mainly because of their uncommonness and lack of economic or medical significance.

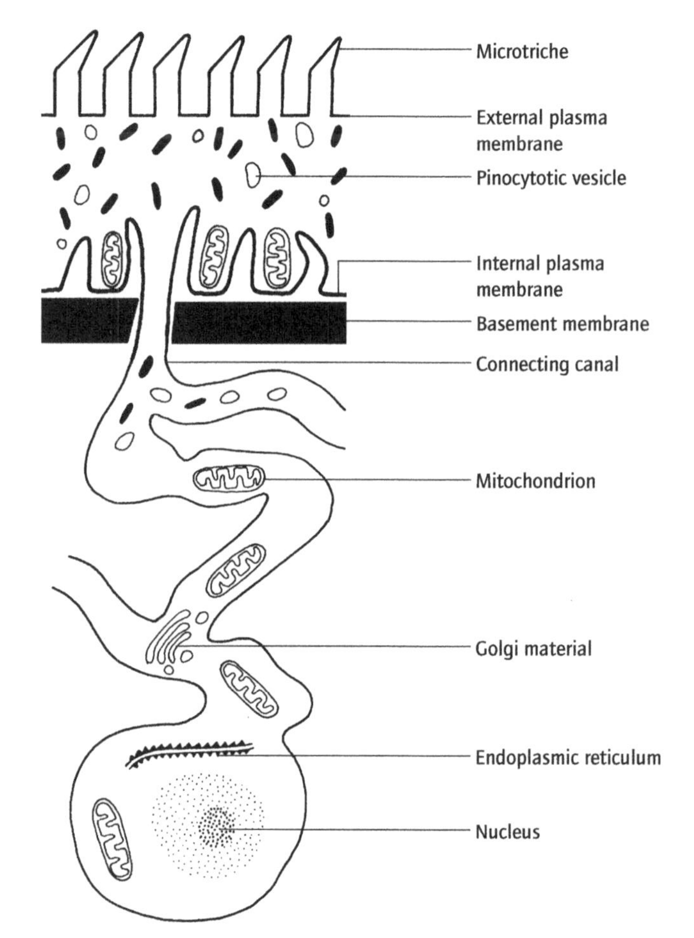

Fig. 3.43 Diagrammatic section of a cestode tegument (after Meglitsch, 1972).

3.7 Further reading

Barnes, R.S.K. 1998. Kingdom Animalia. In: Barnes, R.S.K. (Ed.), *The Diversity of Living Organisms*. Blackwell Science, Oxford.

Bergquist, P.R. 1978. *Sponges*. Hutchinson, London [Porifera].

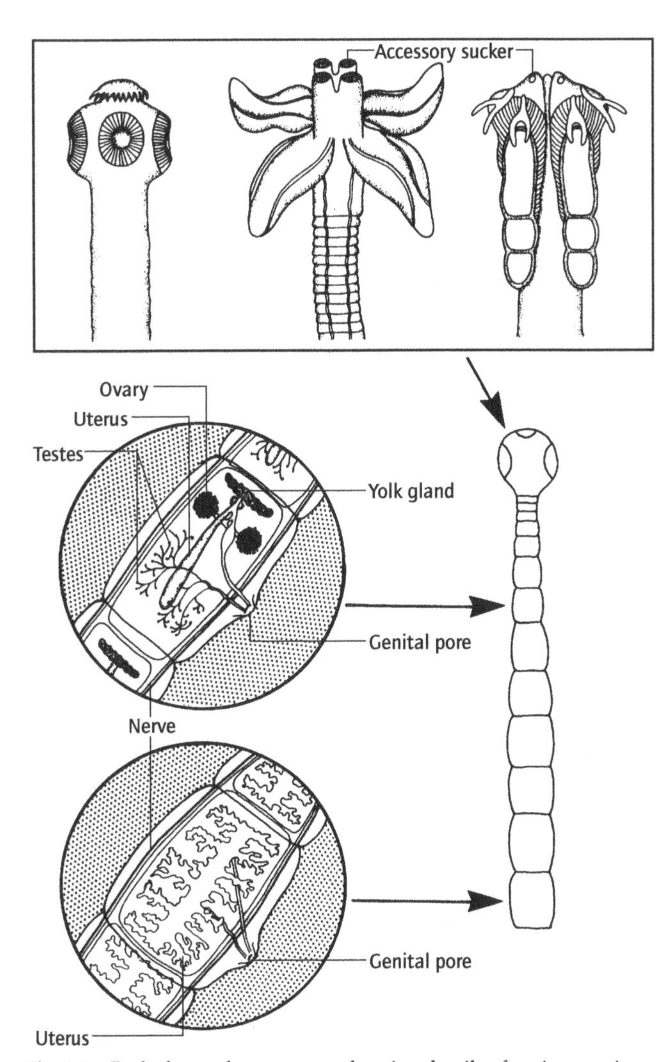

Fig. 3.44 Body form of tapeworms showing details of various regions – three examples of scolexes are given (after various sources).

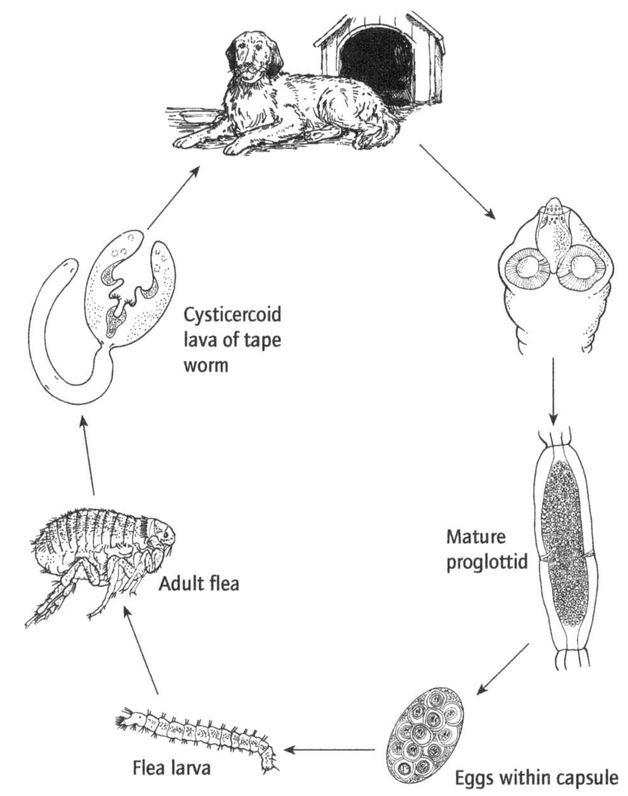

Fig. 3.45 The life cycle of one of the tapeworms infecting the domestic dog (after Joyeux & Baer, 1961).

Corliss, J.O. 1984. The kingdom Protista and its 45 phyla. *Biosystems*, **17**, 87–126 [Protozoa].

Crompton, D.W.T. & Joyner, S.M. 1980. *Parasitic Worms*. Wykeham, London.

Fry, W.G. (Ed.) 1970. *The Biology of Porifera*. Academic Press, New York [Porifera].

Grell, K.G. 1982. Placozoa. In: Parker, S.P. (Ed.) *Synopsis and Classification of Living Organisms*, Vol. 1, p. 639. McGraw-Hill, New York [Placozoa].

Hochberg, F.G. 1982. The 'kidneys' of cephalopods: a unique habitat for parasites. *Malacologia*, **23**, 121–134 [Mesozoa].

Hughes, R.N. 1989. *A Functional Biology of Clonal Animals*. Chapman & Hall, London.

Hyman, L.H. 1940. *The Invertebrates*, Vol. 1. *Protozoa through Ctenophora*. McGraw-Hill, New York [Porifera, Cnidaria, Ctenophora, Mesozoa].

Hyman, L.H. 1951. *The Invertebrates*, Vol. 2. *Platyhelminthes and Rhynchocoela*. McGraw-Hill, New York [Platyhelminthes].

Kaestner, A. 1967. *Invertebrate Zoology*, Vol. 1. Wiley, New York [Porifera, Cnidaria, Ctenophora, Platyhelminthes, Mesozoa].

Lapan, E.A. & Morowitz, H. 1972. The Mesozoa. *Sci. Am.*, **227**, 94–101 [Mesozoa].

Mackie, G.O. (Ed.) 1976. *Nutritional Ecology and Behavior*. Plenum, New York [Cnidaria].

Miller, R.L. 1971. *Trichoplax adhaerens* Schulze 1883: Return of an Enigma. *Biol. Bull. mar. Biol. Lab., Woods Hole*, 141, 374 [Placozoa].

Morris, S.C., George, J.D., Gibson, R. & Platt, H.M. 1985. *The Origins and Relationships of Lower Invertebrates*. The Systematics Association, Special Volume No. 28. Clarendon Press, Oxford.

Parker, S.P. (Ed.) 1982. *Synopsis and Classification of Living Organisms* (2 Vols). McGraw-Hill, New York.

Reeve, M.R. & Walker, M.A. 1978. Nutritional ecology of ctenophores – a review of past research. *Adv. mar. Biol.* **15**, 246–287 [Ctenophora].

Schmidt, H. 1972. Die Nesselkapseln der Anthozoen und ihre Bedeutung für die phylogenetische Systematik. *Helgoländer Wiss. Meeresunters.*, **23**, 422–458 [Platyhelminthes].

Schockaert, E.R. & Ball, I.R. 1981. *The Biology of the Turbellaria*. Junk, The Hague [Platyhelminthes].

Sleigh, M.A. 1989. *Protozoa and other Protists*. Arnold, London [Protozoa].

Smyth, T.D. 1977. *Introduction to Animal Parasitology*, 2nd edn. Wiley, New York [Platyhelminthes].

Sorokin, Y.I. 1993. *Coral Reef Ecology*. Springer, Berlin.

CHAPTER 4

The Worms

Nemertea
Gnathostomula
Gastrotricha
Nematoda
Nematomorpha
Kinorhyncha
Loricifera
Priapula
Rotifera
Acanthocephala
Sipuncula
Echiura
Pogonophora
Annelida
Worms of uncertain phylum

The 14 phyla included in this chapter are all protostome derivatives of the flatworms although, that apart, they share no specific features and do not constitute a natural group of related animals (see Chapter 2); as has been stated, the name 'worm' is an indefinite though suggestive term popularly applied to any animal that is not obviously something else! They are a group of 'convenience', distinguished more by the common absence of the various features which characterize other groups of phyla than by anything else, notwithstanding even their vermiform shape. That is to say that the worms covered here are animals which do *not* possess legs, are *not* covered by a protective shell, are *not* deuterostome, and do *not* bear a lophophore. The equally vermiform phoronans do have a lophophore and are therefore included in Chapter 6; the acorn-worms (enteropneust hemichordates) are deuterostome and are, for that reason, included in Chapter 7, as are the arrow-worms (chaetognaths); whilst the tongue-worms (pentastomans) are considered with their arthropod allies in Chapter 8. In practice, therefore, these 14 groups of worms are simply animals which have retained the general vermiform shape of their ancestors more or less unchanged, having bodies some 2–3 to over 15 000 times longer than wide and flattened or rounded in section.

Although systems of relationship within these worms are not universally agreed, it is nevertheless becoming clear that four clusters of phyla are broadly recognizable. The first such (the Gnathostomula and Gastrotricha) are similar in essential body plan to the supposed ancestral flatworm-like animals and show relatively little within-group diversity; both phyla are characterized by highly individual specializations that prevent them from being accommodated within the Platyhelminthes.

The gastrotrichs may possibly provide a morphological link to the much larger and more diverse second cluster of phyla, variously known as the nemathelminths or aschelminths. It is to these various worms that the arthropods are thought to be related. The group comprises five phyla all with moulted cuticles. The Kinorhyncha, Loricifera and Nematomorpha all share, at some stage of their life history, a distinctive eversible introvert, bearing a stylet-surrounded mouth cone and whorls of scalids; the Priapula possess a similar scalid-bearing introvert, although without oral stylets; and the Nematoda also appear to be linked to this group in that a species from Brazil has an equivalent introvert with spine-like scalids, and the nematodes and nematomorphs share the same highly individual muscular/locomotory system. With the exception of their scalid-bearing introvert, the priapulans are otherwise rather atypical nemathelminths, and could lay claim to be close to the ancestral form and reproductive type of the group.

The third assemblage, the Rotifera and Acanthocephala, lack any such introvert and external cuticle, but share the presence of an intracellular cuticle within the epidermis, as well as a locomotory and/or adhesive proboscis and the epidermal sacs, lemnisci, that may be associated with its operation (the latter two features only occurring in the bdelloid rotifers). The affinities of this group is currently somewhat obscure, although a relationship with the eutrochozoans seems most likely.

The fourth assemblage also contains a diverse series of worm types. The Annelida, Pogonophora and Echiura, however, all share such features as schizocoelic body cavities, chitinous chaetae, closed blood systems and metanephridium-like organs, although the first two phyla are metameric whilst the echiurans are clearly monomeric. Also monomeric are the Sipuncula, of which the 'metanephridia' and schizocoel are often taken to indicate affinity with this fourth group of phyla, although they lack chaetae and any circulatory system. The sipunculans alone possess an introvert; but in contrast to the short, barrel-shaped, spiny organ of the nemathelminths, that of the sipunculans is a long, narrow tube, terminating in detritus-collecting lobes or tentacles. The Nemertea also belong to this group. They are monomeric and non-segmented, but possess a large schizocoelic cavity in which is housed a highly individual prey-catching harpoon. Otherwise their body plan is very similar to that of the flatworms, although they have a circulatory system and a through gut. Where present, the larval stage of all these phyla (and the related molluscs) is a trochophore.

Just as the position of the Priapula within the nemathelminths is aberrant – in respect of some of the peculiarly nemathelminth features they are typical; in respect of others, they are radically unlike all the other phyla – so is that of the Sipuncula with regard to the other schizocoelomate worms. In both these large groups of worms, argument based on some anatomical features can be used to indicate the parallel acquisition of their common grade of organization, whilst other characters can be advanced in support of phylogenetic affinity. What is gland and its duct to one author can be the pharynx and buccal canal to another (this example being cited from two views of the structure of the larval nematomorph). Even some individual phyla, for instance the Nematomorpha amongst the nemathelminths and the Pogonophora amongst the schizocoelomates, can convincingly be argued each to consist of two groups of worms with completely different origins: one subgroup of nematomorphs may have been derived from mermithoid nematodes, whilst the other is related to the kinorhynchs and loriciferans; and one class (or subphylum) of pogonophorans is closely related to the annelids, the other class (or subphylum) may have arisen independently and acquired its schizocoel in parallel!

Clearly it would be premature to amalgamate any of these various groups of worms into a smaller number of larger phyla, with the exception of the union of the Annelida and Pogonophora on phylogenetic grounds and with the possible exception of the nemathelminths. Here we will treat all 14 groups as separate blocks of equal systematic rank.

Also included here (Section 4.15) are four families of worms of more contentious affinity that cannot yet be placed in any existing phylum. One is often given a phylum of its own and it can be argued that new phyla may ultimately have to be erected to accommodate all of them.

4.1 Phylum NEMERTEA (ribbon- or proboscis-worms)

4.1.1 Etymology

Greek: *Nemertes*, a Mediterranean sea-nymph, the daughter of Nereus and Doris.

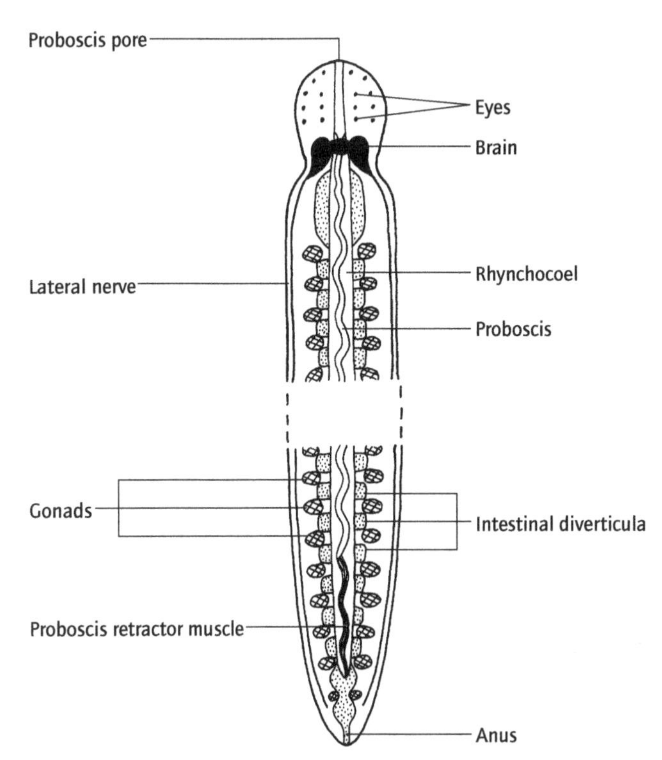

Fig. 4.1 Diagrammatic dorsal view of a nemertine, seen as if transparent (after several sources, especially Pennak, 1978).

4.1.2 Diagnostic and special features (Fig. 4.1)

1 Bilaterally symmetrical, vermiform.
2 Body more than two cell layers thick, with tissues and organs.
3 A through gut and a terminal anus.
4 Without a body cavity, the skeleton being provided by parenchyma.
5 A ciliated outer, cuticle-less epidermis.
6 Body dorsoventrally flattened, often with serially repeated organs (e.g. gut pouches, protonephridial organs, gonads).
7 Without a gaseous-exchange system, but with a circulatory system.
8 Nervous system with a brain and, usually, three longitudinal cords.
9 An eversible and retractable ectodermal proboscis housed in a longitudinal dorsal cavity, the rhynchocoel, which opens to the exterior near the mouth.
10 Eggs cleave spirally.

The <0.5 mm to >30 m long nemerteans or nemertines can be described as 'super-flatworms', the essential features of their bodily organization being basically those of the larger turbellarian platyhelminths (see Section 3.6.2.3.1). They differ from that group, however, in three important respects. First, they possess a more efficient through gut; second, they have a closed blood system, with blood flow (which is irregular) being driven both by contraction of the walls of the blood vessels and by bodily movements; and third, they exhibit a characteristic proboscis,

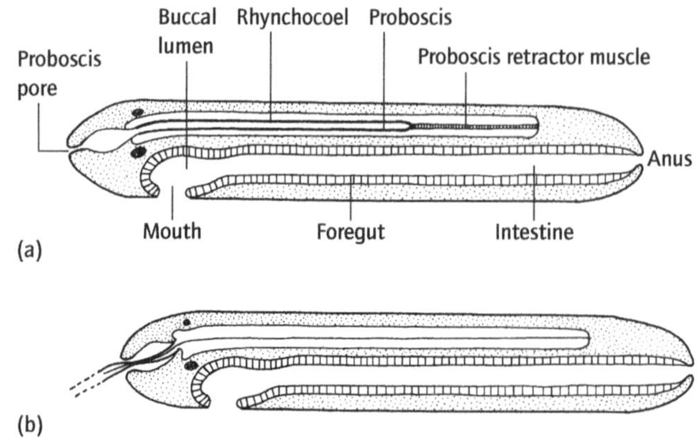

Fig. 4.2 Diagrammatic longitudinal section through a nemertine showing the rhynchocoel and the proboscis (a) retracted and (b) everted (after Gibson, 1982).

Fig. 4.3 Nemertine pilidium larva (after Kershaw, 1983).

which is a completely separate structure from the gut and not a region of it (Fig. 4.2).

This proboscis is housed in a tubular cavity extending almost the whole length of the body, with walls of the same construction as that of the body. Pressure generated within the rhynchocoel by contraction of the wall muscles everts the proboscis, and it is retracted by means of a longitudinal muscle running from its posterior end to the posterior margin of the rhynchocoel. It is used mainly for prey capture, but also, in terrestrial species, for rapid locomotion: the proboscis being everted, attached to the ground ahead of the worm, and the animal then pulling itself forward over its own proboscis.

Nemertine locomotion is otherwise similar to that of the turbellarians, i.e. ciliary gliding, the cilia beating in secreted mucus, or waves of muscular contraction passing along the ventral surface in the larger forms. Some species can swim by means of dorsoventral undulations of the body.

Multiplication occurs both by asexual fragmentation and by sexual reproduction, the gonads being temporary structures formed from aggregations of differentiated mesenchyme cells which become enclosed within membranes and, when mature, connected to the exterior by temporary ducts. Most species are gonochoristic, although some freshwater and terrestrial forms are hermaphrodite, and capable of self-fertilization. Direct development is the norm, but members of one order possess a larval stage (Fig. 4.3); fertilization is usually external.

Almost all species are predatory, capturing organisms ranging in size from protists to molluscs, arthropods and fish. Nemertines are notable for including the longest of known animals: *Lineus longissimus* regularly attains 30 m and some individuals can probably achieve twice this length when fully extended.

4.1.3 Classification

The 900 known species are included in two classes.

Class	Order
Anopla	Palaeonemertea Heteronemertea
Enopla	Hoplonemertea Bdellonemertea

4.1.3.1 Class Anopla

In the anoplans, the central nervous system is located within the body wall, which often has three layers of muscle: either two layers of circular on either side of a single layer of longitudinal, as in the order Palaeonemertea, or two layers of longitudinal on either side of a single layer of circular, as in the order Heteronemertea. The anoplan proboscis is relatively simple, without regional differentiation or stylets, although large numbers of epithelial rhabdoids, of the type frequently found in the turbellarians, may be present; the proboscis pore opens completely separately from the mouth.

The majority of anoplans (Fig. 4.4) are marine and benthic, but three species occur in fresh waters and several inhabit brackish regions. Heteronemertines are the only group to possess a larval stage.

4.1.3.2 Class Enopla

Enoplans possess a central nervous system located internally to the body wall musculature, which is two-layered, and a proboscis differentiated into regions, the central one of which is a short muscular bulb bearing one or more stylets (except in the aberrant and commensal, filter-feeding genus *Malacobdella* which is placed in its own order, the Bdellonemertea). The enoplan gut is complex and opens to the exterior through an aperture in common with the proboscis: in the majority of species (the order Hoplonemertea), the intestine bears numerous pairs of lateral diverticula; whilst in the bdellonemertines, the large pharynx is the organ of filter feeding and it contains ciliated papillae, which trap food particles without the use of mucus.

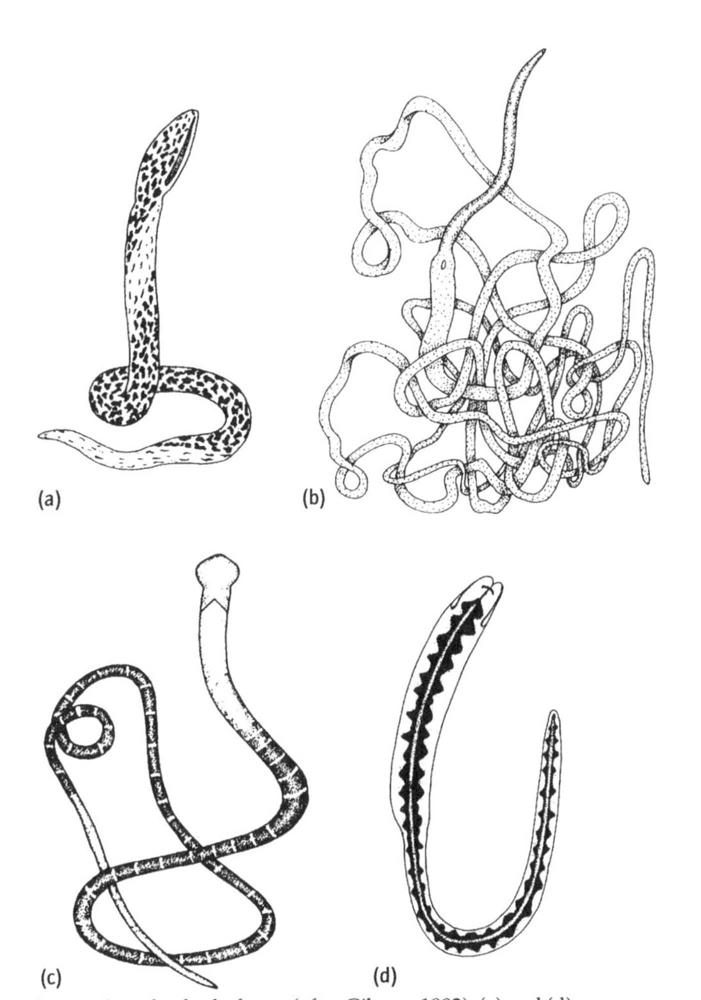

Fig. 4.4 Anoplan body forms (after Gibson, 1982): (a) and (d), heteronemerteans; (b) and (c), palaeonemerteans.

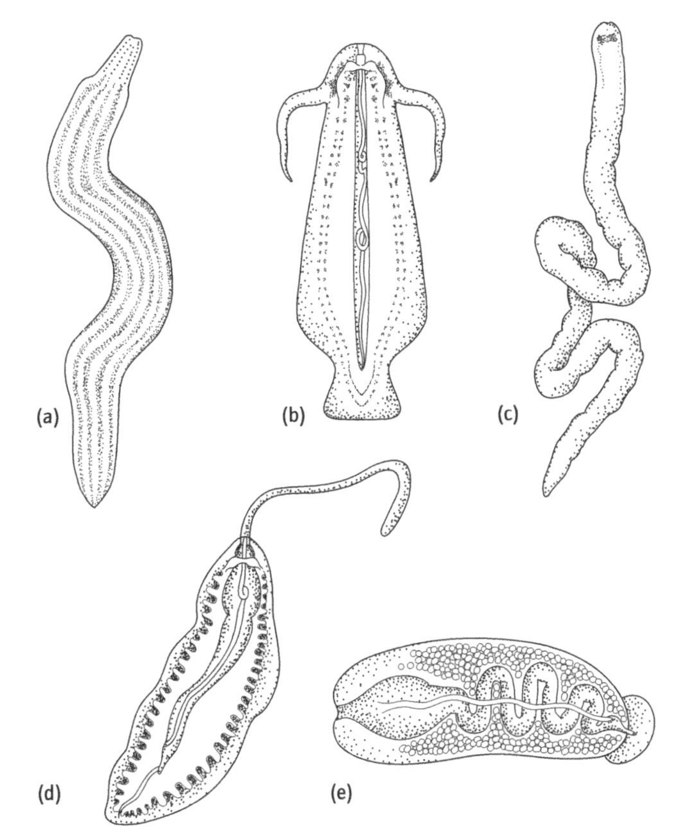

Fig. 4.5 Enoplan body forms (after Gibson, 1982): (a)–(d), hoplonemerteans; (e) a bdellonemertean.

Although most enoplans are marine and benthic, several are pelagic, including both swimming and floating forms (Fig. 4.5), some are freshwater and others terrestrial, and a few are commensal (in, for example, the atrial cavity of tunicates) or parasitic.

4.2 Phylum GNATHOSTOMULA

4.2.1 Etymology

Greek: *gnathos*, jaw; *stoma*, mouth.

4.2.2 Diagnostic and special features (Figs 4.6 and 4.7)

1 Bilaterally symmetrical and vermiform.
2 Body more than two cell layers thick, with tissues and organs.
3 With a blind-ending gut, although a temporary anus may form.
4 Without a body cavity, the skeleton being provided by poorly developed parenchyma.
5 Without circulatory or gaseous-exchange systems.

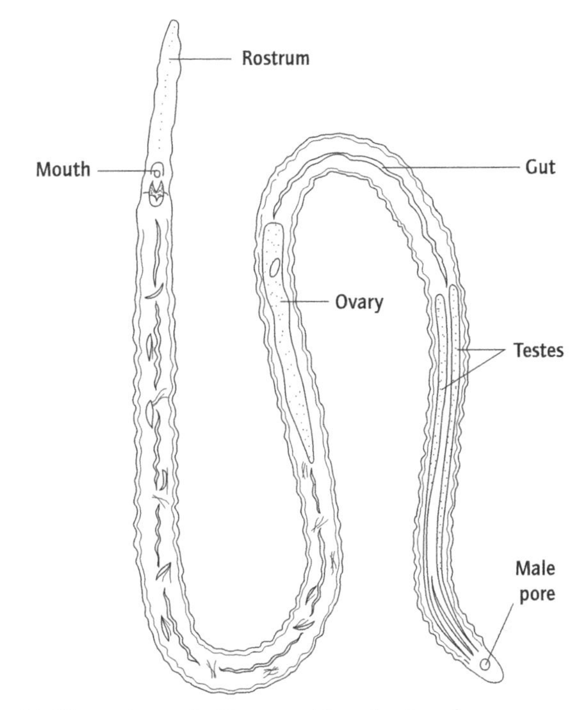

Fig. 4.6 The anatomy of a filospermoid gnathostomulan (after Sterrer, 1982).

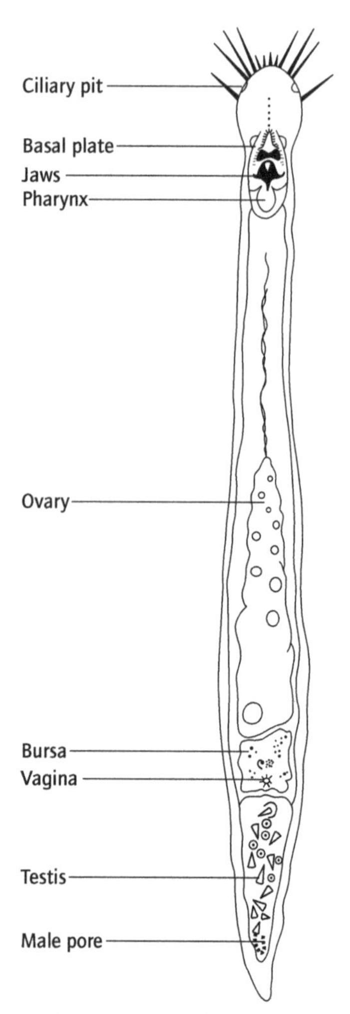

Fig. 4.7 The anatomy of a bursovaginoid gnathostomulan (after Sterrer, 1982).

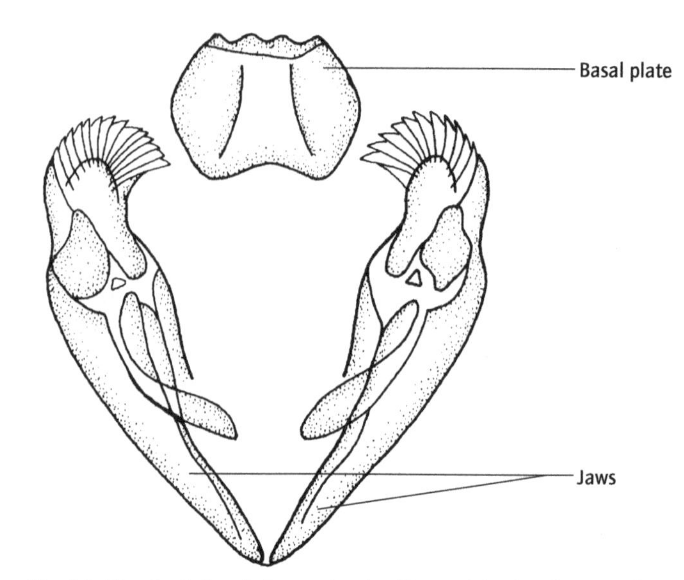

Fig. 4.8 The single basal plate and paired jaws of the Gnathostomula (after Sterrer, 1982).

6 Excretory organs simple, two-celled protonephridia resembling epidermal cells in that the terminal cell bears only a single cilium.
7 With an outer monociliated, cuticle-less epidermis.
8 With a diffuse, epidermal nervous system.
9 With a complex feeding apparatus formed by paired jaws and a single basal plate.
10 Hermaphrodite (simultaneously so).
11 Eggs cleave spirally.
12 Development direct.
13 Marine, interstitial often in anoxic sands.

These minute (<3 mm length), transparent worms possess a bodily organization essentially similar to those of the free-living turbellarian flatworms (Section 3.6.2.3.1) and of the gastrotrichs (Section 4.3). They differ from both these groups, however, particularly in respect of their highly specialized, muscular pharynx which bears a complex jaw system used in grazing bacteria, protists and fungi from the surfaces of sand grains (Fig. 4.8). In search of such prey, gnathostomulans move through sedimentary interstitial spaces by swimming or gliding using their ciliated epidermis, the cilia of which can beat either forwards or backwards, and by means of sudden contraction of the three or four pairs of longitudinal muscles which comprise the musculature of the body wall. The single ovary releases a single, large egg at a time, which is probably always fertilized internally. It is possible that during the life cycle, several gnathostomulans exhibit an alternation of forms between a non-sexual feeding stage and a sexual non-feeding one.

This phylum was not described until 1956 when the first gnathostomulan was discovered. This late discovery was not a result of rarity of the animals – they may achieve densities of 600 000 m^{-3} – but of previous lack of investigation of the anoxic layers of marine sediments, and of the extent to which gnathostomulans distort on attempted preservation.

4.2.3 Classification

The 100 or so described species are placed in two orders within a single class: the extremely elongate filospermoids (Fig. 4.6) with a long anterior rostrum but without anterior paired sense organs, and without a penis or a sperm-storage sac (the bursa); and the squatter bursovaginoids (Fig. 4.7) which lack an elongate rostrum but possess paired sensory bristles, cilia and pits anteriorly, and have a penis and a bursa. Many more species undoubtedly await description.

4.3 Phylum GASTROTRICHA

4.3.1 Etymology

Greek: *gaster*, stomach; *thrix*, hair.

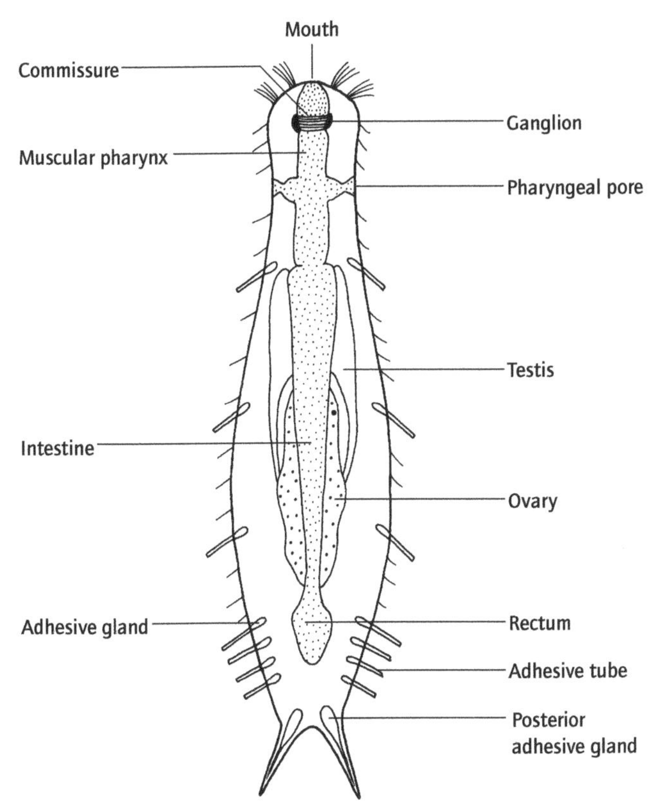

Fig. 4.9 Diagrammatic dorsal view of a gastrotrich, seen as if transparent (after several sources). Note: diagram shows features of both macrodasyidans (lateral adhesive tubes) and chaetonotidans (pharyngeal pores) that are not simultaneously present in any individuals.

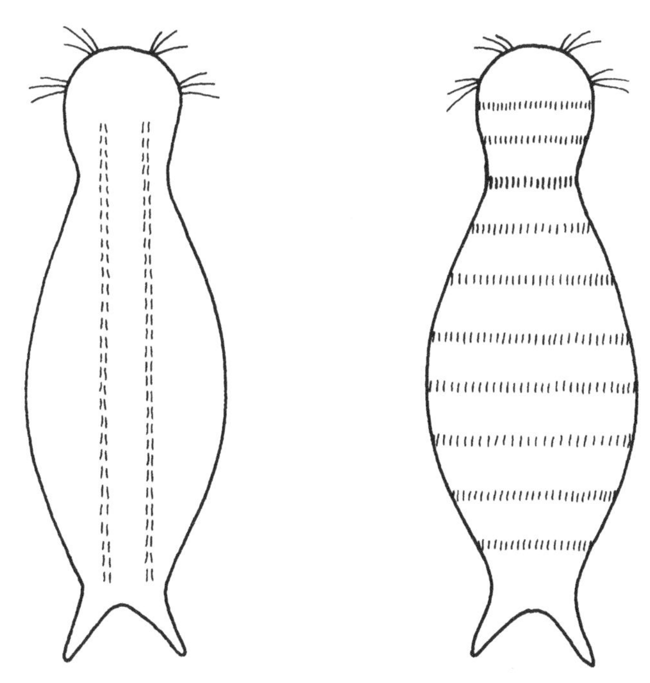

Fig. 4.10 Diagrammatic ventral views of two gastrotrichs showing different patterns of ciliation (after Hyman, 1951).

4.3.2 Diagnostic and special features (Fig. 4.9)

1 Bilaterally symmetrical, vermiform.
2 Body more than two cell layers thick, with tissues and organs.
3 A through gut and a subterminal anus.
4 Without a body cavity (notwithstanding the much quoted early view to the contrary).
5 Without circulatory or gaseous-exchange systems.
6 An outer non-chitinous cuticle and, usually, a monociliated dorsal and multiciliated ventral cellular epidermis; some layers of the cuticle extending as sheaths over each individual cilium.
7 Nervous system with a ganglion on each side of the anterior pharynx, connected dorsally by a commissure, and a pair of longitudinal cords.
8 Body covered by cuticular scales, spines or hooks, and bearing up to 250 adhesive tubes.
9 Hermaphrodite or parthenogenetic.
10 Eggs released by rupture of the body wall and cleave in a bilaterally radial manner, but development is determinate.
11 Development direct.
12 Aquatic; interstitial, surface dwelling or, rarely, planktonic.

Gastrotrichs possess a more or less transparent, dorsoventrally flattened body of up to 4 mm in length (although usually <1 mm). Anteriorly, sense organs of the types seen in bursovaginoid gnathostomulans occur (paired bristles, cilia and sensory pits), together with, in some, eye-spots; posteriorly, the body may end in a stout fork or a thin tail. The adhesive tubes provide a means of temporary attachment to surfaces and, although the body wall contains both circular and longitudinal muscles, movement is by ciliary gliding. The cilia of the ventral surface are often non-uniformly disposed, e.g. in transverse or longitudinal bands, in patterns characteristic of the various genera (Fig. 4.10).

They feed on bacteria, protists and detritus swept into the mouth by the beating of buccal cilia or by pharyngeal pumping. In common with the nematodes, the pharynx is triradiate in section and is lined by a single layer of myoepithelial cells; and, as in rotifers and kinorhynchs, the freshwater species bear a pair of solenocytic protonephridia.

4.3.3 Classification

More than 450 species are known, apportioned between two orders within a single class: the strap-shaped marine macrodasyidans (Fig. 4.11) which possess a pharynx perforated through to the exterior by a pair of pores, probably serving as an exit for water taken in whilst feeding as in the pterobranch hemichordates (Section 7.2.3.2, and see Section 9.2.5); and the marine and freshwater chaetonotidans (Fig. 4.12), which are normally fusiform in shape, lack pharyngeal pores, have adhesive tubes, if at all, only posteriorly and may be parthenogenetic.

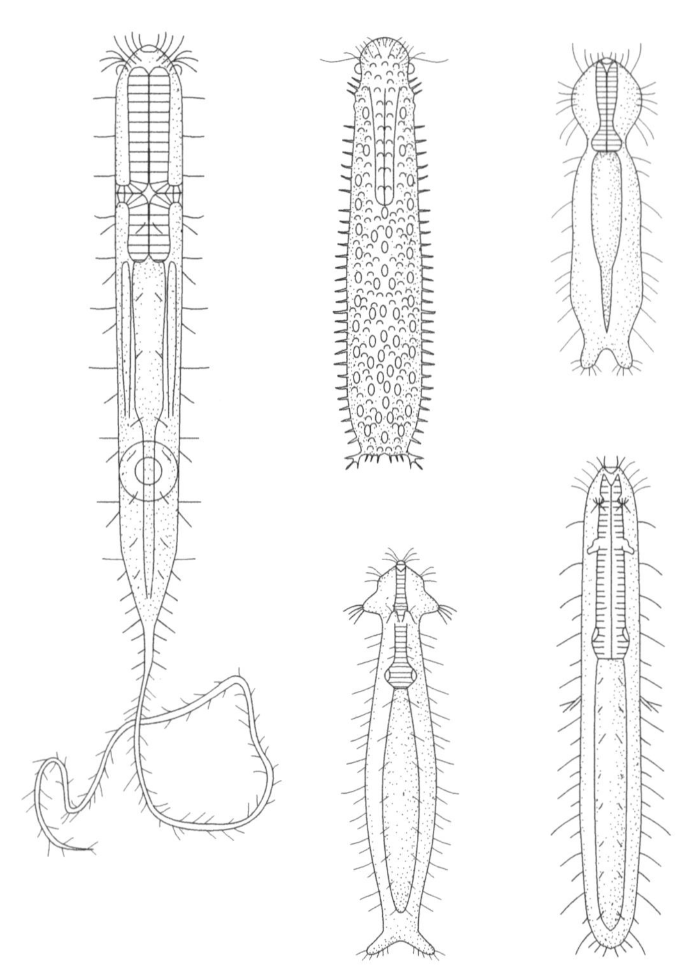

Fig. 4.11 Macrodasyidan body forms (after Grassé, 1965; Hummon, 1982).

4.4 Phylum NEMATODA

4.4.1 Etymology

Greek: *nema*, thread; *eidos*, form.

4.4.2 Diagnostic and special features

1 Bilaterally symmetrical, vermiform, but with a tendency to radial symmetry around the longitudinal axis.
2 Body more than two cell layers thick, with tissues and organs.
3 A complex cuticle.
4 Body wall without circular muscles.
5 Body cavity a pseudocoel, usually derived from the blastocoel.
6 Muscular gut leading from an anterior mouth via a muscular pharynx to the subterminal anus.
7 Longitudinal muscles arranged in four zones; ventral, lateral and two dorsal epidermal chords.
8 Nervous system with longitudinal nerves in the mid-ventral and mid-dorsal epidermal chords and with direct contacts with the muscle cells in the contralateral muscle fields.
9 Cross-sectional area always circular; body fluid always maintained under high pressure.
10 Without a circulatory system.
11 Excretory system without flame cells or nephridia. Excretory tubules in one, or a small number of, renette cells.
12 Embryonic cleavage pattern neither spiral nor radial, but highly determinate with T-shaped arrangement of cells at the four cell stage.
13 Development always direct, but in parasitic forms larvae may be infective stages.

The phylum Nematoda is one of the great success stories of the animal kingdom. More than 15 000 species have been described, of an estimated 1 million living species. In contrast with the other large phyla, this diversity of species is not based upon great diversity of structure. All nematodes are constructed along the same fundamental plan. The many species represent minor modifications of a successful formula. As this includes the ability to withstand potentially harmful environments, many nematodes are parasitic.

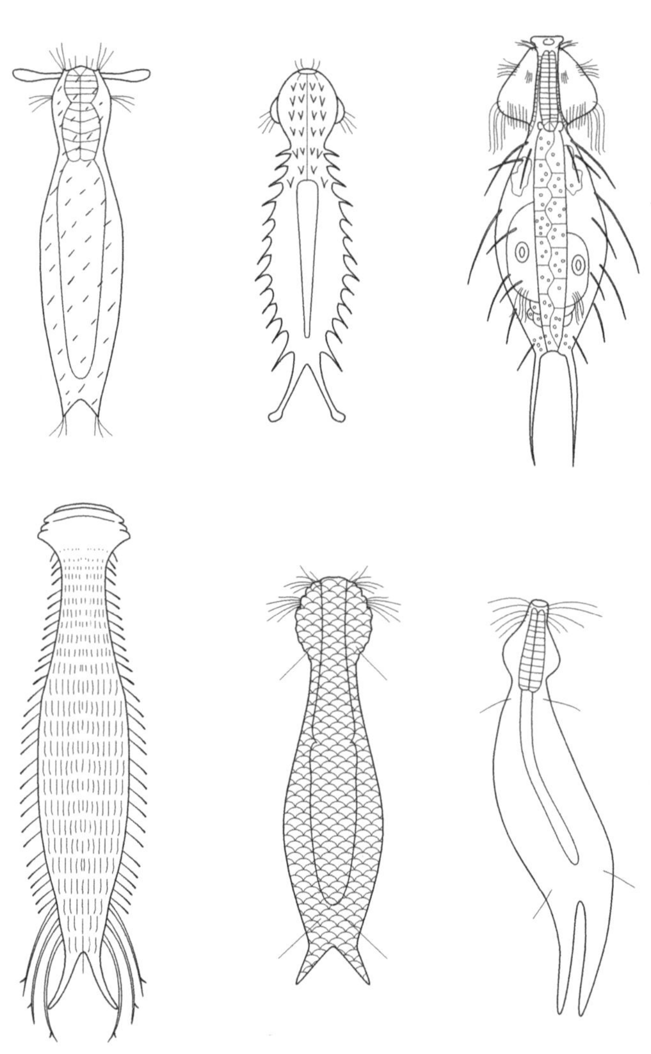

Fig. 4.12 Chaetonotidan body forms (after Grassé, 1965; Hummon, 1982).

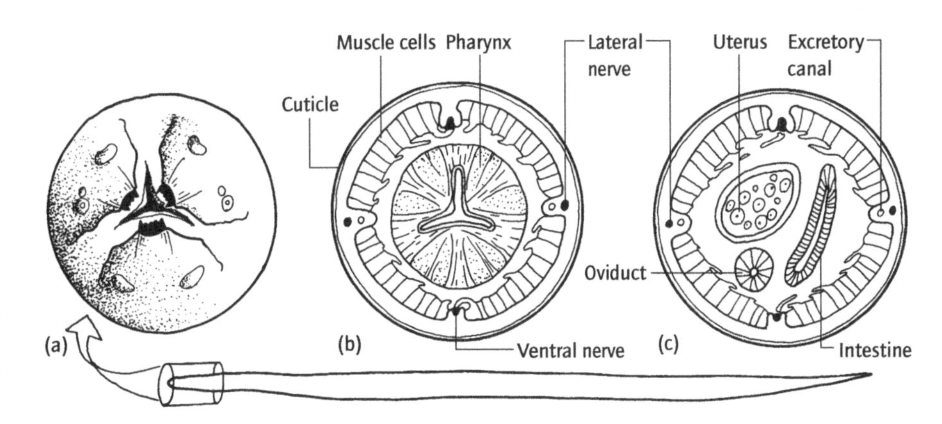

Fig. 4.13 The structure of a nematode as seen in transverse section: (a) stereoscopic view of the head showing the triangular arrangement of the mouth, lips and the associated sense organs; (b) cross-section in the pharyngeal region; (c) cross-section through the mid-body region.

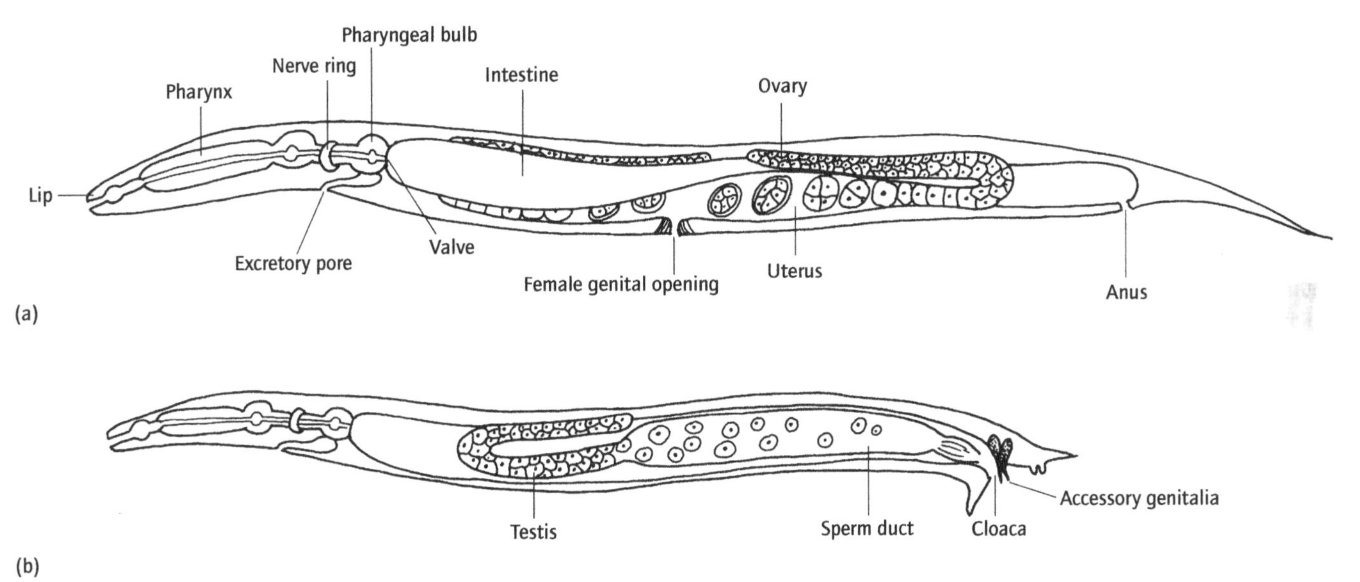

Fig. 4.14 The general structure of a nematode (based on the parasitic genus *Rhabditis*): (a) a female; (b) a male.

A nematode is a 'tube within a tube', pointed at both ends and round in cross-section (Fig. 4.13). Much of the body cavity is filled with the paired reproductive organs, ovaries or testes. These are serially arranged, and often coiled. The positions of the openings are shown in Fig. 4.14a and b.

The structure of the body wall is best seen in cross-section and is unique to the phylum (Fig. 4.13b).

The outermost region is a complex cuticle in which as many as nine layers can be recognized (Fig. 4.15). Among them three bands of crossed fibres forming a spiral network are particularly conspicuous.

The fibres are inelastic and so limit the volume of the worm while they allow changes in shape. The fibres are sufficiently strong to permit high internal pressures to be maintained permanently and the impermeable nature of the cuticle prevents fluid loss. In such a high-pressure system the cross-section of the worm is always circular (see Chapter 10 and Fig. 10.9 for further details). The longitudinal muscles work against the internal pressure and so generate changes in shape. This high-pressure

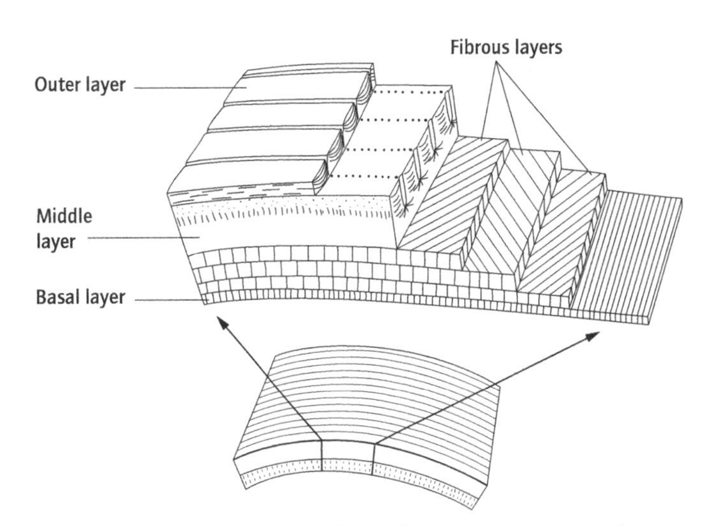

Fig. 4.15 Structure of the nematode cuticle. The cuticle consists of an outer striated layer, an inner homogeneous layer and a complex of fibrous layers (after Clark, 1964).

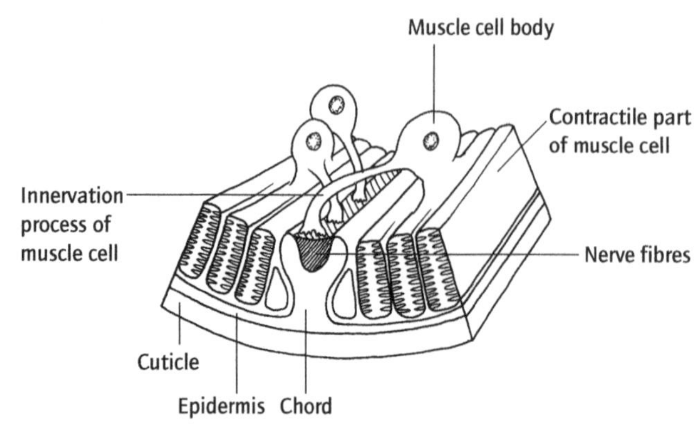

Fig. 4.16 Stereoscopic view of the nematode neuromuscular complex. Note the contractile filament of the muscle cells, the cell bodies and innervation process which make direct synaptic contact with the nerve fibres in the lateral, dorsal or ventral cords.

hydraulic system is of fundamental importance in the biology of the nematodes. It is, for instance, associated with the unusual body plan in which there are no circular muscles.

Beneath the complex cuticle lies the ectodermal epidermis containing the cells of the nervous and excretory systems, the principal elements of which are located in four longitudinal thickened strands or chords. The dorsal and ventral chords contain the prominent dorsal and ventral nerves. In the quadrants between the epidermal chords lie the cells of the longitudinal muscles; these are few and fixed in number. After the earliest stages of embryonic development, bodily growth involves an increase in cell size rather than cell number (see Chapter 15 for details of the experimental use of this feature by contemporary developmental biologists).

The excretory system exemplifies the extreme economy of cell numbers characteristic of the phylum. The primitive condition involves one or two specialized *renette* cells. In more advanced arrangements a system of tubes is developed within the glandular *renette* cells, which in some forms is lost (see Section 12.3.1 and Fig. 12.12).

The muscle cells could be more correctly described as neuromotor units (Fig. 4.16). The contractile fibres of each cell are situated distally and rest on the epidermis. An elongated innervation process leads directly to nerve fibres in the dorsal and ventral chords, according to the position of the muscle cells. As it approaches the nerve fibres, each muscle cell is drawn out into a number of processes, the tips of which form synapses (see Chapter 16) with other muscle fibres and the nerve cells. This may provide a system that permits the simultaneous contraction of all the muscle cells. This arrangement in which muscle cells send fibres to nerve cells is unusual, since in most other animals nerve cells send processes to contact the motor cells (but see Section 16.4.2).

Nematodes are microphagous mostly fluid feeders. Since the body fluids are maintained under a permanent state of high pressure, internal tubules would collapse if allowed to equilibrate with external pressures. There must therefore be some

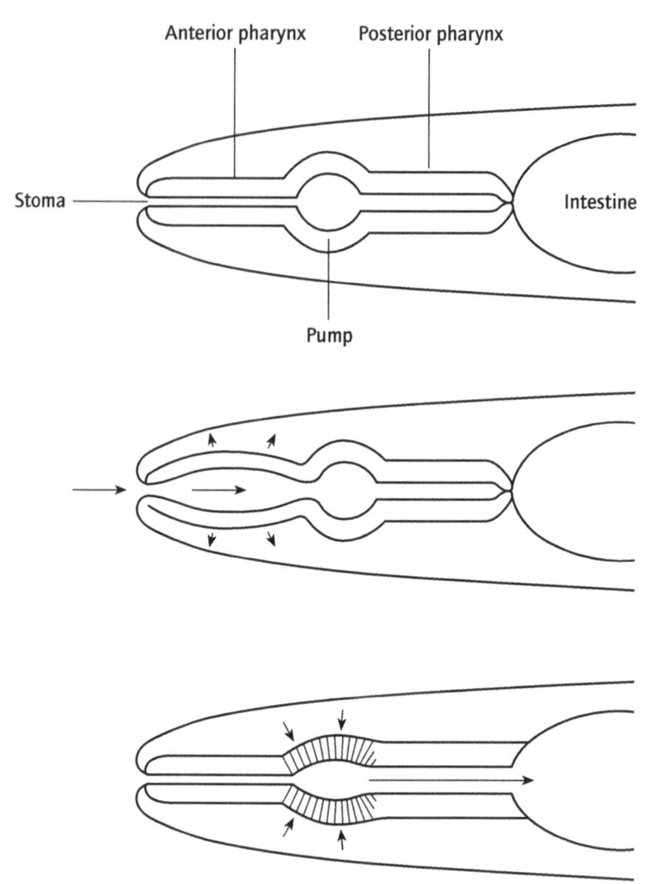

Fig. 4.17 The mechanism of the nematode pharyngeal pump. The radius of the anterior pharynx is much smaller than that of the posterior pharynx. The valve at the posterior is closed when dilation of the pharynx occurs. Consequently liquid flows into the anterior pharynx. When the pharynx contracts the posterior valve is open and liquid flows backward because of the greater radius of the posterior pharynx (after Croll & Matthews, 1977).

mechanism to prevent this happening during feeding when fluids enter the gut against a pressure gradient. The mechanism is illustrated in Fig. 4.17. The anterior opening of the gut (termed the *stoma*) leads into a muscular pharynx, at the other end of which is a valve. The pharynx pumps food against a pressure gradient and the unidirectional flow towards the intestine depends on the relative dimensions of the fore and hind parts of the anterior food duct.

Nematodes are equipped with various sensory devices. These include: *cephalic papillae; amphids* (chemosensory pits often associated with anterior amphid glands); *cephalic setae; somatic setae; caudal papillae* and *spicules; ocelli* and *phasmids* (Fig. 4.18). Phasmids are caudally located sensory pits, not unlike amphids, and sometimes these two are associated with glands. The Nematoda is subdivided into two classes by presence or absence of phasmids.

The rather uniform appearance of the nematodes is a consequence of the unique features of their design which involves high internal pressure and impermeable cuticle. There are

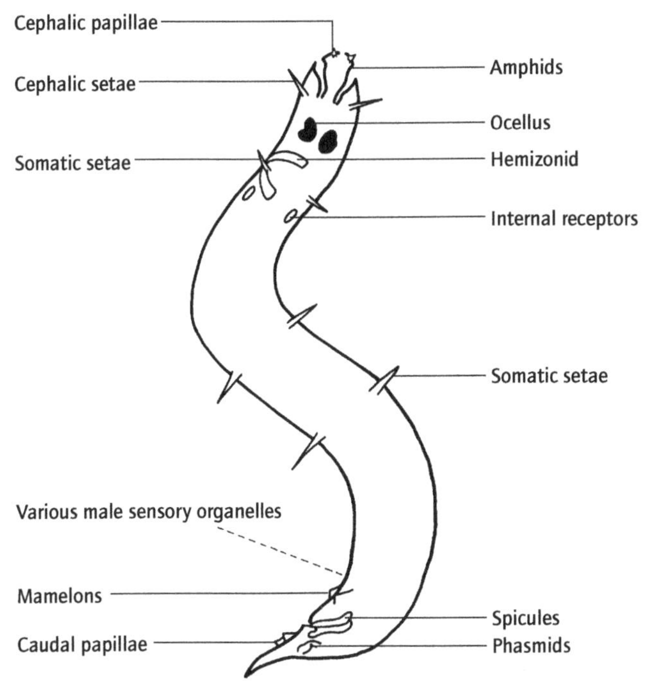

Fig. 4.18 A generalized scheme showing different types of sense organs found in nematodes. No single species is known to have all of these (after Croll & Matthews, 1977).

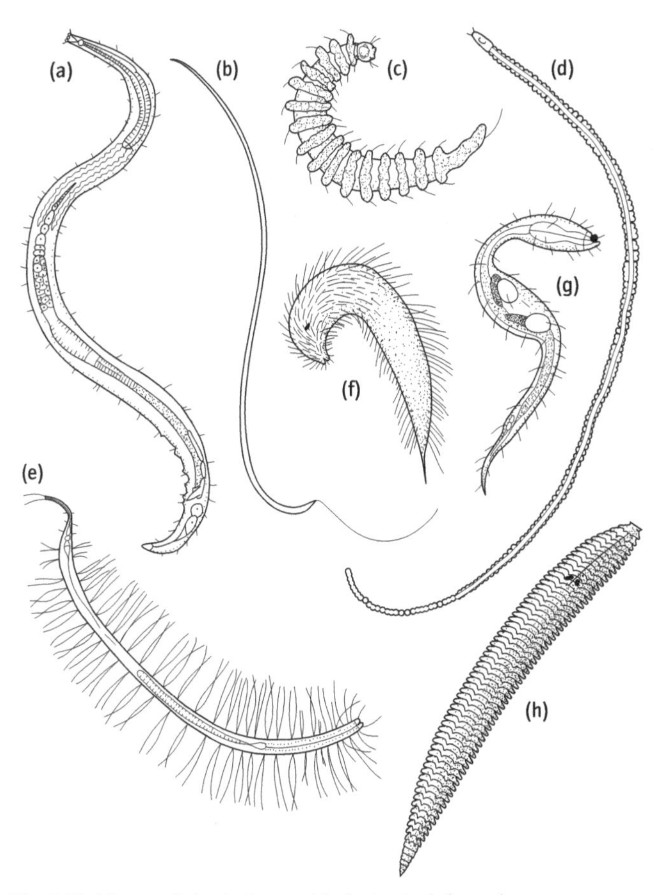

Fig. 4.19 Nematode body forms: (a), the typical shape (a monhysteridan); (b)–(h), more unusual forms (b, an enoplidan; c and f, desmoscolescidans; d and g, desmodoridans; e, a monhysteridan; h, a tylenchidan). (After De Coninck, 1965 and Riemann, 1988).

Class	Order
Adenophorea	Enoplida
	Isolaimida
	Mononchida
	Dorylaimida
	Trichocephalida
	Mermithida
	Muspiceida
	Araeolaimida
	Chromadorida
	Desmoscolecida
	Desmodorida
	Monhysterida
Secernentea	Rhabditida
	Strongylida
	Ascaridida
	Spirurida
	Camallanida
	Diplogasterida
	Tylenchida
	Aphelenchida

variations in overall shape and some species have hair-like spines and other projections from the cuticle. The range of form is illustrated in Fig. 4.19.

Despite their rather uniform appearance the nematodes can be adapted to an enormous diversity of circumstances. This is best illustrated by their feeding mechanisms and by the diversification of their life cycles associated with parasitism. Many free-living nematodes feed on bacteria ingested as a suspension or on detritus. Some, however, are predators, especially of other nematodes and micro-organisms, and they have developed a variety of tooth- or jaw-like plates at the anterior surfaces of the stoma. These plates pierce the prey, the contents of which are then sucked out by the pumping action of the pharynx.

A number of free-living species are phytophagous. Some ingest fungi and yeast cells whole, but most pierce the cells of the plants and suck out the contents. To feed in this way, many phytophagous nematodes possess a special structure in the buccal cavity: spear-like to open up the cells, or hollow to use as a stylet under suction from the pharyngeal pump.

The feeding mechanisms exhibited by free-living nematodes can all be exploited by parasitic forms. The gut cavities of vertebrates, for instance, are filled with a rich mixture of bacteria and cell debris which forms a substrate for nematodes feeding by bacterial ingestion. Similarly, plate-like jaws can be used to tear the linings of the gut releasing the blood corpuscles on which the nematodes feed.

Debilitation caused by nematode infections of the alimentary system of vertebrates depends in part on this feeding biology. The human 'pinworm', *Enterobius vermicularis*, is relatively harmless while feeding on bacteria in the gut and is pathogenic only if the population increases sufficiently to cause blockage of the alimentary system. The human hookworm, *Ankylostoma duodenale*, is a more serious pathogen, causing intestinal bleeding and then ingesting the blood corpuscles. A human patient with 100 such worms could lose 50 ml of blood a day.

All phytoparasitic nematodes have a stroma with a stylet or spear. Many form permanent cysts in which the sedentary females feed on giant cells which they cause the host plant to produce, perhaps by interfering with the production of natural plant growth substances or their inhibitors. The most important genera are *Heteroda* and *Meloidogyne*.

Parasitic habits have arisen in nematode phylogeny on numerous occasions. This reflects a number of features fundamental to the structural organization of the phylum.

1 The complex impermeable and resistant cuticle.
2 Internal fertilization and the capacity to produce highly resistant eggs.
3 Microphagous feeding habits.
4 Small body size.
5 Chemical diversity and the mechanisms necessary to avoid the defence systems of the hosts.

Parasitism in many invertebrates involves extreme modification of adult form and the development of specialized life cycles with asexual multiplication phases, as seen in the many parasitic platyhelminths (Sections 3.6.3.2 and 3.6.3.3). This modification has not, however, occurred with the nematodes. The adults are 'preadapted' to the parasitic mode of life and their pattern of development has remained unchanged. The basic life history is illustrated in Fig. 4.20a. It involves cross-fertilization of separate males and females, and the release of highly protected 'eggs'. Four larval stages occur.

From this basic life cycle it is possible to recognize a series of adaptations to parasitism. In no way, however, should the various stages be conceived as representing steps in an evolutionary sequence.

In the least modified life cycle the larvae are free-living in the soil. Occasionally adults of a species may be either free-living or parasitic. The life cycle of *Strongyloides*, for instance, can be entirely free-living, but some third-stage larvae enter a mammalian host through the skin and move via the trachea to the gut, where females live and reproduce parthenogenetically (see Chapter 14). The eggs either enter the free-living life cycle or give rise to further infective third stage larvae (Fig. 4.20b).

The hookworm *Ancylostoma* has a similar life cycle but with no free-living adults. Both males and females occur in the enterine population (Fig. 4.20c). The first- and second-stage larvae are typical rhabditiform microphagous larvae feeding in the soil. The third-stage, infective larvae are encysted and do not feed.

In the unusually large, human gut nematode *Ascaris*, the free-living larval stage is suppressed. The eggs are voided with the faeces but the first- and second-stage larvae remain encysted within the eggs, which are ingested with contaminated food. The third-stage larvae hatch in the gut lumen but do not remain there. Instead they pass via the circulation to the lungs and thence back to the intestine (Fig. 4.20d). This complex life cycle may reflect an evolutionary history in which there was a secondary host, now lost.

Similar life cycles occur in other human gut parasites, e.g. *Trichurus* and *Enterobius*, but with no signs of the dual circulation prior to establishment in the gut. The females of *Enterobius* can produce severe itching around the anus. This causes scratching and leads to direct reinfection and transfer to other potential hosts via the host's fingers.

Some nematode parasites of vertebrates do have secondary hosts and may exploit food chain relationships. The larvae of mermithids, for example, are parasitic in insects but the adults are free-living (cf. the phylum Nematomorpha, Section 4.5).

4.5 Phylum NEMATOMORPHA (horsehair worms)

4.5.1 Etymology

Greek: *nematos*, thread; *morphe*, form.

4.5.2 Diagnostic and special features (Fig. 4.21)

1 Bilaterally symmetrical, thin, elongate worms (<3 mm diameter, 10 cm to >1 m length).
2 Body more than two cell layers thick, with tissues and organs.
3 With a straight through gut which is often degenerate and probably always non-functional.
4 Body monomeric, with a pseudocoelomic cavity, often occluded by mesenchyme.
5 Body wall with flexible, collagenous cuticle, epidermis, and layer of longitudinal muscle; without circular muscle layer.
6 Without circulatory, excretory or gaseous-exchange organs.
7 Nervous system intraepidermal, with an anterior nerve ring and one or two non-ganglionated longitudinal cords.
8 Gonochoristic, with elongate single or pair of gonads; fertilization internal via spermatophores.
9 Adult free-living but short-lived.
10 Larval stage infects arthropods (or, rarely, leeches), developing in their haemocoels; juveniles with three oral stylets and three whorls of hooked scalids (Fig. 4.22).
11 Freshwater or in moist soil; one genus marine.

The bodily organization of adult nematomorphs is essentially similar to that of the nematodes (Section 4.4) and, like them, they move by means of undulatory waves passing along the body in the dorsoventral plane. The marine species and the male sex in the freshwater/terrestrial group are most mobile; the larger freshwater females usually lead a relatively sedentary life, curled or coiled on the substratum. The most obvious difference

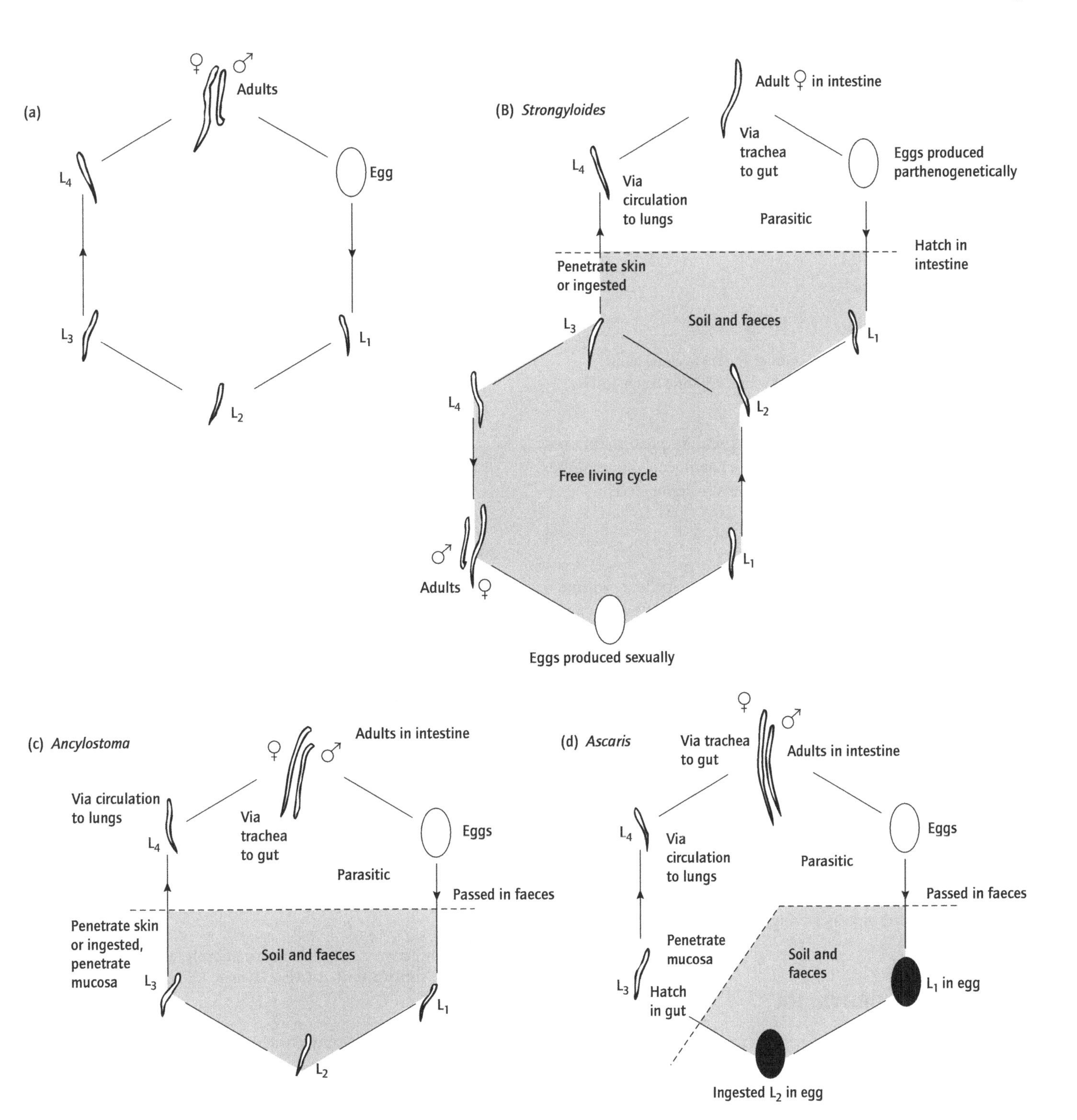

Fig. 4.20 Life cycles of nematodes: (a) the basic life cycle with four larval stages; (b)–(d) modified life cycles of species adapted to an endoparasitic existence. (After Croll & Mathews, 1977.)

between the adults of this group and the nematodes is the degenerate nature of the gut and/or of its orifices, which correlates with the reduction of the adult stage to a brief, dispersive and reproductive phase in the life history.

The feeding phase is the parasitic juvenile that, although it possesses a gut, nevertheless probably obtains most or all of its food requirements by absorption from the host's haemocoel across its body surface. The larvae and juveniles are most unlike nematodes, instead resembling adult kinorhynchs (Section 4.6) and loriciferans (Section 4.7) in their oral stylets, scalids and anterior gut anatomy (Fig. 4.22). As the larval introvert is everted (Fig. 4.22a), aided by a transverse septum isolating the

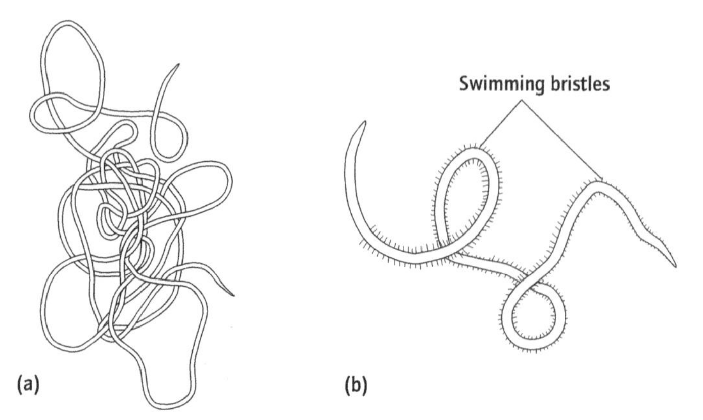

Fig. 4.21 Body form of (a) adult gordioid and (b) nectonematid nematomorphs. (After Margulis & Schwartz, 1982 and Fewkes, 1983.)

anterior pseudocoel, these stylets and scalids presumably aid penetration of the host's tissues into the haemocoel, but whether this is via the integument or the gut is as yet undecided.

4.5.3 Classification

The 250 known species are placed in two orders within a single class. By far the more numerous group, the freshwater gordioids (Fig. 4.21a), parasitize uniramians. They are distinguished by a single (ventral) nerve cord, paired gonads, and, at least when pre-reproductive, a body cavity almost totally filled with mesenchyme cells. The few known marine nectonematids (Fig. 4.21b), on the other hand, which parasitize decapod crustaceans, possess a second (dorsal) nerve cord, a single discrete (males) or diffuse (females) gonad, and a pseudocoel unoccluded by mesenchyme. They also bear a double row of topographically lateral bristles along most of the body which increase the undulatory surface during swimming. Some zoologists consider these two orders to be only distantly related, regarding the nectonematids as being derived from the nematodes, and the gordioids as having affinity with the kinorhynchs and loriciferans.

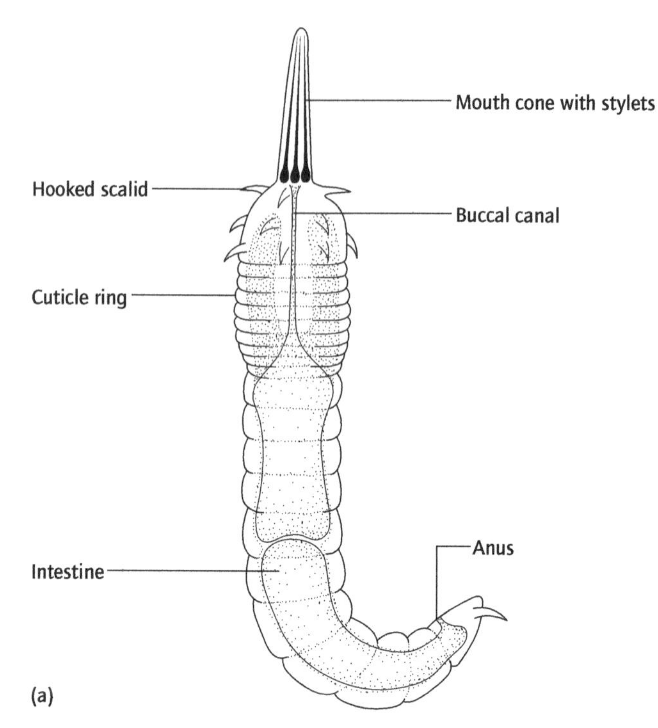

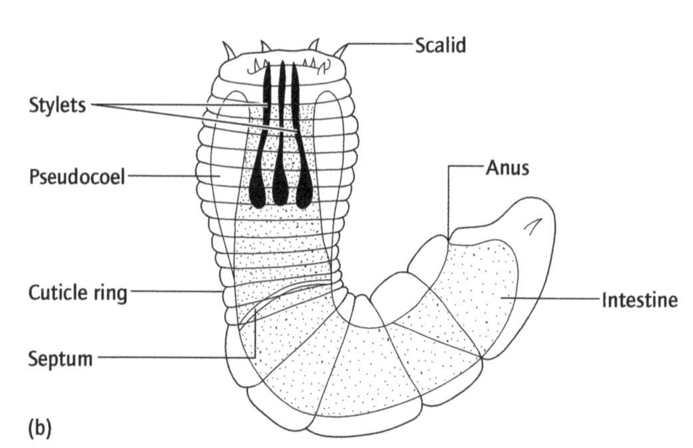

Fig. 4.22 The larva of a gordioid nematomorph: (a) with introvert everted; (b) with it retracted. (After Hyman, 1951 and Pennak, 1978.)

4.6 Phylum KINORHYNCHA

4.6.1 Etymology

Greek: *kinema*, motion; *rynchos*, snout.

4.6.2 Diagnostic and special features (Fig. 4.23)

1 Bilaterally symmetrical; vermiform but rather short.
2 Body more than two cell layers thick, with tissues and organs.
3 Alimentary system tubular with a posterior anus, and a muscular pharynx and protrusible mouth cone containing the buccal cavity.
4 Body divided externally into a fixed number (13 or 14) of segments or 'zonites'.
5 Epidermis with a spiny cuticle comprising single dorsal and paired ventral plates on each zonite.
6 Body cavity a pseudocoel, derived from the persistent blastocoel.
7 Body wall with circular and diagonal muscles.
8 Excretory system of protonephridia in the 11th zonite; the solenocytes with paired cilia and multiple nuclei.
9 Nervous system with anterior circumenteric ring and clusters of ganglion cells in a ventral nerve cord, reflecting the segmental appearance of epidermis and musculature.
10 Invariably small; almost always members of the marine meiofauna.
11 Without external ciliation.

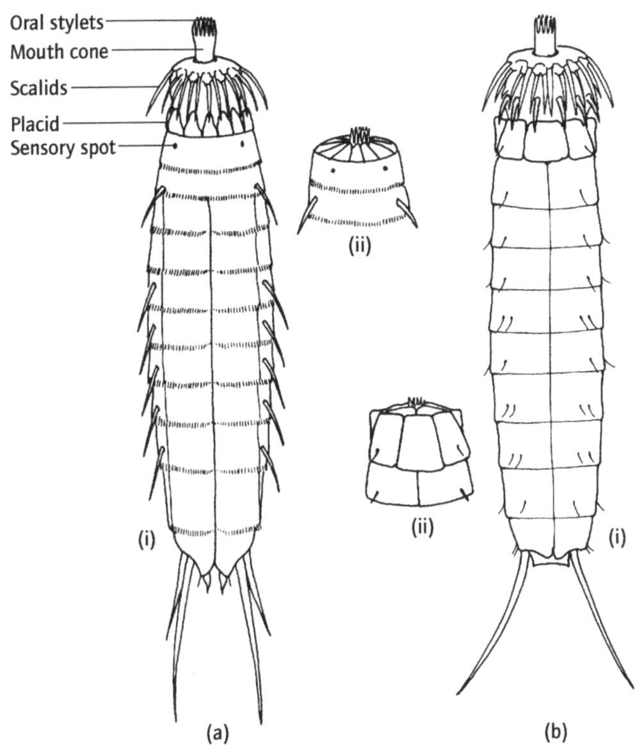

Fig. 4.23 The external morphology of Kinorhyncha: (a) Cyclorhagida: (i) ventral view; (ii) anterior view with introvert withdrawn. (b) Homalorhagida: (i) ventral view; (ii) anterior view with introvert withdrawn and protected by plates of the third zonite. (After Higgins, 1983.)

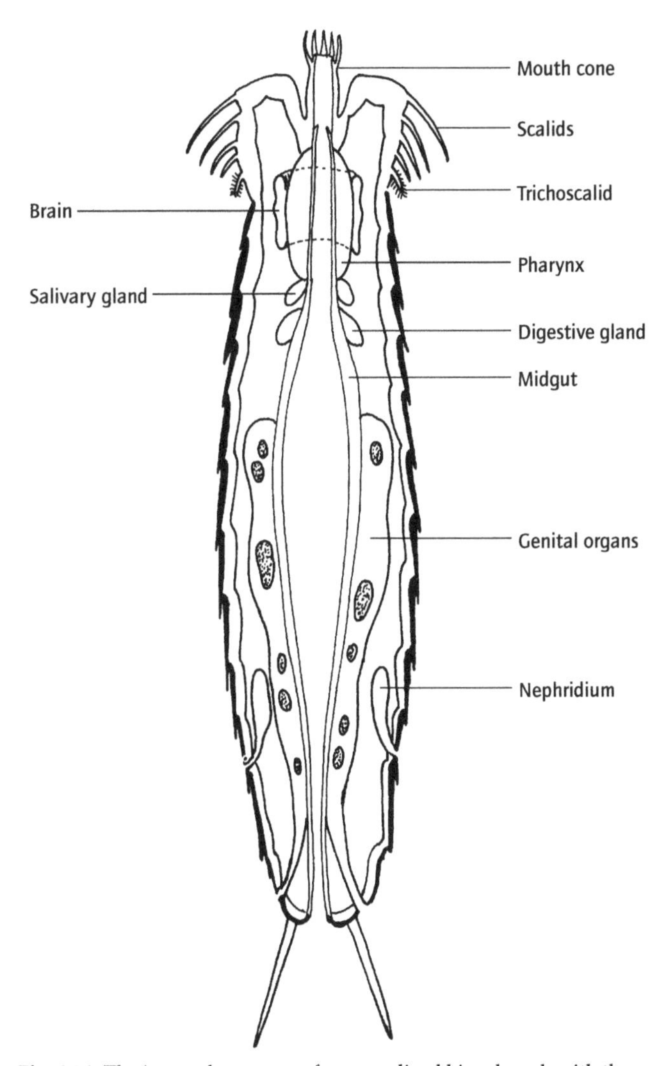

Fig. 4.24 The internal anatomy of a generalized kinorhynch with the introvert extended.

The Kinorhyncha are one of the least-known groups of animals. They have strong affinities with other aschelminth phyla, especially the Loricifera and Nematomorpha.

The most characteristic feature is the spiny cuticle which is clearly subdivided into a fixed number of transverse zonites. The segmentation of the cuticle affects the epidermis, musculature and nervous tissues, giving them a segmented structure. This segmentation is not homologous, however, with that of the segmented coelomate worms, the Annelida (Section 4.14).

The Kinorhyncha are all minute, living as meiofauna in muddy marine substrate. About 100 species have been described, all similar in basic functional design.

The mouth cone with its stylets forms a protractable introvert which can be withdrawn into the second segment or neck (Fig. 4.23). This introvert is armed with a circlet of recurved scalids.

The body region consists of ten unmodified trunk zonites, each with a single dorsal plate, the *tergite*, and paired ventral plates, the *sternites*. The body wall musculature flattens the trunk as dorsal and ventral plates are pulled together and pressure in the pseudocoel everts the introvert. The animals move by thrusting the introvert forwards, the recurved spiny scalids serving as anchors, while the body segments are pulled forward by retractor muscles.

Internally (Fig. 4.24) they share some features with the other aschelminth phyla. The epidermis is somewhat syncytial in structure, forms longitudinal thickenings mid-dorsally and laterally (as in the Nematoda) and enters the major body spines. It also forms cushions projecting into the pseudocoel. There is a muscular pharynx, which, with its triangular cross-section, resembles that of many aschelminths.

4.6.3 Classification

The arrangement whereby the introvert is retracted is used to distinguish the two orders. In the Cyclorhagida, only the first zonite can be withdrawn and it is then protected by large plates on the second zonite (see Fig. 4.23a and b). Both the first and second zonites of the Homalorhagida can be retracted, and they are usually protected by ventral plates on the third zonite (see Fig. 4.23).

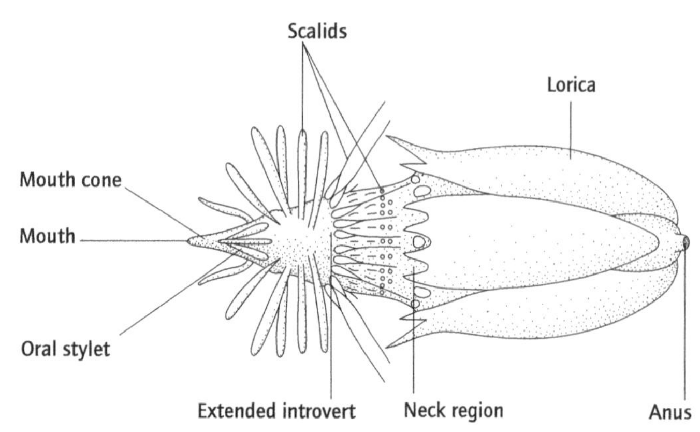

Fig. 4.25 Dorsal view of a loriciferan (simplified, after Kristensen, 1983).

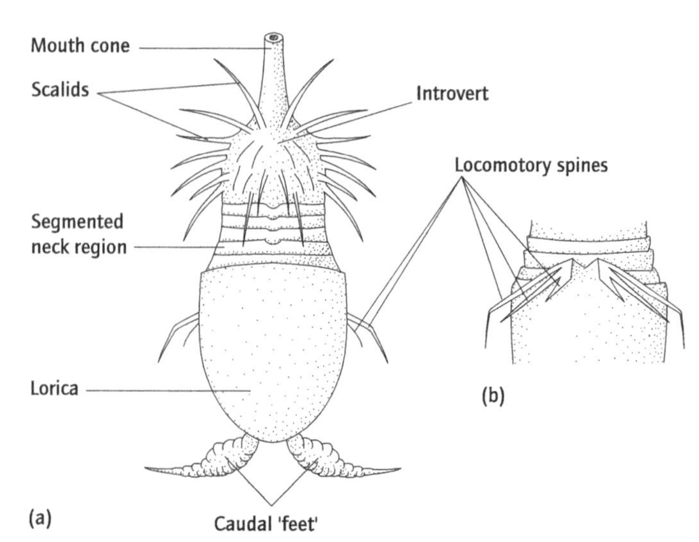

Fig. 4.26 (a) Dorsal view of last-instar larva, and (b) a ventral view of its locomotory spines (simplified, after Kristensen, 1983).

4.7 Phylum LORICIFERA

4.7.1 Etymology

Latin: *lorica*, a breastplate or corselet; *ferre*, to bear.

4.7.2 Diagnostic and special features (Fig. 4.25)

1 Bilaterally symmetrical.
2 Body more than two cell layers thick, with tissues and organs.
3 A straight through gut and a terminal anus.
4 A body cavity.
5 Body with three regions: an anterior, eversible head or introvert bearing up to nine whorls of backwardly directed paddle-, spine- or tooth-like scalids and a protrusible mouth cone surrounded by eight or nine rigid oral stylets; a short, externally segmented neck region, bearing rows of plates; and a trunk covered by a cuticularized lorica that is either divided longitudinally by 22–60 folds or composed of six longitudinal plates and bearing anteriorly hollow, forwardly directed spines. In adults, the introvert and neck can be withdrawn into the loricate trunk, the neck plates probably serving as a protective covering after withdrawal.
6 Gut with a large, muscular pharyngeal bulb and a telescopically extrusible buccal canal; whole gut possibly lined with cuticle.
7 Excretory system of one pair of protonephridia.
8 One pair of gonads; gonochoristic.
9 Well-developed brain and ganglionated ventral nerve cord.
10 A larval stage similar to a small version of the adult (Fig. 4.26), except for the presence of two sets of locomotory organs: (a) two or three pairs of spines on the anteroventral region of the lorica, using which it can crawl; and (b) a pair of movable, foot-like caudal appendages with which it swims. A number of larval instars occur, separated by moults.
11 Interstitial in marine sediments, down to 8000 m.

The structure of these minute animals (<0.4 mm) is as yet known only in outline fashion, and no information is available on their embryology, life style or habits; they were first described only in 1983. One reason for their late discovery would appear to be that they cling tightly to sediment particles, or possibly to other organisms, and are not susceptible to the standard extraction techniques used to collect interstitial marine species; another may be the superficial resemblance of the withdrawn animal to a rotifer or to a priapulan larva.

The most obvious feature of their body plan is its division into an anterior introvert and neck, and a posterior trunk encased by the lorica. In the adult, the whole of the anterior region can be retracted into the loricate trunk, although the larva can only withdraw the introvert into the neck. This system is closely similar to that in the priapulans, larval nematomorphs, the nematode *Kinochulus*, and, especially, the kinorhynchs (Section 4.6). In both kinorhynchs and loriciferans the 'head' is everted by body cavity pressure and withdrawn by specific retractor muscles, whilst the mouth cone with its oral stylets is operated independently, being protracted and retracted muscularly. The function of extension and withdrawal of the anterior end of the body in loriciferans is not known; in other groups it accomplishes locomotion either through sediments or through animal tissues, as well as prey capture and consumption, and the loriciferans may be ectoparasitic.

The gut bears a distinct resemblance to that of the tardigrades (Section 8.1), in that the anterior region (Fig. 4.27) comprises a telescopic buccal canal, a muscular, placoid-bearing pharyngeal bulb, a pair of accessory stylets, and two large salivary glands (the only glands to issue from the gut). Whether this resemblance is due to convergent evolution is not yet known, although it presumably indicates that the loriciferans feed by piercing with the oral stylets and then sucking fluids by means of a pharyngeal pump.

4.7.3 Classification

Although only 14 species have been formally described, some 18

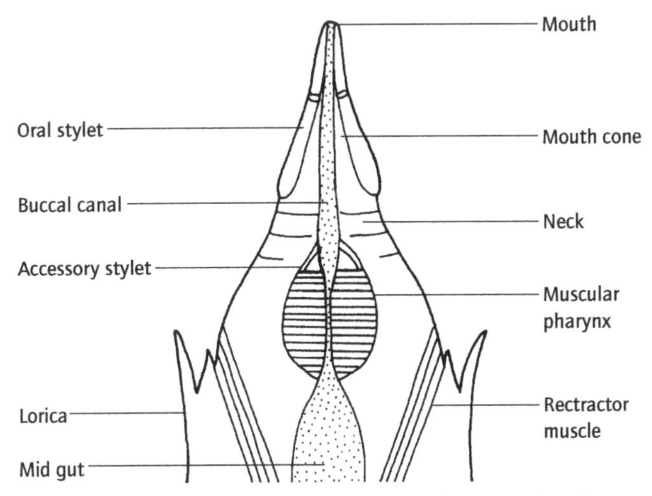

Fig. 4.27 The anterior gut of the Loricifera (mouth cone and neck extended and everted, respectively) (after Kristensen, 1983).

others are known and await description. All can be placed in the one order, Nanaloricida.

4.8 Phylum PRIAPULA

4.8.1 Etymology

Greek: *Priapos*, a phallic deity personifying male generative power.

4.8.2 Diagnostic and special features

1 Bilaterally symmetrical.
2 Body more than two cell layers thick, with tissues and organs.
3 A large retractable introvert or *prosoma* and a spiny or scaly, undivided *trunk*; sometimes a branching caudal appendage.
4 Alimentary system with anterior mouth surrounded by spine- or hook-like scalids, and a posterior anus sometimes surrounded by a crown of spines.
5 Excretory and genital systems closely associated in a urogenital organ with multiple solenocytes.
6 Nervous system with circum-oral ring and ganglionated ventral cord.
7 Body cavity spacious, possibly a true coelom.
8 Separate sexes, with external fertilization.
9 Larva without cilia and loricate (i.e. enclosed within plates).
10 Without a circulatory system but with corpuscles containing hemerythrin in the body cavity.

The Priapula contains a small but characteristic group of marine worms whose affinities are far from clear. They have some resemblance to the aschelminth pseudocoelomate phyla, with which they have often been classed, but functionally they are more similar to the coelomate worms of Sections 4.11–4.14. Some authors have viewed their body cavity as a coelom in that it has a mesodermal lining, but electron microscopical studies

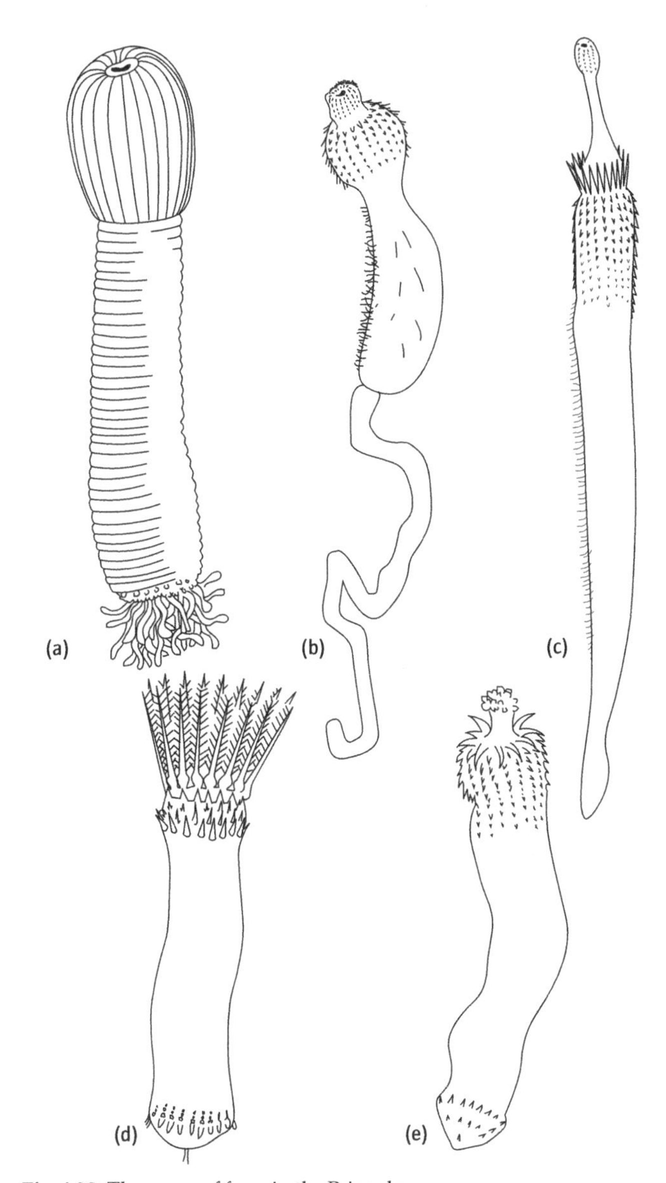

Fig. 4.28 The range of form in the Priapula.

show this lining to be unique and unlike that of the coelomates. More detailed knowledge of the origins of the body cavity is required. What little is known of their embryology suggests that the priapulans show radial not spiral cleavage.

The range of form is shown in Fig. 4.28. The body of most species is composed of a large bulbous prosoma or introvert and a trunk. The introvert terminates in a mouth surrounded by a ring of five circum-oral spine-like scalids. This large barrel-shaped structure with its longitudinal rows of scalids is usually everted, but can be retracted into the trunk, from which it is usually separated by a distinct collar. The often annulate trunk is usually scaly or spiny and bears a chitinous cuticle that is moulted periodically. At the posterior end of *Priapulus* is a curious branching structure, the caudal appendage, which may have a respiratory role (Fig. 4.29a). The cuticle overlies an epidermis in which there are fluid-filled spaces between the cells; it differs

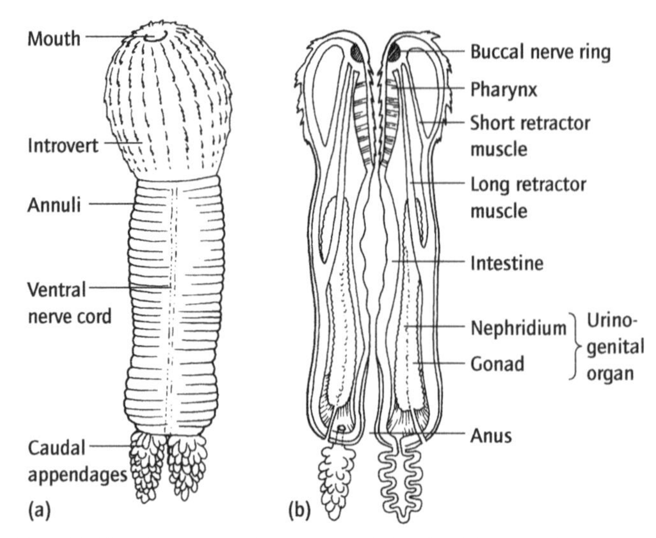

Fig. 4.29 The structure and anatomy of *Priapulus*: (a) external morphology; (b) schematic internal anatomy.

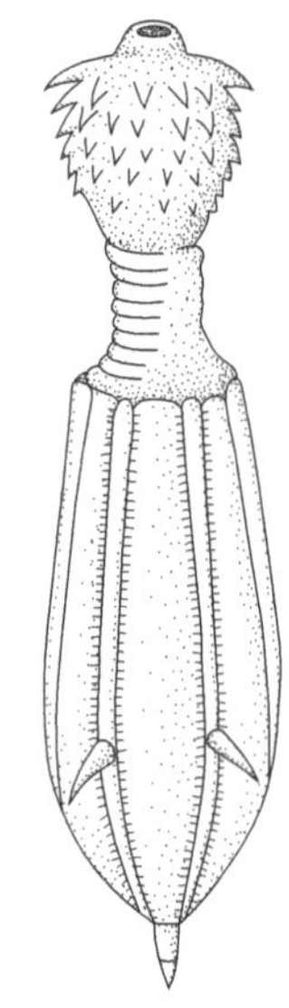

Fig. 4.30 A loricate larva of a priapulan.

in structure from those of both the coelomare and aschelminth worms. The body wall has both circular and longitudinal muscles which exert pressure on the fluid-filled body cavity to evert the prosoma. The gut is a simple, straight tube with a horny, toothed buccal region and a muscular pharynx (Fig. 4.29b); there is no mouth cone, nor any oral stylets.

A unique feature is the conjunction of the genital system and the solenocytes of the protonephridia to form the paired urinogenital organs (Fig. 4.29b).

The loricate larva may be pelagic but is enclosed within cuticular plates, hence its name (Fig. 4.30). The larva already has the retractable scaly prosoma and is able to burrow effectively, whereas the adults find it difficult to re-enter the substratum once displaced. The larva may be the only stage in the life cycle normally to enter the substratum. The superficial resemblances between the spiny introvert of the Priapula and that of the Kinorhyncha, and between the loricate larva and rotifers, are not generally considered sufficient grounds for considering these phyla as being particularly closely related.

4.8.3 Classification

The 16 species of priapulans can be divided between two orders, the Priapulida containing the larger, more common predatory forms (Fig. 4.29a), and the Seticoronaria which have thin scalids in the form of a crown of stiff 'tentacles' (see Fig. 4.28d) and possess a circum-anal ring of hooks. Priapulans have a long history, several being known from the Cambrian.

4.9 Phylum ROTIFERA

4.9.1 Etymology

Latin: *rota*, a wheel; *ferre*, to carry.

4.9.2 Diagnostic and special features

1 Bilaterally symmetrical.
2 Body more than two cell layers thick, with tissues and organs.
3 With a crown of cilia in the anterior part of the body in the form of pre-oral and post-oral bands. Often organized into two wheel-like ciliary organs from which the name of the group is derived.
4 Alimentary system with anterior mouth, complex jaw apparatus, muscular pharynx, and posterior anus opening in a common cloaca with urinogenital system.
5 Epidermis with a fixed small number of nuclei; it includes an intracellular cuticle often thickened to form an encasement or lorica.
6 Body not segmented.
7 Excretory system, protonephridia.
8 No circulatory system or respiratory organs.
9 Body cavity pseudocoelomic.
10 Sexes separate, but males often rare or absent and when present almost always dwarf.
11 Minute, rarely reaching 3 mm in length.
12 Development direct, following modified spiral cleavage. No further nuclear divisions occur after embryonic stages.

The Rotifera are among the smallest animals; their dimensions when adult are similar to the ciliated protists and larvae of many other phyla. Their bodies are composed of a small and strictly determined number of cells, or, strictly speaking, nuclei, since many tissues are syncytial (a feature shared with several other aschelminth phyla).

The Rotifera as adults characteristically pursue a mode of life similar to that of the ciliated larvae of larger invertebrates; indeed they are superficially like the trochophore larvae of marine Annelida, Sipuncula and Mollusca. The characteristic feature of a trochophore is the equatorial ring of cilia, the prototroch,

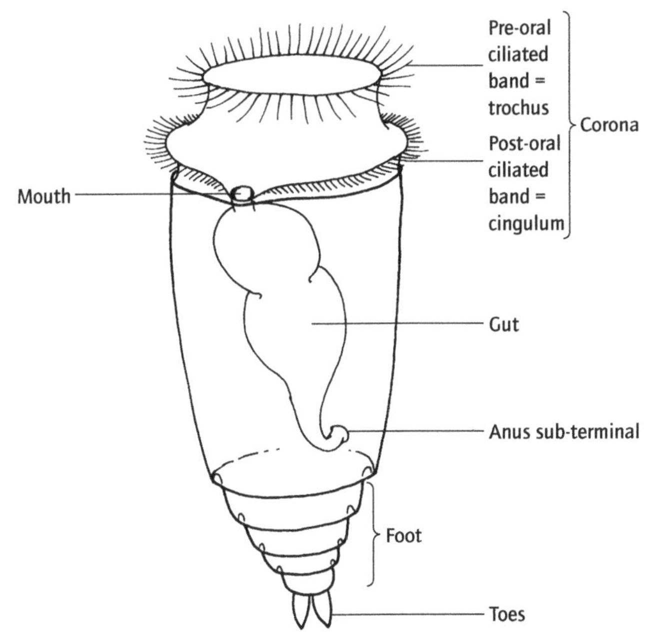

Fig. 4.31 A diagrammatic representation of the rotiferan body plan showing the two ciliated girdles.

which has two bands of cilia – pre-oral and post-oral – beating in contrary motion. A similar arrangement occurs in most planktonic Rotifera where the cilia of the corona form two bands – pre-oral (the trochus) and post-oral (the cingulum) (Fig. 4.31).

These structural similarities, however, do not imply any close phylogenetic relationship and other features suggest closer affinities with the aschelminth phyla, especially with the Acanthocephala. Some authors regard the Rotifera as being reproductively active, permanent larvae, but of what line of descent is unknown. No other aschelminth phylum has ciliated larvae, from which rotiferans could have derived.

Externally the body is often covered by a sculptured cuticle forming a cup-like lorica, the open end of which bears the ciliated corona and the mouth. The corona may be highly specialized and in sessile forms it can be much reduced. In these (see Fig. 4.35), the cilia are often modified as stiff sensory hairs. The corona can usually be withdrawn into the lorica. Posteriorly the body narrows to form the movable narrow 'foot'; this is often ringed or annulate to give the appearance of pseudo-segments. Like the corona the foot can usually be withdrawn, the pseudo-segments sliding telescopically one within another. The foot ends in a pair of toes which serve to anchor the organisms permanently or temporarily to the substratum; it may be absent or reduced in permanently planktonic forms. Internally the animals are relatively simple (Fig. 4.32), one of the consequences of small size. The mouth parts, however, are complex and vary according to the diet. The basic plan of the *trophi*, the hard parts of the *mastax* or feeding apparatus, is illustrated in Fig. 4.33. The fulcrum, which lies below the middle line of the trophi, supports two rami which branch symmetrically from it. Upon these are hinged the paired *unci* and *manubria*. Different trophi are illustrated in Fig. 4.33. They can be adapted for piercing and sucking cell contents, seizing prey in a pincer-like manner, grinding, crushing or tearing it.

Most rotifers have two protonephridia, each composed of a syncytium with a small number of cell nuclei. Inhabiting fresh waters, there is a high rate of fluid flow through the nephridia and the fluid is hypo-osmotic.

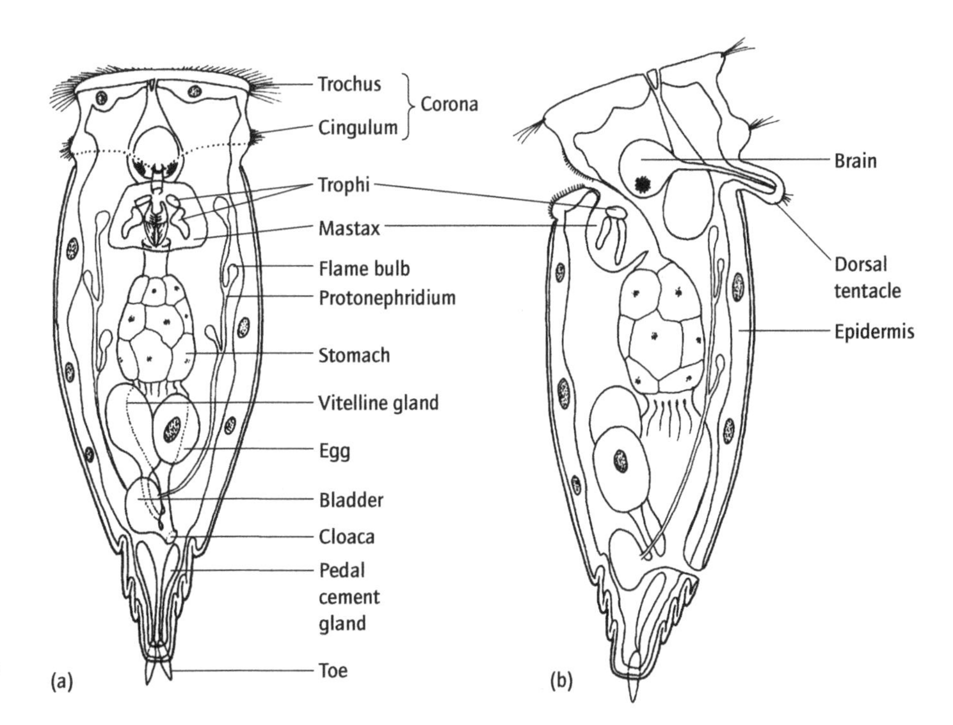

Fig. 4.32 General features of rotiferan anatomy: (a) ventral view; (b) side view.

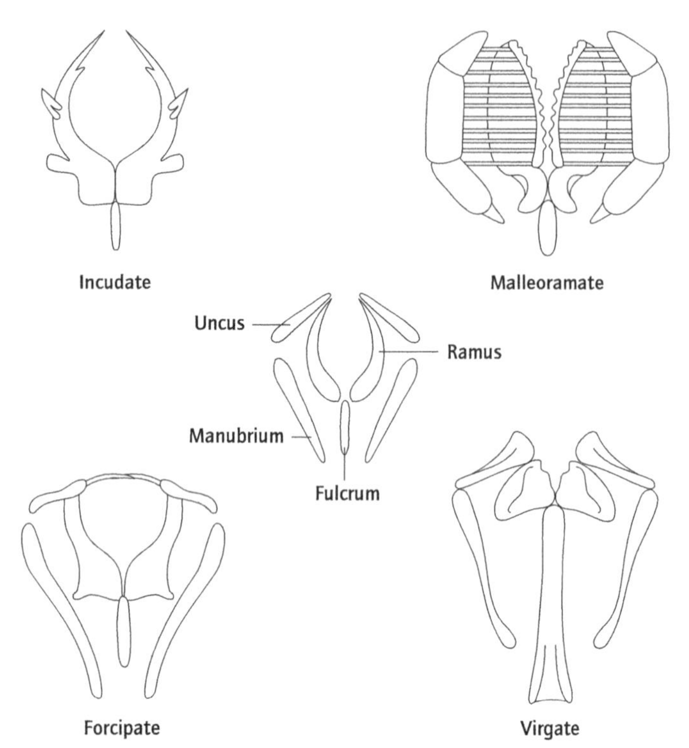

Fig. 4.33 The range of form of the mouth parts of Rotifera (after Donner, 1966).

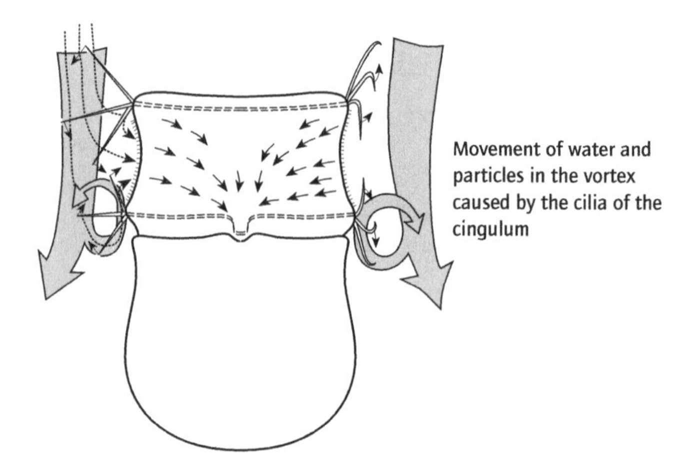

Fig. 4.34 The feeding mechanism of a pelagic rotiferan. The large cilia of the anterior girdle (trochus) draw a current of water in an anterior/posterior direction. This provides support and carries particles. The cilia of the more posterior girdle (cingulum) beat in a counter direction. This creates a vortex which traps food particles which are then transported to the mouth.

4.9.3 Classification

The phylum Rotifera is subdivided into three component classes.

Class	Order
Seisonidea	Seisonida
Bdelloidea	Bdelloida
Monogonata	Ploima Flosculariida Collothecida

4.9.3.1 *Class Bdelloidea*

The class Bdelloidea for the most part have a characteristic 'two-wheeled' corona (Fig. 4.34) and the body is not enclosed within a lorica. Most species creep in a looping manner but also swim well by means of the unfolded corona. The bdelloids are adapted to impermanent and cryptic habitats; they are common in the interstices of the damp sand and soil of lake and river beaches, and in association with mosses in periodically damp terrestrial habitats. They creep rather than swim, although the ciliary discs are well developed, and have massive, powerful, pedal cement glands.

Their reproductive biology is most unusual. The entire group appears to be amictic, i.e. is without meiosis or crossing over in any form. Males have never been found and the whole population comprises obligately parthenogenetic females (see Section 14.2.1). They are also characterized by an ability to pass into a state of anabiosis when they become dried out, and can survive in this desiccated state for many years, then withstanding extremes of temperature, +40 to −200°C.

4.9.3.2 *Class Monogonata*

The majority of rotifers belong to this class, which is distinguished by the single (not paired) ovaries. Several orders can be recognized.

The largest order, the Ploima, contains a wide variety of sessile and free-swimming forms (Fig. 4.35a,b,c). The foot, when present, has two toes, and the mouth usually lies within, not anterior to, the cingulum. These rotifers often have a stiff lorica.

The order Flosculariida (Fig. 4.35f,g) is a second group of either free-swimming or sessile rotifers in which the ciliary bands of the trochus are clearly separated into trochal and cingular bands. The foot, if present, does not have paired toes.

In the Collothecida the trochus is a large funnel often with the cilia forming stiff sensory hairs (see Figs 4.35d,e). The females are always sessile, attached by cement glands, and the body is often encased in a gelatinous mass.

Unlike the bdelloids, monogont rotifers do exhibit sexual reproduction; nevertheless populations are dominated by females for most of the year; the relatively simple, often dwarf but free-swimming, males occur only briefly and cyclically at certain times.

During most of the year diploid amictic females produce eggs that develop without fertilization into young females. On occasion, however, perhaps as a response to environmental conditions, morphologically distinct mictic females appear which lay haploid eggs. These either develop rapidly into haploid

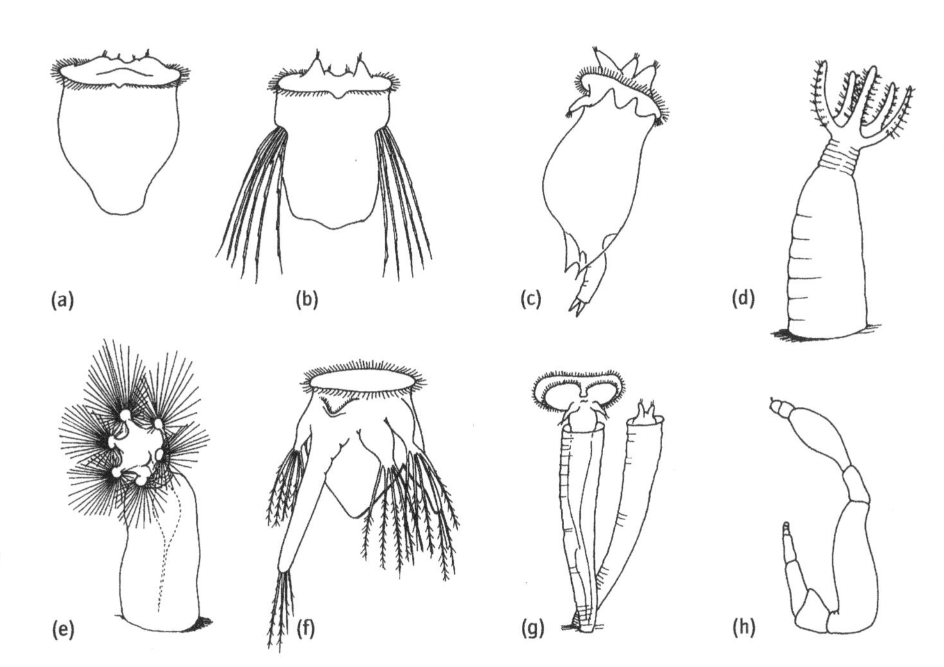

Fig. 4.35 The range of form among rotiferans of the classes Monogonta and Seisonidea: (a–c), Ploima; (d,e), Collothecida; (f,g) Flosculariida; (h) Seisonida. (After Donner, 1966.)

males or, if fertilized, become zygotes that develop into amictic females. The mictic females therefore display facultative parthenogenesis (see Section 14.2.1). The species *Brachionus plicatilis* is now of major economic importance in many systems of marine aquaculture providing an easily cultivated natural food, processing microalgae for the larvae of invertebrates and marine fish.

4.9.3.3 Class Seisonidea

This small class of marine rotifers live on the gills of crustaceans – *Nebalia* and some isopods. There is only one genus, *Seison* (see Fig. 4.35h). Individuals are relatively large (a few millimetres) and have a much reduced corona and prominent mastax. The ovaries, like those of the Bdelloidea, are paired, but unlike that group there is a normal pattern of gonochoristic sexuality in which fully developed males and females are equally common in the population.

4.10 Phylum ACANTHOCEPHALA

4.10.1 Etymology

Greek: *akantha*, prickle; *kephale*, head.

4.10.2 Diagnostic and special features

1 Bilaterally symmetrical, vermiform.
2 Body more than two cell layers thick, with tissues and organs.
3 Alimentary system lacking.
4 Body without segmentation, but superficial transverse annulation sometimes present.
5 A prominent hooked proboscis.
6 Body cavity a pseudocoel.
7 Epidermis syncytial with a small, fixed number of relatively large nuclei.
8 Nervous system a single ventral/anterior ganglion with paired or single nerves to the organs.
9 Nephridia occasionally present.
10 Respiratory and circulatory systems absent.
11 Separate sexes, with internal fertilization and viviparous development.
12 Infective larvae occupy a secondary insect host.
13 Adults always parasitic in the alimentary tract of vertebrates.

The general body form is illustrated in Fig. 4.36. Most species are small, 1 mm to a few centimetres in length, but a few attain lengths up to 1 m. Their most striking feature is the proboscis with its recurved spines. This, together with the neck region, can be inverted but it normally forms a permanent attachment to the host tissues. As in other pseudocoelomates, certain tissues tend to be constructed from a strictly determined small number of cells.

The body wall is composed of a thin cuticle overlying a syncytial epidermis in which there is a further intracellular cuticle and a small number of precisely positioned and rather large nuclei. Indeed the position of the giant nuclei is a useful diagnostic feature. Beneath the epidermis is a thin layer of circular and longitudinal muscles overlying the pseudocoel. Two vertical flaps of tissue (the lemnisci) project into the pseudocoel and in these run fluid-filled channels from the epidermis. The endoderm is reduced to a ligament along which are suspended the reproductive tissues.

As in many parasites the adults show extreme simplification, with loss of many of the organ systems characteristic of free-living animals, but with hypertrophy of the reproductive organs. The Acanthocephala are gonochoristic, an unusual feature for endoparasites (but see Section 4.4); because of this they exhibit

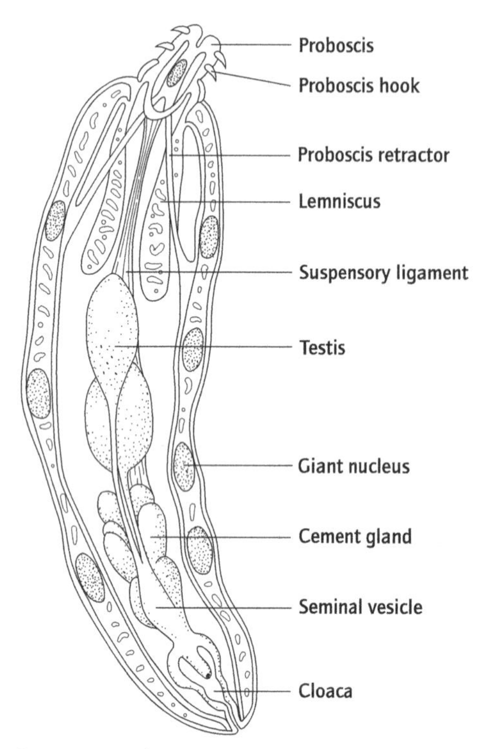

Fig. 4.36 The anatomy of a typical acanthocephalan. Based on *Neoechinorhyncus* (male).

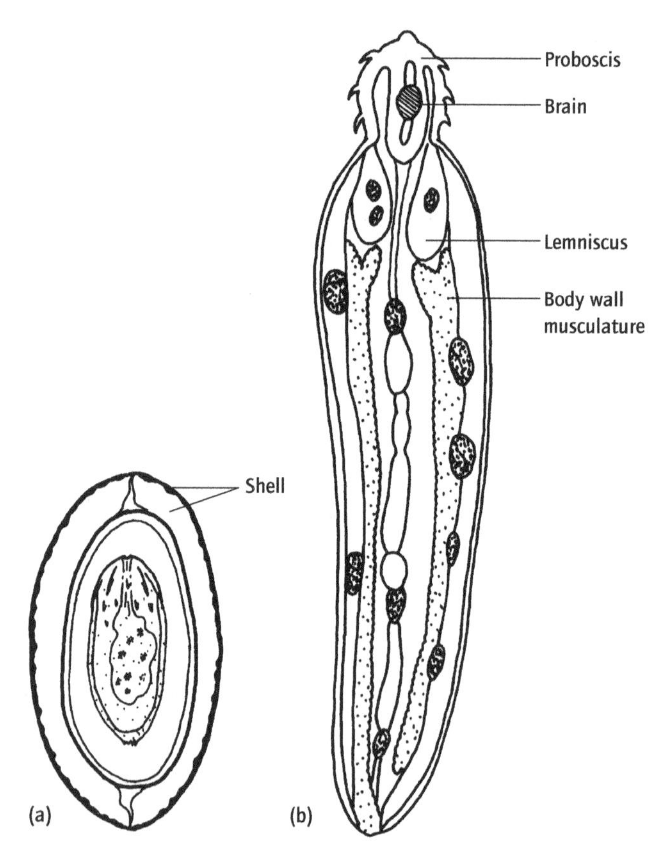

Fig. 4.37 Stages in the life history of the Acanthocephala: (a) the egg containing a larva; (b) the hatched and infective acanthella stage taken from a beetle larva.

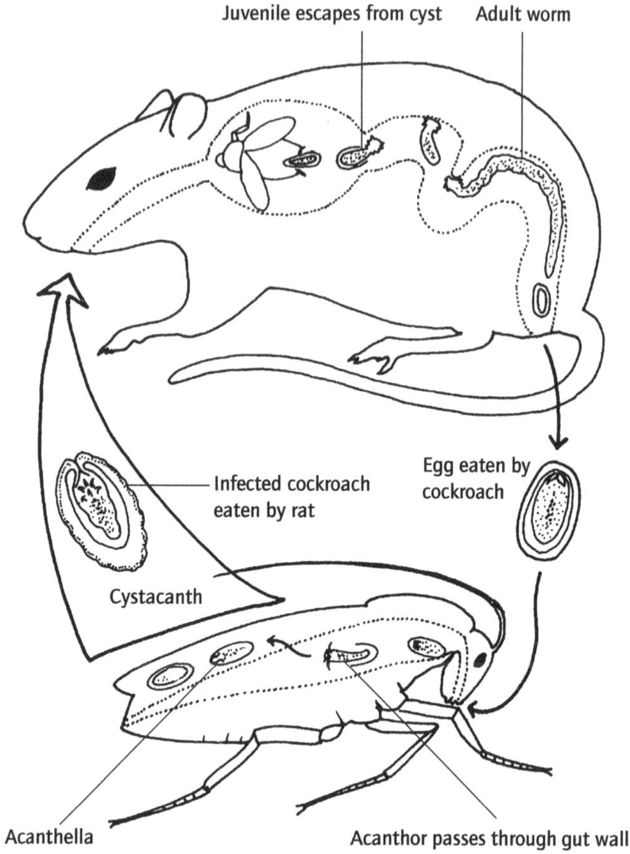

Fig. 4.38 The life cycle of *Moniliformis*, an acanthocephalan with a mammalian primary host (after Noble & Noble, 1976).

complex copulatory behaviour to ensure internal fertilization. Sperm are injected into the female genital tract which is plugged by cement glands after copulation. Fertilization takes place in the pseudocoelom.

The larvae of many specialized parasites betray the phylogenetic relationships of the adults, but this is not the case in this phylum. The larvae too are highly specialized and clearly share the structure of the adult. Following internal fertilization, the early larvae stages are encapsulated as 'eggs' (Fig. 4.37a). These shelled larvae pass out with the faeces and must be eaten by an insect secondary host before further development can take place. In the insect the shelled larva hatches and makes its way to the haemocoel where it develops into a juvenile *acanthella* larva. (Fig. 4.37b). Such a larva may become encysted in the insect and this, when eaten by the vertebrate, re-establishes the adult in its definitive host. Figure 4.38 illustrates the life cycle of *Moniliformis*, a parasite of mice, rats, cats and dogs.

4.10.3 Classification

The 1000 species are distributed between three orders in a single class. Two of the orders contain parasites of freshwater fish, the third those of terrestrial tetrapods.

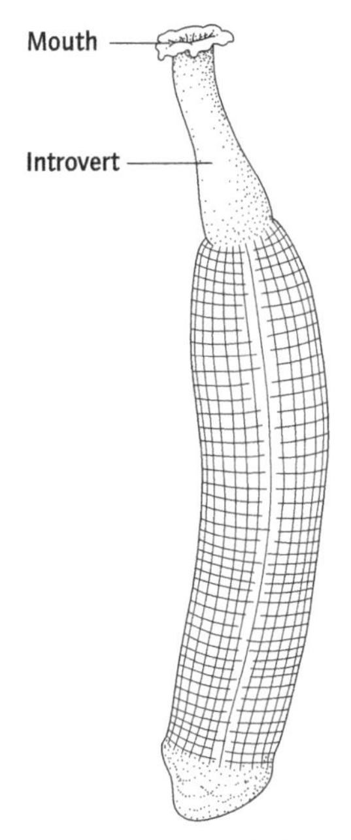

Fig. 4.39 The external appearance of a typical sipunculan.

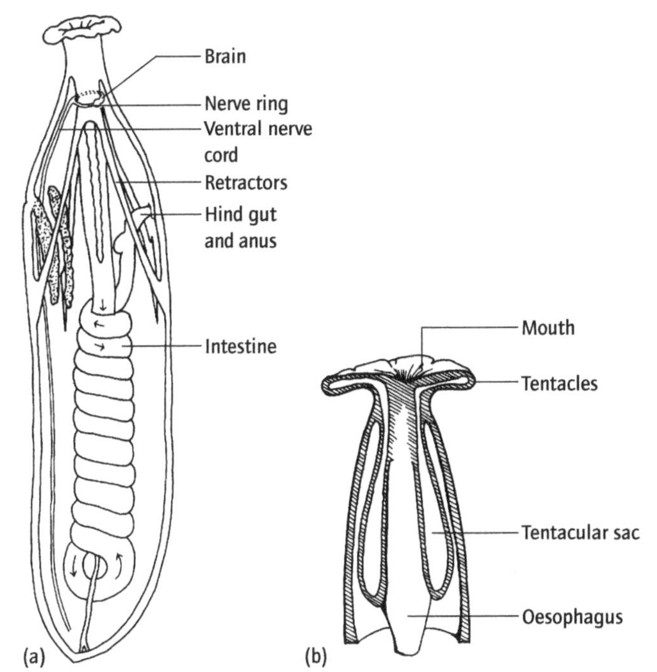

Fig. 4.40 (a) The internal anatomy of a sipunculan shown by dissection from the left side. (b) Detail showing the separate hydraulic system of the tentacles and introvert.

4.11 Phylum SIPUNCULA

4.11.1 Etymology

Latin: *sipunculus*, little pipe.

4.11.2 Diagnostic and special features (Figs 4.39 and 4.40)

1 Bilaterally symmetrical, vermiform.
2 Body more than two cell layers thick, with tissues and organs.
3 A muscular U-shaped gut having both a mouth and an anus, the latter situated dorsally in the anterior part of the body.
4 Body divided into an anterior part, the introvert containing the mouth, and a stouter posterior trunk.
5 Body not divided into segments.
6 Body cavity a schizocoel, but without internal septa.
7 Outer epithelium covered by a cuticle, but without bristles or chaetae.
8 Without circulatory or differentiated respiratory systems.
9 Nervous system with an anterior brain, circum-oesophageal ring and a ventral nerve cord without ganglia.
10 A single nephridium or pair of nephridia.
11 Development with spiral cleavage, typically to a trochophore larva that in some species is superseded by an oceanic 'pelagosphaera' larva unique to the phylum.

The body is divided into two distinctive regions, of which the introvert can be retracted into the posterior trunk by powerful retractor muscles (see Fig. 4.40), and be everted by hydraulic pressure generated by the trunk musculature. Local relaxation of the circular muscles can cause dilations that wedge the worm in the substratum while permitting eversion of the introvert.

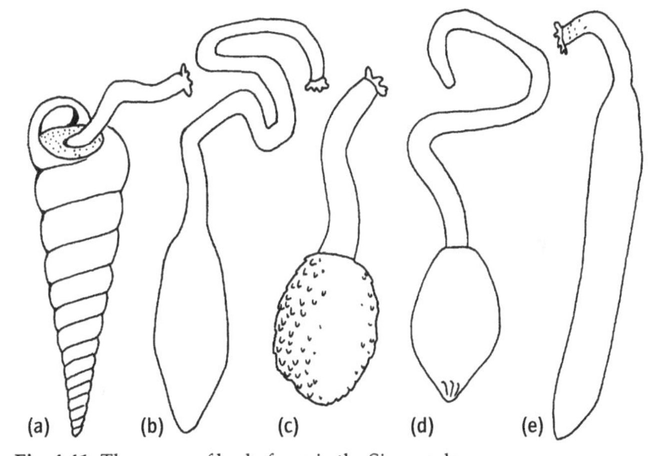

Fig. 4.41 The range of body form in the Sipuncula.

Sipunculans are generally non-selective deposit feeders, using the tentacles which surround the tip of the introvert to gather food. These tentacles are operated by a hydraulic system separate from that of the general body cavity. A pair of ducts lead from each tentacle into a ring-like canal from which extend one or two sacs running parallel to the axis of the introvert. The dorsal and ventral compensation sacs are compressible and serve as hydraulic reservoirs for the canal system (Fig. 4.40b).

Many species live in muddy substrata in a non-permanent burrow but some inhabit empty mollusc shells or worm tubes. Boring forms construct a tube in calcareous coral material. The range of form is illustrated in Fig. 4.41.

The nervous system of the sipunculans resembles that of the annelids and echiurans, with an anterior ganglion (the brain) in the introvert, a circum-oesophageal ring and a ventral nerve cord. There are, however, no signs of segmental ganglia.

The spacious hydraulic coelom also serves for the accumulation of gametocytes, shed into it from the simple gonads at an early stage of differentiation and stored there until large numbers of mature gametocytes have been accumulated. The gametocytes are then shed to the exterior via the nephridia. This pattern is modified in the minute form *Golfingia minuta*, a protandric hermaphrodite producing large eggs which develop directly. This is a consequence of its small size, the pattern of reproduction exhibited by most sipunculans being generally characteristic of large-bodied marine invertebrates.

The majority of Sipuncula hatch into a pelagic trochophore larva essentially similar to that of some annelids. In some sipunculans the trochophore larva metamorphoses directly to the juvenile adult, but usually it metamorphoses to a secondary larva, the pelagosphaera, which may be benthic but is usually pelagic. There are four developmental pathways in Sipuncula as illustrated in Fig. 4.42.

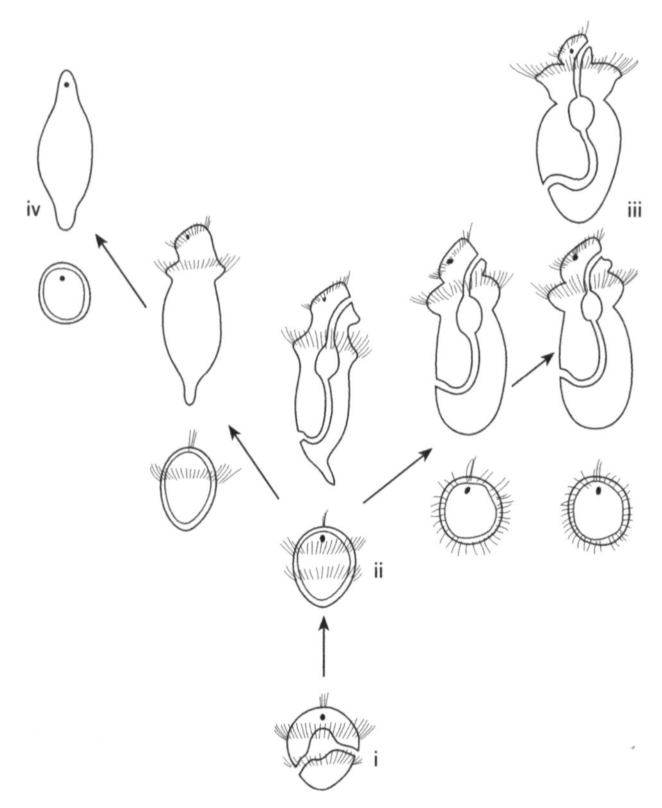

Fig. 4.42 Developmental pathways of the Sipuncula: (i) development via the trochophore larva; (ii) development via a non-feeding trochophore larva and a new phylum-characteristic pelagosphaera; (iii) development with large planktotrophic pelagosphaera larvae characteristic of the open oceans; (iv) development via yolky non-feeding larvae with eventual suppression of the pelagic larval stages. (Modified after Rice, 1985.)

4.11.3 Classification

All 250 known species can be placed in a single class divisible into two orders, the Sipunculida and the Phascolosomatida.

4.12 Phylum ECHIURA

4.12.1 Etymology

Greek: *echis*, viper; *ura*, tail.

4.12.2 Diagnostic and special features

1 Bilaterally symmetrical, vermiform.
2 Body more than two cell layers thick, with tissues and organs.
3 Muscular gut bearing both anterior and posterior openings.
4 Body with a large extensile and contractile anterior projection or 'proboscis' with lateral folds leading to the mouth.
5 A single undivided schizocoelomic body cavity present between muscular components of the body wall and the gut.
6 Body not divided into segments.
7 Chaetae present as a single pair in anterior ventral region and sometimes elsewhere.
8 Closed blood system with dorsal and ventral vessels; open vascular system in one group.
9 Nervous system with circum-oesophageal ring and ventral nerve cord, without definite ganglia.
10 Excretory system with up to 400 nephridia not arranged in a metameric manner.
11 Development via spiral cleavage to a trochophore-like larva.

The Echiura are unsegmented coelomate worms with affinities with the Annelida (Section 4.14) within which they were once classified. Their most distinctive feature is the proboscis, which contains the anterior lobe of the nervous system and which is probably homologous with the annelid prostomium (Fig. 4.43). The paired ventral chaetae or hooks in the anterior region of the body are also annelid-like features, as are the nephridia with a ciliated funnel, the arrangement of muscular layers in the body wall and the structure of the alimentary system. The Echiura differ from the annelids, however, in the complete absence of segmentation or metamerism. They can be regarded as being close to the annelid line of evolution. Indeed there is evidence from their development that there may have been metamerism at an early stage in their evolution.

All are marine, living a sedentary existence in soft substrata. Most are detritus feeders living in permanent burrows, with the proboscis forming a non-selective food-collecting device directing material along its gutter, towards the mouth (see Fig. 9.5). The range of body form is illustrated in Fig. 4.44. The genus *Urechis* exhibits a different mode of feeding. It lives in a deep U-shaped burrow through which it drives water by waves of peristaltic muscular activity.

The reproductive biology of the Echiura is that characteristic of large-bodied coelomate worms in the marine environment.

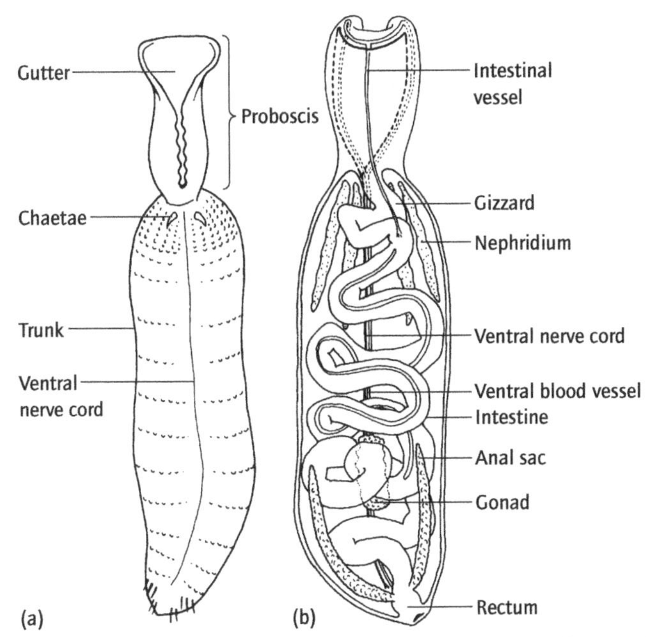

Fig. 4.43 The external morphology and anatomy of Echiura: (a) ventral view (*Echiurus*); (b) dissection from the dorsal side.

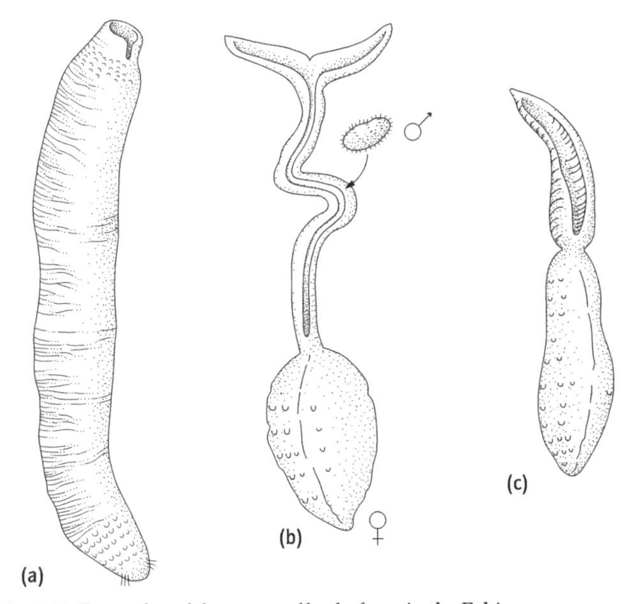

Fig. 4.44 Examples of the range of body form in the Echiura.

Sexes are separate and the coelom forms a cavity in which gametocytes can develop prior to mass discharge via the nephridia, in a spawning crisis.

One genus of Echiura exhibits a specialized mode of sexual reproduction discussed in Section 14.2.3.3. The males are dwarf individuals living on the proboscis of a female (Fig. 4.44b). Dwarf males were formerly thought to be a feature of only the bonellid echiurans, but the phenomenon has been discovered recently in other echiurans, indicating perhaps that it has arisen more than once.

4.12.3 Classification

The 150 species are placed in three orders within a single class dependent on the disposition of the body wall muscle layers, the numbers of nephridia and the open or closed nature of the blood system. Members of one order, the Xenopneusta, have adapted the posterior part of the gut as a respiratory organ.

4.13 Phylum POGONOPHORA

4.13.1 Etymology

Greek: *pogon*, beard; *phoros*, bearer.

4.13.2 Diagnostic and special features (Fig. 4.45)

(Note that the interpretation of various anatomical features of the pogonophorans is still somewhat in a state of flux, not least in respect of the homology, or otherwise, of the structures of the recently discovered vestimentiferans with those of the other members of this phylum. It is, for example, still far from agreed which is the dorsal and which the ventral surface.)

1 Bilaterally symmetrical, elongate, metamerically segmented worms (0.5 mm–3 cm diameter, 5 cm–3 m length) which permanently inhabit chitinous and protein tubes within which they can move.

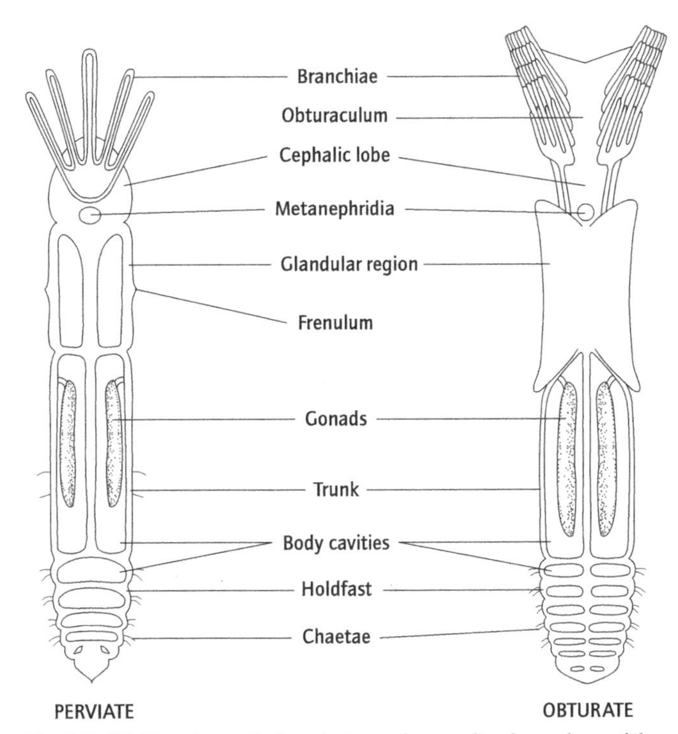

Fig. 4.45 Highly schematic dorsal views of generalized members of the two pogonophoran classes, showing the main regions of the body in diagrammatic form and their body cavities (after Southward, 1980 and Jones, 1985).

2 Body more than two cell layers thick, with tissues and organs.
3 Adults without any mouth or gut. Larval gut, if present, transforms into bacteria-containing 'trophosomal' tissue.
4 With single and/or paired body cavities of uncertain nature (although usually assumed to be schizocoels, they lack a peritoneal lining).
5 Body with four regions: an anterior 'cephalic lobe' bearing 1 to >1000 tentaculate 'branchiae' supported hydrostatically; a short 'glandular region' with its paired cavities sometimes largely occluded by muscular tissue; an extremely elongate 'trunk' with a single pair of large hydrostatic compartments and, often, with various adhesive or secretory papillae; and a short 'holdfast' (forming an attachment or burrowing organ) composed of up to 30 segments each with a single or paired body cavity and each budded off from a terminal proliferation zone.
6 Body wall with cuticle, epidermis, circular and longitudinal muscle layers; the holdfast and, in some, the trunk with chaetae; some regions with tracts of cilia.
7 With a closed blood system including a heart enclosed within a pericardial cavity in the cephalic lobe.
8 With a pair of metanephridium-like excretory organs in the cephalic lobe.
9 Gaseous-exchange organs assumed to be the branchiae.
10 With an epidermal nervous system comprising an anterior nerve ring or mass and usually a single longitudinal cord lacking ganglia (this is generally regarded as being ventral); multiple cords may occur in some regions.
11 Gonochoristic, with a pair of elongate gonads in the trunk, those of the male producing spermatophores.
12 Fertilization assumed to be external; development indirect, eggs and young larvae being brooded within the maternal tube or free-living and ciliary feeding.
13 Marine, usually in deep waters, i.e. 100–4000 m.

Although first discovered in 1900, little is known of these deep-sea worms. It was not until 1964, for example, that complete specimens were obtained (until then the metameric holdfast region was unknown), and up to that date they were generally thought to be related to the lophophorate and deuterostome groups, not least because of their apparently tripartite, oligomeric bodies. The discovery of the segmented and chaetae-bearing terminal section of the body, together with the description of the first vestimentiferan in 1969, however, led to a changed consensus that they had affinities with the annelids. Molecular sequence data have now confirmed this view and it is fairly clear that both groups of pogonophorans are really very highly modified polychaete annelids (Section 4.14.3.1). Nevertheless, we have retained them as a phylum here in recognition of their distinctive anatomy and life style.

Pogonophorans dwell within erect, closely fitting tubes, secreted by the glandular region and thickened by secretions of the trunk, within which they are supported by means of (a) the trunk and/or holdfast chaetae, (b) some of the trunk papillae, and, whilst the branchiae are extended into the surrounding water, (c) by ridges or wings of the glandular region. The

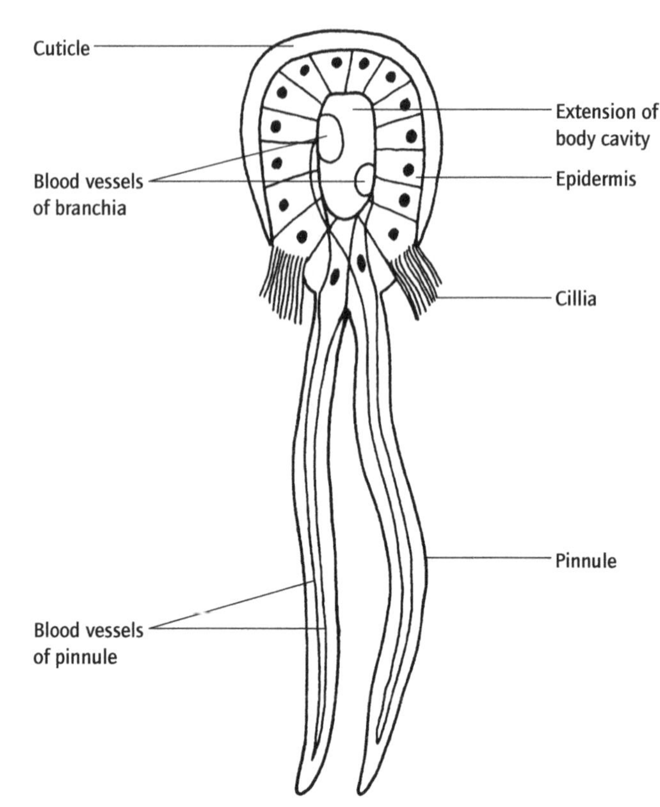

Fig. 4.46 Transverse section of a perviate branchia showing the two pinnules and tracts of cilia (after Ivanov, 1963).

branchiae vary greatly in their number and arrangement; often their bases are fused together or mounted on a tongue-like or ridged structure. In the largest class they are possibly held so as to enclose a central tubular cavity, and, in those species with only a thick single 'tentacle', this is coiled achieving the same effect (see Fig. 4.47c). Into this cavity project paired rows of 'pinnules' – elongate extensions of the epidermal cells – flanked by long epidermal cilia (Fig. 4.46). The action of these cilia is most likely to be the creation of a water current down the intrabranchial cavity, but the function of the pinnules is uncertain.

Lacking a gut, the manner in which adult pogonophorans feed has excited much speculation. It is now certain that those species which live near submarine hydrothermal fumaroles or 'cold seeps' derive most, if not all, their requirements from intracellular chemoautotrophic bacteria in the trophosome, which metabolize the reduced sulphur compounds and methane issuing from such vents, and that equivalent bacteria are also present in such perviate species as have been investigated. The haemoglobins of the worms have the property of binding hydrogen sulphide which is transported to the symbiotic chemotrophic bacteria without toxic effects on the pogonophoran tissues.

4.13.3 Classification

The 140 species of pogonophorans are divided between two classes.

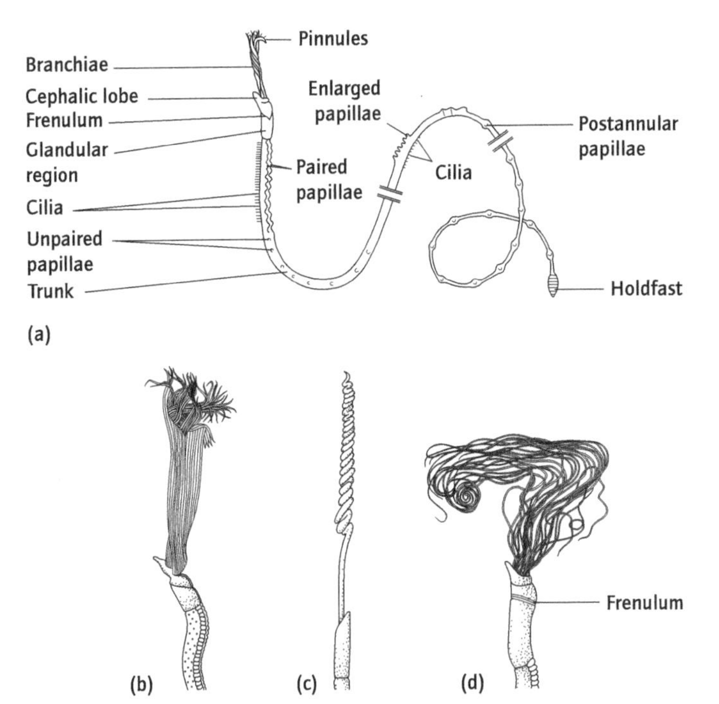

Fig. 4.47 Perviate pogonophorans: (a) a diagrammatic representation of the external appearance of a typical perviate (much shortened), showing its main features (after George & Southward, 1973) (note that in life the body is oriented straight and vertical); (b)–(d) the anterior ends of three perviates showing different numbers and arrangements of branchiae. (After Ivanov, 1963.)

Class	Order
Perviata	Athecanephria Thecanephria
Obturata	Vestimentifera

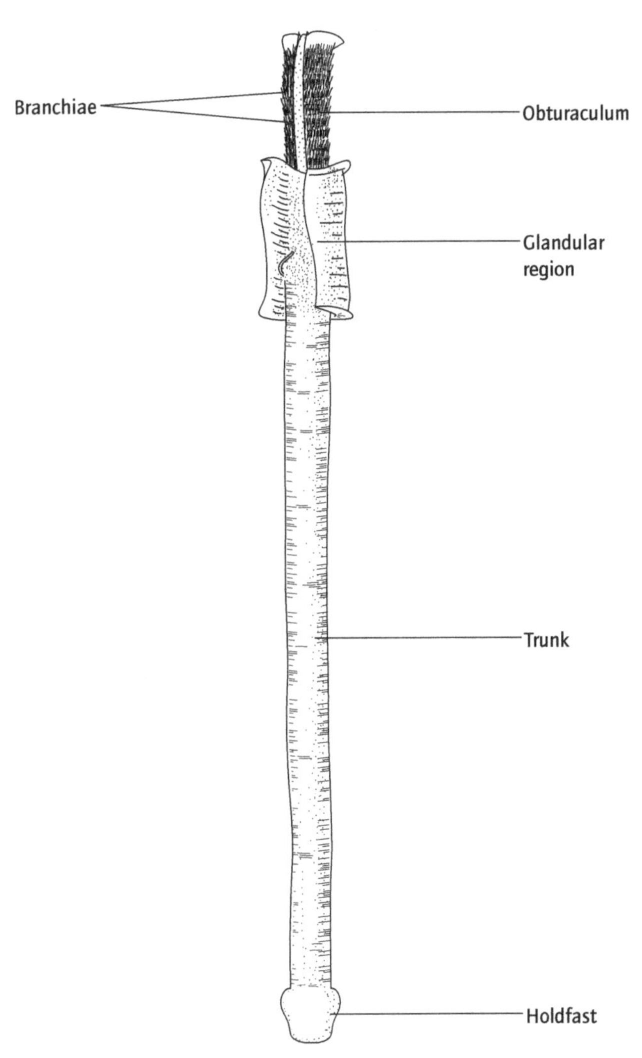

Fig. 4.48 A vestimentiferan (after Gage & Tyler, 1991).

4.13.3.1 Class Perviata

These, the typical pogonophorans, are relatively small species (with diameters of <3 mm and lengths of <85 cm) living in tubes anchored in soft sediments. They possess few (1–250) elongate branchiae, and a raised ridge, the 'frenulum', running obliquely around the glandular region (Fig. 4.47) – this probably supports the anterior part of the body at the mouth of its tube. The 125 species are placed in two orders.

4.13.3.2 Class Obturata

The single order (Vestimentifera) (Fig. 4.48) comprising this class contains much bigger worms (1–3 cm diameter, and lengths of more than 2 m in some) with shorter, more numerous branchiae – in excess of 1000 – mounted on an 'obturaculum', part of which closes the aperture of the tube after the animal has withdrawn. In contrast to the Perviata, vestimentiferans lack chaetae on the trunk and do not possess a frenulum, the latter being replaced by two large wings which meet in the dorsal midline and extend anteriorly over the base of the obturaculum.

As more and more seeps and hydrothermal vents are being explored, so the number of known vestimentiferan species is showing an almost exponential increase. The first species was announced in 1969; the second in 1975; the third in 1981; six new ones in 1986; and the current total of known species is now 15. Although a recent review of the group raises it to the level of phylum, and divides it into two classes and three orders, in view of the continual change in knowledge and opinion which will undoubtedly accompany the discovery of more material it seems advisable at the moment to retain the traditional system of a single class and order until the classification of the group stabilizes somewhat.

4.14 Phylum ANNELIDA

4.14.1 Etymology

Latin: *annellus* or *annelus*, a diminutive of *anulus*, a ring.

4.14.2 Diagnostic and special features

1 Bilaterally symmetrical, vermiform.
2 Body more than two cell layers thick, with tissues and organs.
3 A muscular gut with mouth and anus.
4 Body divided into segments (the segmentation may not be visible externally, but is always evident in the nervous system).
5 A presegmental prostomium containing a nervous ganglion, and a post-segmental pygidium.
6 Body cavity a series of schizocoels, obscured in specimens with anterior and posterior suckers.
7 Body cavity often subdivided by transverse septa, but frequently suppressed or obscured in some or all segments.
8 Outer epithelium covered by a cuticle and with epidermal bristles or chaetae in bundles or singly, except in specimens with anterior and posterior suckers.
9 Body wall muscular, often with complete circular muscle layers and four blocks of longitudinal muscles.
10 A closed blood system.
11 Nervous system with presegmental supra-oesophageal ganglion, circum-oesophageal ring and a ventral nerve cord with segmental ganglia.
12 Segmental ducts of mesodermal and ectodermal origin, which may be combined, restricted to one or a few segments and/or partially suppressed.
13 A varying degree of cephalization.
14 Development with spiral cleavage, but with modification of this and epibolic gastrulation in forms with yolky eggs.
15 Planktonic development in marine forms sometimes via a free-living trochophore larva but this stage frequently encapsulated. Freshwater and terrestrial forms with encapsulated eggs.

The structural grade of organization of the Annelida combines the hydrostatic and functional properties of the coelomate grade of organization with a segmented body plan. At its most primitive level of expression, the segmentation affects the ectoderm and the mesoderm and includes segmental septation of the coelom. Many annelids, however, have suppressed or modified this pattern and are secondarily aseptate or acoelomate.

Segmentation (Fig. 4.49, see also Fig. 4.51), is established during development from paired mesodermal growth zones and a corresponding ectodermal ring in front of the pygidium, the posterior region through which the endoderm opens to the exterior. New segments are formed at the anterior face of the pygidium, so that the last-formed segment is always the most posterior (Fig. 4.49b). In some Polychaeta the first three segments are formed precociously and simultaneously, and may be specialized for planktonic life, but subsequent segments are produced successively. Annelids may continue to add segments throughout adult life, though a definitive number is usually reached. In adult Polychaeta and Oligochaeta, renewed production of segments can be stimulated by transection, and the loss of posterior segments results in the formation of a new pygidium and prepygidial growth zone (Fig. 4.50) (see Section 15.5).

At the most fundamental level, the annelid body can be thought of as comprising a presegmental prostomium containing the supra-oesophageal ganglion, a series of more or less identical structural units, the segments, and a posterior post-segmental region, the pygidium, in the anterior ventral region

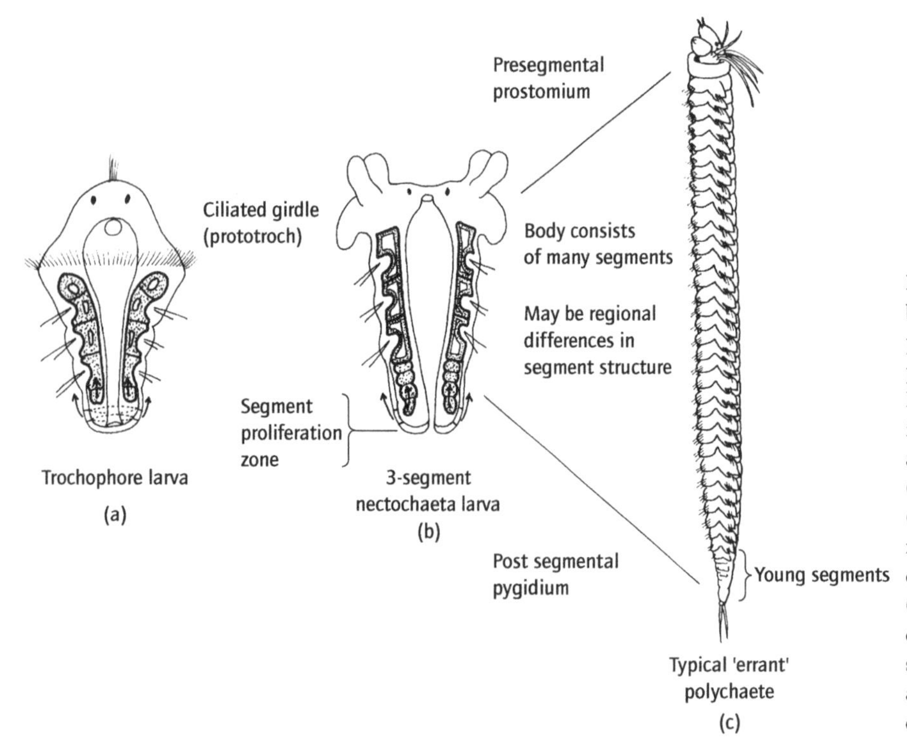

Fig. 4.49 The basic annelid body plan is formed by the intercalation of segments between the presegmental prostomium and a segment proliferation zone at the anterior of the pygidium. In many polychaetes the first three segments are formed precociously and simultaneously; all others are added progressively often to a defined number. (a) Diagrammatic representation of a trochophore. (b) Formation of the first three segments. The most anterior segment may lose the chaetae and contribute to the sense organs of the anterior end. (c) Basic regions of the body. In this case as in an errant polychaete of the order Phyllodocida. In the sedentary orders the segments of a thoracic region and a more posterior abdomen may have rather different architecture (see also Fig. 15.14).

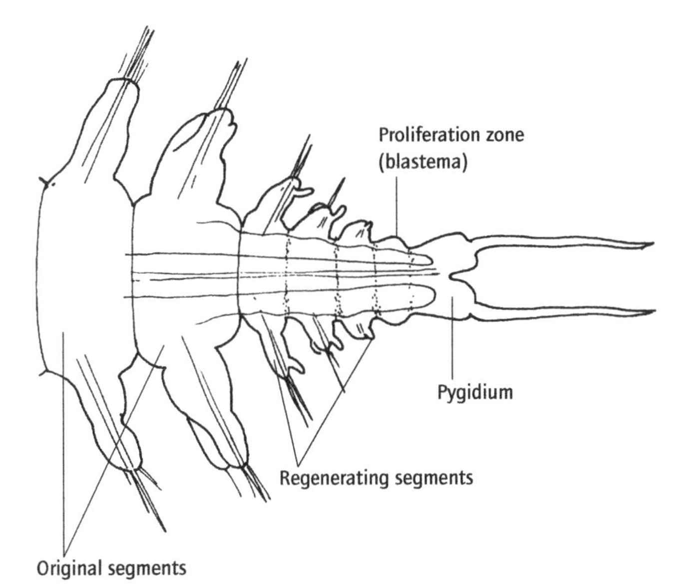

Fig. 4.50 Regeneration in *Nereis*. Loss of segments causes formation of a new segment proliferation zone and the production of a series of segments, the most posterior of which is the youngest.

of which is the segment blastema or growth zone (Fig. 4.51). The fundamental features of the segmentation are as follows:

1 The ectoderm: segmental arrangement of the chaetae; the nervous system with a segmental ganglion and associated nerves; and the ectodermal nephridia.

2 The mesoderm: segmentation of the musculature associated with the chaetal sacs/parapodia, and of the corresponding vessels; septation of the coelomic space, creating isolated body cavities; segmental arrangement of the germinal epithelia and the associated paired coelomoducts. These elements of the underlying segmentation of the annelids are shown diagrammatically in Fig. 4.51.

The proto-annelids probably had complete septa and a complete series of ectodermal protonephridia and mesodermal coelomoducts. This arrangement has not persisted in any living annelid. Protonephridia, with a group of flame cells, are retained in a few families of polychaetes but the nephridia usually have an open ciliated funnel, a nephrostome. Nephridia and coelomoducts have different embryological origins (Section 12.3.1) and are separate in the oligochaetes. In the essentially marine Polychaeta the nephridium and the coelomoduct are usually combined to form a compound structure, the nephromixium (Fig. 12.11), and in those families lacking complete septa and having open coelomic cavities the number of germinal epithelia and segmental ducts is greatly reduced.

4.14.3 Classification

There are at least 75 000 described species of annelid, which are readily subdivided into three major groups, the Polychaeta, Oligochaeta and Hirudinea, and two minor ones. The Hirudinea diverged from an early line of oligochaete evolution, and these two groups, together with the much smaller Branchiobdella, form the class Clitellata.

Class	Subclass	Order
Polychaeta		Orbiniida Ctenodrilida Psammodrilida Cossurida Spionida Questida Capitellida Opheliida Phyllodocida Amphinomida Spintherida Eunicida Sternaspida Oweniida Flabelligerida Poeobiida Terebellida Sabellida Nerillida Dinophilida Polygordiida Protodrilida Myzostomida
Aeolosomata		Aeolosomatida
Clitellata	Oligochaeta	Lumbriculida Moniligastrida Haplotaxida
	Branchiobdella	Branchiobdellida
	Hirudinoidea	Acanthobdellida Rhynchobdellida Arhynchobdellida

4.14.3.1 Class Polychaeta

The mainly marine polychaetes are remarkable for their morphological and anatomical diversity; they take their name from the numerous chaetae or bristles, inserted in two groups rather than singly, in the biramous segmental parapodia (Fig. 4.52). Many exhibit indirect development (Fig. 4.53), though this may be suppressed.

About half the polychaete families and species can be conveniently grouped as 'errant' forms; these belong mainly to two large and easily defined orders, the Phyllodocida and the Eunicida. The Phyllodocida have an axial proboscis (the anterior part of the alimentary canal) everted hydraulically and withdrawn by retractor muscles. This is often armed with a small number of jaws of hardened proteins that may contain a high proportion of heavy metal ions, but which are never calcified (Fig. 4.54).

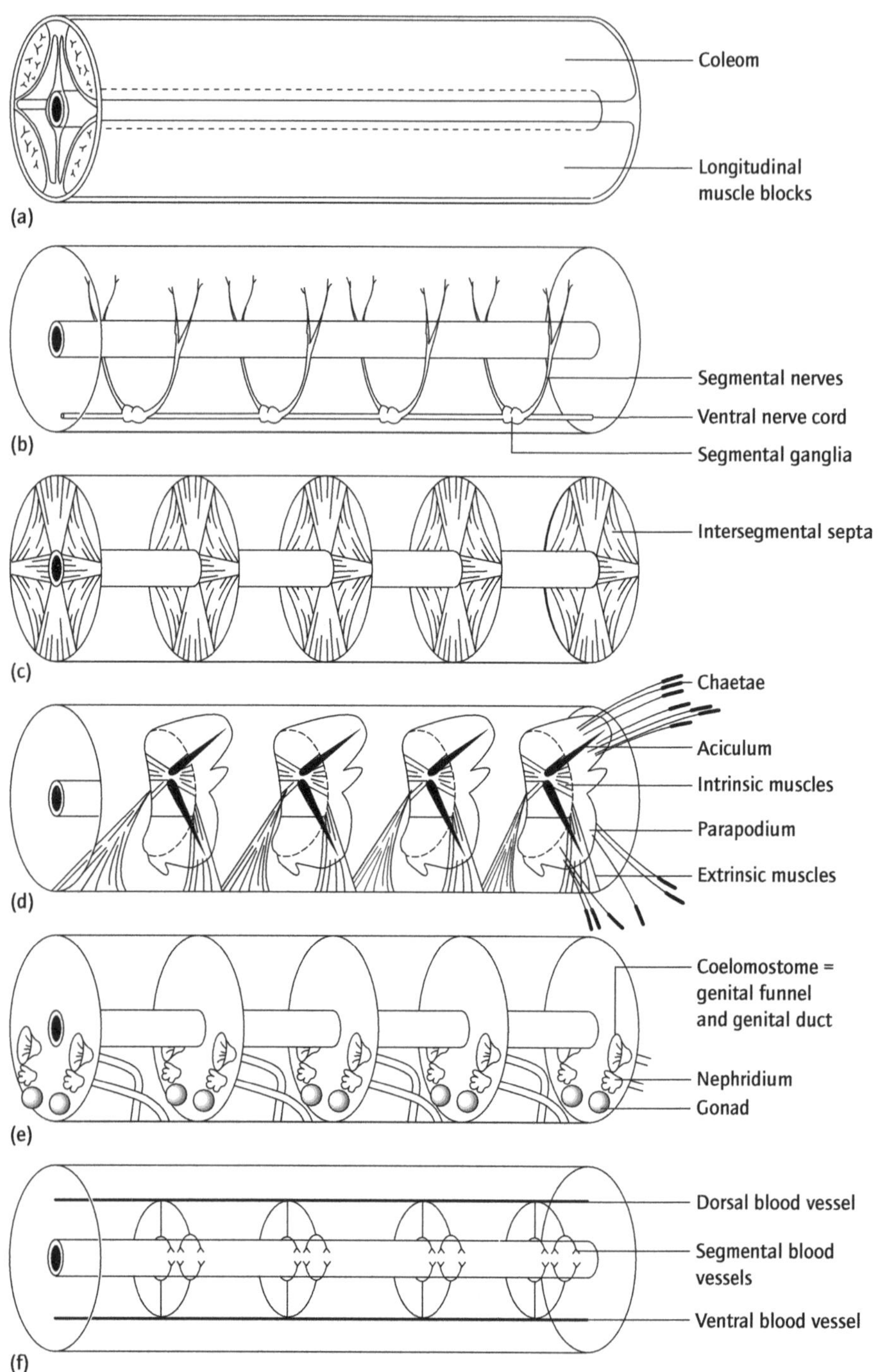

Fig. 4.51 Diagrammatic representation of the components of the segmented body plan of a hypothetical protoannelid ancestor. Living annelids exhibit some, but not necessarily all, of these features. (a) The longitudinal muscle blocks and the gut are not segmented. The body cavity is a true coelom with peritoneal lining. (b) The segmentation of the ventral nervous system, with segmented ganglia and nerves. This is the most conserved element of segmentation in living annelids (see section on Hirudinea below). (c) Subdivision of the coelom by transverse septa. This feature is retained in most Oligochaeta and some Polychaeta, but can be much modified and is lost in Hirudinea. (d) Segmentation of the ectoderm and mesoderm due to the development of parapodia or segmentally arranged chaetae. This feature is prominent in Polychaeta, less prominent in Oligochaeta and suppressed in Hirudinea. (e) Segmentation of genital ducts, excretory ducts and germinal epithelia. Protoannelids are supposed to have had a complete series of mesodermal coelomoducts (for the discharge of gametes) and ectodermal nephridia. They are also supposed to have had a pair of germinal epithelia in each coelomic compartment. These conditions are not found in any living annelids and are frequently much modified. (f) Segmentation of the blood vascular system.

The Eunicida are superficially like the Phyllodocida, but the proboscis is not axial but is an eversible buccal mass, armed with a complex array of calcified jaws, a pair of chisel-like mandibles and several pairs of maxillae (Fig. 4.55). The range of form among the errant polychaetes is illustrated in Fig. 4.56. Many spend most of the adult life in a gallery or permanent burrow system from which they partially emerge to feed. Most have a well-developed head with a variety of sense organs and a complex brain. Many are carnivorous but some are detritivores, filter-feeders or omnivores. They have well-developed parapodia, with the dorsal and ventral lobes supported by a stiff proteinaceous rod – the aciculum.

Some errant polychaetes display *epitoky* or swarming behaviour associated with sexual reproduction. This often involves a complex metamorphosis of the adult worms. Epitoky takes place in two fundamentally different ways (Figs 4.57I,II). In epigamy the whole worm is transformed into the swarming epitoke whereas in schizogamy (stolonization) the posterior segments of sexually mature worms become detached as migratory gamete-bearing stolons.

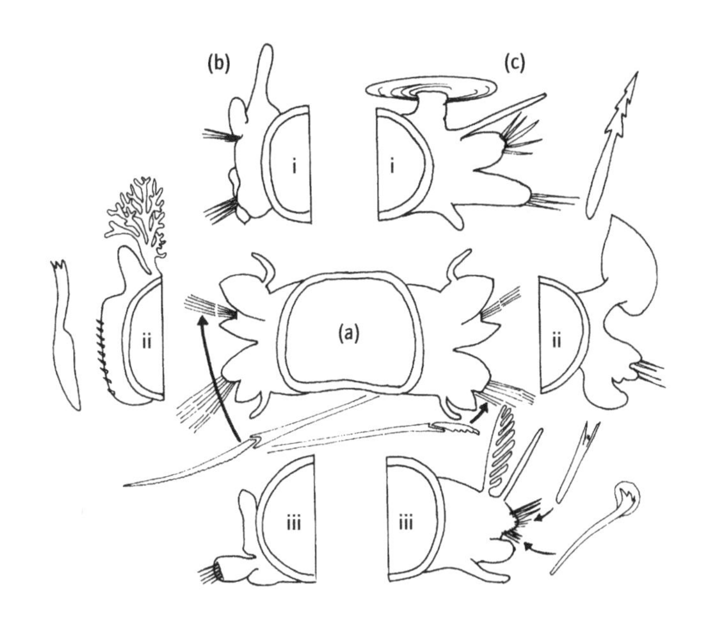

Fig. 4.52 Examples of the range of form among polychaete parapodia. (a) The fundamental biramous type as found in *Nereis*. Details show the morphology of some chaetae. (b) Examples of parapodia among sedentary families: (i) Spionidae; (ii) Arenicolidae; (iii) Sabellidae. (c) Examples of parapodia and chaetae among errant families: (i) Polynoidae; (ii) Phyllodocidae; (iii) Eunicidae.

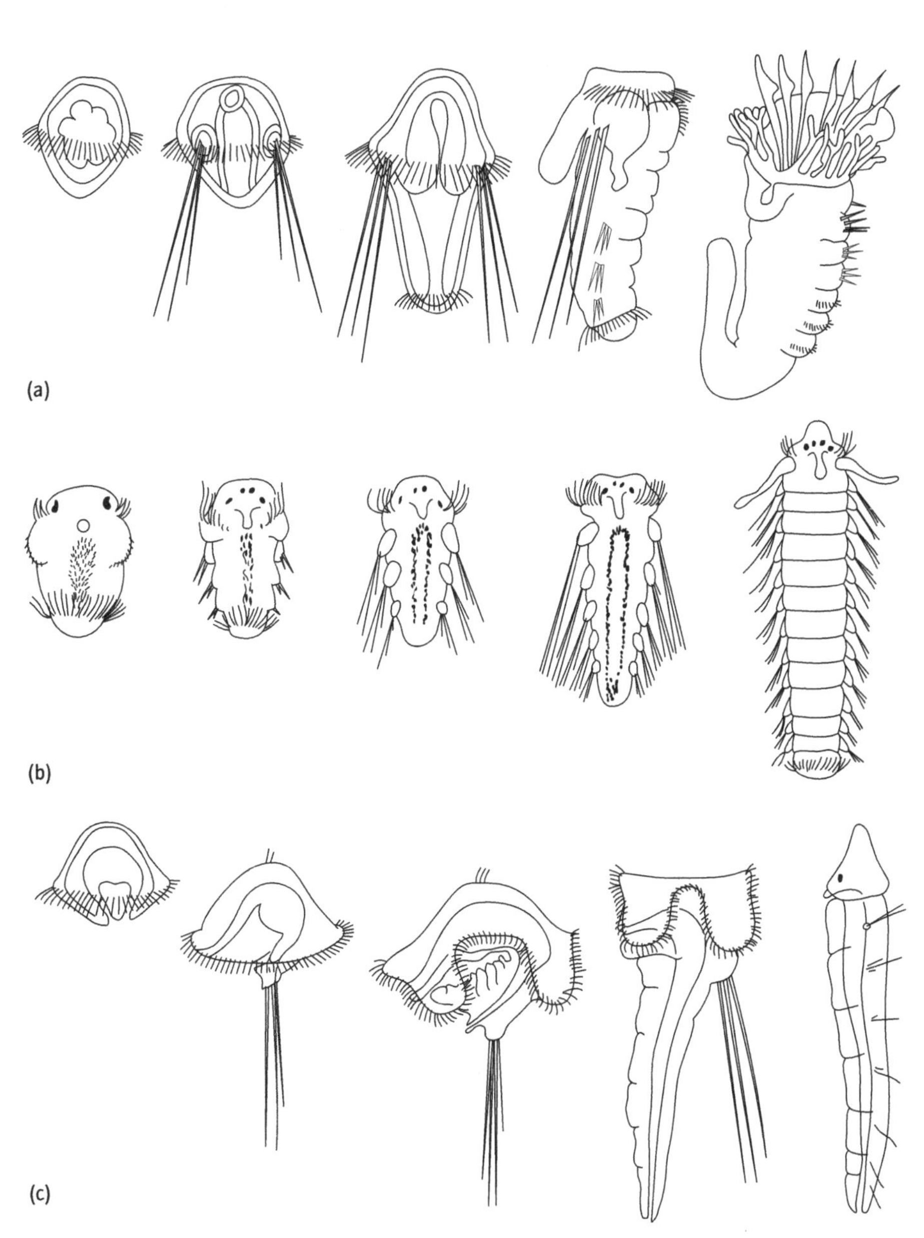

Fig. 4.53 Examples of planktonic development stages among the Polychaeta: (a) *Sabellaria*; (b) *Spio*; (c) *Owenia*.

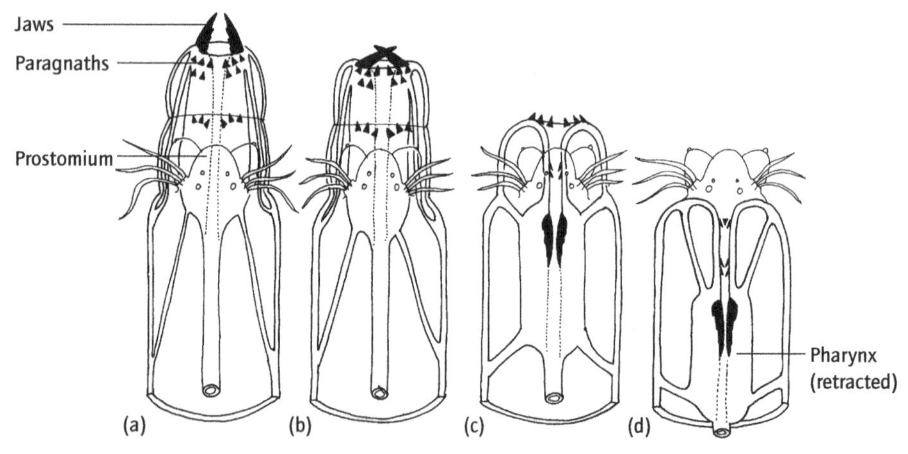

Fig. 4.54 The proboscis of the phyllodocid type. This is a muscular pharynx which is everted by coelomic pressure and withdrawn by the action of retractor muscles as shown in (a)–(d). It is often, as in *Nereis*, armed with proteinaceous jaws.

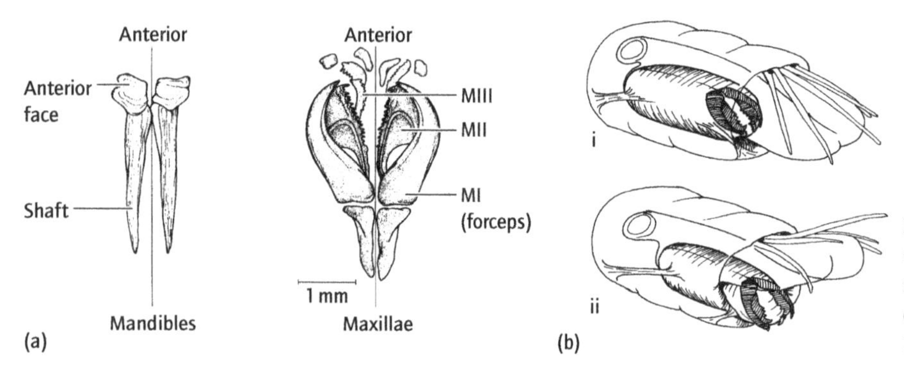

Fig. 4.55 The jaw apparatus of the eunicid type. (a) The isolated jaw complex. (b) (i) The jaws retracted in the muscular ventral tongue-like floor of the pharynx; (ii) the everted jaws. (After Olive, 1980.)

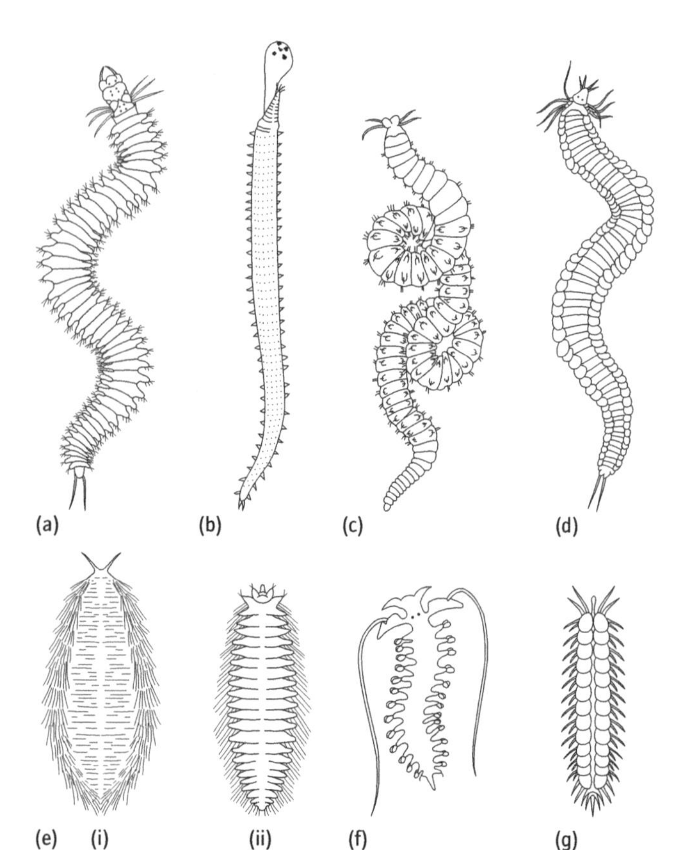

Fig. 4.56 The range of form among the 'errant' Polychaeta; (i) and (ii) are the dorsal and ventral views, respectively, of the animal. All except (c) (Eunicidae) are members of well-known families of the order Phyllodocida. The errant polychaetes are usually identified to the level of family the orders being rarely used. (a) Nereidae, (b) Glyceridae, (c) Eunicidae, (d) Phyllodocidae, (e) Aphroditidae, (f) Tomopteridae, (g) Polynoidae.

The remaining polychaetes are frequently grouped as the sedentary forms and 'archiannelids'. Both are diverse and artificial assemblages, in which it is difficult to recognize relationships. Most biologists refer directly to the rather distinctive families rather than to orders.

In the 'sedentary polychaetes' there is usually regional variation in structure of the segments and distinct thoracic and abdominal regions can be recognized (Fig. 4.58). They are all microphagous deposit or suspension feeders and do not have the wriggling or walking locomotory patterns exhibited by the 'errant' species (see Section 10.5). Their diversity arises from their widely different modes of life. Some live in burrows, others in tubes of secreted parchment-like material or calcium carbonate, or sand grains. Several families exhibit a functional convergence with Oligochaeta. They burrow in soft substrata and are either 'swallowers' or sand-grain 'lickers'. They have a small

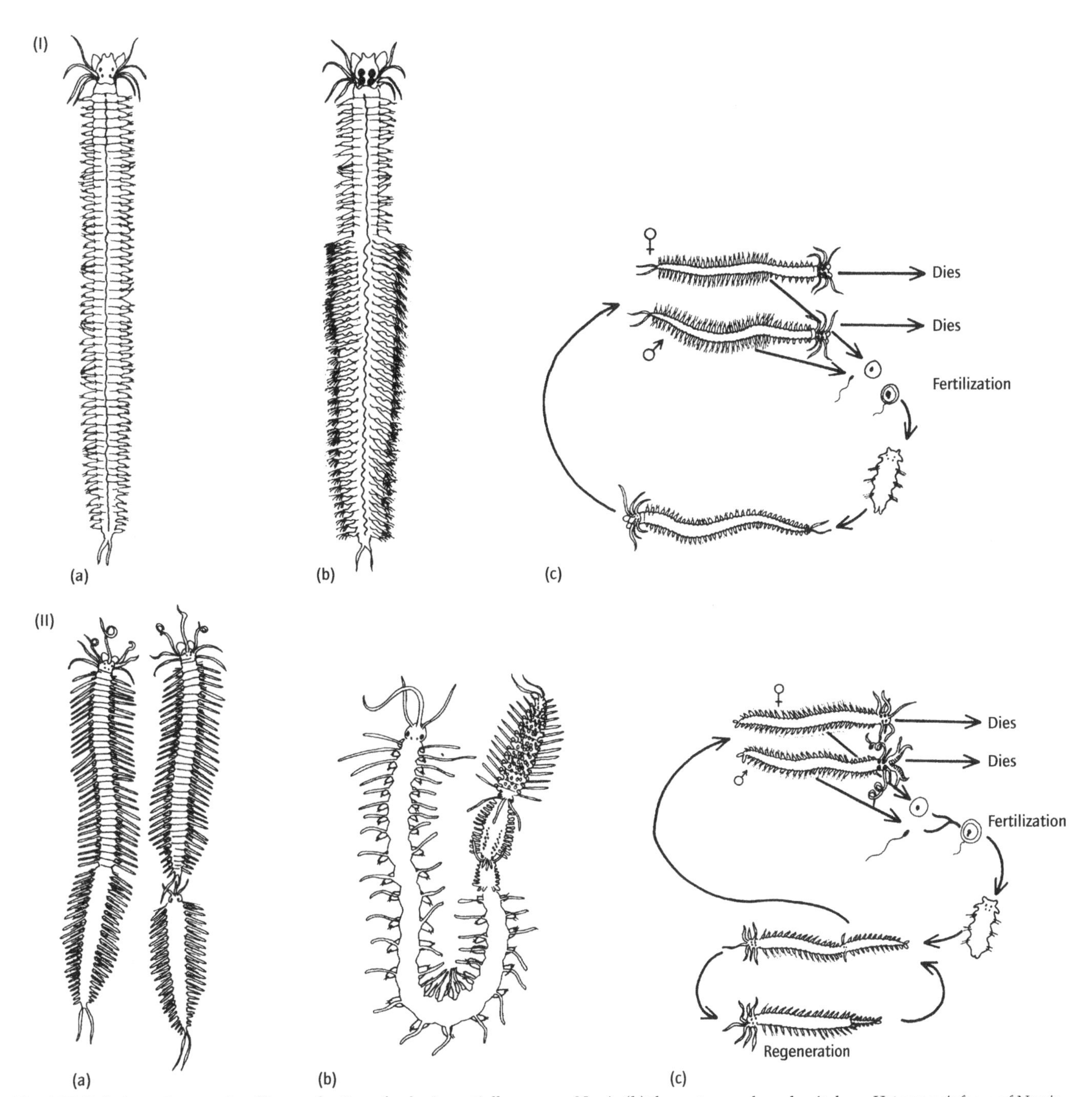

Fig. 4.57 Epitoky and swarming. The production of pelogic partially metamorphosed sexually mature worms has arisen independently several times in the Polychaeta. The process takes place in two fundamentally different ways. I: Epigamy. Mature specimens undergo metamorphosis, swarm and die: (a) the unmodified atokous form of *Nereis*; (b) the metamorphosed epitokous *Heteronereis* form of *Nereis*. (c) Diagrammatic representation of the life cycle. II. Schizogamy: (a) production of a single stolon by modification of posterior segments; (b) production of multiple stolons by terminal budding, the terminal stolon being the oldest; (c) diagrammatic representation of the life cycle.

prostomium without prominent sense organs, simple chaetae in reduced parapodia, well-developed circular muscles and complete septa. Several families have soft prehensile tentacles which collect fine particles of organic sediment and convey them to the mouth (see Fig. 9.5). Such tentacles are often outgrowths of the prostomium. Some rather advanced families of tubiculous worms have a crown of stiff prostomial tentacles which act as a true filter-feeding device (Fig. 9.2). Other feeding specializations are discussed in Chapter 9 (see, e.g. Fig. 9.5).

The archiannelids (Fig. 4.59) are an unrelated assemblage of minute polychaetes independently adapted to the interstitial environment in marine sands.

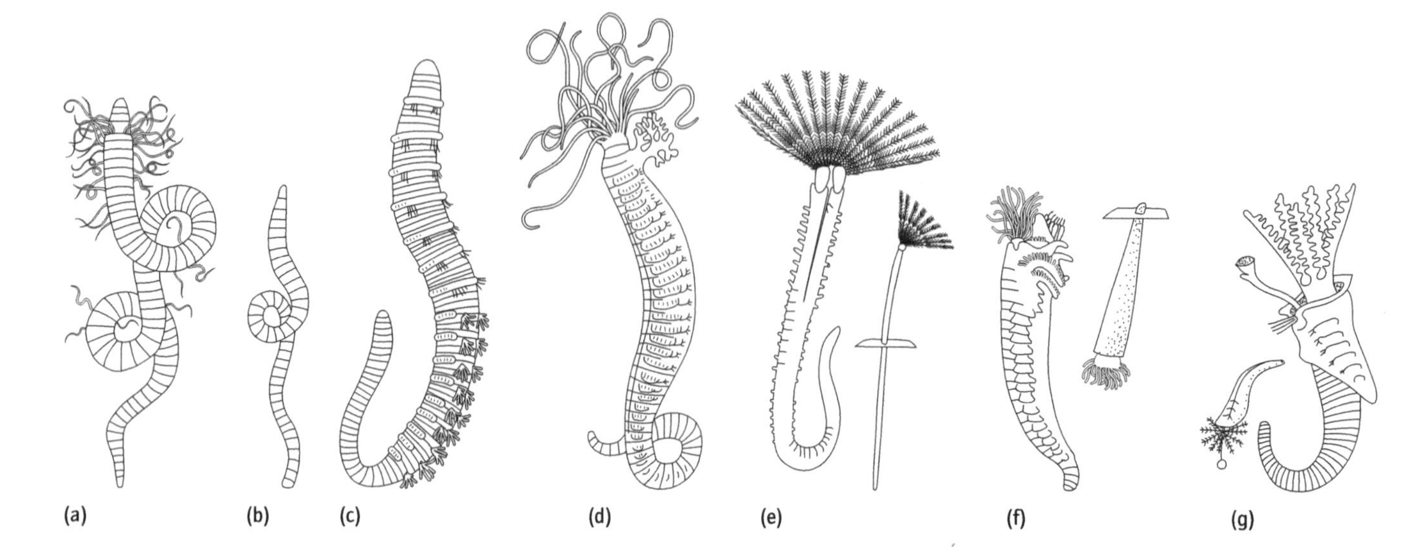

Fig. 4.58 Examples of the range of body form among the sedentary polychaetes. (Smaller figures show the characteristic mode of life. See also Fig. 9.6.) (a) Cirratulidae, (b) Capitellidae, (c) Arenicolidae, (d) Terebellidae, (e) Sabellidae, (f) Pectinariidae, (g) Serpulidae.

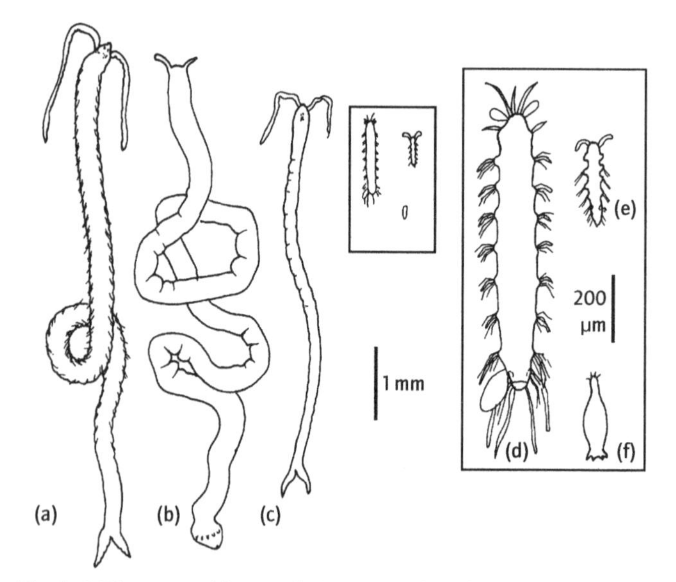

Fig. 4.59 The range of form and size among the polychaete families formerly grouped together as archiannelids. These are now thought to be secondarily simplified as an adaptation to small size (after Jouin, 1971).

4.14.3.2 *Class Clitellata*

Subclass Oligochaeta The oligochaetes are predominantly terrestrial or freshwater annelids, with secondarily marine representatives especially in estuarine and interstitial environments.

In their locomotion, oligochaetes exploit to the full the hydrostatic properties of a completely septate and segmented grade of organization. They have therefore retained a body plan close to that attributed to the proto-annelids (see Fig. 4.51), but they are not primitive. The oligochaetes are specialized in being simultaneous hermaphrodites, in shedding small numbers of large, yolky eggs into a protective nutrient-filled envelope, the cocoon, secreted by the *clitellum*, and in having a greatly reduced number of gonads. The arrangement of the oligochaete genitalia is the basis for their classification into three orders.

Functionally and ecologically, there are essentially two types of oligochaetes: the mainly aquatic *microdrile* species and the terrestrial *megadriles* or earthworms.

The functional anatomy of the earthworms is remarkably constant and can be illustrated with reference to *Lumbricus terrestris* (Fig. 4.60a).

A cross-section (Fig. 4.60b) shows the following features: waterproof cuticle lubricated by secretory goblet cells in the epidermis; the epidermis; a nervous layer; a complete layer of circular muscles; blocks of longitudinal muscles; chaetae inserted singly, or in small groups as in *Lumbricus*; a coelom divided by complete septa and lined with peritoneum; muscles of the gut wall; and the endodermal lining of the alimentary system.

The worm is constructed on a segmental plan, but there is considerable specialization in the anterior region, especially in the structure of the alimentary and reproductive systems (Fig. 4.60a and b, and Fig. 4.61). Earthworms engage in complex pseudocopulatory behaviour during which paired worms become coated in mucus, and sperm are transferred externally to the spermathecae or sperm sacs in the region of the clitellum (Fig. 4.62). Fertilization takes place within the cocoon, and once mated an individual earthworm can release many egg cocoons.

The microdrile oligochaetes are smaller than the earthworms. They comprise several families living mainly in fresh water, although the Enchytraeidae are primarily terrestrial. Some even occur in the deep sea. They are more variable in body form than earthworms, and some have prominent hair-like chaetae which resemble those of some polychaetes (Fig. 4.63). Their reproductive biology is frequently highly specialized. Asexual

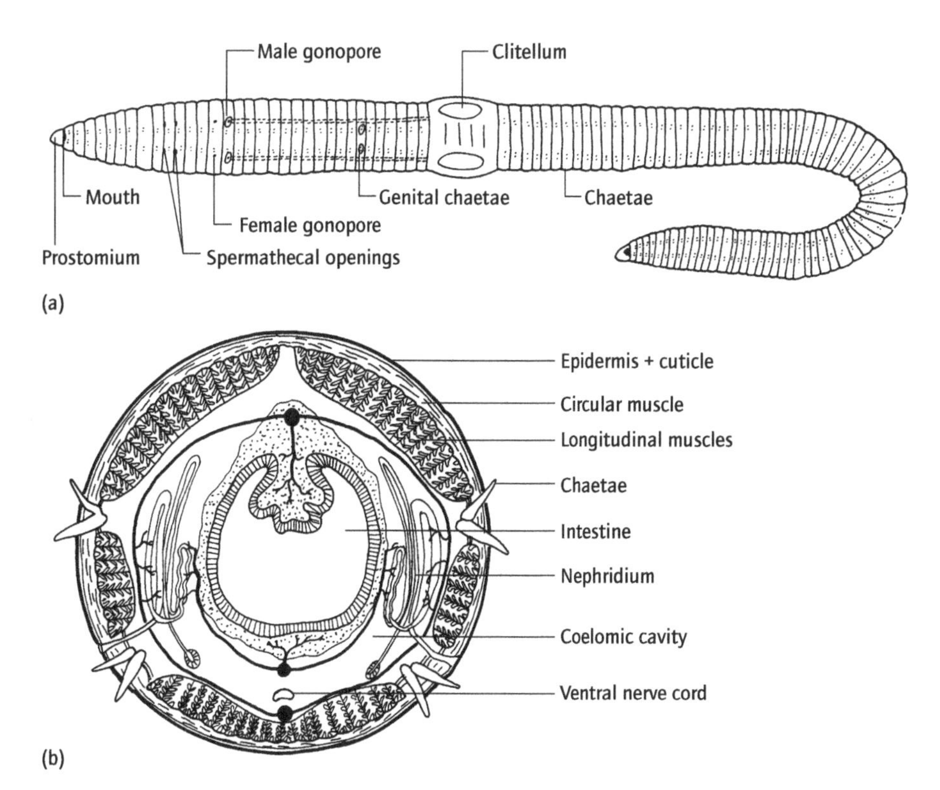

Fig. 4.60 (a) The external morphology of the earthworm *Lumbricus terrestris* as seen in ventral view. (b) Diagrammatic cross-section of the earthworm *Lumbricus terrestris*.

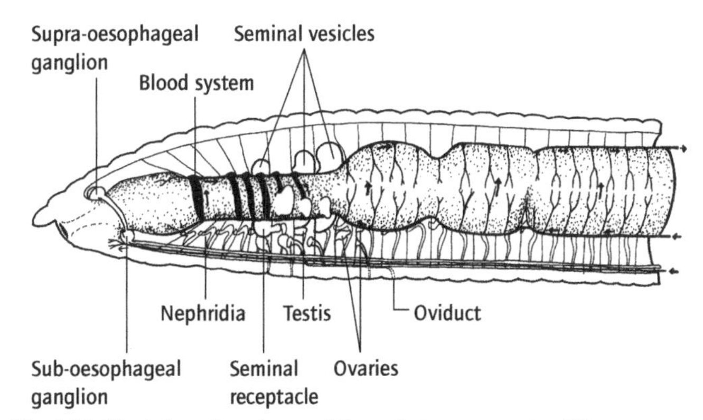

Fig. 4.61 The internal anatomy of the anterior segments of the earthworm *Lumbricus terrestris*.

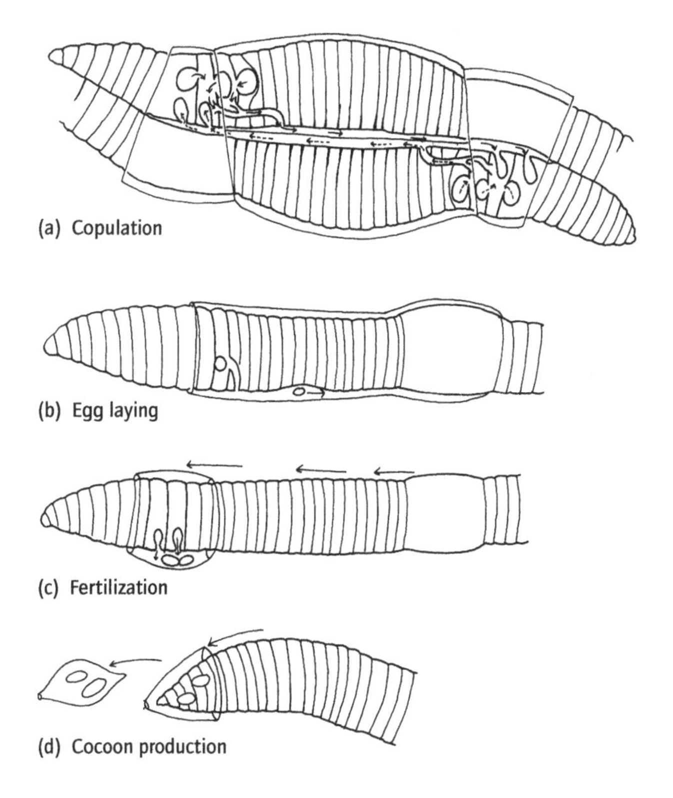

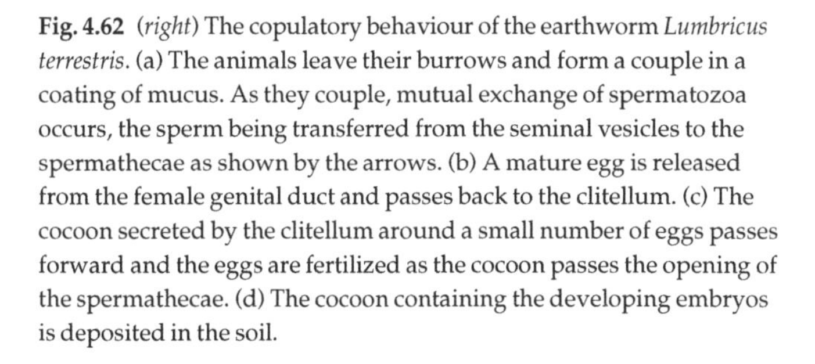

Fig. 4.62 (*right*) The copulatory behaviour of the earthworm *Lumbricus terrestris*. (a) The animals leave their burrows and form a couple in a coating of mucus. As they couple, mutual exchange of spermatozoa occurs, the sperm being transferred from the seminal vesicles to the spermathecae as shown by the arrows. (b) A mature egg is released from the female genital duct and passes back to the clitellum. (c) The cocoon secreted by the clitellum around a small number of eggs passes forward and the eggs are fertilized as the cocoon passes the opening of the spermathecae. (d) The cocoon containing the developing embryos is deposited in the soil.

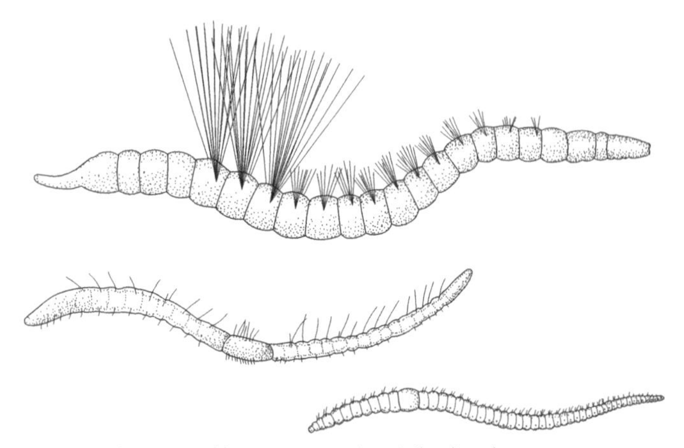

Fig. 4.63 The range of form among microdrile oligochaetes.

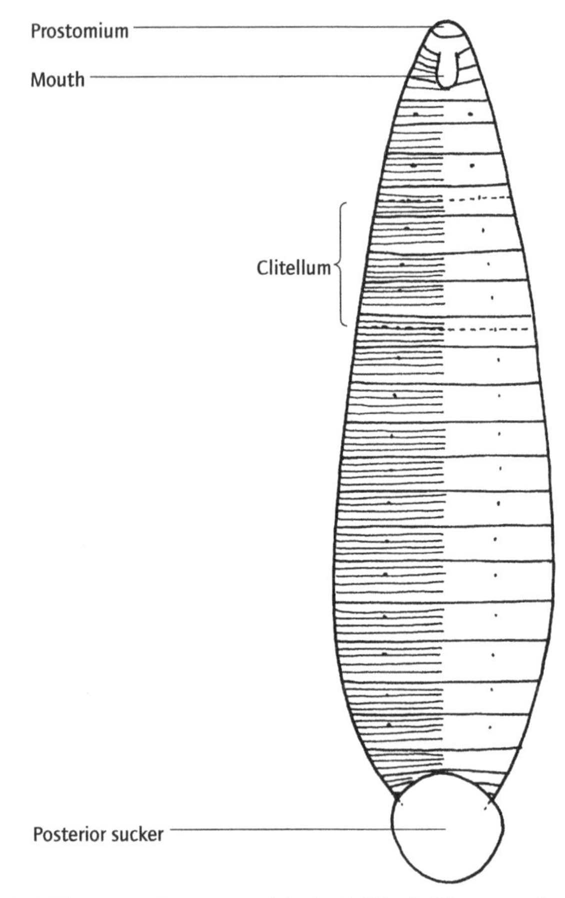

Fig. 4.64 The external anatomy of the leech *Hirudo*. The external annulation is shown on the left and the true segmentation on the right.

reproduction by spontaneous fission (schizogenesis) is common and indeed sexual reproduction has never been observed in some forms.

Subclass Hirudinea (leeches) The special features of leeches follow the development of suckers which enable them to form fixed attachments at the two ends. The body then acts as a single functional hydrostatic unit. Consequently the segmentation of the coelomic spaces, ectoderm and mesoderm has largely been lost. Functionally leeches operate in a similar manner to flatworms, but with the advantages of coelomic sinuses for the transmission of forces, a nervous system with segmental ganglia for the coordination of complex behavioural and locomotory patterns and a closed vascular system with pigments for respiration.

All leeches have precisely 33 segmental ganglia. The head sucker is formed from segments 1–4 and the posterior sucker from segments 25–33 (Fig. 4.64). There are no chaetae and externally segmentation is obscure, although the body is divided transversely by a series of annuli which do not correspond to segmental boundaries. A cross-section shows that the space between the mesoderm and the muscular wall of the gut is filled with a loose 'parenchyma-like' botryoidal tissue (Fig. 4.65). Some internal organs, especially the germinal epithelia and the nephridia, do reflect the ancestral segmental condition (Fig. 4.66).

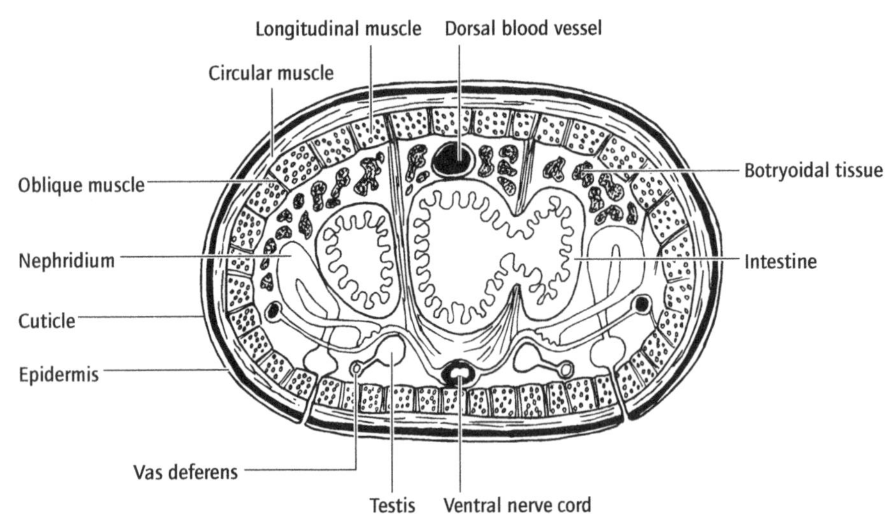

Fig. 4.65 A semi-diagrammatic transverse section of the leech *Hirudo*. The coelomic space is filled with a parenchyma-like tissue. The spaces between these cells may be organized into well-defined sinuses.

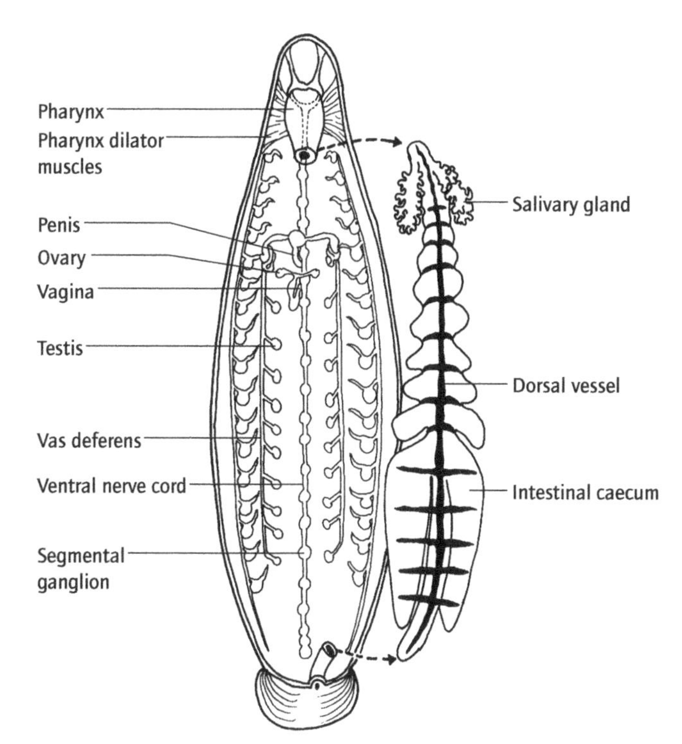

Fig. 4.66 The internal anatomy of the leech *Hirudo*, as revealed by dissection from the dorsal side. The gut is shown displaced.

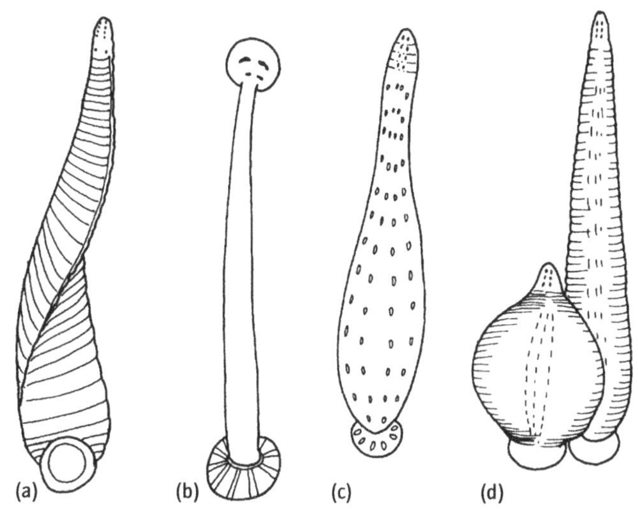

Fig. 4.67 The range of form in the leeches: the form on the extreme right shows the characteristic ability of leeches to show great changes in shape.

Leeches are hermaphrodite, with several pairs of testes, a pair of ovaries, and a single genital opening. Fertilization is internal and reproduction requires transfer of spermatozoa between partners, often with complex copulatory behaviour. In many species the structurally complex spermatozoa are injected hypodermically and make their way to the ovaries, between the cells of the recipient animal. After mating, adult leeches will lay numerous cocoons, each with one or a few eggs. Some of the large species survive for several years but the majority are annual, overwintering as young embryos in the cocoons.

The range of body form is illustrated in Fig. 4.67. All are carnivorous but not all are external parasites or blood-sucking forms. Classification is primarily based upon the structure of the mouthparts and the mode of feeding.

The Acanthobdellida is a peculiar primitive group of ectoparasites of salmonid fish. They lack the anterior sucker, have a few segments with chaetae and have a segmental coelom. They thus provide a link between the oligochaete and leech-like grades of organization.

Subclass Branchiobdella These are minute ectoparasites of freshwater Crustacea. They have 15 or 16 segments, the first four fused to form a cylindrical head with a sucker. These little-known clitellates seem to form an independent line from an oligochaete-like ancestor, paralleling in many ways the evolution of the Hirudinea (Fig. 4.68).

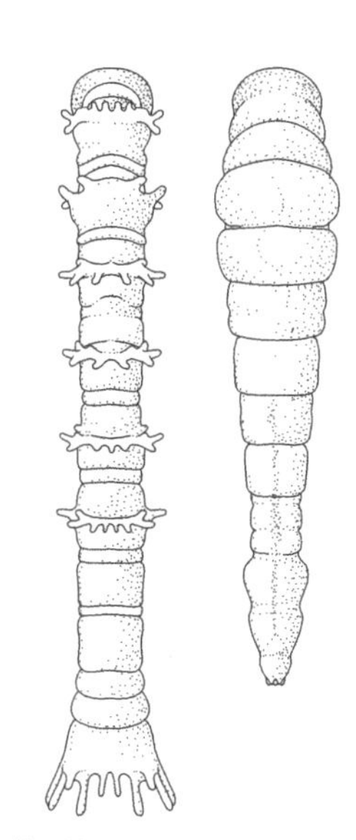

Fig. 4.68 Branchiobdella. The external morphology of typical species.

4.14.3.3 Class Aeolosomata

These minute annelids were formerly thought to be primitive oligochaetes but are now regarded as independent of clitellate evolution. All are small, and most are minute, living in the interstitial environments of fresh and brackish waters. They are hermaphrodite but have a single ovarian segment, and testes in segments both anterior and posterior to this segment. The so-called clitellum is composed of ventral glands and is not homologous to the dorsal clitellum of the clitellate annelids.

A typical species is illustrated in Fig. 4.69. It has a ciliated prostomium forming the main locomotory organ and rather long hair-like chaetae. The Aeolosomata, like the archiannelid families, exhibit a simplified structure due to their minute size, and their relationship to other annelids is unknown. There are only about 25 species in this class.

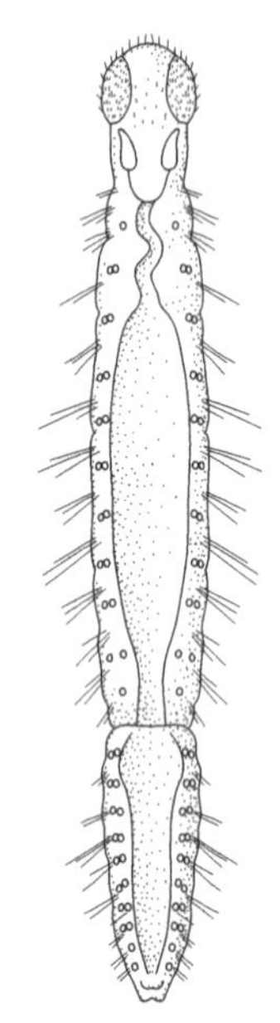

Fig. 4.69 Aeolosomata. A typical species.

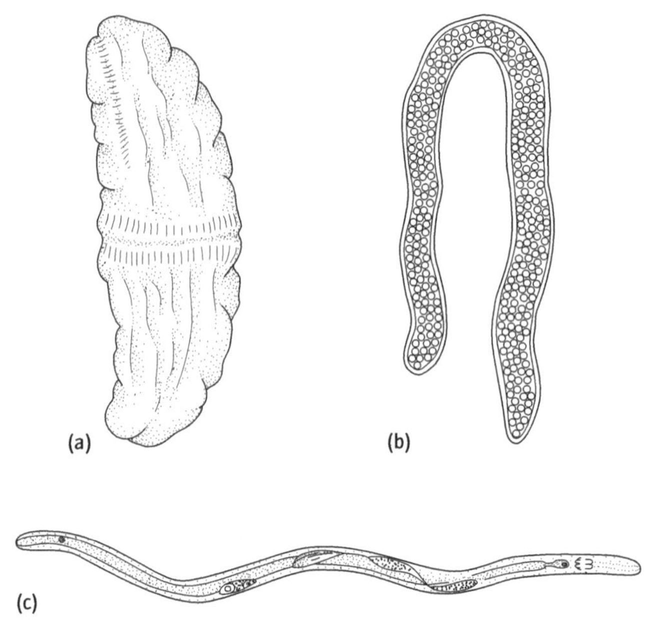

Fig. 4.70 Three worms of uncertain phylum: (a) *Xenoturbella*; (b) *Buddenbrockia*; (c) *Lobatocerebrum*. (After Barnes, 1998.)

4.15 Worms of uncertain phylum

Four types of living worms cannot easily be accommodated in any of the above phyla, largely because they display curious mixtures of features otherwise found in widely separated groups of animals. All four highlight the problem of establishing the relationships of small, acoelomate animals in which simplicity or their peculiarities might be primary and an indication of an evolutionary position near the base of animal diversification, or else be secondary, consequent on a parasitic way of life and/or a paedomorphic origin from some unknown, but more complex ancestor.

4.15.1 Xenoturbellids

The two species of *Xenoturbella* (Fig. 4.70a) are large (up to 30 mm long), hermaphrodite, flatworm-like animals that occur on soft marine sediments in northwestern Europe. Although in several respects similar in their anatomy to the acoelomorph flatworms, in other respects their fine structure shows superficial similarities to the deuterostomes (Chapter 7). *Xenoturbella* has been regarded as a paedomorphic deuterostome (hemichordate?) larva, and as an early offshoot from the line that ultimately gave rise to the turbellarians (Section 3.6.2.3.1); the latest suggestion is that, on the basis of its embryology, it is closely related to the bivalve molluscs, being in effect a metamorphosed trochophore larva that has undergone dedifferentiation.

4.15.2 Buddenbrockids

The 3 mm long *Buddenbrockia plumatellae* (Fig. 4.70b), first described in 1910, lives within the body cavity of the phylactolaeme bryozoan *Plumatella* and those of some other bryozoan genera (Section 6.3.3.1). It lacks a gut and a nervous system, and is somewhat reminiscent of a rhombozoan (Section 3.5.2) or orthonectidan (see below), yet unlike those groups it possesses longitudinal tracts of muscle fibres and its outer layer of cells are not ciliated. It moves rather like a nematode.

4.15.3 Lobatocerebrids

Three or four species of the hermaphrodite interstitial worm *Lobatocerebrum* (Fig. 4.70c), described in 1980, are now known from soft marine sediments in the North Atlantic and Red Sea. These up to 4 mm-long worms are superficially very like free-living flatworms, but the organization of their body wall, gut and male reproductive system suggest an affinity with the annelids (Section 4.14). As with *Xenoturbella* above, there are two contrasting views of its relationships: (i) that it is a paedomorphic larva of some larger, possibly polychaete annelid, worm, or (ii) that it is an early acoelomate offshoot of the line that ultimately gave rise to the schizocoelomate worms.

4.15.4 Orthonectidans

For a long time these parasites which live unattached within the tissues of marine turbellarians, annelids, nemertines, molluscs and echinoderms were considered to be related to the phylum Rhombozoa (Section 3.5.2). (Their name (Greek: *orthos*, straight; *nektos*, swimming) reflects this by its opposition to the rhombozoans which are the 'rotating animals' – a result of their helical symmetry.) They are now sometimes considered to be a phylum in their own right, the Orthonecta, probably allied to the Platyhelminthes, or actually to be flatworms. Two phases, sexual and asexual, occur, the body form of each being extremely simple. The

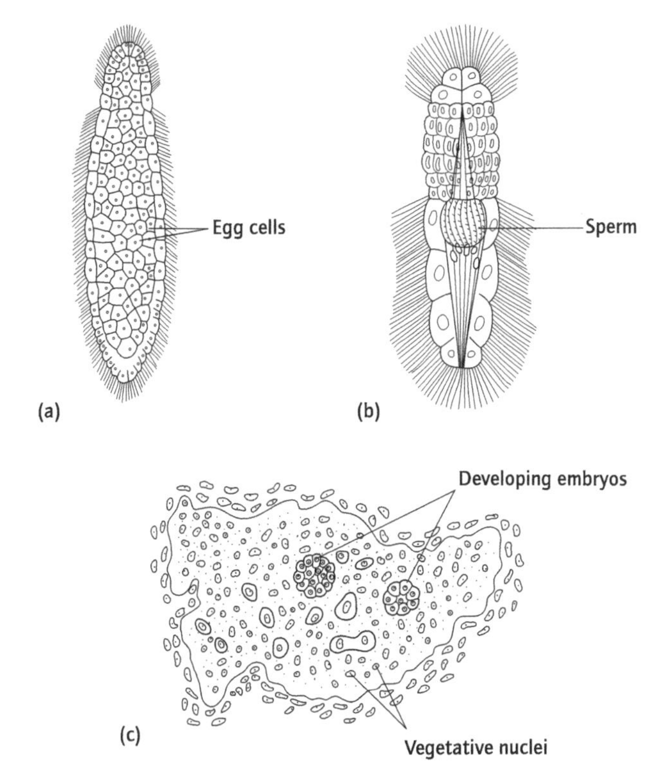

Fig. 4.71 Female (a) and male (b) orthonectidans (after Marshall & Williams, 1972). Orthonectidan plasmodium (c) (after Caullery & Mesnil, 1901).

sexes of the sexual phase are separate and dimorphic, although each is virtually just a single annular layer of ciliated somatic cells surrounding a mass of gametes (Figs 4.71a, b), although in some circular and longitudinal muscles occur beneath the somatic cell layer. Both phases are considerably less than 1 mm long. Sperm shed by the males enter the bodies of the females and after fertilization ciliated larvae – in effect a layer of ciliated cells surrounding a few 'germinal cells' – develop within the females. On their release, they invade the tissues of their host, the ciliated layer of cells is lost and the germinal cells multiply to form a multinucleate plasmodium (Fig. 4.71c). This plasmodium can fragment asexually to give rise to other plasmodia. Eventually, the sexual phase is produced, a single plasmodium producing either those of a single sex or both males and females, and these sexual phases then leave the host and complete the life history.

Although these animals are, in the sexual phase at least, bilaterally symmetrical, they clearly do not possess the other characteristics of the Bilateria (having effectively only a single cell layer and no organ systems, for example): their extreme bodily simplicity, however, is assumed to be an example of parasitic degeneration and they are thought to have evolved from more complex ancestors. The germinal cells, for example, are considered to be mesodermal in origin; and the muscles possessed by some do suggest such an origin. (They also show similarities to a phylum of plasmodial protists, the Myxospora – Box 3.1 – that may well be parasitically degenerate animals in origin.) Only some 10 species are known, all being placed in the single order Orthonectida.

4.16 Further reading

Bird, A.F. 1971. *The Structure of Nematodes*. Academic Press, New York [Nematoda].

Boaden, P.J.S. 1985. Why is a gastrotrich? In: Conway Morris, S. *et al.* (Eds) *The Origins and Relationships of Lower Invertebrates*, pp. 248–260. Clarendon Press, Oxford [Gastrotricha].

Croll, N.A. 1976. *The Organisation of Nematodes*. Academic Press, London [Nematoda].

Croll, N.A. & Mathews, B.G. 1977. *Biology of Nematodes*. Blackie, London [Nematoda].

Dales, R.P. 1963. *Annelids*. Hutchinson, London [Annelida].

De Coninck, L. 1965. In: Grasse P.P. (Ed.), *Traite de Zoologie*, **4** (2), 3–217.

D'Hondt, J.-L. 1971. Gastrotricha. *Oceanogr. Mar. Biol., Ann. Rev.*, **9**, 141–192 [Gastrotricha].

Donner, J. 1966. *Rotifers* (transl. Wright, H.G.S.). Warne, London [Rotifera].

Edwards, C.A. & Lofty, J.R. 1972. *The Biology of Earthworms*. Chapman & Hall, London [Annelida].

Gibson, R. 1972. *Nemerteans*. Hutchinson, London [Nemertea].

Hyman, L.H. 1951. *The Invertebrates*, Vol. 2. *Platyhelminthes and Rhynchocoela*. McGraw-Hill, New York [Nemertea].

Hyman, L.H. 1951. *The Invertebrates*, Vol. 3. *Acanthocephala, Aschelminthes and Entoprocta*. McGraw-Hill, New York [Gastrotricha, Nematoda, Nematomorpha, Kinorhyncha, Priapula, Rotifera, Acanthocephala].

Ivanov, A.V. 1963. *Pogonophora*. Academic Press, London.

Kaestner, A. 1967. *Invertebrate Zoology*, Vol. 1. Wiley, New York [Nemertea, Gastrotricha, Nematoda, Nematomorpha, Kinorhyncha, Priapula, Rotifera, Acanthocephala, Sipuncula, Echiura, Annelida].

Kristensen, R.M. 1983. Loricifera, a new phylum with aschelminthes characters from the meiobenthos. *Z. Zool. Syst. Evolutionsforsch.*, **21**, 163–180 [Loricifera].

Mill, P. 1978. *Physiology of the Annelids*. Academic Press, London [Annelida].

Nørrevang, A. (Ed.) 1975. *The Phylogeny and Systematic Position of Pogonophora*. Parey, Hamburg [Pogonophora].

Rice, M.E. & Todorovic, M. 1975. *Proceedings of the International Symposium on the Biology of Sipuncula and Echiura*. Smithsonian Inst., Washington [Sipuncula & Echiura].

Riemann, F. 1988. In: Higgins, R.P. & Thiel, H. (Eds), *Introduction to the Study of Meiofauna*, pp. 293–301. Smithsonian Institution Press, Washington, DC.

Sterrer, W. 1972. Systematics and evolution within the Gnathostomulida. *Syst. Zool.*, **21**, 151–173 [Gnathostomula].

Sterrer, W., Mainitz, M. & Reiger, R.M. 1985. Gnathostomulida: enigmatic as ever. In: Conway Morris, S. *et al.* (Ed.). *The Origins and Relationships of Lower Invertebrates*, pp. 181–199. Clarendon Press, Oxford [Gnathostomula].

CHAPTER 5

The Molluscs

The molluscs are a distinctive and individual phylum that appears to be related to the eutrochozoan worms, especially to the sipunculans. They were portrayed in Chapter 2 as being effectively chunky flatworms with dorsal protective shields. Nevertheless, the majority of species have diverged very far from flatworm body form and life style. Indeed, the members of this highly successful phylum have radiated into as much morphological diversity as have the various assemblages of phyla described in the other chapters in this section, and as such deserve a chapter to themselves.

5.1 Phylum MOLLUSCA

5.1.1 Etymology

Latin: *molluscus*, a soft nut or soft fungus.

5.1.2 Basic and special features (Fig. 5.1)

1 Bilaterally symmetrical.
2 Body more than two cell layers thick, with tissues and organs.
3 A through gut.
4 Without a body cavity other than that provided by blood sinuses.

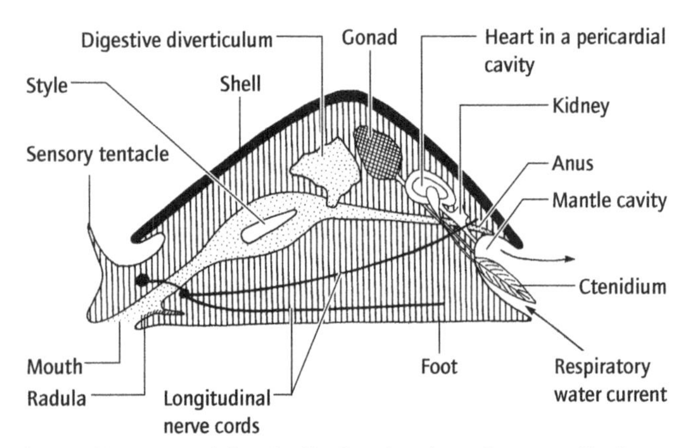

Fig. 5.1 Diagrammatic longitudinal section through a generalized basic mollusc.

5 Body monomeric and highly variable in form, but basically squat and often conical, frequently elongated in the dorsoventral plane to form a 'visceral hump'; essentially with an anterior head bearing eyes and sensory tentacles, a large flat ventral foot, and a posterior mantle cavity, but all of these subject to considerable modification.
6 A protective, external dorsal shell of protein (conchiolin) reinforced by calcareous spicules or from one to eight calcareous plates, secreted by the dorsal and lateral epidermis (the mantle); sometimes shell secondarily reduced, covered by tissue, or lost, and sometimes enlarged so as to cover whole body.
7 A toothed, chitinous, tongue-like ribbon, the radula, can be protracted from the buccal cavity through the mouth; rasped particles are borne in a mucous cord which is winched into the stomach by a style housed in a style sac (see Fig. 9.17 and Section 9.2.5).
8 Gaseous exchange effected by one or more pair/s of ctenidial gills housed in the mantle cavity (sometimes lost).
9 An open blood system and a heart enclosed within a mesodermal cavity, the pericardium, through which the intestine also passes.
10 A pair of sac-like 'kidneys', opening proximally into the pericardium and discharging into the mantle cavity.
11 Nervous system with a circum-oesophageal ring and two pairs of ganglionated longitudinal cords, sometimes highly concentrated.
12 Typically with a single pair of gonads, discharging into the mantle cavity, primitively via the pericardium and kidneys.
13 Eggs cleave spirally.
14 Development indirect, via trochophore and veliger larval stages (Fig. 5.2), or secondarily direct.

With the possible exception of the Nematoda (of which thousands of species probably await discovery and description), the Mollusca, with almost 100 000 species, is the second largest animal phylum. Their success is probably not so much attributable to any particular special anatomical or ecological features of the group as to the extreme plasticity and adaptability of the basic molluscan body plan, the list of features given above having been extensively modified in a wide variety of fashions by

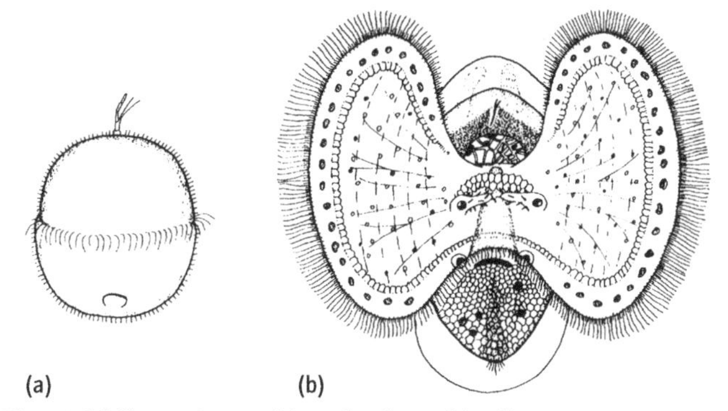

Fig. 5.2 Molluscan larvae: (a) trochophore; (b) veliger. (After Hyman, 1967.)

the different component classes. This plasticity can be illustrated, on the one hand, by the variation displayed in the function of any single molluscan structure (e.g. the shell, besides being protective, may form a buoyancy device, a burrowing organ, or an endoskeletal plate), and, on the other, by the multiplicity of structures which have been adapted to serve the one function (food-catching organs include ciliated tentacles or palps, greatly enlarged gills, sucker-bearing arms, radular teeth, etc.).

In body shape, molluscs range from cylindrical, burrow-dwelling 'worms' lacking both a foot and a shell, and greatly elongated along their anteroposterior axes, to almost spherical, effectively headless clams encased within large bivalved shells; whilst in size they extend from 2 mm long planktonic and interstitial species up to the giant squids which measure, including their arms, more than 20 m in length, in their case as a result of elongation in the ancestral dorsoventral plane. Molluscs occupy all major habitats and include representatives of all known feeding types; they include the most and the least mobile of all free-living invertebrates; and within the one phylum are placed animals with some of the least developed brains and sense organs, together with the most intelligent of invertebrates. ('If the Creator had indeed lavished his best design on the creature he shaped in his own image, creationists would surely have to conclude that God is really a squid' (Diamond, 1985).)

As discussed in Chapter 2, the basic molluscan morphology is really that of a flatworm (Section 3.6) with two distinctive additional features (and a few others consequent on these two): the radula and the dorsal shell. The radula (Fig. 5.3) is housed in a posteroventral diverticulum of the buccal cavity and it comprises a cartilaginous skeletal element, the odontophore, on which is mounted a movable linear radular ribbon bearing

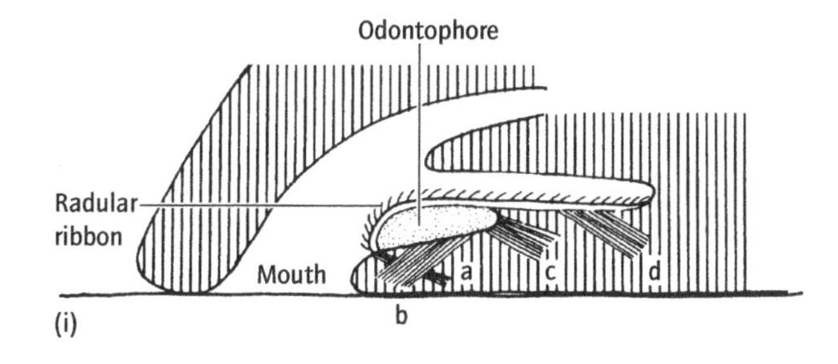

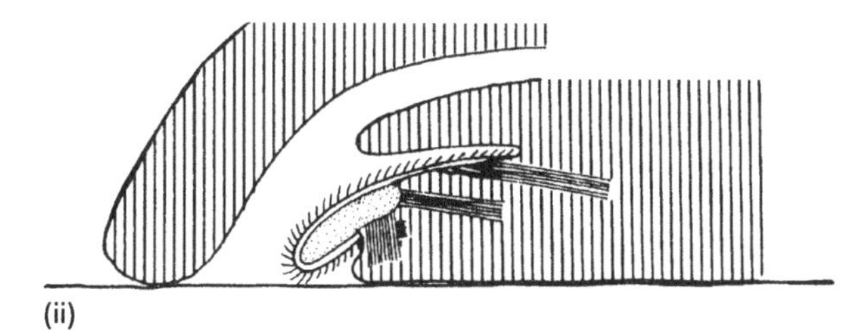

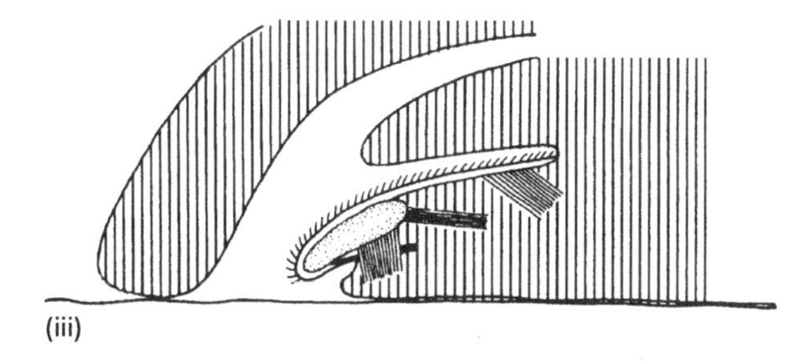

Fig. 5.3 Radular rasping in molluscs (after Russell-Hunter, 1979 and other sources).

transverse rows of backwardly pointing teeth in a longitudinal series. Protractor muscles can cause both the odontophore to be everted through the mouth and a degree of movement of the ribbon over its basal support. This action also causes the individual teeth to be erected. These teeth then rasp any surface to which they are applied, and detached particles are conveyed into the buccal cavity as the retractor muscles bring the odontophore and ribbon back through the mouth. Notwithstanding that they may be mineralized by incorporation of SiO_2 or Fe_3O_4, the chitinous teeth are subject to considerable wear, and therefore the whole ribbon is moved slowly forward over the odontophore, bringing unused tooth rows into play, whilst replacement ribbon is secreted on to the proximal end. The radula is thus essentially an organ of browsing, grazing or boring, although it has been modified in some predatory species for prey capture by enlarging the individual teeth (and correspondingly reducing the number in each row) and, in some, connecting them to poison-secreting glands.

The dorsal protective covering of the ancestral dorsoventrally flattened mollusc was probably a proteinaceous and chitinous cuticle reinforced by calcareous ossicles or scales and secreted by the epidermis. In most descendant forms, however, the calcium carbonate content has been greatly increased and deposited as a large, individual plate or plates covered by a thin conchiolin periostracum. Typically, the calcite or aragonite is laid down in distinct layers, each in an organic matrix: an outer prismatic layer and a variable number of inner nacreous layers. Once a distinct shell had been evolved, the major axis of molluscan growth appears no longer to have been along the anteroposterior axis, as it is in most vermiform animals, but dorsoventral, so that the shell and the animal within it assumed – and in most surviving groups still assumes – a generally conical shape, sometimes elongately so and then coiled into a planar or helical spiral. The mantle which secretes the shell usually extends as a fold over the unshelled regions of the body, forming a covering 'skirt'; so therefore may the shell, and retractor muscles can pull the shell down over the unprotected head and locomotory ventral foot for safety. At the same time, the foot can cling tightly to the substratum by generating suction.

The presence of a dorsal covering has immediate consequences in respect of gaseous exchange across the body surface, and concomitantly with the development of the shell must have proceeded the elaboration of a specialized unshelled region of the body surface into a gill, together with the evolution of a transport system for circulating the respiratory gases through the body (if such was not already present). The overlapping of the molluscan body by the mantle and shell is most marked laterally and/or posteriorly, where a mantle cavity is enclosed (Fig. 5.4). In this cavity are located one or more pairs of distinctive ctenidial gills; each gill is composed of a central flattened longitudinal axis, from each side of which issue flat triangular filaments, each filament supported by a small chitinous rod along the frontal surface. Cilia on the gills drive water between the individual filaments, through which blood diffuses in a counter-current flow (see Fig. 11.9). Water normally enters at the sides of the

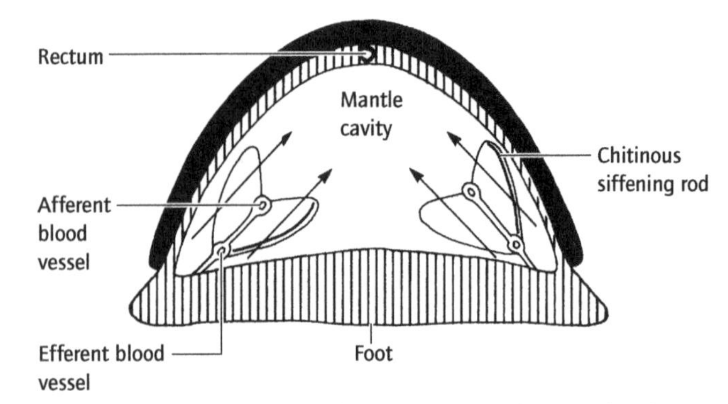

Fig. 5.4 Cross-section through the mantle cavity of a generalized basic mollusc showing the bipectinate ctenidial gills and the direction of the inhalent water currents resulting from ciliary action.

mantle cavity and flows out dorsally along the mid-line, the anus and kidneys discharging into the outgoing current.

As stressed above, however, the nature and development of the radula, shell and ctenidia, together with most other anatomical features of the molluscs, vary considerably in the different component classes, and hence further characteristics of the phylum will be treated separately in the following sections. Because molluscs, more than any other invertebrate group, have been the favoured subject of the attention of neurobiologists, the reader should also consult Chapter 16.

5.1.3 Classification

The phylum Mollusca consists of eight classes, of which two – the gastropods and bivalves – contain between them over 98% of the known living species.

5.1.3.1 Class Chaetodermomorpha

The chaetodermomorphs are a peculiar group of shell-less and vermiform molluscs, much elongated along the anteroposterior axis, which construct and inhabit vertical burrows in soft marine sediments. The cylindrical body, from 2 mm to 14 cm in length, is positioned head downwards in the burrow, the terminal mouth ingesting sediment. The mantle cavity, containing one pair of bipectinate ctenidia, is posteroterminal (Fig. 5.5) and therefore positioned in the mouth of the burrow.

A foot is lacking, the mantle covers the entire body; movement is therefore atypical and is effected by peristaltic contractions of the well-developed body wall musculature. Instead of a shell, the epidermis secretes a chitinous cuticle in which are embedded imbricating scales which all point posteriorly; these also cover the whole body surface. Several characteristically molluscan organ systems are absent: the very poorly marked head is without eyes or sensory tentacles; there are no excretory organs or gonoducts (the gonads discharging via the pericardial cavity); and some species lack a radula. The head does possess a terminal cuticular plate associated with various glands (Fig. 5.6), but as yet its function is uncertain.

Class	Subclass	Superorder	Order
Chaetodermomorpha			Caudofoveata
Neomeniomorpha			Aplotegmentaria
			Pachytegmentaria
Monoplacophora			Tryblidiida
Polyplacophora			Lepidopleurida
			Ischnochitonida
			Acanthochitonida
Gastropoda*	Prosobranchia		Docoglossida
			Pleurotomariida
			Anisobranchida
			Cocculiniformia
			Neritida
			Architaenioglossa
			Ectobranchida
			Neotaenioglossa
			Heteroglossa
			Stenoglossa
	Heterobranchia	Pulmonata	Archaeopulmonata
			Basommatophora
			Stylommatophora
		Gymnomorpha	Onchidiida
			Soleolifera
			Rhodopida
		Opisthobranchia	Cephalaspida
			Anaspida
			Saccoglossa
			Nudibranchia
			Pleurobranchomorpha
			Umbraculomorpha
		Allogastropoda	Pyramidellomorpha
Bivalvia	Protobranchia	Ctenidiobranchia	Nuculida
		Palaeobranchia	Solemyida
	Lamellibranchia	Pteriomorpha	Arcida
			Mytilida
			Pteriida
		Palaeoheterodonta	Trigoniida
			Unioniida
		Heterodonta	Venerida
			Myida
		Anomalodesmata	Pholadomyida
			Poromyida
Scaphopoda			Dentalida
			Siphonodentalida
Cephalopoda	Nautiloidea		Nautilida
	Coleoidea		Sepiida
			Teuthida
			Octopoda
			Vampyromorpha

* The classification of the Gastropoda is still in a state of flux as the old pre-1930 system, based upon different grades of organization, is slowly giving way to more phylogenetically based approaches. That given above should perhaps be regarded as illustrative of modern classifications, in that no individual scheme has yet had time to become generally accepted. For an alternative system, see Haszprunar (1988).

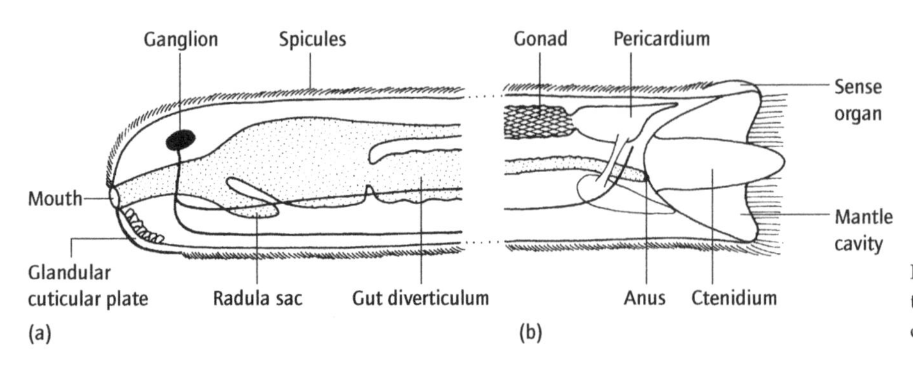

Fig. 5.5 Diagrammatic longitudinal section through the anterior (a) and posterior (b) ends of a chaetodermomorph mollusc. (After Boss, 1982.)

The 70 species of these gonochoristic, deposit-feeding molluscs are grouped in a single order.

5.1.3.2 Class Neomeniomorpha

Members of this class are superficially similar to the chaetodermomorphs in being shell-less, effectively headless, and vermiformly elongated along their anteroposterior axis (Fig. 5.7), and, like them, they lack excretory organs, gonoducts, and, in some, the radula. They differ, however, in numerous other respects, and are generally regarded as being more closely related to the shell-bearing mollusc groups. The 1 mm to 30 cm long body is laterally compressed and possesses a longitudinal ventral groove in which are located one or more small ridges

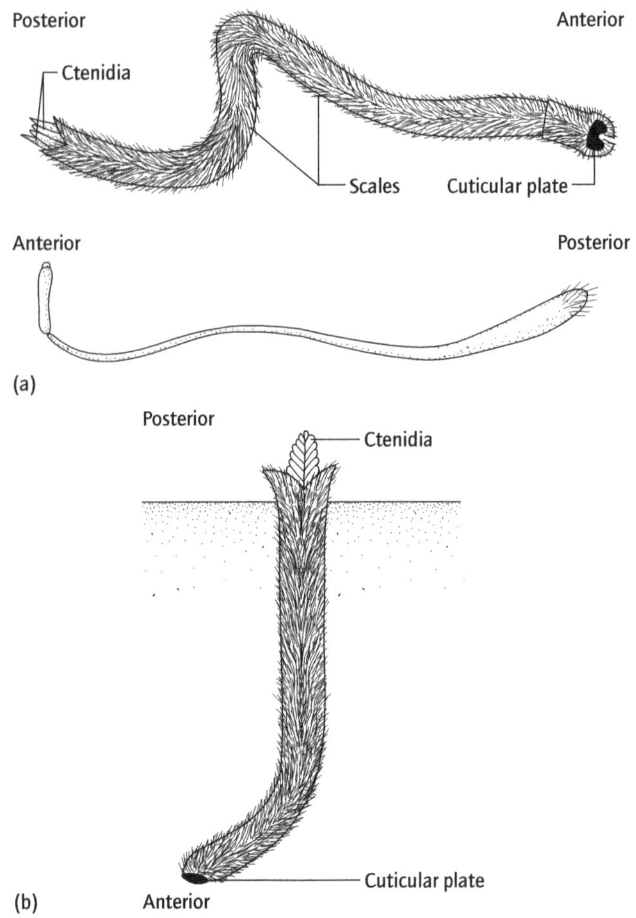

Fig. 5.6 Body form of chaetodermomorphs: (a) external appearance; (b) life style in marine sediment (after Hyman, 1967 and Jones & Baxter, 1987).

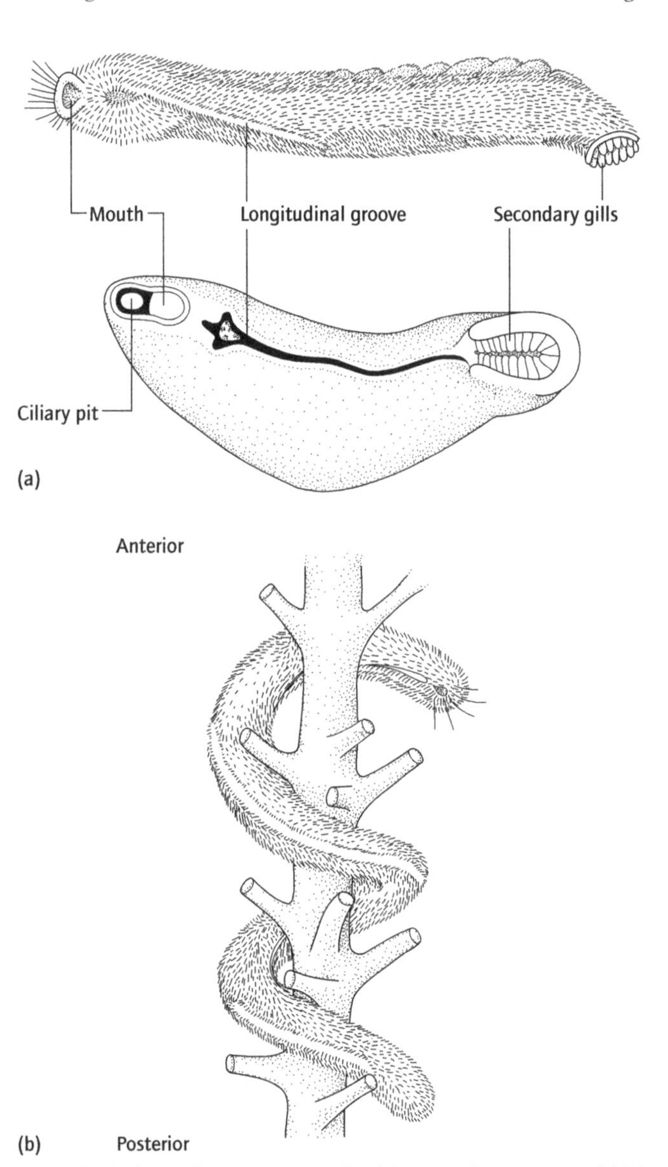

Fig. 5.7 Body form of neomeniomorphs: (a) external appearance; (b) life style, crawling over a colonial hydroid. (After Jones & Baxter, 1987.)

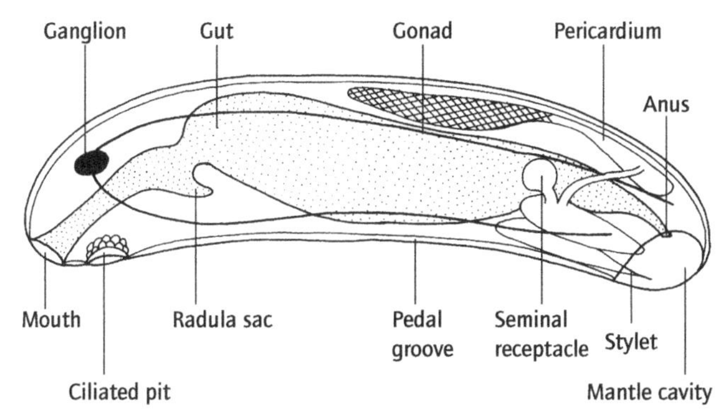

Fig. 5.8 Diagrammatic longitudinal section through a neomeniomorph (after Boss, 1982).

thought to represent a highly reduced foot. The mantle covers the body, apart from this groove, and has embedded in it one or more layers of separate calcareous scales or spicules beneath the cuticle. At the anterior end of the groove are located a ciliary pit and the anteroventral mouth; and at the posterior is the posteroventral mantle cavity which lacks ctenidia, although secondary gills in the form of folds or papillae are often developed (Fig. 5.8).

Neomeniomorphs are carnivores, feeding on the cnidarians on which they are usually to be found; their method of ingestion, however, is somewhat uncertain because of the frequent reduction or absence of the radula – in some species the fore-gut is protruded to engulf the food. Although the body wall musculature is well developed, movement is effected not by muscular means but by gliding, using the cilia on the eversible ridges within the longitudinal ventral groove. All species are hermaphrodite; copulation occurs with the assistance of stylets, in several species the sperm being stored by the recipient in seminal receptacles.

The 180 known species are all marine and are divided between two orders dependent on the number of layers of calcareous bodies in the mantle and on the presence or absence of epidermal papillae.

5.1.3.3 Class Monoplacophora

The monoplacophorans were thought to have become extinct in the Devonian, until, in 1952, living specimens were obtained from an oceanic trench in the Pacific Ocean. Further material from other localities has since been discovered, bringing the number of known surviving species to eight, all referable to a single order.

All are small (3 mm–3 cm), marine, gonochoristic, deposit feeders with a body covered dorsally by a single conical or cap-shaped shell, beneath which is a weakly muscular, circular foot surrounded laterally and posteriorly by an extensive mantle cavity (Fig. 5.9). Within the mantle cavity, around the sides of the foot, are located five or six pairs of monopectinate ctenidia. Other organs also occur in multiple pairs: there are six pairs of lobular kidneys, discharging separately into the lateral regions of the mantle cavity; two pairs of gonads; and eight pairs of foot-retractor muscles (Fig. 5.10). The head is distinct, but poorly developed and without sensory tentacles (except around the mouth) and eyes; the radula, however, is well developed. The anus opens into the posterior section of the mantle cavity.

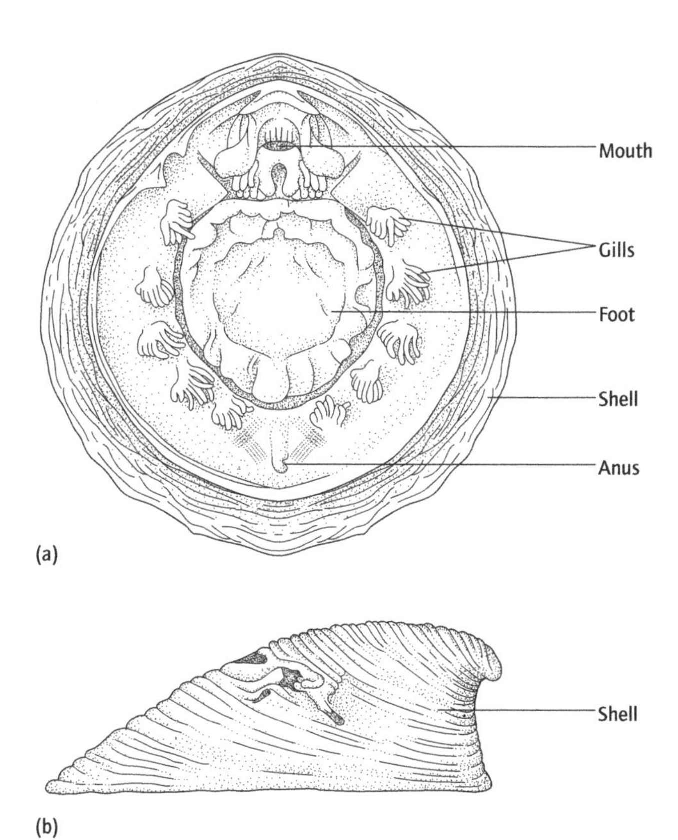

Fig. 5.9 A monoplacophoran mollusc in (a) ventral and (b) lateral view. (After Lemche & Wingstrand, 1959.)

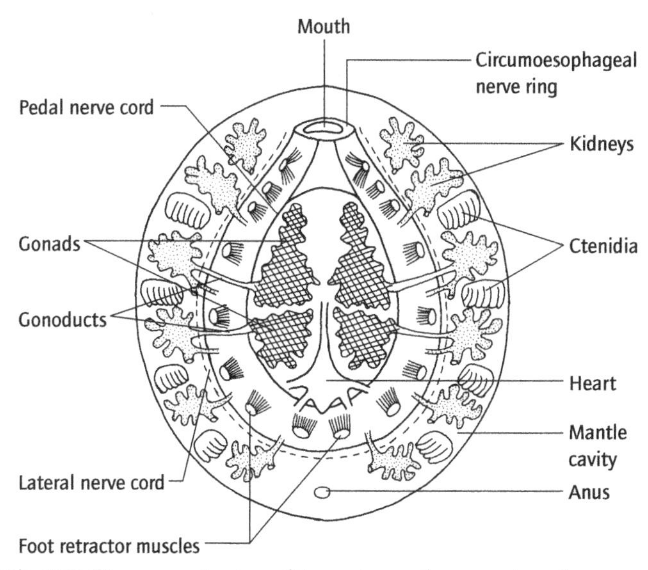

Fig. 5.10 Diagrammatic view of the anatomy of a monoplacophoran, seen from the ventral side with the foot and body wall removed, showing the serially repeated internal organs and ctenidia (after Lemche & Wingstrand, 1959).

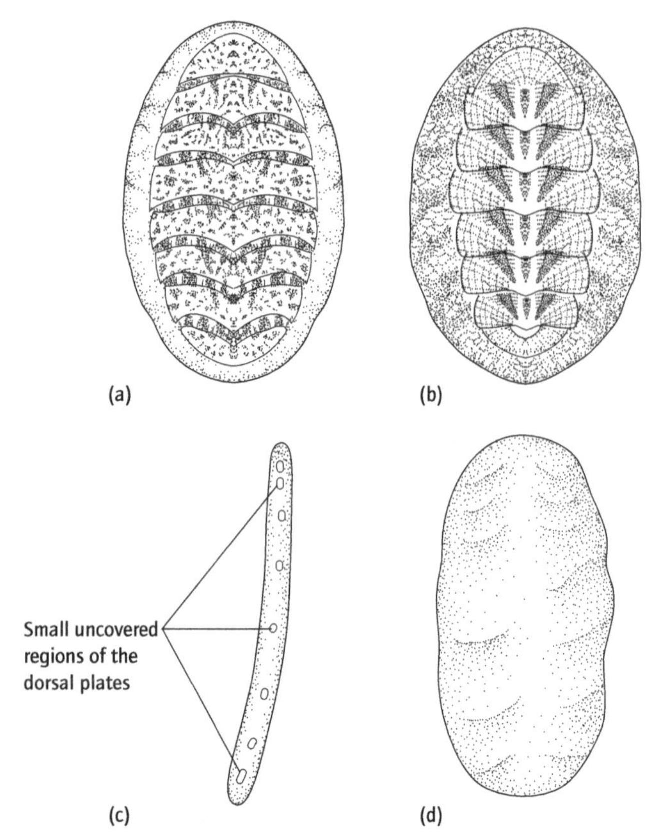

Fig. 5.11 Dorsal views of four polyplacophoran molluscs (a, b – Ischnochitonida. c,d – Acanthochitonida) showing differing degrees of cover of the shell plates by the girdle. (After sources in Hyman, 1967.)

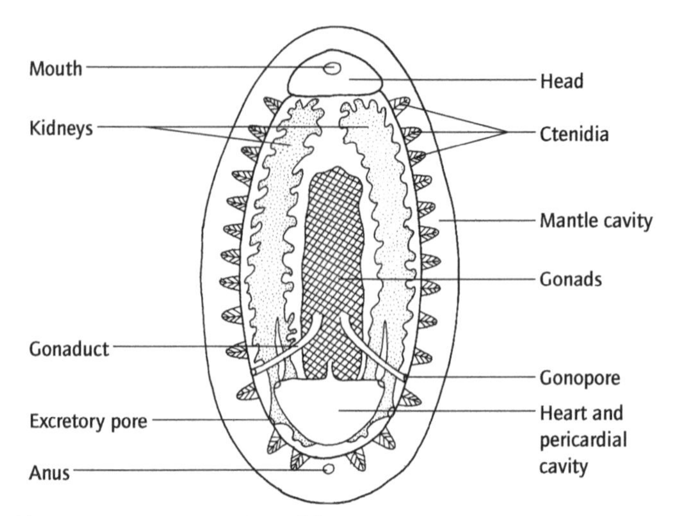

Fig. 5.12 Diagrammatic view of the anatomy of a generalized polyplacophoran, seen from the ventral side with the foot, body wall and gut removed (after Hescheler, 1900).

5.1.3.4 Class Polyplacophora (chitons)

The oval to somewhat elongate, dorsoventrally flattened polyplacophorans are distinguished by their possession of a linear chain of eight, serially overlapping, dorsal shell plates, and by the development of the surrounding mantle into a thick 'girdle', the cuticle of which often bears spines, scales or bristles. In several species this girdle partially or even completely covers the shell plates (Fig. 5.11). The division of the dorsal protective covering into a number of separate elements enables chitons to roll up into a ball in the same fashion as several woodlice and millipedes.

The head is poorly developed and hidden beneath the anterior girdle; eyes and sensory tentacles are lacking, but the radula is large and bears many teeth in each transverse row. Most of the ventral surface is occupied by a large, muscular, elongate foot with which chitons slowly crawl over hard substrata. The foot and girdle can also be used to generate considerable suction, permitting the animals to adhere tightly to surfaces. The narrow mantle cavity surrounds the foot on all sides except immediately anteriorly, and contains from six to 88 pairs of bipectinate ctenidia; posteriorly it receives the anal opening and the discharges of the single pair of kidneys. These are large and, together with the single (fused) gonad (which has paired gonoducts and gonopores), often extend the whole length of the body (Fig. 5.12). The nervous system is relatively simple; the cords, for example, are not ganglionated.

The 550 species of polyplacophorans are all marine and gonochoristic; most are grazers of algae. They range in length from 3 mm up to 40 cm. Three orders are differentiated on the location within the mantle cavity of the gills, on the presence or absence of attachment teeth on the shell plates, and on the extent to which the plates are covered by the girdle.

5.1.3.5 Class Gastropoda

The gastropods are a very large and diverse group sharing the common feature that during development the visceral hump is rotated through some 180° in an anticlockwise direction relative to the head and foot, so that the mantle cavity occupies a forwardly facing position and the anus and kidneys therefore discharge anteriorly: gastropods undergo 'torsion' (Fig. 5.13). This is brought about by the asymmetrical development of the two pedal retractor muscles and/or by differential growth (see Section 15.4.1), and may have had as its selective advantage the ability of the well-developed gastropod head to be accommodated in the mantle cavity on retraction of the animal into its shell, and of the chemoreceptory sense organs in the mantle cavity, the osphradia, to sense the water ahead of, rather than behind, the moving animal. An automatic consequence of this torsion, however, would have been that the exhalent water current from the mantle cavity, bearing the excretory and alimentary discharges, flowed out immediately over the head. Such fouling of the cephalic sense organs has been circumvented by modifying the path of the exhalent stream; for example, by evolving a slit or aperture(s) in the dorsal shell through which the exhalent current can leave, or by the loss of the ctenidium on the right-hand side of the body and the production of a unidirectional through current, passing into the mantle cavity to the left of the head and out to the right (Fig. 5.14). In association with the

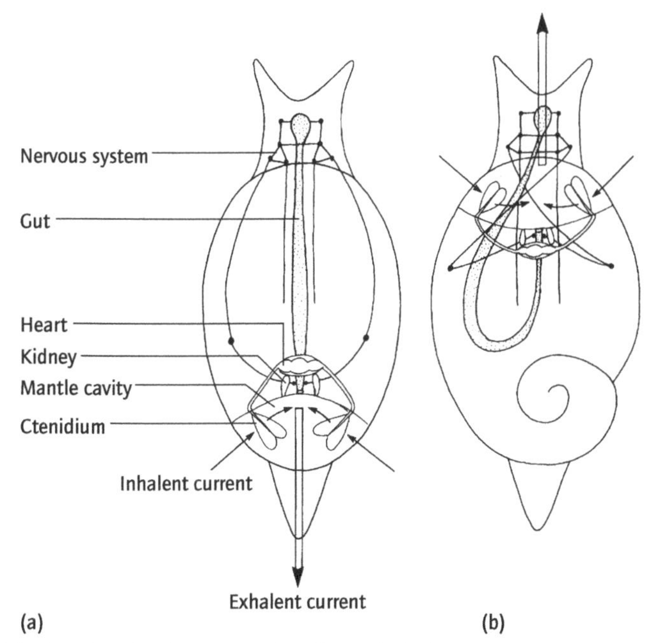

Fig. 5.13 Torsion in gastropod molluscs: (a) before torsion; (b) after torsion.

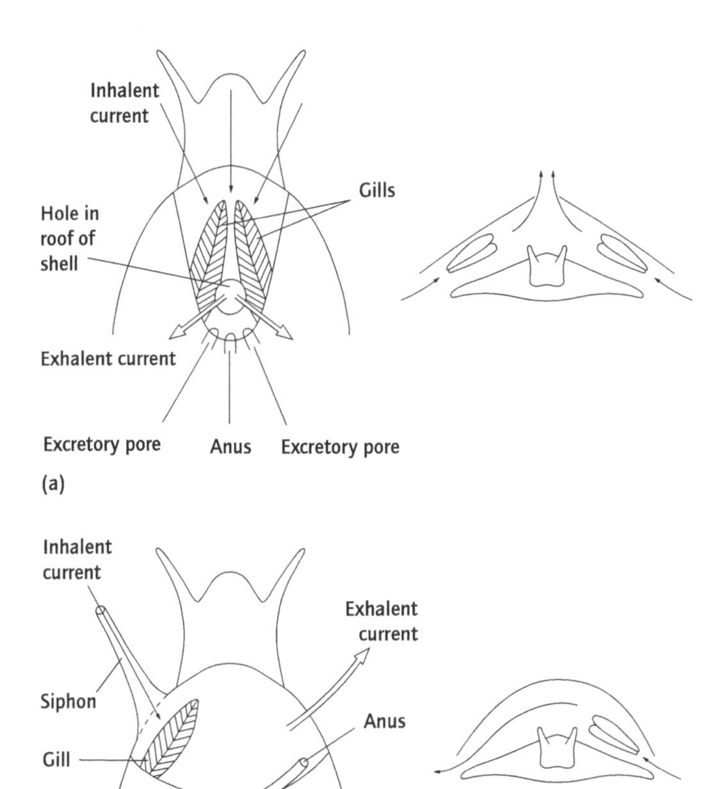

Fig. 5.14 Two modifications of the path of the respiratory water current in gastropods, necessitated by torsion: (a) the exhalent current leaving via a dorsal hole in the shell; (b) loss of the right ctenidium and production of a cross-directed current, the anus being displaced in the direction of the current.

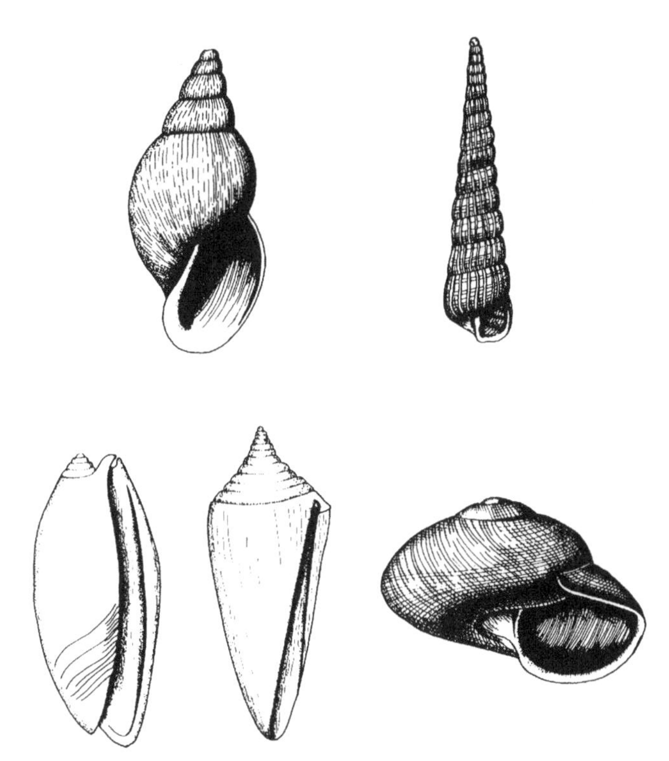

Fig. 5.15 Spiral gastropod shells (after sources in Hyman, 1967).

latter system, the (inhalent) left-hand side of the mantle cavity is often drawn out into a manoeuvrable siphon. Many phylogenetic lines of gastropods, however, have secondarily undergone detorsion, the visceral hump being detorted through some 90° so that the mantle cavity (if retained) lies on the right-hand side of the body, and the anus usually once again becomes posterior.

The basic body plan of the gastropods – from which many species have departed radically – is that of a squat mollusc, with a well-developed crawling foot and a well-defined head bearing a radula, a pair of jaws, a pair of eyes, and one or more pairs of sensory tentacles, both head and foot being completely retractable into a single, thick, helico-spirally coiled shell (Fig. 5.15), of which the aperture can be plugged, after retraction, by a calcareous or organic operculum borne on the posterodorsal part of the foot. Primitively, at least, the mantle cavity contains a single pair of bipectinate ctenidia, and these gonochoristic molluscs develop via trochophore and veliger larval stages.

The 55 000 species of the mainly marine subclass Prosobranchia have largely retained this basic body form, including the shell, operculum, and torsional state, although the ctenidia display a trend towards reduction from the ancestral bipectinate pair, through a single (left) bipectinate ctenidium, to a single (left) monopectinate gill. Correlated with this, the right kidney, right auricle of the heart, and right osphradium are also lost with the right ctenidium. Most prosobranchs remain gonochoristic, although a few are sequential hermaphrodites. The large majority are benthic snails which graze algae (limpets, winkles, topshells, etc.) and sessile animal colonies (e.g. cowries) or deposit feed (e.g. mudsnails); one group (order Stenoglossa) includes

Fig. 5.16 Gastropod body forms: (a), and (b) (*facing page*), representatives of the different groups of prosobranchs (a) and heterobranchs (b). (Note that most approximate to snail, limpet or slug form). (c) (*facing page*) Some atypical body forms associated with unusual gastropod life styles. (After many sources.)

many predatory forms, which characteristically have a radula with only a few, large, fang-like teeth in each radular row; and a few species suspension feed using external mucous nets (see Fig. 9.5) or enlarged ctenidia (e.g. slipper limpets) paralleling the bivalves in this respect. Amongst prosobranchs and indeed all gastropods, the limpets (order Docoglossida) form a very isolated group with several peculiarities of the shell, radula and gut.

From a prosobranch origin evolved the second gastropod subclass, the Heterobranchia: several lines of gastropods which display marked trends towards (a) reduction, internalization or loss of the shell, (b) loss of the operculum, (c) detorsion, (d) loss of the ancestral ctenidium and its replacement by secondary gaseous-exchange surfaces, and (e) simultaneous hermaphroditism. Two major superorders of heterobranchs are the marine opisthobranchs (with some 1000 species) and the largely terrestrial and freshwater pulmonates (20 000 species). The opisthobranchs (the sea-slugs, sea-hares, pteropods, etc.) tend to replace the ctenidium by secondary gills or by gaseous exchange across the often papillate body surface. Most of these

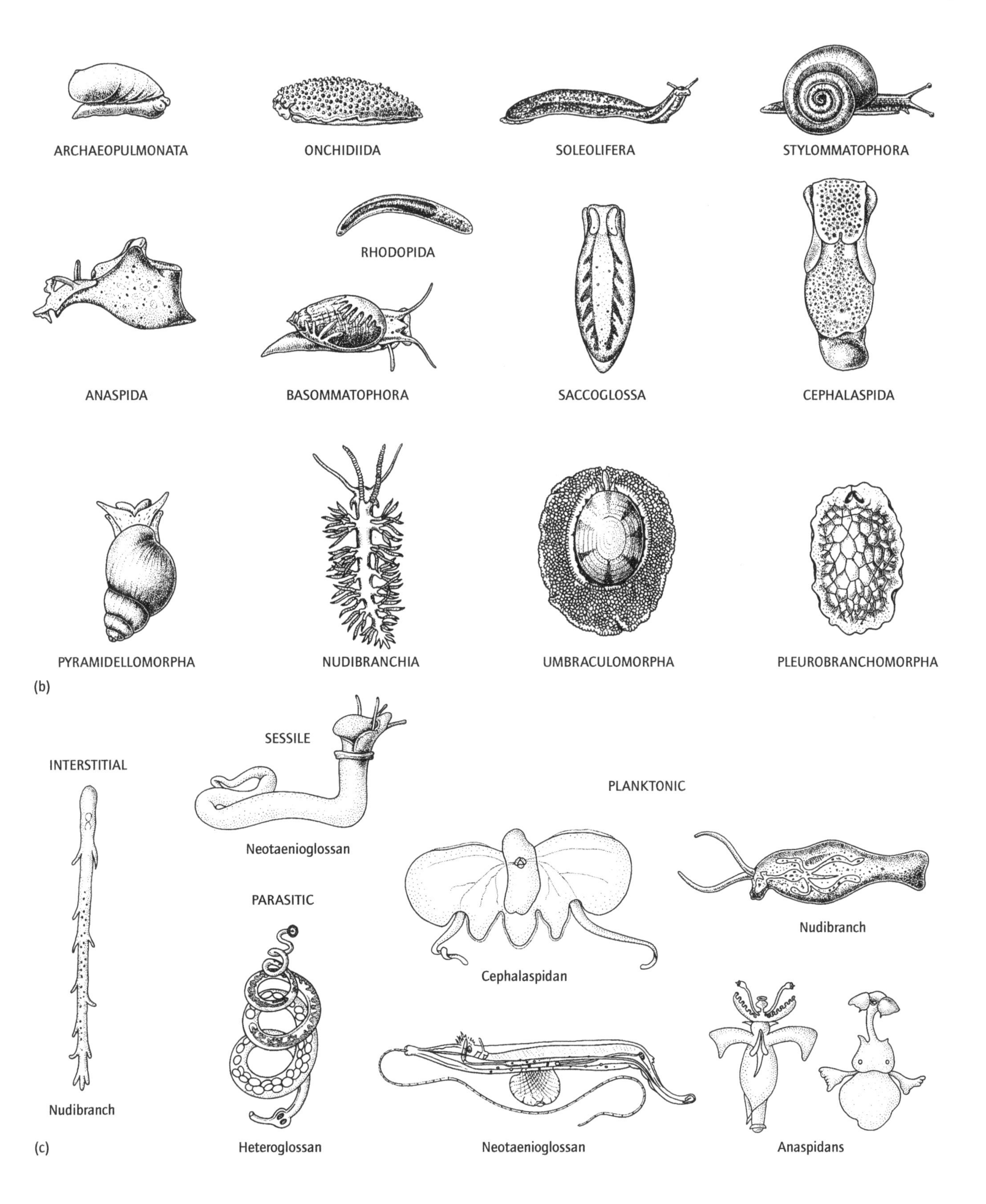
ARCHAEOPULMONATA
ONCHIDIIDA
SOLEOLIFERA
STYLOMMATOPHORA
RHODOPIDA
ANASPIDA
BASOMMATOPHORA
SACCOGLOSSA
CEPHALASPIDA
PYRAMIDELLOMORPHA
NUDIBRANCHIA
UMBRACULOMORPHA
PLEUROBRANCHOMORPHA
(b)
SESSILE
INTERSTITIAL
PLANKTONIC
Neotaenioglossan
PARASITIC
Nudibranch
Cephalaspidan
Nudibranch
(c)
Heteroglossan
Neotaenioglossan
Anaspidans

species are carnivores, including via ecto- and endoparasitism and planktonically as well as on or in the bottom, although several consume algae (often by suctorial feeding) and one planktonic group suspension feeds using secreted mucous nets (Fig. 9.5) and cilia on their broad, wing-like feet. Up to four pairs of cephalic tentacles may be present.

The pulmonates are principally characterized by conversion of the mantle cavity into an airbreathing lung with a contractile opening, the pneumostome. The majority are grazers of land plants, and all have abandoned larval stages, directly developing into young slugs or snails. Many pulmonates have retained the spiral gastropod shell, although it is usually thin, but several members of the terrestrial order Stylommatophora have lost it and thereby form the land-slugs. A third superorder of heterobranchs, the Gymnomorpha, are also slugs, but these three probably unrelated groups of marine or terrestrial gastropods (totalling 200 species) each show a different mixture of opisthobranch and pulmonate features; in much the same way the final superorder, the allogastropod snails (with 500 species), show an amalgam of prosobranch and opisthobranch characteristics. Clearly, the Heterobranchia are gastropods typified mainly by the parallel evolution of many 'progressive' features and it is difficult to disentangle their various interrelationships.

In total, therefore, the Gastropoda includes some 77 000 species of slug- or snail-like molluscs (Fig. 5.16, pp. 126–7) reaching up to 60 cm in height. They can be distributed between 23 orders.

5.1.3.6 Class Bivalvia

The bivalves are essentially laterally compressed molluscs completely enclosed within a pair of shell valves. Being so covered, they are relatively sedentary or even sessile animals, several being cemented or otherwise attached to the substratum, and, as in other animals 'hidden' within a thick, protective casing, the head is greatly reduced (Fig. 5.17). It is without radula, eyes and tentacles, although the missing cephalic sense organs may effectively be replaced by tentacles and, in a few, eyes positioned around the margins of the mantle. The two shell valves, which are lateral and articulate along the dorsal mid-line, open passively by virtue of an elastic dorsal ligament; they must, therefore, be kept closed actively. This is achieved by two adductor muscles (in some, reduced to one) which possess a catch-fibre mechanism so that they can remain contracted without the need for continual expenditure of energy and for rotational relaxation and contraction of the individual fibres.

The body, which in the largest species may exceed 1 m in length, is located dorsally within the shell, the large mantle cavity occupying the lateral and ventral remainder. When the adductor muscles relax, the shell valves gape slightly, permitting a water current to be drawn into and through the cavity, and in forms which burrow into soft sediments allowing the laterally compressed foot to be protruded outside the confines of the shell. Frequently, the ventral and lateral margins of the mantle fuse together, leaving only a gap for the foot and small apertures through which water can flow – in these regions the mantle is

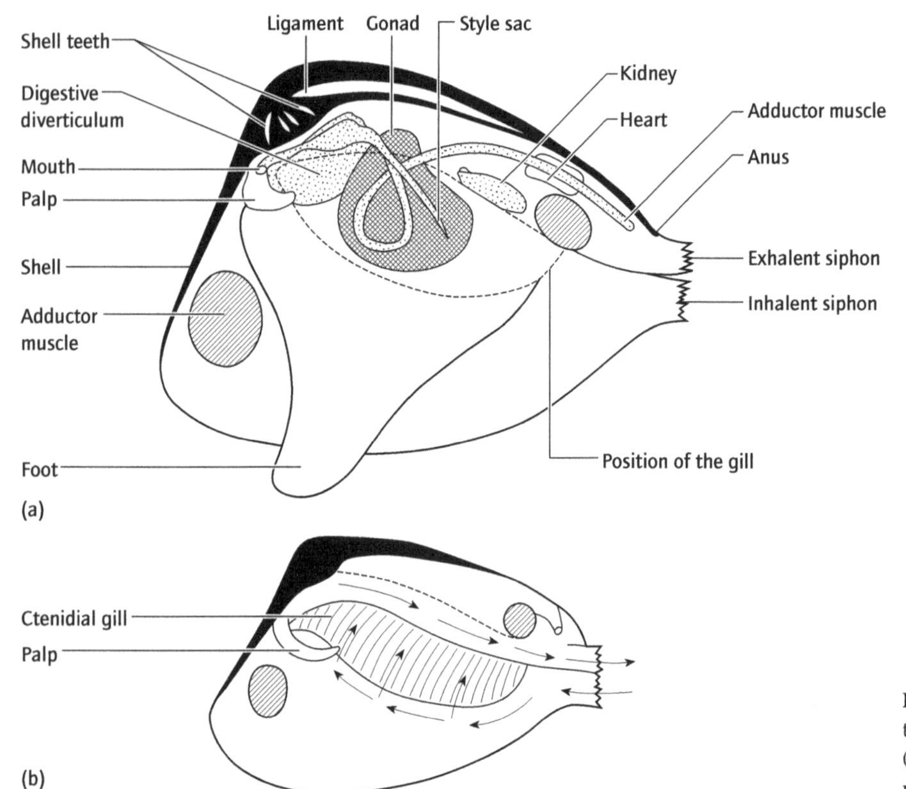

Fig. 5.17 Diagrammatic longitudinal section through a bivalve mollusc: (a) with gill omitted (for clarity); (b) with the gill *in situ*, showing the path of the feeding current. (After several sources.)

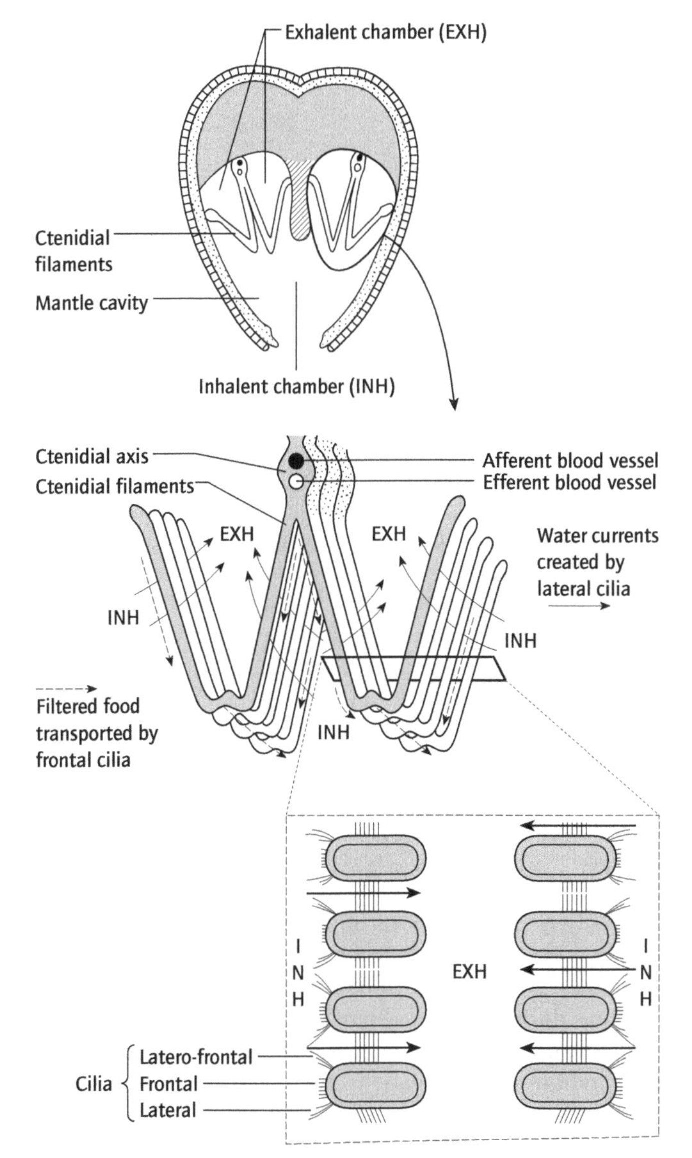

Fig. 5.18 Transverse section through a lamellibranch bivalve, illustrating filter-feeding by means of the ctenidial gills (after Russell-Hunter, 1979).

often drawn out into inhalent and exhalent siphons which may or may not be retractable into the shell.

Three broad categories of bivalves can be distinguished on the basis of the nature and function of the ctenidia. In the primitive forms (the subclass Protobranchia), the paired ctenidia serve mainly the ancestral gaseous-exchange function, and feeding, when not entirely dependent on symbiotic chemoautotrophs, is accomplished by means of labial palps, on either side of the mouth, which in one order (Nuculida) bear long tentacles which roam into or over the surrounding sediment and convey food particles to the palps for sorting before ingestion (see Fig. 9.6). These species are mucociliary deposit feeders. In the large majority of species (the superorders Pteriomorpha, Palaeoheterodonta and Heterodonta), however, the paired ctenidia are

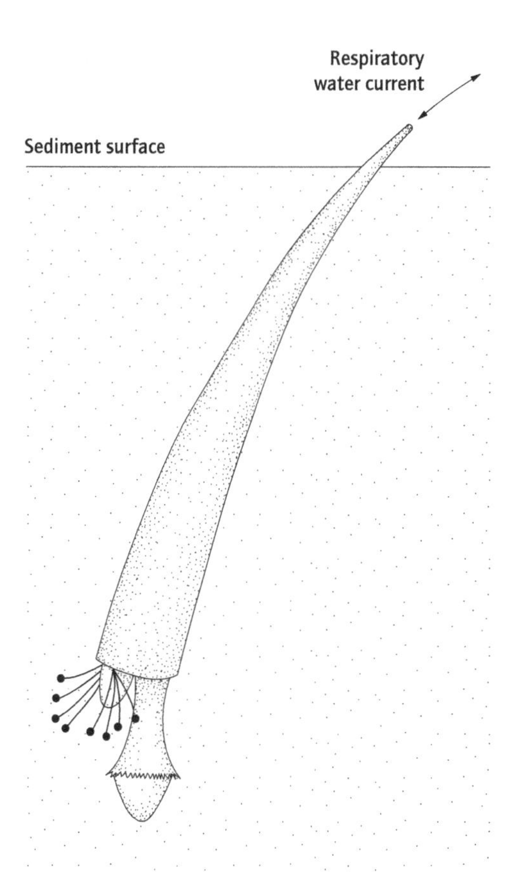

Fig. 5.19 Life style of scaphopod molluscs.

greatly enlarged and folded each into a W-shape in cross-section to form the organs of filter feeding (Fig. 5.18). Surface deposits are sucked into suspension, or material already in suspension is drawn into the mantle cavity with the inhalent current, particles are filtered as the current passes through the gills and then passed to the mouth via the palps, all by (muco)ciliary means. Finally, in the 'septibranch' members of the third group (the poromyidan Anomalodesmata), the gills have been greatly reduced, forming muscular pumping septa by means of which, together with an enlarged and raptorial inhalent siphon, small animals are sucked into the mantle cavity, there to be captured by the muscular palps and ingested.

The 20 000 bivalve species are common benthic animals in the sea and in fresh waters. Most are gonochoristic and undergo indirect development, although the larval stages of some freshwater species are aberrant in being adapted to parasitize fish. Eleven orders may be recognized.

5.1.3.7 Class Scaphopoda (tusk-shells)

The 2–150 mm long scaphopods are elongate, cylindrical molluscs almost completely enclosed by the mantle, which secretes a single, tubular, calcareous shell open at each end. They burrow in soft marine sediments, living with the somewhat narrower end of the shell-tube projecting slightly above the sediment surface (Fig. 5. 19). From the large ventral aperture of the shell

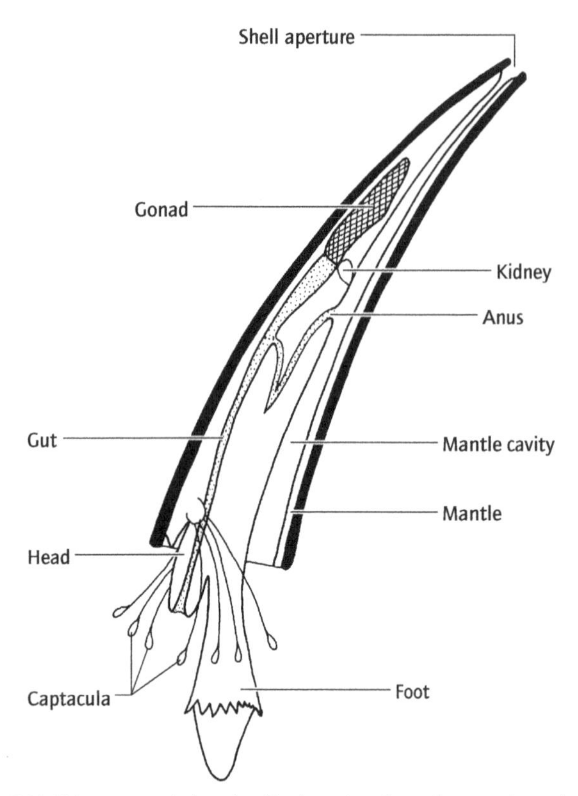

Fig. 5.20 Diagrammatic longitudinal section through a scaphopod (after various sources).

project the conical or cylindrical, burrowing foot and the small proboscis-like head, which lacks eyes and sensory tentacles, but possesses a radula, a single median jaw, and paired clusters of narrow, clubbed, contractile filaments, the captacula. The numerous captacula are the organs of deposit feeding, smaller food particles being conveyed back to the mouth along the filaments by cilia, the larger ones adhering to the sticky clubbed tips and being brought directly to the mouth (see Fig. 9.5).

The plane of elongation of the body is difficult to determine. If the anus and mantle cavity are regarded as being posterior, then, like the gastropods and cephalopods, scaphopods are greatly elongated in the dorsoventral plane (Fig. 5.20); alternatively, the mantle cavity and anus are often considered to be ventral, in which case the scaphopods are lengthened along the anteroposterior axis. The mantle cavity extends the posterior height (or ventral length) of the shell and lacks ctenidia. Water is drawn into it through the dorsal (or posterior) aperture by mantle cilia, and is periodically pumped out through the same aperture by muscular contraction, gaseous exchange being effected across the mantle. All species are gonochoristic and possess a single gonad that discharges via the right kidney.

The 350 species are placed in two orders, dependent on the number and shape of the captacula and on the shape of the foot.

5.1.3.8 Class Cephalopoda

The wholly marine cephalopods are, anatomically and behaviourally, the most sophisticated of the molluscs and arguably of

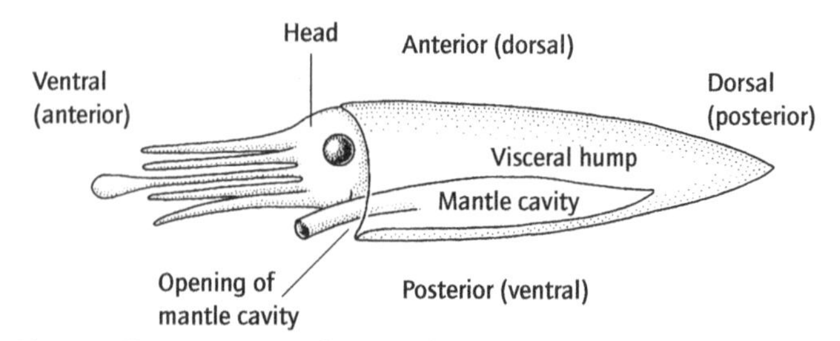

Fig. 5.21 Diagrammatic side view of a cephalopod mollusc showing the changed orientation of the body from that in the ancestral mollusc: the orientation of the living cephalopod is shown in parentheses by the ancestral states.

all invertebrates. They also include the largest species, giant squid achieving total lengths in excess of 20 m. The diagnostic features of cephalopods are the structures produced by what in other molluscs would be the foot, which as such is lacking. The anterior region of the embryonic foot develops into a series of prehensile arms or tentacles around the mouth, and the posterior portion forms a muscular funnel around part of the opening of the mantle cavity. Water can be taken into the mantle cavity around the margins of the head and forcibly expelled through the manoeuvrable funnel by muscular contraction. Cephalopods are essentially pelagic animals which swim by jet propulsion after mobile prey, these being caught by use of the arms. Like the gastropods, cephalopods are elongated in the dorsoventral plane, but in association with their swimming life style, the ancestrally ventral region has functionally become anterior, and the dorsal visceral hump effectively forms the posterior end of the animal (Fig. 5.21). The mantle cavity therefore opens anteriorly and thus during rapid swimming cephalopods move backwards; during slower locomotion, movement of the funnel can permit progression in a variety of directions. Prey having been caught with the arms, it is macerated by a pair of beak-like dorsoventral jaws and then by the radula. Some of the arms are also modified for copulation (spermatophore transfer) in the male, most species being gonochoristic, with a single gonad and direct development. Somewhat surprisingly considering their large size, almost all cephalopods, except the nautiloids, are relatively short-lived animals, which breed once and then die ('semelparous' – see Section 14.3).

Other features of the group differ quite markedly between the two surviving subclasses. The six living species of the Nautiloidea possess a single external planospiral shell, comprising many chambers in a linear series, only the last of which is inhabited, although a thin filament of living tissue, the siphuncle, extends through the other chambers (Fig. 5.22). As the animal grows, new material is laid down so as to add a roughly cylindrical section to the old shell; it then moves forwards, secretes a new partition behind it, and thereby adds another chamber to the existing series. Nautiloids are poor swimmers, water being expelled from the mantle cavity only by contraction of the funnel muscles, and they rely on buoyancy to remain

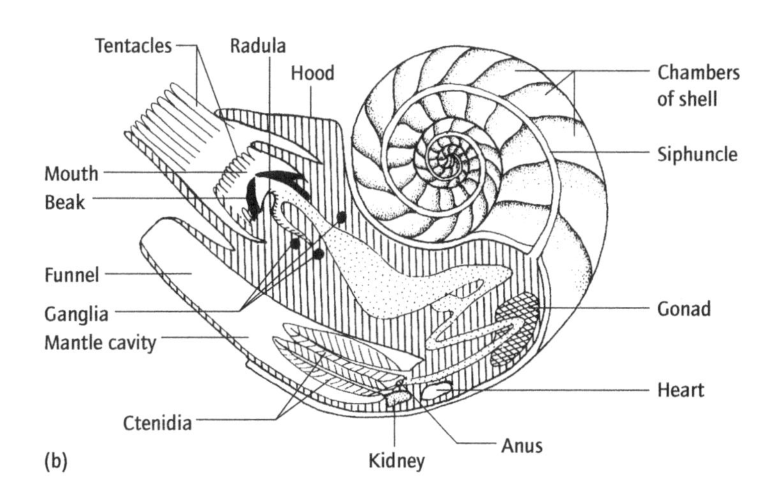

Fig. 5.22 Nautiloid cephalopods: (a) external appearance; (b) diagrammatic longitudinal section. (After Boss, 1982 and others.)

suspended in the water column. The pressure-insensitive buoyancy system is provided by the uninhabited shell chambers, the siphuncle absorbing the water originally contained in a chamber and secreting gas instead, with the result of neutral buoyancy (see Section 12.2.2). Many (80–90) suckerless tentacles surround the mouth, of which four are modified for spermatophore transfer; there are two pairs of ctenidia and of kidneys; and the nervous system and eyes are relatively simple.

In the Coleoidea, however, there is a marked trend of shell reduction correlated with a more active swimming life. In this group, water is expelled from the mantle cavity more forcefully by the simultaneous contraction of powerful circular muscles in the mantle wall, co-ordinated by large stellate ganglia from which issue giant fibres, of increasing diameter the greater the distance between the innervated muscles and the ganglion. The cuttlefish (order Sepiida) have retained the calcareous shell as a buoyancy device (Section 12.2.2), but it is reduced in size and lies within the body; in the squids (orders Teuthida and Vampyromorpha), it is further reduced to a thin internal cartilaginous element with no buoyancy function; whilst in most octopuses (order Octopoda) it has been lost. Many pelagic coleoids are streamlined and torpedo-shaped (Fig. 5.23), and have continually to swim to remain within the water mass, undulations of lateral fins permitting a slow but efficient form of locomotion. Some, however, have evolved alternative buoyancy aids, such as the replacement of heavy divalent cations by ammonia in an expanded pericardial cavity, coupled with low-density gelatinous tissues. The octopuses have adopted a largely benthic existence, although some forms have a web of skin between the arms that can be used as a sail to drift with near-bottom currents. Such species have reduced or vestigial ctenidia. In contrast with the nautiloids, coleoids otherwise have a single pair of sturdy, folded ctenidia, a single pair of kidneys, and eight

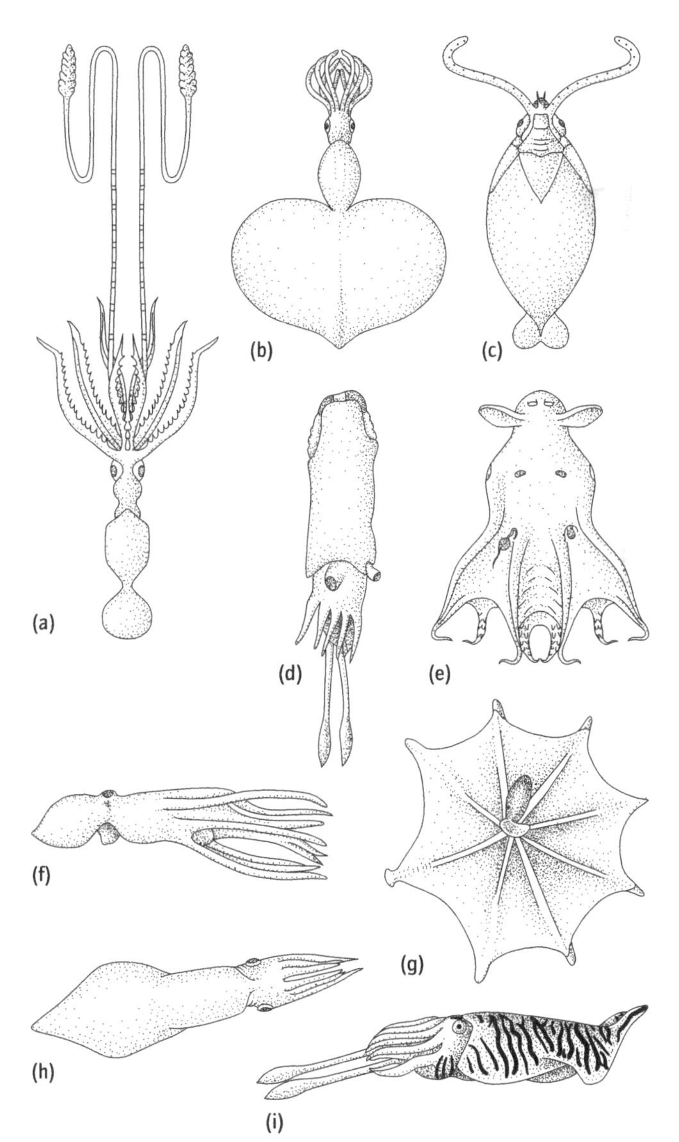

Fig. 5.23 Coleoid cephalopod body forms (after various sources). (a)–(c) and (h), Teuthida; (d) and (i) Sepiida; (e) Vampyromorpha; and (f) and (g) Octopoda.

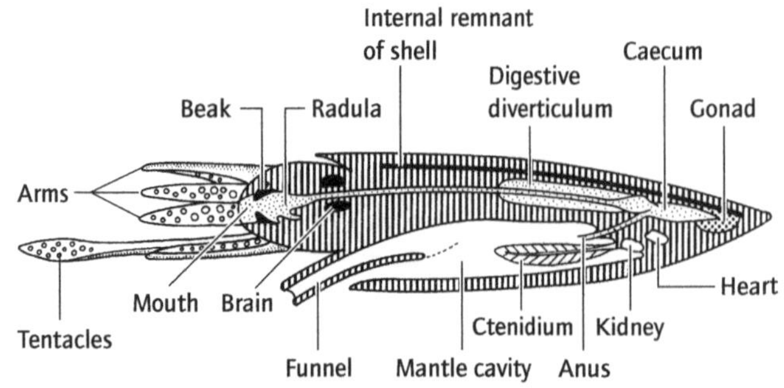

Fig. 5.24 Diagrammatic longitudinal section through a coleoid (after Boss, 1982 and others).

short sucker-bearing arms (two of them usually modified for spermatophore transfer) together, in the squids and cuttlefish, with two elongate, tentaculate, prey-catching arms bearing pads of suckers at their tips (Fig. 5.24).

Perhaps the most distinctive features of the 650 species of coleoids, however, are the closed nature of the blood system and the highly concentrated state of the nervous system which includes well-developed eyes. The ctenidia lack cilia (water being driven through the gills and mantle cavity by the muscular pumping which effects the jet-propelled locomotion), but contain capillaries – not arranged in a counter-current flow – through which blood is pumped by branchial hearts. The circum-oesophageal nerve ring is enlarged into a highly complex brain, to a greater degree than seen in any other invertebrate, and the paired eyes innervated from the brain are of the same general form as possessed by the vertebrates, with cornea, iris diaphragm, lens and retina (Fig. 5.25). In contrast to the vertebrates, however, the photo-receptors of the coleoid eye are directed *towards* the incoming light. The nervous system also controls the numerous surface chromatophores, that unusually are operated by small muscles and hence can respond rapidly, giving rise to almost instantaneous changes in colour pattern.

5.2 Further reading

Fretter, V. & Peake, J. (Eds) 1975–78. *Pulmonates* (2 Vols). Academic Press, London.

Haszprunar, G. 1988. On the origin and evolution of major gastropod groups, with special reference to the Streptoneura. *J. Moll. Stud.*, **54**, 367–441.

Hughes, R.N. 1986. *A Functional Biology of Marine Gastropods*. Croom Helm, Beckenham.

Hyman, L.H. 1967. *The Invertebrates*, Vol. 6: *Mollusca 1*. McGraw-Hill, New York.

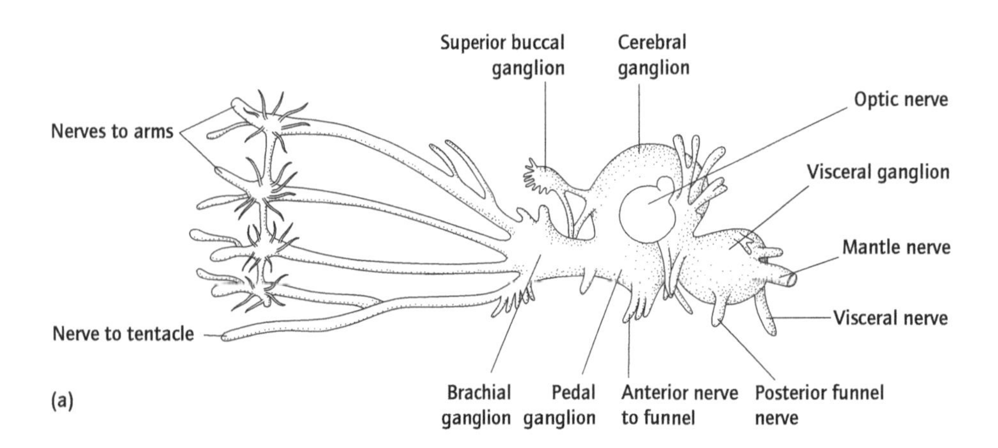

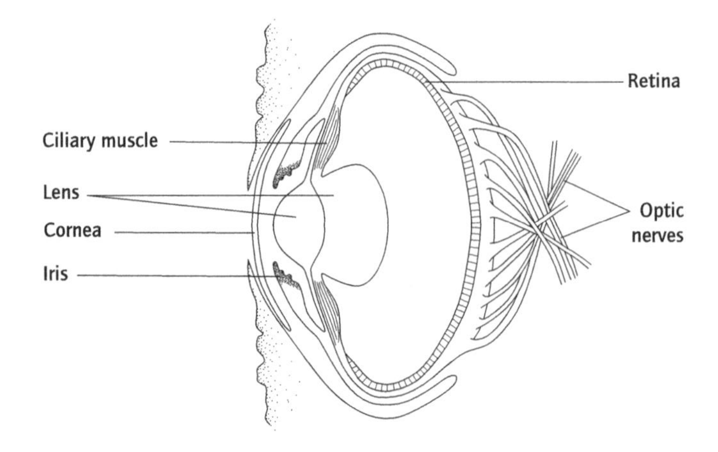

Fig. 5.25 (a) The brain and (b) eye of coleoids. (After Wells, 1962 and Kaestner, 1967.)

Kaestner, A. 1967. *Invertebrate Zoology*, Vol. 1. Wiley, New York.
Morton, J.E. 1979. *Molluscs*, 5th edn. Hutchinson, London.
Purchon, R.D. 1968. *The Biology of the Mollusca*. Pergamon Press, Oxford.
Runham, N.W. & Hunter, P.J. 1970. *Terrestrial Slugs*. Hutchinson, London.
Solem, A. 1974. *The Shell Makers*. Wiley, New York.
Taylor, J.D. (Ed.) 1996. *Origin and Evolutionary Radiation of the Mollusca*. Oxford University Press, Oxford.
Wells, M.J. 1962. *Brain and Behaviour in Cephalopods*. Heinemann, London.
Wells, M.J. 1978. *Octopus*. Chapman & Hall, London.
Wilbur, K.M. (Ed.) 1983–88. *The Mollusca* (12 Vols). Academic Press, New York.

CHAPTER 6

The Lophophorates

Phorona
Brachiopoda
Bryozoa
Entoprocta
Cycliophora

The three phyla usually regarded as comprising the lophophorates, the Phorona, Brachiopoda and Bryozoa, share, as their name suggests, the common feature of a lophophore: a circular, horseshoe-shaped or complexly whorled or folded ring of ciliated hollow tentacles serving as a suspension-feeding apparatus. This lophophore, which encircles the mouth but not the anus, is supported by its own separate hydrostatic body cavity.

Basically, the lophophorate body is tripartite in form, with a minute pre-oral section (the prosome), a small second region (mesosome) bearing the mouth and lophophore, and a much larger third part (metasome), which comprises the major part of the body, contains the other organ systems and bears, near the base of the lophophore, the anus. A further, rather distinctive, common feature of the lophophorates is that their 'gonads' are not discrete organs, merely loose peritoneal aggregations of germ cells. In the majority of lophophorates the prosome has been lost, and even when present it is insignificant.

The lophophorates are clearly protostomes. Nevertheless they have been traditionally allied to the deuterostomes (Chapter 7) on the basis of (i) the shared tripartite, oligomeric, body plan, (ii) the presence of a mesosomal lophophore in the deuterostome pterobranchs (Section 7.2.3.2), and (iii) details of their cleavage pattern and nervous system. Other evidence, however, including from molecular sequence data and the shared possession of chitinous setae, places them with the other protostome groups and particularly close to the eutrochozoans, e.g. molluscs, sipunculans and annelids. Also controversial are the inclusion of the bryozoans (see below) and whether the two groups of the brachiopods are related to each other.

Lophophorates are sedentary or sessile, and their bodies are protected within a secreted external covering. This may take the form of a chitinous tube within which the animal is free to move, or a gelatinous, chitinous or calcareous shell or box to which the epidermis is permanently attached.

The lophophorates, therefore, appear a well-knit group and some authorities have argued that a single phylum Lophophorata should include them all. That this has not become widely accepted is in part due to continuing debate on the relationships of the bryozoans to the other groups, and in part to the affinities of the fourth phylum, the Entoprocta, which are still mysterious. The feeding tentacles of the entoprocts do not conform to the definition of the lophophore, and yet they do show similarities to the bryozoans in particular, although their cleavage pattern is spiral and their development is determinate – markedly non-lophophorate characteristics. Their inclusion in this chapter is therefore largely a matter of convenience. So is that of the most recent phylum to be described, the Cycliophora, which on present knowledge seem possibly related to the Entoprocta and/or to the Bryozoa.

6.1 Phylum PHORONA

6.1.1 Etymology

Derived from the generic name *Phoronis*, one of the epithets of the Egyptian goddess Isis.

6.1.2 Diagnostic and special features (Fig. 6.1)

1 Bilaterally symmetrical, vermiform.
2 Body more than two cell layers thick, with tissues and organs.
3 A through, U-shaped gut.
4 An oligomeric, tripartite body, each region with a single body cavity.
5 Prosome very small; mesosome small but with a large lophophore supported by the mesocoel; metasome large and elongate.
6 Body wall without a cuticle, with muscle layers.
7 A closed blood system, with haemoglobin in corpuscles.

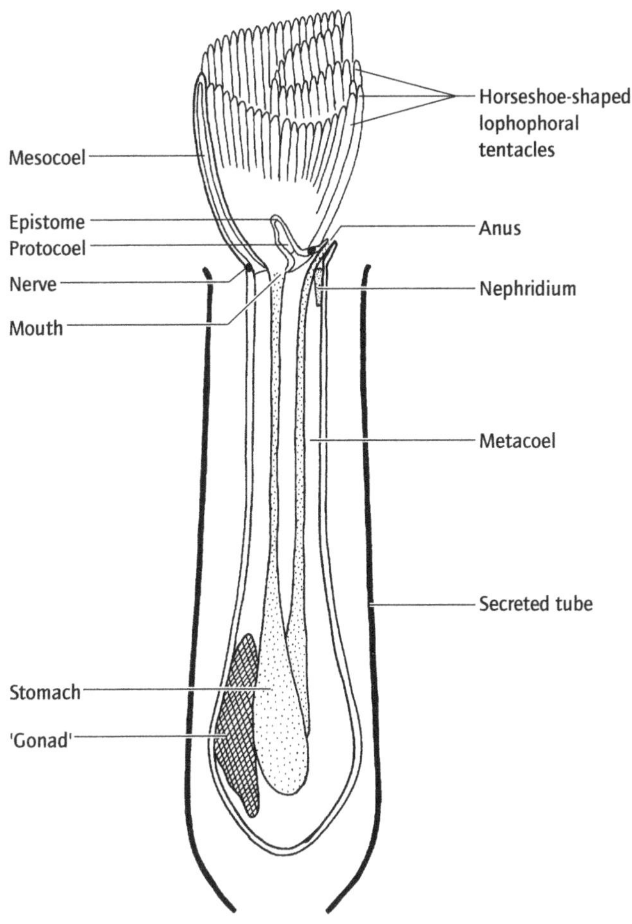

Fig. 6.1 Diagrammatic longitudinal section through a phoronan (after Emig, 1979).

8 One pair of metanephridia-like organs (protonephridia in the larva).
9 Nervous system diffuse, basiepidermal.
10 Eggs cleave radially.
11 Development indirect, usually via an actinotroch larva.
12 Marine.

Phoronans are sedentary worms permanently inhabiting secreted chitinous tubes partially buried in soft sediments or, less commonly, attached to hard substrata. Although exceptionally up to 50 cm in length, most are less than 20 cm long, of which the majority is formed by the metasome or trunk. The only other externally visible structure is the large and topographically terminal, suspension-feeding lophophore, which comprises up to 15 000 tentacles arising from the mesosome in the form of a horseshoe (Fig. 6.2); in species with many tentacles, the free arms of the horseshoe are curled round into spirals.

Most phoronans are hermaphrodite, although a few are gonochoristic. The sex cells originate from the peritoneum and are shed into the metacoel, from which they escape via the nephridia. Fertilization is external although the eggs and developing larvae (Fig. 6.3) may be brooded within the cavity enclosed by the lophophore, or within the tube, for between 40% and 75% of their developmental period. One species is capable of asexual multiplication by fission and budding, the other species by transverse fission only.

6.1.3 Classification

Only 15 species are known, placed in two genera within one family.

6.2 Phylum BRACHIOPODA (lamp-shells)

6.2.1 Etymology

Greek: *brachion*, arm; *pous*, foot.

6.2.2 Diagnostic and special features (Fig. 6.4)

1 Bilaterally symmetrical.
2 Body more than two cell layers thick, with tissues and organs.
3 Either a through, U-shaped gut or one secondarily blind-ending.
4 An oligomeric, bipartite body, each region with a single, essentially enterocoelic body cavity.
5 Prosome absent; mesosome small but with a large complex lophophore supported by the mesocoel (see Fig. 9.2) and, in some by calcium carbonate extensions of the shell; metasome small.

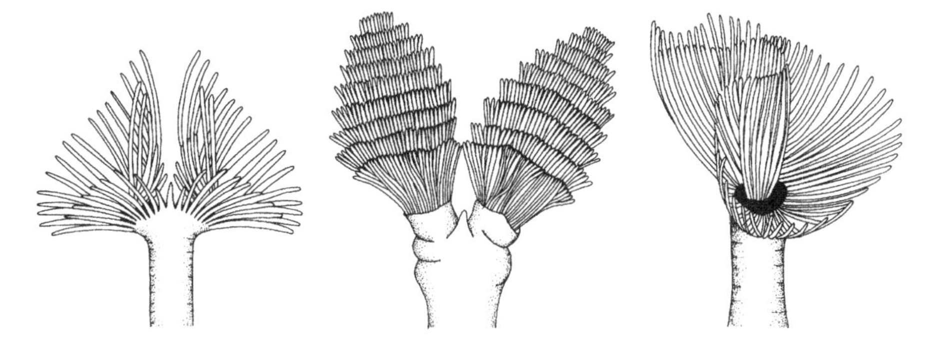

Fig. 6.2 Lophophore form in phoronans (after Emig, 1979).

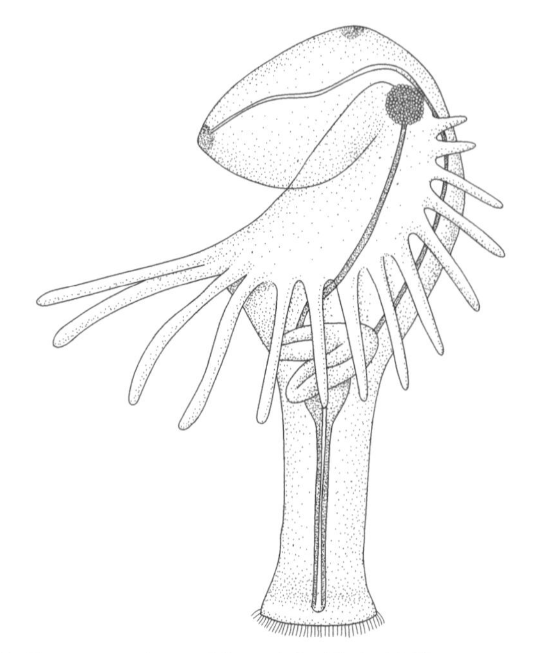

Fig. 6.3 Phoronan actinotroch larva (after Emig, 1979).

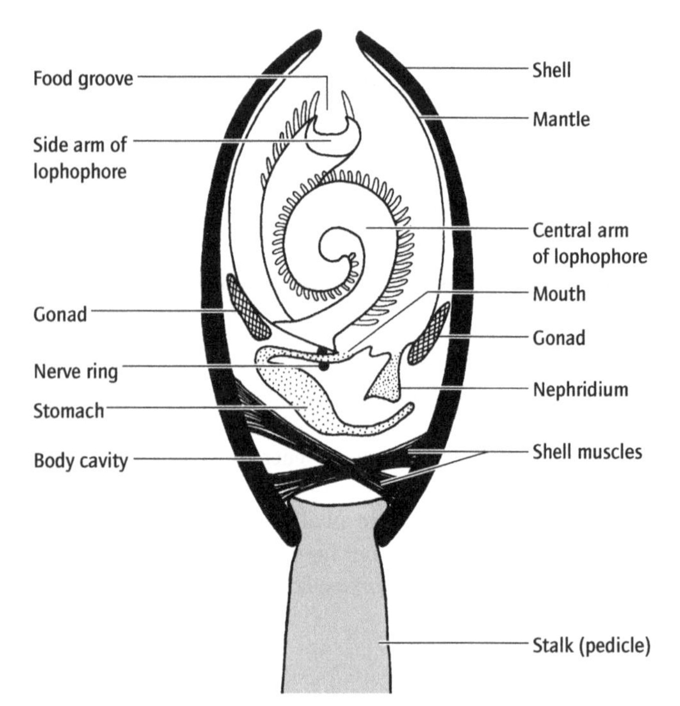

Fig. 6.4 Diagrammatic longitudinal section through a generalized brachiopod.

6 Body completely enclosed, apart from the stalk, within a bivalved shell which may be cemented to the substratum, attached to rock or anchored in soft sediments by a stalk (pedicle), or be unattached; the small body is located posteriorly within the shell, most of the enclosed cavity being occupied by the lophophore.

7 An open blood system with a heart or hearts.

8 With one or two pairs of metanephridia-like organs.

9 Nervous system in the form of a ganglionated circum-oesophageal ring, from which individual nerves issue.

10 Without asexual multiplication; mostly gonochoristic, without discrete gonads but with four aggregations of sex cells associated with the peritoneum, gametes being shed into the metacoel and leaving via the nephridia.

11 Eggs cleave radially.

12 Marine.

The surviving brachiopods, which are but a small remnant of this once important phylum (335 living species as opposed to 26 000 fossil ones), are in their external appearance and general life style remarkably similar to the bivalve molluscs (Section 5.1.3.6). Both groups are sedentary or sessile, suspension-feeding animals enclosed within a bivalved shell, which is secreted by an epidermal mantle and covered by an organic periostracum. The suspension-feeding apparatus of the brachiopods, however, is a circular, looped or spiral lophophore (not ctenidial gills), and the two valves of their shell are dorsal and ventral (not lateral, as in bivalve molluscs). A further point of convergence is the reduction of the ancestral head region (in brachiopods, loss of the lophophorate prosome) consequent on the complete encasement of the body within a thick external shell. The shell of brachiopods is unusual, however, in that it is penetrated by numerous caecae containing body tissue, the caecae functioning as sites of synthesis and storage of materials for use in, e.g. reproduction. Indeed half the total tissue weight may be located within the shell valves themselves.

All living brachiopods are relatively small (<10 cm shell length or width) and benthic in habitat. In evolutionary terms, they may be equivalent to shelled phoronans (Chapter 2) and indeed the brachiopod condition may have arisen polyphyletically from such worms (Fig. 2.11).

6.2.3 Classification

Two classes are recognized which display many divergent features.

6.2.3.1 Class Inarticulata

Inarticulate brachiopods, as their name suggests, are characterized by having shell valves, which are usually composed of calcium phosphate and chitin, held together solely by muscles.

Class	Order
Inarticulata	Lingulida Acrotretida
Articulata	Rhynchonellida Terebratulida Thecideidida

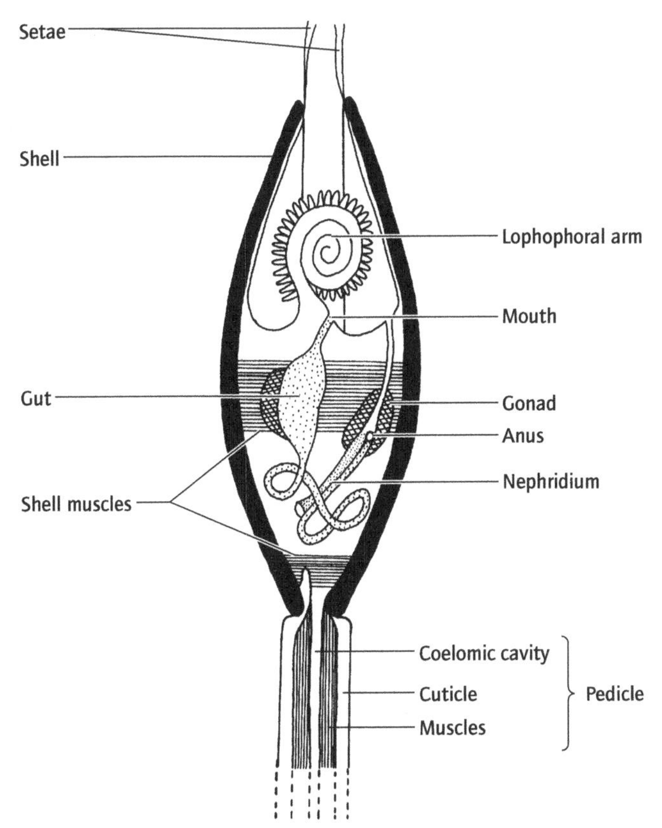

Fig. 6.5 Diagrammatic longitudinal section through an inarticulate brachiopod.

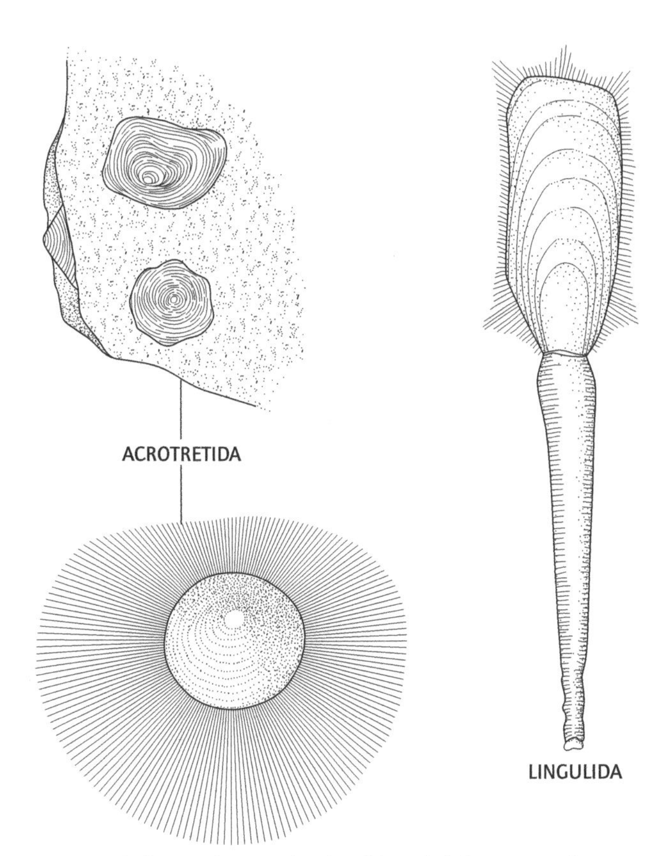

Fig. 6.6 Body forms of inarticulate brachiopods (after sources in Hyman, 1959).

Posteriorly, from between the two valves, emerges the pedicle, which is an integral part of the body formed from the larval mantle; it contains an extension of the metacoel and intrinsic muscles; in some groups it is absent, the ventral valve then being cemented to the substratum. In contrast to other brachiopods, an anus is present (Fig. 6.5) and the lophophore is relatively simple and unsupported by shelly material.

The larva is a small free-swimming version of the adult, complete with shell and lophophore, the cilia on which provide the means of propulsion through the water; correspondingly, there is no metamorphosis on settlement. During development, mesoderm develops from the larval archenteron in the enterocoelic manner, but the body cavities form schizocoelically within the enterocoelic mesodermal cell masses.

The 47 species are divisible into two orders: the burrowing Lingulida and the sessile and limpet-like Acrotretida (Fig. 6.6).

6.2.3.2 Class Articulata

The articulates possess calcium carbonate shells, the valves of which articulate by means of teeth present on the ventral valve that insert into sockets on the dorsal. The ventral valve also bears a slit or notch through which the pedicle, if present, emerges; the dorsal valve often includes inwardly directed supports for the large and complexly looped or spiralled lophophore (Fig. 6.7). Neither intrinsic muscles nor an extension of the metacoel are present in the pedicle, which is derived not from the mantle but from one of the three body regions of the larva. This larval stage (Fig. 6.8) is markedly different in appearance from the adult, and undergoes metamorphosis. In contrast to the inarticulates, the gut is blind-ending and both mesoderm and the contained body cavities develop enterocoelically.

Articulates appeared relatively late in brachiopod evolution (in the Ordovician) and nearly 300 species remain extant. These are placed in three orders on the basis of the structure of the lophophore and of its supporting calcareous system.

6.3 Phylum BRYOZOA

6.3.1 Etymology

Greek: *bryon*, moss; *zoon*, animal.

6.3.2 Diagnostic and special features (Fig. 6.9)

1 Modularly colonial animals; each colony with from a few to a million individuals (zooids) formed by asexual budding from a founding ancestrula; each zooid is in tissue contact with its immediate neighbours.
2 Zooids often polymorphic, with feeding, defensive, brooding types, etc.

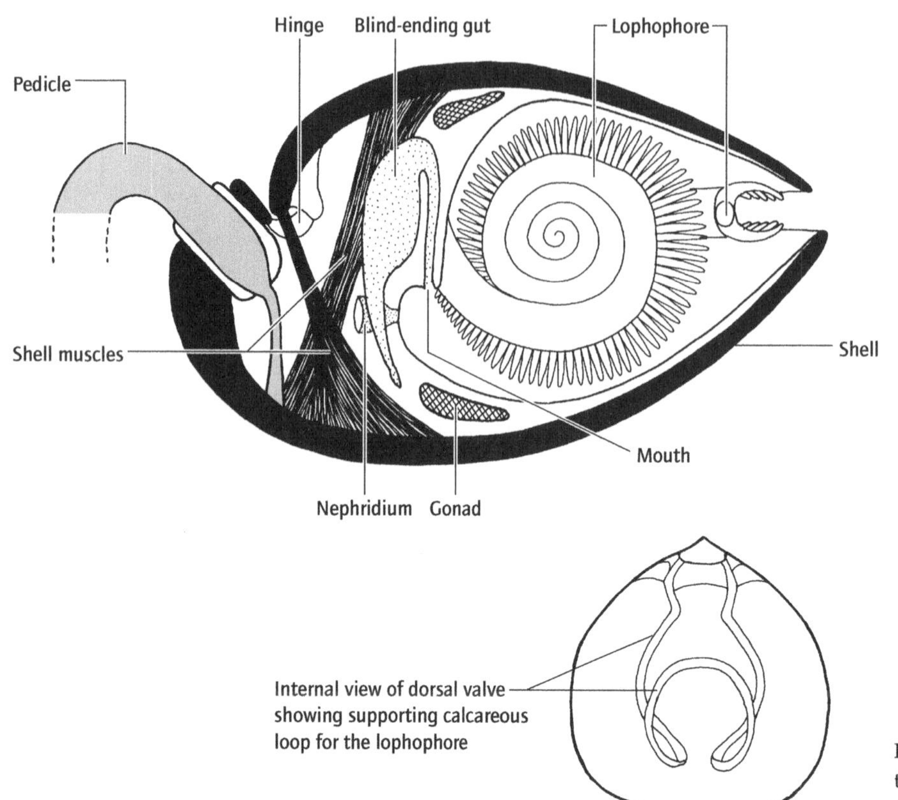

Fig. 6.7 Diagrammatic longitudinal section through an articulate brachiopod (after Moore, 1965).

3 Colonies are hermaphrodite, although zooids may be hermaphrodite or gonochoristic.
(Features below refer to the characteristic individual zooid.)
4 Bilaterally symmetrical.
5 Body more than two cell layers thick, with tissues and organs.
6 A through, U-shaped gut.
7 An oligomeric, bi- or tripartite body, each region with a single body cavity formed *de novo* during metamorphosis; some with an additional metasomal body cavity.
8 Prosome present only in one group, small; mesosome small, and with a relatively small, circular or horseshoe-shaped lophophore supported by the mesocoel; metasome large, sac-like.

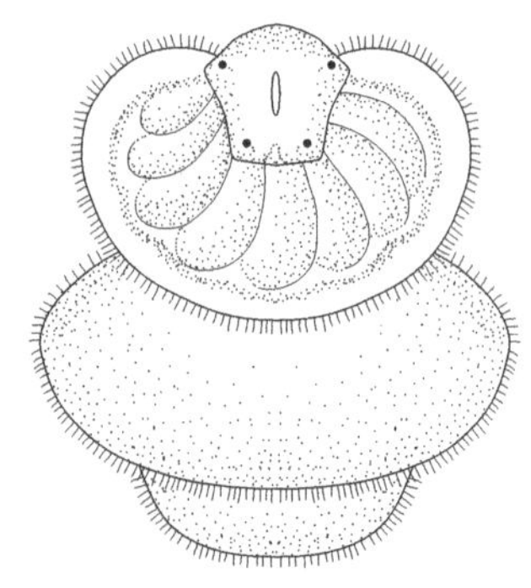

Fig. 6.8 The larva of an articulate brachiopod (after Lacaze-Duthiers, 1861).

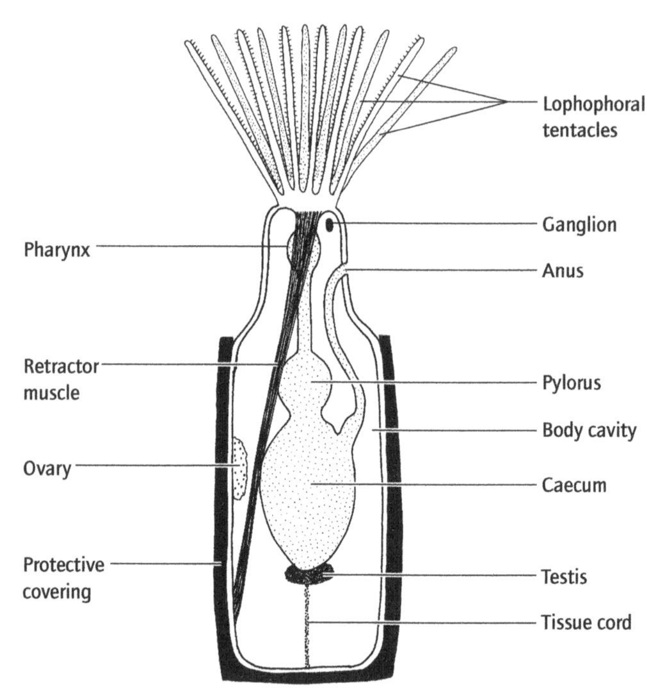

Fig. 6.9 Diagrammatic longitudinal section through a generalized bryozoan.

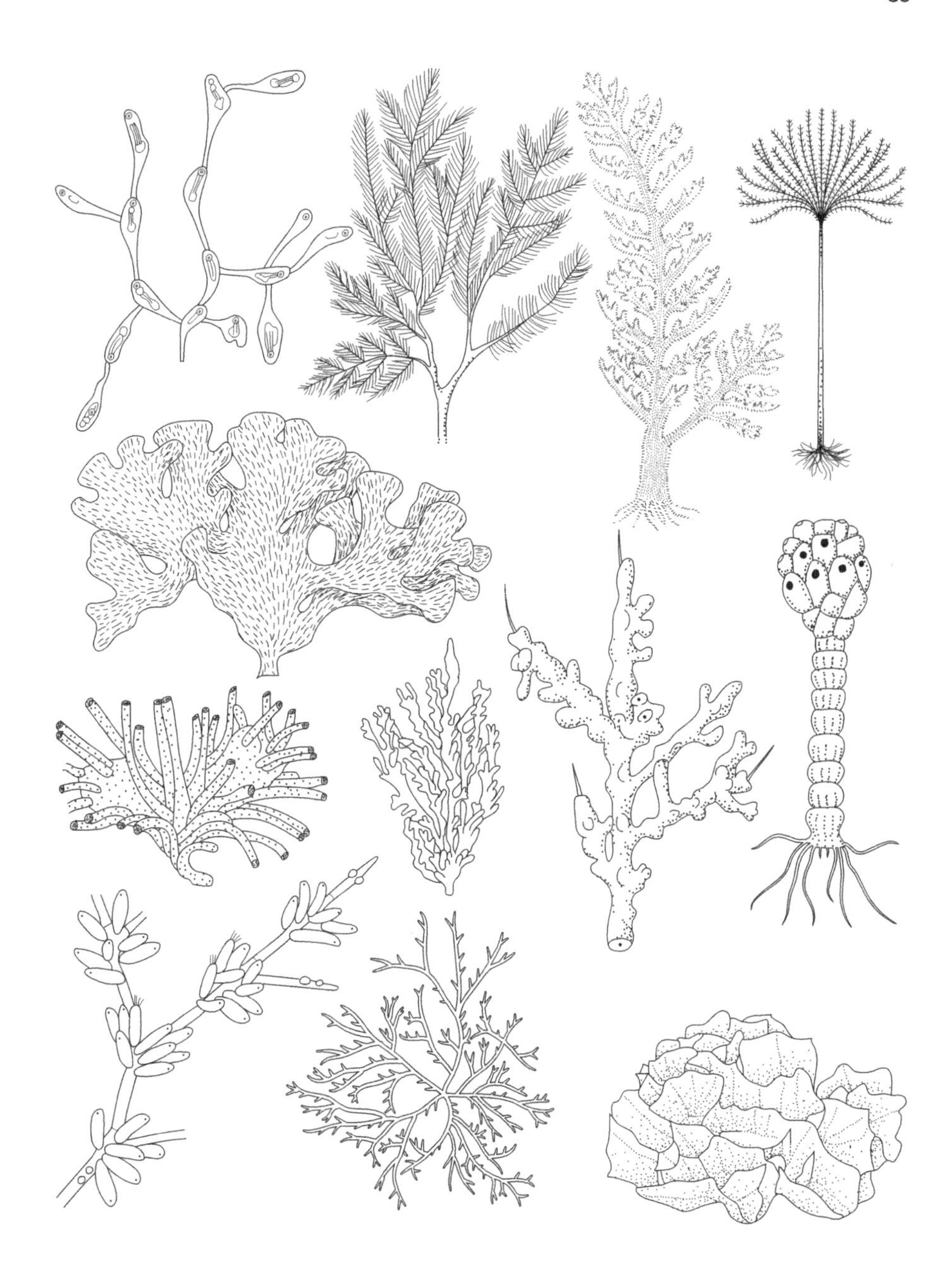

Fig. 6.10 Colony form in bryozoans (after several sources).

9 Body enclosed or embedded in a chitinous, gelatinous or calcareous tube, box or communal matrix, except for an orifice through which the lophophore can be protruded (by hydrostatic pressure) and withdrawn (by retractor muscles).

10 Without circulatory or excretory systems.

11 Nervous system in the form of a ganglion between mouth and anus, from which issue a circumpharyngeal ring and individual nerves.

12 Gonads are 'peritoneal', gametes rupturing into the metacoel and leaving through 'coelomopores' associated with the lophophore.

13 Eggs cleave radially and are usually brooded.

14 Development normally indirect.

Bryozoans appear similar to minute (*c.* 0.5 mm) phoronans (Section 6.1) which form extensive stoloniferous, mat-like, arborescent or foliaceous colonies by asexual budding (Fig. 6.10). Similarly to the other lophophorate phyla (Sections 6.1 and 6.2), they exhibit, besides the suspension-feeding lophophore, an oligomeric body plan, radial cleavage, and an apparently similar gonadal system in that the sex cells develop from cells in the membrane lining the body cavity. The nature of their body cavities is debatable, however, because they have no embryological precursors nor temporal continuity with the embryonic germ layers. During metamorphosis, the tissues of bryozoan larvae are histolysed and then completely reorganized into the adult body. Mesoderm and the body cavities of the other lophophorates are basically enterocoelic, deriving from the embryonic gut; but the larval gut of bryozoans, if present, is broken down at metamorphosis. There is, therefore, no obvious mesodermal origin of any of the adult cavities, and the coelomic status that they are usually given is based mainly on analogy with the phoronans and brachiopods. This is not uncontested: an alternative school of thought regarded them as being equivalent to the supposedly pseudocoelomic cavity of the entoprocts (Section 6.4).

As a group, bryozoans are a very successful line of suspension feeders, more so today than any other lophophorate. With a few exceptions, colonies are sessile, encrusting or attached to relatively firm substrata; one exception, the freshwater *Cristatella*, can creep slowly (*c.* 10 cm day^{-1}) by means of a muscular 'foot'; whilst another, the marine *Selenaria maculata* can walk quite quickly (1 m h^{-1}) on the long setae of its avicularia zooids (see Section 6.3.3.3). *S. maculata* lives in shallow, coral-reef habitats and it has been recorded as moving towards patches of sunlight. The colony is green and it is therefore possible that this behaviour is related to the presence of symbiotic zooxanthellae in its tissue, as seen in many other reef invertebrates (Section 9.1). One recently discovered Antarctic species is planktonic; it forms a 30 mm diameter hollow sphere with the lophophores of the monolayer of zooids projecting outwards all over its external surface.

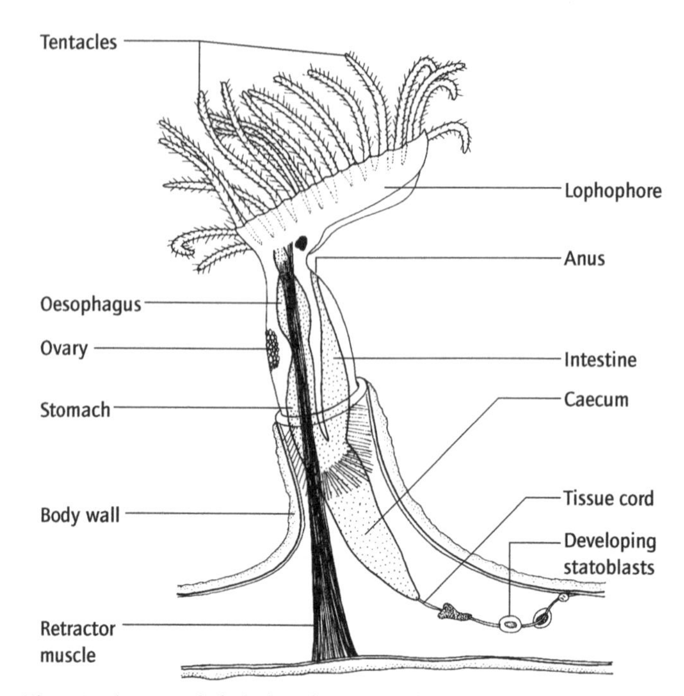

Fig. 6.11 An extended phylactolaeme zooid, seen as if transparent (after Pennak, 1978).

6.3.3 Classification

Most of the 4300 described species are marine, but one class and various individual species in the two other classes are freshwater.

Class	Order
Phylactolaemata	Plumatellida
Stenolaemata	Cyclostomata
Gymnolaemata	Ctenostomata Cheilostomata

6.3.3.1 Class Phylactolaemata

The phylactolaemes are relatively unspecialized bryozoans. The zooidal body has retained the presumed ancestral tripartite form, in that a prosomal flap of tissue, the epistome, containing a hydrostatic protocoel, overhangs the mouth, as in the phoronans. The body wall also bears well-developed layers of circular and longitudinal muscle; contraction of the circular layer generates the pressure required to protract the lophophore, which is large with up to 120 tentacles issuing from a horseshoe-shaped ridge (Fig. 6.11). Individual zooids are monomorphic, cylindrical and among the largest of the bryozoans; the body cavity compartments of adjacent individuals often interconnect.

All species are freshwater and, calcium salts being less abundant in fresh than in salt water, the epidermis secretes a non-calcified, chitinous or gelatinous case, and, in common with other freshwater invertebrates, they have evolved a resting-stage mechanism for dispersal and/or overwintering. This is highly individual in nature: on a cord of tissue running from the gut to the body wall, aggregations of epidermal and 'peritoneal' cells develop into 'statoblasts' (Fig. 6.12) – yolk-rich cells enclosed within (usually) disc-shaped chitinous valves. These may remain attached to the body wall and serve to reinstate the colony *in situ* after die-back, or be liberated during life or after death of the

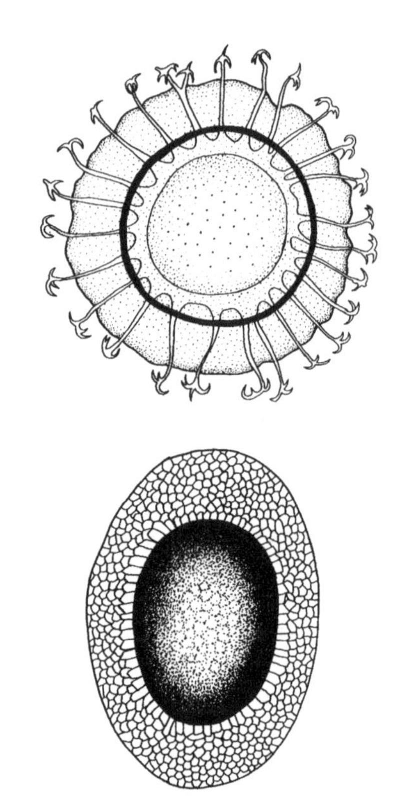

Fig. 6.12 Phylactolaeme statoblasts (after Hyman, 1959).

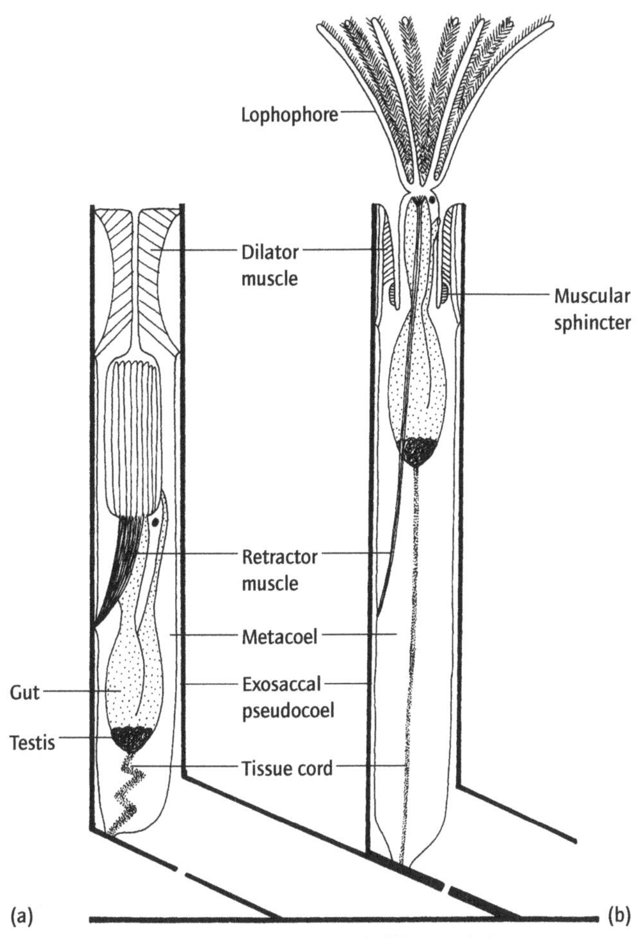

Fig. 6.13 Stenolaeme zooids: (a) retracted; (b) extended. (After Ryland, 1970.)

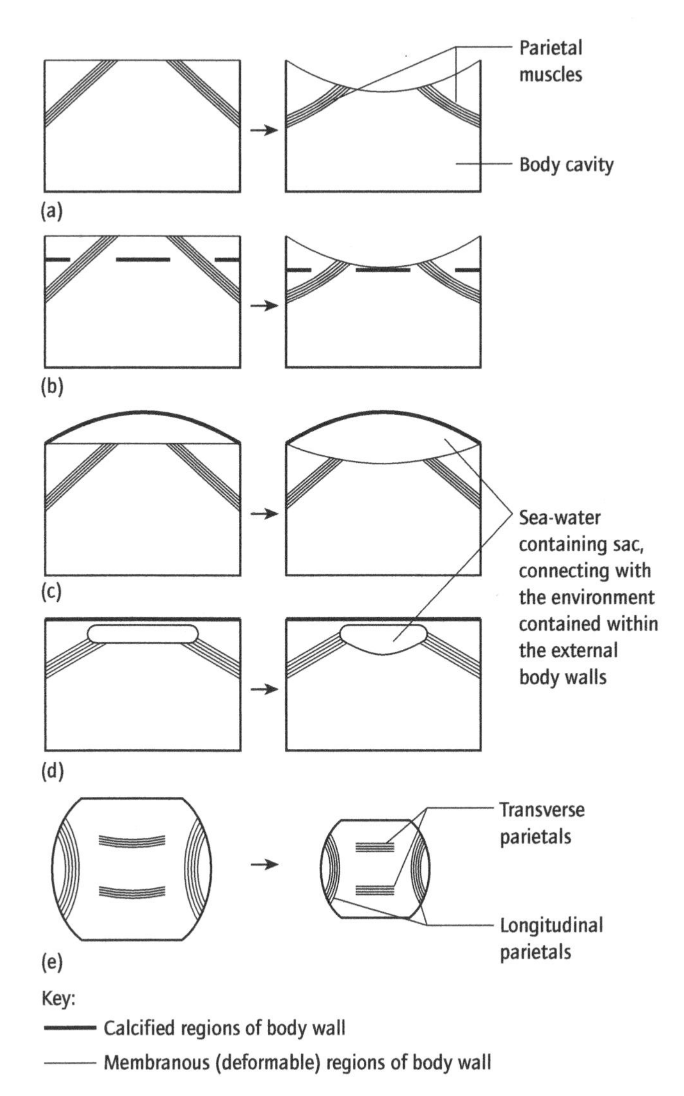

Fig. 6.14 Sections through the zooidal boxes of various gymnolaeme bryozoans, showing how contraction of parietal muscles causes deformation of the body walls and thus generates increased pressure in the body cavity: (a)–(d) cheilostomes; (e) ctenostome. (After Ryland, 1970.)

zooid. The latter type often can float and may be dispersed over large distances. Statoblasts are highly resistant both to freezing and to desiccation, and, in temperate latitudes, after winter they 'germinate' and reform into founding zooids.

A single order containing some 50 species is recognized.

6.3.3.2 *Class Stenolaemata*

The stenolaeme zooids are also cylindrical, but they lack a prosome and muscle layers in the body wall. Their cylindrical tubes, bearing a terminal circular orifice, are heavily calcified beneath an outer hyaline cuticle or cellular layer; the orifice is closed after retraction of the lophophore by a membrane and not an operculum. The pressure required to protract the lophophore, which is circular and comprises only some 30 tentacles at most, is generated by dilator muscles acting on the fluid both in the metacoel and in an exosaccal body cavity generally regarded as being pseudocoelomic (Fig. 6.13). Stenolaeme zooids display a limited degree of polymorphism.

The reproductive system of stenolaemes exhibits several peculiarities, the chief of which is a form of polyembryony. After fertilization, the zygote cleaves to produce a ball of blastomeres, but this ball then buds off a series of secondary embryos, which in turn may bud off tertiary embryos: more than 100 blastulas may be derived from a single zygote.

Most groups of stenolaemes are extinct; a single order containing some 900 marine species survives.

6.3.3.3 *Class Gymnolaemata*

Today, the gymnolaemes are the most abundant and successful of the bryozoans, with over 3000 species, mainly in the sea but also in brackish and fresh waters. Similarly to the stenolaemes, their zooids lack a prosome and muscle layers in the body wall, and possess a relatively small, circular lophophore; but the generally short and squat zooids generate their lophophore-protracting pressure by muscular deformation of the body wall, parietal muscles acting on a specific membranous region (Fig. 6.14). The body wall, therefore, if calcified, is only partially

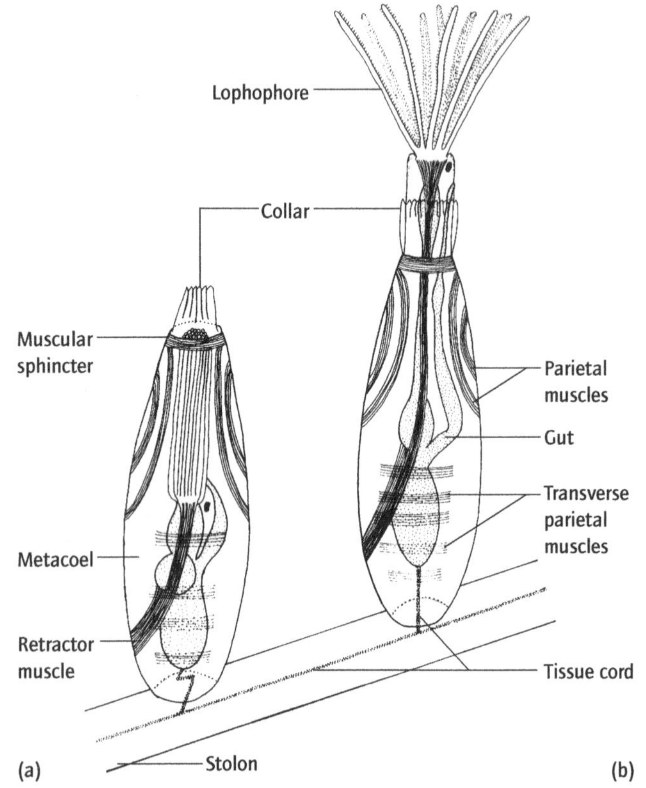

Fig. 6.15 Ctenostome zooids: (a) retracted; (b) extended. (After Ryland, 1970.)

so. After retraction of the lophophore the orifice can, in many species, be closed by an operculum. Zooidal polymorphism reaches its greatest degree in this group, with feeding (autozooids), packing (kenozooids), grasping (avicularia), cleaning (vibracula), and brood-chamber (ooecia) individuals.

Two orders are distinguished: the Ctenostomata, which have non-calcified walls and lack opercula (and are often cylindrical with a terminal orifice) (Fig. 6.15); and the Cheilostomata, which possess flattened, box-shaped zooids with partially – often largely – calcified walls, an operculum, and a frontal orifice (Fig. 6.16).

6.4 Phylum ENTOPROCTA

6.4.1 Etymology

Greek: *entos*, inside; *proktos*, anus.

6.4.2 Diagnostic and special features (Fig. 6.17)

1 Bilaterally symmetrical, goblet-shaped.
2 Body more than two cell layers thick, with tissues and organs.
3 A through, U-shaped gut.
4 Space between body wall and gut filled with a gelatinous 'mesenchyme' – interpreted by some as an occluded pseudocoelomic body cavity.
5 Body in the form of a hemispherical to ovoid calyx, bearing a ring of tentacles around both mouth and anus, and attached to a substratum by a contractile stalk.
6 Body wall with a cuticle but without muscle layers.
7 Without circulatory or gaseous-exchange systems.
8 One pair of protonephridia (or, in the freshwater genus, many protonephridia).
9 Nervous system in the form of a ganglion between mouth and anus, from which individual nerves issue.
10 Tentacles not retractile but they can contract and fold inwards to occlude the intratentacular cavity.
11 Eggs cleave spirally.
12 Development indirect.

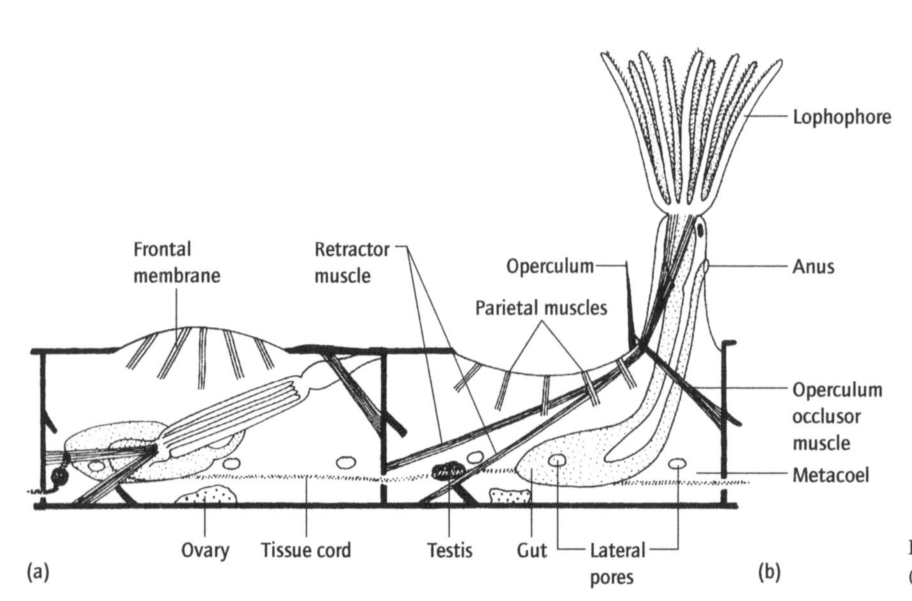

Fig. 6.16 Cheilostome zooids: (a) retracted; (b) extended. (After Ryland, 1970.)

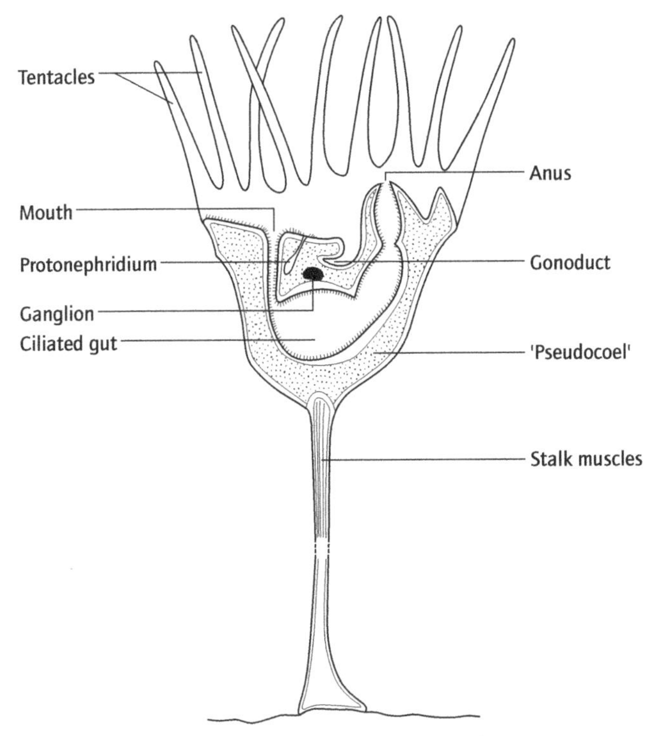

Fig. 6.17 Diagrammatic longitudinal section through an entoproct (after Becker, 1937).

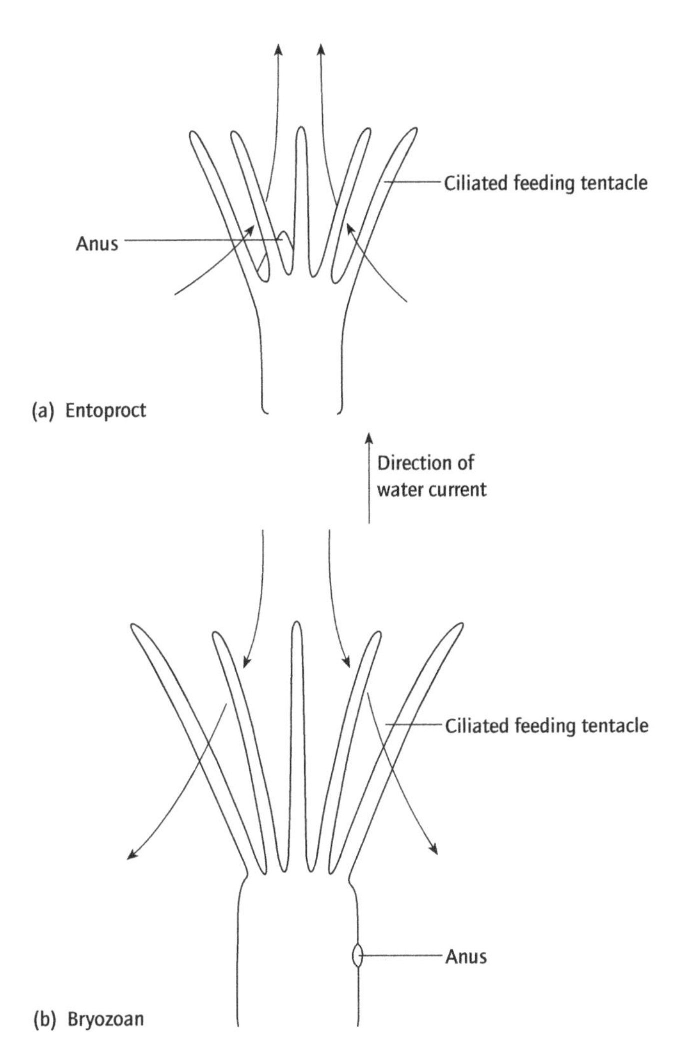

Fig. 6.18 Comparison of the feeding currents of entoprocts (a) and bryozoans (b) in relation to the position of the anus.

Entoprocts are small (0.5–5 mm high), solitary or colonial animals living temporarily or, more commonly, permanently attached to a substratum, including that provided by other organisms. All are suspension feeders, the 6–36 extensions of the body wall which form the tentacles bearing cilia that collect food particles from the water and transport them, in mucus, to the mouth. In forms creating their own feeding current, water passes from outside the tentacular ring, between the tentacles and out via the intratentacular space (Fig. 6.18): the opposite system to that operated by the lophophorates. The anus, which is mounted on a cone, discharges into the central exhalent current.

Asexual multiplication by budding is widespread and can give rise to modular colonies; sexually, entoprocts are probably all hermaphrodite, apparently gonochoristic species being differently aged sequential hermaphrodites. Sperm are shed into the water, but fertilization is likely always to be internal. The larva is a planktotrophic or lecithotrophic trochophore (Fig. 6.19), which in most species metamorphoses into the adult; in some, however, the adult develops from a bud produced by the larval stage.

The relationships of this small phylum are contentious. Some authorities interpret the extensive 'mesenchyme' tissue as a haemolymph-filled pseudocoelom and ally the entoprocts with the other pseudocoelomate phyla (Chapter 4); others regard them as being related to the bryozoans, because of some developmental similarities, the body cavity of the bryozoans also being of a contentious nature (Section 6.3.2).

6.4.3 Classification

The 150 described species are placed in a single class and order. All but one freshwater genus are marine.

6.5 Phylum CYCLIOPHORA

6.5.1 Etymology

Greek: *cyclion*, small wheel; *phoros*, bearing.

6.5.2 Diagnostic and special features (Fig. 6.20)

1 Bilaterally symmetrical, ovoid.
2 Body more than two cell layers thick, with tissues and organs.

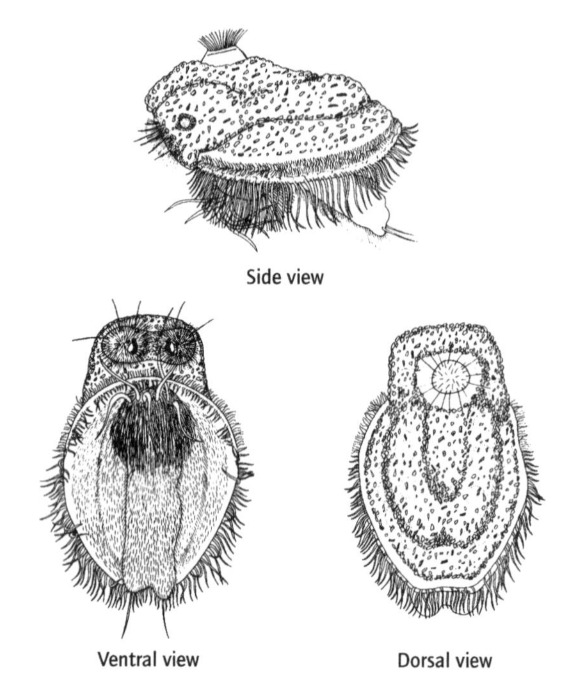

Fig. 6.19 An entoproct larva (after Nielsen, 1971).

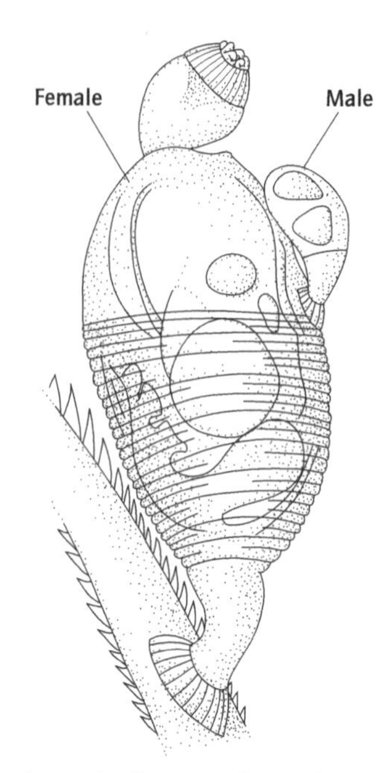

Fig. 6.20 A cycliophoran feeding stage from which a female sexual stage is about to emerge; a small male is attached to the feeding-stage cuticle (after Conway Morris, 1995).

3 A through, U-shaped gut.
4 Space between body wall and gut filled with large mesenchyme cells; without a body cavity.
5 Body in the form of a funnel-shaped feeding organ bearing a crown of compound cilia, a trunk containing the organs, and a cuticular stalk and attachment disc.
6 Body wall with a cuticle.
7 Without circulatory or gaseous-exchange organs.
8 Excretory organs in the form of a pair of protonephridia only in the sexually produced larval stage.
9 Nervous system in the form of a ganglion between oesophagus and rectum in the trunk.
10 Life cycle complex, involving asexual multiplication phases and replacement of internal organs by internal budding, and sexual production of larvae at time of host moulting.

Only described in December 1995, the cycliophorans are minute epibionts on the setae of the mouthparts of decapod crustaceans. Most of the life-cycle appears to be spent as a solitary, 350 μm long 'feeding stage'. This has a body comprising: (i) a funnel-shaped feeding organ constricted at its base and flaring distally into a circular whorl of compound cilia surrounding the mouth; (ii) a trunk housing the brain and a gut which terminates in an anus located on a small papilla near the base of the feeding organ; and (iii) a short cuticular stalk ending in a circular attachment disc. The body is covered by a sculptured cuticle that is similar in construction to that of some nemathelminths (Sections 4.4–4.10) and gastrotrichs (Section 4.3).

The young feeding stage asexually develops an internal bud that grows within it and eventually replaces the feeding organ, gut and nervous system of the parent. Several generations of such buds may occur within the life span of one 'individual' feeding stage which increases in size during the process. Eventually maturity is reached and instead of a new feeding stage a 'pandora larva' is budded internally; this in turn develops further internal buds. On being released, it settles and repeats the feeding-stage asexual cycle. When the host is about to moult, the cycliophoran feeding stage buds (again internally) either a male or female sexual form. The male is a much smaller version (<100 μm long) of the feeding stage, but lacks a gut and feeding organ. On its release, this dwarf male settles onto a feeding stage that contains a developing female. After fertilization, the single egg of the female develops into a 'chordoid larva', still inside the mother. This eventually escapes from the shell of the dead female, disperses by means of locomotory cilia, and colonises a new host. It then initiates a new cycle of asexual feeding stages which continues whilst the host is in intermoult. The life cycle is therefore complex (Fig. 6.21) and based on the asexual production of clonal individuals.

Cycliophorans feed on suspended particles, using the crown of cilia, in a manner essentially similar to rotifers (Section 4.9), although insofar as is known they appear most closely allied to the entoprocts (Section 6.4).

6.5.3 Classification

A single species, living on the Norway lobster, *Nephrops*, is currently known, although other species probably await discovery. It is placed in the order Symbiida.

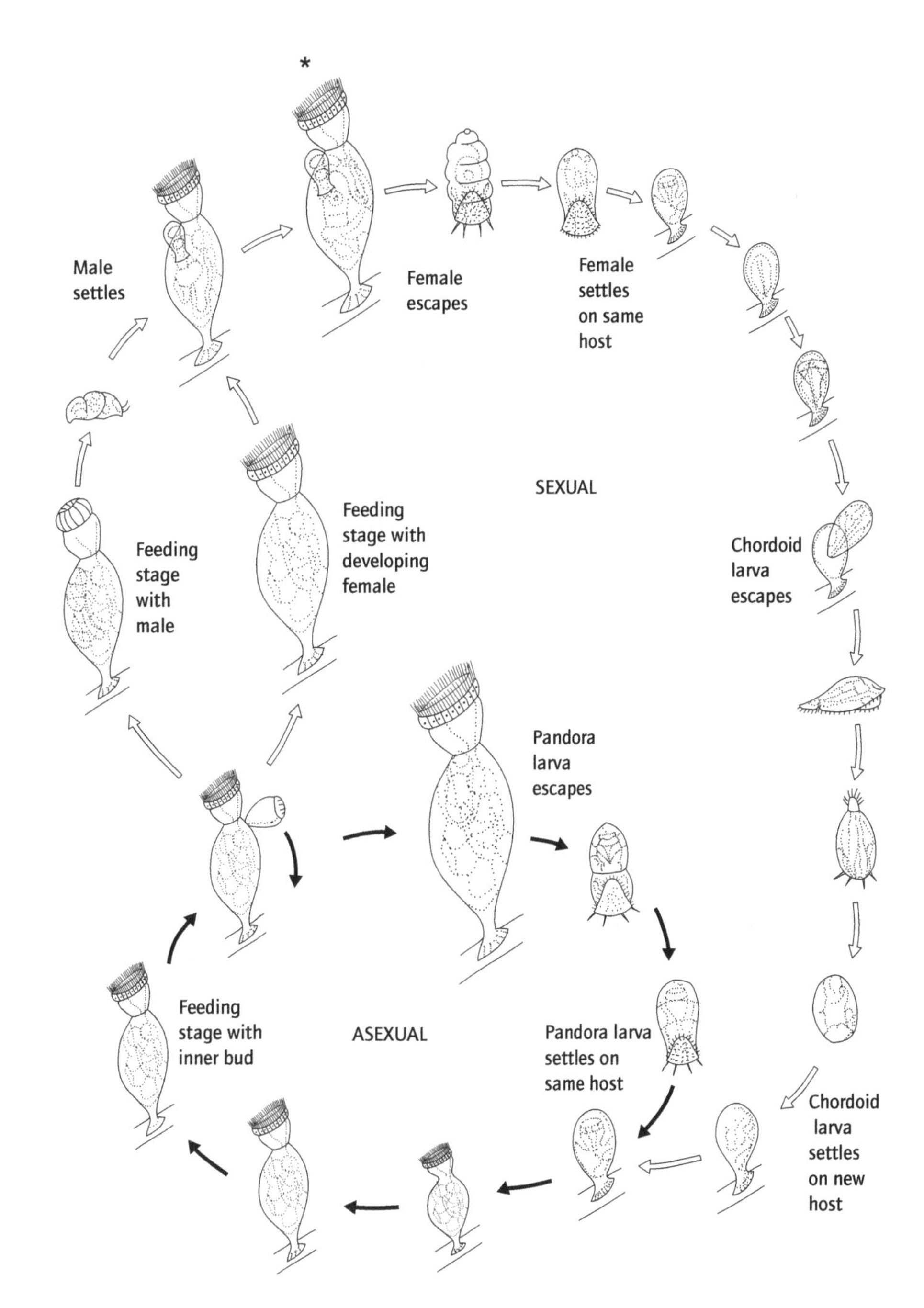

Fig. 6.21 The complex cycliophoran life cycle. The stage in the cycle shown in Fig. 6.20 is asterisked (after Funch & Kristensen, 1995).

6.6 **Further reading**

Emig, C.C. 1979. *British and Other Phoronids.* Academic Press, London [Phorona].

Funch, P. & Kristensen, R.M. 1995. Cycliophora is a new phylum with affinities to Entoprocta and Ectoprocta. *Nature (Lond)*, **378**, 711–714.

Halanych, K.M., Bacheller, J.D., Aguinaldo, A.M.A., Liva, S.M., Hillis, D.M. & Lake, J.A. 1995. Evidence from 18S ribosomal DNA that the lophophorates are protostome animals. *Science (New York)*, **267**, 1641–1643.

Hyman, L.H. 1951. *The Invertebrates*, Vol. 3: *Acanthocephala, Aschelminthes and Entoprocta.* McGraw-Hill, New York [Entoprocta].

Hyman, L.H. 1959. *The Invertebrates*, Vol. 5: *Smaller Coelomate Groups.* McGraw-Hill, New York [Phorona, Brachiopoda & Bryozoa].

Nielsen, C. 1971. Entoproct life-cycles and the entoproct/ectoproct relationship. *Ophelia*, **9**, 209–341 [Entoprocta].

Rudwick, M.J.S. 1970. *Living and Fossil Brachiopods.* Hutchinson, London [Brachiopoda].

Ryland, J.S. 1970. *Bryozoans.* Hutchinson, London [Bryozoa].

Wright, A.D. 1979. Brachiopod radiation. In: House, M.R. (Ed.) *The Origin of Major Invertebrate Groups*, pp. 235–252. Academic Press, London [Brachiopoda.]

CHAPTER 7

The Deuterostomes

Chaetognatha
Hemichordata
Echinodermata
Chordata

That the hemichordates, echinoderms and chordates are a natural group of related phyla is suggested by their common possession of a number of unusual developmental, structural and biochemical features, and of a series of trends unknown in other groups. During the early embryological development of most deuterostomes, for example, (a) the blastopore does not form the mouth, which is therefore a secondary opening into the gut (and hence the name 'deuterostome'), (b) cleavage of the cells of the blastula occurs in a radial pattern, (c) the developmental fate of the cells is not fixed until a relatively late stage of morphogenesis ('indeterminate development'), and (d) their body cavities are formed by outpocketings from the embryonic gut, creating a series of 'enterocoelic pouches' (see, for example, Fig. 2.7). These are typical of deuterostomes particularly in combination, since individually they (especially (b) and (d)) are also known in various other phyla. Further deuterostome characteristics include photoreceptors of the ciliary type (in contrast to the rhabdomeric photoreceptors of the protostomes), the prevalence of monociliated cells in the epidermis, creatine phosphate as the phosphate store (rather than the more usual arginine phosphate of the protostomes), and the virtual absence of chitin.

The ancestral deuterostome was probably not dissimilar to the modern pterobranch hemichordates (see Section 7.2.3.2). That is they would have been short, sedentary, worm-like animals with tripartite oligomeric bodies, the second (mesosomal) region bearing a lophophore, and the third (metasomal) compartment the temporary aggregations of peritoneal cells which probably comprised the gonads; further, their nervous system would have largely been in the form of a diffuse subepidermal plexus concentrated into one or more longitudinal cord-like thickenings.

In contrast to all known lophophorates, however, although similar to some gastrotrichs (Section 4.3.3), the water currents used to convey those food particles trapped by the lophophore into the anterior gut left the body via a number of lateral apertures extending from the pharynx right through to the body surface. The ancestral gut, therefore, was connected to the external environment by more orifices than just mouth and anus.

Most subsequent lines of deuterostome evolution lost or greatly modified the lophophore, lost the original tripartite body plan in association with the evolution of a sessile, attached existence (echinoderms and some chordates) or of a paedomorphic, free-swimming life style (most chordates), and two groups, at least, modified the pharyngeal perforations to serve alternative functions associated with pharyngeal filter feeding (early chordates) or with gaseous exchange (enteropneust hemichordates and many later chordates).

Other trends within the deuterostomes not shown by protostome phyla include: enrolling the subepidermal nerve plexus to form a hollow, dorsal nerve tube (chordates and some hemichordates); developing a propulsive post-anal tail moved in S-shaped undulations by longitudinal muscles (chaetognaths and chordates); and the deposition of protective dermal calcareous plates, which in two separate lines subsequently became used as part of a hard internal skeletal system against which locomotory muscles could act (some echinoderms and some chordates).

The chaetognaths although sharing several features with the other deuterostomes seem most likely to have evolved them by convergence. Their true affinities are unknown and even the molecular sequence data that have resolved a number of controversial issues has failed to shed much light on the issue. They are treated in this chapter for convenience.

7.1 Phylum CHAETOGNATHA (arrow-worms)

7.1.1 Etymology

Greek: *chaite*, long hair; *gnathos*, jaw.

7.1.2 Diagnostic and special features (Fig. 7.1)

1 Bilaterally symmetrical, vermiform.
2 Body more than two cell layers thick, with tissues and organs.

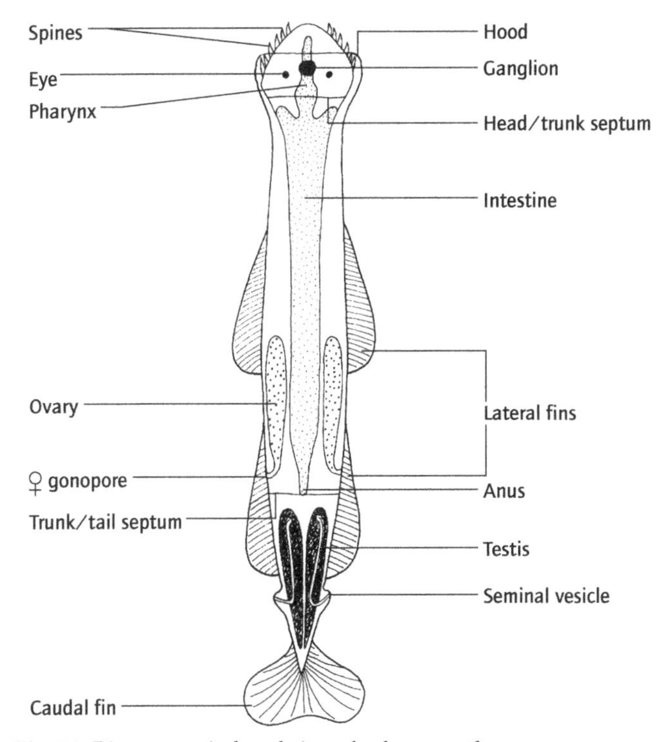

Fig. 7.1 Diagrammatic dorsal view of a chaetognath (after several sources).

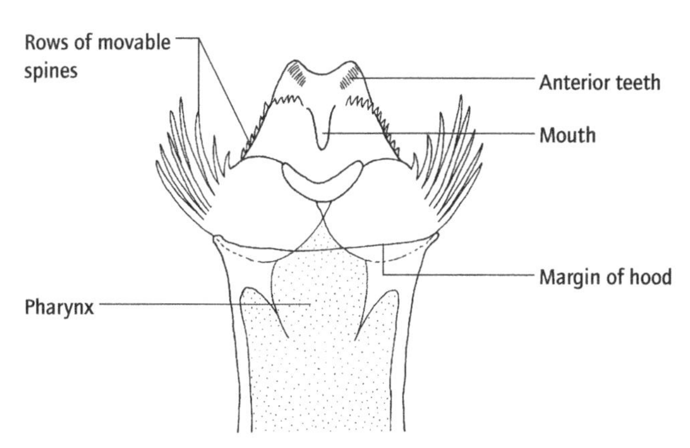

Fig. 7.2 Ventral view of the head of a chaetognath (after Ritter-Zahony, 1911).

3 A straight, through gut and a non-terminal, ventral anus.
4 An oligomeric, tripartite body divided by septa into a head, trunk and post-anal tail; the trunk and tail bearing lateral and caudal non-muscular fins.
5 Each body region with one or two body cavities, which are enterocoelic in origin but in the post-juvenile stages lack a peritoneal lining.
6 Body wall with a non-chitinous cuticle and bands of longitudinal muscle.
7 Without circulatory, gaseous-exchange and excretory systems.
8 Nervous system with a circum-pharyngeal ring ganglionated dorsally and laterally, from which individual nerves issue.
9 Hermaphrodite, with large paired testes and ovaries, the latter with gonoducts.
10 Eggs cleave radially.
11 Development deuterostomatous and direct.
12 Marine.

Chaetognaths are torpedo-shaped carnivores which swim by means of rapid flicks of the post-anal tail in the dorsoventral plane, the lateral fins having a stabilizing function. Prey are grasped by series of movable, non-chitinous spines flanking and in front of the ventral chamber which leads into the mouth (Fig. 7.2), although, when not feeding, the head with its spines is covered by a fold of the body wall, the hood, which probably serves mainly to reduce drag as well as for protection.

A pair of eyes, formed by the fusion of individual ocelli, are present on the head. The photoreceptors in these eyes are of the ciliary type, as in the other groups of deuterostomes, but the chaetognaths are not typical deuterostomes. In particular, none of the adult body cavities possesses a peritoneum (otherwise a diagnostic feature of a coelom), and the tail enterocoels, of which there are one or two, are secondary derivations of the paired trunk cavities and therefore do not originate as separate pouches from the archenteron as in the other tripartite oligomeric groups. The head cavity is single.

Chaetognaths may attain lengths of up to 12 cm and form the dominant group of planktonic predators in the sea, preying on organisms ranging in size from protists to young fish as large as themselves which they subdue with the aid of the neurotoxin tetrodotoxin.

7.1.3 Classification

The 90 known species are placed in a single class containing two orders on the basis of the presence (Phragmophora) or absence (Aphragmophora) of a transverse ventral musculature. The Phragmophora includes, *inter alia*, the only non-planktonic genus, the benthic *Spadella* (Fig. 7.3).

7.2 Phylum HEMICHORDATA

7.2.1 Etymology

Greek: *hemi*, half; Chordata, referring to the phylum of that name (Section 7.4).

7.2.2 Diagnostic and special features

1 Bilaterally symmetrical.
2 Body more than two cell layers thick, with tissues and organs.
3 A through, straight or U-shaped gut.
4 An oligomeric, tripartite body, comprising a large prosomal proboscis, a small mesosomal collar, and a large, elongate or sac-like metasomal trunk.

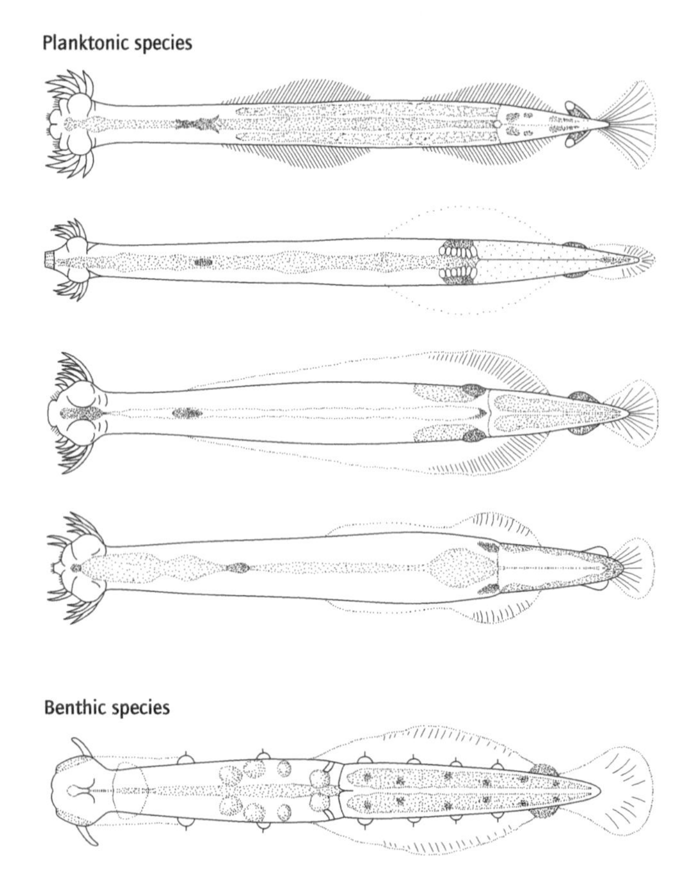

Fig. 7.3 The body form of chaetognaths (after Pierrot-Bults & Chidgey, 1987).

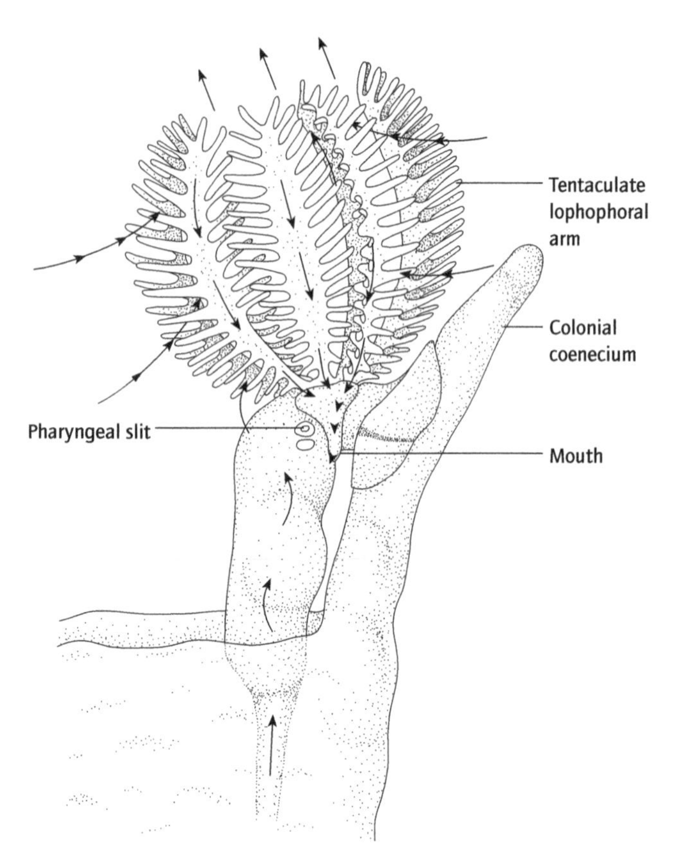

Fig. 7.4 Feeding currents in a pterobranch hemichordate (*Cephalodiscus*). Curved arrows indicate water moving towards lophophoral arms and into the rejection current; upwardly pointing arrows represent the rejection current in the centre of the lophophoral apparatus; arrows on the body show particle movements caused by epidermal cilia; and arrowheads show food moving towards mouth (after Lester, 1985).

5 Each body region with one (proboscis) or two (collar and trunk) enterocoelic body cavities.
6 In some, a lophophoral organ on the collar is supported by the mesocoels.
7 In all but one group, the upper half of the pharyngeal wall is perforated by 1 to >100 pairs of slits through which a water current is discharged; only the lower half of the pharyngeal tube is then alimentary.
8 A ciliated, cuticle-less epidermis, in some secreting an external non-chitinous tube.
9 Circulatory system partially open.
10 Excretory organ a glomerulus formed by peritoneal evaginations into the protocoel.
11 Nervous system diffuse and basiepidermal, in some with a hollow dorsal neurocord in the collar; basiepidermal network concentrated mid-dorsally and mid-ventrally.
12 Gonochoristic.
13 Eggs cleave radially.
14 Development deuterostomatous and indirect.
15 Marine.

Hemichordates are of two main types, which differ considerably in their body forms and life styles: the vermiform, burrowing enteropneusts and the sessile, tubicolous and often colonial pterobranchs. Both, in their different ways, employ mucociliary feeding mechanisms, however. In the pterobranchs, a lophophoral organ is present dorsally on the collar, the arms of which are held so as to form a cone flaring away from the ventral mouth (in the lophophorate phyla and in all other groups feeding by means of a tentacular ring, the tentacles are held so as to surround and encircle the mouth), and water is induced to flow in through the circlet of tentacles and out through the intralophophoral cavity (as also seen in, for example, the sabellid polychaetes and the entoprocts). Particles then pass down the outside faces of the tentacular arms, by ciliary means, towards the mouth (Fig. 7.4). In addition, food particles may be collected by cilia over the general body surface. The latter is the sole method of ciliary feeding of the lophophore-less enteropneusts, the ciliary tracts being especially well developed on the proboscis (Fig. 7.5). The pterobranchs and enteropneusts may be connected via the possibly intermediate 'lophenteropneusts', as yet known only from photographs of the deep-sea floor.

In both forms, ingestion of food particles is probably aided, as in other microphagous feeders, by a water current drawn into the buccal cavity, and, as in some gastrotrichs (Section 4.3.3), this

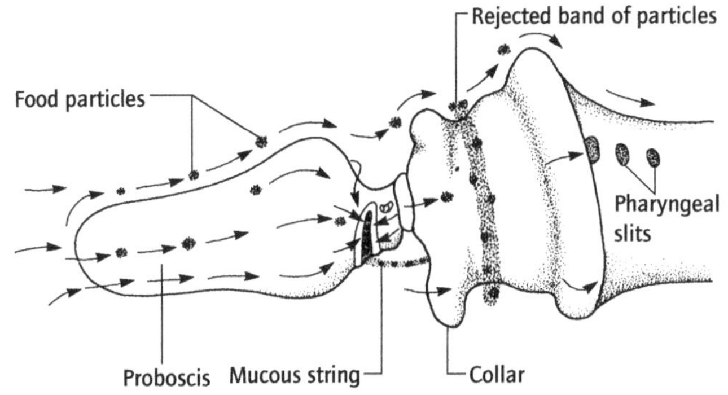

Fig. 7.5 Ciliary feeding currents on the proboscis and collar of an enteropneust hemichordate (after Barrington, 1965).

water current is discharged through pharyngeal pores opening on the body surface. Such a unidirectional flow is a more efficient system than seen in the lophophorate phyla, for example, which periodically have to expel through the mouth the excess water taken in with the food particles.

As their name suggests, hemichordates share a number of special features with the phylum Chordata. Chordates are generally considered to be characterized by four such special features: the presence of a notochord; a pharynx perforated to the exterior by slits or pores; a post-anal tail and a hollow dorsal nerve tube (Section 7.4). Of these, at least the enteropneusts share two (the hollow dorsal nerve tube and the perforated pharynx), and arguably a third in that the young of some species have been described as possessing a post-anal tail. Nevertheless, as in the chaetognaths, the chordate tail is essentially a locomotory structure, which is not the case with any post-anal hemichordate body region (of course, many invertebrates, particularly those with U-shaped guts, possess post-anal body regions, and the juvenile enteropneust 'tail' is probably homologus with the pterobranch stalk). A notochord is definitely lacking in the hemichordates, however, although a forwardly directed diverticulum of the gut, issuing from the buccal cavity, was for many years misinterpreted as a notochord. Some hemichordates do then possess half of the special chordate features, and would thus appear to be aptly named.

7.2.3 Classification

The 100 known species of hemichordate are divided between three classes.

Class	Order
Enteropneusta	Helminthomorpha
Pterobranchia	Rhabdopleurida Cephalodiscida
Planctosphaeroidea	Planctosphaerida

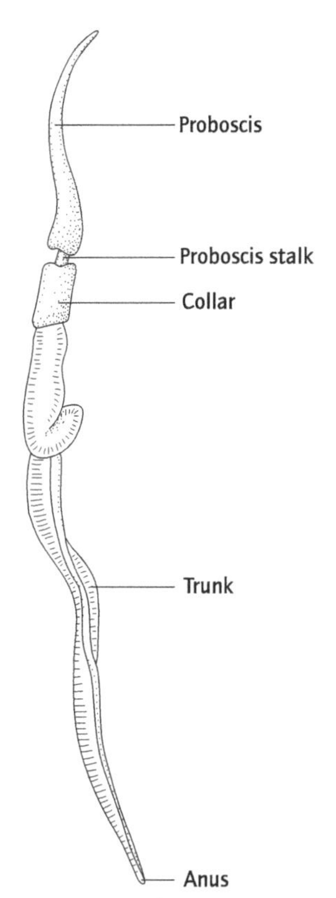

Fig. 7.6 The external appearance of an enteropneust (after Marion, 1886).

7.2.3.1 *Class Enteropneusta (acorn-worms)*

These solitary, mobile, elongate worms, of up to 2.5 m in length, occupy burrows in soft sediments, live under stones, etc., where they deposit- and suspension-feed. Their bodies are in the form of (a) a long proboscis, the cilia on which provide both the main propulsive force and the transport system for food particles; (b) a short, lophophore-less collar; and (c) a very long trunk bearing a terminal anus and many pharyngeal pores (Fig. 7.6), the number increasing throughout life. In the pharyngeal wall these perforations are U-shaped, with a supporting system of skeletal bars, but they open on the body surface as small, circular dorsal pores. The neurocord and glomerulus are well developed in enteropneusts, but the body cavities are largely occluded by muscle fibres and connective tissue formed from the peritoneum, which replace much of the body wall musculature.

Both asexual multiplication by fragmentation and sexual reproduction occur, many pairs of gonads, each discharging by means of a duct, being located in the anterior region of the trunk. Fertilization is external and proceeds through developmental stages closely similar to those of many echinoderms to, in most species, a tornaria larva (Fig. 7.7).

All 70 enteropneust species are placed in the one order.

7.2.3.2 *Class Pterobranchia*

The pterobranch hemichordates are sedentary, tube-dwelling animals, of up to 12 mm in length, with a short body comprising

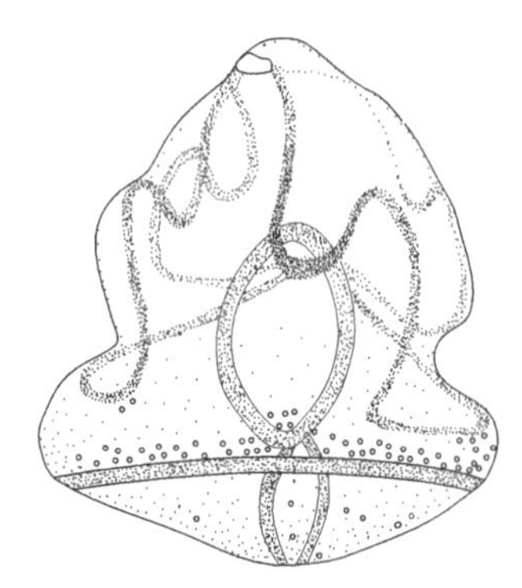

Fig. 7.7 The tornaria larva of an enteropneust (after Stiasny, 1914).

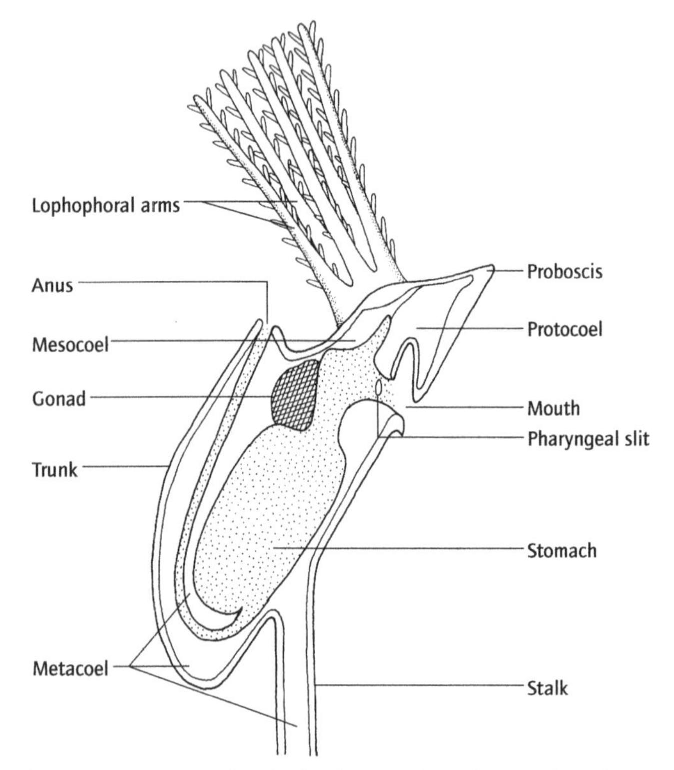

Fig. 7.8 Diagrammatic longitudinal section through a pterobranch (after McFarland *et al.*, 1979 and others).

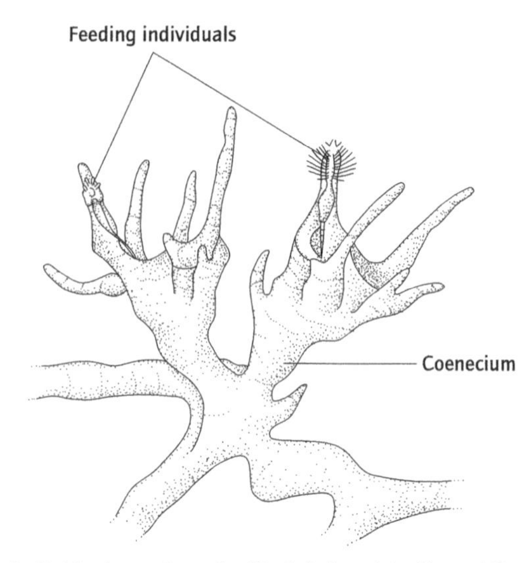

Fig. 7.9 Individual pterobranchs (*Cephalodiscus*) feeding whilst clinging to spines projecting from the common coenecium (after Lester, 1985).

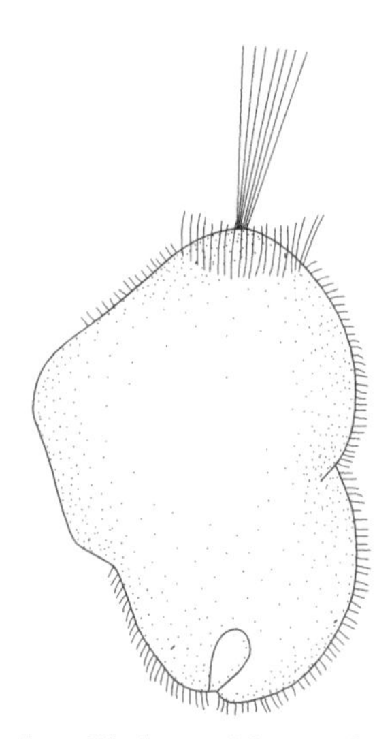

Fig. 7.10 The trochophore-like larva of the pterobranchs (after Schepotieff, 1909).

(a) a shield- or disc-shaped proboscis, which is responsible both for the (ciliary) locomotion within the tube and for secreting this collagenous structure, (b) a collar bearing ventrally the mouth and dorsally from one to nine pairs of lophophoral arms from each of which issues a double row of ciliated and mucus-secreting tentacles, and (c) a short sac-like trunk with near the collar, an anal papilla, and, subterminally, a contractile stalk (Fig. 7.8). This stalk may end in an organ of temporary attachment, may pass into a communal stolon, or be wrapped around a support, like the prehensile tail of many an arboreal mammal. In contrast to the enteropneusts, at most only one pair of pharyngeal perforations are present, the neurocord is absent and the glomerulus is poorly developed, the coeloms are not occluded, and only one or a pair of gonads occur.

Asexual budding is widespread, leading in one group to colonies of individuals joined by their stolonic stalks and, in the other, to clones of separate individuals living within a common 'coenecium' (Fig. 7.9). Little is known of their sexual reproduction, but the larval stage (Fig. 7.10) does not resemble that of the enteropneusts.

The 10 species of pterobranchs are divided between two orders. The Cephalodiscida contains the non-colonial forms which either creep over the surfaces of hydroid colonies or live within a communal coenecium, climbing up to feed near its apertures (Fig. 7.9). They possess one pair of pharyngeal perforations, from four to nine pairs of lophophoral arms, and a pair of gonads. The Rhabdopleurida are the colonial pterobranchs (Fig. 7.11): they have a single gonad, a single pair of lophophoral arms, and no pharyngeal slits.

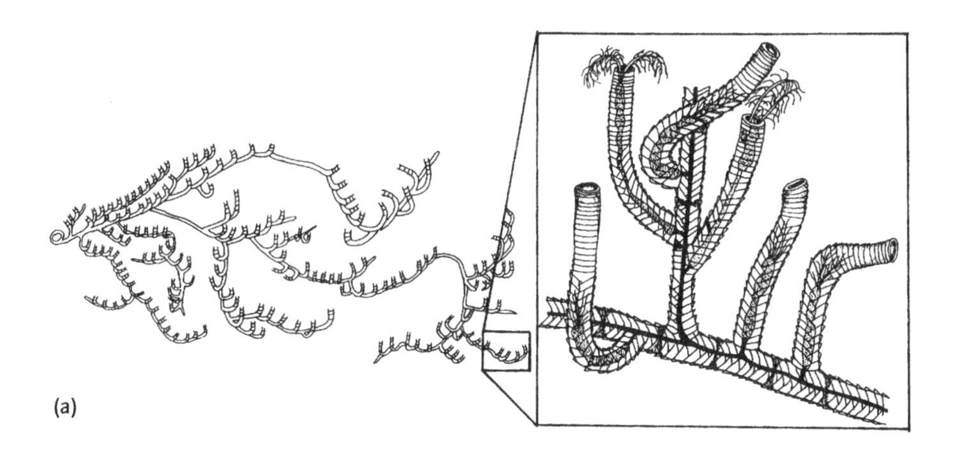

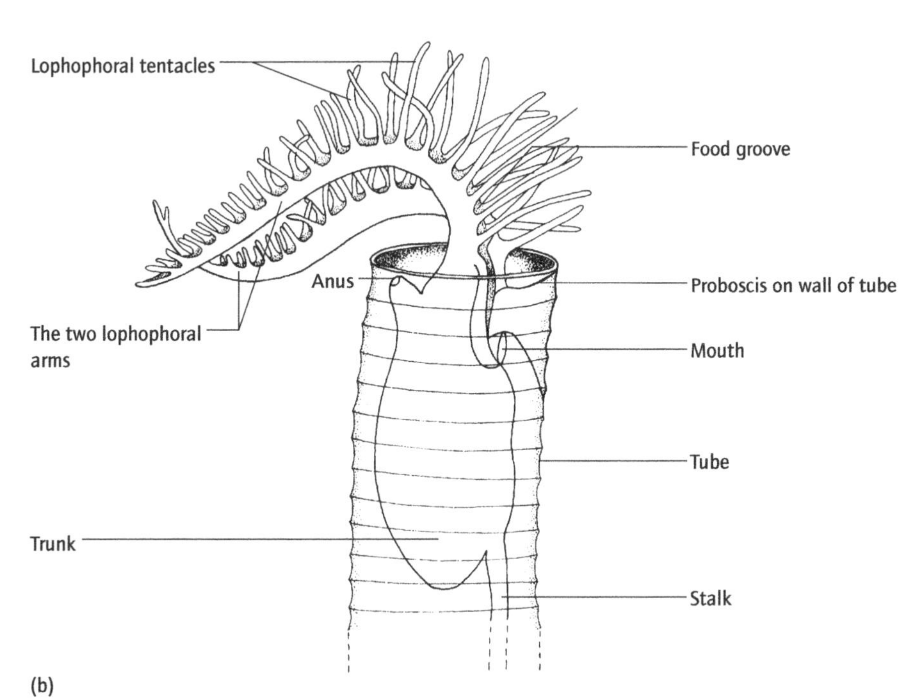

Fig. 7.11 Part of a rhabdopleuran colony (a), with enlarged detail of an individual zooid (b). (After Grassé, 1948 and others.)

7.2.3.3 *Class Planctosphaeroidea*

This class was erected to receive, hopefully only temporarily, some giant larvae (up to 2.2 cm in diameter) resembling in their general form those of enteropneusts (Fig. 7.12), which have been taken by plankton tows in the Atlantic Ocean since the early 1930s. As yet the adult form is still unknown, and their status and affinities therefore remain uncertain.

7.3 Phylum ECHINODERMATA

7.3.1 Etymology

Greek: *echinos*, hedgehog; *derma*, skin.

7.3.2 Diagnostic and special features (Fig. 7.13)

1 Adult with five-rayed symmetry, effectively radial in most, bilateral in some.
2 Body more than two cell layers thick, with tissues and organs.
3 A through gut (secondarily blind-ending in some, lacking in one group).
4 Body shape highly variable: ancestral and sessile forms with spherical or cup-shaped body attached aborally to the substratum by a stalk and with the upwardly directed mouth surrounded by a circle of five arms; derived free-living forms, without stalk, with body oriented so that the mouth is on the lower surface (or, more rarely, is anterior), with or without discrete arms, sometimes bilaterally symmetrical.

Fig. 7.12 Ventral view of *Planctosphaera* (after Spengel, 1932).

5 Without any differentiated head.
6 Basically with oligomeric, tripartite paired enterocoelic body cavities, the metacoels ('somatocoels') forming the main body cavity; in a few forms with direct development, the body cavities are formed schizocoelically.
7 A water vascular system derived mainly from the left mesocoel (left 'hydrocoel') and partly from the left protocoel (left 'axocoel'), which contains and is in indirect communication with sea water and which operates hydraulically the locomotory tube feet and/or feeding tentacles ('podia'); part of the system forms a ring canal encircling the oesophagus, from which issue an ambulacral canal along each arm.
8 A mesodermal, subepidermal system of calcareous ossicles or plates, often bearing projecting tubercles or spines. The dermis also includes a unique 'catch connective tissue' system.
9 Without excretory organs.
10 Circulatory 'haemal' system poorly defined, its function largely being accomplished by coelomic fluids.
11 Nervous system subepidermal, in the form of a circum-oesophageal ring from which issue diffuse nerves along each ambulacrum.
12 Usually gonochoristic.
13 Eggs cleave radially.

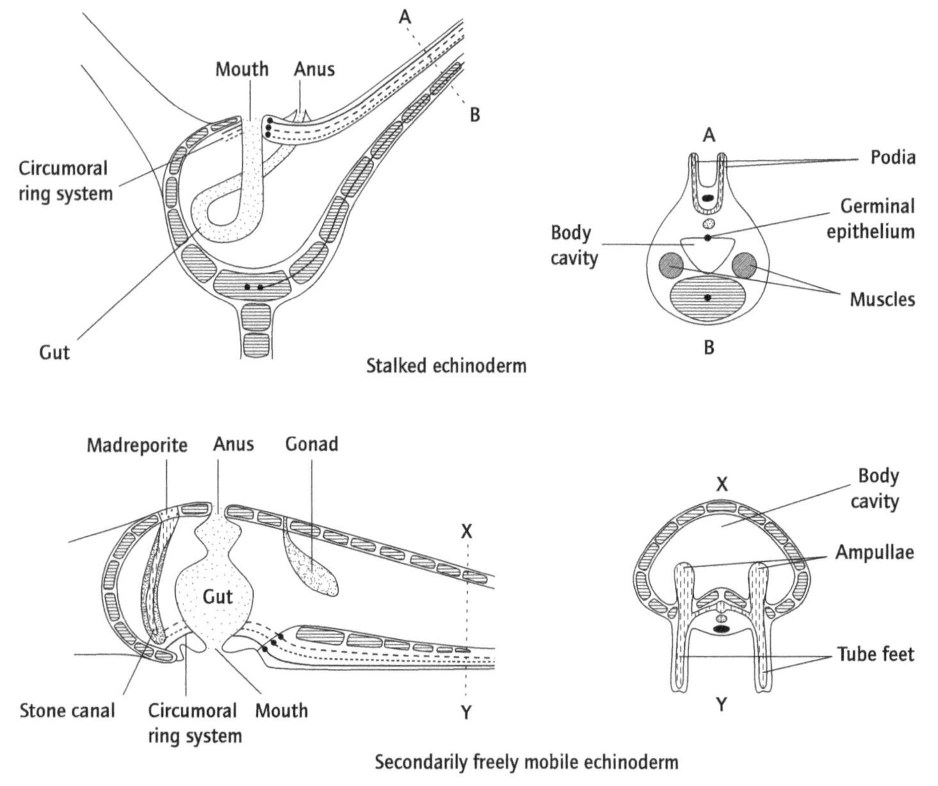

Fig. 7.13 Diagrammatic sections through two generalized echinoderms (after Nichols, 1962).

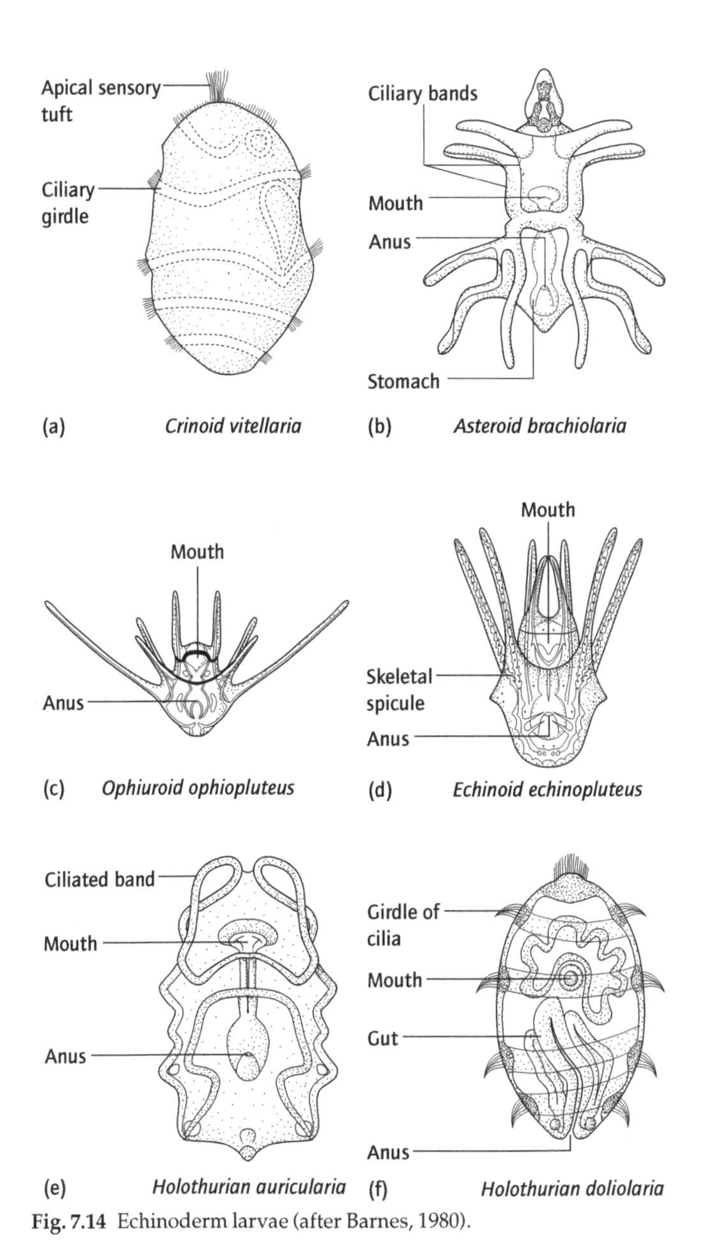

Fig. 7.14 Echinoderm larvae (after Barnes, 1980).

14 Development deuterostomatous, characteristically indirect via ciliated, bilaterally symmetrical larvae (Fig. 7.14).
15 Marine.

The echinoderms are a highly individual group, with three very distinctive features: their symmetry, calcareous mesodermal skeleton, and water vascular and other coelomic systems. All three of these can be related directly to their sessile, suspension-feeding origins. Sessile suspension feeders of whatever ancestry tend towards radial symmetry of their food collection apparatus (Section 9.1), and in the echinoderms this takes the form of a circle of five arms (often dichotomously branching), which in the ancestral forms were equivalent to, and perhaps derived from, the lophophorate arms of animals such as the pterobranch hemichordates (Section 7.2.3.2). Like the latter, the hydraulic system of the arms and of the ciliated tentacles borne on them is provided (mainly) by mesocoelic pressure. A number of echinoderm lines, however, have secondarily become free-living, and in most of these the position of the body axis in life has been turned through 180° so that the erstwhile upper mouth- and arm-bearing surface is in contact with the substratum. The former feeding podia now function in locomotion, with relatively little anatomical change being developed into tube feet (one common adaptation to this changed function being the development of a hydraulic reservoir, the ampulla, for each tube foot, permitting them individually to dilate and contract). The arms have also often become much larger and broader (as in the asteroids) or have been incorporated into – wrapped around – the body (as in the echinoids and holothurians). In any event, the five arms still dominate the body plan, so that in the majority of free-living forms the ancestral, effectively radial symmetry has been retained and consequently the animals can move in any direction. (Not all the early types of echinoderm displayed the pentaradial symmetry characteristic of all surviving groups: some were helicoidally symmetrical, and others showed no basic pattern of symmetry at all – see Fig. 2.12.)

Sessile animals also require means of support and protection from predators and wave action. The echinoderms have achieved these by depositing a system of ossicles in their dermis, each ossicle being a porous lattice of calcium carbonate which behaves as if it was a single crystal of calcite. Individual ossicles are often increased in size by accretion to form plates which may interlock with other such plates, even to the extent of forming a rigid almost external box or test, and in common with other such enclosed animals (e.g. bivalve molluscs and brachiopods) any ancestral head has been lost. The overlying, generally ciliated epidermis may be missing and parts of the test thereby become the outer margin of the body. Further, the calcareous ossicles may secondarily assume a role in locomotion, replacing the tube feet in this respect, with muscles running between different ossicles effecting movement of the arms or of spines against the substratum (see Chapter 10).

Echinoderms have been described as being very small animals in very large bodies. Most of their apparent body is test material and/or coelomic cavities, and their metabolic rate per unit gross weight is minute. In fact, their oxygen consumption per unit live tissue weight is also tiny. This is largely because they do not use muscles to maintain posture, as other animals do; instead, this is carried out by a unique dermal 'catch connective tissue' system. The rigidity of this can be varied rapidly, under nervous control, to an extent almost equivalent to that between a solid and a liquid state. When parts of the animal move, the collagen fibres can slide relative to each other, but when a given desired position has been achieved the fibres can (reversibly) lock together immovably as a result of changes induced in the extracellular matrix. Position can then be maintained without any muscular activity. No other animals have evolved such a system, but then no other mobile animals are so poorly equipped with muscles as a consequence of their sessile ancestry.

Although the body of an echinoderm is monomeric, three pairs of enterocoelic pouches develop (sometimes in modified form), as in the oligomeric, tripartite groups to which echinoderms are generally regarded as being related. And, as in the hemichordates, the metacoels ('somatocoels') form the main body cavity, and (in the echinoderms, *one* of the) mesocoels ('hydrocoels') provide the hydraulic system of the lophophore-like arms; the right mesocoel normally atrophies to a small pulsatile sac. The right protocoel ('axocoel') forms the axial sinus, a space around part of the haemal system of debatable function; whilst the left one is incorporated into the mesocoelic water vascular system. The protocoels and mesocoels of the oligomeric enterocoelic animals communicate with the external environment each by means of a small pore, and this is classically regarded as having been greatly developed by the echinoderms as a means of varying the quantity of liquid in the water vascular system. The common axohydropore is in the form of a porous plate, the madreporite, through which it is suggested water is exchanged with the environment (effectively the water vascular system contains sea water with a few coelomocytes, rather than coelomic fluid), a 'stone canal' running from that plate to the circumoesophageal ring. Actual passage of water across the madreporite has never been observed, however, and it may serve instead to equalize hydrostatic pressure inside and outside the animal.

The peculiarities of the echinoderms are therefore not so much in their anatomy *per se*, but result from the fact that they are the only successful free-living animals to have descended from an attached, sessile group probably capable of moving only their tentaculate arms and of bending their stalk.

7.3.3 Classification

Although some 6750 living species are known, the diversity of echinoderms is much lower now than it was in the Palaeozoic: only six of the 24 component classes survive.

Class	Order
Crinoidea	Isocrinida
	Comatulida
	Millericrinida
	Bourgueticrinida
	Cyrtocrinida
Asteroidea*	Brisingida
	Forcipulatida
	Valvatida
	Notomyotida
	Paxillosida
	Velatida
	Spinulosida
Ophiuroidea	Oegophiurida
	Phrynophiurida
	Ophiurida
Concentricycloidea	Peripodida
Echinoidea	Cidaroida
	Echinothuroida
	Diadematoida
	Pedinoida
	Salenoida
	Phymosomatoida
	Arbacioida
	Temnopleuroida
	Echinoida
	Holectypoida
	Clypeasteroida
	Cassiduloida
	Spatangoida
	Neolampadoida
	Holasteroida
Holothuroidea	Dendrochirotida
	Dactylochirotida
	Aspidochirotida
	Elasipoda
	Apodida
	Molpadiida

* Currently in a state of flux.

7.3.3.1 Class Crinoidea (feather-stars and sea-lilies)

Crinoids are the only surviving echinoderms to have retained the ancestral body posture of an upwardly directed mouth in the centre of a circle of mucociliary, suspension-feeding, podia-bearing arms. Although basically five arms are present, they may divide repeatedly to form from ten to more than 200 apparent arms, each with numerous side branches (pinnules) on to which the upwardly facing ambulacral grooves extend (Fig. 7.15). Particles are trapped by the ciliated and mucus-secreting podia and conveyed along ciliated ambulacral grooves to the mouth, and thence into the U-shaped gut (Fig. 7.16). The water vascular and other coelomic systems are simple, there being no ampullae to operate the podia and no madreporite, the often numerous stone canals opening into the somatocoels. (The podia are extended by contraction of the sphinctered water vascular canals.)

The globose body is permanently or temporarily attached to the substratum by a non-contractile aboral stalk (which can attain a length of 1 m), often bearing whorls of cirri and terminating in an attachment disc or in a system of cirri or rootlets which can grasp or anchor it in the sea bed. In many living species, after a brief period of attachment, the animal becomes free-living, swimming by use of its feather-like arms and temporarily attaching itself to the substratum by aboral cirri: the ancestral, mouth-upwards posture is nevertheless maintained. The arms, body and stalk possess heavily calcified plates, which take up most of the body volume so that the tissue space is small.

Gonadal tissue is diffuse; gametes form from areas of peritoneum within the arms, develop in small proximal 'gonads', and are released by rupture of the pinnule walls. A vitellaria larva (see Fig. 7.14a) is characteristic; this metamorphoses into a miniature stalked version of the adult.

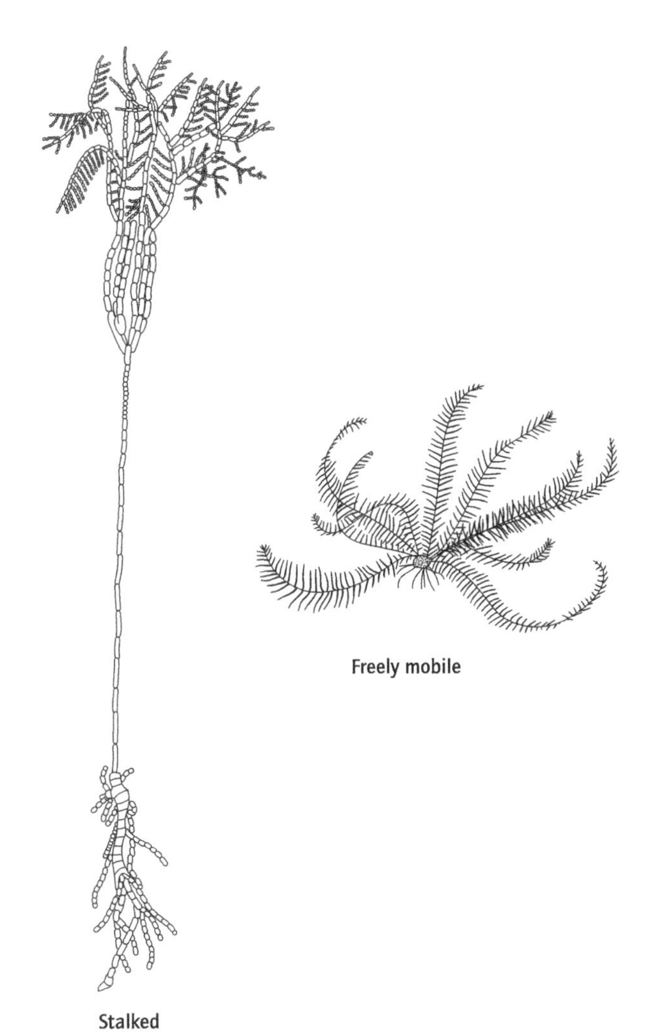

Fig. 7.15 Crinoid body forms (after Danielsson, 1892 and Carpenter, 1866).

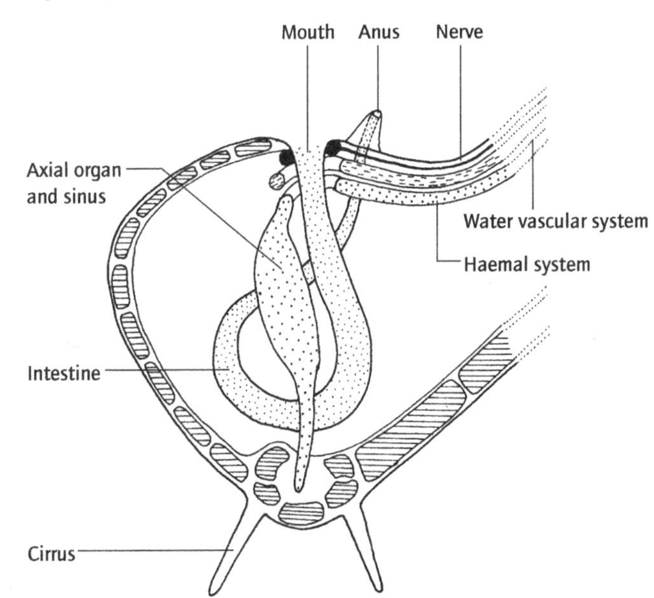

Fig. 7.16 Diagrammatic section through the body and part of an arm of a free-living crinoid (after Nichols, 1962).

Five orders are distinguished, mainly on details of the construction of the stalk and of its system of attachment; they contain a total of 625 species.

7.3.3.2 Class Asteroidea (starfish)

The asteroids are generally characterized by a flattened body which grades imperceptibly into the five, or sometimes more (up to 40), arms. These may be short and broad, giving the animal a pentagonal outline, through to very long and slender (Fig. 7.17). The animals are unattached and freely mobile, the mouth and ambulacral grooves being positioned on the lower surface and the madreporite and anus (if present) aborally. The ambulacral grooves are well developed and protected along each margin by a double row of unfused ossicles, through notches in which pass ducts from the internal ampullae to the locomotory tube feet, which are often suckered. Otherwise, the calcareous skeleton is somewhat loosely organized, although it bears external tubercles and spines, sometimes in definite arrangements. Some of the spines may be modified into pedicellariae (Fig. 7.18): groups of usually three small ossicles which can interact in the manner of scissors or forceps, and thereby remove other organisms attempting to settle on its body surface.

The body cavities, including the water vascular system, are extensive, and include small extensions of the somatocoels, papulae, which project through the body wall aborally and serve a gaseous-exchange function (Fig. 7.19). Some asteroids are hermaphrodite and several forms of multiplication and development occur, including asexual fission, brooding and direct development (especially in high latitudes), and indirect development via bipinnaria and brachiolaria larvae (see Fig. 7.14b). Most species are scavengers or predators on sessile or sedentary prey, but deposit and suspension feeders are also represented. In macrophagous feeding, the prey may be ingested whole or the stomach, which is very large, everted through the mouth on to the prey tissues; from the stomach issue a pair of large caeca into each arm.

Seven orders, containing 1500 species, are recognized on the basis of the type of pedicellariae present, the arrangement of the ambulacral ossicles, and the form of the tube feet. On its discovery in 1962, the genus *Platasterias* was hailed as a surviving somasteroid, an echinoderm class otherwise long extinct. The consensus view today, however, is that it is not a somasteroid but a specialized asteroid.

7.3.3.3 Class Ophiuroidea (brittlestars and basket-stars)

Like the asteroids, to which they are related, the ophiuroids are flat, free-living echinoderms with a mouth located on the lower surface of the body; but in this group, the central body is small, disc-shaped and sharply demarcated from the long, narrow arms (Fig. 7.20). The calcareous skeleton occupies most of the arm volume (and is similarly well developed in the central disc), individual arm ossicles being fused together to form longitudinal series of 'vertebrae' which articulate with each other and which can be moved by intervertebral muscles. By these means,

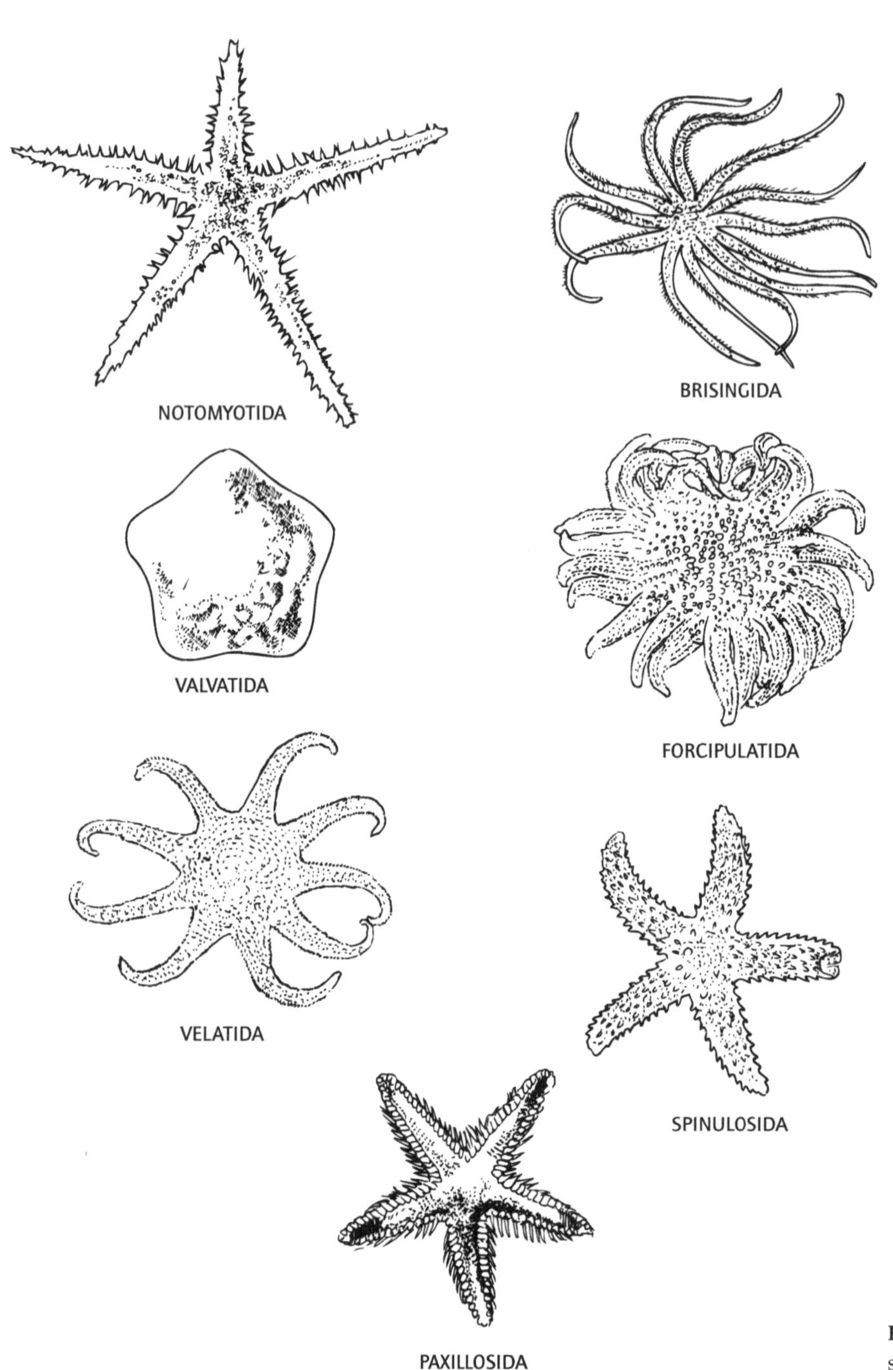

Fig. 7.17 Asteroid body form (after several sources).

ophiuroids walk on their arms, the central disc often being held off the ground; accordingly, the tube feet have no locomotory function, except to help the arms gain purchase on the substratum, and are used mainly for feeding, the ambulacral grooves being enclosed within the skeletal plates. In contrast to the asteroids, the madreporite is on the oral surface.

Many ophiuroids are mucociliary suspension feeders, in several of them the five basic arms repeatedly bifurcating, but deposit feeding and omnivorous scavenging are also common; all possess a large stomach but lack an intestine and an anus. Between the stomach and the oral surface are positioned ten invaginations of that lower surface, the bursae (Fig. 7.21), which function as the specialist organs of gaseous exchange, water being drawn in through slits around the mouth by ciliary beating or by muscular pumping. Into these bursae discharge the gonads, and eggs are often brooded there; in some species the embryo is actually attached to the bursal wall and develops viviparously. Development may be direct or via an ophiopluteus larva (see Fig. 7.14c).

Three orders are distinguished, with a total of 2000 species.

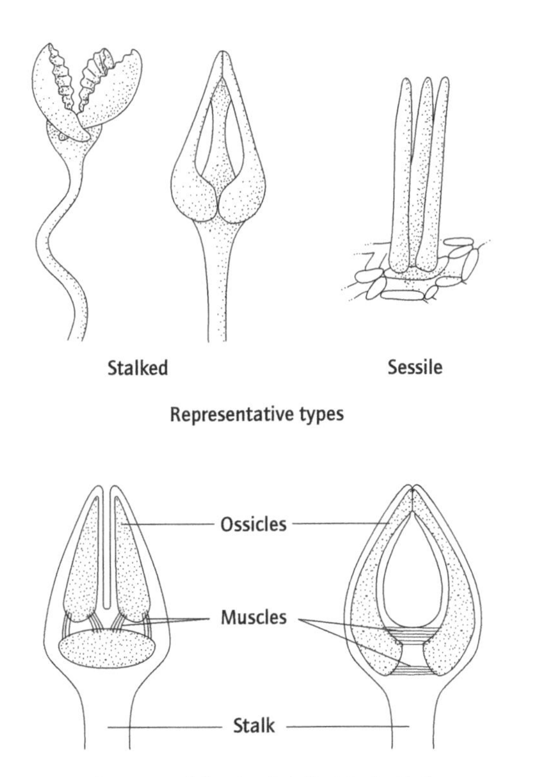
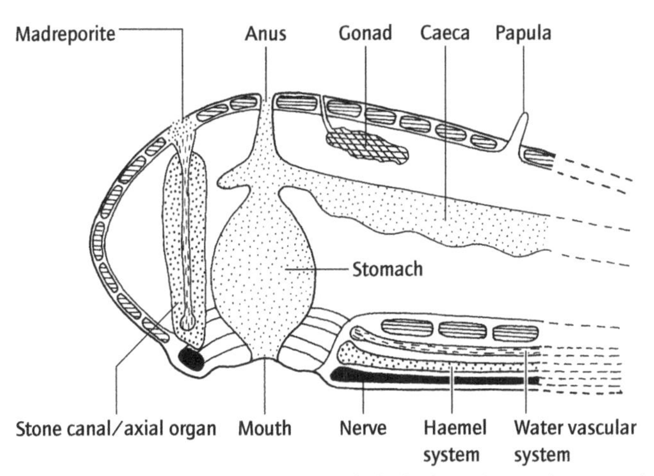
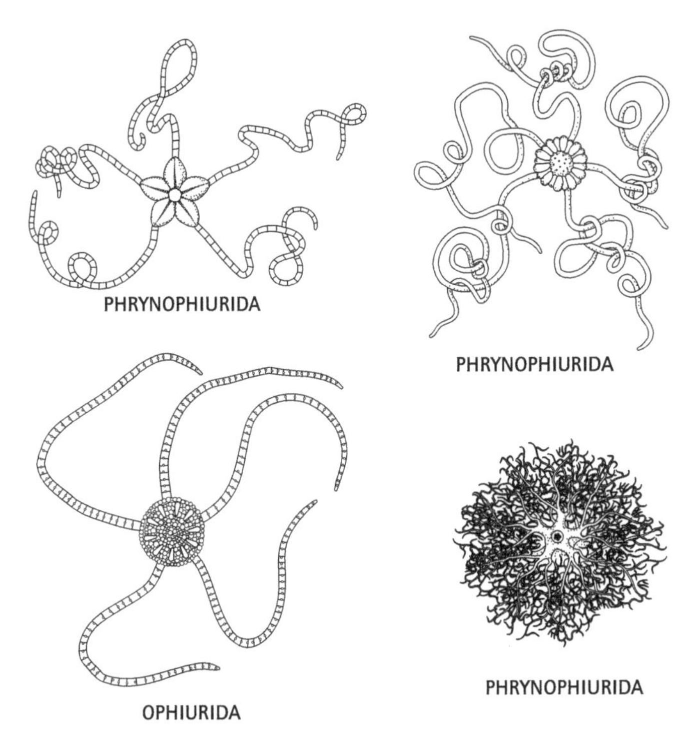

Fig. 7.20 Ophiuroid body forms (after several sources).

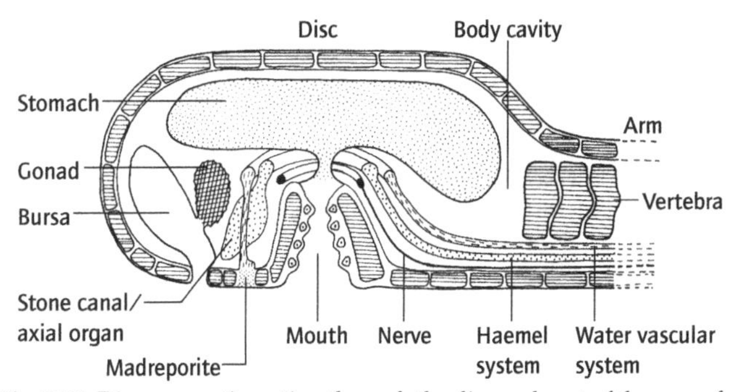

Fig. 7.21 Diagrammatic section through the disc and part of the arm of an ophiuroid (after Nichols, 1962).

Fig. 7.18 Pedicellariae (after Mortensen, 1928–51 and Hyman, 1955).

Fig. 7.19 Diagrammatic section through the body and part of an arm of an asteroid (after Nichols, 1962).

7.3.3.4 Class Concentricycloidea

In 1986 a new class of bizarre, medusa-like echinoderms was described from nine specimens of a single species found at some 1000 m depth off the coast of New Zealand on waterlogged wood. Two years later a second species was described, also from wood but this time from a depth of 2000 m off the Bahamas. The <15 mm diameter body of the sole genus *Xyloplax* is in the form of a flat, armless disc surrounded by a circlet of spines (Fig. 7.22). The upper surface of the disc is covered by scale-like plates, and the lower surface bears two concentric water–vascular ring canals from which issues a single ring of marginal tube feet. Both species are gonochoristic and sexually dimorphic; five pairs of gonads are present. The Bahamian species possesses a wide, centrally located mouth leading into a shallow, blind-ending stomach; the other species is without either mouth or gut, the central part of the lower surface being occupied by a membraneous 'velum'.

7.3.3.5 Class Echinoidea (sea-urchins, sand-dollars, etc.)

Echinoids are spherical or secondarily flattened, free-living echinoderms which lack arms (Fig. 7.23) (in a sense, the arms have been incorporated into the body). They are characterized by the development of their calcareous ossicles into a rigid test of attached plates, each ambulacral and interambulacral zone possessing essentially two vertical columns of such plates which curve around the body aborally from the mouth. This test bears

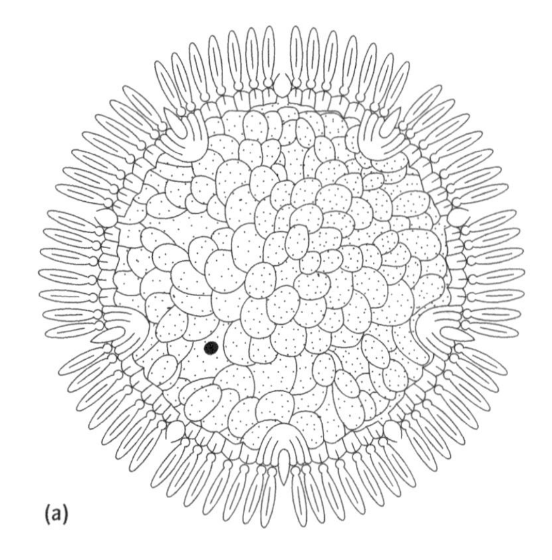

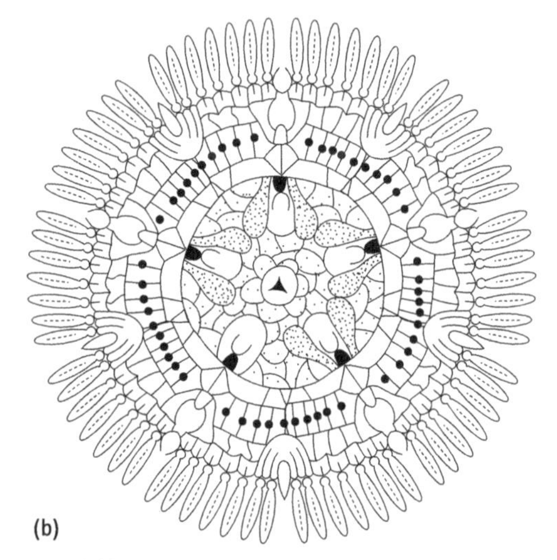

Fig. 7.22 Dorsal (a) and ventral (b) views of the concentricycloid *Xyloplax*. (After Baker *et al.*, 1986.)

pedicellariae (see Section 7.3.3.2) and movable spines on or with which some species can walk or burrow; otherwise locomotion is by means of tube feet that emerge through pores in the ambulacral plates.

The mouth, which is on the lower surface, bears a distinctive grazing apparatus of five large and several small plates, Aristotle's lantern (Fig. 7.24), with which algae and sedentary or sessile animal prey can be chewed, the lantern being protracted through the mouth for this purpose. In the epifaunal 'regular' species, the mouth is in the centre of the lower surface and the anus is mid-aboral. A number of species burrow into soft sediments and have become secondarily bilaterally symmetrical and often greatly flattened (these comprise the 'irregular' species); in these, the anus is displaced markedly towards the 'posterior' end and the mouth may move somewhat 'anteriorly'. Their diet is detrital, in association with which the lantern is often reduced or absent and feeding is effected by specialized, mucociliary tube feet.

The body cavity is large (Fig. 7.25) and bears extensions in the form of yet further modified tube feet (in irregular species) or peristomial gills (in regular forms) for gaseous exchange. The four (irregular) or five (regular) large gonads also project into the body cavity, discharging via aboral gonopores. Brooding occurs in some; in most, however, development is indirect, via an echinopluteus larva (Fig. 7.14d).

The ordinal classification of echinoids is based largely on fossil forms; 15 surviving orders contain a total of 950 living species.

7.3.3.6 Class Holothuroidea (sea-cucumbers)

In comparison with other free-living echinoderms, holothurians lack arms (equivalently to the echinoids, the arms have been incorporated into the body) and possess a bilaterally symmetrical body greatly elongated along the oral/aboral axis, so that they lie on their sides with three ambulacra 'ventrally' and the remaining two in a 'dorsal' position (Fig. 7.26). The body wall is also unusual in being leathery, with well-developed circular and five longitudinal muscles, the calcareous skeleton being reduced to separate microscopic ossicles.

Around the 'anterior' mouth, the water vascular system supports from eight to thirty finger-like, branched or shield-shaped oral tentacles which are used in deposit, or more rarely, suspension feeding (Figs 9.2 and 9.5). The through gut, which is often long, terminates in a 'posterior' cloaca or rectum and anus, the cloaca receiving in many species a pair of large diverticula lying in the body cavity, the respiratory trees, into which water is pumped for gaseous-exchange purposes (Fig. 7.27). Some species eviscerate the posterior gut, including the respiratory trees, through the anus on provocation, whilst a few possess specific sticky or toxin-containing Cuverian tubules associated with the respiratory trees, with the same function. The madreporite lies free in the body cavity.

Locomotion is very slow and is achieved in most species by tube feet, which are often scattered over the body surface, rather than being confined to the ambulacra; some groups, however, completely lack tube feet, and in them movement is effected by peristaltic muscle contractions, aided by small, pointed ossicles projecting through the body wall and anchoring dilated regions of the body against the surrounding sediment. Several are sedentary and use the tube feet more for attachment than for locomotion, and some deep-sea forms have greatly enlarged and elongated tube feet on which they walk like stilts (Fig. 7.28). Most species are epifaunal, although several burrow in soft sediments and one group includes planktonic forms.

A single gonad is normally present, discharging through a gonopore near the oral tentacles. Brooding of the embryos is common, in some actually within the body cavity and in one species within the ovary itself. A few are protandrous hermaphrodites. In non-brooding species, development is indirect, proceeding via vitellaria or auricularia and doliolaria stages (see Fig. 7.14e).

Six orders are differentiated on the basis of the form of the oral tentacles, the nature of the tube feet, and the shape of the body. There are some 1150 living species.

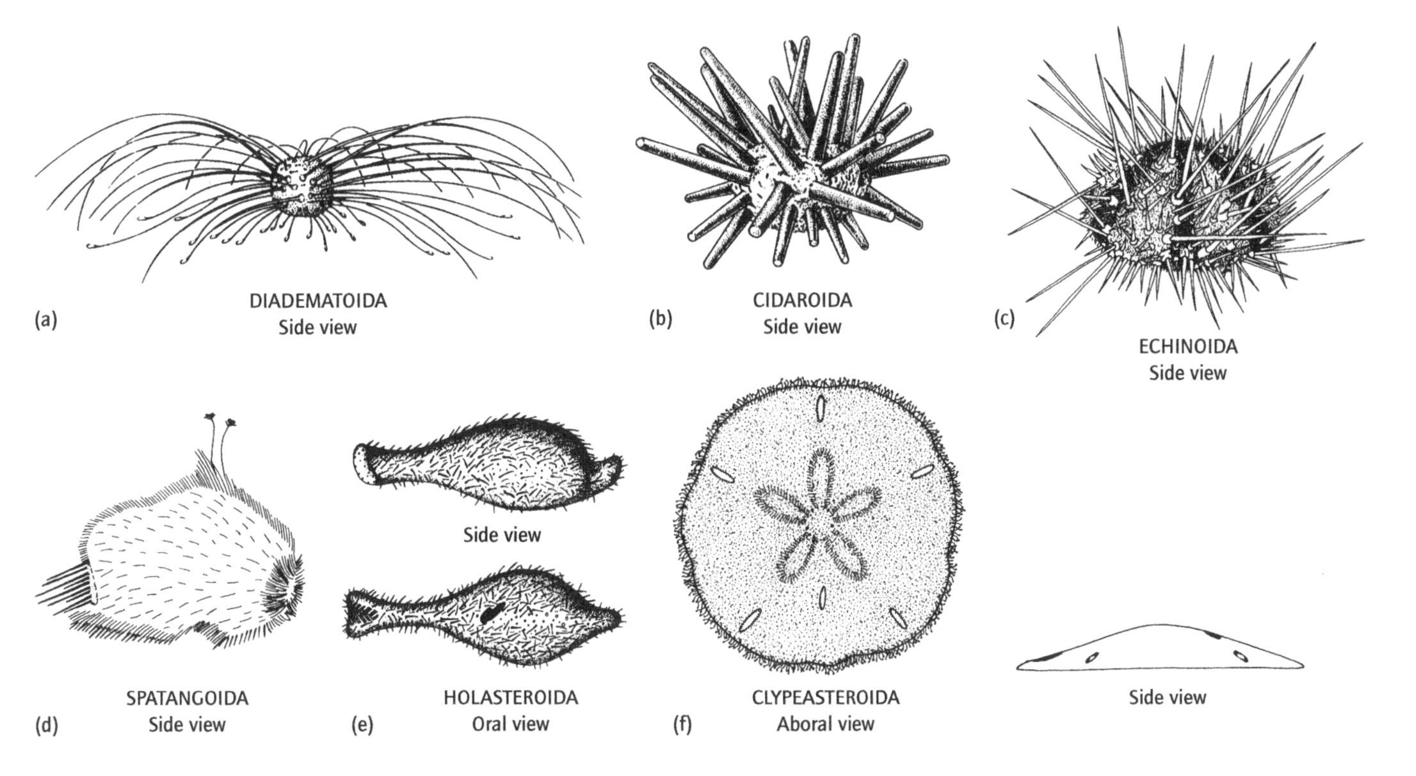

Fig. 7.23 Echinoid body forms: (a)–(c) are 'regular'; (d)–(f) 'irregular'. (After several sources, principally in Hyman, 1955.)

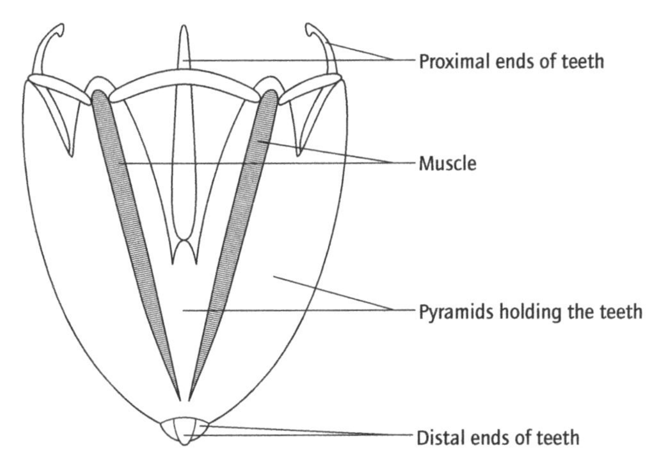

Fig. 7.24 Aristotle's lantern (after Hyman, 1955).

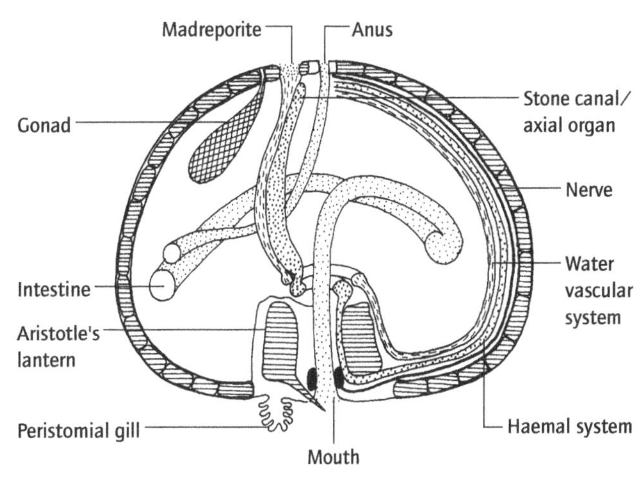

Fig. 7.25 Diagrammatic section through the test of an echinoid (after Nichols, 1962).

7.4 Phylum CHORDATA

7.4.1 Etymology

Latin: *chorda*, a cord (of cat gut).

7.4.2 Diagnostic and special features (Fig. 7.29)

The phylum Chordata comprises three subphyla, of which only two are invertebrate and therefore fall within the compass of this book. These two differ quite markedly, as might be expected from their subphylum status, and hence will be treated separately below. Nevertheless, they share the following characteristics:

1 Bilaterally symmetrical.
2 Body more than two cell layers thick, with tissues and organs.
3 A through gut and a non-terminal anus.
4 Body essentially monomeric, without a distinct head and without appendages or jaws.
5 A large pharynx, the wall of which is perforated through to the exterior by from a few to very many pharyngeal slits.
6 Mucociliary suspension feeders (as described in Section 9.2.5), the feeding water current leaving the body through the pharyngeal slits.

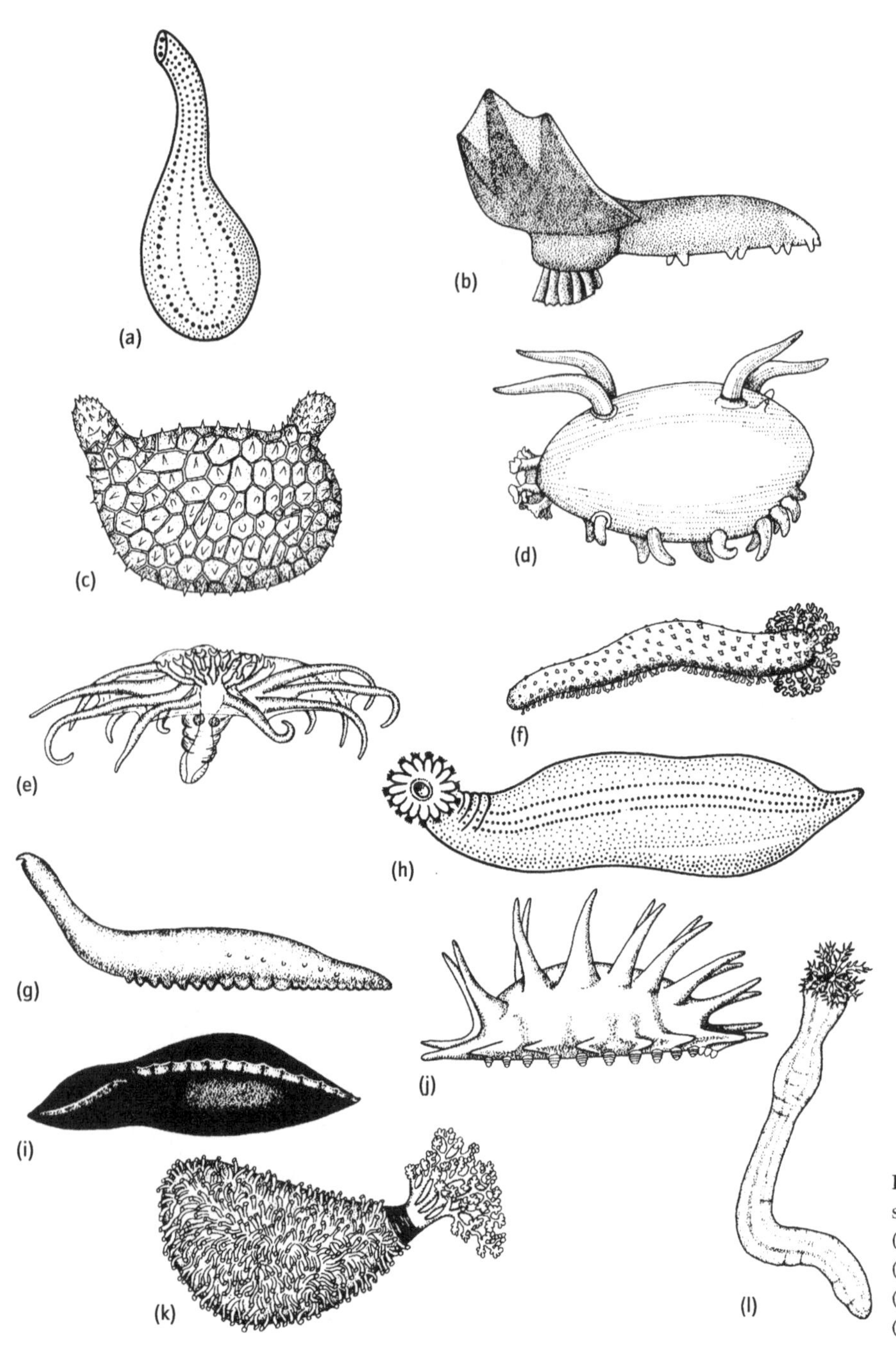

Fig. 7.26 Holothurian body forms (after several sources, principally in Hyman, 1955). (a) and (c) Dactylochirotida; (b), (d), (e), (g), (i) and (j), Elasipoda; (k) Dendrochirotida; (f) Aspidochirotida; (h) Molpadiida; and (l) Apodida.

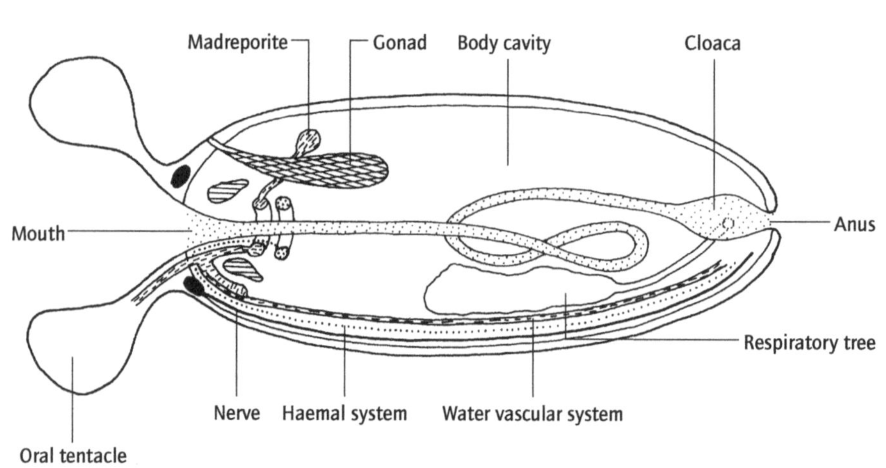

Fig. 7.27 Diagrammatic section through a holothurian (after Nichols, 1962).

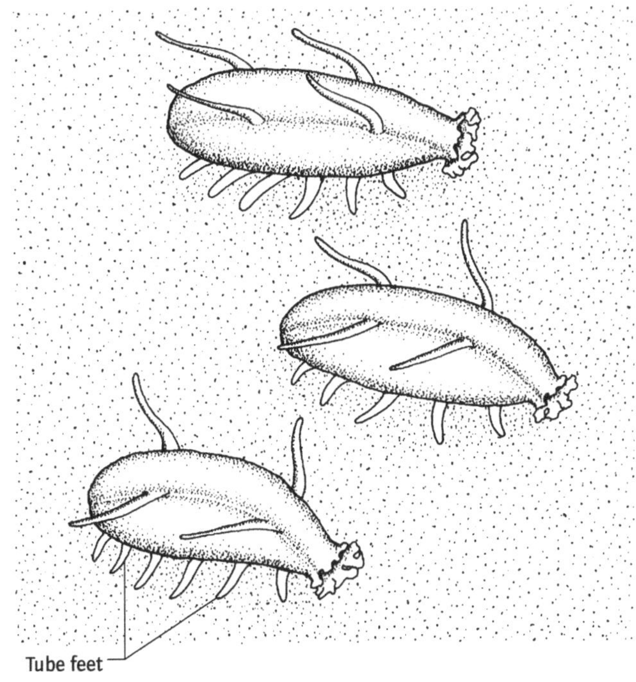

Fig. 7.28 Holothurians walking across the ocean floor by means of large, elongate tube feet.

7 Usually the pharyngeal region surrounded by a secondary body wall which encloses an atrial cavity connecting to the exterior by a single pore (see Fig. 9.14).
8 Epidermis not secreting an external cuticle and without cilia.
9 At some or all stages of the life history, an internal, dorsal skeletal rod, the notochord.
10 At some or all stages of the life history, a hollow nerve tube running dorsally to the notochord.
11 At some or all stages of the life history, a muscular post-anal tail serving as the organ effecting swimming.
12 Circulatory system partially open.
13 Eggs cleave radially.
14 Development deuterostomatous, usually indeterminate and indirect.
15 Marine.

7.4A Subphylum UROCHORDATA (tunicates)

7.4A.1 Diagnostic and special features

1 A notochord, hollow nerve cord and post-anal tail only in the larval stage (if present) and in one permanently larval group.
2 Without any coelomic body cavities.
3 Without excretory organs.
4 Without segmentation of muscular or other structures.

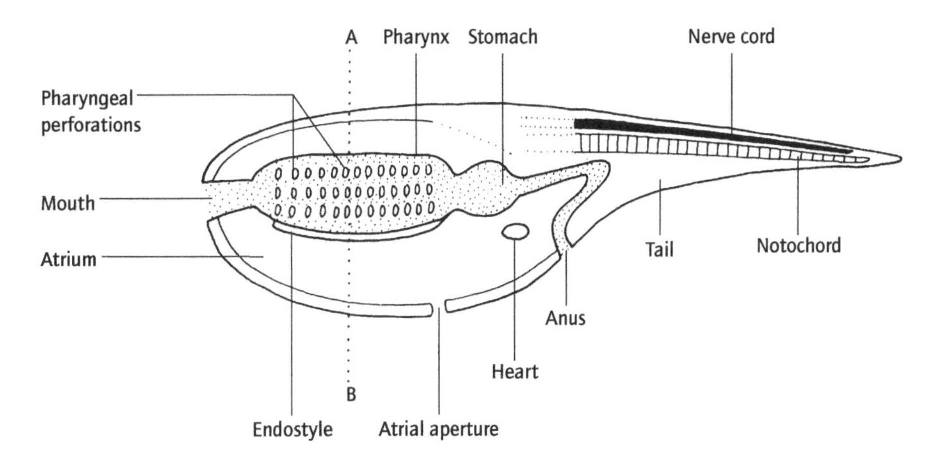

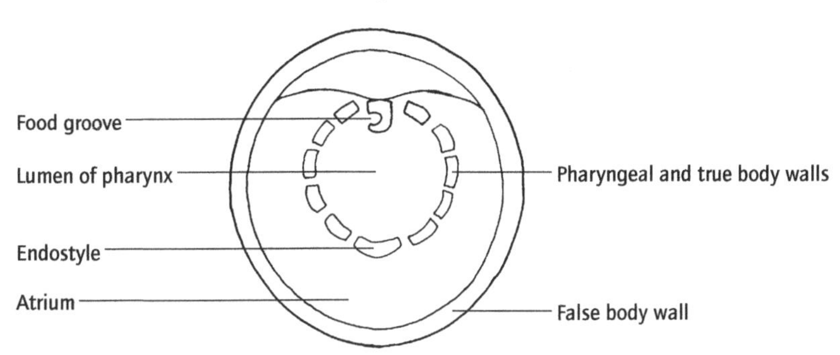

Fig. 7.29 The basic body plan of a generalized invertebrate chordate.

5 Body wholly enclosed within a secreted tunic, test or 'house', usually composed of cellulose and protein, and containing cells which have migrated from the body and, in some, with extra-corporeal blood vessels.
6 A U-shaped gut and a large pharynx usually occupying most of the body volume.
7 Nervous system in the form of a ganglion between mouth and atrial apertures, from which individual nerves issue.
8 Hermaphrodite, usually with a single ovary and testis.

The urochordates are sessile or free-living, solitary or colonial filter feeders which display their chordate affinities only in the free-swimming larval stage.

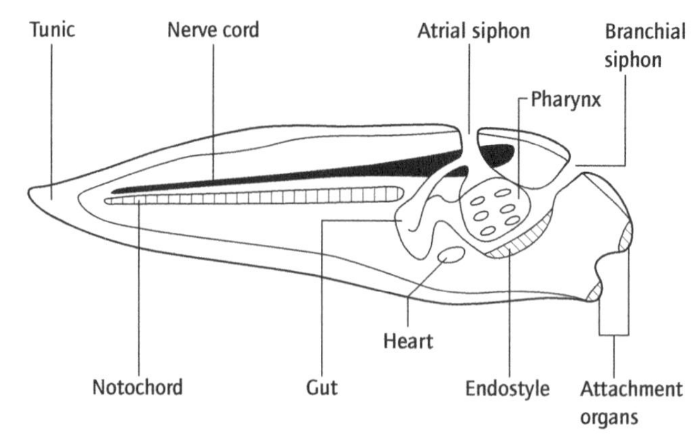

Fig. 7.30 Anatomy of the ascidian tadpole larva (after McFarland *et al.*, 1979 and others).

7.4A.2 Classification

The 2000 species of urochordates are divided between three classes.

Class	Order
Ascidiacea	Aplousobranchia Phlebobranchia Stolidobranchia Aspiraculata
Thaliacea	Pyrosomida Doliolida Salpida
Larvacea	Copelata

7.4A.2.1 Class Ascidiacea (sea-squirts)

This class contains the 1850 sessile, benthic species. A free-living tadpole larva (Fig. 7.30) is normally possessed and this eventually attaches itself by its head end to the substratum. Differential growth during metamorphosis then leads to an effective rotation of the anterior/posterior axis of the body through 180° so that the mouth and the associated branchial siphon of the tunic or test move to lie at the opposite end of the body to the point of attachment (Fig. 7.31). This branchial siphon leads via a tentaculated orifice into a huge pharynx, normally perforated by very many small slits (except in four deep-sea species, comprising the order Aspiraculata, in which the pharynx is reduced, the pharyngeal slits are absent, and the branchial siphon is modified into a prey-capturing series of prehensile lobes; Fig. 7.33h). The tentacles surrounding the opening to the pharynx serve to prevent the entry of large particles. The atrial aperture is also drawn out into a siphon which is sited near to the branchial one (Fig. 7.32).

A basal heart is present which, unusually, pumps blood in two directions, the heartbeat reversing periodically. The blood is peculiar in that some of its cells contain high concentrations of heavy metals, particularly vanadium, niobium or iron, in association with sulphuric acid. Vanadium concentrations in these cells may attain 10^5–10^6 times the background levels in the external sea water.

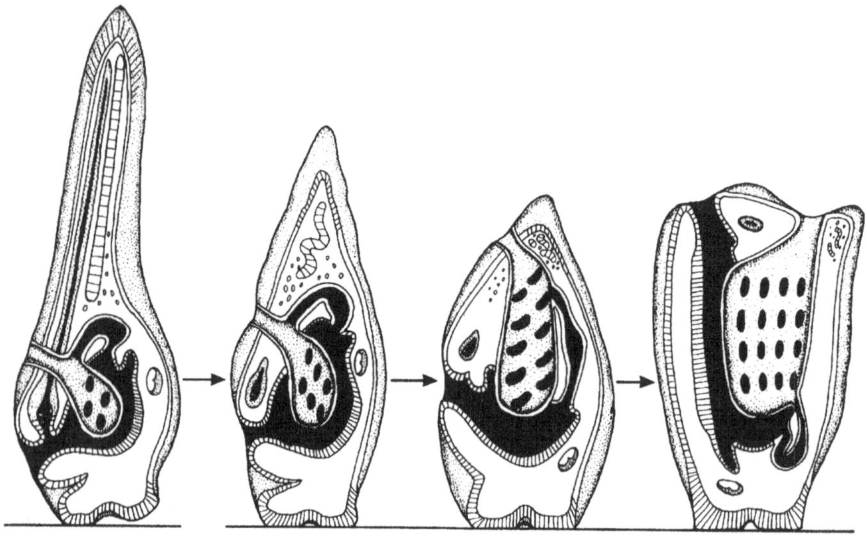

Fig. 7.31 Metamorphosis of the tadpole larva (after Barnes, 1980).

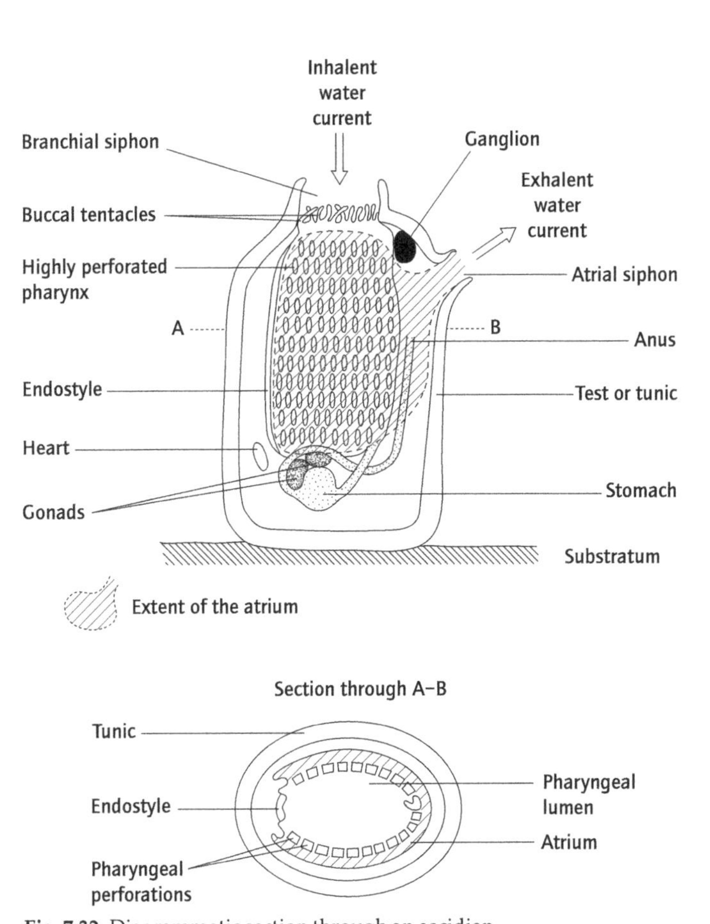

Fig. 7.32 Diagrammatic section through an ascidian (after several sources).

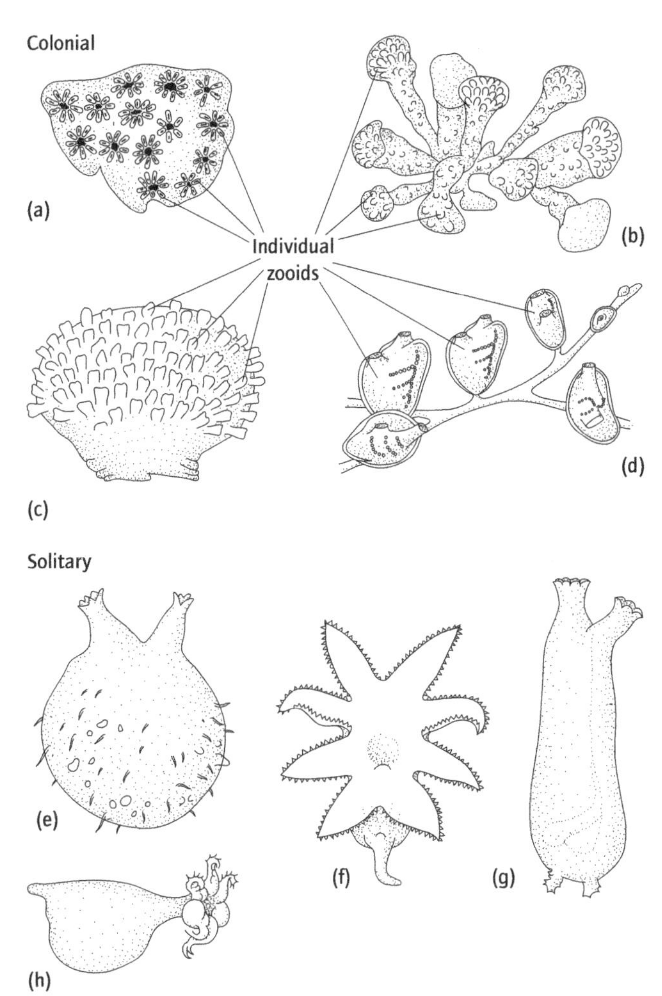

Fig. 7.33 Colony and body form of ascidians (after Millar, 1970 and others). (a) and (e) are stolidobranchs; (b) is an aplousobranch; (c), (d), (f), and (g) are phlebobranchs; and (h) is an aspiraculatan.

Asexual multiplication by budding is common, proceeding in a variety of manners from different regions of the ascidian body. As in the unrelated cnidarians and bryozoans, budded individuals may remain associated with each other in colonies, although there is no zooidal polymorphism. Some colonial forms are embedded in a communal test, many sharing a single atrial opening; others are connected by basal stolons; and many colonies adopt characteristic shapes. In some, budding begins whilst the ascidian is still larval. The majority of ascidians, however, are solitary and these non-colonial forms are often much larger than the individual colonial zooids, up to 15 cm in height or more (Fig. 7.33).

Besides the aberrant Aspiraculata, the sea-squirts are included in three orders dependent mainly on the location of the gonads and on the structure of the pharyngeal wall: in the Aplousobranchia, which are all colonial, the gonads are positioned in the loop of the intestine and the pharynx possesses a simple wall; the mainly solitary Stolidobranchia have their gonads embedded in the body wall alongside the pharynx which is folded longitudinally and bears internal bars; whilst the mostly solitary Phlebobranchia possess gonads sited as in the Aplousobranchia but their pharynx, although unfolded, has raised internal longitudinal bars formed by bifurcating papillae.

7.4A.2.2 *Class Thaliacea (the pelagic tunicates)*

The 70 thaliacean species are all planktonic and use their feeding current as a means of jet propulsion through the water, the branchial and atrial apertures being located at opposite ends of their fusiform or barrel-shaped bodies. Asexual budding also occurs in all forms, buds forming on a ventral stolon which arises immediately behind the pharyngeal endostyle.

Two different life styles occur. The order Pyrosomida includes colonial thaliaceans which occupy a common cylindrical test with a central lumen opening to the environment at only one end. Each zooid is positioned such that its branchial aperture is on the outside face of the communal cylinder and its atrial aperture is on the inside face. All the feeding currents therefore discharge into the colonial lumen and the whole tubular colony, which may be several metres long (giving rise, it has been suggested, to several sightings of great sea-serpents), moves as a single unit through the water (Fig. 7.34). The individual zooids possess many pharyngeal perforations, similar to those of the ascidians to which they are closely related.

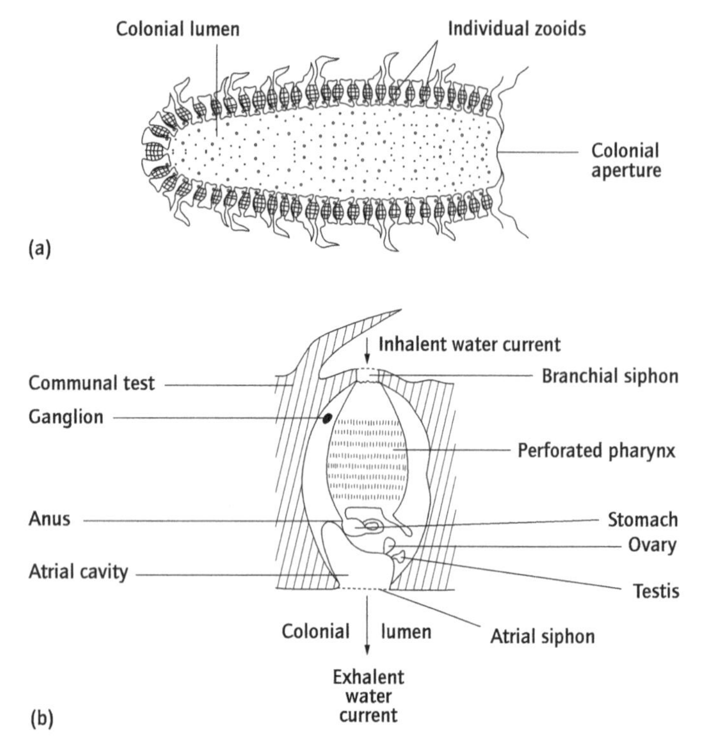

Fig. 7.34 Diagrammatic sections (a) through the colony and (b) through the individual zooid of a pyrosome. (After Grassé, 1948 and Fraser, 1982.)

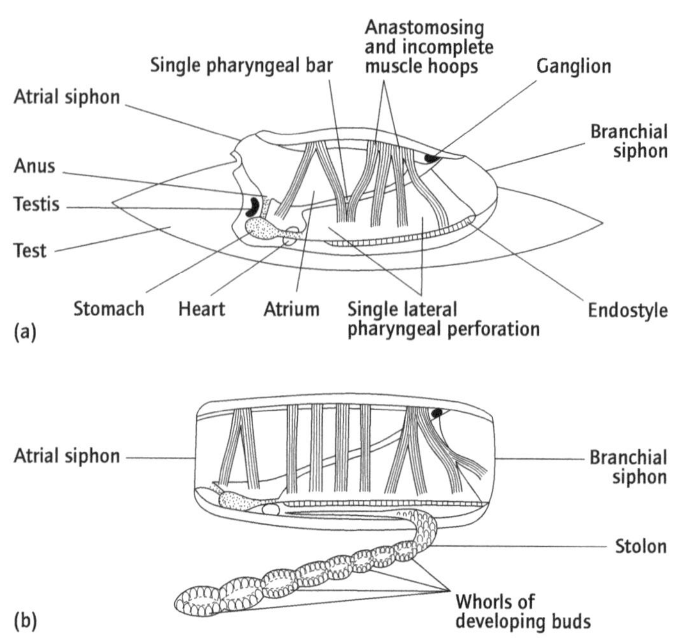

Fig. 7.35 The anatomy (diagrammatic) of the two stages in the life cycle of a salp: (a) the sexually reproducing aggregate 'generation'; (b) the asexually multiplying solitary 'generation'. (After Berrill, 1950, Fraser, 1982 and others.)

The orders Salpida and Doliolida, however, are mainly solitary, although they exhibit an alternation between solitary and aggregate forms of different sexuality. In the salps (Fig. 7.35), the solitary phase multiplies asexually, giving rise to chains of individuals which eventually separate to yield the solitary sexual 'generation'; in the doliolids, on the other hand, it is the solitary phase which reproduces sexually to produce the asexually multiplying aggregate 'generation' (Fig. 7.36). Both groups exhibit zooidal polymorphism (during different stages of their life history), possess bands or hoops of muscle around the body which in the salps drives water through (instead of ciliary power), have more or less transparent gelatinous tests, and have reduced numbers of individually larger pharyngeal perforations – a development which reaches its extreme form in the salps which possess only two relatively enormous slits (see Fig. 9.14). The test of the salps is a cellulose-containing tunic essentially similar to that of the ascidians and pyrosomes; that of the doliolids, however, is a cellulose-less cuticle.

Only the doliolids have a larval stage in their life history, and this, it is believed by some, gave rise to the following class by paedomorphosis.

7.4A.2.3 Class Larvacea (appendicularians)

The planktonic larvaceans possess a morphology basically that of the characteristic urochordate larva (Fig. 7.37), the small body bearing a large persistent tail, complete with hollow nerve

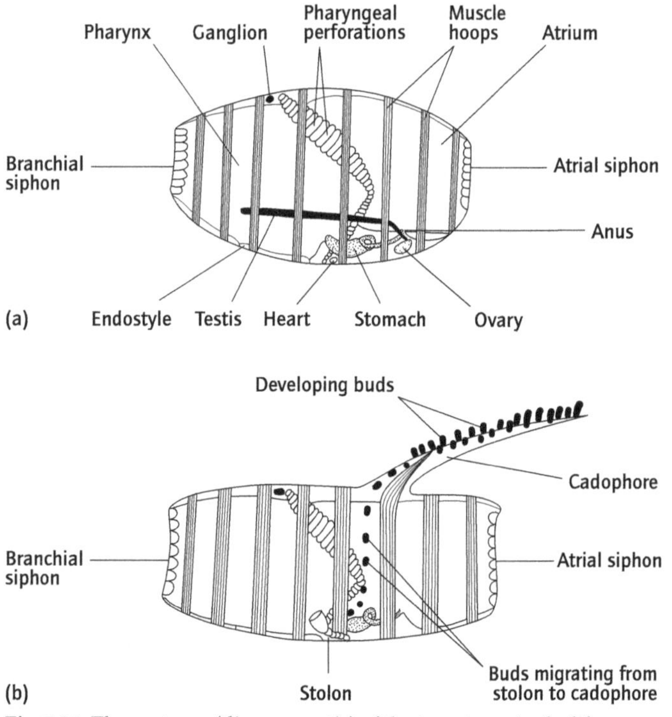

Fig. 7.36 The anatomy (diagrammatic) of the two stages in the life cycle of a doliolid: (a) the sexually reproducing solitary 'generation'; (b) the asexually multiplying aggregate 'generation'. (After Berrill, 1950, Fraser, 1982 and others.)

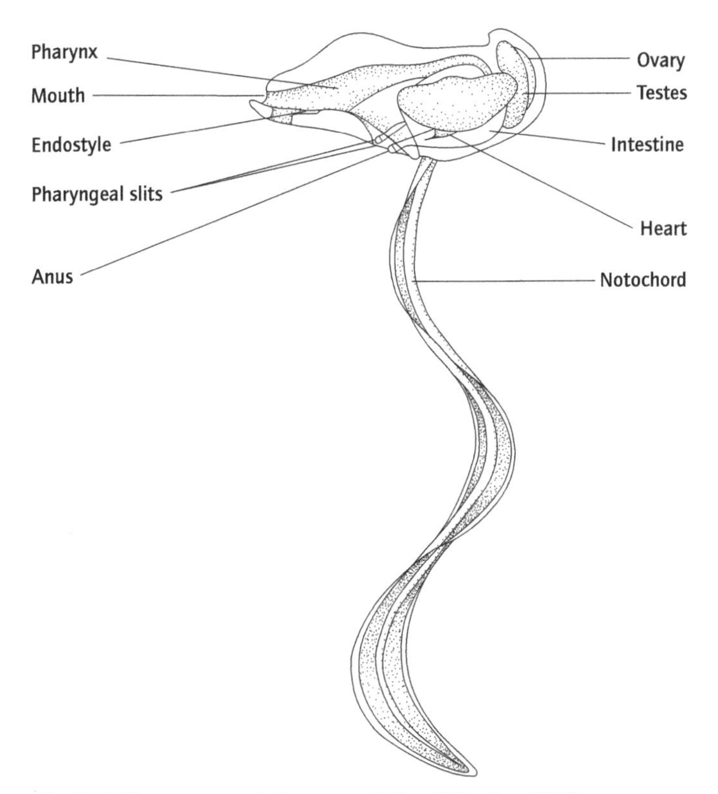

Fig. 7.37 The anatomy of a larvacean (after Alldredge, 1976).

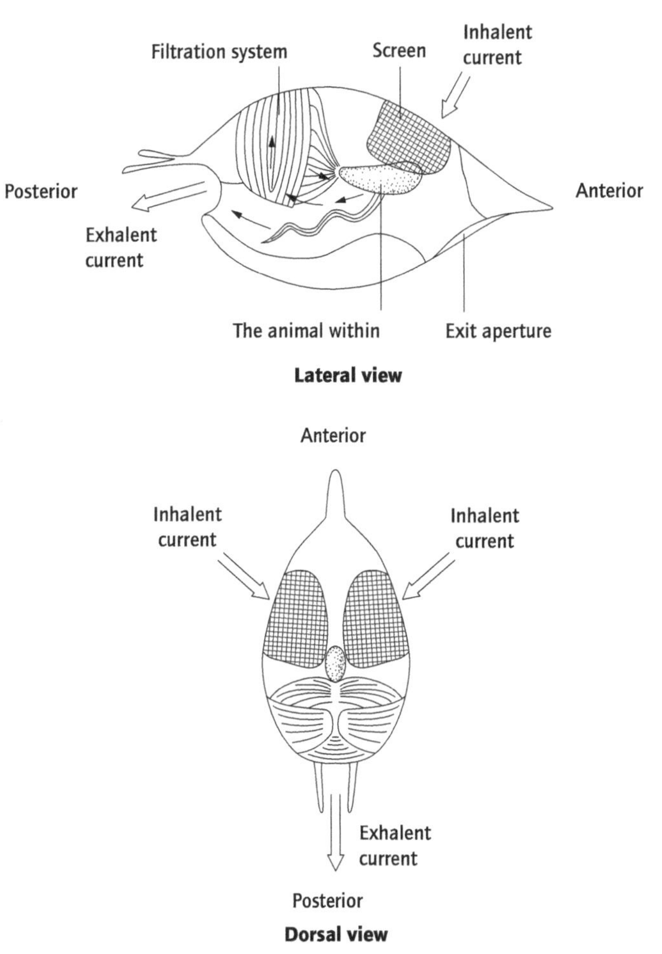

Fig. 7.38 The larvacean 'house' and the feeding currents induced through it (after Hardy, 1956 and others).

cord and notochord, and totalling only some 5 mm in length. Epidermal glands secrete not a cellulose test but a thin gelatinous cuticular 'house' which forms an external filtration apparatus (Fig. 7.38), water being drawn into this 'house' through screens by the beating of the tail and food particles, especially nanoplanktonic algae, being filtered by fine meshworks within the gelatinous structure (see Fig. 9.4). Filtered material is then ingested with the aid of a further water current, created by pharyngeal cilia, which leaves the gut through a single pair of pharyngeal slits opening individually on the body surface. The atrial cavity is therefore the area within the 'house'. When the screens and filters of the 'house' become irreversibly clogged, larvaceans can abandon their tests through a 'door' (which is normally in a closed position) and inflate a new pre-formed 'house'.

Larvaceans are all solitary, and reproduction is solely sexual. Unusually amongst urochordates, one species is gonochoristic. A single order is recognized for the 70 known species.

7.4B Subphylum CEPHALOCHORDATA (lancelets)

7.4B.1 Diagnostic and special features (Fig. 7.39)

1 Body laterally compressed, fish-like.
2 Notochord extends the whole length of the body.
3 The hollow dorsal nerve cord extends almost the whole length of the body, but is not dilated anteriorly to form a brain.
4 A persistent post-anal tail.
5 Body with serially repeated muscle blocks, nerves, excretory organs and gonads.
6 Body cavities formed enterocoelically from many serially repeated pouches; the ventral parts of the separate coelomic pouches merge together and the dorsal parts become obliterated by muscle.
7 Excretory system closely resembles protonephridia, but is formed by peritoneal cells.
8 Pharyngeal region covered by a secondary body wall formed by a pair of metapleural folds growing ventrally and fusing along the mid-ventral line.
9 Pharynx large, occupying half the body length.
10 Gonochoristic, with many single or paired gonads.

Cephalochordates are small (up to 10 cm long), free-living animals which are sedentary and benthic whilst feeding, but which are capable of swimming to change feeding location or to escape from predators, the notochord acting as an incompressible but flexible longitudinal strut. Contraction of the longitudinally arranged muscles therefore bends the body into a series of S-shaped wriggles, rather than shortening it (an equivalent system of movement to that seen in the chaetognaths and nematodes).

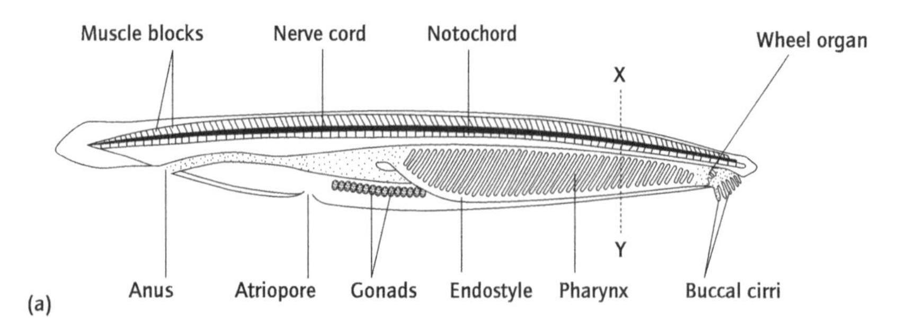

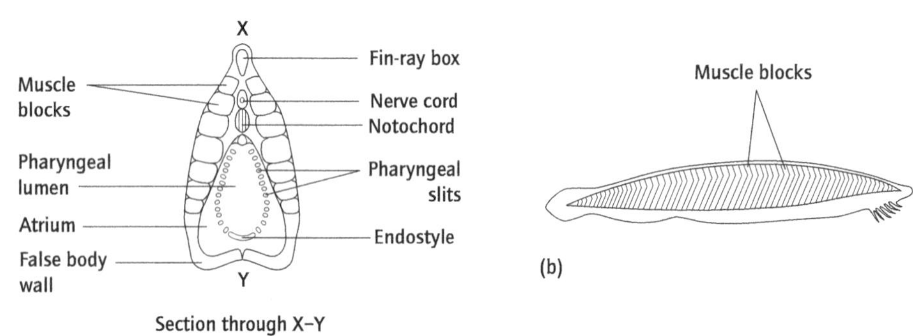

Fig. 7.39 Cephalochordates: (a) diagrammatic longitudinal section through the body; (b) appearance in life. (After Young, 1962.)

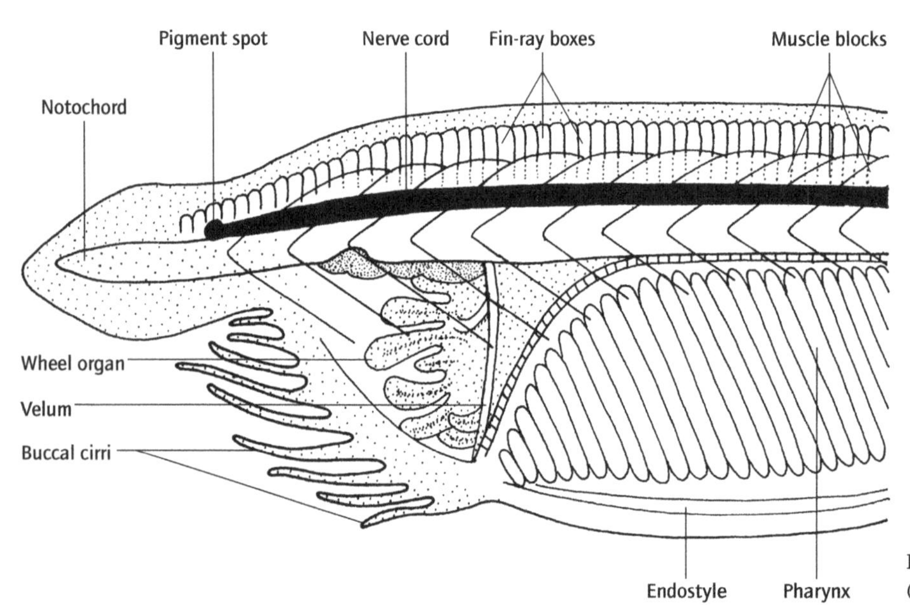

Fig. 7.40 The anterior end of a cephalochordate (after Young 1962).

The cephalochordate feeding system is essentially the same as that employed by the urochordates, and in this group too the buccal region is guarded by a series of tentacles (the buccal cirri and velar tentacles – Fig. 7.40) which prevent the entry of undesired particles. The pharynx, however, is elongate rather than barrel-shaped, and the atriopore is in the mid-ventral line just anterior to the anus. No surrounding test drawn out into siphons occurs.

To some degree, the cephalochordates provide a link between the invertebrates and the third chordate subphylum, the Vertebrata, especially the agnathans which also lack appendages and jaws, and the larval stage of one group of which (the lampreys) possesses a mucociliary filtering pharynx similarly to the lancelets (although water is driven through its pharynx by muscular pumping). Most obviously, however, the cephalochordates differ from the vertebrates in their lack of a differentiated head with paired sense organs, a brain and a protective cranium, in possessing a nerve cord to which the muscles send processes (forming the apparent ventral roots) and peripheral nerve fibres which lack myelin sheaths, and in their very thin monolayered epidermis. Nevertheless, in outline fashion they do perhaps indicate the general nature of the likely bodily organization of the 'protovertebrate'. Equally, their basic similarity to an enlarged and neotenous version of the larval stage of the otherwise most unvertebrate-like urochordates is also apparent.

7.4B.2 Classification

The 25 known species are placed in a single class and order.

7.5 Further reading

Alvarino, A. 1965. Chaetognaths. *Oceangr. Mar. Biol., Ann. Rev.*, **3**, 115–194 [Chaetognatha].

Barrington, E.J.W. 1965. *The Biology of Hemichordata and Protochordata*. Oliver & Boyd, Edinburgh [Hemichordata and Chordata].

Berrill, N.J. 1950. *The Tunicata*. Ray Society, London [Urochordates].

Hyman, L.H. 1955. *The Invertebrates*, Vol. 4: *Echinodermata*. McGraw-Hill, New York [Echinodermata].

Hyman, L.H. 1959. *The Invertebrates*, Vol. 5: *Smaller Coelomate Groups*. McGraw-Hill, New York [Hemichordata and Chaetognatha].

Nichols, D. 1969. *Echinoderms*, 4th edn. Hutchinson, London [Echinodermata].

Rowe, F.W.E., Baker, A.N. & Clark, H.E.S. 1988. The morphology, development and taxonomic status of *Xyloplax*, Baker, Rowe & Clark (1986) (Echinodermata: Concentricycloidea) with the description of a new species. *Proc. Roy. Soc. Lond.* (*B*), **233**, 431–459.

Young, J.Z. 1981. *The Life of Vertebrates*, 3rd edn. Clarendon press, Oxford [Chordata].

CHAPTER 8

Invertebrates with Legs: the Arthropods and Similar Groups

Tardigrada
Pentastoma
Onychophora
Chelicerata
Uniramia
Crustacea

Besides their general protostomatous condition, the six phyla included in this chapter really share only two characteristic anatomical features. They bear pairs of legs along all or part of the length of the body, each pair usually being served by ganglionic swellings on the longitudinal nerve cord/s; and they possess pseudocoelomic body cavities, often filled with blood and then termed haemocoels.

Three of the phyla (the Tardigrada, Pentastoma and Onychophora) are soft-bodied animals which use their body cavities as hydrostatic skeletons: in effect, these are worms with soft fleshy, unjointed, claw-bearing legs formed by finger-like outgrowths of the body and capable of being moved by extrinsic muscles. We suggested in Chapter 2 that the Onychophora and Tardigrada are survivors of the lobopod 'proto-arthropods' of the Cambrian. The tardigrades in particular share a number of anatomical and life-style features with the vermiform relatives of the arthropods, the nemathelminth ecdysozoans. Although lobopod in its structure, the Pentastoma seems most likely to have acquired this state secondarily, in that the concensus view is that it is a degenerate branchiuran crustacean, although affinity with the chelicerate mites has also been argued. Here we treat it as a separate phylum, not least because of its peculiar anatomy.

In contrast, the 'true' arthropods (the Crustacea, Chelicerata and Uniramia) possess a hard, jointed, sclerotized cuticular exoskeleton composed of chitin and protein, sometimes impregnated with calcium carbonate (Fig. 8.1). This covers the whole body including the legs, which are therefore also jointed (the word arthropod is derived from the Greek: *arthron*, joint; *podos*, feet) (Fig. 8.2). As in various other animals, being covered by an external cuticular system imposes constraints on growth, and necessitates a series of moults. This problem is particularly acute in respect of the arthropods where the cuticle is also the skeleton. During moulting, therefore, the old skeleton is partly resorbed and then shed, and a new soft skeleton, which has been developed beneath the old, is inflated (by the intake of air or water into the body) and hardened. The animal is especially vulnerable during this period and moulting often takes place whilst in hiding. In origin, the exoskeleton was probably a series of protective plates or hoops of cuticle, as seen in the kinorhynchs (Section 4.6) and in some tardigrades (Section 8.1), for example, and this was later adapted to serve a skeletal function in partial replacement of the ancestral hydrostatic pseudocoel/haemocoel. In some, replacement has remained only partial since a number of arthropods still extend their legs by hydraulic pressure and only flex them using their exoskeletal–muscular system.

The body of an arthropod is in origin fundamentally monomeric, although extensive metamerism of the body wall, exoskeleton and some internal structures has occurred in association with each pair of legs (see, e.g. Fig. 2.8). In essence, therefore, the arthropod body plan is that of a small anterior region (acron) and an equivalent posterior portion (telson) without legs, and number of intervening sections (the segments) each with one pair of legs, and with serial repetition of leg-based organs (muscular, nervous, skeletal, etc.). Non-leg-associated organs, for example the excretory and reproductive systems, are not serially repeated, however. In many lines of arthropods, there has been considerable fusion of leg-bearing segments, loss of appendages, and/or differentiation of various body regions; specialization of the anterior legs into feeding organs is particularly widespread. Apart from the common modification of their cuticle to serve a wholly or partially exoskeletal function (and any feature consequent on this, e.g. the jointed condition of the legs), the absence of cilia, and the tendency to develop compound eyes, the arthropods share few features, however, and hence it is possible that the arthropod state represents a grade of organization rather than the distinguishing characteristic of a single phylogenetic line.

Together the arthropods comprise the large majority of animal species, mainly as a consequence of their successful

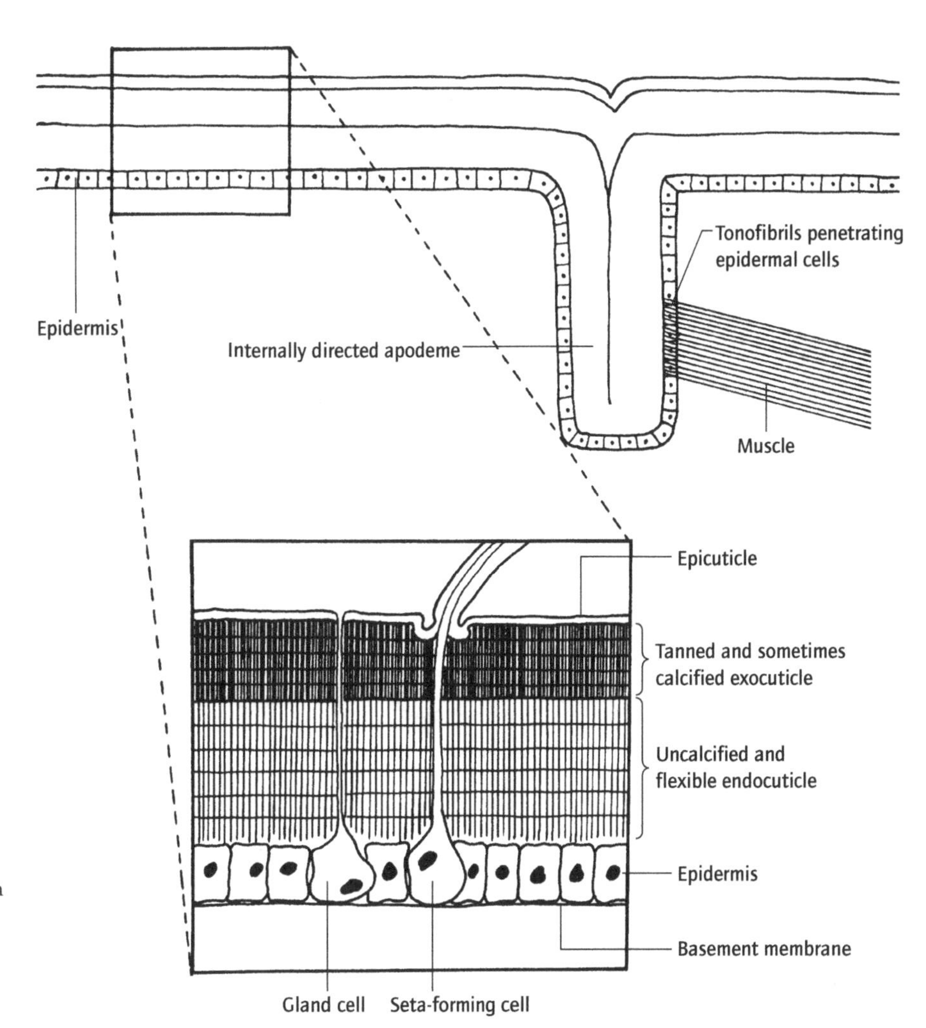

Fig. 8.1 Section through the cuticular exoskeleton of an arthropod showing the various layers and an internal projection (an apodeme) serving an internal skeletal function (after Hackman, 1971 and others).

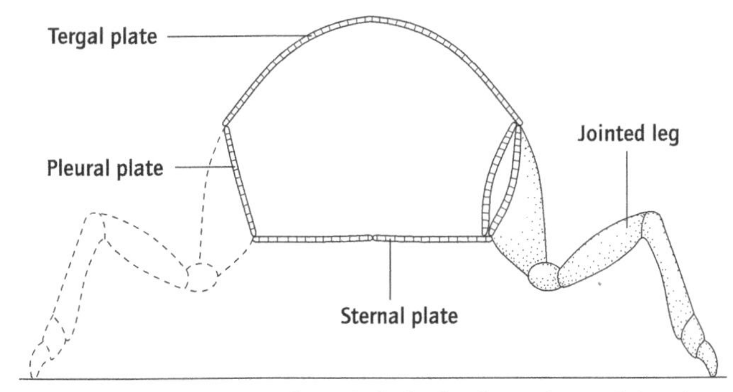

Fig. 8.2 Transverse section through the body of an arthropod showing the various exoskeletal plates encircling the body and a jointed limb. The number of separate articles of which the limb is composed varies from group to group, each phylum having its own nomenclature.

conquest of the land and of the ease with which small terrestrial organisms can speciate. Their success as terrestrial animals, in marked contrast to most groups of invertebrates, probably owes much to the evolution of water-conserving excretory systems and gasous-exchange organs, and the development of a desiccation-resistant impermeable epicuticle.

8.1 Phylum TARDIGRADA (water bears)

8.1.1 Etymology

Latin: *tardus* slow; *gradu* step.

8.1.2 Diagnostic and special features (Fig. 8.3)

1 Bilaterally symmetrical; minute, squat.
2 Body more than two cell layers thick, with tissues and organs.
3 A through, straight gut.
4 Body monomeric, although with four pairs of short, unjointed, claw-bearing legs on which the animals crawl (Fig. 8.3a) using extrinsic muscles; leg pairs served by serially repeated nerve ganglia.
5 Well-developed pseudocoelomic body cavity forming the hydrostatic skeleton.
6 Body wall with cuticle-covered epidermis, but without muscle layers; essentially non-chitinous cuticle moulted and often bearing spines and/or thickened into plates; network of individual smooth muscle cells criss-cross the body.
7 Body has a fixed number of cells (eutelic).

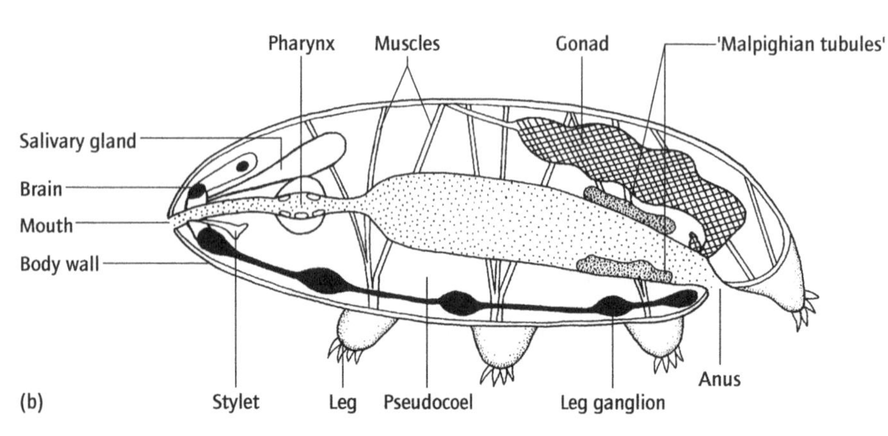

Fig. 8.3 Tardigrade structure: (a) external appearance of animals crawling on a filamentous alga (after Marcus, 1929); (b) a diagrammatic longitudinal section through a generalized tardigrade (after Cuénot, 1949).

8 A muscular, pumping pharynx bearing chitinous plates (placoids); a pair of buccal stylets can be protracted through the mouth to pierce the prey.
9 Without circulatory system or gaseous-exchange organs.
10 Three 'Malpighian tubules' possibly form an excretory system in some species.
11 Nervous system with a brain and paired longitudinal ventral cords bearing leg-associated ganglia.
12 Gonochoristic, sometimes parthenogenetic; with a single gonad.
13 Development direct.
14 Free-living, inhabiting water films interstitially or associated with vegetation on land, in fresh water and in the sea; cryptobiotic.

The tardigrades display a peculiar amalgam of features: like the other protoarthropods, they possess paired claw-bearing legs; like the lophophorates and deuterostomes, they show during their development, albeit only transitorily, paired enterocoelic pouches (five pairs); and with the pseudocoelomates they share a general level of bodily organization and life style.

Although several species occur in relatively permanent habitats, such as marine sands and shell gravels, many characterize temporary water films and labile water bodies. These latter tardigrades have evolved a variety of resistant stages. When the film of water around moss leaves evaporates, for example, the tardigrades too lose water through their permeable cuticles. Most of their bodily fluids can be lost as they shrivel to small barrel-shaped 'tuns' (Fig. 8.4). Tuns can survive for up to 10 years in the dry state (and probably for longer), their oxygen consumption falling to one six-hundredth of normal. In this state, they can withstand temperatures of −272°C for over 8 hours, and of up to 150°C. Less severe adverse circumstances can be avoided (a) by encystment, the animal pulling in its legs, detaching itself from its cuticle, as during moulting, and curling up into a ball within the cuticular shell; or (b) by the production of thick-walled resting eggs. It is possible that including such phases of suspended animation, the tardigrade lifespan may exceed 60 years.

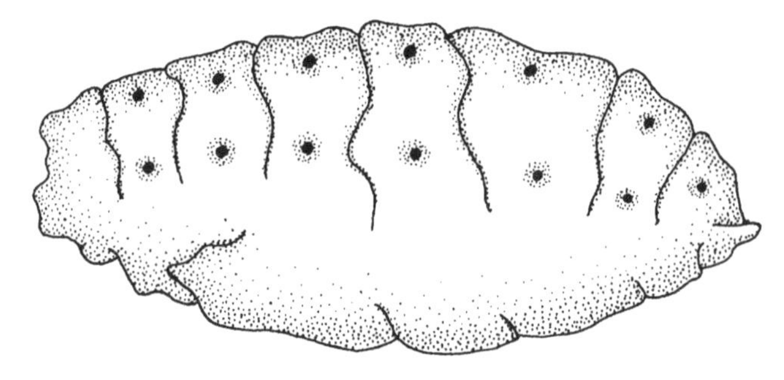

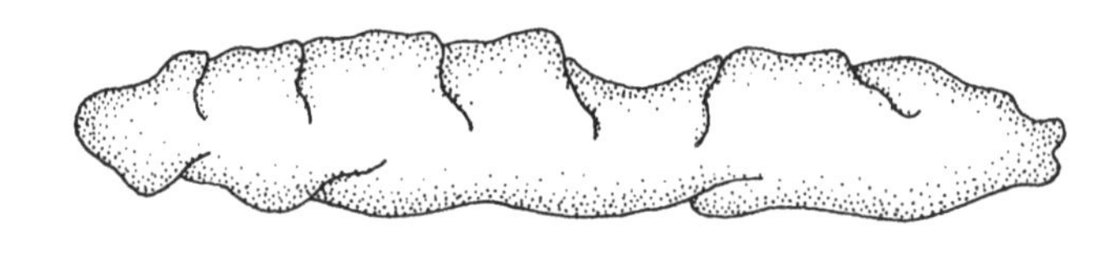

Fig. 8.4 A tardigrade tun (after Morgan, 1982).

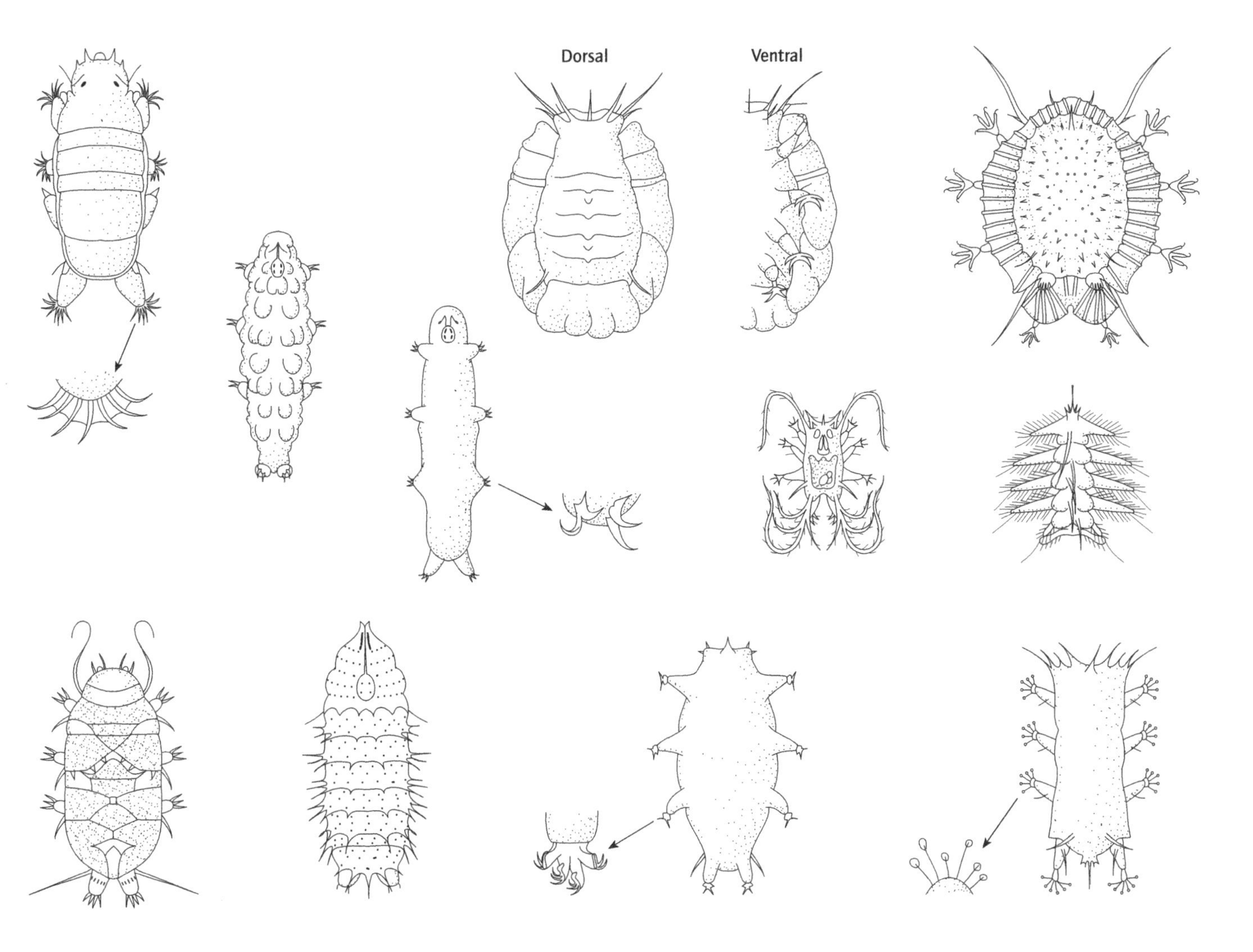

Fig. 8.5 Variation in cuticular form and ornamentation of tardigrades, and in shape and number of their claws (after Morgan & King, 1976 and others.)

All species are suctorial feeders, although the prey consumed may be algal or plant cells, or associated interstitial or cryptobiotic animals such as rotifers, nematodes and other tardigrades. A few are parasitic, one within the gut of gastropod molluscs. The mouth is applied to the food and the two stylets are then protracted by muscles through the mouth, piercing the prey. Fluids and organelles can thereafter be sucked into the gut by the pumping action of the pharynx, the pharyngeal placoids probably serving to macerate any solid particles ingested.

8.1.3 Classification

The 400 living species, which range in size from 0.05 to 1.2 mm in length, vary little in their anatomy, except in respect of ornamentation of the cuticle and in the form and number of their claws (Fig. 8.5); all are contained within a single class.

8.2 Phylum PENTASTOMA (tongue-worms)

8.2.1 Etymology

Greek: *pente*, five; *stoma*, mouth.

8.2.2 Diagnostic and special features (Fig. 8.6)

1 Bilaterally symmetrical; flattened, vermiform.
2 Body more than two cell layers thick, with tissues and organs.
3 A through, straight gut.
4 Body monomeric, although annular; with, anteriorly, two pairs of claw-bearing 'legs' or with only two pairs of claws (Fig. 8.7).
5 A pseudocoelomic hydrostatic skeleton.
6 Body wall with cuticle-covered epidermis and layers of striated circular and longitudinal muscles; cuticle chitinous, porous and moulted.
7 Without excretory, circulatory or gaseous-exchange organs.
8 Nervous system with a brain, a ganglion associated with each 'leg' (or all five ganglia fused into a single mass), and a ventral nerve cord.

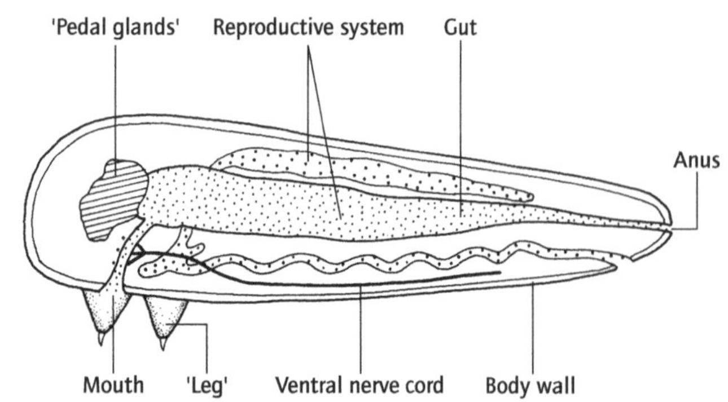

Fig. 8.6 Diagrammatic longitudinal section of a female pentastoman (principally after Cuénot, 1949).

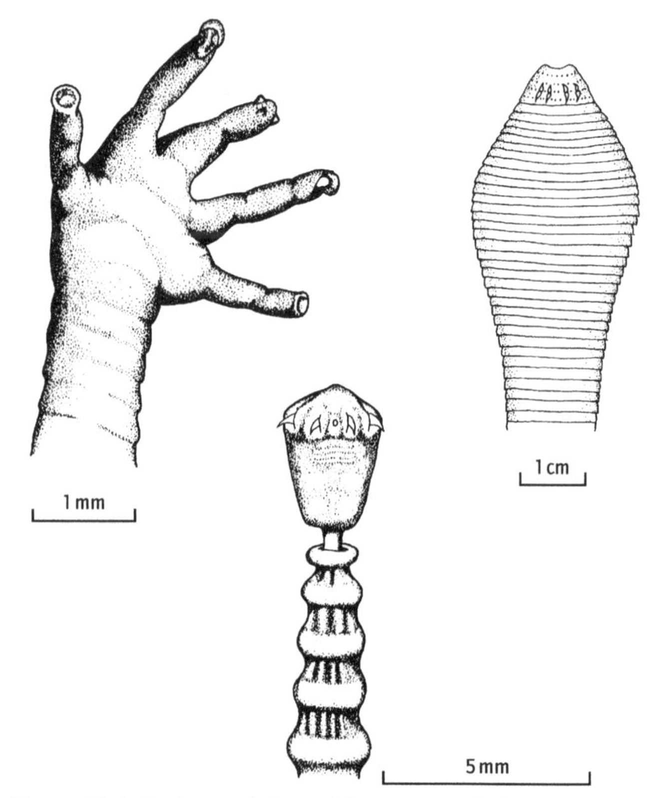

Fig. 8.7 Variation in morphology of the anterior end of the body of pentastomans, showing the extent of development or reduction of the 'legs' (ventral views) (after Kaestner, 1968).

9 Gonochoristic; with one or two gonads.
10 Fertilization internal, via copulation.
11 Development via three 'larval stages', the first with three pairs of lobed, leg-like, unjointed appendages.
12 Blood-consuming parasites of the naso-pulmonary system of vertebrates.

Like many endoparasites, the bodies of pentastomes are dominated by the reproductive system and very many, small eggs are produced. Fertilization is internal, and the fertilized eggs, numbering up to half a million at a time, are retained within the maternal body for a period before being released. The uterus which accommodates them then effectively occupies the whole of the body of the female worm, enlarging over 100 times. Three 'larval' or juvenile stages are passed through sequentially, the first while still within the egg shell. This passes out of the host's body through its alimentary canal (all hosts are predatory terrestrial or freshwater vertebrates, 90% of pentastoman species parasitizing reptiles). If this larva, still within its egg capsule, is swallowed by an intermediate host (an omnivorous or herbivorous insect, fish or tetrapod), the larva emerges and bores its way into the intermediate host's tissues by means of three chitinous stylets, moving on its short, stumpy legs. On reaching a specific region (the host's liver, etc.), the larva encysts and develops into the secondary larval stage. If the infected intermediate host then becomes the prey of the definitive reptile, bird or mammal host, the tertiary larva, which resembles a small version of the adult, emerges and migrates to the lungs or nasal passages. Some species lose their legs during one of the final larval moults, retaining only the terminal hook-shaped claws; many larvae also originally possess a pair of claws per leg, this number reducing to one per leg by the adult stage.

8.2.3 Classification

Some 100 species of these small (2–16 cm long) worms have been described; all can be placed in the one class.

8.3 Phylum ONYCHOPHORA

8.3.1 Etymology

Greek: *onychos*, claws; *-phoros*, bearer.

8.3.2 Diagnostic and special features (Fig. 8.8)

1 Bilaterally symmetrical; elongately and cylindrically vermiform.
2 Body more than two cell layers thick, with tissues and organs.
3 A through, straight gut, bearing, anteriorly, a pair of mouthparts, each with two claw-like mandibles (forming an inner and outer jaw blade); fore- and hind-guts lined with cuticle; without digestive diverticula.
4 Body with 14–43 pairs of short, unjointed, fleshy legs along its length; each leg a hollow evagination of the body bearing a terminal pad, pairs of claws and intrinsic muscles (although leg movements are effected by extrinsic muscles); each leg pair has associated with it a pair of heart ostia and of excretory organs.
5 A well-developed haemocoelic body cavity forming the hydrostatic skeleton; with a tubular heart but without other blood vessels.
6 Body wall with a cuticle-covered epidermis and layers of circular, oblique and longitudinal smooth muscle; cuticle very thin, flexible and chitinous.
7 Excretory organs serially repeated pairs of sac-like glands, the anterior ones forming salivary glands and the posterior ones gonoducts.

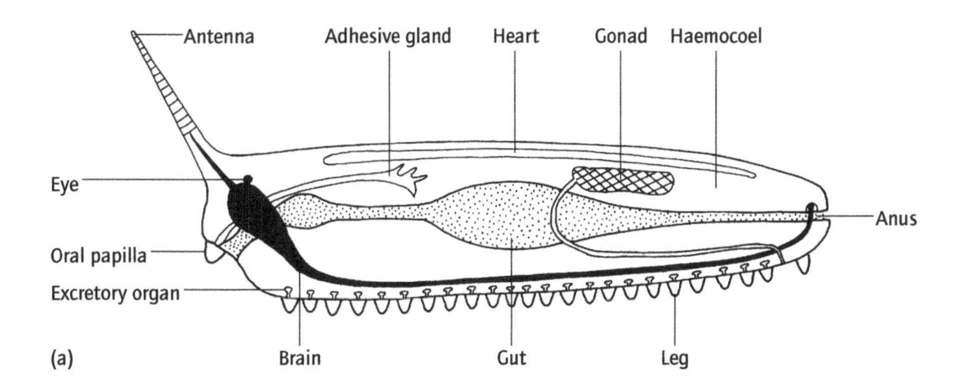

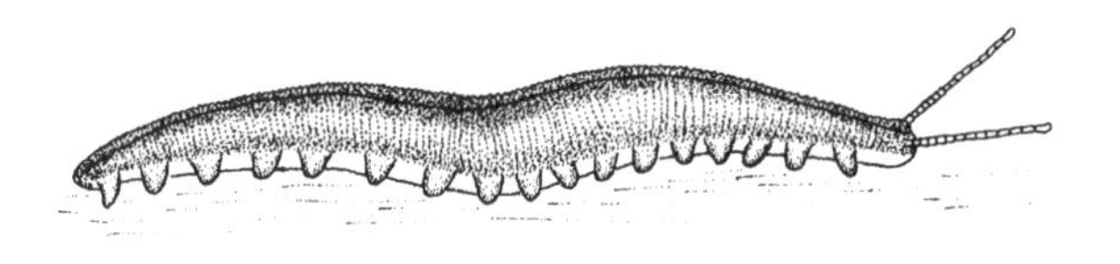

Fig. 8.8 Onychophoran structure: (a) diagrammatic longitudinal section; (b) external appearance. (After Sedgwick, 1888 and Cuénot, 1949.)

8 Gaseous-exchange organs simple tubular tracheae issuing in tufts from numerous small spiracles.
9 Nervous system with brain and a pair of very widely separated ventrolateral cords joined by nine or ten rung-like cross-connectives in each leg 'segment' but without distinct ganglia; sense organs include a pair of annular antennae, each with a small, simple eye at its base.
10 Gonochoristic, with paired gonads; fertilization internal, via spermatophores.
11 Development direct.
12 Free-living, terrestrial.

For many years, the onychophorans have been of scientific interest chiefly as living examples of a half-way stage between the worm and arthropod grades of organization. Like worms, they are soft-bodied and possess hydrostatic skeletons, ciliated excretory ducts and smooth muscle layers in the body wall; and, similarly to the arthropods, they bear legs, tracheae, a heart with ostia, longitudinally partitioned blood sinuses, and jaws derived from the appendages, in this case from the claws that terminate the walking legs. The precise structure of their arthropod-like features, however, strongly indicates that they have achieved them in parallel, and that whereas they may illustrate what the early uniramians, for example, may have looked like, the onychophorans cannot be ancestral to any known arthropod group. Their tracheae, for instance, are simple, mostly unbranched tubes issuing many at a time from the many (up to 75) spiracles that are scattered over each leg-bearing 'segment' (Fig. 8.9), and their jaws, which move in the anterior/posterior plane, act independently of each other and function as ripping organs by virtue of their pointed tips rather than as chewing appendages. Other onychophoran peculiarities include the structure of the ventrolateral nerve cords, with their numerous connectives but without 'segmental' ganglia.

The onychophorans are terrestrial animals, largely confined to humid microhabitats and environments. Their cuticle is only

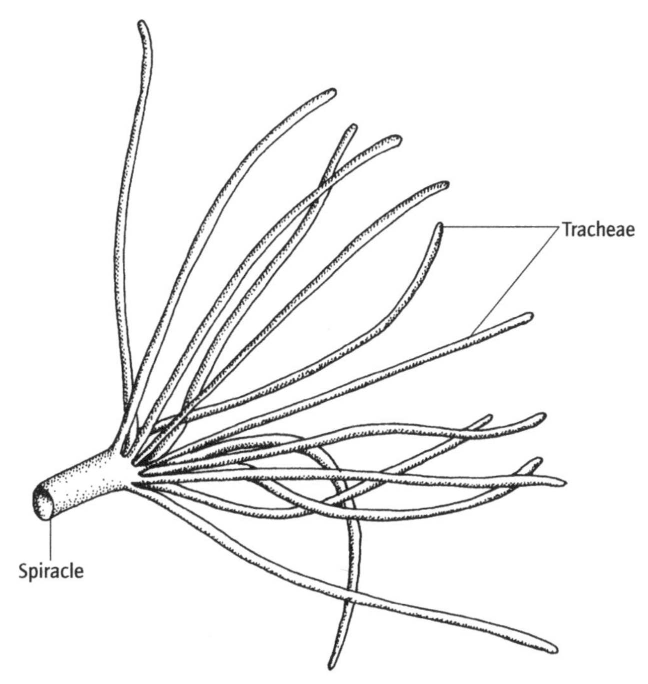

Fig. 8.9 A spiracle and tuft of tracheae (after Clarke, 1973).

1 μm thick (thinner even than the thin arthropod epicuticle, see Fig. 8.1) and is permeable to water, whilst the spiracles have no closing mechanism. Lost water can, however, be replaced by evaginating thin-walled vesicles on to damp surfaces through slits in the cuticle. Onychophorans are mainly nocturnal predators, detecting prey with their antennae and capturing even quite active animals such as grasshoppers by spraying a mucus-like substance distances of up to 0.5 m from adhesive glands which open on paired oral papillae which flank the mouth. This mucus hardens almost immediately on exposure to the air to

form an extremely sticky meshwork entangling potential prey. The same technique serves as a method of defence.

8.3.3 Classification

The 70 species of onychophoran achieve lengths of up to 15 cm; all are placed in a single class and order.

8.4 Phylum CHELICERATA

8.4.1 Etymology

Greek: *chele*, talon; *cerata*, horns.

8.4.2 Diagnostic and special features (Fig. 8.10)

NB, the Pycnogona differ in numerous respects – including in many of those listed below – from the other chelicerates.

1 Bilaterally symmetrical, <1 mm–60 cm long arthropods varying in body shape from elongate to almost spherical.
2 Body more than two cell layers thick, with tissues and organs.
3 A through, straight gut, from the mid-gut region of which issue from two to many pairs of digestive diverticula which secrete enzymes and intracellularly digest and absorb food (these diverticula arise not as outgrowths from the embryonic gut but by partitioning of the embryonic yolk masses before the gut becomes differentiated); mouth anteroventral.
4 Body divided into two regions, an anterior 'prosoma' formed by the acron and six appendage-bearing segments, and wholly or partly covered by a dorsal carapace, and a posterior 'opisthosoma' without legs and with only highly modified appendages, if any.
5 Appendages uniramous; prosomal appendages comprising one pair of chelate, subchelate or stylet-like 'chelicerae', one pair of chelate, leg-like or feeler-like 'pedipalps', and four pairs of walking legs, all attached near to the ventral mid-line and, in some, extended by haemocoelic pressure; without antennae or mandibles.
6 Only one pair of appendages (the chelicerae) form mouthparts, although medially directed processes of the basal article of one or more other limbs ('coxal endites') may crush food or spoon it into the mouth.
7 Usually (unless secondarily lacking) with direct median and indirect lateral ocelli on the prosoma; in one group aggregations of the lateral ocelli form compound eyes.
8 Opisthosoma sometimes externally segmented and then with up to twelve segments, in some divided into a broad anterior 'mesosoma' and a narrow posterior 'metasoma', and in several with a projecting post-anal spine, sting or flagellum.
9 A prosomal excretory system of blind-ending coxal glands, and/or an opisthosomal system of branched endodermal Malpighian tubules arising from the mid-gut and discharging mainly guanine.
10 A non-calcareous exoskeleton and sometimes also with a plate-like mesodermal endoskeleton in the prosoma.
11 Gaseous-exchange organs associated with the opisthosomal appendages or with their embryological primordia; in marine forms, these are external gill-books, in terrestrial forms, the internal lung-books and the sieve- or tube-tracheae derived from them.
12 Blood system involved in the circulation of respiratory gases and usually containing haemocyanin.
13 Nervous system with separate ganglia along the length of the body or, more usually, concentrated into a single prosomal mass.
14 Gonochoristic, with external fertilization in solely marine classes (although the two partners associate closely in pseudocopulation during mating) and internal fertilization via copulation or spermatophores in the primarily terrestrial class; gonopores on the second opisthosomal segment.
15 Juvenile stages small versions of the adult, usually hatching with the full complement of limbs.
16 Originally benthic marine, one class has colonized the land and fresh water highly successfully.

Chelicerates differ from the two other arthropod phyla in a number of major respects, as can be seen from the listing above, paralleling the uniramians in their evolution of Malpighian tubules and tracheae. The most obvious distinctive feature concerns their appendages. All lack mandibles or any other limbs

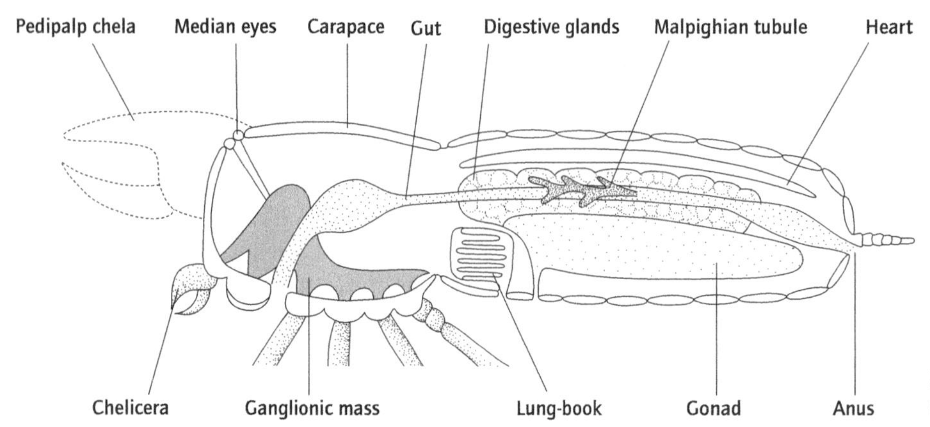

Fig. 8.10 Diagrammatic longitudinal section through a generalized chelicerate.

which are capable of working against each other to bite or chew. Food may be caught and ripped apart by chelate or subchelate limbs (the pedipalps and/or chelicerae); nevertheless, except in a very few cases, only very finely particulate or, characteristically, liquid food can be ingested. Indeed, the mouth itself is usually screened by setae to prevent the entry of large particles. Even so, the chelicerates are almost entirely predatory. Their feeding speciality is to hold the prey close to the mouth, pour digestive enzymes into it, and then imbibe the products of this external predigestion. If enzymes are not actually injected into the prey, digestion or mechanical breakdown occurs in a pre-oral space enclosed by the coxal endites of some or all of the prosomal limbs, which are often arranged almost radially around the mouth. In association with this method of feeding, the fore-gut is adapted to form a pump or pumps, and the mid-gut, especially its many diverticula (which can occupy most of the body volume), is the site of final digestion and absorption. Further, the chelicerates lack antennae, although the pedipalps or first or second pairs of walking legs may be modified to serve a similar function.

Within the chelicerates, there is a morphological series indicating how a tracheal system could evolve from external gills. In the merostomatans, which are marine and almost certainly the ancestral stock of the chelicerates, the organs of gaseous-exchange are – as in most other marine animals – external gills. These arise from the posterior margin of the flap-like opisthosomal appendages, the beating of which drives water over the gill lamellae. Appropriately, the terrestrial arachnids are air-breathing but, in contrast to the uniramians (see Section 8.5), their characteristic gaseous-exchange organs have retained an embryological association with the posterior margin of the opisthosomal limb primordia (in the arachnids, these opisthosomal limb primordia do not develop into legs, although the spinnerets of spiders and the pectines of scorpions are highly modified opisthosomal appendages). The lung-books (Fig. 8.11), for example, form in this way and are equivalent to a series of gill lamellae housed in pockets sunk into the body. Each lung-book is an invagination with, dangling into it, a series of parallel lamellae held apart by struts, between which air moves by diffusion and into which blood flows within a sinus. In some lung-books, the lamellae are elongate and tube- rather than plate-like, and dependent on the number of such tubes these are termed sieve-tracheae (many closely associated tubes) or tube-tracheae (few tubes). These tracheal systems are therefore essentially elongate, internalized, leg-associated gills. In some spiders, additional secondary tracheae have developed from an alternative origin: from hollow, internal projections of the exoskeleton ('apodemes' – see Fig. 8.1).

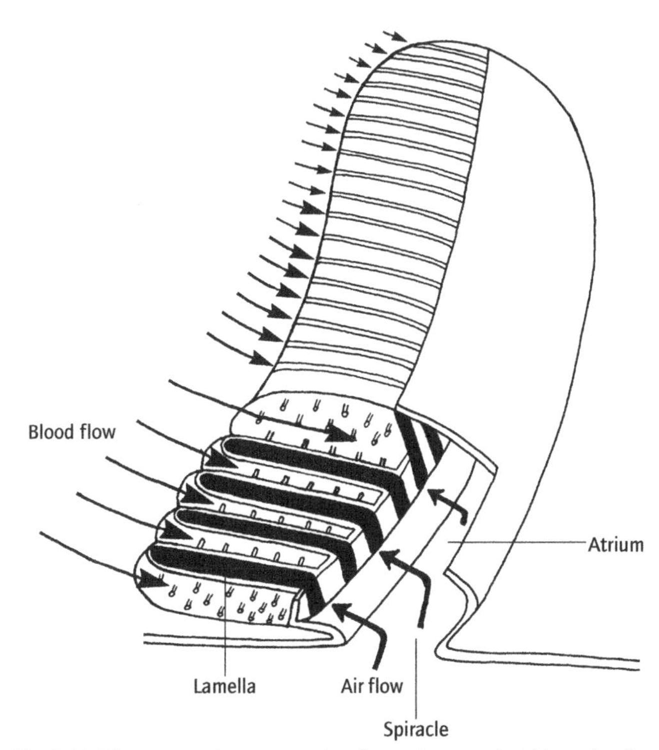

Fig. 8.11 Diagrammatic stereosection through an arachnid lung-book, showing the circulation of blood and air (after Barnes, 1980).

Class	Order
Merostomata	Xiphosura
Arachnida	Scorpiones Uropygi Schizomida Amblypygi Palpigradi Araneae Ricinulei Pseudoscorpiones Solpugida Opiliones Notostigmata Parasitiformes Acariformes
Pycnogona	Pycnogonida

8.4.3 Classification

The 63 000 described species of chelicerates are placed in three rather disparate classes, of which one is arguably unrelated to the other two.

8.4.3.1 Class Merostomata (horseshoe crabs)

Although a dominant group of invertebrates until the Permian, the merostomatans are today represented by only four species in one order (Xiphosura). The horseshoe crabs are large marine chelicerates with a thick, horseshoe-shaped carapace covering the large prosoma and extending both anteriorly, so that the mouth is mid-ventral, and laterally, hiding the appendages from dorsal view. The small, hinged opisthosoma is a flat plate partly inset into a notch in the carapace and fringed laterally with stout spines; terminally, it bears a long, post-anal, caudal spine

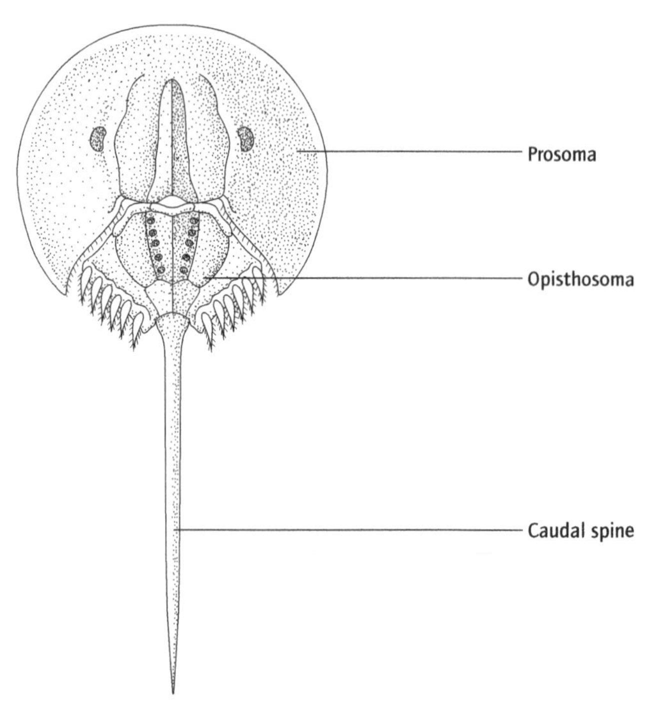

Fig. 8.12 A merostomatan in dorsal view (after Kaestner, 1968).

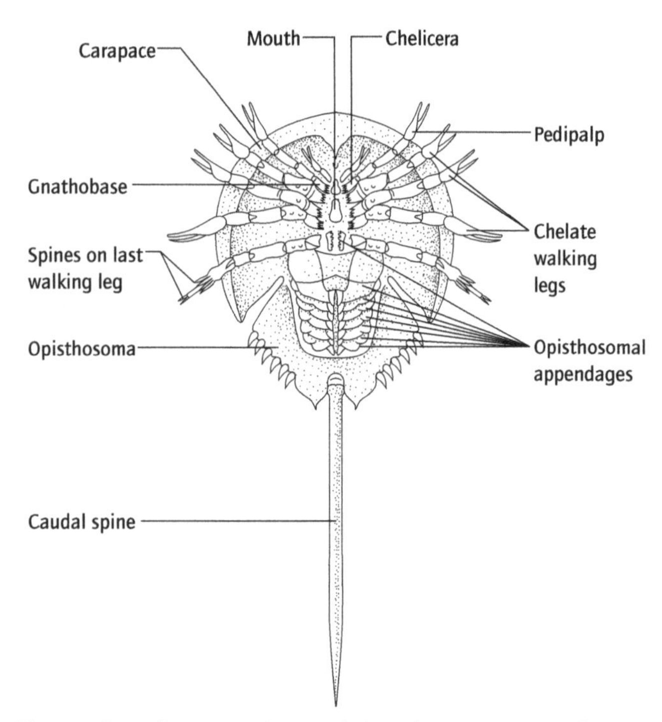

Fig. 8.13 Semi-diagrammatic ventral view of a merostomatan showing the various appendages (after Savory, 1935).

(Fig. 8.12). The chelicerae, pedipalps and all but the last pair of walking legs are chelate, this last pair bearing instead several terminal spines or lamellae used during burrowing (Fig. 8.13).

Unlike the other surviving chelicerates, horseshoe crabs possess almost a full complement of opisthosomal appendages: the most anterior pair are small and tubular, and the remaining six pairs form flat plates which propel the animals during their upside-down swimming. The first of these plates also serves as a genital operculum, whilst the posterior five in addition bear the external gills. Other features not inappropriate in a marine animal but atypical amongst the chelicerates in general include the lack of Malpighian tubules and the occurrence of external fertilization. Alone amongst living members of this phylum merostomatans also have large, diffuse, lateral compound eyes of a rather individual type; in contrast to the arachnids, they have their gonads and digestive diverticula in the prosoma.

The horseshoe crabs are nocturnal benthic animals found in shallow coastal waters, where they burrow in soft sediments and crawl over the surface, preying on large molluscs and polychaetes which are crushed by the coxal endites ('gnathobases') and in a specialized gizzard.

8.4.3.2 Class Arachnida

The almost entirely land-based arachnids comprise over 98% of living chelicerate species and display a number of marked adaptations to terrestrial existence, e.g. a Malpighian-tubule excretory system, internal air-breathing gaseous-exchange organs, a cuticle waterproofed by a wax layer, and internal fertilization. In morphology, they range from elongate, well-armoured forms with large raptorial pedipalps, conspicuous external segmentation, and opisthosomas divided into meso- and metasomes (scorpions, whipscorpions, etc.) to almost spherical species with thin exoskeletons, no externally visible segmentation and less evident pedipalps (most spiders and mites, etc.); this diversity is reflected in their division into 13 orders (Fig. 8.14). The scorpion-like body form is almost certainly the primitive condition as it is little removed from that of some of the (now extinct) eurypterid merostomatans, which were freshwater and probably amphibious in habitat. Several modern arachnids have recolonized fresh waters, and some mites inhabit the sea.

In contrast to the xiphosurans, the arachnids are characterized by the absence of compound eyes and of opisthosomal ambulatory appendages, by the presence of long, non-chelate walking legs, and by the location of the gonads and digestive diverticula in the opisthosoma. In several groups of arachnids, the general reduction in massiveness of the exoskeleton has extended to the freeing of the last two prosomal segments from the carapace, although these may have separate dorsal plates, and in the evolution of a soft, flexible opisthosoma. In most mites these last two prosomal segments have been incorporated into the opisthosoma to form a two-regioned body differing from the standard chelicerate pattern. More generally, however, the prosoma and opisthosoma are retained as the two fundamental divisions of the body, the two being separated by an internal diaphragm or by a narrow stalk ('pedicel'); this separation may permit the prosomal body pressure to be raised (to extend the legs) without affecting that in the opisthosoma.

The arachnids are a most successful group, with a life style based mainly on preying on the equally successful insects.

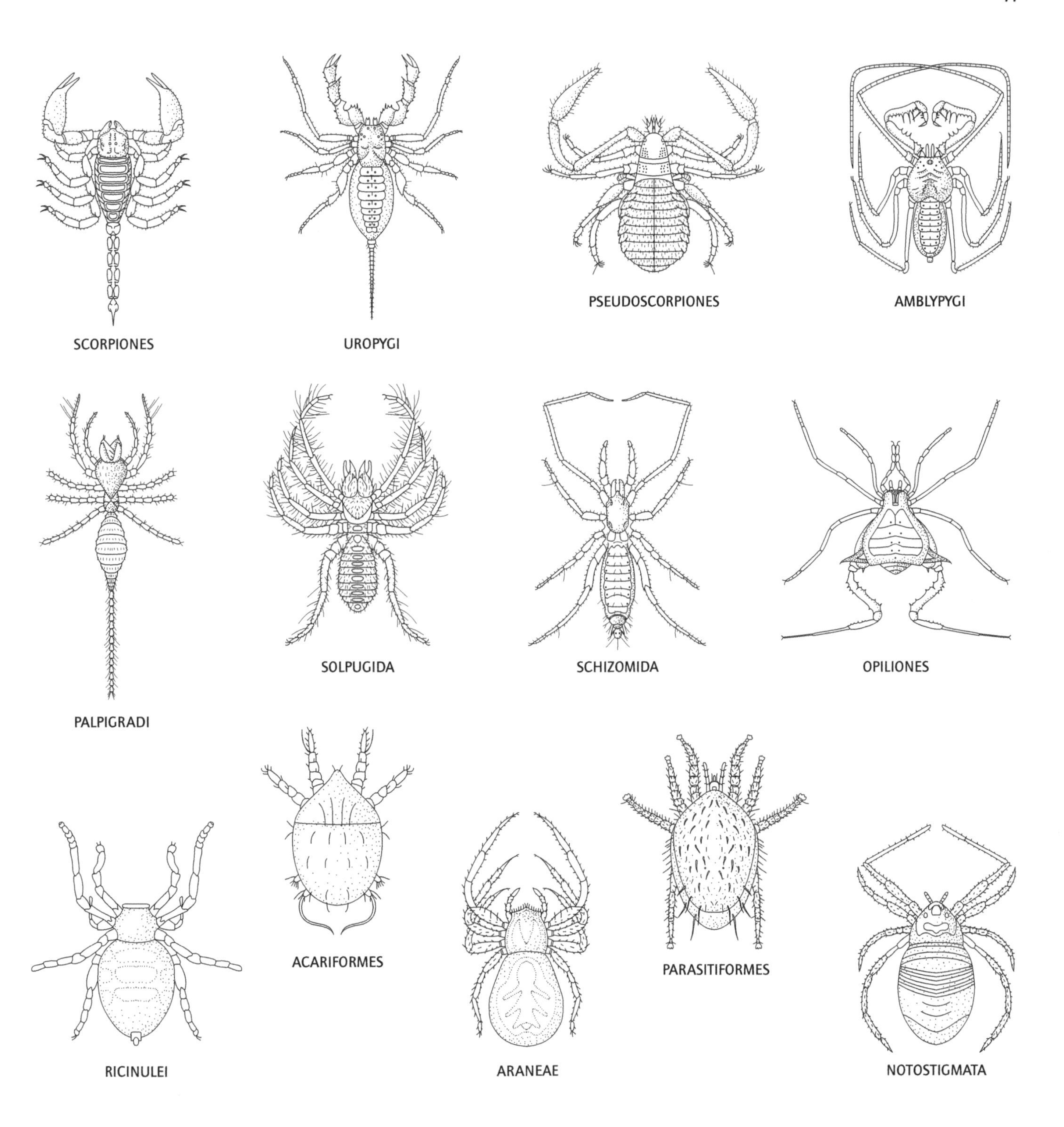

Fig. 8.14 Body form of the various arachnid orders (after Savory, 1935; Hughes, 1959 and others).

Notable adaptations to permit the capture of such highly mobile organisms include the possession of poison glands, the secretions of which can be injected into their prey, and the production of silk, a phenomenon culminating in the construction of webs by spiders. They range in size from <0.1 mm to 18 cm, and include over 62 000 species. Unusually, some mites are non-predatory, feeding on plant substances and on detritus; several are parasites.

8.4.3.3 Class Pycnogona (sea spiders)

These small benthic marine arthropods (body length <6 cm) are considered by several zoologists to be unrelated to the other chelicerates, and possibly to the other arthropods. Their first

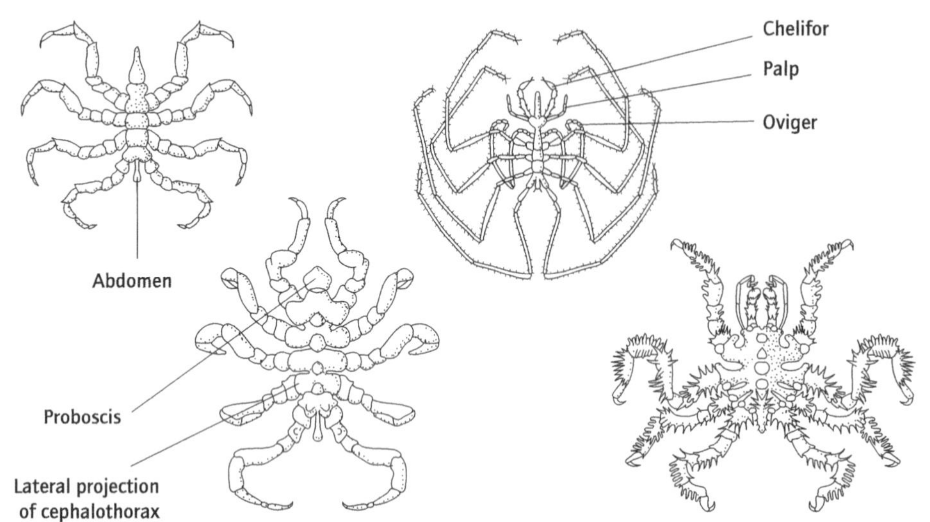

Fig. 8.15 Pycnogonan body forms (principally after Hedgpeth, 1982).

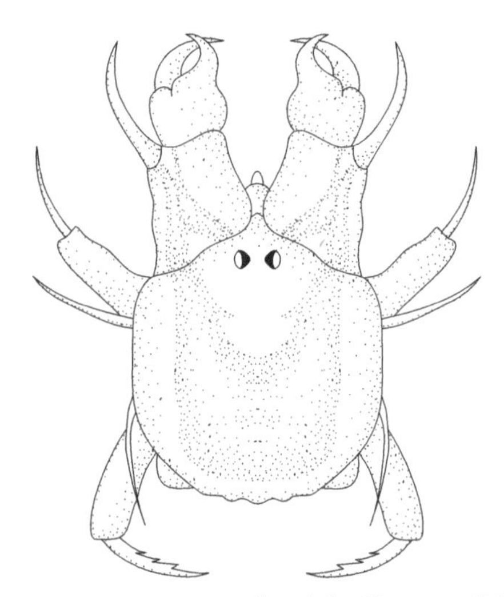

Fig. 8.16 A pycnogonan protonymphon (after Kaestner, 1968).

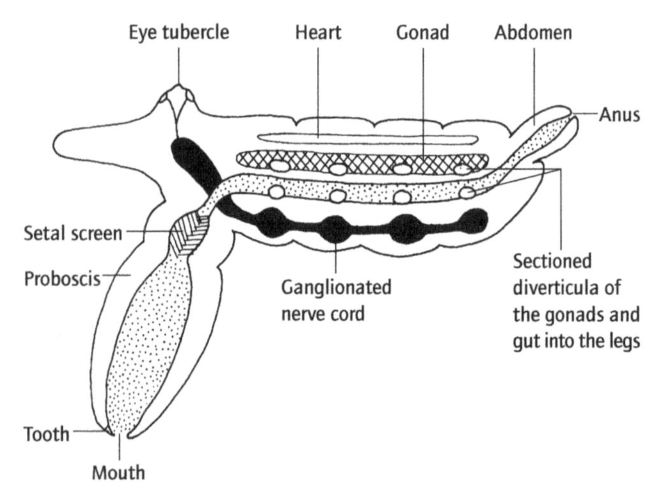

Fig. 8.17 Diagrammatic section through the body and proboscis of a pycnogonan (after King, 1973).

appendage, the chelifor, *is* chelate like a chelicera; the second one, the palp, *may* be the homologue of the pedipalp; and most species *do* possess four pairs of walking legs. Some species, however, have five pairs of walking legs and a few have six pairs, whilst an additional appendage, the oviger, is located in between the palp and the first walking leg. This is attached ventrally, not laterally as in the other appendages – the walking legs articulating with stout lateral projections from the body (Fig. 8.15). In several species, the chelifores, palps and ovigers are vestigial or absent; the walking legs, however, are always well developed and may be very long (up to 75 cm span has been recorded).

The cephalothorax (prosoma?), which is not covered by any carapace, bears a central prominence, with two pairs of simple eyes, and anteriorly or ventrally, a large, tubular, non-retractable proboscis on which the mouth is terminal. The abdomen (opisthosoma?) is a minute, unsegmented papilla lacking any appendages. No excretory or gaseous-exchange organs occur. Peculiarly, the gonads extend well down into the legs, as indeed do the digestive diverticula, and the ova ripen in the legs. During pseudocopulation, the gametes are discharged from multiple gonopores positioned at the base of each walking leg or at those of the last two pairs, and, after fertilization, the male gathers up the eggs and carries them around in balls attached to its ovigers. Eventually, a distinctive larval stage with three pairs of appendages, the protonymphon (Fig. 8.16), hatches and, most commonly, begins a semiparasitic life on or in cnidarians or molluscs.

The adults mostly feed on sponges, cnidarians or bryozoans, part of the prey being grasped by the chelifores (if present) whilst powerful suction generated by the pharynx results in ingestion of tissues, aided by the gnawing action of three teeth occurring at the tip of the proboscis. Within the pharynx, further teeth or strong setae macerate the food, and a collar of setae at the pharyngeal base prevents the intake of anything other than minute particles into the oesophagus (Fig. 8.17).

More than 1000 species have been described, all referable to a single order.

8.5 Phylum UNIRAMIA

8.5.1 Etymology

Latin: *unus*, one; *ramus*, branch.

8.5.2 Diagnostic and special features (Fig. 8.18)

1 Bilaterally symmetrical; <1 mm–35 cm long arthropods varying in body shape from extremely elongate to almost spherical.
2 Body more than two cell layers thick, with tissues and organs.
3 A through, straight gut lacking any digestive diverticula.
4 Body divided into two regions, a 'head' formed by the acron and three or four appendage-bearing segments, and a 'trunk' bearing pairs of walking legs; in one subphylum, the trunk comprises a series of up to 350 relatively uniform segments, the great majority of which bear walking legs; in the other subphylum, the trunk is differentiated into a 'thorax' with three pairs of legs, and an 'abdomen' of up to eleven segments with only highly modified appendages, if any.
5 Appendages uniramous, those of the head comprising one pair each of 'antennae', 'mandibles', and maxillae, and in some groups a second pair of maxillae, those of the trunk all form functional or modified walking legs; without chelicerae or chelate limbs.
6 Two or three pairs of mouthparts (the mandibles and maxillae), members of each pair working with or against the other; the basal articles of the maxillae, or second maxillae in those groups having two pairs, fuse to form a plate flooring the pre-oral cavity (the labium or 'lower lip').
7 Head with lateral ocelli, frequently organized into compound eyes; sometimes also with median ocelli.
8 Trunk, but not head, externally segmented.
9 Most members of one subphylum with one or two pairs of wings on the thorax.
10 With a fat body in the haemocoel, often closely associated with the gut.
11 Excretory system in the form of zero to two pairs of maxillary glands in the head, and one to 75 pairs of unbranched ectodermal Malpighian tubules arising from the hind-gut near its junction with the mid-gut and discharging mainly ammonia and/or uric acid.
12 Exoskeleton calcareous or, more commonly, non-calcareous.

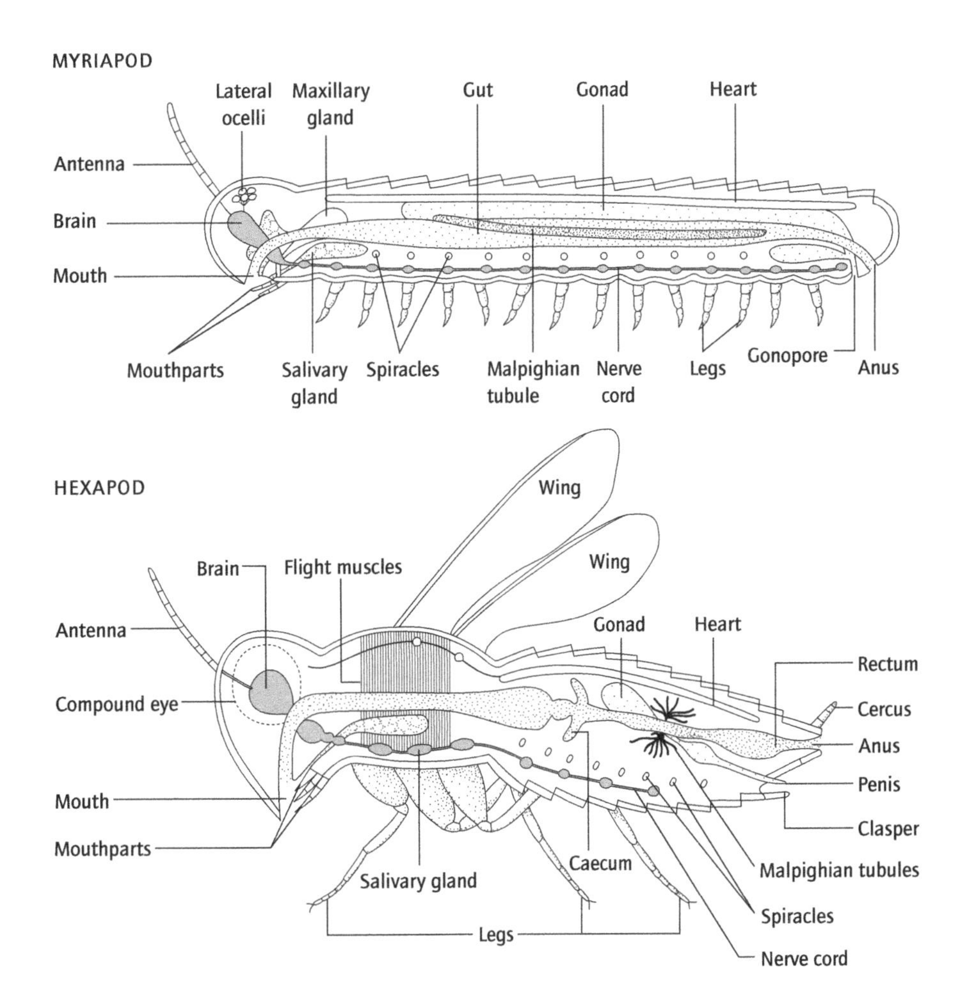

Fig. 8.18 Diagrammatic longitudinal sections through two generalized uniramians.

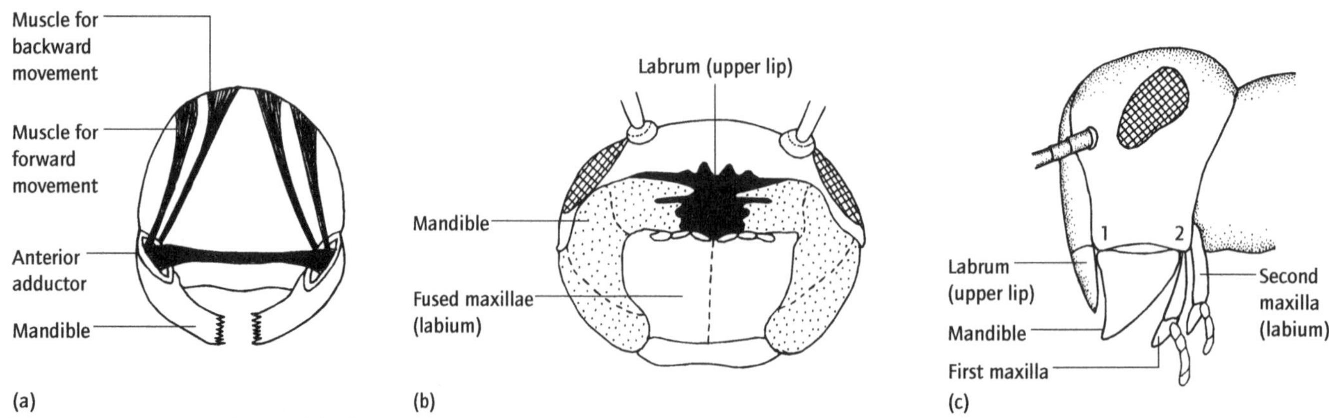

Fig. 8.19 Uniramian mouthparts: (a) a diagrammatic transverse section through the head showing the primitive state in which the mandibles are aligned vertically downwards and operated by a similar series of muscles to those of the walking legs; (b) an anteroventral view into the pre-oral cavity of a millipede, showing the upper lip (labrum) formed by sclerites of the head capsule, and the lower lip (labium) formed by fusion of the maxillae (note the horizontal orientation of the mandibles); (c) side view of an insect head with three pairs of mouthparts and a mandible articulating with the head capsule at two points (labelled 1 and 2). (After Kaestner, 1968.)

13 Gaseous-exchange organs paired, branched tracheal tubes through which air diffuses; tracheae open (primitively) via one pair of spiracles on each leg-bearing segment and terminate, internally, in tracheoles; tracheal system not associated embryologically with the appendages.
14 Blood without any circulatory function in respect of the respiratory gases, and without respiratory pigments (except in a few larval stages).
15 Gonochoristic, with internal fertilization via spermatophores or copulation; ancestrally with gonopores on last trunk segment, modified in several forms.
16 Several groups with a distinct larval stage quite unlike the adult form; others hatching with less than the full complement of segments (and legs), additional segments being added at each moult, even, in some, after reproductive maturity.
17 Primarily terrestrial, several in fresh water but very few in the sea.

The Uniramia probably originated on land and, with the arachnid chelicerates which they outnumber in both numbers of species and numbers of individuals, they are the dominant land invertebrates. Part of the reason for their terrestrial success is held in common with the arachnids. Both groups, independently, have acquired a gaseous-exchange system which relies on diffusion of the gases from the environment to the tissues and vice versa and is therefore more efficient in preventing water loss than, for example, is that of the tetrapod vertebrates, in which much water vapour is expired during forced ventilation of their lungs. The shared Malpighian-tubule excretory system is, by virtue of its association with the gut, also at least potentially capable of reducing water loss through the reabsorption of water from the nitrogenous wastes after their discharge into the hind-gut. And the presence of the cuticular exoskeleton around the external surface of the body is a greater barrier to water loss than are the soft, moist integuments of annelids and molluscs. Nevertheless, in the majority of the uniramian classes, such prevention of water loss is only partially successful, and they remain largely tied to moist microhabitats in and near the soil. Their spiracles cannot be closed, for example; little water is reclaimed from the nitrogenous waste, which is largely in the form of ammonia; and the cuticle is relatively permeable. It is in only one class that the water loss has been further, and greatly, reduced by the evolution of spiracular closing mechanisms, of an impermeable waxy epicuticle, and of a water reclamation system in the rectum. This class, that of the pterygote insects, is also by far the largest and most diverse of the Uniramia.

Although the above water-retaining mechanisms may explain why the uniramians are at least as successful as the arachnids, they do not appear able to account for their much greater apparent success. This is probably largely attributable to the possession by the uniramians of jaws capable of biting and chewing. They can take solid foods into the gut, and are not confined to a liquid diet and therefore to prey which can be externally predigested. In effect, this means that plant materials are available to the uniramians, and the land is, above all, characterized by an abundance of relatively tough plant tissues which need to be cropped and chewed before they can be ingested. In marked contrast to the almost entirely predatory arachnids, only one major group of the Uniramia, the centipedes, are exclusively carnivorous; although, for the reasons outlined in Chapter 9, most species are consumers of dead and decaying plant substances rather than the living plants themselves.

In origin, the uniramian jaw, the mandible, is the first walking leg, of which only the basal portion develops. Its cutting edge is a large, medially directed projection immovably attached to the remainder of the jaw in all but the millipedes and symphylans. Primitively, the mandible articulates with the body in the same manner as the other walking legs, being oriented perpendicularly to the head capsule (Fig. 8.19a). In most uniramians, however, its alignment has changed so that it lies parallel to the ventral surface of the head, having undergone a rotation towards the anterior. Its erstwhile dorsal articulation with the body has therefore become a posterior one and, in some, a second point of articulation is developed anteriorly (see Fig. 8.19c). The freedom of movement of the jaws is then mainly limited to an opening and closing motion. The floor or posterior margin of the space in front of the mouth in which the mouthparts function is formed by the fused basal parts of the maxillae, or second maxillae in groups having two pairs, which have also been derived from walking legs (Fig. 8.19b). This contrasts with the position in the two other arthropod phyla, in which this labium is part of a ventral exoskeletal sternite of the body. In the myriapods, mouthpart anatomy varies little in its overall structure, but, in the pterygote insects, the morphology of the mandible and maxillae is extremely diverse, correlated with their radiation into numerous feeding modes (see Section 8.5.3B.2).

8.5.3 Classification

The more than one million uniramian species can be distributed between two subphyla, the one with four component classes and the other with six. The uniramian body plan is relatively conservative, however, and the differences between most of these classes are not nearly so marked as between those of many other large phyla, e.g. of the Mollusca, Chelicerata and Crustacea.

Class	Order
Chilopoda	Scutigerida
	Lithobiida
	Scolopendrida
	Geophilida
Symphyla	Scolopendrellida
Diplopoda	Polyxenida
	Glomeridesmida
	Oniscomorpha
	Polyzoniida
	Stemmiulida
	Spirobolida
	Iuliformida
	Typhlogena
	Chordeumatida
	Polydesmida
Pauropoda	Pauropodida

8.5.3A Subphylum MYRIAPODA

The myriapod classes probably represent the first major radiation of this phylum (or the surviving members of it), one of these classes then providing the starting point of the second major radiation in which the classes of the second subphylum arose. Nevertheless, in spite of the great anatomical similarity between the Myriapoda and the Hexapoda, and the interlinking apterygote insects, molecular sequence evidence suggests that the two groups are more closely related to other arthropods than they are to each other, the myriapods to the chelicerates and the hexapods to the crustaceans. Until this conflict between molecular and anatomical data is resolved it seems most appropriate to retain the traditional view here.

Their primary distinguishing feature is that the trunk comprises a series of more-or-less identical segments, each of which, except for one or two terminal ones and sometimes for the first, bears a pair of walking legs; there is no differentiation into a thorax and abdomen. They do share a number of other common features, although these are not exclusively myriapod characters (since several are shared with the apterygote insects – see Section 8.5.3B.1). There is only a single pair of (usually elongate) Malpighian tubules, for example; the head lacks median ocelli but bears the 'organs of Tömösvary', which are probably sensitive to air-borne chemicals and maybe also to humidity; the individual articles of the antennae possess their own muscles; the nervous system is not concentrated, having ganglia in each trunk segment; and, as indicated above, their powers of water retention are relatively poor. The various classes differ mainly in the numbers of pairs of maxillae, the position of the gonopores, the numbers of dorsal tergal plates, and in which segments lack legs.

Although an exclusively terrestrial subphylum, several species in each class inhabit the intertidal zone of the sea.

8.5.3A.1 Class Chilopoda (centipedes)

Centipedes are elongate to very elongate, dorsoventrally flattened myriapods with a head possessing two pairs of maxillae, and a trunk comprising from 15 to more than 181 leg-bearing segments (always an odd number) together with two terminal legless ones, the pregenital and genital (Fig. 8.20). All their legs are similar, although sometimes increasing in length posteriorly, except for the first pair which, diagnostically, are modified into prey-catching organs – large, claw-like, poison-gland containing fangs (Fig. 8.21).

Although most species have quite large numbers of lateral ocelli and one group (the order Scutigerida) – alone amongst the myriapods – possesses large compound eyes, the carnivorous centipedes are nocturnal or live beneath the soil surface and locate their arthropod and oligochaete prey with the antennae or, more rarely, with the legs. Some groups lack eyes.

As in most myriapods, sperm transfer is effected by means of an external spermatophore, which is extruded from the terminal gonopore only after considerable courtship behaviour. In all but one group, the male also protrudes a spinneret from the same orifice and lays down a series of silk threads in which the spermatophore may be suspended and which may guide the female to it. Several species guard the eggs and even the young after they have hatched.

Unusually in this subphylum, however, the young of many species hatch with the full complement of segments and legs

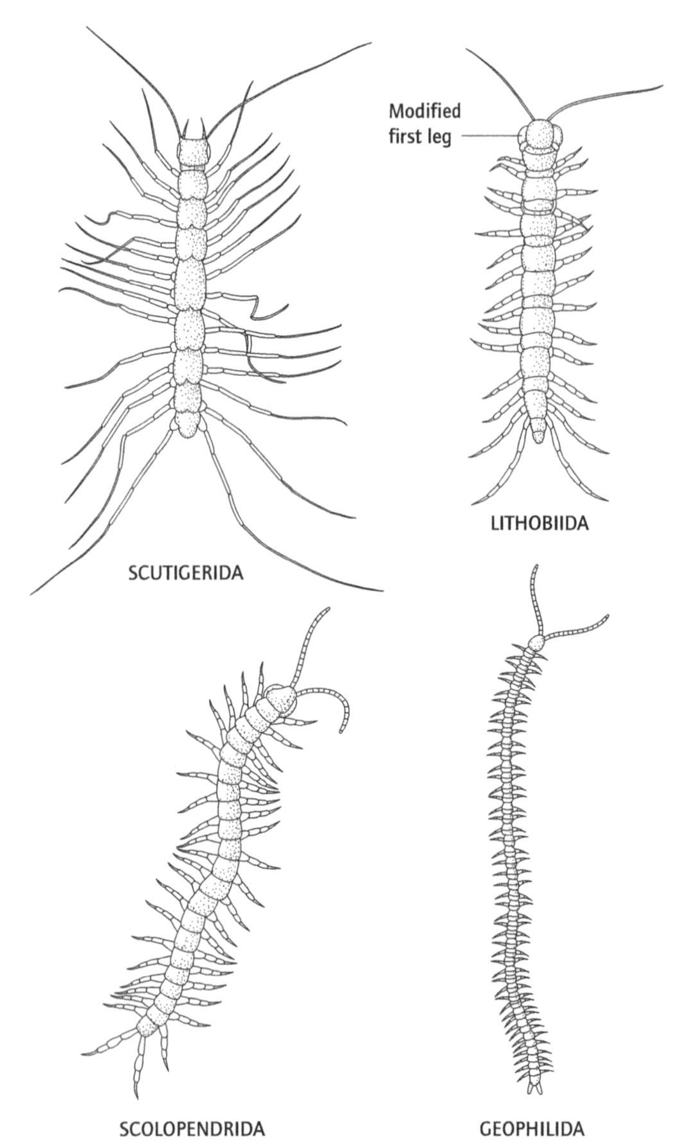

Fig. 8.20 The body form of centipedes (after Lewis, 1981).

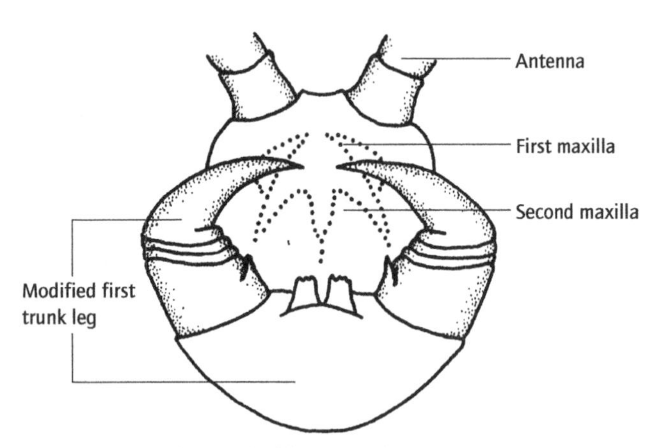

Fig. 8.21 The highly modified first pair of centipede trunk limbs (after Borror *et al.*, 1976).

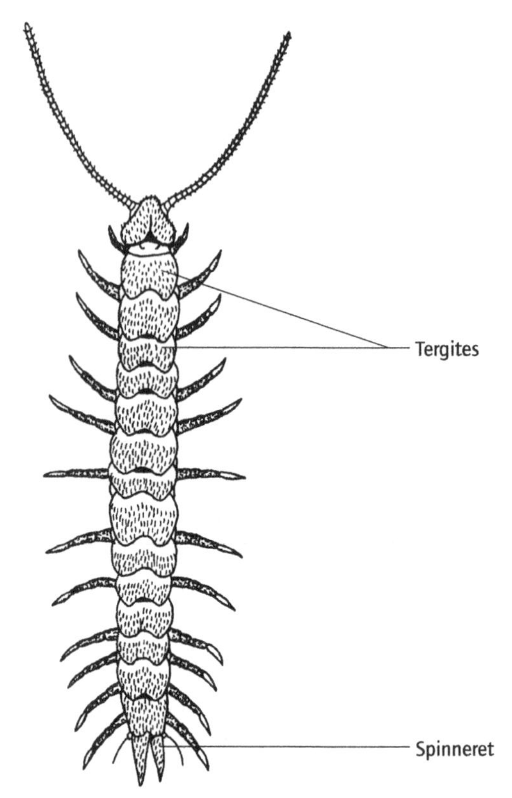

Fig. 8.22 The external appearance of symphylans (after Kaestner, 1968). Note that the number of tergites exceeds that of the leg-pairs.

(including, somewhat paradoxically, those centipedes with the largest number of segments), whilst, more typically, others hatch with less than the adult number, some with only four.

The 3000 species, which attain lengths of up to 27 cm, are divided between four orders.

8.5.3A.2 Class Symphyla

The symphylans are small myriapods (<8 mm long) sharing the same general body plan as the centipedes, i.e. there is a head with two pairs of maxillae, and a trunk with, in this group, twelve leg-bearing segments and two terminal leg-less ones, the last fused with the telson (Fig. 8.22). Spinnerets are also located on the last free body segment. In contrast to the centipedes, however, there are many more dorsal tergal plates than there are segments – up to 24 – permitting increased flexibility of the body; the gonopores secondarily open anteriorly (on the third segment); and the first pair of walking legs do not form fangs, although they are distinctive in being smaller than the following ones, sometimes only half the size. Also, unusually amongst myriapods, the single pair of short tracheae open via spiracles on the head.

In a number of other respects, however, this class resembles the members of the subphylum Hexapoda, particularly the apterygote insects. Their mouthparts are essentially similar, for example, and small styli and eversible coxal sacs are present in association with most legs; the coxal sacs can be everted by

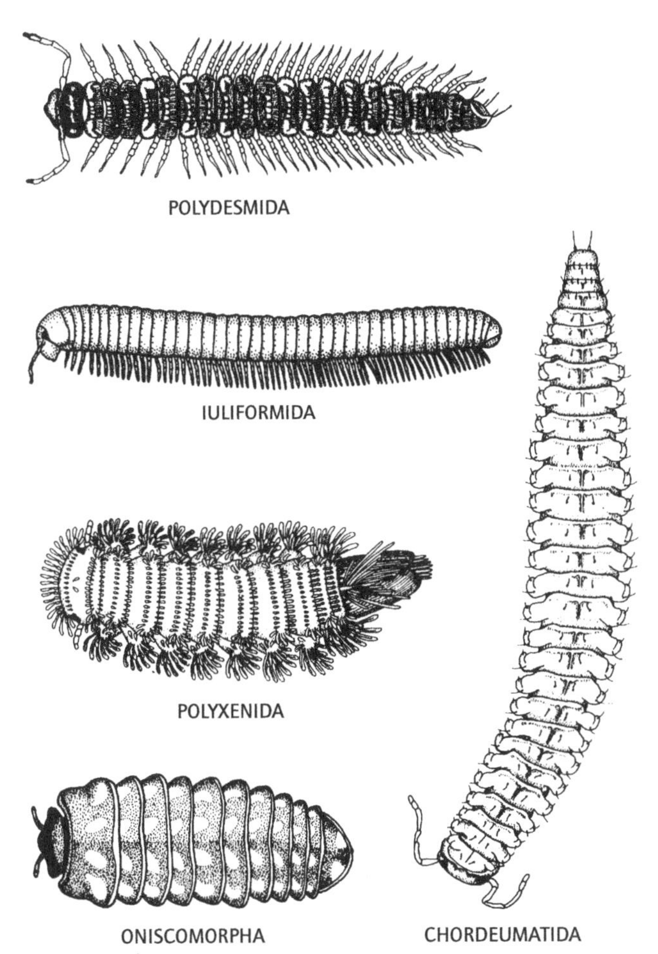

Fig. 8.23 Millipede diversity (after several authors, principally Blower, 1985).

haemocoelic pressure and used to take up environmental water. Since, like most myriapods, symphylans hatch with less than the full complement of segments, it is possible that the hexapods evolved from symphylan-like ancestors by paedomorphosis (some millipedes and pauropodans also hatch with only three pairs of legs).

The 160 species of the small, blind, soft and pallid symphylans are mainly herbivorous, feeding especially on live rootlets and living in soil, leaf litter and rotting wood, under stones, etc. All species are placed in a single order.

8.5.3A.3 Class Diplopoda (millipedes)

The short to very elongate millipedes (Fig. 8.23) have as their most conspicuous distinguishing feature the occurrence of 'diplosegments'. The diplopod trunk is composed of a leg-less first segment, three following segments each with the typical myriapod single pair of legs, and then a series of from five to more than 85 'rings' each with two pairs of legs, ganglia, heart ostia, etc., in front of the one or more leg-less segments which comprise the terminal segment-proliferation zone. Each diplosegmental ring, which together comprise most of the almost cylindrical or, rarely, flattened trunk, is formed by the partial or complete fusion of the trunk segments in pairs; at the very least, such diplosegments share a common dorsal tergite. In many species, one or two leg pairs (those of the seventh apparent segment, or the posterior pair of the seventh and the anterior pair of the eighth) are, in the male, modified to varying extents to form copulatory 'gonopods', which collect sperm from the gonopores on the third segment and transfer it to the corresponding gonopores of the female. In other species, the mandibles are used to transfer the sperm, or the gonopores of the two sexes may be brought into close proximity, or, in classic myriapod fashion, sperm packets and silk guide threads are produced. The eggs of most millipedes hatch with only seven trunk 'segments', new ones being added throughout life, and long after the onset of sexual maturity.

Other millipede characteristics include the presence of only one pair of maxillae, of large numbers of lateral ocelli arranged in blocks which superficially resemble compound eyes (some species, however, lack eyes), and, except in one order, of a calcified exoskeleton – millipedes being the only uniramians to have such.

The 10 000 species are disposed in ten orders. All are slow-moving feeders on plant material, usually only after it has started to decompose, which they obtain in the same microhabitat types as the symphylans (rotting logs, leaf litter, etc.). The largest achieve a length of 28 cm.

8.5.3A.4 Class Pauropoda

The minute pauropods (<2 mm length) live in the same habitats as the symphylans and millipedes, where they bite into fungal hyphae and suck out their contents by means of a pumping foregut. Although inconspicuous, they are often abundant. In their body plan, they bear a similar relationship to the millipedes as do the symphylans to the centipedes. The head, for example, bears a single pair of maxillae; the first trunk segment is leg-less, as are the last two; the gonopores are anterior, on the third segment; the trunk segments are arranged in incipient diplosegments, a single tergal plate partially or wholly covering segments 1 and 2, 3 and 4, 5 and 6, and so on (Fig. 8.24), up to the total of eleven or twelve segments, the last fused with the telson; and the young hatch with few segments and most with three pairs of legs.

Pauropods are not just incipient millipedes, however, as they show a number of distinctive features. They lack a heart, for instance, and most lack tracheae; even when present, the tracheae are very small. Further, they possess branched antennae.

The 500 species, all of which are blind, soft and colourless, are placed in a single order.

8.5.3B Subphylum HEXAPODA (insects)

Only one characteristic separates all the members of this subphylum from the Myriapoda: the hexapod trunk is subdivided into a thorax of three leg-bearing segments, and an abdomen of eleven segments without walking legs; although abdominal

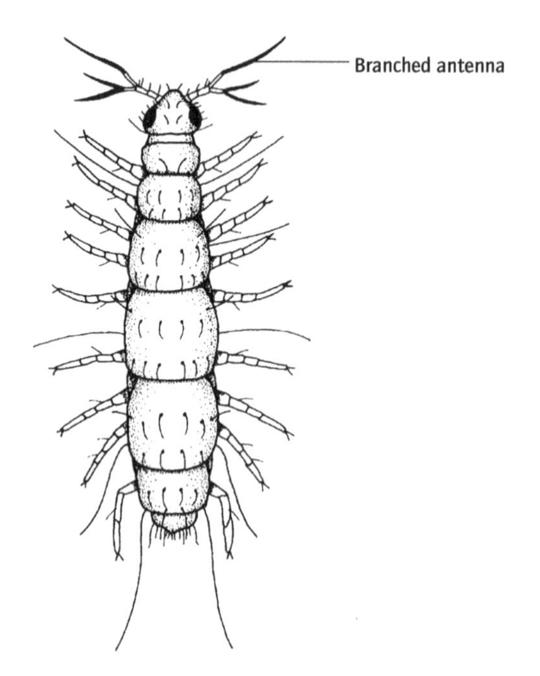

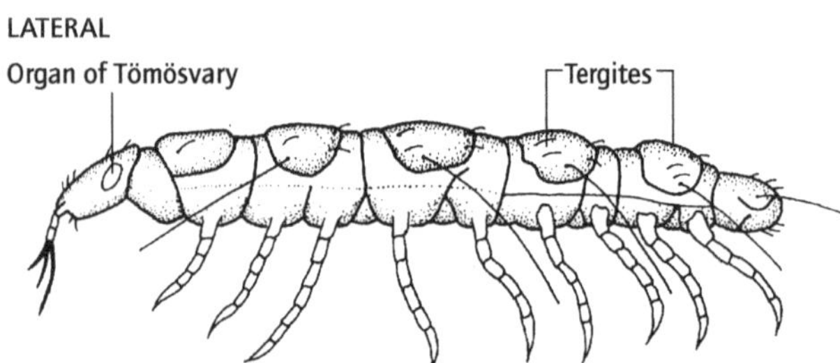

Fig. 8.24 The external appearance of pauropods (after Kaestner, 1968 and Borror *et al.*, 1976). Note that the tergites are less numerous than the leg-pairs.

appendages of some form may be present and loss or fusion may reduce the number of abdominal segments to less than eleven. All hexapods also have two pairs of maxillae and terminal or subterminal gonopores, and most possess median ocelli as well as lateral ocelli or compound eyes.

Class	Superorder	Order
Diplurata		Diplura
Oligoentomata		Collembola
Myrientomata		Protura
Zygoentomata		Thysanura
Archaeognathata		Microcoryphia
Pterygota	Palaeoptera	Ephemeroptera Odonata
	Orthopteroidea	Blattaria Mantodea Isoptera Zoraptera Grylloblattaria Dermaptera Orthoptera Phasmida Embioptera Plecoptera
	Hemipteroidea	Psocoptera Mallophaga Anoplura Thysanoptera Homoptera Heteroptera
	Endopterygota	Coleoptera Strepsiptera Hymenoptera Raphidioida Neuroptera Megaloptera Mecoptera Diptera Siphonaptera Trichoptera Lepidoptera

8.5.3B.1 The Apterygote classes (wingless insects)

Five primitively wingless groups of insects are recognized, each of which must have diverged from the others at a very early stage of insect evolution (if indeed they did not derive separately from their myriapod ancestor or ancestors), and hence each can be regarded as constituting a separate class. Only one of these classes shows any clear affinity with the winged insects, which comprise the sixth hexapod class. All the apterygote classes are small, with only a single order each, and they are most conveniently treated together.

The five classes are broadly separable into three groups, of which the first (the diplurans, Class Diplurata; the spring-tails or collembolans, Class Oligoentomata; and the proturans, Class Myrientomata) all possess mouthparts partially enclosed within the head capsule, and share, between them, a number of other features in common with the myriapods rather than with the other insects. Their habits, including reproductive behaviour, and habitats are also very myriapod-like; not surprisingly, therefore, many entomologists would redefine the Myriapoda so as to include these groups within that subphylum. Amongst the presumed ancestral characters which they have retained are: mandibles with a single articulation with the head; a three-lobed hypopharynx (a median tongue-like organ associated with the salivary glands); eversible coxal sacs and abdominal styli; antennal articles with their own musculature; organs of Tömösvary (or structures very similar to them); and abdominal appendages of some form. All are small (generally with lengths of less than 7 mm) and, perhaps for that reason, lack Malpighian tubules,

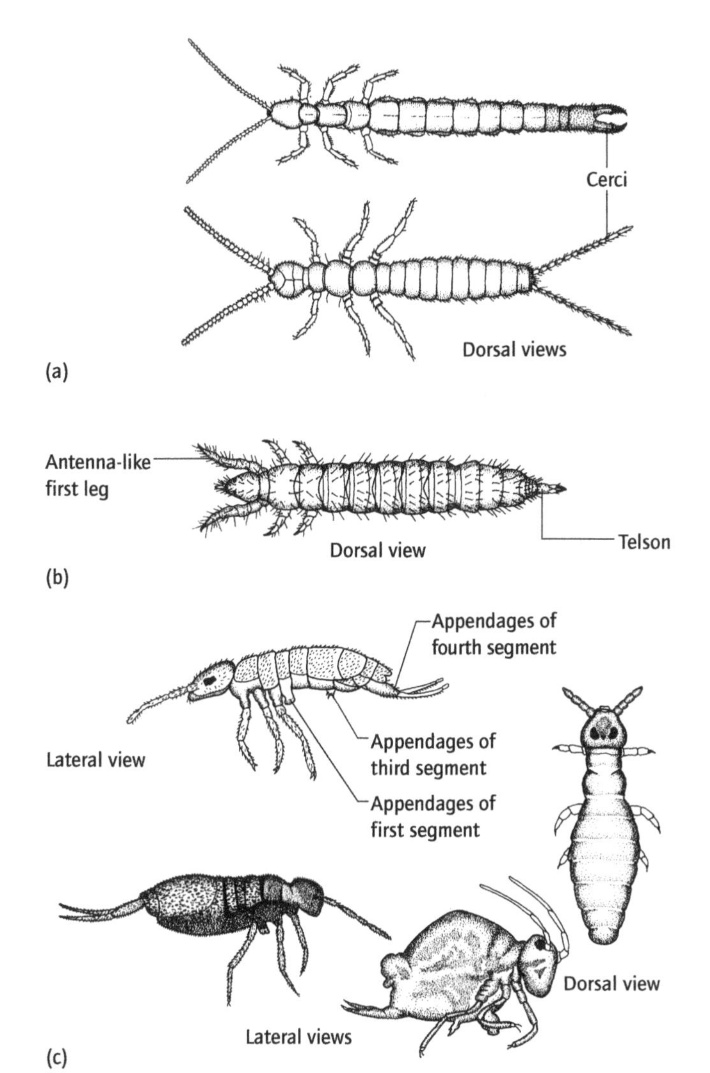

Fig. 8.25 The myriapod-like apterygote classes: (a) two diplurans; (b) a proturan; (c) various spring-tails, one showing the modified abdominal appendages. (After Imms, 1964 and Wallace & Mackerras, 1970.)

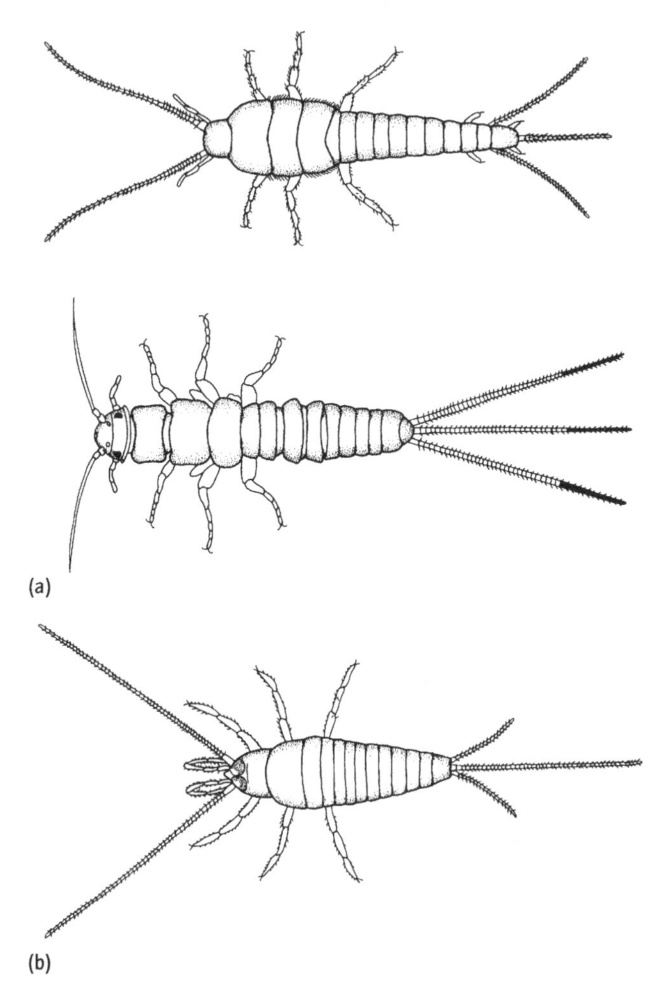

Fig. 8.26 The hexapod-like apterygote classes: (a) two zygoentomatans; (b) an archaeognathan. (After Imms, 1964 and Daly *et al.*, 1978).

nitrogenous waste being eliminated through the mid-gut epithelium; two of the three classes also lack eyes.

The Diplurata (Fig. 8.25a) are without particular specializations such as distinguish the other two classes. The Myrientomata (Fig. 8.25b), however, have vestigial antennae, their function having been taken over by the first pair of thoracic legs, and piercing, stylet-like mandibles. Like most myriapods, the number of their trunk segments increases after hatching until the full adult complement of eleven is attained. The Oligoentomata (Fig. 8.25c), on the other hand, have retained only five abdominal segments (plus the telson) and have adapted two pairs of abdominal appendages to form the organ which gives them their common name of spring-tails. The limbs of the fourth segment form a spring which can be forcibly extended by haemocoelic pressure. When not in use it is held by a catch developed from the appendages of the third segment. Further, the limbs of the first abdominal segment are involved in the formation of a large ventral tube which contains eversible, coxal-sac-like vesicles and is of debated function. It may be used to take up environmental water. In none of these three classes are the gonopores on the eighth (female) and tenth (male) abdominal segments as in all other insects, nor do they display low rates of water loss.

The second grouping, containing the silverfish and firebrats (Class Zygoentomata) (Fig. 8.26a), differs from the pterygote insects to a much smaller extent. Apart from their primitive winglessness (and the thoracic structure associated with this state), they differ essentially only in the retention of three ancestral (and general apterygote) characteristics: the presence of abdominal styli; the absence of copulatory organs, sperm transfer being by externally deposited spermatophores; and in continuing to moult after attaining sexual maturity. For this reason, they are generally regarded as being close to the ancestry of the pterygotes.

The bristle-tails (Class Archaeognathata) (Fig. 8.26b), which form the third general group, straddle the interface between the Zygoentomata and the other apterygotes, possessing some of the advanced features of the silverfish and firebrats but retaining

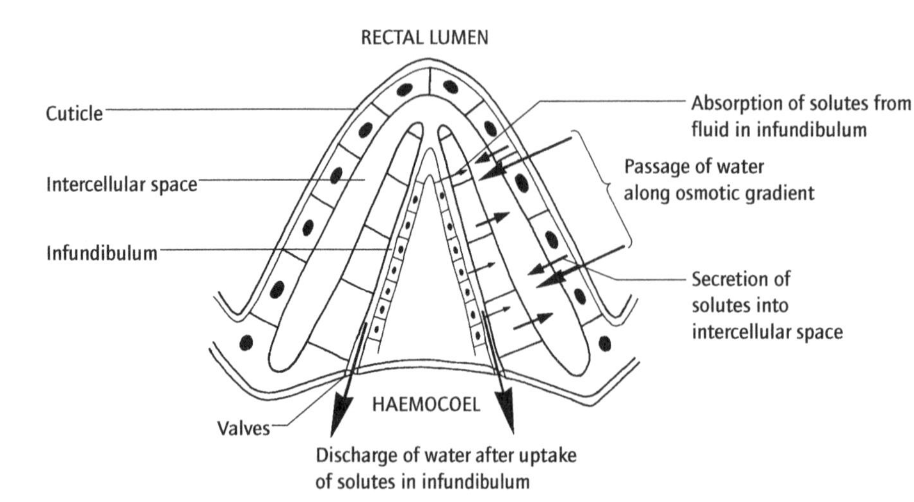

Fig. 8.27 A diagrammatic section through a rectal papilla of a blowfly, showing the system of intercellular spaces into and from which solutes can be moved, and into which water from the rectal lumen is induced to pass, ultimately to be discharged into the haemocoel (greatly simplified after Gupta & Berridge, 1966).

many of the ancestral ones of the myriapod-like groups, often in respect of the same appendage or organ system. Thus the mouthparts are not enclosed within the head, but the mandibular articulation, hypopharynx and maxillae are all otherwise in the ancestral myriapod state. They also display all the primitive characters retained by the zygoentomatans. Just as it is possible that the winged insects were descended from the Zygoentomata, so it is possible that the latter were derived from the Archaeognathata, and so both classes are usually included within the same grouping as the 'true insects' by entomologists. They are both also generally larger (up to 2 cm in length) and less tied to humid litter and soil microhabitats than the other apterygotes.

The five classes together total somewhat over 3100 species, the majority of which are scavengers and feeders on decaying plant material, although some capture other small hexapods.

8.5.3B.2 Class Pterygota (winged insects)

The 29 orders of winged insects, including secondarily wingless forms, comprise the third major radiation of the Uniramia, and it is this class which currently dominates the phylum with some 98% of its living species. Above all else, their success has been made possible by emancipation from the humid types of microhabitats to which all the other uniramian classes are confined. And this, in turn, is largely a result of their ability to restrict the rate at which water leaves their bodies. The evolution of this state probably occurred in a number of stages, beginning with the properties of the cuticle and of the tracheae.

It is possible that an epicuticular wax layer was originally of adaptive value as a water-repelling hydrofuge layer: in environments liable to become waterlogged, small animals can become trapped in water films, and water uptake is a much more serious problem than water loss; the maxillary glands of apterygotes may have to pump out considerable volumes of surplus water. Some spring-tails have developed a series of wax-covered tubercles which serve a hydrofuge function, and the ancestral pterygote can be envisaged as being similarly equipped. It also may well have lacked a tracheal system, as do most of the myriapod-like apterygotes, this perhaps largely being a consequence of the small size of any animal originating by paedomorphosis, especially when its ancestors were already small. Once tracheae developed (were redeveloped), the waxy hydrofuge layer could spread to cover the whole body surface without impeding gaseous exchange.

Terrestrial arthropods may lose more water through their tracheal system, however, than through the cuticle, and so a waterproofed integument would by itself be ineffective.

Characteristically, pterygotes possess muscles which close the spiracles (and may have others to open them) and, for example, insects living in dry habitats may open the spiracles for only a small fraction of the time. Other means by which water loss through the spiracles is reduced may also be present. One such is to store the carbon dioxide generated in a non-gaseous phase for long periods. As oxygen in the tracheal air is consumed and not replaced by an equivalent volume of carbon dioxide, so a partial vacuum develops and serves to draw more air in through the spiracles in a one-way flow, counteracting any tendency of water vapour to diffuse out.

A final refinement of the water-conservation system is located in the hind-gut in the form of a mechanism whereby water can be resorbed from the faeces and excretory discharges as they pass through. Again, this may take several forms, of which one widespread one involves the active secretion of inorganic and/or organic solutes into a system of intercellular spaces within pads or papillae in the rectum wall. Water then diffuses passively from the gut lumen into these spaces along the osmotic gradient, and as the fluid is transported into the haemocoel, the solutes are absorbed and recycled (Fig. 8.27).

The colonization of non-humid habitats made possible by these heightened powers of water economy would have had many repercussions on the biology of the early pterygotes, and these account for many of the characteristic features of the group. The deposition of external sperm droplets or spermatophores, for example, is not a viable strategy in relatively dry conditions, and pterygotes transfer sperm directly, by copulating, although the spermatozoa are still usually contained within

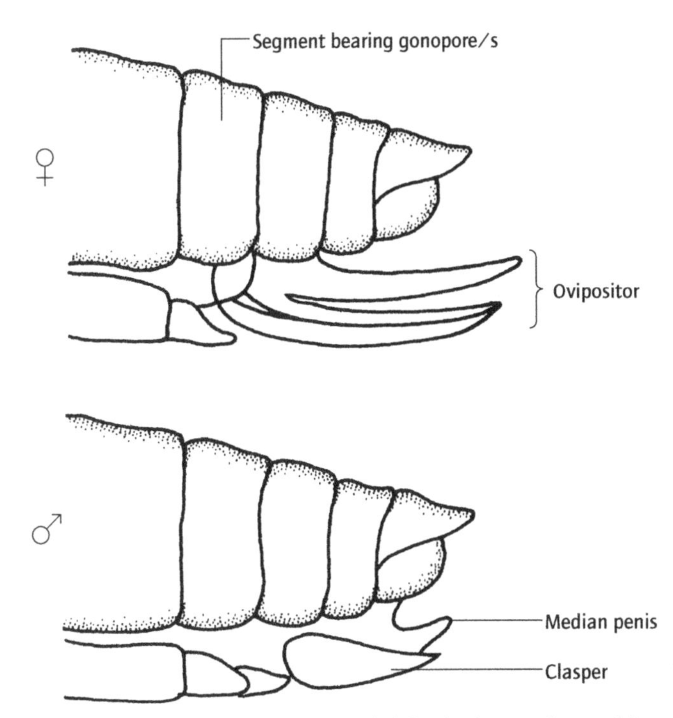

Fig. 8.28 Modification of the segmental abdominal appendages of the ancestral uniramians to serve egglaying and copulatory functions in the pterygotes (lateral view) (after Snodgrass, 1935). The ovipositor of the female represents the basal portions of the appendages of the eighth and ninth abdominal segments; the male gonopore/s belongs to the tenth segment, although in most insects it is displaced anteriorly so as to appear to be on the ninth. Note, however, that many pterygotes have evolved secondary copulatory organs and means of oviposition.

a spermatophore. Primitively, the females retain the ancestral paired gonopores and the male possesses paired copulatory organs, although in most groups only a single, median female pore occurs and correspondingly the two male organs fuse into a single structure. Males also possess a pair of claspers which hold the female whilst the eversible distal section of the male duct or distinct gonopod is inserted through her gonopore to deposit one or more spermatophores or sperm packets into the bursa copulatrix or some other section of the female reproductive tract. On release from the spermatophore, the sperm are usually stored – sometimes for considerable periods of time – in a spermatheca, until eventually they fertilize the eggs as these are laid, entering through minute pores in the shell which characteristically protects pterygote eggs. A number of female insects mate only once, although the sperm received may fertilize several batches of eggs; no species moults once it has achieved sexual maturity. Many possess an ovipositor, permitting the eggs to be laid in specific, usually concealed sites; in others, the terminal portion of the abdomen can be extended in the form of a tube to achieve the same end, whilst in several hymenopterans (bees, wasps, etc.) the ovipositor has been adapted to form a sting and is no longer associated with egg laying. The male claspers and gonopod, and the female ovipositor (Fig. 8.28), are derived from the ancestral abdominal appendages of the genital segments; apart from the widespread occurrence of a pair of terminal cerci, which are the appendages of the eleventh segment, these copulatory and egg-laying organs are the only remnants of the series of these segmental appendages to remain functional in the living adult pterygotes. Abdominal limbs, with styli, are well developed in some Palaeozoic pterygotes, however, and they may even have been functional legs.

More than any other feature, however, the Pterygota are diagnosed by their possession of wings – another development only possible after release from confinement to humid habitat systems. One pair of wings, in addition to the walking legs, occurs on each of the second and third thoracic segments of the adult, although one or both pairs may be reduced or lost. Some of the Palaeozoic insects referred to above had functional, albeit small, wings on the first thoracic segment as well. The origin of these wings, and the original selective advantages of the wing precursors, are unknown, although they may have developed from the trachea-rich, plate-like lateral expansions of the thoracic tergites that occur, for example, in some zygoentomatans and in various fossil hexapods (see Section 10.6.2). Certainly in living pterygotes, they form from evaginations of the body wall, four buds of epidermis growing out dorsolaterally. The upper and lower surfaces of these buds fuse together except along a series of channels in which blood flows and tracheae and sensory nerves are located. Eventually, the epidermal cells secrete a thin encasing cuticle, and when the moult into the adult occurs, the small, somewhat fleshy wing buds are inflated by haemocoelic pressure, the lining of the channels is reinforced with cuticle (to become the final 'wing veins'), the epidermis degenerates, and the wing comprises a thin double layer of cuticle. Of necessity, therefore, the wings are operated by extrinsic muscles, some, and only some, of which are attached to the wing base. The wings must also be hinged to permit them to describe the elliptical or figure-of-eight path required for flight. In small insects, the wings may beat at rates of up to one thousand times per second and, unusually amongst animals, there are several contractions of the flight muscles consequent on each received nerve impulse (see Section 16.10.5). The wings of primitive insects are usually relatively large and contain numerous wing veins, although the size of wing and number of veins have been greatly reduced in many evolutionary lines (Fig. 8.29); this correlates with a progression from long-bodied animals with a stable but unmanoeuvrable flight to short-bodied forms with unstable but highly manoeuvrable flight characteristics. These same lines have also evolved the ability completely to fold up the wings when they are not in use, a development which reaches a climax when all or part of the fore pair are modified to serve as protective coverings for the folded hind wings. Insect flight is considered in Section 10.6.2.

The final factor in the evolutionary success of the winged insects is probably the diversity of the food sources which they can exploit. This is reflected in an almost bewildering variety of mouthpart structure, different orders modifying different appendages, although the pterygotes are united, and contrast

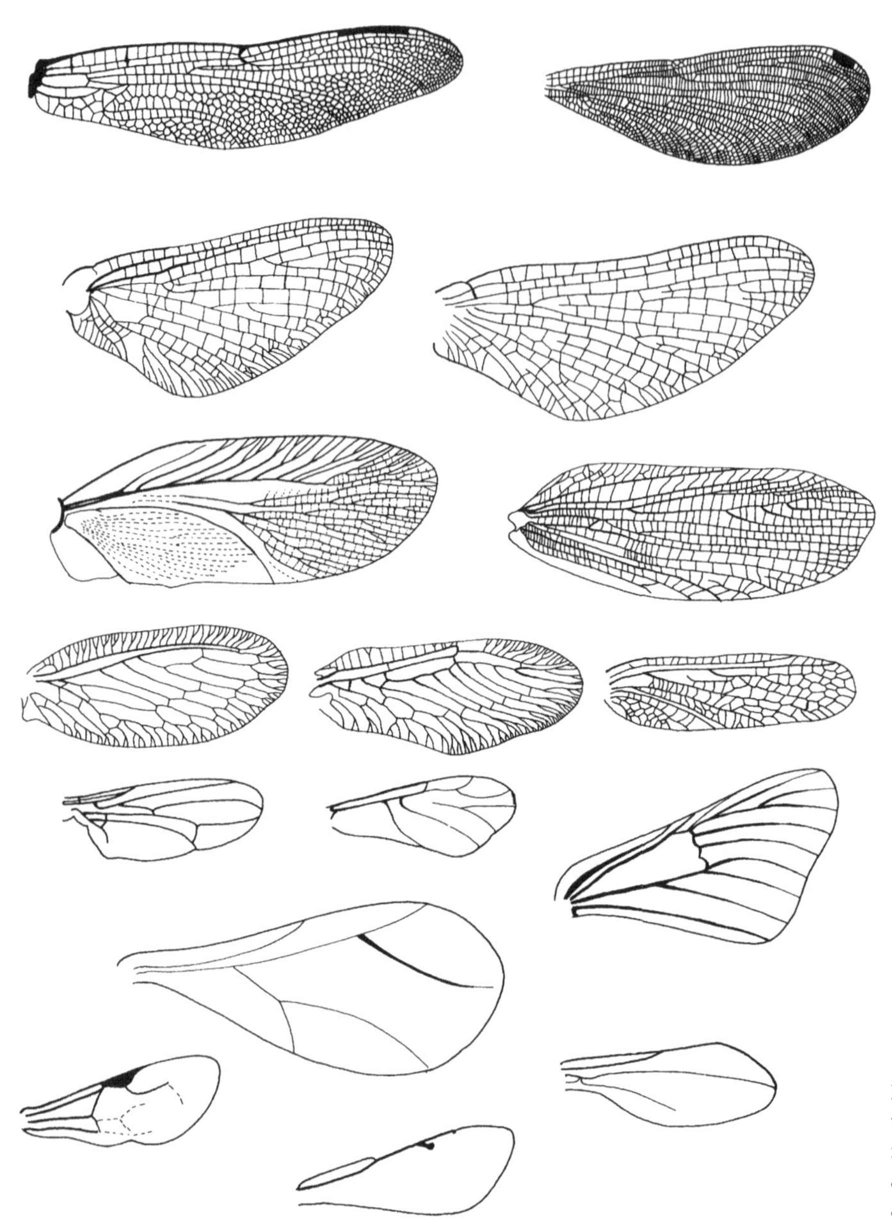

Fig. 8.29 Variation in the venation of pterygote wings (after several authors). Primitive wings with numerous veins are shown at the top of the figure, and down the figure is arranged a series displaying a progressive reduction in the number of veins.

with most apterygotes (except the Zygoentomata), in having mouthparts which are not enclosed within the head capsule, a mandible with two points of articulation, and a single-lobed hypopharynx. The ancestral form of these appendages was almost certainly the standard myriapod system adapted for biting and chewing, and several modern species have retained this configuration (Fig. 8.30a). From this, however, have evolved many lines of piercing and sucking mouthparts, of which only a few examples can be given here.

The bugs, for example, can, dependent on group, feed on plant or animal fluids, by means of a beak formed from the labium (the second maxillae) within which are located two pairs of stylets derived from parts of the mandibles and maxillae (Fig. 8.30b). The mandibular stylets pierce the prey's tissues and allow the maxillary stylets to enter the wound. The structure of the maxillae is such as to enclose a pair of canals, down one of which passes saliva, and up the other the tapped fluids. The mosquitoes (Fig. 8.30c), on the other hand, have two additional stylets formed by the labrum (upper lip), which encloses the food canal, and the hypopharynx, which contains the salivary duct. In both cases, however, the labium serves solely as a guard and it telescopes or folds back as the other mouthparts enter the tissues.

Other insects suck more readily available liquids, such as nectar or fluids oozing from wounds. Butterflies and moths, for example, imbibe nectar through an elongate proboscis which can be extended by haemocoelic pressure (reduction in this pressure leads to re-coiling under the proboscis's own elasticity). This is formed from part of the maxillae; all the other mouthparts are reduced or missing (Fig. 8.30d). In some moths, the tip

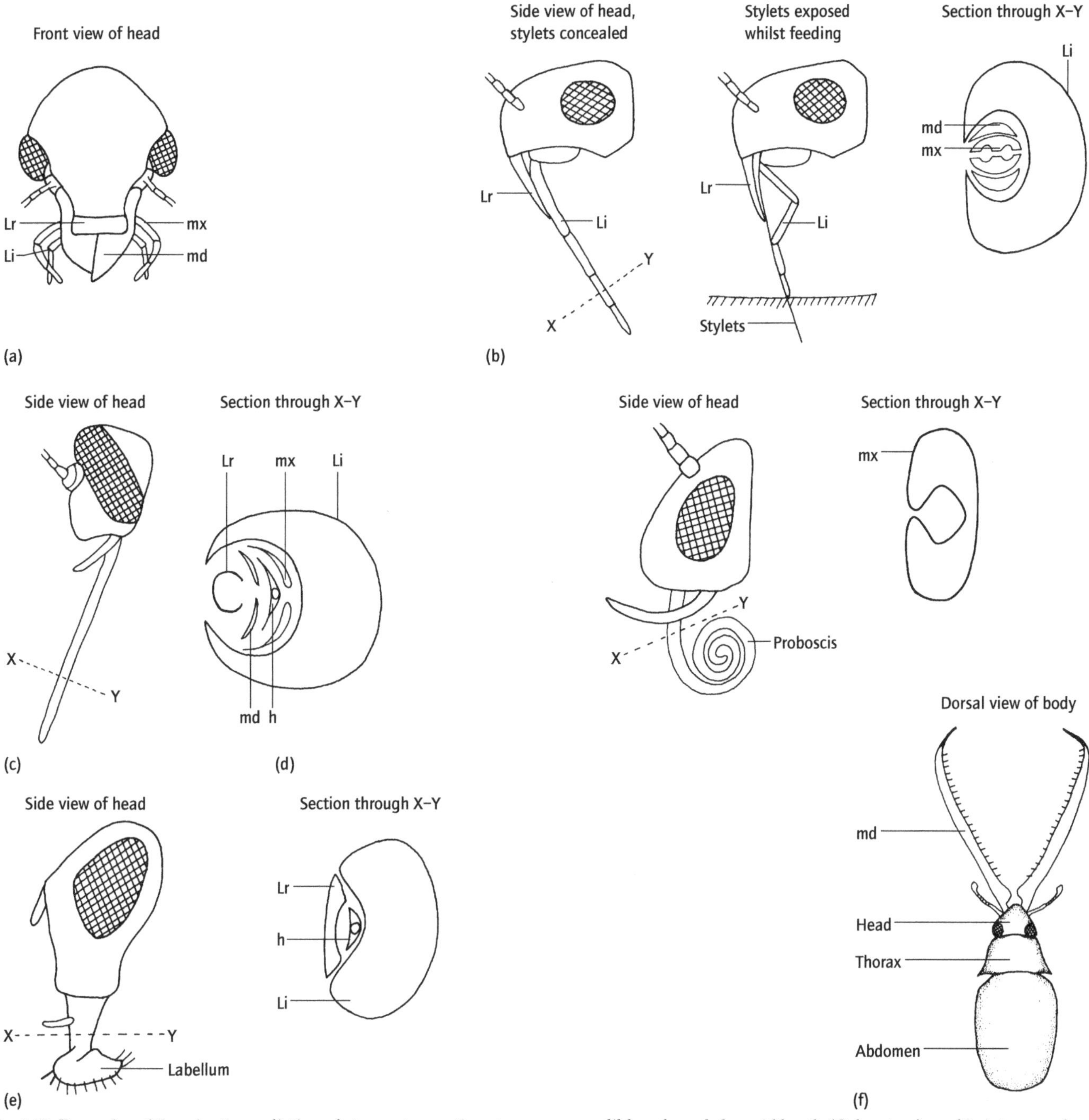

Fig. 8.30 Examples of the adaptive radiation of pterygote mouthparts (see text): (a) biting mandibles of a beetle (Coleoptera); (b) the piercing/sucking beak of a bug (Heteroptera); (c) the mouthparts of a mosquito (Diptera); (d) proboscis of a nectar-sucking butterfly (Lepidoptera); (e) sponging mouthparts of a housefly (Diptera); (f) the greatly enlarged mandibles of a male lucanid beetle (Coleoptera) used in intrasexual combat and not for feeding. (After several authors, principally Borror *et al.*, 1976.) Key to labels: Lr = labrum; md = mandible; mx = maxilla; Li = labium; h = hypopharynx.

of the proboscis is sharp and barbed, and can be used as a piercing organ. In the houseflies, it is the labium which functions as the organ of uptake. Again, the mandibles are absent, but in this group the labrum and hypopharynx are enclosed within a groove in the large labium, which bears terminally a large, soft, bi-lobed 'labellum' which acts like a sponge (Fig. 8.30e).

In yet further species, the mandibles although huge are not used in feeding at all, but rather in defence or in fighting for mates (Fig. 8.30f); and some members of at least eight groups do not feed when adult and have vestigial mouthparts. In a number of pterygotes, especially those with a distinct larval stage or with aquatic juveniles, the main part of the life cycle and the one

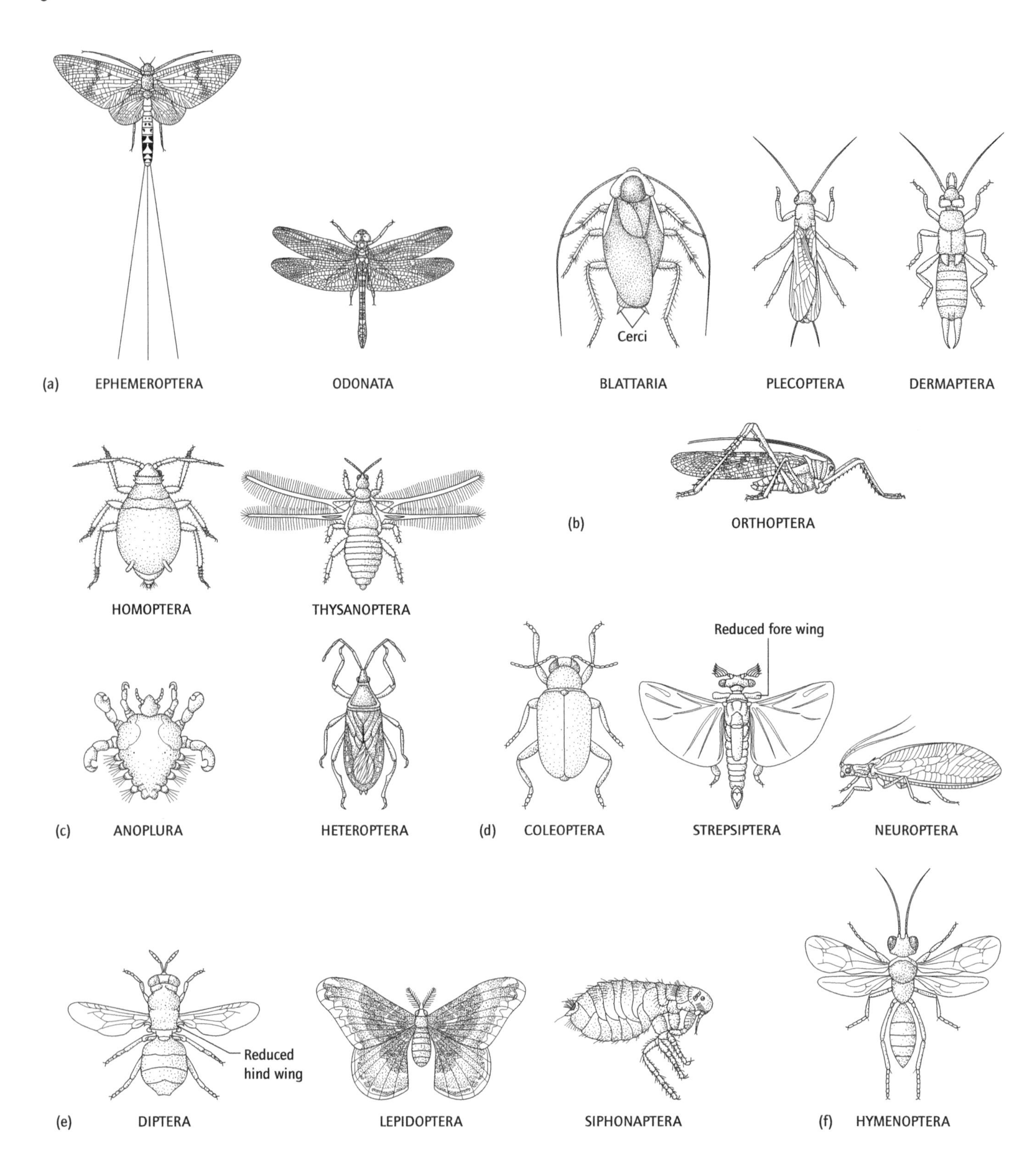

Fig. 8.31 Diversity of pterygote adult body form: (a) representative palaeopterans; (b) orthopteroids; (c) hemipteroids; (d) neuropteroid endopterygotes; (e) panorpoid endopterygotes; (f) hymenopteroid endopterygotes. (After various authors.)

concerned with feeding occurs before sexual maturity; the adult is a short-lived dispersal and reproductive 'machine'.

The one million species of winged insects are broadly divisible into four clusters of orders (Fig. 8.31). The palaeopterans (two orders) are chiefly distinguished by a (primitive) inability to flex their wings, which are large and have numerous longitudinal and cross veins, back over their abdomens. Their juvenile stages (naiads), which are long-lived and pass through many moults, are always aquatic and hence differ in their adaptations from the adults. Many, for example, have external pairs of tracheal gills: leaf-like abdominal plates which resemble true gills except that instead of containing a blood supply (the blood system of uniramians not being concerned with the distribution of respiratory gases), they possess closed tracheal tubes. These paired structures may be derivatives of the ancestral abdominal appendages. All other pterygotes are 'neopteran', i.e. can lay the wings flat over the abdomen when at rest.

The second cluster is the orthopteroids (10 orders). This basal group of the Neoptera, typically, have biting mouthparts, and an unconcentrated nervous system, simple elongate multiarticulate antennae, a pair of terminal cerci, numerous Malpighian tubules, eversible copulatory organs, and a series of several nymphal instars which change slowly and progressively towards the adult bodily form. All the adult organ systems, including wings, compound eyes, etc., gradually develop through the various juvenile stages. To a considerable extent, therefore, the ecology of the young is the same as that of the adult.

The six hemipteroid orders display a similar progressive change of the nymphs towards the adult form, although in most the changes are concentrated in the final instar or instars. The majority are fluid feeders and have parts of the maxillae adapted to form sclerotized stylets. Characteristically they possess few (four or less) Malpighian tubules, lack cerci, have a much-reduced wing venation, bear short antennae with few articles, and display a distinct gonopod. None of their nymphs possess ocelli.

In several respects, the remaining cluster, the endopterygotes (11 orders), resemble the hemipteroids, although they do have short cerci and may bear many Malpighian tubules. But whereas all the other neopterans display the gradual transition from the first nymphal stage to the adult referred to above, during which the wing buds develop externally to the body, in this group larval stages occur which differ profoundly from the adult form in their anatomy, diet and, often, habitat – a characteristic unique amongst the terrestrial arthropods. Many flies and hymenopterans, for example, have limbless worm-like larvae, whilst, conversely, lepidopterans and some other hymenopterans have larval stages with additional, secondary abdominal walking legs ('prolegs') (Fig. 8.32). In all these larvae, the wing buds develop beneath the integument and hence are not visible externally. Since the larvae differ so markedly from the adult, there is a complete metamorphosis separating the two (see Section 15.5.1). This occurs during a specialized inactive phase, the 'pupa' (Fig. 8.33); in it, all or many of the larval tissues are broken down by histolysis, the adult form is assembled *de novo*, and the developing wings appear outside the integument for the first time. Some pupae retain the moulted integument of the last larval instar as a protective casing around them (Fig. 8.33b). This strategy of separate larval and adult ecologies appears, evolutionarily, to have been highly successful in that 85% of all species of winged insects are endopterygote; some larval stages can even occur in what, to a basically terrestrial animal, are most inhospitable habitats, such as intertidal marine sediments and temporary pools of water – the larva of one African midge has powers of cryptobiosis, after dehydration, rivalling those of the tardigrades (see p. 170).

The pterygotes and other hexapods are all relatively small, and several attempts have been made to show that the largest living species (at a live weight of 70 g) are at the theoretical maximum limit of insect size for physical reasons (related to their possession of tracheal systems that rely on diffusion, to the gravitational problems associated with moulting exoskeletons on land, and so on). Some fossil species, however, were the size of crows and hawks – well above the alleged maximum size permitted by physical constraints. These pre-date the successful vertebrate invasion of the land, and it is more likely that the observed small degree of overlap in size between surviving insects and the terrestrial vertebrates indicates a degree of subdivision of the size spectrum between the two groups. In essence, it is competition and/or predation from vertebrates that keep the modern insects small.

8.6 **Phylum CRUSTACEA**

8.6.1 Etymology

Latin: *crusta*, a rind or crust.

8.6.2 Diagnostic and special features (Fig. 8.34)

1 Bilaterally symmetrical, <0.1 mm–60 cm long arthropods (some achieving a leg span of up to 3.5 m, and others a weight of >20 kg) varying in body shape from elongate to spherical.
2 Body more than two cell layers thick, with tissues and organs.
3 With a through, straight gut, from the small mid-gut region of which issue two digestive diverticula in which digestion and absorption of food take place; those species not already feeding on very small particles or fluids possess a grinding mechanism in the fore-gut; the digestive diverticula arise as outgrowths from the embryonic gut.
4 Different groups of crustaceans vary markedly in the manner in which the body is subdivided and in the number of segments comprising each division; nevertheless, basically a head, formed by the acron and five appendage-bearing segments, can be distinguished, as can a trunk of from two to >65 segments and a terminal telson which often bears a pair of processes, the 'furca'; the first and in various groups up to seven other trunk segments are often fused with the head to form a 'cephalothorax'; and the trunk is usually subdivided into a thorax and abdomen on the basis of its appendages, the thoracic segments not incorporated into the cephalothorax comprising the 'pereon' and the abdominal ones, the 'pleon'; the cephalothorax and, in some groups, most or the whole of the body is enclosed within an outgrowth

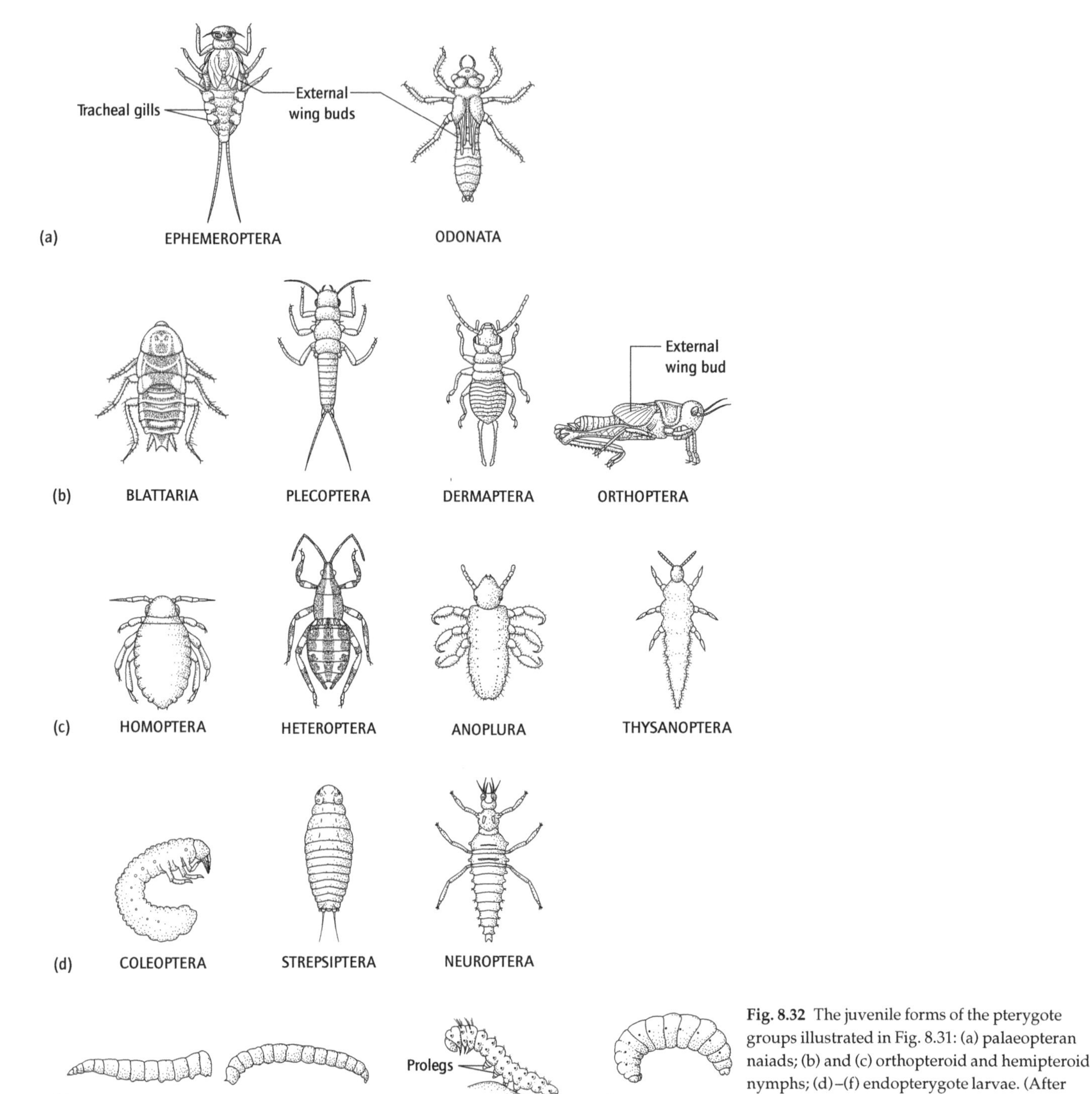

Fig. 8.32 The juvenile forms of the pterygote groups illustrated in Fig. 8.31: (a) palaeopteran naiads; (b) and (c) orthopteroid and hemipteroid nymphs; (d)–(f) endopterygote larvae. (After various authors.)

of the head, the 'carapace', which extends laterally to overhang the sides of the body.

5 The cylindrical or leaf-shaped appendages are all basically biramous, the two branches normally being of different size and shape, and often bearing further secondary branches; the appendages of the head comprise two pairs of 'antennae' (the first pair being the 'antennules'), one pair of 'mandibles', and two pairs of 'maxillae'; those of the trunk vary greatly in their number, shape and regional differentiation; primitively each segment bears a pair of limbs, although those of the abdomen are often absent; without chelicerae but some limbs may be chelate.

6 With three pairs of primary mouthparts (mandibles and the two maxillae) and with, in many groups, one to three pairs of accessory mouthparts, 'maxillipeds', arising from those thoracic segments incorporated into the cephalothorax; members of each pair of mouthparts work with, or against, each other.

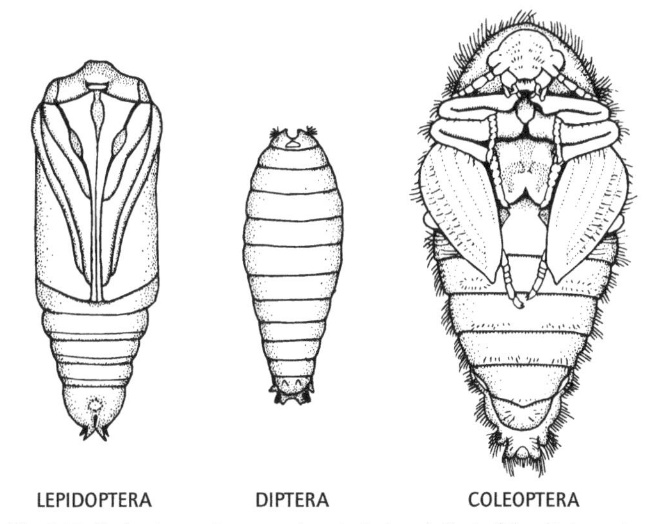

Fig. 8.33 Endopterygote pupae (ventral views), that of the dipteran is enclosed within a 'puparium', the cast integument of the last larval instar (after Jeannel, 1960 and Wallace & Mackerras, 1970).

7 Head with median ocelli or lateral compound eyes, the latter sometimes located on movable stalks.
8 Trunk, but not head, normally with evident external segmentation, often concealed beneath the carapace, and sometimes lost.
9 Excretory system of blind-ending antennal and/or maxillary glands.
10 Exoskeleton often calcareous.
11 Gaseous exchange effected across the inner wall of the carapace, across the general body surface, or by means of gills developed from parts of the thoracic or abdominal limbs.
12 Blood system with haemocyanin and, rarely, other pigments.
13 Nervous system with paired ganglia in each segment; primitively with separate ventral cords and ganglia, more often fused together; all thoracic ganglia sometimes fused into a single mass.
14 Gonochoristic or, rarely, hermaphrodite, with internal fertilization via copulation by means of gonopods or penes; gonopore location variable, often thoracic.
15 Eggs usually carried by the female or brooded within specialized pouches; some hatch with the full adult complement of segments, most do so as a 'nauplius' larva with only three segments (Fig. 8.35) (some highly specialized parasitic forms are recognizable as crustaceans only by virtue of their nauplius larvae).
16 Essentially marine, although several (13%) are freshwater and a few (3%) are terrestrial.

The crustaceans are *the* marine arthropods, with all but 3% of the arthropod species known from the sea. They dominate the plankton (as they also do in fresh waters) and they are one of the three or four most important members of the benthos, in respect both of interstitial and of macroscopic species. In addition, several are parasitic.

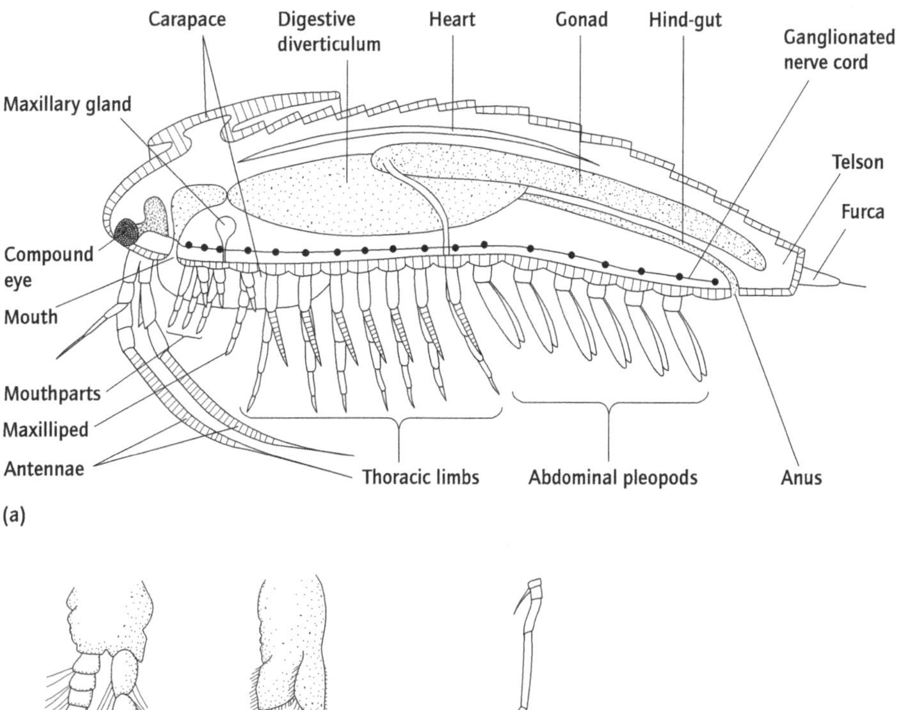

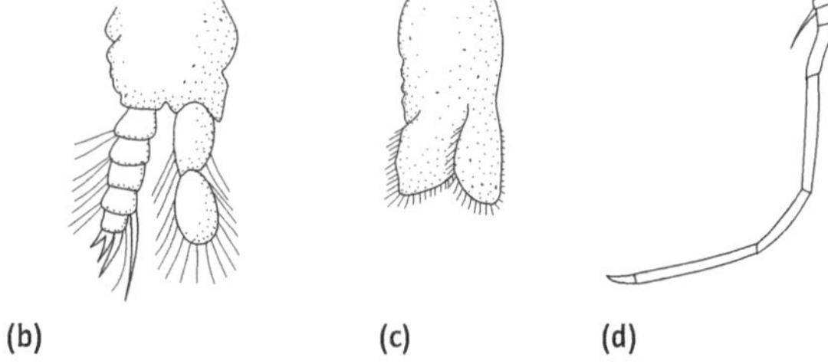

Fig. 8.34 Crustacean morphology. (a) A diagrammatic longitudinal section through the body of a generalized crustacean. (b)–(d) Characteristic biramous limbs (greatly simplified) (partly after McLaughlin, 1980): (b) with one branch tubular, the other leaf-like; (c) with both branches leaf-like (a typical swimming limb); (d) with both branches tubular, one much reduced (a typical walking leg).

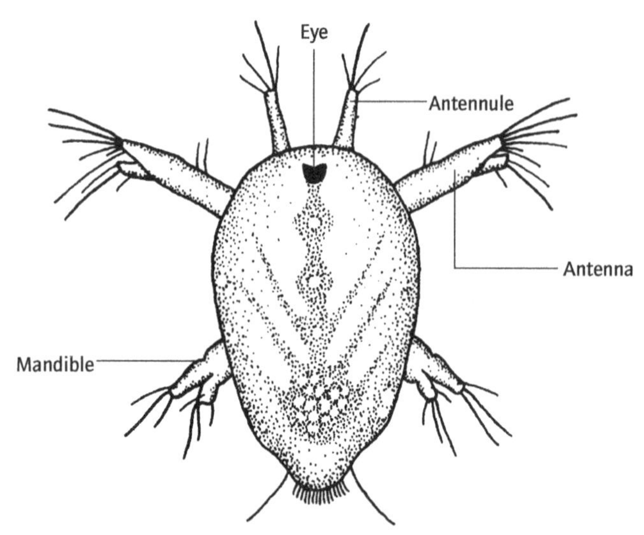

Fig. 8.35 A nauplius larva (after Green, 1961).

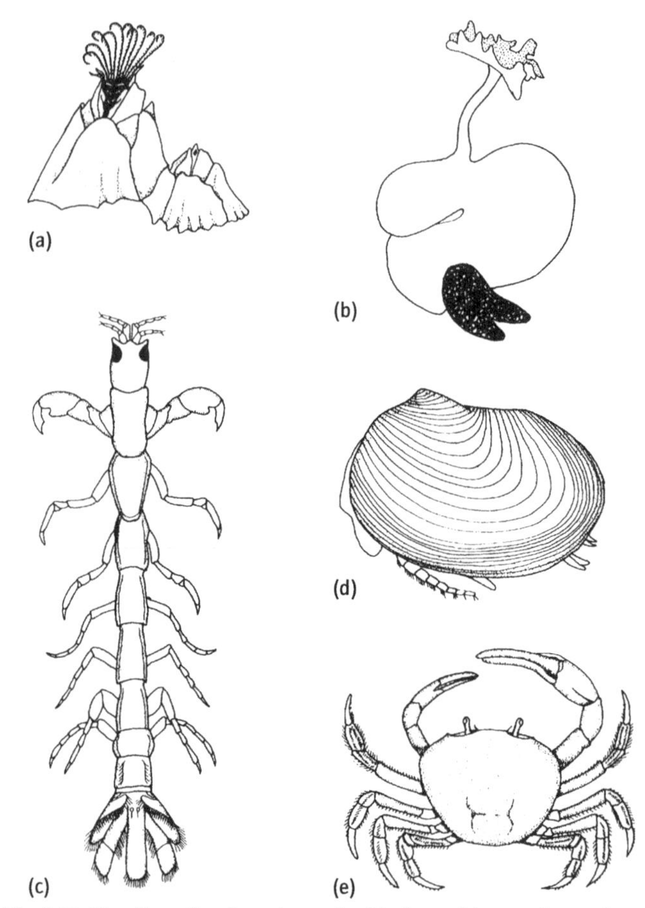

Fig. 8.36 The diversity of crustacean bodily form: (a) acorn barnacles (Cirripedia); (b) a parasitic copepod (Copepoda); (c) an isopod (Malacostraca); (d) a conchostracan (Branchiopoda); (e) a crab (Malacostraca). (After various authors.)

To a considerable extent, their success must be attributable to the general arthropod characteristic of jointed limbs. Legs and/or swimming paddles permit rapid locomotion and provide effective means of moving from one habitat patch to another. Doubtless, the other arthropod feature of an exoskeleton also contributes; certainly the larger, relatively heavily armoured forms possess a significant measure of immunity from casual predation. But perhaps the most important reason for the virtual ubiquity of the aquatic Crustacea is the variety of their body forms, even though their internal anatomy is relatively uniform.

Whereas the chelicerate (Section 8.4) and hexapod (Section 8.5.3B) bodies conform to a standard pattern, and the myriapods (Section 8.5.3A), although displaying much variation in the number of segments, nevertheless possess a very conservative body form, such is far from the position in the crustaceans; there is no such thing as the typical crustacean body plan. Some have a head and trunk, others head, thorax and abdomen, several possess a cephalothorax, pereon and abdomen, and one major group only a cephalothorax and abdomen; in a few, the abdomen is missing, and, in some, a differentiated thorax or head is effectively absent too – further, the number of segments comprising these different bodily blocks also varies from group to group, and even within a single class. Similarly, the form of the limbs can range from walking legs, equivalent to those of the chelicerates and uniramians, to foliose paddles; the antennae, for example, may be sensory, the organs of propulsion or of food collection, the means of attachment to a host, claspers used during copulation, and so on. An acorn barnacle, parasitic copepod, conchostracan, crab and anthurid isopod (Fig. 8.36) display few, if any, obvious signs that they belong to the same group of animals.

This structural plasticity has permitted crustaceans to swim, burrow, crawl, bore into wood, live cemented to rock, hunt, browse, suspension and deposit feed (see Fig. 9.3), parasitize most phyla of animals including their own, and it is probably no exaggeration to conclude, occupy every possible type of marine niche. They are, however, principally aquatic animals. Members of several groups are terrestrial, but usually only marginally so. In particular, their gaseous-exchange systems remain those of their aquatic relatives, restricting them to humid habitats. Moreover, some, e.g. the land crabs, must return to water to breed since their pattern of reproduction and development still includes an aquatic larval phase. Only the sowbugs or woodlice include species with specific terrestrial adaptations, and they are the most widely distributed of the land Crustacea. In addition to the pleopodal gills typifying their order (Isopoda), several woodlice have developed trachea-like invaginations of the integument of the pleopods (the abdominal appendages) which extend into the haemocoel within these limbs. Other woodlice, however, have only evolved mechanisms for keeping the surfaces of their gills moist, for example by channelling water droplets from the body surface on to the pleopods.

Class	Superorder	Order
Cephalocarida		Brachypoda
Branchiopoda		Anostraca Notostraca Conchostraca Cladocera
Remipedia		Nectiopoda
Mystacocarida		Derocheilocarida
Branchiura		Arguloida
Copepoda		Platycopioida Calanoida Misophrioida Cyclopoida Gelyelloida Mormonilloida Harpacticoida Monstrilloida Siphonostomatoida Poecilostomatoida
Tantulocarida		Tantulocaridida
Cirripedia		Rhizocephala Ascothoracica Thoracica Acrothoracica Facetotecta
Ostracoda		Myodocopida Cladocopida Podocopida Platycopida Palaeocopida
Malacostraca	Phyllocarida	Leptostraca
	Hoplocarida	Stomatopoda
	Syncarida	Anaspidacea Stygocaridacea Bathynellacea
	Pancarida	Thermosbaenacea
	Peracarida	Mysidacea Cumacea Spelaeogriphacea Tanaidacea Mictacea Isopoda Amphipoda
	Eucarida	Euphausiacea Amphionidacea Decapoda

8.6.3 Classification

The almost 40 000 species of this phylum are divided between ten classes. It is becoming increasingly customary to apportion these ten to four larger groupings (e.g. subphyla), of which the relatively unspecialized classes Remipedia and Malacostraca are placed in separate monotypic subphyla, and the two remaining subphyla, both of which probably originated by paedomorphosis, contain the Cephalocarida and Branchiopoda (subphylum Phyllopoda) and Mystacocarida, Branchiura, Copepoda, Tantulocarida, Cirripedia and Ostracoda (subphylum Maxillopoda). It is possible, however, that both these subphyla represent grades of organization rather than natural phylogenetic clades and hence the subphylum classification is not adopted here. Although the Malacostraca are often termed the 'higher Crustacea', they are a basal, rather primitive group; it is the Phyllopoda and Maxillopoda which show the most advanced features.

8.6.3.1 Class Cephalocarida

The small, blind cephalocarids (<4 mm in length) are detritus-feeding, bottom-dwelling marine animals which were only discovered in 1955 although, locally, they can be abundant. They display some features which are usually regarded as being primitive. The body (Fig. 8.37) is divided into a head, thorax

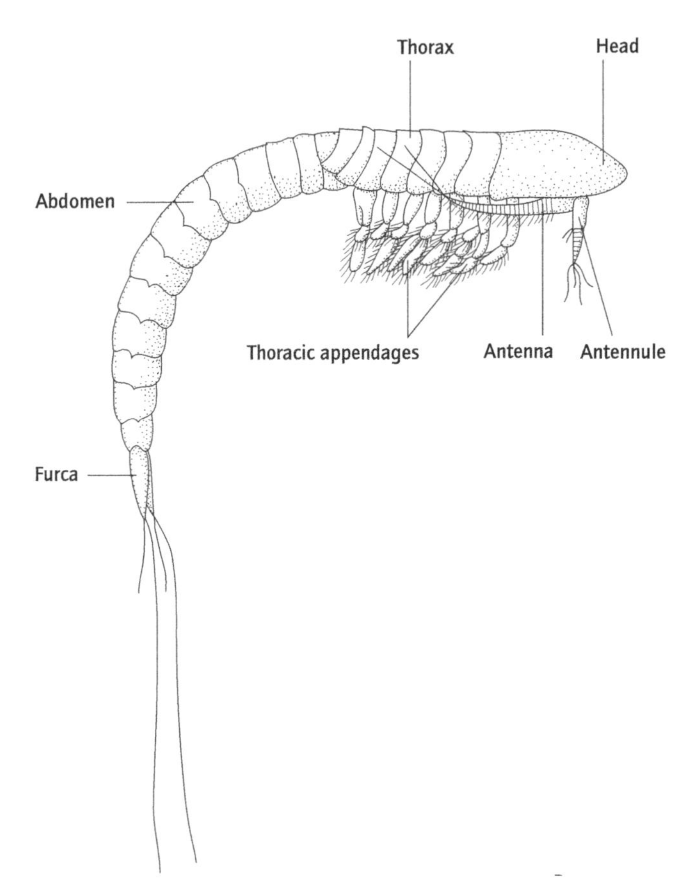

Fig. 8.37 A cephalocarid (lateral view) (after Sanders, 1957).

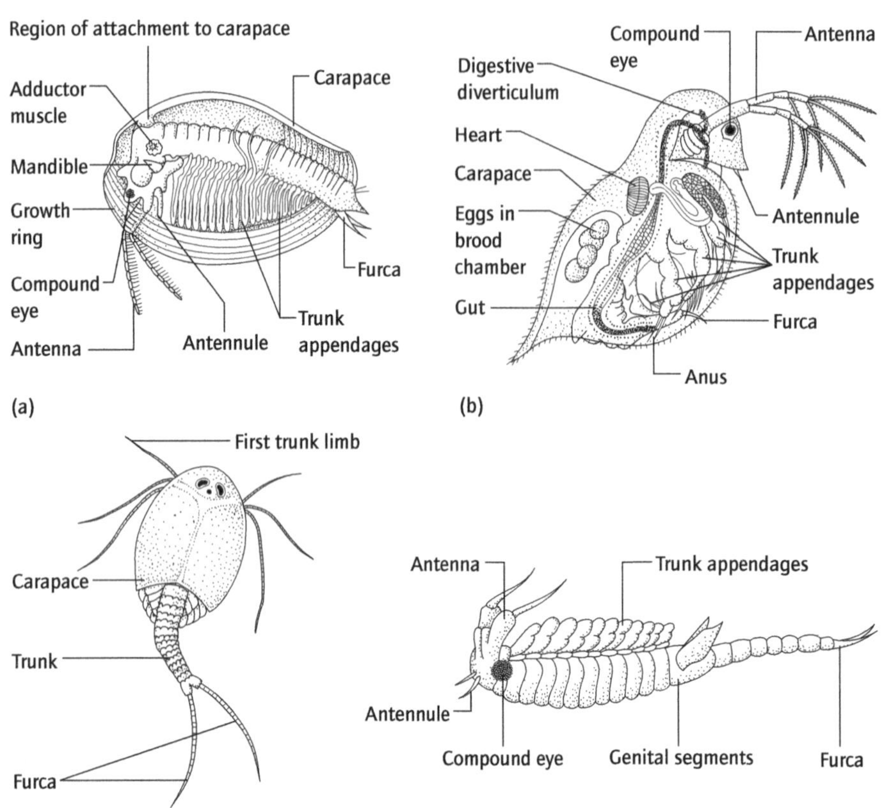

Fig. 8.38 Branchiopod morphology: (a) Conchostraca, lateral view with the left-hand valve of the carapace removed (see also Fig. 8.36d) (after Kaestner, 1970); (b) Cladocera, lateral view as seen through the transparent carapace (after Belk, 1982); (c) Notostraca, external appearance in dorsal view (after Kaestner, 1970); (d) Anostraca, lateral view of its upside-down swimming position (after McLaughlin, 1980).

and abdomen, without any development of a cephalothorax or carapace, and all eight pairs of thoracic limbs are similar and effectively identical to the second maxillae, each one having both leaf-like and tubular elements. The eleven-segmented abdomen lacks appendages, however, except for the first segment which retains reduced limbs to which the egg sacs are attached. Unusually, cephalocarids are hermaphrodite with the paired ovaries and testes sharing a common duct, and development is a rather gradual process through the many larval stages.

The ten known species are all elongate and cylindrical, their bodies terminating in a telson with a long furca of which each ramus bears a bristle. Only one order is included.

8.6.3.2 Class Branchiopoda

The branchiopods are a diverse group (Fig. 8.38) of mainly freshwater crustaceans characterized by small to vestigial head appendages (except, usually, the antennae), by not having any trunk segments fused to the head, and by the trunk bearing a series of similar limbs which usually decrease in size posteriorly, the last few segments lacking limbs altogether. These limbs are typically leaf-shaped swimming and/or filter-feeding organs supported more by haemocoelic pressure than by the cuticle; they also bear gills. Many species reproduce parthenogenetically and brood their eggs; several produce resting stages.

The four orders differ considerably in their body forms. The conchostracans (clam shrimps) and cladocerans (water fleas) both have short, sometimes almost circular, bodies, locomotory antennae, a claw-like furca, and a dorsal brood chamber within the laterally compressed carapace. But whereas the clam shrimps possess a series of up to 30 or more trunk segments and their carapace encloses the whole body including the head, the carapace of the water fleas, although often large, never encloses the head and is, in some, reduced to a small dorsal brood chamber, whilst there are never more than six pairs of trunk limbs (Fig. 8.38a and b). The conchostracan carapace is not moulted but grows by the addition of concentric rings, like those of the bivalve molluscs, and like them it is held shut by an adductor muscle which acts against an elastic hinge ligament.

In the Notostraca (tadpole shrimps), on the other hand, the carapace is a wide, dorsoventrally flattened, somewhat horseshoe-shaped shield from out of which projects the narrow cylindrical posterior end of the body with its two long, annulate furcal rami (Fig. 8.38c). Notably, the posterior trunk segments are only partially differentiated, so that one apparent segment may bear up to six pairs of limbs. Up to 70 pairs of trunk limbs can occur, of which the eleventh carry the brood chambers and the first are larger than the rest. The fourth group, the Anostraca (brine or fairy shrimps), lack any carapace (Fig. 8.38d) and form their brood chamber within the body, from the dilated vagina,

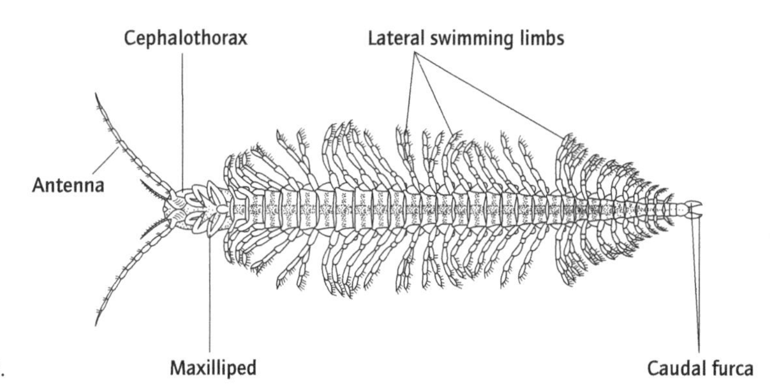

Fig. 8.39 A remipedian (ventral view) (after Yager & Schram, 1986).

although when full of eggs the genital segments form a large projecting bulge, the ovisac. Both the notostracans and anostracans characteristically occur in harsh environments, e.g. temporary pools or saline lakes, and have developed resting stages in extreme form. Their eggs are tolerant of temperature extremes and desiccation; some can remain dormant for up to 10 years.

None of the 850 living species exceeds 10 cm in length, most are less than 3 cm.

8.6.3.3 Class Remipedia

This class is represented by nine species known from marine caves in the North Atlantic and Caribbean (Fig. 8.39), and as yet very little is known of its biology. The smallish, elongate, translucent body (<1–4 cm in length) comprises a short, carapaceless cephalothorax of the head and first trunk segment, and a long trunk of over 30 similar segments each with a pair of leaf-like, lateral limbs used in upside-down swimming. The head appendages include a pair of peculiar rod-like processes in front of the antennules and prehensile mouthparts, including the maxilliped. No species possesses eyes; they probably locate their animal (?) food by chemosensory means.

8.6.3.4 Class Mystacocarida

Mystacocarids (Fig. 8.40) are minute (<1 mm in length), elongate, pigmentless, interstitial marine crustaceans distinguished chiefly by a head which is divided into a small anterior and a large posterior portion, and a trunk of ten segments of which the first bears a maxilliped even though it is not fused to the head. Although the head appendages are large (and are used in locomotion), those of the trunk are either reduced to small, single-articled structures, as on segments 2–5, or are missing; the telson, however, bears a large, pincer-like furca. A primitive feature of this group is that the members of each opposing pair of trunk ganglia touch but do not fuse; neither are compound eyes or digestive diverticula present, possibly as a consequence of their small size. One peculiar feature, otherwise unknown in the Crustacea, is that the posterior portion of the head and each trunk segment bears a pair of lateral, toothed grooves of which the function remains to be discovered.

The twelve species are all placed in a single order.

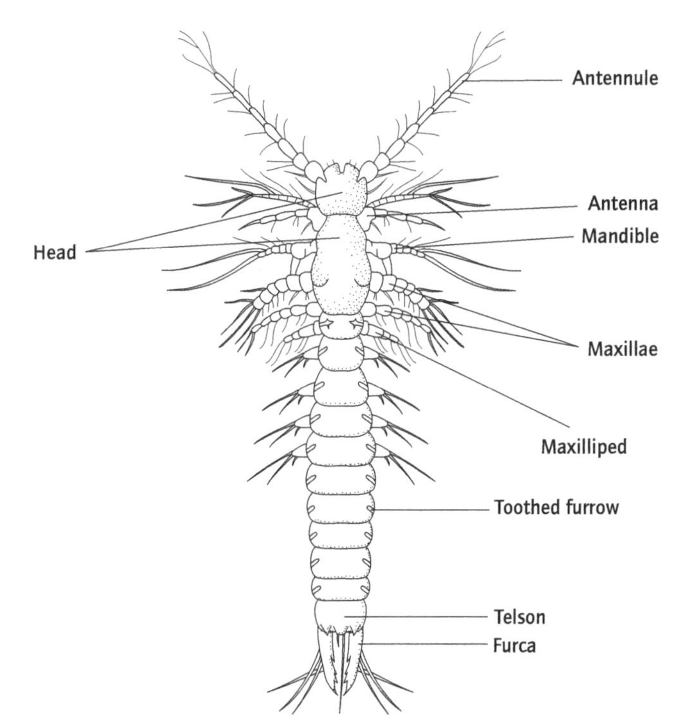

Fig. 8.40 A mystacocarid (dorsal view) (after Kaestner, 1970).

8.6.3.5 Class Branchiura (fish lice)

The branchiurans are smallish (<3 cm in length) periodic ectoparasites of marine and freshwater fish, piercing their much larger prey with the small mandibles and ingesting a fluid meal, usually blood, like marine fleas or mosquitoes. Their markedly dorsoventrally flattened body comprises a cephalothorax of head and first thoracic segment, a pereon of three segments, and a bi-lobed unsegmented abdomen; the cephalothorax and, in some, much of the pereon is covered by a large, flat carapace, circular, bi-lobed or arrowhead in shape, which extends laterally or postero-laterally (Fig. 8.41), and bears a pair of compound eyes.

Their head appendages are either minute or modified into organs of attachment to the fish, ending in hooks or, as in the case of the first maxillae, often in large, stalked suckers. All four

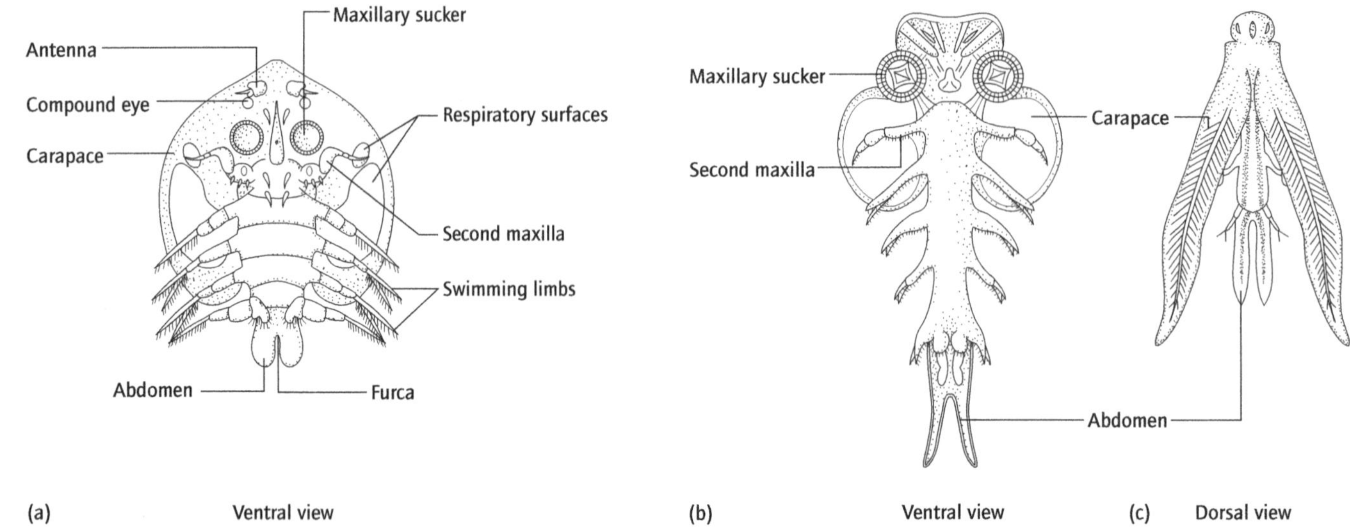

Fig. 8.41 Branchiuran morphology and diversity (after Kaestner, 1970).

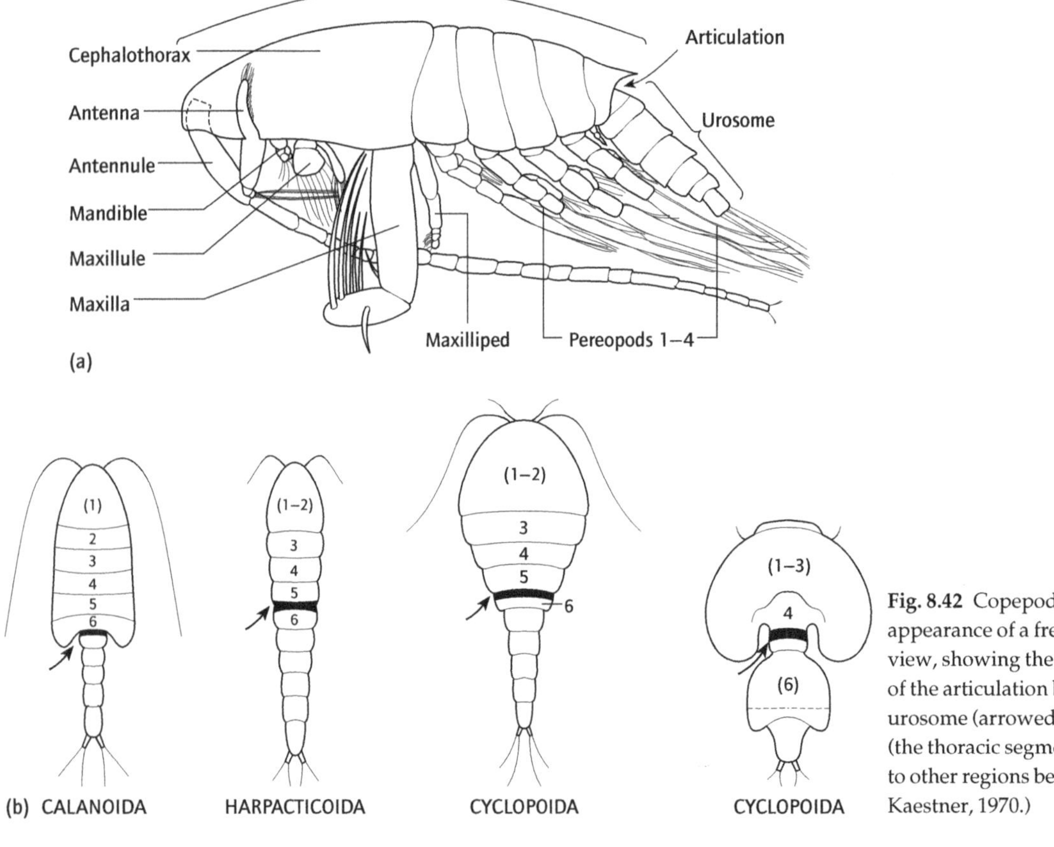

Fig. 8.42 Copepod morphology: (a) external appearance of a free-living copepod, in lateral view, showing the appendages; (b) the location of the articulation between the prosome and the urosome (arrowed) in various types of copepod (the thoracic segments are numbered, those fused to other regions being in parentheses). (After Kaestner, 1970.)

pairs of thoracic appendages, including that of the segment wholly or partially incorporated into the cephalothorax, form swimming limbs; the abdomen, however, lacks any appendages. Somewhat unusually, the eggs are not brooded or carried by the female but are attached to the substratum or to benthic vegetation.

The 150 species are contained within a single order.

8.6.3.6 Class Copepoda

The copepods are the dominant members of the marine plankton and to a slightly lesser extent of that in fresh water as well; there are numerous interstitial benthic species; and about one-quarter of all species are parasitic, attacking animals ranging from sponges to whales. Although most species are small (<2 mm in length), exceptionally one free-living species approaches a length of 2 cm and an ectoparasitic form achieves 0.3 m.

Basically, the copepod body comprises a head, with well-developed mouthparts and antennae, a six-segmented thorax bearing swimming limbs, and a five-segmented appendage-less abdomen, but various segments may be fused together in a wide range of fashions (one thoracic segment, at least, always being fused to the head, and, in many, a second is also incorporated into the cephalothorax), and the primary division into cephalothorax, pereon and abdomen does not necessarily reflect the manner in which, in practice, the body is subdivided, if indeed there is any regional differentiation. The parasitic species, for example, show various degrees of bodily degeneration, in extreme form including loss of all apparent segmentation and of appendages. In the free-living (and some parasitic) species, there is a major functional subdivision of the body, marked by a point at which the posterior region articulates with the anterior. Except in the cylindrical interstitial species, the portion anterior to this articulation (the 'prosome') is oval, sometimes elongately so, whilst the posterior part of the body (the 'urosome') is narrowly tubular (Fig. 8.42). Although in one group, this articulation is indeed sited at the division between cephalothorax + pereon and abdomen, in many it occurs between the third and fourth segments of the pereon (in these groups corresponding to the fifth and sixth thoracic segments) so that the last segment of the pereon (and thorax) forms part of the urosome. Copepods always lack a carapace and compound eyes.

Parasitic species display a wide range of body forms (Fig. 8.43), the sac- or worm-like bodies of the most degenerate types being unrecognizably crustacean except during their development. These often carry their eggs in long strings, as opposed to the one or two oval egg sacs of the free-living species.

The 8400 species can be divided between ten orders, although a further two have recently been proposed.

8.6.3.7 Class Tantulocarida

Tantulocarids are minute (usually <0.2 mm, always <0.75 mm) ectoparasites of marine crustaceans (copepods, ostracods and peracaridan malacostracans), except for the adult males and the sexual adult female phase, living permanently attached to their

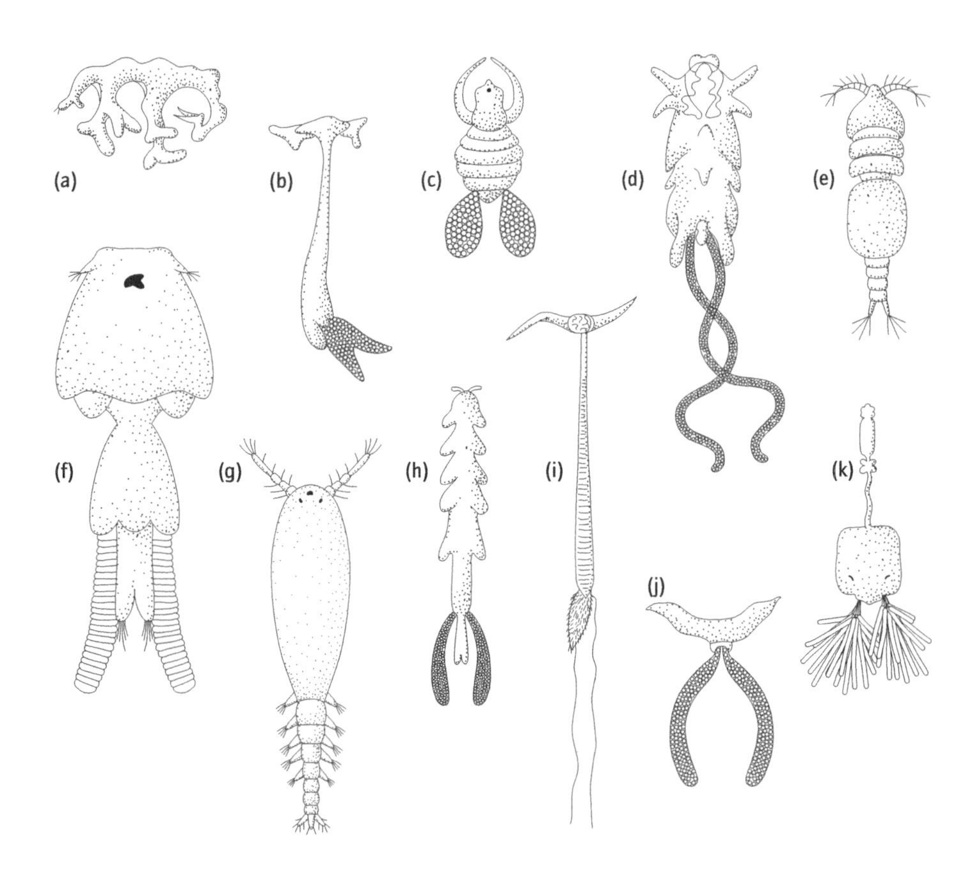

Fig. 8.43 The range of bodily form and degeneration in parasitic copepods (after various authors). (a), (d) and (h) are poecilostomatoids; (b) and (e) are cyclopoids; (c), (f) and (i)–(k) are siphonostomatoids; and (g) is a monstrilloid.

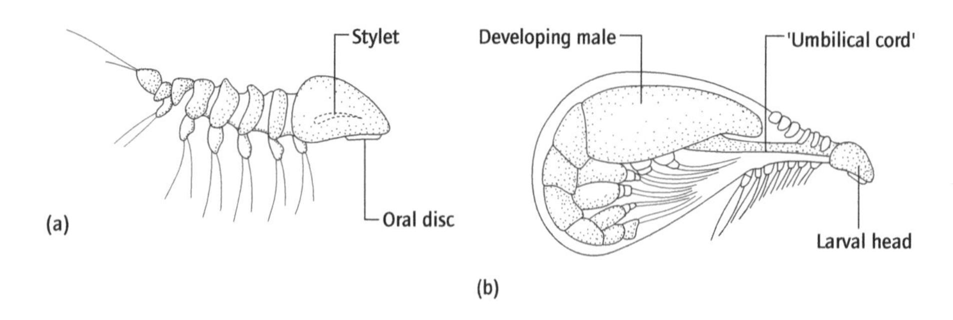

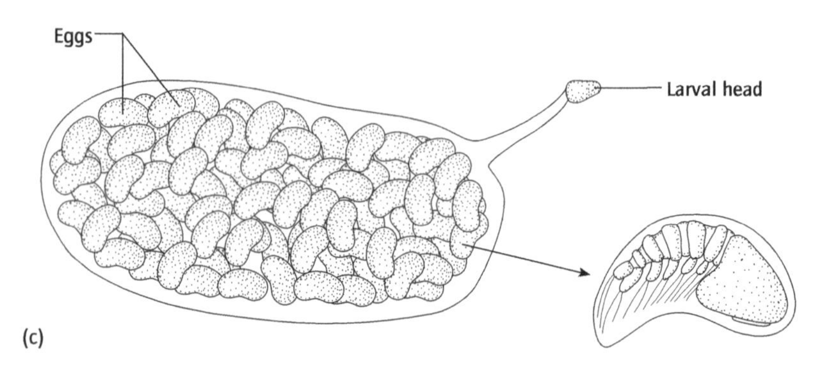

Fig. 8.44 Body form of tantulocarids: (a), tantulus larva; (b), male developing within larval body and attached to the host via the larval head and 'umbilical cord'; (c), adult female filled with eggs, together with detail of a larva developing within an egg. (After Boxshall & Lincoln, 1987.)

host by an oral disc and imbibing host liquids through a hole made by a median ventral stylet, there being no primary or accessory mouthparts. Their anatomy (Fig. 8.44) basically comprises an eyeless and appendage-less head, a six-segmented thorax bearing five similar pairs of biramous limbs and a posterior pair of uniramous ones, and a two- to six-segmented limbless abdomen (including the telson); the first two thoracic segments in the adult male being incorporated into a cephalothorax covered by a cephalic shield. The 'tantulus' larvae hatch at an advanced stage with six free thoracic limb-bearing segments and up to seven abdominal ones; they do not appear to moult. Metamorphosis into the adult form is of a very peculiar type: the presumed parthenogenetic and asexually multiplying type of female develops as a large appendage-less egg sac which erupts from the larva dorsally and remains attached to the host via the larval head; whilst the non-feeding and free living male develops from a mass of dedifferentiated tissue within the larva, remaining attached to the host during this process by an 'umbilical cord' which similarly passes through the larval head. The adult male thoracic appendages are well developed, presumably for swimming. A non-feeding, free-living type of female has recently been described, suggesting that two possible life cycles occur. This bears paired antennules (the only well-defined cephalic appendages known in tantulocarids) and a five-segmented trunk, of which only two segments bear limbs. Little is known of their biology, although they occur from 20 down to 5000 m depth.

Since they lack almost all of the diagnostic features listed in Section 8.6.2, their assignment to the Crustacea must be somewhat conjectural. Nevertheless they do share certain features with the copepods and cirripedes – also groups in which major departures from the basic crustacean body plan occur in association with a parasitic life style (see Sections 8.6.3.6 and 8.6.3.8). The dozen or so known species are placed in a single order.

8.6.3.8 Class Cirripedia

As a group, the cirripedes are the most highly modified of the Crustacea, being either sessile or dwellers in other organisms in a parasitic manner. They are effectively headless, most lack an abdomen, and there is little or no evident segmentation. In the extreme form of the order Rhizocephala, they resemble nothing so much as a bracket fungus, with a network of fine tubes spreading through all the tissues of the host (almost invariably a decapod crustacean) and an external sac containing the gonads (Fig. 8.45a).

The order Ascothoracica (Fig. 8.45b), which parasitize cnidarians and echinoderms, are the least specialized anatomically. In a few of these, there are the rudiments of a head bearing chelate antennules (a most bizarre feature for a crustacean), and a thorax with six pairs of swimming limbs, all enclosed within a bivalved carapace, from out of which projects a free abdomen of five segments. The more familiar barnacles (order Thoracica) can be visualized as ascothoracicans which have become sessile, the six pairs of thoracic legs – the 'cirri' – forming filtering or food-collecting organs, and the bag-like carapace being reinforced with from few to many calcareous plates (Fig. 8.45c and d) that are not moulted but grow around their margins. Like the other cirripedes, initial attachment to, in this case, the substratum (rock, shell or seaweeds) is effected by the antennules; after successful attachment, this pre-oral region may enlarge to form an elongate column, as in the goose- or stalked barnacles, or it may form only a thin attachment disc, as in the acorn forms. The

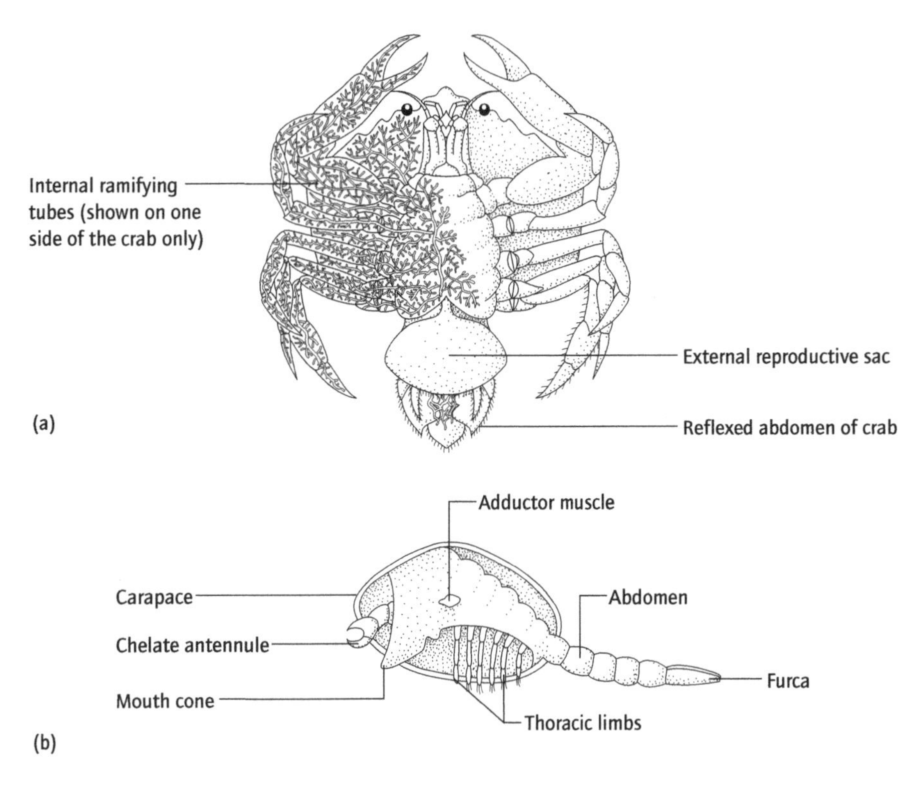

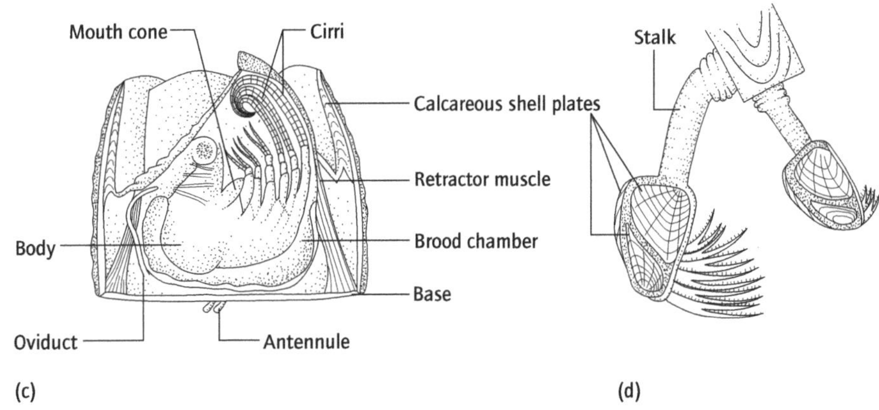

Fig. 8.45 Cirripedia: (a) a rhizocephalan infesting a crab, the latter shown in ventral view as if transparent; (b) a diagrammatic lateral view of an ascothoracican shown with the left-hand carapace valve removed; (c) an acorn barnacle in lateral view shown with some of the carapace and its plates removed (see also Fig. 8.36); (d) external appearance of two stalked barnacles attached to a block of wood (a and b after Kaestner, 1970; c and d after Zullo, 1982).

fourth group, the order Acrothoracica, are essentially similar to the barnacles, although they lack calcareous plates and bore into corals or, more rarely, mollusc shells. In both barnacle-like groups, the penis may be very long in order to reach into nearby attached (and therefore sessile) individuals. A fifth order, the Facetotecta, is known only from larvae (the so-called Y-nauplii and Y-cyprids).

Within the class as a whole, there are clear trends towards hermaphroditism or reduction of the male sex to minute proportions, and towards reduction of the gut – in many it is blind-ending and, in the rhizocephalans, absent. All have an individual type of second larval stage after the nauplius, the 'cypris', which locates the host or settlement site and thereafter metamorphoses into the young adult stage. The 1000 species are all marine.

8.6.3.9 Class Ostracoda

Ostracods are very small crustaceans (mostly <1 mm long, although rarely approaching 2 cm) with a short oval body enclosed within the bivalved and often calcarous shell formed by the carapace (Fig. 8.46). Like the conchostracans (and the bivalve molluscs), the two carapace valves are provided with transverse adductor muscles which act against an elastic hinge ligament; hinge teeth may even be present. Unlike those groups, however, the shell is shed and reconstituted at each moult.

Their sac-like bodies have no visible signs of segmentation but, judging from the appendages, what was ancestrally the head forms half of the body volume. Only a total of five, six or seven pairs of appendages are present, of which the first four certainly belong to the 'head' (the antennules, antennae, mandibles and first maxillae), and the last two, one or both of

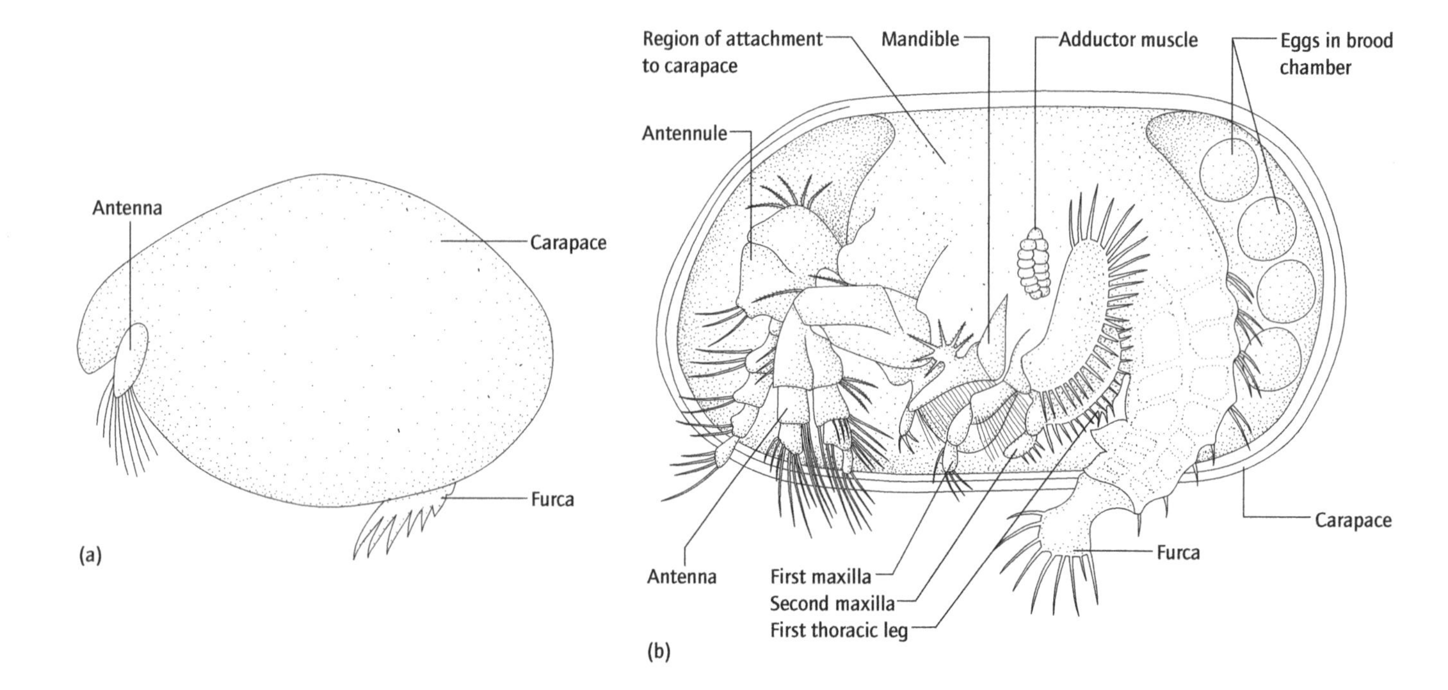

Fig. 8.46 Ostracoda: (a) external appearance in lateral view; (b) shown with the left-hand valve of the carapace removed. (After Cohen, 1982.)

which may be missing, are 'thoracic'; but authorities differ on the nature of the intervening fifth pair. Most regard it as the cephalic second maxilla, but some consider it to be thoracic and term it either the maxilliped or first thoracic leg. If it is the second maxilla, then some ostracods are the only crustaceans to have retained only the appendages of the head. The function of these various limbs varies from group to group, but the antennae are usually the main swimming organs, and the first thoracic limbs, if present, are often the walking legs. The eggs are usually brooded beneath the carapace, although they may simply be attached to the substratum or to submerged vegetation.

The five orders (one of them as yet known only from empty shells) contain a total of 5700 species, most of them marine. A few species are terrestrial, having invaded the humus/litter layer in forests via fresh water.

8.6.3.10 Class Malacostraca

The Malacostraca is by far the largest class of the Crustacea, with some 23 000 species, and, arguably, it contains a greater diversity of body forms than any other class in the animal kingdom; just one of its 16 orders, the Decapoda, includes such varied organisms as crabs, crayfish, shrimps and hermit-crabs. The main feature which unites the class is that the body fundamentally comprises a head, an eight-segmented thorax, and a six- (or rarely seven-) segmented abdomen, all these regions being equipped with a full complement of segmental appendages, including the abdomen. Their diversity, however, can be gauged from the fact that from none to all eight thoracic segments may be incorporated with the head into a cephalothorax; that from none to three pairs of thoracic appendages may form maxillipeds; and that a carapace may be present or absent (either primitively or secondarily), whilst, if present, it may cover some or all of the anterior region of the body (from only the first two thoracic segments up to all the thoracic and several abdominal ones).

Typical malacostracan specializations are the presence in the fore-gut of a 'stomach', in which the food is ground into finely particulate form and any coarse particles remaining are filtered out of the material which then passes into the digestive diverticula (Fig. 8.47), and the development of the appendages to serve a variety of functions, the posterior thoracic limbs being walking legs (pereiopods), the first five pairs of abdominal ones forming swimming organs (pleopods) and the last pair of abdominal appendages (the 'uropods') comprising a tail fan with the telson, which terminates the body instead of the more usual furca. Unless secondarily lost, a well-developed pair of compound eyes are also characteristic. Many species are large and well calcified, and several show marked degrees of concentration of the nervous system and complex behaviour patterns. They are important members of the marine nekton and benthos; many occur in freshwater streams, rivers and lakes; and several are terrestrial, including all those crustaceans which can survive in other than permanently humid habitats.

Six major groups of orders can be distinguished. Two of them have retained the basic body plan of head, thorax and abdomen, i.e. without any development of a cephalothorax or of accessory mouthparts. One, the Phyllocarida (Fig. 8.48a), are the only malacostracan group to have retained a seventh abdominal segment and the furca, and to lack uropods. Their most obvious feature is a large, laterally compressed, bivalved carapace which

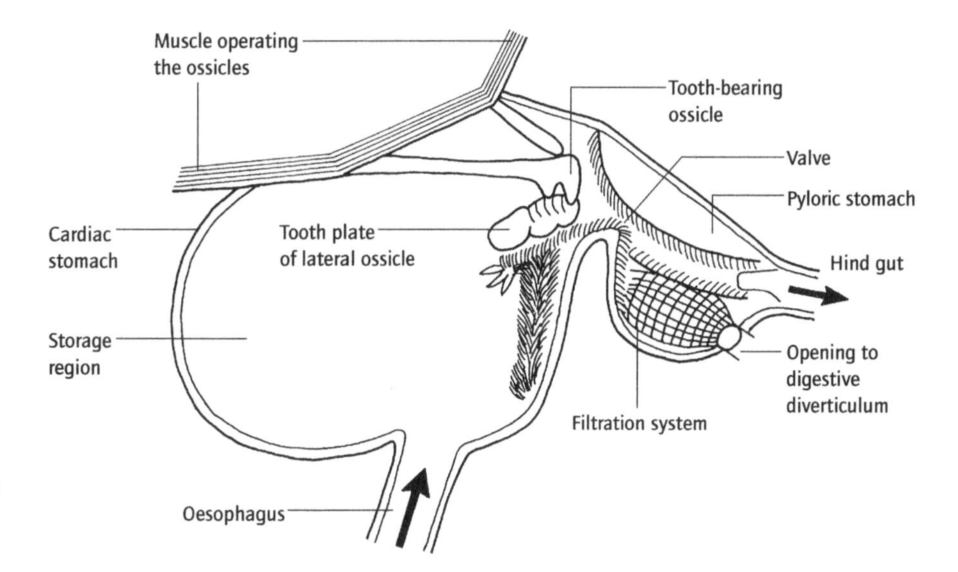

Fig. 8.47 A diagrammatic longitudinal section through the fore-gut stomach of a decapod malacostracan showing the ossicles of the 'gastric mill' which grind the food ingested, and the filtration system in the pylorus guarding the opening into one of the digestive diverticula (after Warner, 1977).

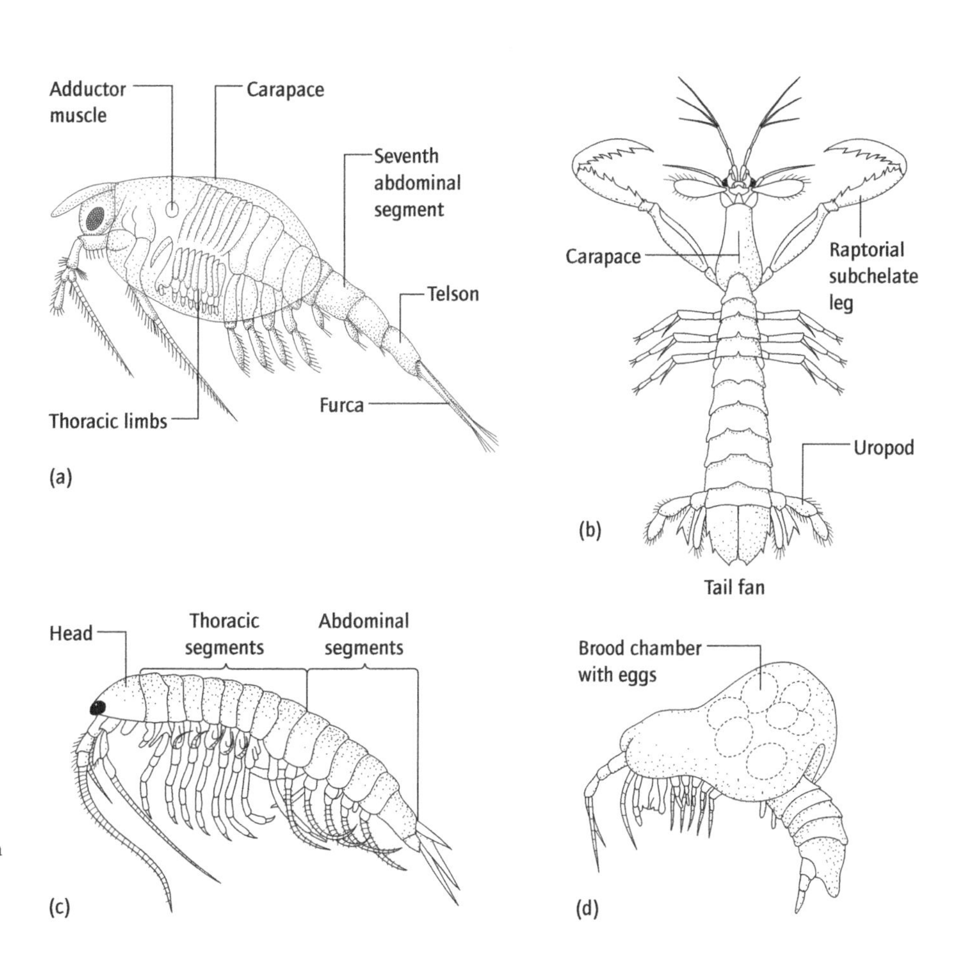

Fig. 8.48 Malacostracan diversity I. (a) Phyllocarida, seen in lateral view as if the carapace was transparent; (b) Hoplocarida, dorsal view; (c) Syncarida, lateral view; (d) Pancarida, lateral view (the carapace of the female depicted is greatly enlarged and swollen to act as the brood chamber). (After various authors.)

covers the thorax and its eight pairs of leaf-like limbs, setae on which enclose the brood cavity. The second group, the Hoplocarida (mantis shrimps) (Fig. 8.48b), also have a carapace but it is much smaller, covering only half of the dorsoventrally flattened thorax. Particularly characteristically, their antennules are triramous and the first five pairs of their thoracic limbs are subchelate and are not concerned with locomotion; the second pair are large and raptorial, whilst the next three pairs, in

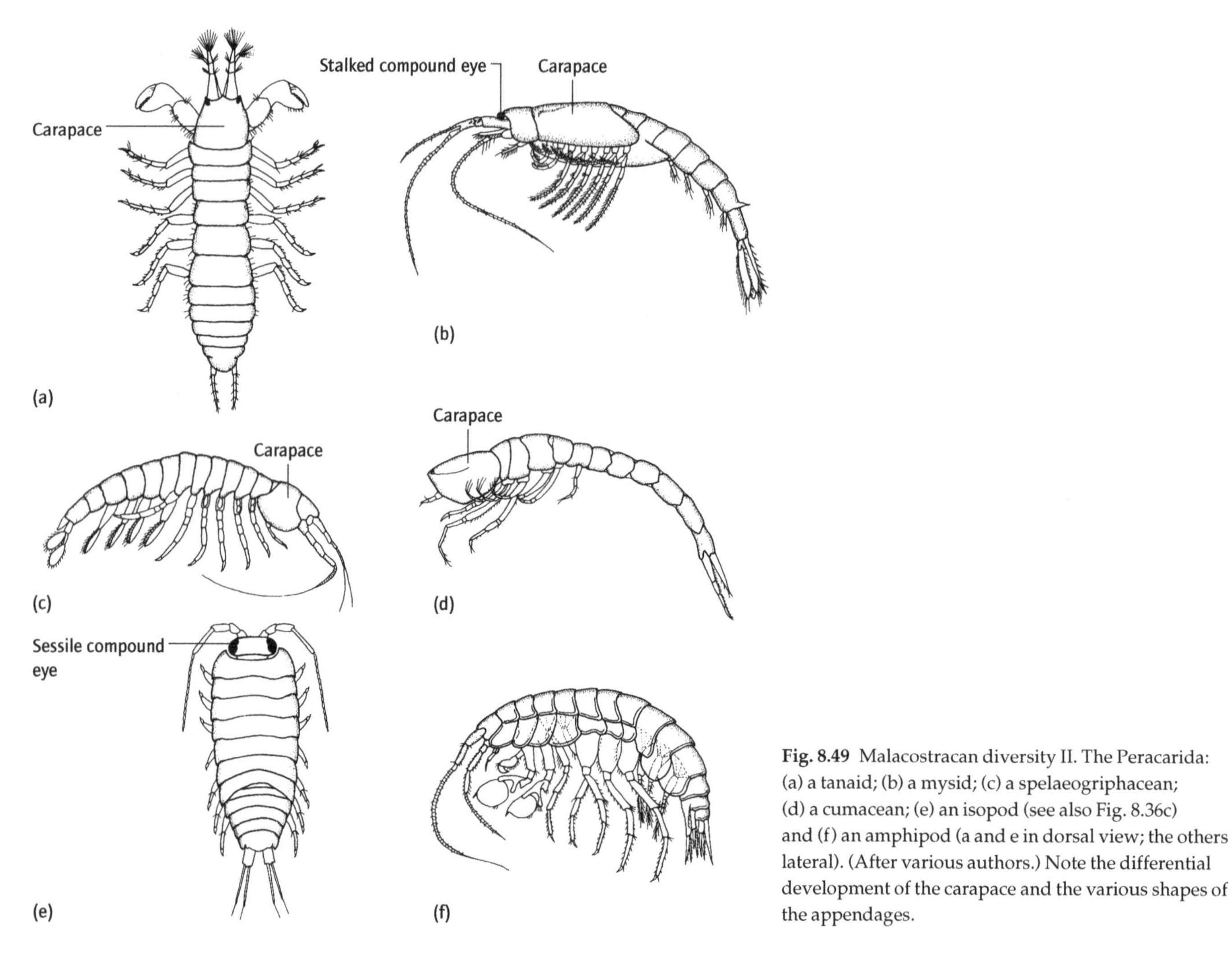

Fig. 8.49 Malacostracan diversity II. The Peracarida: (a) a tanaid; (b) a mysid; (c) a spelaeogriphacean; (d) a cumacean; (e) an isopod (see also Fig. 8.36c) and (f) an amphipod (a and e in dorsal view; the others lateral). (After various authors.) Note the differential development of the carapace and the various shapes of the appendages.

females, carry the egg masses. In some species, the second pair form 'fists' which can be shot out at 1 cm ms^{-1} to deliver an impact equivalent to a 0.22-calibre bullet!

In contrast to these two groups, the next three superorders have at least one (and usually only one, but three at the most) thoracic segment fused to the head in a cephalothorax, and therefore usually possess a pereon of seven segments and an abdomen of six. Carapace development, however, is very variable. The solely freshwater Syncarida (Fig. 8.48c) lack any carapace (primitively, it is generally thought) and tend towards the loss of appendages from the rather uniform trunk segments. The blind Pancarida (Fig. 8.48d), which inhabit caves and interstitial habitats, sometimes in hot springs of up to 45°C, are also mainly freshwater; they are distinguished by a short carapace which serves as the brood chamber (as well as the centre of gaseous exchange). The much larger third grouping, the Peracarida (Fig. 8.49), however, form their brood chamber from outgrowths of the pereiopods. A carapace and stalked compound eyes were probably present in the ancestral members of this superorder, but both are subject to marked reduction in various lines so that their compound eyes are mostly sessile and in the members of two important orders (the Amphipoda and Isopoda) a carapace is absent. Peracarids are very variable in form and habitat: some swim, several burrow, and many crawl; a number are parasitic; their bodies may be elongate or squat, dorsoventrally or laterally compressed or shrimp-like. The woodlice have successfully colonized the land (see Section 8.6.2).

The final group, the Eucarida (Fig. 8.50), have all their thoracic segments fused and incorporated into the cephalothorax, so that the body comprises only a cephalothorax and abdomen; and the whole cephalothorax is enclosed by the carapace which usually extends down laterally to cover the thoracic gills within a protective chamber. In general, the abdominal pleopods are used in swimming, and, in decapod females, for carrying the egg masses, whilst the thoracic limbs either serve as feeding appendages, or for prey-capture and walking. In the largest order, the Decapoda, there are three pairs of maxillipeds and an evolutionary trend towards reduction of the abdomen, which culminates in the crabs in which the small abdomen is folded underneath the large and often broad carapace, so as not to be visible from above. Alone amongst crustaceans, some crabs are much broader than they are long.

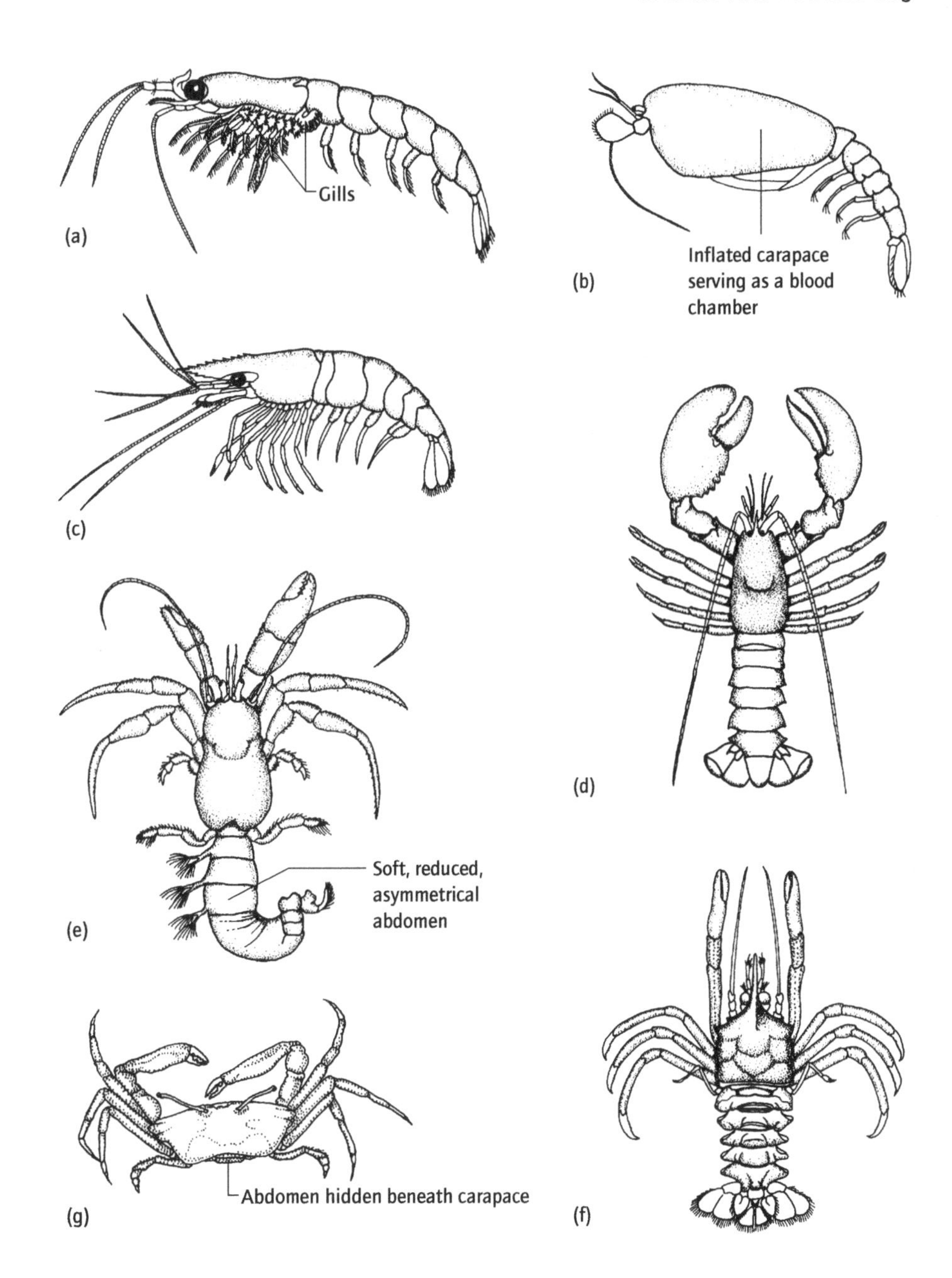

Fig. 8.50 Malacostracan diversity III. The Eucarida: (a) a euphausiid; (b) an amphionid; (c)–(g) decapods; (a–c in lateral view; d–g dorsal). (After various authors.)

8.7 Further reading

Aguinaldo, A.M.A., Turbeville, J.M., Linford, L.S., Rivera, M.C., Garey, J.R., Raff, R.A. & Lake, J.A. 1997. Evidence for a clade of nematodes, arthropods and other moulting animals. *Nature (Lond)* **387**, 489–493.

Arnaud, F. & Bamber, R.N. 1987. The biology of Pycnogonida. *Adv. Mar. Biol.*, **24**, 1–96.

Borror, D.J., De Long, D.M. & Triplehorn, C.A. 1976. *An Introduction to the Study of Insects*, 4th edn. Holt, Reinhart & Winston, New York.

Boudreaux, H.B. 1979. *Arthropod Phylogeny with Special Reference to Insects*. Wiley, New York.

Chapman, R.F. 1969. *The Insects: Structure and Function*. English Universities Press, London.

Clarke, K.U. 1973. *The Biology of the Arthropoda*. Edward Arnold, London.

Cloudsley-Thompson, J.L. 1968. *Spiders, Scorpions, Centipedes and Mites*. Pergamon Press, Oxford [Chelicerata & Myriapoda].

Daly, H.V., Doyen, J.T. & Ehrlich, P.R. 1978. *Introduction to Insect Biology and Diversity*. McGraw-Hill, New York.

Gupta, A.P. (Ed.). 1979. *Arthropod Phylogeny*. Van Nostrand Reinhold, New York.

Kaestner, A. 1968. *Invertebrate Zoology*, Vol. 2. Wiley, New York [Onychophora, Tardigrada, Penstastoma, Chelicerata & Myriapoda].

Kaestner, A. 1970. *Invertebrate Zoology*, Vol. 3. Wiley, New York [Crustacea].

King, P.E. 1973. *Pycnogonids*. Hutchinson, London.

Lewis, J.G.E. 1981. *The Biology of Centipedes*. Cambridge University Press, Cambridge.
Little, C. 1983. *The Colonisation of Land*. Cambridge University Press, Cambridge [All terrestrial arthropods].
Manton, S.M. 1977. *The Arthropoda*. Oxford University Press, Oxford.
McLaughlin, P.A. 1980. *Comparative Morphology of Recent Crustacea*. Freeman, San Francisco.
Ramazzotti, G. 1972. *Il Phylum Tardigrada*, 2nd edn. Istituto Italiano di Idrobiologia, Pallanza.
Rosa, R. de, Grenier, J.K., Andreeva, T., Cook, C.E., Adoutte, A., Akam, M., Carroll, S.B. & Balavoine, G. 1999. *Hox* genes in brachiopods and priapulids and protostome evolution. *Nature (Lond)*, **399**, 772–776.
Savory, T.H. 1977. *Arachnida*. Academic Press, New York.
Schram, F.R. 1986. *Crustacea*. Oxford University Press, New York.
Sedgwick, A. 1888. *A Monograph of the Development* of Peripatus capensis, *and of the Species and Distribution of the genus* Peripatus. Clay, London.

PART 3

Invertebrate Functional Biology

Whereas Part 2 described the diversity of invertebrate body plans and biologies, this Part concentrates on the unifying features of their functional anatomy, physiology and behaviour. It is clear that, whatever their structure, evolutionary history or ecology, all animals have certain common requirements needed to achieve, at least potentially, their own individual survival and that, in the longer term, of their genes. Hence they possess equivalent suites of functional systems permitting the acquisition of the necessary resources and information, and the processing and ordering of these inputs. The selective advantages associated with different body plans, life styles and habitats, however, have favoured different solutions to many of these common problems; and, as in Part 2, the following chapters present invertebrate functional biology against a background of the various selection pressures and of optimal solutions to interacting pressures.

Some requirements are necessary for immediate survival, and selection here may often act powerfully on living individuals. Animals, for example, need to find, consume and assimilate energy- and chemical-containing food materials, often in the face of considerable competition for these resources (Chapter 9), whilst at the same time avoiding becoming the food of other consumers (Chapter 13). Other necessities – those permitting any animal to be capable of functioning at all – can be envisaged as having been faced and overcome relatively early in the evolutionary history of most lineages, and thereafter as being subject largely to stabilizing selection. Thus most animals require locomotory systems and all need some parts of their bodies to be capable of movement to obtain food, escape from consumers, avoid unfavourable environmental conditions, and so on (Chapter 10); they must also exchange respiratory gases with their environment and carry out energy-yielding metabolic reactions (Chapter 11); being of a different chemical composition to their environment, regulation of their internal composition and/or concentration, including the elimination of metabolic waste products, will be required (Chapter 12); and information on both the internal and external environments must be obtained, evaluated and, if appropriate, acted on, and the various levels of development or activity of the different functional systems of the body have to be timed and co-ordinated if an individual multicellular organism is to act as a unitary whole in its behaviour and physiology (Chapter 16). Finally (Chapters 14 and 15), animals must adopt reproductive and life-cycle strategies that will maximize their genetic contribution to future generations; and the zygotes or other propagules formed must, in turn, develop into organisms capable themselves of reproducing and acquiring and processing their own resources, sometimes via distinct larval stages adapted for dispersal, the finding of specific host organisms, or feeding before metamorphosis into the reproductive adult form.

Since research into all these various fields has been carried out on only a very limited number of invertebrates, the animals upon which this section is based will be a minor fraction of those described in Part 2. For understandable reasons, this experimental material has been drawn mainly from the more numerous and larger invertebrate groups (e.g. arthropods, molluscs). This, however, will serve to redress the apparent bias against these so-called 'major phyla' created by the approach adopted in Part 2.

CHAPTER 9

Feeding

To some extent, animal feeding is a neglected field of enquiry. True, there is a wealth of information available on, for example, the mechanics of suspension feeding and on how predators catch their prey, and much is known of gut anatomy and digestive physiology, but why animals feed on what they do consume has largely been assumed to be self evident: that is what they are adapted to capture and digest. But why are the most primitive types of animals almost exclusively carnivorous? And why is so much plant material left unconsumed? Why are some species more generalist in their diets than others?

In this chapter, we look at the phylogenetic constraints which have channelled animal feeding (the evolutionary past), at the different types of feeding mechanism possessed by animals (the heritage of the past), and at the rapidly growing field which investigates the pros and cons of taking different individual items within the broad range of a diet and how the prey can influence consumer choice (the ecological present).

9.1 Introduction: the evolution of animal modes of feeding

All the phyla of animals, with the possible exception of the Uniramia, evolved in the sea, a relatively stable and uniform habitat. Except in the marginal intertidal and immediately coastal zones, physical variables such as temperature, salinity, ionic composition, O_2 saturation, rarely vary enough to pose any threats to the survival of marine animals, and physiological limitation of biological activity rarely occurs. Marine species therefore only require two – or possibly three – requisites for survival, both or all of them relating to feeding. They need to obtain sufficient food; they must avoid becoming the food of other organisms; and they may require an exclusive area of space in which to achieve the other two. Survival of the individual requires only these. If, of course, their genes are to survive through a longer period of time than the lifespan of any individual organism, they will also need to produce the maximum possible number of surviving and reproducing offspring (see Chapter 14). All other biological attributes, whether anatomical, physiological, biochemical or developmental, are simply the mechanistic ways of maximizing the chances of obtaining and processing these fundamental requisites.

The first animals inhabited the surface of the Precambrian sea bed (Chapter 2): what would they have had available to them as potential food? The answer can only be colonial or unicellular bacteria and protists, which, in all but the shallowest of waters, would mainly have also been heterotrophic. Sufficient light to permit photosynthesis penetrates only some 100 m into open sea water, and usually much less than this in shallow regions (some 20–30 m into coastal seas, and sometimes only a few centimetres in highly turbid silt-laden inshore waters). Most of the continental shelf and all of the ocean bed receive no sunlight. The primary production of organic materials by the photosynthetic protists of the sea is therefore a surface phenomenon, far removed from the vast majority of the benthic habitat. Today, most benthic animals are almost entirely dependent, ultimately, on the rain of dead and already partially decayed material from above, and on the organisms responsible for its decomposition both during its descent through the water and after it has settled on the sea bed.

Only in the marginal shallow waters would living photosynthesizers have been available to any animal which could either filter them from suspension in the water or else graze or browse the attached forms. Apart from the phylogenetically isolated poriferans, filter feeding was a much later specialization in animal evolution, as were browsing and grazing (Chapter 2). Bacterial chemosynthesis on the sea bed was, however, and in some areas still is, an important source of primary production; both bacterial photosynthesis and chemosynthesis are characteristic of anoxic or microaerobic subhabitats or reduced substrata.

It is therefore not surprising to find that the simplest surviving flatworms, and presumably the ancestral forms, are and were essentially consumers of bacteria and protists associated in some manner with the bottom sediments and rocks. What is perhaps more surprising is that in overt or covert ways this ancestral diet has dominated the nutritional lives of all the flatworm descendants, even including the terrestrial ones. Bacteria and protists are abundant in marine sediments, but they are individually small and are often relatively widely dispersed in space. They rarely occur in dense clumps, although we will consider some exceptions later. This has had two fundamental repercussions on animal life styles.

First, animals must be mobile to find new supplies when local stocks have been exhausted – and mobility is one of the hallmarks of the animal condition – and, second, a consumer of small, widely scattered organisms must itself be relatively small. The larger an animal is, the greater will be its metabolic requirements and the more food in total it must obtain per unit time. This can be illustrated by a plot of body mass against the corresponding energy expenditure on maintenance (Fig. 11.16a). A large animal could not subsist on a diet of bacteria or small protists because it could not find and consume enough of them per unit time.

But to increase in size is itself likely to be selectively advantageous, for three reasons:

1 Larger animals tend to be able to produce more offspring than smaller animals (i.e. differential reproductive output).
2 Larger size can confer greater immunity from consumption by other organisms (i.e. differential survival).
3 Larger animals tend to be able to displace smaller animals from limited shared resources (i.e. differential survival again).

In addition, assuming equal digestibility, it is always energetically more efficient to ingest larger individual items of any given type of food than it is smaller ones, and to achieve this in turn requires larger size.

So considering these selective benefits of increase in size, it is also not surprising to find that most surviving flatworms are larger than the bacteria-consuming species and that they have turned to the consumption of larger food items – other animals – rather than bacteria and protists. Today, the two more significant lines of structurally simplest animals, the bilateral flatworms and the radial cnidarians, are both essentially carnivorous. Nevertheless, there is a limit imposed on flatworm body size by the constraint of diffusion distance (Section 11.4.1), and so the continued selective advantage of increased size would render advantageous any morphological changes which permitted this to occur. This, together with escape from flatworm predation, probably provided two of the main impetuses behind the wide diversification of vermiform animals which occurred early in animal evolution.

One notable concentration of protist tissue does occur, however, in the shallowest of marine regions, in the form of filamentous or thallose, colonial or multicellular algae: seaweeds. Although the larger seaweeds pose particular problems (see below), the filamentous forms, and the juvenile stages of the macroscopic types, can relatively easily be removed from their attachment to the substratum by a rasping organ (see Fig. 5.3) making available for digestion this abundant and concentrated source of photosynthetic tissue. One of the early flatworm derivatives, the Mollusca, evolved in association with this specialized feeding mode, safe both from the vigorous water movements typical of shallow waters and from predation beneath their dorsal protective shells. Radular rasping is also an effective means of obtaining food from sessile or sedentary animal prey otherwise protected with external shells, sheaths or boxes, and indeed from terrestrial plants, and during their subsequent evolution the molluscs extended their basic feeding method to utilize most of the other available types of food materials.

The other early flatworm derivatives either retained the ancestral flatworm diet or evolved mechanisms for concentrating the rain of particles carried by, or sedimenting out of, the overlying water. Under many marine conditions, the fall-out rate and the suspended organic load of sea water are effectively constant, and so, provided that a consumer positions itself appropriately to intercept and concentrate this supply, it need not move much, if at all, or only one organ system need be moved, and being sessile or sedentary will reduce the basal metabolic requirements for energy and therefore the minimum quantity of food needed per unit of body weight. A diet largely based on bacteria, protists and dead organic matter is then still possible in a relatively large animal.

Several of these other early flatworm descendants extracted particles from suspension in the water by creating a current through some form of filtration device ('suspension feeding'); others collected those particles and their microbial associations which had already sedimented out on to the substratum surface ('deposit feeding'); and yet further forms intercepted the passive fall-out before it reached the bottom ('sedimentation interceptors').

Suspension-feeding groups include the poriferans, the lophophorate phyla and the invertebrate chordates, and several members of the Mollusca, the Annelida and the arthropod groups, including some which later in phylogeny exported suspension feeding into the water column and up into the zone of photosynthetic production. The alimentary adaptations of suspension feeders will be covered in a later section of this chapter (Section 9.2.5) but here we can note some generalities of the feeding process. All suspension feeders have a filter of some form and either a feeding location positioned so as to intercept a natural (and persistent) water current or else means of creating their own current through the filter.

In sponges, the same cells – the choanocytes – both create the water current by the beating of their flagella and trap the food particles contained in it on the collar of microvilli surrounding the flagellum (Fig. 9.1). The small mesh size of this microvillar filter enables particles down to the size of bacteria to be retained. In all suspension feeders, it would clearly be advantageous not to filter the same volume of water more than once, and poriferans avoid this by drawing water in through many small pores scattered over the general body surface, and by expelling it through only one or a few larger exit apertures (see Section 3.2). The more powerful water current which results drives the filtered water well away from the sponge body. (The colonial and planktonic pyrosome tunicates have evolved an equivalent system and the single colonial exhalent current further serves to drive the colony through the water by jet propulsion; Fig. 7.34.)

In contrast, the suspension-feeding products of the radiation of flatworm descendants have adapted or evolved specific

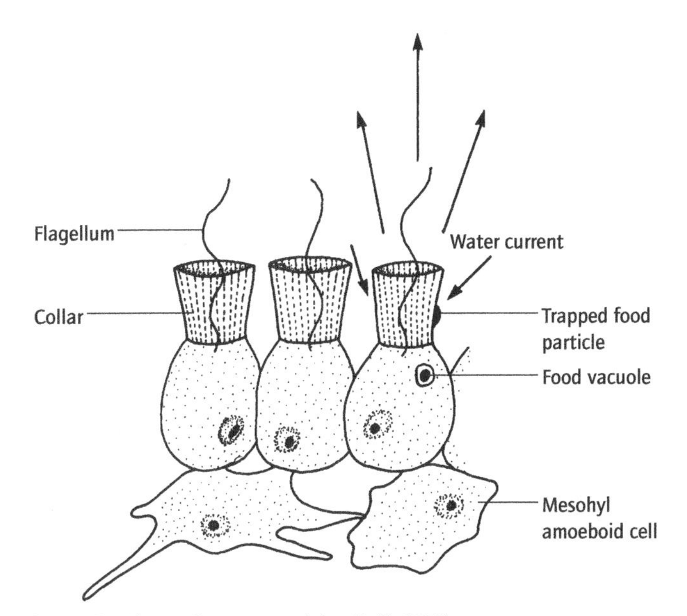

Fig. 9.1 Poriferan choanocytes (after Brill, 1973).

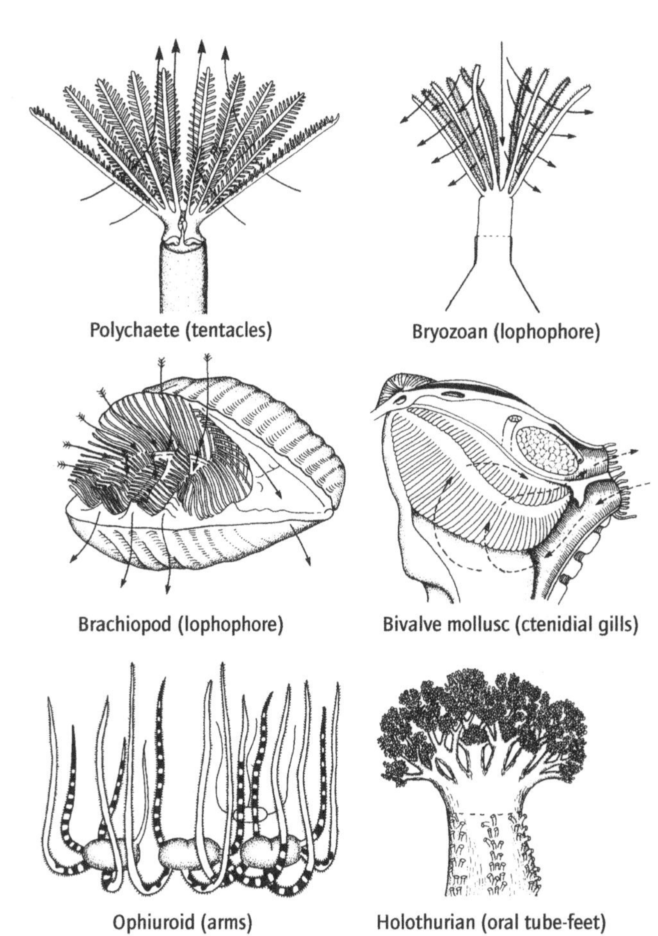

Fig. 9.2 External organs used for mucociliary suspension feeding (after several sources).

organs for filter feeding: tentacles on or near the head in the annelids, the mesosomal lophophore in the phoronans and their relatives, greatly enlarged ctenidial gills in bivalve molluscs, and so on (Fig. 9.2). All, however, operate in the same basic fashion. The beating of cilia in tracts on some region of the filtration organ create the water current, whilst other ciliary tracts trap and convey the food particles, with or without the aid of mucus, to the mouth. Particle capture is largely size specific, but yet further tracts of cilia, either on the filter itself or on some accessory organ, sort the trapped particles into those to be ingested and those to be rejected, often in the form of a pseudofaecal pellet. The invertebrate chordate system is exactly equivalent (Section 9.2.5).

The arthropod filter feeders, however, lack cilia by virtue of their possession of an exoskeleton, and instead use setae on various of their appendages to form the filter (Fig. 9.3). Their feeding current is created by swimming-like movements of the filter-bearing or of other limbs. Setose filters are generally coarser than ciliary ones and therefore can trap only larger particles. Small planktonic crustaceans like copepods (Section 8.6.3.6) have traditionally been regarded as setose filter feeders. This, however, is doubtful: copepods operate at such low Reynold's numbers (see Section 10.1) that movement of a setose filter through the water – even of a coarsely meshed one – would act as a paddle, not as a sieve. Copepods and similar crustaceans are, therefore, more likely to capture minute particles raptorially, including by a 'fling-and-clap' method that is the reverse of the clap/fling generation of flight described in Box 10.8. In other cases, particles are likely to stick to the alleged sieve by virtue of its adhesive properties or of electrostatic forces.

All the animals using tentacle-like structures to suspension feed, and many other filter feeders as well, are sessile or sedentary, and the tentacular feeding organ is effectively radially symmetrical with the mouth usually being located in the centre of the feeding apparatus – the same functional system as shown by the primarily radial cnidarians (see Section 3.4.2). The body is generally protected by an external shell or tube or is situated within a burrow in the substratum. The animal and/or its delicate feeding apparatus can therefore be withdrawn for safety from predators into cover. Under such semi-enclosed circumstances, a terminal anus would of necessity result in the passage of faecal material along the whole length of the external body surface before it could escape to the environment. Accordingly, there is a marked tendency in tubiculous and equivalent species for the gut to be U-shaped, with the anus, and often the excretory organs, discharging near the anterior end of the body in the path of the exhalent water current (e.g. Fig. 7.8). The same anatomical system is also found in many sedentary deposit feeders for the same reasons (e.g. Fig. 4.40). (Alternatively, the burrow may possess (at least) two openings at the surface and the occupant maintain a unidirectional current of water through it, or, more rarely, the animal may live doubled up within its tube so that both mouth and anus are at the single aperture.) Studies of models have shown that a wide porous crown of tentacles atop a cylindrical body, as in most tubiculous worms, creates a

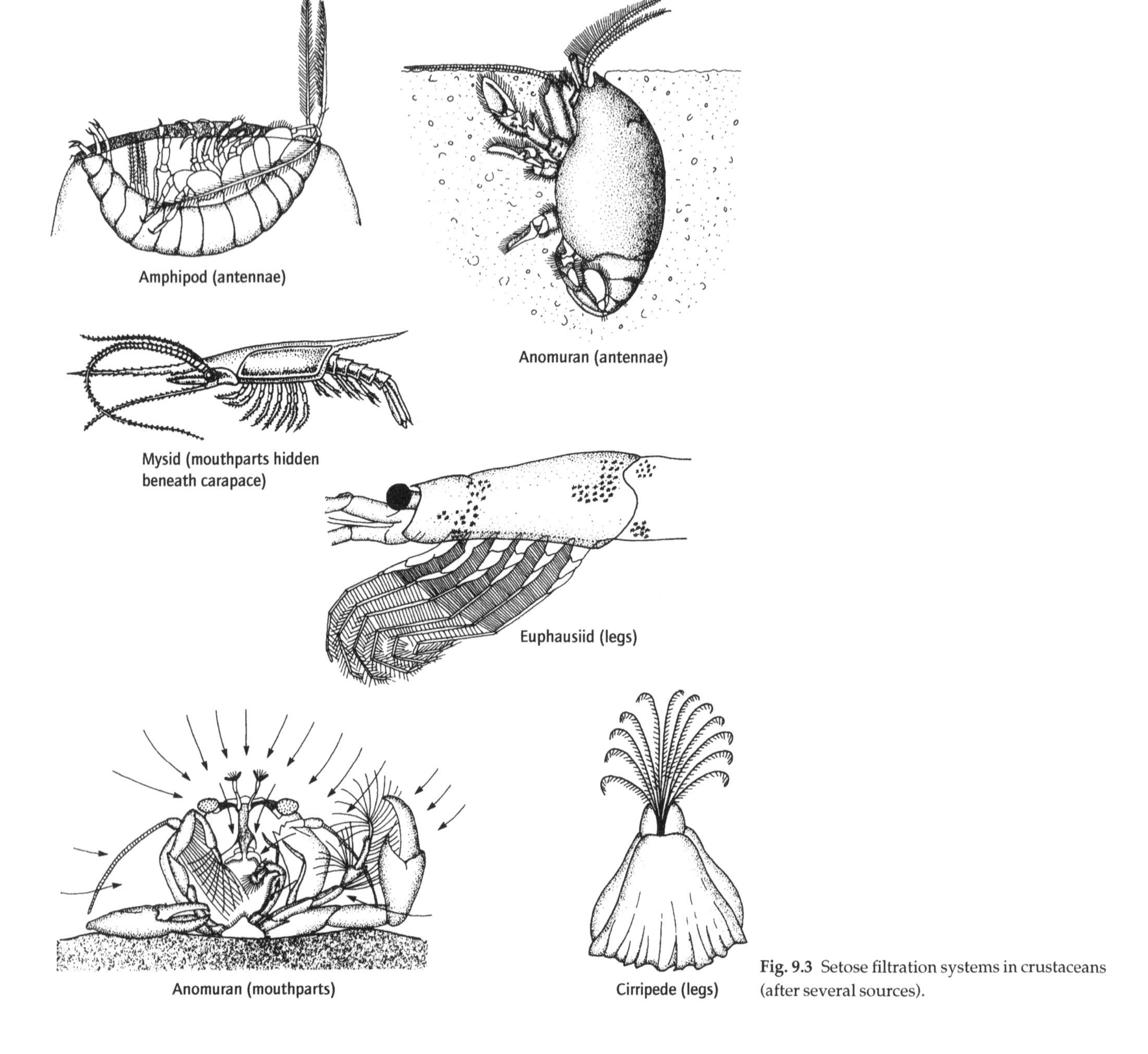

Fig. 9.3 Setose filtration systems in crustaceans (after several sources).

special hydrodynamic effect in a water current: material in a circular zone around the worm is lifted off the substratum and moved upwards through the tentacular ring; and potential food particles are slowed down and recirculated around the tentacles by turbulent flow – all without any active feeding current being maintained.

Some annelids, including the ragworm *Hediste* (*Nereis*) *diversicolor* and *Chaetopterus*, appendicularian tunicates, and some gastropod molluscs and echiurans, have evolved a filtration apparatus which is not part of the body at all, but is a secreted structure. A meshwork of mucous threads is formed in the shape of a net or bag, and a current of water is drawn through it by means of movements of all or part of the body equivalent to those used for locomotion. After an interval, the mucous net is consumed, together with such material as it has collected, and a new filter is secreted (Fig. 9.4). Some corals, other gastropod molluscs and insect larvae have independently evolved a similar feeding system, although largely dependent on natural water currents rather than self-generated ones.

The collection of surface material by deposit feeding requires no such elaborate organs as those of suspension feeders, and indeed some worms simply consume the surface layer with a completely unspecialized mouth region. Nevertheless, in a number of cases, notably in the sipunculans, holothurians and echiurans, and in several annelids and some hemichordates, a specialized series of lobes or tentacles around the mouth or an

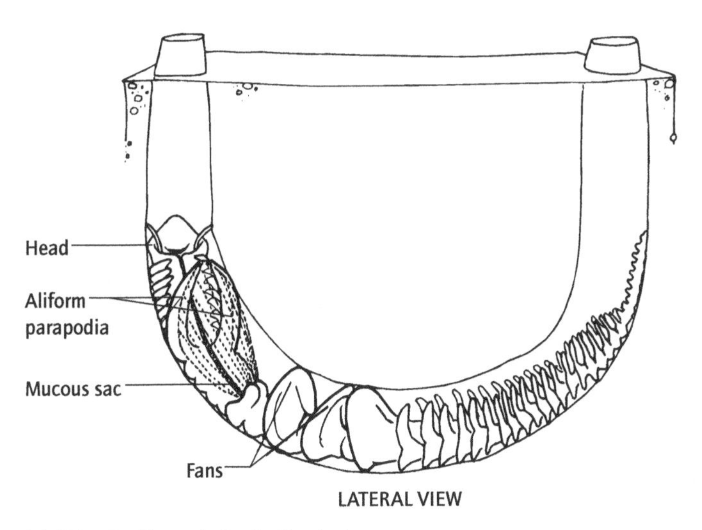

(a) Tube-dwelling polychaete *Chaetopterus*

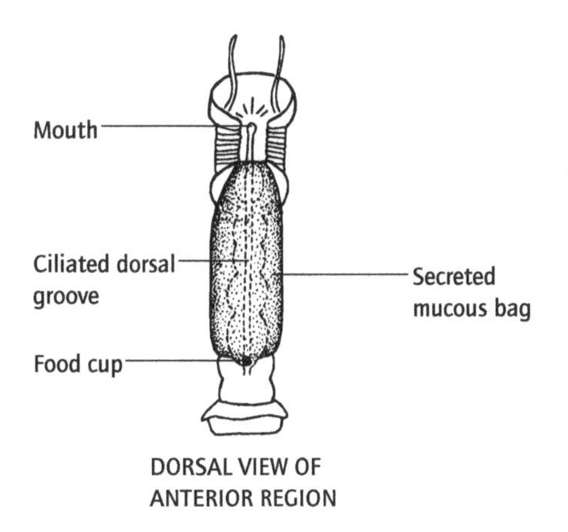

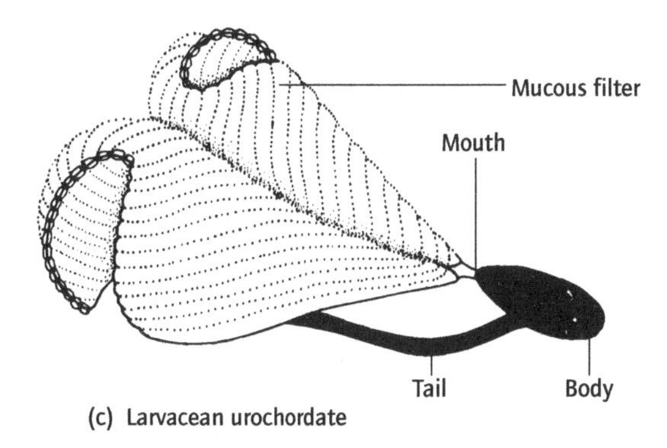

(c) Larvacean urochordate

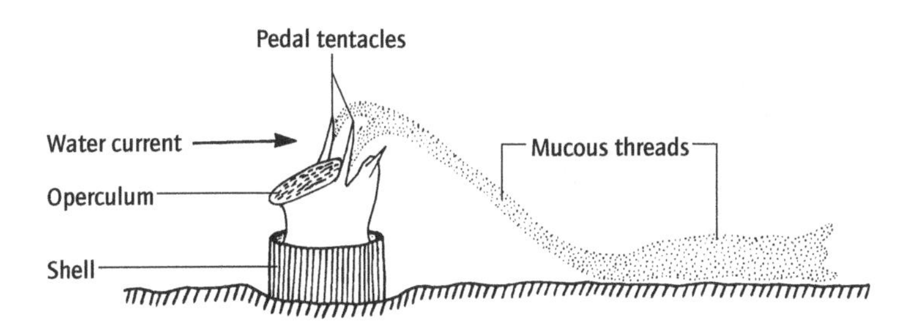

(b) Vermetid gastropod mollusc

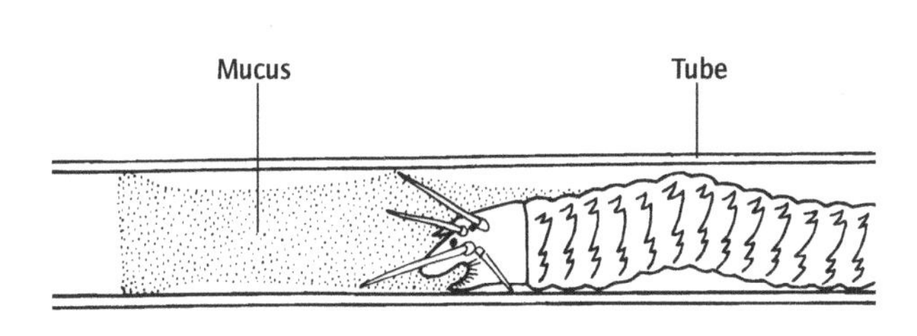

(d) *Nereis (Hediste) diversicolor* (polychaete)

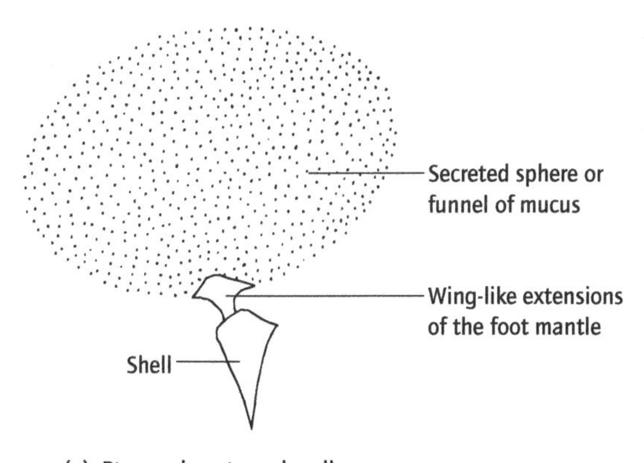

(e) Pteropod gastropod mollusc or 'sea butterfly'

Fig. 9.4 Filter feeding by the use of a secreted mucous meshwork (after several sources).

extensible proboscis, all covered by ciliary tracts, may be moved through or across the sediment, the more effectively to collect food particles in the vicinity of the burrow system (Fig. 9.5). Deposit feeders with lobes or tentacles around the mouth share several features in common with suspension feeders (as we noted above in respect of the U-shaped gut) and several of them can feed in both modes. Two problems peculiar to deposit feeders, however, are (a) that organic material may comprise only a small proportion of the background sediment, and (b) that much of the organic matter present may be the relatively refractory and indigestible residues which have accumulated precisely because they cannot be used by animal consumers, only by bacteria (see Section 9.2.6).

Therefore, except in shallow regions where benthic photosynthetic protists can live on the sediment surface, deposit feeders may be dependent on such bacteria as can convert these refractory organic residues into digestible materials, and on those protists and interstitial animals as are themselves dependent on the

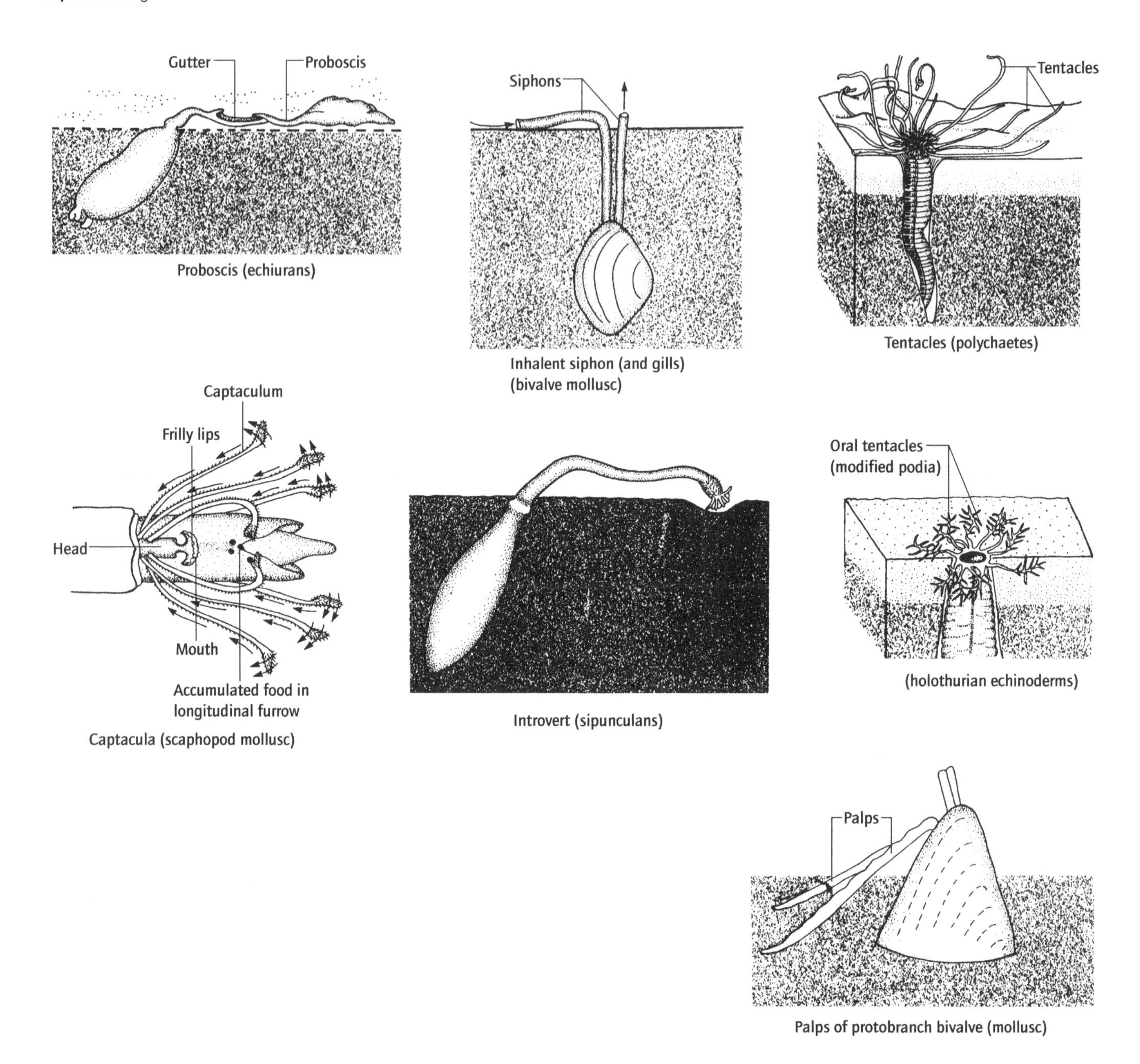

Fig. 9.5 Organs of mucociliary deposit feeding in burrowing invertebrates (after several sources).

bacteria. In any event, large quantities of sediment may have to be ingested (and deposit feeders may rework benthic sediments and so act as powerful agents of bioturbation) or large quantities must be processed by sorting organs in order to obtain sufficient digestible food. The tentacles of many deposit-feeding worms are capable, for example, of extending large distances away from the burrow in which the body is housed. Except in particularly rich areas, deposit feeders cannot therefore achieve the densities attained by suspension feeders, for which space rather than food is often the limiting resource. Because many are ultimately reliant on bacterial productivity rather than on the rate of sedimentation of detrital particles, and because bacterial productivity may itself be limited by factors other than the supply of carbon (such as nutrient shortage in the interstitial water), the growth rate of deposit feeders is also often slower than that of suspension feeders.

The final category of consumers of materials originating in the overlying water, the sedimentation interceptors, such as several of the stalked echinoderms, have a body attached to the sea bed but positioned well above it so as to intercept the supply of detritus before it becomes diluted by incorporation into the sediment (Fig. 9.6). A series of radially arranged arms bearing hydraulically operated and mucus-laden papillae collect the particles, which are then conveyed to the mouth, located in the centre of the circle of arms, along ciliated grooves.

Judging from the surviving members of potentially ancestral animal groups, there was probably one other, and completely different, mode of nutrition present very early in the history of animal multicellularity. Indeed in the Precambrian, it may have been the predominant type. Several marine animals living in very shallow waters contain in their surface tissues symbiotic oxyphotobacteria (prochlorophytes and cyanobacteria) or photosynthetic protists (unicellular dinoflagellates or chlorophytes), and some flatworms, nudibranchs and cnidarians appear to be entirely dependent on these symbionts for their nutrition, although most are not capable of digesting them. Such acoels, soft corals and a jellyfish live with their bodies permanently (in the case of sessile species) or temporarily (in mobile forms) exposed to full sunlight. Their symbionts photosynthesize in the normal manner, but in part using inorganic nitrogen and phosphorus, stripped of their organic binding during the animal's metabolism, as a source of nutrients, and using the host's respiratory carbon dioxide as a carbon source. The animal partner appears in some way to render the symbiont's cell walls thinner and/or more leaky than normal, and some of the photosynthesized products, including sugars, lipids and amino acids, therefore diffuse out into the animal's tissues where they are assimilated, and the breakdown products then diffuse back to the symbionts. In effect, the animal milks its captive populations of photosynthetic bacteria and algae, which can attain densities of 30 000 mm^{-3} of host tissue (Fig. 9.7). Other marine species are at least partially dependent on symbiotic photosynthesis.

Some deep-sea pogonophorans are equivalently dependent, in whole or in part, on symbiotic chemoautotrophic bacteria which utilize the reduced sulphur compounds and methane that issue from vents, seeps and fumaroles in the sea floor; and various oligochaetes, other worms and bivalve molluscs associated

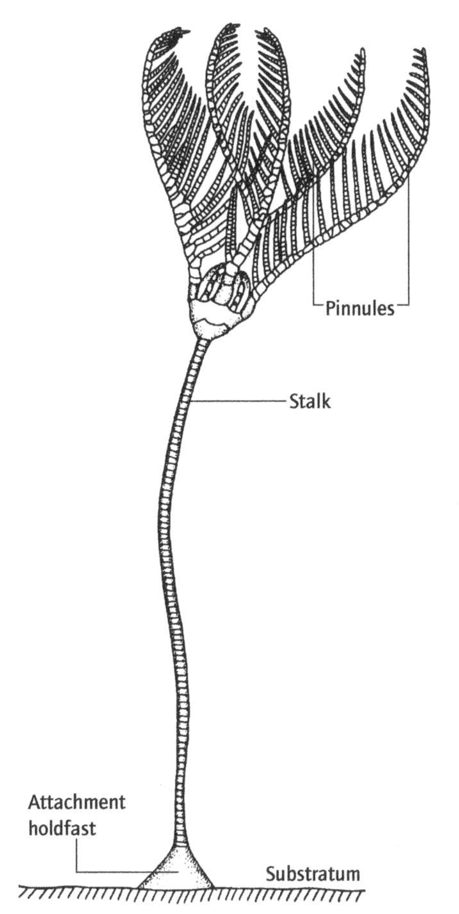

Fig. 9.6 A sediment-intercepting stalked echinoderm (crinoid) (after Clark, 1915).

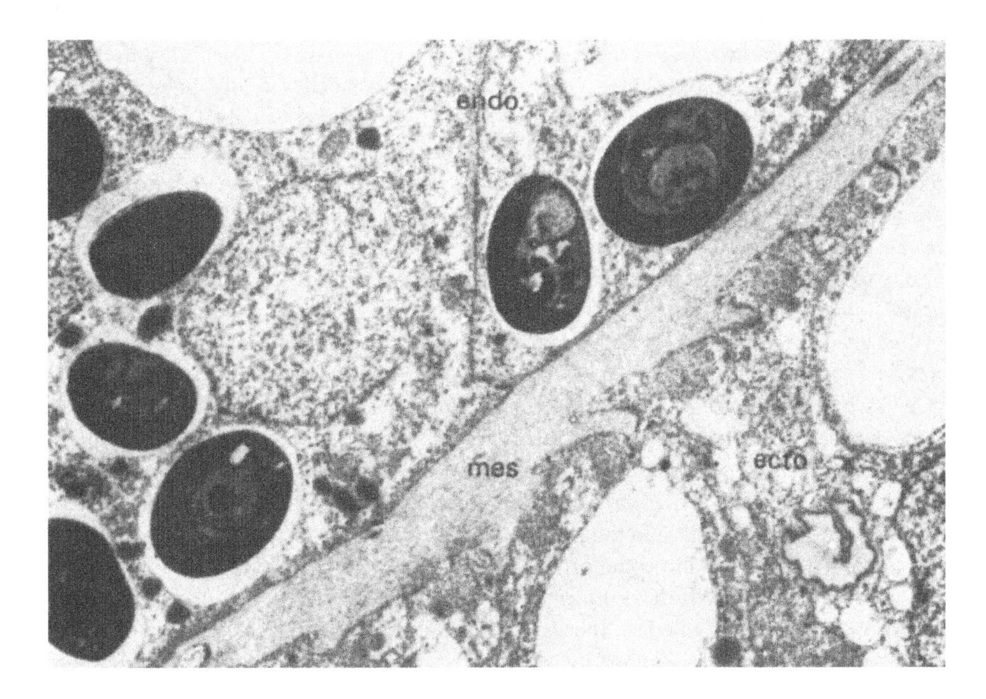

Fig. 9.7 An electronmicrograph of algal symbionts in the tissues of a cnidarian: ecto = ectoderm; endo = endoderm; mes = mesoglea (from Muscatine *et al.*, 1975, with permission).

with other sulphide-rich marine habitats have also been shown to obtain their nutrition in a similar fashion, receiving some 50% of the carbon fixed by their sulphur-oxidizing bacteria. That this symbiotic contribution to animal nutrition occurs in representatives of most animal phyla, and particularly widely in the groups thought to be closest to the ancestral ones, suggests that although multicellularity itself may not have had a symbiotic origin (Chapter 2), the origin and success of the early groups of animals may well have been aided by it, especially perhaps where food availability was limiting. (In fact, the chloroplasts of some chromophytan algae may have been derived from other, one-time endosymbiotic, eukaryote algae, and, if this is so, then some eukaryotes have evolved from symbiotic unions of separate organisms which were themselves eukaryote.)

The feeding methods and food items reviewed above must for hundreds of millions of years have been the only ones found in the animal kingdom amongst free-living species, and marine animals today still basically subsist on a diet of living or dead bacteria, protists and each other. Eventually, however, animals of marine ancestry colonized the land and although terrestrial bacteria, protists and other animals could continue to permit essentially unchanged ancestral diets for many (except for the ineffectiveness of suspension feeding on land), in their new habitat they would have encountered a new, plentiful, but very different, food source – the terrestrial bryophytes and tracheophytes. These pose a considerable problem to consumers in that a high proportion of their biomass is in the form of tough, indigestible cell walls of cellulose and almost inert, supporting, structural polymers such as lignin. Ancestrally, most animals had never needed cellulases or similar enzymes to digest their food, and the early terrestrial species must have found themselves completely ill-equipped to tackle this abundant but refractory source of potential food. Not until the Cretaceous did feeding on plants become widespread on the land.

To some extent, however, a few marine species had already overcome a very similar problem. The larger seaweeds of shallow marine habitats also contain complex carbohydrates, in this case to provide the strength required to resist vigorous water movements. Once their macroscopic form has developed, they are as potentially usable as is the majority of terrestrial plant tissue. Only two groups of marine animals have evolved the ability to cope with this generally unattractive material whilst it is still living.

One group is certain sea-urchins. Most consumers of seaweed material can do so only after bacteria have converted refractory polysaccharides into digestible form; these include the classic deposit feeders. Bacteria, especially the anaerobic fermenters, can break down a very wide range of organic molecules, including those in petrol and many plastics, but instead of relying on this activity taking place in the external environment, some sea-urchins have internalized the process. They have a special region in their gut in which a culture of anaerobic fermenting bacteria is maintained. In effect, the sea-urchins take into their guts the coarse seaweed tissue and by so doing feed the bacteria; the bacteria digest this material, and the echinoderms then subsist on a diet of the products of bacterial digestion and on the bacteria themselves. A further problem associated with seaweed material is its low nitrogen status per unit weight, but, under anaerobic conditions, several bacteria can fix atmospheric nitrogen dissolved in the sea water and hence the gut bacteria can boost the nitrogen status of the overwhelmingly carbohydrate sea-urchin diet. Although in this sense pre-adapted to terrestrial herbivorous diets, no echinoderm has become terrestrial, but many of the most successful terrestrial herbivores have solved the problems in a parallel manner.

The other group is the gastropod molluscs, and not surprisingly in view of their browsing and grazing origins in shallow waters, they are one of the few animal groups to have evolved the required enzymes to break down a large number of carbohydrate polymers including, in some, cellulases. In contrast to the echinoderms, the gastropods have proved a successful group of land herbivores, as any gardener will testify.

Although a few terrestrial arthropods have also evolved cellulases, the strategy of a symbiotic gut microbiota is the one adopted by many of the more successful terrestrial herbivores, including cockroaches, termites, several beetles (and the mammalian vertebrates). These can only subsist on their plant diet through the intermediary of their symbiotic bacteria and protists, either via the fermentation of the carbohydrate chains and/or via the provision of the required additional levels of organic nitrogen. Such so-called herbivores are as dependent on bacteria and protists as are the acoel flatworms, the sipunculans and the deposit-feeding annelids.

Not all parts of a given plant are equally refractory, however; the young photosynthetic leaves, in particular, pose less problems to animal digestion. Even so, few types of consumer are able to break down unbroken cell walls to get at the cell contents by their own enzymatic repertoire. Land invertebrates lacking the ability to decompose cellulose, symbiotically or enzymatically, have evolved two techniques for releasing the contents of plant cells or otherwise obtaining plant fluids. Caterpillars, grasshoppers and various other insect consumers of leaves bite small pieces or thin strips from the plant and utilize the contents of such cells as were ruptured during detachment and subsequent chewing of the leaf fragment. Intact cells, however, are unavailable and hence relatively little food (one-third, on average) is obtained from each fragment ingested. Large quantities of material must therefore be consumed to offset the inefficiency of utilization, and this inefficient system can only be maintained because of the large biomass of the raw material available. The second technique is to tap into the plant's fluid-transport systems by means of piercing and sucking mouthparts (Fig. 9.8; see also Fig. 8.30), as seen in all homopteran bugs, e.g. aphids. They avoid the refractory structural carbohydrates but may still require symbiotic gut bacteria to boost the nitrogen status of this dilute and protein-deficient liquid.

The exploited plants, however, would clearly be advantaged if they could deter consumption of their photosynthetic tissues

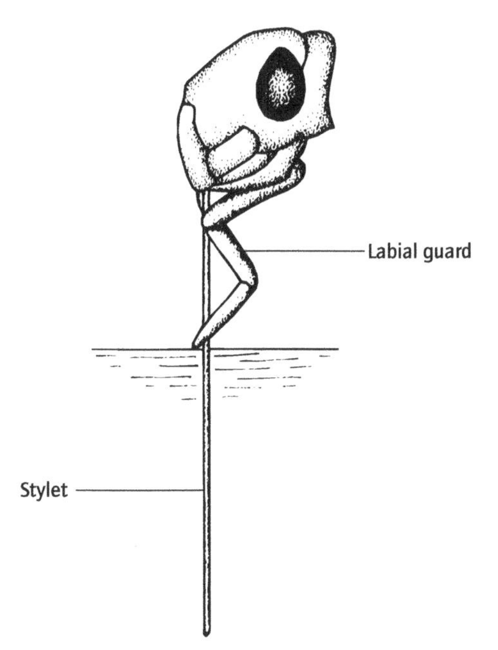

Fig. 9.8 The head of a plant-sucking bug, with its stylet inserted into the tissues of the plant (after Barnes, 1980).

and in common with other sessile organisms (colonial marine animals, for example), i.e. those least able to defend themselves by other means, they have evolved mechanical and chemical defences of great subtlety (Section 9.2.4). As is usual in such systems, this has led to an arms race between the consumed and consumer in terms of toxin production and detoxification or resistance.

Only one category of plant consumer efficiently utilizes plant materials without the aid of symbiotic bacteria and protists. These are species feeding on substances or structures produced by the plants specifically to attract animal consumption – nectar, fruits, nuts, etc. – in order to achieve animal-effected pollination and/or dispersal. Although fruits and nuts are generally targeted on vertebrates, nevertheless arthropods and molluscs may pirate this resource before it is consumed by birds or mammals.

Fungi, with their chitinous cell walls and lack of refractory supporting tissues, present fewer basic problems for animal digestive systems and are widely consumed. Indeed, some insects go so far as to collect plant material, chew it into a pulp, and then use that pulp as a substrate for fungal growth, ultimately consuming the fungi. To a degree, fungi replace the ancestral algal component in terrestrial animal diets.

Despite the widespread popular notion that terrestrial food chains are predominantly of the form plant → herbivore → carnivore, an impression which is reinforced by the amount of scientific attention which has been devoted to herbivorous animals (especially grazing mammals), the base of most terrestrial food webs is not living plant tissue and the grazing down of this by herbivores, but the decomposer food chain. It is through the detritus- and litter-feeding animals that most energy flows; a pathway mediated by bacteria, protists and fungi. Terrestrial animals have therefore not escaped from the consequences of their marine ancestry in terms of what they can efficiently process, and even today less than 3% of forest productivity is consumed, whilst it is still alive, by herbivores.

9.2 Types of animal feeding: patterns of acquisition and processing

9.2.1 Classification of feeding type

The influence of early assumptions of the overriding importance of the grazing food chain in ecological interactions is also reflected in the names given to the classical trophic levels. For many years, animal feeding was classified on the basis of this food chain and of the systematic affinities of the food species consumed, such that 'herbivores', 'carnivores' and the intermediate 'omnivores' were recognized. This classification of feeding possesses many disadvantages: relatively few species are exclusively herbivorous or carnivorous, especially when all the stages in their life history are taken into account, so that most animals fall into the catch-all omnivorous category; secondly, being based on the old two-kingdom approach to classification (organisms being either 'plant' or 'animal'), it leaves open to question the placing of animals dependent on bacteria and protists, which as we have seen is an important group of consumers; and thirdly, the phylogenetic relationships of the prey species are not necessarily relevant to the feeding ecology of the consumer.

It has become customary to distinguish categories of consumer on the basis of their general feeding methods. Thus we have hunters, parasites, grazers and browsers, suspension feeders, deposit feeders, and those deriving their nutrition symbiotically. These divisions cut right across the boundaries of the systematic position of the prey species: grazers and browsers, suspension feeders and deposit feeders, for example, may consume any or all of bacteria, protists, fungi, plants and animals. If the feeding techniques of the animal consumers do not necessarily distinguish between the various kingdoms of organisms, then neither should we when analysing feeding biology, and these categories are those which were introduced in the foregoing section. Here we will cover them in rather more detail, including an introduction to the alimentary and digestive features of each type.

9.2.2 Common features

Most invertebrates possess a through gut with a more or less anterior opening to the exterior, the mouth, into which food is taken, and a more posterior opening, the anus, from which indigestible residues and excretory products discharged into the gut can be voided. Some, however, principally the cnidarians and the flatworms but also a number of other types most of which have secondarily evolved this condition, have a blind-ending gut with a single opening serving both for ingestion and egestion; and others, including several parasitic forms but also a few

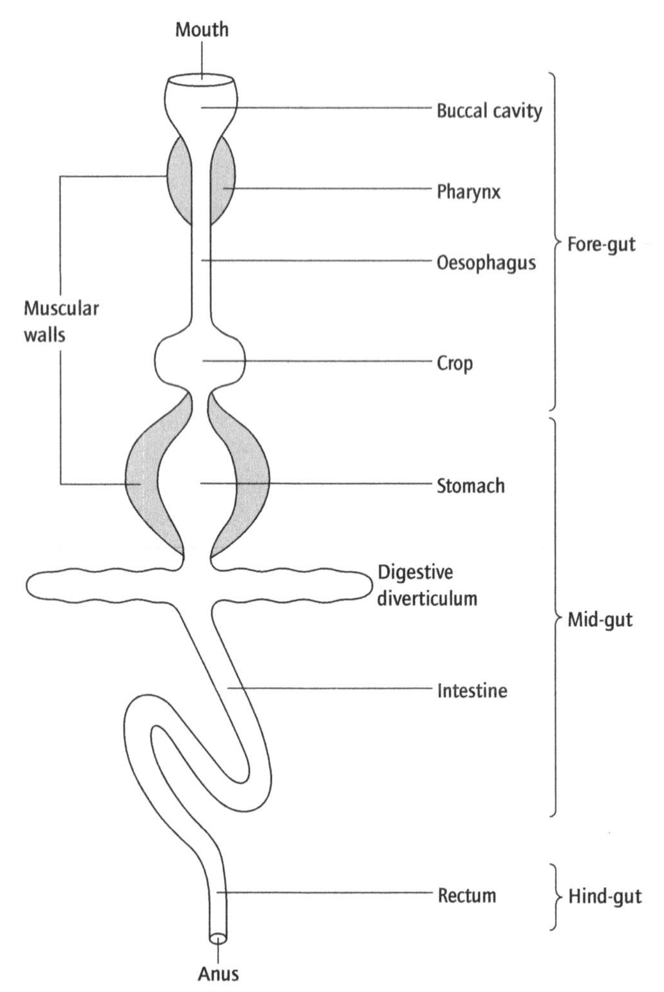

Fig. 9.9 The regions of a generalized invertebrate gut (diagrammatic).

free-living species, lack an alimentary canal and absorb food materials from internal symbionts or directly across the outer body surface. Variations on these themes are numerous. In sessile species, for example, the mouth is often located in the centre of the upper surface, rather than being anterior; in some free-living forms with a recent sessile ancestry, e.g. the free-moving echinoderms, the mouth is located in the centre of the lower surface; and the chordates and some others have evolved a gut with multiple openings to the external environment (Section 9.2.5).

Developmentally, the gut comprises three regions: an anterior fore-gut of ectodermal origin and lining (an inpushing of the outer body surface); a similarly ectodermal hind-gut; and an endodermal digestive and absorptive mid-gut, often bearing blind-ending diverticula. These three basic regions are further subdivided into different functional sections (Fig. 9.9). The fore-gut may comprise (a) a buccal cavity into which the mouth opens and into which 'salivary glands' discharge (sometimes specialized to produce sticky secretions, anticoagulants or toxins); (b) a muscular pharynx which may aid in food ingestion by acting as a pump, by forming an eversible organ of prey capture, or, in chordates, by serving as a filter for suspension feeding; (c) a short region of conduction, the oesophagus; and/or (d) a storage organ, the crop, especially well developed in consumers taking large meals at infrequent intervals which are released only gradually into the mid-gut.

The mid-gut is typically differentiated into (a) a muscular stomach, (b) various blind-ending secretory and/or absorptive outgrowths, and (c) an intestine. The stomach is the site of mechanical breakdown and sorting of the ingested material, sometimes possessing a distinct region – the gizzard – concerned with trituration, and it may also be partially the region of digestion. More often, however, either or both of digestion and absorption are effected in the various diverticula or caeca which issue from the gut just posterior to the stomach. In molluscs and crustaceans these caeca may be elaborated into large complex organs – the hepatopancreas.

Primitively, digestion is largely intracellular, food particles being taken up by phagocytosis by the cells lining the absorptive region and being digested within vacuoles. This is prevalent in the poriferans, cnidarians, flatworms and in various other animals in which the food is in finely particulate form by the time that it reaches the mid-gut. In more complex animals, and particularly in those species ingesting large individual masses of food, much digestion proceeds via enzymes secreted into the lumen of the gut and the products of digestion are then absorbed by the cells of the gut wall. In such species, the diverticula of the gut are normally purely secretory, and the intestine is the site of absorption. Extracellular digestion permits enzymatic specialization both of the individual cells and of different regions of the gut, but it also necessitates the production of a greater total quantity of enzymes since the lumen of the gut is a large open system in which it is difficult to maintain optimum concentrations of digestive activity.

Finally, the hind-gut (if present) is formed by the rectum, in which water may be absorbed (in some terrestrial animals; see Fig. 8.27) and faecal pellets may be formed prior to discharge through the anus.

From their food, all animals require energy-yielding compounds for immediate and later use, amino acids for synthesis into structural and metabolic proteins, and various other elements and compounds, such as vitamins, for use as catalysts or in other biochemical reactions. That is to say that they require to ingest carbohydrates, lipids, proteins and vitamins, and to digest them into a form suitable for absorption, if they cannot be absorbed directly. Animals with a diet low in proteins and/or vitamins normally require symbiotic gut bacteria for the synthesis of these compounds, and others ingesting complex carbohydrate polymers may also require symbiotic bacteria and protists to ferment them into simpler organic molecules. Much of the mass of animal faeces is composed of gut bacteria.

9.2.3 Hunters and parasites

Hunters are mobile animals which attack, kill and consume individual prey items one at a time, almost invariably other mobile animals. Three broad types can be distinguished: pursuit hunters, such as squid, which chase after, catch and subdue highly mobile prey; searchers, such as many gastropods and arthropods, which actively forage, seeking out prey items with less-developed powers of movement than themselves; and ambushers, such as spiders and preying mantises, which, apart from the final dart or spring, may be relatively sedentary.

Pursuers and ambushers (and to a lesser extent searchers) possess weapons of prey capture and immobilization. Most characteristically, these are either organs surrounding the mouth region, e.g. chelate or subchelate appendages in arthropods, sucker- or hook-bearing arms in cephalopods, and spines in chaetognaths (Fig. 9.10), or are powerful jaws associated with the anterior gut, in annelids mounted in an eversible pharyngeal region which can be forcibly and rapidly shot out to catch a prey item. Some ambushers attract their prey to them by mimicry: a number of siphonophores, for example, possess tentacles bearing copepod mimics, and they prey largely on copepod feeders (mostly other crustaceans). Once caught, the prey can be ingested whole, torn apart by appendages and consumed piecemeal, or have its body fluids sucked out. If ingested whole, the anterior region of the gut is capable of being distended to accommodate the meal.

Many searchers, on the other hand, feed on relatively sedentary prey which may protect themselves from attack by external coverings of calcium carbonate, cellulose, chitin, etc. (Sections 9.3.3 and 13.2.1). Searching hunters, if they have specialized predatory techniques, are therefore adapted: to bore through protective casings (e.g. by use of the radula in molluscs); to prise apart elements of the casing sufficiently to evert the stomach through the gap, then to secrete enzymes on to the unprotected tissues and to absorb the products of the extracorporeal digestion; to swallow the prey whole and crush the shell in the gizzard; or to suck out individual polyps or zooids from the communal matrix by means of a proboscis or stylets and a pharyngeal pump (e.g. several opisthobranch molluscs, pycnogonids).

Several types of hunter are suctorial feeders. In addition to the groups just mentioned, several predatory insects and above all the spiders either suck the fluids directly from their prey or inject salivary proteolytic enzymes into captured individuals (together with paralysing toxins) which liquefy the tissues so that they may be pumped into the predator. It is clearly but a small step from suctorial feeding of this type to an ectoparasitic life style, feeding on a host's fluids without killing it. The categories of hunter and parasite merge cleanly into one another, and in large measure their differentiation is simply a matter of relative sizes of consumed and consumer. A species of leech sucking the blood of a large mammal is unlikely to cause the death of the prey attacked by withdrawing what to the mammal would be an insignificant volume of blood; a different species of leech, on the other hand, sucking the blood of a small pond-snail would, by so doing, kill it and would therefore qualify as a hunter. Even amongst equally sized animals there are problems of categorization. Some planktonic polychaetes attack and consume the head ends of arrow-worms, an attack which the prey survives, later regenerating its missing head. Is the polychaete a predator or a parasite?

A similar argument applies to many endoparasites. Several hymenopteran insects, for example, complete part of their life cycle within other animals and kill them in the process. Such 'parasitoids' consume their prey slowly from the inside towards the surface, rather than more rapidly the other way around as in the classic hunter. In these insects, the adult injects an egg or eggs into the prey individual (usually another insect), and the developing larval stage/s consume the host's tissues before pupating and metamorphosing. The adult stage is a typical hunter except in the sense that it does not do the consuming, only the attacking; its progeny are the consumers. Again, however, it is really only a question of relative size. A hymenopteran or dipteran larva is large relative to the size of the parasitized insect prey and therefore the host may only be just large enough to support from one to a few larvae through to pupation. But the smaller adult stage of, say, a trematode fluke or a nematode, although feeding on the host's tissues in an essentially identical fashion, is very small compared to a mammal host, for example, and it does not cause its host's death. Because the two feeding patterns are not distinguishable, except in extreme form, it is not surprising that many groups of small predatory invertebrates also contain parasitic species.

A rather clearer distinction can be drawn, at least in respect of feeding biology, between hunters and parasites which consume the fluids and/or tissues of their prey on the one hand, and gutless endoparasites inhabiting a host's alimentary tract on the other, although again the evolutionary transition between an endoparasitic consumer and an endoparasitic absorber is not a very major one. Several parasitic worms, most notably the tapeworms and the acanthocephalans, lack guts but possess an outer cuticle which is disposed in a series of microvilli (termed microtriches) equivalent to those possessed by the absorptive cells of the guts of other animals. Within the alimentary canal of their host, these worms absorb the products of their host's digestive processes, their integument also protecting them from this enzymatic activity. Whereas the other types of 'parasite' considered above are in effect micropredators, tape-worms and acanthocephalans are genuinely parasitic in the normal English usage of that word in that they subsist on the resources obtained by another species. The same is true of some other gut parasites (including several nematodes which do have fully functional alimentary systems) that consume only the intestinal contents of their host. Yet other gut-inhabiting animals, however, consume the tissues of the gut wall and the host's blood and are therefore predatory.

Feeders on animal tissues and fluids ingest a readily digestible, protein-rich material, and accordingly typically they secrete proteases and possess relatively short, simple guts, specialized only – if at all – anteriorly, in respect of a crushing gizzard (if whole prey are ingested) or a pumping or bulk-storage crop. The

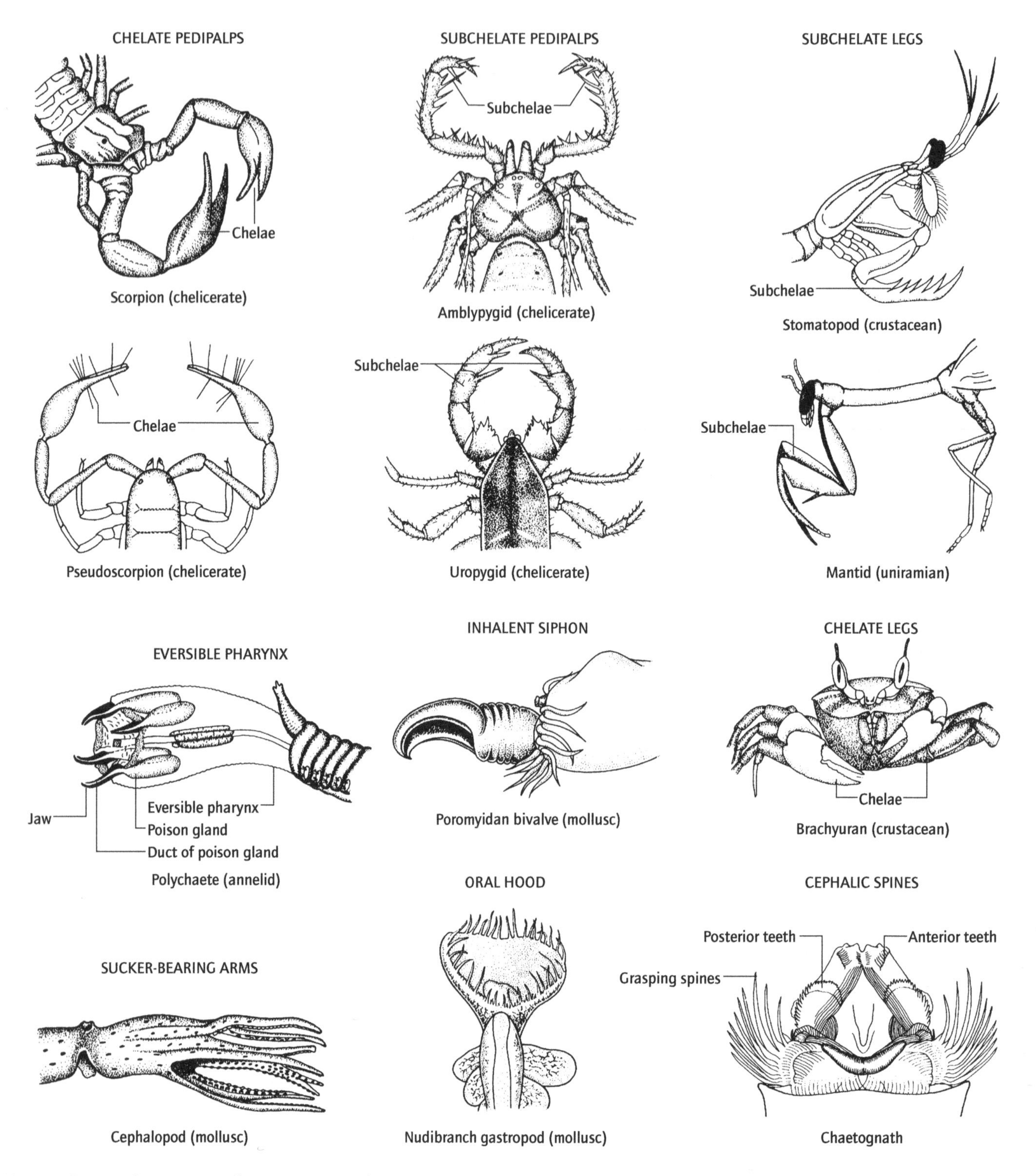

Fig. 9.10 Organs of prey capture (from many sources).

high protein content of animal material also makes it attractive to some essentially non-carnivorous animals at certain critical stages of their life histories. Some female dipteran insects, mosquitoes for example, require to take a blood meal to provide the protein to invest in eggs, although the males and the larval stages may not consume any animal food.

9.2.4 Grazers and browsers

Grazers and browsers are mobile consumers of sessile prey, cropping exposed tissues without, usually, killing the prey individual or colony. On land, the food sources are plants and fungi, but, in the sea, colonial animals (such as cnidarians, bryozoans or tunicates), bacterial colonies and multicellular algae can be grazed in an equivalent manner. Removal of the food materials requires the possession of hard biting or rasping mouthparts, e.g. the radular ribbon of molluscs (Fig. 5.3), Aristotle's lantern in sea-urchins (Fig. 7.24), the sclerotized jaws of insects (Fig. 8.30), etc. – although finding and acquiring the resources are not the major problems faced by grazers and browsers (in marked contrast to most types of hunter), since consumable material is often plentiful. The difficulties encountered by this category of consumer are: (a) the chemical defence systems evolved by the otherwise defenceless sessile prey species; and (b) the small proportion of digestible material in every unit weight ingested as a result of the abundant structural or protective refractory compounds in their prey, and, often, the protein-deficient nature of the utilizable organics.

The problem of the high proportion of indigestible carbohydrate polymers in the bodies of seaweeds and plants, e.g. agar, algin, laminarin, cellulose, has, as we have seen (Section 9.1), partly been solved enzymatically, in association with the evolution of elongate alimentary canals (especially the mid-gut region) to increase the area of those regions responsible for the digestion of refractory materials, and partly with the aid of symbiotic bacteria and protists housed in specialized gut compartments, often the hind-gut but sometimes the crop or stomach. A large storage crop is often also present.

The alimentary symbionts ferment the polysaccharides anaerobically, releasing fatty acids and other simple carbohydrates which can be absorbed. This system reaches its greatest development in those termites which consume the most refractory of all natural organic materials, wood. In these, the hind-gut is large, larger than the whole of the rest of the gut (Fig. 9.11), and it contains a dense culture of hypermastiginan flagellates. These ingest the wood particles phagocytically and themselves contain symbiotic bacteria which are probably mainly responsible for digesting some of the cellulose content of the wood. The lignin component is probably indigestible. The isopod crustacean *Limnoria* (the gribble) is another wood feeder, but it appears to lack gut symbionts, digesting some of the cellulose and hemicellulose components of the wood with its own enzymes. Structural carbohydrates, if they can be digested, provide an abundant source of energy-yielding materials, but relatively little else. In *Limnoria*, the required proteins must be derived from fungi infecting the wood consumed; wood without any such decomposer organisms already colonizing it cannot sustain the animal.

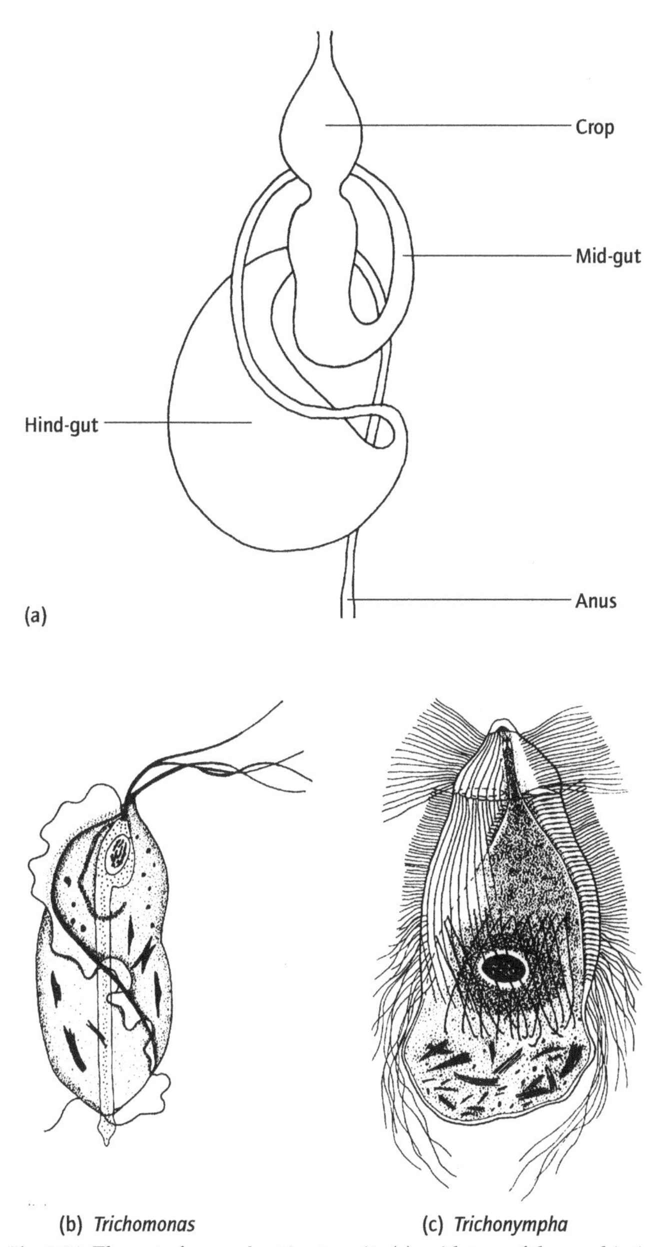

Fig. 9.11 The gut of a wood-eating termite (a), with two of the symbiotic flagellates (b) and (c) found in its hind-gut (a after Morton, 1979; b and c after MacKinnon & Hawes, 1961).

Scale effects also operate in the consumption of relatively non-nutritious plant materials. Large animals, i.e. vertebrates of rabbit size or larger, have a low specific metabolic rate, reducing the requirement for energy per unit body weight, and are able to store larger quantities of herbage in their gut fermentation chamber because of their large body volume; being homoiothermic, mammals are also able to maintain efficient fermentation temperatures. They are therefore able to subsist on coarse grasses, etc., ingested in bulk. Small animals, such as herbivorous invertebrates, must, however, feed on higher quality material and must gain entry to the plant cell contents by piercing the cell, rasping away the cellulose wall with a radula, or biting through it with their mouthparts. Bulk ingestion is generally not possible, simply by virtue of their small size. Even small birds and mammals cannot subsist on low-grade materials: if herbivorous, the diet must be restricted to energy-rich seeds and similar items.

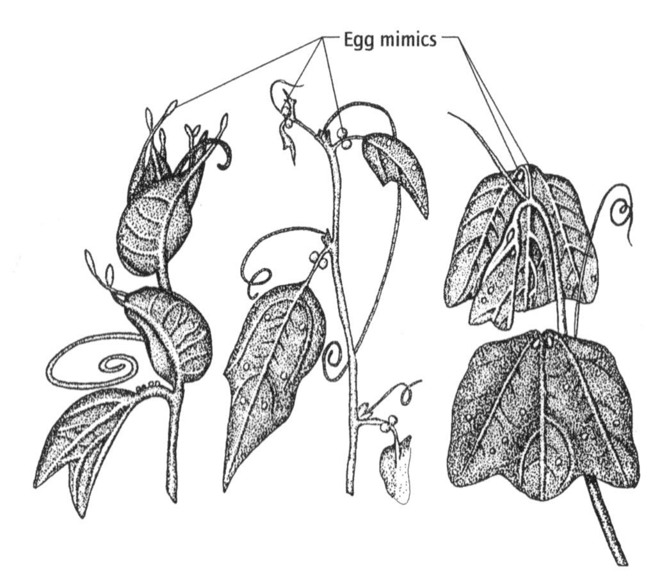

Fig. 9.12 The leaves and stalks of various passion-flower vines bear egg mimics to deter female *Heliconius* butterflies from depositing eggs on them (leaves already bearing eggs are avoided by the butterflies, probably because newly hatched larvae are cannibalistic). (After Gilbert, 1982.)

The chemical defences of plants are many and varied, including alkaloids (e.g. nicotine, cocaine, quinine, morphine and caffeine), glucosinolates, cyanogenic glycosides and tannins, and these substances would appear to be deployed specifically to deter consumption of their susceptible tissues. Some of the chemicals are straightforwardly toxic – natural insecticides, for example derris and pyrethrum. Others, e.g. the tannins, bind with proteins when released, rendering them indigestible and further reducing the protein status of the food ingested. Yet others mimic closely the consumer's own hormones or pheromones, adversely affecting growth, development or reproduction, and initiating inappropriate behavioural reactions. In some plants, these chemical defences are mobilized only when grazing pressure starts, so as to avoid the diversion of resources away from growth and reproduction when this is unnecessary, and it has recently been shown that in at least one case adjacent plant individuals of the same species can react, by chemical mobilization, when a nearby individual is attacked by grazers, even though they themselves have not yet been grazed.

Other plant defences are structural, for example hair-like projections which release sticky secretions or noxious chemicals when triggered, and even anatomical mimetic resemblances to leaves already infested with consumers (Fig. 9.12), or symbiotic, especially as seen in the associations of a variety of angiosperms and other tracheophytes with ants. The ants remove browsing insects from the plant by interference competition 'in exchange' for the provision of nectar and of nest sites. Comparable structural, chemical and other defence systems are displayed by marine animals subject to grazing pressure, although as yet these have been little studied.

Nevertheless, consumers have been coevolving with their sessile prey for many millennia and individual grazers and browsers have evolved the ability to detoxify, avoid, sequester or excrete the specific defensive chemicals of their particular prey species, to bypass the structural defences, and to escape the attention of the defending ants by chemical mimicry of these insects. In spite of this, however, the huge biomass of living plant tissue present in most terrestrial habitats suitable for plant growth, and of seaweeds in kelp-forests and algal beds, must indicate that much macrophyte tissue remains effectively unavailable and/or unusable by grazers and browsers, and is only an adequate diet after death and decomposition.

Grazing and browsing are not the only feeding techniques appropriate to the consumption of macrophyte material. Some herbivores in both the sea (e.g. some opisthobranch molluscs) and the land (e.g. many hemipteroid insects) have circumvented the structural carbohydrate problem by sucking out the contents of individual cells or by inserting cannula-like mouthparts into the xylem or phloem vessels of tracheophytes (see Figs 9.8 and 8.30). In the latter case, hydraulic pressure in the transport vessel may be sufficient to pump the fluid directly into the gut of, for example, an aphid, and aphids parasitize the host plant in much the same way as does a tick or a female mosquito its animal host. Indeed, as noted above with respect to hunting carnivores, it is but a small step from ectoparasitically feeding on a larger host to becoming endoparasitic, and several nematodes, for example, are endoparasites of plants in the same manner as are other nematodes of animals, and several insect larvae live within plants, there feeding on plant tissues by endoparasitic grazing.

Even when equipped with cellulases and equivalent enzymes or with symbiotic microorganisms in the gut – and even more so in the absence of both – digestion and assimilation of macrophyte tissue are typically very inefficient, and in the case of liquid feeders, through-put of carbohydrate may be greatly in excess of requirements. Faecal production by these categories of consumer is therefore copious, and the faeces contain much unassimilated organic matter. They therefore provide an important ecological pathway by which materials photosynthetically fixed by living macrophytes are made available to other categories of animal consumer, especially to the deposit feeder.

9.2.5 Suspension feeding

The essential features of the means by which suspension feeders filter their food from water have been outlined in Section 9.1. In groups utilizing a non-anatomical external secretion (see Fig. 9.4), there appear to be no further specifically suspension-feeding adaptations. Such animals are perhaps more nearly equivalent to ambushing hunters trapping relatively small individual prey in nets, at least in so far as their gut anatomy is concerned. Certainly the two feeding modes do intergrade: a web-building spider could equally well be argued to be an aerial suspension feeder or an ambushing hunter; and the cnidarians straddle the interface between the two in the sea. They possess radially symmetrical rings of tentacles around the centrally located mouth and the polyp phase, at least, is sessile – both characteristic suspension-feeding adaptations – and indeed their prey are mainly zooplanktonic animals suspended in the water. The separate prey items are not so much passively trapped, however, as attacked individually with nematocysts, albeit that the prey themselves discharge the nematocysts by accidentally touching them rather than the cnidarian initiating the attack.

In contrast, groups with a ciliary or mucociliary filter-feeding system also possess a distinct series of alimentary specializations. In most such species, those in which the filtration apparatus is either freely projecting into the water or else is protected within an external shell (as in the brachiopods and the bivalve molluscs) (see Fig. 9.2), the fore-gut is simply a short region connecting the external filter to the stomach. Most unusually, however, in the chordates, the fore-gut is the site of the filtration process itself and hence is highly specialized. The sides of the pharyngeal wall of these animals are perforated by numerous small openings, 'stigmata', which extend right through the body wall to open on the body surface. Water taken in through the mouth can then pass into the pharynx, through the stigmata and back to the environment in a unidirectional flow. Although unusual in that the pharynx is the site of filtration, this system has some parallels amongst other animals which probably indicate its evolutionary origins. Feeders on small particles often use a water current to drive collected particles into the mouth, and this transport stream has to be expelled in some way. In the lophophorates, it appears that the water is simply 'regurgitated' at intervals through the mouth, and the ingestion of particles is then temporarily suspended. In the macrodasyid gastrotrichs (Fig. 4.9) and the cephalodiscid hemichordates (Section 7.2.3.2), however, the pharynx bears a pair of perforations extending through to the body surface through which such ingested water can be discharged without interrupting food intake. The early chordates would appear to have adapted a similar current for direct filter feeding, thereby doing away with the necessity of withdrawing an external lophophore-like organ at times of danger from predation, etc. (which also interrupts feeding), whilst the enteropneust hemichordates developed the same current for gaseous-exchange purposes, as indeed did the later aquatic chordates.

The pharynx of suspension-feeding chordates is very large, comprising the majority of the body volume. Being perforated by thousands of stigmata in many species, the body wall in the pharyngeal region is almost non-existent and would be completely ineffective. Accordingly, it has been replaced by a secondary or false body wall, formed by folds of tissue in the cephalochordates and by a secreted cellulose test in the tunicates. The body proper is therefore wholly or partially surrounded by a morphologically external cavity, the atrium, between the true and false body walls, into which water passes after flowing through the stigmata, and from which water is discharged to the

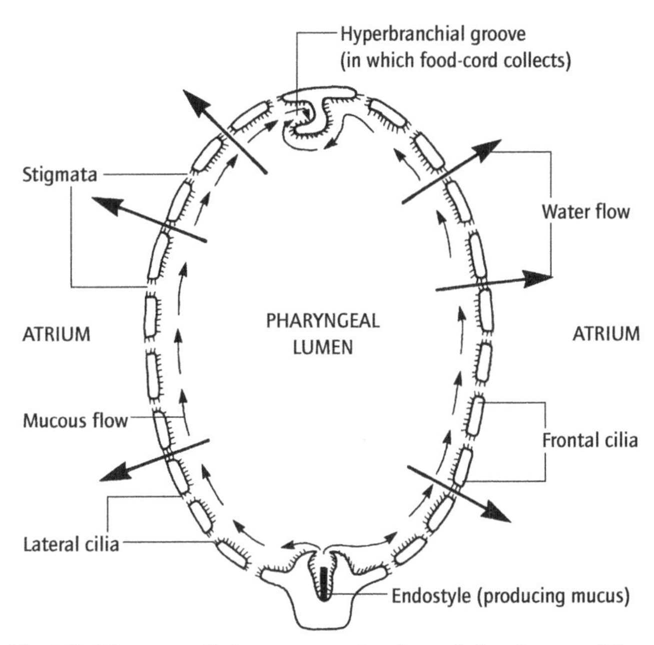

Fig. 9.13 Diagrammatic transverse section through the pharynx of the invertebrate chordates showing the nature of the filter-feeding system (after several sources).

environment via a single aperture, the atriopore or atrial siphon (Fig. 9.13).*

Mucus is produced by a gland, the endostyle, running along the ventral mid-line of the pharynx, and this mucus is induced to move up the perforated sides of the pharynx in thin sheets. The mucous sheets intercept potential food particles, in some cases down to 0.5 μm in size, as the water flows out of the pharynx through the stigmata. Eventually, the food-laden sheets meet dorsally where they are formed into a longitudinal cord in the mid-dorsal hyperbranchial groove. All the motive power in this system, whether of water or of mucus, is provided by tracts of cilia in the pharynx, except in the pelagic salps.

In the pelagic tunicates generally, the feeding current also provides the means of propulsion, the oral and atrial apertures being located at opposite ends of the body. In association with this jet propulsion in the salps, the numbers of stigmata have been reduced down to only two, the mucous sheets forming a single, internal, conical net slung across the lumen of the pharynx (Fig. 9.14) with the motive power of the water being provided by hoops of muscle around the body (Fig. 7.35).

Not only in these invertebrate chordates, but also in the lophophorates and in the filter-feeding molluscs, food passes from the filtration organ (of whatever type) into the stomach

* Essentially the same system is, of course, present in vertebrate chordates. In fish, the gill slits (≡ stigmata) perforating the body wall from gut to exterior are fewer, however, and the body wall is thicker, so that an additional false body wall is unnecessary, and the water current is used for gaseous-exchange purposes, not primarily for suspension feeding.

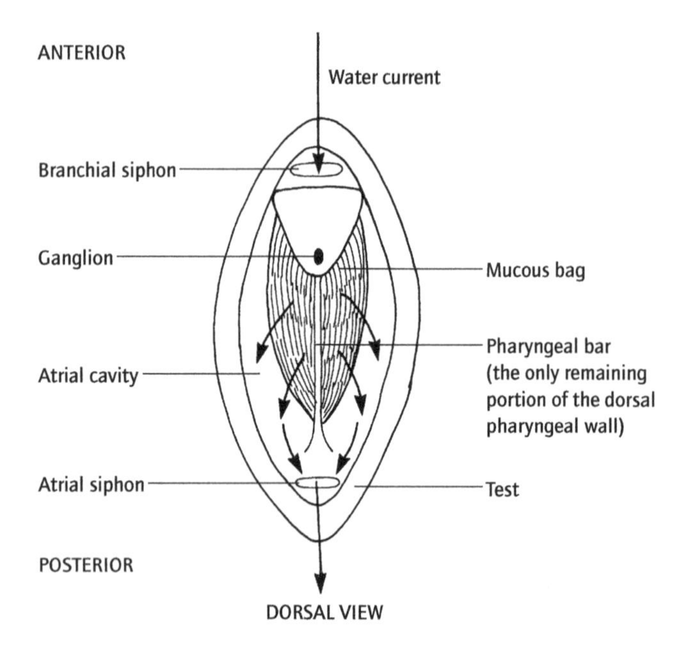

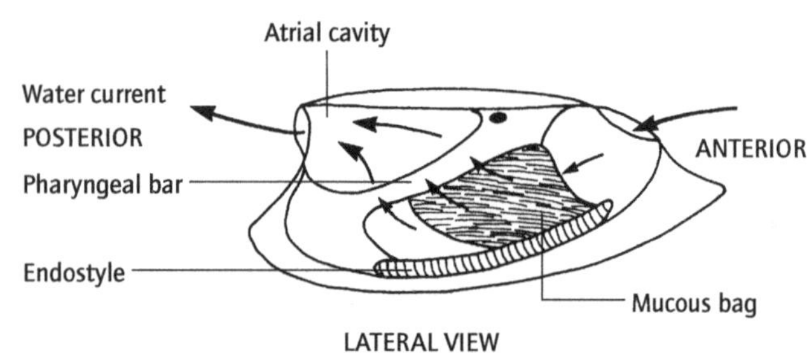

Fig. 9.14 A salp showing the mucous bag suspended across the two large pharyngeal openings (after Berrill, 1950 and others).

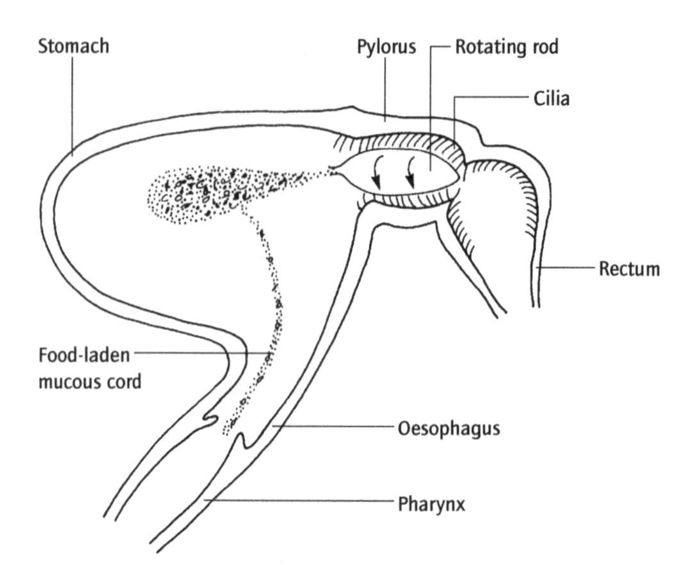

Fig. 9.15 The stomach of a bryozoan lophophorate showing the mucofaecal rod used to winch food into the gut (after Gordon, 1975).

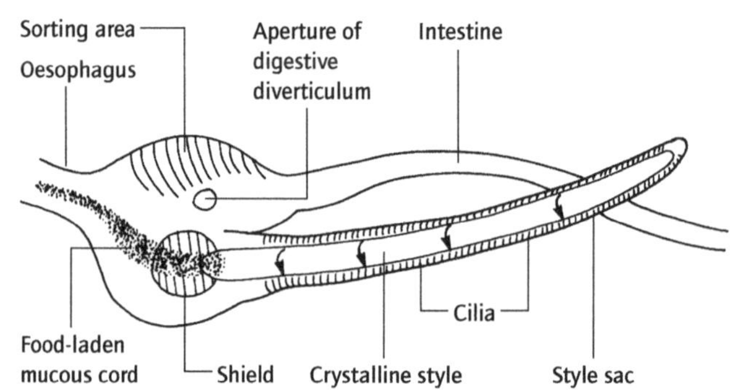

Fig. 9.16 The stomach of a filter-feeding mollusc showing the crystalline style used to winch food into the gut (after Morton, 1979).

in the form of a cord of food-laden mucus. In all mucociliary feeders, the pH of the stomach lumen is acid. This reduces the viscosity of the mucous cord, permitting the trapped food particles to be released. Digestion of the food and absorption of the products takes place within the stomach and/or in the mid-gut diverticula issuing near it. Undigested material passes through into a generally short intestine, in which the pH is alkaline, increasing the viscosity of the mucus again and rendering more easy the production of faecal pellets or strings.

The motive power to transport the mucous cord into the stomach is also provided by cilia, acting directly on the cord in the chordates, but indirectly in the lophophorates and molluscs. In both these latter groups, a rotating rod projecting into the stomach winches in the cord. The rod of the lophophorates is formed of mucus and faecal material and is housed in the pylorus, the cilia of which rotate it (Fig. 9.15). Every so often, the used rod is passed into the intestine, converted into a faecal pellet and voided, a new rod being then formed.

In the molluscs, the rod, known as the crystalline style, is much larger, more permanent, and is composed of a hyaline mucoprotein. It also projects into the stomach, but in the molluscs from a special diverticulum, the style sac, the lining sac cilia rotating the style. Not only does the tip of the style act as a winch (Fig. 9.16), but the tip gradually dissolves in the stomach releasing the enzyme amylase. As the tip is lost, the style is slowly moved forward into the stomach (again by ciliary action) and more style material is secreted proximally so as to maintain constant stylar length. The tip is therefore maintained in contact with the stomach wall against which it rotates: this has the effect of a rotary pestle and serves to free particles from the now less viscous cord and distribute them over a gastric sorting region in which cilia can further subdivide the particles collected.

9.2.6 Deposit feeding

Probably the majority of living animal species consume detritus (organic material of small particle size) or litter (material of larger dimensions), but, in spite of this, the precise nature of their diet is still uncertain. A leaf ingested by, say, an earthworm is not just a dead leaf; it is a whole ecosystem in microcosm. On and in the leaf tissue will occur the bacteria and fungi responsible for its decay, together with various protistan consumers of the decomposer organisms (e.g. amoebae, ciliates, heterotrophic flagellates) and some only slightly less microscopic animals such as nematodes and mites feeding on the smaller organisms. On the surface of the leaf may well occur photosynthetic algae and cyanobacteria, and the dead and decomposing remains of other organisms, not least those derived from animal faeces. All of these living and dead components may be ingested by the consumer with the leaf, and the problem has been to determine which are digested and assimilated, and which meet most of the metabolic requirements of the consumer.

Nor is it necessarily the case that the consumer simply ingests background material and digests out whatever it can. Several deposit feeders are now known to be capable of much greater degrees of selective ingestion than was once thought to be possible; although few species in total have yet been re-examined in this light.

Selective ingestion would certainly appear to be advantageous, if possible, since the nutritive values of different elements in the aggregate 'detritus' vary very widely. Comparison of the organic content of the sediment ingested by some unselective marine deposit feeders and of the faeces resulting after consumption, and exposure of samples of the sediment to enzymes known to be present in their guts, have both indicated that in such species the majority of what little organic matter is present in the sediment is not available to them. Less than 5–10% of the organic detrital pool may be digestible. (In clean sandy beaches, organic matter of *any* sort may comprise a very small proportion – <1% – of the ingestible material. The litter layer in a forest may be richer in organic matter, but still relatively little of it may be nutritious.)

By the time any piece of organic debris becomes incorporated into the surface layer of litter or detritus on the substratum, it will have already lost most of its original food value. If a leaf, then the parent plant may have translocated all the soluble compounds into its perenniating tissues before shedding that leaf. Leaching will rapidly cause the loss of other organics in the first few hours after shedding. Any fragments of organic matter may only arrive on the substratum after passage through the gut of a consumer, during which most of the utilizable substances may have been removed: faecal material is one of the major sources of organic matter in sediments or soils. Therefore, the organic material remaining in an aged item of organic debris is likely to be in the form of those refractory structural, skeletal or protective substances that defy most animal digestive systems – detritus feeders do not generally possess cellulases, for example.

If, however, an item of debris is recent in origin, deposit feeders may be able to digest some of its contained organics, although even so its protein content is liable to be either very low or unavailable as a result of the presence of tannins (see Section 9.2.4). In general, though, it would appear most likely that consumers of dead organic matter (excepting scavengers of animal carcases) are reliant on the living associates of the debris, not on the debris itself which may merely be a suitable vehicle to transport the micro-organisms into the gut. On the land and in fresh water, those fungi which are responsible for litter degradation are probably of primary importance as the real food source of deposit feeders, whilst, in the shallow fringes of the sea, photosynthetic unicellular or mat-forming protists are probably the elements of greatest nutritional import. Selective feeders capable of ingesting solely diatoms have been shown to be able to achieve a 70% assimilation efficiency, in comparison to less than 4% in related species feeding unselectively on the background organic pool. Throughout most of the sea, however, under circumstances in which photosynthesis is impossible, in which the fallout is sparse, and in which by the time debris reaches the sea bed only refractory substances remain, animal consumers must be dependent on bacteria, both for energy-yielding and proteinaceous materials.

The guts of deposit feeders are typically unspecialized, with a tendency towards elongate intestines (see Figs 4.40 and 4.43), for the same reason as in the grazing and browsing category. Like the latter too, faecal production may be copious although of much less nutritive value to other consumers.

9.2.7 Food from symbionts

Although relatively few animal species are totally dependent on the photosynthesis of endosymbionts for their nutrition in the manner displayed in Fig. 9.17 (see Section 9.1), many species, ranging phylogenetically from sponges to molluscs and from cnidarians to chordates, derive some nutritional benefit from the symbiotic photosynthesis of cyanobacteria (e.g. some sponges and echiurans), prochlorophytes (tunicates), or the more widespread dinoflagellates (present in the form of zooxanthellae in many marine invertebrates) or chlorophytes (in the form of zoochlorellae in mostly freshwater species). Some stony corals, for example, obtain two-thirds of their metabolic requirements

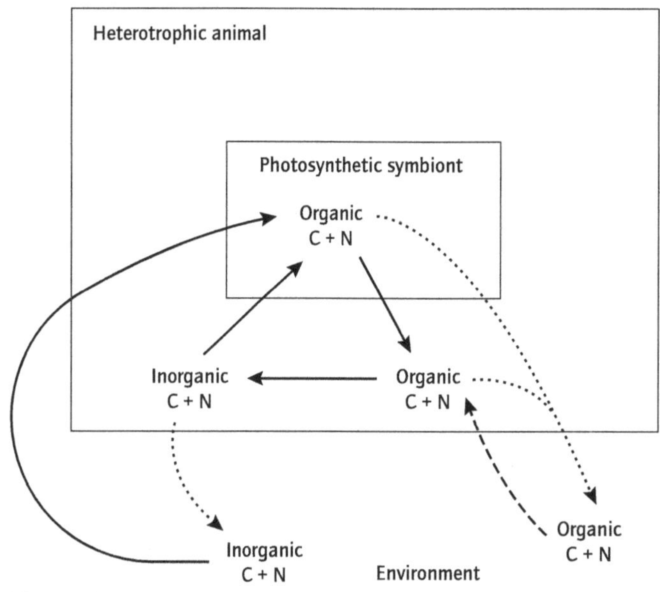

Fig. 9.17 Diagrammatic representation of carbon (C) and nutrient (N) fluxes in symbiotic interactions (after Barnes & Hughes, 1982).

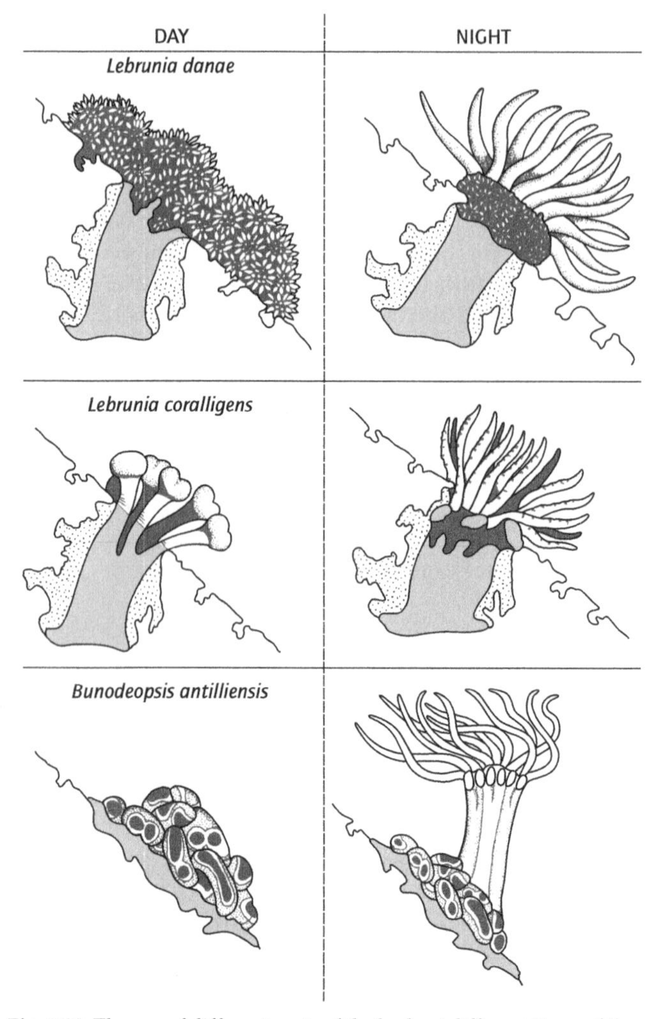

Fig. 9.18 The use of different parts of the body at different times of the day by three coral-reef anemones: zooxanthellae-containing structures during daylight (left-hand column); and nematocyst-bearing tentacles at night (right-hand column) (after Sebens & De Riemer, 1977).

from the intracellular zooxanthellae, and only one-third from external sources including the typically cnidarian capture of zooplankton. In a number of sea-anemones, the symbiotic zooxanthellae and the organelles of prey capture are located in different parts of the body: the symbiont-bearing 'organs' are extended during the hours of daylight and the nematocyst-bearing tentacles at night (Fig. 9.18). Symbiosis can also supplement herbivory. Sacoglossan molluscs feed suctorially on the cells of macroscopic green (and other) seaweeds, and some species can sequester intact and still functional chloroplasts from their prey into their digestive mid-gut diverticulum (Fig. 9.19). The chloroplasts become engulfed by the phagocytic digestive cells but in them continue to function photosynthetically for more than 2 months in some cases. Up to half of the carbon fixed passes to the mollusc and this may be sufficient to satisfy its respiratory requirements. (The chloroplasts of some *Euglena* may have had an evolutionary origin in a similar process.)

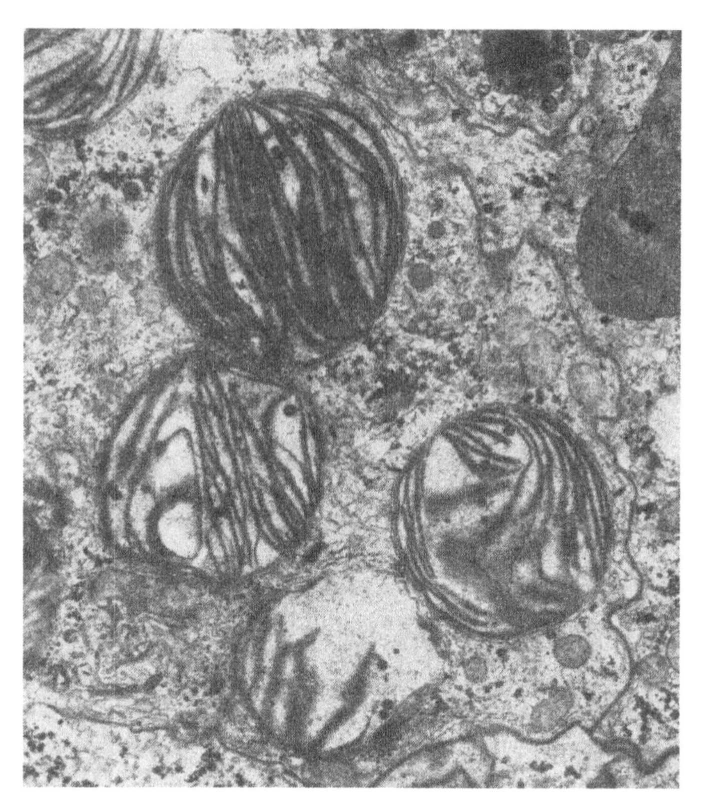

Fig. 9.19 An electronmicrograph of functional chloroplasts in the digestive diverticulum of a sacoglossan mollusc (from Trench, 1975).

A rather different symbiotic relationship occurs between plants which secrete nectar and insects which gather to consume this sugar-rich fluid and thereby effect pollination. Several adult insects consume only nectar (to provide the energy required for flight); their larval stages are the feeding phase of the life cycle and they accumulate sufficient resources to permit the assembly of the adult body which then requires only energy-yielding substances during its short existence.

Finally, some relatively large, free-living animals associated with hydrothermal vents, cold seeps and regions of hydrocarbon discharge in the sea bed lack guts at all stages of their life history (e.g. a few bivalve molluscs, oligochaete and polychaete annelids and all pogonophorans) and their mode of feeding is dependent on symbiotic chemoautotrophic bacteria often located in endodermal tissues that would in related groups have formed the gut.

9.3 Costs and benefits of feeding: optimal foraging

9.3.1 Introduction

We have seen that animals are clearly constrained in their types of diet by their evolutionary pasts: an animal in a lineage of suspension feeders could not easily adapt to pursuit hunting nor vice versa. Nevertheless, in many species there is a degree of flexibility possible, and in some generalist feeders this may

Table 9.1 Feeding preferences of *Hediste diversicolor* and rank order of assimilable energy per food item.

Feeding mode and food type	Preference order	Rank order of assimilable energy per item
Scavenging on dead *Macoma* (bivalve mollusc)	1	1
Hunting live *Tubifex* (oligochaete annelid)	2	2
Hunting live *Corophium* (amphipod crustacean)	3	4
Hunting live *Erioptera* larvae (dipteran insect)	4	3
Deposit feeding on surface sediment particles	5	5
Suspension feeding on particles in the overlying water	6	6
Browsing live *Enteromorpha* (green alga)	7	7
Browsing live *Ulva* (green alga)	8	8
Hunting live *Hydrobia* (gastropod mollusc)	9	9

even permit switching from one feeding mode to another, and from one prey type to a completely different sort of food. The estuarine polychaete *Hediste diversicolor*, for example, can suspension feed (Fig. 9.4d), deposit feed, behave as a hunter or a scavenger, or browse pieces from macroscopic algae (Table 9.1).

All animals encounter a variety of potential food items during their daily lives and they are therefore faced with alternative food sources. For the most specialist of consumer, this may only be different individuals of the single prey species taken, but more generally the alternatives will at least include individuals or material of more than one food species. Hence, within limits set by their evolutionary pasts, animals face alternative courses of action, or 'choices', in ecological time – from minute to minute, hour to hour or day to day – in terms of whether to catch and consume a given item or to reject it in favour of another in space or time. Such 'decisions' are not without important potential consequences, since the benefit to be derived from consuming different items will vary. Some will be more nutritious than others. The most nutritious are also likely to be favoured by many consumers, and this is likely to give rise to intense utilization and resultant scarcity of that resource. Others are low in energy and nutrient content, and may therefore be generally less favoured foods but correspondingly relatively abundant and more readily available in any given habitat.

Nutritive value (per unit weight consumed) is only one element in any potential choice of food item, however, since besides yielding the obvious benefits, feeding activity also incurs costs. Whatever the form of the feeding process, time and energy have to be expended on finding, consuming, processing and digesting the food materials, and this time and energy could have been devoted to other purposes. The precise nature of feeding costs will vary from feeding type to feeding type, but in all cases one would expect selection to act in favour of a feeding strategy that maximizes the net gain (benefits minus costs) obtained. Food is one of the prime requirements for survival and hence any individual consumer maximizing this net gain would be at a selective advantage over individuals behaving differently, in respect of both survival and reproduction, since it would be more likely: (a) to be able to devote less time in total to foraging (thereby decreasing predation risk); (b) to grow faster or to a larger overall size (with consequent reproductive advantages); (c) to be healthier and therefore the more able to withstand parasitic infection, to escape from predators, etc.; and (d) to derive a greater benefit from fewer resources at times of food shortage.

How then can consumers maximize their net gain per unit time? This has been investigated via simple models of optimal foraging strategy and by testing the predictions of these models through experiment and observation. As is usual in the design of experiments, this has proceeded by isolating one particular variable, in this case prey choice. It should be remembered, however, that in the real world, an animal's activity at any one time is a compromise between many conflicting pressures. Maximization of net gain from feeding will in itself be advantageous, but so will be avoidance of being eaten by predators, breeding behaviour, and so on. These other desiderata may well place constraints on feeding and render consumption of prey less than optimally efficient: a consumer may have little option but to consume anything it finds in the brief period in which feeding is possible.

Further, the living food of animal consumers does not simply wait for its chance to nourish consuming species! Indeed, living prey have a greater vested interest in not being consumed than the consumer has in feeding on them; this has been termed the 'life/dinner principle'. In any encounter between, say, a cuttlefish and a shrimp, the cuttlefish has only its dinner at stake whilst the shrimp is fighting for its life, and selective pressures will vary accordingly.

In the following paragraphs, we will discuss how consumers could best maximize their net gains and how the potentially consumed could minimize that gain, thereby decreasing their own risk of being eaten. Mobile consumers, free to accept or reject discrete individual prey items, form the simplest case and we will devote most attention to that category; suspension and deposit feeders face rather different problems and so we will treat them separately.

9.3.2 The theory of optimal foraging

Any potential food item will have a particular 'food value' to a consumer. Animals require both energy-yielding compounds from their diet and those that will permit growth of somatic and reproductive tissues and cells, and so this overall food value should take both of these requirements into account. Unfortunately, the two may be required at different times of the year or during different stages of the life history, so that at some point energy may be the overriding factor, at another point organic nitrogen may be in particular demand, whilst for some species (e.g. land snails) there may be periods in which inorganic elements or compounds (such as calcium) may override all other requirements.

In practice, most work has used the energy content of the food as a convenient measure of its food value, and, in these simplified terms, the food value can be expressed as the energy which would be gained by consuming a given item (E_g) minus all the energetic costs associated with catching, subduing, consuming and digesting it (the 'handling cost', E_h). In addition, different potential food items will occur in any given habitat at different frequencies, and, if a consumer has a preference for some particular type of item, it may have actively to search for it and thereby incur a 'searching cost' (E_s). The net energy gain will therefore be:

$$E_g - E_h - E_s.$$

Alternatively, time, which can be more easily measured, may be substituted for energy on the debit side of this expression, and the food value given as energy gain per unit time. This will be E/T_h, if a prey item has already been encountered, or $E/(T_h + T_s)$ if a search-time element is included (where T_h is the 'handling time', T_s is the 'search time', and E is the available energy content $\equiv E_g$ above).

Let us take the simpler case first, in which there is no search-time element. A consumer encounters different potential food items at random in its habitat, and we are asking which of them it should eat in order to maximize its energy gain per unit time. The answer is that we would expect it to take preferentially those with the highest food value as measured by E/T_h (or, should nutrients, N, be the dietary requirement at that time, those with the highest value of N/T_h): consumers should choose the most profitable prey. This simple model gives rise to a number of testable and tested predictions:

1 A consumer should eat *only* the prey type with the highest food value, if the encounter rate with that prey type is sufficiently high that inclusion of other prey types (with lower food values) would decrease the average rate of energy intake.

2 If, however, the prey type with the highest food value is encountered at less than this rate, a consumer should expand its range of prey types consumed to include the next most valuable prey items, and so on.

3 If it takes a considerable time to distinguish different prey types (as, for example, when discrimination is by touch), lower value prey should be eaten if they are encountered frequently, even though higher prey may be still plentiful; whereas if prey recognition is instantaneous (as when discrimination is by sight), lower value prey should not be eaten if higher value foods are available instead. (Recognition of prey food value is simply an additional component of handling time.)

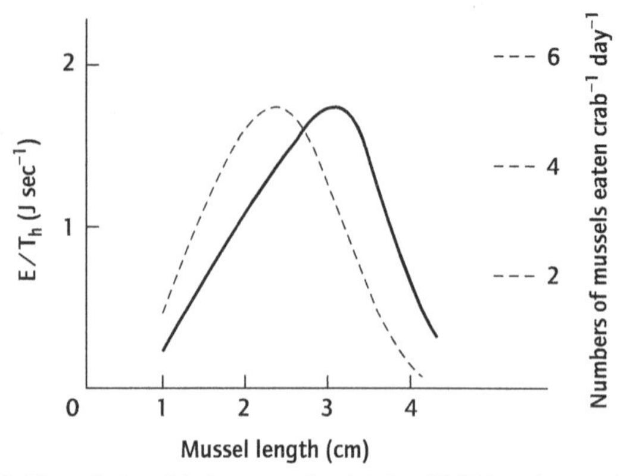

Fig. 9.20 The relationship between food value (E/T_h) and prey choice in 6.0–6.5 cm broad crabs (*Carcinus maenas*) feeding on mussels (*Mytilus edulis*) – see text and Box 9.1. (After Elner & Hughes, 1978.)

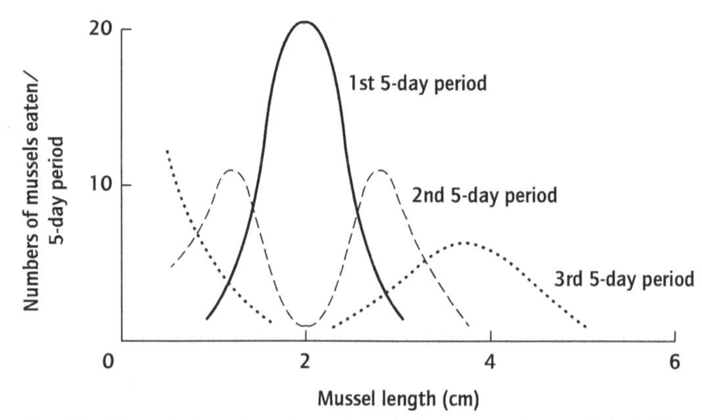

Fig. 9.21 Prey choice when the preferred size range of mussels (*Mytilus edulis*) could be depleted as a result of consumption by crabs (*Carcinus maenas*) – see text and Box 9.1. (From data in Elner & Hughes, 1978.)

All these predictions have been found to hold true, for example, in the shore crab, *Carcinus*, when feeding on mussels, *Mytilus*, of different individual sizes (see Figs 9.20–9.22; Box 9.1).

Tests involving a consumer and a variety of different prey species have been less frequently undertaken, but the polychaete *Hediste diversicolor*, for example, cited above as consuming a very wide range of different foodstuffs, does exhibit a preference hierarchy which corresponds well with the amount of assimilable energy in each food type (see also Box 9.1).

The basic model can now be made slightly more complex by the incorporation of different search times for different types of food. Let us assume that T_h is constant for items of given prey type and size, and vary the magnitude of T_s dependent on the frequency of various potential prey items in a habitat. Let there also be two types of prey, x and y, such that the food value of x is greater than that of y, i.e.

$$\frac{E(x)}{T_h(x)} > \frac{E(y)}{T_h(y)}$$

Clearly, if a consumer encounters prey of type x it should always eat it; it would never do better by rejecting it in favour of prey of

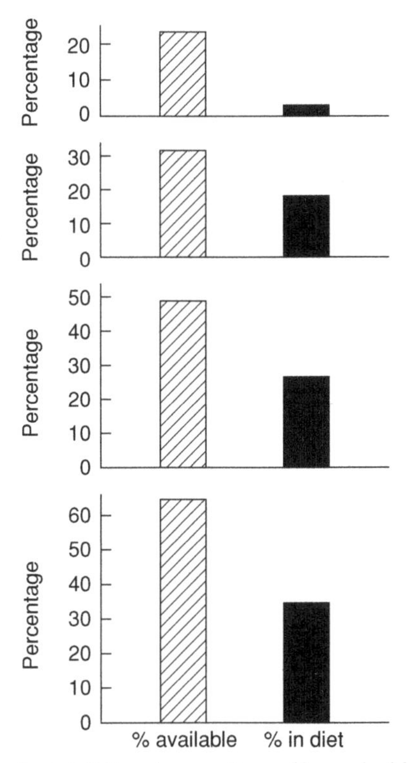

Fig. 9.22 The effect of different proportions of less valuable mussels (*Mytilus edulis*) on the preference displayed by crabs (*Carcinus maenas*) – see text and Box 9.1. (From data in Elner & Hughes, 1978.)

type y (all other things being equal). But what if the first prey encountered by the consumer is of type y: should it eat it, or reject it and continue to search for the more profitable prey of type x? To maximize the net rate of energy gain, the decision to accept or reject y is dependent solely on the frequency of type x prey, i.e. on the magnitude of $T_s(x)$. Prey of type y encountered should be consumed if

$$\frac{E(y)}{T_h(y)} > \frac{E(x)}{T_h(x) + T_s(x)}$$

and it should be ignored if

$$\frac{E(y)}{T_h(y)} < \frac{E(x)}{T_h(x) + T_s(x)}$$

In other words, a given prey item encountered should be consumed if, during the handling time concerned, the consumer would not be able to obtain a more profitable item.

A further complication is that many potential foods occur in highly clumped distribution patterns – in the form of discrete patches – for example, encrusting organisms on marine rocky substrata, ants in anthills, mysids in swarms, nettles in clumps, etc. Hence consumers face another type of problem: any given patch, region or area of local concentration will contain a finite amount of food, and consumption of that food may result in a decrease in its abundance and diminishing returns to the consumer. For how long, then, should a consumer remain in a patch before moving on to the next? Should it exhaust the patch completely first, or would it be advantageous to move before that point, and, if so, when?

Although superficially dissimilar, exactly similar problems are faced by other types of consumer. In high latitudes, for example, the amount of food available is dependent on the seasonal climate, and consumers have to decide when to migrate out of a geographical area of declining resources. When should the swallows leave? Several individual predator/prey interactions also fall into a similar category, especially where the consumer takes a considerable period of time to consume all the tissues of a captured item, and including those cases in which the consumer sucks the juices from its prey. With bees imbibing nectar from a flower, water-bugs sucking the fluids from mosquito larvae, and even lions feeding on an antelope, at first the rate of food intake from the prey is high but eventually it will fall when much of the easily obtainable and more nutritious materials have been ingested. When should the consumer leave the old item and obtain a new one? When all the food materials have been extracted, or before this point?

Box 9.1 Choice of mussel prey by crab consumers

By observation, it is relatively easy to quantify the time taken for a crab to break open a mussel and consume the flesh (T_h); and the energetic content of the flesh obtained (E) can be determined by bomb calorimetry. Crabs crush mussel shells with their chelae to open them, and there is therefore an appreciable handling time which increases more or less exponentially with mussel size. Large mussels thus take a very long time to open but have large values of E, whereas small mussels can be opened very quickly but contain little energy. For a given size of crab, E/T_h varies with mussel size as shown in Fig. 9.20; the same figure also displays the size range of mussels actually consumed by that size of crab. Clearly, the agreement between the mussels consumed and those with the greatest predicted food value is good, and prediction **1** (p. 228) holds true.

In this experiment any mussel eaten was replaced immediately by one of the same size, so that the highest value mussels were always maintained at a frequency high enough to result in depression of the average gain rate should a lower value prey by included in the crab diet. In a second experiment, however, consumed mussels were not replaced and the highest value mussels became reduced in frequency. The reaction of the crabs is shown in Fig. 9.21: as in prediction **2** (p. 228), they expanded their intake range to include the next most profitable sizes of mussels.

Crabs distinguish the sizes of different mussels by touch, and in a third experiment the relative proportions of different size mussels were varied. Figure 9.22 shows that, in conformity with prediction **3** (p. 228), less valuable mussels were taken in small numbers when they were abundant, even though optimally sized mussels were also plentiful.

When presented with different sized prey individuals of one prey type, *Carcinus* therefore appears able to select those sizes of mussels which maximize its intake of energy per unit time. In some way it must be able to assess the food value of different sized mussels.

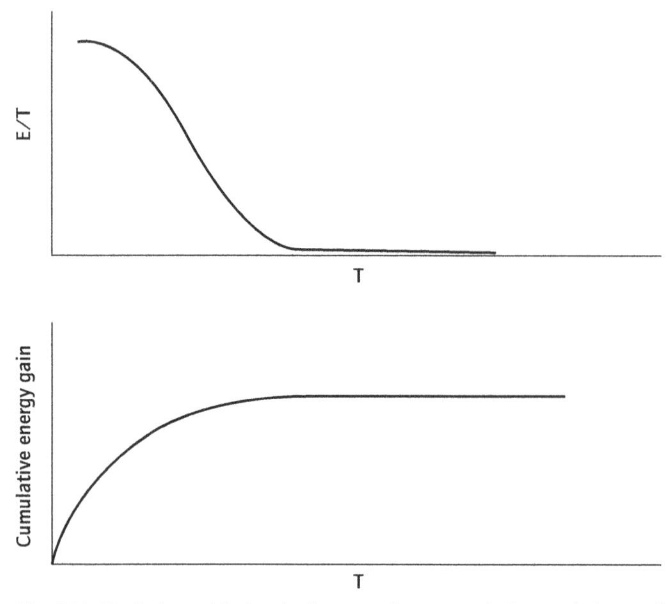

Fig. 9.23 Variation with time in the rate of energy gain from a finite pool – see text.

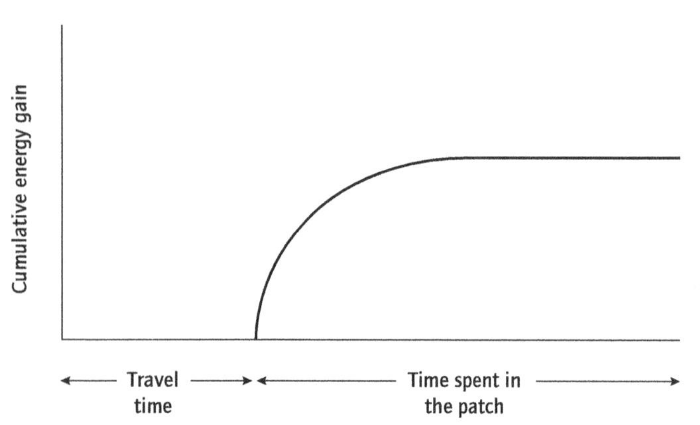

Fig. 9.24 Energy gain from an exploited patch in relation to the time spent within each patch and the time taken to travel to that patch – see text.

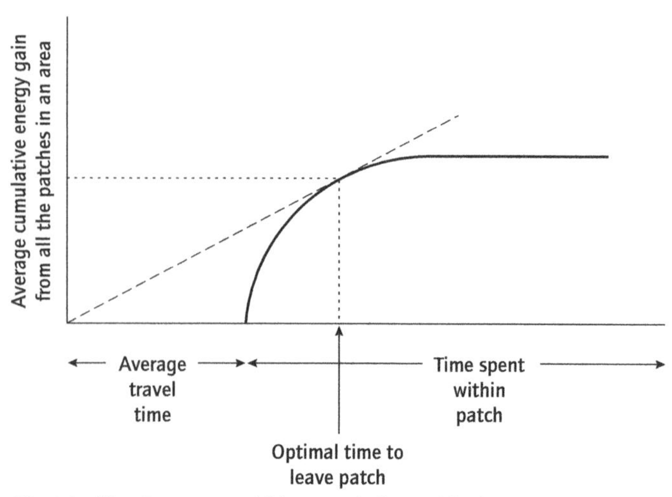

Fig. 9.25 The time spent within a patch that yields the maximum cumulative gain per unit total foraging time is given by the tangent from the origin to the gain curve – see text.

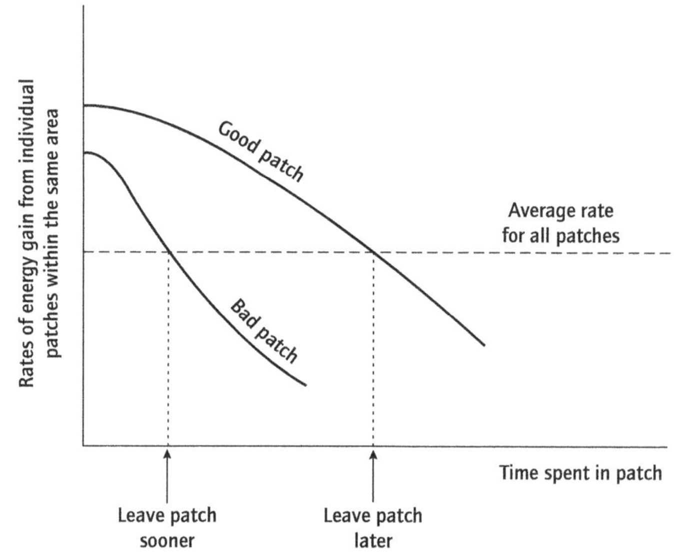

Fig. 9.26 The effect of patch quality on the optimal time to be spent in different patches within a given area – see text.

When a consumer begins to exploit a hitherto unexploited patch, geographical region or captured prey individual, the rate of energy gain per unit time will show a relationship with time of the form displayed in Fig. 9.23. Sooner or later the consumer will have to leave and move elsewhere or catch another item, and movement between patches may be expensive in time and energy. Even such a slow moving animal as the winkle, *Littorina*, uses twelve times more energy per unit time when crawling than when grazing in one spot, and clearly long distance migration is very expensive indeed. So it is necessary to allow for the time (corrected for the different energy demands) taken to travel between, as well as within, patches.

If then, we plot the average net rate of cumulative gain in energy from all the patches in a given habitat against time, and include the average time required to travel between these patches (Fig. 9.24), we can ask the question 'what period of time spent in a patch will yield the maximum cumulative gain per unit total time (i.e. travelling time plus time within the patches)?' The answer will be given by the tangent from the origin to the cumulative gain curve (Fig. 9.25) – the line from origin to curve which has the steepest slope. Hence a consumer which leaves a patch or ceases consuming a given prey item when its rate of gain falls to this average value will maximize its energy intake per unit total foraging time. Consumers should give up feeding on a diminishing resource when they would do better by leaving it!

This being the case, then it follows that: (a) where patches vary in quality within any given habitat, consumers should remain longer in the better patches (Fig. 9.26); and (b) where patches are

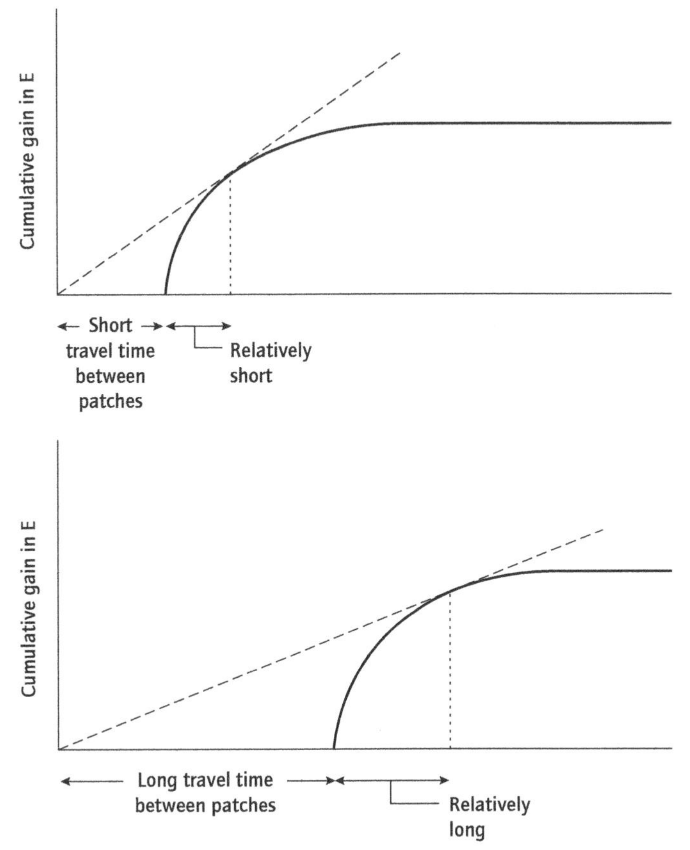

Fig. 9.27 The effect of differing average travel times on the optimal time to be spent in patches of the same quality – see text.

of approximately the same quality but vary in their distances apart, it would be advantageous to remain in a patch longer if it took a longer travel time to reach it (Fig. 9.27) (i.e. when travel costs are higher, the returns from moving will be lower, and hence a consumer should remain longer in each patch).

Patches may not be good or bad, however, solely on the basis of their intrinsic merit. The food that can be obtained per unit time or per unit effort may well also depend on how many other consumers are already exploiting any patch in question. In other words, a consumer may face a choice between, on the one hand, seeking or continuing to obtain a small share of the resources in some high-quality patch that is already being heavily utilized or, at the other extreme, a larger share of a relatively unused low-quality patch. Many animals, ranging from flies on piles of dung to intertidal crabs on mudflats, have been shown to distribute themselves in such circumstances according to 'ideal free theory': patches of variable quality become filled differentially by consumers in such a manner that the reward that unit competitor gains from each patch is the same. This assumes that all individuals are free to exploit whatever patch they wish. In practice, of course, individuals can differ in their competitive abilities and inferiors may be displaced onto poorer quality patches, complicating the ideal distribution (see, for example, Parker & Sutherland, 1986).

9.3.3 Mobile consumers and their prey

How might the general considerations outlined above affect the feeding biology of mobile consumers feeding on individual items, and how can their prey minimize the chance of being consumed?

A given quantity of food can clearly be composed of many small prey items or a few large ones, and these two extreme cases bear different costs and different benefits. The gain from a single large item is great, but large prey are likely to be highly mobile and it may therefore require the expenditure of much time and energy to capture any single such item. Conversely, a diet of many small items may not require much time to be devoted to each individual capture but there will probably be a considerable search-time element. These differences will have repercussions on consumer and prey alike.

Pursuit hunters typically spend a large proportion of their total feeding time chasing large individual prey items, not always successfully. There is much energy expended on each potential prey individual, therefore, and correspondingly the gain from it must be great. It will be to the advantage of the hunter to minimize this pursuit cost as much as possible, and to the advantage of the prey to maximize it. This is likely to lead to an arms race between predator and prey, with the predator concentrating on that limited range of potential prey species that are most susceptible to its form of pursuit and capture whilst still being maximally rewarding energetically, and adapting more and more closely to the flight patterns of these prey species. The effect of maximizing net gain is for pursuit hunters to become specialist in their diet.

Besides the obvious pressure in favour of increased speed and agility, the most general means by which the prey species can decrease capture success rates of pursuit (and other) hunters is living in groups (swarms, schools, clumps, herds). This derives from a number of effects.

1 One commonly recorded observation is that a large group can detect and react to a predator's approach more quickly than can a small group or a solitary individual. The small, flightless, marine hemipteroid *Halobates*, for example, skates on the surface film of the sea in a similar fashion to pond-skaters on inland waters. When not feeding, it aggregates in 'flotillas', and the flotillas display clearly defined responses to the approach of predators (birds, fish, predator models). The distance away at which an experimental predator model produced a behavioural avoidance response in the flotilla varied with flotilla size as shown in Fig. 9.28: detection distance was greater in larger flotillas, although clearly there is a critical size of flotilla after which there is no further increase in detection range (the maximum having been reached).

2 A second effect of living in a group is confusion of the potential predator. Pursuit hunters attack one potential prey item at a time, and we have already seen that it will be to the advantage of the consumer to capture a prey individual of high food value. Groups may react to the close proximity of a predator by the

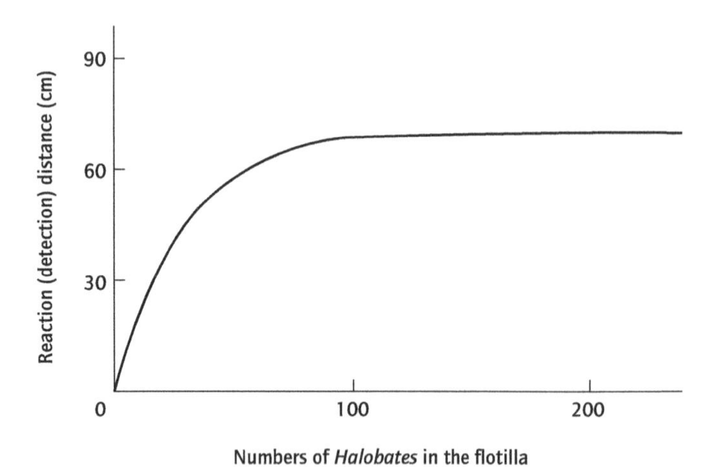

Fig. 9.28 Variation in the distance away at which an experimental predator model evoked a behavioural response in the sea-skater *Halobates* in relation to the number of individuals in the flotilla (after Treherne & Foster, 1980).

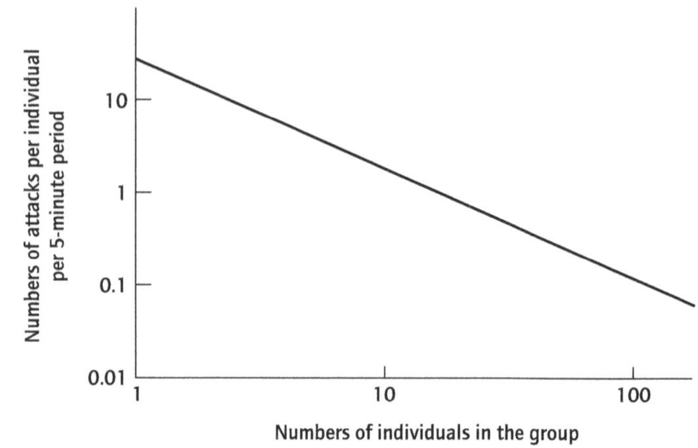

Fig. 9.29 Number of attacks by fish on individual sea-skaters, *Halobates*, in relation to the number of individuals in the group (after Treherne & Foster, 1980).

individual members scattering in all directions. This increases the difficulty of 'homing in' on any one target individual and results in considerable target switching as different prey move across the sensory field.

3 Predators may be deterred from attacking a group by group-defence behaviour, whereas they suffer no such inhibition when encountering a solitary member of the same species. Most of the classic examples of group defence are from within the vertebrates, but invertebrate examples are also known, most obviously in the social hymenopterans. In addition, some sawfly larvae exude a sticky resin from their mouthparts when attacked, e.g. by predatory insects. This defence does not deter pentatomid bugs from attacking solitary larvae, but bugs which attack larvae in aggregations rapidly become covered in the resin and may barely be able to move afterwards.

4 The three effects of group living described above all have the effect of increasing the time taken for a predator to make a successful capture; it may, on average, be advantageous for the predator to switch its attention to another, non-social prey. A fourth effect relates to the probability of any given individual within a group being the subject of a successful predatory attack; the larger the group, the smaller the probability that any particular individual will be the one caught during a given attack. This has been shown in the bug *Halobates*, cited above in respect of the detection effect, when preyed on by fish (Fig. 9.29). An individual in a group of ten receives only one-tenth of the number of attacks directed at a solitary individual, and a bug in a group of 100 receives only one-hundredth of that number.

The effect is not necessarily purely statistical. The probability of being the individual taken from a group is often highest for those on the periphery, and least for those in the centre of the aggregation. In a number of instances, individuals have been described as constantly moving from the periphery to the centre, leaving at the margins of the group those individuals least able to maintain a relatively safe position within the 'selfish herd'. In this way, individuals are actively minimizing their own chances of being consumed.

In contrast, hunters of the searching and ambushing categories prey on individuals which are easily caught by virtue of their small size (relative to the consumer). All and any easily captured prey may be consumed in order to ingest sufficient food in total, and searchers can most readily maximize their net gain by being generalist in diet. Further, because search time will be inversely proportional to the overall abundance of suitable prey, searchers are likely to have particularly generalist diets in food-poor habitats, and at times of food scarcity. At times of peak abundance of any one particular prey species, however, they are likely to spend a disproportionate amount of time utilizing that prey type (become temporarily specialist), switching to other food species if and when their numbers also increase.

Prey individuals subject to predation by searching hunters will be advantaged by maximizing a consumer's search time. This can be achieved in many different ways, which increase the difficulty of either finding the prey or of recognizing the prey as potential food. These include (Fig. 9.30):

1 Inhabiting microhabitats which cannot easily be investigated by most potential predators, e.g. dwelling in crevices, under stones, in burrows.

2 Being cryptic, i.e. by a combination of shape, posture and surface pattern (or odour) blend in with the background and thereby increase recognition time (some species can change colour or pattern to merge with more than one background type; others can modify their local environment to conceal themselves more effectively).

3 Being mimetic, i.e. similarly to crypsis, increases recognition time by mimicking an object of no or low food value – a dead leaf, twig, faecal pellet, etc. – or a distasteful species or one capable of aggressive retaliatory defence.

Increase in the necessary search time is not the only means of decreasing an individual's food value to a consumer; handling time can also be increased and the energy content per unit

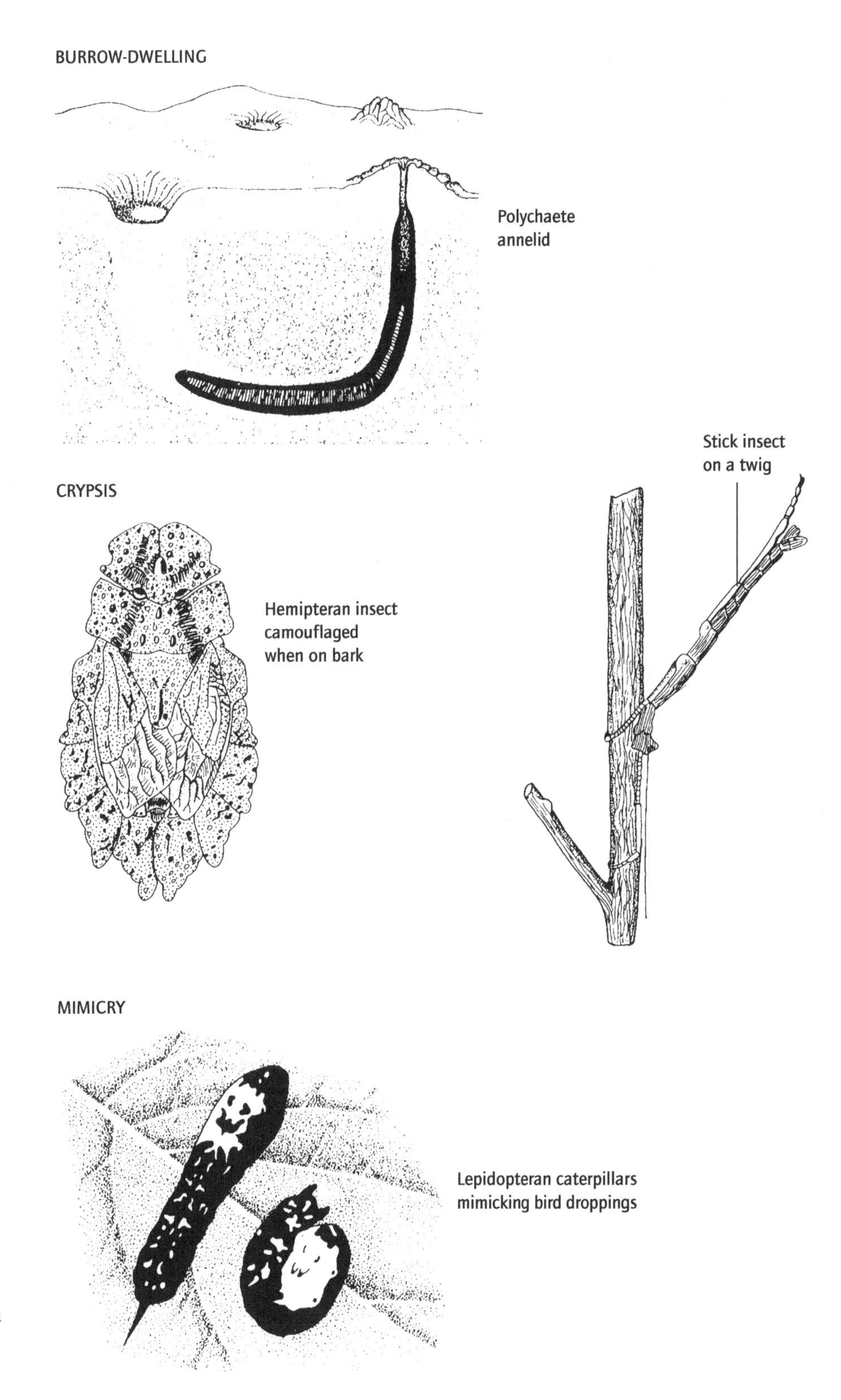

Fig. 9.30 Mechanisms for increasing the search times of potential predators (from various sources).

weight can be decreased. Handling time is often increased by the evolution of (Fig. 9.31):

1 Defensive weaponry, e.g. stings, jaws, etc.

2 Defensive armour, e.g. shells, calcareous or chitinous plates.

3 Shapes which increase the difficulty of grasping, manipulating and/or ingesting the food item, e.g. spines. Spines and similar protuberances on the body surface are particularly effective in this respect, for not only do they make handling time-consuming,

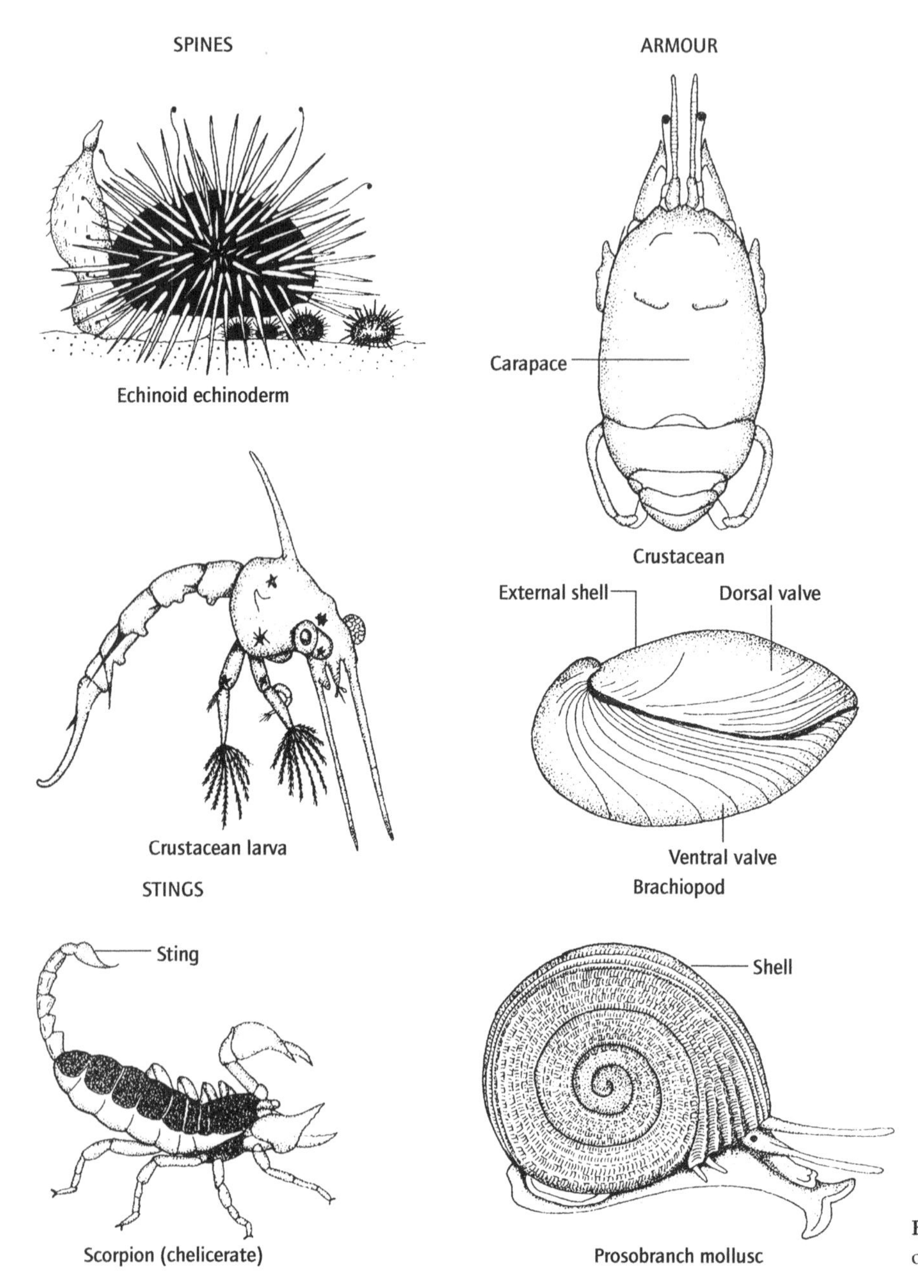

Fig. 9.31 Structures increasing the handling times of potential predators (from various sources).

but they also increase the effective body size – thereby taking the prey out of the size range of at least some predators – for little increase in body tissue.

Energy and/or nutrient content per unit body weight or volume can be reduced by diluting the quantity of living tissue with inert, nutrient-poor or inedible substances. Palatability can be decreased by the presence of noxious substances in specific organs or generally in the tissues. In this case, a level of trial and error learning is necessary on the part of the consumer, and the prey often make this process as rapid and effective as possible by coupling their distastefulness with bright warning coloration.

Increase in handling time, decrease in energy content per unit mass and decrease in palatability (or outright toxicity) are also precisely the defensive tactics utilized by the prey of grazing and browsing consumers, which functionally are equivalent to searching hunters except in respect of the sessile and frequently photosynthetic nature of their prey. Searchers are normally larger and longer lived than their animal prey, but the relationship is usually the other way around in grazers/browsers and their food species: the prey are the larger and longer lived. Hence whereas it is difficult for the prey of searching hunters to do anything other than 'teach' their long-lived consumer to avoid them by use of noxious chemicals, it is easier and more advantageous for long-lived plants to kill their short-lived insect browsers with toxins and thereby cause selection for individuals which ignore that plant as food.

All of these defensive systems (and see also Section 13.2.1) make it advantageous for a generalist consumer to search for other prey types – more obvious, less protected, more palatable species – but by virtue of their abundance resulting from successful avoidance of generalist consumption, sessile or sedentary species relying mainly on passive mechanical or chemical defences are open to attack from specialist consumers which have managed evolutionarily to crack the defensive codes. Grazers and browsers, and specialist consumers of cryptic, mimetic and noxious animal prey have participated in similar evolutionary arms races with the prey species, as have pursuit hunters and their highly mobile prey. Indeed, once specialist consumers have evolved the mechanisms whereby the toxic or noxious defences of their prey organisms can be rendered ineffective, the consumer may even use the chemicals to their own advantage by sequestering them in their bodies as a defence against their own predators.

9.3.4 Suspension and deposit feeders

Consumers which capture particles by filtering a current of water clearly have no pursuit or search costs. Instead they bear the costs of filtration and those associated with the rejection of unwanted particles. The mesh sizes of their filters can, in general, only be altered in evolutionary time and hence there is little or no possibility of differentially selecting out particles of high food value from the water: filters are size not edibility specific. A sessile filter feeder can only maximize its net rate of energy gain by altering:

1 The filtration rate in relation to the relative abundance of different types of particles in the water.

2 The rejection rate in relation to the relative nutritional values of the particles and to the costs of rejection.

Filtration rate should increase as the concentration of high-value particles in the water increases (Fig. 9.32), but there may be a limit determined by the ability of the gut to process the collected material. In mucociliary feeders, this could be the ciliary movement of the mucous cord within the gut; and in the arthropods sieving out particles with their setae, it could be the packing of the gut lumen with ingested particles. Once the gut becomes filled, the rate of energy intake will be limited by the passage time through the gut and the speed with which digestion can proceed. Hence filtration rate should decrease, once the gut is full, to that value which maintains the gut in a full state. Equivalently to the position with predators consuming only the highest value parts of a prey item before moving to the next (Section 9.3.2), it may pay suspension feeders, dependent on the quality and abundance of their food, to pass food rapidly through the gut digesting only some of it, rather than pass it through more slowly and digesting a greater proportion.

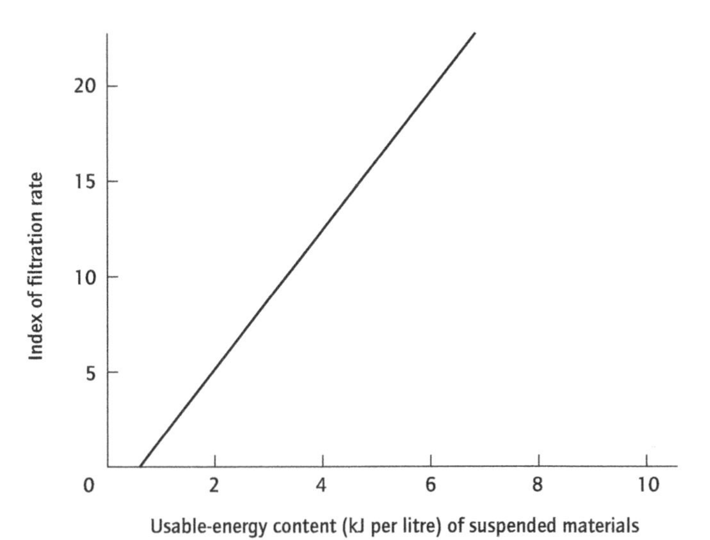

Fig. 9.32 The relationship between concentration of edible particles in the overlying water and the filtration rate of the polychaete *Hediste diversicolor* when behaving as a suspension feeder (after Pashley, 1985). (See Fig. 9.4d)

The cost of rejecting unwanted particles may be high, and therefore as the cost of rejecting a given particle type increases so the ingestion rate of that particle type may also be expected to increase. When rejection costs are very high, it may be less costly to reject even completely indigestible particles through the gut than by means of sorting organs. Under such circumstances, a suspension feeder could be in energetic deficit and it would then be advantageous to cease feeding completely until the suspended load of the water changed.

With such a food-collection system, there is very little, if any, opportunity to become a specialist consumer except on an evolutionary time scale and then only on certain size ranges of particles. Pelagic tunicates, for example, possess very fine filters capable of retaining bacteria; whales, on the other hand, sieve out the large planktonic euphausiids. These differences are simply functions of the mesh sizes of their filters, and also of the relative sizes of the filterers. Whales could not operate a filter capable of collecting bacteria and protists: they would be unable both to move forward through the water and to filter, for exactly the same reason that it is not possible to tow a fine-meshed plankton net through water rapidly and still permit it to catch anything. Water is simply pushed ahead of the mesh and little passes through. Such selectivity as suspension feeders show is therefore not exercised at the level of the diet, but is displayed mainly at the stage of the free-living larval phase possessed by most species, when it settles in an area likely to provide the sessile adult with an adequate supply of appropriate suspended food particles.

Deposit feeders are somewhat intermediate in character between the suspension-feeding category and those of the hunters and grazers/browsers considered above. In as much as they are consuming non-living organic matter, they are feeding on material of low nutritional status which is correspondingly plentiful and which, being dead, does not protect itself from being consumed. Some may simply pass the inorganic sediment, together with such organic matter as it may contain, through their guts, and they will therefore be subject to the same gut-packing restrictions as apply to some suspension feeders. When the potential ingesta contains less than a certain threshold digestible content, it may also be energetically advantageous not to process it at all but to move and resume feeding elsewhere. Many deposit feeders, however, are capable of selective ingestion (Section 9.2.6), and such species may behave like searching hunters, except, instead of the whole animal moving, the 'searching' is conducted by highly mobile feeding tentacles or similar organs.

In general, the living components of the diet of suspension and deposit feeders are minute in relation to the size of the consumer, and, although some of the consumed species do have chemical defence systems, most appear to lack any form of defence against consumption. Instead, they have rapid rates of sexual and/or asexual multiplication which are more than capable of making good predatory losses. In this they are somewhat equivalent to the prey of grazers and browsers, which characteristically possess sufficient potential for growth or asexual division to replace lost structures, such as leaves, polyps or zooids, so that only exceptionally severe rates of consumption cause the death of the individual or of the colony.

9.4 Conclusions

We have seen that animals originated as small consumers of bacteria, protists and each other, and that these ancestral diets not only persisted relatively unchanged for millions of years but have also influenced animal feeding ever since. The selective advantage of large size favoured the evolution of mechanisms for increasing the quantities of these small and widely scattered items that could be ingested, but did not lead to any radically different new diets. Consumption of macrophyte material, whether algal or plant, evolved relatively late and, unless mediated by intestinal micro-organisms, remains an inefficient process, not least because of the poor nutritional quality of structural carbohydrate tissues, but also because of the defensive repertoires of these otherwise easily available materials.

The categorization of animals as 'herbivores', 'carnivores' or 'omnivores' is not helpful in understanding their feeding biology: on such a basis, most species are omnivores. Of more value is classification by feeding mechanism – hunter/parasites, grazers/browsers, deposit feeders, suspension feeders, absorbers of the products of internal symbionts, etc. – a system which cuts right across the taxonomic position of the prey species, and is concerned more with the consumer than with the consumed.

Although constrained to varying extents by their evolutionary pasts, all animals nevertheless face choices in ecological time: whether to attempt to consume this individual prey item rather than another (in space or time); whether to remain feeding in this area rather than move to another; and so on. Consumers do not appear to react to such 'decisions' randomly, and optimality theory provides one model with some success in predicting foraging behaviour. Within the constraints imposed by factors other than food quality and supply (e.g. predation, breeding biology), several animals do feed in such a manner as to maximize their gain (benefits derived minus costs borne). Prey species on the other hand have been selected so as to maximize the relative cost of consuming them, by their behaviour, their morphology or their biochemistry. As might be expected, the different feeding modes will have inherently different potential solutions to the maximization of gain, with consequent selective advantages in favour of a generalist or a specialist diet.

Feeding behaviour would seem to be just as much under a day-by-day selective control as other, more familiar aspects of animal biology.

9.5 Further reading

Barnard, C.J. (Ed.) 1985. *Producers and Scroungers*. Croom Helm, London.

Begon, M., Harper, J.L. & Townsend, C.R. 1996. *Ecology*, 3rd edn. Blackwell Science, Oxford.

Bennett, V.A., Kukal, O. & Lee, R.E. 1999. Metabolic opportunists: feeding and temperature influence the rate and pattern of respiration in the high arctic woollybear caterpillar *Gynaephora groenlandica* (Lymantriidae). *J. exp. Biol.*, **202**, 47–53.

Crawley, M.J. 1983. *Herbivory*. Blackwell Scientific Publications, Oxford.

Doeller, J.E., Gaschen, B.K., Parrino, V. & Kraus, D.W. 1999. Chemolithoheterotrophy in a metazoan tissue: sulfide supports cellular work in ciliated mussel gills. *J. exp. Biol.*, **202**, 1953–1961.

Esch, G.W. & Fernandez, J. 1992. *Functional Biology of Parasitism: Ecological and Evolutionary Implications*. Chapman & Hall, New York.

Fabricius, K.E., Benayahu, Y. & Genin, A. 1995. Herbivory in asymbiotic corals. *Science*, **268**, 90.

Hodkinson, I.D. & Hughes, M.K. 1982. *Insect Herbivory*. Chapman & Hall, New York.

Hughes, R.N. (Ed.) 1993. *Diet Selection*. Blackwell, Oxford.

Jennings, D.H. & Lee, D.L. (Ed.) 1975. *Symbiosis*. Cambridge University Press, Cambridge.

Jennings, J.B. 1972. *Feeding, Digestion and Assimilation in Animals*, 2nd edn. Macmillan, London.

Jørgensen, C.B. 1975. Comparative physiology of suspension feeding. *Annu. Rev. Physiol.*, **37**, 57–79.

Julian, D., Gaill, F., Wood, E., Arp, A.J. & Fisher, C.R. 1999. Roots as a site of hydrogen sulfide uptake in the hydrocarbon seep vestimentiferan *Lamellibrachia* sp. *J. exp. Biol.*, **202**, 2245–2257.

Lee, R.W., Robinson, J.J. & Cavanaugh, C.M. 1999. Pathways of inorganic nitrogen assimilation in chemoautotrophic bacteria–marine invertebrate symbioses: expression of host and symbiont glutamine synthetase. *J. exp. Biol.*, **202**, 289–300.

Mason, C.F. 1977. *Decomposition*. Edward Arnold, London.

McNeill, A.R. 1996. *Optima for Animals*, revised edition. Princeton University Press, Princeton.

Morton, J. 1979. *Guts*, 2nd edn. Edward Arnold, London.

Owen, J. 1980. *Feeding Strategy*. Oxford University Press, Oxford.

Parker, G.A. & Sutherland, W.J. 1986. Ideal free distributions when individuals differ in competitive ability: phenotype-limited ideal free models. *Anim. Behav.*, **34**, 1222–1242.

Randall, D., Burggren, W.W. & French, K. 1997. *Animal Physiology. Mechanisms and Adaptations*, 4th edn. W.H. Freeman, New York.

Schmidt-Nielsen, K. 1997. *Animal Physiology. Adaptation and Environment*, 5th edn. Cambridge University Press, Cambridge.

Smith, D.C. & Douglas, A.E. 1987. *The Biology of Symbiosis*. Edward Arnold, London.

Southward, E.C. 1987. Contribution of symbiotic chemoautotrophs to the nutrition of benthic invertebrates. In: Sleigh, M.A. (Ed.) *Microbes in the Sea*, pp. 83–118. Wiley, New York.

Taylor, R.J. 1984. *Predation*. Chapman & Hall, New York.

Townsend, C.R. & Calow, P. (Ed.) 1981. *Physiological Ecology*. Blackwell Scientific Publications, Oxford.

Tunnicliffe, V. 1992. Hydrothermal-vent communities of the deep sea. *Am. Sci.*, **80**, 336–349.

Vacelet, J. & Boury-Esnault, N. 1995. Carnivorous sponges. *Nature (Lond.)*, **373**, 333–335.

Vermeij, G. 1987. *Evolution and Escalation. An Ecological History of Life*. Princeton University Press, Princeton.

Weibel, E.R., Taylor, C.R. & Bolis, L. (Eds) 1998. *Principles of Animal Design. The Optimisation and Symmorphosis Debate*. Cambridge University Press, Cambridge.

Wildish, D. & Kristmanson, D. 1997. *Benthic Suspension Feeders and Flow*. Cambridge University Press, Cambridge.

Wright, S.H. & Manahan, D.T. 1989. Integumental nutrient uptake by aquatic organisms. *Annu. Rev. Physiol.*, **51**, 585–600.

Wright, S.H. & Ahearn, G.A. 1997. Nutrient absorption in invertebrates. In: Dantzler, W.H. (Ed.) *Handbook of Physiology*. Section 13 *Comparative Physiology*, Vol. II, Chapter 16, pp. 1137–1206. Oxford University Press, Oxford.

CHAPTER 10

Mechanics and Movement (Locomotion)

Some invertebrate animals are able to fly, others use six, eight or indeed many legs to walk or run. In aquatic environments some can swim, others crawl or creep over the surface. Many invertebrates live in more or less permanent tubes and burrows and tubiculous invertebrates may move around in their tubes or expend energy to drive water through their tubes to provide oxygen and a source of food. As adults they are often sessile and fixed permanently to a substrate but have pelagic freely moving larvae. Even sedentary animals expend energy moving water and the mechanics of this is similar to that of locomotory movement. As discussed in Chapter 14 the larvae of many sessile invertebrates are motile and their movements during the pelagic phase are important in establishing contact with an appropriate environment for sessile life. Sexual reproduction requires contact between male and female gametes and this is often achieved by the movements of the male gametes (Spermatozoa) which are typically motile especially those of marine invertebrates. Whatever the pattern of locomotion there must be the expenditure of energy and the mechanical principles required for locomotion or to move the environment relative to the body are the same. The opening section of this chapter will deal with the basics of mechanics and define some terms. This will be followed by a consideration of the mechanisms by which animal cells are able to generate forces. The harnessing of these forces in systems of locomotion makes possible the variety of locomotory modes among invertebrate animals. A common feature is the utilization of respiratory energy as discussed in Chapter 11. We will learn that the underlying principles are the same for flagella, cilia and muscle cells.

A comparative study of invertebrate locomotion reveals the diversity of locomotory modes and provides the basis for a discussion of the influence of scale and ratio between body mass and surface area. The surface area increases by the square of the linear dimensions while the body mass increases in proportion to the cube, consequently large animals are not able to move by means of ciliary locomotion and must rely on muscular contraction to develop the forces required for movement. Many aquatic organisms are soft bodied and do not have a hard mechanical skeleton. They have what may be termed a 'hydrostatic skeleton' and we will learn how this enables worms and similar animals to move in a variety of ways. The transition from aquatic to terrestrial environments brings enormous mechanical challenges. Buoyancy is reduced but so is frictional drag. Jointed hardened legs with an external skeleton are used by many terrestrial animals as well as by the predominantly marine crustaceans. The opportunities presented by the oxygen-rich terrestrial environment has enabled the insects to develop a quite remarkable range of flying techniques. Scientists are only now developing an understanding of the complex aerodynamics of insect flight which is certainly not always what it seems. The largest invertebrates, the marine squids, use aquatic jet propulsion and complete the range of locomotory modes that will be discussed in this chapter.

10.1 Introduction

Invertebrates show a diverse range of structure. The basal animal groups (Chapter 3) and the many worm-like animals (Chapter 4) are soft-bodied animals. They move by flagella, cilia (Section 10.2) or by the use of a hydrostatic skeleton (Section 10.5). Some invertebrates, however, have developed hardened tissues that permit the use of a mechanical skeleton. The most adept are the animals with jointed legs, the Arthropods and similar groups (Chapter 8). These animals are able to exploit the principles of a mechanical lever and, in the case of the insects, have developed a truly remarkable array of flying and jumping mechanisms.

Virtually all classes of invertebrates are able to perform mechanical work with the result that the body moves with respect to the environment or the environment is moved past the fixed body of the organism. The energy which is invested in such activities must result in a return to the animal. This return on the energy invested can be thought of as being an enhanced rate of feeding or other resource acquisition, a reduced rate of predation, the avoidance of harmful environmental changes, enhanced dispersal and/or contact with potential partners in sexual reproduction.

When animals move they obey the same basic principles as all moving objects (Newton's laws of motion). These principles state that:

1 If a body is at rest relative to its environment, it can be set in motion only by the application of an external force.

2 A body moving in a straight line will continue to do so, unless acted upon by an external force.

3 The application of an unbalanced force to a mass in motion results in an acceleration or deceleration of the mass in the direction of the force.
4 For every action there must be an equal and opposite reaction.
The First Law of Thermodynamics states that the energy within a closed system remains constant, though it can be changed from one form to another.

These concepts apply universally and will be illustrated in general terms before the movement and activities of the different groups of invertebrates are explained. Box 10.1 summarizes the general principles governing the locomotion of animals and defines some units.

In order to move, an animal must expend energy that would otherwise have to be spent and there is therefore a net cost of movement (Schmidt-Nielsen, 1984). It can usually be estimated from measurements of the metabolic rate (oxygen consumption) during movement minus the resting levels of oxygen consumption, although the cost of passive movement may not be measured in this way (see below).

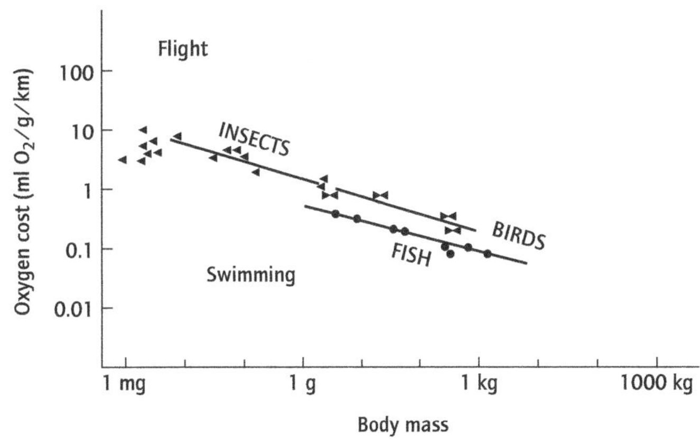

Fig. 10.1 The cost of locomotion expressed as the amount of O_2 needed to transport a 1 g animal over a distance of 1 km. Note that flying insects and flying birds lie on the same regression line. Most data are available for mammals; how would invertebrates compare? ● = Swimming, ◀ = flight. (After Schmidt-Nielsen 1984.)

Box 10.1 Mechanical Terms and Definitions

1 Force
A **force** is detected by its effect on a **mass**.

The effect of a force on a mass is to change the direction or rate of movement of a mass, the value of a force is equal to the product of the mass and the acceleration, i.e.

Force is the **mass** × **acceleration**

$F = M\,a$

In SI units a force of **1 newton** (N) is required to impart to a mass of 1 kilogram (1 kg) an acceleration of 1 metre per second per second.

2 Reaction Force
When a body exerts a force on a second body, it is as though the second body exerts a force of equal magnitude but in the opposite direction. This force is called the **reaction force**.

3 Work
Mechanical work is done when a force imparts an acceleration to a mass. For a stationary mass the work done is the product of the force and the distance moved.

$W = F\,d$

and

$W = (M\,a)\,d$

The international unit of work is the **joule** (J).

4 Power
Power may be thought of as the rate at which work is done and therefore **power** is force × velocity

The international unit of power is the **watt** (W) where

1 watt = 1 joule per second.

The locomotory patterns of animals can be classified as: passive transport; swimming; walking; running; jumping; flying. Invertebrates exhibit all of them. The net cost of movement per unit size increases in the order: passive transport, swimming, walking, running, jumping. Flying therefore appears as a means of escaping the high metabolic cost of rapid movement in the terrestrial environment. Each of the active modes of locomotion listed above has a progressively higher net cost of movement per unit size.

Comparisons between invertebrates and vertebrates are difficult because of the differences in scales and their modes of movement, but there is a comparable set of data for flying insects and birds. Among the vertebrates there is, for each of the modes of locomotion listed above, a negative relationship between the logarithm of the net cost of movement and the logarithm of the body mass. The data for insects show the same negative relationship and the insects and birds fit the same negative regression line remarkably well despite differences of scale up to seven orders of magnitude (Fig. 10.1).

These relationships demonstrate a dramatic reduction in the net cost of movement per unit mass with increasing size.

The locomotory patterns exhibited by organisms are intimately linked to the supply of energy and thus to the feeding biology. Organisms that exploit low-concentration food resources, such as deposit feeders, will also be limited to low-energy locomotory systems. At the scale of most invertebrates, these food resources will be exploited only by sedentary benthonic animals or those which move relatively slowly. In an evolutionary sense, the emergence of multicellular animals and plants (see Chapter 2) established dense pockets of energy which could be exploited by animals with greater powers of movement. Many evolutionary advances can be interpreted as advances in animal mechanisms that permitted the exploitation of these dispersed but dense aggregations of energy.

Paradoxically the relationships set out in Fig. 10.1 also suggest that for the very largest animals, the net cost of movement per unit weight might be expected to be very low, especially for a swimming animal with neutral buoyancy. Indeed the world's largest animal, the blue whale, meets these requirements and feeds on a low-energy dispersed food resource by filtering plankton.

Animals that exhibit passive transport need not do work on the environment in order to move. Instead they rely on buoyancy and natural movements of the medium in which they live. The aquatic and aerial plankton are assemblages of species that make extensive use of passive transport. Most are also capable of swimming and must do so to overcome downward drift if they are not neutrally buoyant. The physical conditions governing passive transport in aquatic environments (i.e. marine and freshwater habitats) are very different from those of the aerial plankton due to the great differences in the viscosity of water and air. Because of this, members of the aquatic plankton can achieve much greater sizes than can the aerial plankton. Most of the primary production in the seas and oceans is due to the phytoplanktonic algae suspended in the surface waters, and there are vast numbers of permanent and temporary zooplankton exploiting this source of production as the food resource. The costs of movement for these animals may be partly obscured as 'the cost of achieving neutral buoyancy'. This could be the cost of secretion of spines or the cost of deployment of energetically expensive lipids.

When an animal moves it will have to overcome forces which tend to resist its movement. This is referred to as 'friction' or 'drag'. Since the resistance of the medium through which an object moves tends to change its rate of movement (by causing deceleration), the resistance is a force. It can be expressed in the units defined in Box 10.1. In any system, energy is conserved and mechanical energy can be converted into kinetic energy which is stored in the movement of a mass, but which will eventually be lost as heat. The units are defined in Box 10.2.

The mechanical work performed in locomotion is ultimately derived from energy generated by chemical reactions in the cells of which the animal is composed (see Chapter 11), but the chemical work is not necessarily equal to mechanical work. A weightlifter may be exhausted in trying to lift a weight which does not move, but he has not, by these definitions, performed mechanical work. He has, of course, done metabolic work, his muscles use ATP at metabolic cost, and through his efforts he has produced heat. We therefore have to define the efficiency of locomotion in recognition of the fact that energy may be expended which does not contribute to the movement of the body. We can determine the mechanical work done by measuring mass and distance moved. This work we could describe as the 'useful work'. The total energy output of the animal in motion is more difficult to measure (but can be estimated by measuring the rate of oxygen consumption), and the efficiency of the animal (like any machine) can then be expressed as the ratio:

$$\frac{\text{output of useful work}}{\text{input of energy}} \qquad (1)$$

We shall see that some systems of animal locomotion may be more efficient than others, but that efficient systems may have limited power, i.e. the rate at which they can work may be limited. As elsewhere in biology we find that there are potential trade-offs, one important one being that between efficiency and absolute power.

Organisms in motion have inertia and will experience drag, i.e. forces acting contrary to the direction of locomotion due to the resistance of the medium through which they are moving (see Box 10.2). According to the First Law of Thermodynamics, a body that is moving in a straight line will continue to do so unless acted upon by an external force. In a viscous medium the body will experience 'drag' a force acting contrary to the direction of locomotion due to the resistance of the medium through which it is moving. The extent to which such forces are important is profoundly influenced by the relative size of the organism, its absolute velocity, and the viscosity (or friction) of the medium over or through which it is moving. These interrelationships can be expressed as the *Reynolds number (Re)*. This dimensionless number is the ratio of inertial to viscous forces and is given by the relationship:

Box 10.2 Kinetic energy and friction

1 Kinetic Energy
In the absence of friction a mass will accelerate during the application of a force (Box 1). The work done will be defined by the distance moved by the mass during the time that the force was applied and after the application of the force the mass would continue to move in a straight line at constant speed for ever.

$F = M\,d$ and $W = F\,d$ (Box 10.1).

The energy is stored in the form of kinetic energy where,

$KE = \frac{1}{2} m\,V^2$

and V is the velocity of the mass. The SI units are metres per second ($m\ s^{-1}$).

2 Friction
In the real world a moving mass comes to rest because it is subject to a **frictional force** which causes deceleration. This force is often called **drag**.

3 Dissipation of kinetic energy.
Kinetic energy may be dissipated as heat energy as frictional force causes a mass to become stationary.

4 Constant velocity
In order to move at a constant velocity a mass must be subject to a constant force equal and opposite to the frictional force opposing its movement.

$$\text{Re} = \frac{\text{velocity} \times \text{dimension of the system}}{\text{kinematic viscosity}} \qquad (2)$$

$$\text{Re} = ud/v \qquad (3)$$

The dimension may be chosen as the length of the entire organism or the length of a component part such as a limb. The velocity is the maximum velocity achieved and the viscosity is that of the medium. The viscocity may be thought of as the 'stickiness' of the medium and when the Reynolds number is large the stickiness of the medium may be ignored. When the Reynolds number is small, this is no longer true. For such animals, viscous forces will be important and may dominate their lives. The range of Reynolds numbers for invertebrates is through several orders of magnitude. For a large flying insect such as a dragonfly moving in air at 2–7 m s^{-1}, the Reynolds number is $> 10^4$ and may reach 30 000; in contrast, the Reynolds number of an invertebrate larvae swimming in sea water by ciliary locomotion at speeds up to 1 mm s^{-1} is about 0.3.

From observations on the movement of a whole range of organisms it has been shown that the Reynolds number increases according to the following equation:

$$\text{Re} = \text{approximately } 1.4 \times 10^6 \times d^{1.86} \qquad (4)$$

where d is the linear dimension of the object. Note that the Reynolds number increases by a power relationship of the linear dimension, so large animals have much larger Reynolds numbers.

In aquatic environments, Reynolds numbers are much lower for the same relative velocities because of the higher viscosity (see equation 2 above). What this means is that larger animals are more likely to be influenced by viscous forces in aquatic environments and are therefore able to exploit passive movement and to float. If these organisms are not neutrally buoyant, there will be a net downward drift due to gravitational forces. The rate of downward movement can be reduced by increasing drag (friction). This can be achieved by a variety of devices.

Devices to reduce the range of downward movement may be spines or threads. Examples include the elongate larval chaetae of some polychaete larvae, sculptured spines on the carapace of larval crustacea and on the larval shells of some molluscs and byssus threads secreted by some larval bivalves (Fig. 10.2).

Many small organisms in aquatic environments capture food particles, or locomote through the water by using appendages bearing arrays of bristles. The way in which these bristle-bearing appendages work depends on the Reynolds number associated with their movement. They may behave effectively as paddles (i.e. the arrays of bristles are not leaky) or as rakes when water can pass between the bristles in an array. In aquatic organisms these setose paddles mostly operate at Reynolds numbers less than 1 at a range of values in the order 10^{-5} to 1. A copepod *Centropages typicus*, for instance, has been found to operate its maxilliped setae at a Reynolds number of 0.1. When the Reynolds number is very low, viscous forces predominate. A characteristic of this type of movement will be marked changes in velocity. Imagine that you were swimming through treacle! If you watch planktonic crustacea swimming, you will see that they too show this jerky type of movement.

Aerial planktonic organisms are always very small and it has been suggested that they may be so small, that even in the less viscous medium of the air, the Reynolds number may be very low. This now seems unlikely and the so-called 'smoke midges' that have setose rather than membranous wings are, like other insects, flying (see Fig. 10.28 below). The wing operates at Reynolds numbers such that the plumose wing acts as an aerofoil shedding vortices and the flight can be analysed in the same way as other insects (see Section 10.6.2).

Passive movement may still be possible for larger invertebrates if neutral buoyancy is achieved. The density of an organism can be reduced by:

1 Reducing the concentration of heavy elements in the body tissues. This may involve making the whole body rather dilute as in the Cnidaria, especially jellyfish, which are isosmotic with sea water and neutrally buoyant. Mollusca typically have a calcareous shell but some mollusca are pelagic swimming animals. These animals often have shells in which the calcareous components of the shell are replaced with tanned proteins which are metabolically more expensive. Similarly, some pelagic organisms also reduce the concentration of denser ions such as sulphate or magnesium. The squids have a high ammonium ion concentration. All these adaptations must carry a metabolic cost which ought to be counted as part of the cost of movement.

2 Increasing the concentration of stored oils and fats; these storage products have a specific gravity of about 0.9. Most planktonic organisms consequently have a high concentration of fats and oils.

3 Development of gas-filled flotation chambers. These are characteristic of the siphonophores (see pp. 58–60) and pelagic cephalopod molluscs such as the Nautilus.

10.2 The generation of forces by animal cells

The locomotion of all animals is due to changes in cell shape resulting from movements of elements of the cytoskeleton requiring metabolic energy. These movements enable a cell to perform work (see Box 10.1). The movements of protista include movements associated with amoeboid creeping, the use of a flagellum and ciliary movement (see Chapter 3). Multicellular animals may also use flagella or ciliary mechanisms to perform work on the environment but larger animals use the contraction of specialized muscle cells as shown in Fig. 10.3. The cellular mechanism of these different types of movement are based on similar mechanisms by which cells can change shape. Eukaryotic cells are able to change shape because of the properties of the protein filaments that make up the cytoskeleton. The movements of micro-organelles within the cell is also controlled by the cytoskeleton. Microtubules are stiff hollow cylinders of the protein tubulin. The movement of specific 'cargo' molecules through a cell is brought about by the movement of the proteins

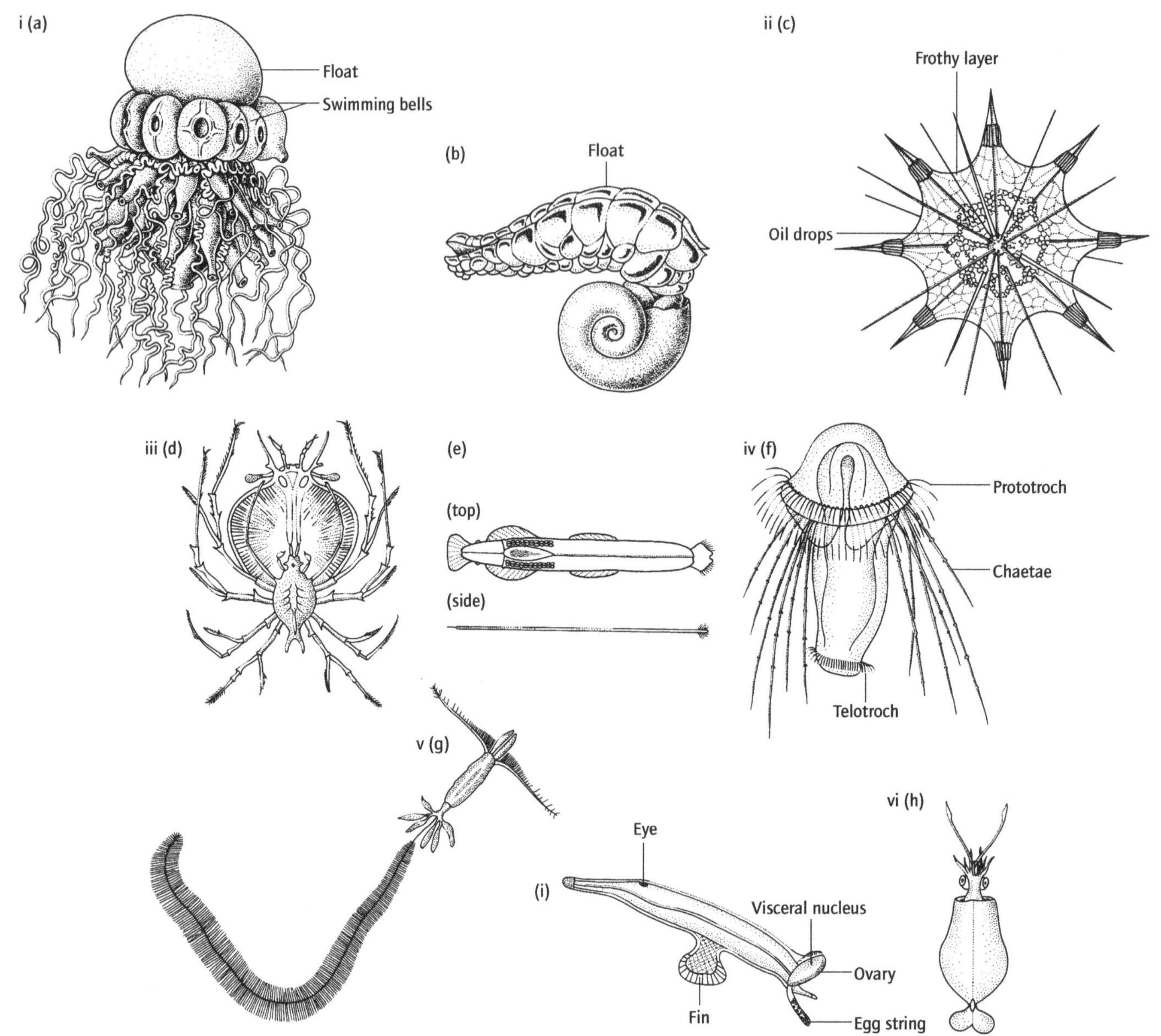

Fig. 10.2 Examples of methods to increase buoyancy or reduce the rate of sinking in invertebrates. (i) Development of gas-filled floats: (a) a siphonophoran colony (see also Fig. 3.20a), (b) *Janthina*, a pelagic gastropod mollusc. (ii) Secretion of gas bubbles and oil droplets in the cytoplasm: (c) a radiolarian (Protista) with oil droplets in the inner cytoplasm and CO_2 gas in the outer cytoplasm. (iii) Flattened body shape: (d) the flattened phyllosome larvae of a spiny lobster (Crustacea: Decapoda), (e) *Sagitta elegans*: a chaetognath (see also Fig. 7.5). (iv) Spines and elongate chaetae: (f) Trochophore larvae of *Sabellaria*, a Polychaete worm. The long erectile chaetae may also reduce predation by increasing the effective size. (v) Trailing threads: (g) *Calocalanus plumulosus* (Crustacea: Copepoda). Trailing threads are also secreted by some mollusc (bivalve) larvae. (vi) Reduction of mass by the exclusion of heavy ions. Squids (h) and Heteropod molluscs are among the organisms that increase buoyancy in this way. (From various sources and after Nybakken, 1988. (f) after Anderson, 1973.)

kinesin and dynein along the microtubules. This migratory movement requires the breaking of linkages between the proteins and the microtubule and is dependent on ATP hydrolysis. Microtubules and tubulin are also involved in the function of cilia and flagellae.

Cilia and flagellae are tubular projections of the cell surface bound by a cell membrane continuous with that of the cell. The core structure of the cilium or flagellum is remarkably similar across the whole of the animal kingdom and is referred to as the axoneme. The axoneme is made up of a longitudinal array of microtubules arranged in a very distinctive pattern. There is an inner pair of microtubules and an outer ring of nine incomplete pairs of microtubules, the outer doublets. Each doublet is composed of one complete tubule and one incomplete tubule. This typical '9 + 2' array is illustrated in Fig. 10.4 and occurs in the cilia and flagellae of all eukaryotic organisms. The array

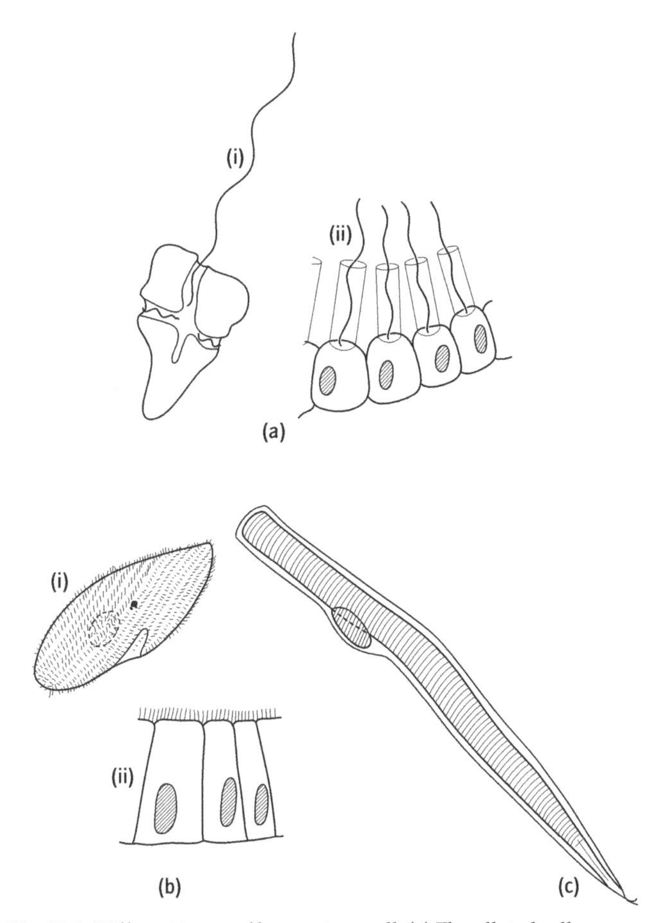

Fig. 10.3 Different types of locomotory cell. (a) Flagellated cells: (i) free-living flagellate; (ii) choanocytes of Porifera. (b) Ciliated cells: (i) free-living ciliate; (ii) ciliated cells of an animal. (c) Muscle cells.

of tubules is held in place by regularly spaced proteins and there are also dynein proteins which are involved in the bending process that enables the cilium and flagellum to function in a locomotory or fluid moving system.

The ciliary dynein protein is a complex of several polypeptides. One foot-like pole of the protein binds permanently to microtubule A and the larger head-like protein binds to microtubule B but this binding depends on the hydrolysis of ATP and is therefore energy dependent. When the ATP is hydrolysed, the dynein protein head moves along the B tubule and, since the microtubules are tied to each other by cross-binding proteins, the tubule will bend. In this way, chemical energy is used up in the creation of a bending movement in the axoneme and this can be converted into useful work by the progressive movement of the cilium of flagellum.

The main differences between these structures relates to their absolute length and the relationship between the length and the wavelength of the cycle of bending movements (Fig. 10.5). The wavelength of the locomotory wave of a flagellum is less than the absolute length (Fig. 10.5a). As a consequence, a flagellum generates a force parallel to the longitudinal axis. A cilium, in contrast, is relatively short and the wavelength is greater than its length. It therefore appears to make a cycle of movements. These movements typically involve a rather stiff power stroke (Fig. 10.5b) and a more flexed recovery stroke. The net effect is the generation of a force at right angles to the length. Propulsive forces are generated by movements of large numbers of cilia on a surface which are beating in a co-ordinated manner. These cilia are frequently arranged in bands or girdles. The total power is related to the number of cilia beating in the same direction and this depends on the length of the band or the area of the ciliary surface.

Muscle cells make use of a second class of cytoskeleton fibres, the actin filaments. These are composed of two helical strings made up of many units of the protein actin. The actin filaments are involved in the control of the shape and polarity of cells and these protein filaments also permit the changes in cell shape which can be used to generate locomotory forces. In muscle cells, actin filaments are associated with filaments of another protein myosin. These two proteins form the muscle fibres, or myofibrils, which typically contain two types of fibre: the thin and thick fibres. Actin filaments form the thin fibres and myosin filaments form the thick fibres. A muscle cell is composed of masses of myofibrils which together may constitute more than 65% of the total cell mass. In many muscle cells a functional unit can be recognised, the sarcomere. This unit is composed of bands of the thick (myosin) and thin (actin) filaments. The structure of a muscle cell is illustrated in Fig. 10.6 at various levels of magnification.

The muscle is able to perform work because of sliding movements between the actin and myosin filaments. These sliding movements require energy in the form of ATP. The thick myosin filaments can be seen at the highest levels of electron-microscopic imaging to have a large number of lateral projections, the myosin heads, which can bind to actin protein in the actin fibrils. During muscular contraction, the myosin fibril in effect 'walks along the actin fibrils' in such a direction that the myofibril shortens.

In the absence of ATP, the myosin heads are locked to the actin filaments and the muscle is rigid (rigor mortis). In living animals, this locked state is short-lived. In the presence of ATP which binds to the myosin head, the two filaments are unlocked and the myosin head is freed from the actin filament. Hydrolysis of the ATP releases a phosphate ion and ADP still bound to the myosin. In this state, the shape of the myosin head is changed and it approaches another actin-binding site. Tight binding of the myosin head to the new actin site generates a force moving and locking the mysosin head to its new actin site and releases the ADP and inorganic phosphate. The muscle filaments are once again in a locked state but in a new position. Energy has been consumed and ATP reduced to ADP. Further movement requires more energy once again in the form of an ATP molecule. A diagrammatic representation of this cycle is shown in Fig. 10.7.

The properties of a muscular system are determined by the rate at which the sarcomeres shorten and the number of

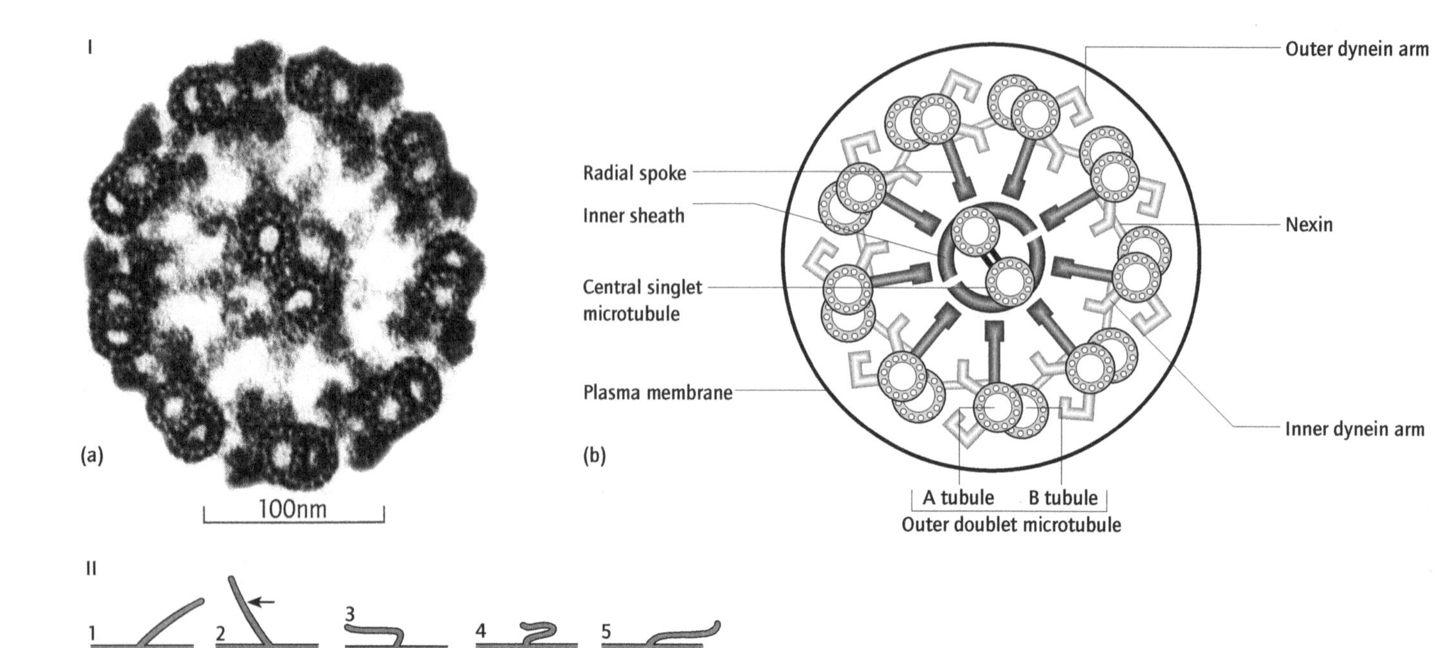

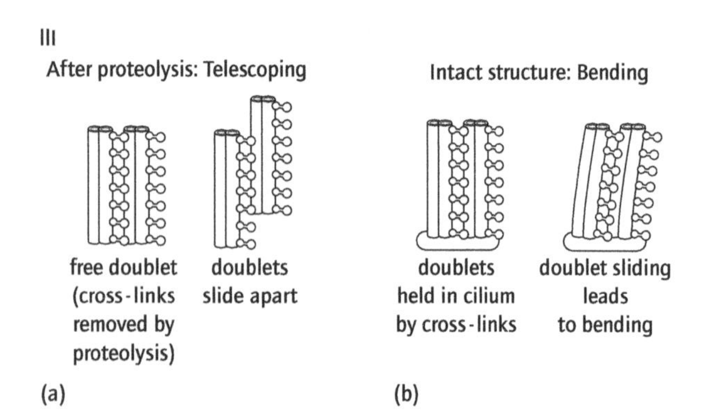

Fig. 10.4 The arrangements of microtubules and bending mechanism of the axoneme of a flagellum and cilium. (I) A diagrammatic representation of the component parts of the axoneme as visualized by electron microscopy in transverse section: (a) electron micrograph image of the axoneme; (b) diagram of the component parts of the axoneme. (II) The sequence of bending movements of a cilium: the stiff 'power stroke' (1,2) and the flexed 'recovery stroke' (3,4,5). (III) The molecular mechanism of the bending of the axoneme. (a) Proteolysis has removed cross linking proteins. The outer microtubule doublets can slide against each other and this causes axoneme to elongate. (b) In the intact cilium the doublets are tied to each other by protein cross linkages and the sliding of the doublets forces the axoneme to bend. (After Alberts, B., Bray, D., Lewis, J., Raff, M., Roberts, K. & Watson, J.D. 1994. *Molecular Biology of the Cell*. Garland Publishing, New York.)

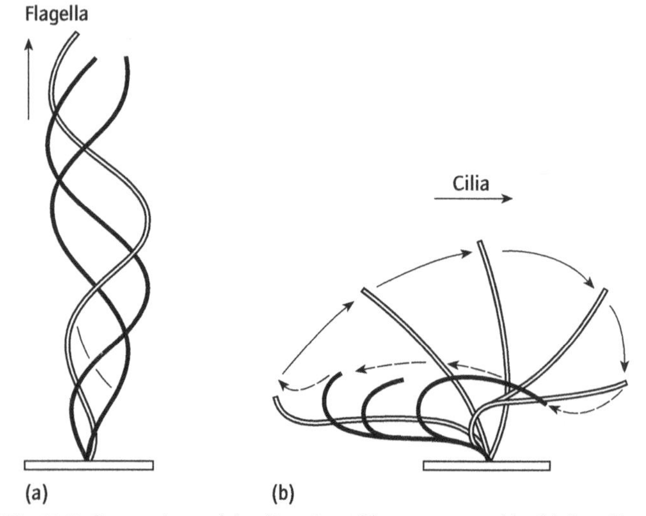

Fig. 10.5 Comparison of the direction of force generated by (a) flagella and (b) cilia.

sarcomeres in series and in parallel. The number of sarcomeres in series determines the intrinsic rate of shortening. The power of the muscle depends on the number of filaments in parallel. The force generated per unit cross-sectional area is remarkably constant through the animal kingdom. A feature of a muscle cell is that it can only generate power during shortening and there must be some opposing force to extend a shortened muscle. Once all the sarcomeres have reached their fullest state of contraction and the degree of overlap between the thin actin filaments and thick mysosin filaments cannot be increased, the potential power of the muscle is zero.

10.3 Ciliary locomotion

Invertebrates (other than the very simple placozoans; Section 3.3) do not utilize flagella in their locomotory activities, although this method of locomotion is retained as the almost universal mechanism for the locomotion of their spermatozoa. Flagellated choanocyte cells are, however, used by the Porifera

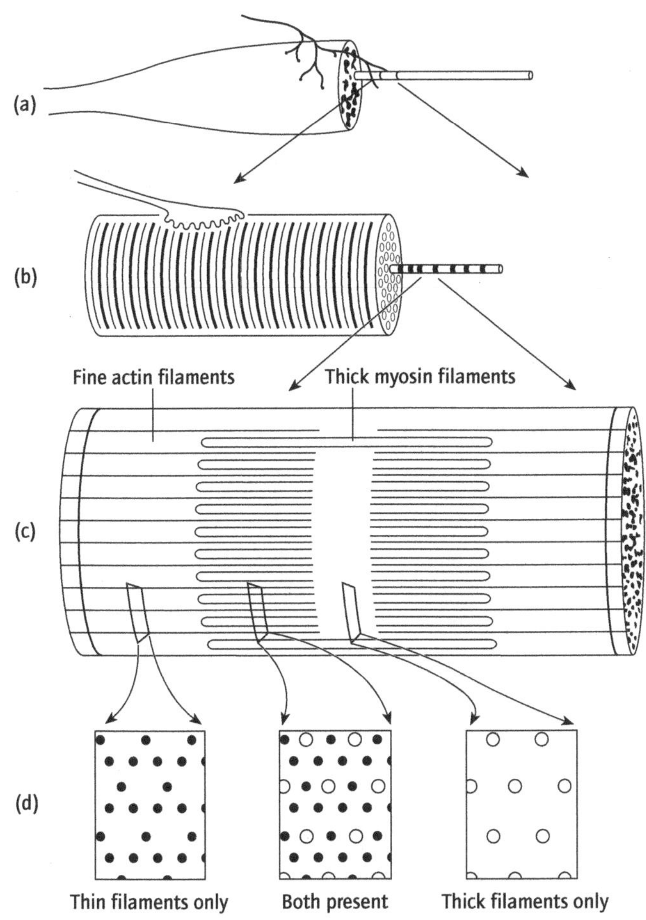

Fig. 10.6 The structure of a muscle at different levels of magnification. (a) The entire muscle with innervation; (b) a single muscle fibril with nerve ending (synapse); (c) the functional unit, the sarcomere; (d) diagrammatic representation of an electron-microscopic image of a fibril cross-section in a region of thin filaments only, mixed and thin filaments, thick filaments only.

to drive water through their chambers and canals (see Chapter 3).

The deployment of large numbers of cilia over the entire body surface or in distinct ciliated bands or girdles, however, is a common locomotory device, but is generally restricted to animals less than 10^3 μm in length. This is because the individual cilium has a fixed mode of beating and the only way in which the sum of the locomotory forces can be increased is by an increase in the number of cilia. This is limited, however, by the surface area of the organism, whereas the mass of organism increases with the volume. Moreover, hydrodynamic efficiency of ciliated organisms decreases with increasing size; consequently, larger organisms moving in this way are required to expend much greater energies to maintain the same relative velocities.

The larvae of many invertebrates such as the polychaetes, molluscs and echinoderms have their cilia arranged in bands or girdles. As these larvae grow they need more power for feeding and movement. This requires either an increase in the number of bands or an increase in the relative length of the bands. Unless the larva is neutrally buoyant, its mass will increase with the cube of the linear dimensions. Its surface area will only increase by the square of the linear dimension, hence unless there is a change in shape, the power per unit mass will decrease. Because of this, there is a disproportional increase in the length of the ciliary bands of invertebrate larvae as they grow and the locomotory bands will be thrown into massive folds and projections as illustrated in Fig. 10.8.

Most organisms that use cilia as the only means of locomotion are relatively small. The Ctenophores, however, are relatively large. They are near to neutral buoyancy and have cilia fused into compound structures in the ciliary comb rows. These animals can achieve speeds of up to 15 mm s^{-1} and may be near to the limits of the locomotory potential of cilia-based locomotory systems.

The activity of cilia is co-ordinated in such a way that each cilium has the same pattern and frequency of beating, but cilia in successive rows are at a different phase of this cycle (Fig. 10.9a). The co-ordination is due to viscous coupling of the hydrodynamic system. The movement of an individual cilium involves a power stroke and a more flexed recovery stroke that results in a net flow of water in the outer boundary layer (Fig. 10.9b). For a free-swimming organism, the reaction to the force generating this flow of water will drive it in the opposite direction.

The wave of activity may pass in the same direction of the power stroke or in the opposite direction so that each cilium makes its power stroke slightly after the one immediately in front of it. There will then be no interference between cilia during the execution of the power stroke. Cilia co-ordinated in this way are said to exhibit *metachronal rhythmicity*. The ciliated larvae of marine invertebrates most frequently serve the functions of dispersal and feeding. The ciliary bands provide the power for both. The movements of the cilia generate a mass flow of water (relative to the animal surface) resulting in locomotion or prevention of downward drift due to gravitational forces. The moving-water mass contains small particles which represent a food source. However, in order to feed, these particles must be trapped and this requires a localized zero flow rate for the water. The larvae are found to be of two functionally different types which trap particles in different ways. The trochophore larvae of Polychaetes and Molluscs have a complex ciliary girdle, the prototroch, which is composed of a longer anterior row of cilia and a shorter posterior row, separated by a food groove with very short cilia leading to the mouth (Fig. 10.10). The two rows of cilia beat in opposite directions. The longer anterior cilia being the more powerful, there is a net resultant force in the outer boundary layer. The inner cilia beat in the opposite direction resulting in localized eddies where the net movement of water is zero. This delivers small particles passing in the inner parts of the moving boundary layer into the food groove with zero velocity, so that they may then be driven towards the mouth.

Echinoderms trap particles in a different way. There is only a single band of cilia and all the cilia beat in the same direction. Food particles are trapped by localized reversal of power stroke

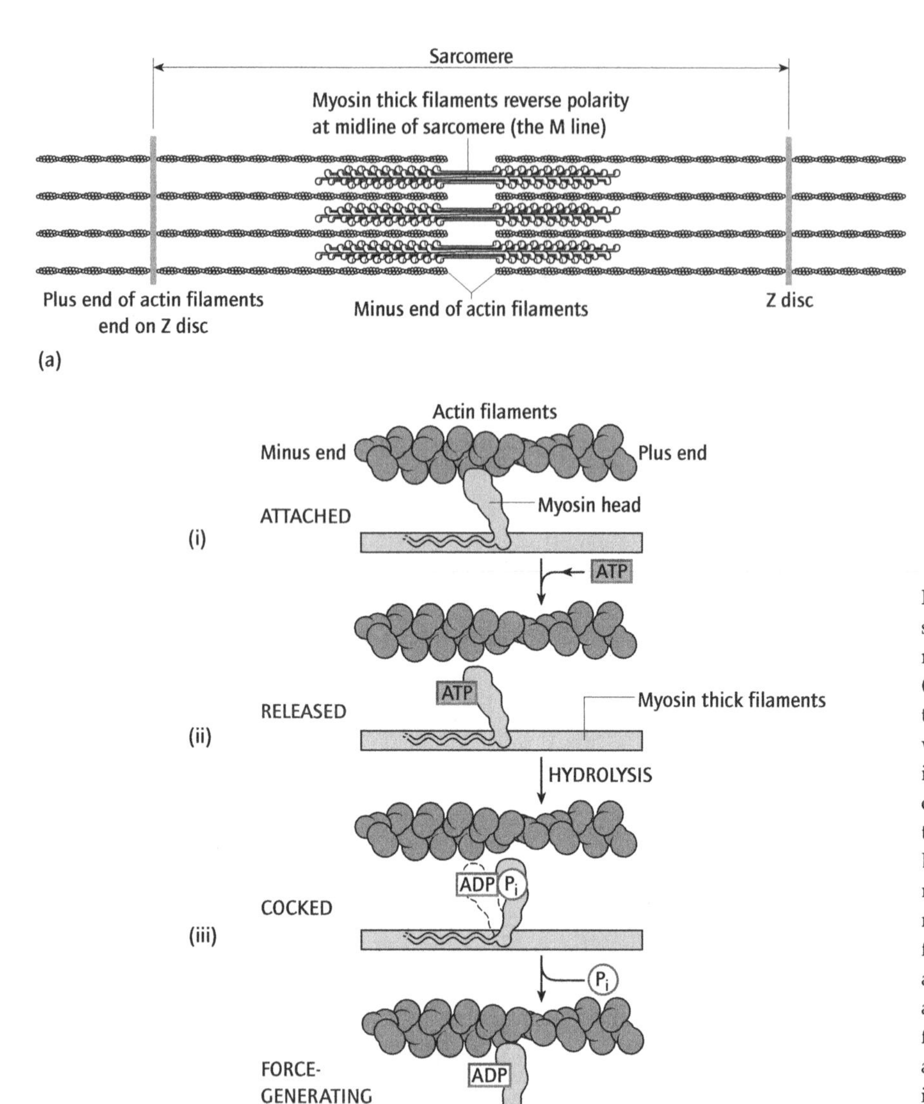

Fig. 10.7 (a) Representation of the longitudinal structure of the protein filaments in a striated muscle. The thin actin filaments have a polarity (+) and (–) the plus ends oriented and anchored to the proteins of the Z discs. Thick myosin filaments which have a midpoint polarity reversal are inserted between the thin filaments and when energy is available 'walk' inwards between the thin filaments towards their plus ends. (b) (i–v) Diagrammatic representation of the molecular mechanisms involved in the energy-consuming migration of myosin filaments along the actin filaments. In (i) the myosin is 'locked' onto the actin filament. In living cells this is a transient state as a free ATP molecule will bind to the mysosin filament (ii) head causing it to unlock from the actin. Hydrolysis of the ATP to ADP plus an inorganic phosphate ion triggers a large change of shape, the so-called 'cocked' position (iii). A force is generated as the cocked myosin head binds tightly to a new position (iv) releasing the ADP. The two filaments are again tightly bound (v) and a new cycle requires the supply of an energy-rich ATP molecule. (Based on a figure in Rayment *et al.* 1993. *Science* 261: 5058, after Alberts *et al.* 1994 – cited Fig. 10.4)

induced in groups of cilia when particles in the moving boundary layer approach close to the cilia. The reversal of power stroke causes local eddies leading to zero velocity and again the trapping of food particles.

The relationship between linear dimensions, surface area and volume in animals means that only relatively small animals are able to move by simple cilia alone. For any organism of constant shape, the surface area increases in proportion to the square of the linear dimension whereas the volume (hence mass) increases in proportion to its cube. These geometric relationships also have profound functional implications affecting the functional design of respiratory and excretory systems (see also Chapters 11 and 12), as well as the mechanics of locomotion and support.

The transition from ciliary to muscular locomotion is likely to have taken place among the flatworm-like ancestors of the Bilateria, and the two modes of locomotion are found to coexist in some free-living flatworms and nemertines.

The epidermis of free-living flatworms (turbellarians) and nemertines is abundantly ciliated and the smallest specimens, which are about 1-mm long, lie at the upper end of the size range for efficient locomotion using ciliary mechanisms. Larger flatworms (triclads and polyclads), which retain ciliary creeping as the principal means of locomotion, do so by becoming flat;

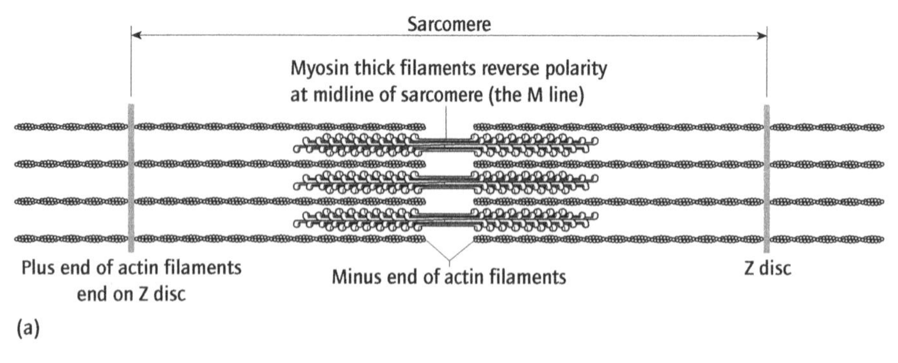

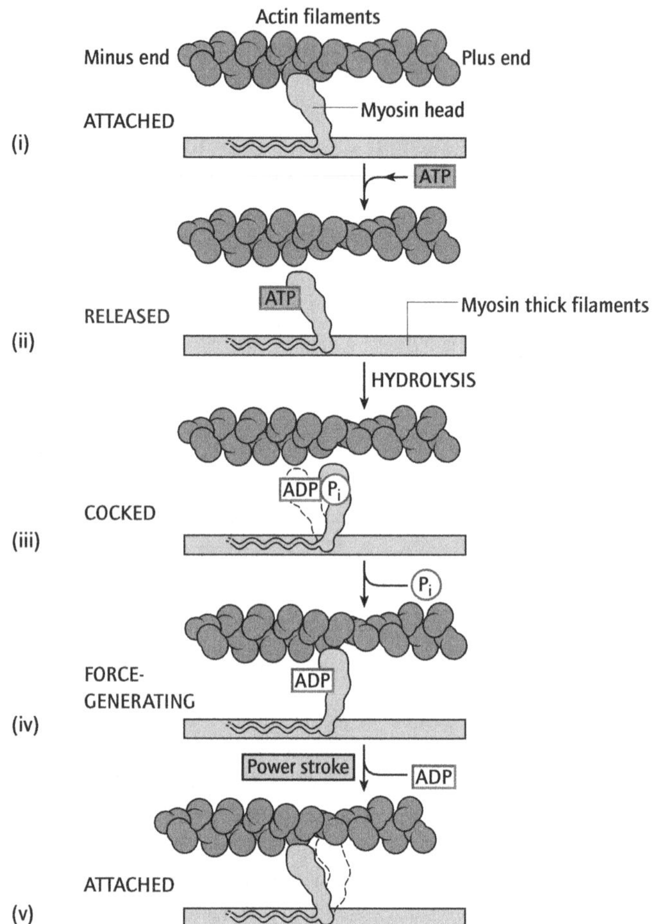

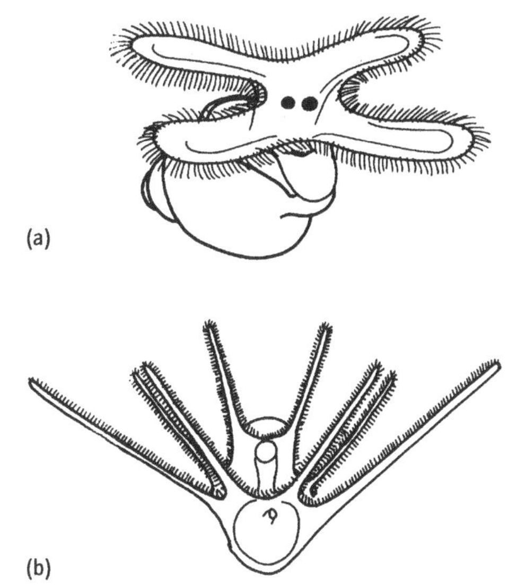

Fig. 10.8 Larvae with elongated ciliary bands: (a) gastropod veliger larva; (b) echinoderm pluteus larva.

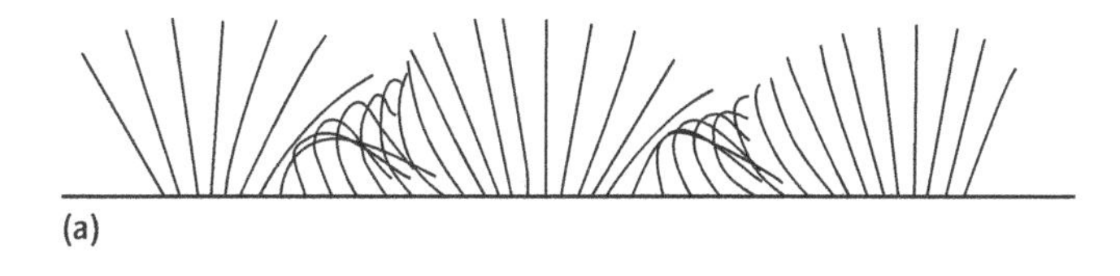

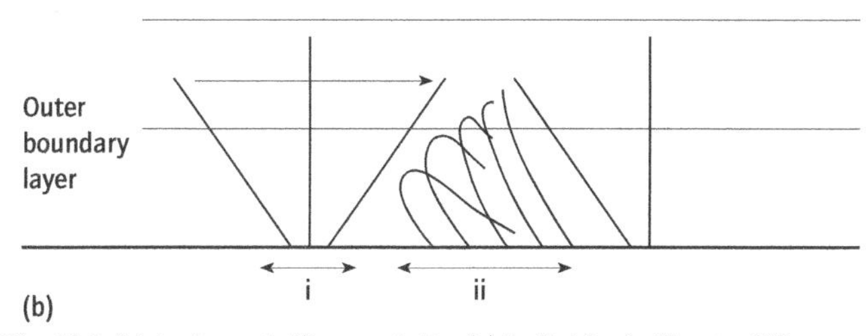

Fig. 10.9 Metachronal ciliary activity. (a) Individual cilia at a different phase of the cycle of activity; (b) the power stroke (i) and recovery stroke (ii) produce a net force in the outer boundary layer.

consequently, their surface area increases by more than the square of the linear dimensions.

The largest organisms to move by ciliary creeping are presumably nemertines, such as *Lineus longissimus*, which can achieve lengths of many metres yet have a maximum width of only *c*. 1–2 mm. This gives the required high surface area, but it is a solution open only to aquatic species. In terrestrial environments, the high rates of water loss which would be associated with such a high surface area are not acceptable. The muscular activities of flatworms and nemertines are varied and involve pedal locomotory waves, peristalsis or looping movements with anterior and posterior adhesion. These have often been developed to a much greater degree by other phyla and the principles will be described in the following sections.

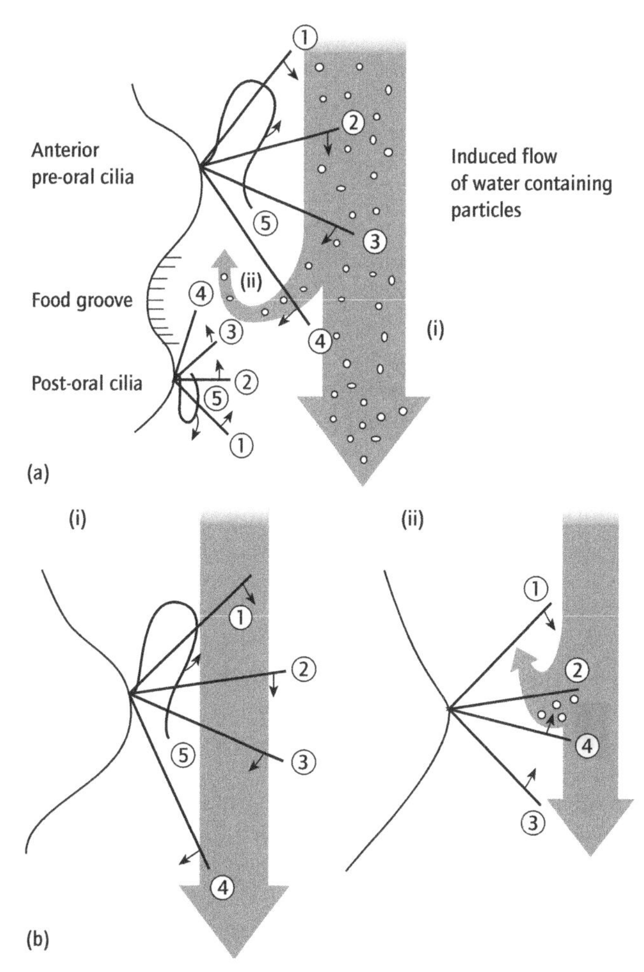

Fig. 10.10 Two modes of using ciliary bands for locomotion and for food capture. (a) Opposed ciliary bands of trochophore and similar larvae of annelids and molluscs. The prototroch consists of two bands of cilia. A pre-oral band of larger more powerful cilia (i) and a post-oral band of weaker cilia (ii) beating in a counterdirection. The powerful cilia result in the current of particle-containing water as shown. The counterbeating cilia set up local eddies that reduce the velocity locally resulting in particles being trapped and being transferred to the mouth via the ciliated food groove. (b) Echinoderm larvae and others have only a single band of cilia. These beat to produce a water current and particles are trapped by the response of cilia that respond to the presence of small particles and reverse their beat with a stiff recovery stroke. This localized reaction also traps food particles from the moving water mass.

10.4 Muscular activity and skeletal systems

As explained above, chemical energy can be used by muscle cells to cause shortening of the muscle fibres. This provides a means by which the power of an organism can be related to the mass of the muscle cells rather than to the surface area. Consequently, most larger invertebrates use muscular rather than ciliary locomotory systems. Muscle can only be part of a locomotory system if there is a skeletal system to transmit the forces generated.

The skeletons of animals are of two fundamentally different types. Fluid or *hydrostatic skeletons* are used by soft-bodied animals. They function because of the non-compressibility and fixed volume of a liquid which can therefore be used to transmit pressure. Locomotory systems using hydrostatic skeletons are especially characteristic of the worm-like animals (see Chapter 4). Animals with hardened tissues may also use *rigid skeletons*. Rigid skeletons are used by a number of invertebrate groups including some echinoderms, larval chordates but most importantly by the various arthropod phyla, the invertebrates with jointed legs (see Chapter 8). The distinction between animals with rigid skeletons and fluid skeletons is far from being absolute. Many of the non-arthropod phyla have some rigid skeletal parts; the spines of echinoid echinoderms, the 'vertebrae' of ophiuroid arms and the aciculae in the parapodia of polychaete worms are examples.

Similarly in the arthropods, a hydrostatic skeleton is often used for the transmission of forces as in the movements of the gut and in arachnids for the extension of jointed limbs. The rigid skeletons of the arthropods are exoskeletons in which the hardened skeletal elements enclose the soft tissues. In the chordates an elastic skeletal rod, the notochord, has been developed as a simple internal skeleton. From this origin, the vertebrates with their characteristic cartilaginous or bony skeletons have evolved.

In all locomotory systems involving muscular activity, the muscles can only generate power when they contract, and a force must be applied to restore them to their original length. This is most frequently accomplished by antagonism between different sets of muscle fibres, but the restorative force can also be provided by the action of the cilia (as in some cnidarians), or by energy stored in elastic tissues. The fluid skeletons of soft-bodied invertebrates can be employed in a wide variety of ways, but the basic principles are always the same. These are explained in Box 10.3.

Many soft-bodied animals are capable of very great changes of shape. Some of the most deformable are the nemertines, but there are physical limits to deformability since the volumes of the tissues and usually of the body cavities of animals are fixed. In the nemertines, there are inelastic fibres in the body wall inclined at an angle to the longitudinal axis. If the fibres were to be stretched out until they were parallel to the longitudinal axis, the enclosed volume would be zero; similarly if the fibres had an angle of 90° to the axis, the volume would again be zero. Between these two impossible extremes is a position where the enclosed volume is maximal; it occurs when the fibres have an angle of approximately 55° to the longitudinal axis. Nemertine worms conform very closely to the limits set by this system. They are spherical in cross-section when maximally contracted and when maximally extended. In between, the volume of the worm is less than that would be enclosed by the fibre system and the worm has a flattened or elliptical cross section. Figure 10.11 shows the theoretical and observed limits for the nemertine worm *Amphiporus*. This worm conforms very closely to the limits set by the fibre system, being longest when the fibres have

Box 10.3 The hydrostatic skeleton

1 The contraction of muscles surrounding a fluid-filled cavity will increase the pressure of the fluid:

$$\textbf{Pressure} = \frac{\textbf{force}}{\textbf{area}}$$

The SI unit of pressure is newtons per square metre N m^{-2}.

2 The **force** acting on any surface is therefore F = pressure × area.

3 Pressure acts equally at all points of the surface and acts at right angles to the surface (see below).

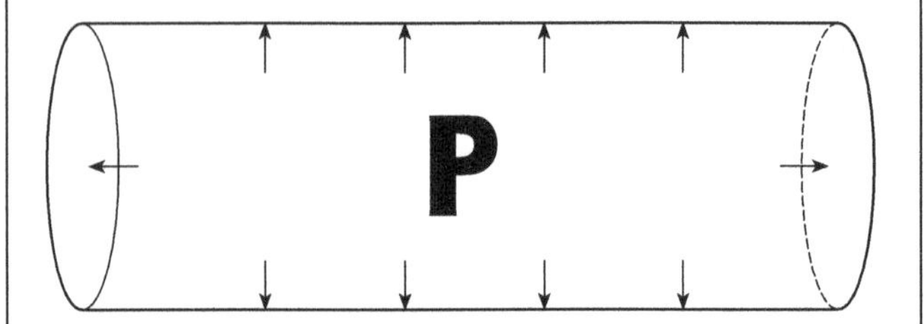

Pressure acts in all directions in a fluid.

4 The internal pressure in the hydraulic system can be resisted by the boundary surface.

If the resistance of the boundary surface is lower in any region than that necessary to contain the pressure, fluid movements will cause a change in shape. This is the basis of locomotion in all animals that use a fluid skeleton.

As shown below, muscles in the right- and left-hand parts of a cylinder oppose each other.

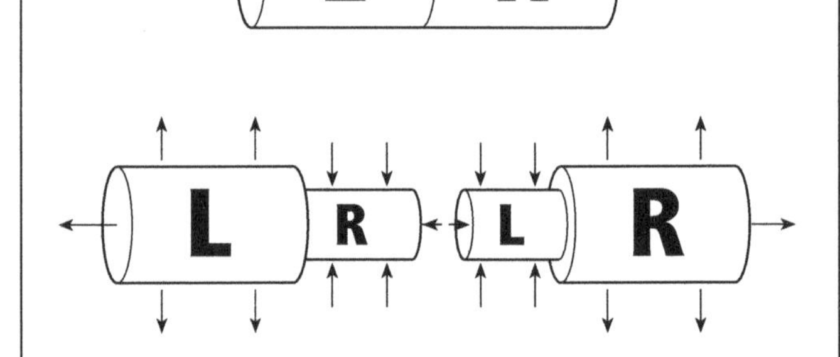

Antagonism between connected compartments in the system, in this case left and right.

5 A very versatile system results if muscles are arranged around the body wall as circular and longitudinal sets.

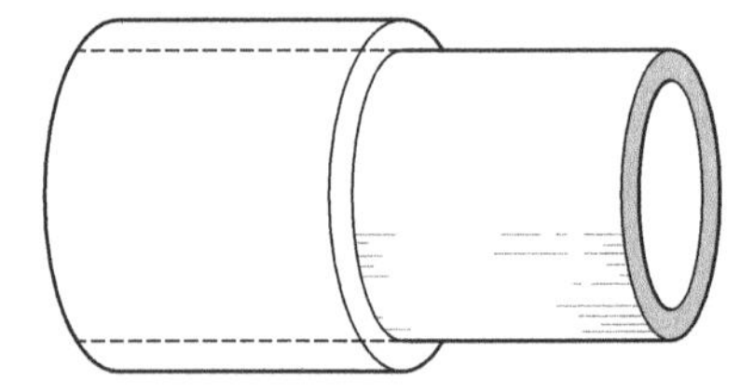

Opposing circular and longitudinal muscles, circular muscles on the outside.

Continued

Box 10.3 *(cont'd)*

6 Any section of an undivided cylindrical animal can then perform a wide variety of changes in shape.

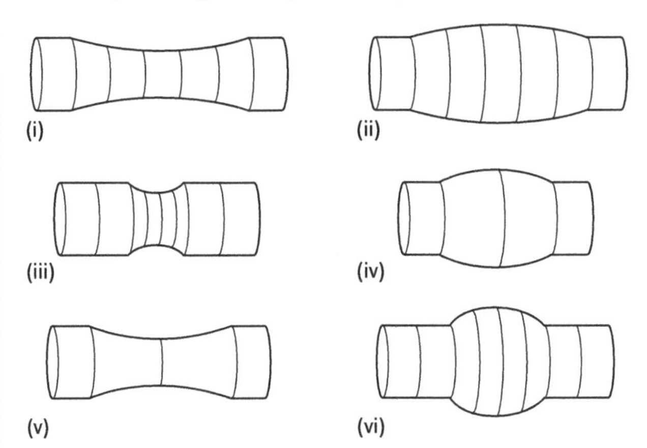

Body regions not of fixed volume. (i) Circular muscles contract, fluid exported; (ii) circular muscles relax, fluid imported; (iii) longitudinal and circular muscles contract, fluid exported; (iv) longitudinal and circular muscles relax, fluid imported; (v) (vi) body regions of fixed volume; (v) longitudinal muscles relax, volume constant; (vi) circular muscles relax, volume constant.

7 Changes in shape can perform mechanical work. The work done will be the product of the force applied (=pressure × area) and the distance moved.

In the example below a proboscis is extended while much of the body contracts.

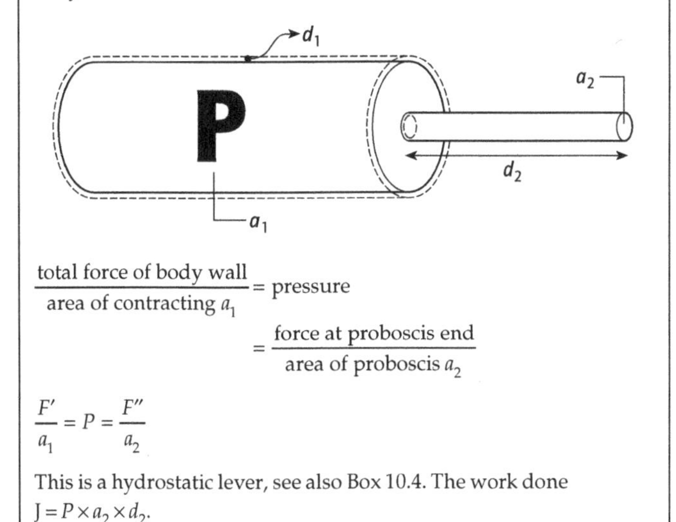

$$\frac{\text{total force of body wall}}{\text{area of contracting } a_1} = \text{pressure}$$

$$= \frac{\text{force at proboscis end}}{\text{area of proboscis } a_2}$$

$$\frac{F'}{a_1} = P = \frac{F''}{a_2}$$

This is a hydrostatic lever, see also Box 10.4. The work done $J = P \times a_2 \times d_2$.

an angle of about 80°. Some nemertines are rather less variable in shape because of inelasticity of the tissues.

Nematode worms have a high internal pressure and the cross-sectional shape is always circular – they are always, in effect, at a position of maximum contraction with a high angle between the fibre system and the longitudinal axis. Contraction of the longitudinal muscles will tend to reduce the volume, and since fluids are not compressible, such contractions will raise the internal pressure which acts against the muscles. Contralateral contractions cause a change in shape (sinusoidal flexing) and the internal pressure will return the muscle to its original length when it relaxes. The nematode worms therefore do not couple circular and longitudinal muscles as illustrated for a low internal pressure worm in Box 10.3.5 and there are no circular muscles in the body wall (see Fig. 4.13). The high-pressure system of the nematodes is another aspect of the design stemming from the complex fibre and cuticle system of the body wall (Fig. 4.15) that permits high internal pressure to be maintained without fluid loss.

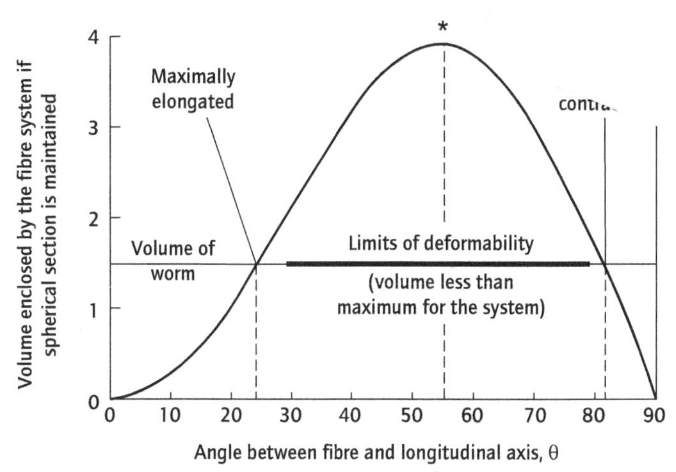

Fig. 10.11 The physical properties of an inelastic spiral fibre system and the shape of worm-like animals. The curve shows the theoretical limits of the volume enclosed by such a fibre system. Nemertine worms have volumes much less than the maximum and consequently can show great changes in shape. The heavy line shows recorded data for the nemertine Amphiporous. A worm whose volume was close to the maximum, as marked *, would not be capable of changes in shape without increasing the internal pressure. (Redrawn from Clark, 1964.)

Some soft-bodied animals with oblique muscle fibres such as leeches are able to set the fibre angle at 55° and to become in effect rigid. Invertebrates that have rigid skeletons are conveniently grouped together as the Arthropods (see Chapter 8 for systematic discussion) and these organisms undoubtedly include the most successful groups of invertebrates in the marine and terrestrial environments.

Among the many advantages of the functional design of the arthropods are the mechanical advantages that accrue when muscles acting on rigid or jointed skeletons are also able to transfer forces from the site of generation to the point where they are applied to the environment. A rigid skeletal element can act as a mechanical lever, the principles of which are explained in Box 10.4. Note, however, that the example illustrated in Box 10.3.7, is also a lever system which depends not on relative length but on relative area.

Box 10.4 Rigid skeletons – the principle of a lever

1 A force F can be made to move a load M when it causes rotation about a fulcrum. If F is the force applied and F_R is the force of reaction, they are related by the relationships

$$F \times d_1 = F_R \times d_2$$

where d_1 is the distance moved by F and d_2 is the distance moved by F_R.
Note that the pivot reverses the direction of the work done.

2 A small force can move a large load but energy will be conserved such that:

$$\frac{F_R}{F} = \textbf{mechanical advantage} = \frac{d_1}{d_2} = \textbf{velocity ratio}$$

3 A large effort can also be made to move a small load over a large distance.

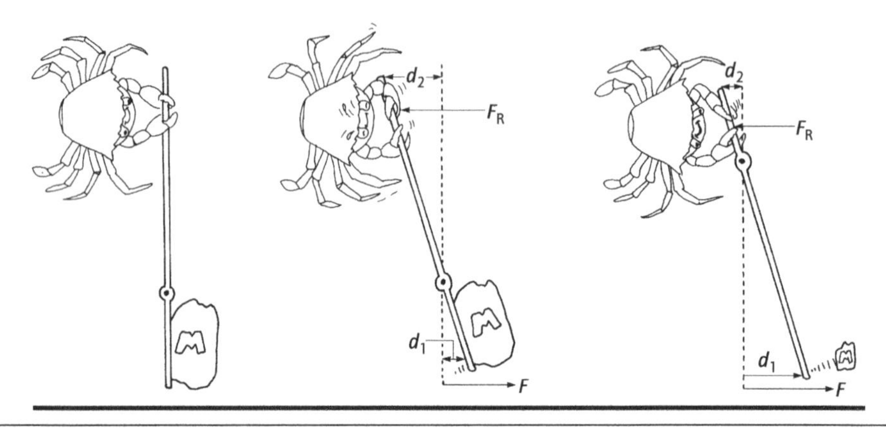

10.5 Burrowing, creeping, crawling, walking and running: locomotion over and through a solid substrate

10.5.1 The locomotion of soft-bodied invertebrates

Many soft-bodied invertebrates are able to move over a firm substrate that is not substantially deformed by it. In order to move, the animal must transmit a force to the substratum through a fixed point, often referred to by the French term 'point d'appui'.

In soft-bodied animals, the locomotory system often involves the propagation of waves of contraction and relaxation in muscles that have their longitudinal axes (fibre orientations) parallel to the direction of locomotion. Flat worms, some cnidarians and above all the gastropod molluscs move by means of waves of activity in the muscular surfaces applied to the substratum. These waves of contraction are called *pedal locomotory waves*. Pedal locomotory waves can be seen quite easily by examining the undersurface of a planarian or a snail while it crawls along a glass plate. In species of the land snail *Helix*, several waves crossing the whole of the foot will be seen simultaneously. The waves are seen through the glass as dark and pale bands passing over the surface of the foot. Each wave moves in the same direction as the snail but at a greater rate. A locomotory wave like this that moves in the same direction as the animal is called a *direct locomotory wave*. The velocity of the animal relative to the ground is often represented by the symbol V and the velocity of the locomotory wave relative to the ground is represented by the symbol U. In the case of *Helix* the velocity V is less than the velocity U but they have the same direction.

In other mollusc species, for example the marine limpet *Patella*, a smaller number of waves can be seen on the foot at any time and the waves pass in the opposite direction to the movement of the animal. This type of wave form is described as a *retrograde wave*. In limpets, the waves on the left- and right-hand side of the foot appear to be a half wavelength out of phase. You will notice this because the light and dark bands that you see moving backwards along the foot if you watch it through a glass plate, only extend half way across the foot. A wave crossing the whole of the foot, like in *Helix* is called a *monotaxic wave*, a wave that only crosses half the foot and where left- and right-hand sides are out of phase with each other are called *ditaxic waves*.

The different types of activity have different characteristics. The monotaxic, direct locomotory waves of *Helix*, in which the wavelength is much less than the length of the foot, can be considered to be a low-geared system, giving relatively low speed and poor manoeuvrability but the ability to move a relatively large mass or overcome large resistive forces. Ditaxic systems with a wavelength as long as or longer than the foot give higher relative speeds and much greater manoeuvrability.

The molluscan foot is an organ of considerable complexity. Its principal skeletal element is a complex system of blood sinuses. Although the animals are acoelomate, there is a hydrostatic skeletal system within the foot. The principal muscle systems are not longitudinal muscle fibres, but opposing systems of anterior and posterior oblique muscles (Fig. 10.12). It used to be thought

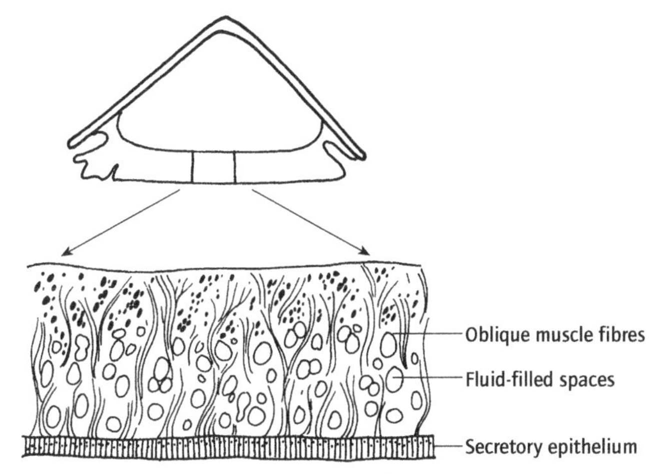

Fig. 10.12 The structure of the molluscan foot.

that each segment of the foot actually become detached (as shown in Box 10.5a,b) during the passage of a locomotory wave, but this is unlikely due to the huge forces which would be necessary to separate two closely adhering wet surfaces. The formation of points d'appui and zones of movement depend on the properties of the mucous layer beneath the foot. The mucus acts as an elastic solid under conditions of low lateral stress but as a viscous liquid under higher lateral stress. At the points d'appui it acts as an elastic solid but is fluid under the wavefront where the lateral forces are greater.

Some snails, including *Helix*, can exhibit an alternative 'galloping' movement, in which retrograde waves of muscular activity, with a wavelength about equal to the length of the foot, are generated. This type of locomotion can further be modified to the stage where there is attachment to the substratum alternatively at anterior and posterior extremities – a type of locomotion which is more characteristic of the looping movements of leeches described below.

Many large flatworms and most nemertine worms exhibit a muscular component to their locomotion in which the alternating waves of contraction of the circular and longitudinal muscles generate retrograde peristaltic waves which enhance the locomotory activity of the surface cilia. This system is most highly developed in the septate coelomate worms and is particularly characteristic of the earthworms.

Figure 10.13a illustrates the movement of an earthworm. Note that segments are stationary when the circular muscles are relaxed and longitudinal muscles contracted (see Box 10.5). The segments in front of the shortest segment are elongating and exert a backthrust against the ground through the points d'appui while the posterior segments are contracting and exert tension through this point (Fig. 10.13b). There will naturally be a tendency for slip to occur, but this is prevented by the erection of stout chaetae.

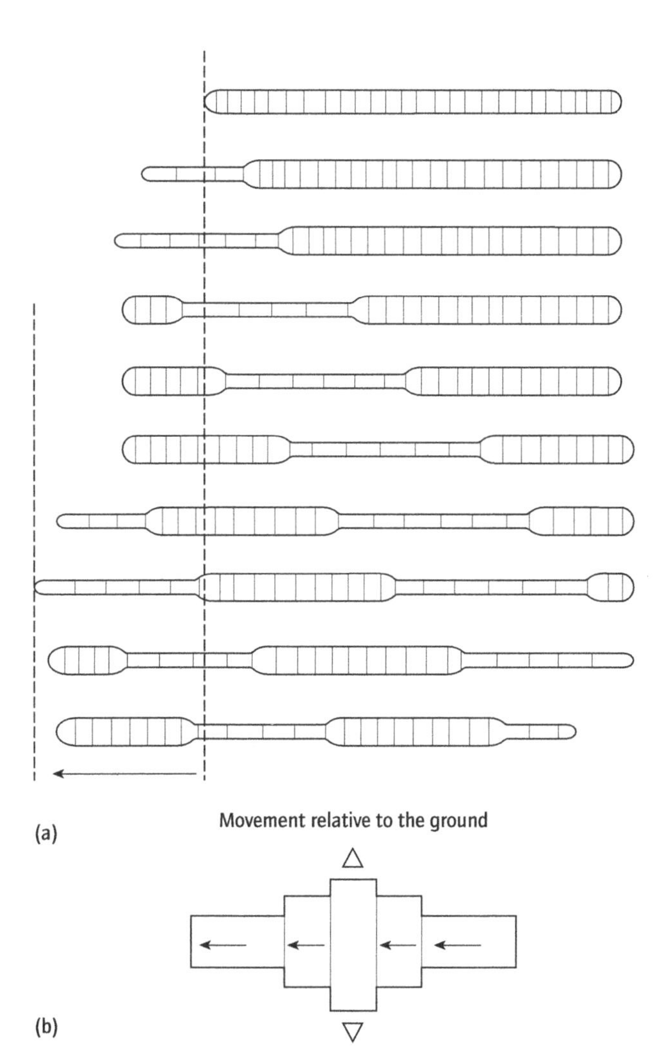

Fig. 10.13 (a) Successive stages in the movement of an earthworm. Segment volume is fixed as in Box 10.5, Fig. a(ii). The segments are stationary with respect to the ground when the longitudinal muscles are contracted, and circular muscles relaxed; segment volume is fixed and the locomotory wave is retrograde. (b) Forces acting on the fixed segments with erected chaetae.

The pressure waves associated with the contractions of circular and longitudinal muscles are separated (Fig. 10.14) and the intersegmental septae effectively isolate pressure changes in individual segments. This is not possible in aseptate animals. The highest pressures are recorded when the circular muscles contract and the segments would be penetrating the substratum. This is thought to be one of the major advantages of the septate condition responsible for the evolution of metamerism in annelid worms. Each segment is a separate hydrostatic element with its own fixed volume (see Box 10.4). Although we have considered earthworms in the context of movement over a flat surface, there is little doubt that they are primarily adapted for movement between the interstices of the soil.

Box 10.5

The illustrations show the body of an animal in which five regions are identified (they may or may not be true segments) and which are attached to the ground when extended in (a), when contracted in (b). The waves of muscular activity can be in a muscular foot as shown in a(i) and b(i) or in a cylindrical worm as shown in a(ii) and b(ii).

The result is always the same. If the points d'appui are formed in regions where the longitudinal dimensions of the body are maximal, the animal will move in the same direction as the wave – the wave is direct. If, on the other hand, the points d'appui are in regions of longitudinal muscle contraction so that the fixed body regions are at their shortest, the animal will move in the opposite direction to the wave, the wave is said to be retrograde.

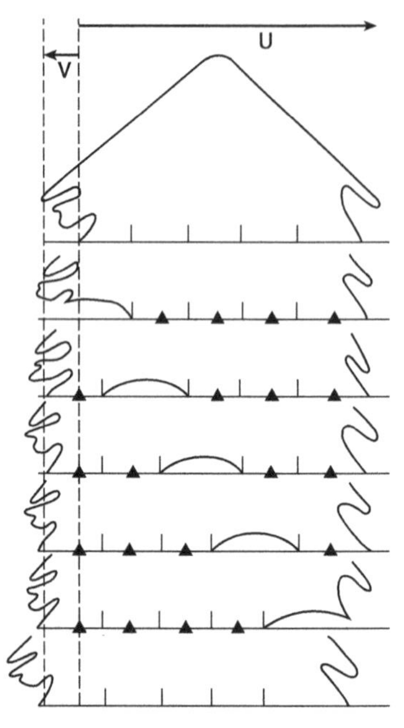

(a) (i) Points d'appui: short segments wave retrograde

(b) (i) Points d'appui: long segments wave direct

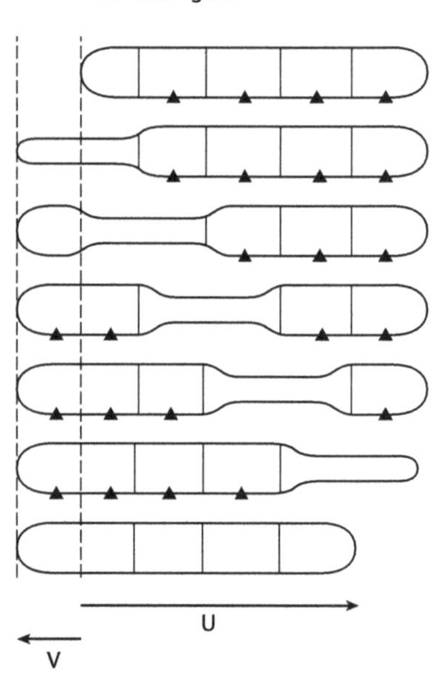

(a) (ii) Points d'appui: short segments wave retrograde (cf. earthworms)

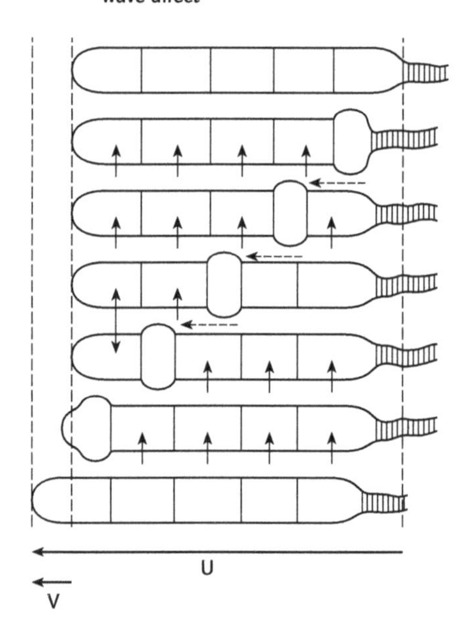

(b) (ii) Points d'appui: long segments wave direct. For a cylindrical worm in a tube this will require the erection of chaetae, but see also nereidiform locomotion.

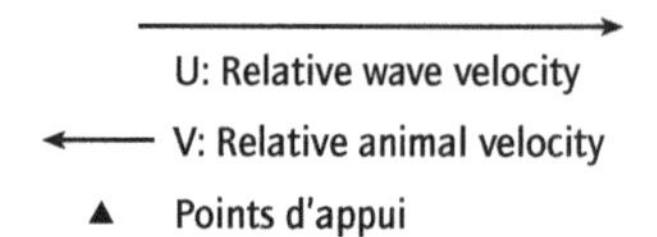

U: Relative wave velocity

V: Relative animal velocity

▲ Points d'appui

↑ Erectile chaetae to form points d'appui (see also Fig. 10.18)

←---- Slippage

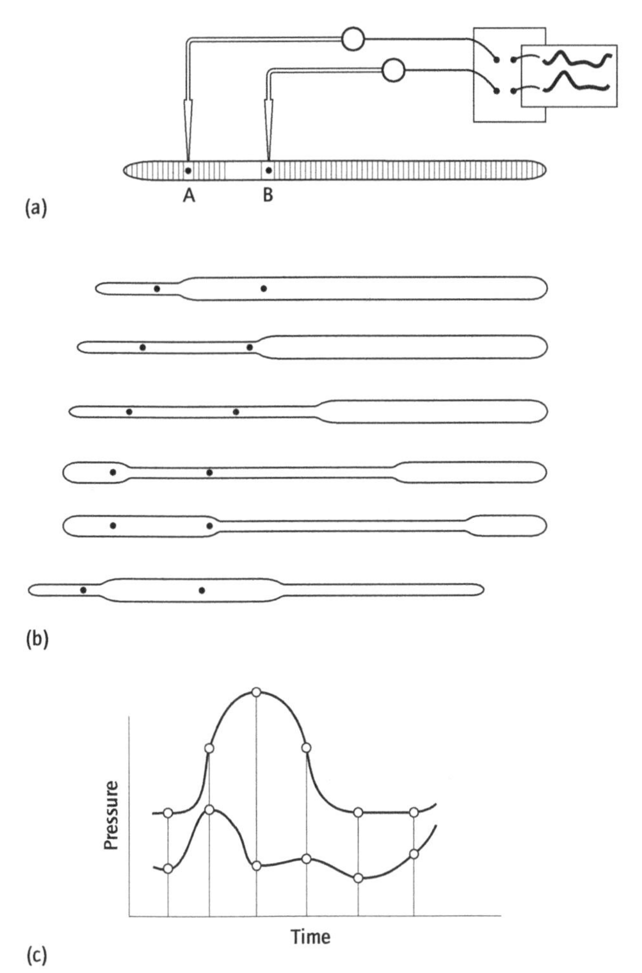

Fig. 10.14 Studies of locomotion in the earthworm. (a) System in which pressures in individual segments A and B can be recorded by the insertion of plastic tubes directly linked to a pressure transducer. (b) Successive position of the two segments A and B during recording. (c) The recorded pressure waves. (After Seymour, 1969.)

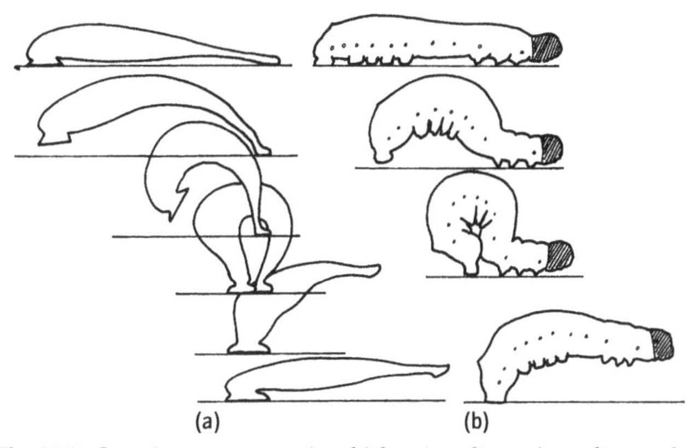

Fig. 10.15 Looping movements in which points d'appui are alternately at anterior and posterior ends. (a) A leech; (b) a caterpillar (lepidopteran larva).

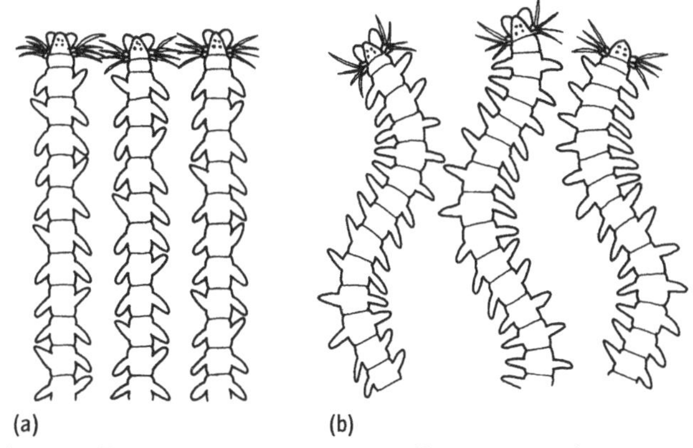

Fig. 10.16 Diagrammatic representation of locomotion in the polychaete *Nereis*. (a) Slow crawling; (b) rapid crawling. For reasons of simplification the movements of the chaetae are not shown. Chaetae can be protruded and retracted to increase step length.

A dramatically different means of locomotion has been developed by the leeches though it can be derived from that shown by the earthworms. These animals have posterior and anterior suckers which provide particularly effective points d'appui. The whole body acts as a single hydrostatic system, the gut providing a single fluid-filled cavity, and attachment occurs alternatively by anterior and posterior suckers (Fig. 10.15a).

Similar movements are also exhibited by some insect larvae, e.g. lepidopteran caterpillars, in which arching movements are equivalent to the contraction of longitudinal muscles (Fig. 10.15b). In leeches the body cavity (coelom) is virtually obliterated by a deformable tissue, the botryoidal tissue, and the animals are effectively aseptate and acoelomate from a design point of view. They are undoubtedly annelid worms which betray their segmented origins in the details of their anatomy, especially in the nervous system and in the expression of typical segmentation Hox genes.

The errant polychaete worms exhibit a mode of locomotion which is very different from both that of the oligochaetes and leeches. It involves the movement of multiple limbs, the tips of which are made to move backwards relative to the body. Because the tips are attached to the ground, forces are generated that cause the body to move forwards.

When a *Nereis*, for example, is crawling slowly (Fig. 10.16a) there is a metachronal wave of activity in the parapodia passing forwards from the tail to the head and with the left and right parapodia being exactly one half wavelength out of phase. This direct wave ensures that each limb executes its power stroke just beforehand. As a *Nereis* crawls more quickly or swims, the same ditactic direct locomotory wave is exhibited by the longitudinal muscles of the body wall are also used to generate propulsive force (Fig. 10.16b). A synchronous wave of activity in the left and right longitudinal muscle is initiated so that each parapodium executes its power stroke as it is passed by the crest of the sinusoidal wave in the body wall which moves forward synchronously with the wave of parapodial activity.

The forces transmitted to the ground are now not only those generated by the parapodial musculature, but also ones

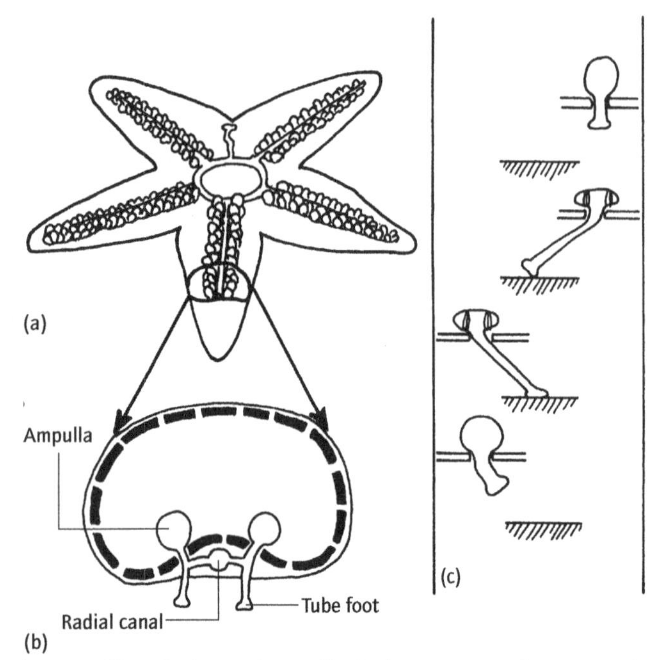

Fig. 10.17 The tube foot and ampullae in the water vascular system of an echinoderm: (a) the general arrangement; (b) a diagrammatic cross-section of an arm to show the water vascular system; (c) stepping cycle of a single tube foot (retractor muscles in the tube foot are not shown).

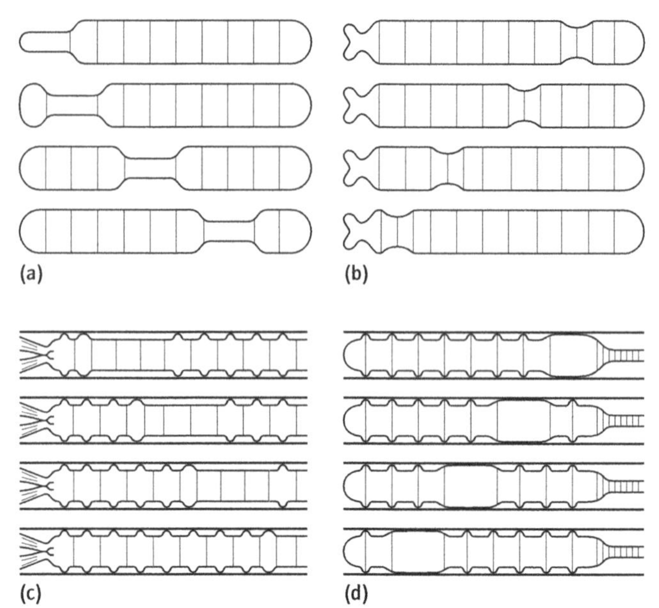

Fig. 10.18 Four modes of peristalsis which may be exhibited by tubiculous worms. In (a), (b) and (d) segment volume is not fixed. In (c), which represents the fanworm *Sabella*, segment volume is fixed due to complete septa. Locomotion during peristalsis is prevented by the formation of points d'appui in segments of intermediate length by the erection of chetae. The inside of the tube is slippy to allow the shortened segments; which do not form the points d'appui in this case, to move inside the tube as a kind of plunger.

generated by the contraction of the contralateral longitudinal muscles.

The locomotory patterns of snails and of soft-bodied animals like the earthworm and *Nereis* follow some general rules. It seems odd at first sight that the locomotory waves may pass either in the same direction as the animals movement or in the opposite direction. It is found that direct locomotory waves are found when points d'appui are formed in regions where the longitudinal muscles have maximum length. Retrograde locomotory waves are found where points d'appui are formed when the longitudinal muscles have minimum length. This general principle is illustrated for snails and burrowing worms in Box 10.5.

The water vascular system of echinoderms provides a unique ambulatory system. In starfish, for example, there are typically five arms and in each one there is a water vascular canal radiating from the circum-oral ring. Along each radial water vascular canal is a series of reservoir-like ampullae and tube feet (Fig. 10.17a,b). The ampullae and tube feet have muscular components which work against each other. Contraction of the muscles compressing the ampullae drives water into the tube feet whereas contraction of the tube feet must involve a mass flow of water into the ampullae. Each ampulla does not act as a simple reservoir for its own tube foot, as there is mass movement of fluid between regions of the system, and a tube foot can be extended to have a volume greater than that of its ampulla when fully dilated. In general, however, muscles which compress the ampullae act antagonistically to those which shorten the tube feet.

The tube feet are extended by hydraulic pressure, can perform single step-like motions and may be provided with adhesive suckers at their tip (Fig. 10.17c).

A peculiar feature of starfish locomotion is that the tube feet do not exhibit any detectable metachronal rhythmicity.

10.5.2 Burrowing and movements within tubes

Burrowing worms tend to have large body cavities and to be at least partially aseptate. Animals with this structure can exhibit a wide variety of movements within the burrow (Fig. 10.18); four types of peristaltic locomotion are possible because any region of the body can simultaneously increase in length and girth. These movements may be used for irrigation, locomotion, or irrigation and locomotion together.

Unsegmented, and segmented but aseptate, animals with large body cavities can also perform major changes in shape which can be used when burrowing into the substratum; and the sipunculans, for example, are capable of rapid re-entry into the substratum. The associated movements are illustrated in Fig. 10.19, together with a trace of the pressures generated internally. These animals have the capacity to exhibit a high rate of work, but such activity would not normally be maintained for long periods of time. During phases of high coelomic pressure, all the muscles of the body must do metabolic work to maintain constant length. Mechanical work is performed by the

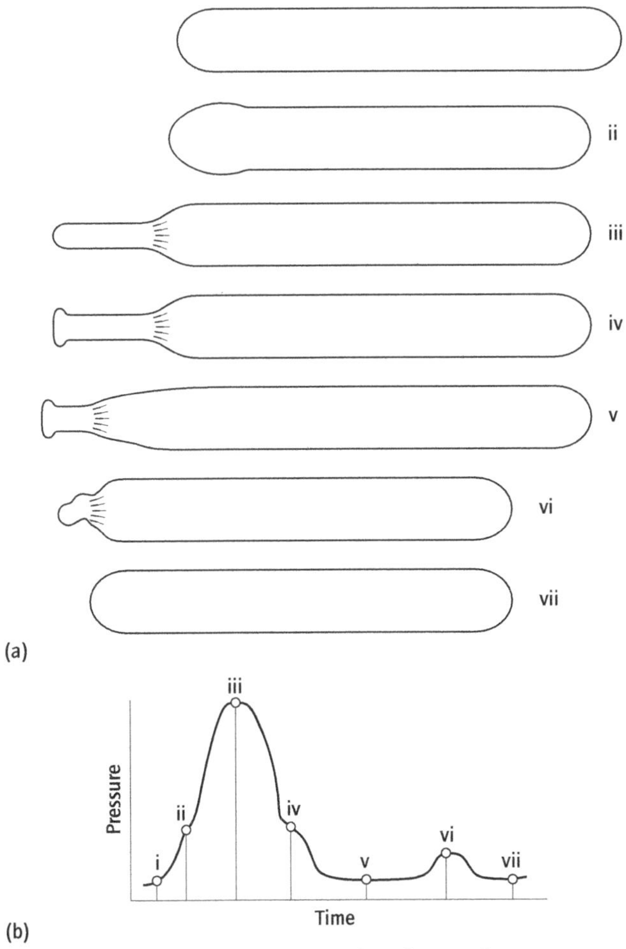

Fig. 10.19 (a) Diagrammatic representation of successive movements of a sipunculan during burrowing into or through the substratum. (b) Internal pressures recorded during the burrowing cycle. Note that highest pressures are recorded during penetration into the substratum (as in earthworms, cf. Fig. 10.14) but all regions of the body have the same pressure. (After Trueman & Foster-Smith, 1976.)

controlled relaxation of specific muscles allowing certain regions of the body to be extended (see Box 10.3). The rate of work can be high, but efficiency is low due to the requirements for continuous muscle tone at all times of high coelumic pressure. It could be argued that the evolution of transverse septa in the annelids occurred at least partly as a means of resisting the outward movement of the body wall without muscular activity, and which thus increased the mechanical efficiency (cf. Section 10.1).

The polychaete worm *Arenicola marina* is a segmented worm which is perhaps secondarily adapted to a burrowing existence. Its segmented structure confers a number of advantages. Nervous co-ordination is enhanced by the segmented ganglia; there are chaetae with parapodia for formation of points d'appui and a well developed vascular system with vascular gills. Mechanically, *Arenicola* has an undivided trunk coelom which gives it the wide repertoire of locomotory movements needed and it is capable of high rates of working. This gives it the ability to re-

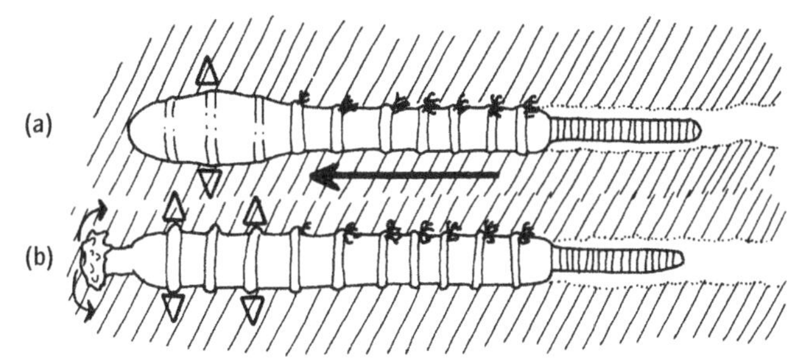

Fig. 10.20 Different stages in the burrowing of the lugworm. (a) Terminal anchor. The anterior part of the worm is dilated while the posterior is drawn forwards. (b) Penetration anchor. The anterior segments form 'flanged' anchor points while the proboscis excavates the sand and the head is thrust forwards. (After Trueman, 1975.)

enter the substratum by the alternate formation of penetration and terminal anchor points (Fig. 10.20), which are also to be found in burrowing cnidarians and in the later stages of burrowing by bivalves (Fig. 10.21).

Arenticola normally lives in an open J-shaped tube which is part of a U-shaped burrow system (Fig. 9.30). The animal excavates and eats sand at the base of the head shaft using scraping and ingesting movements by the proboscis. The rate of working is low, and for this activity the anterior coelom is isolated by the pharyngeal septa. These have valves and can either isolate the anterior coelom for sustained activity or, when open, allow the high pressures generated in the trunk coelom to do work via the extruded proboscis when burrowing more actively.

Tubulous worms which retain the intersegmental septa, such as the fanworm *Sabella*, have a special problem. It is necessary for them to irrigate the tube by peristalsis. With fixed segment volume they can only do so by employing a peristaltic wave of type (c) in Fig. 10.18: consequently they tend to move in a direction opposite to the waves. This is prevented in *Sabella* by the formation of points d'appui when the segments have intermediate length through the erection of chaetae (see Box 10.5b(ii)). Other segments slip inside the very smooth lining of the animal's tube.

10.5.3 The locomotion of invertebrates with jointed limbs

In many errant polychaetes, the role of the longitudinal muscles in crawling is slight, most of the tractive power being developed by the parapodial muscles, which cross segmental boundaries. This tendency is especially marked in the scale worms and their relatives, and in these conditions the transverse septa are reduced or lost. The locomotion of these animals involves intrinsic and extrinsic muscles of the parapodium. The chaetae can be protracted and retracted by intrinsic muscles in the chaetal sac, and both intrinsic and extrinsic muscles operate the parapodium which can be lifted and depressed, and moved forwards and backwards causing stepping movements. This walking movement is developed further by animals with jointed exoskeletons.

The Crustacea, Chelicerata and Uniramia are three diverse and successful groups of invertebrates. They are all characterized by

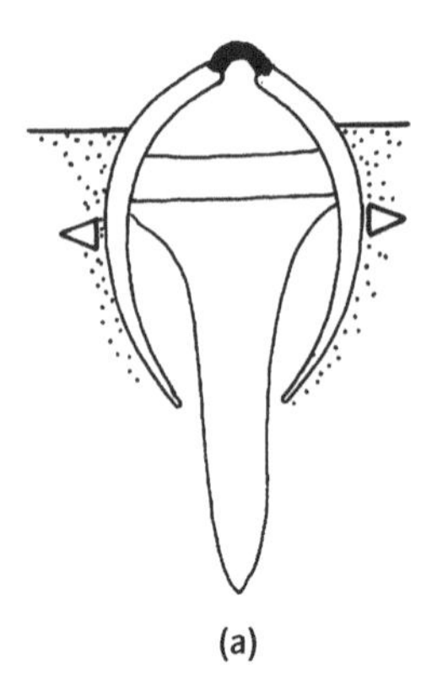

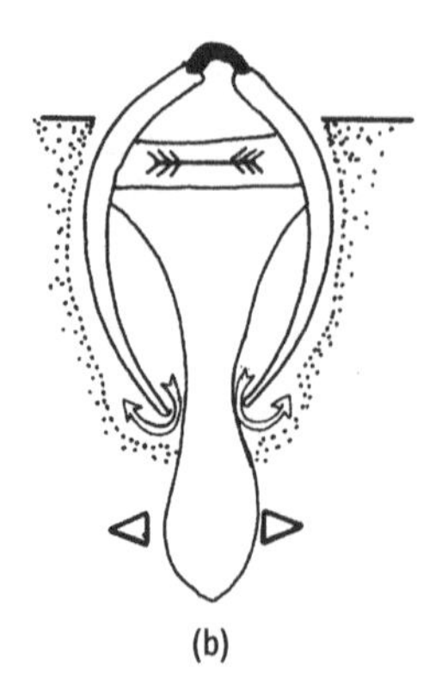

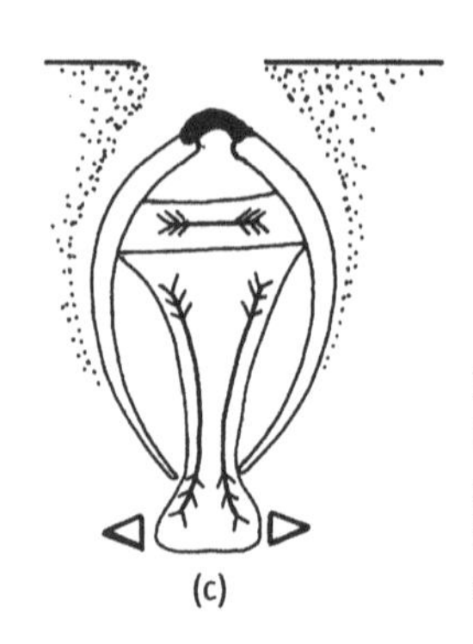

Fig. 10.21 Burrowing of a bivalve mollusc. (a) The shell gapes and forms an anchor during penetration of the foot. (b), (c) The tip of the foot is dilated while the shell is drawn down into the sand. (After Trueman, 1975.)

their hardened exoskeletons. This enables them to have jointed limbs, and the success of the three groups must in part be attributed to the locomotory abilities which this confers. In addition, pterygote Hexapoda (insects) are the only vertebrates to have evolved wings; this gives them the power of true flight and is one of the factors that has allowed their dominance of terrestrial environments. Despite their polyphyletic origins, the walking limbs of the most highly evolved members of the Crustacea, Chelicerata and Uniramia are remarkably uniform in structure as a result of convergent evolution. The limbs are composed of a series of jointed elements becoming progressively less massive towards the tip (Fig. 10.22a). Each joint is articulated to allow movement in only one plane. These limb joints allow extension and flexion of the limb; rotation of the limb plane at the basal joint with the body is also possible and this is often responsible for forward movement. The body is typically carried slung between the laterally projecting limbs (Fig. 10.22b) and walking movements do not involve any raising or lowering of the centre of gravity.

As the base of the limb is rotated, the tip of the foot can be made to trace a linear path parallel to the axis of motion by extension and flexion of the limb joints. The limb acts as a mechanical lever (see Box 10.4) and in long-legged arthropods the velocity ratio is such that a large force can be made to move the relatively small load through a large distance giving a rapid walking gait.

The bulkiest of the muscles providing the locomotory power are located not within the limb but within the body, and since the animals do not bob up and down during forward locomotion, only the limb undergoes major changes in momentum. This is kept to a minimum by the common structural design, in which the mass of the limb is reduced from base to tip. Lateral undulations of the body such as those exhibited by polychaete worms (see Fig. 10.16) and some centipedes (see below) do involve changes in momentum, and suppression of this is likely to have been a powerful influence in arthropod evolution.

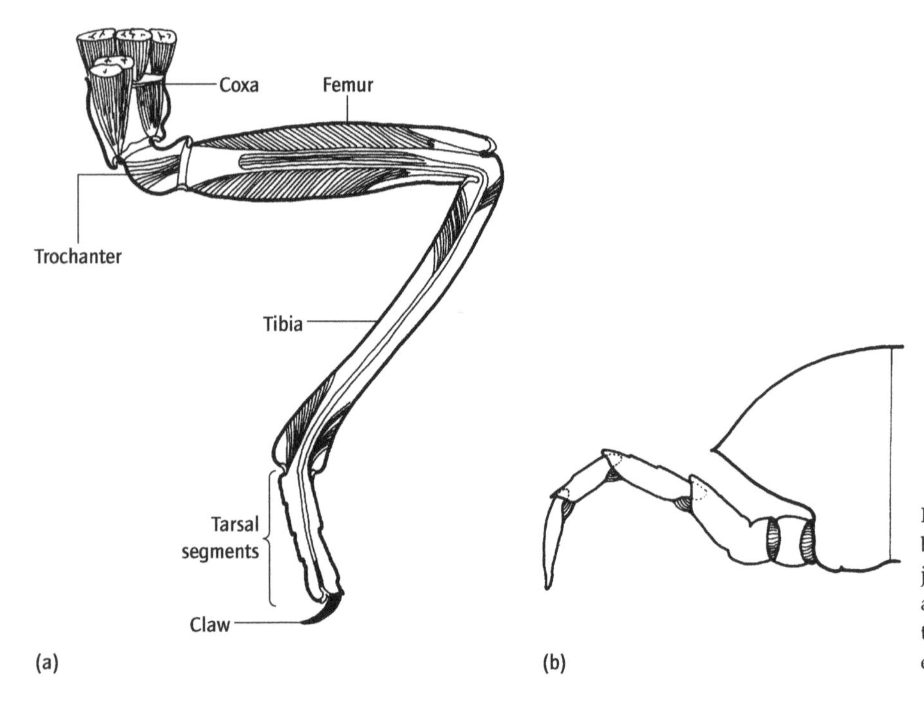

Fig. 10.22 (a) A typical arthropod limb. Note the bulk of muscles in basal sections, tendons and joints. (b) The characteristic attitude of the arthropod limb. Protraction involves rotation of the proximate joint in a horizontal plane and extension of the more distal joints.

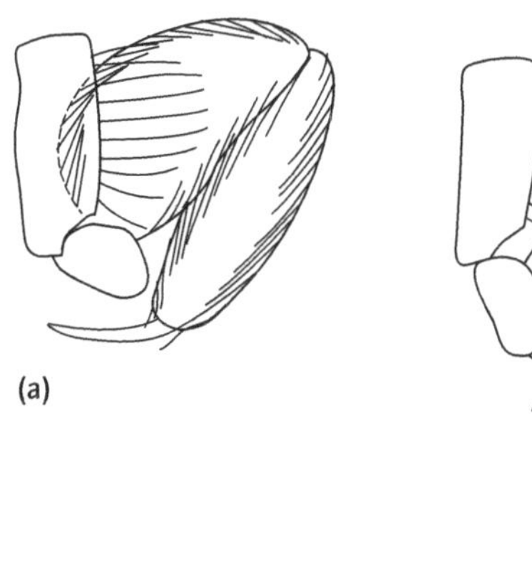

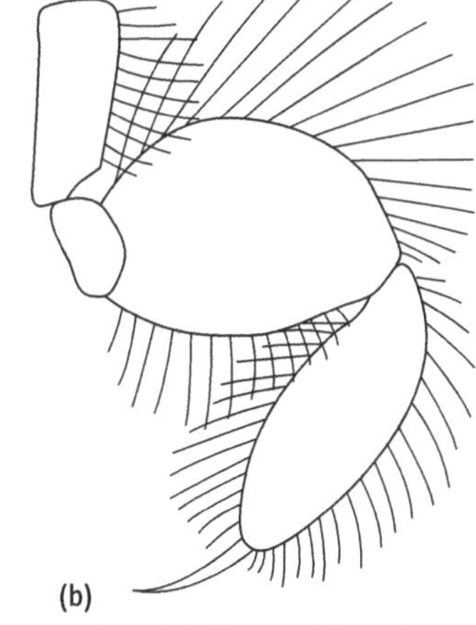

Fig. 10.23 The role of setae in the generation of differential drag forces in the swimming of Crustacea with relatively low Re numbers. (a) Attitude during recovery stroke; (b) attitude during effector stroke. (After Hesseid & Fowtner, 1981.)

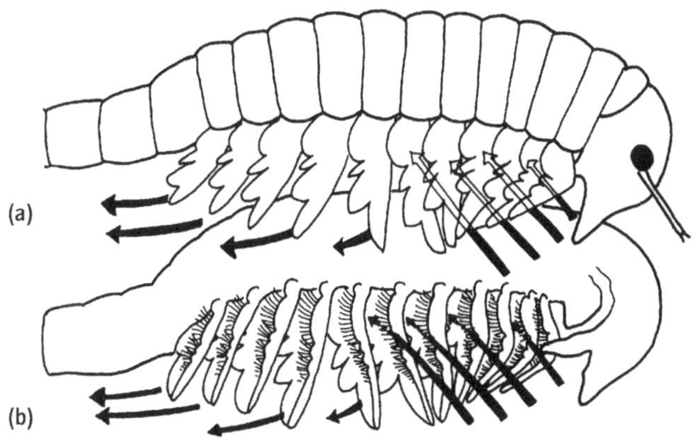

Fig. 10.24 The swimming/feeding currents produced by the limbs of phyllopod Crustacea. (a) Outer view of the right-hand side; (b) inner view of the left-hand side, showing the bristles which filter the food particles which are then passed forwards to the mouth. Water is drawn into the mid-ventral line and then forced backwards when it passes through the filter on the endopod or inner limb part.

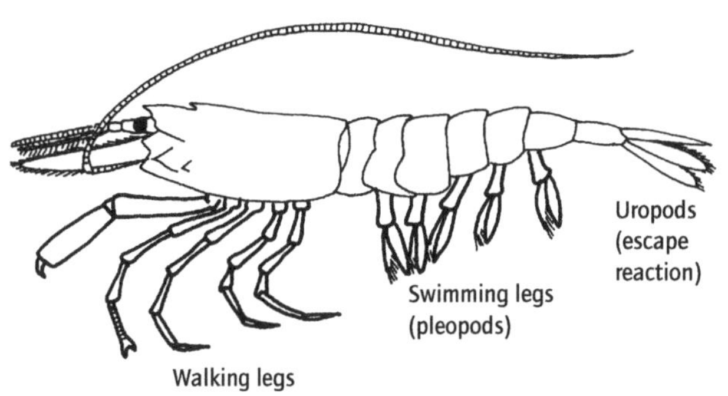

Fig. 10.25 Shrimp-like malacostracan with both walking and swimming limbs.

10.5.3.1 Locomotion of crustaceans (walking in the sea can easily turn to swimming)

The crustacean limb is based on a plan which has several component parts: two unpaired basal segments, a foliaceous epipod, and paired distal parts – the exopod and the endopod (see Figs 8.34b–d). In the living phyllopod Crustacea, such as the Cephalocarida and the Branchiopoda (fairy shrimps) (see Figs 8.37–8.39), there are similar limbs on each non-cephalized thoracic segment, which serve the functions of both locomotion and feeding.

The crustacean limb can act as a paddle in which the surface area presented to the water is at the maximum during the power stroke, but is much reduced due to folding and the movements of the setae during the recovery stroke (Fig. 10.23). There is differential drag between effector and recovery strokes. The swimming of the primitive crustaceans, however, is also due to the expansion and contraction of the spaces formed between the limbs. Water is drawn into the mid-line between the limbs (Fig. 10.24) and passes into the interlimb space on each side and exits laterally. The unidirectional flow is due to the valve-like action of the outer exopod. Food particles are trapped by setae on the endopod and are passed from seta to seta along a mid-ventral food groove forwards towards the mouth. The metachronal rhythm of the limb movement is therefore responsible for both locomotion and feeding and probably also for respiration.

A special feature of crustacean locomotion is the relative ease with which the transition between walking and swimming may be made. Many of the shrimp-like malacostracans have thoracic walking limbs and abdominal swimming limbs (pleopods) (Fig. 10.25). There are many lines of evolution leading to the loss of walking ability (see Fig. 8.50a) or reduction in the role of the swimming pleopods, as in the lobsters and crabs (see Fig. 8.50d–g). The lobsters also have a well-developed escape reaction involving the last pair of abdominal appendages, the uropods (see Fig. 8.50d, g). The powerful muscles of the abdomen can be contracted to result in its sudden flexion, which causes the uropods to exert a force on the water and the lobster moves rapidly backwards (see Box 16.8).

In the true crabs (Brachyura), the abdomen is reduced and most walk on five pairs of thoracic limbs. Walking in an aquatic medium poses special problems. Water movements such as waves or tidal flows will tend to lift and dislodge the animal and walking can only be resumed by re-establishing the connection with the substratum. An animal such as a crab walking in water will be subject to substantial forces other than weight. Its movements will generate drag and in some circumstances lift and the animal may be buoyant. This complex of forces is shown in Fig 10.26a. The proximity of the animal to the susbstratum will alter the fluid movements over the animal and cause it to be dislodged. For this reason an organism walking in water, unlike an animal walking on land, may need to grip the substrate to avoid being dislodged. When walking in a fluid medium, the water will tend to impede the movement and rapid movements may generate lift. It is therefore relatively easy to make a transition

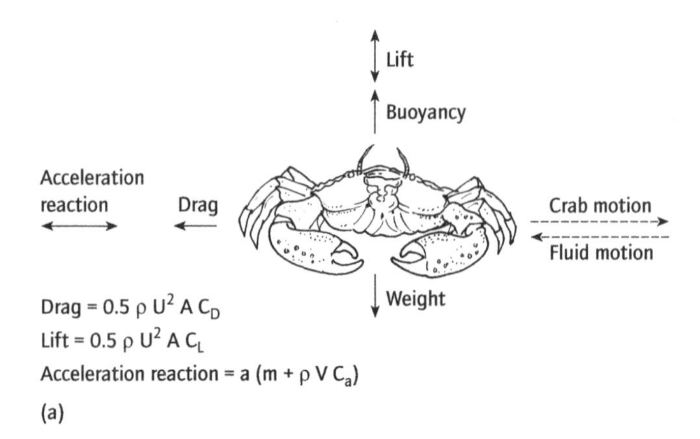

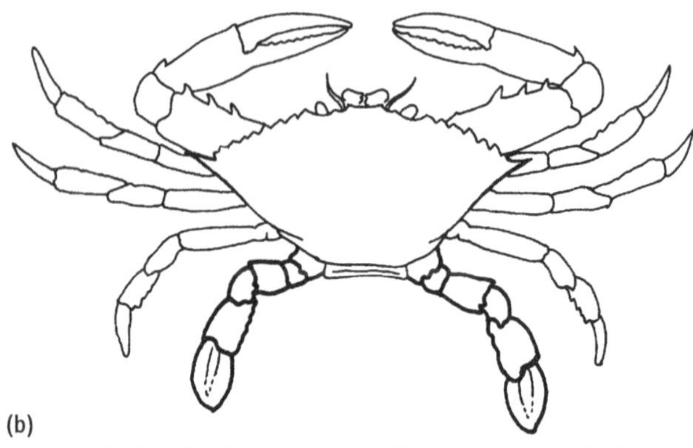

Fig. 10.26 (a) Possible forces generated by a sideways ambulatory crab in moving water. (b) The swimming crab *Callinectes*. The hind limbs act as hydrofoils during swimming.

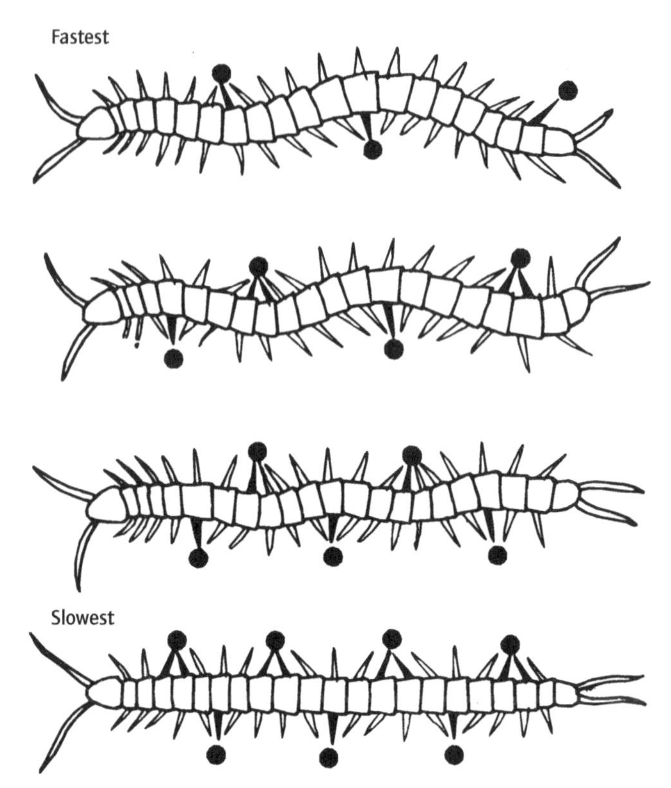

Fig. 10.27 Tracings of a centipede running progressively faster. The black dots mark limb tips which are stationary and in contact with the ground. (After Manton, 1965.)

to swimming. Some crabs are specialized for swimming as well as walking. In the portunid *Callinectes* (Fig. 10.26b), the last thoracic limbs are flattened and modified for sustained swimming. This crab swims by rotational movements of the legs which act as hydrofoils.

10.5.3.2 Terrestrial locomotion: walking and running

Animals living in terrestrial environments are much denser than the air in which they live; they therefore require structural support, and those that move frequently, or quickly, make use of rigid skeletal elements which interact with the ground. The skeletal elements may have to resist considerable deformative forces. These include bending forces, normal to the axis of the limb, and torsion forces acting about the axis of the limb.

The structure which gives this property with the minimum mass is a hollow cylinder. Rapid locomotion also requires flexible links (joints) between the rigid elements, tendons for the transmission of the forces generated by muscles and energy-storing devices. The arthropod cuticle is ideally suited to meeting all of these demands. The basic material is a carbohydrate–protein complex, chitin. This is a tough and flexible substance, but is not itself suitable for forming the rigid skeletal elements which are required. The cuticle can be hardened, however, by the formation of linkages between the protein elements. This tanned protein is called sclerotin. It can give to the cuticle a very high degree of rigidity comparable with bone or it can be relatively soft. This enables the cuticle to form flexible hinges characteristic of the articulation of arthropod limbs. Each limb joint usually permits movement in one plane only, this being achieved by the development of internal buttresses.

The simple flexible hinge joint allows for flexion by muscular activity while extension of the joint often involves the elasticity of the joint membrane or involves a hydraulic mechanism in which the forces causing extension of the joint are transmitted by the haemocoelic fluid in the limb.

The stepping locomotion of terrestrial arthropods typically involves rotation of the laterally directed limb axis. In the Diplopoda (millipedes) there are a very large number of short limbs (Fig. 8.23). A metachronal rhythm (see cilia movement in Section 10.2) of leg movements passes forwards, consequently each limb makes its power stroke shortly after the limb immediately behind. This could be referred to as a 'low-geared' locomotory system in which a relatively large force can be applied to the anterior end of the animal, but maximum speed is low. Diplopoda are herbivorous animals mostly living in rotten wood and similar substrata, and this is a suitable mode of movement for that mode of life.

Many of the Chilopoda (centipedes) however, are more active predatory animals capable of greater speed. Figure 10.27 shows tracing of a centipede running progressively faster, as the speed

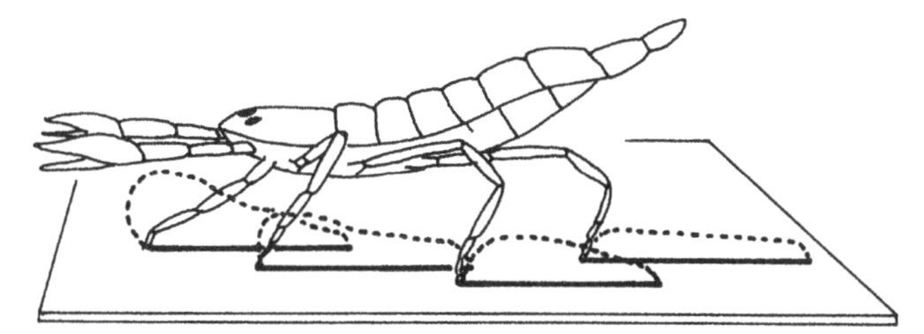

Fig. 10.28 Side view and trajectories of the limb movements of a scorpion. Note that the paths of the individual limbs do not cross each other. (After Hesseid & Fowtner, 1981.)

increases the number of limbs in contact with the ground, i.e. points d'appui, is reduced. The low speed movements of Chilopoda are very similar to those of the nereid polychaetes (Fig. 10.16). There is, however, a tendency for the number of limbs in contact with the ground to be fewer and the flexion of the body to be greater with increasing speed.

Lateral undulations of the body serve to increase step length in arthropods with short limbs, but they introduce significant changes in momentum. This can be circumvented by increasing the length of each limb and reducing the number of limbs. Interference between limbs is prevented by deployment of the legs in different longitudinal planes (Fig. 10.28). During the evolutionary history of the arthropod phyla, there has been a marked tendency for a reduction in the number of walking limbs. In Crustacea, the Decapoda have five pairs, but often as few as three or four pairs are actually used in walking. The arachnids have four pairs and the insects three pairs of limbs. Walking requires the lifting of the limb (elevation) and its forward movement (protraction) followed by lowering of the limb (depression) and its backward movement relative to the body (retraction). During retraction, the limb tip will be anchored to the ground and the centre of gravity of the animal moved forward relative to that of the limb tip. Movements in the various hinges in the limb can maintain the body at a constant height above the ground during such a step.

Figure 10.28 shows the side view and trajectories of the limbs of a scorpion as it walks forwards. Interference between the limbs is prevented by the different lateral position of each limb tip during the movement. The trajectory of each limb is rather different and this must involve complex co-ordination of the movements of the joints in each limb. The trajectories of individual limbs are generally non-overlapping in this way as indicated in Fig. 10.29. Most arthropods walk forwards rotating the basal joint of the limb relative to the body (Fig. 10.29b–d) but crabs walk in a sideways fashion, protraction being achieved by extension of the lower limb joints.

With a small number of limbs, there are problems of stability if too many limbs are in motion at once. In insects which have only three pairs of limbs, stability requires that the limbs are moved in such a sequence that at least three limbs are in contact with the ground at all times, and that the centre of gravity falls within a triangle drawn between the limb tips. The most frequently observed stepping pattern in insects meets this requirement. Two sets of limbs are moved alternatively, each set forming a stable base while they are in contact with the ground. The limbs move in metachronal sequence from rear to front with legs on opposite sides of a single segment being completely out of phase with each other. These features are also exhibited by multi-limbed animals such as nereid worms (see Fig. 10.16) and centipedes (Fig. 10.27). Different walking speeds can be achieved within the alternating triangle gait by alterations of the relative duration of promotion (p) and retraction (r), there being a marked tendency for the ratio p/r to increase as the rate of forward progression is increased.

Some insects show variation on the alternating triangle pattern, especially where one pair of limbs is specialized for a different function, as in the mantids where the first pair of limbs rarely support the body.

10.6 Swimming and flying

The movement of animals through the fluid media – water and air – when swimming or flying, does not rely on the formation of points d'appui. To move in water and in air it is necessary to generate forces that set the medium in motion. Action and reaction

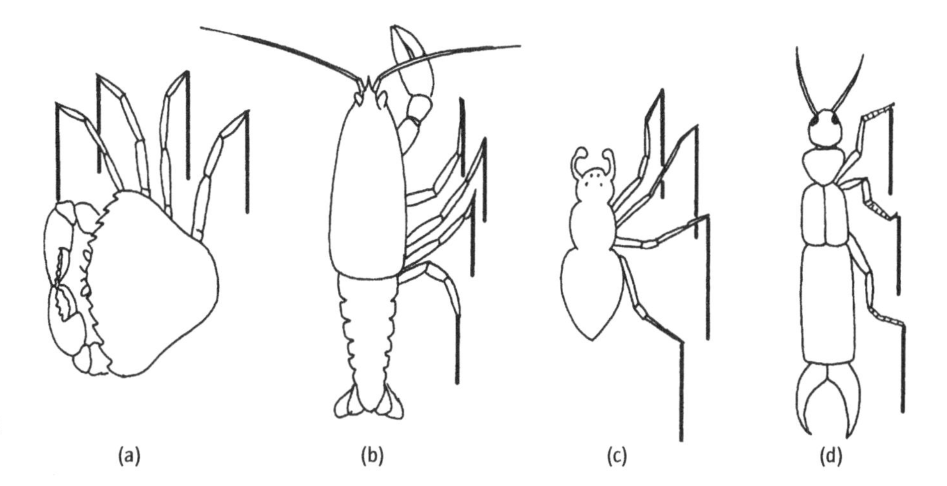

Fig. 10.29 The limb trajectories of a variety of arthropods: (a) crab; (b) lobster; (c) arachnid; and (d) insect. (After Manton, 1952.)

are equal and opposite, therefore a force moving the organism in an opposite direction to the medium will be generated. In fluids, this type of movement is called 'swimming' in water and 'flying' in the air. The analysis of these movements is technically difficult and the analyses and descriptions that follow are only first approximations. Flight, in particular, is notoriously difficult to analyse in a satisfactory way, though great advances are being made through modern recording techniques. This is leading to understanding of the non-steady-state dynamics involved.

10.6.1 Swimming

The smallest soft-bodied invertebrates swim using the locomotive power of bands of cilia as described in Section 10.3. Apart from the Ctenophora, however, which have comb-rows formed from fused cilia, larger invertebrates use muscular power in swimming. In order to do so, the forces generated by the muscles must be transmitted to the medium (usually water) through which the animal is moving. For larger animals the Reynolds numbers (see Section 10.1) will be greater than unity and the inertia of the body cannot be ignored.

A large animal whose Reynolds number is greater than 1 can only swim if it sets in motion a mass of water. In other words, it must create a wake. An ocean going liner cannot possibly move without creating a wake, and the same is true in swimming and in flying (see Section 10.6.2). The water movements generated by swimming will consist of a series of rotating masses of water known as 'vortices'. The analysis of the energy in these moving bodies of water provides a means of understanding the forces generated by the animal. If you watch an expert swimmer swimming breaststroke you will see these vortices. If you are a less than expert swimmer yourself you will realise that you move through water at relatively low Reynolds numbers and your movement will be very much a stop-start affair. The swimming of many small crustaceans may be like this, and the water is to them a viscous medium. Large expert swimmers however achieve moderately high Reynolds numbers and their locomotion is much smoother without a stop-start or saw-toothed velocity pattern.

The vortices created by expert swimmers provides a record of the work done. Swimming of this kind is particularly characteristic of vertebrate animals, but there are many soft-bodied invertebrates, which are not neutrally buoyant in water, that are able to generate lift and propel themselves forward. The principal mechanisms are: (a) the backward propagation of waves in smooth-bodied animals; (b) paddling and rowing; and (c) jet propulsion.

Smaller animals for which the Reynolds number is less than 1 cannot exploit vorticity. They may still be able to swim, but the forces that result in locomotion are due to differential drag.

Much of the mechanical analysis of undulatory swimming is based on a study of the movements of the common eel, but comparable locomotory waves can be generated by smooth-bodied worm-like invertebrates; a good example is the leech *Hirudo*. When it swims, *Hirudo* is flattened dorsoventrally and retrograde waves are generated. The wave passes backwards at a velocity (U) relative to the ground which is greater than the forward velocity (V) of the body. The principles of this type of swimming are outlined in Box. 10.6.

The swimming of the rough-bodied polychaete worms is rather different. The wave passes in the same direction as the locomotion. This is partly due to the complex fluid dynamics that result from the oscillation of a flanged body. It is also because most of the power derives from the actions of the parapodia each of which executes a power stroke as they are passed by the crest of the wave of longitudinal muscle contraction. Because this wave passes forwards each parapodium makes its backward power stroke just after the parapodium behind it and this avoids interference. (The only other way of achieving this is for each parapodium to execute its power stroke at the same time as the others. Worms cannot do this but it is the solution adopted by a university rowing eight) (see also Fig. 10.16b).

Many large vertebrates exhibit a thrust-and-glide mode of swimming in which kinetic energy is stored in the body following intermittent power strokes. Large-bodied animals are able to generate forces that accelerate relatively large masses of water at low velocities in what has been described as 'an exchange of effort for momentum'. The relationship can be represented as

$$m_w u_w = mu$$

where m_w and u_w represent the mass and velocity of the water. The mass and velocity of the animal are represented by m and u and the equation defines the momentum achieved by the animal, expressed in kg m s^{-1}.

Some animals adopt a different approach and accelerate a small mass of water giving a high velocity. This mechanism is known as *jet propulsion*. Typically, a small body of water will be forced through a narrow aperture to acquire a high velocity. This requires the generation of a large force that can result in massive acceleration. The metabolic costs of this are very high and this technique is mostly used in escape mechanisms. The benefit of escape (your life) being sufficient to justify the high cost. Examples of escape mechanisms are the tail flip of some crustaceans, such as lobsters, and the flapping of a scallop. When threatened by a starfish predator, a scallop will open the two valves, then clap them suddenly shut; the mantle edge directs the enclosed mass of water towards a direction behind the hinge. This causes the scallop to shoot 'gape forwards'.

The energy cost of accelerating a mass of water is given by the relationship $0.5 m_w u_w^2$. To achieve high water velocities therefore requires a large amount of energy. There is no requirement to maintain an escape response indefinitely, the change in velocity being such as to take the prey safely out of harm's way – an 'escape into the third dimension' – and it can justifiably be powered by anaerobic respiration.

A few animals, however, use jet propulsion for sustained locomotion over large distances. This is typically seen in medusae where water is driven out of the undulating bell, though at relatively low velocity, and among the cephalopod molluscs.

Box 10.6 The swimming of smooth-bodied worms

1 The swimming of smooth-bodied animals can be analysed by considering the motions of a single length of the body.

2 Such a unit describes a figure-of-eight motion in which it has an angle θ to the transverse axis of motion as it crosses the axis of motion. In this analysis only the forces acting at the mid-line are considered, although this is a simplification.

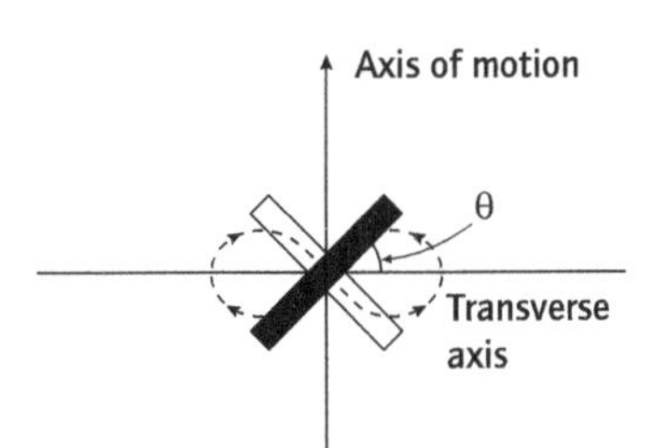

An understanding of the forces involved requires a knowledge of the resolution of forces (see Box 10.3).

3 If the body were at rest the motion of the body as it crosses the longitudinal axis of motion would exert a force and set in motion a mass of water (see Box 10.2 – kinetic energy and friction).

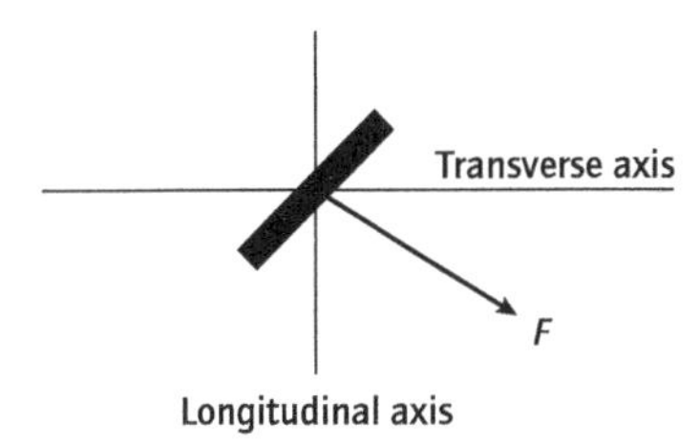

One component of the force F, F'', will cause water to move along the surface of the body, but the component F' acts at right angles to the body and causes water to be set in motion.

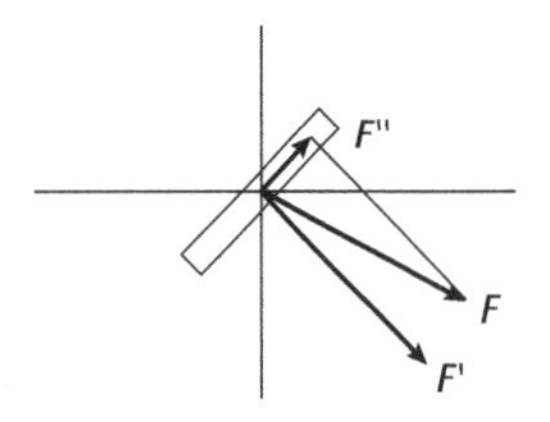

The reaction force RF' equal and opposite to F' can be resolved into components parallel to and normal to the axis of motion of the animal.

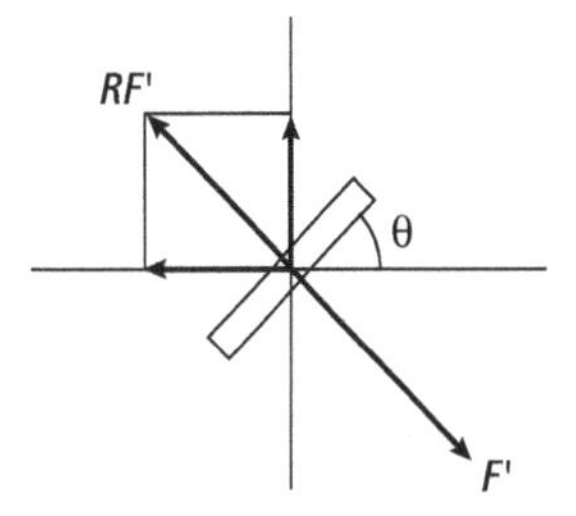

During the passage of a complete wave the transverse forces cancel each other out but the components acting along the axis of motions are summed in effect.

As the animal gathers speed the motion of the body is such that the effective angle between the body and the axis of movement is reduced. This angle is called the angle of attack (α).

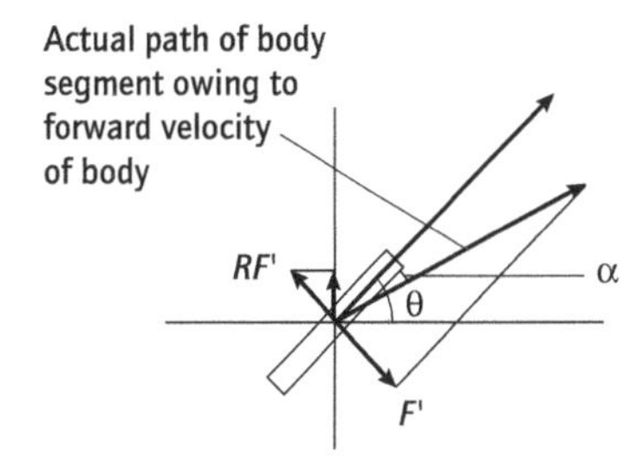

This effectively reduces the reaction force acting along the axis of motion. The animal will accelerate until this force is equal to the drag.

Cephalopods, like *Octopus*, use jet propulsion as other animals do, only as an escape response. There are cephalopods, however, which sustain jet propulsive swimming over long periods of time. The principles are explained in Box 10.7. The jet pressures generated and the speeds attained by different cephalopods vary considerably and *Nautilus*, a living example of a more primitive cephalopod type, uses low-pressure jetting. The cost of transport is energy per unit mass per unit distance and can be expressed in the units J kg^{-1} m^{-1}. This has been calculated for the cephalopods shown in Fig. 10.30 where it is compared with values for a medusa jetting at low velocity but at low cost and with a representative teleost such as the salmon. The cost of transport of the cephalopods is much higher than that of the teleost across the range of typical speeds although it is clearly much less than that recorded for an escaping scallop. The cephalopods appear to be the true athletes among marine invertebrates. How they sustain these energy costs and how they minimize them by increasing the efficiency of swimming by riding ocean current systems remains a challenge for the future.

10.6.2 Flying: invertebrates that have conquered the air

10.6.2.1 The possible origin of flight

Of all the means by which invertebrates move none can be more remarkable than the flight of insects. By flying, insects are able to escape from predators, move from one resource to another or find a mate. The flight of insects is so well known that we almost take it for granted. We all know that insects can fly, but how did flight evolve and how does it work?

Box 10.7 Jet propulsion

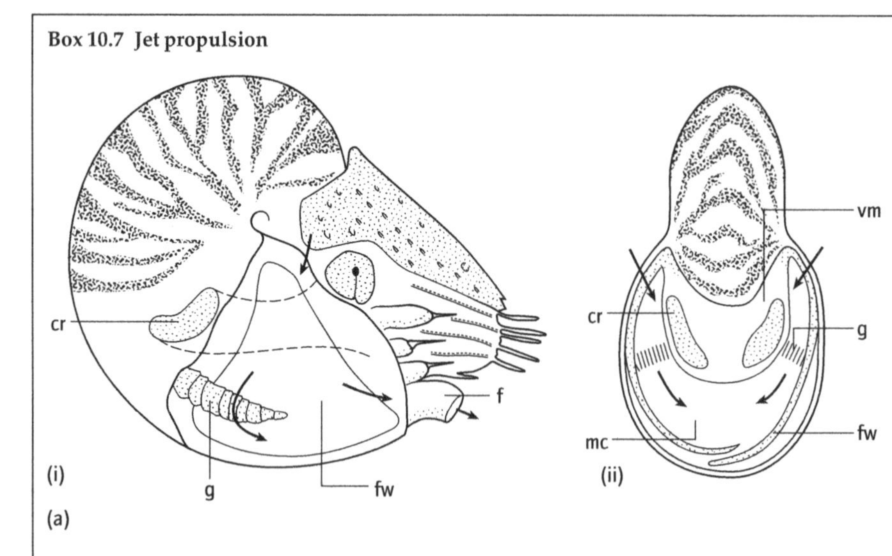

(a) The locomotor anatomy of *Nautilus*. (i) As seen in lateral view. (ii) In cross-section. *mc* mantle cavity, *fw* funnel wing muscles, *g* gills, *cr* cephalic retractor muscles. (From Chamberlain, 1990.)

1 In jet propulsion the animal imparts a velocity to a small body of initially enclosed water. The force against the water and the reaction force act along the line of movement.

2 Jet propulsion is seen in the primitive cephalopods such as *Nautilus* involves low mantle pressure and relatively low propellant velocity.

3 In most living squids, the pressures generated and the propellant volumes achieved are much greater than in *Nautilus*. In squids, the pressure in the mantle cavity is raised by the contraction of muscles surrounding it. There is therefore a force acting at right angles to the body wall in all places.

Force = pressure × area

The jet thrust (kg m s^{-2}) is equal to the product of jet velocity u_j (m s^{-1}), propellant flow Q (m^3 s^{-1}) and density d_w. This thrust depends on the pressure p (Pa) generated in the mantle cavity and the area A (m^2) of the propellant opening.

The equation for this relationship is:

$$u_j Q d_w = 2Ap$$

4 Because the mass of water expelled is smaller than the mass of the body of the animal, the jet is expelled at higher speed than the animal is propelled in the opposite direction. Jet engines with this characteristic have low efficiency.

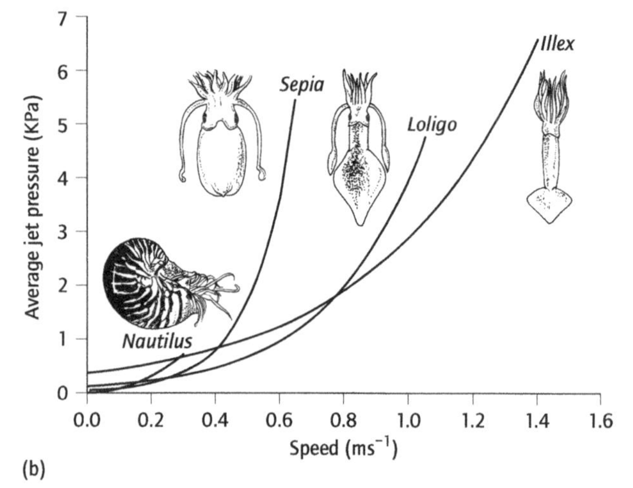

(b) Jet pressures and speed achieved by a variety of cephalopods.

5 Part of the work done by the muscles in contracting the mantle cavity is used to distort the elastic body wall, but there are also antagonistic muscles which restore the mantle cavity to its original volume. These are the cephalic retractor muscles shown in (a) above.

The fossil evidence for the origin of winged insects is crucially missing, so the evolution of flight among the insect ancestors about 300 million years ago has to be derived from other sources of information. The observation that the pattern veins on insect wings follows a common pattern suggests that flight may have arisen only once. However, the evolution of flight initiated an incredible radiation and diversification among insects and there are now many different flight patterns and mechanisms. The wings of insects are adult structures that become fully functional only in the adult or imago state. The pre-adult stages do not have functional wings. In holo-metabolous endopterygote insects the developing wings are internalized imaginal discs, but in hemi-metabolous exopterygote insects the developing wings are external.

Powered flight must have arisen after the evolution of proto-wings performing some other function. One possibility is that powered flight evolved from the movement of insects that exhibited aerial gliding. This is not, however, the only possibility. An alternative hypothesis is that wings evolved from the articulated gill plates that are still used for ventilation and locomotion by some aquatic insect larvae. A plausible model is provided by the stonefly *Allocapnia vivipara*. This stonefly is not

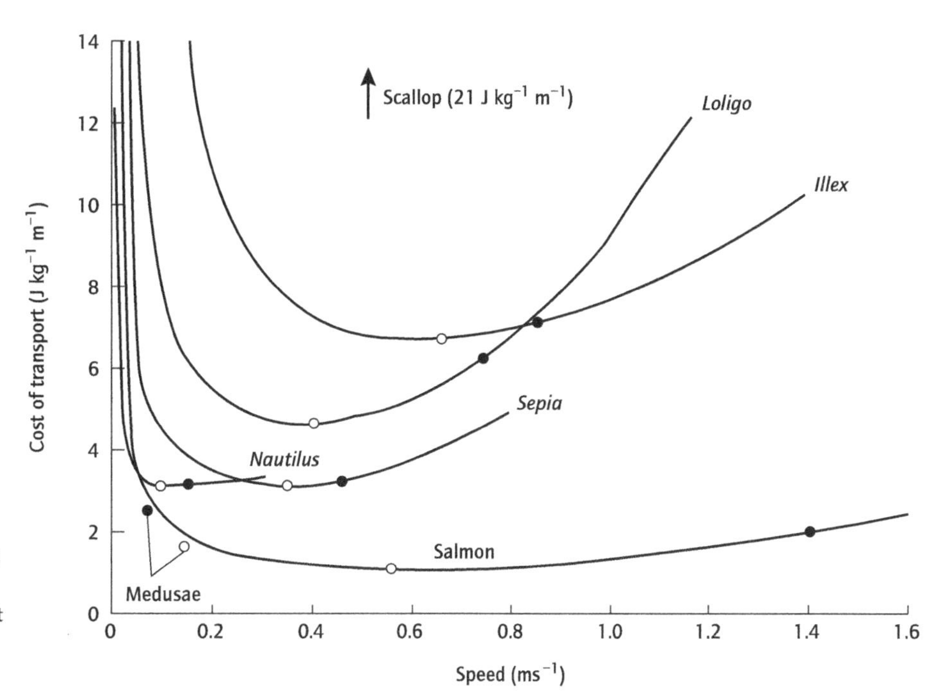

Fig. 10.30 A comparison of the cost of transport over a range of velocities for the cephalopods illustrated in Box 10.7 compared with the teleost fish, the salmon. (After O'Dor, R.K. & Weber, D.M. 1991. *J. exp. Biol.*, **160**, 93–112.)

able to fly but uses its wings to sail across the water surface. The wings are raised in wind and this provides a non-flapping means of aerodynamic movement. Some other stoneflies, e.g. *Taeniopteryx burksi*, are able to flap their wings to assist surface sailing but are not capable of free, surface-independent, flight. This suggests that insect wings may have evolved 'from articulated gill plates of aquatic ancestors through an intermediate semi-aquatic stage'.

Whatever the means by which insect wings first arose, during the subsequent 300 million years or so, the basic mechanism has been modified and refined many times. Consequently present-day insects exhibit a very wide range of structural and aerodynamic adaptations. The modes of flying among insects are much more diverse than those of birds or bats. Figure 10.31 shows a selection of wing shapes for insects which are here drawn all the same size but which, in reality, differ markedly in absolute size (see also Figs 8.29 and 8.31).

The wings that made gliding or flying possible in terrestrial insects may also have been involved in temperature regulation or respiration. Indeed, many living insects use their wings as do the Lepidoptera (e.g. butterflies) to raise the body temperature by absorbing solar energy, while others such as the Hymenoptera (e.g. bees and wasps) will raise body temperature prior to flight by flapping movements. It has been suggested that changes in relative scale could bring about a changed situation in which wings, which were adaptative elements developing functions related to temperature regulation or respiration acquired new aerodynamic properties. We have already seen, however, that this is not the only way in which wings may have arisen.

Primitive flying insects probably used relatively slow wing movements and may have been capable of gliding flight during which stable or steady-state aerodynamic forces operated. The aerodynamics of the flapping flight of living insects, however, can be only be understood if non-steady state conditions and unstable air flows are taken into account.

10.6.2.2 The mechanics and control of wing movement

The dragonflies, locusts and relatives (Odonata) are relatively large insects that use slow wing movements. The muscles that power the flight are direct flight muscles attached to the wing bases as shown in Fig. 10.32. Contraction of the inner muscle pairs which are attached directly to the proximate ends of the wing base will lift the wings and contraction of the outer pairs of muscles which are attached more distally will lower the wings. These flight muscles arranged like this are termed 'direct flight muscles'. In the dragonflies, there is a one-to-one relationship between the frequency of the nerve impulses which stimulate the contractions of the direct flight muscles and the frequency of contraction as shown in Fig. 10.32a(iii). Insects with this one-to-one relationship between the frequency of flight muscle contraction and spiking in flight muscle innervation are said to be 'synchronous'.

Dragonflies and locusts also have flight muscles that attach to the walls of the thorax. The contraction of these muscles cause the thorax to change in shape and this has the effect of moving the wings. Muscles arranged in this way are referred to as 'indirect flight muscles'. In the Odonata, the indirect flight muscles are relatively small but in the majority of insect species the indirect flight muscles are much larger and provide most of the power in flight.

Indirect flight muscles act by deforming the thoracic segmental box to which the wings are attached and exploit the

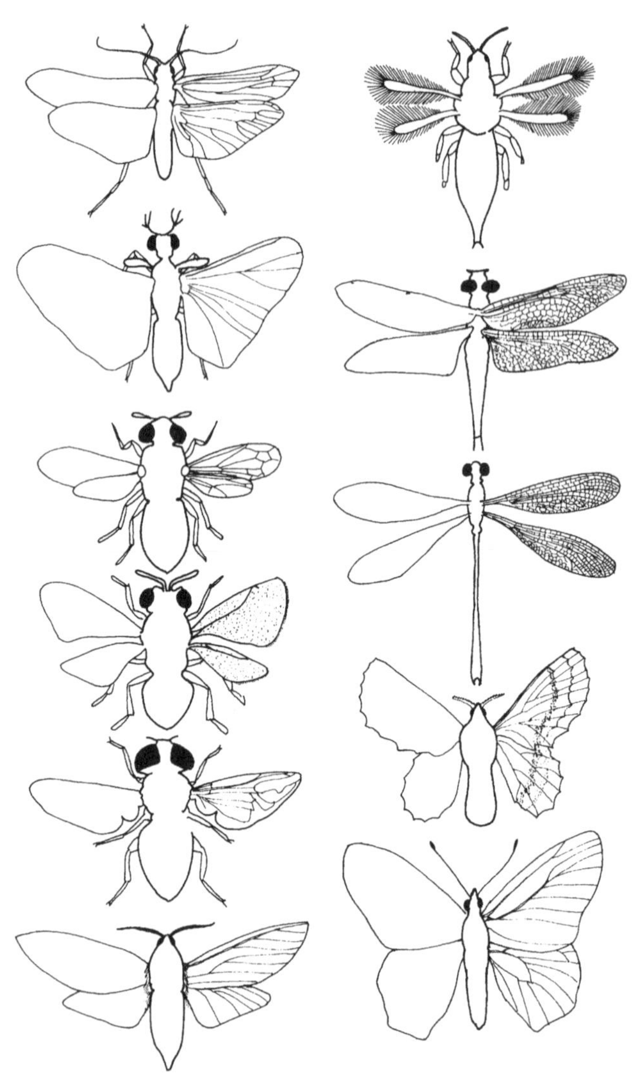

Fig. 10.31 Wing shapes in insects. These animals are here drawn as approximately the same size; in reality they have markedly different dimensions. (See also Fig. 8.29.) (After O'Dor and Weber, 1991.)

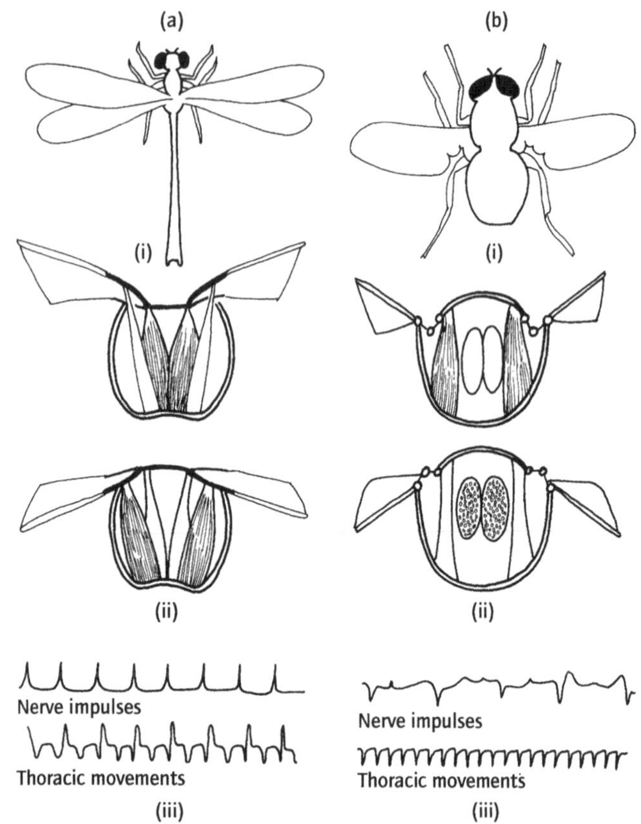

Fig. 10.32 Structural and neuromuscular adaptations for flight in insects. (a) Direct flight musculature, e.g. Odonata: (i) a representative dragonfly; (ii) cross-section of the thorax of a dragonfly showing the direct flight muscles attaching to the wing bases; (iii) traces of the synchronous nerve impulses and thoracic movements recorded in insects of this type. (b) Indirect flight musculature, e.g. Diptera: (i) a representative fly; (ii) cross-section of the thorax showing the main antagonistic flight muscles; (iii) traces of the asynchronous nerve impulses and thoracic movements recorded in insects of this type. (After Pringle, 1975.)

energy storing or elastic properties of the insect cuticle. The opposing muscles are pairs of vertical and longitudinal indirect flight muscles as shown in Figure 10.32b. The thorax is made up of three articulated curved plates on each side, the tergal, pleural and sternal plates (Fig. 10.33). The wing is a lateral extension from the tergal plate that rests on and articulates with a projection of the pleural plate, the wing process.

When the longitudinal indirect flight muscles contract, the tergal plate bends upwards and the wing is forced downwards articulating on the wing process. When the vertical muscles contract, the tergal plate is pulled downwards and flattened which causes depression of the wing. The wing is a lever (see Box 10.4) and the small movements of the tergal plate induce relatively massive movements of the wing tip which must therefore achieve high velocities. This makes possible the acrobatic and agile flight of so many insects.

This system has many special properties. The tergal plate moves from one stable state to another and in doing so stores elastic energy. The contractions of the flight muscles force the tergal plate to move against an elastic strain by deforming the pleural plate. As the wing passes past the midpoint it will move suddenly to the new stable position. This allows for the rapid release of energy stored in the deformed pleural plate. It was formerly thought that the wing in effect 'clicks' between two stable positions (see Fig. 10.34). More detailed studies of the kinematics of the wing beat, however, do not support this mechanism. It predicts that the velocity of the wing tip would slow as it approached the midpoint of the movement followed by acceleration towards the stable position. This is not observed and the precise way in which the thoracic box stores and releases energy remains a subject of active investigation.

The work done by a flying insect includes the work required to accelerate the wing and the mass of air which moves with the

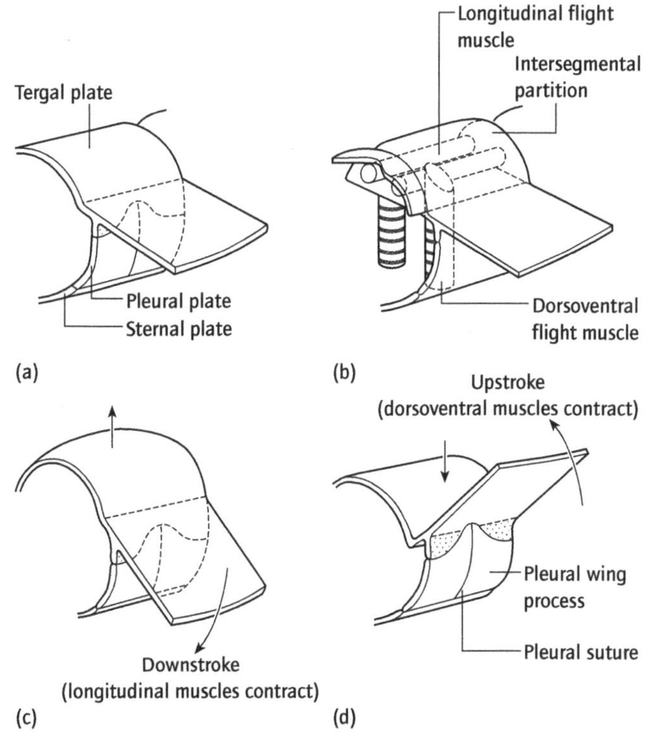

Fig. 10.33 The generation of wing movements by the action of indirect flight muscles on the insect 'thoracic box'. (a) The three components of the thoracic box: the tergal, pleural and sternal plates. (b) The dorsoventral and longitutudinal indirect flight muscles which oppose each other through the elastic properties of the plates. (c) Contraction of the longitudinal muscles causes the tergal plate to arch upwards forcing the attached wings to move downwards, articulating with the wing process of the pleural plate. (d) Contraction of the dorsoventral flight muscles causes flattening of the tergal plate and the wings move upwards.

wing. As the wing decelerates, a great deal of the inertial energy is stored in an elastic system and is released during a subsequent power stroke. In locusts and dragonflies which, as we have seen above, have direct flight muscles, the energy is stored in the elastic resilin protein in the cuticle. Results using the bumble bee as a model suggest that a substantial amount of the energy is stored in the elastic indirect flight muscles. According to this model the wing of the bumblebee behaves like a pendulum that is being pulled upon by antagonistic muscles that themselves behave like moderately stiff springs (Fig. 10.35).

In some insects with indirect flight muscles, there may be a one-to-one relationship between the period of wing beating and nerve spike frequency in the nerves driving the muscle contractions. These insects therefore are also synchronous. Much more often, however, the frequency of muscle contraction and wing beating is much greater than the spike frequency in the flight muscle innervation as in Fig. 10.32b(iii). Insects showing this pattern are said to be 'asynchronous'. Asynchrony is an adaptation that enables muscles to contract and relax many times per second. The frequency in some tiny flies may be as high as several hundred beats per second. This is because the muscles are highly sensitive to stretch. When one set of indirect flight muscles contracts, it stretches the antagonists. This acts directly on the contractile apparatus to initiate a new cycle of contraction (see Fig. 10.35). Asynchronous flight muscles are found usually in small insects such as flies, bees, wasps and beetles. Moths and butterflies have large synchronous but indirect flight muscles.

As we shall see below, the aerodynamics of insect flight involves much more than a simple up-and-down movement. Fine adjustments and twisting of the wings are involved and these movements are controlled, in insects with large indirect flight muscles, by the smaller direct flight muscles. This allows for steering and quite remarkable aerobatic ability.

10.6.2.3 The wing as an aerofoil and generator of air vortices

The evolution of flight in insects has been associated with functional developments that permitted the insects to fold the wings. The wings of the earliest insects like those of the living Odonata could only be held out rigidly to the side. Subsequent developments have made it possible for some insects to fold the wings vertically, as in mayflies and butterflies. Even greater advantages accrue from the development of truly folding wings that can be closed over the abdomen. This allows the adult insect to creep into cryptic habitats and most modern insect orders

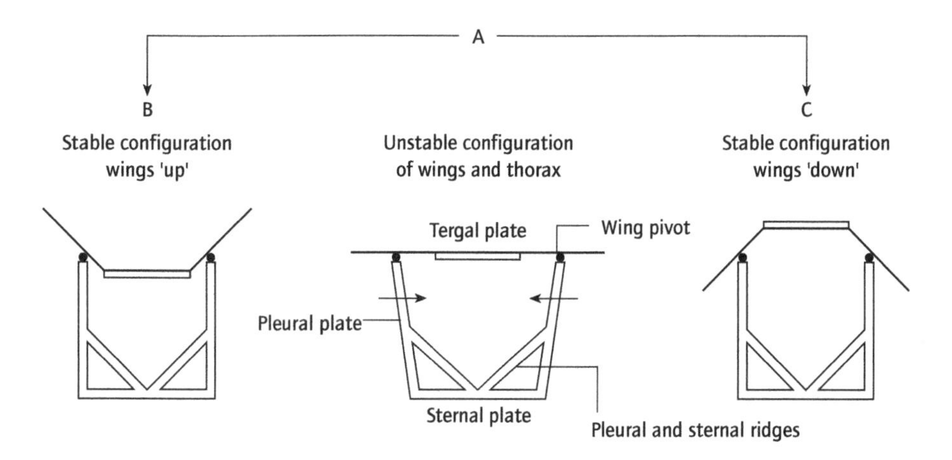

Fig. 10.34 Diagrammatic representation of the 'click mechanism' that plays a role in insect flight. The 'up' and 'down' positions are stable but the midpoint where the wing passes through the horizontal plane is unstable because the sides of the elastic pleural plate are pushed outwards in this position. Analysis of the kinematics of wing movement shows that the wings do not reach minimum velocity at this point as predicted by this model and the click mechanism is not the only determinant of wing movement. (After Brackenbury, J. 1995. *Insects in Flight*. Blandford, London.)

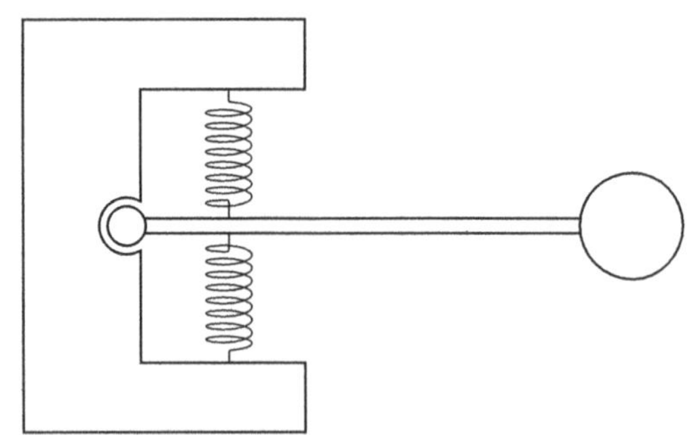

Fig. 10.35 A mechanical model of the insect wing system. Stretching of a spring stores energy and causes a resonant response. The spring stresses and inertial forces are sufficiently large that the gravitational force on the wing can be ignored. (After Josephon, R.K. 1997. *J. exp. Biol.*, **200**, 1227–1239.)

can fold the wings in this way. This in turn has permitted the evolution of protective wing cases as in the Coleoptera (beetles) (Fig. 8.31).

The paired wings of insects are frequently locked together in a single aerodynamic unit. Butterflies and moths (Lepidoptera) and bees and wasps (Hymenoptera) are well-known examples. In Coleoptera, the anterior pair of wings are modified as wing cases but may have an important aerodynamic role in the rapid flight of these insects. In the Diptera (true flies, hover flies, mosquitoes, etc.), which can have the most aerobatic flight capability, the posterior wings have been modified as specialized sense organs – the halteres. It is interesting that the wing-like origin of the halteres can be seen in some genetic mutations which cause halteres to develop as wings (see Chapter 15).

The control of flight in the advanced insects is highly developed. The asynchronous nerve impulses that characterize the flight of the dipterans (Fig. 10.32a) maintain the flight muscles in the state of excitation which is characterized by high levels of cytoplasmic calcium ions. The frequency of wing beating is then a function of the physical properties of the thoracic box and spring-like quality of the flight muscles. One of the effects of the contraction of one set of flight muscles is to cause a stretching of the opposing set which sets in train a sequence of alternate contractions which will be maintained so long as the flight muscles are in the excited state.

The wings of insects are not simple rigid paddles but flexible, often pleated structures whose shape during the movements of flight is determined by the pattern of venation and the air pressures to which they are subjected. Many have microscopic hairs or scales, and these affect the way in which air will adhere to the surface of flow, and be shed from the wing surface. There is no simple model of insect flight and the mechanisms in different types of insect are remarkably varied. Some general principles, however, can be established. An insect flying through the air at constant velocity will be subject to two forces. A downward force, weight, due to the action of gravity on the mass and a force in the opposite direction to that of the locomotion due

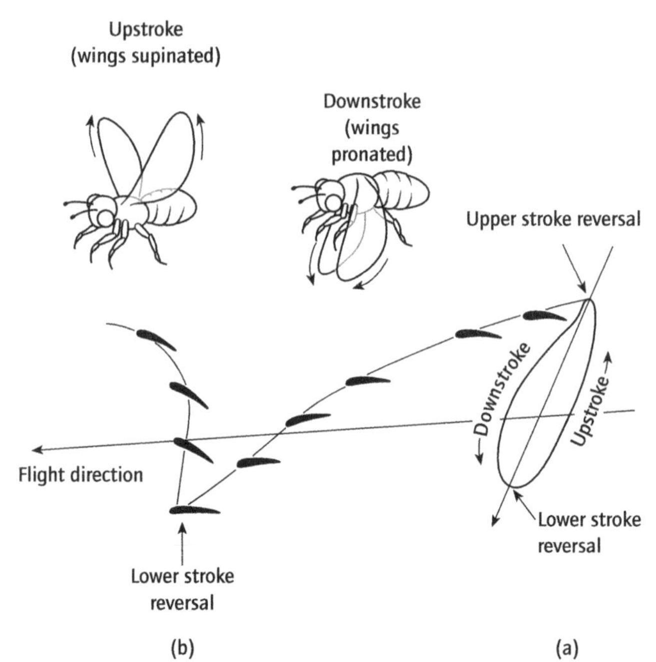

Fig. 10.36 A diagram of the path traced by the tip of a flying insect such as *Drosophila*. (a) The closed loop represents the path of the wing tip relative to the insect. (b) The saw-tooth pattern is the path of the wing tip relative to a fixed point of observation. The dark sections of the wing shows the changing angle of attack of the aerofoil. The angle of attack is the angle between the leading edge of the wing and the air flow (see also Box 10.6). Rotation of the wing to face downwards occurs prior to the downstroke and upwards before the upstroke. Lift is generated during both wing strokes. (After Brackenbury, J. 1995. *Insects in Flight*. Blandford, London.)

to the viscoscity of the air referred to as 'drag'. The wings, therefore, must generate forces which have components equal and opposite to these forces. These forces are referred to as the 'thrust' and 'lift'. If the thrust is greater than the drag the insect will accelerate and if the lift is greater than the weight the insect will rise.

The air is a fluid medium and the movements of the wing which generate thrust and lift can only do so by generating movements of the air just as a moving ship leaves behind a wake of moving water. The movements of flying insects take place at high Reynolds numbers and inertial forces are important. The wings of insects are rotated during flight and generate lift and thrust during both the downstroke and the upstroke. The wing is generally rotated leading edge up during the upstroke and down during the downstroke so that the 'angle of attack' changes (Fig. 10.36). The wing tip describes an elliptical or figure-of-eight closed path relative to the insect which, for an insect with a forward velocity, would be seen as a saw-toothed pattern relative to a stationary point of reference external to the fly. This pattern is also shown in Fig. 10.36.

The coefficient of lift produced by most insects is greater than would be expected for a conventional aerofoil and it has proved impossible to model most insect flight as a steady-state aerofoil. The key to understanding insect flight is the recognition that non-steady state, unstable forces are generated by the wing movements and that the insect leaves in its wake vortices of moving air that provide a record of those forces. As the wing moves through the air, a circular movement of air is induced around the wing. During flight, this tube of moving air is shed from the tips of the wing. The principle is explained in Box 10.8. In order to set a mass of air moving and to leave that mass of moving air behind, it is clearly necessary to generate a force. The

Box 10.8 Vortex generation and flight

1 The flight of insects is a consequence of non-steady-state airflow generated by the movements of the wings. An important feature is the production and shedding of vortices of moving air. In order to generate a vortex in a fluid, work must be done.

2 In flight the movements of the wings generate circulation about the aerofoil surface which is shed from the wing as a vortex when the wings reverse direction or change shape and inclination (see below). The mechanism has been visualized in a fly.

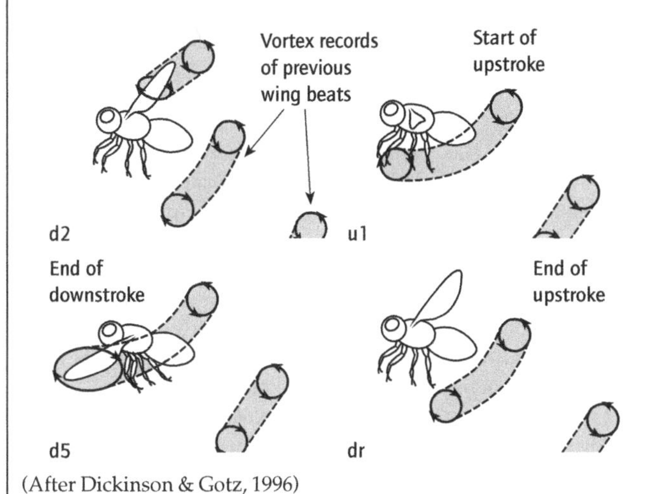

(After Dickinson & Gotz, 1996)

3 As the wings move apart during the downstroke (d1–d4), circulating air is constantly lost from the tips of the wings and this forms a vortex loop of circulating air attached to the wings.

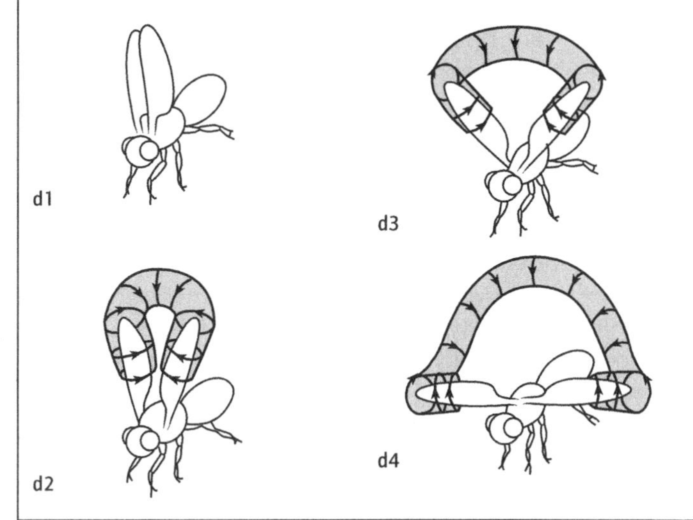

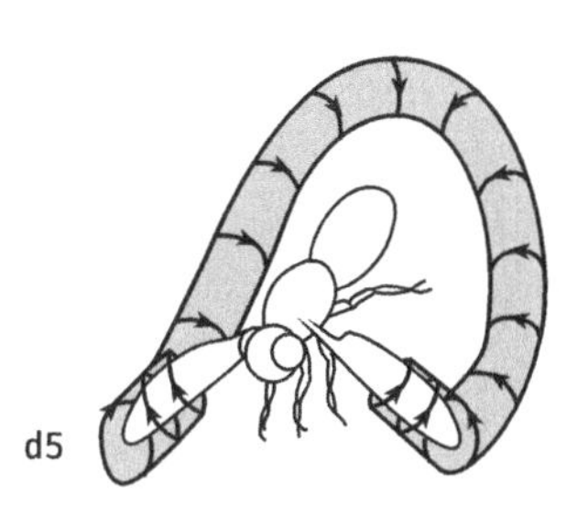

(After Dickinson & Gotz, 1996)

After the end of the downstroke d5, during the ventral flip (vf), the wings shed this vortex loop which moves backwards along the ventral surface of the fly during the upstroke u1–u3.

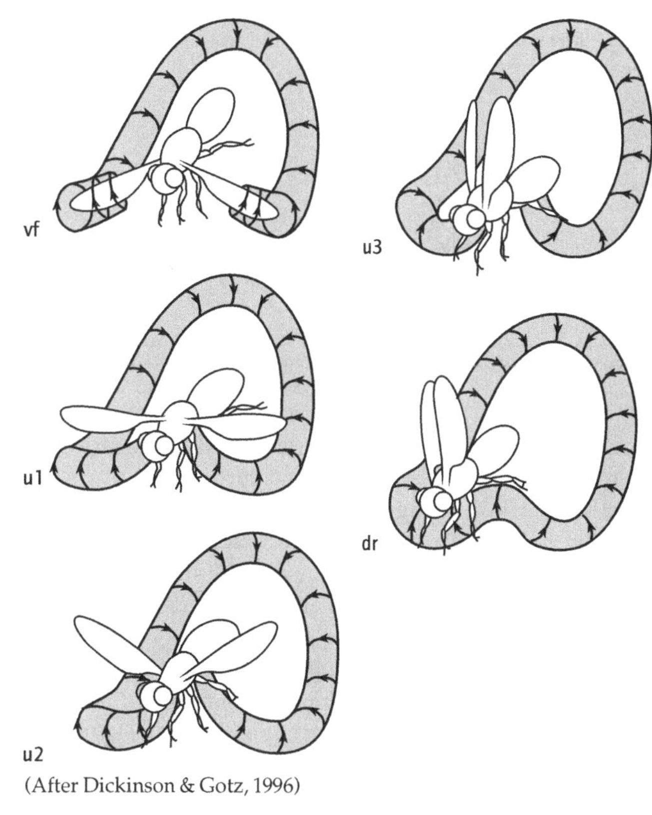

(After Dickinson & Gotz, 1996)

As the wings reach the top of the up-movements (dr), they squeeze together and the vortex loop is shed from the posterior of the fly.

A flying insect can be supposed to be supported and propelled forwards by the reaction force Q resulting from the downward-shed vortex rings (see over).

Continued p. 268

Box 10.8 *(cont'd)*

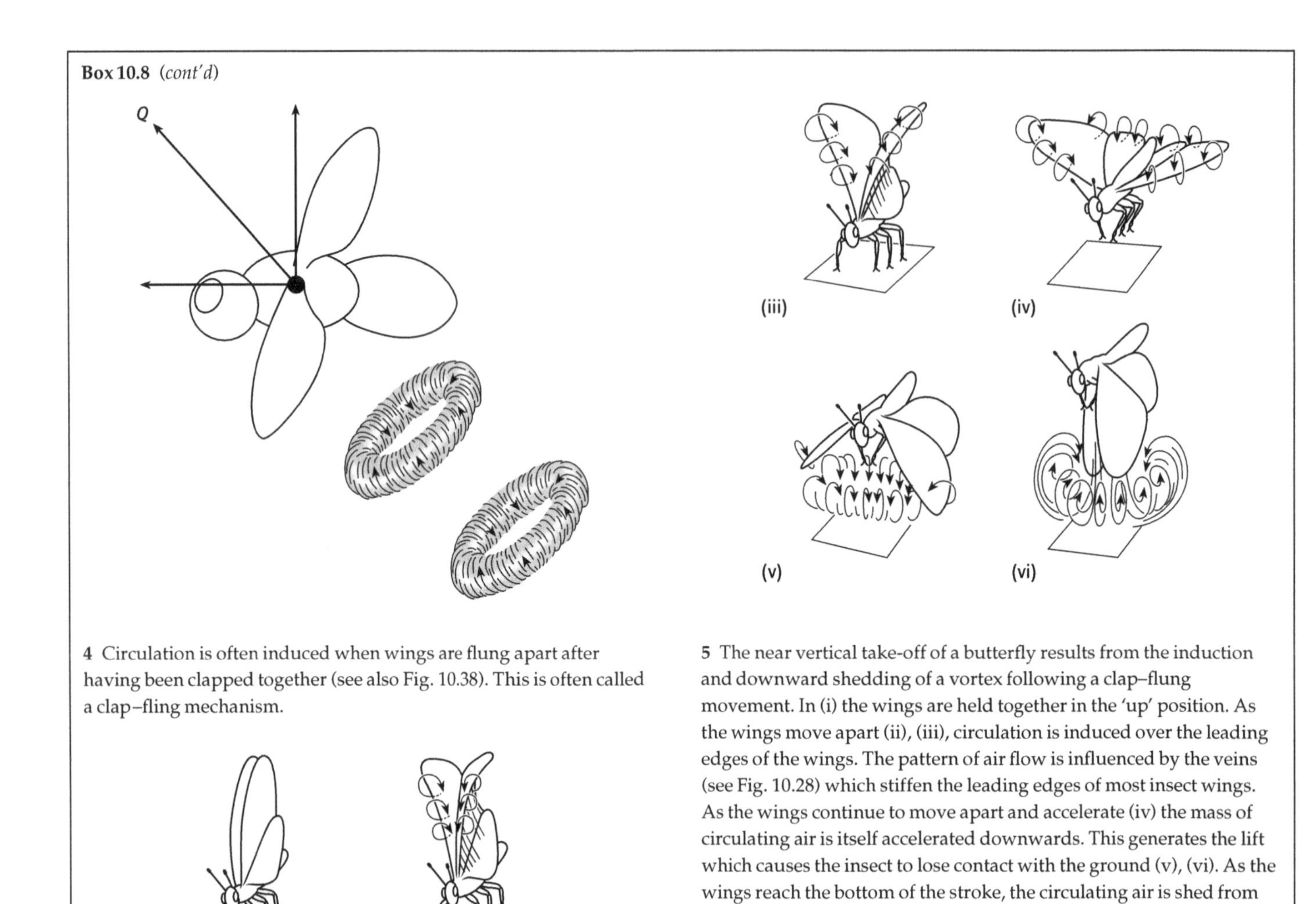

4 Circulation is often induced when wings are flung apart after having been clapped together (see also Fig. 10.38). This is often called a clap–fling mechanism.

5 The near vertical take-off of a butterfly results from the induction and downward shedding of a vortex following a clap–flung movement. In (i) the wings are held together in the 'up' position. As the wings move apart (ii), (iii), circulation is induced over the leading edges of the wings. The pattern of air flow is influenced by the veins (see Fig. 10.28) which stiffen the leading edges of most insect wings. As the wings continue to move apart and accelerate (iv) the mass of circulating air is itself accelerated downwards. This generates the lift which causes the insect to lose contact with the ground (v), (vi). As the wings reach the bottom of the stroke, the circulating air is shed from the wing surface as a vortex ring with downward momentum. (After Kingsolver, 1985.)

reaction to that force will have the forward and downward components that we have termed thrust and lift.

One way of visualizing the movements of the air that are caused by flying is through the observation of an insect in a stream of smoke particles in moving air. The shape and nature of the wake structure is highly complex and depends on the detailed morphology of the wing as it moves through the air. Figure 10.37 shows a stereophotograph taken from a series of images of the wake generated by the tethered flight of the moth *Manduca*. The components of this wake are shown in Fig. 10.37b. When the wings of an insect are clapped together then flung apart, the anterior stiffened edges are the first to be pulled away from each other and air rushes into the low-pressure space developing between the wings (Fig. 10.38). Once the moving air has maximum velocity, no further work can be done by the wings until the vortex is shed. This often happens when the wings change their direction of movement at the bottom of the wing stroke. The venation pattern of the wing influences the changes of shape that occur and this will also influence the pattern of vortex generation and shedding. Some further details are given in Box 10.8.

Large insects like the butterfly use a clap-and-fling mechanism to generate the large lifts required at take off. This is also described in Box. 10.8.

The adaptations of insects for flight are diverse. The larger Coleoptera achieve flight characteristics to give Re values as high as 23 000 and when these insects are flying the stiffened forewings behave as conventional aerofoils. The pleating of the wings of some insects and the surface structure of hairs or scales will all determine the precise pattern of air flow over the wing surface and thus influence the aerodynamic properties of the wing during flight.

10.6.2.4 Jumping insects: flight without wings

Some insects have the ability to jump; in most of them, it is an important escape reaction, and anyone who has tried to catch

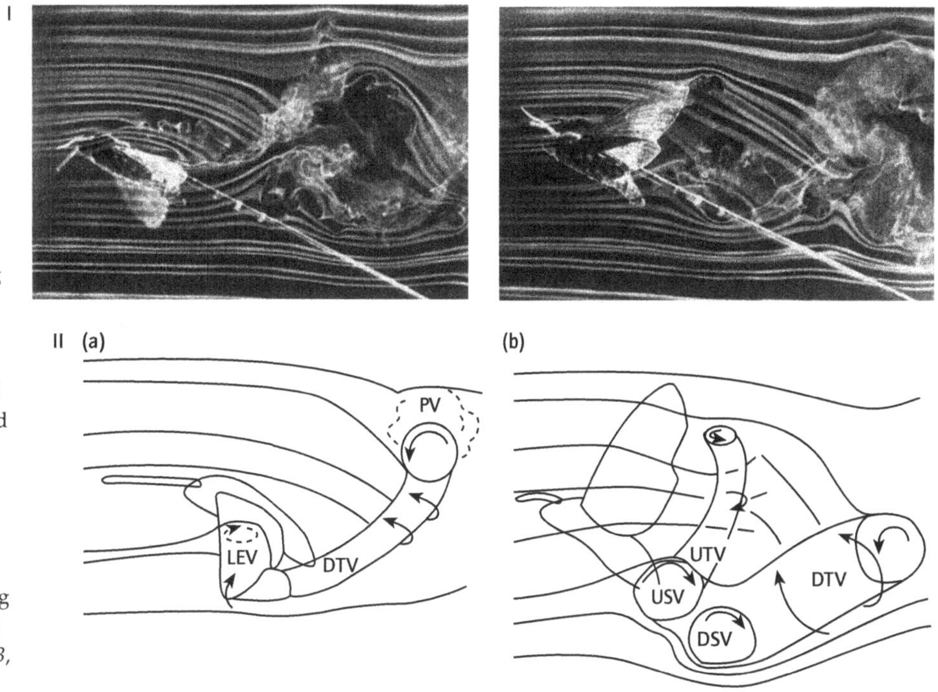

Fig. 10.37 Visualization of the wake of a flying insect is made possible by the use of smoke particles in moving air. (I) Photographs of a tethered Hawkmoth *Manduca* in an airflow visualized with smoke: (a) the wake generated during the downstroke; (b) the wake generated during the upstroke. (II) A schematic diagram of the wake structure: (a) the downstroke; (b) the upstroke. LEV: leading edge vortex. DTV: downstroke tip vortex. PV: pronation vortex. USV: upstroke starting vortex. UTV: upstroke tip vortex. DSV: downstroke stopping vortex. (From Willmott, A.P., Ellington, C.P. & Thomas, A.L.R. 1997. *Phil. Trans. R. Soc. Lond. B*, **352**, 303–316.)

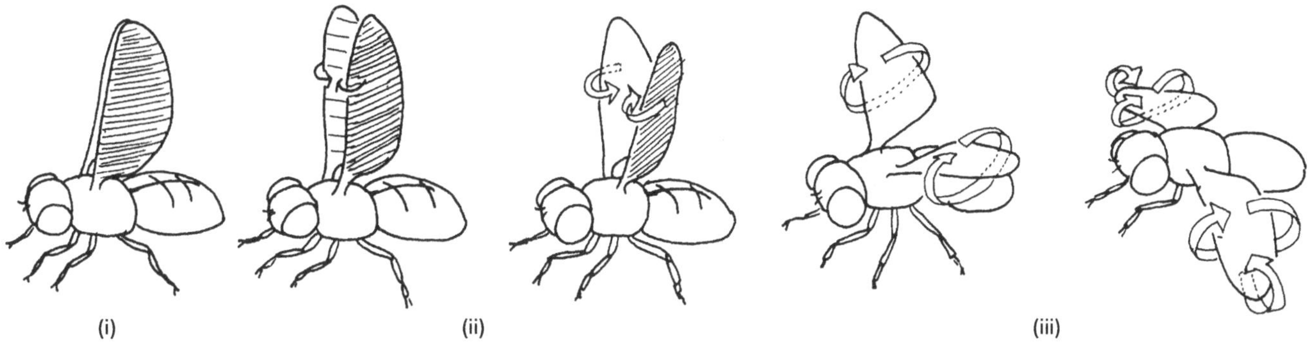

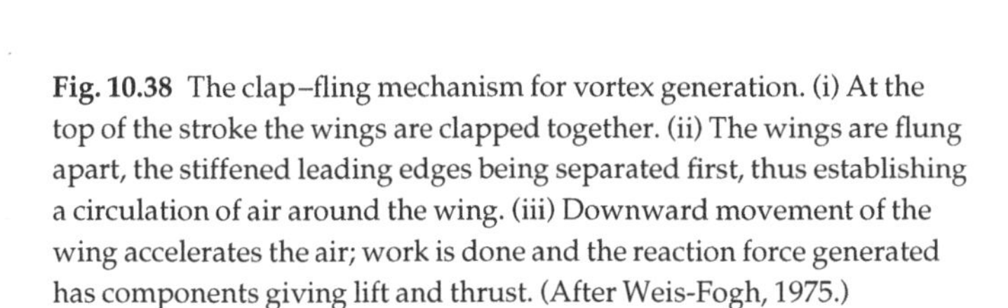

Fig. 10.38 The clap–fling mechanism for vortex generation. (i) At the top of the stroke the wings are clapped together. (ii) The wings are flung apart, the stiffened leading edges being separated first, thus establishing a circulation of air around the wing. (iii) Downward movement of the wing accelerates the air; work is done and the reaction force generated has components giving lift and thrust. (After Weis-Fogh, 1975.)

a flea will be aware how effective this unpredictable behaviour is. The ability to jump is particularly well developed in the fleas, grasshoppers and leafhoppers.

In order to jump, an insect must exert a force against the ground sufficient to impart a take-off velocity to its mass. The height of the jump will be defined by the relationship:

$$\tfrac{1}{2}mV^2 = mgh$$

kinetic energy = potential energy at the peak of the jump

where m is the mass of the insect, V is the velocity at take-off, g is the acceleration due to gravity and h is the height of the jump.

It follows that:

$$h = \frac{V^2}{2g}$$

and that

$$h = \frac{\text{kinetic energy}}{mg}$$

The force exerted against the ground by the tip of the leg of a jumping insect will have vertical and horizontal components as shown for a grasshopper in Fig. 10.39, and the vertical component is equal to $F \sin \theta$.

A jumping insect will only continue to accelerate while its feet are in contact with the ground, and the take-off velocity will be determined by the scale of the force and the time over which it

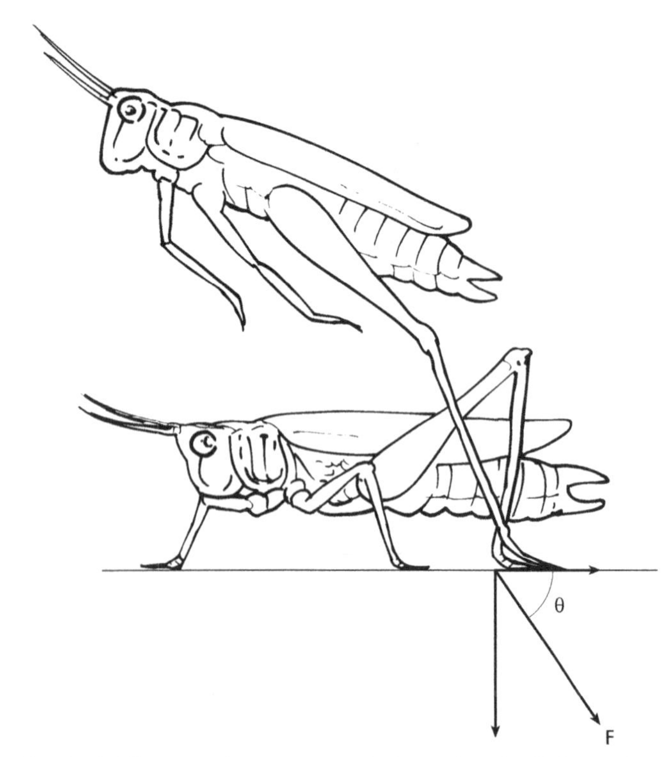

Fig. 10.39 Diagrams of a grasshopper before and during its jump. A force is transmitted to the ground via the articulation of the hind feet (see text). The long legs increase the time during which the force can act and so contribute to the acceleration achieved, but the higher the jump the less time the foot pushes. The jump is the result of an explosive release of energy, and an adult grasshopper requires more than $\frac{1}{2}$s to build up sufficient tension for the jump.

acts. This in turn will be determined by the length of the legs. Long legs also increase the mechanical advantage of the extensor muscles (see Box 10.4). For these reasons all insects which jump have relatively long legs. The limit to this line of evolution is probably set by the mechanical strength of the insect cuticle acting as a lever in this system. In some jumping insects such as the fleas, energy is stored in the elastic cuticle by movements which in effect 'cock' the legs for the jump and this stored energy is released during the relaxation of muscles which occurs when the insect jumps (Fig. 10.40).

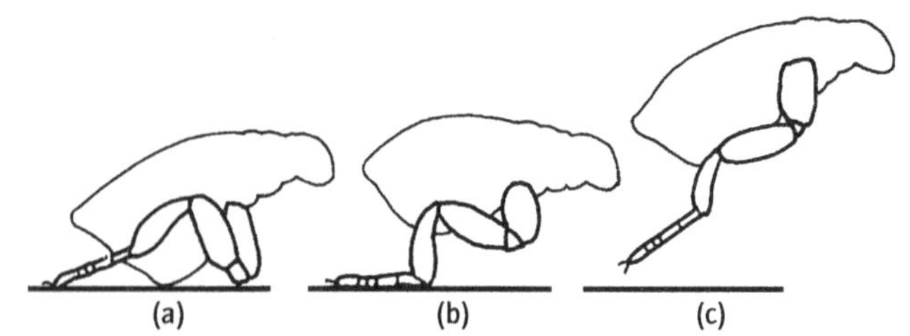

Fig. 10.40 The jump of a flea. (a) The femora are raised and energy is stored in deformed elastic proteins of the cuticle. (b) Locked femora are released by the relaxation of muscles. The force exerted against the ground by the tibia gives the flea a specific velocity which will determine the height of its jump.

10.7 Conclusions

This chapter has provided a general introduction to the locomotory systems of invertebrates. You should understand the effects of scale and the reasons why larger animals use muscle cells to develop force whereas the smallest animals are able to use cilia or flagella.

You should understand that all systems of animal movement obey the same mechanical laws and that a skeletal system is always involved in the application of a force. The skeletal system may transmit forces through a liquid in an enclosed space or through a rigid system of levers.

Some invertebrates are sluggish, slow-moving animals, others are highly mobile and agile; the cephalopod molluscs and the insects, for instance, illustrate very advanced locomotory techniques involving on the one hand jet propulsion and on the other, flapping flight. The analyses given here are expressed in simple terms, but many readers may wish to proceed to a more rigorous analysis. These readers are referred to the list of further reading below.

An understanding of the mechanics of animal locomotion is crucial to an understanding of the evolutionary origins of animal groups, since any proposed ancestor must be structurally sound. It must have worked and obeyed fundamental physical laws just as all the living invertebrates must do.

10.8 Further reading

Anderson, D.T. 1973. *Embryology and Phylogeny in Annelids and Arthropods*. Pergamon Press, Oxford.
Alexander, R.McN. 1982. *Locomotion of Animals*. Tertiary Level Biology. Blackie, Glasgow.
Brackenbury, J. 1995. *Insects in Flight*. Blandford, London.
Clark, R.B. 1964. *Dynamics in Metazoan Evolution*. Clarendon Press, Oxford.
Chamberlain, J.A. 1990. Jet propulsion of *Nautilus*: a surviving example of early Palaeozoic cephalopod locomotor design. *Can J. Zool.*, **68**, 806–814.
Dickinson, M.H. & Gotz, K.G. 1996. The wake dynamics and flight forces of the fruit fly *Drosophila melanogaster*. *J. exp. Biol.*, **199**, 2085–2104.
Elder, H.Y. & Trueman, E.R. 1980. *Aspects of Animal Movement*. Society for Experimental Biology Seminar Series. Cambridge University Press, Cambridge.
Hesseid, C.F. & Fowtner, C.R. (Eds) 1981. *Locomotion and Energetics in Arthropods*. Plenum Press, New York.
Kingsolver, J.G. 1985. Butterfly engineering. *Scient. Am.*, **253** (2), 90–97.
Marden, J.H. & Kramer, M.G. 1995. Locomotory performance in insects with rudimentary wings. *Nature*, **377**, 332–334.
Nybakken, J.W. 1988. *Marine Biology: An Ecological Approach*. Harper & Row, New York.

O'Dor, R.K. & Weber, D.M. 1991. Invertebrate athletes: Trade-offs between transport efficiency and power density in cephalopod evolution. *J. exp. Biol.*, **160**, 93–112.

Rainey, R.C. (Ed.) 1984. *Insect Flight*. Blackwell Scientific Publications, Oxford.

Schmidt-Nielsen, K. 1984. *Scaling: Why is Animal Size so Important?* Cambridge University Press, Cambridge.

Trueman, E.R. 1975. *The Locomotion of Soft-Bodied Animals*. Edward Arnold, London.

Weis-Fogh, T. 1975. Unusual mechanisms for the generation of lift in flying animals. *Sci. Am.*, **233** (5), 81–87.

Willmott, A.P., Ellington, C.P. & Thomas, A.L.R. 1997. Flow visualisation and unsteady dynamics in the flight of the Hawkmoth *Manduca sexta*. *Phil. Trans. R. Soc. Lond. B*, **352**, 303–316.

CHAPTER 11

Respiration

The need for O_2 used to be thought of as a fundamental property of all living things. Oxygen is taken up over respiratory gas exchange surfaces such as gills and lungs, and once inspired it is used to oxidize organic substances – largely but not exclusively carbohydrates – to yield the energy needed to power all active body processes. Life originated without O_2, though, so aerobic respiration is not an essential feature of organisms and indeed some can and do still function anaerobically. This chapter reviews processes of aerobic and anaerobic respiration in invertebrates. We begin by considering the biochemical basis of respiratory processes and then, focusing on aerobic respiration, we consider how O_2 is obtained from the environment and transferred to the tissues, and how the uptake of O_2 is influenced by intrinsic and extrinsic factors.

11.1 Central importance of ATP in respiration

Phosphorylated nucleotides, and particularly *adenosine triphosphate*, play an important part as intermediaries in the transfer of energy from the fuel (foodstuffs) to the power-consuming processes of metabolism. Potential energy in the food is transferred to so-called high-energy phosphate bonds (designated ~P) viz:

$$\text{energy from absorbed food} + A - P \sim P + Pi \rightarrow A - P \sim P \sim P$$

where Pi is inorganic phosphate, and this stored energy can then be yielded up to metabolism giving ADP + Pi. Note, however, that the term 'high-energy phosphate bond' is used rather loosely. The energy is *not* stored in the covalent linkage between the phosphate and the rest of the molecule. Rather, the phosphate-bond energy is a reflection of the energy content of the whole triphosphate molecule, before and after its conversion to the diphosphate.

11.2 Backbone of catabolism

Glycolysis and the tricarboxylic acid (TCA) cycle form the backbone for catabolism in all the invertebrate phyla. These pathways are familiar and are well covered in other textbooks and so only a brief description will be given here.

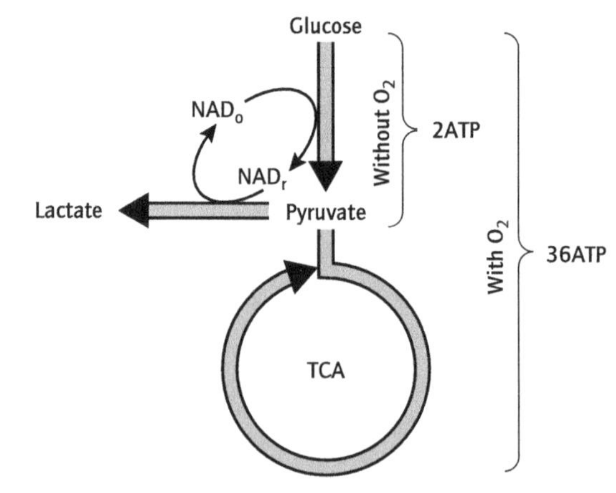

Fig. 11.1 An extremely stylized version of the metabolic pathways concerned with generating ATP.

Figure 11.1 gives a *very stylized* summary of these two pathways. The fuel is glucose, derived directly from food or by enzyme-mediated transformations of other molecules from food or from body stores.

Glycolysis takes place in the cytoplasm of cells and can occur without O_2. It generates ATP from the direct involvement of substrate passing down this pathway – *substrate level phosphorylation.*

The *TCA cycle* occurs in mitochondria and requires O_2. It generates reduced nicotinamide adenine dinucleotide (NAD_r) which is used, in turn, to generate ATP by donating electrons (becoming oxidized; NAD_O) to an electron transport system of cytochromes in which the final electron acceptor is O_2. This is *electron transport phosphorylation*. Notice from Fig. 11.1 that the involvement of O_2 allows the generation of more molecules of ATP per molecule of glucose than the part of the process that does not involve O_2.

11.3 Generating ATP without O_2

Before the evolutionary origin of photoautotrophs, organisms had to function without O_2. Moreover, environmental anoxia (no

O_2) can occur now: for littoral animals during exposure to air at low tides; during burrowing in reducing substrata; in many parasitic habitats. Additionally, specific tissues may become anoxic though the organism as a whole has ample O_2. For example, the muscles concerned with escape reactions in bivalve molluscs typically undergo anaerobic metabolism.

Hence the 'first' organisms depended upon anaerobiosis, and some continue to do so. In theory there are a large number of possible anaerobic pathways but four main ones have been used by invertebrates. Different pathways have evolved to meet particular energy requirements.

Lactate pathway – the best known pathway. It is illustrated in Fig. 11.1; the NAD_r, from glycolysis, being reoxidized by the reduction of the terminal pyruvate to lactate under the catalytic action of the enzyme lactate dehydrogenase. As already noted, it is not efficient in ATP production, but it can generate ATP at a high rate without O_2 and is commonly used to support burst work when muscle tissue can temporarily run out of O_2. It is not, however, of universal occurrence in invertebrates. It probably occurs in insect leg muscles but not in flight muscles which are so well served by tracheae and tracheoles (see p. 276) that they rarely become anoxic.

Opine pathway – this is similar to the lactate pathway and is adapted for burst work – the rapid but not necessarily efficient generation of ATP. In it the carbohydrate is catabolized by glycolysis but the reduction of pyruvate is replaced by its reductive condensation with an amino acid to form an opine – an amino acid derivative:

$$\text{glucose unit} + 2\ \text{amino acid} + 3\ \text{ADP} + 3\ \text{Pi} \rightarrow 2\ H_2O + 2\ \text{opine} + 3\ \text{ATP}$$

Several pathways can be identified depending upon the amino acid and enzyme used. For example, a substance called octopine is formed in the mantle muscles of cephalopods during burst work. Strombine, another opine, has been recorded from some bivalve molluscs.

Succinate pathway – used by organisms such as bivalves inhabiting anoxic muds and endoparasites inhabiting anaerobic sites in their hosts, like the vertebrate gut. Whilst not being capable of generating ATP rapidly, the succinate pathway can generate more per glucose input than the lactate and opine pathways. The succinate pathway retrieves energy from the NAD_r, generated in the initial glycolytic pathway, by an electron transport phosphorylation with fumarate rather than O_2 as a final electron acceptor. The basic framework of the process is illustrated in Fig. 11.2. Phosphoenolpyruvate (PEP), the molecule before pyruvate in glycolysis, is converted to oxaloacetate (by the addition of CO_2 – carboxylation – catalysed by PEP carboxykinase) then to fumarate. Fumarate is oxidized to succinate by an electron transport system. Succinate may then be further metabolized to proprionate and other volatile fatty acids. Oxaloacetate, fumarate and succinate are all intermediaries in the TCA cycle, but occur there in exactly the reverse sequence to the succinate system just described. Hence this pathway can be considered as

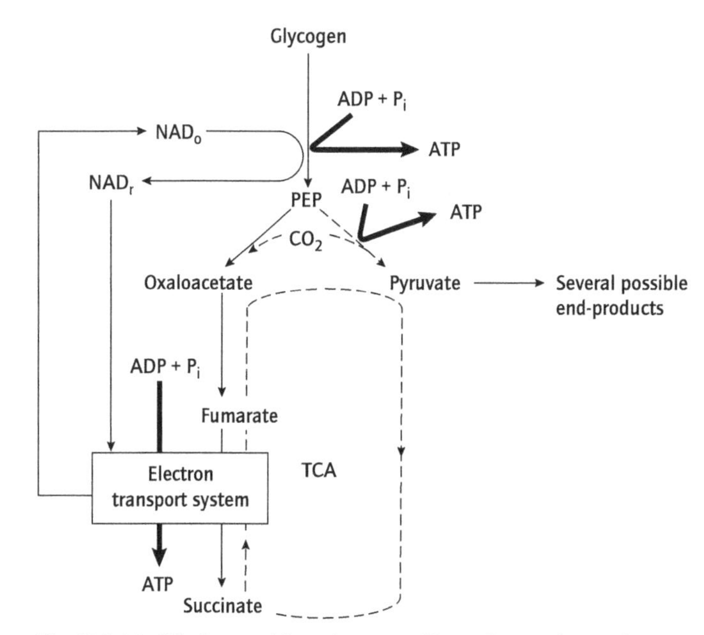

Fig. 11.2 Modified anaerobic pathway used by endoparasites and bivalves living in conditions of hypoxia for long periods (after Calow & Townsend, 1981).

the TCA put in reverse. A part of the pyruvate in this succinate system can also be converted to lactate, acetate, alanine and ethanol. A range of end products is therefore produced. Overall, the process leads to the production of about 4–6 ATP molecules per glucose-unit input.

A very similar reaction pathway, described for bivalves, produces succinate by reduction of oxaloacetate; however, the latter is not produced from PEP but from an amino acid, asparate, by transamination reactions.

Phosphagens – these are important in burst work and act as stores, accepting ~P from ATP in periods of relaxation and delivering it up under periods of hard work and anoxia:

$$\text{Phosphoarginine} + \text{ADP} = \text{Arginine} + \text{ATP}$$

The above system is common in the invertebrates whereas phosphocreatine is used by vertebrates. Exceptions to this rule are echinoderms that may have both phosphagens and the annelids that contain four other phosphagens in addition to phosphoarginine.

Figure 11.3 maps the phyletic distribution of the major pathways (not including phosphagens). All pathways are widespread. Thus the succinate pathway might either have evolved from the glycolysis–TCA system or vice versa, i.e. by the reversal of one or the other (above). It is generally thought that amino acids were prominent components of the early biotic environment and it has been suggested that the earliest system for the generation of ATP involved amino acids as both electron donors and acceptors. Hence, opine pathways could be primitive. Clearly all anaerobic pathways were present at an early stage in the evolution of the invertebrates and their distribution now is probably attributable to selection pressures – specific adaptations

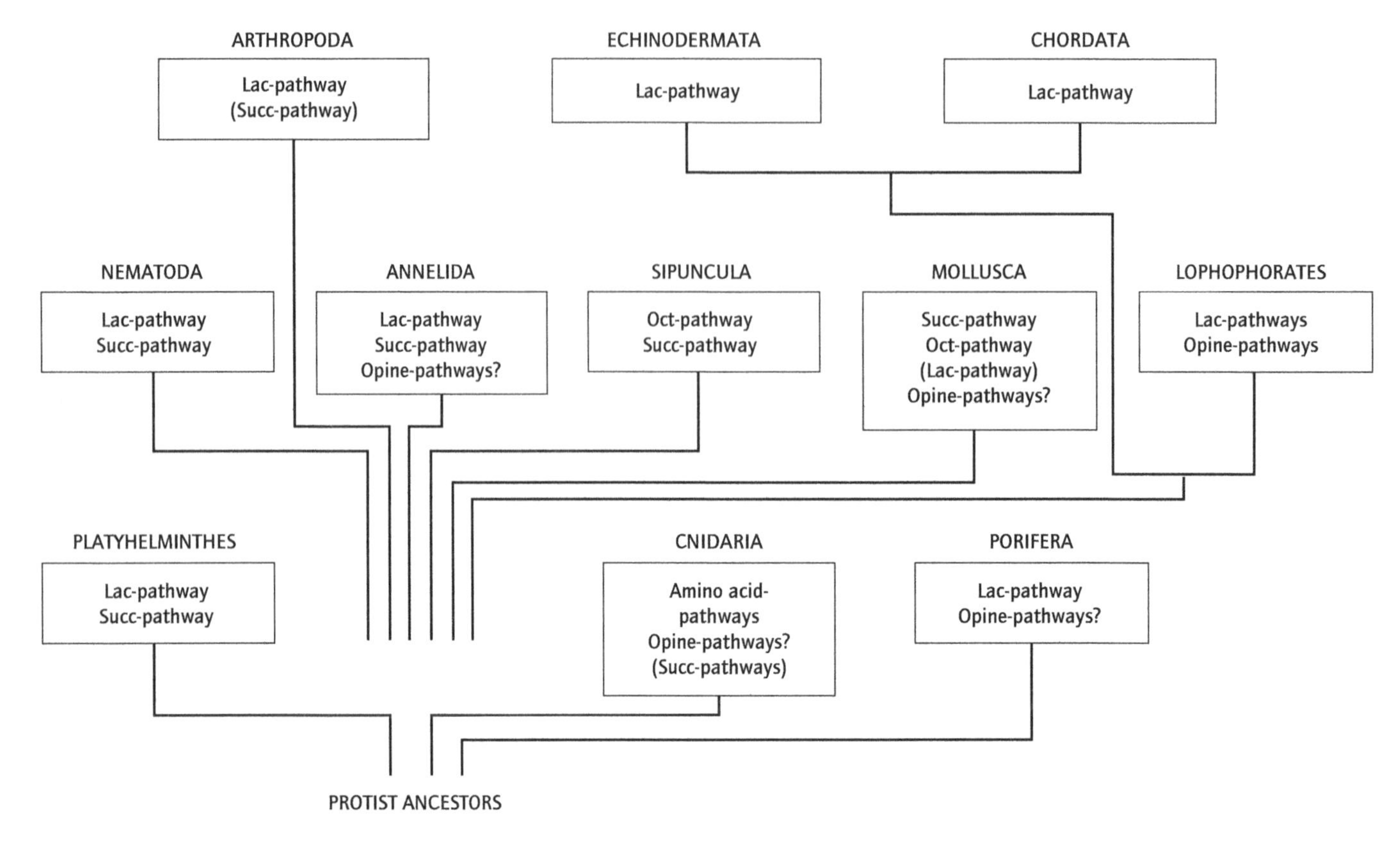

Fig. 11.3 Phyletic distribution of major pathways involved in ATP production. Modified from Livingstone (1983). Key: Lac = lactate; Succ = succinate; Oct = octopine; pathways in parentheses represent special circumstances.

to the ecological circumstances in which they have to work. Some systems have evolved to meet the physiological needs of burst work (produce lots of ATP but not necessarily very efficiently because it is not sustained), others have evolved to supply ATP more efficiently, if less quickly, in response to long-term and sustained anoxia.

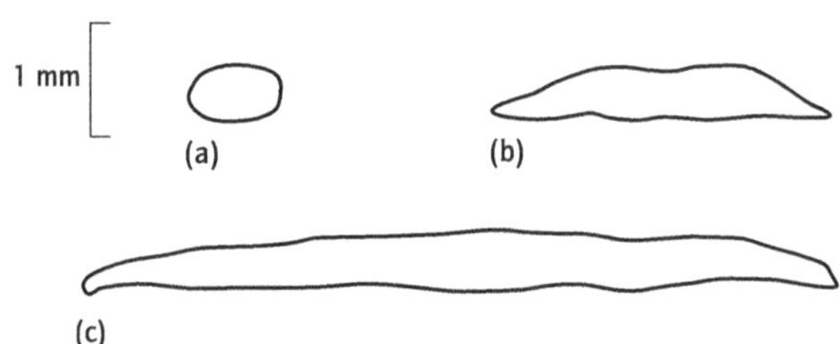

Fig. 11.4 Transverse sections, drawn to the same scale, through various turbellarians: (a) rhabdocoel, (b) triclad, (c) polyclad. (After Alexander, 1971.)

11.4 Uptake of O_2

11.4.1 Diffusion is paramount

Aerobic metabolism depends on O_2 being made available to respiring tissues. Fundamentally, this depends upon diffusion – O_2 molecules moving from high to low partial pressure (Po_2) according to the dictates of Fick's Law. The rate of diffusion of O_2 through tissues depends on Po_2 gradients and also on the properties of the tissues – the latter are often expressed as diffusion coefficients. On the basis of reasonable assumptions about these coefficients and the O_2 demands of tissues it can be calculated that the distance between metabolizing tissue and a respiratory surface can be no more than 1 mm. This is one of the reasons why large, solid-bodied turbellarians have to be flat whereas smaller acoels and rhabdocoels do not (Fig. 11.4). Many jellyfish and anemones, however, do grow very large despite the fact that they are also solid. Here, though, the outer and inner layers of tissue are thin and in direct contact with water in the external environment or the gastrovascular cavity. The thicker mesoglea is largely devoid of cells and has a low metabolic demand. In anemones, the more cellular mesoglea is usually less thick and, because of the complex folding of the gastrodermis (p. 63), always within 1 mm of water circulating in the gastrovascular cavity.

11.4.2 Circulatory systems

One solution to the limitations imposed by diffusion is the evolution of circulatory systems. These increase the capacity for transporting O_2, so that the 1 mm limitation is escaped, and by rapidly removing O_2 from respiratory surfaces maintain a steep

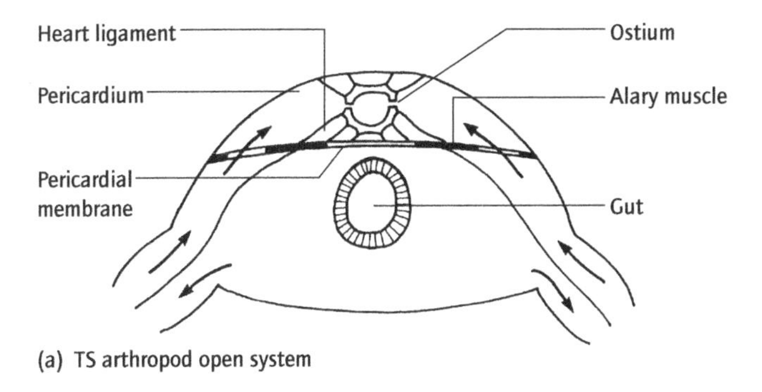

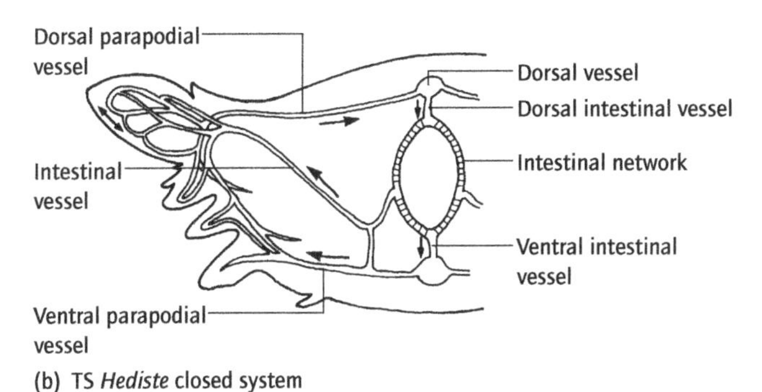

Fig. 11.5 Two kinds of blood system: (a) open system of an arthropod; (b) closed system of a polychaete annelid (*Hediste*).

Po_2 gradient and so increase the rate of O_2 intake. Two main kinds of system have evolved (Fig. 11.5): an open system with large haemocoel (persistent blastocoel or expanded blood vessel) typical of arthropods and molluscs and a closed system of arteries and veins well developed in the Annelida. Both require muscular pumps, e.g. contractile tubes in arthropods, accessed by pairs of valved ostia (Fig. 11.5a), and muscular, pulsating blood vessels (lateral 'hearts') in annelids. Echinoderms and hemichordates have somewhat intermediate systems consisting of small vessels connected to larger sinuses. Holothurian echinoderms, though, have well-developed closed systems. In association with the invasion of land, isopod crustaceans have evolved larger, more muscular hearts than other crustaceans and vessel-like lacunae that form an almost closed system. Insects, though having an open system, do not make much use of it for O_2 transport. Instead they have evolved an elaborate and extensive system of tubes, tracheae and tracheoles (Fig. 11.6), that ensure all actively metabolizing tissues are provided for by almost direct supply of gaseous O_2. (The way these work is explained in the legend to the figure.)

11.4.3 Blood

The blood of many invertebrates (e.g. some mollusc and echinoderm, and all urochordate and cephalochordate species) is colourless. It is similar in composition to sea water. All of the O_2 present is contained in physical solution (typically <0.3 mmol O_2 l^{-1}). One way of increasing the amount of O_2 carried by a given volume of blood, as exploited by a wide variety of different species, is to evolve a respiratory pigment. These pigments are specialized proteins capable of binding reversibly with O_2. Their presence in blood can result in increases (×2–×30) in the amount of O_2 carried. All respiratory pigments resemble each other in that they consist of conjugated proteins linked to prosthetic groups normally containing one of two metals, either iron or copper. The names and some of the properties of the four pigments that can be found across the animal kingdom are summarized in Table 11.1.

If we examine the distribution of respiratory pigments across the major invertebrate phyla (Fig. 11.7) a number of general observations can be made. Respiratory proteins are found in about one third of all invertebrate phyla. In some groups they occur in the blood, either packaged in cells (e.g. haemoglobins of some annelids, some molluscs and some echinoderms and haemerythrins of some brachiopods and some polychaetes) or dissolved in the blood itself (e.g. haemoglobins of some molluscs and some crustaceans, arthropod and mollusc haemocyanins and annelid chlorocruorin). In other groups respiratory pigments can be found within the tissues. Haemoglobins can be found in the body wall of some nematodes, the muscles of some annelids and the pharynx of some Platyhelminthes. There is a general lack of pattern in any one pigment's distribution (perhaps with the exception of chlorocruorin), suggesting the independent evolution of many of these respiratory pigments. Both haemoglobin and haemerythrin are widely distributed across the invertebrate phyla compared to haemocyanin (some arthropods and some crustaceans only) and chlorocruorin (restricted to just four families of polychaetes). Sometimes different pigments even occur within the same individual. For instance in some molluscs haemocyanin is present in the blood but haemoglobin-like molecules are used as non-circulating O_2 carriers in tissues such as gill, muscle and nerve. In only a few phyla does the occurrence of pigment seem to be ubiquitous, e.g. blood cell haemoglobins in the Phoronida and Echiura, blood cell haemerthyrins in the Sipuncula and Priapulida. It is currently thought that the primitive pigment was a haemoglobin contained within red blood cells and that this evolved around the same time as the first circulatory systems, from a non-circulating proto-heme protein.

Despite tremendous structural diversity the respiratory pigments have much in common with regards their function. Many (but not all) of them show cooperative O_2 binding, i.e. binding O_2 molecules makes it successively easier to bind further O_2 molecules. This results in a sigmoid curve for the relationship between the amount of O_2 bound and the O_2 tension the blood is equilibrated with. One such a curve is illustrated in Figure 11.8. This curve shows that the respiratory pigment, in this case haemocyanin from the snail *Helix*, is fully saturated with O_2 >7 kPa. Even if we increase O_2 tension further the haemocyanin cannot bind any more O_2. The total O_2 contained in the blood does, however, increase a little due to an increase in the amount of dissolved O_2. At lower O_2 tensions the pigment gives up its O_2

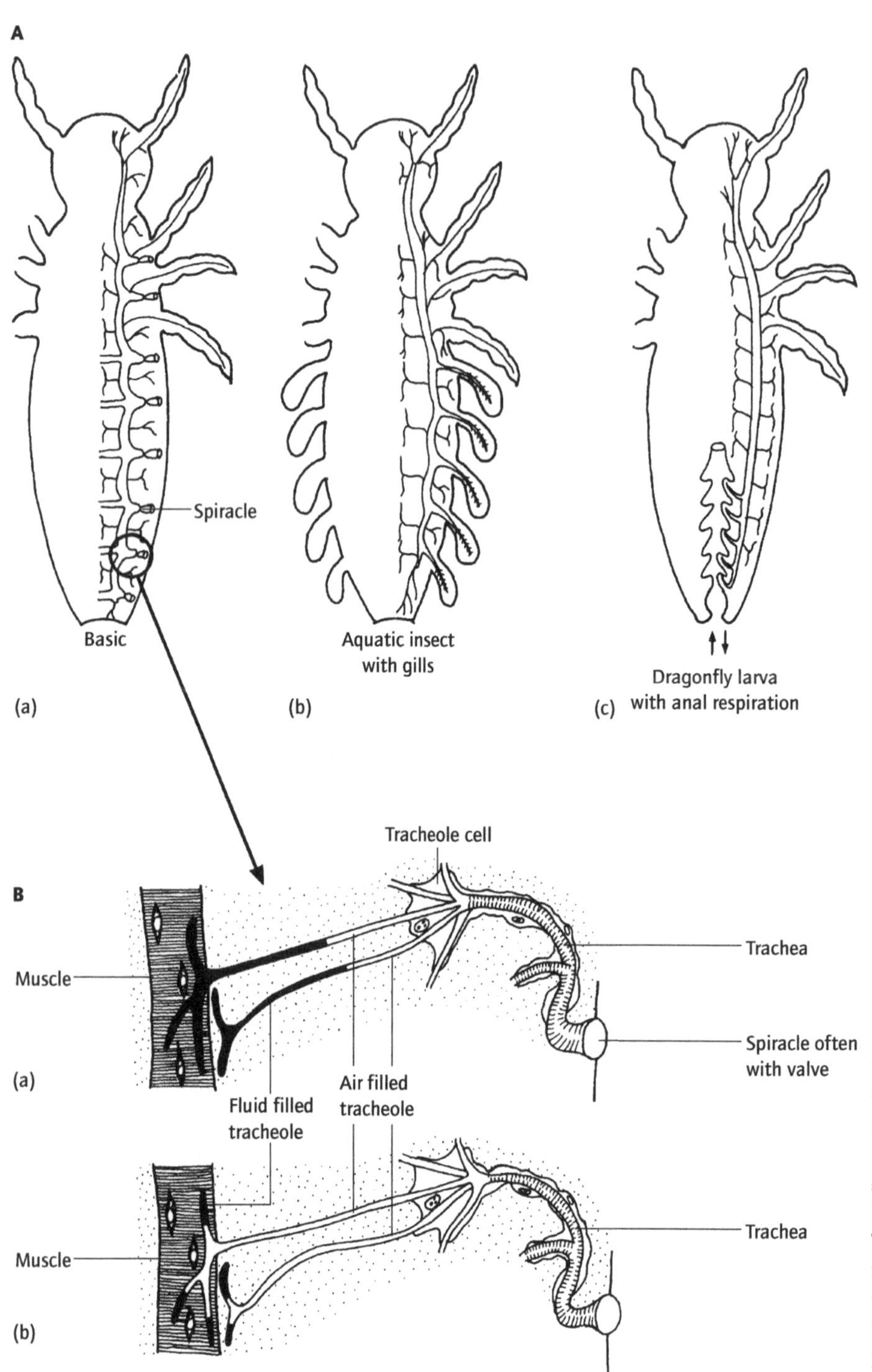

Fig. 11.6 **A** Tracheal systems: (a) basic; (b) aquatic insect with gills; (c) dragonfly naiad with anal respiration. **B** Enlargement of tracheal system (enlarged from **A** (a)) – air enters through spiracles, which in most insects have valves. Large tubes (tracheae) lead to small (tracheoles). The latter can contain fluid (a) but during active metabolism the osmotic pressure of surrounding tissues increases and the fluid is withdrawn (b). Air is then drawn into the tracheoles, and passes by diffusion through their walls into the tissues. (Redrawn from various sources.)

and at an O_2 tension of around 2.2 kPa it is 50% saturated with O_2. This point is referred to as the half saturation pressure, or P_{50}. Values for P_{50} give a useful measure of the affinity of a pigment for O_2. If the pigment is characterized by a high P_{50} value this means it has a low affinity for O_2. On the other hand, if it has a low P_{50} this means that it has a high affinity for O_2. The amount of O_2 (bound and physically dissolved) delivered to the tissues per unit blood is represented by the rectangle situated to the left in Fig. 11.8. This is determined by the concentration of the respiratory pigment, its affinity for O_2 (and how this can be altered by changes in blood chemistry), its cooperativity and the O_2 tensions prevailing in arterialized (P_a) and venous blood (P_v).

The position of an O_2 binding curve can be altered by changes in blood chemistry. For instance for a large number of respiratory pigments a decrease in pH (or an increase in CO_2 tension) results in the curve being shifted to the right. This is known as the Bohr effect. It is assumed that this will aid unloading of O_2 at the tissues at times of heightened energy demand since CO_2

Table 11.1 Structure and function of respiratory pigments.

Name	Structure	Molecular weight range	Function
Haemoglobin	Prosthetic group is haem (a porphyrin), linked to one atom of ferrous iron. Found either in solution or in cells	17 000–3 000 000	Cooperative O_2 binding. Red when oxygenated, blue when deoxygenated
Haemocyanin	Prosthetic group is a polypeptide linked to 2 atoms of copper Always found in solution, never in cells	25 000–6 680 000	Cooperative O_2 binding. Blue when oxygenated, colourless when deoxygenated
Chlorocruorin	Like haemoglobin the prosthetic group is haem linked to one atom of ferrous iron. Always found in solution	3 400 000	Cooperative O_2 binding. Green in dilute solution, red in concentrated
Haemerythrin	Non-porphyrin prosthetic group though it is attached to iron. Always found in cells	17 000–120 000	Violet when oxygenated, almost colourless when deoxygenated

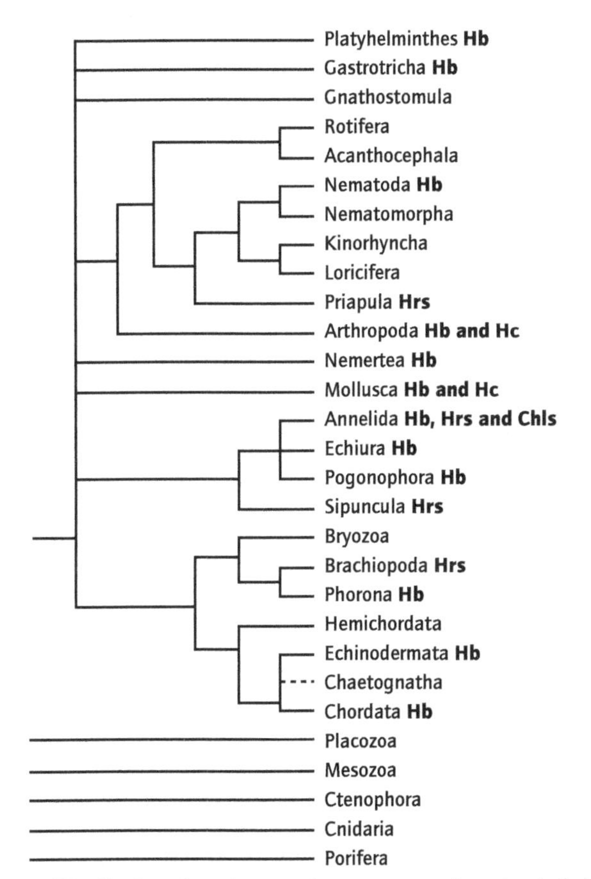

Fig. 11.7 Distribution of respiratory pigments across the animal phyla. Hb = haemoglobins; Hrs = haemerythrins; Hc = haemocyanins; Chls = chlorocruorins. Relationships between the phyla based on information presented in Figs 2.8 and 2.20.

is likely to be highest in tissues with elevated metabolic rates. However, some species, e.g. the horseshoe crab *Limulus* and some molluscs, actually show reverse Bohr effects. That is to say an increase in CO_2 shifts the curve to the left. While in most cases it is not difficult to suggest reasons why animals might display such responses, the extent to which normal and reverse Bohr effects are adaptive is currently a matter of some debate. A number of other substances present in the blood have been shown to alter the position of the O_2 binding curve (at least for a few species). They include divalent ions, such as magnesium and calcium, lactic acid (produced as a result of anaerobic metabolism in some invertebrates), uric acid and the catecholamine dopamine.

External factors such as temperature have also been shown to alter the position of the O_2 binding curve. In general an increase in temperature tends to shift the curve to the right, i.e. there is a decrease in O_2 affinity (increase in P_{50}). However, this is not invariant as there does appear to be a tendency for animals that inhabit thermally variable or unstable environments to possess pigments which are insensitive to temperature, e.g. intertidal hermit crabs.

It has been relatively common in the past to 'provide' functional explanations for every characteristic of the O_2 binding curve. These are based on the assumption that as a result of natural selection there is an optimal set of O_2 binding characteristics for the particular needs and circumstances of each species. This has resulted in some confusion, in no small part due to all of the 'exceptions to the rule' discovered. Given the large number of interrelated factors that can and do influence the O_2 binding curve and how it acts *in vivo* to transport O_2 from the gas exchange surface to the tissues, we should perhaps take added care in so readily ascribing adaptive significance to all of the features we examine. Even keeping all of this in mind we do seem able to identify some broad patterns, where we can relate physiological variation both between and within pigments to the environments (internal and external) in which they operate.

Invertebrates that inhabit O_2 poor environments generally possess a respiratory pigment with a high affinity for O_2. Some of the lowest P_{50} values recorded are for the haemoglobins of parasitic animals, such as the nematode *Ascaris*, which inhabit chronically low O_2 environments. The lugworm *Arenicola* is a marine intertidal annelid that often constructs burrows in deoxygenated muds. It possesses haemoglobins with a considerably greater affinity for O_2 than *Eudistylia*, a worm species which requires good O_2 conditions for survival (Fig. 11.9). The same is

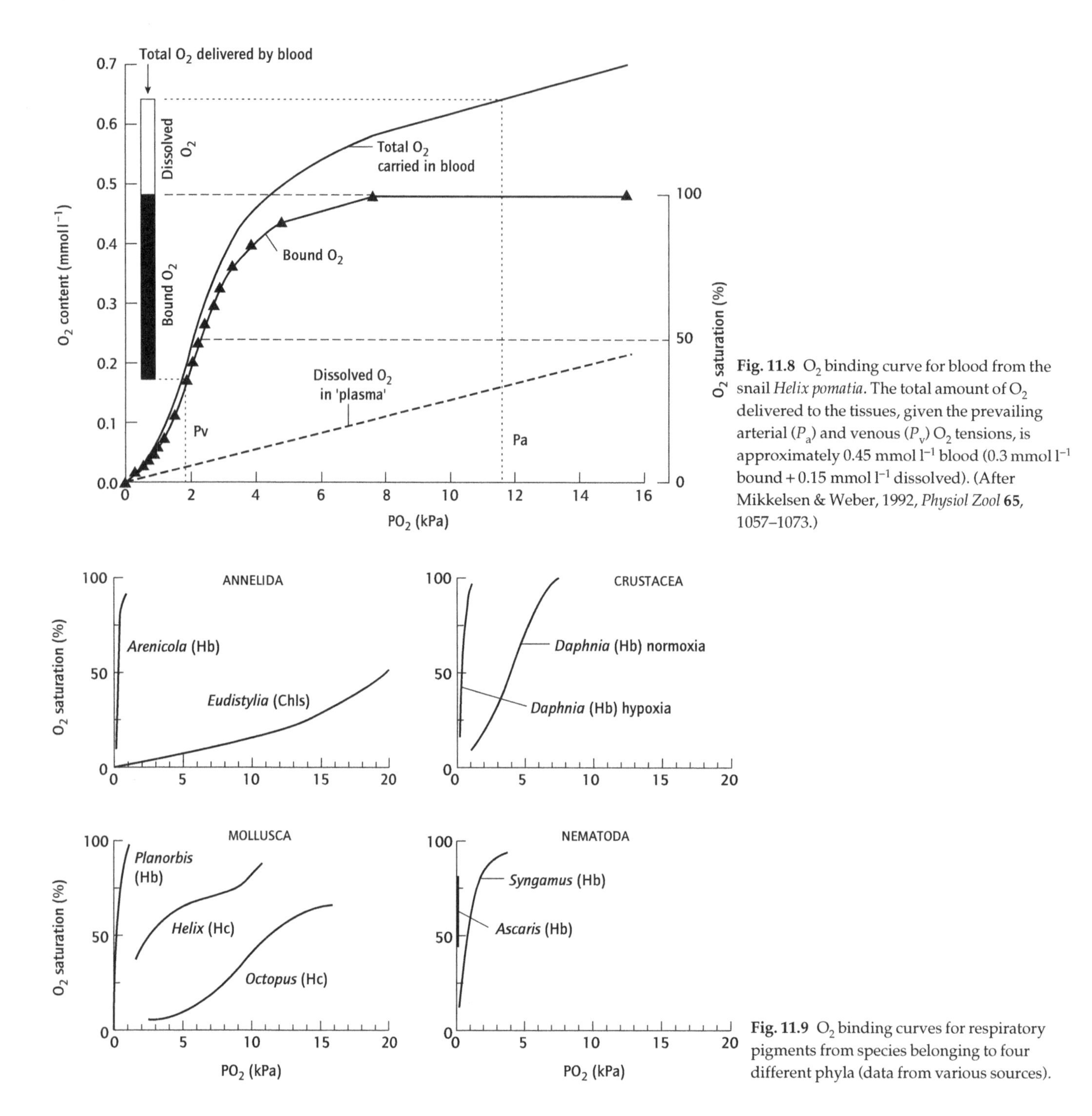

Fig. 11.8 O_2 binding curve for blood from the snail *Helix pomatia*. The total amount of O_2 delivered to the tissues, given the prevailing arterial (P_a) and venous (P_v) O_2 tensions, is approximately 0.45 $mmol\,l^{-1}$ blood (0.3 $mmol\,l^{-1}$ bound + 0.15 $mmol\,l^{-1}$ dissolved). (After Mikkelsen & Weber, 1992, *Physiol Zool* **65**, 1057–1073.)

Fig. 11.9 O_2 binding curves for respiratory pigments from species belonging to four different phyla (data from various sources).

true when we compare haemocyanin from the snail *Planorbis* (which can live in stagnant waters) with a mollusc that does not really tolerate low environmental O_2, in this case the octopus. In some species at least, exposure to low environmental O_2 can induce a change in the molecular structure of the respiratory pigment, changing it, within the course of a few days, from a low to a high O_2 affinity. This is the case in the waterflea *Daphnia* where not only does the O_2 affinity of the haemoglobin change but the total amount of pigment present increases to such an extent that the individual turns bright red.

Conversely, species that normally inhabit well-oxygenated waters, and have well-developed gas exchange surfaces (i.e. they present only a modest diffusion barrier and thus arterialized O_2 tensions are high), tend to have low-affinity pigments; species such as *Eudistylia* and *Octopus*. It is possible, however, that a species inhabiting a well-oxygenated environment may possess a high-affinity pigment. Usually in such cases the respiratory gas exchange surfaces constitute a considerable barrier to diffusion and so arterialized O_2 tensions are much lower relative to ambient, e.g. some crustaceans.

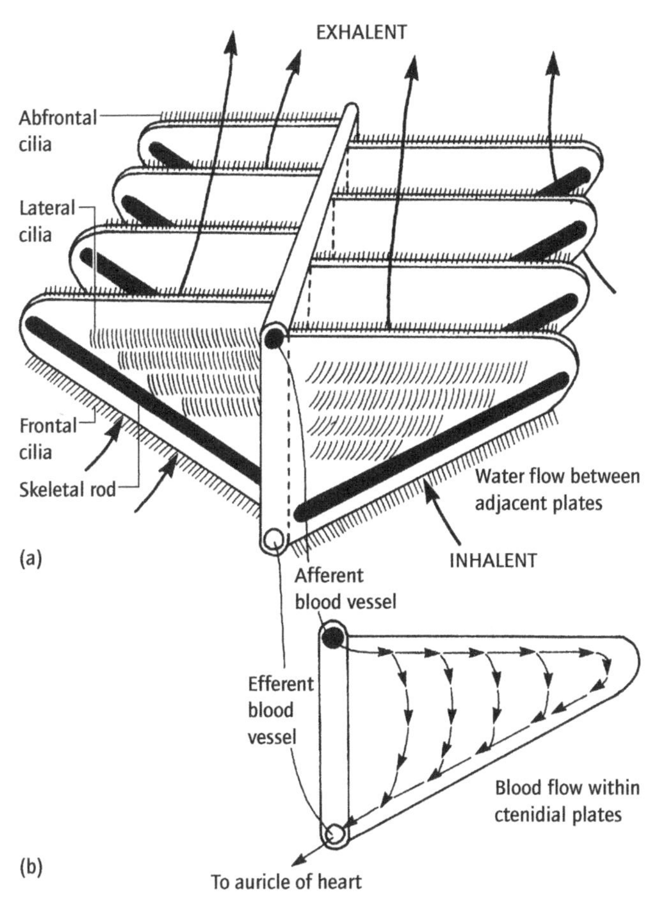

Fig. 11.10 Ctenidial system of a gastropod: (a) shows flow of water over 'gills'; (b) shows flow of blood through them. (After Russell-Hunter, 1979.) See also Fig. 5.18.

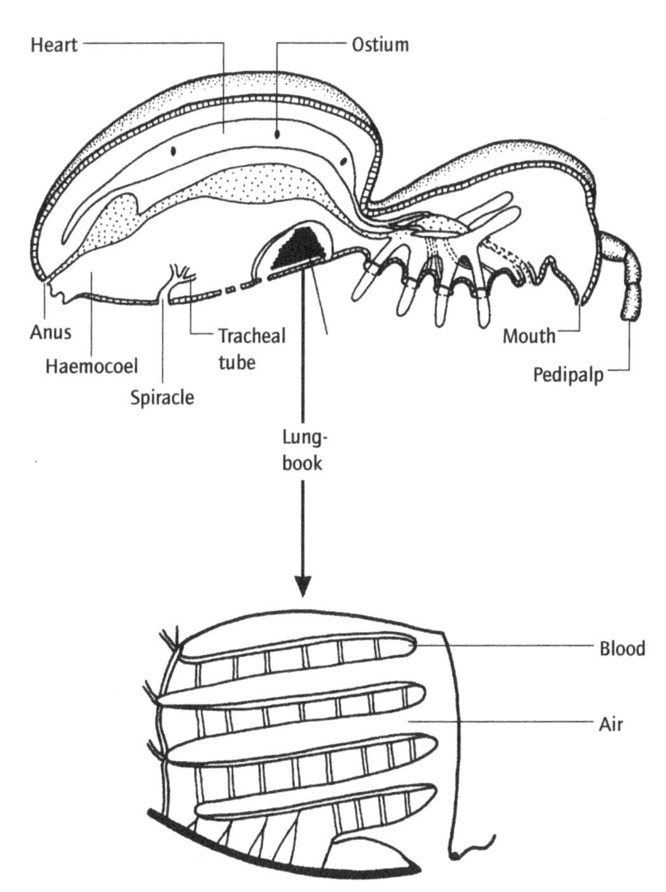

Fig. 11.11 Lung-book of a spider (after Snow, 1970). See also Fig. 8.11.

Related to this issue of O_2 availability there do seem to be consistent differences between the pigments of terrestrial compared with those of aquatic crustaceans. Despite claims to the contrary there appears to be a decrease in O_2 affinity with increasing terrestrial adaptation in the haemocyanins of both crabs and amphipods.

Finally, the respiratory pigments of species that inhabit cold waters do tend to have lower intrinsic affinities for O_2 (higher P_{50}) than their close relatives from warmer, more tropical waters. This means that although not identical O_2 affinities are more similar at prevailing environmental temperatures than they would have been if intrinsic affinities were the same.

11.4.4 Respiratory gas exchange organs

Another limitation on O_2 uptake is the surface area (and thickness) of the gas exchange organs. An unspecialized surface is adequate for flatworms and long, thin nemertines, nematodes and some annelids. However, further increases in size or activity and the evolution of outer protective and relatively impermeable coverings, or shells, required the evolution of vascularized respiratory surfaces. Evaginated surfaces are common in aquatic forms, e.g. 'gills' of some polychaetes and arthropods (Fig. 11.6), ctenidia of molluscs, gill books of *Limulus* and podia of some echinoderms. Figure 11.10 shows the ctenidial system, in diagrammatic form, of an aquatic gastropod. Notice that blood is pumped through the 'gill', by the heart, in the opposite direction to the flow of water (effected mainly by lateral cilia). This countercurrent flow is typical of gills (though not those of the lugworm *Arenicola* or cephalopods) and ensures that water with lowest P_{O_2} is in contact with the least oxygenated blood, hence maximizing the efficiency of transfer of O_2. By contrast, invaginations are common in terrestrial groups, e.g. tracheae (Fig. 11.6), lung-books and sieve and tube tracheae of arachnids (Fig. 11.11) and 'lungs' of pulmonate molluscs. Some aquatic forms have invaginated surfaces – the respiratory trees of holothurian echinoderms (Fig. 11.12) and the anal structures of some insect larvae (Fig. 11.6). Some aquatic insect larvae have physical gills and plastrons (Fig. 11.13), and the way these work is explained briefly in the legend to Fig. 11.13.

11.4.5 Ventilation

One other mechanism that assists in O_2 uptake is the ventilation of the respiratory surfaces. Again this ensures that the supply of O_2 to the respiratory surface is continuously replenished so

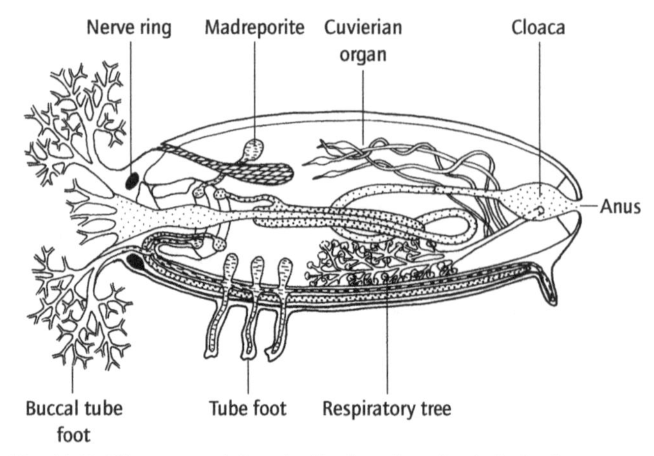

Fig. 11.12 Diagrammatic longitudinal section of an holothurian echinoderm showing respiratory trees (after Nichols, 1969). See also Fig. 7.27.

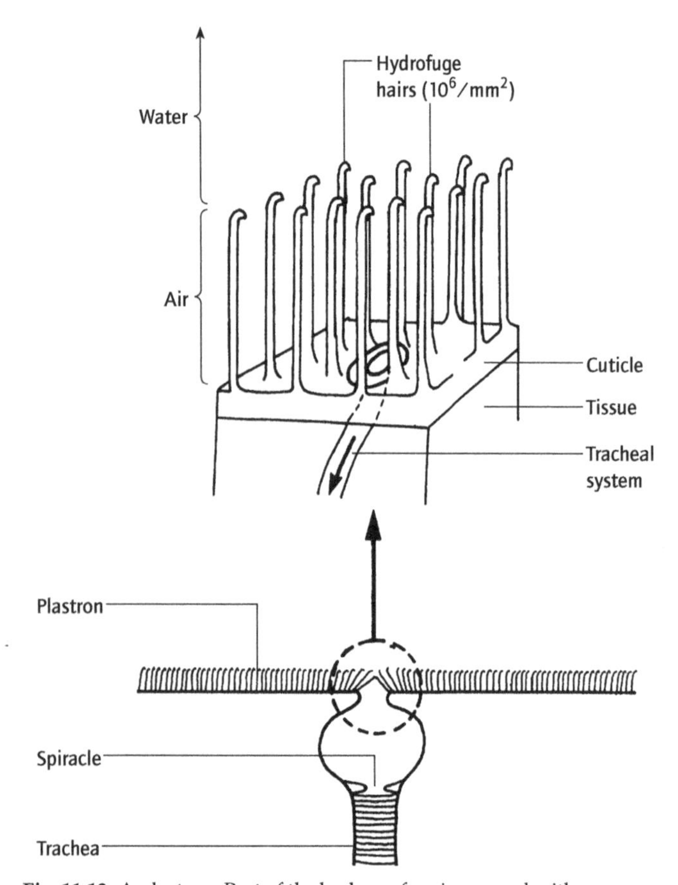

Fig. 11.13 A plastron. Part of the body surface is covered with hydrophobic hairs providing a non-wettable surface, where air remains permanently. It thus acts as a non-compressible gill into which O_2 diffuses from the water; i.e. as O_2 is withdrawn the thick pile prevents intrusion of water, so the volume remains constant and Po_2 must fall causing O_2 to diffuse in from the aquatic environment. In the aquatic insect *Aphelocheirus* the plastron can withstand pressures of several atmospheres before collapse. (After Ramsay, 1962 and Randall *et al.*, 1997).

that the gradient across the surface is maximized. This occurs in animals both with and without modified respiratory surfaces, but not in the tracheal systems of insects. Arrangements range from lateral cilia on the gill lamellae of bivalves to the muscular pumps servicing the respiratory trees of holothurians and anal structures of dragonfly naiads. Tubiculous polychaetes use ciliary currents and/or peristaltic movements. Crustacea commonly depend upon undulating appendages. Ventilation often increases as environmental Po_2 is reduced (see Section 11.6.5).

11.5 Measuring metabolism

Most of the energy that goes into respiration ultimately appears as heat (Fig. 11.14). Hence, if the heat emanating from animals could be measured, it would provide a useful and accurate assessment of metabolism. Devices that are sensitive enough to record the small quantities of heat emanating from invertebrates, involving very sensitive thermopiles, are available – they are called direct microcalorimeters – but they are not commonly used. A trace from a calorimetric experiment, involving one of them, with an aquatic oligochaete, *Lumbriculus variegatus*, as experimental subject, is illustrated in Fig. 11.15. The traces show: (I) the aerobic state; (II) the anoxic state; (III) what happens after the animals are poisoned. Notice that the rate of aerobic metabolism is about four times that of anaerobic metabolism.

It is easier to measure O_2 uptake than heat production. Various techniques are available; these range from chemical titration of O_2 in aqueous solution to physical measures of the volume or pressure of the gas. Several kinds of very accurate and sensitive electrodes are also available. Hence O_2 levels can be measured in the environment of sealed systems containing animals, or in the inflows and outflows of open systems. By definition, this kind of technique will only monitor aerobic processes and so is likely to underestimate respiratory metabolism because it is possible that invertebrates use anaerobic processes in conjunction with aerobic ones even when there is O_2 available.

Nevertheless, most studies of respiration have used O_2 uptake as a measure of metabolism. This kind of information is used below to describe how various factors influence respiratory metabolism. However, there are still complications since the metabolism of invertebrates is so sensitive to the condition of the individual or the surroundings in which it is being kept that what is being measured in one experiment need not be comparable with that being measured in others. The following classification is helpful. Total respiratory metabolism comprises: *standard metabolism*, recorded when the organism is at rest; *routine metabolism*, recorded in routinely active animals; *feeding metabolism*, recorded in animals that have just fed; and *active metabolism*, recorded in animals undergoing substantial activity. Experimentalists often aim at standard metabolism as a repeatable measure of metabolism.

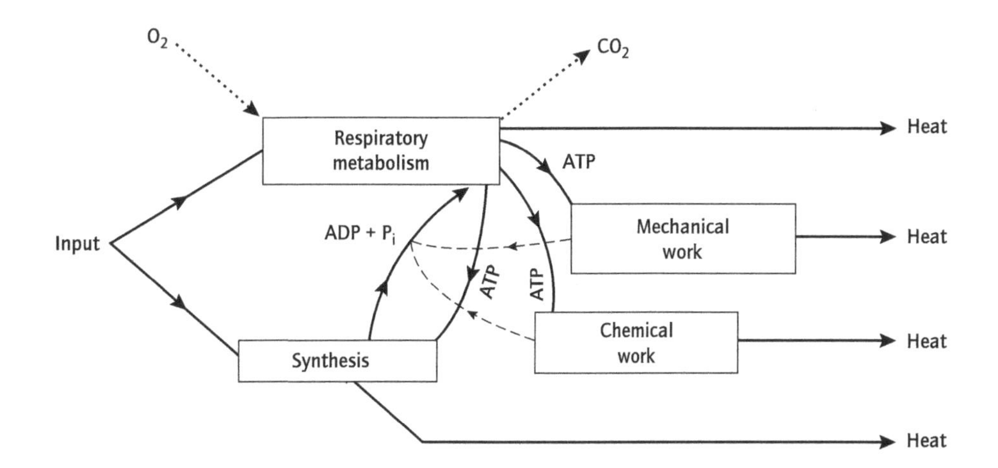

Fig. 11.14 Flow diagram of use of food energy. Respiratory processes generate ATP inefficiently, with heat loss, and most of the energy carried by ATP appears as heat after doing work.

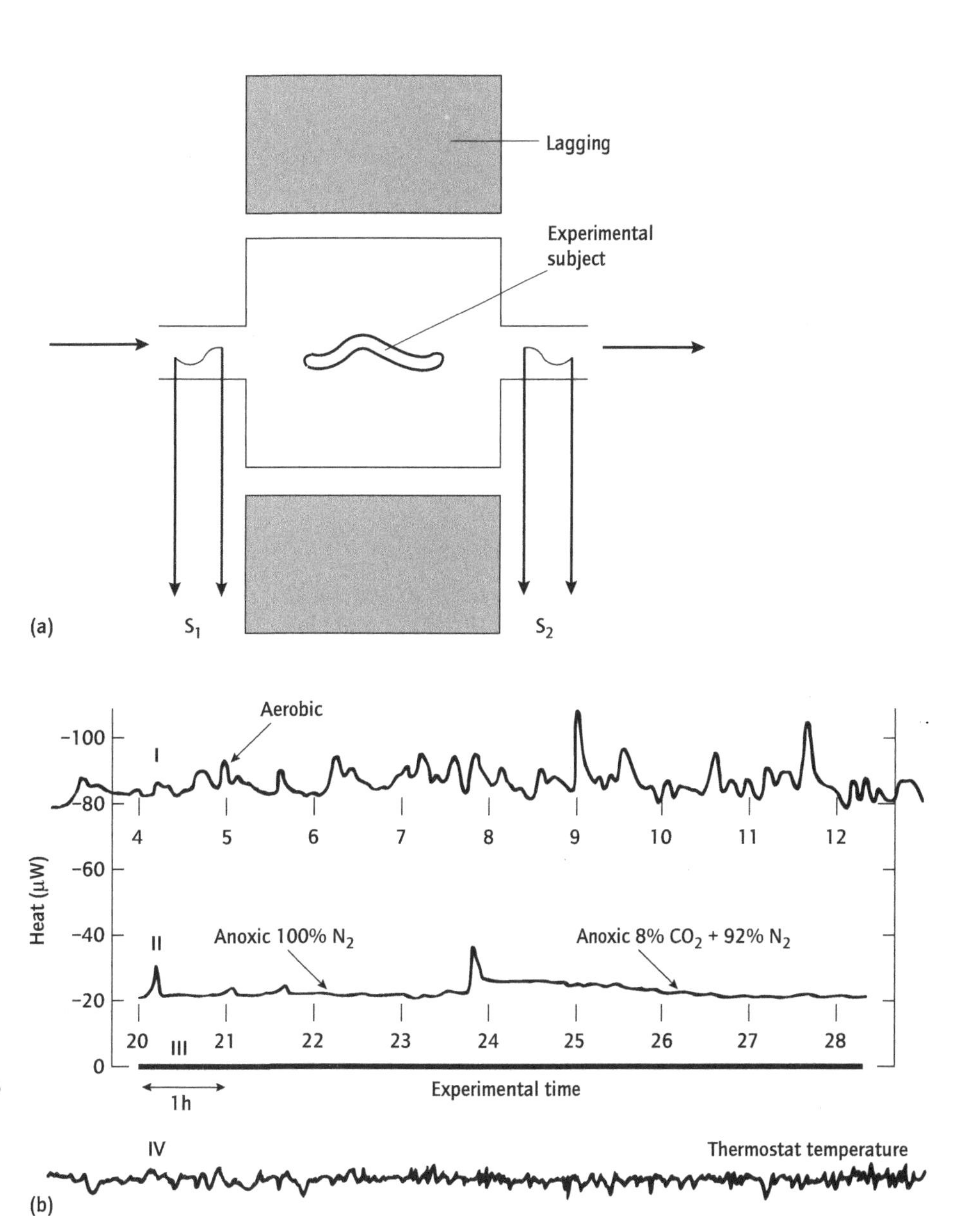

Fig. 11.15 Direct microcalorimeter. (a) The apparatus: very sensitive sensors (S_1 and S_2) measure the temperature of water flowing into and out of a well-lagged chamber containing the experimental subject. (b) Some results described more fully in the text. (After Gnaiger, 1983.)

11.6 Factors influencing respiration

The following is a list of some of the better studied factors that influence the O_2 uptake (as a measure of respiration) by invertebrates. They are ordered in decreasing intimacy of association with organisms, i.e. from what have been described as intrinsic to extrinsic factors.

11.6.1 Body size

It is not unreasonable to expect large invertebrates to respire more than small. However, since O_2 uptake is surface dependent (above) and body surfaces increase in two dimensions whereas body mass increases in three dimensions, it is also expected that the relationship between body mass (as an index of size) and O_2 uptake will not be one of simple proportionality. Comparisons of different-sized individuals within species and of species with different body sizes indicate that standard respiratory rate increases, but at a reducing rate, with body mass represented by the following equation:

$$\text{Resp. rate} = a\,(\text{mass})^b \qquad 11.1$$

where a and b are constants and b is usually less than 1. Taking logarithms gives:

$$\text{log. Resp. rate} = K + b(\text{log mass}) \qquad 11.2$$

where K = log a. Hence plotting the logarithms of O_2 uptake against the logarithms of body mass should give a linear relationship with a slope equivalent to b (Fig. 11.16).

Dividing equations 11.1 and 11.2 throughout by weight gives:

$$\text{Resp. rate per unit mass} = a(\text{mass})^{b-1}$$

$$\log(\text{Resp. rate per unit mass}) = K + (b-1)\log \text{mass}$$

Since b is less than 1, b − 1 will be negative, so plotting log O_2 uptake per unit mass against log mass should give a straight line with a negative slope (see Fig. 11.18). Hence, as expected above, O_2 uptake per unit mass reduces with mass. However, if these relationships were simply a matter of surface area not keeping pace with body mass as mass increases, then b should be 0.67 and b − 1 should be −0.33. The rationale behind this is straightforward. Respiratory gas exchange, being surface dependent (above), should be proportional to body surface area which, in geometric-ally similar bodies, is in turn proportional to body length squared (l^2). Mass is equivalent to volume and hence, again assuming geometrical similarities, to l^3. So respiration should be proportional to the cubed root of mass ($\sqrt[3]{l^3} = l$) squared ($= l^2$), i.e. to ($\sqrt[3]{M^2} = M^{0.67}$). However, rarely are values of b precisely 0.67 and often they approximate more to a value between 0.67 and 1 (Fig. 11.16). Hence, it is usually assumed that though geometrical factors play a part in determining the size dependency of O_2 uptake, they cannot be the only or perhaps even the main basis for this relationship. Just exactly what factors are involved is still not clear.

11.6.2 Activity

A mobile limpet consumes about 1.4 times more O_2 than an inactive one; this figure is about 2 for *Gammarus*, an amphipod, 3 to 4 for *Palaemonetes*, an estuarine shrimp and more than 100 for a flying locust. Again these observations are not unexpected; movement involves the beating of cilia and flagella and the contraction of muscles and so leads to an enhancement of metabolism over the standard rate. To say that an active individual has a greater O_2 uptake than an inactive one is to state the obvious. What is not so obvious, however, is that at comparable levels of activity (e.g. same running or swimming speed) a given mass of a small individual has a greater O_2 uptake than the same mass of a large individual. This means that activity is more costly for a small individual (or species) per unit mass, than it is for a large one.

11.6.3 Feeding

Oxygen uptake often increases immediately after a meal, only to subside again shortly afterwards. This response is referred to as specific dynamic action or SDA (also specific dynamic effect or calorigenic effect) (Fig. 11.17). Both the intensity and duration of SDA vary between species, and even within the same species, for example, as a result of changes in environmental temperature or diet composition. The large Antarctic nemertean *Parborlasia* displays an SDA that last for 30 days and at its peak is ×1.5–2.6 that of resting metabolism. Terrestrial crabs living at tropical temperatures by contrast have an SDA that lasts around 50 h and is × 3 that of resting metabolism. Even within a species the same pattern seems to hold. The predatory leach *Nephelopsis* has an SDA that last 19 hours at 5°C but only 11 h at 25°C.

There are at least three, non-exclusive, explanations of SDA. It may constitute the cost of procuring food and processing it in the gut. Generally speaking it is thought that these costs are a relatively small proportion of SDA, but this is not always so, e.g. about 80% of a 25% increase in postprandial (i.e. after feeding) O_2 uptake by the mussel *Mytilus edulis* could be attributed to the cost of filtering food. Secondly it has been suggested that SDA largely represents the cost of using raw materials obtained from the food to synthesize new tissues. Finally a substantial component of SDA may be due to the cost of degrading and excreting proteins absorbed from food (particularly when protein-rich) that are in excess to requirements. In this connection it is interesting that, almost irrespective of taxon, SDA is generally greater in animals fed on a high protein diet compared with those fed on carbohydrates or fats.

While feeding may lead to an acute increase in O_2 uptake, continuous exposure to low rations is often associated with a reduced O_2 uptake. Part of this may be due to the absence of SDA. This said in many cases it is predominantly due to reductions in levels of activity and even economization in maintenance metabolism. Such reduced rates of O_2 uptake are

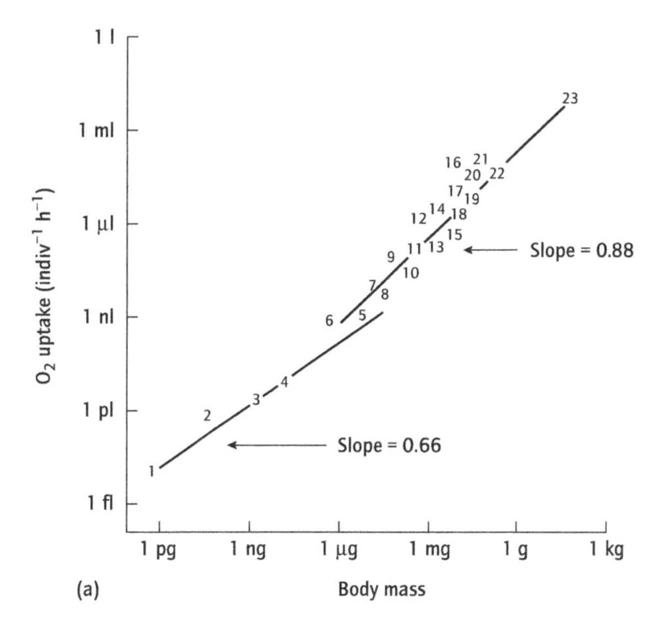

Code number	Group	*n*	b
Unicellular ectotherms			
1	Bacteria	5	0.68
2	Fungi	2	
3	Flagellates	4	1.33
4	Ciliates	5	0.28
5	Rhizopoda	5	0.93
Multicellular ectotherms			
6	Nematoda	24	0.82
7	Microcrustacea	12	0.91
8	Mites	71	0.61
9	Collembola	29	0.74
10	Isoptera (larvae)	4	0.75
11	Enchytraeidae	61	0.87
12	Coleoptera (larvae)	17	0.67
13	Isoptera (larvae)	21	0.94
14	Formicidae (workers)	23	1.14
15	Lumbricidae (cocoons)	3	1.00
16	Opiliones	30	0.69
Multicellular ectotherms cont.			
17	Diplopoda	77	0.79
18	Aranea	6	0.81
19	Isopoda	40	0.69
20	Mollusca	6	0.76
21	Coleoptera (adults)	14	0.81
22	Lumbricidae (adults)	18	0.76
23	Macrocrustacea	3	0.81

(b)

Fig. 11.16 (a) Double-logarithmic plots of O_2 uptake against body mass for a variety of taxa. (b) List of taxa used in compiling a, each with the slopes b of log–log plots of the data for the taxon. (After Phillipson, 1981.)

characteristic of some deep-sea and some cave-dwelling species as well as some intertidal species that occur at different levels of the shore. For example, the subtidal barnacles *Balanus crenatus* and *B. rostratus* have an uninterrupted supply of food and are characterized by comparative high rates of O_2 uptake, compared with lower and mid-shore species *B. glandula* and *B. cariosa*, which cannot feed during periods of low tide (Fig. 11.18). The upper shore *Chthamalus* species can only feed at the periods of highest tide, and during neap tides may not be even covered by water at all. Thus not surprisingly the upper shore species is characterized by the shortest feeding time and corresponding lowest rates of O_2 uptake (Fig. 11.18).

11.6.4 Temperature

Changes in ambient temperatures can only influence metabolism if they influence body temperatures – something which is *generally* true for invertebrates but not for mammals and birds. Invertebrates are therefore described as *poikilothermic* (*poikilo* = Gk varied) and mammals and birds as *homeothermic* or *homoiothermic* (*homoio* = Gk keeping same). However, many invertebrates, such as those in the tropical open oceans and the deep seas, live at constant temperatures and are therefore not strictly poikilothermic. Hence a more general term to describe the metabolic properties of invertebrates is *ectothermic*, referring to the source of the heat, and mammals and birds are

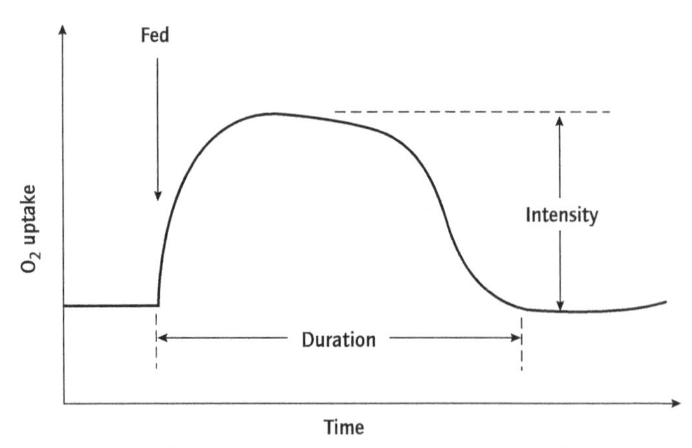

Fig. 11.17 Specific dynamic action – an increase in metabolism following a meal. Both the intensity and duration of the effect vary between and within species.

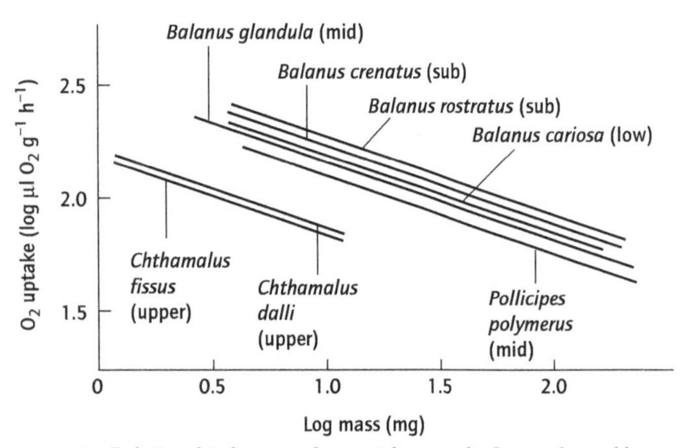

Fig. 11.18 Relationship between log weight-specific O_2 uptake and log body mass, for several species of barnacles. (After Newell & Branch, 1979.) See text for details.

endothermic. But even this classification is not truly general, for some insects generate large amounts of heat in their flight muscles (by endothermy) and this can be used to maintain a constant thoracic temperature (i.e. homoiothermy). For example, bumblebees cannot fly if their flight muscle temperature drops below 30°C or rises above 44°C. At least 90% of energy expended by a flying bee is released as heat within the thorax and, during vigorous flight, the temperature here can rise several degrees within seconds. Moreover, there is evidence that in free flight bumblebees can maintain a fixed internal temperature at different external temperatures. Large bees (primarily queens) can remain in free flight at 0°C, and under these conditions maintain a thoracic temperature at 30°C and above. Such regulation is maintained in a small part passively – by the thick pile of insulating hair on the thorax – but mainly by regulating flight effort.

In conclusion, then, no one term is strictly applicable to all the thermal characteristics of invertebrates but most are poikilothermic and ectothermic.

Within the normal temperature range of a poikilotherm, respiration is increased by increasing ambient temperature. However, as might be expected from the way chemical reactions respond to temperature, the relationship is non-linear – each additive increment of temperature causing a multiplicative increment in respiration. A widely used index of the effect of temperature on metabolism, that takes this into account, is the Q_{10} value, defined as follows:

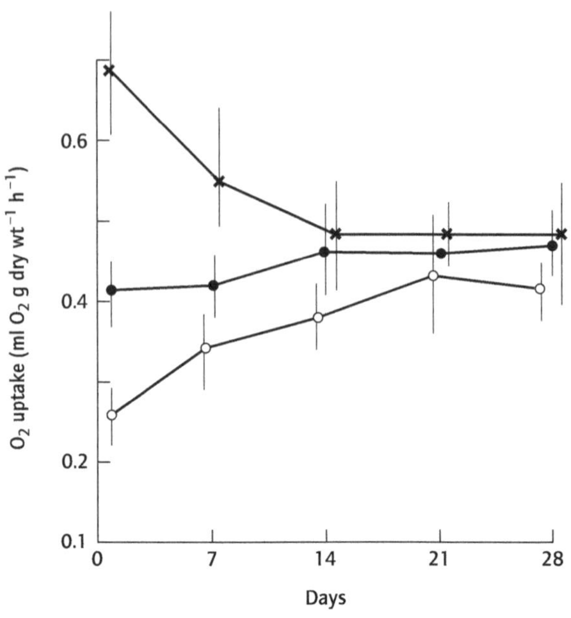

Fig. 11.19 Acclimation by *Mytilus*. ● = continuously at 10°C; ✖ = 10°C to 15°C; ○ = 10°C to 5°C, see text. (After Widdows & Bayne, 1971.)

$$Q_{10} = \frac{R_2^{10(t_2-t_1)}}{R_1}$$

$$\log Q_{10} = \frac{(\log R_2 - \log R_1)10}{t_2 - t_1}$$

where R_1 and R_2 are metabolic rates (e.g. of O_2 uptake) at temperatures t_1 and t_2 °C respectively. Note that if $t_2 - t_1 = 10$°C then Q_{10} is given by the ratio of R_2 and R_1, so Q_{10} indicates the factor by which R is multiplied up for each 10°C increase in temperature. Following an acute change in temperature Q_{10}s of 2 and more are often recorded, but Q_{10} can change dependent on the range of temperatures over which it is measured.

Moreover, the immediate response following a temperature change need not be a lasting one. Figure 11.19 shows what happens when mussels, *Mytilus edulis*, are transferred from an ambient temperature (10°C) to 5°C and 15°C. Oxygen uptake increases dramatically for the 15°C group and then reduces to a lower, steady level, whereas it first reduces for the 5°C group and then shifts with time to a higher steady level. This process is known as *acclimation*; note that Q_{10} values for acclimated values (after the adjustments) have to be lower than for the acute response and for perfect acclimation tend to 1. Such a response is thought to be adaptive because it allows conservation of energy at high temperatures and the maintenance of a high level of ATP production at low temperatures so that body maintenance and vital activity can be kept going. Acclimation responses have been found for nearly all of the major invertebrate groups examined to date. Having said this, there is now a great deal of debate surrounding the adaptive significance of acclimation generally.

Acclimation can also operate in the reverse direction; i.e. O_2 uptake reduces further at low temperatures and increases further at high temperatures with acclimation time, so Q_{10}s between acclimated rates increase from the acute rates. This occurs in some limpets, both freshwater (*Ancylus fluviatilis*) and marine (*Patella aspera*). At low temperature this kind of acclimation may be adaptive, operating like aestivation, and allowing the conservation of energy under winter food shortage or oxygen depletion, perhaps associated with ice cover, in freshwater habitats. Elevated metabolism at high temperature is more difficult to explain and subtle rate compensations are probably involved.

The acclimatory response illustrated in Fig. 11.19 is called *positive acclimation*; that of the limpets is called *negative/reverse acclimation*. The extent of these processes varies from species to species and might also depend upon the state of the animal. Thus in some intertidal invertebrates standard metabolism acclimates (positively) quickly and almost completely to temperature change ($Q_{10} \simeq 1$) whereas the Q_{10} values of routine metabolism are always greater than 1. This response is possibly adaptive because littoral animals are subject to considerable fluctuations in temperature between tides when they are inactive and cannot feed. Some ecological patterns are discernible in the occurrence of low Q_{10}s for the O_2 uptake of quiescent organisms: subtidal organisms, such as the sea urchin *Strongylocentrotus franciscanus* and the sea anenome *Anemonia natalensis*, as well as lower-shore, burrowing animals such as *Diopatia cuprea* and some bivalves, that do not experience marked, short-term variations in ambient temperature, have $Q_{10}s > 1$, whereas intertidal organisms such as *Littorina littorea*, *Strongylocentrotus purpuratus*, *Macoma balthica*, *Actinia equina* and *Bullia digitalis* have $Q_{10}s \simeq 1$. But there are also many contradictions, with some intertidal organisms such as *Patella vulgata* showing no suppression of Q_{10} and some subtidal organisms, such as the polychaete *Hyalinoecia*, having low Q_{10}s.

11.6.5 Oxygen tension (Po_2)

Some invertebrates when exposed to low Po_2 (hypoxia) are unable to maintain their rate of metabolism similar to that present when they are in 'normal' O_2 conditions, i.e. air or air saturated water (normoxia). The rate of O_2 uptake decreases with a decrease in environmental Po_2. Consequently they are referred to as oxyconformers (Fig. 11.20). Generally speaking conformers are found in environments that either do not normally experience pronounced hypoxia (e.g. some mayfly and stonefly larvae in fresh waters, sponges, cnidarians, some arthropods and molluscs in marine waters and many, if not all, terrestrial invertebrates) or are chronically exposed to very severe hypoxia (e.g. facultative anaerobes such as the nematode *Ascaris* and the

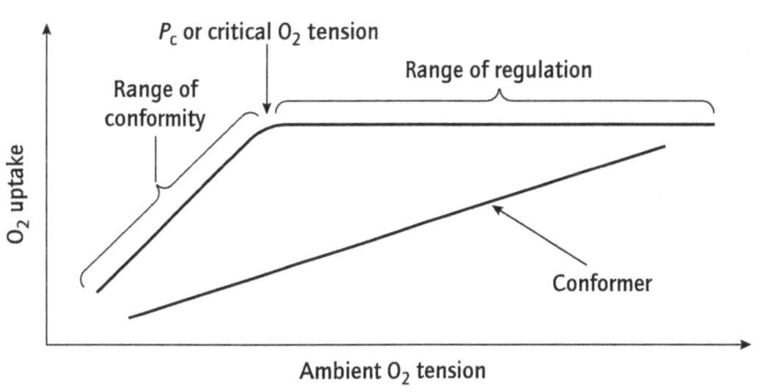

Fig. 11.20 Oxygen uptake as a function of ambient O_2 tension for an oxyregulatory and for an oxyconforming species.

platyhelminth *Fasciola*). In the case of the latter a well-developed capacity for anaerobic metabolism as the primary source of energy production makes any regulatory ability unnecessary.

Many, predominantly aquatic, invertebrates do display varying degrees of ability to maintain a constant O_2 uptake over a wide range of environmental Po_2 values (Fig. 11.20). They are referred to as oxyregulators. However very few species can regulate their metabolism completely. Invariably the Po_2 falls to a level where an individual can no longer maintain such respiratory independence and the regulator becomes a conformer. The point at which this switch takes place is commonly referred to as the critical Po_2 or P_c. The lower the P_c value the greater the regulatory ability. There are significant differences in P_c both between and within species.

Low P_c values are characteristic of species that live in periodically or chronically hypoxic environments. Some species inhabit burrows, often constructed in severely hypoxic muds, on the sea bottom or intertidally in estuaries. They tend to have low P_c values, e.g. the bivalve *Arctica*, 5 kPa and the shrimp *Calocaris*, 2 kPa. Similarly species belonging to a wide range of taxa (including ctenophores, chaetognaths, polychaetes, crustaceans and molluscs) that inhabit O_2 minimum layers found at depths of 400–1000 m in many of the world's oceans also tend to have quite low P_c values (e.g. they can be as low as 0.4 kPa). Intertidal rockpools are exposed to periodic hypoxia, i.e. when the tide is out, during the night when Po_2 falls as a result of the respiration of both animals and plants. Consequently many of their inhabitants are characterized by good oxyregulatory ability, e.g. the intertidal prawn *Palaemon elegans* ($P_c = 1$ kPa).

Hypoxia is also symptomatic of the organic loading associated with pollution in fresh waters, and increasingly also in estuarine and coastal marine environments. In such cases when the Po_2 decreases generally speaking those species which have a good regulatory ability survive while oxyconformers tend to be hit hardest. Thus in organically polluted waterways mayflies and stoneflies (both oxyconformers) are usually absent while species that are good oxyregulators such as the waterlouse *Asellus* and the oligochaete *Tubifex* tend to persist.

A number of intrinsic and extrinsic factors can dramatically affect the P_c value, and so comparisons of actual values must

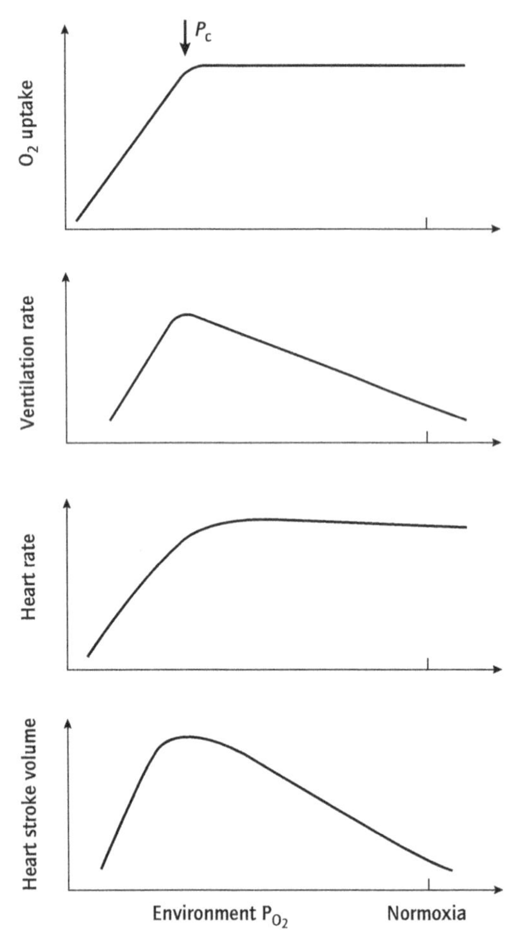

Fig. 11.21 Stylized respiratory response to progressive hypoxia in an oxyregulating crustacean.

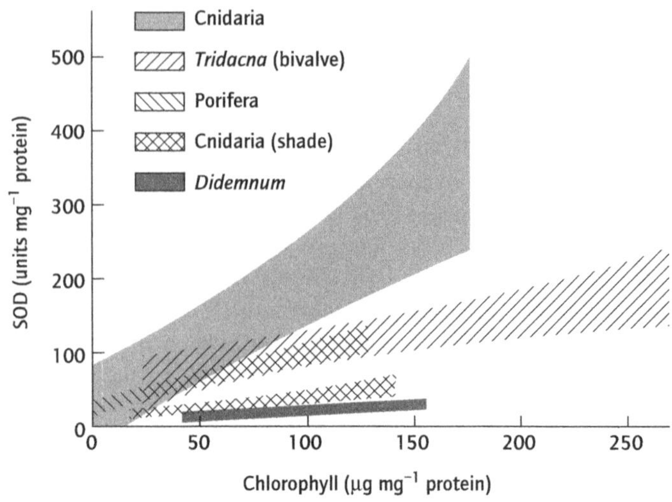

Fig. 11.22 Relationship between SOD activity and chlorophyll concentration for major groups of coral-reef invertebrates (after Shick & Dykens, 1985).

always be made with some caution. Although not invariable, an increase in (a) temperature, (b) metabolic rate, (c) activity and (d) holding time in the laboratory, all seem to decrease regulatory ability (i.e. P_c increases). Pre-exposure to hypoxia, on the other hand, in at least some invertebrate species, can result in an improvement in regulatory ability, e.g. the brine shrimp *Artemia* and the waterflea *Daphnia*.

The mechanisms underlying oxyregulation are relatively well understood at least for a few invertebrate species. In the short-term regulation is achieved by a number of different physiological responses. Most commonly ventilation will increase with decreasing Po_2, at least up until the P_c is reached. Thereafter ventilation rate declines precipitously. For some species an increase in the amount of blood supplied to the tissues by the heart can be linked with oxyregulation. Here there may be a tendency for the stroke volume (the amount of blood expelled by the heart in one beat) to increase with decreasing Po_2 rather than there be an increase in heart rate. However, the metabolic costs of increased ventilation and perfusion are high and largely unsustainable in the long term. Consequently more long-term 'solutions' may involve enhanced O_2 transport in the blood (via induction of respiratory protein synthesis and the manufacture of a pigment with an intrinsically higher affinity for O_2), and overall reduction of metabolic demand. Some of these mechanisms are illustrated in Fig. 11.21.

Before leaving the effect of Po_2 on O_2 uptake, it should be noted that too much O_2 can be as harmful as too little. Free O_2 (superoxide) radicals can denature macromolecules, particularly proteins. This is important when O_2 concentrations are very high, as they can be in the tissues of invertebrates harbouring symbiotic algae (such as many coral-reef dwelling organisms), for O_2 is an end product of photosynthesis. In response, these organisms have evolved cellular defences against O_2 toxicity, i.e. the enzyme superoxide dismutase (SOD), which can eliminate damaging superoxide radicals, and catalase which eliminates the H_2O_2 generated by SOD. Figure 11.22 shows that for a number of invertebrate groups, SOD activity increases with chlorophyll concentration in their tissues, i.e. the potential for the photosynthetic production of O_2 by the symbionts. However, the relationship is not the same for all groups and these differences can be ascribed to differences in the localization of the symbionts: SOD activity is highest in Cnidaria where zooxanthellae are exclusively or predominantly intracellular (cytoplasm is directly exposed to O_2) whereas for the urochordate *Didemnum*, where the symbiotic prochlorophytes are not only extracellular but also extraorganismal (occurring in the lining of the common cloacal chamber), O_2 can be quickly removed in the outgoing current without coming into contact with animal cytoplasm. Some of the lower values in the cnidarian range are from species that are shaded in caves and under ledges.

11.6.6 Salinity

When marine and estuarine invertebrates are challenged with changes in salinity the overall response is a complex mixture of

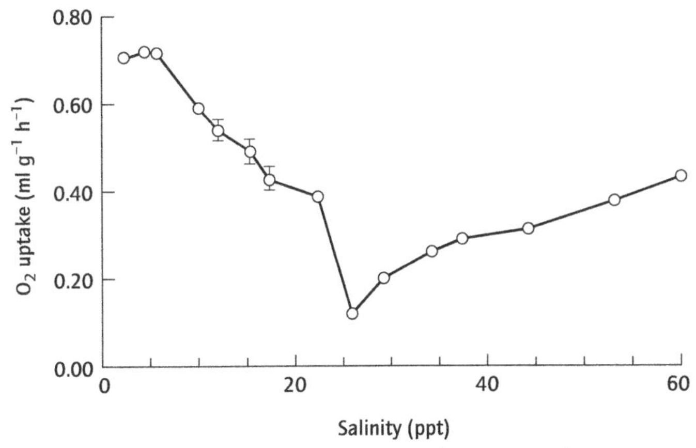

Fig. 11.23 Rates of O_2 uptake of the prawn *Palaemonetes varians*, at different salinities (values are means ± 1SD; n = 3 or 4). (Data from Lofts, 1956.)

physiological alteration and changes in behaviour (i.e. initially they tend to be more active after the change). Many fully marine species are unable to tolerate, or compensate for, the physiological disturbance induced by exposure to low salinities. Consequently they show a decrease in metabolism with a decrease in salinity. The opposite is true for fully freshwater species. Increasing the salinity of their surrounding medium leads to a decrease in metabolism and ultimately death. However, the metabolic response to salinity change can be complicated by a concomitant increase in activity which is itself an energy-demanding process. Thus it is not uncommon in the literature to find an increase, a decrease, or no change in metabolism reported, even for the same species, as a result of changing the salinity of the surrounding medium.

In the case of estuarine species (although there are exceptions) metabolism is lowest at the salinity where the animal is roughly iso-osmotic. Presumably at this salinity the cost of maintaining ionic and osmoregulatory processes is at a minimum. However metabolism tends to increase with either a concentration or dilution of the surrounding medium. This could be due to the increased costs of osmoregulatory work and/or an increase in locomotor activity. Such a pattern can be clearly seen for the brackishwater prawn *Palaemonetes varians* (Fig. 11.23). It displayed a minimum O_2 uptake at an external salinity of 26 ppt, the salinity at which the animal is isotonic.

It is widely held that the costs of osmoregulation and active transport are small and that these costs are 'swamped' by the cost of activity often incurred at the same time. However, there is as yet little agreement on what the actual osmoregulatory cost is, with calculated estimates ranging from around 0.01 to 25% of total metabolism.

11.7 Conclusions

'Respiration' refers to processes that release biologically useful energy from food materials, particularly carbohydrates. These processes are largely, but not exclusively, aerobic, and so 'respiration' is also used to refer to the process of obtaining O_2 and getting rid of waste CO_2. The biochemical aspects of the process are sometimes called 'internal' or 'tissue respiration' and the physiological aspects of O_2 uptake as 'external respiration'. The reader should now be familiar with the mechanism behind both, the systems that obtain O_2 from the environment and deliver it to the tissues, and the intrinsic and extrinsic factors that influence metabolic rates.

11.8 Further reading

Bayne, B.L. & Scullard, C. 1977. An apparent specific dynamic action in *Mytilus edulis*. *J. Mar. Biol. Assoc., UK*, **57**, 371–378.

Bryant, C. & Behm, C. 1989. *Biochemical Adaptation in Parasites*. Chapman and Hall, London.

Bryant, C. (Ed.) 1991. *Metazoan Life Without Oxygen*. Chapman & Hall, London.

Calow, P. & Townsend, C.R. 1981. Resource utilization in growth. In: Townsend, C.R. & Calow, P. (Eds) *Physiological Ecology: an Evolutionary Approach to Resource Utilization*. Blackwell Scientific Publications, Oxford.

Cameron, J.N. 1989. *The Respiratory Physiology of Animals*. Oxford University Press, New York.

Childress, J.J. 1995. Are there physiological and biochemical adaptations of metabolism in deep-sea animals? *Trends Ecol. Evol.*, **10**, 30–36.

Chown, S.L. & Gaston, K.J. 1999. Exploring links between physiology and ecology at macro-scales: the role of respiratory metabolism in insects. *Biol. Rev.*, **74**, 87–120.

Cossins, A.R. & Bowler, K. 1987. *Temperature Biology of Animals*. Chapman and Hall, London.

Fothergill-Gilmore, L.A. 1986. The evolution of the glycolytic pathway. *Trends Biochem. Sci.*, **11**, 47–51.

Heatwole, H. & Cloudsley-Thompson, R.J.L. 1995. *Energetics of Desert Invertebrates*. Springer-Verlag, New York.

Heinrich, B. 1979. *Bumble-bee Economics*. Harvard University Press, Cambridge, Massachusetts.

Heinrich, B. 1993. *The Hot-blooded Insects*. Harvard University Press, Cambridge, Massachusetts.

Huey, R.B., Berrigan, D., Gilchrist, G.W. & Herron, J.C. 1999. Testing the adaptive significance of acclimation: A strong inference approach. *Amer. Zool.*, **39**, 323–336.

Livingstone, D.R. 1983. Invertebrate and vertebrate pathways of anaerobic metabolism: evolutionary considerations. *J. Geol. Soc.* **140**, 27–38.

Lutz, P.L. & Storey, K.B. 1997. Adaptations to variations in oxygen tension by vertebrates and invertebrates. In: Dantzler, W.H. (Ed.) *Handbook of Physiology*. Section 13. *Comparative Physiology*, Vol II, Chapter 21, pp. 1479–1522. Oxford University Press, Oxford.

Mangum, C.P. 1994. Multiple sites of gas exchange. *Amer. Zool.*, **34**, 184–193.

Mangum, C.P. 1997. Invertebrate blood oxygen carriers. In: Dantzler, W.H. (Ed.) *Handbook of Physiology*. Section 13. *Comparative Physiology*, Vol II, Chapter 15, pp. 1097–1136. Oxford University Press, Oxford.

Mangum, C.P. 1998. Major events in the evolution of the oxygen carriers. *Amer. Zool.*, **38**, 1–13.

McMahon, B.R., Wilkens, J.L. & Smith, P.J.S. 1997. Invertebrate circulatory systems. In: Dantzler, W.H. (Ed.) *Handbook of Physiology*. Section 13. *Comparative Physiology*, Vol II, Chapter 13, pp. 931–1008. Oxford University Press, Oxford.

Mill, P.J. 1997. Invertebrate respiratory systems. In: Dantzler, W.H. (Ed.) *Handbook of Physiology*. Section 13. *Comparative Physiology*, Vol II, Chapter 14, pp. 1009–1098. Oxford University Press, Oxford.

Newell, R.C. 1979. *Biology of Intertidal Animals*, 3rd edn. Marine Ecological Surveys Ltd, Kent.

Phillipson, J. 1981. Bioenergetic options and phylogeny. In: Townsend, C.R. & Calow, P. (Eds) *Physiological Ecology: an Evolutionary Approach to Resource Utilization*. Blackwell Scientific Publications, Oxford.

Randall, D., Burggren, W.W. & French, K. 1997. *Animal Physiology. Mechanisms and Adaptations*, 4th edn. W.H. Freeman, New York.

Schmidt-Nielsen, K. 1997. *Animal Physiology. Adaptation and Environment*, 5th edn. Cambridge University Press, Cambridge.

Somero, G.N. 1997. Temperature relationships: From molecules to biogeography. In: Dantzler, W.H. (Ed.) *Handbook of Physiology*. Section 13. *Comparative Physiology*, Vol II, Chapter 19, pp. 1391–1444. Oxford University Press, Oxford.

Spicer, J.I. & Gaston, K.J. 1999. *Physiological Diversity and its Ecological Implications*. Blackwell Science, Oxford.

Wasserthal, L.T. 1997. Interaction of circulation and tracheal ventilation in holometabolous insects. *Adv. Insect Physiol.*, **26**, 298–351.

Willmer, P., Stone, G. & Johnston, I. 2000. *Environmental Physiology of Animals*. Blackwell Science, Oxford.

CHAPTER 12

Excretion, Ionic and Osmotic Regulation, and Buoyancy

Materials in excess of metabolic requirements have to be removed from the bodies of animals. They include indigestible residues that are lost in the faeces (Chapter 9) and CO_2 that is lost in respiration (Chapter 11). There are other excess materials, though, and these include water, various ions and the breakdown products of excess proteins and amino acids. It is the latter, nitrogen-containing substances, that are normally referred to by physiologists as excretory products. Nevertheless, the processes involved in the removal of excess ions, water and nitrogen-containing substances from the bodies of invertebrates are often so intimately related that it is sensible to treat them together. In this chapter we therefore start with an account of excretion, then consider ion–water problems, and finally describe the structure and functioning of so-called excretory systems that are invariably associated with ionic and osmotic regulation but not always with nitrogen excretion.

12.1 Excretion

Waste amino acids derive from excesses absorbed across the gut wall and from the catabolism of proteins. These are most usually broken down further, in an oxidative process that yields ketoacids and ammonia:

$$\underset{\text{amino acid}}{NH_2\text{–}\overset{\displaystyle R}{\overset{|}{C}}HCOOH} + {}^1/_2O_2 = O{=}\underset{\text{ketoacid}}{\overset{\displaystyle R}{\overset{|}{C}}COOH} + \underset{\text{ammonia}}{NH_3}$$

The ketoacids can easily be used in other metabolic pathways, but because ammonia is extremely toxic it must either be removed rapidly from the body or be stored in a harmless form (below). Ammonia is very soluble and is easily lost from aquatic animals by a process of diffusion across the total-body and respiratory gas exchange surfaces. Excretion where ammonia is the principal end product is referred to as *ammoniotelism* and is very common in aquatic invertebrates, including aquatic insect larvae (cf. terrestrial adults, below).

In some species, ammonia can, however, be converted to form less toxic urea:

$$\begin{array}{c} NH_2 \\ | \\ C{=}O \\ | \\ NH_2 \end{array}$$

The use of this substance in excretion is referred to as *ureotelism* and is typical of mammals. Its use is not widespread in invertebrates, possibly because it has the disadvantage of being toxic whilst still entailing the loss of considerable amounts of water in its excretion. Ureotelism is, however, found in some Platyhelminthes, Annelida and Mollusca.

A dominant excretory product of terrestrial invertebrates (and non-mammalian, terrestrial vertebrates) is uric acid – *uricotelism* – and this is a member of a general class of molecules, the purines, others of which are used by some invertebrates – *purinotelism*.

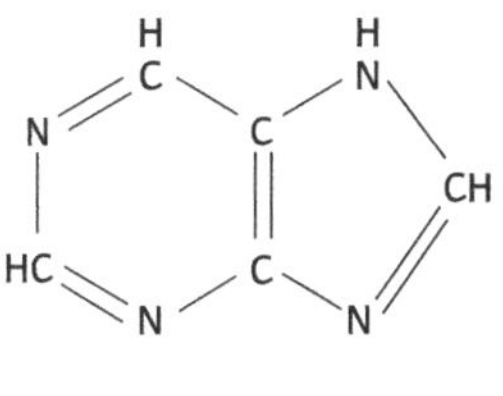

General purine

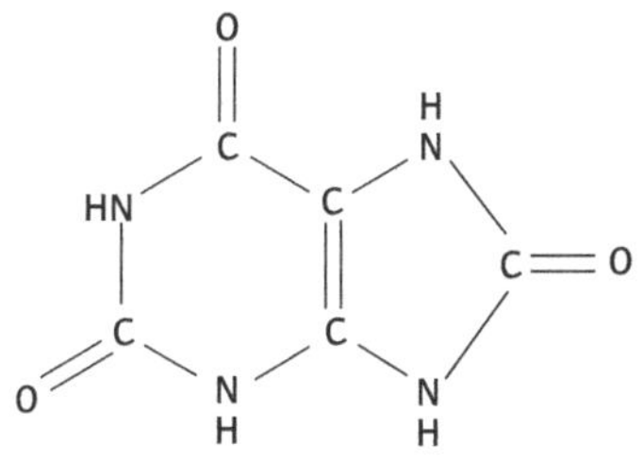

Uric acid

The main reason for the evolution of these excretory products is that they have low toxicity and because of their low solubilities are excreted as solids that require little water for removal. Uricotelism is of importance in the Onychophora, Uniramia and to a lesser extent in the Crustacea and Mollusca. The main excretory product of Arachnida is guanine, another purine.

Guanine

An exception to these trends is found within the terrestrial crustaceans (crabs, woodlice and amphipods) where ammonotelism predominates. Their cuticular surfaces lack the epicuticular wax found in insect cuticles (Section 12.2.5). Consequently, particularly in the smaller forms ammonia can be eliminated from the body by gaseous diffusion. This said uric acid has recently been found to be an important excretory product in some fully terrestrial crabs such as the robber crab *Birgus latro*. Other terrestrial invertebrates that excrete ammonia directly are some oligochaetes and myriapods, but these, like the isopods, live in humid microhabitats where the risks of desiccation are not serious.

Metabolic economy must also be another factor that has played a part in the evolution of excretory systems. Thus purines might be less toxic than ammonia, but they are associated with carbon and hence potential energy. Urea is intermediate in these respects. Table 12.1 compares and contrasts the toxicity, water requirements and energy/carbon losses associated with each of the three main products. Just how important these losses are in the economy of the whole organism is, however, problematical, and of more importance might be the energy lost (mostly as heat) in the metabolic processes leading to the formation of the excretory products.

It should also be realized that excretory substances, particularly less toxic ones such as the purines, need not leave the body at all. This is certainly true in terrestrial and aquatic pulmonate snails in which uric acid can accumulate to considerable extents as snails get larger and older. Urea has also been found to accumulate in the tissues of the tropical pulmonate, *Bulimulus*, during aestivation. The significance of this is uncertain but, as well as serving an excretory function, the enhanced urea concentration in the tissues and body fluids of this snail may cause an increased osmotic pressure and thus a reduced evaporative water loss. And this might increase survival during prolonged drought. Uric acid also accumulates during development in cleidoic eggs. Again pulmonate snails provide good examples – the uric acid content of the egg of *Lymnaea* increases from about 0.5% fresh weight at cleavage to about 4.5% fresh weight at hatching. And it has been argued that the evolution of the cleidoic condition might have been the prime impetus for the evolution of purinotelism. Certainly both traits have been of considerable importance in the conquest of land.

Some species of ascidians possess what have been termed renal sacs because solid uric acid concretions can be found within them. However, these sacs do not open onto the surface. Instead the uric acid accumulates within the sacs throughout the life of that individual. Consequently some doubt has been cast on the excretory function although alternative functions are as yet unclear.

Table 12.1 'Economics' of excretory products.

	C/N	Heat combustion		Toxicity	Water requirements
		kJ mol^{-1}	kJ mol^{-1} N^{-1}		
Ammonia	0	378	378	***	***
Urea	0.5	638	319	**	**
Uric acid	1.25	1932	483	*	*

* Rough quantitative index. Data from Pilgrim (1954).

12.2 Osmotic and ionic regulation

The body fluids of animals are dilute saline solutions with sodium chloride as the predominant electrolyte – in other words they resemble sea water. Indeed it is generally considered that this reflects an origin of life within the sea. Yet there are appreciable differences between the body fluids and sea water (Table 12.2) and Macallum in the 1920s saw in this evidence for

Table 12.2 Concentrations of ions in plasma or body-cavity fluid as a percentage of the concentration in body fluid dialysed against sea water. After Schmidt-Nielsen (1997).

	Na	K	Ca	Mg	Cl	SO_4
Cnidaria						
Aurelia aurita	99	106	96	97	104	47
Echinodermata						
Marthasterias glacialis	100	111	101	98	101	100
Urochordata						
Salpa maxima	100	113	96	95	102	65
Annelida						
Arenicola marina	100	104	100	100	100	92
Sipuncula						
Phascolosoma vulgare	104	110	104	69	99	91
Crustacea						
Maia squinado	100	125	122	81	102	66
Dromia vulgaris	97	120	84	99	103	53
Carcinus maenas	110	118	108	34	104	61
*Pachygrapsus marmoratus**	94	95	92	24	87	46
Nephrops norvegicus	113	77	124	17	99	69
Mollusca						
Pecten maximus	100	130	103	97	100	97
Neptunea antiqua	101	114	102	101	101	98
Sepia officinalis	93	205	91	98	105	22

* This grapsoid crab is the only animal in the table which is hypo-osmotic (ionic concentration 86% that of sea water).

Box 12.1 Some definitions

When solutions are separated by a barrier (membrane) permeable to both solvent and solute:

1 Solutes (ions) pass from the more concentrated to the less concentrated solution by *diffusion* (in fact these ionic movements can be complicated by differences in electrical potential across the membrane and this will be considered in more detail in Chapter 16).

2 Solvent passes from the less to the more concentrated solution by *osmosis*, and *osmotic pressure* is said to force solvent to pass from the low to the high concentration – osmotic pressure is a colligative property, i.e. dependent on the number of solute particles and not on their kind.

Processes 1 and 2 continue until the concentration gradient is abolished.

Membranes associated with organisms are usually permeable to both water and solutes *to some extent* and are subject to simultaneous movement of both, for example:

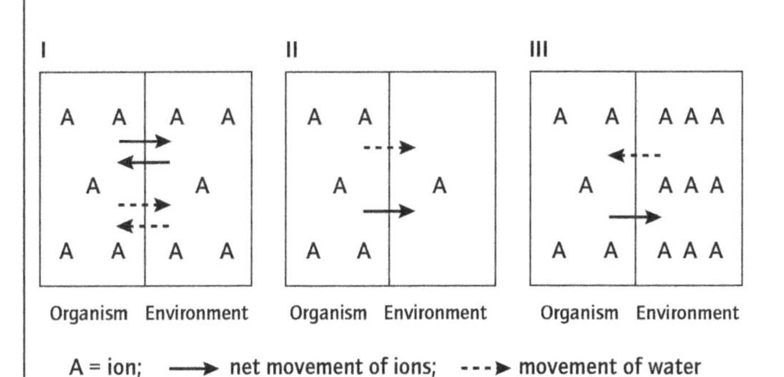

In I the organism is *iso*-osmotic with its environment
In II the organism is *hyper*osmotic with its environment
In III the organism is *hypo*-osmotic with its environment

Isotonic is often used synonymously with iso-osmotic, but this is not strictly correct – since tonicity describes the response of the cell/organism (in terms of volume) and organisms iso-osmotic with surrounding ions are usually, but not always, *isotonic*, e.g. sea urchin eggs are isotonic in iso-osmotic NaCl but not in iso-osmotic $CaCl_2$.

Osmoconformers are animals in which osmotic concentrations of body fluids conform with the environment, i.e. remain iso-osmotic with it.

Osmoregulators are animals that maintain osmotic concentrations of body medium different from the environment.

Ionic regulation is the regulation of concentration of solutes in body fluids that usually differ markedly from their concentration in the environment.

Euryhaline: animals tolerate wide variation in salt concentration of water in which they live.

Stenohaline: animals have limited tolerance to variations in salt concentrations in water in which they live.

Concentrations: it is easy to understand what is meant by a weight of substance but *moles* (=molecular weight in grams or the Avogadro's number [6.023×10^{23}] of molecules of an element or compound) are more elusive. They are, nevertheless, more useful since they measure the number of solute particles – and the osmotic pressure of a solution depends directly upon the number of particles that it contains.

Molarity=moles/litre of solution.

Molality=moles/kg of pure solvent.

a gradual change in the composition of the oceans from what it was when life forms originated to what it is now. However, recent palaeochemical research seems to indicate that early oceans did not differ appreciably in composition from present-day ones.

An implication of Table 12.2 is, therefore, that even animals living in sea water have to regulate the internal composition of their body fluids. This intensifies in more dilute aquatic environments – estuaries and fresh waters – where the inhabitants have to maintain body fluids more concentrated than their surroundings. The opposite is the case in animals (a) that having occurred in fresh waters, evolved more dilute body fluids than sea water but then returned to the sea, and (b) that live in very concentrated media, such as salt lakes. In terrestrial situations, of course, the main problem is the conservation of water.

We now consider the ionic and osmotic challenges associated with these four main ecological circumstances in more detail. A brief definition of various terms, important in understanding ionic and osmotic regulation, is given in Box 12.1.

12.2.1 Marine environment

Most marine invertebrates have body fluids iso-osmotic to sea water and are osmoconformers. Nevertheless, as we have already seen, the ionic composition of body fluids can differ quite appreciably from that of sea water and so there has to be ionic regulation. Something of this is illustrated in the data in Table 12.2. The following points are worthy of note: (a) *Aurelia* regulates mainly sulphate ions, keeping them below the concentration in sea water – this may be to do with buoyancy, since sulphate ions are heavy and their replacement by chloride might decrease the density of this jellyfish and prevent it from sinking – see also the cuttlefish, *Sepia* (and Section 12.2.2); (b) apart from potassium, echinoderms do not regulate their body fluids appreciably and this may be one feature of their physiology that has largely restricted them to the sea; (c) magnesium levels are generally low in the body fluids of arthropods and this might be to do with the fact that the arthropods are generally very active and magnesium can act as an anaesthetic; note in contrast, though, that magnesium concentration is not particularly depressed in the fast-moving *Sepia*.

12.2.2 Aside on buoyancy

Aquatic animals denser than water would tend to sink, so it is advantageous for swimming animals to maintain a density equal to or less than that of water, otherwise they spend energy in keeping themselves from sinking. We have already seen that

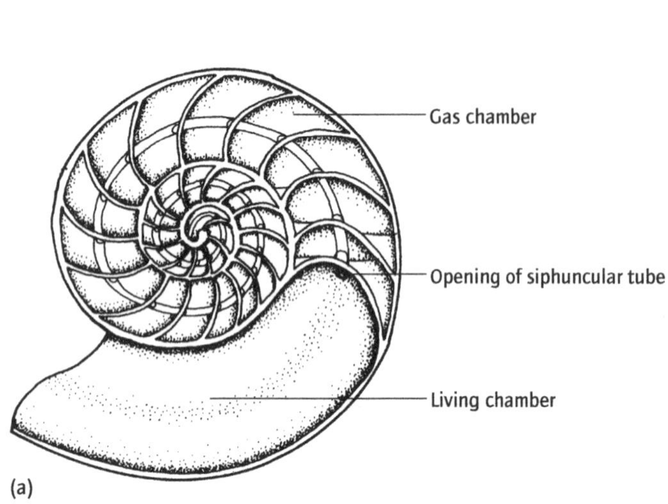

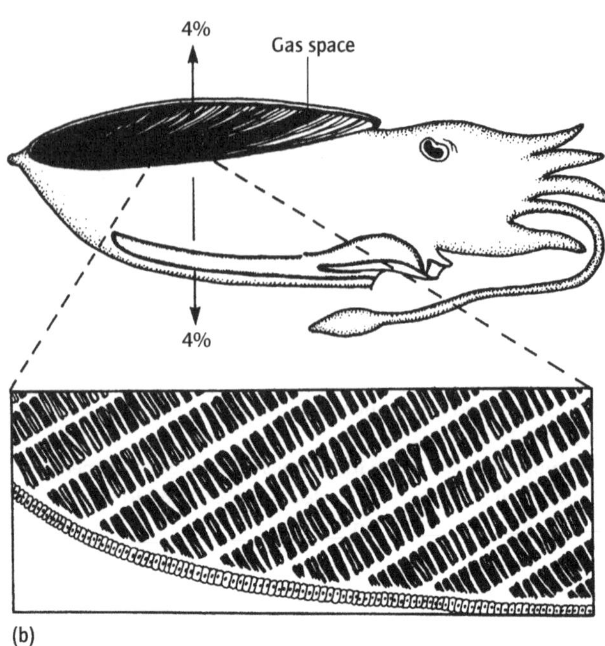

Fig. 12.1 (a) The chambered shell of *Nautilus*. As it grows this animal adds new gas chambers to its shell, one by one. Initially these are filled with a liquid in which NaCl is the major solute. Sodium is removed from this by active transport. The fluid is therefore hypo-osmotic to sea water and water is removed leaving a gas space. Gas diffuses in: N_2 reaches 0.8 atm, the same as in the water and the animal's own tissues, but O_2 is at lower pressure. Because of the rigid system the gas pressure remains at *c.* 0.9 atm, irrespective of the depth at which the animal swims. (b) Cuttlebone is a laminar structure containing gas of approximately the same composition and pressure as in *Nautilus*. However, the oldest and most posterior chambers (marked black) contain fluid – again a solution of largely NaCl. The quantity of gas present can be varied by altering the volume of liquid. In this way the cuttlefish varies its own density and hence can vary its buoyancy and thus depth. This is done by altering the concentration of the NaCl solution by ionic movements; e.g. near the surface the fluid approximates to sea water in its osmotic concentration, but at greater depths the concentration is lowered, the fluid becomes hypo-osmotic to sea water and the animal's own blood, and this generates an osmotic force that balances the tendency of the increased hydrostatic pressure to push water into the cuttlebone. (After Schmidt-Nielsen, 1997.)

this can be achieved, at least in part, by ionic regulation. Here we illustrate this further but also draw attention to other adaptations involved with buoyancy.

Heliocranchia is a deep-sea squid. Here the fluid-filled pericardial cavity is very large. The fluid is less dense than sea water, containing far less sodium but a very high concentration of ammonium ions. The latter form as end products of protein metabolism and diffuse into the acidic pericardial fluid and become trapped there. Finally, as noted for *Aurelia* and *Sepia*, the anions in the pericardial fluid of this squid are almost exclusively chloride, with the heavy sulphate ions being excluded.

Other methods of increasing buoyancy involve: (a) the removal of ions without replacement, but then the body fluids would be hypo-osmotic to sea water and this therefore involves osmotic costs; (b) a reduction in heavy substances, for example, some pelagic gastropods such as heteropods and pteropods have reduced or no shells; (c) an increase in substances such as fats and oils, that are less dense than water, is very common in planktonic crustaceans from both marine and freshwater environments; (d) use of gas floats, such as the soft-walled floats of the Portuguese man-of-war *Physalia* and the rigid-walled floats formed from the gas-filled chambers of nautiloid cephalopods and the cuttlebone of the cuttlefish (Fig. 12.1).

12.2.3 Freshwater environment

If fully marine animals are transferred to dilute sea water, their body fluids either follow the osmotic conditions of the environment faithfully (osmoconformers) or they resist the dilution of their body fluids (osmoregulators). Examples of both are given in Fig. 12.2. Estuarine, and to some extent littoral, animals are naturally subjected to periodic dilution, between tides and after heavy rain, and show both patterns of response.

Osmoconformers, though, are not necessarily more stenohaline than osmoregulators. Thus the osmotic pressure of the blood of the mussel, *Mytilus edulis*, closely follows that of the surrounding water whether it be in the North Sea (normal salinity) or in the Baltic (<half normal salinity). One reason for this tolerance is that though it does not regulate extracellular body fluids it does regulate intracellular ones, and amino acids play an important part in this. These can increase in concentration during dilution, so increasing the osmotic pressure of the intracellular fluids. Accordingly, excretion of ammonia is accelerated

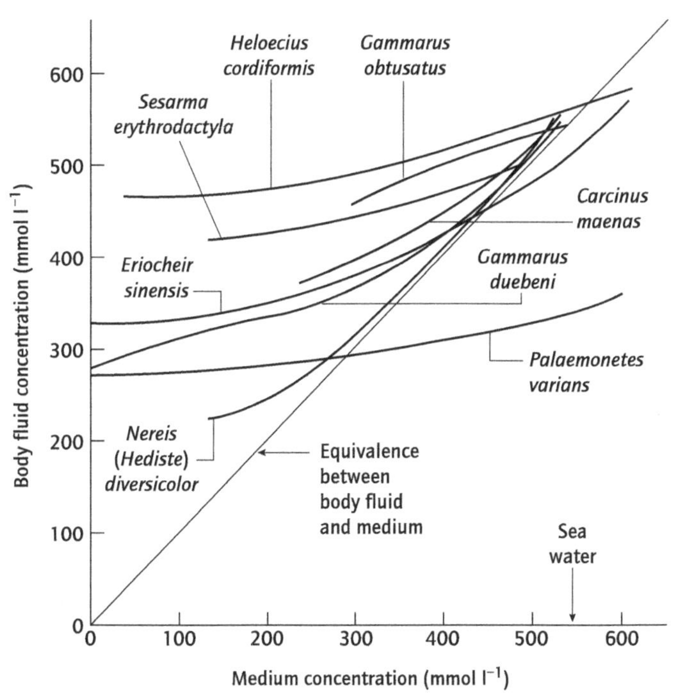

Fig. 12.2 Relation between concentrations of body fluids and medium for several brackish water invertebrates (after Schmidt-Nielsen, 1997).

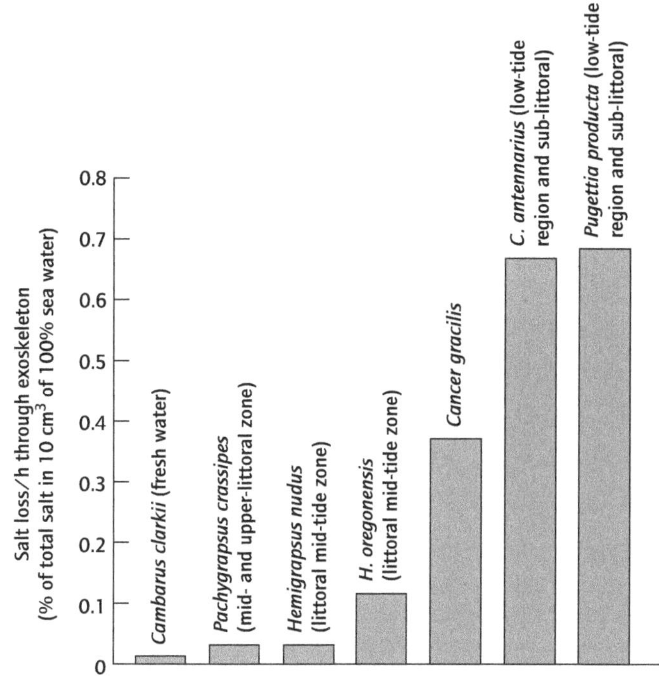

Fig. 12.3 Permeability of the exoskeleton of crustaceans from different environments (after Hoar, 1966).

in *M. edulis* when transferred from high to low salinity. This presumably signals the greater catabolism of proteins and hence animo acid production, intracellularly, in conditions of osmotic stress. Indeed, there appears to be genetic variation in leucine aminopeptidase (LAP), an important enzyme for protein catabolism, and variants with higher catabolic rates are found in higher frequencies in low-salinity, estuarine conditions.

Osmoregulators can control water influx and ionic efflux by evolving a less permeable body surface and/or by actively pumping water out and regulating the influx and efflux of ions to body fluids. Figure 12.3 indicates how the permeability of the exoskeletons of crustaceans is lower in species from more dilute or littoral habitats.

Animals living in fresh water are similar to the osmoregulators in brackish waters, but have to regulate throughout life. Figure 12.4 indicates how the body fluids of some of these animals respond to increasing salinity. Notice that there are great differences in the concentration at which they maintain the body fluids in low salinity (i.e. fresh water) – thus the bivalve, *Anodonta*, and the water flea, *Daphnia*, have lower concentrations than the amphipod, *Gammarus*, and the corixid bug, *Sigara* – but these are all lower than the concentration of the body fluids of marine equivalents, presumably to reduce the costs of maintaining a large differential in osmotic pressure between internal and external environments.

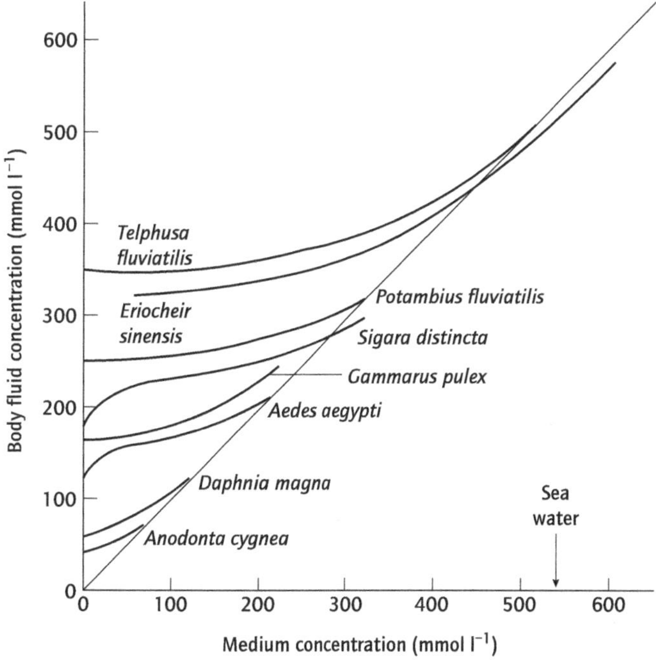

Fig. 12.4 As Fig. 12.2 but for various freshwater animals (after Schmidt-Nielsen, 1997).

Most freshwater animals regulate by having an epidermis of low permeability (Fig. 12.3) and by producing copious amounts of urine. Thus if urine production per day is expressed as a percentage of the body weight of the producer, it is usually considerably less than 10% for marine invertebrates, but considerably more than this for freshwater invertebrates. For example, *Astacus* and *Gammarus* produce around 40% body weight per day, *Daphnia* >200% and the freshwater bivalve *Anodonta* >400%. Usually the urine of these freshwater invertebrates is hypo-osmotic relative to body fluids, useful ions having been selectively

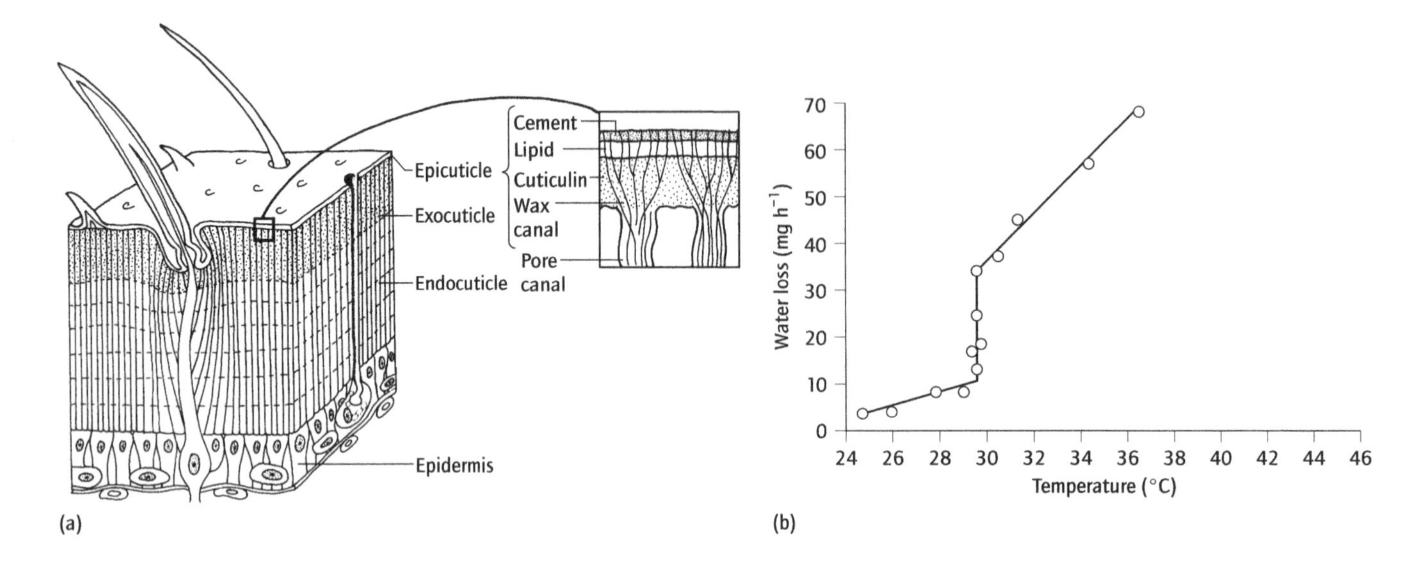

Fig. 12.5 (a) Section of the cuticle of an insect, showing the wax layer (after Edney, 1974); (b) water loss from the cuticle of the cockroach with increasing air temperature. (After Beament, 1958.)

removed in appropriate parts of 'excretory' organs. Nevertheless, despite this and the impermeability of the epidermis, some ions are lost and must be replaced. Some of this occurs from the food, but many freshwater or brackish water invertebrates are also capable of the direct, active uptake of ions from the surrounding medium. This can be demonstrated by first soaking them in distilled water, to reduce the concentration of ions in their body fluids, and then returning them to ordinary fresh water. After a short time, the ionic composition of the body fluids returns to normal even though at all times the freshwater medium is more dilute than the body fluids. This implies uptake by active transport. These processes have been most thoroughly researched in freshwater crustaceans such as the crayfish, *Austropotamobius*, the freshwater isopod, *Asellus*, and freshwater species of the amphipod, *Gammarus*. It is probable that the general body surface plays some part in this process, but there is convincing evidence that for crustaceans the gills are the organs of active ion transport.

Larvae of the mosquito *Aedes aegypti*, with blood containing extremely small amounts of sodium (<30% normal), can be produced by rearing them in distilled water. This can be rectified within several hours of transfer to natural fresh waters. Experiments that involve damaging the anal papillae and blocking the guts of these animals have shown that 90% of this uptake occurs through the papillae and most of the rest through the gut. In terms of ion uptake, the rectal gills of dragonfly larvae play a similar role as the anal papillae of *Aedes*.

12.2.4 Hyperosmotic environments

A few marine and many brackish water invertebrates, such as the shrimp *Palaemonetes*, are hypo-osmotic with respect to full strength sea water (Fig. 12.2). It is generally agreed that these animals have secondarily returned to salt waters from an original freshwater existence. There are also a number of grapsoid crabs that have body fluids hypo-osmotic with respect to sea water (see footnote to Table 12.2) but there is no question that the ancestors of these ever spent time in fresh water and the reason for the hypo-osmotic state here is unclear.

The general problem for organisms in hyperosmotic circumstances, though, is to keep salts out and water in. Invertebrates living in saline waters more concentrated than sea water face the same problem. *Artemia*, the brine shrimp from salt lakes, is an excellent example. This animal maintains its body fluids at a lower osmotic concentration than its surroundings by active regulation; water is taken in by drinking and excess ions are removed through the gills in adults and a specialized neck organ in larvae.

12.2.5 Terrestrial environment

Potentially, the greatest physiological problem for terrestrial animals is dehydration. However, some terrestrial animals, such as earthworms, slugs and snails, have moist surfaces and are restricted to moist habitats in the soil and leaf litter and are not truly terrestrial. After rainfall, conditions in the soil might even become hypo-osmotic and under these conditions some nematodes actively remove water from their own tissues, possibly directly through the intestine. At the other extreme are the insects that can live in very dry environments and have an impervious exoskeleton. The latter is achieved not by the cuticle itself, but by the epicuticular layer of wax (Fig. 12.5a) (see also Section 8.5.3B.2). For example, if this is abraded the evaporative water loss from the body is increased greatly. Similarly, the importance of the wax layer for the retention of water has been demonstrated by measuring the rate of water loss at different temperatures. In this experiment there is a sudden jump in the rate of water loss at a temperature coincident with the melting point of the wax coating (Fig. 12.5b).

As already noted, terrestrial crustaceans lack an epicuticular wax layer and are mostly restricted to moist habitats. Terrestrial amphipods inhabit leaf litter and they carry around with them, on the underside of their (grooved) body, a small amount of water. Immersed in or covered by this water are the gills, and in females the eggs/newly-hatched individuals are also kept in the groove. This exosomatic water is derived both from standing bodies of water and from urine. Its composition is under strict physiological control such that these amphipods carry their own 'ocean' with them onto land. Although terrestrial isopods are often found in micro-habitats of high humidity they are sometimes exposed to drier air; water lost by evaporation can be replaced, in part, from the moist food these animals eat – decaying plant material – and some are capable of drinking and taking up water through the anus. The desert isopod, *Hemilepistus*, avoids excess desiccation behaviourally, by living in burrows during the heat of the day. These may be up to 30 cm deep, are much cooler than the surface desert and their relative humidity can be as high as 95%.

Insects also take up water by drinking, from food and also by the metabolic oxidation of organic molecules, e.g.

$$C_6H_{12}O_6 + 6O_2 \rightarrow 6CO_2 + 6H_2O$$

Moreover, some terrestrial insects and arachnids are able to absorb water vapour directly from the atmosphere. Just how this is achieved is not clear but, in different groups, rectal and buccal epithelia and possibly the tracheal system are involved.

12.2.6 'Invasions' of land and freshwater habitats

It seems indisputable that life evolved and radiated in the marine environment (Chapter 2). Invasions of other major habitats might either have occurred directly (sea to land; sea to fresh water) or indirectly (sea to land via fresh water; sea to fresh water via land). Additionally the sea/land transition might have occurred across the surface of the littoral region or through its interstices. These transitions are illustrated schematically in Fig. 12.6. We have already discussed and dismissed Macallum's theory (p. 290) but it is not unreasonable to presume that ancestral habits and habitats have left some mark on the composition of the body fluids of extant animals. Thus it is possible to hypothesize that terrestrial animals derived directly from marine forebears are likely to have body fluids with higher osmotic pressures than those with freshwater forebears. And this is largely supported by the data summarized in Table 12.3. As might be anticipated, though, there are complications. For example, the terrestrial decapod, *Holthuisana transversa*, is thought to be derived from a freshwater ancestor, and yet has blood with a relatively high osmotic pressure. However, though small freshwater crustaceans do have body fluids with low osmotic pressures (<300 mOsmol kg^{-1}), freshwater decapods, possibly because of their large sizes, have retained much higher values (>500 mOsmol kg^{-1}), so the higher values in terrestrial species are not incompatible with freshwater origin.

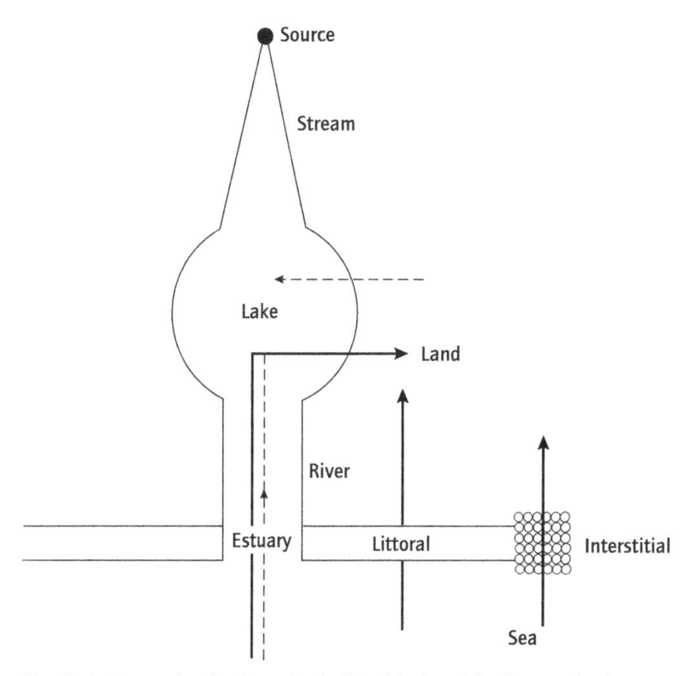

Fig. 12.6 Routes for the 'invasion' of land (—) and freshwater (– –) habitats.

Animals that followed the interstitial route to land, such as some nemertines, nematodes and annelids, are likely to have experienced conditions somewhat intermediate to those that pertain in fully terrestrial and freshwater situations. This is because sediments act as buffers to changes in salinity of the overlying water. Nevertheless, the transition from sea to land, interstitial water will become more dilute so that animals adapted for life in the soil are likely to have evolved good mechanisms for ionic regulation. Hence, very possibly these animals, like those with freshwater ancestors, should have dilute body fluids and the terrestrial nemertine, *Argonemertes dendyi*, that is thought to have invaded land through the interstitial route, does indeed have body fluids with relatively low osmotic pressures (Table 12.3).

From the very limited survey in Table 12.3, it would appear that most invasions of land by invertebrates have occurred directly rather than indirectly. Also two major groups of terrestrial invertebrates, the insects and arachnids, are not mentioned in the table but very probably derived directly from marine ancestors. By contrast to this emphasis in the invertebrates, terrestrial vertebrates have evolved exclusively from freshwater ancestors. Both the terrestrial vertebrates and the terrestrial arthropods, particularly the insects (Section 8.5.3B), have body surfaces that are adapted to resist desiccation. More soft-bodied terrestrial invertebrates avoid desiccation by living in humid habitats, but probably have to suffer some desiccation from time to time, and hence have to tolerate some osmotic concentration. The evolution of dilute body fluids, associated with freshwater origins, also appears to be associated with some inability to tolerate an increase in the concentration of body fluids

Table 12.3 The osmotic pressures of the blood of some terrestrial animals (modified from Little, 1983).

Species	Taxonomic position	Osmotic pressure (mOsmol)	Possible route to land
NEMERTEA			
Argonemertes dendyi	(Hoplonemertea)	145	Marine littoral
ANNELIDA			
Lumbricus terrestris	(Oligochaeta)	165	Fresh water
MOLLUSCA			
Eutrochatella tankervillei	(Prosobranchia)	67	Fresh water
Poteria lineata	(Prosobranchia)	74	Fresh water
Pseudocyclotus laetus	(Prosobranchia)	103	Brackish water
Helix pomatia	(Pulmonata)	183	Salt marshes
Pomatias elegans	(Prosobranchia)	254	Marine littoral
Agriolimax reticulatus	(Pulmonata)	345	Salt marshes
CRUSTACEA			
Undescribed landhopper species	(Amphipoda)	400	Marine littoral
Holthuisana transversa	(Decapoda)	517	Fresh water
Porcellio scaber	(Isopoda)	700	Marine littoral
Cardisoma armatum	(Decapoda)	744	Marine littoral
Coenobita brevimanus	(Decapoda)	800	Marine littoral

All figures are averages for active animals in damp conditions on land, or for equilibrium with fresh water

– possibly proteins evolved for dilute conditions are more easily denatured at higher osmotic pressure. Hence, soft-bodied invertebrates with freshwater ancestors are less well-fitted for terrestrial life than invertebrates with direct marine ancestry.

Having invaded land, some invertebrates secondarily invaded freshwater habitats, for example, some nemertines, myriapods, insects and gastropod molluscs. The insects have been particularly successful in fresh water; out of *c.* 1 million described living species, 25 000 to 35 000 are freshwater for at least one stage of their life cycle. Osmotic problems are probably less important for adult stages because of their impervious cuticles. But for larvae, with hydrophilic and poorly chitinized cuticles, complex mechanisms of regulation have evolved, and ion pumps, located in special structures (see below), are important in this. These soft-bodied larvae and other soft-bodied invertebrates almost certainly experience the same selection pressures, in terms of osmotic conditions, as forms that have accessed fresh waters directly and hence are likely to have evolved body fluids with lower osmotic pressures. Because of this they are probably indistinguishable in these terms from invertebrates with direct origins.

12.3 Excretory systems

'Excretory systems' occur in all major invertebrate phyla, except the Cnidaria and Echinodermata. Porifera do not have 'excretory systems' as such, but freshwater species do possess contractile vacuoles. 'Excretory systems' occur in ammoniotelic animals and hence cannot all be concerned with nitrogen excretion. Instead, their prime function is almost certainly concerned with osmotic and ionic regulation, and occasionally with true excretion, and the same is probably the case for the contractile vacuoles of freshwater sponges (hence the use of quotation marks). Here we first describe the structure and then consider the function of 'excretory systems'.

12.3.1 Structure

Two major categories of system can be distinguished on the basis of development: nephridia, tubules of ectodermal origin that develop from external surfaces and grow in; coelomoducts, tubules of mesodermal origin that develop from internal tissues and grow out.

Nephridia consist of two main types: (i) protonephridia – the closed system of tubules that end in flame cells in platyhelminths (Section 3.6.2) and nemerteans, in flame bulbs in rotifers, and in solenocytes in *Priapulus*, some gastrotrichs, polychaetes and archiannelids (Fig. 12.7); and (ii) metanephridia with ciliated funnels, as in oligochaete annelids (Fig. 12.8) and, arguably, in some other groups.

Coelomoducts are tubular, sometimes ciliated, excretory structures which, before their embryological origin was fully understood, were often labelled 'nephridia'. They occur in onychophorans, arthropods and molluscs. *Peripatus* bears a pair, known as coxal glands, in almost every segment (Fig. 12.9); but they are fewer in number in the other groups. In the crustaceans, glands open to the base of the second antenna (= antennal gland; Fig. 12.9) and/or second maxilla (= maxillary or shell gland; Fig. 12.9), some uniramians have them associated with their maxillae, and in arachnids a pair open in the 6th segment

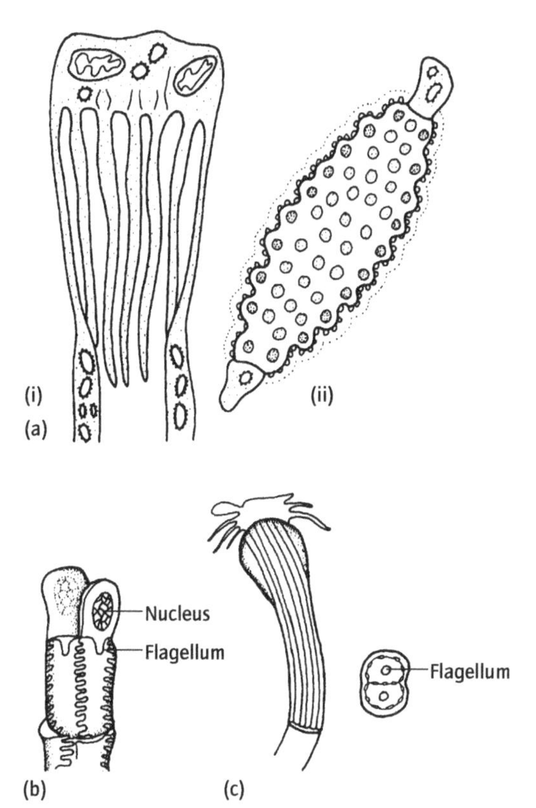

Fig. 12.7 (a) Flame bulb of a rotifer: (i) longitudinal section; (ii) transverse section. (b) and (c) solenocytes of respectively a priapulan and a gastrotrich. (After Barrington, 1979.)

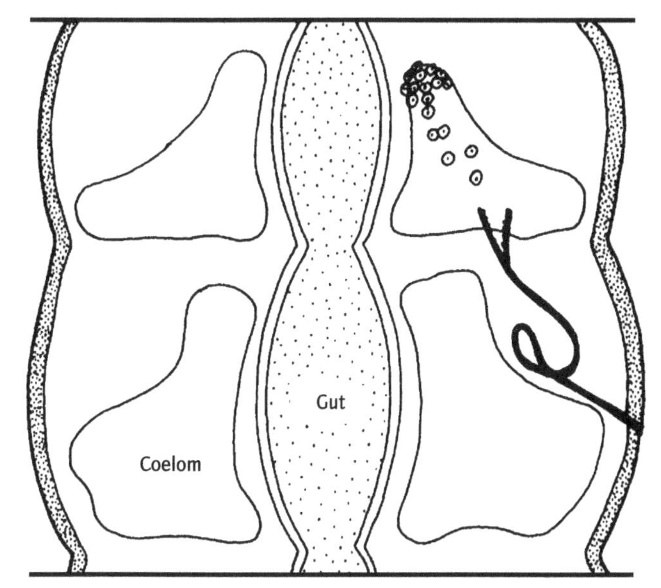

Fig. 12.8 A metanephridium, as in an annelid. (See also Fig. 12.15.)

(= coxal gland; Fig. 12.9). For molluscs the 'excretory' coelomoduct 'kidneys' have their origin in close association with the heart and gonads but in the course of evolution have become more or less distinct, in all the major classes, as illustrated in Fig. 12.10. [It should be noted that although these coelomoducts are by definition 'coelomic', in that their tubular cavities are bounded by a mesodermally derived membrane, they are *not* associated with coelomic body cavities. In fact, animals with excretory coelomoducts do *not* possess coelomic body cavities, and whilst coelomate invertebrates often possess 'coelomoduct' connections between their coeloms and the external environments, these are *not* used for excretion!]

Some systems have both ectodermal and mesodermal components and are referred to as mixonephridia or nephromixia (Fig. 12.11). They are particularly common in polychaetes, but the 'nephridia' of phoronans, sipunculans, echiurans and brachiopods are probably also of this type.

Not related to the above series of organs at all are the 'excretory systems' of nematodes and pterygote insects. *For nematodes*, the system invariably consists of a ventral gland cell, situated in the pseudocoelom, with a terminal ampulla opening to the exterior on the ventral surface by a pore. This may or may not be associated with a tubular system – the so-called H system (Fig. 12.12). The lateral canals are intracellular and the whole system has only one nucleus. *For insects*, Malpighian tubules are the characteristic excretory organs. Tubules, one to many in number, open into the intestine between the mid-gut and hindgut as illustrated in Fig. 12.13. Tubules in the same general position as these, and probably with excretory functions, also occur in myriapods, some arachnids and even tardigrades, but have almost certainly evolved independently at least in the arachnids and tardigrades, so this is an example of convergence. Finally, another peculiar system is the so-called glomerulus of enteropneusts. This is formed by evagination of peritoneum into the proboscis coelom.

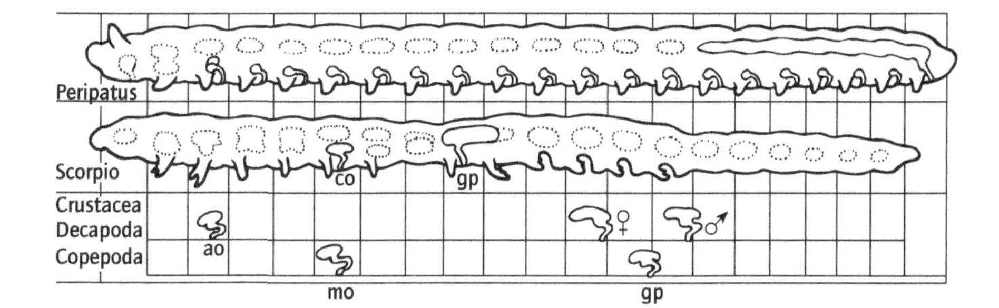

Fig. 12.9 Excretory and genital coelomoducts in the Onychophora and various arthropods (after Goodrich, 1945). ao = antennal gland; co = coxal gland; gp = genital pore; mo = maxillary gland.

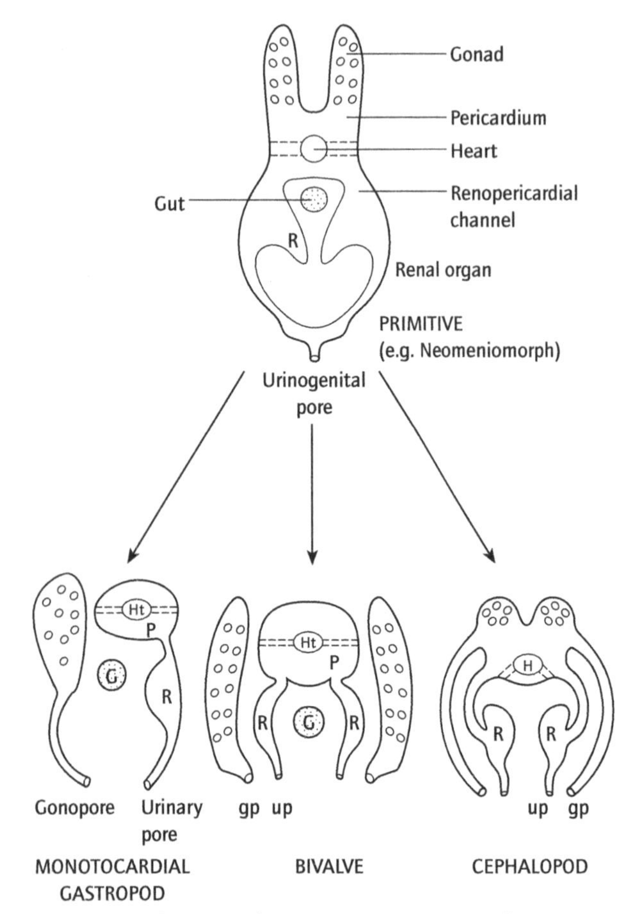

Fig. 12.10 Reproductive and excretory structures in molluscs (after Goodrich, 1945).

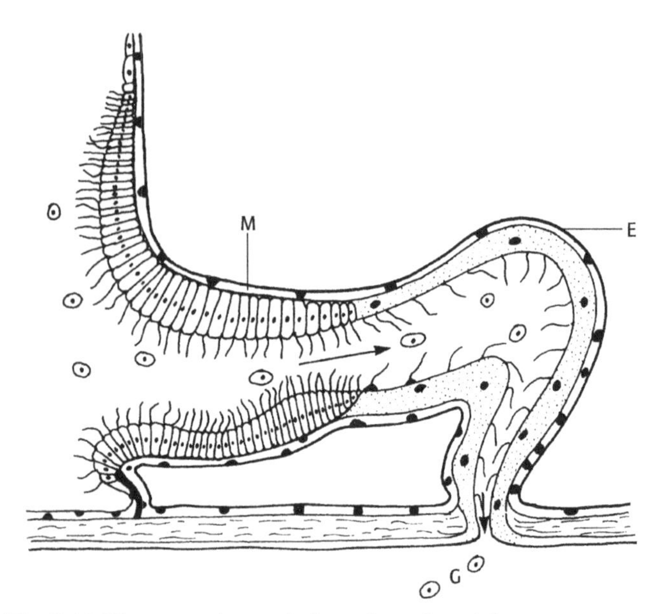

Fig. 12.11 Diagrammatic morphology of a nephromixium (= mixonephridium). M = mesodermal (= coelomoduct component); E = ectodermal (= nephridial component); G = gamete.

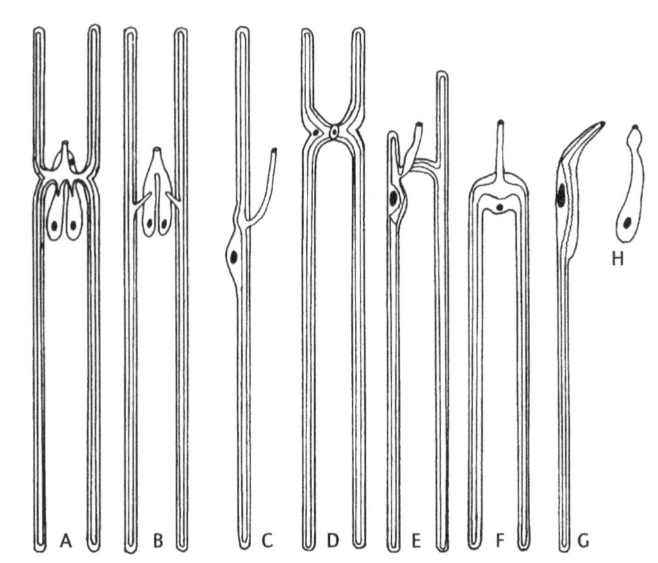

Fig. 12.12 Excretory systems in nematodes. A and B = H-type system with two ventral gland cells; C = asymmetric system, representing one arm of H system; D = H system without gland cells; E and F = shortened H system; G = shortened and reduced H system; H = single, ventral gland cell only present. (After Lee & Atkinson, 1976.)

12.3.2 Function

The previous section indicated that there are a large number of different types of 'excretory systems' with different embryological origins. Despite this diversity of structure, however, there is commonality of function in that only two basic processes are responsible for the formation of excreted fluid.

1 Ultrafiltration: fluid is passed, under pressure, through a semipermeable membrane that holds proteins and similar larger molecules back but allows water and smaller solutes to pass through. The proteins and macromolecules in fact contribute to the osmotic pressure of the fluids in which they occur, and this is known as the colloid osmotic pressure. In situations where there is ultrafiltration this will tend to draw fluid back and a pressure difference is then needed to overcome it. It is thought that the pressure difference set up by the flame may well suffice to do this in protonephridia.

2 Active transport: movement of solutes against a concentration gradient by a process that uses up energy. It can either occur into the excretory system (active secretion) or out of the excretory system (active reabsorption/resorption).

12.3.2.1 Systems thought to involve ultrafiltration

Pressure for ultrafiltration is thought to be generated by the beating flagella or cilia ('flame') in protonephridia, and by 'blood pressure' in coelomoducts, e.g. at the end sac of the antennal (green) gland of crustaceans (Fig. 12.14) and in the heart of the mollusc. Ultrafiltrates pass from these structures into tubules, and in the molluscs, via the pericardium, into the kidneys. All the low-molecular-weight constituents of the body

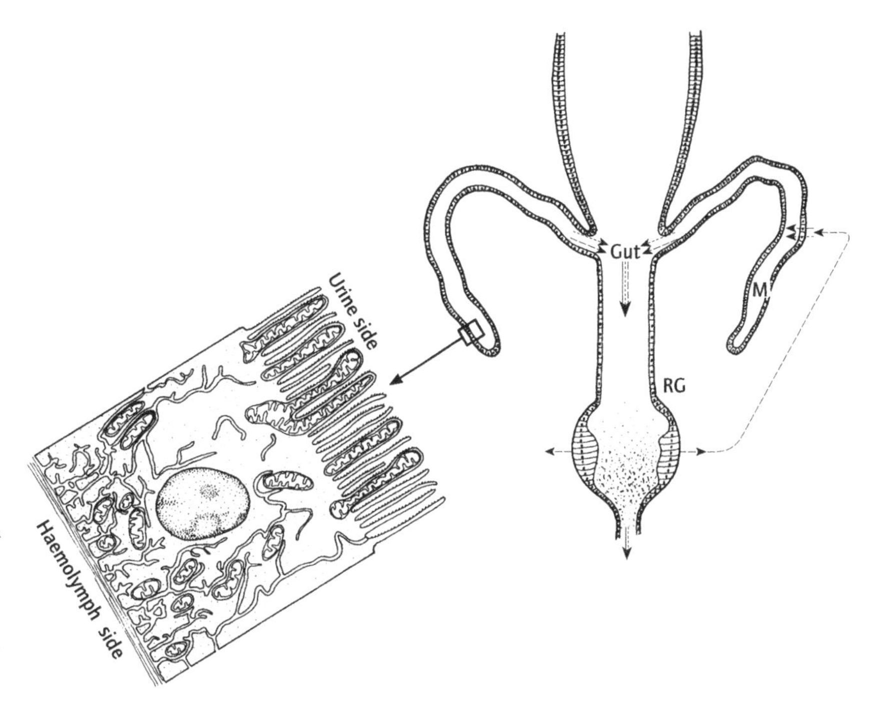

Fig. 12.13 Malpighian tubules (M) and rectal gland (RG) of insects. ⟶ = food; --➤ = water and some ions (notice that the spatial organization of Malpighian tubules and rectal gland is such that water and ions can cycle); ···➤ = uric acid (after Potts & Parry, 1964). The ultrastructure of the Malpighian tube is also shown (after Oschman & Berridge, 1971). Notice that it is rich in mitochondria, suggesting that it is involved in active transport.

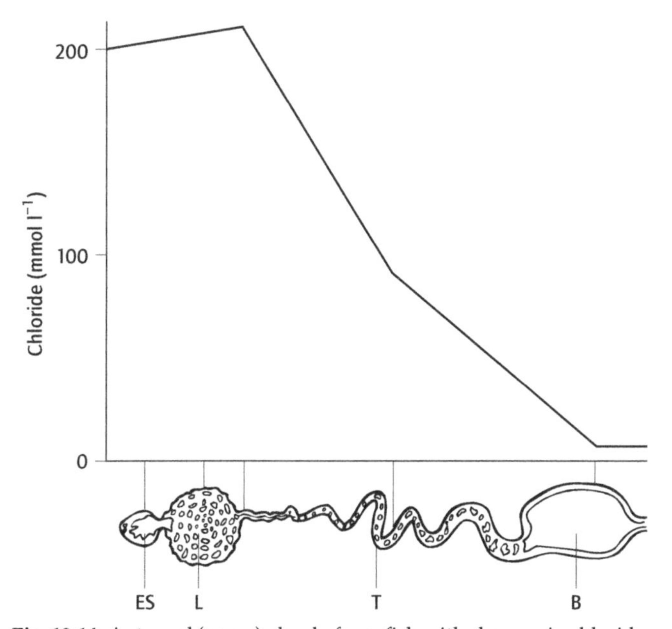

Fig. 12.14 Antennal (green) gland of crayfish with changes in chloride composition of urine as it passes along the organ. ES = end sac; L = labyrinth; T = renal tubule; B = bladder. (After Potts & Parry, 1964.)

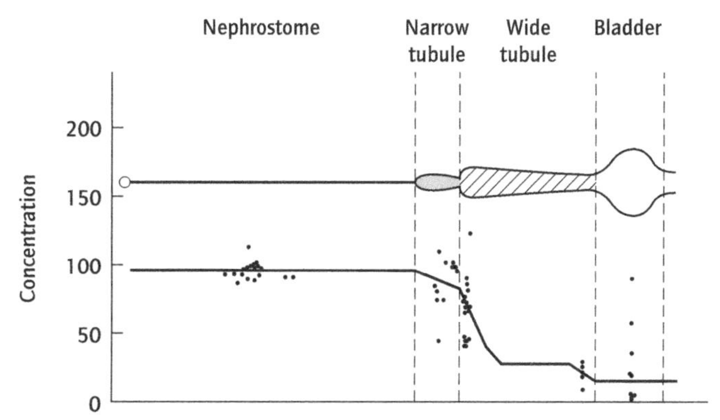

Fig. 12.15 Osmotic pressure along the length of a metanephridium of an earthworm (after Potts & Parry, 1964).

fluids are filtered into the ultrafiltrate in proportion to their concentration in the fluids. Physiologically important molecules such as glucose and, in freshwater invertebrates, ions such as Na^+, K^+, Cl^- and Ca^{2+} are removed in the tubules of the system (see, for example, Fig. 12.14) leaving toxic substances and unimportant molecules behind to be excreted. As already noted, this selective reabsorption involves active transport.

12.3.2.2 Systems thought not to involve ultrafiltration

Ciliary filtration Metanephridia and nephromixia direct fluids from the coelomic body cavity into the 'excretory' tubules by ciliary action. These fluids are again modified in composition by active transport in the tubule of the system (Fig. 12.15).

Excretory materials may also exit via this route. For example, in earthworms chloragogenous tissue is important in excretion. This tissue is of epithelial origin and lies in the coelom against the gut wall. Chloragogen cells store fat and carbohydrate, but they are also sites of deamination, and ammonia and urea pass from them into the coelom, and are swept into the nephridial funnels. Waste particles, released when the chloragogen cells disintegrate, may also be passed out through the nephridial system. In leeches, botryoidal tissue takes over the function of the chloragogenous tissue and in some groups, such as *Theromyzon*, waste products from it pass out through an open nephridial system. In other groups, though, such as *Glossiphonia* and *Hirudo*, the nephridial funnel is sealed from its tubules. Here cilia beat from the metanephridial funnels, but *into* coelomic cavities surrounding the funnels, and serve to distribute amoebocytes that are actually formed in the funnels. The tubules still open to the outside and probably effect some ionic and osmotic regulation.

The Malpighian tubules of insects The direct supply of O_2 to the tissues by a tracheal system has diminished the need for an efficient, pressurized blood system by insects. Hence the Malpighian tubules do not receive a pressurized blood supply, and are surrounded with blood at approximately the same pressure as the contents of the tubes. So ultrafiltration cannot be involved here. Instead potassium ions, and possibly other solutes, are actively secreted into the lumen of the tubule and water follows by osmosis. Urates are also secreted into the lumen and these are kept in solution by the relatively high pH at the distal end of the tubules. This fluid enters the hind-gut and water is resorbed, particularly by rectal glands (see Fig. 8.27), the pH falls and the urate is precipitated as uric acid. The latter is lost in the faeces (see Fig. 12.13).

12.3.2.3 *The nematode system*

Not much is known about the functioning of this system; however, the tubes almost certainly differ in function from the gland cells. The former appear to be involved in ionic and osmotic regulation, but there are no cilia to create flows; it is usually assumed, by default, that ultrafiltration and active transport are important. It seems likely that the system plays a minor role in nitrogen excretion but some nematodes release secretions from the system that appear to have enzymatic activity. These probably derive from the gland cell.

12.4 Conclusions

You should now appreciate that, within the invertebrates, it is difficult, and even unhelpful, to disentangle an account of excretion (loss of nitrogenous wastes) from one of ionic and osmotic regulation and even of buoyancy. The formation, transport and removal of nitrogenous wastes, at some stage, inevitably involves the filtration and flow of body fluids, even if the excreta end up, like uric acid, to be dry. So excretion is intimately associated with the movement of liquids and ions and hence with processes of osmosis, diffusion, active transport and ultrafiltration – and the reader should now have a clear understanding of these. 'Excretory systems' are invariably associated with ionic and osmotic regulation but not always with excretion. Despite their diversity of structure, this chapter focused on the similarities in function of 'excretory systems' in terms of the key processes just noted and, again, it is these principles that the reader should take from the chapter. Finally, the replacement of ions of one mass with those of another, and the replacement of liquid with gas, influence the density of tissue and are, as will now be appreciated, important factors in the control of buoyancy in aquatic invertebrates.

12.5 Further reading

Burton, R.F. 1973. The significance of ionic concentrations in the internal media of animals. *Biol. Rev.*, **48**, 195–231.

Denton, E.J. & Gilpin-Brown, J.B. 1961. The distribution of gas and liquid within the cuttlebone. *J. Mar. Biol. Assoc., U.K.*, **41**, 365–381.

Denton, E.J. & Gilpin-Brown, J.B. 1966. On the buoyancy of the pearly *Nautilus*. *J. Mar. Biol. Assoc., U.K.*, **46**, 723–759.

Durand, F., Chausson, F. & Regnault, M. 1999. Increases in tissue free amino acid levels in response to prolonged emersion in marine crabs: an ammonia-detoxifying process efficient in the intertidal *Carcinus maenas* but not in the subtidal *Necora puber*. *J. exp. Biol.*, **202**, 2191–2202.

Eddy, B.E., Flik, G., Potts, W.T., Hazon, N. & Dimitrijevic, M.R. (Eds) 1997. *Ionic Regulation in Animals: A Tribute to W.T.W. Potts*. Springer-Verlag, New York.

Edney, E.B. 1957. *The Water Relations of Terrestrial Arthropods*. Cambridge University Press, Cambridge.

Edney, E.B. 1974. Desert arthropods. In: Brown, G.W. (Ed.) *Desert Biology*, Vol. 2. Academic Press, New York.

Gilles, R. & Delpire, E. 1997. Variations in salinity, osmolarity, and water availability: vertebrates and invertebrates. In: Dantzler, W.H. (Ed.) *Handbook of Physiology*. Section 13. *Comparative Physiology*, Vol II, Chapter 22, pp. 1523–1586. Oxford University Press, Oxford.

Gordon, M.S. & Olson, E.C. 1995. *Invasions of the Land*. Columbia University Press, New York.

Hadley, N.F. 1994. *Water Relations of Terrestrial Arthropods*. Academic Press, San Diego, CA.

Horne, F.R. 1971. Accumulation of urea by a pulmonate snail during aestivation. *Comp. Biochem. Physiol.*, **38A**, 565–570.

Koehn, R.K. 1983. Biochemical genetics and adaptations in molluscs. In: Hochachka, P.W. (Ed.) *The Mollusca*, Vol. 2, pp. 305–330. Academic Press, New York.

Lee, D.L. & Atkinson, H.J. 1976. *Physiology of Nematodes*, 2nd edn. Macmillan, London.

Little, C. 1983. *The Colonisation of Land*. Cambridge University Press, Cambridge.

Little, C. 1990. *The Terrestrial Invasion: An Ecophysiological Approach to the Origin of Land Animals*. Cambridge University Press, Cambridge.

Morritt, D. & Spicer, J.I. 1993. A brief re-examination of the function and regulation of extracellular magnesium and its relationship to activity in crustacean arthropods. *Comp. Biochem. Physiol.*, **106A**, 19–23.

Morritt, D. & Spicer, J.I. 1998. Physiological ecology of talitrid amphipods: an update. *Can. J. Zool.*, **76**, 1965–1982.

Potts, W.F.W. & Parry, G. 1964. *Osmotic and Ionic Regulation in Animals*. Oxford University Press, London.

Randall, D., Burggren, W.W. & French, K. 1997. *Animal Physiology. Mechanisms and Adaptations*, 4th edn. W.H. Freeman, New York.
Rankin, J.C. & Davenport, J. 1981. *Animal Osmoregulation*. Wiley, New York.
Schmidt-Nielsen, K. 1972. *How Animals Work*. Cambridge University Press, Cambridge.
Schmidt-Nielsen, K. 1997. *Animal Physiology. Adaptation and Environment*, 5th edn. Cambridge University Press, Cambridge.
Spicer, J.I. & Gaston, K.J. 1999. *Physiological Diversity and its Ecological Implications*. Blackwell Science, Oxford.
Willmer, P., Stone, G. & Johnston, I. 2000. *Environmental Physiology of Animals*. Blackwell Science, Oxford.
Wright, P.A. 1995. Nitrogen excretion: three end products, many physiological roles. *J. exp. Biol.*, **198**, 273–281.
Zerbst-Boroffka, I., Bazin, B. & Wenning, A. 1997. Chloride secretion drives urine formation in leech nephridia. *J. exp. Biol.*, **200**, 2217–2227.

CHAPTER 13

Defence

This chapter first of all classifies the various threats to which invertebrates are exposed and then considers how these animals defend themselves against each type of threat. It therefore ranges widely from defence against predators, to defence against pathogens and even to defence, if a defence is possible, against ageing processes.

13.1 Classification of threats

13.1.1 There are two major classes

Figure 13.1 gives some examples of survivorship curves – numbers (or proportions) of individuals all born at roughly the same time (cohorts) alive at different ages thereafter. Curves in (a) are for field populations and here mortality is either focused on juveniles or occurs at a roughly constant rate at each age. Mortality in these natural populations is probably mainly due to ecological or extrinsic factors such as accidents, diseases and predation. Juveniles are often more vulnerable to these than adults, or all age classes are roughly equally susceptible.

Curves in (b) are for laboratory cultures. Here most extrinsic mortality factors can be excluded. However death still occurs but is focused on old-aged individuals. There is an increase in vulnerability with age, possibly due to intrinsic factors, i.e. to ageing processes or senescence.

13.1.2 Ecological (extrinsic) causes of mortality

Ecological causes of mortality are many and varied, but there are four main classes: accidents, diseases, predation and environmental stresses. The first three of these are self-explanatory. Environmental stress can either occur due to absence of an essential factor or to the presence of a stressor – a natural toxin or an artificial pollutant.

13.1.3 Ageing

Ageing apparently occurs when extrinsic mortality factors are excluded, and can therefore be ascribed to internal degeneration of systems, cells and molecules. These intrinsic effects can probably be traced ultimately to the denaturation of important biological molecules – nucleic acids and proteins – under the influences of such processes as thermal noise, cross-linkage of side-chains in macromolecules, autoxidation and so on. Yet it should not be imagined that these intrinsic processes cannot be influenced by extrinsic factors. Figure 13.2 shows that whole-body irradiation of *Drosophila* can shorten life to an extent that depends upon dose. However, despite these effects the shapes of the survivorship curves remain the same, and some gerontologists suggest that lifeshortening here is due to accelerated ageing, possibly due in turn to increased damage to macromolecules – particularly DNA – by the high-energy radiation. Alternatively, addition of certain chemicals, such as vitamin E, to the culture media of nematodes or food of *Drosophila* extends their lives. These chemicals probably work by protecting macromolecules from damage; for example, vitamin E is probably an antioxidant.

13.1.4 Classification

All mortality can therefore be influenced by the external environment; it is just that some causes of mortality are more intimately associated with organisms than others. Table 13.1 classifies mortality factors according to their intimacy of association with a recipient and results in a series ranging from predators at one end to ageing processes at the other. Conversely, the ease of exclusion and hence experimental manipulation of mortality factors reduces along the continuum in the reverse direction, from predators to ageing factors.

13.2 Defence

13.2.1 Against predators

All animals are potential food for other animals (see Chapter 9). They can protect themselves against being eaten in a variety of ways but these can be grouped into one of three main classes of responses: avoiding potential predators; dissuading them; actively repelling them.

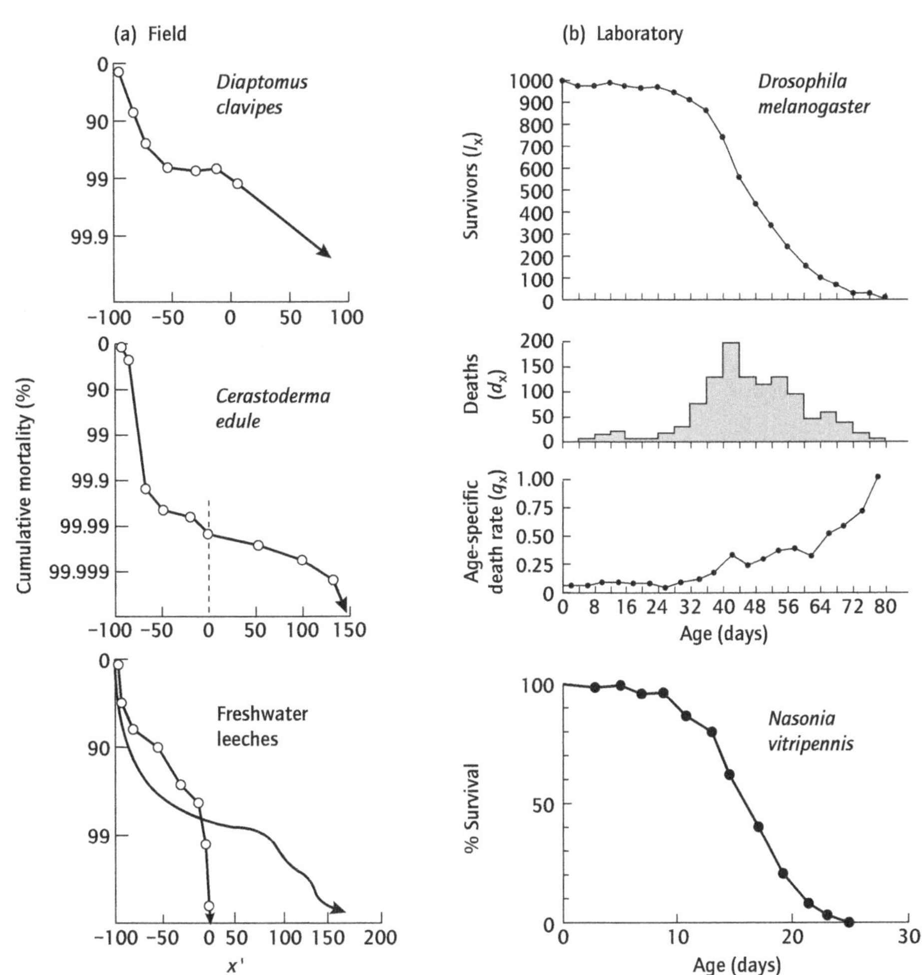

Fig. 13.1 (a) Survivorship curves for field populations of invertebrates. Note x' = % deviation from mean life-span. (After Ito, 1980.) (b) Survivorship curves, distribution of ages at death and age-specific death rate from a laboratory population of *Drosophila melanogaster* (after Lamb, 1977) and survivorship curve for a laboratory population of the hymenopteran, *Nasonia vitripennis* (after Davies, 1983).

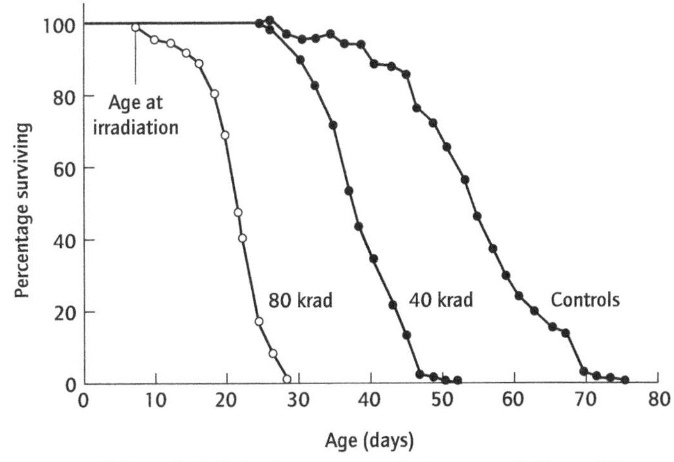

Fig. 13.2 Effect of whole-body gamma-radiation on male *Drosophila melanogaster* (after Lamb, 1977).

13.2.1.1 *Avoidance*

This can involve keeping out of the way of predators and/or being inconspicuous. A possible example, involving both kinds of responses, is the extensive vertical migration exhibited by some marine and freshwater planktonic animals. Though these patterns are generally complex, they often involve downward migration, away from sunlight, during the day and upward migration during the night. Thus, animals avoid being conspicuous to predators in the light and only come to the surface at night when they are likely to be less conspicuous to visual predators (Fig. 13.3a,b). This behaviour is most marked in species and age classes that are particularly vulnerable to predation. Moreover, in a unique study of a copepod in a series of isolated mountain lakes, Gliwicz (1986) has shown that vertical migrations only occur in lakes that contain planktivorous fishes. Also he was able to make observations on the migratory patterns of the copepod in one lake at several different stages in stocking with planktivorous fishes. Twelve years after stocking there was little evidence of vertical migration, but some 23 years later the copepods showed strong migration away from the water surface during the day. However, predation cannot be the whole explanation for the evolution of all vertical migration, for in some invertebrates the daytime descent is often to great depths, far in excess of what would be needed to avoid the light, and some migrating zooplankters luminesce at night so increasing their conspicuousness. Other possible explanations involve optimal exploitation of patchy food (Chapter 9), energy economies and improved horizontal migration.

Table 13.1 Classification of mortality factors.

Mortality agent	Accident	Predators	Disease	External stressor (e.g. pollutant)	Internal stressor (e.g. system degeneration)
Intimacy of association with recipient	X	X	XX	XX	XXX
Ease of exclusion by artificial means	XXX	XXX	XX	XX	X
Response of recipient		Defence	Immunological defence	Tolerance, resistance, repair	Repair

X = low, XX = high.

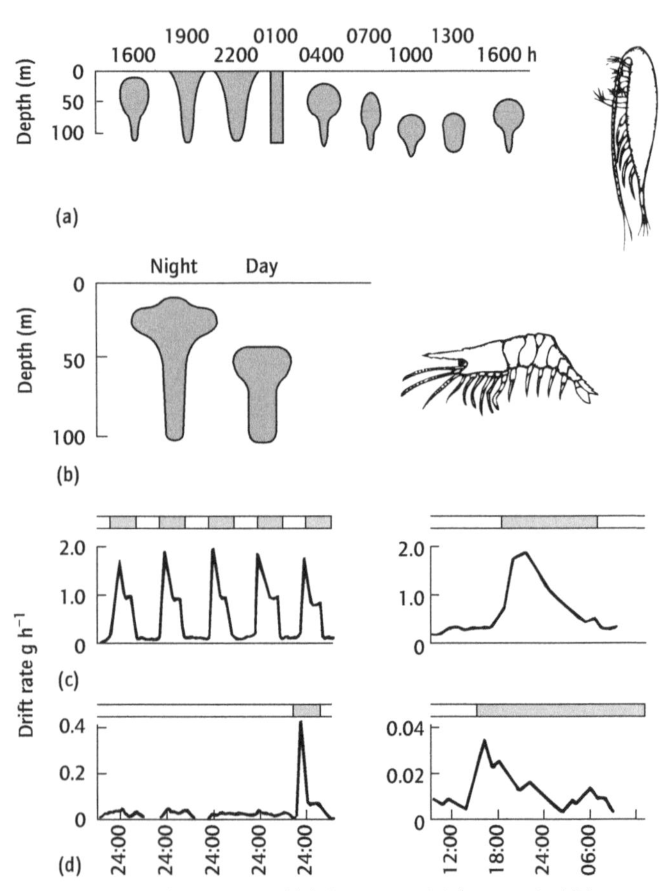

Fig. 13.3 Vertical migration of (a) the copepod *Calanus* and of (b) a deeper-dwelling acanthephyrid prawn (after Barnes & Hughes, 1982). The influence of light and dark on drift of some freshwater invertebrates: (c) open stream; (d) in experimentally manipulated systems (after Holt & Waters, 1967).

Similar vertical migrations, but of a more limited kind, are practised by freshwater invertebrates that live on submerged stones: during the day they are under stones and are inactive, whereas during the night they often emerge on to upper surfaces and become more active. Large numbers of these normally benthic organisms can be collected floating free in running-water systems. This invertebrate drift, as it is called, is particularly abundant during nightfall (Fig. 13.3b,c,d), possibly because it is then that invertebrates crawl on to exposed, upper-stone surfaces and become vulnerable to being washed away.

Escape reactions are an extreme form of locomotory avoidance reactions. They involve use of either normal locomotory responses or the deployment of specialized behaviour. The cuttlefish, *Sepia*, has an 'ink sac' which contains fluid composed of granules of melanin. When attacked, it ejects an ink screen and immediately becomes very pale and swims at right angles to its original path of flight. Bivalve molluscs, such as the usually sessile cockle, *Cerastoderma*, can achieve escape reactions by sudden and rapid foot and shell contractions. Extremely powerful escape movements are made by *Cerastoderma* in response to the tube feet of starfish – probably evoked by a substance released into the water.

Cryptic (concealing) coloration is another widespread method of avoidance in invertebrates (see Fig. 9.32). Examples probably occur in all phyla but have been particularly thoroughly studied in some snails and insects.

1 Banding in *Cepaea nemoralis*. This species of land snail produces a wide range of shell colours and patterns by varying the underlying hue of its whole shell and the number, width and intensity of bands. These variations are genetically controlled. Thrushes feed on *Cepaea*, finding shells by sight and breaking them open on rocks known as thrush anvils. In the mid-1950s A.J. Cain and P.M. Sheppard showed that, on average, shells broken by thrushes have colours more easily seen than those carried by snails still living in the same area. Different patterns are more difficult to see in different places and at different times of the year. Banded light-coloured shells, for example, are difficult to see in lush vegetation (grass fields, and hedgerows) when sharp contrasts of light and dark are produced by the interplay of entering light and narrow shadows. In dark woodlands, uniformly dark and unbanded shells are more difficult to see (Fig. 13.4).

2 Melanism in *Biston betularia*. A large number of moths and other insects have evolved wing and body patterns that make them inconspicuous on lichen-covered trees. Industrial pollutants have killed lichens and sooted up tree trunks. In these areas the previously adapted *typical* moths were replaced by black,

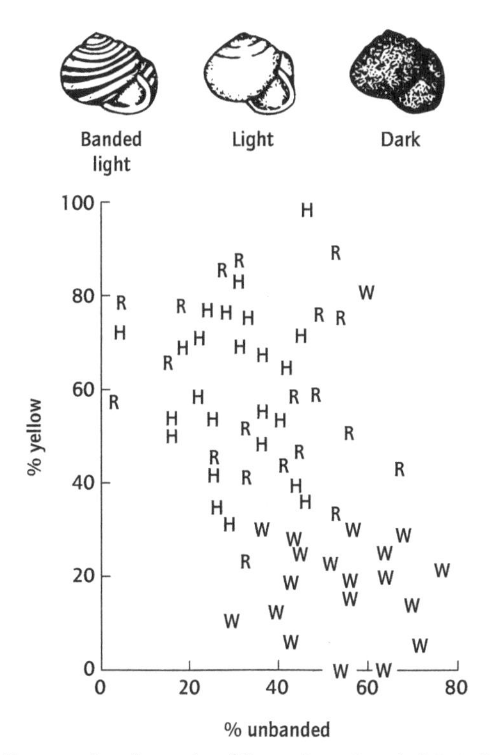

Fig. 13.4 Frequencies of morphs of *Cepaea* in various habitats. Banded light shells are difficult to see in low vegetation (herbage, R; hedgerows, H) whereas dark unbanded shells are more difficult to see in dark woodland (W) (after Calow, 1983; original work by Cain & Sheppard, 1954).

melanic, forms. This is known as industrial melanism (Fig. 13.5). Again in the 1950s H.B.D. Kettlewell showed that birds could be responsible for this in the peppered moth (*Biston betularia*); by direct observation, typical forms were found to be more vulnerable to bird predation than melanics at polluted sites and vice-versa at non-polluted sites. Kettlewell also released peppered moths into an old cider barrel lined with vertical black and white stripes. Sixty-five per cent of the moths took up a position on a background they matched (typicals on a white and melanics on a dark background), so moths appear to have appropriate behaviour to allow them to take up resting positions on backgrounds upon which they are least conspicuous.

(As might be expected there is more to the evolution of banding in shells and melanism in the peppered moth than indicated by the short descriptions given above. For a more thorough description the reader should refer to genetical texts, e.g. Berry, 1977).

Finally, camouflage need not make the bearers inconspicuous but can make them resemble objects not usually associated with food. Many insects resemble parts of plants: twigs and leaves. The young larvae of some of the swallowtail butterflies are conspicuous but escape attack because they are black with a white saddle on their backs and thus resemble bird droppings (see Fig. 9.32)! They change their colour pattern dramatically when they become too large for this method of concealment.

Erichsen *et al.* (1980) have carried out some novel experiments to test the influence of this kind of camouflage on predation by birds. They offered great tits (*Parus major*) a choice between large and small mealworms (*Tenebrio molitor*) in straws; but the large ones were in opaque straws that resembled artificial twigs whereas the small ones were in clear straws and were easily visible. The large worms gave more energy per mouthful than the small ones (about twice as much) and were therefore a more profitable choice, but the small worms were more instantaneously recognizable than the large ones that required the bird to take time inspecting and picking up 'twigs' in the search for worms.

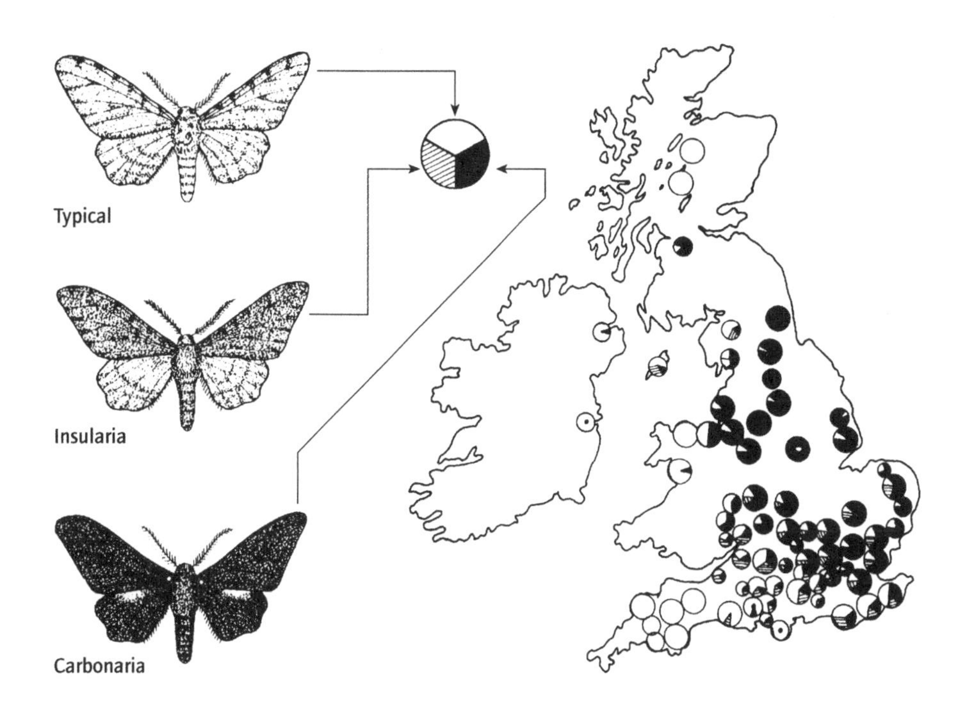

Fig. 13.5 Frequencies of three morphs of *Biston betularia*. Insularia and carbonaria are both melanics. Although looking intermediate between typical and carbonaria, insularia is in fact controlled at a different locus (after Sheppard, 1958).

In choice experiments the birds consistently selected the small mealworms where 'twigs' were abundant, but switched to larger ones where 'twigs' were rare.

Other animals achieve camouflage by attaching to themselves material from the external environment. Spider crabs pick up pieces of algae and other materials which are attached to patches of hooks on their exoskeletons. The cases of caddis larvae might also serve a similar function. Thus larvae of *Potamophylax cingulatus* with leaf cases are less likely to be eaten by trout when on a leafy background than on a sandy one. During larval growth, this animal switches from a leaf to a sand case, but sandgrain cases are no more likely to be eaten when on a leafy as compared with a sandy background. This lack of difference might be due to the low palatability of sandgrain cases (see below). (For more details, see Hansell, 1984.)

13.2.1.2 Dissuasion

Animals can dissuade predators from eating them by either physical or chemical defences.

Calcareous fortifications are used widely – spicules of sponges, calcareous skeletons of corals, tubes of annelids, shells of molluscs and brachiopods, calcified boxes of some lophophorates and tests of echinoderms. Chitinized skeletons of arthropods also form fortifications and are sometimes reinforced with calcium. The latter is particularly true of cirripedes, sessile crustaceans which become covered externally with thick, calcareous plates (see Fig. 8.36a).

Sometimes inorganic inclusions may simply so dilute the nutritious tissues that they make the animals poor-quality food. Such may be the function, at least in part, of the calcareous or siliceous skeletons of sponges, the calcareous skeletons of cnidarian corals, the tubes of grit and sandgrains that surround some polychaete annelids and the stony cases of some freshwater insect larvae (see above). Surprisingly, a small number of marine turbellarians have been found to bear calcareous scales or rods embedded in their body walls (Chapter 3). These probably have a supportive function but it is not unreasonable to assume that they also operate as tissue diluents. Moreover, it is also plausible that the molluscan shell evolved from this kind of origin.

Shells are a particularly obvious form of physical fortification (see Fig 9.33). Strong shell sculpture, occluded shell apertures and low spires coupled with a thick shell are effective devices amongst many marine, benthic gastropods in thwarting the attacks of fishes, crabs, lobsters and other shell-breaking predators (Fig. 13.6). Occluded teeth, inside the apertures of land snails, exclude shell-entering predacious beetles. Strong sculpture in the form of various shell protuberances, ribs, knobs and spines strengthen shells and make them effectively larger and hence more difficult or time-consuming for a predator to handle. Lack of specialized molluscivores may in part explain the general absence of thick shells and elaborate architecture in the freshwater molluscan fauna. Poor calcium

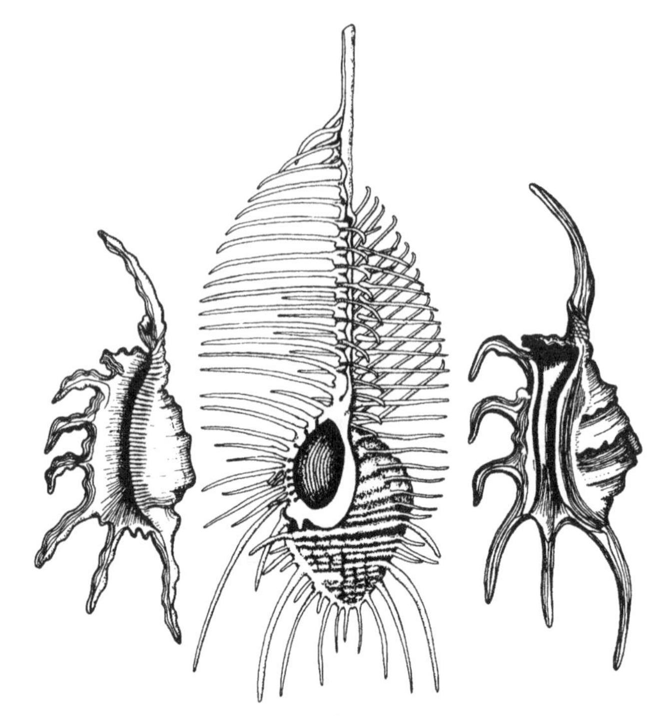

Fig. 13.6 Defensive sculpturing in shells.

availability in fresh water may also be partially responsible for these differences.

An unusual, protective body covering occurs in the ascidians. Their bodies are covered by an epithelium, one cell thick; but this is not the outer covering of the body. Instead it is surrounded by a tunic (the group are sometimes referred to as tunicates – Section 7.4A). This is usually quite thick but varies considerably in texture from being soft and delicate to being tough and similar to cartilage. It consists of a fibrous material, a principal constituent of which, in many, though not all species, is a kind of cellulose called tunicin. Also present are proteins and inorganic inclusions, such as calcium. The tunic can be vascularized and contains amoeboid cells, so it is not just a dead covering.

Not all physical fortifications are secreted by the defenders. Again particularly good examples of this are the cases of caddis larvae. These are built from materials from the surrounding environment. A similar example, but where the defences are formed from materials rejected by the defender, is the so-called 'faecal shield' of the beetle larvae *Cassida rubiginosa*. This consists of a compressed packet of cast skins and faeces carried on a fork-like organ held over its back. The shield is manoeuvrable and is used by larvae to protect themselves against attack from other insects such as ants. Hermit crabs move into empty gastropod shells and thus save on the need to invest in the production of a thick exoskeleton. Their uropods are modified and the larger, left one is used for hooking on to the columella of the shell (see Fig. 8.49e).

Chemical methods of dissuasion are also common in the invertebrates. Many invertebrates lace their tissues with toxins.

As much as 0.3% of the body weight of a nemertine can be made up of neurotoxins. Numerous shell-less gastropod molluscs, e.g. opisthobranchs and pulmonate slugs, use various toxins, including sulphuric acid. Some sponges produce irritating substances and an extract, prepared from freshwater sponges, has been shown to be fatal when injected into mice. Other animals may also make use of these toxins. Some crabs decorate themselves with sponge, perhaps for camouflage or perhaps to 'cash in' on protection derived from the toxins produced by the sponge. Similarly, some hermit crabs occupy snail shells with attached sponges and anemones and derive similar protection.

Some pelagic antarctic amphipods, that cannot defend themselves chemically, dissuade fish predators from feeding upon them by carrying, in their pereiopods, a pteropod that can. In laboratory experiments carried out by McClintock & Janssen (1990) it was shown that amphipods not carrying pteropods were invariably eaten by fishes, whereas those with them were avoided, apparently actively. The benefits to the amphipod must be greater than the costs; e.g. swimming speed reduced by nearly 50%. The pteropod appears to gain nothing, since it does not eat while being carried. The nature of the toxin secreted by the pteropod is not currently known.

Turbellarian flatworms probably do not secrete toxins directly into their tissues but produce rhabdoids – rod-shaped epidermal bodies, arranged at right angles to the surface and secreted by epidermal gland cells (Section 3.6.3.1). They are discharged when the worms are irritated, and that they have a defensive role is suggested by the following easily carried out experiment. Stickleback fishes readily eat the oligochaete, *Tubifex tubifex*, and may take it from forceps. If, however, the *Tubifex* is first smeared with mucus produced by prodding a flatworm then it is rejected by the fish. If, finally, *Tubifex* is coated in mucus from the trails of normally moving worms it will be eaten. The mucus from the disturbed worms contained many rhabdoids whereas that from the non-disturbed worm contained few if any rhabdoids. Rhabdoids probably also have other functions such as in the rapid formation of mucus itself, or as antimicrobial agents.

Toxins are common in insects and they can 'borrow' toxic compounds from the plants upon which they feed. For example, the grasshopper *Poekilocerus bufonius* feeds on milkweeds that contain a number of complex toxins that can disrupt cardiac function – so-called cardenolides. The grasshopper extracts these from its food and stores them in a poison gland. When attacked by predators it defends itself by ejecting a poison spray rich in the plant-derived toxins. When these grasshoppers are maintained on a diet without milkweeds, the cardenolide content of the spray is reduced ten-fold. Monarch butterflies also feed on milkweeds and lace their tissues with cardenolides, making themselves distasteful to avian predators. Butterflies grown from larvae not fed on milkweeds again have no harmful effects on predators. Toxic chemicals such as this, not produced in the insects but received from plants, are sometimes called *kairomones* to contrast them with *allomones*, toxins that confer advantages on the organisms (i.e. plants) that produce them. (For more details see Nordlund and Lewis, 1976.)

Chemical toxins are often associated with warning coloration. Correlations of this kind can be found in many invertebrate phyla, from vividly coloured nemertines and slugs to brightly coloured insects. Such coloration tends to be associated with simple patterns and the colours frequently include red, yellow, or black and white. Everyone will be familiar with the black and yellow bands of bees and wasps.

The evolution of toxins and warning colorations is not straightforward. Their only evolutionary virtue is if they inhibit predation, and yet the only way a predator can be aware of them is by 'having a go'. Kin selection is a possible explanation, where the sacrifice of one individual carrying a gene for the warning toxin can protect relatives in the same group that carries the same gene. Similarly, such a gene could also spread if the bearer were not easily damaged, so that it could survive detection by a predator or if predators were repelled by an obnoxious stimulus before attempting a strike. The majority of warningly coloured insects are tough and not easily damaged and, as with slugs, frequently emit strong odours. Nemertines can regenerate tissues lost to predators.

Warning colours can be mimicked by non-toxic animals and this resemblance is referred to as Batesian mimicry, after the man who first made it explicit. Since the warning colour of the *mimic* is false whereas that of the *model* is not, it follows that: (a) the model must be noxious and bright coloured; (b) the model must be more common than the mimic for if the model were rare the predator would not learn that it is protected and the whole relationship would fail; (c) mimics must occur in close association with models and closely resemble them. Mullerian mimicry is another form of mimicry whereby noxious species converge on the same pattern because they get advantages from each other. Criterion (b) does not apply here and resemblance, criterion (c), need not be as precise. Wasps and bee species carry the same pattern of banding and this is Mullerian mimicry. Many dipterans, particularly hoverflies and some lepidopterans, have evolved wasp/bee-like appearances and this is Batesian mimicry. Figure 13.7 gives some lepidopteran examples.

It is not too difficult to appreciate how Mullerian mimicry might have evolved, but a major problem with Batesian mimicry is that the mimic must resemble the model closely enough to gain protection from it (criterion (c)), so how could forms intermediate between non-mimetic ancestor and mimic have been favoured? One possibility is a two-phase evolutionary process: the establishment of an approximate but, nevertheless, adequate resemblance by means of a major mutation, followed by a gradual improvement by natural selection of more usual small-scale genetic variance.

13.2.1.3 *Repulsion*

Organs that are used to capture and kill prey can often be used actively to repel predators; for example cnidoblasts of

Fig. 13.7 Batesian mimicry between the moths *Alcidis agarthyrsus* (a) and its mimic *Papilio lag* (b) and between the North American viceroy *Limenitis archippus* (c) and its model a monarch *Danaus plexippus* (d). Mullerian mimicry between *Podotricha telesiphe* (e) and *Heliconius telesiphe* (f).

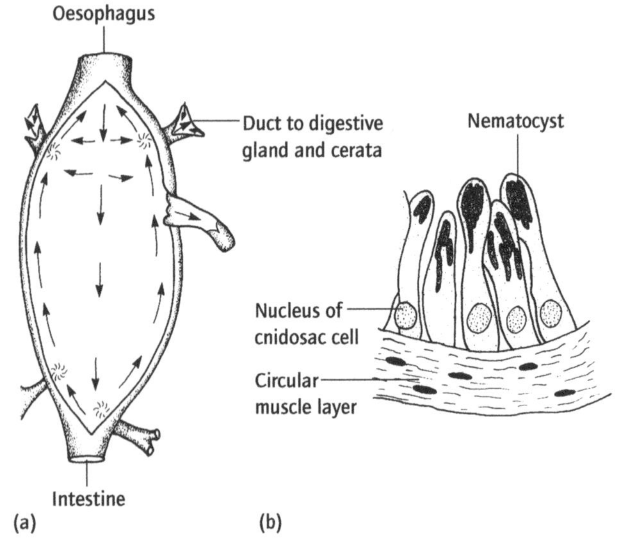

Fig. 13.8 Stomach of nudibranch opened dorsally (a), and cnidosacs (b). (After Barnes, 1980).

cnidarians are used for both attack and defence. Paradoxically, the cnidoblasts of cnidarians that fall prey to nudibranch molluscs can also be borrowed by the predators themselves as defences. Ciliary tracts in the stomachs of the nudibranchs carry undischarged nematocysts to projections on their dorsal surfaces (cerata) in which they are engulfed but not digested (Fig. 13.8). They are moved to the distal tips of the cerata, cnidosacs, which open to the exterior. Discharge from these may be effected by contraction of circular muscles around the cnidosacs. Nematocysts can be replaced in about 10 days and most nudibranchs use only certain types of nematocysts present in their prey. A small number of turbellarians utilize nematocysts of the hydroids they eat in the same way, as does the ctenophore, *E. rubra*, those of its medusa prey.

The chitinized jaws and stings of arthropods are also examples of aggressive structures that are used defensively. On the other hand, some stings are specifically defensive, for example the sting of honey bees. This is formed from the modified ovipositor – no longer used for laying eggs – and consists of paired barbed lancets and the unpaired stylet (Fig. 13.9). At rest this lies in a pocket within the seventh abdominal segment. The mechanism of stinging is described in the legend of the figure. The poison is secreted by a pair of long glands in the abdomen and in the honey bee contains certain enzymes which cause the tissue of the victim to produce histamine.

Beetles of the genus *Brachinus*, commonly known as bombardier beetles, use a defensive spray to ward off predators, such as spiders, preying mantids and even frogs. When disturbed they release this from a pair of glands at the tips of their abdomens. The latter can be rotated so that they can spray accurately in virtually any direction. The active principals of the secretion are benzoquinones, which are synthesized explosively by oxidation of phenols at the moment of discharge. An audible detonation accompanies the emission and the spray is ejected at 100°C!

Most millipedes are relatively slow moving, and as well as having thick calcareous exoskeletons for protection, also carry a battery of repugnatorial glands (Fig. 13.10). The openings are located on the sides of tergal plates or on the margins of tergal lobes – usually a pair per segment, though they are entirely absent from some segments. The composition of the secretion varies with species but can include aldehydes, quinones, phenols and hydrogen cyanide. The HCN is liberated just before use when a precursor and enzyme are mixed from a two-chambered gland. This fluid, which is toxic or repellent to other small animals and from large tropical species seems to be caustic to human skin, is usually released slowly but some species can eject it as a high-pressure jet or spray for 10–30 cm. Here, ejection is probably caused by the contraction of trunk muscles adjacent to the secretory sac. The carnivorous and faster running centipedes are less well endowed with repugnatorial mechanisms. They rely more for protection on speed and the use of poison claws that are also concerned with the capture of prey (Fig. 8.21). Some species do, nevertheless, carry repugnatorial glands and some lithobiomorphs bear large numbers of unicellular repugnatorial glands on the last four pairs of legs, that they can kick out at an aggressor, throwing off adhesive droplets.

The pedicellariae of echinoderms, found in Asteroidea and Echinoidea, are also organs that have evolved specifically for defensive purposes. These are specialized, jaw-like appendages and are used for protection, especially against larvae that might settle on the body surface (Fig. 7.18). There are three main kinds: pedunculate (stalked), sessile (attached directly to the test) and alveolar (somewhat insunk). One of the several kinds of echinoid pedicellaria contains glands that secrete a poison capable of rapidly paralysing small animals and driving larger predators away. The avicularia zooids of some bryozoans (Section 6.3.3.3) have the same function as pedicellariae.

An interesting form of repulsion is by startling potential predators. Some lepidopterans and other insects have large markings on their wings which appear to be imitations of the vertebrate eye. These animals normally rest with 'eyes' concealed but suddenly expose the spots when disturbed. Another possible explanation for them is that they deflect predators to less vulnerable parts of the body or even to defence organs; for example, some wasps have white abdominal spots near their stings. Experimental work with captive birds has produced evidence for both functions of eye-spots. Flash coloration also serves a startling function and the rapid colour changes that occur in cuttlefishes after disturbance provide a particularly vivid example of this.

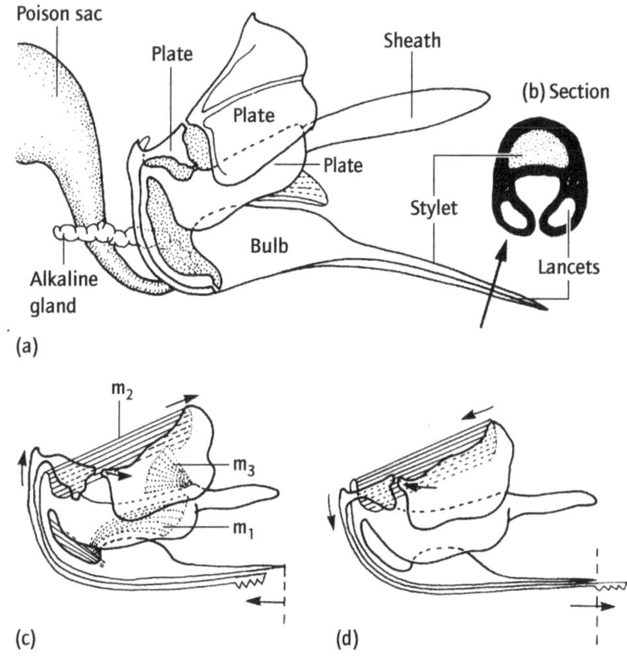

Fig. 13.9 Sting of bee, *Apis mellifera*. The shaft (stylet + barbed lancets) is depressed by contraction of m_1, then the contraction of the powerful m_2 muscles, running from quadrate plate to anterior part of the oblong plate, causes rotation of the triangular plate so as to push out the lancet. Retraction of the lancet is brought about by contraction of m_3. The muscles on the two sides of the sting work alternately, and by successive acts of protraction and retraction drive the lancets more deeply into the body of the victim. Poison is secreted by a pair of thread-like glands in the abdomen. Their secretion accumulates in the poison sac that opens at the base of the sting into the poison canal. The function of the so-called alkaline gland is uncertain. (After Imms, 1964.)

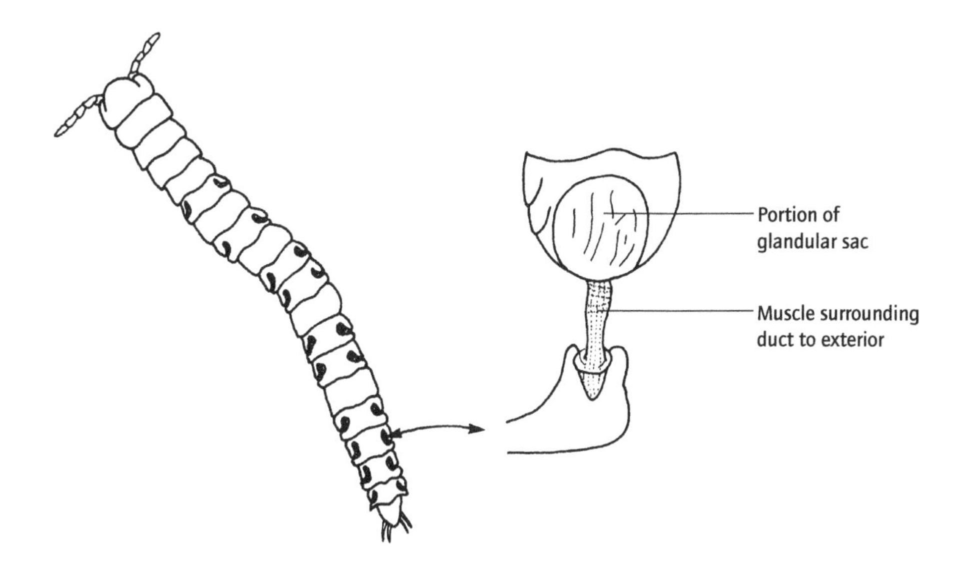

Fig. 13.10 Repugnatorial glands of a millipede (after Cloudsley-Thompson, 1958).

13.2.2 Against internal invaders

Some pathogens can penetrate outer defences but then all organisms have an inner line of defence of one form or another to combat this. In the vertebrates this is achieved by an immune system that involves antibodies capable of neutralizing specific foreign agents. Invertebrates do not have such a specific immunological system, but they do have an inner line of defence – based generally on phagocytic cells capable of recognizing and eliminating foreign material.

13.2.2.1 Recognition of self and non-self are fundamental requirements

A classic example of self-recognition is given by sponge cell reaggregation. By squeezing cells through cloth or by putting them in a solution of ethylene diamine tetraacetic acid (EDTA) it is possible to dissociate whole sponges into slurries of individual cells. Mixtures of cells from different species and even clones within the same species will reaggregate in species-specific and clone-specific fashion.

Grafting experiments also make the same point. Gorgonians (colonial anthozoans), for example, reject grafts from different species (xenografts) and from genetically different individuals of the same species (allografts) but accept grafts from different parts of the same colony (autografts) which consistently fuse.

13.2.2.2 Phagocytic or amoeboid cells are of general importance in invertebrate self-defence

All the reactions noted above were for colonial and encrusting organisms. These are often found in situations where living space is limited and competition for it is keen, so that evolution of self-recognition might have been favoured as a means of maintaining self-integrity. Hence these kinds of systems of self-recognition may be a peculiar consequence of this kind of evolutionary pressure and not a common basis for the evolution of an immunological system within the animal kingdom.

Immunological studies began when, in the early 1900s, Elie Metchnikoff introduced rose spines beneath the epidermis of bipinnaria larvae of starfish and found that within a short time these were 'attacked' by amoeboid cells. He obtained similar

Table 13.2 The invertebrate phyla in which it has been recorded that amoeboid cells remove foreign material.

		Response	
Animal phylum	**Particle or substance injected**	**Phagocytosis**	**Encapsulation**
Porifera	India ink, carmine	+	
	Erythrocytes	+	
	Trematode redia, cercaria		+
Annelida	India ink, carmine, iron	+	
	particles, erythrocytes	+	
	Foreign spermatozoa	+	
Sipuncula	Latex beads, bacteria	+	
Mollusca	Carmine	+	
	India ink	+	
	Erythrocytes, yeast, bacteria	+	
	Thorium dioxide	+	
Crustacea	Bacteria, carmine	+	
Uniramia	Bacteria	+	
	Latex beads	+	+
	Iron, saccharide	+	
	Araldite implants		+
	Bacillus thuringiensis	+	
	India ink, carmine	+	
	Erythrocytes, bacteria	+	
Echinodermata	Bovine serum albumin	+	
	Bovine gamma globulin	+	
	Sea urchin cells (into a sea star)	+	
Urochordata	Carmine	+	
	Glass fragments		+
	Trypan blue	+	
	Thorium dioxide	+	

results when he injected anthrax bacilli into the larvae of the rhinoceros beetle, *Oryctes nasicornis*. From observations such as these, Metchnikoff proposed the idea that amoeboid cells, which are involved in intracellular digestion in many primitive invertebrates (Section 9.2.2), had been retained in the evolution of more advanced animals as an inner defence system. Phagocyte cells are certainly widespread throughout invertebrate animals and experiments involving the introduction of foreign material into living animals have indicated that they are capable of removing a variety of foreign particles (Table 13.2).

13.2.2.3 How do phagocytes discriminate between self and non-self?
Wandering phagocytes must be able to 'ignore' normal tissue of *self* but engulf *non-self* particles. They may also be involved in the removal of damaged 'self' due to the presence of xenobiotic, tissue-damaging substances such as pollutants.

Knowledge of the mechanism is limited. A priori it would seem likely that recognition occurs when a phagocyte makes contact with its target and that since foreign particles are unlikely to produce specific 'kill' signals, it is more likely that self cells produce specific 'don't kill' signals. There is no evidence for an intermediate antibody system as subtle as in the vertebrates, but there is evidence for opsonin (molecules that coat foreign particles so that they adhere to phagocytes and facilitate their phagocytosis) activity in the body fluids of those invertebrates that have fluid-filled cavities. Thus, amoebocytes from the chelicerate horseshoe crabs exhibit no significant bactericidal effect in the absence of serum, but killed *Escherichia* when serum was present. Similarly, phagocytosis of human erythrocytes by haemocytes (blood amoebocytes) from the octopus *Eledone cirrosa* occurred only after they had been exposed to *Eledone* serum. Extracts from many invertebrates act as agglutinins – cross-linking and binding various cells and bacteria *in vitro* (Table 13.2) – and these might have opsonic properties, binding foreign particles to the surfaces of phagocytes. These mechanisms are discussed further in Coombe *et al.* (1984).

Models showing how phagocytes might effect self-recognition are summarized in Fig. 13.11. Direct self-recognition (Fig. 13.11a) probably occurs in the phagocytes that 'patrol' the tissues of solid-bodied invertebrates, whereas the involvement of intermediate factors, opsonins (b), can occur in invertebrates with fluid-filled cavities – the fluids containing the opsonins.

13.2.2.4 Reproduction presents some complications
Where there is internal fertilization (Chapter 14), sperm with a foreign genotype are transferred to the tissues of another organism. Similarly, fertilized eggs and embryos, when they reside within the tissues of a mother (Chapter 15), are genetically semi-foreign. However, under normal circumstances these must not be destroyed by the immunological system of the parent – the 'mother' must be a willing host for the sperm or offspring. Just how the reproductive cells evade immunological destruction is not understood, but the following experiments are illuminating.

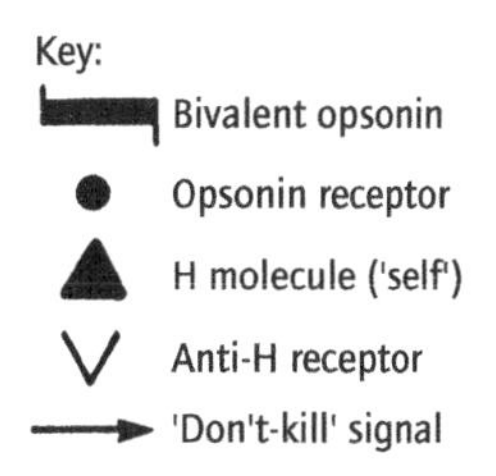

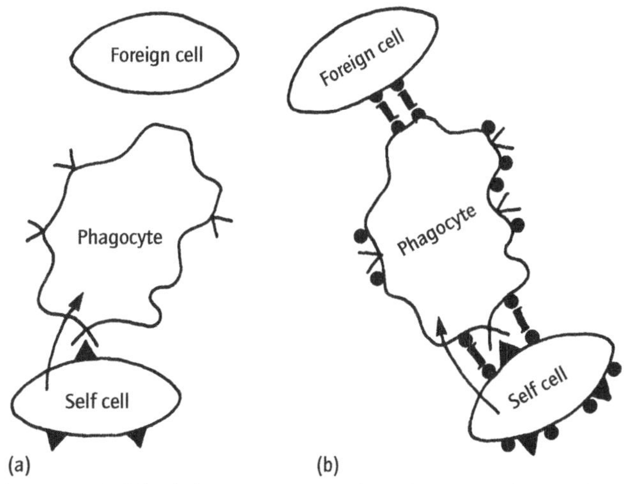

Fig. 13.11 Models of phagocyte immunological systems of invertebrates – see text for further explanation (after Coombe *et al.*, 1984).

Spermatozoa from allogenic donors are not phagocytosed when injected into the coelomic cavity of earthworms, whereas mammalian spermatozoa and spermatozoa from other species of earthworms are phagocytosed. Male sipunculan worms failed to encapsulate homologous eggs when they were injected into their coeloms, and this suggests that propagules have some generally evasive mechanism to deal with host immune systems, for eggs do not naturally occur within the male worm! Eggs damaged by staining, heating or sonication were rapidly encapsulated whereas frozen eggs, though apparently dead, were not. For more information, see Coombe *et al.* (1984).

13.2.2.5 Evasion of host response
Successful parasites have to be able to defend *themselves* against the host immune response. This has been carefully studied in some trematodes where a variety of defences has been discovered. Thus schistosomes coat themselves with antigens identical or similar to those of the host, these either being synthesized by the parasite (molecular mimicry) or derived from the host and bound to the surface of the parasite. Whatever their origin, these antigens mask parasite antigens so they are no longer recognized as foreign. Fasciolid flukes, on the other hand, produce substances that are toxic to lymphocytes and other immunological cells. Also, the glycocalyx of the parasite tegument appears to turn over at a high rate, and by this means flukes may be able to slough off host antibody.

13.2.3 Responses to stressors

Environmental stressors and the responses that they provoke are so varied that it is not possible to treat them all in a comprehensive way. Some 'physiological defences', e.g. to stress from oxygen, salinity, have already been treated in previous chapters (e.g. Sections 11.6.5 and 12.2). Here we describe some general responses that are elicited in invertebrates by stress generally, and by two groups of pollutants specifically: xenobiotics (organic toxicants) and heavy metals.

13.2.3.1 Heat shock proteins

A number of different stresses, including excessive heat, exposure to toxins and low O_2, can act to destabilize the structure of protein molecules. Many organisms in response to such stresses produce a special class of proteins that act as molecular chaperones within cells. These proteins (termed heat shock proteins or *hsps* – but only because they were originally investigated using heat stress) bind to other proteins that are damaged and/or non-functioning. Heat shock proteins either (a) help them (re)attain their native state or (b) minimize the accumulation of non-functioning or toxic aggregations of protein molecules.

Nearly all organisms investigated have genes that encode and express *hsps*. These molecules are very highly conserved. They are commonly assigned to 'families' on the basis of their molecular weight, structure and function (e.g. *hsp110*, *hsp70*).

Such molecular chaperones play numerous roles in the unstressed cell, but they are perhaps best known for being induced by, or coping with, almost every type of stress studied to date (see also Section 13.2.4). It appears that different species have different thresholds for *hsp* expression. Generally thresholds can be correlated with the levels of stress experienced in nature. For example, the cold-water mussel *Mytilus trossulus* has a lower threshold for *hsp70* expression than the closely-related warm-water species *M. galloprovincialis*. Similarly, the fruitfly *Drosophila melanogaster* has a higher threshold for *hsp70* expression than a related species *D. ambigua* which has a more northern distribution. Even modest increases in temperature (1–2°C) can induce *hsp* production (*hsp70*) in some tropical coral species such as *Goniopora djiboutiensis*. Many insect species express *hsps* in response to cold shock or while overwintering in diapause. Not only does *hsp* expression vary between species but it can also vary within a species. Furthermore this variation can be correlated with resistance to stress. Centipedes collected from the immediate vicinity of a smelter had greater levels of *hsp70* than individuals of the same species collected from an uncontaminated site. Individuals of *Drosophila*, derived from a single wild population, showed variation in *hsp70* expression that could be correlated with temperature tolerance and was heritable.

There is evidence that the induction of *hsps* is a costly process. When blue mussels *Mytilus edulis* were exposed to a range of copper concentrations (0–100 μg l^{-1}) for 7 days, there was found to be a positive relationship between external metal dose and the amount of *hsp60* present in tissues (Fig. 13.12a). What is more, for individuals exposed to >32 μg l^{-1} copper there was a decrease in scope for growth (Fig. 13.12b). At even higher concentrations the value was actually negative. Scope for growth is an estimate of the energy available for growth (or reproduction) after the energy excreted and lost through respiration is taken into account. In many cases it is a good measure of the physiological condition of animals. The negative scope for growth accompanying pronounced *hsp60* accumulation in *M. edulis* indicates that at these high copper concentrations the mussels are not even generating enough energy to meet normal metabolic demands.

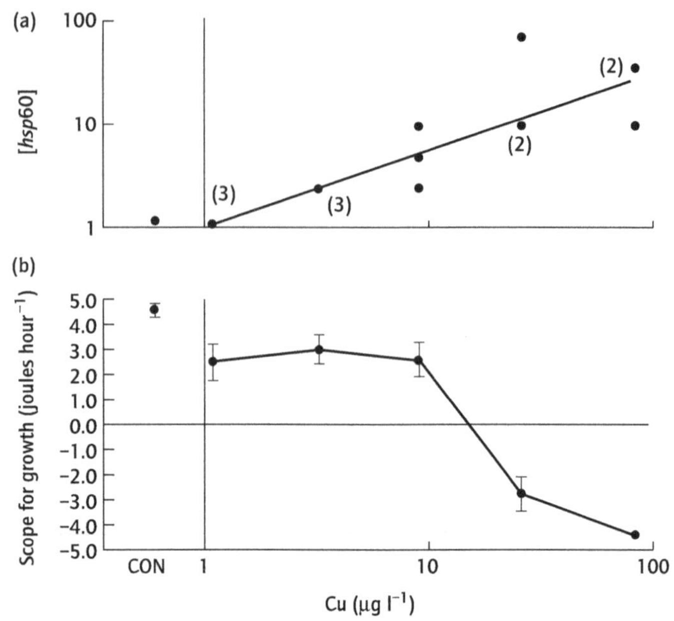

Fig. 13.12 Relative *hsp60* concentrations in mantle tissue (a) and scope for growth (b) expressed as a function of the copper concentration blue mussels were kept at for 7 days. CON = controls. Scope for growth values are given as means ±1 standard error.

While obviously important in stress responses, it should be kept in mind that *hsps* are still only one of many molecular mechanisms of stress tolerance.

13.2.3.2 Mixed function oxygenase (MFO) and xenobiotics

Organic pollutants, such as hydrocarbon compounds from oil spills, can penetrate the tissues of marine invertebrates. They are lipophilic and as such not easily metabolized. Instead they can accumulate in lipid depots and the lipid components of cell membranes until they reach concentrations at which they cause biochemical problems. However, some marine invertebrates, namely polychaetes and some molluscs and crustaceans, contain an enzyme system capable of oxidizing the toxicant by literally adding oxygen atoms, making it more hydrophilic and thus more easily metabolized. The system is fairly non-specific in terms of the substrates that it will attack. It consists of several

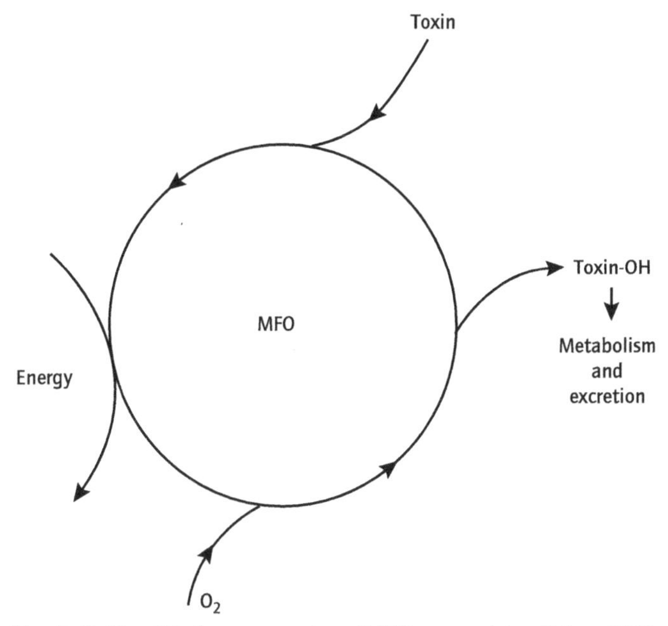

Fig. 13.13 Simplified representation of MFO system (after Calow, 1985).

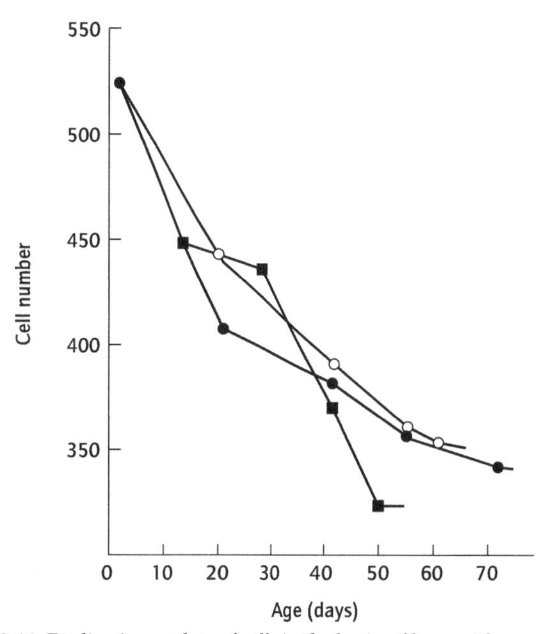

Fig. 13.14 Decline in number of cells in the brain of bees with age (after Rockstein, 1950).

enzymes, and some cytochromes are present. It is primarily membrane-bound in the microsomal fraction of the endoplasmic reticulum of certain tissues (e.g. digestive gland or hepatopancreas). The process of oxygenation is energy expensive and is summarized in a very general way in Fig. 13.13; it involves expoxidation, hydroxylation and dealkylation. It is also an inducible system, i.e. the enzymes associated with it are being produced only when a xenobiotic challenge exists; for example, a cytochrome specifically associated with MFO increases in the tissue of mussels after 1 day of exposure to diesel oil and returns to control concentrations after 8 days' recovery.

Mixed function oxygenases also occur in herbivorous insects, where they are probably involved in dealing with natural organic toxins produced by plants as a defence against these herbivores. For example, polyphagous insects generally have higher MFO activities than stenophagous (more specialist) ones, probably because they are exposed to a wider variety of toxic compounds such as phenolics, quinones, terpenoids and alkaloids.

13.2.3.3 *Metallothioneins*

Heavy metals, such as mercury, cadmium, copper, silver and tin, can be extremely toxic to aquatic invertebrates. For example, they cause denaturation of enzymes by interacting with them and altering their tertiary configuration. However, many invertebrates can detoxify heavy metals by binding them to specialized proteins called metallothioneins. These are low molecular weight compounds, rich in sulphydryl (SH) groups due to high levels of the amino acid cysteine within them. The SH group is capable of combining with or chelating the metal and rendering it less toxic.

13.2.4 Repair–protection against ageing?

In Section 13.1.3 it was hinted that ageing of whole organisms is due to the accumulation of suborganismic damage. Evidence for this is as follows.

Tissue disruption – ageing in dipteran insects is associated with a declining flight capacity and this is correlated with degenerative changes in the structure of the flight muscle. Cell numbers in the brain of worker bees have been shown to decline from a mean of 522 at eclosion to 350 at 10 weeks (Fig. 13.14).

Lipofuscin (known as age-pigment) – probably a product of lipid peroxidation and derived from the breakdown of membranes – has been found to accumulate with age in the tissues of nematodes and insects (Fig. 13.15).

Enzyme fidelity – there is evidence, particularly from studies of nematodes, that enzyme structure and function become impaired with age: enzymes become more sensitive to heat denaturation (indicative of changes in molecular organization), develop different immunological properties and have reduced catalytic capacities. However, these observations do not apply to all the enzymes that have been studied in nematodes, nor to many enzymes studied from other animals.

The root cause of all this damage is likely to emanate from molecular processes such as thermal noise, mistakes being made in the synthesis of proteins and a variety of other processes. At the same time, damaged protein molecules can, in principle, be replaced according to genetic instructions and whole cells can be replaced, again according to genetic instruction, by mitosis (Chapter 1). Indeed evidence from work on nematodes suggests that proteins containing abnormal amino acids have a more

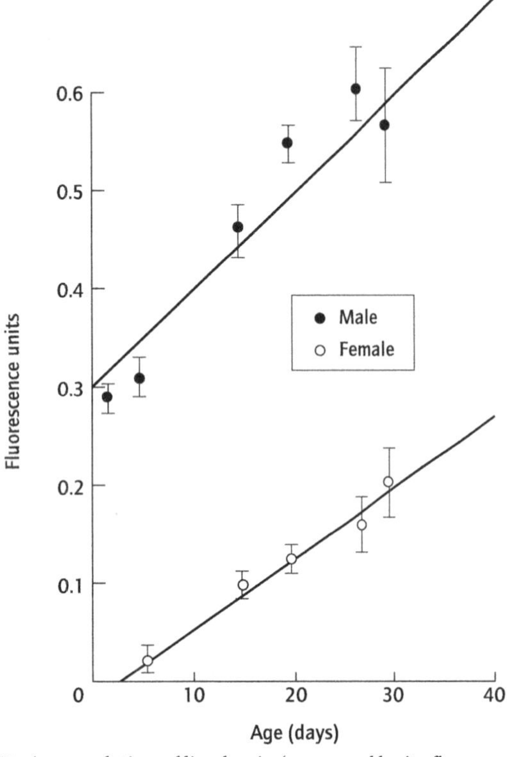

Fig. 13.15 Accumulation of lipofuscin (measured by its fluorescence) in *Drosophila melanogaster* (after Biscardi & Webster, 1977).

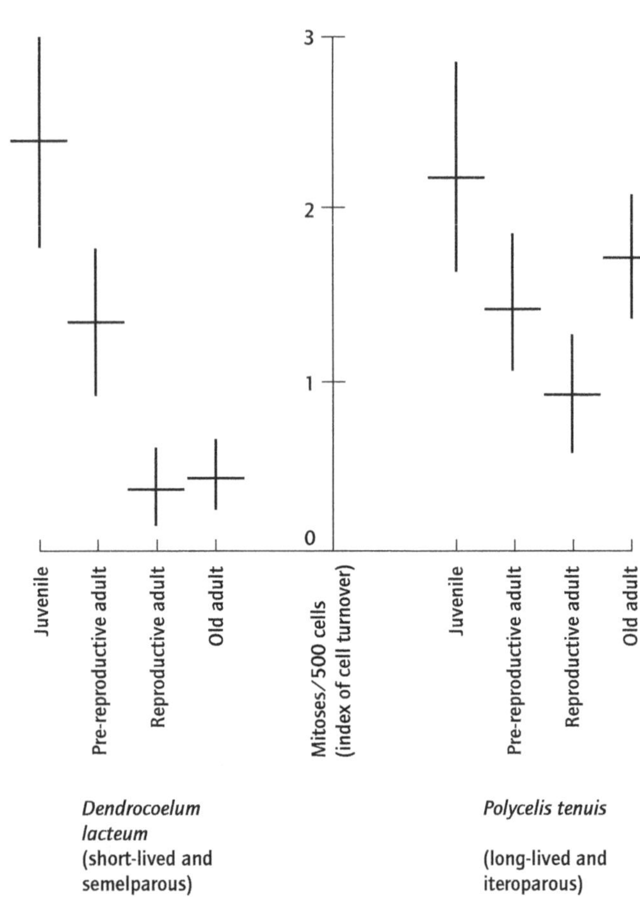

Fig. 13.16 Cell turnover at different stages in the life cycles of two triclads, one long-lived, the other short-lived. The latter invests considerably more in reproduction than the former. In both, cell division reduces from juvenile to adult. However, whereas there is a big further dip in this, following the onset of reproduction in the short-lived form, there is a smaller, non-significant dip in the long-lived form. It is thought that reduced turnover is *caused* by high investment in reproduction and shortens life by acceleration of senescence. (After Calow & Read, 1986.)

rapid turnover than normal proteins, but that this slows down as worms get older. Cell division occurs extensively in the Cnidaria, Platyhelminthes, Annelida and Mollusca but is very restricted in adult nematodes and insects. In freshwater triclad turbellarians there is a reduction in cell turnover with age (Fig. 13.16), which is accentuated with the onset of reproduction in semelparous forms (see Section 14.5). The latter is associated with accelerated ageing. The age-state, vitality, of an organism might therefore depend upon a balance between the generation of damage and its replacement or repair. Not surprisingly, therefore, ageing processes are least obvious in organisms such as cnidarians where tissue turnover is continuous and are most obvious in organisms such as nematodes and insects, where tissue turnover is limited (Table 13.3). Interestingly, mitosis is more extensive and persists for longer in those insects with the greatest longevity, namely coleopterans.

More recently, it has been suggested that heat shock proteins (see Section 13.2.3.1) are able to modulate senescence. It

Table 13.3 Distribution of senescence.

Animal phylum	Suspected lack of ageing in some species	Suspected presence of ageing in some species	Definite ageing in some species
Cnidaria	+	+	
Platyhelminthes	+	+	+
Mollusca	+	+	
Nematoda			+
Annelida	+	+	+
Rotifera			+
Arthropod phyla			+

is thought that proteins get converted to non-functioning states in the cells of ageing animals, where their accumulation may contribute to age-dependent morbidity and mortality. The expression of heat shock proteins has been shown to mitigate such damage, at least in fruit flies and nematode worms.

13.3 Conclusions

The threats to which organisms are exposed are diverse, as, not surprisingly, are the responses that they provoke. Yet it is possible to produce a coherent classification of threats that puts them on a continuum, varying from more to less intimacy of association with the organism that is threatened. Similarly, behind the diversity of defence responses it is possible to perceive some common features. For example, defence mechanisms, whether they be against predators, parasites, microbes, or wear and tear, are all expensive in material and energy. Harvell (1990) measured lifetime costs of defence in the marine bryozoan *Membranipora membranacea*. Here colonies grow spines in the presence of their predators – nudibranchs – and extracts from them; these defences are said to be inducible and such responses are common in other invertebrates as well. Growth rates were reduced and senescence accelerated in colonies that were caused to defend by the regular application of the predator extract. These kinds of costs lead to thoughts about optimal investment in defence mechanisms (of all kinds) that have to be traded off against other components of Darwinian fitness (Sibly & Calow, 1989). They also make sense of the evolution of inducible defences as a way of minimizing costs.

It is these general principles that readers should aim to take away from this and, indeed, all the chapters in this part of the book.

13.4 Further reading

Arking, R. 1991. *Biology of Aging: Observations and Principles*. Prentice-Hall, Englewood Cliffs, New Jersey.

Cooper, E.L. (Ed.) 1996. *Invertebrate Immune Responses*. Springer-Verlag, New York.

Berry, R.J. 1977. *Inheritance and Natural Selection. New Naturalist*, No. 61. William Collins & Co., Glasgow.

Blest, A.D. 1957. The function of eyespot patterns in the Lepidoptera. *Behaviour*, **11**, 209–256.

Coombe, D.R., Ey, P.L. & Jenkin, C.R. 1984. Self/non-self recognition. *Q. Rev. Biol.*, **59**, 231–255.

Davies, I. 1983. *Ageing*. Edward Arnold, London.

Dunn, P.E. 1990. Humoral immunity in insects. *BioScience*, **40**, 738.

Eisner, T., Van Tassell, E. & Carrel, J.E. 1967. Defensive use of a 'faecal shield' by a beetle larva. *Science, N.Y.*, **158**, 1471–1473.

Erichsen, J.T., Krebs, J.R. & Houston, A.I. 1980. Optimal foraging and cryptic prey. *J. Anim. Ecol.*, **49**, 271–276.

Esch, G.W. & Fernandez, J. 1992. *Functional Biology of Parasitism: Ecological and Evolutionary Implications*. Chapman & Hall, New York.

Fainzilber, M., Napchi, I., Gordon, D. & Zlotkin, D. 1994. Marine warning via peptide toxin. *Nature*, **369**, 192.

Feder, M.E. & Hofmann, G.E. 1999. Heat-shock proteins, molecular chaperones, and the stress response: Evolutionary and ecological physiology. *Annu. Rev. Physiol.*, **61**, 243–282.

Klaassen, C.D., Liu, J. & Choudhuri, S. 1999. Metallothionein: An intracellular protein to protect against cadmium toxicity. *Annu. Rev. Pharmacol. Toxicol.*, **39**, 267–294.

Finch, C.E. 1990. *Longevity, Senescence and the Genome*. University of Chicago Press, Chicago.

Gliwicz, M.Z. 1986. Predation and the evolution of vertical migration in zooplankton. *Nature (London)*, **320**, 746–748.

Hansell, M.H. 1984. *Animal Architecture and Building Behaviour*. Longman, London.

Harvell, C.D. 1990. The ecology and evolution of inducible defenses. *Q. Rev. Biol.*, **65**, 323–340.

Livingstone, D.R., Moore, M.N., Lowe, D.M., Nasci, C. & Farrar, S.V. 1985. Responses of the cytochrome P-450 monoxygenase system to diesel oil in the common mussel, *Mytilus edulis* L., and the periwinkle, *Littorina littorea* L. *Aquat. Toxicol.*, **7**, 79–81.

McClintock, J.B. & Janssen, J. 1990. Pteropod abduction as a chemical defence in a pelagic Antarctic amphipod. *Nature*, **346**, 462–464.

McClintock, J.B. & Baker, B.J. 1997. A review of the chemical ecology of Antarctic marine invertebrates. *Amer. Zool.*, **32**, 329–342.

Neill, W.E. 1990. Induced vertical migration in copepods as a defense against invertebrate predation. *Nature*, **345**, 524.

Nordlund, D.A. & Lewis, W.J. 1976. Terminology of chemical releasing stimuli in intraspecific and interspecific interactions. *J. Chem. Ecol.*, **2**, 211–220.

Parker, A.R. 1998. The diversity and implications of animal structural colours. *J. exp. Biol.*, **201**, 2343–2347.

Rainbow, P.S. & Dallinger, R. (Eds) 1993. *Ecotoxicology of Metals in Invertebrates*. Lewis Publishers, Boca Raton.

Rockstein, M. 1950. The relation of cholinesterase activity to change in cell number with age in the brain of the adult worker bee. *J. cell. comp. Physiol.*, **35**, 11–23.

Rutherford, S.L. & Lindquist, S. 1998. Hsp90 as a capacitor for morphological evolution. *Nature*, **396**, 336–342.

Sanders, B.M., Martin, L.S., Nelson, W.G., Phelps, D.K. & Welch, W. 1991. Relationships between accumulation of a 60 kDa stress protein and scope for growth in *Mytilus edulis* exposed to a range of copper concentrations. *Mar. environ. Res.*, **31**, 81–97.

Schmidt-Nielsen, K. 1997. *Animal Physiology. Adaptation and Environment*, 5th edn. Cambridge University Press, Cambridge.

Sibly, R.M. & Calow, P. 1989. A life-cycle theory of responses to stress. In: Calow, P. & Berry, R.J. (Eds) *Evolution, Ecology and Environmental Stress*, pp. 101–116. Academic Press, London.

Tatar, M. 1999. Evolution of senescence: Longevity and the expression of heat shock proteins. *Amer. Zool.*, **39**, 920–927.

Theodor, J.L. 1976. Histo-incompatibility in a natural population of gorgonians. *Zool. J. Linn. Soc.*, **58**, 173–176.
Turner, J.R.G. 1984. Darwin's coffin and Dr. Pangloss – do adaptationist models explain mimicry? In: Shorrocks B. (Ed.) *Evolutionary Ecology*, pp. 313–361. Blackwell Scientific Publications. Oxford.
Turon, X., Becerro, M.A. & Uriz, M.J. 1996. Seasonal patterns of toxicity in benthic invertebrates: The encrusting sponge *Crambe crambe* (Poecilosclerida). *Oikos*, **75**, 33–40.
Willmer, P., Stone, G. & Johnston, I. 2000. *Environmental Physiology of Animals*. Blackwell Science, Oxford.

CHAPTER 14

Reproduction and Life Cycles

The creation of new individuals is a fundamental property of living things. Two aspects of the process can be recognized: the setting aside by adult organisms of material for the purpose of reproduction and the development of new individuals from these materials. At some stage in the life cycle of almost all, but not quite all, animals these two processes involve the production of haploid gametes – eggs or spermatozoa. Fusion of the gametes creates a zygote (fertilized egg) and through subsequent development (see Chapter 15) the zygote becomes a fully differentiated, spatially complex, multicellular organism, like, but not identical, to the parent animals. A single adult of course can set aside materials for a large number of potential offspring and, when the offspring arise from the fusion of haploid gametes resulting from meiosis, each one will have its own unique genetic constitution. This process is called sexual reproduction.

This chapter will be concerned with both sexual and asexual reproduction in a wide range of invertebrate animals. Chapter 15 will deal with the processes of development and differentiation by which the zygote formed as a consequence of sexual reproduction becomes a new individual. The two aspects of reproduction are of course closely interlinked. Sexual reproduction in animals always involves the fusion of small mobile gametes – the spermatozoa and larger, cytologically rich, non-motile gametes – the eggs or ova. Despite this uniformity our survey will also show that animals have evolved a great diversity of other means of propagation (as have plants) that do not involve sexual reproduction via gametes. One of the great challenges of evolutionary theory is to understand why the process of sexual reproduction is so dominant.

We will also see that although the life cycles of animals are similar in the sense that each individual develops from a single fertilized egg, there is tremendous variation in the pattern of allocation of resources to egg production. We will look briefly at the apparently simple questions – 'when, where and by how much to reproduce?' These prove to be far from simple questions to answer, and lie at the heart of evolutionary theory.

In order to understand this diversity, we will also review its control, and the adaptive significance of the different variants, including the selective advantages of sexual reproduction itself. It will become apparent in the following pages that different combinations of reproductive and life-cycle traits are characteristic of different phylogenetic lines of animals, grades of animal organization, and ecological situations. This will make it possible to discuss the richness of invertebrate reproductive strategies and the contribution this has made to current theories of life-cycle evolution.

14.1 Introduction

In Chapter 1, it was suggested that evolution is a natural consequence of systems that persist by a semi-conservative process of replication. All living things possess in common the same type of genetic programme, variation in which arises from mis-copying of base sequences during replication of the DNA molecule. The genome projects are now establishing complete genomic sequences for a range of organisms including the insect *Drosophila* and the nematode *Caenorhabditis* to complement the human genome project. The entire chromosome sequences are being established to reveal the complexity of gene structure and the degree of duplication that has occurred during evolutionary time. These astounding projects confirm the commonality of genetic structure in widely diverged organisms. The evolution of multicellular organisms, however, requires not just the semi-conservative replication of the genome and construction of differential cells, but the replication of organisms, a process which we call 'reproduction'. In the vast majority of organisms reproduction allows genetic recombination. Several systems of sexuality are possible, but in all of them there is a mechanism for the exchange of genetic material between parental types. In animals this requires the exchange of genetic information derived from the two parents. The rise of molecular biology has revealed just how powerful and complex the process of meiosis is for the generation of novelty. The ultimate source of novelty is the process of mutation. This occurs when the information encoded within the DNA base sequence is changed by (i) a substitution of a base in the sequence, (ii) duplication of a sequence of bases – which may be extremely important for the subsequent evolution of diversity – or (iii) deletion. Over evolutionary time the genetic material has increased in complexity. The size of the genomes for a number of organisms is now being established by the various genome projects. One of the organisms chosen for this vast undertaking is the fruit fly *Drosophila melanogaster* and it has been estimated to have a genome of some 1.2×10^8 base pairs and to have in the order of 10 000 genes.

All of this information may be re-sorted (though certainly not at random) during meiosis and the re-sorted information from two individuals is recombined when gametes fuse during the process of fertilization. Thus sexual reproduction creates a new and unique genome each time it occurs. The capacity for novelty generation is enormous. The number of independent maternal/paternal chromosome combinations for a diploid animal is a function of the number of chromosomes i.e. 2^n where n is the haploid chromosome number. The haploid number (n) may be small (4 in *Drosophila melanogaster*) but is more often in the region of 20–30 chromosome pairs. In addition to this source of diversity, the recombination of genetic material through meiosis in incalculable.

Genomes are highly structured linear sequences of base pairs that encode sequences that appear in transcription products (these sequences are termed 'exons') and in addition sequences that do not appear in the mRNA transcripts ('introns') (see also Chapter 15). Meiosis is thus a means of creating enormous novelty and diversity through new assortments of genes, new combinations of regulatory genes and of regulatory components within the DNA sequence as well as new arrangements of base pairs within the protein encoding sequences.

Although sexual reproduction is characteristic of eukaryotes, genetic exchange also occurs in bacteria (by conjugation, for instance) and among viruses as for instance when two different strains infect a single host. Three different systems which allow genetic recombination are illustrated diagrammatically in Box 14.1. All multicellular eukaryotes have adopted the third of the systems illustrated in Box 14.1, i.e. they exhibit sexual reproduction involving meiosis and fertilization.

The evolution of sexual reproduction as it occurs in animals has involved a number of independent steps. If we suppose as an original state one in which there was no sexual reproduction and no genetic recombination, the evolutionary sequence has involved:

1 The acquisition of mechanisms for limited recombination.
2 Meiosis.
3 The evolution of separate mating types (usually two).
4 Anisogamy – the adoption of small mobile (male) gametes and larger immobile (female) gametes.
5 The appearance of gender, i.e. the separation of male and female functions in different individuals.

Understanding the evolutionary processes that have resulted in the universal adoption of sexual reproduction is one of the great intellectual challenges of biology. The study of the invertebrates has a special role to play because of the diversity of expressions of sexuality to be found among them. Figure 14.1 represents a possible outline of key steps in the evolution of continuous obligate sexuality. Living invertebrates provide examples that span the adaptive shifts from the acquisition of multicellularity to continuous obligate sexuality and the involvement of parental

Box 14.1 Systems for sexual reproduction (Myxis)

Sexual reproduction is a system of self-replication which permits the exchange of genetic information (embodied in a DNA molecule) and the creation of an individual whose genetic information is derived by crossing over from different parental sequences.

1 Viruses

(a) The life cycle of bacteriophage viruses. Infection of a bacterial host by injecting their DNA core. The protein coat of the virus remains outside the cell. Inside the bacterium the viral DNA molecules provide the information for the creation of new protein coats and the lysis of the host.

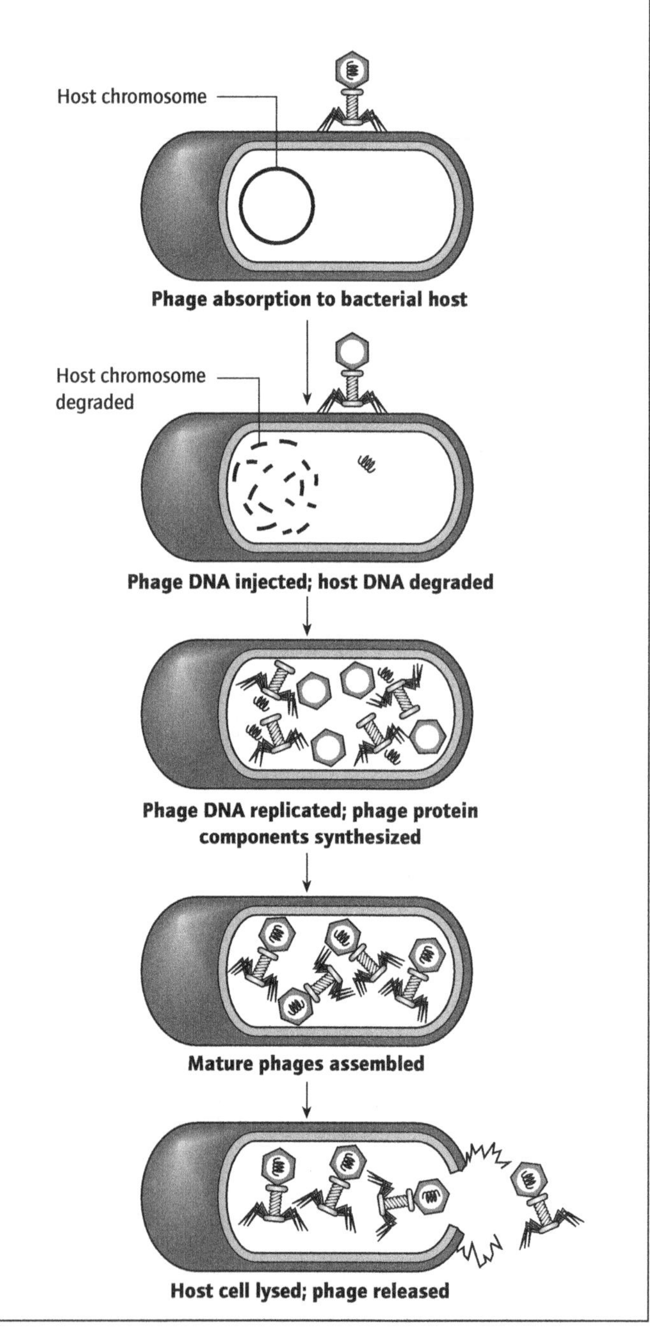

Continued

Box 14.1 *(cont'd)*

(b) Genetic exchange and recombination.

(i) A representation to show two genes A and B and their protein products.

(ii) Mixed infections of different mutants for the same structural genes (A or B) are inactive in certain strains of bacterium, but mixed infections for different structural genes are active.

(iii) The mutant virus particles are able to grow in strain k of the host bacterium.

(iv) Genetic crossing over can occur from mixed infection of the non-complementary type in strain k, and these new viral particles can infect strain B, i.e. sexual reproduction has occurred.

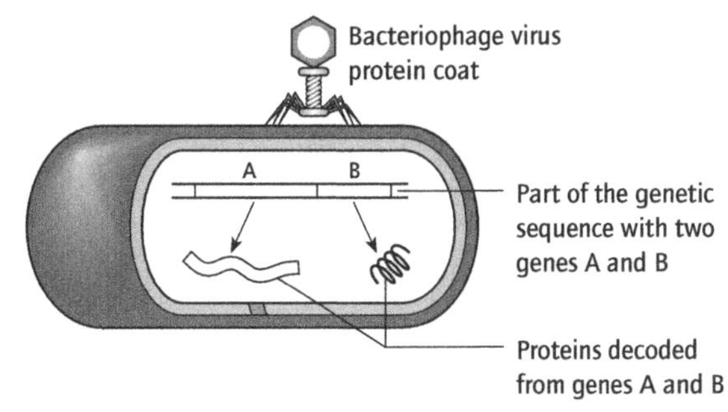

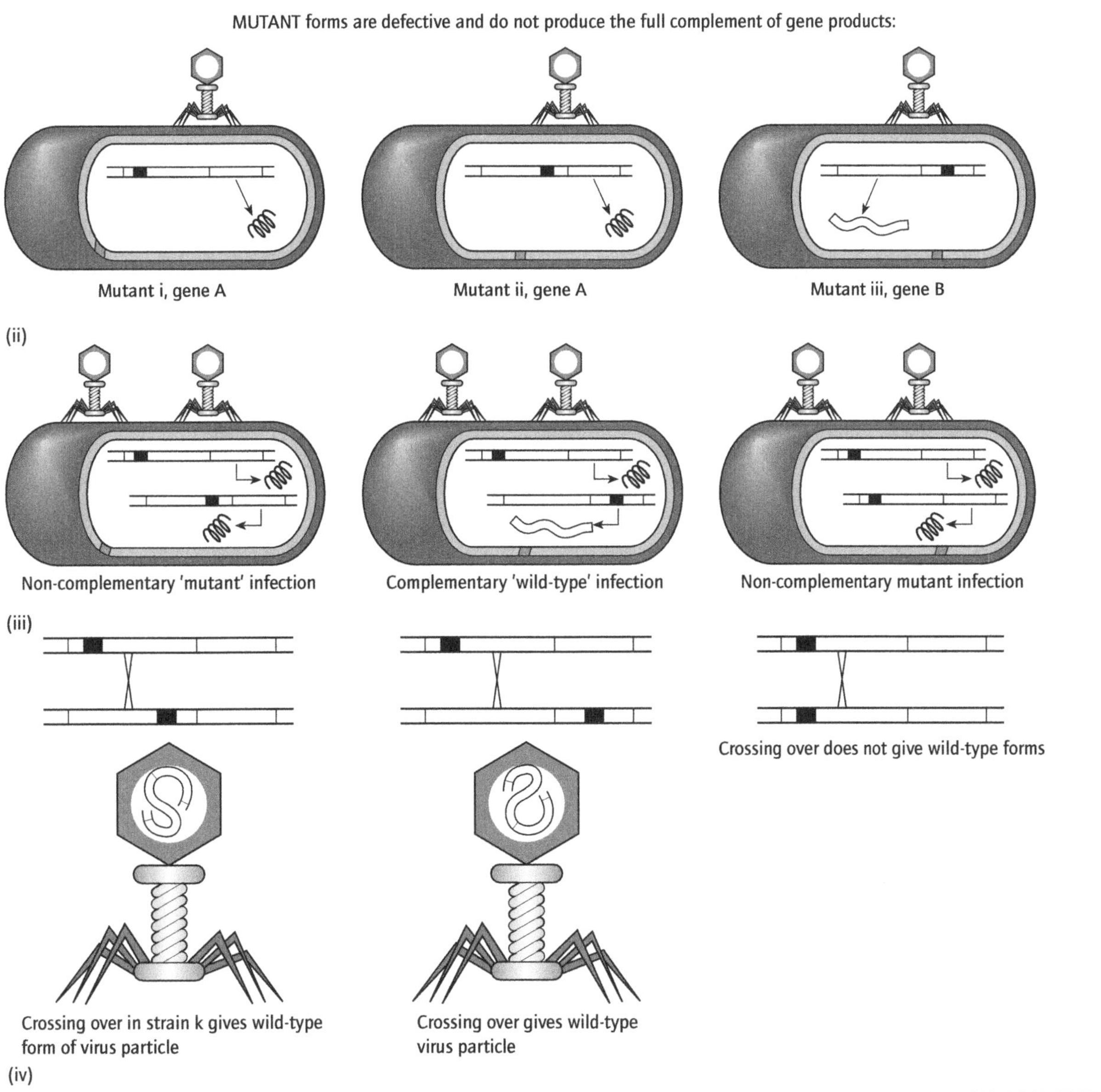

Continued p. 320

Box 14.1 *(cont'd)*

2 Bacteria

(i) Different strains of bacteria are able to conjugate. During the conjugation process an attachment stalk or pilus is formed and the DNA molecule (bacterial chromosome) from the donor (male-like cell) is passed to the recipient.

(ii), (iii) The length of DNA exchanged is time-dependent and can be interrupted by vigorous shaking of a culture.

(iv) Genetic exchanges can occur between the duplicated lengths of chromosome material. In this way bacteria can be produced which combine the genetic characteristics of the two parental types.

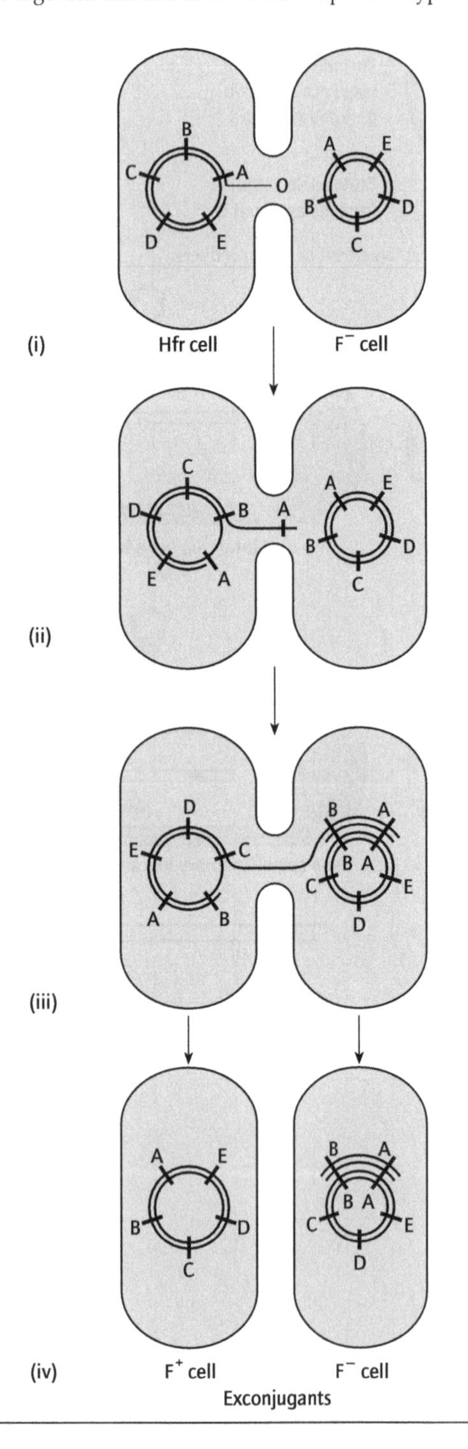

3 Sexual reproduction in higher organisms

Diploid cells have two sets of chromosomes, each containing coded genetic information from one parent. The diploid cell will give rise to haploid gametes by a special cell division called 'meiosis'. The genetic information in the haploid germ cells may be identical to that of the parent or may be different because of crossing over.

Haploid germ cells from different parents (or sometimes form one parent) fuse (a process called 'fertilization') and form a diploid zygote. The genetic constitution of the zygote is not the same as that of the diploid cells which give rise to the haploid gametes. It combines genetic traits from the two parental genomes in each set of chromosomes. The number of unique material/paternal combinations is 2^n where n is the haploid number and the potential for genetic reassessment is incalculable (see text).

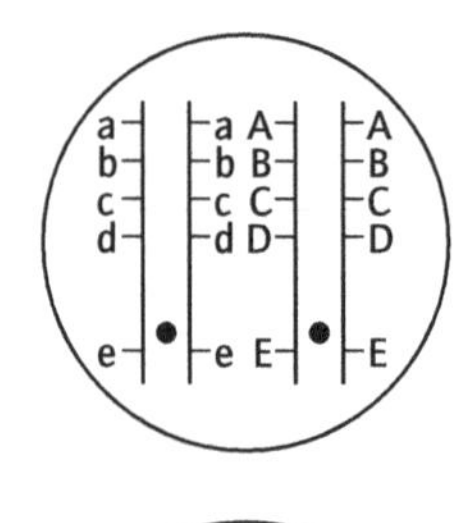

A diploid animal cell. The genes represented by the sequence a–e and A–E occur on chomosomes inherited from two parents. This cell is heterozygous

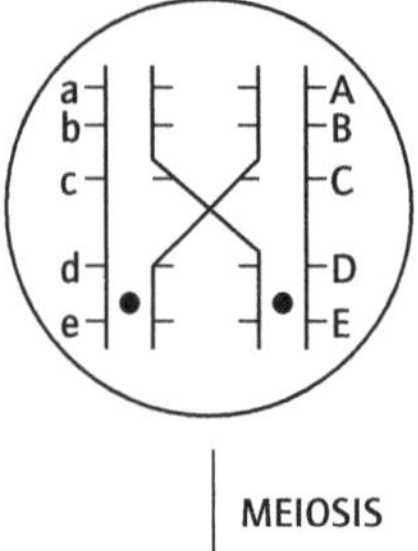

During MEIOSIS genetic exchange (crossing over) can occur between chromatids of the paired chromosomes

MEIOSIS

The gamete may have parental or recombinant sequences of genes

GERM CELLS

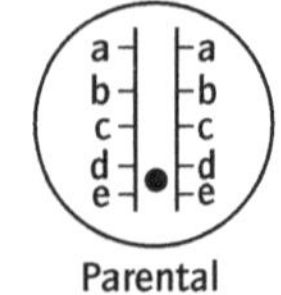

Parental Recombinant Recombinant Parental

The frequency of crossing over is a function of the distance between the genes in the DNA sequence of the chromosome

care. It is interesting that many invertebrate phyla have retained a degree of plasticity and show frequent reversals to supposedly more primitive states of sexuality.

In animals, two very different types of germ cells are usually produced, the spermatozoan and the egg (Section 14.4.2). These are often derived from specialized cells set aside early in embryonic development, but this is not universal and the germ cells are sometimes derived from de-differentiated somatic cells.

The early segregation of the germ cell lineage is an important factor. It ensures that only those genetic sequences that have survived a complete life cycle are included in the new genetic combinations of the new zygote. This new genetic constitution is set aside in the germ line and will not contribute to future generations unless the developing organism that carries it survives to reproduce. In other words 'the adult organism tests the efficacy of the gene combinations it has set aside for the germ cells'. Somatic mutations are excluded. Sexual reproduction therefore requires a life cycle with a periodic return to the single-cell undifferentiated form.

Asexual reproduction does not have this constraint and may involve multicellular propagules (see Section 14.2.1). Nevertheless, even when meiosis does not occur, asexual reproduction sometimes involves the periodic return to a single cell. This is called 'parthenogenesis'.

Multicellular organisms can have different life histories; many plants, for instance, have alternate generations of haploid and diploid individuals (Box 14.2), which may be morphologically different or may be identical. An alternation of generations of this type never occurs in animals, although there may be a striking alternation between diploid phases which reproduce sexually and asexually (see Section 14.2.1).

Alternation of sexual/asexual forms is common in several clades of Cnidaria (see Chapter 3, Figs 3.15, 3.19 and 3.21) but is also common is parasitic flatworms (Chapter 3, Box 3.3) and many annelids where the production of satellite individuals may be a precursor of sexual reproduction (Fig. 4.57). The implications of this alternation between asexual propagation and sexual reproduction within the life cycle of a diploid organism is discussed further in Section 14.2.

Although animals are always anisogamic and do not have alternating haploid and diploid generations, the conditions of sexuality and the organization of the life cycle are extremely varied. Invertebrate animals differ from each other in a great many of what may be termed their 'life-cycle traits' and this is, somewhat anthropomorphically, described as the 'reproductive strategy'. Some of the alternative ones are listed in Table 14.1; many of these traits are covariable so that not all possible combinations are observed.

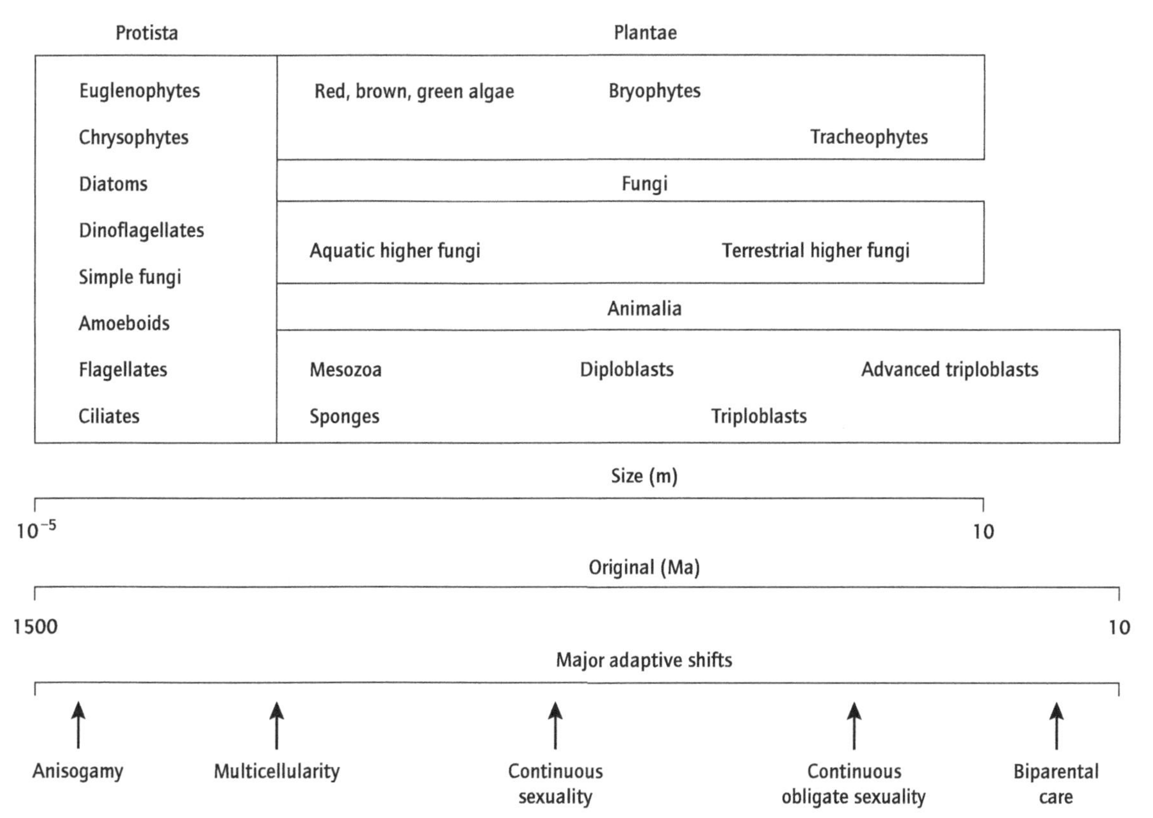

Fig. 14.1 Key steps in the evolution of continuous sexuality – a possible outline. (After Lewis, 1987 in Stearns, 1987.)

Box 14.2 Life cycles and sexual reproduction

1 Sexual reproduction in most protists. Example, *Paramecium*. The life cycle involves conjugation during which genetic exchange occurs. The major events in the conjugation process are illustrated. Note that the *Paramecium* has macro- and micronuclei. The macronucleus is formed by fusion of two micronuclei and replication of DNA. The macronucleus does not take part in the conjugation process. (After Klug and Cummings, 1997.)

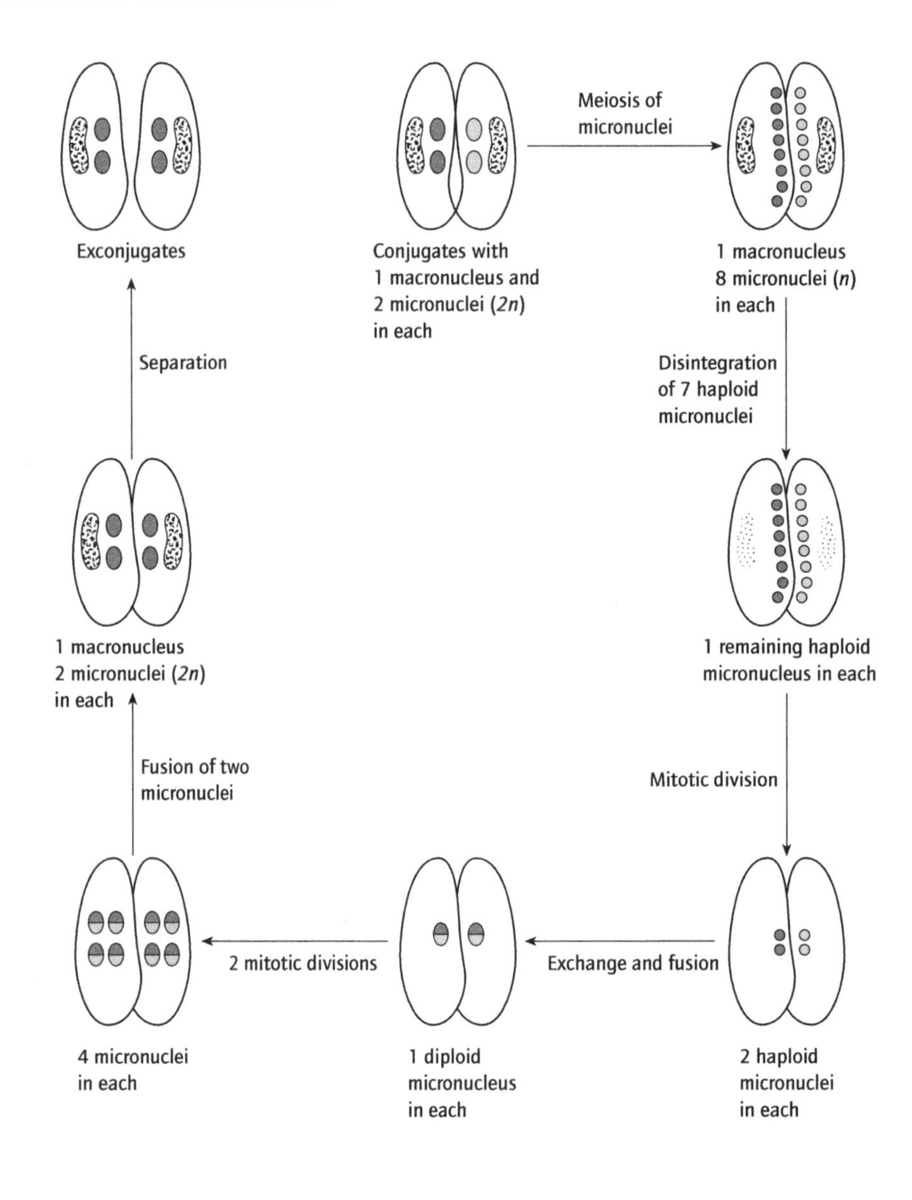

Continued

Box 14.2 *(cont'd)*

2 The life cycle of all multicellular animals (and some algal protists). The multicellular adult is composed of several or many (often different) diploid cells. The multicellular body gives rise to haploid gametes. Fusion creates a diploid zygote. By a series of mitotic divisions a new diploid adult develops.

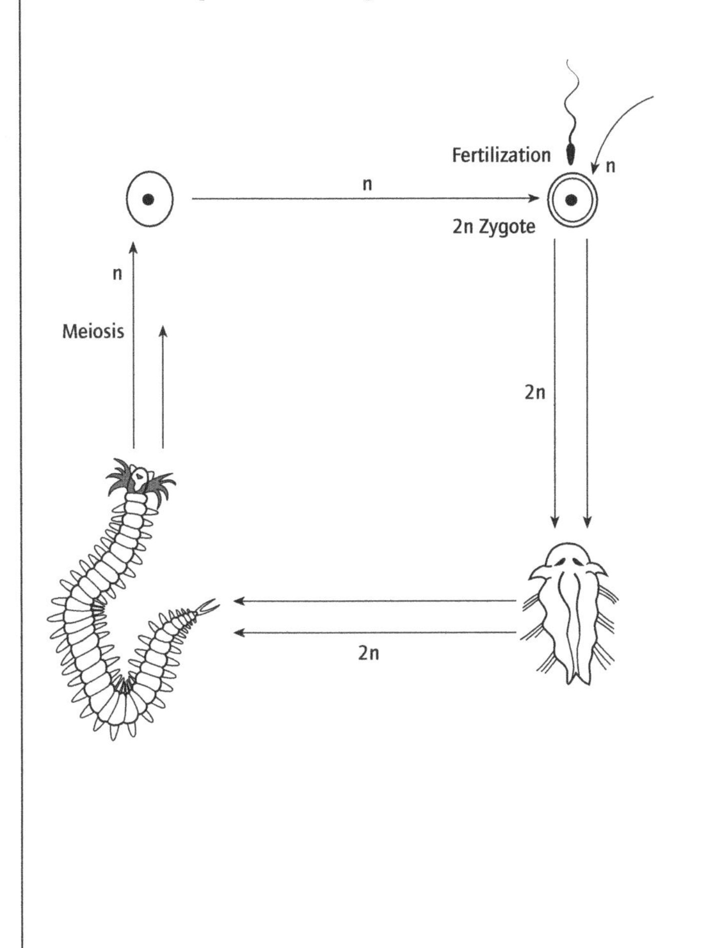

3 The life cycle of many algae and all higher plants involves an alternation between multicellular haploid and diploid phases. The haploid phase is called the 'gametophyte'; it produces (by mitosis) sperm cells and/or egg cells.

Fusion of gametes creates a zygote.

By mitosis this develops into a multicellular body called the 'sporophyte'.

Meiosis creates haploid spores.

Without fusion they proceed through mitotic divisions to create the haploid gametophyte.

The figure below illustrates this process in ferns,

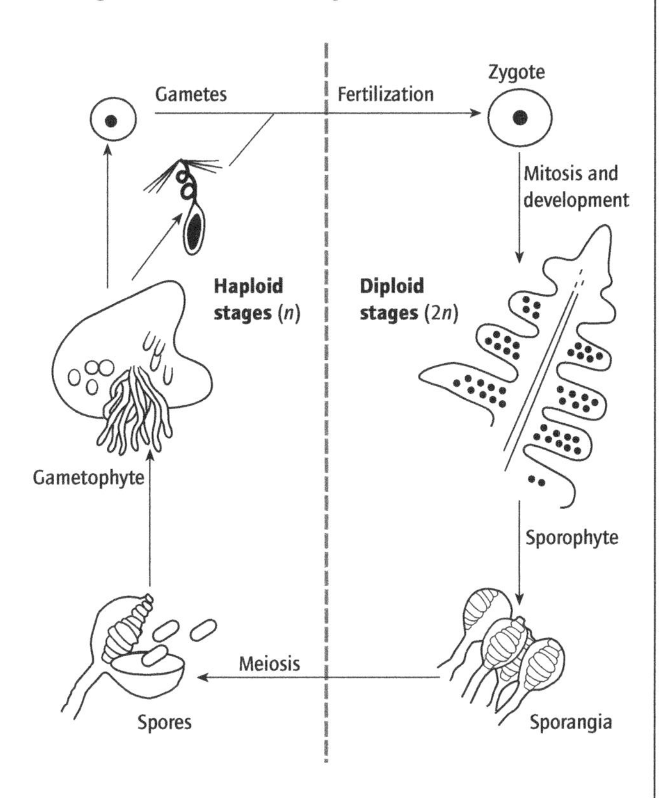

In higher plants the gametophyte is often reduced to a very small number of cells. Animals never show this alternation of haploid and diploid generations. The alternation of sexually and asexually reproducing individuals (see Fig. 14.5) is not equivalent.

14.2 The significance of sexual and asexual reproduction

14.2.1 Asexual reproduction in invertebrate life cycles

Sexual reproduction in animals is almost universal but many organisms can, in addition to reproducing sexually, also reproduce asexually, i.e. produce offspring without recombination of genetic material. Offspring produced in this way will have a genetic constitution almost identical to that of the parental organism. Asexual reproduction can take place either by subdivision of an existing body into two or more multicellular parts ('budding' and 'fission') or by the production of diploid eggs (parthenogenesis). Figure 14.2 illustrates these two basic mechanisms. Both are widespread among the invertebrates.

Fission is particularly common in the soft-bodied phyla such as the Porifera, Cnidaria, Platyhelminthes, Nemertea, Annelida and some Echinodermata. It is not often found in those that have an external casing and is unknown in the Mollusca and the arthropod phyla. Fission may involve simple transection into two fragments, each of which regenerates the missing parts (Fig. 14.2a) or it may give rise to multiple fragments each of which

Table 14.1 Reproductive traits of marine invertebrates.

Trait		
Development	Pelagic { planktotrophic lecithotrophic mixed	Non-pelagic
Egg size	Small *c.* 50 µm ⟷	Large >1000 µm
Fecundity	High 10^6 ⟷	Low 1
Brood frequency	Low 1 per annum ⟷	High Many per annum almost continuous
Broods per lifetime	One	Many
Longevity (Generation time)	Perennial Annual Many years ⟷	Subannual A few days or weeks
Body size	Large Length > 1000 mm ⟷	Small <1 mm
Spermatozoa	Simple	Advanced
Fertilization	External without sperm storage	Internal or with sperm transport or storage
Reproductive effort*	Large Made later	Small Made early

* Reproductive effort can be defined for animals that breed only once as:

$$\frac{E_g}{E_s + E_g}$$

but for animals that breed several times the following equation is preferred:

$$\frac{\Delta E_g}{\Delta E_s + \Delta E_g}$$

where ΔE_s and ΔE_g describe the instantaneous effort per unit time. The term E_g is the energy allocated to germinal tissues; E_s the energy allocated to somatic tissues.

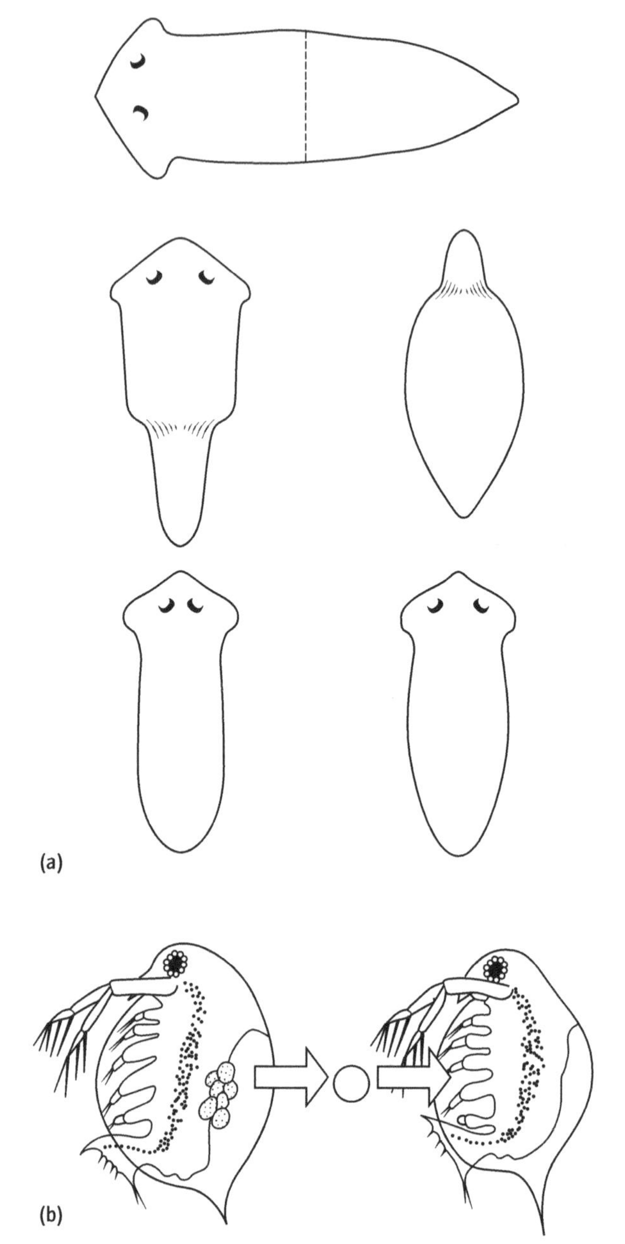

Fig. 14.2 Diagrammatic representation of two types of asexual (amictic) reproduction. (a) Fission: the multicellular body subdivides into one or more multicellular propagules each of which is able to re-constitute the initial body plan. The illustration shows simple fission in a planarian in which the body divides into anterior and posterior parts. Each is capable of regenerating the missing components to re-establish the entire body plan (see also Fig. 14.3). (b) Parthogenesis. The process of meiosis is suppressed, the ovary produces diploid 'eggs' which develop into a new organism without the fusion of a female pronucleus with a male gamete nucleus. The illustration shows a cladoceran crustacean, these animals reproduce by parthenogenesis for a number of generations but with the onset of adverse conditions may revert to sexual reproduction (see also Fig. 14.5b). Continuous parthenogenesis is rare.

can reconstitute a complete animal (Fig. 14.3). Fission is usually combined with the capacity for sexual reproduction in complex life cycles with asexual and sexual individuals (see Fig. 14.5a). The difficulties that fission presents to our concept of 'the individual' are underlined by those invertebrates in which fission is incomplete and thus gives rise to colonial organisms. Colonial organisms are made up from a large number of structural units that can be identified with the 'individuals' of non-colonial relatives. In some cases, all the subunits are similar but in others the subunits have specialized roles and the colony is clearly functioning as the 'individual', perhaps best defined as the unit on which natural selection acts. This seems to be the case, for instance, in complex siphonophoran jellyfish (see Fig. 3.18 and Section 3.4.2.3.1).

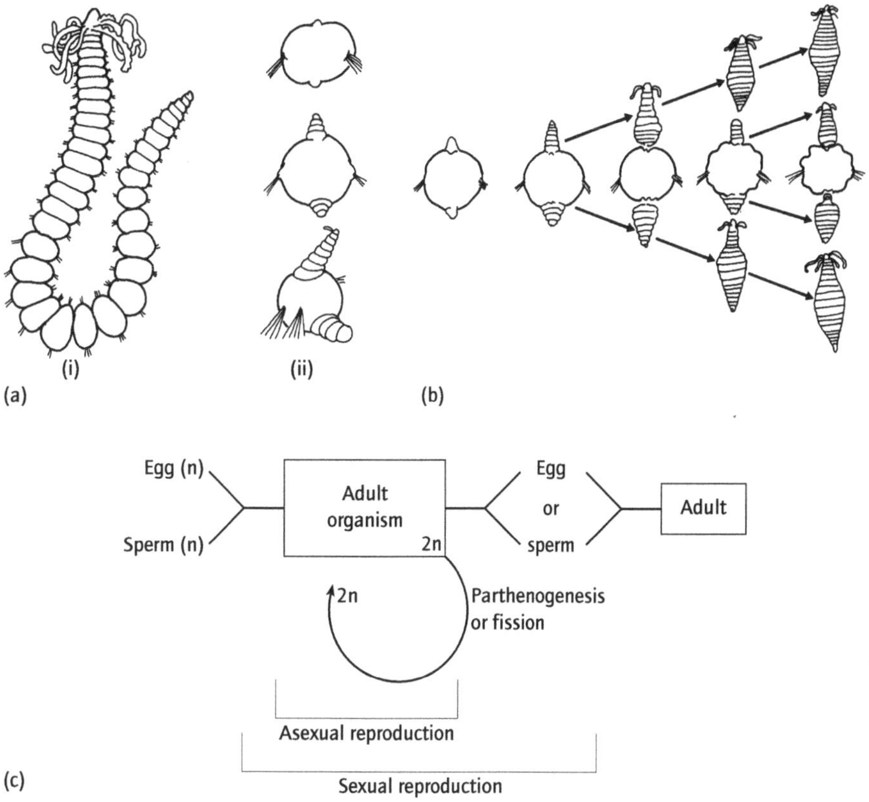

Fig. 14.3 An example of multiple fission: the polychaete *Dodecaceria* (a) (i) The adult prior to fragmentation into individual segments; (ii) stages in regeneration as each segment reconstitutes a new head and a new tail by renewed segment proliferation. (b) The production of primary then secondary individuals from a single segment. Eventually the single segment is incorporated into an individual in which fragmentation has stopped. (c) Diagrammatic representation of the life cycle incorporating asexual and sexual reproduction. Asexual reproduction is followed by a transition to sexual reproduction and all fragments which have reproduction sexually die. (After Dehorne, 1933 and Gibson & Clark 1976).

Colonial structures are encountered particularly frequently in the Cnidaria, the bryozoans and the urochordates. Some compound or colonial invertebrates are illustrated in Fig. 14.4. These animals are more realistically interpreted as being made up of a series of modular units, not individuals.

Parthenogenesis is also widespread among the invertebrates. In this book, 'parthogenesis' is taken to mean asexual reproduction via eggs, but some authors use the term 'parthenogenesis' to include fission. Table 14.2 summarizes some of the terms used to describe the different forms of sexual reproduction that are found amongst the diverse groups of invertebrates. In parthenogenesis, meiosis is suppressed so that the eggs are diploid and do not fuse with male germ cells. The term 'arrhenotoky' is used to describe the related phenomenon in which unfertilized haploid eggs develop into males and fertilized diploid eggs give rise to females. Obligate parthenogenesis, in which sexual reproduction never occurs, is extremely rare, but is found in the bdelloid rotifers (see Section 4.9.3.1) in which males have never been observed and in a few other taxonomic groups. The existence of ancient asexual clades comprising several putative species as in the Bdelloidea (363 species) presents a major challenge to the many theories of sexual reproduction (Section 14.2.4). More frequently, parthenogenesis occurs cyclically together with episodes of sexual reproduction. One or several generations of asexually reproducing individuals are followed by a generation of sexual individuals which usually give rise to resistant resting eggs (Fig. 14.5a). There are some 1000 animal species thought to exhibit obligate parthenogenesis. These are widely scattered at low incidence in taxa that more typically exhibit sexual reproduction. Most are thought to have been recently derived from sexually reproducing forms. Cyclic parthenogenesis, by contrast, is restricted to a much smaller range of just seven taxonomic groups (Table 14.3). This method of reproduction is, however, strikingly successful where it is found; at least 15 000 species exhibit this trait. Monogont rotifers, many small freshwater Crustacea and aphids all characteristically exhibit this type of life history in which the diploid individual that emerges from the resting egg gives rise to a clone of genetically identical descendants. In these circumstances, the growth rate of the population is at a maximum in the parthenogenic phase and the life cycle can be thought of as being a means for the maximum exploitation of a temporarily underexploited food resource where other biological constraints limit body size. At some stage individuals appear in the population which have two types of offspring, sexually reproducing females and males. The transition may be endogenously determined, but more often it is a response to changing environmental conditions such as crowding, food quality or shortening day lengths (see also Section 14.4.4). There may be morphological changes associated with this transition to sexual reproduction (Fig. 14.5b,c) and in the aphids complex changes of morphology also occur during the asexual phase (see Fig. 14.5d).

These different patterns of life history can be understood better if we distinguish between two aspects of individuality. We

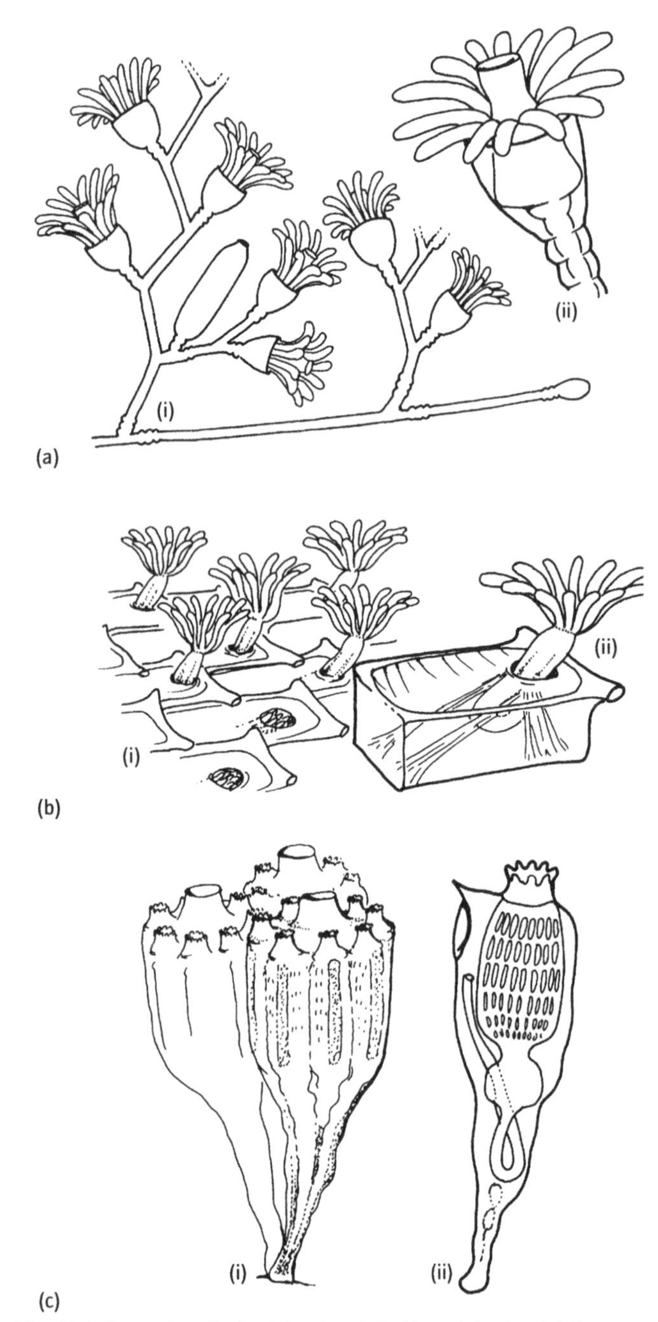

Fig. 14.4 Examples of colonial or 'modular' invertebrates: (a) the hydroid *Obelia*: (i) the branching colony; (ii) an individual polyp; (b) the bryozoan *Membranipara*: (i) part of the mat of zooids; (ii) an individual zooid; (c) the urochordate *Sydnium*: (i) three colonies showing individual inhalent siphons and a shared exhalent siphon, (ii) an individual zooid.

can recognise the 'individual' organism but also the unique genome created by each act of sexual reproduction. Among large organisms, such as vertebrates, the terms are usually synonymous since each individual (apart from 'identical twins') has its own distinctive genome. The development of technologies that make 'cloning' of vertebrates possible, e.g. the creation of 'Dolly' the sheep, has, for many people, raised the spectre of artificially changing the pattern of reproduction in human beings. Among invertebrate species, cloning may be a normal feature of the life cycle. The range of animal forms is such that it has been necessary to develop a terminology to describe the two aspects of individuality. The terms 'ramet' and 'genet' are useful to distinguish these concepts. We intuitively recognize individual ramets by the observation of a single-body component, most often defining an individual by the provision of the head. In the many different species of animal with an asexual reproductive phase, it is also possible to distinguish many individuals in a population that share the same genome, each having been

Table 14.2 Terminology for discussion of sexual and asexual reproduction (after Judson & Normack, 1996).

Asexual reproduction (amixis)
The term 'asexual reproduction' is sometimes used to describe processes in which new individuals are derived from only one parent and without exchange of genetic material. Because there is no genetic recombination the term 'amixis' may also be used. Two forms may be recognized:

Apomixis: reproduction by single cells that are derived by mitosis.

Vegetative reproduction or fission: reproduction by detachment and subsequent differentiation of groups of mitotically derived cells or tissues.

Sexual reproduction
Reproductive processes in which the recombination of genetic material derived from more than one parent is possible. In animals sexual reproduction always involves meiosis and the formation of a haploid egg.

Sexual reproduction with fertilization: processes of reproduction in which meiotically derived haploid eggs and spermatozoa fuse to create a diploid zygote.

Automixis: reproduction by single cells that are derived by meiosis from a single parent with some mechanism other than fertilization to restore the diploid condition.

Parthenogenesis
Development of a new individual (male or female) from an unfertilized 'egg' (apomixis or automixis). Parthenogenesis may alternate with sexual reproduction via eggs in a compound life cycle.

Fig. 14.5 (*opposite*) Alternation of parthogenesis and sexual reproduction in the life cycles of invertebrates. (a) The elements of a generalized life cycle; (b) the life cycle of the freshwater cladoceran *Daphnia*; (c) the life cycle of the rotifer *Brachionus*; the asexual phase is followed by a sexual one in which unfertilized females lay small eggs which hatch as males and fertilized females produce larger eggs which are the overwintering stage in the life cycle; (d) the life cycle of the bird-cherry aphid. All females, other than the oviparous females of autumn, are parthogenetic and viviparous. There is marked polymorphism especially in the production of winged and wingless forms. (b after Bell, 1982; d after Dixon, 1973.)

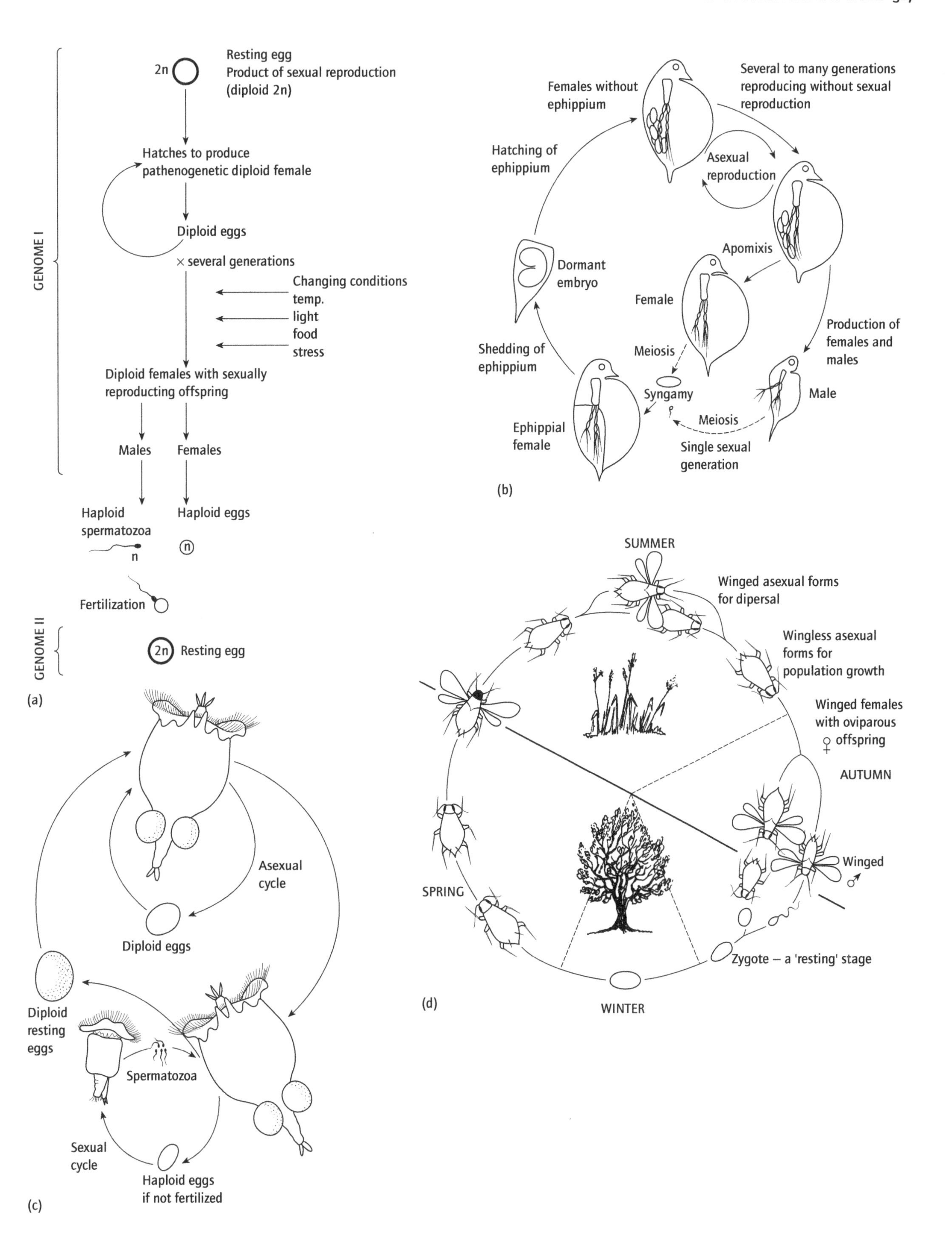
2n
Resting egg
Product of sexual reproduction
(diploid 2n)
Hatches to produce
pathenogenetic diploid female
Diploid eggs
× several generations
Changing conditions
temp.
light
food
stress
Diploid females with sexually
reproducting offspring
Males
Females
Haploid
spermatozoa
n
Haploid eggs
n
Fertilization
GENOME I
GENOME II
2n
Resting egg
(a)
Females without
ephippium
Several to many generations
reproducing without sexual
reproduction
Hatching of
ephippium
Asexual
reproduction
Apomixis
Dormant
embryo
Female
Production of
females and
males
Shedding of
ephippium
Meiosis
Syngamy
Male
Ephippial
female
Meiosis
Single sexual
generation
(b)
Asexual
cycle
Diploid eggs
Diploid
resting
eggs
Spermatozoa
Sexual
cycle
Haploid eggs
if not fertilized
(c)
SUMMER
Winged asexual forms
for dipersal
Wingless asexual
forms for
population growth
Winged females
with oviparous
♀ offspring
AUTUMN
Winged
♂
SPRING
Zygote – a 'resting' stage
WINTER
(d)

Taxon	Parthenogenetic stage	Sex determination	Duration of parthenogenesis
Rotifera	Adult	Haplodiploid	Unlimited
Cladocera	Adult	Environmental	Unlimited
Digenea	Larva	Hermaphroditic, ZW	2–5 generations
Aphidoidea	Adult	XO	Unlimited
Cynipinae	Adult	Haplodiploid	1 generation
Cecidomyiidae	Larva, pupa	Chromosome exclusion	Unlimited
Micromalthidae	Larva	Haplodiploid	Unlimited

Table 14.3 Sex determination system, parthenogenetic stage and duration of parthenogenesis in groups reproducing by cyclic parthenogenesis.

produced by parthenogenesis or fission. A collection of individuals sharing the same genome may be referred to as a 'genet'. The number of individuals sharing the same genome in a population can of course only be determined by analysis of their genetic identity, since because of different effects of the environment, they may not have exactly the same appearance or phenotype.

For those animals without any form of asexual reproduction each individual in a population has its own unique genetic identity and the ramet and genet are synonymous. In the more complex life cycles described above however each unique 'genet' may give rise to many ramets through asexual replication early in the life cycle (polyembryony) or later in the life cycle by larval or adult replication (fission) or by suppression of meiosis (apomictic parthenogenesis) as shown diagrammatically in Fig. 14.6. There are some animals that are both clonal and colonial but since the colonies themselves may subdivide, the physical extent of any one genet is not at all obvious.

To put things simply: asexual reproduction alternating with sexual reproduction provides a means for the genet to grow (i.e. acquire a greater number of mitotically replicated copies) while at the same time the functional unit remains small (as in the aphids) or alternatively allows the same genet to be in different places at the same time (which occurs with cnidarian medusae). It also commonly observed that invertebrate endoparasites have an asexual phase that creates many ramets each with the same genotype. This may be an adaptation for increasing the likelihood that the invasive stages in the life cycle of the genet come into contact with a potential host (see also Box 3.3).

14.2.2 Patterns of sexuality

Over 99% of all invertebrates exhibit sexual reproduction at some stage in their lives, and it is, for most, the only means by which reproduction can take place. This section will describe the diverse ways in which sexual reproduction can occur. Sexual reproduction in animals always involves fusion of relatively large, immobile gametes, the ova, and much smaller, mobile male gametes, the spermatozoa. This is referred to as 'anisogamy'. Whenever male and female gametes occur it is possible to recognise male and female function, but these are not necessarily assigned to separate individuals in the population.

(a) NON-CLONAL. Replication and Duplication only occur as a consequence of multiple gamete formation. Two forms of non-clonal reproduction may be recognized (i) Semelparous and (ii) Iteroparous reproduction.

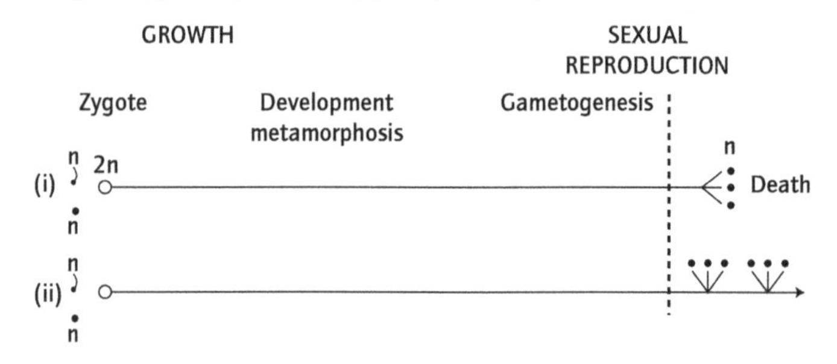

(b) CLONAL. Replication and multiplication occurs prior to gametogenesis and gamete production. The replication may occur at various stages in the developmental cycle to give rise to the following patterns that may be described as: (iii) Polyembryony; (iv) Larval replication; (v) Sexual satellite production; (vi) Parthogenic production of dipoid eggs.

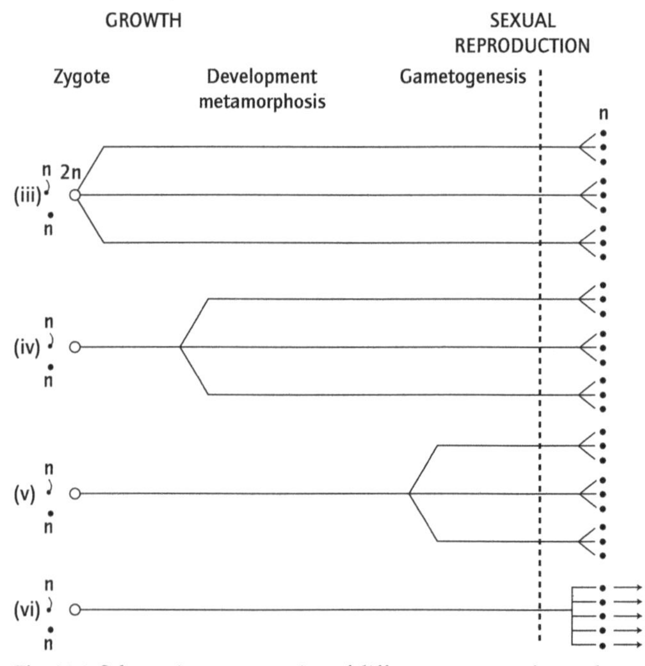

Fig. 14.6 Schematic representation of different patterns of growth, replication and multiplication of body forms. In the case of non-clonal reproduction (a) the number of ramets = number of genets. In clonal reproduction (b) the number of ramets greatly exceed the number of genets. All these patterns of sexuality may be found among the invertebrates. (After, J.S. Pearce, Pearse, V.B. and Newberry, A.T. 1989. *Bull. Marine Sci.*, **45**, 433–446.)

The phenomenon in which the sexes are separated and individuals are either exclusively male or exclusively female is described as 'gonochorism', or alternatively, 'dioecy' (the corresponding adjectives being 'gonochoristic' and 'dioecious'). Many species of animals are known, however, in which the same individual can function both as a male and as a female. This situation is described as 'hermaphroditism' (adj. 'hermaphroditic'), or alternatively, 'monoecy' (adj. Monoecious). When the male and female gametes are present at the same time the animal is said to be a 'simultaneous hermaphrodite' but in other cases there is a functional sex reversal during the life cycle and the animals are said to be 'sequential hermaphrodites'. A summary of their distribution among the major groups of invertebrates is presented in Table 14.4. In the case of sequential hermaphrodites, the sex change may occur at a species-specific size or age, no individuals being pure male or females; alternatively, the sex change may not be so precisely fixed, taking place at a variable age and size. In the latter case some individuals may be purely male or female. The two situations are illustrated in Fig. 14.7a and b.

14.2.3 Mechanisms of sex determination

The sex-determining mechanisms of invertebrates fall into three basic types: (a) maternal, (b) genetic and (c) environmental. The variation in sex-determining methods is somewhat surprising given the highly conserved nature of meiosis in all animals. The method of sex determination varies even within taxonomic groups suggesting that there is some advantage in the retention of plasticity in this feature of the life cycle (see also Section 14.2.3.2 below).

14.2.3.1 Maternal sex determination

The sex of the offspring of some animals is determined by the mother through the production of different types of eggs. An example is illustrated in Box 14.3(1). In this system, inbreeding is inevitable and all members of the adult populations are females. However, despite the maternal determination of egg type, male and female sperm are also found. The female-specific sperm has a chromosome not present in male sperm. It seems that male- and female-determining sperm select the appropriate type of

Table 14.4 Conditions of sexuality in invertebrates.

Phylum	Class	Notes
Porifera		All sponges have the ability to reproduce sexually although many also produce asexual fragments called gemmules
Mesozoa		Functional self-fertilizing hermaphrodites with alternation of sexual and asexual generations
Cnidaria		Gonochoristic but with frequent asexual reproduction by fission. Sometimes a complex life cycle with pelagic medusae producing gametes and asexually reproducing benthic hydroid phase
Ctenophora		Simultaneous hermaphrodites, probably not self-fertilizing
Platyhelminthes	Turbellaria } Monogenea }	Simultaneous hermaphrodites, probably cross-fertilizing Frequent asexual reproduction by fission Sexual reproduction sometimes unknown
	Trematoda	Simultaneous hermaphrodites, probably usually cross-fertilizing in multiple infections. Occasionally self-fertilizing Larval stages exhibit frequent asexual reproduction by multiple fission
	Cestoda	Simultaneous hermaphrodites, usually self-fertilizing. One genus gonochoristic
Gnathostomula		Simultaneous hermaphrodites
Nemertea		Virtually all gonochoristic, occasionally hermaphrodite in fresh waters, some cyclic parthenogenesis
Gastrotricha		Simultaneous hermaphrodites, but parthenogenesis common
Mollusca	Chaetodermomorpha } Monoplacophora } Polyplacophora } Scaphopoda }	All gonochoristic
	Gastropoda	Highly variable sexual conditions Prosobranchia – mostly gonochoristic, but often sequential (protandric) hermaphrodites Opisthobranchia – mostly simultaneous hermaphrodites Pulmonata – all simultaneous hermaphrodites with cross-fertilization
	Neomeniomorpha	Hermaphroditic
	Bivalvia	Gonochoristic
	Cephalopoda	Gonochoristic

Continued p. 330

Table 14.4 (*cont'd*)

Phylum	Class	Notes
Rotifera	Bdelloidea	Obligate parthenogenesis, males unknown. Ancient asexual clades
	Monogonata	Cyclic parthenogenesis, gonochoristic at sexual phase with dwarf males
	Seisonidea	Gonochoristic
Kinorhyncha		Gonochoristic
Acanthocephala		Gonochoristic
Loricifera		Gonochoristic
Nematomorpha		Gonochoristic
Nematoda		Gonochoristic
Priapula		Gonochoristic
Sipuncula		Gonochoristic
Echiura		Gonochoristic, sometimes with dwarf males
Annelida	Polychaeta	Usually gonochoristic, sometimes sequential protandric hermaphrodites, occasionally simultaneous. Asexual reproduction by fission in some species
	Clitellata	Usually cross-fertilizing simultaneous hermaphrodites, sometimes parthenogenic with non-functional males
Pogonophora		Gonochoristic
Phorona, Bryozoa		Simultaneous hermaphrodites, occasional gonochoristic species
Brachiopoda		Mostly gonochoristic, occasionally hermaphrodite
Entoprocta		Simultaneous or sequential hermaphrodites
Hemichordata		Gonochoristic
Echinodermata	Echinoidea, Holothuroidea, Crinoidea	Gonochoristic
	Asteroidea, Ophiuroidea	Mostly gonochoristic, occasionally simultaneous hermaphrodites. Asexual reproduction by fission in some
Chordata	Urochordata	Simultaneous hermaphrodites
Crustacea	Branchiopoda, Ostracoda	Gonochoristic, but with frequent parthenogenesis; males sometimes unknown
	Copepoda	Gonochoristic
	Cirripedia	Thoracicans usually functional hermaphrodites, but in some species gonochoristic with dwarf males. Others gonochoristic with dwarf males
	Malacostraca	Usually gonochoristic, but not infrequently sequential (protandric) hermaphrodites
Chelicerata		Almost always gonochoristic
Onychophora		Almost always gonochoristic
Tardigrada		Almost always gonochoristic
Pentastoma		Almost always gonochoristic
Uniramia		Almost always gonochoristic, occasionally with parthenogenesis or arrhenotoky. Cyclic parthenogenesis in some taxa

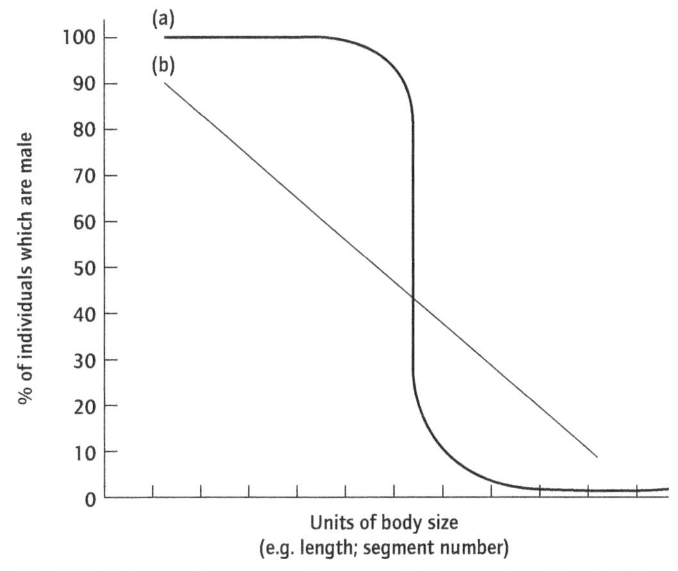

Fig. 14.7 Patterns of sex reversal in sequential hermaphrodites: (a) false gonochorism; (b) unbalanced hermaphrotitism.

egg, although the mechanism by which this achieved is not yet understood. The X sperms select large (female generating) eggs to fertilize while the O sperms select small (male forming) eggs (see Section 14.2.3.2). The honey bees (Insecta Hymenoptera) present a particularly well-known case where the sex of the offspring can be controlled by the female. The queen honey bee (*Apis mellifera*) is able to control the fertilization of her eggs. Fertilized diploid eggs will normally develop as sterile female workers, but have the potential to become functional females if exposed to appropriate conditions during their development, whereas unfertilized eggs develop into haploid males (see Table 14.3).

14.2.3.2 Genetic sex determination

In some animals males and females have visibly different chromosomal complements. Most often females have two identical sex chromosomes, the so-called X chromosomes, and the males have unlike chromosomes, one X chromosome and one Y chromosome. The sex of the zygote is here determined by the sex chromosome complement of the fertilizing sperm. Chromosome sex-determining mechanisms occur among the insects in a number of different forms. Chromosomal sex determination is known only spasmodically in other groups; it is well known in the nematode *Caenorhabditis elegans*. The molecular mechanism of this mode of sex determination have been investigated in *C. elegans* and in the fruit fly *D. melanogaster*. The mechanics appear to be rather different and do not share homologous molecular machinery. As shown in Box 14.3, female *Drosophila* have two X chromosomes and males have an odd pair: the X and Y chromosomes. At meiosis, all the eggs receive an X chromosome but exactly half of the spermatozoa receive a Y chromosome. Random fusion of gametes therefore establishes a sex ratio of 1 : 1. In many invertebrates, heterogamy has not been observed; nevertheless the sex ratio may be fixed. Heterogamy has been described in insects, arachnids and nematodes and recently in a polychaete worm, but is otherwise not widespread. Theoretical considerations suggest that the sex ratio, and investment in male and female functions among the offspring of sexually reproducing organisms, will normally be equal. This is because male and female offspring offer equal routes to reproductive success in diploid organisms with haploid gametes. The circumstances where this general rule does not apply are only partly understood, and the exceptions to genetic mechanisms of sex determination, may have evolved primarily as means of controlling the sex ratio among offspring.

14.2.3.3 Environmental sex determination

The sex of an individual is not always determined at or before fertilization; it sometimes depends on the environmental conditions experienced by the developing embryo or larva. One of the best known invertebrate examples is that of the echiuran worm *Bonellia viridis*, the males of which are dwarf and parasitic on the females (see Section 4.12). It was demonstrated at the beginning of the century that the sex of the free-swimming planktonic larvae is not fixed, those that settle onto mud become females and only those that settle on, or very close to the large female proboscis, will become males (Box 14.3(3)).

Careful experiments have shown that the females release a substance which has a profoundly masculinizing influence on the developing larvae. In the absence of this pheromone almost all larvae become females. In most other echiurans sex determination is genetical, but dwarf males have recently been discovered in some other families not closely related to *Bonellia*, and this may indicate independent evolution of environmental sex determination in other members of the phylum. Echiurans are gonochoristic, but the process of sexualization in many sequential hermaphrodites is also profoundly influenced by environmental conditions. A pheromonal mechanism is thought to be involved in the sex-determining mechanism of the slipper limpet *Crepidula fornicata*, as illustrated in Fig. 14.8. These limpets form themselves into stacks: the lowermost individual is always a female and the upper ones are males, while intermediate animals in the stack may be hermaphrodites. Environmental sex determination has also been investigated in some nematodes, polychaetes and amphipod crustaceans. In the amphipods the temperature of the water in which the larvae develop determines their sex, a phenomenon which is echoed among vertebrates in lizards, crocodiles and turtles where temperature-dependent sex determination has been particularly well studied.

14.2.3.4 Plasticity and contingency in sex determination

Recent studies of the molecular mechanisms of sex determination suggests that genetic plasticity in the means of sex determination tends to persist in evolutionary time. It has been suggested that switching between environmental and genetic

Box 14.3 Systems of sex determination

1 An example of maternal sex determination; *Dinophilus gyrociliatus*, a minute polychaete worm.

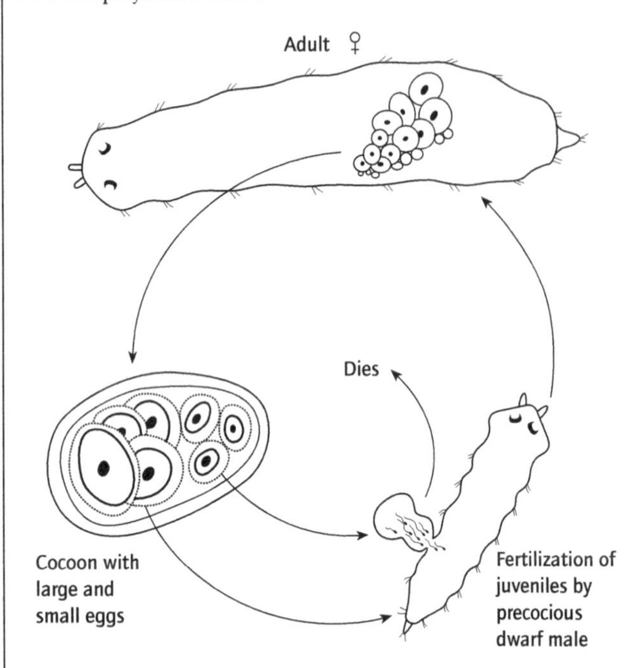

Two types of egg are produced in the ovaries.
Large eggs become females; smaller eggs become precociously mature dwarf males. Two types of sperm nevertheless occur. It is now known that genetically different types of sperm select the appropriate type of egg to give XX females and XO males. The sex ratio will be determined by the relative number of large and small eggs and is thus determined by the female.
Insemination of the female embryos occurs in the cocoon.
All adults are inseminated females.
2 An example of genetic size determination in the fruit fly *Drosophila*.

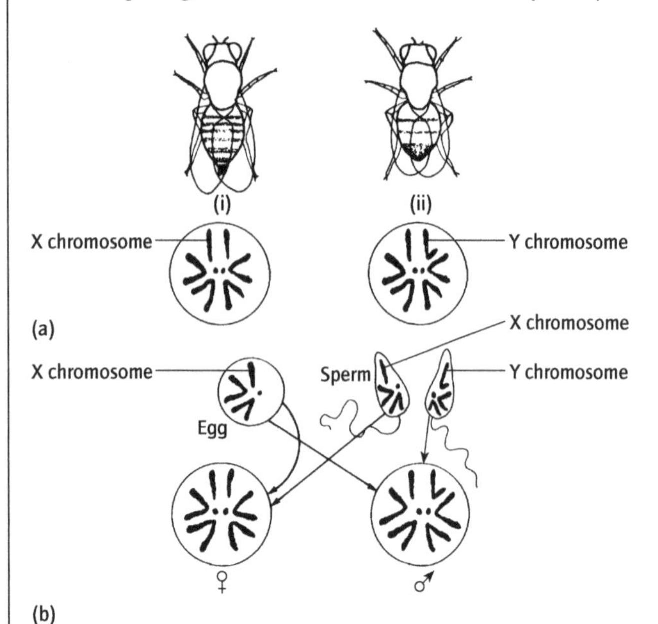

(a) (i) Females have a diploid genetic constitution with equal sex chromosomes XX. All the offspring have equal-forming capability.
(ii) Males have a diploid genetic constitution with unequal sex chromosomes, X and Y. Males produce sperm with two different sex-forming capabilities. Sperm with a Y chromosome will father females; sperms with an X chromosome will father males.
(b) The sex of the offspring is determined by which of the two possible sperm types happen to fertilize the eggs. The sex ratio will normally be 1 : 1 with this system of sex determination.
(c) High levels of the X to autosomal chromosome ratio in females (two X chromosomes) lead to high level of translation of the products of the sex lethal genes (*sx1*). This in turn leads via a sequence of gene switches to the production of the female form of the double-sex gene product, dsx^{f}. The default pathway occurs when the X to autosomal chromosome ratio is low and leads to the production of the male form of the double-sex gene product, dsx^{m}.
3 Environmental sex determination.
(a) In some organisms the sex of an individual is determined after fertilization. The example shows the echiurid worm *Bonellia viridis*. The ciliated planktonic larvae have the potential to develop as a female (all the large adults are female) or as a dwarf male. Larvae settling on an uninhabited area of the sea floor tend to develop as females. Larvae settling in the vicinity of a female proboscis are induced by secretions of the female to develop into dwarf males. Careful experiments show that there is also some element of genetic determination of sex but this is normally overridden by the environmental factor.
(b) The well-studied nematode *Caenorhabditis elegans* has a chromosomal mechanism of sex determination but many other Nematoda exhibit environmental sex determination, conditions such as the degree of crowding are found to determine the sex of an individual.

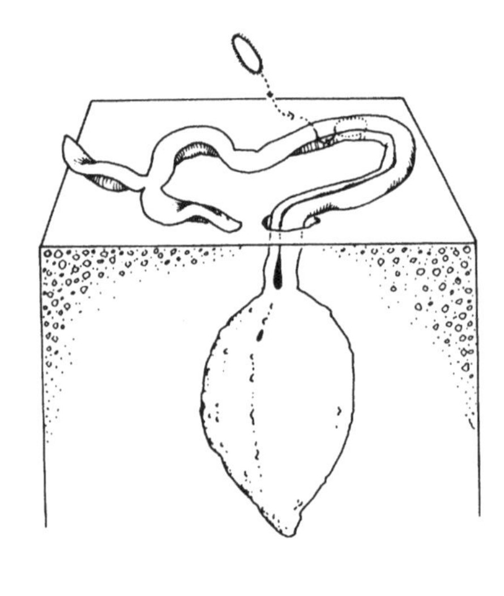

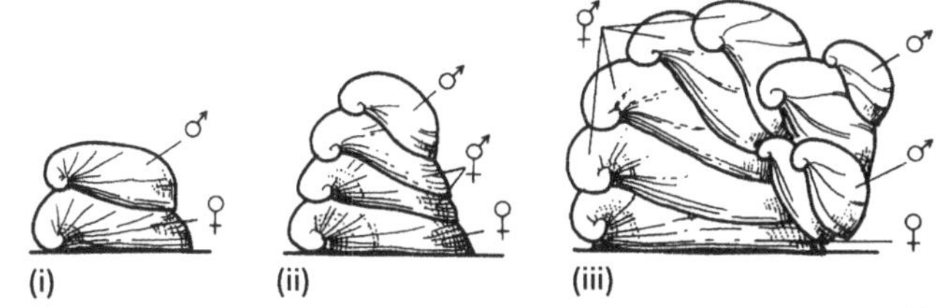

Fig. 14.8 Sex determination in the slipper limpet *Crepidula fornicata*; (i) early pair formation; (ii), (iii) late and more complex associations with males, females and intersexes.

mechanisms of sex determination may occur frequently within a clade during its evolutionary history.

In *Drosophila*, as explained in Box 14.3(1) and illustrated further in Fig. 14.9 sex is determined by a balance between genes on the X chromosomes and on the other autosomal chromosomes ultimately leading to activation or suppression of male or female specific products of the *Dsx* gene. An intermediary in this chain reaction is the transformer gene *tra*. Temperature-sensitive alleles of the transformer gene *tra* are known which lead to situations in which the female phenotype is expressed at colder temperatures and the male phenotype is expressed at warmer temperatures. There is therefore environmental input in a process that otherwise appears to be a classic chromosome sex-determination mechanism.

A different mechanism has been found in *Caenorhabditis elegans*. This nematode normally has two sexual types, hermaphroditic individuals and males. The genetic sex-determining mechanisms control the production of the hermaphrodite form and the male form. However mutations are known which transform XX or XO hermaphrodite individuals into females and some closely related species are strictly gonochoristic.

It seems that while sex is a highly conserved process, its expression and control has remained highly labile. Evolutionary and selective forces can then easily lead to changes in the modes of sexuality and in the factors which regulate the expression of male and female characters in individuals.

14.2.4.1 *Why reproduce sexually?*

One of the outstanding challenges to evolutionary biology is to understand why different patterns of reproduction are observed. Why, for instance, is sexual reproduction so dominant? The answer is not as self-evident as might first appear. Sex has been described as an 'enigma within a mystery' (Hurst & Peck, 1996). Sexual reproduction creates diversity among offspring and it reduces the expression of potentially harmful genes, but these advantages must be offset against the disadvantages or costs. Sexual reproduction reduces the efficiency with which genes are transmitted between generations, since only half of the genome of the offspring of sexually reproducing animals is contributed by each parent. Since fitness is measured by the effective contribution of genes to subsequent generations this is a major cost.

Other costs are associated with the need to devote resources to sexual display and courtship, and with the higher rates of mortality incurred whilst finding a mate. There are temporal costs associated with the production of gametes and their subsequent fusion, and the need to ensure fertilization introduces yet more costs associated with the greater exposure to risk through predation or disease, waste of gametes and the failure to find a suitable mate. A general theory of sexual reproduction must explain why sexual reproduction is universal when it is apparently so costly. The enormous apparent cost of sex, sometimes referred to as the 'Maynard–Smith' paradox stems from the cost of male reproduction.

Many of the theories suggest that sexual reproduction has a long-term advantage to the species or group but concede a short-term advantage to individuals with the capacity for asexual reproduction; they are therefore group selection theories and many biologists believe that as such they cannot be sustained. Recently, more acceptable theories have been developed in which the selective advantage of sexual reproduction is perceived as a property of the individual, but not of the group or species. It would be best to conclude, however, that the problem is far from resolved.

It has been suggested that sexually reproducing individuals have an advantage because their rather variable offspring, which together have many different genetic constitutions, will have an average fitness in the changed world of the future which is greater than that of the identical offspring of an asexual organism. Such a theory suggests that sexual reproduction will have its greatest advantage in unstable environments and will tend to lose its short-term advantage in relatively stable ones. A survey of the distribution of asexual reproduction among invertebrates in different habitats, however, does not confirm that prediction. Asexual reproduction is particularly frequently encountered in organisms exploiting unstable, fluctuating resources while sexual reproduction is almost universal in those exploiting the most stable environments. Thus asexual reproduction is much more commonly encountered among freshwater than among marine representatives of the same taxonomic groups.

An alternative theory supposes that the selective advantage of sexual organisms arises from the ability of their diverse offspring to compete in the structurally complex world of a saturated environment where there is intense competition. All sexually produced offspring of an asexually reproducing animal will not be able to supplant them from all the different habitats which they have come to occupy. The theory was called by its proposer (Bell, 1982) 'the theory of the tangled bank'. It suggests that sexual reproduction would predominate in stable complex environments, but that asexual reproduction might be expected to occur where species are maintained below the carrying capacity of the environment and opportunities for ecological opportunism exist. In these circumstances animals which have the highest possible rates of growth or growth in potential numbers of descendants are favoured.

(a)

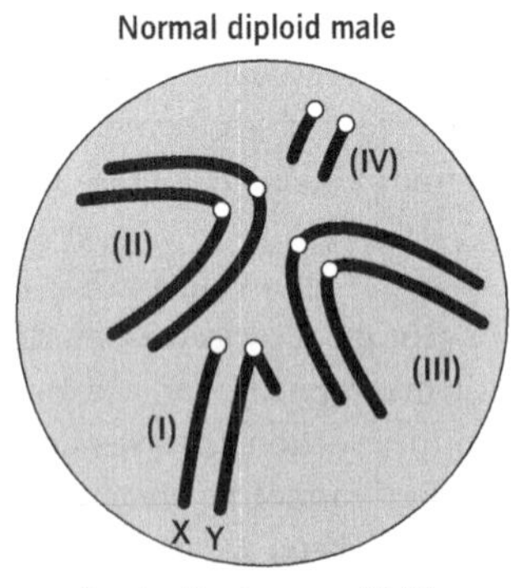

Chromosome composition	Chromosome formulation	Ratio of X chromosomes to autosome sets	Sexual morphology
	3X/2A	1.5	Metafemale
	3X/3A	1.0	Female
	2X/2A	1.0	Female
	2X/3A	0.67	Intersex
	3X/4A	0.75	Intersex
	X/2A	0.50	Male
	XY/2A	0.50	Male
	XY/3A	0.33	Metamale

(b)

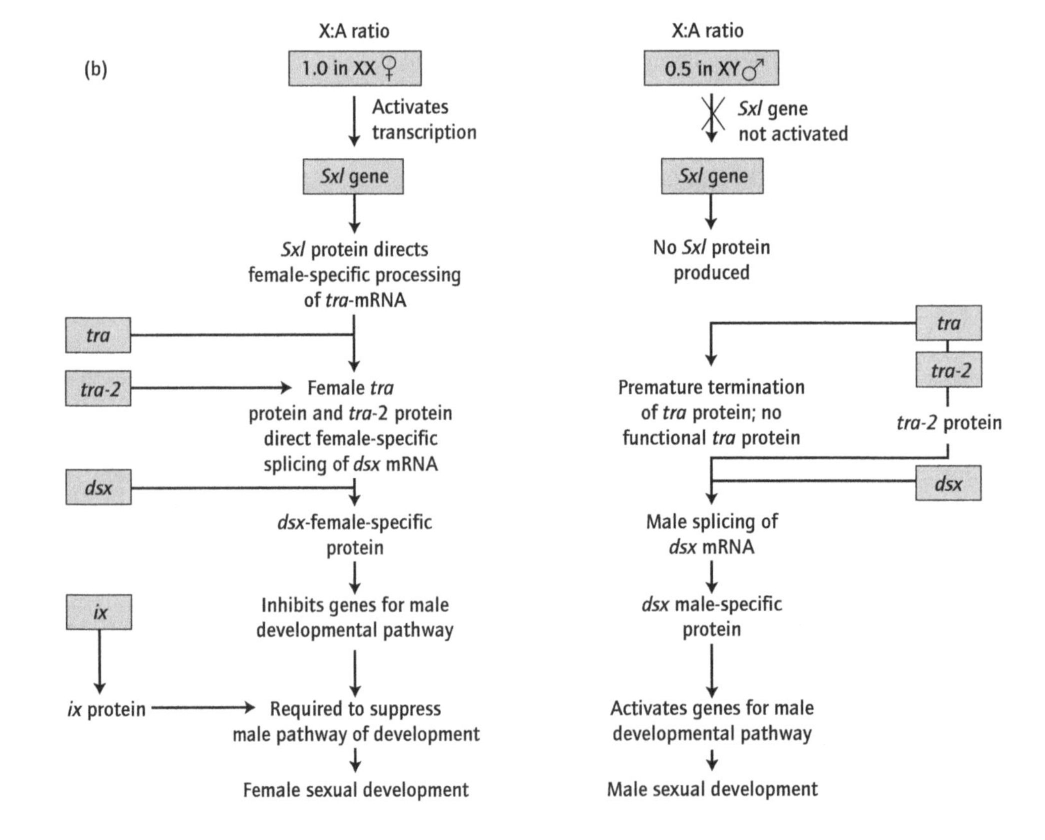

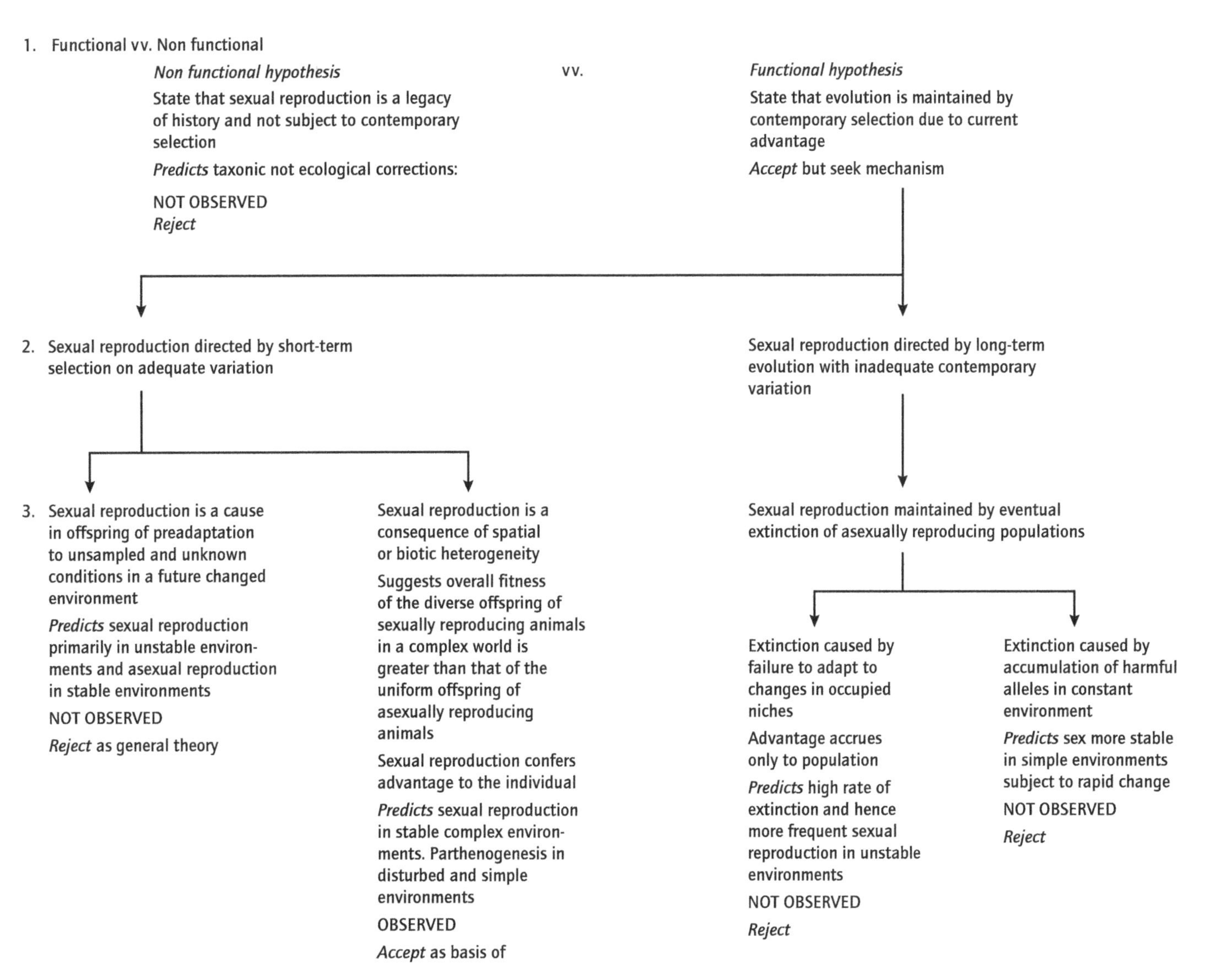

Fig. 14.10 A scenario for distinguishing between alternative theories of the selective advantage of sexual reproduction. (After Bell, 1982.)

Fig. 14.9 (*opposite*) The genetic basis of heterochromosomal sex determination in *Drosophila melanogaster*. (a) The ratio of X chromosome to autosomal chromosome numbers determines the sexual state of an individual. (After Klug, W.S. and Cummings, M.R. 1997. *Concepts of Genetics*, 5th edn., fig. 9.7.) (b) The method of sex determination involves a hierarchy of genes and the process of 'mRNA splicing'. This mechanism can give rise to a situation where the expression of single gene can give rise to a family of proteins. The X chromosome/autosomal chromosome ratio controls the expression of the *Sxl* gene which is not transcribed when the X : A ratio is low (males). In females *Sxl* is transcribed and in turn controls the expression of the *tra* (transformer) gene leading to the accumulation of transformer gene proteins in females which directs female specific gene splicing of the *dxl* (doublesex) gene mRNA. The cascade of genetic interactions caused by the female and male forms of the *dxl* (doublesex) gene mRNA involves suppression of male pathway in females and activation of the male developmental genes in males. (After Klug, W.S. and Cummings, M.R. 1997. *Concepts of Genetics*, 5th edn., Prentice Hall Int. Inc., New Jersey. Fig. 19.24.)

There are other theories to account for the dominance of sexual reproduction (Fig. 14.10). The most powerful of these alternative theories, sometimes referred to as the theory of the Red Queen, suggests that sexual reproduction confers an important short-term advantage over asexual competitors because organisms are in effect engaged in a 'co-evolutionary arms race'. Herbivores, predators, disease organisms and parasites are major causes of death. Sexual reproduction produces diverse offspring that are less easily identified targets for predatory and perhaps even more importantly parasitic organisms.

The costs of sexual reproduction are disproportionately large for smaller organisms and obligate continual sexuality without asexual reproduction appears to have become stable in virtually all of the largest animals, e.g. the vertebrates. Invertebrates are, because of mechanical constraints, smaller than most vertebrates. As a consequence there is much great diversity of sexual activities and life cycles. With their larger generation time, larger, long-lived animals may be more susceptible to invasion by parasitic organisms, and larger long-lived organisms may thus be forced into continuous obligate sexuality, whereas those animals with shorter generation times may be free to exhibit a much wider set of sexual conditions. It may also be in the interests of microorganisms to inhibit the sexuality of potential hosts, and there is some evidence that this does occur.

14.2.4.2 *Hermaphroditism versus gonochorism*

Just as we can ask 'what are the selective advantages of sexual as opposed to asexual reproduction?' we can also ask what determines whether animals which reproduce sexually should be gonochoristic or hermaphrodite. There are both taxonomic and ecological components of the variation. Table 14.4 shows that certain taxonomic groups are predominately hermaphrodite, others almost exclusively gonochoristic. It is also observed that hermaphroditism is more frequently encountered in some environments than others. Freshwater and terrestrial annelids and molluscs, for instance, are most often hermaphrodite, whereas their marine relatives are predominately gonochoristic. Similarly, deep-sea crustaceans are more often hermaphroditic than their shallow-water relatives. Both observations suggest a functional link and that it is possible to seek functional explanations of the observed patterns. Natural selection determines patterns of sexuality. A number of models to account for the evolution of herma-phroditism have been proposed. Three of these are summarized below:

1 *Low-density model*: when organisms exist in low densities or are immobile or sedentary then simultaneous hermaphroditism increases the probability that rare encounters between individuals will be fecund, and if other individuals are not encountered self-fertilization is possible.

2 *Size advantage model*: when one of the sexual functions has an advantage related to size but the other does not, then sequential hermaphroditism will be adopted.

3 *Gene dispersal model, low-density version*: when population numbers are low, inbreeding and random genetic drift may occur leading to reduced offspring fitness; in these circumstances hermaphroditism increases the effective population size.

A more general selectionist theory, which encompasses these different models, can be proposed. At the heart of the theory is the simple but profound observation that all sexually reproduced animals have precisely one mother and one father. Exactly half the zygote genome is contributed by the mother and exactly half by the father. Consequently both male and female reproductive functions are equal means to reproductive success.

An organism has limited resources at its disposal and these must be allocated to maintenance and reproduction in such a way that the overall fitness of the organism is at a maximum. Some of the costs associated with sexual reproduction arise from the need to construct accessory structures, e.g. the glands and ducts which are necessary to discharge the gametes satisfactorily, and from increased mortality associated with the need to find a mate. Hermaphrodites must carry both types of cost, and their total fixed costs, in relation to any finite resource to be invested in reproduction, are likely to be higher than would be the case for an individual expressing only one sexual function. We would thus expect evolution to favour gonochorism.

This conclusion is modified, however, if the return on investment in either sexual function declines; in these circumstances it can be argued that hermaphroditism will be favoured. A declining return on energy allocated to female reproductive function, for instance, might be expected in animals which brood their offspring in a chamber within which space is limited. These concepts are illustrated in more detail in Box 14.4.

Box 14.4 Hermaphroditism versus gonochorism: an investment trade-off

1 Investment and benefit

An organism must invest resources in sexual function to gain some fitness benefit. The fitness benefit can be thought of as the expectation of future offspring and can be formally represented by the Euler–Lotka equation (see also Section 14.5.1).

$$1 = \tfrac{1}{2}\Sigma e^{-Ft} s^t n^t$$

or

$$1 = \tfrac{1}{2}\int s e^{-Ft} s^t n^t$$

2 Male and female function

Imagine that a limited resource (such as energy) may be invested in two different investment accounts that we will call (a) male function and (b) female function. The investments in these two functions include:

Male function

- production of male germ cells
- copulatory organs
- mate searching
- defence of territory, etc.

This investment may be represented as V_m.

Female function

- production of female germ cells
- female accessory gland investment levels
- mate searching
- offspring provisioning (may also be a male function)

This investment may be represented as V_f.

3 Investment gain curves

If zero investment is made in male function there can be no benefit from this function and male function contributes nothing to fitness. Similarly if zero investment is made in female function there can be no benefit from that. A sexually reproducing organism that invests nothing in either male or female function therefore has zero fitness. But as investment may be made in either one or both functions it is possible to imagine (and in some circumstances to measure) the fitness

Continued

Box 14.4 *(cont'd)*

benefit that is derived from an investment in either function. The pattern of benefit in relation to investment may be termed a 'fitness gain curve'.

As investment increases so the benefit that is derived from that investment increases. The investment–benefit response for the two types of investment is called the male and female gain curve. This may be a linear response (i) or saturating response (ii).

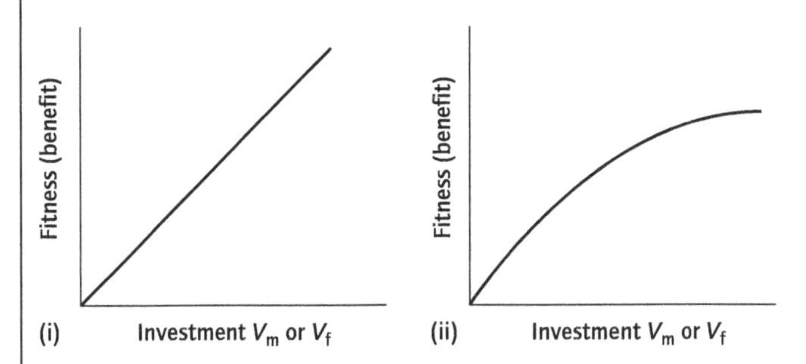

4 Trade-off between investments in male and female function
We can also imagine simultaneous investment in the two alternative means of producing offspring: male function V_m and female function V_f. We can represent the response in relation to lines of equal fitness or so called fitness isoclines. These are represented as parallel lines in (iii). These isoclines represent the benefit of investment and of course fitness is zero when the investment in both functions is zero.

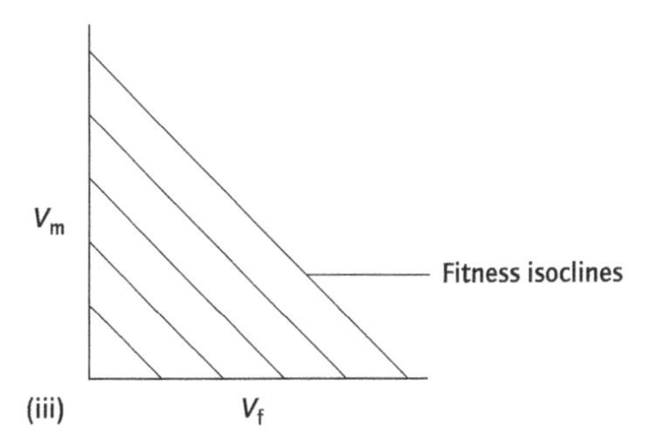

Three possible trade-off curves that may result from investment in the functions V_m and V_f are:

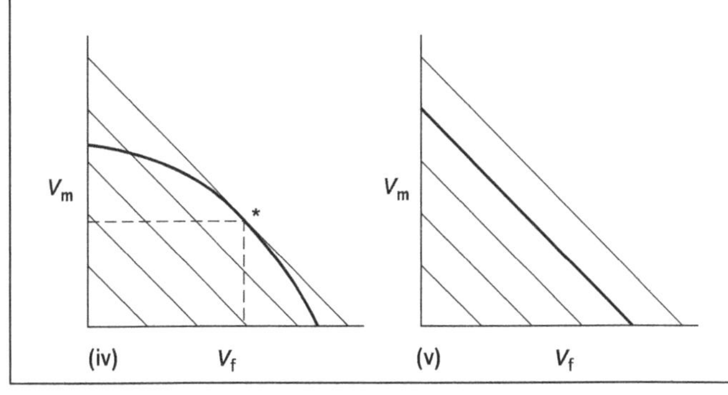

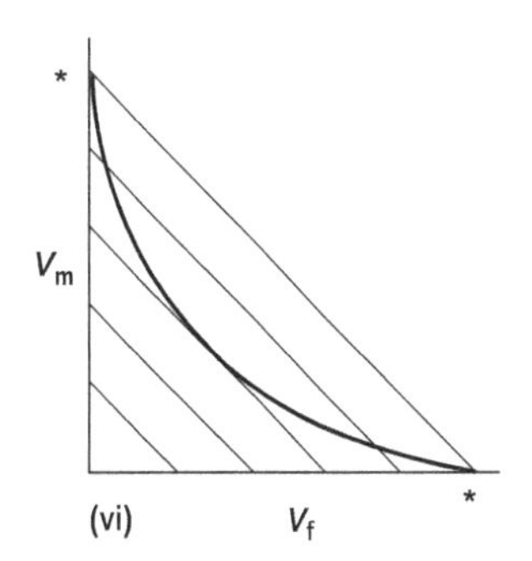

In (iv) the fitness curve is convex and maximum fitness occurs at the point* when there is a balance between investment in both male and female function. When the fitness trade-off curve is shaped like this, the optimum strategy to maximize fitness is hermaphroditism. Maximum fitness is associated with intermediate levels of investment in both V_m and V_f.

In (v) there is a linear trade-off curve and all values of relative investment confer equal fitness.

In (vi) the trade-off curve is concave and maximum fitness is associated with the maximum level of investment in either V_m or V_f. The optimum strategy is then for a female to produce offspring that will be male or female, i.e. gonochorism (dioecy). By taking the analogy with economic investment further it is possible to show that the optimum investment is always when the slope of the trade off curve between the logarithm of male and female investment has a slope of –1.

5 Interpreting the trade-off curve
Conditions leading to the trade off curve as in (iv) will be those that favour hermaphroditism whereas conditions resulting in a trade-off curve as in (vi) will be those that favour gonochorism. Situations leading to (v) may never occur, it would imply that gonochorism and hermaphroditism have equal fitness.

But what could be determining the shape of the trade-off curve? Eric Charnov has suggested that the fixed costs of male and female functions will normally cause the trade-off curve to be concave as in (vi) because a simultaneous hermaphrodite has to carry the fixed costs of investment in both male and female secondary sexual characteristics and a sequential hermaphrodite has to carry the cost of sex reversal. Therefore gonochorism will usually maximize fitness. Under some conditions, however, one or both gain curves may saturate as in (ii) and in these circumstance a form of hermaphroditism will be the Evolutionary Stable Strategy (ESS).

One or both gain curves may saturate in brooding animals at low population density and when the number of partners that may be inseminated by a male is limited. These situations are also discussed in the text.

14.3 The organization of sexual reproduction and life histories: reproductive traits and functions

14.3.1 Introduction

There are many different patterns of sexual reproduction, and the most important of the variable traits which together constitute the reproductive strategy are:

1 The maximum potential lifespan.
2 The number of breeding episodes per lifetime.
3 The pattern of gamete discharge which may involve storage and mass release or alternatively progressive discharge of gametes as they are produced.
4 The degree of synchronization between members of the population.
5 The pattern of mating and degree of outbreeding in the population.
6 The relative size and costs of the gametes.
7 The mode of development and the extent to which juveniles and adults are exposed to different selective pressures.
8 The relative proportion of total available resources allocated to reproduction.

The breeding patterns of animals are frequently classified according to the number of breeding episodes per lifetime and the duration of adult life. Several schemes, which are not mutually exclusive, are used: some of them are compared in Table 14.5.

Table 14.5 Systems for the classification of breeding patterns.

Breeding occurs once per lifetime		
Nouns:	Semelparity/monotely	Virtually synonymous terms;
Adjectives:	Semelparous/monotelic	monotely used primarily by annelid biologists
All insects fall into this category; their life cycles are further described as follows:		
Univoltine	one generation per year, i.e. annuals	
Multivoltine	many generations per year	
Bivoltine	two generations per year	
Semivoltine	one generation every second year	
Breeding occurs several times per lifetime		
Nouns:	Iteroparity/polytely	
Adjectives:	Iteroparous/polytelic	
Annual iteroparity (polytely)	Breeding occurs in discrete episodes separated by periods of usually 1 year	
Continuous iteroparity (polytely)	Breeding more or less continuous during an extended breeding season. When total lifespan is 1 year or less may be indistinguishable from univoltine	

14.3.2 Marine invertebrates

The environments in which animals live profoundly influence the patterns of reproduction. Marine invertebrates are able to discharge unprotected gametes without waterproof coatings into the surrounding medium where fertilization may take place. Freshwater and terrestrial invertebrates, however, are not able to do this. The osmotic stress of exposure to fresh water, or the tendency to lose water on the land, prevents it. In these environments the eggs must be protected and this is one of the reasons why the patterns of reproduction in fresh water and terrestrial environments tend to be rather different from those seen by larger organisms in the marine environment.

The externally fertilized eggs of marine invertebrates frequently develop into mobile planktonic larvae (see the individual systematic sections and Chapter 15). These two factors have profound influences on the whole pattern of reproduction.

The majority of the animal phyla with marine representatives have most, or at least some, species with pelagic larvae and external fertilization; they exhibit a pelago-benthic life cycle as illustrated in Box 14.5. It is also true, however, that all the phyla with species having pelagic larvae have some, or many, species which have non-pelagic, benthic larvae. Furthermore closely related species in the same genus are sometimes found to exhibit contrasting modes of development. We can argue therefore that the conditions required for the invasion of freshwater and terrestrial environments may well have existed already in marine organisms prior to the invasion of that habitat during evolutionary time. In the major phyla with abundant marine representatives, e.g. Annelida, Mollusca, Echinodermata and Crustacea, the majority of species have pelagic larvae that feed and disperse during the planktonic phase. These are often called 'planktotrophic larvae'. A smaller number of species have larvae which are planktonic but which do not feed during the relatively short pelagic phase, being supplied instead with sufficient yolk to reach metamorphosis without feeding. This requires higher investment per egg by the parental organism. Such larvae are called 'lecithotrophic larvae'. It is also found that the development of some marine invertebrates is completed without a distinct pelagic larval phase. Animals with this type of development are said to exhibit 'direct development'. This subdivision of types is not absolute and many larvae exhibit mixed development. They may develop to a relatively advanced phase, feeding on yolk supplies provided by the parents, before eventually beginning to feed in the planktonic larval phase prior to the completion of metamorphosis. It has been estimated that over 70% of all marine invertebrates in temperate zones have planktotrophic pelagic development which implies that in most circumstances this pattern of reproduction has clear advantages. These are thought to be due to some or all of the following:

1 Exploitation of the temporary food resource provided by phytoplanktonic blooms.
2 Colonization of new habitats.
3 Expansion of geographical range.

Box 14.5 The life cycle of marine invertebrates

The eggs and spermatozoa of marine invertebrates may be released into sea water where fertilization can take place.

This has profound implications for their reproductive biology.

Below are examples of the characteristic larvae of the major groups of marine invertebrates. The sketches show the adult form and the corresponding larvae. Note the adults are drawn to a much smaller scale than the larvae.

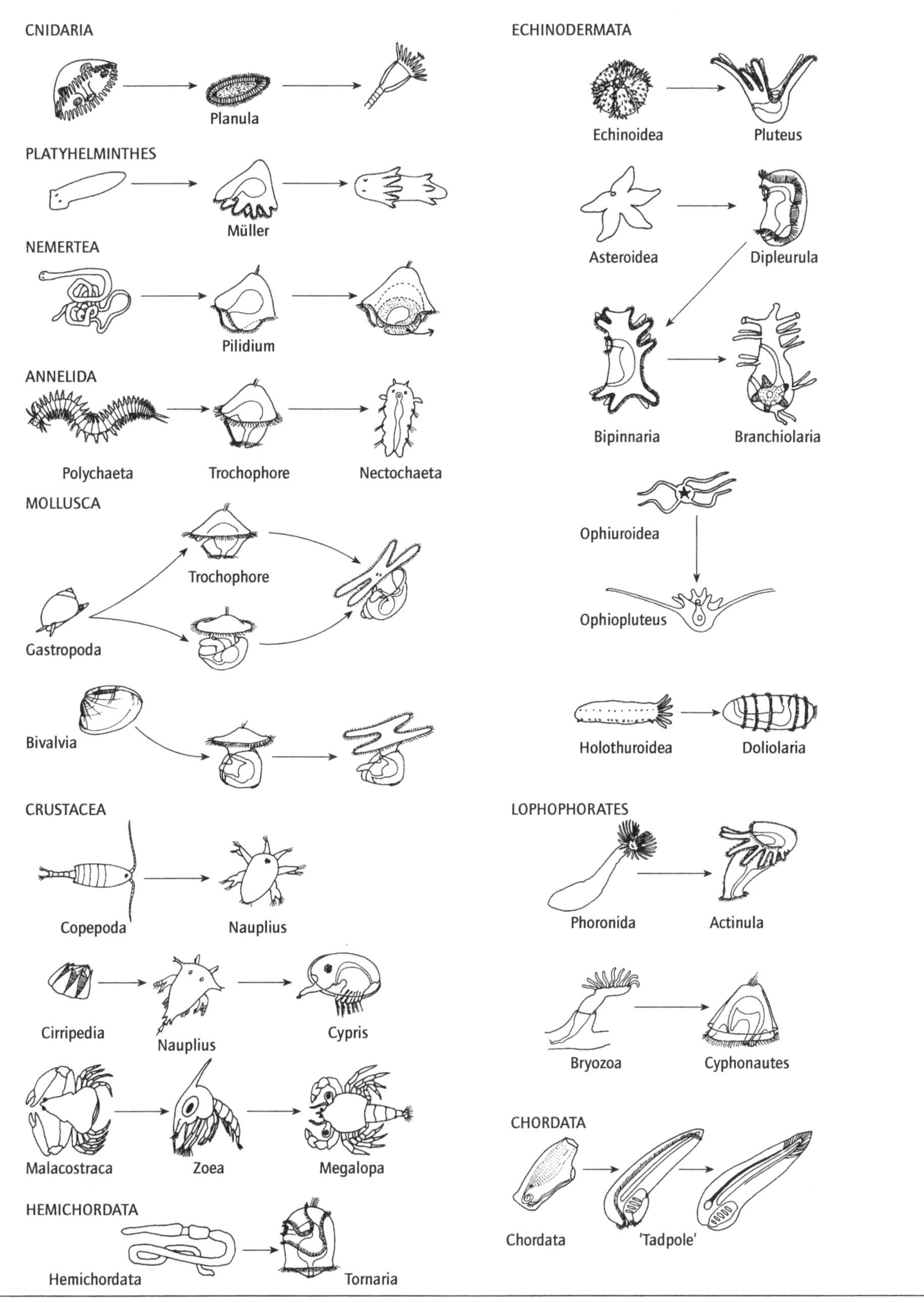

Continued p. 340

Box 14.5 *(cont'd)*

The figure below shows characteristic life cycles: (i) pelago-benthic – with pelagic larval stage; (ii) holo-benthic – suppression of the primitive pelagic larval stage; (iii) holo-pelagic – with entirely pelagic history.

All three patterns can be found in many groups. The examples given here are all gastropod molluscs.

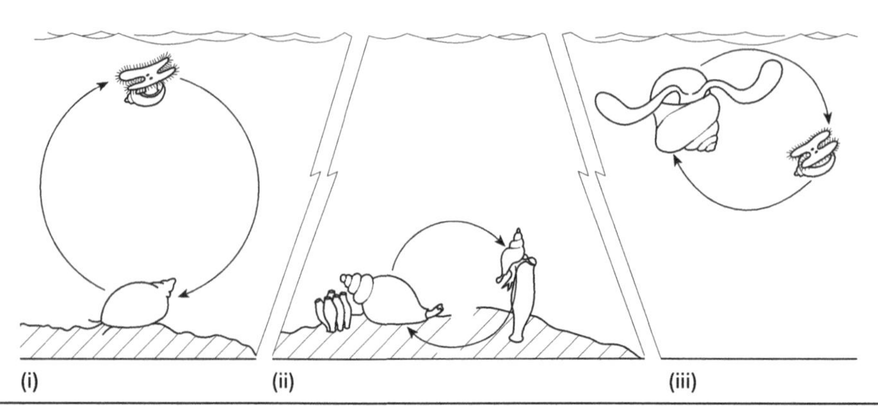

4 Avoidance of catastrophe associated with local habitat failure.
5 Avoidance of local and sib-competition.
6 Exposure of diverse offspring to the maximum degree of habitat diversity.

Pelagic development is frequently (but not always) associated with external fertilization in which the sperm contacts the egg following dispersion free in the sea water. This in turn is associated with the possession of a round-headed spermatozoon of the type illustrated in Fig. 14.12 (see below) which is best called by the functional name 'ect-aquasperm' to avoid unwanted phylogenetic implications. A simple round-headed free-swimming sperm type is found throughout the animal kingdom when fertilization occurs externally in sea water. These two traits help to define a pattern of reproduction which is typically, though not universally, found in a diverse range of marine invertebrates but which should not be regarded as being primitive as some have suggested.

As well as these two traits, there are several others that together constitute a syndrome of co-variable traits characteristic of the life histories of marine invertebrates (Table 14.6a). This syndrome of traits however is not always found. The smallest representatives of all the main classes of marine invertebrates, such as those inhabiting interstitial spaces in sands, usually exhibit a rather different set of reproductive traits as summarized in Table 14.6b. It is interesting to note that virtually the same set of traits could be listed as being characteristic of invertebrates inhabiting non-marine environments – in freshwater, soil and terrestrial conditions. This implies that there are strong functional constraints on the reproductive traits that may be adopted by animals living under different environmental conditions.

For many years it was implicity assumed that certain of these reproductive traits could be interpreted as being 'primitive' or plesiomorphic within a given clade and, to a certain extent,

Table 14.6 Co-variable traits in marine invertebrates.

(a) Co-variable traits set I: frequently exhibited by larger marine invertebrates	(b) Co-variable traits set II: frequently exhibited by smaller (often minute) marine invertebrates
Eggs freely discharged into water, free pelagic development	Eggs not freely discharged (frequently brooded)
External fertilization (ect-aquasperm)	Internal fertilization (often with sperm storage in spermathecae)
Small quantity of yolk in the egg	Sperm type with diverse specialization and usually filiform morphology
Total equal cleavage	Large quantities of yolk in the egg
Blastula with blastocoel	Very unusual or superficial cleavage (see Chapter 15)
Gastrulation by invagination	Blastocoel obliterated
Planktotrophy in the larva	Gastrulation by epiboly not invagination (see Chapter 15)
Discrete, once per year reproduction with strong seasonality, between and within individual synchrony of gametogenesis	Lecithotrophy in the larvae
Long-term storage of accumulated germ cells	Frequent episodes of egg production often during an extended breeding season
Large body size, with capacious body cavity or intercellular spaces	This set of traits is observed in interstitial and minute marine animals but is not necessarily associated with small body size in freshwater and terrestrial organisms

earlier editions of this text reflected this generally accepted view. In particular, possession of a single round-headed sperm and development via a pelagic larval phase tended to be regarded as 'primitive' traits among invertebrates. We would wish to caution against the interpretation of any reproductive traits being in this sense 'primitive' unless that interpretation is supported by a body of independent evidence from cladistic or molecular phylogenies. Hypotheses about phylogenic relatedness may be tested through the formal analysis of a large range of characters (cladistic analysis) or from the data provided by investigations of gene sequences (molecular phylogeny).

Without implying any phylogenetic or evolutionary sequence however it is nevertheless possible to recognize some functional co-variable associations among reproductive traits as set out below.

1 Individual egg size is co-variable with developmental mode. Smaller eggs frequently develop through a free-swimming pelagic larval stage; larger eggs more frequently develop as a short-lived lecithotrophic larvae or are brooded or have direct development.

2 Sperm morphology is co-variable with the site and mode of fertilization. Free-swimming round-headed sperm with simple mitochondria are associated with fertilization in the external medium (sea water) and filiform spermatozoa with elongated nuclei, long acrosomes and modified mitochondria are associated with brooding and where various forms of internal fertilization occur.

3 Spawning mode is, to some extent, co-variable with body size among marine invertebrates: organisms with small body size rarely exhibit broadcast spawning but large-bodied invertebrates frequently (but not invariably) do so.

No doubt other sets of similarly co-variable traits could be identified but the assumption that any one state is primitive (within a clade) without supporting evidence should always be avoided.

Formal cladistic analysis suggests that a high degree of plasticity with respect to reproductive traits may be retained within a clade. We have already seen from an analysis of the molecular mechanisms controlling sex determination that such fundamental traits as chromosomes versus environmental determination and gonochorism versus hermaphroditism have retained a degree of plasticity and are unlikely to have become fixed by previous evolutionary events. The potential to make an evolutionary response to changed conditions seems to be an important feature of reproduction among animals. Care must be taken in accepting this argument, however, for it is in itself a group selection theory.

The mass discharge of gametocytes in an annual spawning crisis is very common among large-bodied marine invertebrates that have the capacity to store the gametes in body cavities (often the coelom) and then release them in a single seasonal spawning crisis. One of the reasons why minute organisms do not exhibit broadcast spawning may be because it is not possible for them to store sufficient quantities of gametocytes in this way. There are two quite different life cycles in which mass spawning occurs and in which a clear annual spawning season may be expressed at the population level. In the more common pattern, spawning occurs at annual intervals during a lifespan of two or more years in an iteroparous life history (Table 14.5). Some marine animals, most notably all members of the polychaete family Nereidae and all cephalopod molluscs other than the primitive nautiloids breed only once per lifetime and are said to have a semelparous life cycle (Table 14.5). In such animals spawning may be synchronized but occurs only once per lifetime. Mass spawning in animals of this type is followed by the genetically determined death of the individual though the age at reproduction, and hence death, may be variable according to environmental conditions experienced during early life.

14.3.3 Freshwater and terrestrial invertebrates

The choice between planktotrophic or lecithotrophic pelagic development and direct development, which is such a feature of reproduction in the sea, is not open to those animals inhabiting freshwater and terrestrial environments. The osmotic and other stresses in these habitats preclude release of naked unprotected spermatozoa and eggs, and fertilization must be internal. Similarly developing embryos must be protected against water loss or osmotic stress, e.g. they must be enveloped in a waterproof coat or in a cocoon. Consequently, a pelagic larval phase rarely occurs and the larger, soft-bodied non-marine invertebrates usually exhibit most of the following reproductive traits:

1 Viviparity or deposition of eggs in impermeable membranes or cocoons.

2 Internal fertilization, requiring direct pairing between partners.

3 Structurally complex, often filiform spermatozoa.

4 Investment of relatively high levels of maternal resources in each egg, and consequently lower fecundities.

5 Brood care or provisioning of offspring.

6 Repeated or episodic egg laying, exploiting the potential that an internal sperm store provides for continuous reproduction.

7 Hermaphroditism. (This is especially true for the soft-bodied invertebrates but is not the case for the arthropods.)

You will notice how similar this set of traits is to those listed in Table 14.6b for smaller marine invertebrates.

Some freshwater animals seem to betray a recent origin from marine ancestors; the freshwater bivalves for instance, e.g. *Anodonta*, are structurally very similar to their marine relatives and they do release relatively large numbers of what are in effect modified veliger larvae called 'glochidia'. The larvae are not free-living, however, but are external hitch-hikers on the gills or skin of freshwater fish.

The life cycles of the freshwater platyhelminthes, clitellate annelids and pulmonate molluscs are remarkably similar. They are all simultaneous hermaphrodites (unless sexual reproduction has been suppressed), have complex sexual behaviour, highly specialized accessory glands and the capacity to protect their embryos. Their life cycles also often involve extended periods of egg-laying activity rather than the sychronized mass release of gametes so characteristic of animals that live in the sea. The principal trade offs that are made in such species are those between fecundity and longevity; some species live for only one year or less but others, usually those that achieve a larger body size, live for several years.

Freshwater and terrestrial environments are more extreme in physical variables than are marine ones, and most of the invertebrates inhabiting them have the capacity to enter a physiological resting state known as 'diapause'. In this state the organism can lose water and withstand extreme conditions without harm; the metabolic demand is lowered and the requirement for an external energy source can be nil so that the periodic absence of food which occurs in temperate and polar regions can be withstood. Long-lived species such as some pulmonates and clitellates will enter the diapause state as adults, but many smaller forms do so as eggs. Thus many populations of leeches only exist as embryos in cocoons during the winter months. The habit of overwintering as resting eggs is particularly characteristic of minute forms such as the rotifers and several groups of crustaceans such as the cladoceran water-fleas. As explained above, these organisms exhibit life histories in which there is an alternation of asexual and sexual generations (see Fig. 14.5); the diapause eggs, which arise from asexual reproduction, hatch in the spring to give rise to new clones of offspring which can exploit the new resources. The terminology already introduced (see Fig. 14.6) is useful here – the clone of aphids sharing the same genome may be referred to as the genet which is the unit of evolution whereas the many individuals produced parthenogenetically that share this genome may be distinguished as multiple ramets within that genet.

The bdelloid rotifers exhibit an extreme form of diapause; their encysted eggs can remain in what is virtually a state of suspended animation for years until they are blown into suitable conditions for growth. This capacity may be related to the absence of sexual reproduction; by avoiding unfavourable conditions they live to enjoy what are in effect permanent optimal conditions. They will be less likely to be targeted by parasites and disease organisms and will thus escape the conditions that are thought to favour sexual reproduction (Section 14.2.4.1).

The insects (Hexapoda) and the spiders and mites (Arachnida) have most successfully adapted to terrestrial environments (see Sections 8.4.3.2 and 8.5.3B.2) and particularly in the case of the insects, some have secondarily adapted to fresh waters as adults or as larval or nymphal forms. The reproductive biology of these groups differs from that of the soft-bodied invertebrates in one important respect. They are very rarely hermaphrodite; the vast majority of species reproduce only sexually and the sexes are always separate. Their success is due in large part to the development of a waterproof covering – the integument or cuticle – but is also due to their ability to lay waterproof eggs. Such eggs must be fertilized before they are laid; consequently, all of them exhibit internal fertilization, often associated with complex copulatory behaviour. This latter requires contact between males and females and thus leads to the possibility of sexual selection. The prevalence of gonochorism is thought to be a consequence of their high mobility.

It is traditional to subdivide the life cycles of insects according to modes of development. Two major types may be recognized. In many insects, external wing buds begin to develop prior to the final instar moult to the sexually mature adult condition these are said to show the exopterygote ('hemimetabolous') condition whereas in many others the wing buds develop internally from imaginal discs that do not begin to differentiate until the final moult, the endopterygote ('holometabolous') condition. The development of the imaginal discs of holometabolous insects is described in Chapter 15 and the endocrine control in Chapter 16. Some wingless insects continue to grow and moult after they have become adult and may said to be 'ametabolous'. A different, more functional approach will be used here. The life histories of insects can be classified according to the different patterns of activity and function during the pre-adult and adult stages. Insects grow through a series of moults or instars, and it is possible therefore for the different instars to have different functions. The principal ones are:

1 Development and differentiation.
2 Food and other resource acquisition.
3 Dispersal and resource tracking.
4 Mating and mate selection.
5 Allocation of resources to offspring.
6 Selection of sites for offspring growth.
7 Oviposition.

A number of different insect life histories are analysed in this way in Box 14.6. Note the marked differences in the allocation of dispersal and resource acquisition functions and the contrasts which exist between most insects and the marine invertebrates. In the latter, dispersal is often a function of the juvenile phase and resource acquisition a function of the adult phase. In the Insecta these functions are frequently reversed as illustrated in Box 14.6.

The simplest life history is one in which there is a gradual transition during development to the adult condition and both adults and larvae exploit the same food resource. Both juveniles and adults have the additional roles of dispersal, resource tracking, mating and egg laying. This simple life history is exhibited by the Orthoptera (e.g. the locusts) and is illustrated in Box 14.6(1).

Not infrequently, however, adults and larvae feed in different ways, and are consequently subject to quite different selection pressures. Dipteran blowflies, for instance, have larvae feeding on a rich but temporary food resource, dead meat. This must be tracked by the parents which often feed in a quite different way (Box 14.6(2)). Several insect larvae have an aquatic existence and the niche differentiation between adults and juveniles is then even more marked.

From this situation, life histories may have arisen in which the

Box 14.6 A functional analysis of the life histories of insects

1 Juveniles and adults exploit a similar food resource: Juvenile functions – development and differentiation, resource acquisition. Adult functions – resource acquisition, dispersal and resource tracking, mating and oviposition. For example the Orthoptera (locusts and grasshoppers).

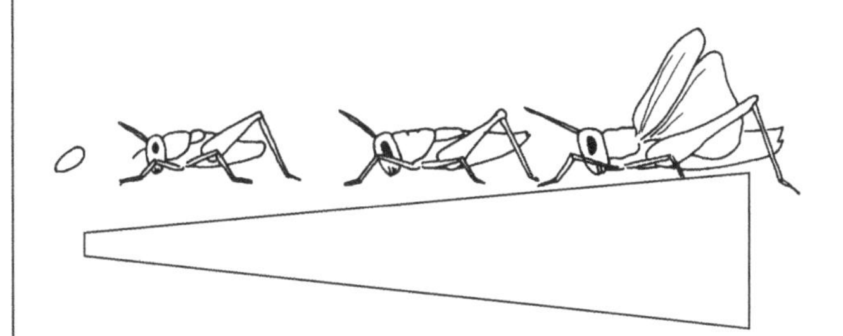

2 Juvenile and adults exploit a different food resource:
Juvenile functions – development and resource acquisition.
Pupal functions – differentiation and development.
Adult functions – resource acquisition, resource tracking, mating, and selection of sites for offspring.
For example blowflies and other Diptera.

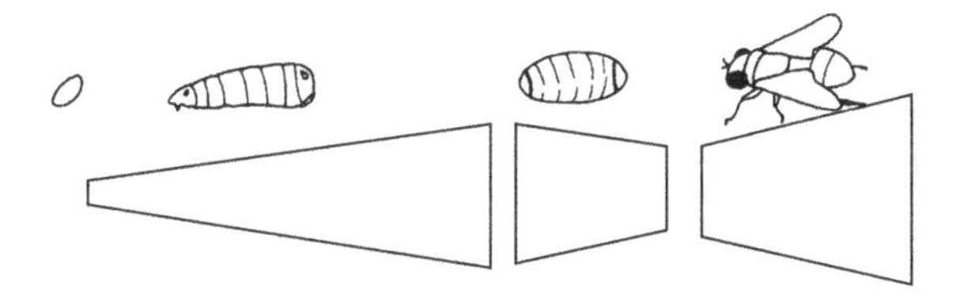

3 Adults ephemeral and not involved in resource acquisition; juvenile functions, development and resource acquisition, adult functions, mating, resource tracking and dispersal.
(i) Larvae aquatic, e.g. Ephemeroptera.
(ii) Larvae terrestrial, e.g. Lepidoptera. In this example the adults are longer-lived and have mouthparts specialized for nectar collection as a fuel for flight.

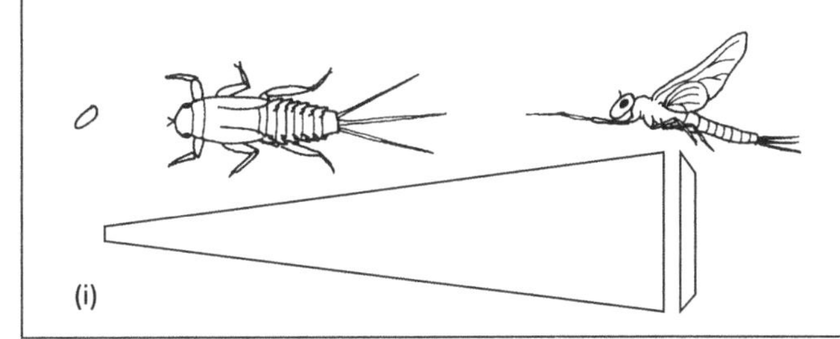

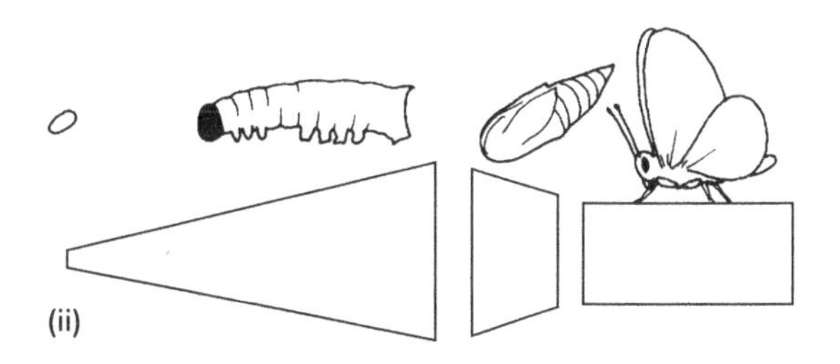

4 Adults solely responsible for resource acquisition – provisioning:
Larval roles – resource acquisition, dispersal and resource tracking, mating, allocation to offspring, selection of sites for offspring development, oviposition.
For example a solitary hymenopteran wasp.
The adult provides all resources available to the offspring.

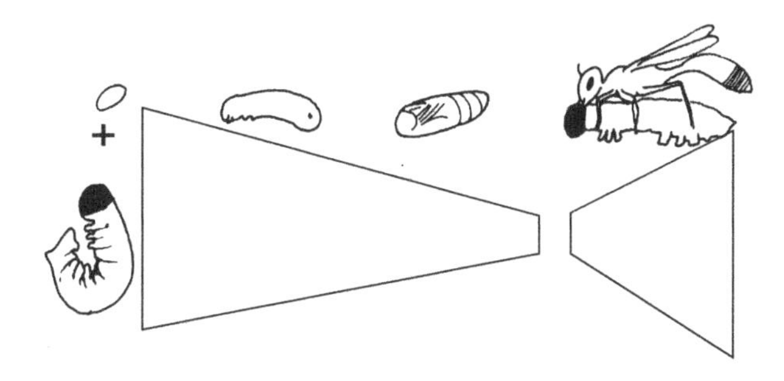

5 Resource acquisition, development, mating and oviposition separated by caste differentiation:
Juveniles – resource utilization and growth.
Sterile workers – resource acquisition.
Queens (fertile adults) – mating and mate selection, oviposition, dispersal and resource tracking.
Males (fertile adults) – mating – no role in resource acquisition.
For example honey bees.

roles of dispersal and resource tracking have been separated from those of resource acquisition, and the adults feed either very little or not at all. This separation of function may have arisen independently many times in the evolution of the insects. The most extreme examples are the Ephemeroptera, Plecoptera and Trichoptera, the larvae of which are aquatic carnivores. The adults or imagos of these groups do not feed and only live for a few hours. During the brief adult life, they mate and deposit eggs in an environment suitable for their subsequent survival and development (Box 14.6(3)).

The well-known life history of the Lepidoptera is similar, although the adults do obtain some energy from nectar sipping. The caterpillar larvae have one basic function, which is to acquire as much of the available resource as quickly as possible. The transition to the adult phase, which has the roles of mating and resource tracking, involves a non-feeding pupal stage (Box 14.6(3)). The food resource may be seasonal and it can be tracked through time as well as space by the intervention of diapause as egg, pupa or adult, or by migration.

The adults of some insects have acquired not only the functions of dispersal, mating and resource tracking, but also resource acquisition. They are solely responsible for finding the food and making it available to their offspring, this phenomenon is described as 'provisioning'; it is observed in some Orthoptera, Coleoptera and Diptera. It is most characteristic of the Hymenoptera (Box 14.6(4)) and is thought to have played a crucial role in the evolution of eusocial behaviour among the bees and wasps (Box 14.6(5)).

14.4 The control of reproductive processes

14.4.1 Ultimate and proximate factors

The life cycles discussed in the previous sections involve complex sequences of cellular activity, and these must be coordinated in order to bring about an orderly progression of events and a proper relationship with outside factors, and where necessary to maintain the proper degree of synchrony between different members of the population. It is usually supposed that populations of animals that show clear cycles of reproductive activity do so in response to cycles of environmental change. Environmental conditions are not constant and it follows that certain times will be more favourable for reproductive activity than others. Evolutionary forces which select for reproduction to occur at the most favourable times are referred to as the 'ultimate factors' controlling reproduction. They are not necessarily the same ones as those used to control the reproductive cycle. Gametogenesis and particularly oogenesis may take several months for completion, and the environmental signals that regulate the progression of the cellular events culminating in reproduction may be quite different from those that confer a selective advantage on individuals that breed at a particular time.

Environmental events that regulate the progression of gametogenesis and which thus control the time and season of reproduction are referred to as the 'proximate factors'. In order to control reproductive processes environmental changes must be detected and the information integrated within the central nervous system of the reacting individual before being transduced in the form of a change in nervous, neuroendocrine or endocrine activity (Section 16.12). There is, in effect, a chain of command which results in a highly structured species-specific reproductive cycle (Fig. 14.11). This controls the differentiation of the germ cells, the flow of energy, and its relative allocation to reproductive processes, maintenance and growth.

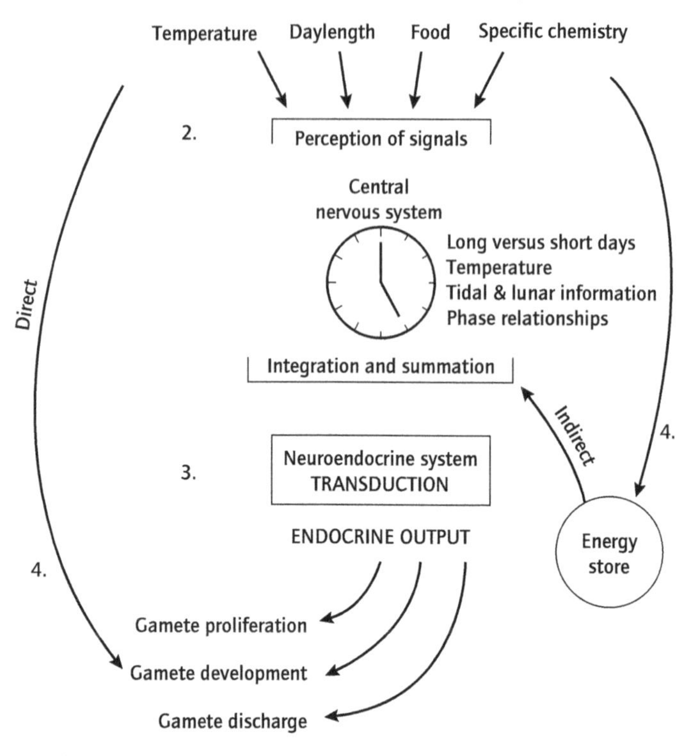

Fig. 14.11 A diagrammatic representation of the chain of control elements involved in regulating an externally synchronized reproductive cycle. (After Olive, 1985a.)

14.4.2 Component processes: gametogenesis

Spermatogenesis results in the formation of the male germ cells and oogenesis in the formation of the female germ cells; they are rather different processes and are illustrated in Fig. 14.12 and described in a little more detail below.

14.4.2.1 Spermatogenesis

Spermatogenesis in most invertebrates is completed quickly; it often involves frequent mitotic divisions prior to the onset of meiosis to yield vast numbers of germ cells. Each spermatogonium which transforms to a primary spermatocyte will give rise to four spermatids which, by a process of differentiation, give rise to the spermatozoa (Fig. 14.13a). Marine invertebrates that exhibit broadcast spawning, i.e. in which fertilization takes place in the sea water, typically have a round-headed sperm type known as an ect-aquasperm. There is a terminal acrosome, a large spherical nucleus, a short middle piece with simple unmodified mitochondria and a long flagellum with typical

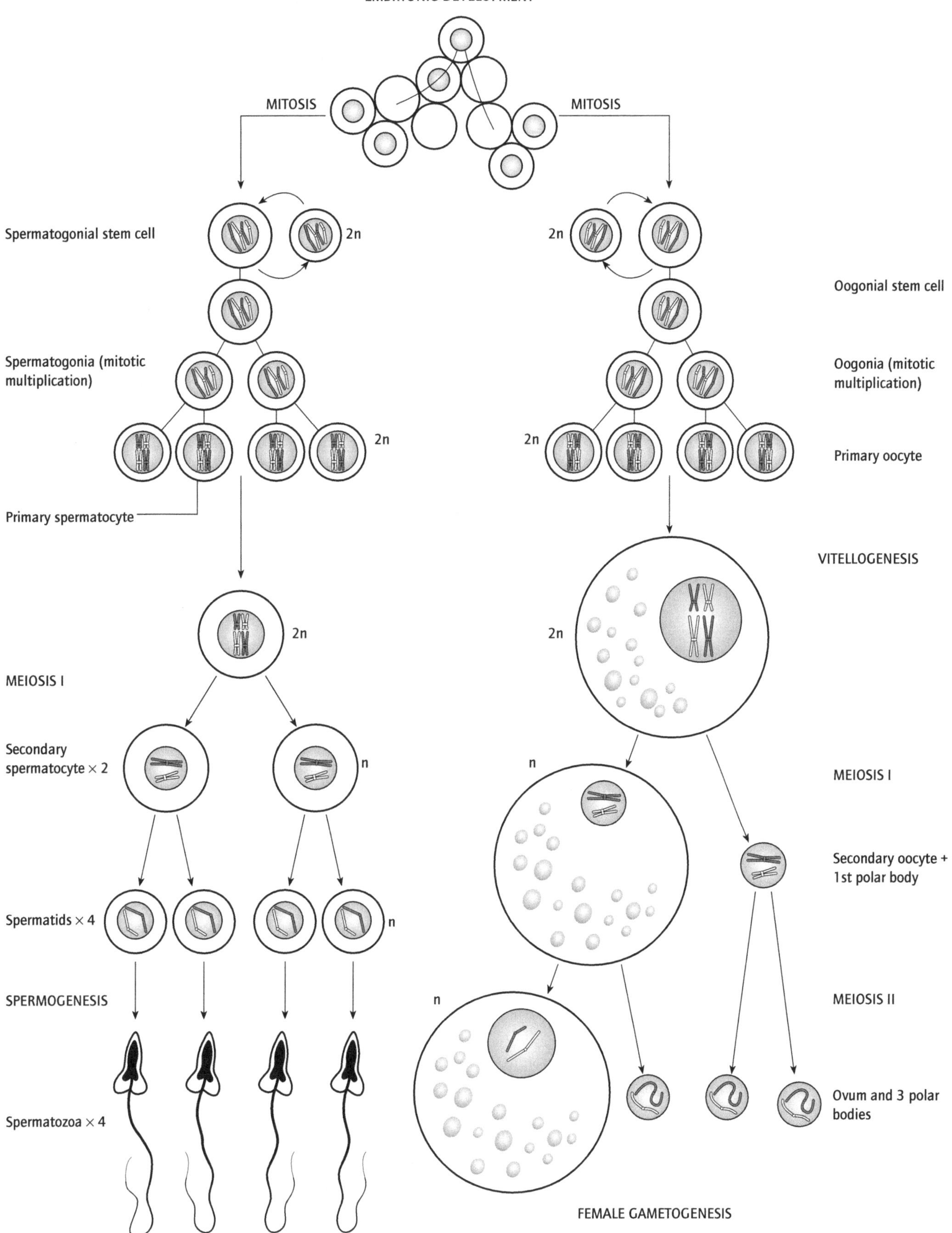

Fig. 14.12 A diagrammatic representation of the principal cellular events of gametogenesis.

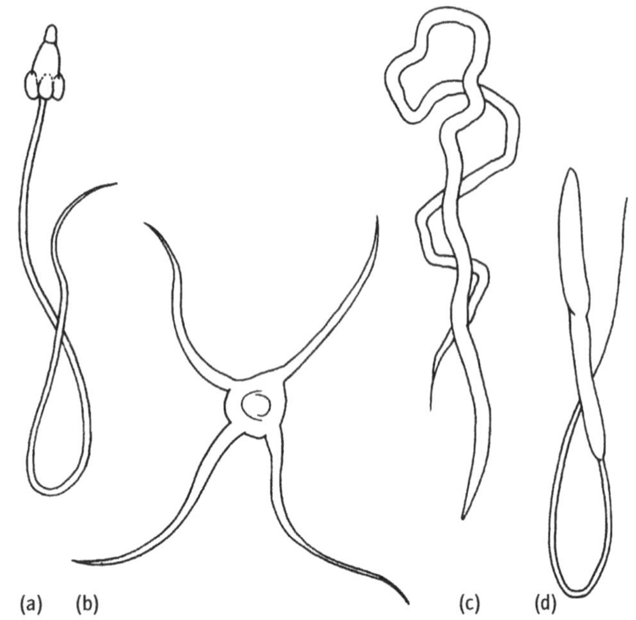

Fig. 14.13 (a) Spermatozoon of the primitive round-headed type. (b)–(d) Spermatozoa of a more advanced filiform type. Spermatozoa with a similar filiform shape may have very different internal structure and organization.

microfilament arrangements (see also Chapter 10, Fig. 10.4). A round-headed sperm is shown in Fig. 14.13a. Animals with different modes of fertilization often have rather different sperm. The mode of fertilization may involve copulation and internal fertilization, or a system for sperm transfer via spermatophores or the sperm may be collected from sea water by females and stored prior to use in special vesicles – the spermathecae. Marine invertebrates that exhibit these modes of fertilization usually have sperm that are elongated. Ultrastructural studies have revealed a wide diversity in the structure of such elongate sperm (Fig. 14.13b,c,d).

The round-headed sperm type should not be regarded as being 'primitive' within any clade and the terms 'ect-aquasperm', 'ent-aquasperm' and 'introsperm', which do not carry phytogenetic implications, must be used in the absence of a clearly tested phylogeny.

14.4.2.2 *Oogenesis*

Oogenesis is a protracted process during which food reserves are deposited in the developing egg, usually during an extended prophase, and meiosis is not completed until shortly before or after fertilization (Fig. 14.12).

In some marine invertebrates, e.g. Sipuncula, Echiura and some Polychaeta, the developing oocytes are solitary cells suspended freely in the coelomic fluid. Such a pattern is described as 'solitary oogenesis' (Box 14.7(1a)); in these cases the metabolites to be stored in the oocyte cytoplasm may be taken up by the egg from the surrounding body fluids as low-molecular-weight precursors (amino acids, simple sugars and monoglycerides) and built up into complex storage products, collectively called 'yolk', by synthetic organelles in the ooctye cytoplasm. This pattern is described as 'autosynthetic' and is particularly associated with solitary oogenesis, but the association is not absolute (see below).

Most invertebrates, however, exhibit either 'follicular' or 'nutrimentary oogenesis' (Box 14.7(1b,c)). In follicular oogenesis the developing oocytes are intimately associated with an epithelium of somatic cells – the follicle cells – which form a box-like covering around the oocyte. Nutrimentary oogenesis involves sibling cells of the oocyte derived by incomplete cytokinesis during the mitotic divisions of the mother cell or oogonium, prior to the onset of meiototic prophase. There is an oocyte–nurse cell complex of this type in the eunicid polychaetes *Diopatra* as shown in Box 14.7(1ci), but more often the oocyte–nurse cell complex is also surrounded by a layer of follicle cells as in Box 14.7(1cii). The ovary may then consist of a series of ovarioles each of which is a string like series of developing follicles. Studies of the developing oocytes of the fruit fly *Drosophila*, which are of this type, are now providing important information into the origin of localized molecular information in the cytoplasm of the developing egg that creates the regional organization of the future embryo. The follicle cells and nurse cells have an important role in establishing gradients which profoundly influence the body architecture and this is discussed further in Chapter 15.

In the more complex types of oogenesis, most of the high-molecular-weight materials deposited in the oocyte cytoplasm are synthesized not in the oocyte cytoplasm but by somatic cells elsewhere in the body. The high-molecular-weight yolk precursors (vitellogenins) and other materials are then transported to the oocytes by the body fluids or blood vascular system. This pattern of oogenesis is described as 'heterosynthetic'. It was first discovered in insects but has subsequently been shown to be widespread. This pattern of yolk synthesis is characteristic of the Crustacea and the Mollusca and is known in some Annelida.

In the Polychaeta in which autosynthesis and solitary oogenesis were thought to be typical a remarkably wide range of patterns of oogenesis can be found, rivalling those of the insects in structural complexity. Some are now known to exhibit follicular and nutrimentary oogenesis and in the Nereidae, where oogenesis is solitary, there is evidence of heterosynthesis. The different patterns of oogenesis are explained further and illustrated in Box 14.7.

In turbellarians the 'egg' is often a complex structure composed of a relatively yolk-free oocyte, combined with nurse cells which are packed with a yolk-like cytoplasm produced by the vitellarium (Fig. 14.14) all packaged together within a tanned protein coat. This type of egg formation is termed 'ectolecithal' and is thought to be an advanced trait in platyhelminthes. Some free-living turbellarians have a more primitive endolecithal mode of egg development in which yolk is stored in the oocyte cytoplasm. Freshwater and terrestrial annelids and molluscs show similar reproductive adaptations from this point of view;

Box 14.7 Patterns of egg formation and oogenesis invertebrates

1 *Solitary, follicular and nutrimentary oogensis.*
One way of classifying the diverse patterns of oogenesis among invertebrates is by reference to the degree to which other cells are intimately involved in this process.

Three basic patterns can be recognized:
(a) Solitary: The oocytes develop without any close association with other cells. Especially in animals with large spacious body cavities. The oocytes float freely in the coelomic cavity of echiuroids, sipunculans and some, but not all, polychaete annelids and in such cases that is where most of the vitellogenic growth takes place.

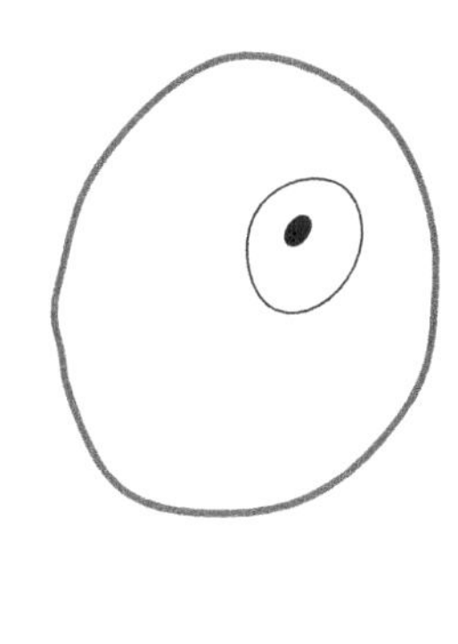

(a)

(b) Follicular: The oocytes are intimately associated with somatic cells. These cells may play an important role in the transport of macromolecules to the germ cell cytoplasm and may cover the surface of the germ cells.

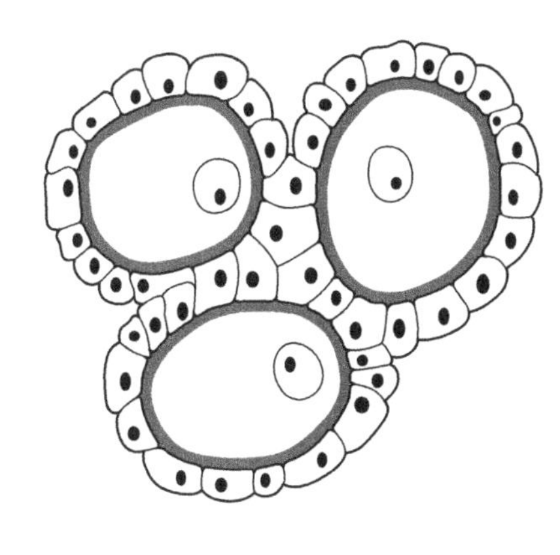

(b)

(c) Nutrimentary: The oocytes retain a close relationship with other cells of the germ cell line not destined to become germ cells. Incomplete cytokinesis results in the formation of a syncytial complex with cytoplasmic connections between cells of the complex. Usually only one cell of the complex becomes an egg; the other cells are referred to as nurse cells and contribute in some way to the development of the egg. Two types of association may be found.

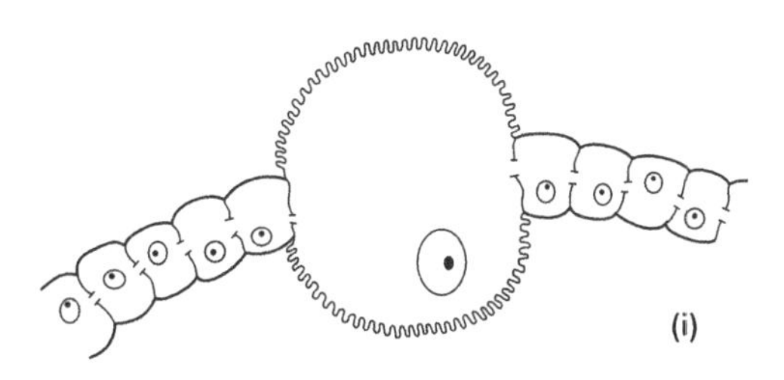

(i)

(i) Without follicle cells. The illustrated complex is from the polychaete annelid *Diopatra*. The oocyte is one of a chain of cells, many such chains being produced by the ovaries. There are no follicle cells.
(ii) With follicle cells. The ovaries of insects are always follicular. Sometimes, as in the dipteran insects, the follicle contains a syncytial complex of up to 16 cells arranged in a syncytium with cytoplasmic connections as shown. The nurse cell–oocyte complex is completely enclosed by the follicular epithelium.

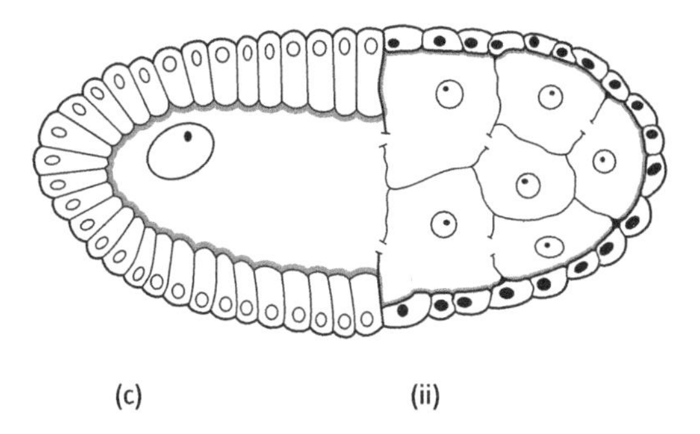

(c) (ii)

2 *Nutrient transfer and biosynthesis.*
During oogenesis the developing oocyte accumulates massive reserves of yolk, RNA molecules and other substances. These may be synthesized by the oocyte itself or by the cytoplasm of other non-germ cells.
(i) Autosynthesis: The site of synthesis of macromolecules and storage products is the oocyte cytoplasm using genetic products of the primary oocyte nucleus.

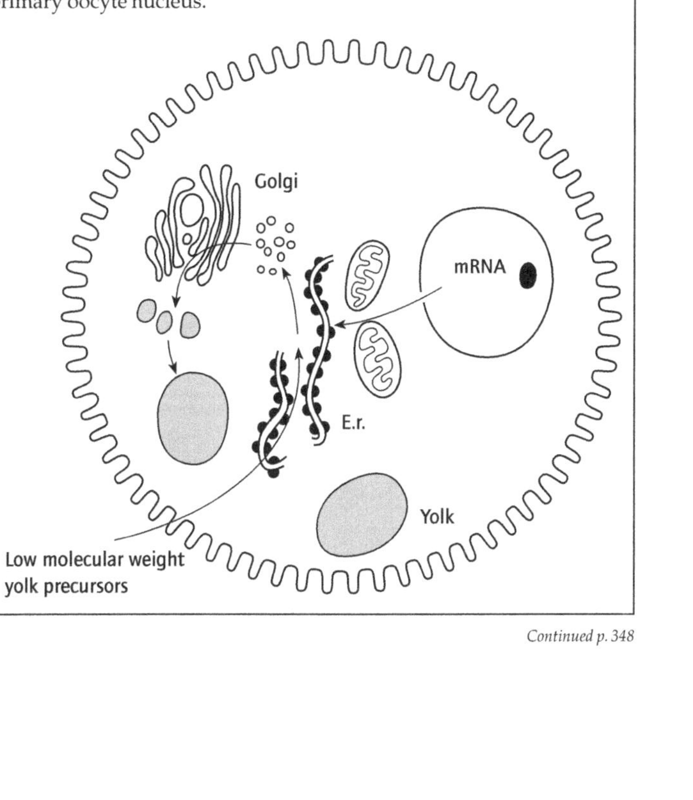

Continued p. 348

Box 14.7 (*cont'd*)

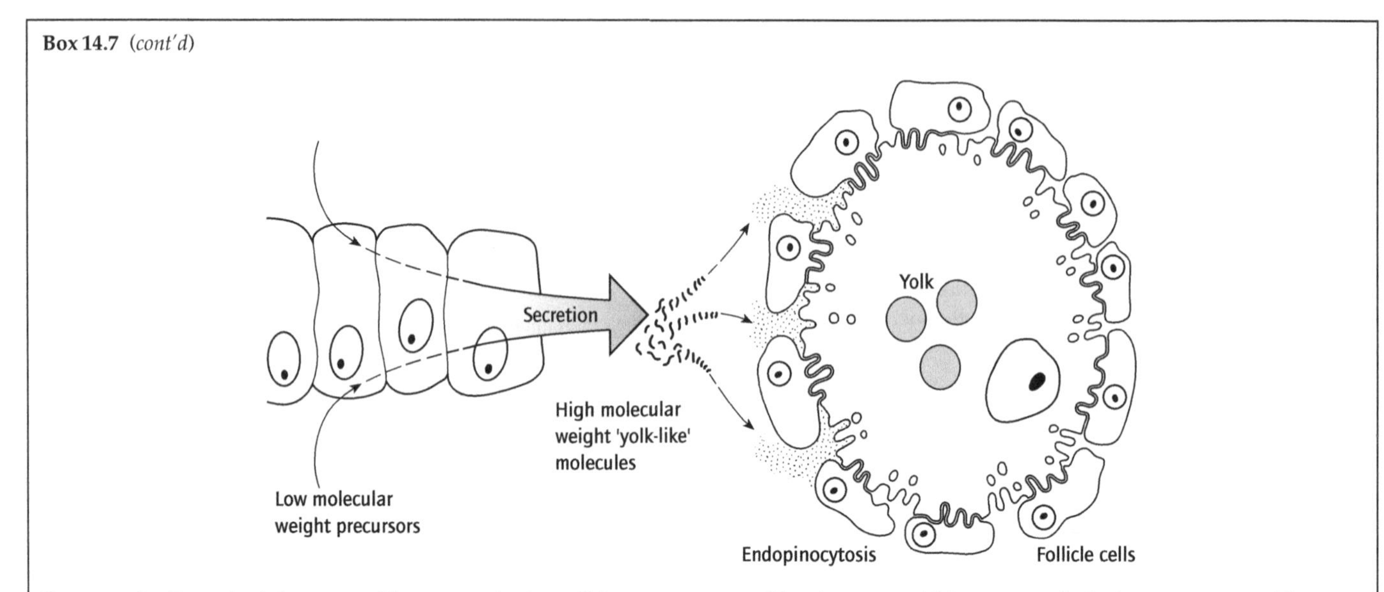

Patterns of yolk synthesis in oocytes. The autosynthetic condition usually associated with solitary oocyte development. Low molecular weight precursors are absorbed by the microvillus surface of the oocyte, and the synthesis of high molecular weight storage products takes place in the oocyte cytoplasm using informational molecules derived from the oocyte nucleus.

(*ii*) *Heterosynthesis*: The storage products of the oocyte are first synthesized by other cells. Accessory cells may have an important role in the transfer and uptake of such materials, and in many cases the heterosynthetic production of yolk may be thought of as an amplification system. This may permit, for instance, very rapid oocyte growth.

The heterosynthetic condition, usually associated with a follicular mode of oocyte development but not necessarily so. Complex yolk precursors of high molecular weight are transported to the oocyte in the body fluids (blood system, haemolymph, coelomic fluid, etc.). These high molecular weight substances are manufactured from low molecular weight precursors by non-germ cells using informational molecules of the accessory cells. The diagram represents the situation in some insects.

in both, the eggs are laid within cocoons containing a nutritive albumen which supplements the storage materials deposited in the egg cytoplasm. The provision of large amounts of yolk or albumen, within a waterproof tanned protein cocoon will enable the offspring to develop to such a stage that they are able to withstand the rigours of exposure to the stresses of the terrestrial and freshwater environments.

14.4.3 Synchronous reproduction of marine invertebrates

The dominant pattern of reproduction among marine invertebrates (Section 14.3.2) requires a high degree of synchronization of reproductive events within individuals and between different members of a population. The degree of synchronization can be very dramatic indeed. Table 14.7 records some of the dates and times when the breeding of the Pacific palolo worms (Polychaeta, Eunicida) has occurred during the last 100 years: spawning has a precise and fixed relationship to the time of the third lunar quarter which first occurs after a date in early October. The timing is accurate to within one day and spawning also occurs at precisely the same time of each day. A similarly precise timing of reproduction has been observed in the Japanese crinoid *Comanthus japonicus* (Echinodermata, Crinoidea) and Fig. 14.15 shows that the whole cycle of gametogenesis spanning almost the whole year is constrained into this pattern. These are extreme examples perhaps, but the reproduction of most marine invertebrates in virtually all phyla involves to some degree this kind of synchronization.

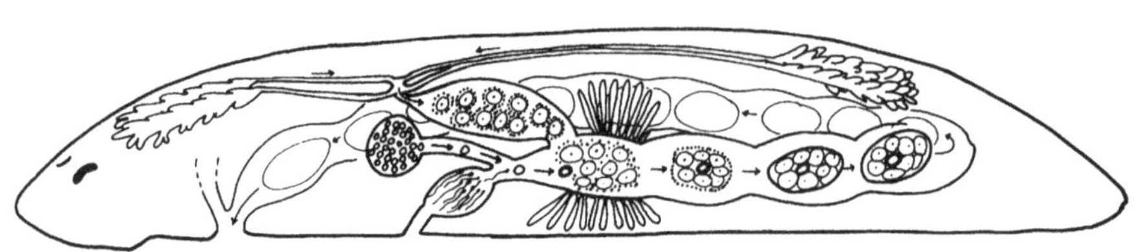

Fig. 14.14 Diagrammatic representation of a turbellarian showing the complex arrangement of glands associated with the production of the complex 'egg' which is made up of a fertilized ovum, extra-ovarian yolk cells and a protein coat. The arrows show the path of the egg as it passes from the ovary and is fertilized, and incorporated into the complex with protein and nutrient cells before it is finally released to the exterior (see also Fig. 3.38).

Table 14.7 A sample of data from a compilation showing the synchrony of emergence of sexually mature palolo worms (*Eunice viridis*) in the Samoan Islands.

Year		Third quarter of the moon		Dates of emergence	
		Oct.	Nov.	Oct.	Nov.
19 years	1843	16	14	15/16	
	1862	15	14	15/16	14/15
19 years	1874	31	30	31	1
	1893	31	29	31	1
19 years	1926	27	26	28	
	1927	17	15	17	
	1928	5	4		4
	1929	24	23	25	
	1930	14	13	14/15	
	1943	20	19	20	
	1944	8	7		7/9
	1945	27	26	28	

The data show that: (a) spawning never occurs before 8 October; (b) the worms spawn in October if the third lunar quarter falls after 18 October; (c) each 19th year spawning occurs on the same date. In addition, the time of day of emergence is also determined with precision.

One of the most striking discoveries of marine biologists in recent years has been the discovery that a multitude of species in a given locality may spawn all at the same time. This phenomenon of simultaneous epidemic spawning was first described for animals of the Australian Great Barrier Reef in 1981. As many as 86 species were observed to spawn at the same time on just one or two days of the year. This creates a massive 'slick' composed of billions of eggs and sperm of many species in the water at the same time. Fertilization takes place and this slick is the source of all the young recruits to maintain the coral community structure. Similar phenomenon have now been observed at higher latitudes and these observations pose a number of questions that are difficult to answer:

- How are the large numbers of germ cells produced by the individual animals all brought to maturity at precisely the same time?
- How is the high level of synchrony maintained among the different members of the population?
- How is the reproductive state of the many different species involved in mass spawnings synchronized?
- What is the selective advantage of highly synchronized and multiple species spawning?
- What prevents cross-fertilization and hybrid production when closely related species spawn together?

It was formerly thought that variations in environmental temperature might provide the most important timing signals, but this is clearly not sufficient to account for the extreme examples cited above. The temperature cycle is subject to too much random variation (i.e. is too noisy a signal) to be solely responsible for the observed patterns.

It is perhaps surprising that the waters of the tropical seas provide some of the best examples of synchronized breeding among marine invertebrates because, in a general sense, the tropical seas are thought to be less subject to seasonal variation that those of temperate regions. The deepest regions of the oceans are even more stable. Indeed the deep sea was for a long time thought to represent an unchanging environment. The waters in the depths of the ocean are cold, around 5 °C, and with almost no seasonal variation in temperature, there is no incident light and food was thought to be sparse. It was very surprising therefore when scientists began to collate evidence during the 1980s that even in the depths of the ocean there were invertebrates, molluscs, crustacea and echinoderms, that produced small eggs, had ect-aquasperm and released small eggs attributes normally associated with seasonal discontinuous breeding. It was also found that these probably developed via planktotrophic larval stages despite living at such great depths.

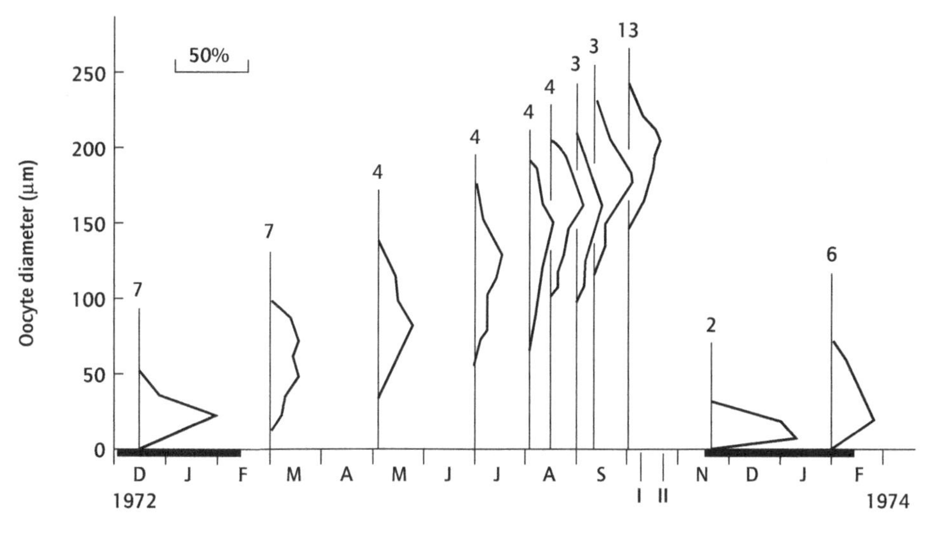

Fig. 14.15 The pattern of egg development in the crinoid *Comanthus japonicus*. The sequence of gametogenesis occupies 12 months and culminates in highly synchronized breeding on two days of each year with a precise correlation to the phases of the moon. The horizontal bar represents the time of gametocyte proliferation. (After Holland *et al.* 1975.)

Evidence pointed towards discrete synchronized breeding in many deep-sea animals just as in the shallow-water temperate zone cousins of these animals. How could this be brought about? One theory to emerge was that the seasonal fluxes of 'marine snow', a floculum of planktonic detritus that descends into the depths of the oceans, may provide both the selective advantage and the environmental clues to permit seasonal reproduction at these depths. More recently, a team working from Japan have shown that the deep-water bivalve mollusc *Calyptera soyoae* shows highly synchronized bursts of reproduction. These giant clams live in the cracks between blocks of basalt lava at great depths in the oceans in this quasi-constant world of cold and darkness. Occasionally the males would release clouds of sperm and this would be followed shortly by mass release of eggs by females – evidence of pheromone action? Observations and experiments conducted more than 1000 m below the surface demonstrated that the spawnings tend to occur when very small changes in temperature occur (in the range 0.1–0.2 °C) and that spawnings can be triggered by artificially raising the temperature by such amounts within experimental domes on the ocean floor.

Outside the deep ocean regions the sea is a complex rhythmic environment (Table 14.8) and reproductive cycles of marine invertebrates can have fixed phase relationships to all of these different cycles. It has been shown that species of Polychaeta, Echinodermata and Crustacea can all show clear-cut responses to the relative duration of day length, a phenomenon known as 'photoperiodism'. These responses are as complex as the photoperiodic responses of the terrestrial insects to be described below. Marine animals can also show direct responses to cycles of moonlight, and can exhibit endogenous rhythms of circa-lunar and circa-tidal periodicity which can be entrained by exposure to appropriate external time-setting programmes (called 'zeitgeber') to run at exactly tidal or lunar rates. It is also becoming evident that underlying the overt annual reproductive cycles of some marine species are endogenous rhythms of circa-annual periodicity. There is clearly a strong selective

Table 14.8 Geophysical cycles in the marine environment.

Name		Periodicity
Metonic	Recurrence of phases of the sun and moon	19 years
Annual	Cycle of earth about sun	1 year
Lunar	Cycle of moon about earth	29.5 days
Semi-lunar	Recurrence of phases of tidal and solar cycles (neap/spring tide cycle)	15 days
Tidal	1 lunar day	24.8 hours
Daily	1 solar day	24 hours
Semi-lunar	Recurrence of high or low tides with semi-diurnal tides	12.4 hours

Observations show that the reproductive activities of at least some marine invertebrates are correlated with each one of these cycles. This does not establish what is the causal factor (see text for discussion).

pressure to achieve reproductive synchrony and the animals will consequently have adapted and respond to many of the complex geophysical signals that occur in the marine environment. Terrestrial organisms, on the other hand, are exposed to more extreme signals relating to the annual and solar day cycles due to the motions of the earth relative to the sun and responses to these have come to predominate (see below).

Repeated cycles of gamete production and release are clearly key elements in the reproductive biology of marine animals. Such cycles can be controlled at a cellular level by the cyclic production of hormones. Such hormones can be described by reference to their functions as being of two basic types: those that have a gonadotrophic role and those that induce spawning or gametocyte maturation and/or activation, as illustrated in Fig. 14.16. The substances involved in controlling these basic functions may be of quite a different molecular structure but there is also evidence of a high degree of conservatism. In the echinoderms, the gonadotrophic function is associated with changes in the relative levels of the 'vertebrate' hormones

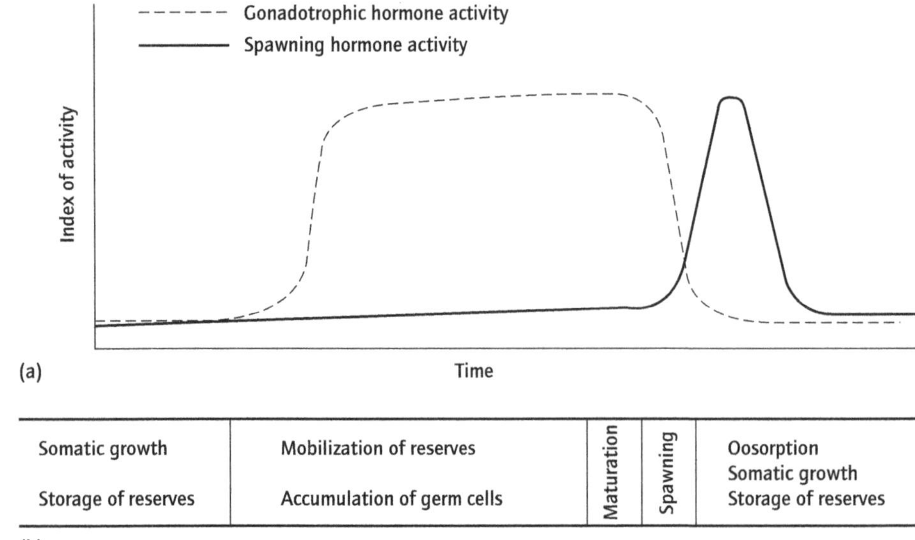

Fig. 14.16 Control of annual reproductive cycles through the production and release of gonadotrophic and spawning-inducing hormones: (a) a generalized scheme; (b) cellular activities associated with different phases of the hormonal cycle.

progesterone and oestrone, while spawning is initiated by a cascade reaction starting with the release of a neurosecretory peptide from the radial nerves, which in turn induces production of the simple substance 1-methyladenine by the ovary (see also Section 16.11.4). 1-Methyladenine causes contraction of the ovarian muscles and thus initiates spawning, but it is also involved in gamete maturation. Its release by the follicle cells of the ovary induces the production of a third messenger, maturation promoting factor, MPF, from the inner surface of the oocyte membrane which ultimately leads to germinal vesicle breakdown and renders the oocyte fertilizable. The receptor protein transducing the signal is a 39 kDa G-protein. The G-protein signalling pathway is an example of a highly conserved mechanism which is involved in the regulation of cell functions in a wide variety of organisms (Fig. 14.17a).

In the starfish the G-protein signalling regulates renewed cell cycling involving pulses of production of the kinase Cdc2 interacting with the proteins cyclin B and cyclin A (Fig. 14.17b) (Kishimoto, 1998) in the developing egg. The maturation-inducing hormone response leads to the completion of meiosis I and II but the female pronucleus cannot proceed further unless fertilization occurs some time between re-initiation of meiosis and the G1 arrest following completion of meiosis II (details in Fig. 14.17b). Mechanism such as this ensure a simultaneous maturation response by all the thousands of oocytes in an individual starfish. Similar mechanisms are known in Polychaeta (e.g. *Arenicola*) and mechanisms of this type can be linked to external signals so that all members of the population may complete meiosis at the same time. The signalling pathways appear to have been highly conserved but the inputs that are detected and transduced in this way are highly diverse. The input signals may be associated with the geophysical cycles in the environment (temperature/photoperiod) but may also include chemical signals released by other organisms called 'pheromones'.

The observation that the spawning of female deep-sea clams (referred to above) occurs just after the release of sperm by the males as first triggered by a slight rise in temperature points to the operation of a pheromonal mechanism. It would, of course, be extremely difficult to isolate the chemicals involved at that great depth so this must, for the time being, be pure conjecture. Progress has been made, however, in the isolation and chemical characterization of pheromones co-ordinating the reproductive behaviour of several shallow-water invertebrates. Nereid polychaetes provide good models for work on the chemical nature of marine pheromones. They take part in the characteristic 'nuptial dance', in which ripe animals leave their burrows in the sea floor and swarm to the sea surface for mating. This behaviour can be induced by a chemical bouquet. In the species *Platynereis dumerilli*, the nuptial dance is induced by release of the ketone 5-methyl-3-heptanone while male gamete release is triggered by uric acid release. Species specificity seems to be partly due to different thresholds of response but may also involve responses to complex suites of pheromones giving a characteristic species-specific 'bouquet'.

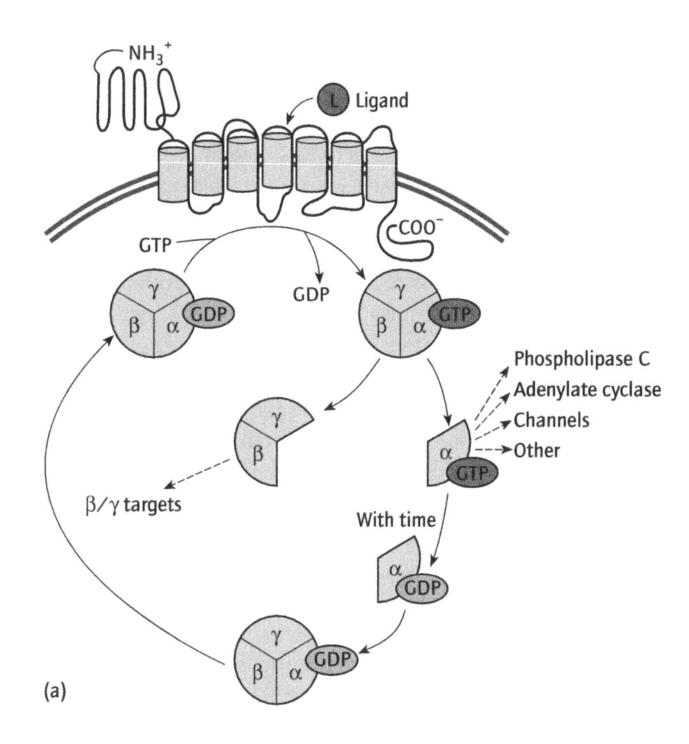

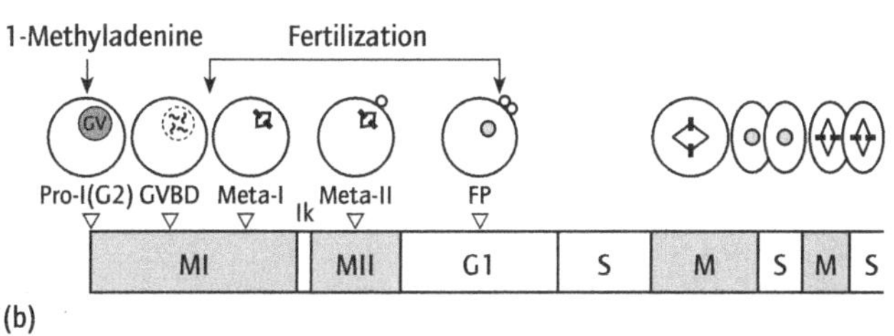

Fig. 14.17 Molecular mechanism controlling renewed cell cycling during gamete maturation. (a) The G-protein signalling pathway. Extracellular signals may bind membrane bound protein receptor molecules. On activation a trimeric G-protein with GDP binds to the inner surface of the receptor, exchanges GDP for GTP and is released from the receptor as an α-unit with GTP and as a γβ dimer. The α GTP is highly unstable and rapidly reverts α GDP and binds again to the γβ dimer. (b) Cell division responses to an external signal (1-methyladenine) and G-protein transduction in the starfish *Asterina*. Fully developed immature oocytes are arrested at the prophase (Pro-1) of meiosis I (MI). At this stage, the germinal vesicle (oocyte nucleus) is intact. Maturation hormone (1-methyladenine) activates a G-protein receptor cascade and re-initiates meiosis leading to germinal vesicle breakdown to be followed by two successive cell divisions (M-phases). The oocyte will then arrest again unless fertilization has occurred. 1-Methyladenine stimulation activates cyclin β/Cdc2 kinase and cycles of kinase activity are associated with successive cell divisions. (After Kishimoto, T., 1998.)

The nereid polychaetes together with the cephalopod molluscs also provide interesting examples of another variation in the pattern of seasonal reproduction. All members of these taxonomic groups breed only once in their lifetime. In the Nereidae the total life span is greater than the modular interval between spawning events at the population level so that in the population a cyclic pattern with an annual, lunar or semilunar periodicity is observed. Systems of this kind are associated with endocrine control systems with a strong element of positive feedback. Sexual reproduction in semelparous organisms involves a massive and rapid redeployment of reserves during which stored metabolites are transferred to the developing germ cells since the animals do not survive breeding. In Cephalopoda, the transition is stimulated by secretions of the optic glands, which could thus be said to have a gonadotrophic role, but the optic glands themselves are inhibited by the activity of the optic nerves. Isolation of the glands or cutting the optic nerves provokes sexual maturation and ultimately death. The transition is normally an irreversible one. Sexual reproduction in the Nereidae (Polychaeta) also involves a programmed death at the time of reproduction. This is a consequence of the adoption during their evolutionary history of epitoky, the process whereby sexually mature individuals undergo a somatic metamorphosis, leave the benthic habitat in which they have been living and take part in a breeding swarm – the 'nuptial dance' (see above and also Chapter 4, Fig. 4.57). The co-ordination of this behavioural change by the programmed release of pheromones has been described and it seems likely that the change in behaviour accompanying the discharge of the gametes was associated with a greatly increased risk. In these circumstances reproduction is best delayed until the animal has sufficient resources to make the risks 'worthwhile'.

14.4.4 Reproductive cycles and diapause in terrestrial and freshwater environments

The reproductive biology of terrestrial and freshwater invertebrates is strongly influenced by the need to fertilize ova internally and to lay well-protected eggs. Linked to this is the requirement to store spermatozoa. In temperate, boreal and polar latitudes, most freshwater and terrestrial invertebrates show seasonal reproductive patterns with prolonged periods of egg laying, interspersed with periods when sexual reproduction does not take place. Such a reproductive cycle is characterized not so much by the extreme synchrony of reproductive events as by the controlled transition from a state of reproductive activity to one of reproductive inactivity. Three states of reproductive inactivity can be recognized:

- *Quiescence*: a direct and temporary response to adverse conditions which is reversed as soon as favourable conditions return.
- *Facultative diapause*: a direct response to unfavourable conditions which once initiated will not reverse until some fixed period has elapsed.
- *Obligate diapause*: a phase of reproductive activity which recurs at specified times each year irrespective of the onset of adverse conditions.

Quiescence and facultative diapause are associated with the adult stages of pulmonate molluscs, earthworms and some insects; obligate diapause, on the other hand, is more frequently associated with earlier stages of development – eggs, larvae or pupae. Many populations of terrestrial and freshwater invertebrates may exist only as eggs during certain periods of the year. This may be during the winter months in temperate and boreal regions or, in tropical regions may enable the organisms to avoid exposure to periods of drought or extreme wet.

Insects which are active in the summer and which enter diapause in the autumn are called 'long-day insects', whereas others such as the silk worm *Bombyx mori* which are winter active are described as 'short-day insects'. The terms are descriptive, but they are also apt, for it is indeed the day length which controls the transition from one physiological state to the other.

14.4.5 Reproductive cycles, biorhythmicity, photoperiodism and the biological clock

Virtually all organisms have biological clocks although the nature of the biological clocks was for many years clothed in mystery. Now it seems that virtually all organisms have the ability to measure the passage of time, exhibit cycles of activity in virtually all their cellular and physiological activities linked to the ticking of an internal clock. They are able to use external 'environmental information' to keep the internal clock in step with real time and the movements of the celestial bodies. They are also able to use the process of daily time measurement to respond to changes in relative duration of light and dark in the solar day to regulate seasonal activities. As usual many an important breakthrough leading to a dramatic increase in understanding came from the study of invertebrate animals especially the fruit fly *Drosophila*.

We now know that the molecular basis of the biological clock is highly conserved and that similar proteins are involved in insects and mammals. The fundamental features of the clock arise from the properties of a number of auto-regulatory proteins. The 'biological clock' is constructed from gene transcription factors that feedback and inhibit their own transcription (see also Box 16.7). The timing components are due to the time required for gene transcription, movement of mRNA to the cytoplasm, transcription and dimer formation. The core clock proteins are the transcripts of the period gene *per* and the timeless gene *tim*. These proteins are only able to move into the nucleus in dimer form. Kinase proteins are involved in dimer formation and also form part of the clock mechanism (Fig. 14.18). One or several of the gene transcription factors from which the biological clock is constructed are also photoreceptor molecules and are sensitive to specific light wavelengths. Thus a common mechanism that helps to explain the basic properties of all circadian biological clocks is emerging. To be effective a biological clock must have the following properties: (i) time keeping (i.e. endogenous cycling), (ii) an entrainment mechanism. For terrestrial organisms, entrainment is brought about by input of information about the light–dark cycle. In marine invertebrates, photoperiodic inputs have many functions but in addition

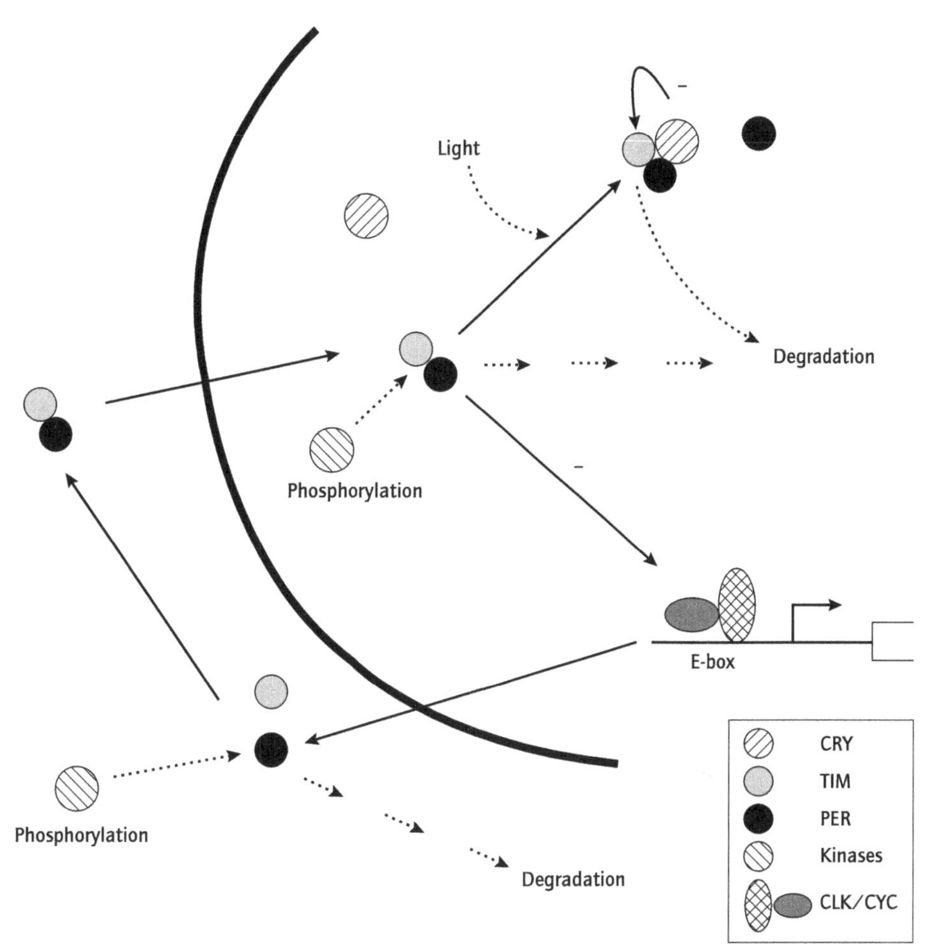

Fig. 14.18 A conceptual model of the molecular components of the circadian clock in *Drosophila*. The clock involves two proteins period (PER) and timeless (TIM) that are autoregulatory via a negative-feedback loop. In addition a number of other gene products are involved. The two positive transcription factors clock (CLK) and cycle (CYC) bind to specific sequences on the *per* and *tim* promoters (Ebox). In this way they drive *per* and *tim* transcription, which reaches peak levels at the beginning of the night. The increase in mRNA is not immediately followed by a substantial increase in protein levels, since phosphorylation of PER, probably through the action of several kinases, targets the protein for degradation. One of these kinases is the product of the gene double-time (*dbt*). TIM is not phosphorylated by DBT and in the cytoplasm is more stable than PER. The interaction with TIM stabilizes PER allowing accumulation of the two proteins. It is also essential for nuclear entry of the complex. The PER–TIM dimer enters the nucleus where it represses CLK/CYC switching off transcription. In the nucleus PER and TIM are further phosphorylated and targetted for degradation. In this subcellular compartment PER is now more stable than TIM, so it persists for longer and can be found as a monomer towards the end of the night and the beginning of the day. Light drives the degradation of TIM, which explains why TIM is not found during the day. Also without TIM, new PER cannot accumulate. Furthermore light promotes the interaction of the cryptochrome (CRY) protein with the PER–TIM complex. As a result PER–TIM inhibitory effect on CLK/CYC is repressed. The combined effects of light, low levels of TIM and repression of the residual PER–TIM complex through the action of CRY, allow the start of a new cycle of transcription. In constant darkness the cycling of *per* and *tim* mRNA and protein levels continues. Therefore additional regulatory steps are likely to be required. (Figure and accompanying text by kind permission of Professor C.P. Kyirakou and Dr E. Rosato, Departments of Genetics and Biology, Leicester University.)

inputs relating to the lunar cycle of light at night and the related tidal cycles may be equally important so that marine organisms appear to be rhythmically more complex than terrestrial ones. It remains to be discovered whether the greater diversity of clocks found among marine organisms share the same basic molecular structure.

In order to be effective, many physiological and behavioural activities of organisms are best performed at certain phases of the day or night. The duration of day and night however changes seasonally except in the deep tropics where there is a constant 12-hour day and night and in the deepest regions of the oceans where light does not penetrate. Consequently a simple, fixed period clock, no matter how accurate it is, cannot keep track of the changing time of dawn and dusk as it changes seasonally. This is perhaps the reason why the daily clock has a 'circadian', i.e. near 24-hour periodicity not an absolute 24-hour periodicity.

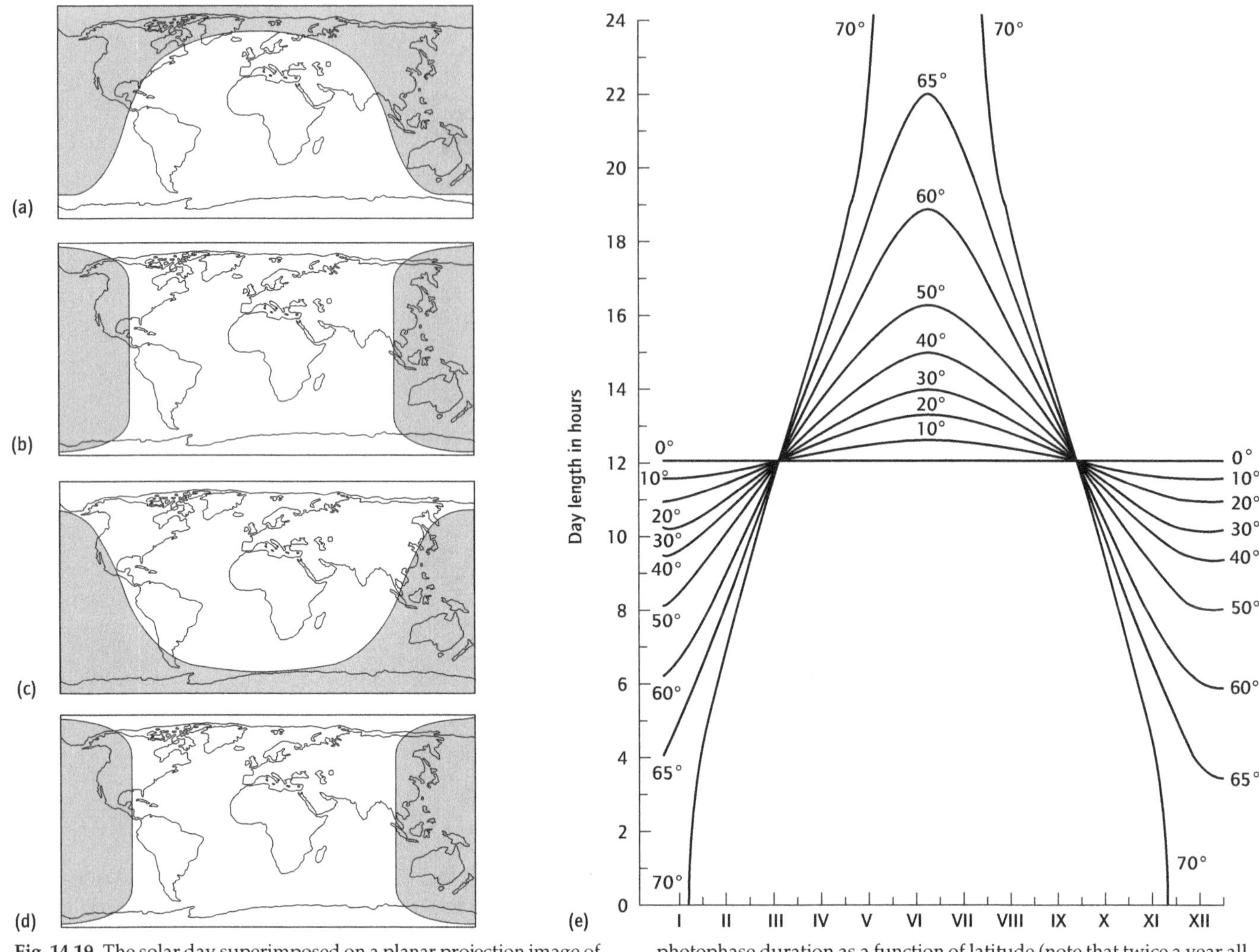

Fig. 14.19 The solar day superimposed on a planar projection image of the world at four different epochs: (a) the winter solstice – 21 December; (b) the spring equinox – 22 March; (c) the summer solstice – 21 June; (d) the autumn equinox – 23 September. (e) Seasonal variation in photophase duration as a function of latitude (note that twice a year all points on the earth's surface experience 12 hours between sunset and dawn and that the amplitude of the photoperiodic signal increases with latitude).

Exposure of *Drosophila* kept in the dark to brief burst of light can have one of three different effects according to the time of day when the light-exposure occurs. A light pulse in what would be the early evening causes a phase delay whereas a light pulse late at night causes a phase advance. If the light pulse is received during what would be the day (the subjective day) it has little effect on the setting of the internal clock.

The molecular mechanism of the biological clock can therefore respond to the relative duration of night and day so that daily activities can occur say 'at dawn' or 'shortly before dawn'. The ability to track the changing phase relationship between the length of the night and day also provides a mechanism by which organisms can achieve a high degree of synchrony and control of annual events and we have seen that this is very important in the regulation of reproductive activities.

The changing cycle of day length associated with the progression of the seasons is a precise source of seasonal information in non-tropical regions of the world. The relative day length throughout the year at different latitudes is illustrated in Fig. 14.19a–d. Note that twice a year, on 21 March and 23 September, the day length is exactly 12 hours light and the night length is 12 hours dark at every point on the earth's surface (Fig. 14.19e). Marine, freshwater and terrestrial animals can all respond to the changing cycle of day lengths and this information can be used to control a yearly progression. The phenomenon is known as 'photoperiodism'. The responses of many animals to the relative day length (or the relative night length) are usually non-linear and light periods less than some critical length are interpreted as being 'short' and those greater than the critical length are interpreted as 'long'. This can be demonstrated by exposing colonies of insects to different light–dark regimes and recording the frequency with which diapause is induced (Fig. 14.20). The accuracy of the time measurement can be within 30 minutes or less. Photoperiodism is now known to control the seasonal activities

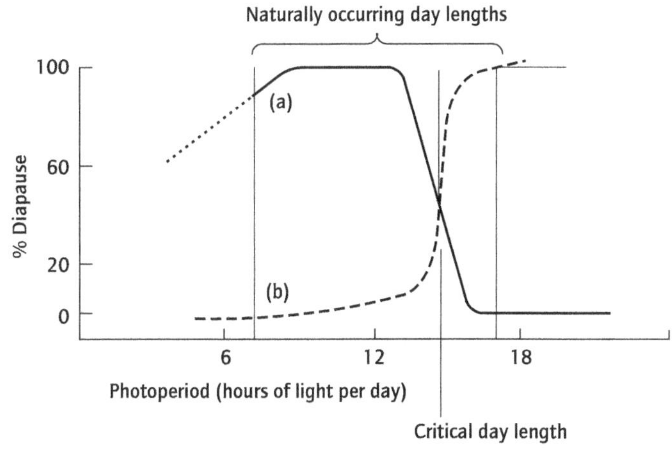

Fig. 14.20 Schematic representation of the photoperiodic response of (a) long-day and (b) short-day insects, where the response is measured as the percentage of individuals entering diapause when exposed to fixed photoperiods as shown.

of a wide variety of organisms but is best known for several groups of insects. Experimental analysis of photoperiodism in insects suggested that in some, notably the Aphidae, the photoperiodic response involves a mechanism with the properties of an interval time. In others the circadian system is clearly involved in photoperiodism. The experimental evidence for this has been obtained from ingenious and painstaking experiments in which insects were exposed to different combinations of light and dark cycles such that the total duration of the light–dark cycle was not necessarily 24 hours. In aphids the clock appears to have the property of an interval timer measuring the deviation of night length (Fig. 14.21a). In most insects however the photoperiodic clock shows resonance indicating the involvement of a continuously running circadian clock. Evidence for this is summarized in Fig. 14.21b.

Restricting the time of reproduction to a particular time of year may be presumed to maximize fitness but the reasons for this are difficult to establish. In the marine environment, larval production can be timed to occur just prior to the appearance of the temporarily rich food source of the phytoplankton bloom. In the terrestrial domain, diapause, or a switch from parthogenetic growth to sexual reproduction and then to diapause, can be timed so that it occurs before the onset of poor food conditions during the winter period. Food resources are therefore prevented from becoming limiting (see Fig. 14.5).

The ability to synchronize spawning brings other advantages:
- Highly synchronized spawning maximizes fertilization rate when fertilization is external.
- The synchronous mass discharge of gametes leads to predator swamping so that the larvae survival is the greater.
- Synchronized breeding may enhance the reproductive value of adult animals by permitting the effective partitioning of activities in a seasonally changing environment.

Photoperiodic mechanisms may predominate in the terrestrial environment and are certainly a component of the reproductive cycles of marine animals but the sea is a complex rhythmic environment in which several geophysical cycles interact (see Table 14.8). The most extreme examples of reproductive synchrony reflect this and, as explained in Table 14.7, have components of annual, lunar, tidal and solar periodicity.

14.5 Reproduction and resource allocation

The variety of reproductive patterns that may be found among the invertebrates makes them ideal subjects for the study of evolution and selection of reproductive traits. Through investigations of the invertebrates it may be possible to gain greater insight into some fundamental questions relating to the evolution of animals, and to find the experimental material that will permit the testing and further development of life-history theory. This section will consider the life histories of invertebrates from a theoretical point of view, to explain the patterns of resource allocation between adult organisms and their offspring.

It is assumed that for most organisms one or more resource is limited; there is then a fundamental dichotomy between the allocation of the limited resource(s) to enhance either (a) adult survival and growth or (b) offspring production and their survival.
- How much of the limited resource(s) should be allocated to reproduction and when?
- How much of the limited resource(s) should be allocated to each new offspring?

In recent years a rather unified life-history theory has begun to emerge that might be termed the 'demographic theory of life-history evolution'. At the core of this theory is the concept that natural selection always acts to maximize fitness but that genes relating to reproductive traits can act in two rather different ways (see Section 14.5.1):
- by enhancing survival
- by enhancing fecundity

The trade-offs between these two aspects of gene action may explain much of the diversity in the patterns of reproduction and resource allocation exhibited by organisms under different ecological circumstances. Life-history theory should explain why different patterns of resource allocation are favoured in different circumstances, and the theory should be tested by critical testing of its predictions.

14.5.1 An introduction to demography

A population of organisms reproducing by sexual reproduction is often composed of individuals of different ages, The youngest have recently been born and the oldest are near to death. Such a population may be described not only by the total number of individuals but also in terms of their age distribution. There is, of course, an intimate relationship between the age distribution, survival and birth rate. Such a population will be in steady state when, in any defined period:

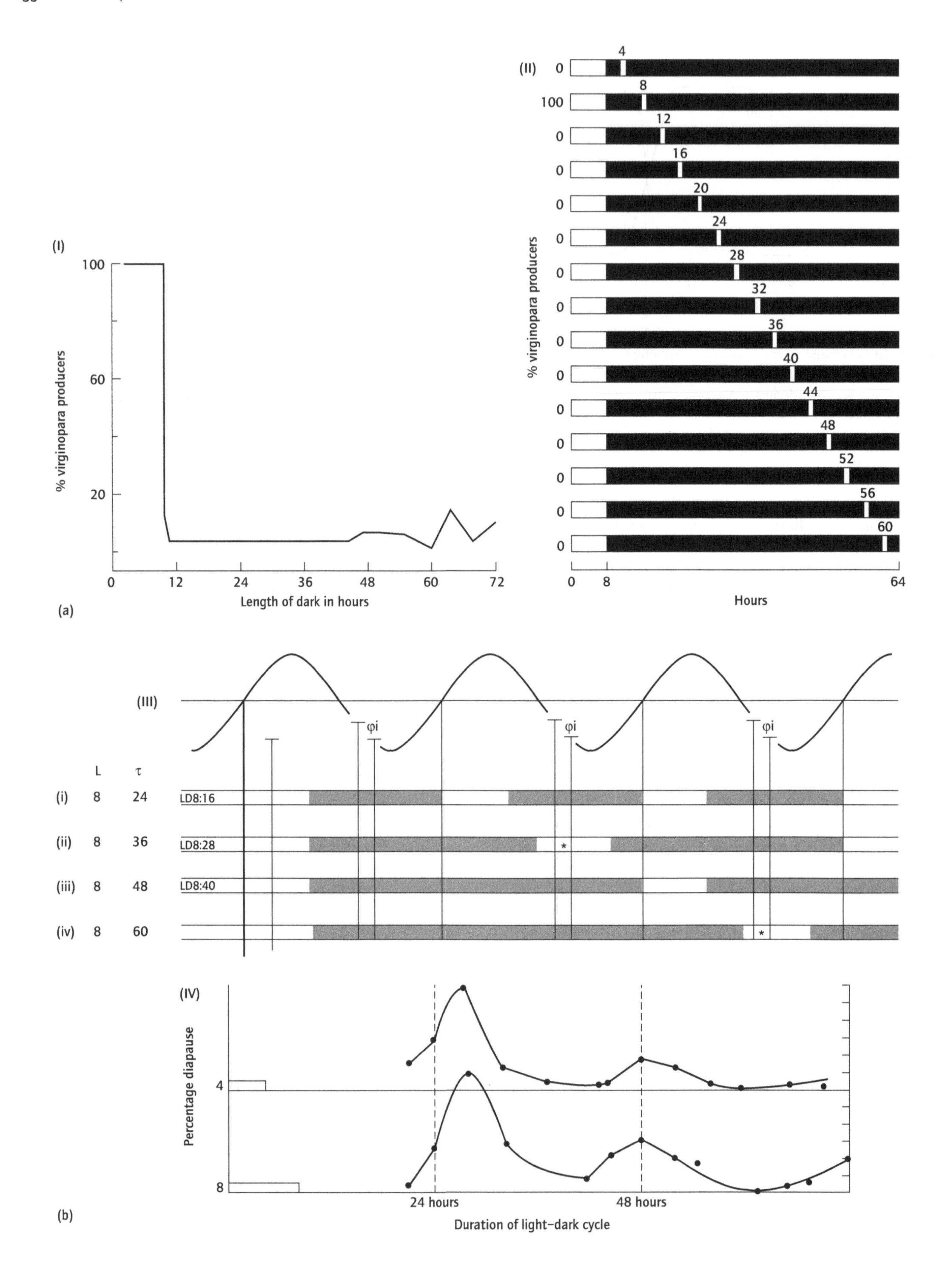
(I)
% virginopara producers
Length of dark in hours
(a)
(II)
% virginopara producers
Hours
(III)
φi
L τ
(i) 8 24 LD8:16
(ii) 8 36 LD8:28
(iii) 8 48 LD8:40
(iv) 8 60
(IV)
Percentage diapause
24 hours
48 hours
Duration of light–dark cycle
(b)

Total numbers of births + immigration = Total deaths + emigration

In a closed system without immigration and emigration, the population will be increasing if the birth rate exceeds the death rate and, conversely, will be declining if the death rate exceeds the birth rate. In these circumstances, the rate of change in the population can be described by the simple relationship:

$$dN/dt = rN \quad \text{or} \quad N_t = N_0 e^{rt} \tag{1}$$

where N is the number of individuals in a population, N_0 is the number of individuals at some time 0, N_t is the number of individuals at some later time t. The exponent r is of particular interest as it determines the rate of population increase with time. The factors that determine the value of r are crucial and will be discussed further below. No population can grow exponentially for ever, and most will fluctuate about a mean level which represents the notional carrying capacity of the environment. One way of representing this in mathematical terms is to introduce a factor K into the growth equation such that:

$$dN/dt = rN\frac{(K - N)}{K} \tag{2}$$

Fig. 14.21 (*opposite*) Evidence of interval timer and circadian clock functions in photoperiodic regulation of insect diapause. (a) Interval timer mechanisms in aphids. (I) Percentage of wingless parthogenic females (characteristic of the summer months) as a function of night length (scotophase) when the duration of light (photophase) is 8 hours – note lack of resonance cf. (IV). (II) Response to a 1-hour light break after various periods of dark following an 8-hour photophase. If the light pulse occurs after 4 hours it is interpreted as a new dusk and the following long night suppresses virginoparae production. An 8-hour initial dark period suppresses virginoparae production, i.e. it is 'interpreted' as a short night less than critical night length and is therefore characteristic of the summer. All other treatments suppress virginoparae production and simulate autumn/winter conditions. (b) Evidence for the involvement of a free-running circadian clock in the regulation of photoperiodic phenomena. (III) The design of a resonance experiment. It is supposed that there might be an endogenous oscillator (represented by the sine wave) in which there is some photo-responsive phase φi. This endogenous oscillator represents the molecular clock with properties explained as in Fig. 14.18. Different light–dark cycles are represented as the light and dark bands. In the examples (i)–(iv) the photophase (lights are on) is constant at 8 hours. The dark period is variable to give different cycle times τ as shown. Note that in (i) and (iii) φi always falls in the dark but in (ii) and (iv) φi sometimes falls in the light. If this is interpreted by the circadian clock as a long day a resonance effect would be expected. (IV) Experimental demonstration of 'resonance' indicating the involvement of an endogenous circadian oscillator. The experiment shows the percentage of flesh flies *Sarcophaga* which enter pupal diapause when they are reared in photoperiodic regimen with 4 and 8 hours of light. In these, and in other photophases not shown here, there is a clear resonance effect.

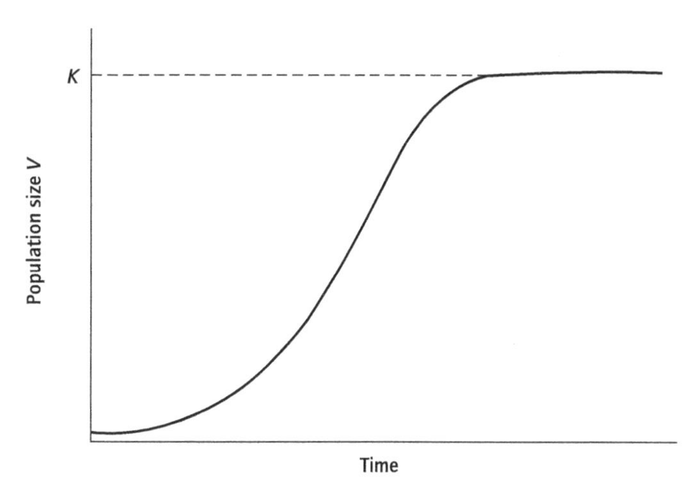

Fig. 14.22 A graphical representation of the logistic equation $dN/dt = rN(K - n)/K$ from which the terms r and K may be derived. (From Pianka, 1978.)

This equation is referred to as the logistic equation. It describes a population in which the rate of population growth declines as N tends to K (Fig. 14.22). The logistic equation is a simple one and no real population of animals follows it exactly. Most populations will be found to fluctuate about an average level, but there are marked contrasts in overall stability. Some are characterized by very large fluctuations in number, being highly unstable and subject to population crashes but with a high rate of recovery due to exceptionally high maximum rate of population growth. Others are more stable but show slower intrinsic rates of population growth. The two extremes were formerly referred to as r-selected and K-selected species with reference to the terms in the logistic equation above.

This idea was embodied in an influential theory of life-history evolution, the so-called r–K-selection theory. In this theory, it was supposed that natural selection operates in two fundamentally different ways by maximizing either r or K. In a later development of the theory, it was also supposed that species could be classified as being r- or K-selected by reference to their reproductive traits. The theory in this simple form is no longer generally accepted as the sole basis for understanding the evolution of life histories and patterns of sexual reproduction, but it was important as an early example of a testable general theory.

One major objection to the r–K-selection theory is the realization that natural selection always operates in the same way. If a new rare allele is to be incorporated into a population, it must maximize fitness and this means that the rate of growth of the population of organisms that inherit the allele must be greater than that of the population that does not. In other words 'fitness' is measured in the same terms as r above whatever the circumstances. Moreover, the concept of the carrying capacity K is a purely abstract concept. It is not possible to say how any particular allele could affect K. For reasons such as these, life-history theory has moved on and contemporary life-history theory supposes that, just as the growth characteristics of a population are determined by the average survival and average fecundity of individuals in a population (see life table data below), so the fitness of individuals within a population can also be described in terms of survival and fecundity of their offspring.

A mathematical representation of the fitness of an individual carrying a rare dominant gene and reproducing sexually is defined for an animal with discrete reproductive events in each year of life to age ω, which is the last age at which reproduction occurs, by:

$$1 = \frac{1}{2}\sum_{t=1}^{t=\omega} s_t n_t e^{-Ft} \qquad (3)$$

or for an organism reproducing continuously

$$1 = \tfrac{1}{2}\int e^{-Ft} s_t n_t \, dt \qquad (4)$$

Equation (3) represents the discrete reproduction form and (4) the continuous reproduction form of the Euler–Lotka equation where t is any specified age, s_t is survival to age t, n_t is fecundity at age t, and F is an estimate of fitness.

Accordingly the fitness of an individual and the rate of increase of a population are defined in similar terms – the fitness of a genetically defined individual in a population being the exponential rate of increase of its offspring among the population of competing organisms.

There are, therefore, alternative routes to enhanced fitness. These are: increased survival and increased fecundity. This dichotomy lies at the heart of the demographic theory of life-history evolution.

The key parameters of average survival and fecundity during the entire lifetime of an organism are not easy to establish, but this information is essential if the life history is to be understood in dynamic terms. The appropriate information is most easily understood when presented in the formal manner of a life table. A life table sets out the average survivorship and average fecundity of an average female in a population over the entire lifespan. Such a table can be constructed from observations on a population if (i) it can be assumed to be approximately in steady state and (ii) the ages of individuals can be determined. Such a life table is called a 'static life table'. If this is not possible it may be better to record the survival and fecundity of an identified or marked group of organisms in a population over the entire life time of the group and to make the assumption that the marked group is typical of the population as a whole. A life table constructed in this way is referred to as a 'cohort life table'.

It is not easy to determine the exact age of invertebrates but some can yield the required information. It happens that many invertebrates show age-related striations in skeletal tissues and skeletal growth lines can be used to estimate the age of individuals. In these circumstances a life table can usually be constructed. The best known growth lines are the striations in the shells of some bivalve molluscs, but others include growth bands in corals, in the calcareous plates of echinoderms and in the proteinaceous jaws of some polychaetes.

Whatever skeletal record is used, some independent validation of the age interpretation is required and this may be difficult to achieve. The most useful skeletal records have lines to indicate both annual and daily events and such records can provide important information about the history of the earth. A major problem for the construction of life tables for organisms with mobile or pelagic larvae, as is the case for many marine invertebrates, but which are sessile as adults, is that of defining the population.

An example is illustrated in Table 14.9; this shows a life table for a North Sea barnacle (Crustacea, Cirrepedia – see Section 8.6.3.8). These are sessile animals as adults, and individuals can be marked and observed over several years. The life table shows the average survival s^t and the average fecundity n^t at each age t.

The sum of the products of average survival and average fecundity over the entire life span, $\sum s^t n^t$, is the average lifetime reproductive output of a female born into that population. It is often referred to as R_0 the 'net reproductive rate'. The generation time can be estimated as the age of the parent of an average newborn offspring. This can be calculated from the life table by the relationship:

$$T = \sum t s_t n_t \qquad (5)$$

The intrinsic rate of population increase r is, as we have seen above, a key parameter for population growth. It is formally defined by the Euler–Lotka equation which we have also used to describe fitness (equations (3) and (4) above) and which in a discrete form to describe the rate of population growth could be written using the same notation as in equation (3) as:

$$1 = \frac{1}{2}\sum_{t=1}^{t=\omega} s_t n_t e^{-rt} \qquad (6)$$

This equation is difficult to solve, but when a population is near to steady state, R_0 is not greatly different from unity and in these circumstances r may be estimated from the approximation

$$r \cong \frac{\ln R_0}{T} \qquad (7)$$

A positive value for r, as in Table 14.9, suggests that the population is increasing in size and if r is negative the population is decreasing.

Another important concept for life-history theory is that of the 'residual reproductive value' of an organism. At any time when reproduction takes place, the reproductive output of the organism can be subdivided into two components:

Present reproductive output + Future reproductive output

Assuming near steady-state conditions, at any age x, the reproductive value is given by

$$V_x = \sum_{x}^{\omega} \frac{s_t}{s_x} n_t \qquad (8)$$

and when the population is not in steady state,

Table 14.9 Life table of *Balanus glandula* (from Hines, 1979).

	Age (t) months	Survivorship to age s_t	Average eggs at age (t) n_t	$s_t n_t$	$t s_t n_t$
Estimated mortality	0	1.000	0	0	0
prior to settlement	3	1.17×10^{-4}	0	0	0
	12	2.04×10^{-5}	20 504	0.418	5.016
	24	3.84×10^{-6}	66 814	0.256	6.153
	36	2.05×10^{-6}	113 125	0.231	8.335
	48	1.28×10^{-6}	140 742	0.180	8.641
	60	7.33×10^{-7}	159 435	0.117	7.008
	72	4.18×10^{-7}	170 892	0.075	5.151
	84	2.44×10^{-7}	176 922	0.043	3.629
	96	1.42×10^{-7}	180 540	0.026	2.469

$R_0 = \Sigma s_t n_t = 1.346$

Generation time $T = \Sigma t s_t n_t = 46.402$ months

Intrinsic rate of population increase for time in months

$$r = \frac{\ln R_0}{T} = 0.006$$

i.e. population would be growing slowly if this were to continue.
In this case, survivorship was determined by observing the number of survivors for each 1000 eggs from the ratio between egg production and numbers of spat settling. This is an estimate. Subsequently survivorship was observed directly. These data give column s_t.
Fecundity was estimated by measuring the mean basal diameter of the barnacles at each age and measuring the relationship between basal diameter and number of eggs per brood.
Finally the number of broods per year was estimated. The product gives an estimate of the mean fecundity at each age. These data give column n_t.

$$V_x = \sum_x^{\omega} \frac{s_t}{s_x} n_t e^{-r} \qquad (9)$$

where t has all values from $t = x$ to $t = \omega$ the last age at reproduction.

Note that in equations (8) and (9), slightly different notation has been used. When $t = x$, $s_t / s_x = 1$ and consequently the expansions and the equations (8) and (9) define present fecundity n_x at age x and future fecundity offset by the probability of surviving from age x to all future ages. When the population is not in steady state, the value of future offspring must also be adjusted by a factor to take account of the rate of population growth (or decline) (equation 9). This is because a given number of offspring will contribute relatively less to the total population at some future data if the population is growing. The more complex mathematics of equation (9) are necessary to allow for the effects of population change, the exponential term adjusts the average survival according to the expected change in population number. In an intuitive sense, the value of one individual offspring is less the larger size of the population.

The reproductive value of an individual (equations 8 and 9) changes with age. In many invertebrate populations the reproductive value of a new-born propagule is low because only a very small number of newly produced offspring will survive to breed. The reproductive value of an individual at the age of first reproduction will be higher and this provides a means of estimating the fitness of an individual independently of the survival through the larval phase. If an individual is born into a population in which there is a high risk of mortality prior to the attainment of reproductive condition, then the reproductive value of that individual will increase with time, will peak and then decline. Note that in Table 14.9 the reproductive value of a barnacle aged from 12 months onwards is very much greater than the equivalent value of a new-born barnacle egg.

14.5.2 Assumptions of a general theory of life-history evolution

A general theory of life-history evolution will incorporate certain assumptions, e.g.:

- That natural selection acts on individual life-history traits.
- That natural selection tends to maximize fitness of individuals (fitness is defined in equations (1) and (2) but see also Section 14.5.3).
- That individual life-history traits can evolve independently.

In the general theory touched on below it is also supposed that:

- The resources available to an organism are limited.
- An increase in reproductive effort (i.e. allocation of resources to reproductive activities) results in an increase in reproductive output and a reduction in somatic investment.

- Increased reproductive effort will result in increased fecundity, increased offspring survival, increased rates of offspring growth and maturation, or some combination of these.
- A reduction in somatic investment will result in either reduced adult survival or reduced growth and future fecundity, and hence reduced residual reproductive value.

A key feature of these assumptions is the possibility of a trade-off between current reproductive output and residual reproductive value as defined by equations (8) and (9). There are clearly different routes to long-term reproductive success and hence many different patterns of reproduction.

14.5.3 The demographic theory and invertebrate life histories

A theory of life-history evolution must, if it is to be useful, make predictions possible, and these predictions must be testable. One way of testing the theory is to consider whether the observed patterns of reproduction in different circumstances are explained by the theory. There may of course be factors which modify the predictions of the theory, and these include the influence of the evolutionary history of a taxon which may impose constraints on the possible routes to reproductive success.

The life-history characteristics exhibited by organisms are supposed, like other traits, to be determined by natural selection. Characters will be selected if they increase the fitness of an individual, i.e. the population of organisms with the character grows more quickly than the population without.

In equations (3) and (4) above, fitness is defined in terms equivalent to r, the intrinsic rate of population increase in equation (6). The concept that there is some global measures of fitness that can be maximized is itself the subject of discussion. More rigorously, r is an appropriate measure of global fitness only under conditions of an unlimited homogeneous and constant environment, i.e. when density-dependent factors do not influence the outcome of selection. The net reproductive rate R_0 may also be used as a measure of global fitness when the environment is stationary. In many circumstances neither r nor R_0 can be computed with accuracy but the residual reproductive value for organisms of known age can be calculated, i.e. where larval survival rate is unknown but adult survival rate is known. In these cases reproductive value may be easier to work with and it may be assumed that natural selection will tend to maximize r by maximizing 'reproductive value relative to reproductive effort' at all ages. If there is sufficient genetic variation to permit the attainment of an optimum combination of reproductive traits, we can expect global fitness to be maximized by the selection of traits subject to the constraints and trade-offs which limit the set of possible life histories which can be exhibited by any organism.

This dichotomy between alternative traits can be interpreted within a framework which considers the relationship between global fitness and any two life-history traits. We may consider for example egg size and egg number. If some resource, e.g. energy, is limiting it follows that the resource can be allocated in three different ways: (i) to present reproduction; (ii) to survival and maintenance; or (iii) to growth. It is also assumed that any allocation of resources to 'present reproduction' can be made to provide a larger number of relatively cost-free offspring or, a smaller number of more costly offspring (see Table 14.6).

In many classes of marine invertebrates, it has been found that within individual species egg size is rather invariant, but closely related species exhibit very different egg sizes. This is generally associated with the mode of development so that smaller eggs develop via a pelagic larval phase and larger eggs do not. There are some exceptions, but in general it is the smaller of two closely related species, or the smaller of the two closely related coexisting members of similar taxa, that have the larger eggs and have the higher investment per egg. This seems paradoxical at first but the larger species is more likely to produce eggs with a lower investment per egg. The absolute levels of reproductive effort defined as the relative proportion p of the energy available that is invested into present reproduction may, however, be the same whatever the size of an individual egg (see also Box 14.4).

A study of reproductive effort in two species of small littoral gastropod molluscs, showed hardly any difference in the relative allocation of energy to reproduction despite striking differences in the relative cost of individual eggs. *Lacuna vincta* has a long pelagic larvae phase and relatively small eggs whereas *Lacuna pallidula* has much larger eggs and direct development. The proportion of the overall energy budget allocated to egg production in these two species, however, differed by less than 4%. The dichotomy between alternative traits may be referred to as a trade-off situation.

The problem for life-history theory is the prediction of the circumstances under which one or another optimized trade-off confers maximum fitness. The formidable problem that this represents for invertebrate biologists can be envisaged by considering the very simple model represented in Fig. 14.23. This represents the life history of an iteroparous animal in which it is supposed that there is a finite probability of survival to adulthood and a different finite probability of survival by an adult between successive reproductive episodes (S_a). A fixed number (n) offspring are produced at each time of breeding. Even with this very simple life history, no less than 10 independent two-trait trade-offs have been identified (see Fig. 14.23). Of these trade-offs, some are already familiar to us. The trade-off between egg number and egg size (pelagic versus non-pelagic) development for marine invertebrates, for instance, is represented by the second trade-off in the list.

Given the rich diversity of life cycles among the invertebrates, it is tempting to make comparisons between taxa, but this approach is not very constructive. It is much better to analyse variation within individual taxa.

The observed set of life-history traits of any individual represents the response of that individual with its own (often unique) genotype to the specific set of environmental conditions experienced. The term 'reaction norm' is used to express the 'full set of phenotypes that a specific genotype could express in interaction with the full set of environments in which it can survive'.

Box 14.8 Reproduction now or in the future?

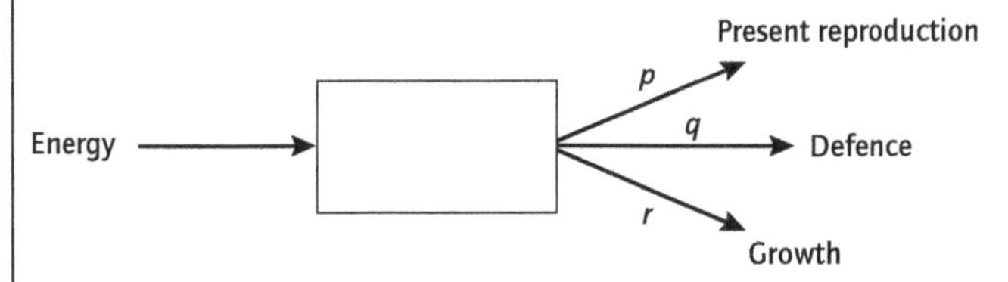

The global fitness term which may be represented by the reproductive value of an individual aged $1(V_1)$, where the age of first reproduction is 1. This reproductive value can be represented as having two components, present fecundity and future expectation of fecundity. This can be represented diagrammatically:

Reproductive value at age 1 $\quad V_1 \quad = \quad$ [Present fecundity] $\quad + \quad$ [Expected future fecundity]

Or mathematically:
When near steady state

$$V_i = \left[\frac{s_{i[t=1]}}{s_1} m_1\right] + \left[\sum_{2}^{\omega} \frac{s_i}{s_1} m_i\right] \qquad (1)$$

where i represents the age at breeding at each age (t) from $t = 1$, the first age of reproduction, to $t = \omega$, the last age of reproduction. When the population is subject to growth at some instantaneous rate r, in a non-stable population, the value of future reproductive effort should be offset by a term to reflect the rate of change in population size. This is necessary because, in an expanding population future offspring are of less value than present offspring. This can be represented as follows:

$$V_i = \left[\frac{s_{i[t=1]}}{s_1} m_1\right] + \left[\sum_{2}^{\omega} \frac{s_i}{s_1} m_i\right] e^{-rt} \qquad (2)$$

Resources allocated to present reproduction contribute to reproductive effort at the present age. Resources allocated to future reproduction may be used to maximize survival, defence or growth, and hence the prospect of offspring at some future time. The trade-off between present reproductive effort and adult survival can be represented in a fitness diagram (recall Box 14.4 i,ii). When the trade-off relationship is concave as in the left-hand diagram (i), then optimal fitness accrues from maximum reproductive effort and minimal adult survival, i.e. semelparity, whereas when the trade-off curve is convex (ii), maximum fitness is conferred by some intermediate reproductive effort and finite adult survival, i.e. iteroparity. (After Steams, 1992.)

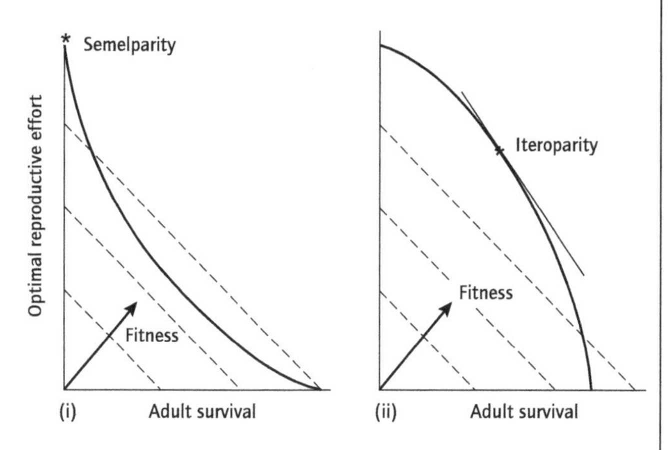

Many eggs or few eggs?

- Resources allocated to present reproduction may be allocated to a larger number of small eggs or to a smaller number of large eggs.
- The number of eggs may be limited either by the energy available (p) or by the volume available for egg storage. Then:

$$\text{Number of eggs} \propto \frac{\text{Energy allocated to eggs}}{\text{Energy per egg}}$$

$$\text{Number of eggs} \propto \frac{\text{Volume of mother}}{\text{Volume of egg}}$$

(see also Fig. 14.23).

Some life-history traits are relatively static and fixed. Others may be variable according to the environmental conditions. It is the set of 'reaction norms' that has arisen through natural selection.

Large-scale long-time experiments are now underway which seek to explore the rapidity with which 'reaction norms' can change under directional selection. This is important because the longer-term effects of pollution can be expected to have genetic consequences. To understand them it is necessary to have a sound understanding of ways in which natural selection affects reproductive processes. We have seen that in order to understand why individual animals reproduce as they do it is necessary to understand: (i) the constraints and opportunities set by the environments in which they live; and to (ii) understand how the evolutionary history of an organism may limit the reproductive options open to it. Experimental programmes must address these key questions at a level below that of the species, so that life-history theory is not tested solely by comparisons made between species each with its own long evolutionary history. The invertebrates are a highly diverse assembly of organisms with diverse life histories. Many have short generation times and this will favour the testing of ideas through experiment.

14.6 Conclusions

Life-history theory is a rapidly expanding field which seeks a functional explanation of the diverse sets of reproductive traits that may be found among organisms. Particularly rapid progress has been made in the development of a general theory of sexual reproduction, and observations on the invertebrates have contributed a great deal to the database that underpins the developing theory. It is clear that sexual reproduction is the dominant mode of reproduction amongst all living things, although it may be combined with episodes of asexual reproduction in complex life histories such as those

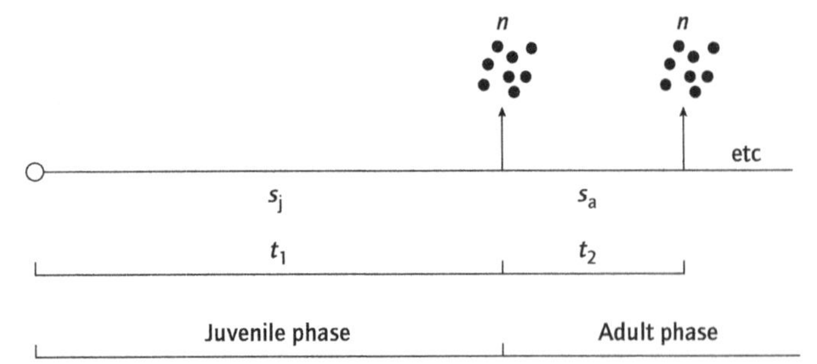

Fig. 14.23 A very simple life-history model for an iteroparous animal with the following parameters: survival through the juvenile phase s_j, the number of gametes produced at each reproductive event n, the survival between successive reproductive events s_a and the time intervals t_1 between zygote formation and the onset of reproduction and t_2 between successive breeding events. This very simple model defines 10 two-parameter trade-off situations. These include:

- n versus s_a — Reproduction is risky, s_a is low, maximize n at first reproduction.
- n versus s_j — High fecundity leads to poor survival, trade-off many small versus few large offspring.
- s_j versus s_a — Parents defend or provide resources for the young, increased parental survival decreases offspring survival.
- s_a versus t_1 — Increased parental investment increases the rate of development.
- s_a versus t_1 — Feeding is risky for juveniles.
- s_j versus t_2 — Feeding is risky for adults.
- n versus t_1 — A small number of offspring may be produced early or a larger number later.
- t_2 versus s_j — Investment in offspring to increase their survival increases the interval between broods.

(After Sibly, R.M. 1991. The life history approach to physiological ecology. *Funct Ecol.*, **5**, 184–191.)

of many cnidarians, flatworms, annelids, small crustaceans and some insects.

The widespread occurrence of sexual reproduction is associated with a bewildering variety of reproductive states and patterns of allocation to reproductive processes. This variety has been examined from a taxonomic, environmental and functional viewpoint. It is clear that all possible combinations of reproductive traits are not found equally in all environments, nor in all taxonomic groups. Reproduction in the sea is often associated with the release of free-living pelagic larvae, and this trait is associated with a number of others, including external fertilization, mass epidemic spawning and the production of simple spermatozoa and energy-poor eggs. Under special circumstances in the sea, and in all terrestrial and freshwater environments, pelagic larval development is suppressed. This is usually associated with internal fertilization, sperm storage and the production of energy-rich eggs.

All patterns of reproduction which differ from random are the product of the controlled allocation of resources to specific reproductive functions. If there is a strong element of synchronization among members of the population there must be the input and neuroendocrine transduction of environmental information. This subject has been briefly touched on in this chapter, and will be discussed more fully in Chapter 16.

We have examined the recent development of theories relating to the allocation of limiting resources between the conflicting demands of parents and offspring, and seen how it is possible to classify environments to make sense of the patterns of allocation that are observed. We look forward to the increasing role of experiment and observation at the subspecies level to test and refine the theory, with practical applications in relation to longer-term effects of sublethal levels of environmental change (pollution, global warming, etc.) which may be detected in changed reproductive characteristics.

The analysis and description of reproductive processes in invertebrates provides the background to studies of ecology and community biology, as well as providing an introduction to the study of invertebrate development which follows (Chapter 15).

14.7 Further reading

Detailed background to the reproduction of invertebrates can be obtained by reference to two multi-treatises and a continuing series of review volumes:

Adiyodi, K.G. & Adiyodi, R.G. (Eds) 1993. *Reproductive Biology of Invertebrates*. Wiley, New York.
Vol. 1. *Oogenesis, Oviposition and Oosorption*.
Vol. 2. *Spermatogenesis and Sperm Function*.
Vol. 3. *Accessory Glands*.
Vol. 4. *Fertilisation, Development and Parental Care*.
Vol. 5. *Sexual Differentiation and Behaviour*.
Vol. 6. *Asexual Propagation and Reproductive Strategies*. Parts A and B.
Further volumes pending.

Giese, A.G. & Pearse, J.S. (Eds). *Reproduction of Marine Invertebrates*. Academic Press, New York.
Vol. 1 (1974) *General Introduction, Acoelomate and Pseudocoelomate Metazoans*.
Vol. 2 (1975) *Entoprocts and Lesser Coelomates*.
Vol. 3 (1975) *Annelids and Echiurans*.
Vol. 4 (1977) *Molluscs: Gastropods and Cephalopods*.
Vol. 5 (1979) *Molluscs: Pelecypeds and Lesser Classes*.
Vol. 6 (1991) *Echinoderms and Lophophorates*.

Giese, A.G., Pearse, J.S. & Pearse, V.B. (Eds).
Vol. 9 (1987) *General Aspects: Seeking Unity in Diversity*.

Advances in Invertebrate Reproduction. Elsevier Science, Amsterdam.
Vol. 2. Clark, W. & Adams, T.S. (Eds) 1981.
Vol. 3. Engels, W. (Ed.) 1984.
Vol. 4. Porchet, M. (Ed.) 1986.
Vol. 5. Hashi, M. (Ed.) 1990.

Further continuations of this series have continued as special editions of the journal *Invertebrate Reproduction and Development*

The following are monographs which address different aspects of invertebrate reproduction:

Begon, M., Harper, J.L. & Townsend, C.R. 1986. *Ecology: Individuals, Populations and Communities*. Blackwell Scientific Publications, Oxford.
Bell, G. 1982. *The Masterpiece of Nature: The Evolution and Genetics of Sexuality*. University of California Press, Berkeley.
Brady, J. 1979. *Biological Clocks* (Studies in Biology, 104), Edward Arnold, London.
Calow, P. 1978. *Life Cycles*. Chapman & Hall, London.
Charnov, E. 1982. *The Theory of Sex Allocation*. Princeton University Press, Princeton, New Jersey.
Cohen, J. 1977. *Reproduction*. Butterworth, London.
Grahame, J. & Branch, G.M. 1985. Reproductive patterns of marine invertebrates. *Oceanography Marine Biology Annual Review*, **23**, 373–398.
Greenwood, P.J. & Adams, J. 1987. *The Ecology of Sex*. Edward Arnold, London.
Hurst, L.D. & Peck, J.R. 1996. Recent advances in understanding of the evolution and maintenance of sex. *Trends in Evolution and Ecology*, **11**, 46–52.
Maynard-Smith, J. 1978. *The Evolution of Sex*. Cambridge University Press, Cambridge.
Pianka, E.R. 1978. *Evolutionary Ecology*. Harper & Row, New York.
Roff, D.A. 1992. *The Evolution of Life-Histories: Theory and Analysis*. Chapman & Hall, New York, London.
Saunders, D.S. 1977. *The Introduction to Biological Rhythms*. Blackie, Glasgow.
Sibly, R.M. & Calow, P. 1986. *Physiological Ecology of Animals*. Blackwell Scientific Publications, Oxford.
Stearns, S.C. 1992. *The Evolution of Life-Histories*. Oxford University Press, Oxford.

CHAPTER 15

Development

Felix qui potuit cognoscere causas
Motto: Churchill College Cambridge

Chapter 14 described the patterns of reproduction in animals and in many ways the story of animal development begins in the adult organism with the formation of the gametes – oocytes or spermatozoa (see Section 14.2). In sexually reproducing organisms the haploid gametes fuse to form a new diploid zygote and it is from this rather specialized cell that the adult multicellular organism is derived. The sequence of early cell divisions is referred to as 'cleavage', a precisely organized process of cell division not associated with cell growth. Cleavage subdivides the cytoplasm of the zygote in a predetermined way into a larger number of smaller cells which retain the spatial organization of the fertilized egg. The cytoplasm of the egg is characterized by the storage of materials which support the future development of the embryo. Among these materials are stored messenger RNA (mRNA) molecules and proteins (mRNA transcripts) which will have a profound influence on the subsequent development of the embryo. For this reason there are strong effects of the cytoplasm inherited from the egg on the developmental fate of the cells that develop from the egg. These are called maternal effects. Some time before the first cleavage occurs a three-dimensional spatial arrangement of the cytoplasmic structure of the egg is created by cytoplasmic movements. This spatial arrangement establishes the primary axes of the future embryo. Among invertebrates with a grade of organization more complex than the cnidarians this will involve definition of anterioposterior and dorsoventral axes. This spatial organization may become established during the development of the egg (as in the insect Drosophila *– see below) or, may be set up by cytoplasmic movements in the fertilized egg that are initiated at the time of fertilization. Whatever the case, the importance of this spatial organization of oocyte cytoplasm can hardly be overemphasized. It creates localized distributions of maternal gene products which have profound influences in the determination of the fate of the cells produced during early cleavage.*

The term 'cleavage' is used to describe the highly conserved patterns of cell division in early animal development when the fertilized egg first becomes subdivided into a larger number of small cells. A relatively few, rather different, patterns of cleavage occur in invertebrates. The spiral cleavage pattern of molluscs, annelids and the other so-called protostome phyla and the radial pattern of the deuterostomes (e.g. echinoderms) are well known. In recent years, the superficial cleavage pattern of the insect Drosophila *and the cleavage pattern of the nematode* Caenorhabditis *have assumed greater importance as studies in developmental genetics, using these organisms as models, have become a major focus of developmental research.*

The subsequent development of the embryo involves spatial reorganization to give the adult body plan, and, at a later stage, differentiation of cells to those found in the functional larva or adult. The developmental sequence thus involves:

1 *Gamete formation and storage of developmental information*
2 *Fertilization*
3 *Activations of zygote metabolism and translation of maternal messenger molecules (mRNA)*
4 *Cleavage*
5 *Activation of zygote nucleus and transcription of new zygote-specific information molecules (mRNA)*
6 *Organogenesis*
7 *Differentiation*

Development also frequently involves a metamorphosis when a differentiated larva adapted to one set of environmental conditions changes suddenly into an adult with rather different morphology adapted to quite different conditions, and which may have very different functions (see Chapter 14). Invertebrates have provided particularly valuable models for the analysis of the biochemical and molecular mechanisms involved in cell differentiation and regional organization. This trend has continued and the insect Drosophila melanogaster *and the nematode* Caenorhabditis elegans *are now amongst the most important models for the investigation of the molecular basis of regional organization and the determination of developmental fate. Moreover the isolation and determination of the structure of genes controlling early development in insects and in vertebrates such as ourselves is showing that their early development is based on processes regulated by the same highly conserved genes.*

Developmental studies have always been important for the light they shed on the relationships between animals and the development of precise molecular tools has revolutionized this approach. A new scientific discipline that might be termed 'evolution of development' based on genomic techniques is now emerging.

This chapter will examine the development of invertebrates with particular reference to experimental investigations of the underlying processes involved in the determination of cell fate. Wherever possible an insight will be given into the gene regulatory mechanisms that provide a functional explanation of cellular interactions discovered by developmental biologists earlier in the century, before the discovery of the molecular basis for the control of cell differentiation. The review includes an analysis of fertilization, cleavage, regional organization and the determination of cell fate. The emphasis will be on experimental studies and a historical approach to the study of animal development will be retained. Progress is now more rapid than at any time but the science of developmental biology is based on a single quest – to understand the re-creation of a fully differentiated adult organism from what appears to be (but in reality is not), an unstructured egg.

The chapter will conclude with a discussion of regeneration among the invertebrates. During regeneration, a complete pattern is reconstituted from a fragment of a pattern. The processes involved therefore mimic those of normal development.

15.1 Oogenesis: the storage of developmental information

The title of this section is taken from the prophetic title of a book on development by the embryologist Raven. It encapsulates precisely the importance that events during oogenesis have for the subsequent development of a new organism.

15.1.1 The wheel of development

In Chapter 14, we learned of the diverse processes of asexual reproduction which may be used to 'reconstruct' a fully differentiated organism. Some of these, such as parthenogenesis, involve differentiation from a single egg-like cell, while others, such as fragmentation, budding and gemule formation, involve regeneration from multicellular components. Despite the great diversity of asexual processes that can be used to reconstruct an organism, in the vast majority of animals the adult individuals (or colony or clone) has arisen from a single cell – the fertilized egg.

It is an almost universal feature of animal life that each evolutionary line periodically passes through the single-celled state and subsequently redevelops a multicellular differentiated state. A number of common features in this developmental cycle have been recognised and can be set out as in Fig. 15.1 as a 'developmental wheel'. The process of development is a single continuous process but a number of common steps can

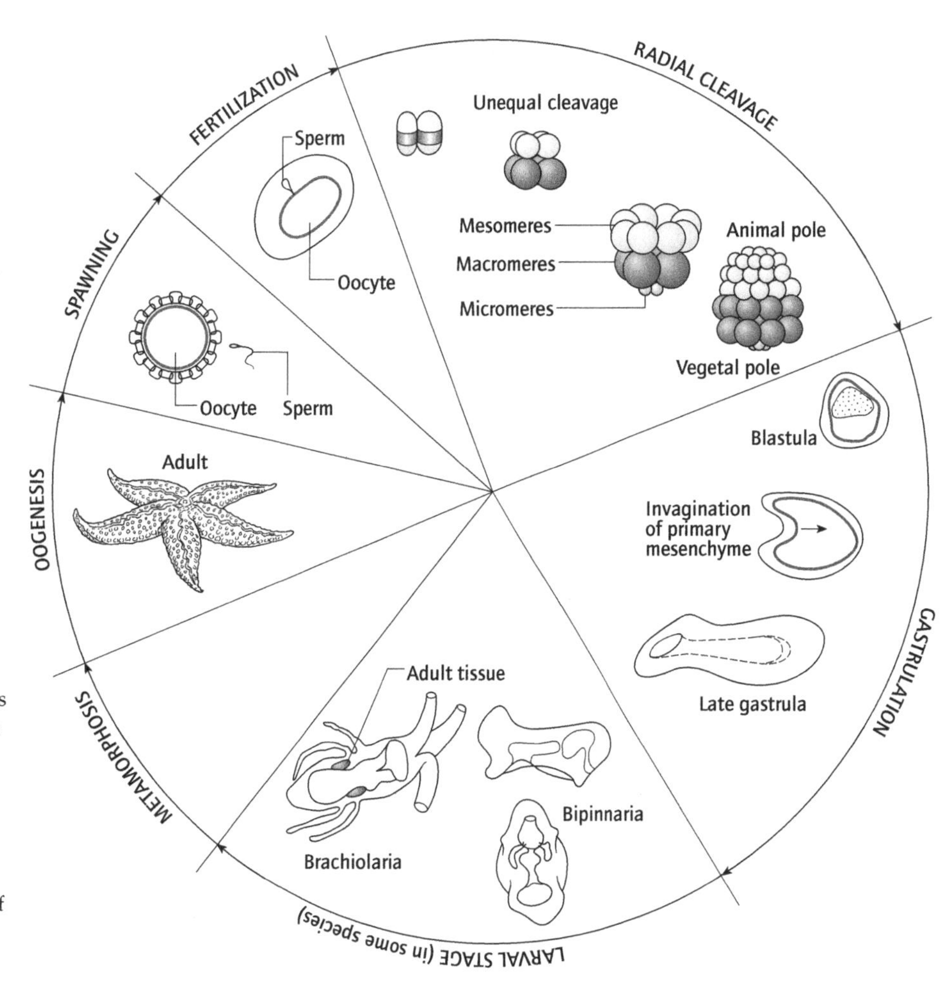

Fig. 15.1 The developmental wheel. Animal development can be likened to a cyclic process during which an adult organism creates germ cells which at fertilization fuse to form a zygote which must then reconstitute an adult. The key processes are: fertilization, cleavage, gastrulation, larval development, metamorphosis and organogenesis. (This diagram is based partly on the observations of Rebecca Platt, a second-year zoology student at Newcastle University.)

be recognized. It should be remembered, however, that all invertebrates do not have the organ grade of organization. Metamorphosis is particularly striking in invertebrates such as molluscs, annelids, insects and tunicates.

The formation of a diploid zygote is achieved at fertilization when two haploid gamete cells meet and fuse (egg n + spermatozoan n = zygote $2n$). It may be considered that animal development starts here but it is instructive to move back around the developmental wheel and start our discussion of animal development with the formation of the egg.

15.1.2 Oogenesis

The basic patterns of egg formation were described in Chapter 14 (see Fig. 14.12 and Box 14.7). The fully developed eggs are large cells that contain within the cytoplasm all the materials needed for the subsequent development of a new individual at least up to the stage at which feeding can begin. In addition there are materials involved in the reaction of the egg after contact with a sperm (see Section 15.2). These materials (cortical granules) are in the outer, cortical, region of the egg and after the intial fertilization reaction, help to keep out further sperm. The egg must fuse with only one!

Stored materials in the egg include:
- Protein and lipoprotein yolk spheres (see Box 14.7) often containing a large molecular weight protein – vitellin;
- lipid droplets;
- mitochondria;
- abundant ribosomes (rRNA);
- cortical structures – granules and/or alveoli;
- stored heterogeneous gene transcripts (mRNA).

These materials may be arranged in a concentric radial pattern. If so the concentric pattern may change at the time of fertilization prior to the first cell division (cleavage – see Fig. 15.8).

In many invertebrates a clear axial polarity is established during oogenesis and it is possible to distinguish a relatively yolk free 'animal' pole and a more yolky 'vegetal' pole. In the insect *Drosophila*, the polarity arises from the polarized position of the egg in its follicle (see Box 14.7c.ii). In this case the polarization has an important role in defining the future anterioposterior axis of the embryo. Oocyte cytoplasm contains transcribed gene products – stored mRNA. These gene transcripts are 'masked' in some way and are not immediately translated into functional proteins. Their position in the egg cytoplasm can therefore have a profound influence on developmental pathways later in development as explained in Box 15.1.

Box 15.1 Cytoplasmic localization of mRNA and the establishment of the anterior–posterior axis in *Drosophila*

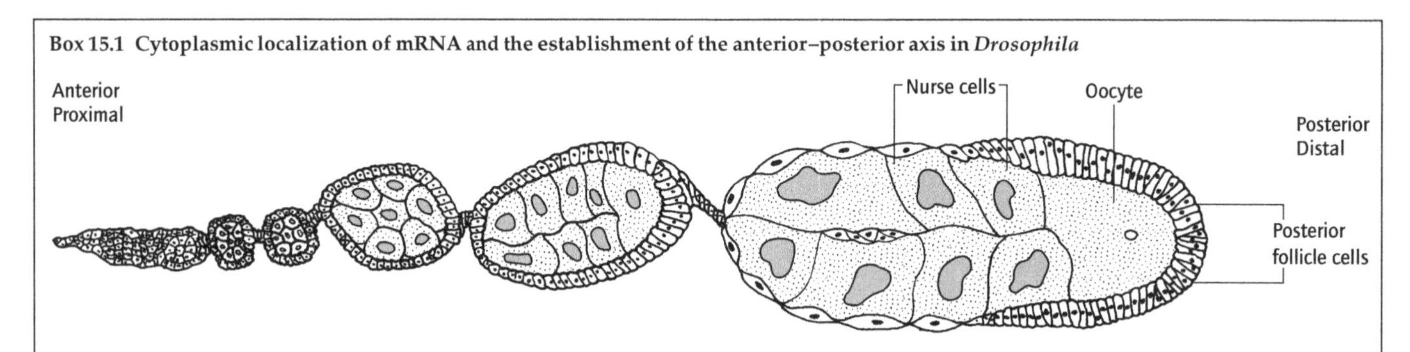

1 The fly *Drosophila melanogaster* is an important model for the genetic analysis of animal development. It has provided a starting point for the study of the molecular basis for regional organization.

2 The eggs of *Drosophila* develop within string-like ovarioles grouped together to form the ovaries, as shown above.

Each ovariole consists of a string of developing oocyte/follicles. The germ cell stem cells are located at the proximal end of each ovariole and the ovariole can be thought of as a production line with mature oocytes emerging from the distal ends of the string ready for fertilization and egg laying.

3 A stem cell divides and produces a daughter cell which begins to move down the ovariole production line. In *Drosophila* (but not all insects) the nucleus of each daughter cell, or primary oocyte, divides again four times but cytokinesis or cell division is incomplete so that a syncytium of 16 cells is produced. In this syncytium the cells are interconnected by cytoplasmic bridges (ring canals) in a very precise pattern as shown below:

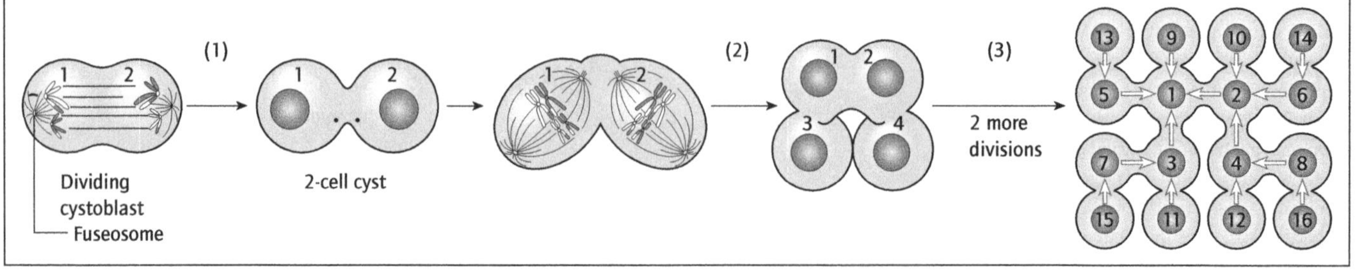

Continued

Box 15.1 *(cont'd)*

Note: two of the interconnected cells have more ring canal connections than the others and one of these two cells will become the oocyte; the other 15 cells will become nurse cells.

4 The nurse cells form a kind of 'amplification' system for the production of RNA (mRNA and ribsosomes). Substantial quantities of RNA are synthesized within the nurse cell nuclei and transported out first to the nurse cell cytoplasm and then, through the ring canals, into the oocyte cytoplasm.

This picture published by Bier provided a dramatic first illustration of the transfer of mRNA from the nurse cells to the oocyte.

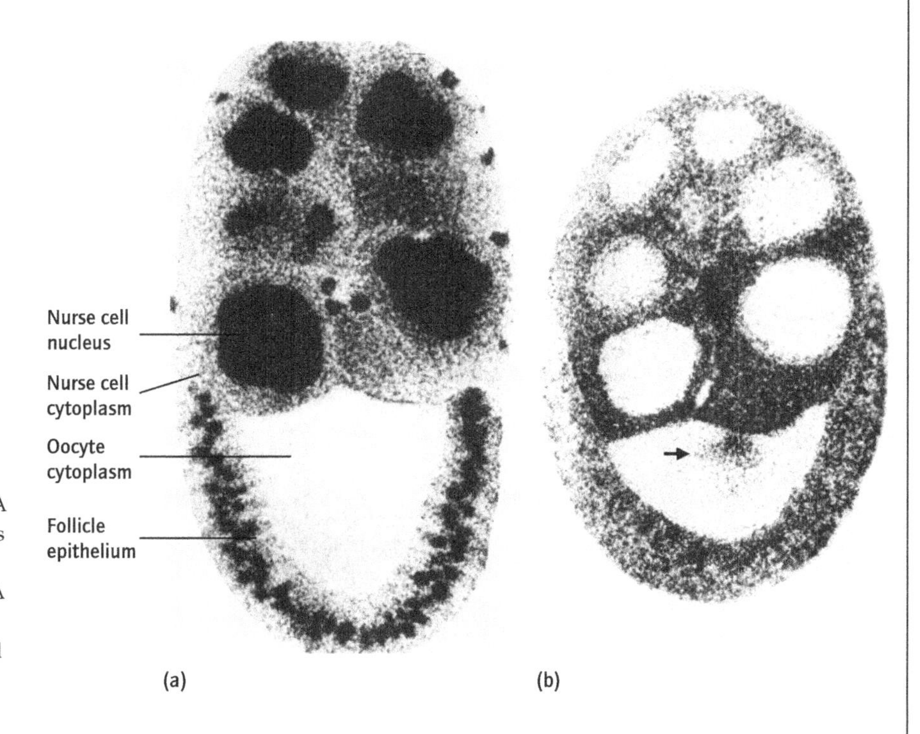

Transport of mRNA from nurse cells into fly oocytes. (a,b) Autoradiographs of the follicle cell of the housefly, *Musca domestica*, after incubation with [^{3}H]cytidine. (a) Egg chamber fixed immediately after label was introduced. The nuclei of the nurse cells are heavily labelled, indicating that they are synthesizing new RNA. The oocyte remains unlabelled. (b) A similar egg chamber fixed 5 hours later. Label is gone from the nurse cell nuclei but has moved into the cytoplasm. Moreover, radioactive RNA can be seen passing into the oocyte cytoplasm through the channel between the nurse cell and the oocyte (arrow).

5 The oocyte–nurse cell complex becomes surrounded by somatic (follicle cells) to form the ovoid maturing egg follicle.

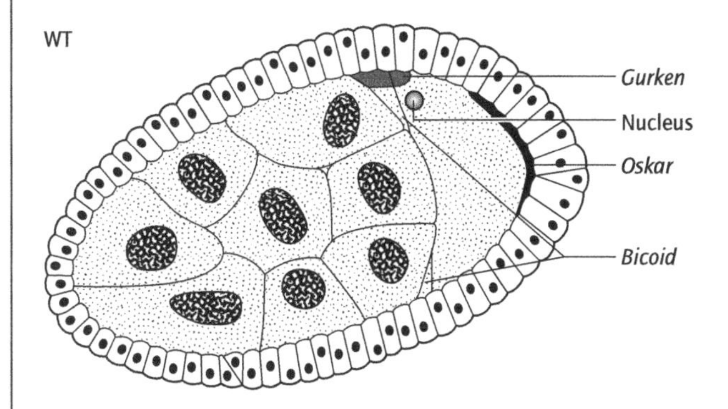

6 Polarization of the egg cytoplasm. The egg is in contact with the follicle cells at one pole and with nurse cells at the other. It is therefore in a polarized position and this spatial pattern becomes embodied as the future anterior–posterior axis of the embryo and the molecular steps involved are now becoming understood.

The oocyte nucleus transcribes only a few genes itself, one of these is the gene *gurken*. This is a regulatory gene that has effects on the adjacent follicle cells. Those follicle cells that have been influenced by gurken gene products are called the posterior polar follicle cells (PPFC). These cells respond by translation and transcription of a variety of other gene products. These in turn influence the microtubulin proteins in those parts of the egg cortex that they contact. Thus signals are being bounced back and forth between the oocyte and the neighbouring follicle cells.

7 Regional signal information. As the definitive oocyte grows the overlying follicle cells eventually collapse but the underlying egg carries the imprint of their previous position and molecular activity. This then controls the spatial distribution of two very important gene products that have moved into the oocyte cytoplasm from the nurse cells (see above). The regional identifiers are bicoid mRNA the product of the gene *bicoid* and nanos mRNA the product of the gene *nanos*. The gene product nanos mRNA becomes restricted to those regions of the oocyte cytoplasm that have been influenced by the PPFC. A second gene product bicoid mRNA in contrast is sequestered at the opposite end of the egg, i.e. nearer to the nurse cells away from the cortical region of the egg influenced by PPFC. The egg is thus polarized in molecular terms.

The two gene products bicoid and nanos mRNA subsequently play an important role in establishing the proper sequence of axial structures – head, thorax and abdomen, by regulating the regional transcription and/or translation of other gene products. The details of this story are now emerging and are discussed further in Section 15.4.2 Box 15.9.

Continued p. 368

Box 15.1 (*cont'd*)

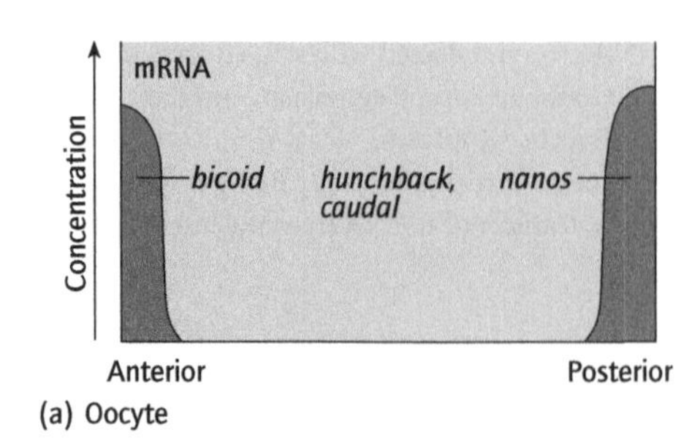

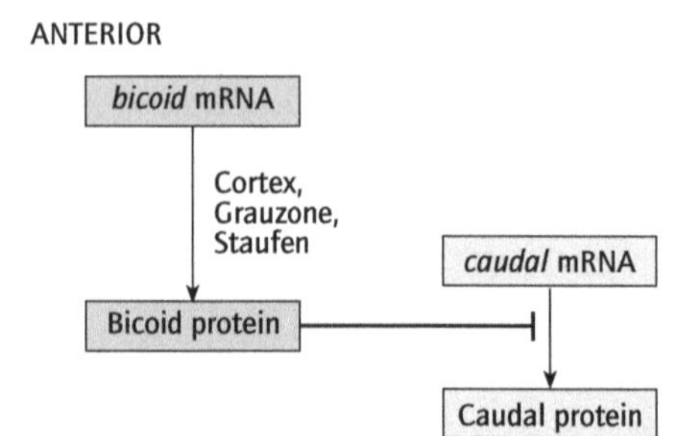

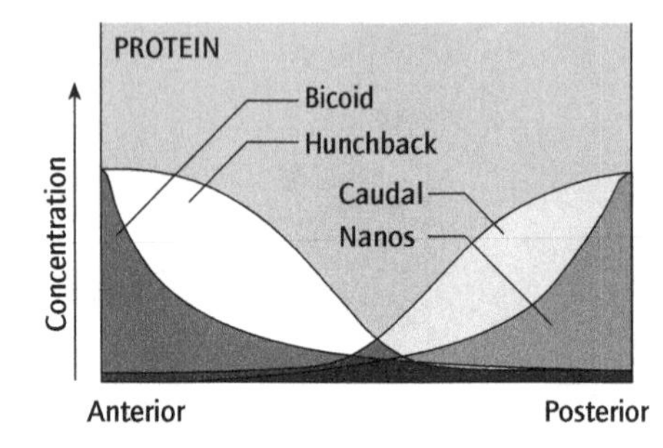

(b) Early cleavage embryo

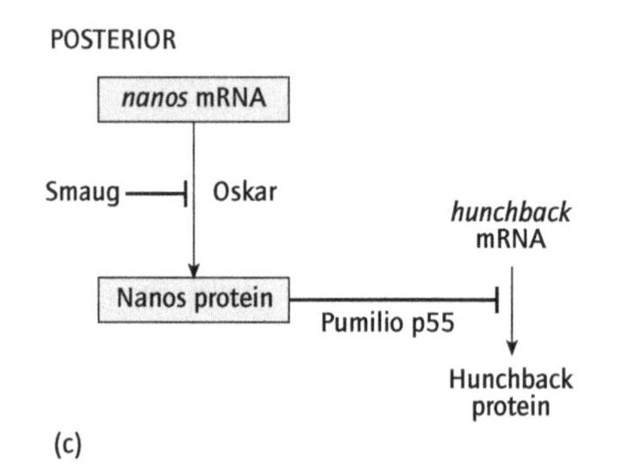

(c)

A model of anterior-posterior pattern generation by the maternal effect genes. (a) The *bicoid*, *nanos*, *hunchback* and *caudal* messenger RNAs are placed into the oocyte by the ovarian nurse cells. The *bicoid* message is sequestered anteriorly. The *nanos* message is sent to the posterior pole. (b) Upon translation, the bicoid protein gradient extends from anterior to posterior, and the nanos protein gradient extends from posterior to anterior. Nanos inhibits the translation of the *hunchback* message (in the posterior), while bicoid prevents the translation of *caudal* message (in the anterior). This results in opposing caudal and hunchback gradients. The hunchback gradient is secondarily strengthened by the transcription of the *hunchback* gene from the anterior nuclei (since bicoid acts as a transcription factor to activate *hunchback* transcription). (c) Parallel interactions whereby translational gene regulation establishes the anterior–posterior patterning of the *Drosophila* embryo. In the anterior of the embryo, *bicoid* mRNA is bound to the anterior cytoskeleton and is inhibited from translation by having a small polyadenylate tail. Upon fertilization, the tail is extended in a manner dependent on the cortex, grauzone, and staufen proteins, and the *bicoid* mRNA is translated. The bicoid protein suppresses the *caudal* mRNA from being translated. In the posterior region of the embryo, *nanos* mRNA is suppressed in the oocyte by the smaug protein (which binds to its 3′ UTR). At fertilization, oscar aids its translation, and the nanos protein acts as a translational suppressor of *hunchback* mRNA. (After Gilbert, 1997; Macdonald and Smibert, 1996, *Current Opinion Genetics & Development* **6**, 403–407.)

Experiments with sea-urchins first established that early developmental morphology is controlled by maternal genes and not by the zygote genome alone (Fig. 15.2). This provided an important early insight into the role played by stored mRNA in animal development.

15.1.3 Fertilization and the initiation of development

Fertilization is a complex process. Its principal components are:

1 Physical juxtaposition of the gametes.

2 Surface membrane interaction leading to the union of the sperm and the egg.

3 A physiological reaction at the surface of the egg leading to a block to further sperm entry, usually called the 'block to polyspermy'.

4 Activation of the oocyte metabolism.

5 Fusion of the pronuclei to form the new diploid genome of the zygote.

6 The initiation of early cleavage.

Marine invertebrates with external fertilization have been especially important in establishing the general principles of this process. The precise sequence of events depends on the state of maturation of the 'egg' when gamete fusion occurs (Table 15.1). The eggs and sperm of many marine invertebrates are shed into sea water and it is in this medium that fertilization takes place. Spermatozoa may be activated by the pH change that occurs when they are mixed with sea water and may show random movements that will tend to increase the frequency of contact with eggs if they are also suspended in sea water. There is increasing evidence that at close range chemical interactions may guide the spermatozoa towards the egg surface.

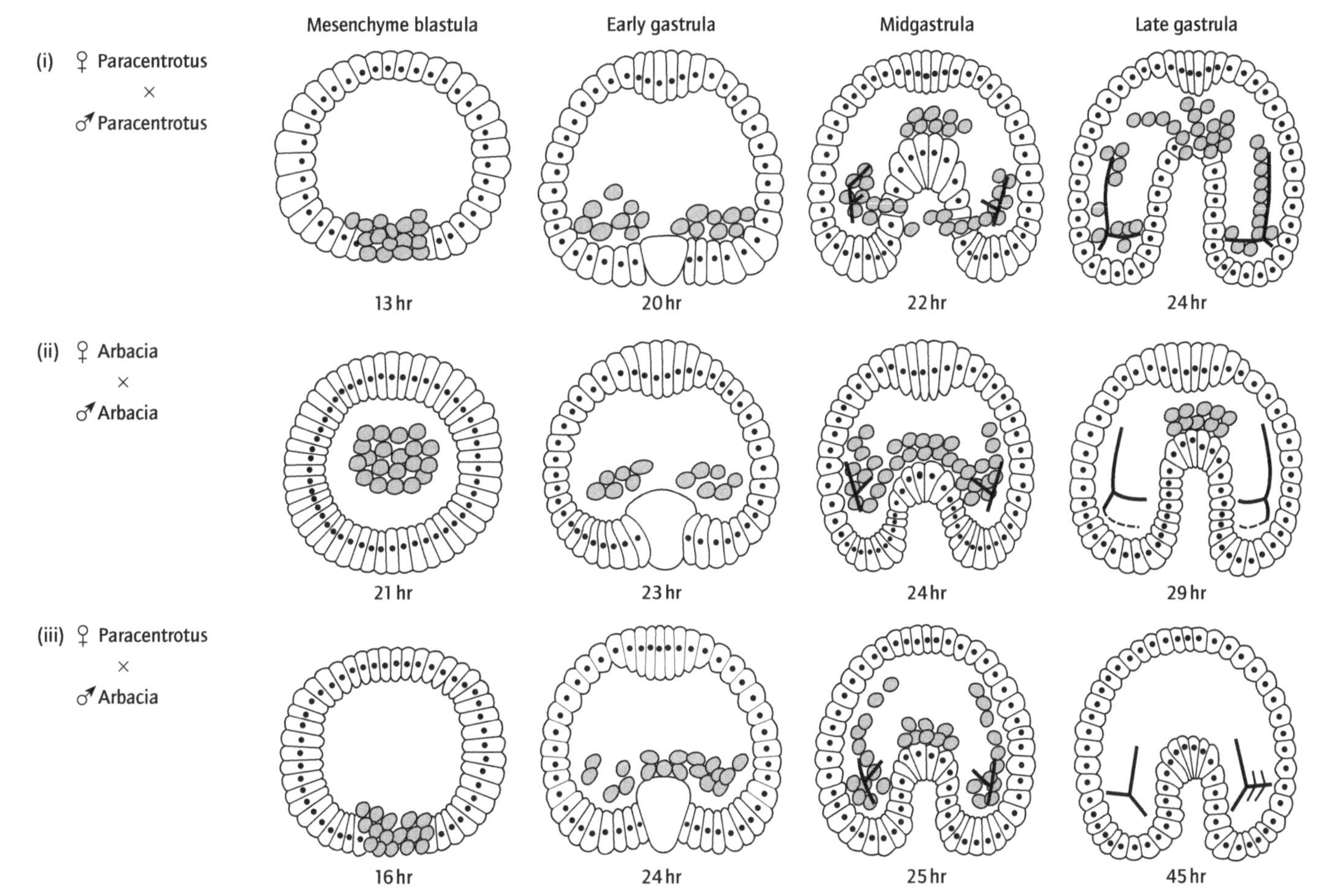

Fig. 15.2 Diagrammatic representation of the early development of two species of echinoderm: (i) *Paracentrotus lividis* and (ii) *Arbacia lixula*. (iii) The hybrids formed by fertilization of eggs of *Paracentrotus* with the sperm of *Arbacia*. The hybrid embryos develop for about 45 hours but follow the maternal forms. Based on A.H. Whitley and F. Baltzer (1958). (After Davidson, 1968, *Gene Action in Early Development*, Academic Press, N.Y.)

Table 15.1 Diversity of the stages in meiosis at which cell division arrests pending fertilization. Examples are given.

Pre-vitellogenesis: primary oocyte	**Post-vitellogenesis**			
	Prophase I	**Metaphase I**	**Metaphase II**	**Post-metaphase II**
Planaria – *Otomesostoma* Polychaeta – *Dinophilus, Saccocirrus* Onycophora – *Periopatopsis*	Nematoda – *Ascaris* Mesozoa – *Dicyema* Porifera – *Grantia* Polychaeta – *Nereis* Mollusca – *Spisula* Echiura – *Urechis*	Nemertea – *Cerebratulus* Polychaeta – *Chaetopterus, Arenicola, Pectinaria* Mollusca – *Dentalium* Echinodermata Asteroidea – *Asterias, Asterina*	Chordata – *Branchiostoma*	Cnidaria Echinodermata Echinoidea – *Psamechinus, Echinus, Arbacia*
Not common associated with modified sexuality, see Box 14.3	Sperm fusion reactivates meiosis. Problem of linking receptivity of sperm and developmental competence	Hormone-signalling links spawning and progress from prophase to metaphase I. See Chapter 16, Section 16.11.4. In starfishes 1MeAd signals. GVBD and unfertilized eggs go to G1 after metaphase II	Not common among invertebrates other than Chordata. Pattern shared with most vertebrates	In echinoids has been used as a convenient model for studying the biochemical consequences of fertilization

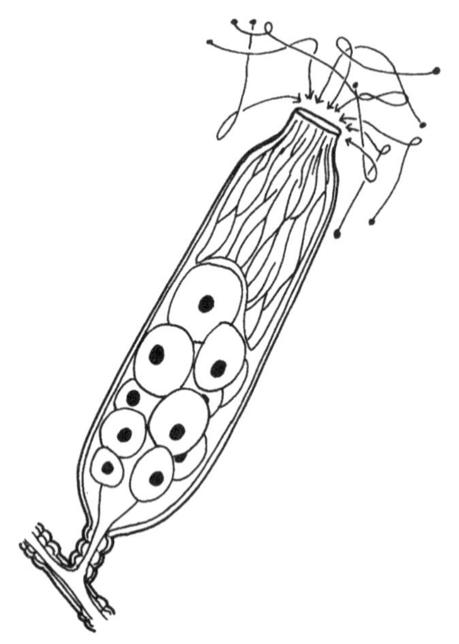

Fig. 15.3 A gonangium of the hydroid *Campanularia* and tracings of the tracks of spermatozoa as recorded by cinematography. This was one of the first cases where chemotaxis between eggs and spermatozoa could be demonstrated. It is certainly not unique; chemotaxis between sperm and eggs has now been demonstrated in urochordates, molluscs and annelids. It is probably a widespread phenomenon. (After Miller, 1966.)

15.1.4 Interaction of sperm and egg at a distance

The eggs and sperm of the colonial hydroid *Campanularia* provided the first clear experimental evidence of chemotaxis between sperm and eggs. In this species the eggs are not released freely into sea water but are retained in the flask-shaped gonangium, and the sperm must reach the eggs through the narrow opening. Cine film showed that the tracks of individual sperm were not random but directed towards the mouth of the gonangium (Fig. 15.3). A substance was extracted from the mouth of the gonangium that would attract spermatozoa.

The close range attraction of sperm towards oocytes is now believed to be not uncommon and has been observed in species where the eggs are freely discharged as well as in those where the eggs are retained in a protective capsule. Convincing evidence has been obtained using annelids, molluscs, tunicates and echinoderm eggs as well as hydroids and the chemical nature of the substances involved is being elucidated. In echinoderms a short-chain (1.4 kDa) peptide containing just 14 amino acids has been isolated from the eggs of the sea urchin *Arbacia* that has sperm attracting properties. This peptide – resact – is found in the jelly surrounding the egg at the time of egg laying. The sperm are able to move towards a concentration centre in sea water of nanomolar quantities. The reaction of *Arbacia* sperm to the resact molecule is species-specific but this is not a property of all sperm chemoattractants. A larger 12.5-kDa 32 amino acid peptide termed Startrac (STARfish a TRACtant) has been isolated from starfish eggs but the reactions of sperm caused by this substance have not been found to be so precisely species-specific.

Sperm chemotaxis is probably not effective at distances of more than one or two egg diameters (0.2–0.5 mm). Nevertheless substances diffusing from the eggs or egg jelly layer may elicit enhanced sperm motility. The peptide resact, for instance, is one of many substances in the jelly layer of sea-urchin eggs which stimulate sperm mobility and cause an increase in sperm oxygen consumption rate.

In the lugworm *Arenicola marina*, a different mechanism operates. Male spawning is initiated by the release of the fatty acid 8.11.14-eicosatrienoic acid. Pools of sperm are discharged onto the surface of the sand, in which these animals live, during low tide. The sperm are disaggregated from the syncytial cell masses in which they developed within the body cavity at the time of spawning, but the disaggregated sperm remain relatively

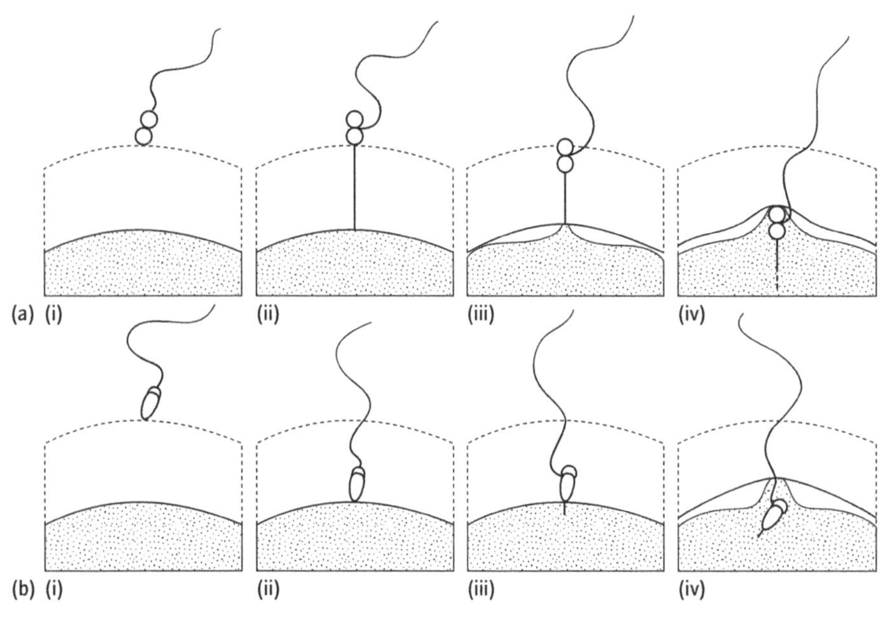

Fig. 15.4 Observations on the changes visible with the light microscope during the fertilization process of (a) a starfish (*Asterias*) eggs and sperm and (b) a sea-urchin (*Arbacia*) egg and sperm. (a) *Asterias*. (i) Approach of the spermatozoon to the egg which is surrounded by a jelly coat. (ii) The acrosome at the front (in the sense of direction of forward movement) of the sperm. (iii) The fertilization response occurs spreading away from the initial point of contact. (iv) The spermatozoon is engulfed by a cone-like eruption of oocyte cytoplasm. (b) *Arbacia*. (i) Approach of the spermatozoon to the jelly coat. (ii) Passage of the spermatozoon through the jelly coat. (iii) Contact of the sperm head with the vitelline membrane. A very small acrosome filament may be visible as shown. (iv) Fertilization response and incorporation of sperm nucleus. (After Austin, 1965.)

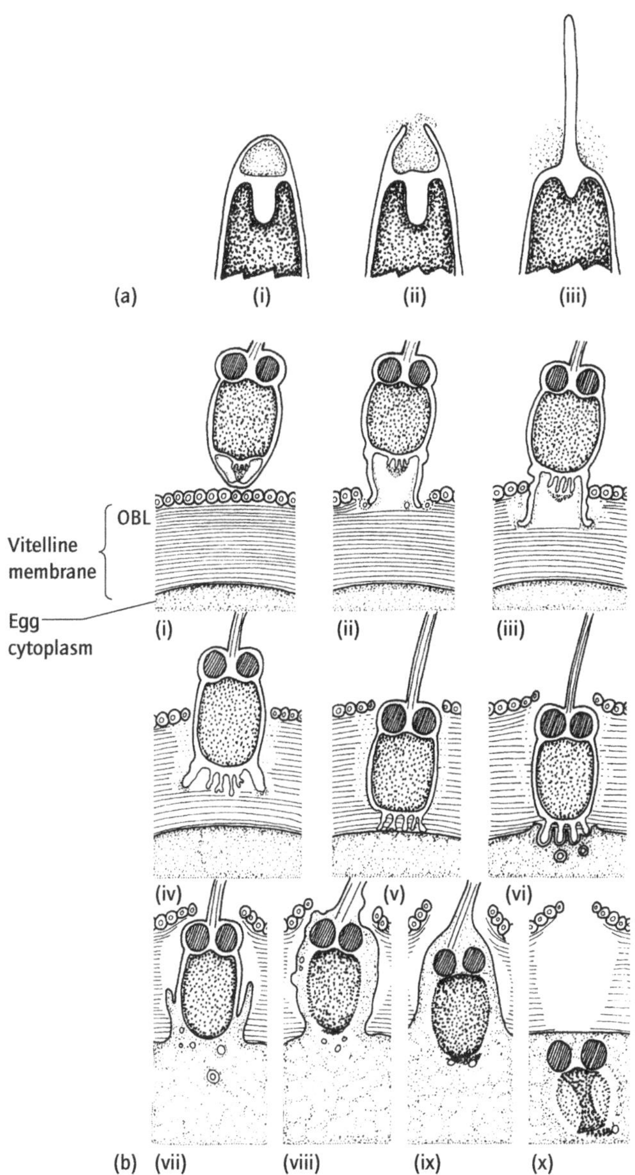

Fig. 15.5 Fertilization as seen with the electron microscope. (a) The acrosome reaction of the sea-urchin *Arbacia*. The acrosome filament of this species is too small to be seen easily with the light microscope. (i) The front part of the sperm head as seen with the electron microscope. (ii) Fusion between acrosome membrane and sperm membrane to release acrosome contents. (iii) Elongation of the acrosome filament. (b) Detailed analysis of the fertilization process of the polychaete *Hydroides*, as revealed by the electron microscope. (i) Approach of the spermatozoon to the outer border layer (OBL) of the vitelline membrane. The OBL is formed from the tips of the microvilli at the oocyte surface. (ii), (iii) Fusion of acrosome vesicle membrane causing release of the acrosome contents. Beginning of vitelline membrane penetration. (iv) Continuing penetration of sperm head and elongation of multiple acrosome tubules. (v), (vi) Contact and eventual fusion of acrosome tubules and oocyte plasma membrane. (vii), (viii) Incorporation of the sperm as oocyte cytoplasm moves into the joint oocyte/sperm membrane (caused by membrane fusion). (ix) Further incorporation. (x) Sperm nucleus and mitochondria incorporated into the egg cytoplasm. Note the hole left in the vitelline membrane. (Redrawn from electron micrographs of Colwin & Colwin, 1961.)

immobile in the sperm pools until the next tidal inundation causes them to be mixed with sea water. The pH change that occurs then (sea water has a pH of 8.2) is responsible for initiating changes that result in sperm activation and acquisition of forward mobility.

15.1.5 Contact between sperm and egg: the acrosome reaction

Observations on the interactions between invertebrate sperm and eggs have been important in leading to a better understanding in this important process that, in effect, begins the process of animal development. At the beginning of the century, it was observed that substances would diffuse from sea-urchin eggs into sea water that had the property of coagulating or binding spermatozoa. It is now known that the diffusing substances were protein-bound oligosaccharides. Most eggs have a species-specific sperm coagulant at the surface, but only in some cases will this diffuse into the surrounding medium, as it does from sea-urchin eggs.

The binding of protein bound oligosaccharides to species-specific receptor molecules on the sperm head initiates the first in a series of membrane fusions that are involved in the fertilization reaction. The head of most spermatozoa of the ectaquasperm type has, at its forward tip, a more or less complex vesicle, the acrosome vesicle. This is where the first changes involved in the fertilization reaction occur. It is only just possible to see what happens with a light microscope (Fig. 15.4) but the electron microscope provides a powerful tool which has made the sequence of events easier to visualize. As a sperm approaches the egg surface, but before direct contact is made, there is a fusion of the acrosome vesicle membrane and the sperm plasma membrane (Fig. 15.5 a,i,ii; b,i,ii). This releases the enzymic contents of the acrosome vesicle and the subsequent lysis permits the expanding acrosome tubules or filament to penetrate the egg coats and contact the egg plasma membrane.

The next stage in the fertilization process is the fusion of the sperm plasma membrane with that of the oocyte; this effectively creates a single new hybrid cell. Again the echinoderms provided excellent material for the initial investigations of what happens. The acrosome reaction exposes the most distal part of the spermatozoon, the acrosome tubule. In echinoderms and the enteropneust *Saccoglossus* (Fig. 15.5a) the acrosome tubule is a single filament which rapidly elongates. In some annelids there are multiple tubules (Fig. 15.5b), but the mechanism is otherwise similar.

Acrosome–sperm membrane fusion is signalled by an influx of Ca^{2+} ions and an efflux of H^+ following the species-specific binding of sperm-receptor sites to the sperm coagulant substances at the surface of the egg or in the jelly layer. The ionic signal initiates the processes of action polymerization that causes the tubule extension and activates ATPase activity. The increased ATPase causes the characteristic increase in sperm motility. The next stage of fertilization is the fusion of sperm and egg membranes.

The acrosome reaction exposes proteins that will bind to specific receptor sites at the egg surface. In sea urchins the protein is a species-specific 30 500 Da protein – bindin – capable of binding to isolated vitelline membranes or dejellied eggs. Following this

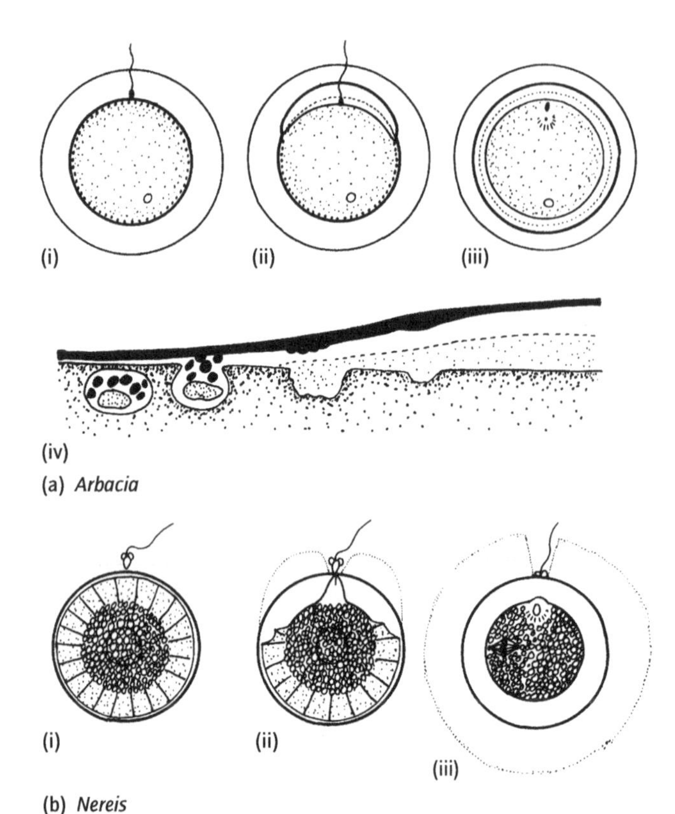

Fig. 15.6 The fertilization reaction: elevation of the fertilization membrane. (a) The sea-urchin *Arbacia*. (i) The sperm contacts the egg plasma membrane, cortical granules intact. (ii) Cortical granules break down and fertilization membrane is raised; the reaction moving progressively away from the point of contact. (iii) The fertilized egg. Note that the jelly coat is present prior to fertilization. (iv) Diagrammatical representation of the chain of events as seen in the electron microscope. (b) The polychaete *Nereis*. (i) The sperm approaches the egg. There is no jelly layer at this stage. (ii) Progressive spread of fertilization reaction from the point of contact. The outer part of the egg is composed of large, cortical alveoli which break down in this process. (iii) The fertilized eggs. In most species the contents of the cortical alveoli extrude through the fertilization membrane (ex-vitelline membrane) and form a new jelly coat. (After Austin, 1965.)

species-specific binding, fusion between egg and sperm membranes is initiated.

15.1.6 The reaction of the egg to fertilization

15.1.6.1 The block to polyspermy

The initial reactions to bindin–antibindin reaction and the fusion of egg and sperm are sometimes referred to as the 'block to polyspermy'. There are two components: the first is not visible but involves a change in electrical potential at the surface of the egg. A few seconds later a visible cortical reaction spreads radially away from the point of sperm–egg contact. This reaction involves fusion of the membranes of the cortical granules, formed during late oogenesis, with the egg membrane and the release of their contents into the perivitelline space. The result is the raising of the fertilization membrane, as illustrated in Fig. 15.6. A reaction of this type is almost universal; it can lead to the emission of a jelly coat (Fig. 15.6b), and an egg which has reached this stage of development cannot fuse with other sperm cells.

The eggs now shows profound metabolic changes. The unfertilized egg is in some ways in a state of suspended animation. The binding reaction of bindin protein to the oocyte membrane antibindin receptor site initiates a sequence of events (summarized in Fig. 15.7). There is a major increase in oxygen consumption which peaks during the first minute when the fertilization membrane is formed, but remains much higher than it was in the unfertilized egg. Subsequently, there is an increase in the rate of protein synthesis, as masked messenger RNA (mRNA) molecules, which were stored in the egg cytoplasm during oogenesis, are made available. The first cleavage will follow some predetermined time after fertilization, but important changes in the distribution of the cytoplasmic constituents may be observed prior to the first division.

In bilaterally symmetrical animals these cytoplasmic movements establish the primary axes of the future embryo by defining anterior, posterior, dorsal and ventral quadrants in the undivided egg. Such movements were first described in an ascidian and these and more recent developments are described below.

15.1.6.2 Activation of the egg cytoplasm and establishment of the major embryonic axes

Recent developments in our understanding of the molecular basis of regional organization have confirmed the importance of these pioneering observations. Understanding of the molecular basis for the establishment of the major embryonic axes through the creation of anterior–posterior and dorsal–ventral gradients has developed rapidly through studies of *Drosophila*. This is explained in Box 15.1.

In *Drosophila*, these axes are established prior to fertilization but in most other organisms the embryonic axes are established, or fixed, at the time of fertilization. Random events such as the point on the egg surface where sperm binding occurs can have the most profound consequences. Prior to fertilization a tunicate egg is radially symmetrical with a gradient in the distribution of yolk along the animal–vegetal axis (Fig. 15.8) but it develops an asymmetry that defines the dorsal–ventral axis shortly after fertilization and before completion of the first cleavage.

When the sperm fuses with the egg cytoplasm a calcium spike (net increase in free Ca^{2+} ions) is initiated beginning from the point of sperm entry. This causes a contraction of subsurface actin filaments away from the animal pole and making the animal–vegetal pole gradient of yolk distribution even more extreme. However the sperm entry point has created an element of asymmetry and a sperm aster (i.e. the microtubular structure to be involved in egg pronucleus and sperm pronucleus fusion) begins to form. This establishes a second axis at right angles to the animal–vegetal axis which defines the future anteroposterior organization of the body plan. It is interesting to notice a difference between the time of signalling this secondary axis in the tunicate and *Drosophila*. In the case of the tunicate there is no signalling of the dorsal–ventral axis until after fertilization whereas

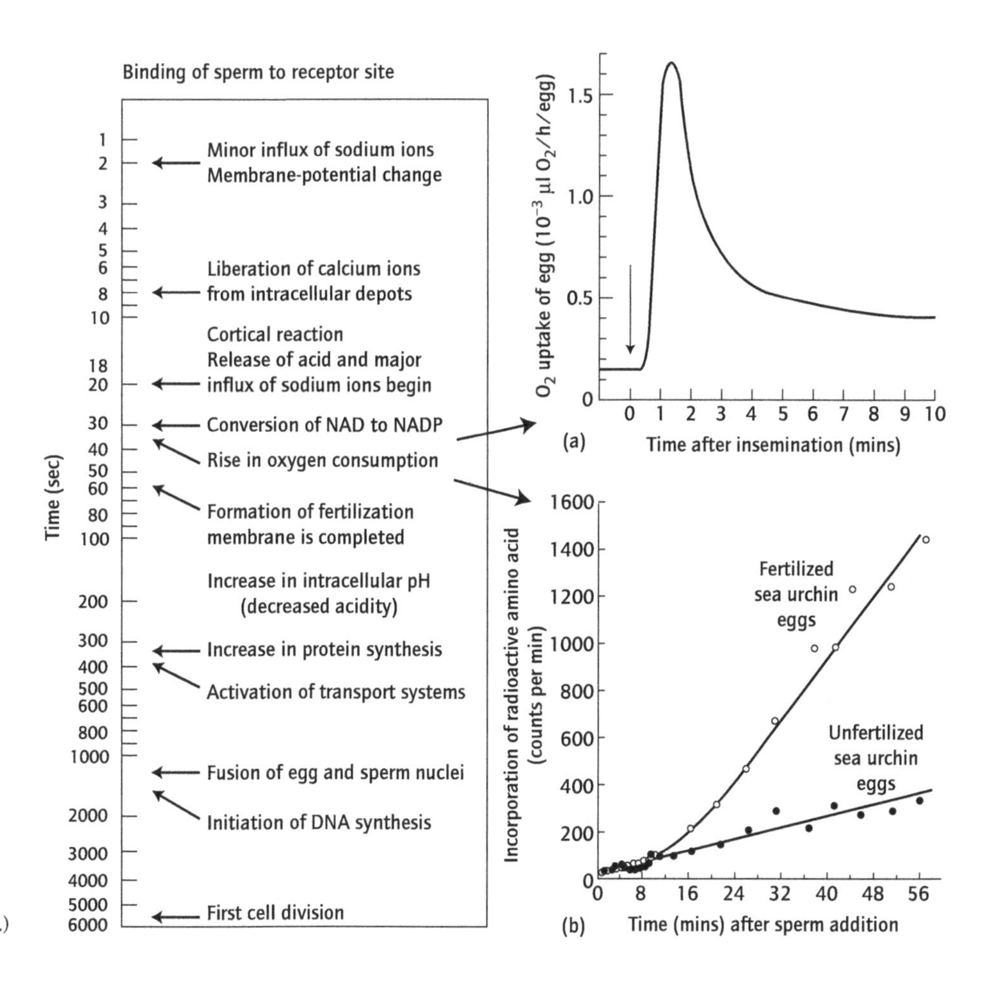

Fig. 15.7 The sequence of events following contact between sperm and egg membranes during sea-urchin fertilization (after Epel, 1977). Inset are examples of experimental data in support of this scenario. (a) Oxygen consumption (after Ohnishi & Sugiyama, 1963); (b) protein synthesis. (After Epel, 1967.)

in *Drosophila* the dorsal–ventral axis is established during oogenesis (Box 15.1). The nematode *Caenorhabditis* provides a third example where the processes of axis formation are becoming understood in molecular terms (Fig. 15.9). Again it is found that the axes are established at the moment of fertilization.

15.1.6.3 *The unmasking of stored mRNA*

A wide variety of mRNAs are stored in the oocyte cytoplasm and they are translated shortly after fertilization. The masked mRNAs include a wide diversity of transcripts encoding proteins that will function early in development. These include the cyclin proteins which regulate the cleavage pattern, tubulin proteins, involved in cytokinesis or cell division as well as 'maternal effect' gene products, such as bicoid and nanos genes, the role of which in setting up the anterior–posterior gradient in *Drosophila* was explained in Box 15.1.

The unmasking of mRNA is an important consequence of fertilization. Studies of sea urchins provided an initial indication of the importance of stored mRNA and some early evidence was presented in Fig. 15.2 and Fig. 15.7. Fertilization causes an increase in the rate of uptake and incorporation into protein of radioactive amino acids (indicating *de novo* protein synthesis). There is a characteristic delay from the moment of fertilization to the onset of increased protein synthesis (about 9 mins in Fig. 15.7b) and this led David Epel to suggest that during this time unmasking of the stored mRNAs in the egg cytoplasm occurred. Details of the molecular mechanisms are now emerging. The oocytes of the mollusc *Spisula* contain mRNAs encoding the protein cyclin-A. The mRNAs for this protein and for other mRNA species are bound to an 82-kDa masking protein in unfertilized eggs. Fertilization causes ionic changes leading to activation of cdc2 kinase and phosphorylation of the cdc-2 releases cyclin-A mRNA and thus initiates cyclin-A translation. Other mechanisms for regulating the translation of stored mRNA messages have been found – cutting/construction of polyadenyl tails, regulation of the internal pH, capping/uncapping of the ends of the mRNA sequences – and the process is clearly of great importance during the early stages of animal development.

15.2 **Patterns of early development**

15.2.1 Cleavage

Once mitosis has been initiated, embryos of most animals follow a precisely determined pattern of cleavage during which the cytoplasm of the relatively massive egg is subdivided into smaller cellular units. These cleavage patterns are such that any spatial organization in the fertilized eggs is retained into the multicellular embryo and, most give rise to a hollow ball of cells called a 'blastula'.

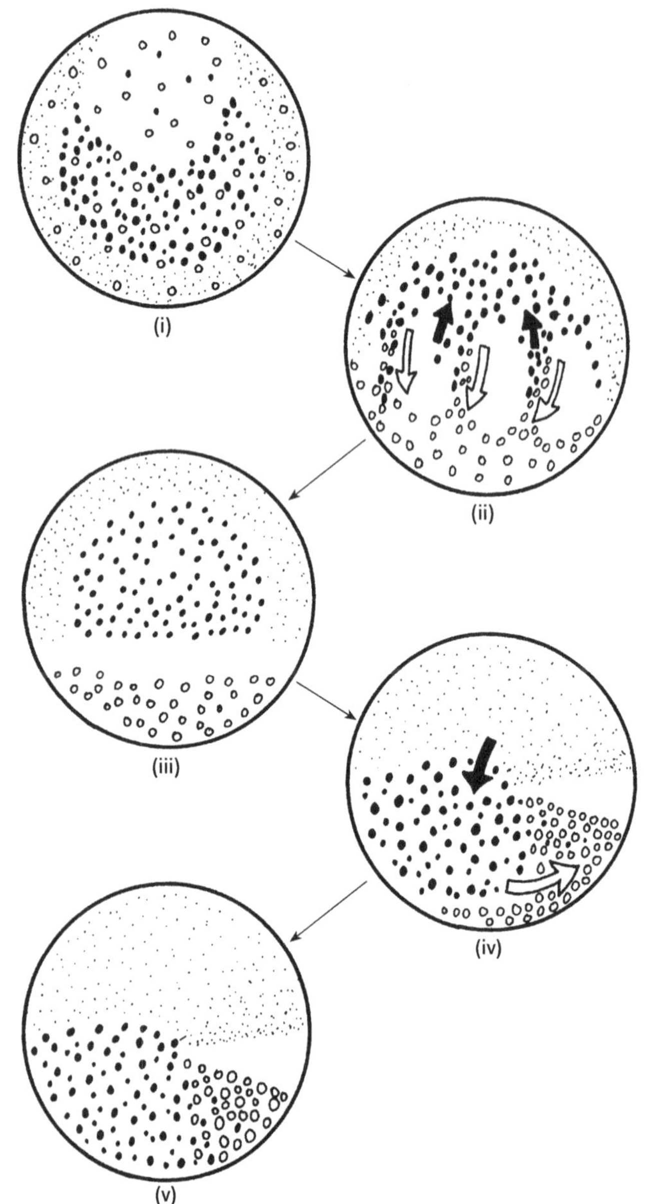

Fig. 15.8 Movements of the visibly different types of cytoplasm described in the early twentieth century in the ascidian *Styela partita*. These observations were important in demonstrating the significance of cytoplasmic factors in determining cell fate. (i) The radially symmetrical unfertilized egg. (ii), (iii) Migration of 'yellow' cytoplasm to the vegetal pole and segregation of 'grey' cytoplasm in the central mass of the egg. (iv) Segregation and migration of yellow cytoplasm and clear cytoplasm to one side of the egg. The egg becomes bilaterally symmetrical. (v) The bilaterally symmetrical egg with cytoplasm in positions occupied at the time of the first cleavage. (See also Box. 15.4 and Box 15.5.)

Many of the protosome phyla (e.g. Nemertea, Annelida, Sipuncula, Echiura, Mollusca and perhaps Pogonophora) exhibit spiral cleavage described in some detail in Box 15.2. The spiral cleavage exhibited by these phyla has sometimes been said to indicate a relatively close phylogenetic link between them. This thesis, however, has to be interpreted with care. The observation that several phyla have a spiral pattern of cleavage in common need imply only that this has been a conservative feature of their evolution, and it cannot be taken to indicate a relatively recent divergence from a common line of descent. If the ancient flatworm pattern of cleavage was spiral then the observation that some other pattern is common among several phyla may suggest a more recent evolutionary divergence. This is shown, for instance, by the deuterostome phyla. Many of these exhibit a different pattern of cleavage, termed 'radial', in which the products of the first transverse divisions lie directly over each other. A relatively simple example, found in the holothurian *Synapta* is illustrated in Fig. 15.10. All the cells are virtually identical in size and the embryo develops progressively to the hollow-ball blastula stage.

A more complex radial pattern is exhibited by the sea-urchins, which have been extensively used in experimental embryological research. Individual cells cannot be identified since the first two cleavages are equal, but different layers of cells can be identified in the 64-cell stage embryo, as explained in Box 15.3.

The yolky eggs of the arthropods, especially the insects, show yet another pattern of cleavage described as 'superficial' or 'endolecithal' because of the distribution of the yolk as a central mass. The cytoplasm is restricted to a superficial layer and the initial nuclear divisions are not accompanied by cell division. The nuclei eventually move into the superficial layers and cell boundaries are formed around them (Fig. 15.11). The cytoplasm at the posterior pole is specialized, and those nuceli that enter the pole plasm have special properties and are the only ones that can form germ cells.

The nematode *Caenorhabditis elegans* has become another key species for the analysis of animal development. It is especially useful for the study of cell differentiation because the adult has a very small number of precisely determined cells. The initial polarity of the egg is established at the time of fertilization. The first division separates two cells, a larger A cell and a smaller P cell. The pattern of cleavage is precisely controlled and the A cell divides before the P cell to give a characteristic three-cell stage (see Fig. 15.9). Interestingly the P cell contains a *nanos*-like mRNA and is the only cell from which future germ cells can be derived. Remember that *nanos* mRNA is sequestered at the future posterior end of the insect egg (Box 15.1) and that the cells which inherit the posterior pole plasm are the only ones that will form germ cells (Fig. 15.11). Common molecular pathways involved in early specification of cell fate are now known in organisms that are not thought to be closely related. The conclusion to be made is that these genetic mechanisms were present in early ancestors of the diverse animal phyla and that the mechanisms have been highly conserved. This in turn is giving rise to a new scientific discipline – the evolution of developmental mechanisms.

The subsequent pattern of development in *Caenorhabditis* is very much dependent on the spatial arrangements of the cells and the contacts that are made between them as they move within the tight confinement of the tough shell membrane typical of nematode worms.

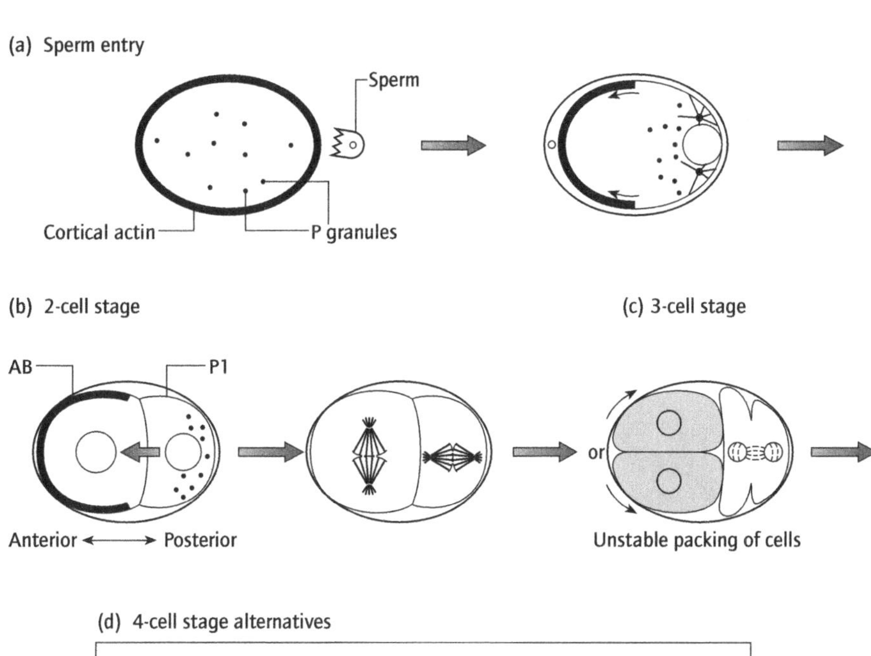

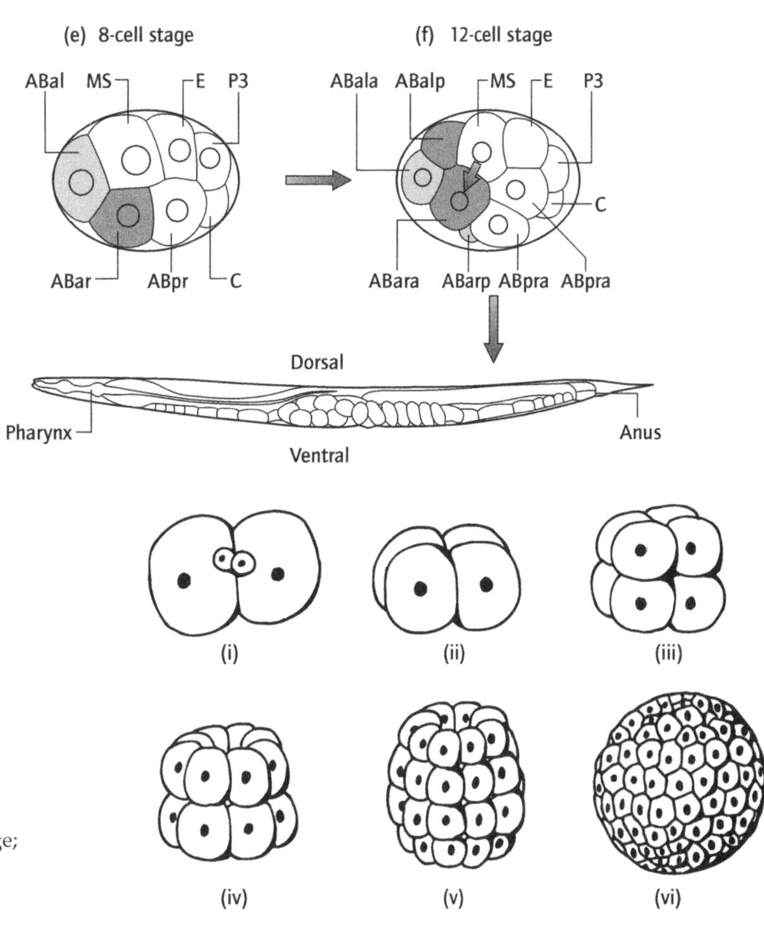

Fig. 15.9 Specification of the anterior–posterior axis in the nematode *Caenorhabditis elegans*. (a) The point of sperm entry establishes a polarized cytoplasmic architecture. (b) The first cell division creates two cells AB and P arranged along the anterior–posterior axis. (c) A three-cell stage is formed with alternative stable packaging arrangements at the four-cell stage (d). Subsequent divisions lead to the arrangements of cells shown in (e) and (f) and the future adult body plan is established.

Fig. 15.10 Cleavage in the echinoderm *Synapta*: (i) two-cell stage; (ii) four-cell stage; (iii) eight-cell stage; (iv) 16-cell stage; (v) late cleavage; (vi) blastula change.

Box 15.2 Spiral cleavage

This pattern of cleavage is characteristic of several protosome phyla including the Nemertea, Annelida, Sipuncula, Echiura and Mollusca. It is most easily seen in embryos with relatively little yolk such as those of the marine polychaetes and molluscs.

1 The unfertilized egg is radially symmetrical about an axis from the animal (less yolky) pole to the vegetal pole (Fig. i). All segments of such an egg cut along the animal–vegetal axis are equivalent.

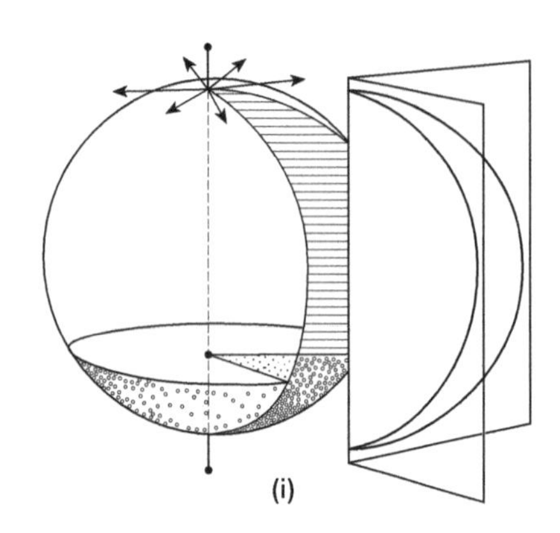

2 After fertilization the egg is bilaterally symmetrical.
The principal planes of the embryo can then be recognized (i–iii):

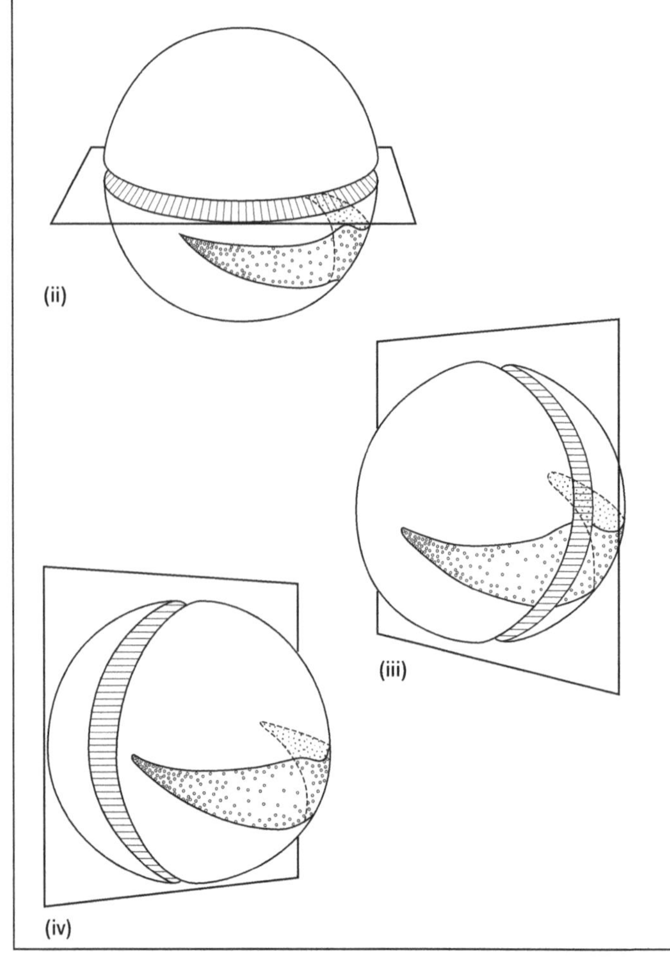

(i) Transverse – plane of symmetry separates anterior from posterior.
(ii) Sagittal – plane of symmetry separates left from right.
(iii) Frontal – plane of symmetry separates dorsal from ventral.

3 The first two cleavage planes are longitudinal and bisect the angles between frontal and sagittal.

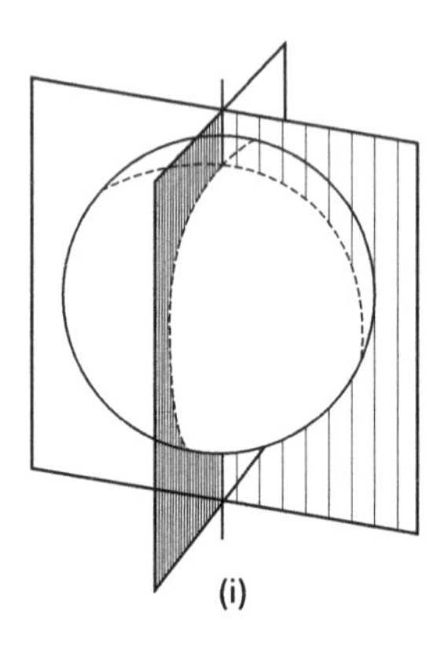

The AB cell is usually larger than the BC cell. The first separates two cells, AB and CD (ii). The second separates four cells called A, B, C and D (iii).

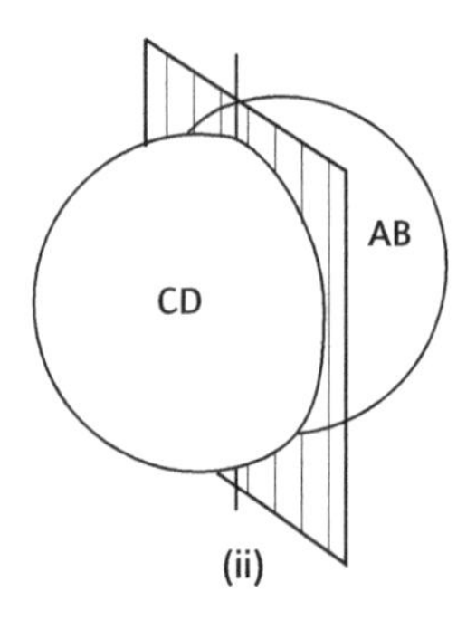

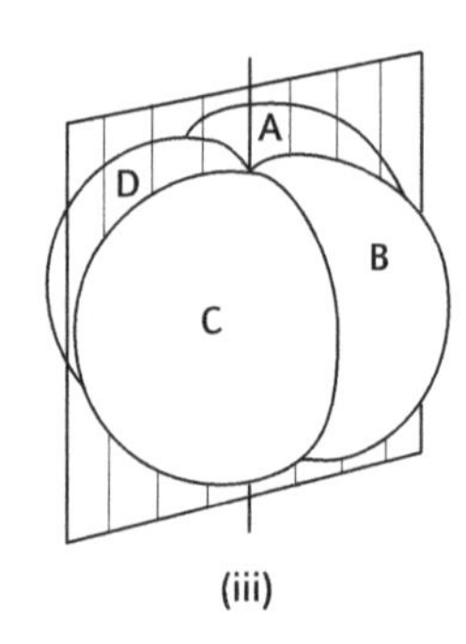

The D cell is usually larger than the other three and by reference to this cell all subsequent cleavage cells can be identified and named individually.

4 The third cleavage plane is transverse and passes above the equator. It separates four smaller cells at the animal pole from four larger ones at the vegetal (i).

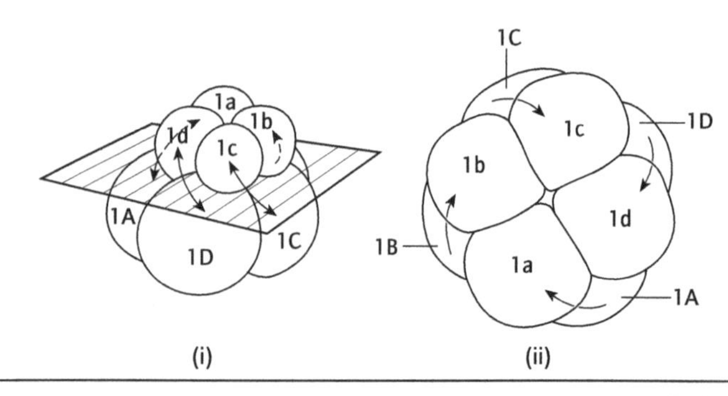

Continued

Box 15.2 *(cont'd)*

The four smaller cells are referred to as the first quartet of micromeres. Because the cleavage plane is inclined to the long axis of the embryo, the micromeres lie over the intercell boundaries between the macromeres. When seen from the animal pole, the cells appear to be rotated in a clockwise direction.

5 The eight cells of the embryo can now be individually identified as:

1a	1b	1c	1d
1A	1B	1C	1D

All subsequent divisions are transverse and result in subdivision of existing micromeres and the generation of new quartets of micromeres by unequal cleavage of the macromeres.

The fourth cleavage produces an embryo with 16 cells. At this stage they can be individually identified as:

Subdivision of	$1a^1$	$1b^1$	$1c^1$	$1d^1$
First quartet	↕	↕	↕	↕
	$1a^2$	$1b^2$	$1c^2$	$1d^2$
Second quartet	2a	2b	2c	2d
	↕	↕	↕	↕
Macromeres	2A	2B	2C	2D

Individual cleavage planes are inclined to the long axis such that there is alternately clockwise and anti-clockwise rotation of the micromeres.

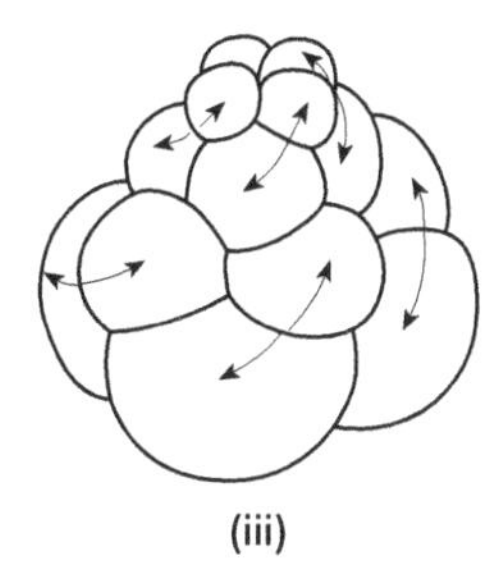

(iii)

6 A conventional scheme of notation has been developed which allows each cell to be identified up to the 64-cell stage. The formal scheme (completed only for the important D-cell lineage) is set out opposite.

As explained further in the text and in Boxes 15.3 and 15.4 the D-cell lineage is particularly important.

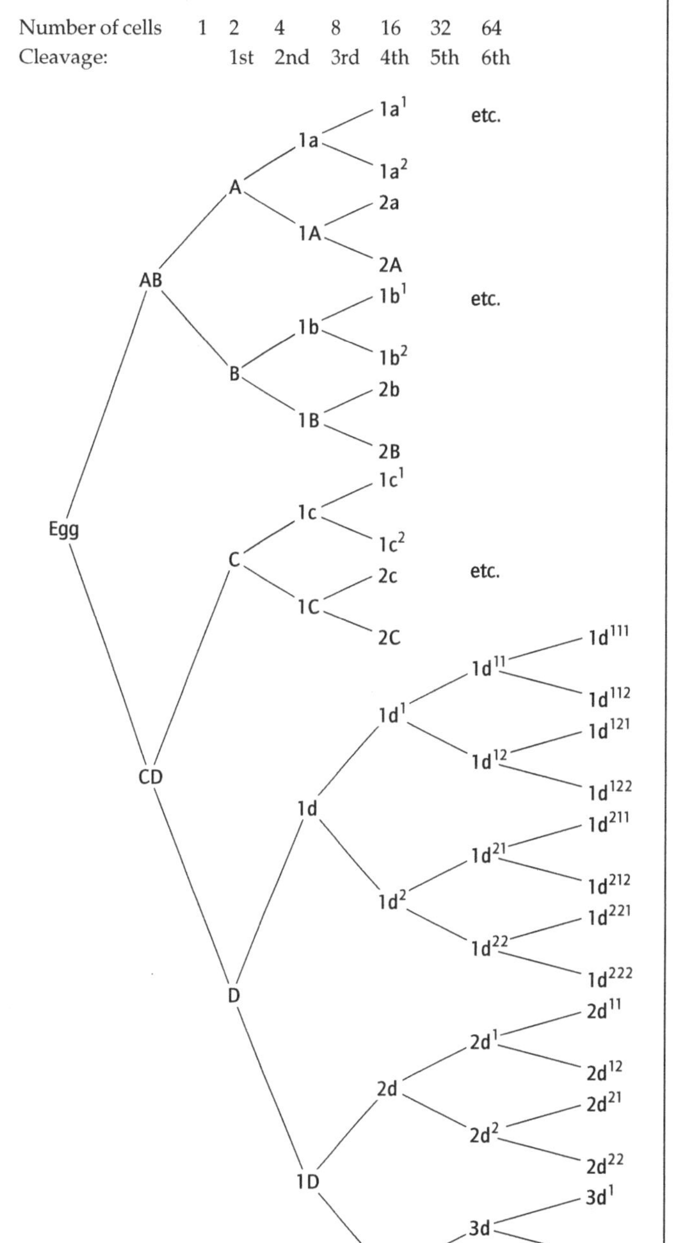

15.2.2 Gastrulation

With exceptions of the Placozoa, Mesozoa, Porifera, and the two coelenterate phyla, the invertebrates are triploblastic animals. That is to say, their bodies are derived from three embryologically distinct cell layers, the ectoderm, mesoderm and endoderm (some, for example the Pogonophora, now have no trace of endoderm in the adult). The embryo which results from early cleavage, the blastula, however, is only composed of a single layer of cells. The process of gastrulation creates the three-layered embryo in which ectoderm, mesoderm and endoderm tissue layers can be recognized. During gastrulation, not only do cells move to new positions to create a more physically and visibly complex embryo but the cells also become more restricted in their subsequent developmental fate. They may be said to have become more 'determined' in their developmental fate. Gastrulation frequently involves invagination of cells. The precise details vary between organisms and, being so diverse, the

Box 15.3 Radial cleavage

A pattern of cleavage which contrasts with that of the annelids and molluscs is seen, for example, in the echinoderms and in modified form in the chordates. In the sea-urchins (Echinoidea) a very precise pattern of cleavage is observed.

1 *The cleavage pattern*. The fertilized egg is bilaterally symmetrical (i). The first two cleavage planes are along the animal/vegetal axis, and they separate two, then four, equal cells (ii and iii).

The third cleavage is transverse and unequal, passing just above the equator (iv). It separates four *'mesomeres'* from slightly larger *'macromeres'*.

The fourth cleavage is different in the upper and lower layers of cells. In the upper half it is longitudinal to give a ring of three cells. In the lower tier it is transverse and very unequal to give four *macromeres* and four *micromeres*. In this example the micromeres are at the vegetal pole (v).

The next two divisions are transverse in the upper half, and alternately longitudinal and transverse in the lower half, to give a 64-cell stage embryo as seen in (vi).

2 *Cell identification*. Because the first four cells are structurally identical, individual cells cannot be identified. Cell layers, however, can be identified. These are conventionally identified as in the following scheme at the 64-cell stage:

Animal 1 layer = an1 = upper 16 cells.
Animal 2 layer = an2 = lower 16 cells of the animal hemisphere.
Vegetal 1 layer = veg1 = ring of eight macromere cells beneath the equator.
Vegetal 2 layer = veg2 = lower ring of eight macromere cells in the vegetal hemisphere.
Micromeres = mic = group of much smaller cells at the vegetal pole.

3 *Blastula*. The cleavage sequence produces a blastula as illustrated in (vii): a hollow ball of ciliated cells.

The initiation of gastrulation is marked by the invagination of the micromeres (viii).

Gastrulation is further described in Section 15.2.2 and in Fig. 15.9.

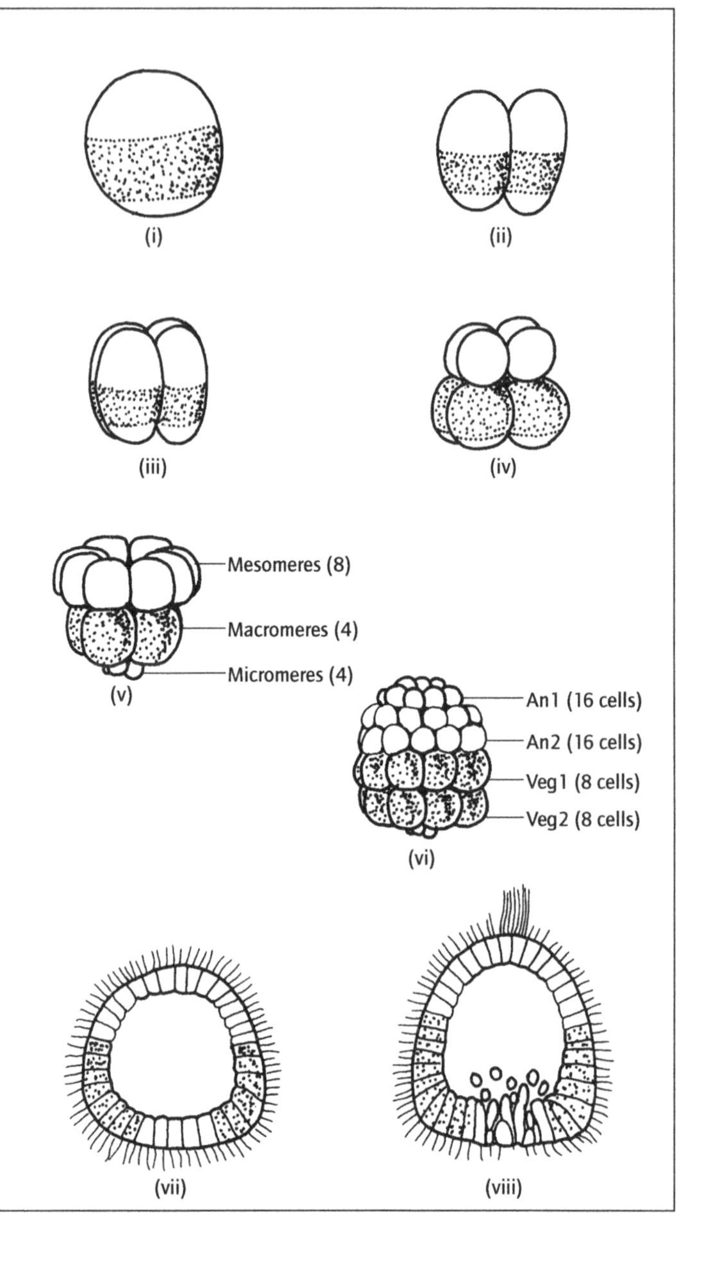

invertebrates have provided some of the best known models for the investigation of the processes involved. Three different patterns of gastrulation will be described here: in echinoderms; in protostomes, such as annelids and molluscs; and in insects. It is important to remember however that even in these groups the precise pattern may be modified especially among the echinoderms and protostomes when there is a large amount of yolk in the egg.

Gastrulation is particularly easily studied in the transparent relatively yolk-free embryos of many echinoderms such as sea-urchins (Fig. 15.12). Primary mesenchyme cells derived from the micromeres of the 64-cell embryo (see Box 15.3) invaginate towards the blastocoele but retain contact with the inner wall. The invading primary micromeres take up a specific position where they form clusters and secrete the silicate spicules which support the arms of the echinoid pluteus larva.

The next stage of gastrulation involves an inward 'buckling' of cells at the vegetal pole. This can be easily seen in living sea-urchin and starfish embryos about 48 hours after fertilization took place and is illustrated in Fig. 15.12(i–iv). The first phase of inward invagination may be caused by forces generated within the clear transparent 'hyaline' layer covering the outer surface of the blastula. This layer is a bilaminate structure, and like all such stuctures it will develop curvature if there is differential expansion of the layers. Differential expansion of the layers is thought to be caused by the secretion of chondroitin sulfate-proteoglycan (cspg for short) from the vegetal plate cells. As this substance diffuses into the inner lamella of the hyaline layer it absorbs water,

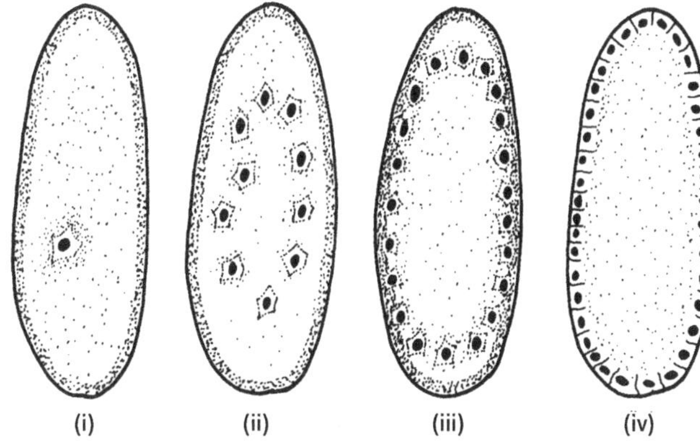

Fig. 15.11 Superficial cleavage of an arthropod (insect) egg. (i) The single nucleus lies in the central yolky part of the egg. (ii) Nuclear division results in the appearance of several nuclei but no cell boundaries. (iii) The nuclei move out to the peripheral ooplasm which is relatively yolk free. (iv) Radial cell walls appear between the nuclei. The number of nuclei is not represented exactly. Note the formation of pole cells in (iv).

causes a localized expansion, and the generation of tensions that are released by the observed inward bending of the layer, and so initiating the inward movement of the cells at the vegetal pole.

Later phases of gastrulation involve spreading and migration of the partially invaginated cells at the vegetal pole. This results eventually in the formation of the larval hind gut or 'archenteron'. In sea-urchin embryos, this inward movement may be further aided by the direct contacts of filopodia formed between the secondary mesenchyme cells (see Fig. 15.12iv) and the inner surface of the blastocoel. Such filopodia are not essential for gastrulation, however, and no such contacts are made in starfish embryos.

A specific cell-surface protein can be detected for the first time at the precise moment when primary mesenchyme cells begin to leave the blastula wall. Permanent contact between the secondary mesenchyme filopodia and the blastula wall occurs with the appearance of specific proteins in localized areas of the blastula wall.

At the end of gastrulation, the embryo has an outer layer of ectoderm, an inner tube opening posteriorly and a number of mesenchymal cells. Such an embryo is called a 'gastrula'. At this stage the definitive adult mesoderm has not yet been delaminated from the archenteron. The mesenchyme cells do not form the adult mesoderm but may contribute to skeletal and muscular elements in the larva which are lost at metamorphosis. The formation of the mesoderm is described in Section 15.4.

In deuterostomes, the blastophore becomes the anus of the functional larva. The mouth will form as an ectodermal invagination.

Gastrulation in the molluscs and annelids is rather different. It also involves the internalization of the presumptive endoderm and of the small number of cells from which the adult mesoderm will be derived. In forms with relatively yolk-free eggs, gastrulation also involves invagination as in Fig. 15.13i, but in more yolky eggs the gastrulation movements involve an amoeboid-like migration of cells over the relatively inert yolk-filled endoderm (Fig. 15.13ii,iii).

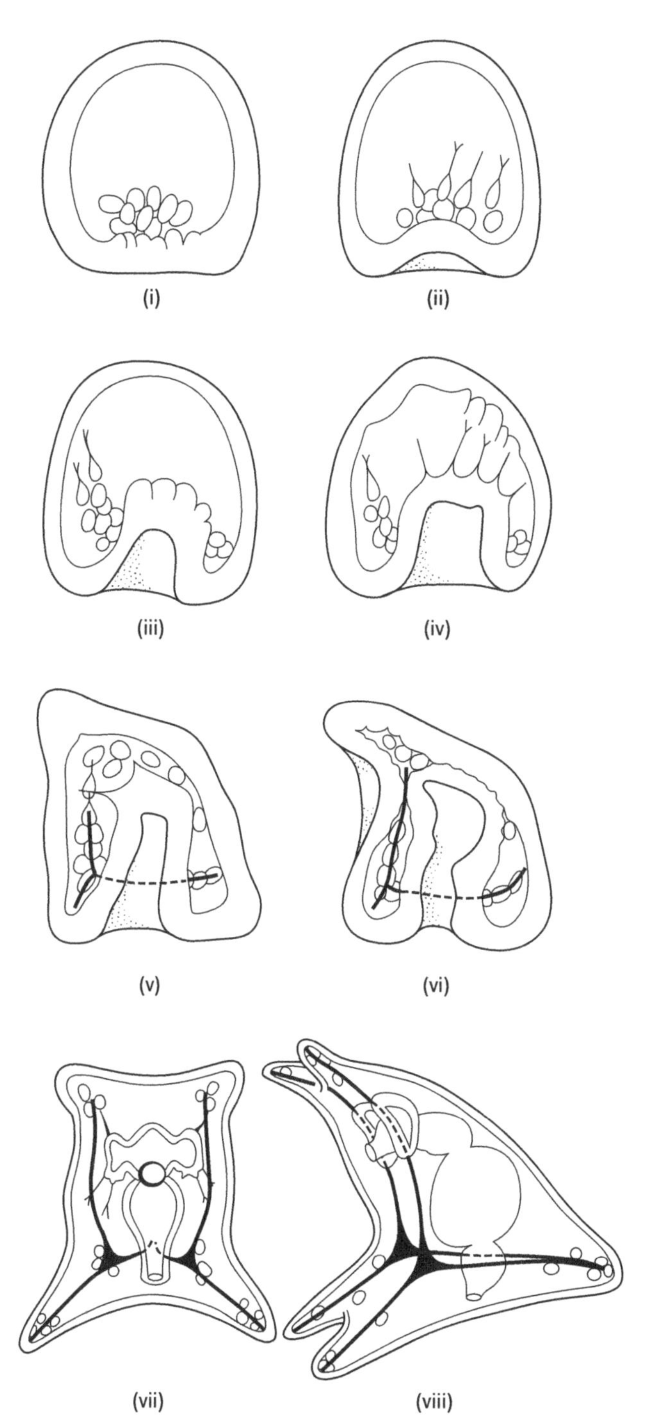

Fig. 15.12 Gastrulation in a sea-urchin. (i) Early gastrulation, invagination of primary mesenchyme. (ii), (iii) Migration of mesenchyme and the initiation of invagination. (iv) Filopodial contacts by secondary mesenchyme. (v) Late gastrula, appearance of skeletal spicules secreted by primary mesenchyme. (vi) Complete gastrula (prism stage embryo). Initiation of the oral field. (vii) Early pluteus seen from the oral surface. (viii) Early pluteus seen from the site. (After Trinkaus, 1969.)

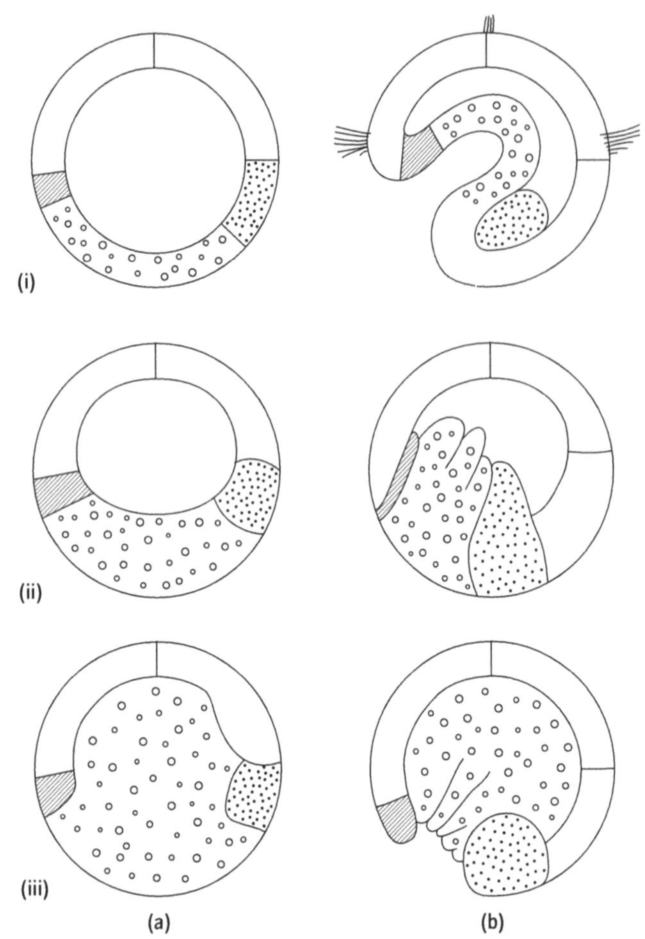

Fig. 15.13 Gastrulation in protosomes (these examples are based on studies of polychaetes with different amounts of yolk. (a) The structure of the blastula; (b) mid-gastrula. The morphologically important cells derived from 4d which form the adult mesoderm are shown with light stipple, the mouth-forming cells are cross-hatched and the gut cells are shown with open stipple. (i) Relatively yolk-free form; (ii) moderately yolky form; (iii) yolky form in which gastrulation is by epiboly.

The anterior lip of the blastopore includes the cells which form the mouth of the larva, and it is this feature which was considered to be such a fundamental one of early development that several phyla were grouped together as the protosomes.

The underlying cleavage pattern is obscured in yolky eggs and gastrulation is less easily observed. The blastocoel is frequently obliterated. The presumptive endoderm cells at the vegetal pole are internalized by a process of overgrowth by the animal pole cells. This process of spreading of the animal pole cells is referred to as 'epiboly'.

In the annelid larva illustrated in Fig. 15.14, all the segmented structures will be derived from the products of the paired mesodermal bands and the ectoteloblast ring. Structures derived from the other parts of the trocophore larva are the prostomium and the pygidium (see Section 4.14) and these can be considered to be asegmental.

A rather different mode of gastrulation is seen in the arthropods (Fig. 15.15). In insects, for example, the 'blastopore' is a furrow through which a ventral strip of prospective mesoderm cells destined to form body musculature, gut wall, gonads, etc. is invaginated. The interior of this embryo is not a hollow space but is an enclosed mass of acellular yolk. The ganglionic cells of the segmented insect nervous system also invaginate at this time.

Gastrulation in insects has assumed greater importance in studies of animal development because of the great strides being made in understanding the processes of cell determination that accompany gastrulation. It is now possible to define the primary tissue layers – ectoderm, mesoderm and endoderm – in terms of the genes that are expressed during gastrulation.

The first cells to invaginate and move towards the centre of the yolk-filled interior of the *Drosophila* embryo lie along the mid-ventral gastrulation furrow and these cells will form the central nervous system (Fig. 15.15iii,iv). Gastrulation then involves internalization of the mesoderm forming cells found in the mid-embryonic region of the ventral furrow (Fig. 15.15v–vii) and lastly invagination of the presumptive endoderm cells at the two ends of the furrow. The invaginating endoderm components of the gut are then followed by ectoderm cells that form the foregut and the hindgut of the future insect. The mesoderm cells form a flattened tube (Fig. 15.15vi,vii). The cross-sectional structure of the *Drosophila* embryo is made more complex by the folding of the embryo back on itself so that in a transverse cross-section the major axis is sectioned twice.

15.3 Experimental embryology of invertebrates: the determination of cell fate

15.3.1 Introduction

A complex multicellular animal is derived from a single undifferentiated totipotent structure, the fertilized egg. During its development it proceeds to a state where it is composed of a larger number of cells, groups of which have different morphologies, chemical composition and function. Experimental embryology is the study of how populations of differentiated cells with specific functions arise in developing embryos, and some of the most favourable model systems for the analysis of the processes in-volved have been provided by the embryos and larvae of invertebrates. More recently, two systems have gained particular importance. *Caenorhabditis elegans*, a nematode, has been chosen for major study, the object of which is to identify and characterize each gene involved in development and to establish the lineage of every cell. This is feasible because of the small, fixed number of cells that make up the adult and the very determinate pattern of development. The insect *Drosophila* has for nearly a century been the model for genetic studies of eukaryotes. Much more is known of the genetics of this animal than any other. Spectacular advances are being made in the unravelling of the genetic control of developmental processes,

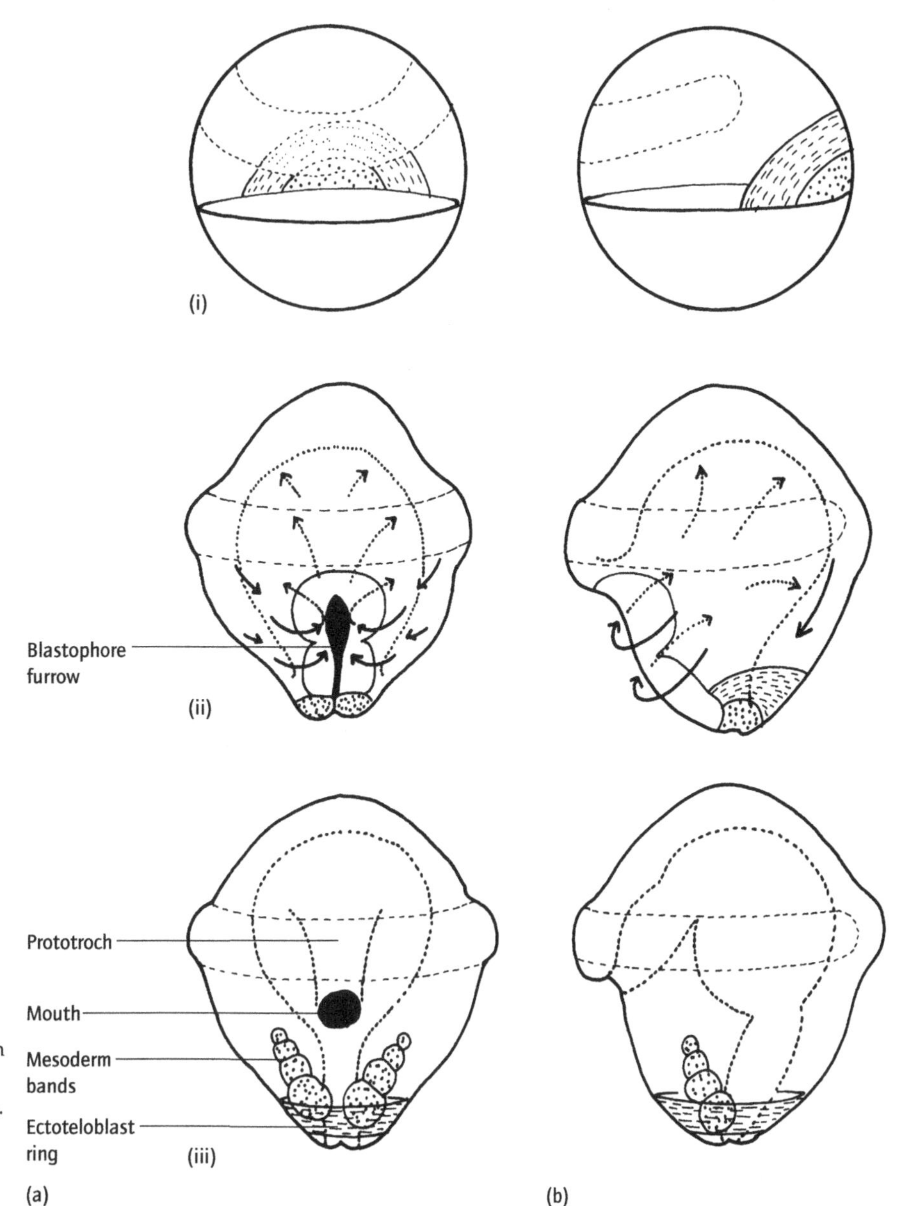

Fig. 15.14 Gastrulation in the protosomes, e.g. Polychaeta. (a) Ventral view. (b) Left-hand view. (i) Diagrammatic representation of a simple fate map. (ii) Gastrulation movements: heavy lines show the path of cells to the blastophore, broken lines the migration of invaginated cells. Note that the stomadaeum or mouth is created by invagination of cells through the anterior blastophore. The mesoderm bands are derived from paired mesentoblast cells migrating through the posterior margin of the blastophores. (iii) The definitive trochophore larva with paired mesoderm bands and ectoteloblast blastema which remains at the anterior face of the pygidium through adult life. The prostomium, peristomium and pygidium are therefore asegmental structures. (Partly after Anderson, 1964.)

especially in the control of regional organization, as will be explained briefly below.

During early embryonic development, maternal genes which have been stored in the oocyte cytoplasm are used to support the activities of the developing embryo.

Eventually, however, the zygote's own novel genome is called into play, and since many different types of cell eventually develop, there must be control over the information contained in the zygote nucleus. The potential control points are many and include each or several of the steps in the following sequence:

1 Genetic information in the zygote nucleus.
2 Allocation of information to daughter cells.
3 Transcription of genetic information.
4 Population of mRNA molecules.
5 Export of mRNA from nuclei to cytoplasm.
6 Population of informational molecules in the daughter cell cytoplasm.
7 Translation of information in the mRNA molecules.
8 Formation of cell-specific proteins.
9 Function of specific substances.

In the majority of organisms the nuclei of differentiated cells contain the same genetic information as the nucleus of the fertilized egg. The evidence is based on the results of:

- Experimental transplantation of nuclei from differentiated cells into enucleated egg cytoplasm.
- Studies of the cytology of chromosomes during early cleavage.
- Comparison of the banding patterns of the giant chromosomes during early cleavage.

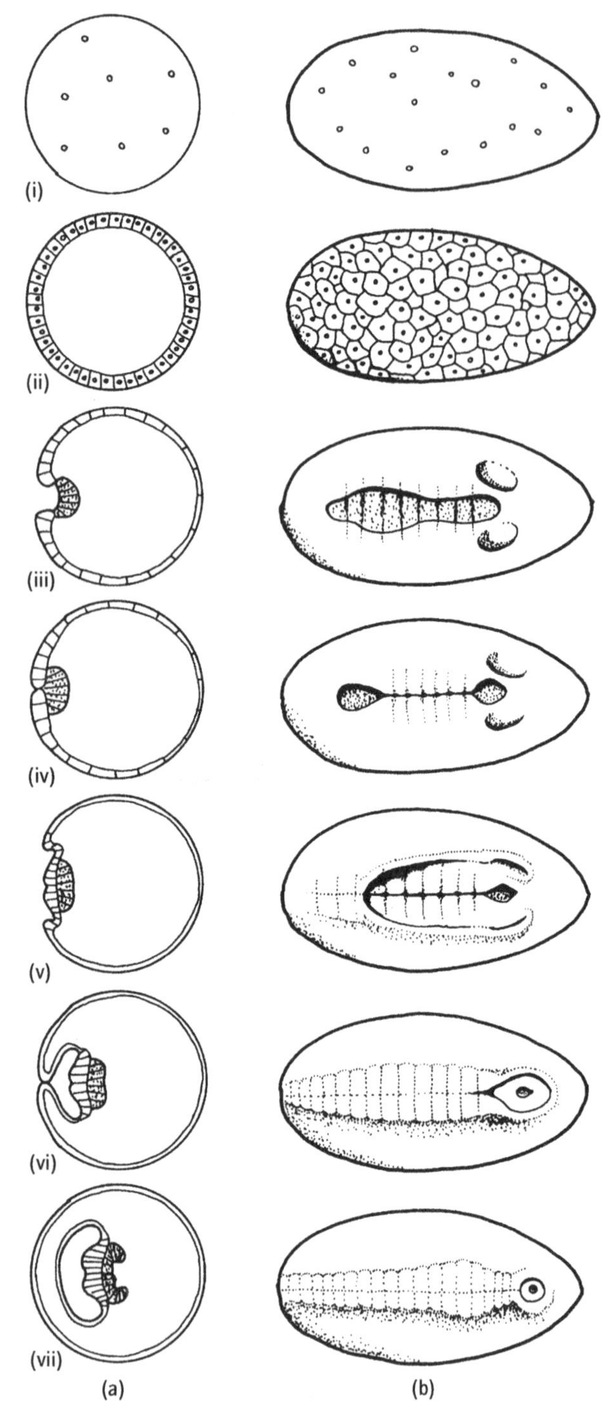

Fig. 15.15 Gastrulation in insects. (a) In cross-section. (b) Ventral view. (i) Early nucleus proliferation; (ii) superficial cleavage; (iii), (iv) invagination of mesoderm; (v), (vi) segmentation of the neuroectoderm, formation of extra-embryonic membranes; (vii) the completed gastrula. (After Slack, 1983.)

- Studies of differentiation and de-differentiation during regeneration (see also Section 15.6).
- Analysis of DNA 'fingerprints' from different tissues.

There are, however, exceptions to this general rule. The primordial germ cells of the nematode *Ascaris* are, for instance, the only cells of the embryo to receive a full complement of chromosomes. In cells of other lineages chromosomes are lost during early cleavage, as explained in Fig. 15.16. The somatic cells all have the same genetic complement so the loss of chromatin is not part of a differentiation mechanism. Similar observations have been made for a very small number of other animals. We can conclude that chromatin elimination is not a mechanism for the control of differentiation in animals.

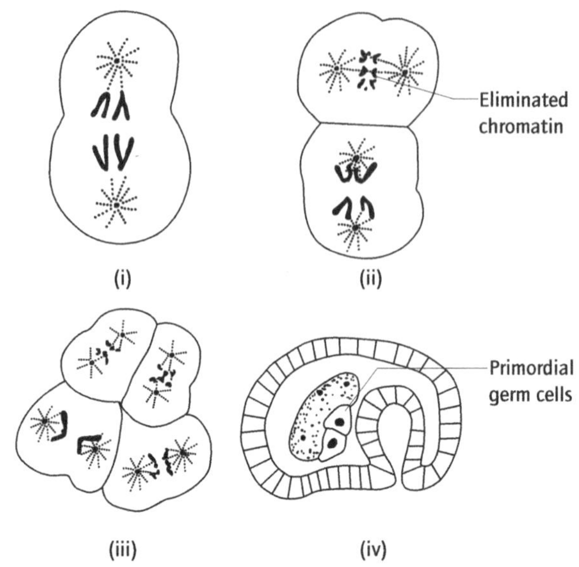

Fig. 15.16 An example of chromosome diminution during early embryonic development (*Ascaris*). (i) First cleavage, no chromosome loss. (ii) Elimination of chromosome material during cleavage of the anterior cell. (iii) Elimination of chromosome material and rearrangement of cells, at the four cell stage. Only one cell retains the complete set of chromosomes and is therefore totipotent. (iv) The germ cell cytoplasm and cells with the full chromosomal complement can be traced to the two primordial germ cells of the gastrula.

The specification of posterior function is associated with the expression of the *nanos* gene (see Box 15.1) implying genetic similarity with the regional specification of other organisms. This apparent conservation of mechanisms of regional organization is discussed further below.

If all the nuclei of the developing embryo have the same genetic constitution, then differentiation must be an expression of an interaction between cytoplasmic information in the cell and that derived from the genetic information in the nucleus.

15.3.2 Mosaic versus regulative development

By the early part of the twentieth century experimental embryologists had investigated the ability of isolated cells of invertebrate embryos to develop normally. In some cases, as in the sea-urchins, each one of the blastomeres at the four-cell stage would, in isolation, give rise to a complete and perfectly normal pluteus larva (see Box 15.4(1)). Such an embryo was said to be capable of

Box 15.4 Regulative and mosaic development I

In the late nineteenth and early twentieth centuries, it became possible to study the developmental capacities of individual cells isolated from early cleavage stages of the embryos of marine invertebrates at the two or four cell stage. The results could be strikingly different.

1 Echinoderm embryos. Isolated cells of the four cell stage can each give rise to a normal (but small) pluteus. The small embryos were said to be regulative.

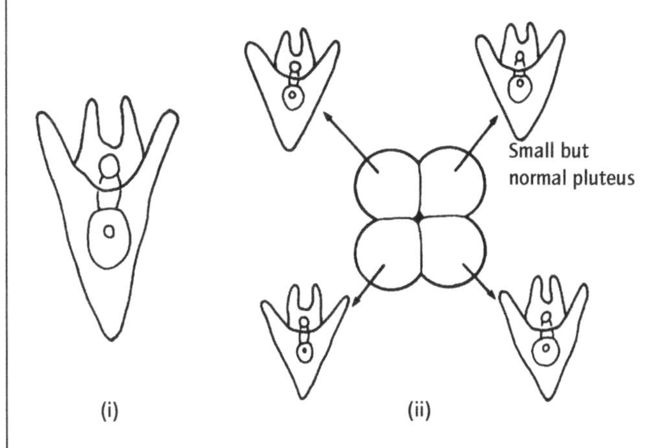

(i) The normal pluteus.
(ii) The results of cell isolation.

An isolated blastomere of the four cell stage divides to give two mesomeres, one macromere and one micromere, i.e. it follows the cleavage pattern of a typical embryo (see Box 15.3), and would go on to produce a normal larva.

2 Mollusc embryos. Isolated cells at the two and four cell stage give rise to unbalanced or deficient embryos. Embryos developing from isolated D cells usually give rise to more normal embryos than those derived from A, B or C cells.

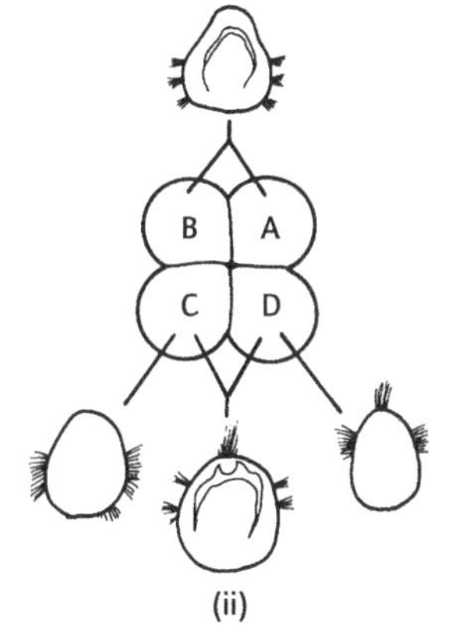

(i) A normal trochophore of *Patella*.
(ii) Abnormal trochophores from isolated cells as indicated.

3 Ascidian embryos. The eggs of ascidians often have visibly different regions of cytoplasm (see Fig. 15.6); these are segregated to different cells.

Embryos developing from isolated blastomeres are very different and their development reflects the structures they would have given rise to in an intact embryo.

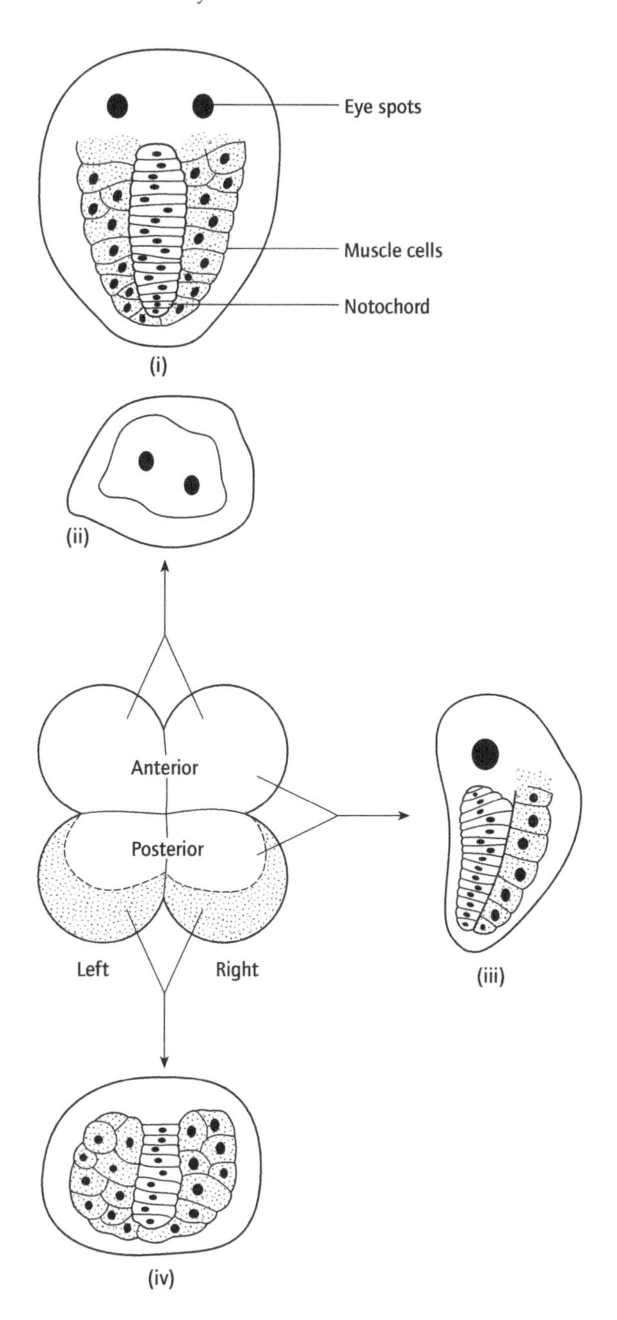

(i) The normal 'tadpole' embryo with paired eyes, notochord and paired muscle blocks;
(ii) Isolated anterior cells;
(iii) Isolated right-hand cells;
(iv) Isolated posterior cells.

Box 15.5 Cell fate, fate maps and cytoplasmic localization: ascidian embryos

1 *Natural markers.* After fertilization some embryos have clearly different regions of cytoplasm and bilateral symmetry. The figure shows how the different cytoplasmic regions of an ascidian egg (i) are segregated into the cells of the vegetal pole, the anterior and posterior cells at the vegetal pole are further distinguished by the localization of different-coloured yolk – the yellow cytoplasm identifies the vegetal pole cells that will form the muscle cells of the tadpole larva (ii); during gastrulation these cells are invaginated, (iii) and (iv). After gastrulation, cells containing the visibly different cytoplasm will have given rise to different embryonic structures, (v) and (vi). The question arises in this case, 'do the natural cytoplasmic markers determine the fate of the cells?'

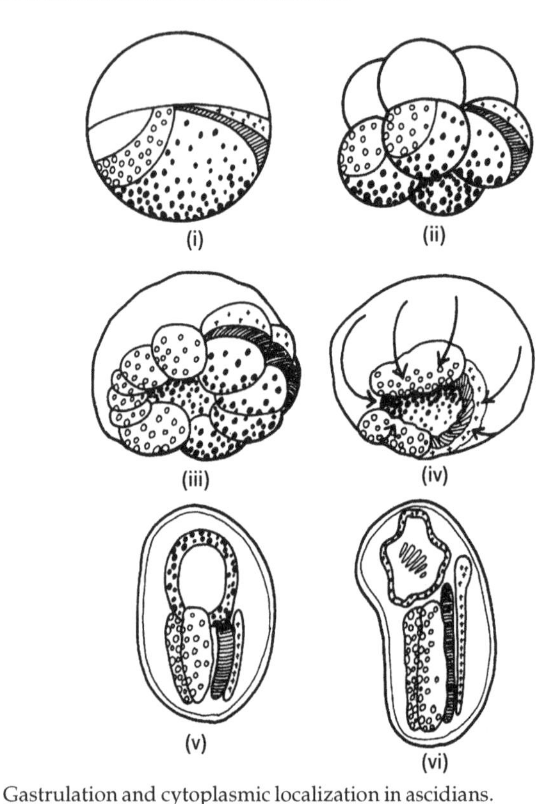

Gastrulation and cytoplasmic localization in ascidians.

2 *In the ascidian, cytoplasmic factors do determine cell fate.* A region of yellowish cytoplasm acts as a natural marker for the muscle cells of the tadpole larva (Box 15.3(3)).

3 The enzyme *acetylcholinesterase* is specific to the developing muscle cells and can be stained in the cytoplasm.

The following experimental results have been observed:

(a) The enzyme normally appears after 8 hours of development;

(b) The enzyme appears at the appropriate time even if cell division has been inhibited by the application of drugs;

(c) The appearance of the enzyme is inhibited by the application of actinomycin D up to 5 hours or the application of puromycin up to 7 hours.

(d) Compression of the egg during cleavage can lead to yellow cytoplasmic allocation to cells which do not normally receive it. The cells then show acetylcholinesterase activity. These experiments suggest that the genes which code for the enzyme acetylcholinesterase are transcribed only in cells which inherit some factors normally associated with the yellow cytoplasm.

The timing of the gene transcription and translation is under the control of a cellular clock which operates independently of cell division.

4 When larval nuclei are transplanted into enucleated fragments the subsequent cells show the structures typical of the cytoplasm, not of the implanted nuclei, confirming the importance of cytoplasmic determinants in determining cell fate. There is, of course, a genetic basis to the localization of cytoplasmic determinants and this is now being studied in insect larvae, as explained in Section 15.4.2.

5 The maternal factor stored in the 'yellow cytoplasm' of the egg (see Fig. 15.8), which has an important function in muscle cell determination, has now been identified. This cytoplasm contains the mRNA gene of the gene *macho-1* which has been sequenced. The *macho-1* mRNA encodes a zinc-finger protein. Depletion of the mRNA results in loss of primary muscle cells in the larva. At the late gastrula stage this mRNA is localized in just two cells. This work confirms the existence of a muscle cell determining factor first postulated 100 years ago.

regulation. In other cases the embryos developing from isolated blastomeres at the four-cell stage were very deficient (see Box 15.4(2) and (3)). It appeared that some cells inherited a particular type of cytoplasm and that this cytoplasm determined their developmental fate. Further analysis of these results requires knowledge of the normal developmental fate of the individual cells, as expressed in a fate map of the uncleaved cytoplasm of the egg or of the blastula. This can be painstakingly built up by marking specific regions of the cytoplasm and tracing the fate of the marked cells. A natural fate map of the ascidian *Styela partita*, however, was established by careful observation of the egg after fertilization (see Fig. 15.8). After fertilization, the egg becomes visibly bilaterally symmetrical due to the movements of the cytoplasm. Prominent among its regions is a crescent of yellowish cytoplasm. This can be traced through the early cleavage stages to the muscle blocks of the tadpole larva. All the presumptive muscle cells inherit some of the yellow cytoplasm (see Box 15.5).

The experiments with ascidians provide compelling evidence for the importance of localized cytoplasmic substances in the determination of cell fate and these early insights are now being thoroughly substantiated.

Experiments with insect cells also show that the developmental fate of the nuclei can be determined by the cytoplasm into which they move. The cleavage pattern is described in Fig. 15.11. There is often a distinct cytoplasm at the posterior pole of the egg

– the pole plasm. Only nuclei which become incorporated into cells with pole plasm have the capacity to form germ cells (i.e. remain totipotent) and embryos from which the pole plasm has been removed are sterile. Pole plasm of a genetically identified stock of *Drosophila*, when injected into the anterior part of an egg of a different strain of fly, causes the cells which form in this anterior region also to become germ cells.

Development however, is not to be understood as being solely the result of pattern of distribution of cytoplasmic determinants. It also involves interactions between cells. Detailed experiments with the mud snail *Nassarius obsoletus* (often referred to by the earlier name of *Ilyanassa*) reveal how cytoplasmic determinants interact during development. They are explained briefly below and in more detail in Box 15.6. *Nassarius obsoletus* exhibits spiral cleavage (see Box 15.2) in which a prominent polar lobe appears at the first and second cell divisions. This special pole plasm is allocated specifically to the D cell. Embryos from which the polar lobe is removed are symmetrical, the special timing of cleavage in the D-cell lineage is disturbed and the embryos develop into very deficient 'lobeless' larvae (see Box 15.6). 'Lobeless' larvae lack eyes, statocysts, a foot, velum, shell, heart and organized intestine.

It is possible, however, to recognize some products of cell differentiation. The cell lineage chart in Box 15.6 reveals the significance of these observations. Many of the lobe-dependent structures are not derived from the cells which inherit the polar lobe material. The developmental fate of the cells derived from 1a and 1c is to form the eyes, and 2d forms the shell. These developmental fates are determined not only by their cytoplasmic inheritance but also by their interactions with other cells of the D-cell lineage during the critical period from the appearance of cell 2d to the formation of 4d. Similarly the only cells which are *competent* to give rise to the shell gland are derived from the single cell 2d, but they will also do so if cell 2d has been influenced by contacts with other cells in the D lineage.

In sea-urchin embryos, the developmental fate of the different layers of cells is to an even greater extent determined by their relative position in the embryo. Their developmental fate is not

Box 15.6 Regulative and mosaic development II: the development of the marine snail *Nassarius obsoletus*

1 The organization of the veliger larva of *Nassarius*. Different regions of the larvae and the associated structures are derived from specific cell lines. The general relationship between the larval structures and the cells derived from the four quartettes of micromeres (1a, 1b, 1c and 1d) is illustrated in Figure (i).

2 The cell lineage of the veliger larva. Cell lineage studies show that specific larval structures develop from the products of individual cells which can be identified and named during early cleavage.

The cell lineage diagram shows tentative assignments for larval structures at the 29-cell stage using the nomenclature presented in Box 15.1.

Note that cells 2d and 4d have particularly important roles in the formation of the veliger. Cell 4d is called the mesentoblast cell.

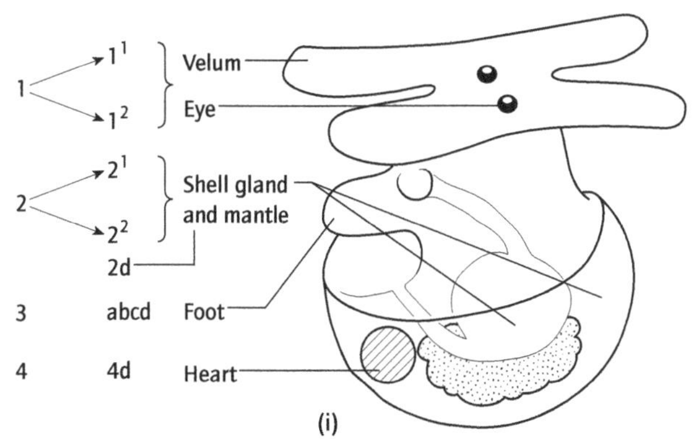

The organization of the veliger larva.

3 The organizing role of the pole plasm. In this snail a prominent polar lobe of specialized cytoplasm appears shortly before the first and second cleavage, see (ii) and (iii). The cytoplasm of the pole plasm is inherited by the D-cell lineage. The polar lobe can be removed quite simply during early cleavage by microsurgery. The results are dramatic.

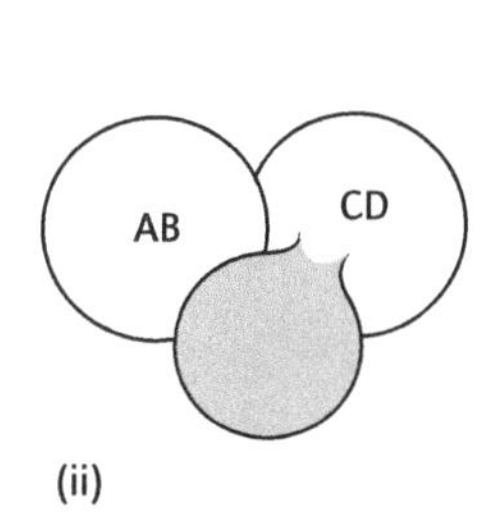

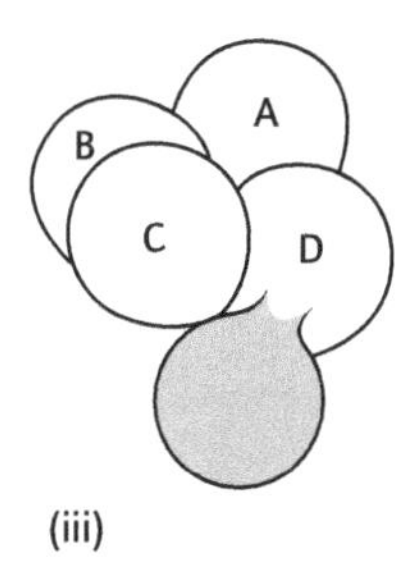

Lobeless larvae produced in this way lack all structural organization but show some signs of cell differentiation. An example is illustrated below (iv). Compare this with the fully developed larva above (i).

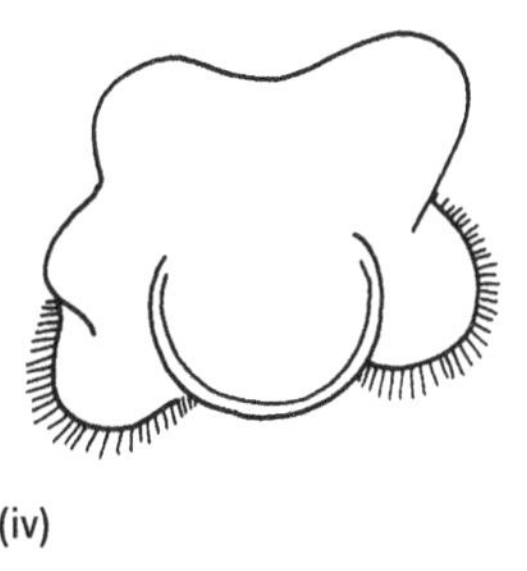

In a classic series of experiments, A.C. Clement tested the effects of removal of the D cell progressively during development. The results are summarized in Table B.1 (overleaf).

Continued p. 386

Box 15.6 *(cont'd)*

Table B.1 Summary of the effects of progressive ablation of the D-cell lineage in *Nassarius* (from data of A.C. Clement, 1962).

Cell destroyed	Embryo	Defects found
1 D cell (see Box 15.1)	ABC	As lobeless embryos Lacks: intestine, heart, shell, foot, statocysts, eyes
2 1D cell	ABC + 1d	As for ABC
3 2D cell	ABC + 1d + 2d	As for ABC
4 3D cell	ABC + 1d + 2d + 3d	Shell variable, lacks intestine, lacks heart
5 4D cell	ABC + 1d + 2d + 3d + 4d	None

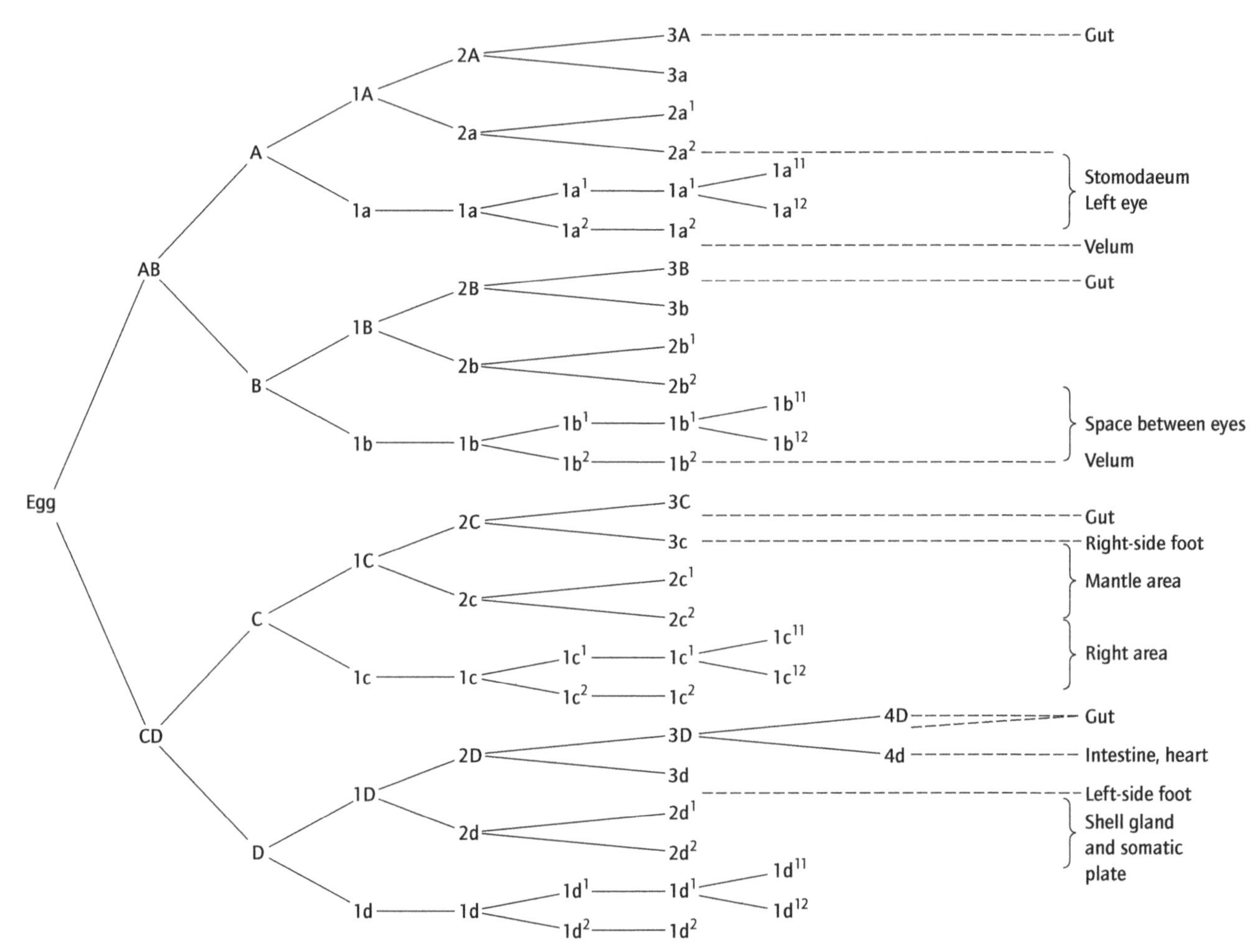

Diagram to show cell lineage.

4 Interpretation: evidence of inductive interrelationships. Refer to the cell lineage diagram and the table above; note that: eyes form from derivatives of cells 1a and 1c, and shell gland from derivatives of cell 2d. Cells 1a and 1c are present in all cells of the series of experiments shown in the table but eyes form only in operations (4) and (5). Why? Answer: eyes only form in the presence of D-lineage cells at least to the formation of 3d.

The shell gland is only normal if 2d is in the presence of D-lineage cells at least until the formation of 3d.

Evidence of mosaic development: eyes normally form from 1a and 1c, and destruction of 1a or 1c results in the failure to develop eyes.

Conclusion: only derivatives of 1a and 1c are competent to form eyes, but these cells require the inductive influence of the D-cell lineage in order to become determined in that fate.

Heart normally forms from cell 4d, and only cell 4d is ever competent to produce the differentiated heart (see text for further discussion).

fixed by their cytoplasmic inheritance. The principal components of the sea-urchin larva can be traced to the cell layers of the late cleavage stage embryo according to the following scheme:

an1 cells	apical ciliary tuft
an2 cells	oral field stomodaeum
veg1 cells	ectoderm
veg2 cells	archentron and coelomic pouches
micromeres	primary and secondary mesenchyme

This relationship is explained further in Box 15.7. The transverse divisions clearly separate material of different developmental potential, but the embryo does not develop as a simple mosaic of parts. Each cell layer develops according to its position relative to two gradients in the embryo, one declining from the animal towards the vegetal pole and another declining from the vegetal to the animal pole. Some of the crucial experiments which lead to this conceptual framework are summarized in Box 15.7.

Box 15.7 Regulative and mosaic development III: Experimental analysis of sea-urchin development

1 *Cell layers.* Cleavage of the sea-urchin egg is described in Box 15.2. The principal layers of the 64-cell embryos are:

An1 – 16 cells
An2 – 16 cells
Veg1 – eight cells
Veg2 – eight cells
Micromeres.

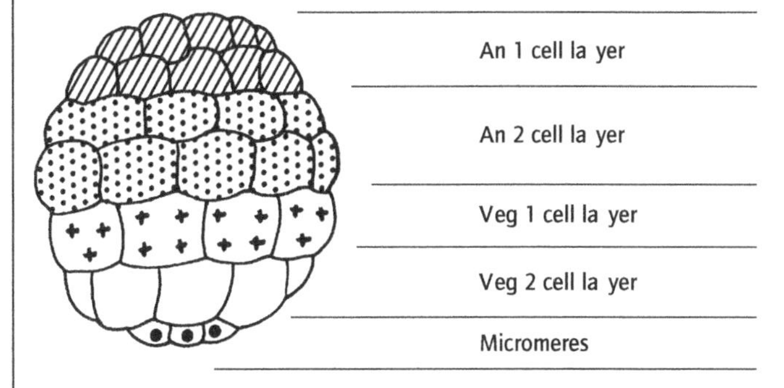

Relationship of the cell layers to the gastrula and early pluteus.

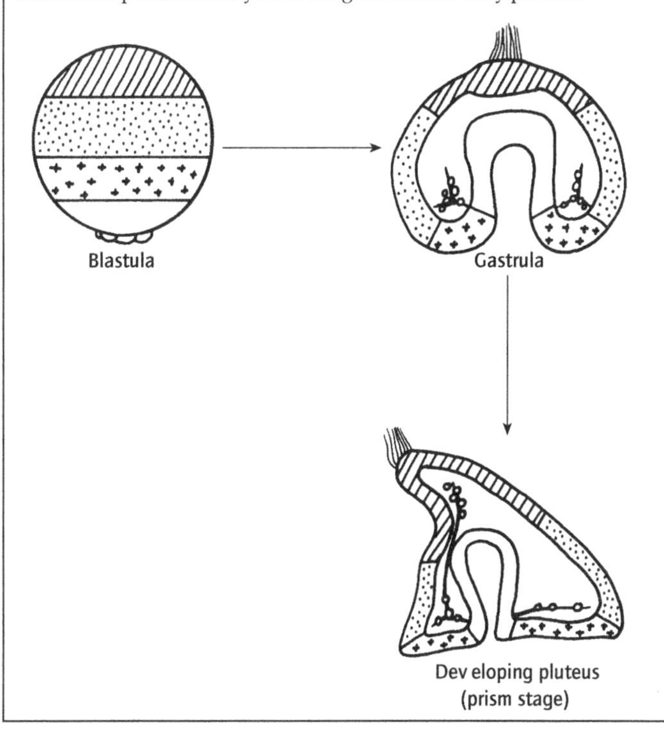

2 *The development of isolated half-embryos.* Isolated animal half-embryos give rise to permanent blastulas with an over-developed apical tuft. Isolated vegetal half-embryos give rise to more or less normal embryos with over-development of the gut. They are said to be vegetalized.

These observations could be compatible with self-differentiation of the primary cell layers, a 'mosaic' theory of development, but further experiments show this to be incorrect.

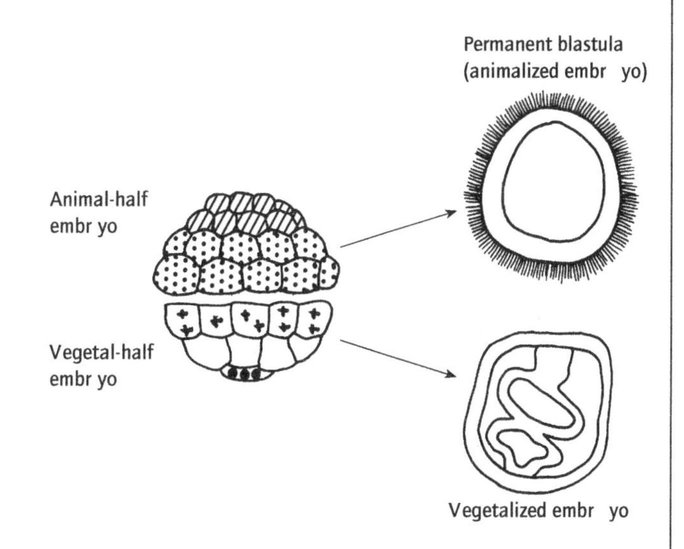

3 *Evidence for the gradient theory of development.* An isolated animal half-embryo will develop into a normal pluteus if combined with four micromeres.

(i) Subdivision of the embryo.
(ii) Combination of animal hemisphere with four micromeres.
(iii) Gastrulation.
(iv) A normal pluteus.

Continued p. 388

Box 15.7 *(cont'd)*

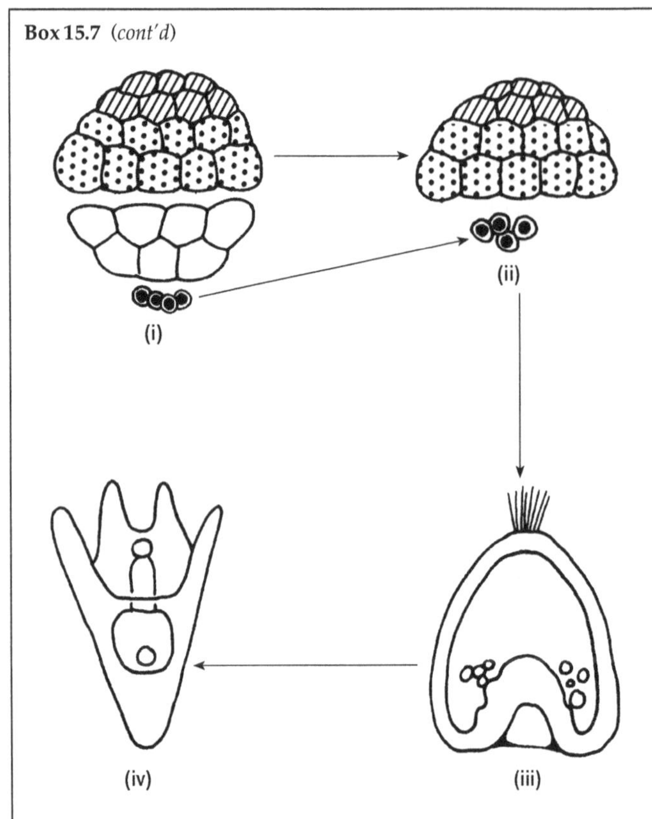

In the experiment described in **3** the four micromeres could be said to 'induce' archenteron-forming capability in the cells of the an-2 layer.

Cells of the an-2 layer are competent to differentiate as archenteron, although they would not normally do so.

4 A larger number of experiments have been performed on the development of isolated layers of the embryo in isolation and in combination.

The results of one such series are summarized below. Normal embryos can be produced by 'balancing' animal and vegetal tendencies:

an1 + 4 micromeres give a normal pluteus
an2 + 2 micromeres give a normal pluteus
veg1 + 1 micromere give a vegetalized embryo

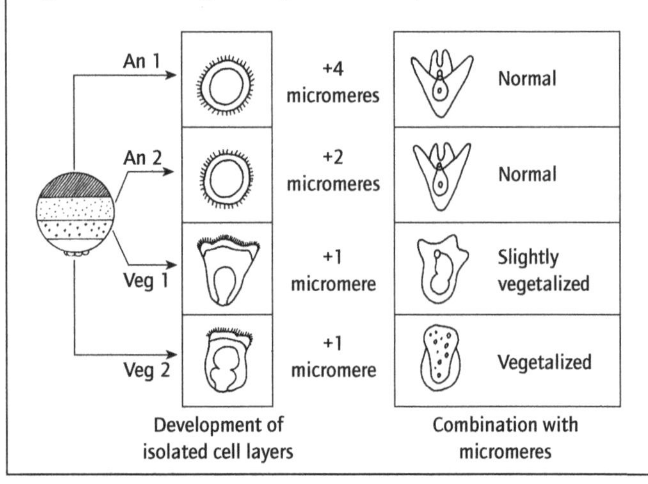

5 As shown, these results can be interpreted as the expression of a two-gradient system. It is supposed that in the normal embryo the relative position of a cell layer can be identified in relation to its position in the two gradients.

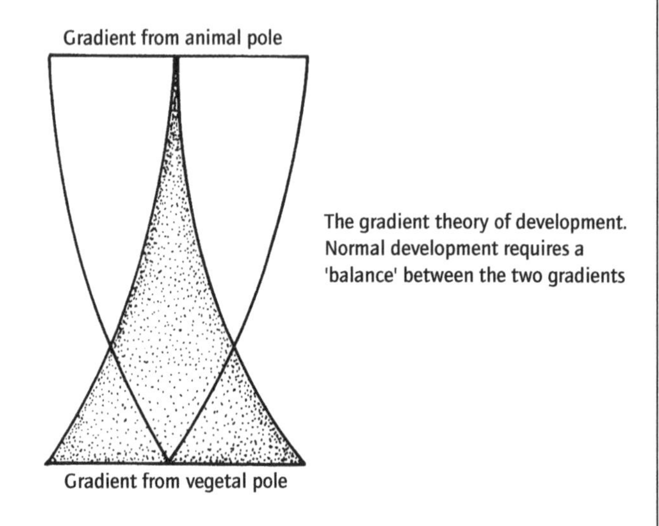

The gradient theory of development. Normal development requires a 'balance' between the two gradients

6 *Visualization of the gradient system.* Animalizing and vegetalizing agents. Some substances cause the development of abnormal animalized or vegetalized embryos like those developing from isolated animal or vegetal hemispheres.

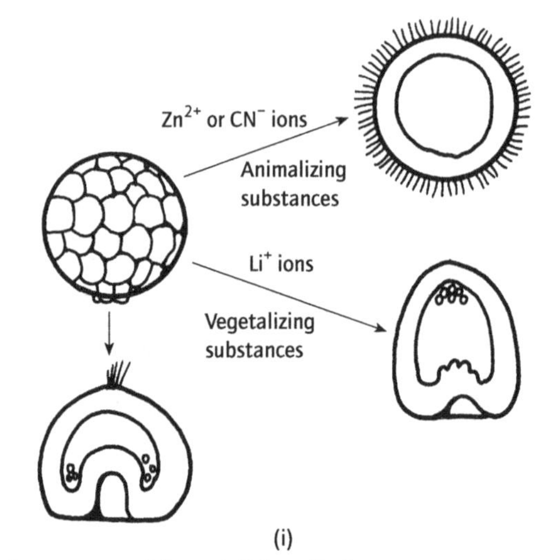

Li^+ ions have a powerful vegetalizing effect.
Zn^{2+} and CN^- ions cause an animalizing effect.

Continued

Box 15.7 *(cont'd)*

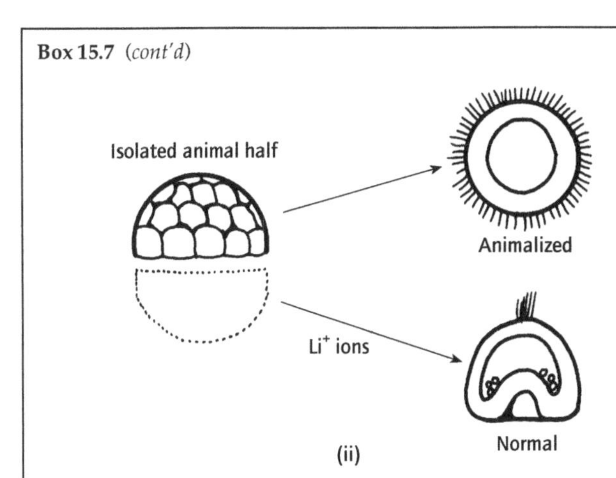

Li^+ ions can cause an isolated animal hemisphere to gastrulate normally, however.

7 *Visualization of gradients of metabolic activity.* Direct evidence for gradients of metabolic activity at the poles of the early gastrula is obtained by staining with the vital dye Janus green. This dye becomes pink, then colourless.

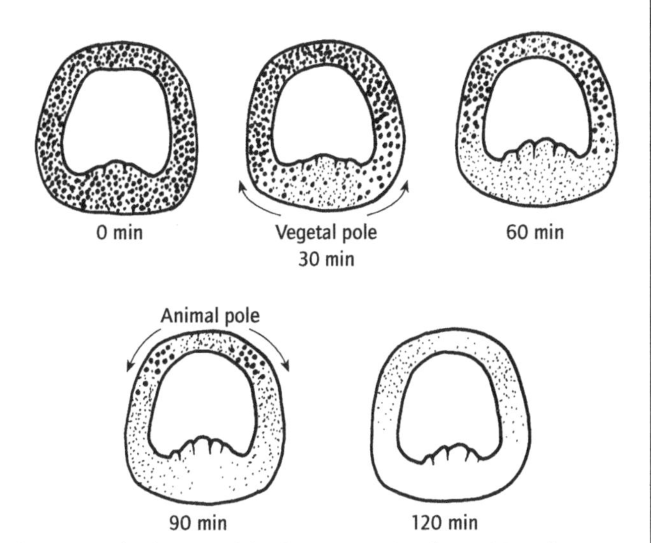

The course of reduction of the dye in normal embryos. Note the centres of reduction radiating from animal and vegetal poles.

Li^+ ions cause animal hemispheres to develop normally.
Li^+ ions suppress the animal pole centre of Janus green acidification.

8 The gradient system reflects the molecular basis of regional organization. Although the echinoderm system is not fully understood it shares similarities with other systems and rapid developments are now being made in the analysis of genetic components of regional differentiation in insect embryos. See Section 15.4 and Box 15.9.

The molecular mechanisms that lie behind the rather abstract concepts that emerged from the classical phase of experimental developmental biology are being spectacularly revealed by studies of *Drosophila*.

15.4 The developmental genetics of *Drosophila melanogaster*

The fruit fly *Drosophila melanogaster* has come to have a very special role in contemporary studies of animal development. The reasons for this are many; in large part they arise from the unequalled wealth of genetic information that exists for this organism. It also arises from the structural organization of the genetic material in giant polytene chromosomes in larval and some adult cells, and from the unique properties of the imaginal disc system.

In the holometabolous insects, groups of cells are set apart during early embryogenesis and they differentiate only during metamorphosis. At this time the groups of imaginal cells, which are organized either into rather diffuse nests or more commonly into well-organized 'imaginal discs', give rise to the entire ectoderm, to the salivary glands, and to the other internal organs. A plan of the imaginal disc system of a dipteran larva and its relationship to the adult structure is shown in Fig. 15.17.

15.4.1 The imaginal disc system: determination of cell fate and transdetermination

Early cleavage in the dipteran insect *Drosophila melanogaster* gives rise to two distinct populations of cells. Some 10 000 cells give rise to the structures of the larva; they have polytene chromosomes in which there are up to 1000 parallel strands of homologous DNA. These cells are large and functional in the larva and pupa stages (see also Section 15.5). The cells of the future adult, however, do not, for the most part, have polytene chromosomes and they are not functional in the larva. Instead about 1000 cells are set aside as groups of undifferentiated cells in the imaginal discs. The differentiation of these cells will be initiated by changes in hormonal milieu at pupation. At this time the hormone 20-hydroxyecdysone causes the larval cells to degenerate and the cells of the imaginal discs to differentiate into the adult structures and tissues (see Section 16.12.4).

Although undifferentiated, the imaginal disc is a highly determined structure rather more like a mosaic embryo than a

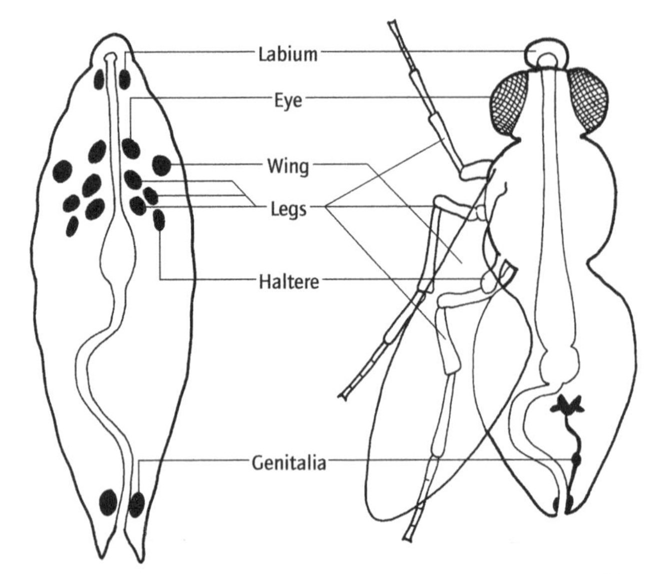

Fig. 15.17 The imaginal disc system of a dipteran fly (e.g. *Drosophila*). The position of the discs in the larva and the structures to which they give rise in the adult are indicated. Note that the alimentary system is not replaced at metamorphosis.

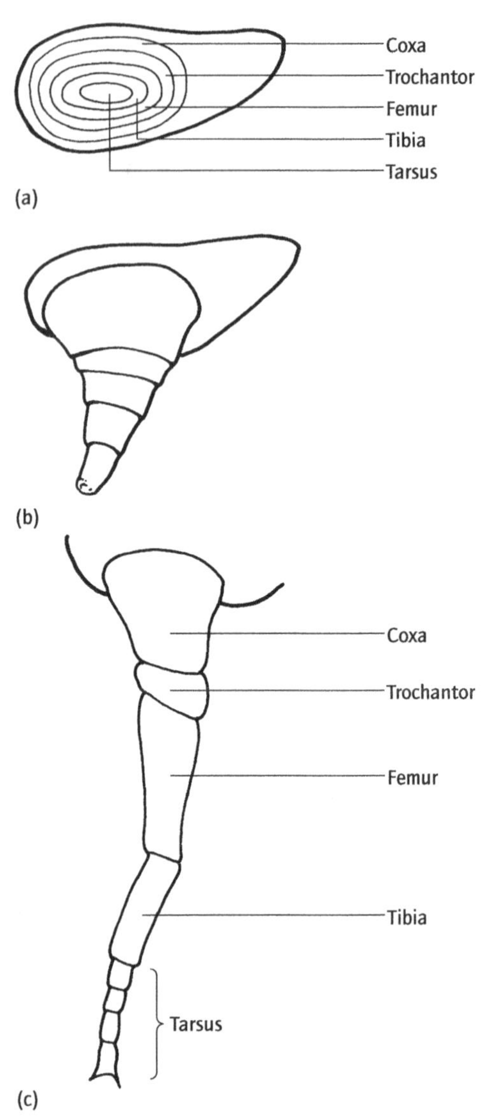

Fig. 15.18 The development of an insect limb from its imaginal disc. (a) Fate map of the imaginal disc as determined by selective ablation; (b) progessive development of the limb by outgrowth of the imaginal disc; (c) the completed limb with its component parts.

regulative one in its properties. There is, however, evidence for a gradient system underlying the determination of the disc. A fate map of a disc can be constructed by selective ablation of parts of it followed by implantation into a late larva. At metamorphosis an incomplete imaginal structure will develop according to which parts of the disc were removed. The fate map of the leg disc, for instance, has a concentric pattern (Fig. 15.18a). The outer zones give rise to proximal leg structures while the inner zones give rise to distal ones (Fig. 15.18b,c). The leg disc can be dissected and implanted into a young larva. The limb bud then regenerates by compensatory growth. Some fragments reconstitute an entire limb disc but others show deficiencies depending on the position in the disc from which the fragment was derived. This indicates a structural hierarchy within the disc.

An undifferentiated cell is often said to be 'determined' in its developmental fate when experiments show that the pattern of its future differentiation is fixed and it is not influenced by the cellular environment in which it is placed. The cells of the imaginal disc of insects have this property. Rapid progress is now being made towards an understanding of the molecular basis for the determination of cell fate and cell identity in imaginal discs.

A disc can be isolated from an embryo and transplanted to a different region of a host embryo (Fig. 15.19). It will continue (usually) to develop according to its original regional identity. In other words, if it is a prospective leg disc, the cells will continue to develop into a leg. Two elements of the determination can be recognized:

- The overall regional identity of the disc.
- The identity of individual cells within the disc, which permits the differentiation of an organized integrated structure.

The cells of the imaginal disc are capable of proliferation (mitotic division) and disc growth is a normal part of development. The disc is also capable of compensatory growth and a disc fragment can, as explained above, reconstitute an entire disc.

The hormone milieu of an adult insect allows proliferation of imaginal disc cells but does not permit or cause differentiation of the disc. This will only take place if the disc is placed in the hormonal milieu of a late-stage larva and pupa. Using these facts, experimental biologists have conducted many intriguing experiments, the results of which were often quite unexpected.

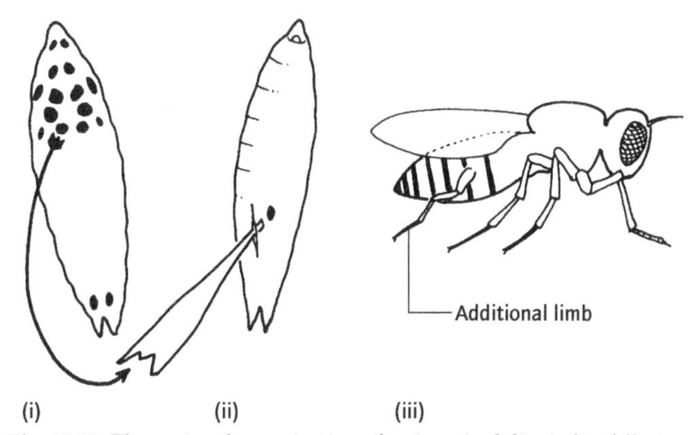

Fig. 15.19 The regional organization of an imaginal disc is fixed (but see Box 15.8). The leg disc of a dipteran larva is (i) isolated by dissection and (ii) implanted into the abdominal region of a host embryo. Eventually the larva will complete metamorphosis; (iii) the resulting fly has an additional limb in the abdomen.

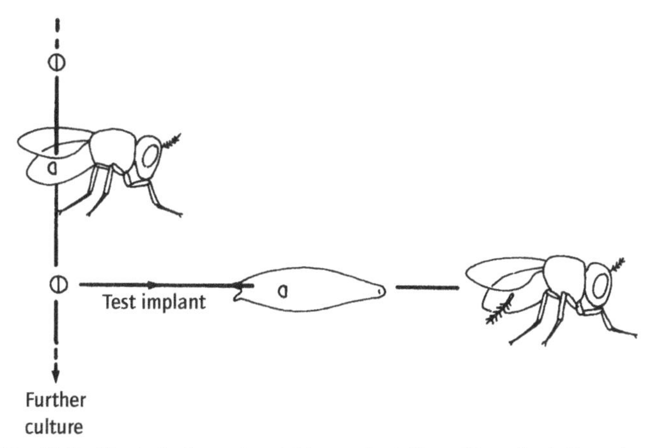

Fig. 15.20 The technique of serial transplantation of imaginal discs. A disc is removed from a donor larva, dissected and placed into an adult fly where it will proliferate and regenerate. The regenerated disc can be recovered from the host fly, dissected again and one component tested for developmental capacity in an old larva, the other fragment being transferred to a host adult for further culture. See also Box 15.8.

A disc can be maintained in its undifferentiated state long after it would normally have metamorphized by serial culture in adult flies. The low levels of moulting hormone (hydroxy-ecdysone) in the adult haemolymph do not allow the disc cells to undergo metamorphosis, although they can proliferate.

Disc fragments can be serially transplanted through many generations and their stage of determination checked at any time by transplanting test implants into old larvae. The technique is illustrated in Fig. 15.20.

The cells of imaginal discs tested in this way normally retain their original regional identity and will continue to undergo metamorphosis into that structure, be it an eye, an antenna or a wing, etc. according to the developmental fate of the original disc. The rule of fixed regional determination, however, is not absolute; sometimes a disc fragment gives rise to a structure which is appropriate to a different region of the embryo (Box 15.8). This phenomenon is called 'transdetermination'. What is particularly striking is that the various cells of the disc still behave in a co-ordinated determined manner: the cells behaving as if they were now part of a complete but different fate map.

Many observations were made and certain transdeterminations found to be more frequently observed than others (see Box 15.8 for more detail). When all the cells of the disc suddenly change their state of determination in this way it is as if there were some master control system and the genetical basis for this is discussed in the next section.

15.4.2 Genetic control of regional development in *Drosophila*

The imaginal discs of *Drosophila* larvae have a regional identity and, as we have seen, the original identity can change spontaneously in predictable ways. Similarly, many mutants of *Drosophila* have been isolated that cause monstrous abnormalities because they change the regional organization of the embryo. Mutations that caused bizarre changes in morphology, termed 'homeotic mutants', were described by Bateson at the end of the nineteenth century. In *Drosophila* they often have the effect of changing the regional nature of an adult appendage or sensory structure. There are homeotic mutations, for instance, that cause an antenna to grow as a leg (dominant mutation *Antennapedia, Antp*) or cause halteres to develop as wings (dominant mutation *Ultrabithorax, Ubx*). These homeotic regulator genes are now known to form part of a hierarchical pattern-forming system. This pattern can be traced back in *Drosophila* to the pattern formation that is initiated by genes that control the cytoplasmic polarity of the egg (see Box 15.1).

15.4.2.1 *Homeotic mutations, the homeobox and the homeodomain*

Understanding of the control of development in *Drosophila* and elucidation of the molecular basis for the regulation of DNA transcription and mRNA translation have gone hand in hand. The *homeotic* mutants of *Drosophila* have such far-reaching effects because they are part of a very large class of genes which encode 'transcription factors' i.e. proteins that regulate the transcription of other genes, sometimes by working in tandem with co-regulator genes which enhance their activity.

A large class of genes encoding transcription factors have in common a sequence of DNA known as the 'homeobox' incorporating the name given to this class of mutations by Bateson more than a century ago. The homeobox is a 180 base pair DNA sequence that has been highly conserved. The homeobox encodes a sequence of 60 amino acids, the 'homeodomain' and structural features of this region of regulatory protein enable it to bind to DNA and so control the transcription of other genes. Highly precise binding occurs when two genes act together as in the precise binding of the Hox gene *Ubx* (see below) to target

Box 15.8 Serial transplantation of imaginal discs and the discovery of transdetermination

1 Using the technique of serial transplantation, the state of determination of a disc fragment derived from a single donor disc can be tested many times.

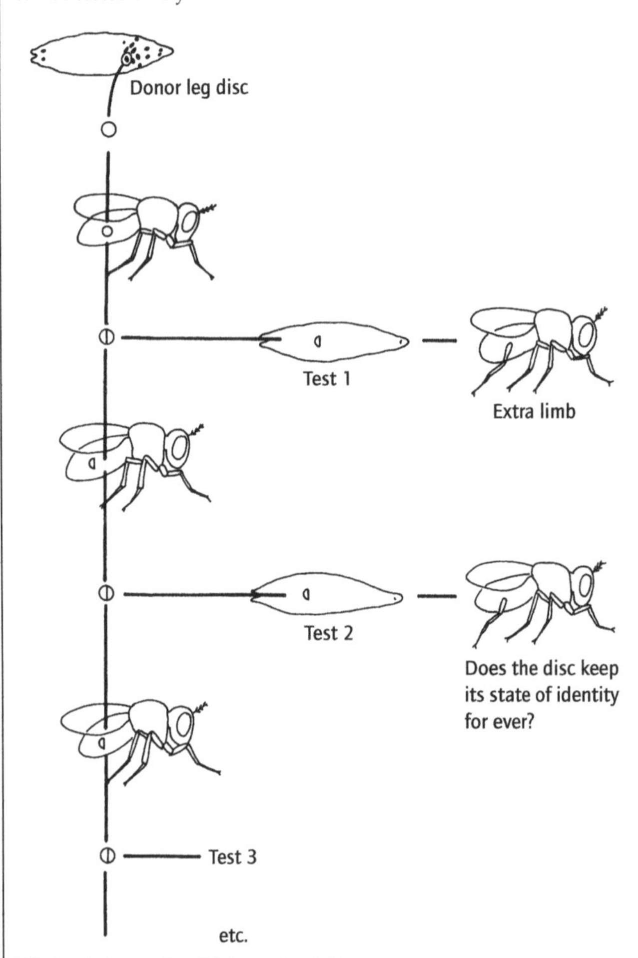

The technique of serial transplantation.

If the state of determination of the original disc is maintained, each test implant will reflect the character of the original disc. This is usually the case and the state of determination is stable.

2 If a large number of test implants are made, however, spontaneous changes in the state of determination may be observed. Not all spontaneous changes are possible. Some of those that have been observed in *Drosophila* are shown below.

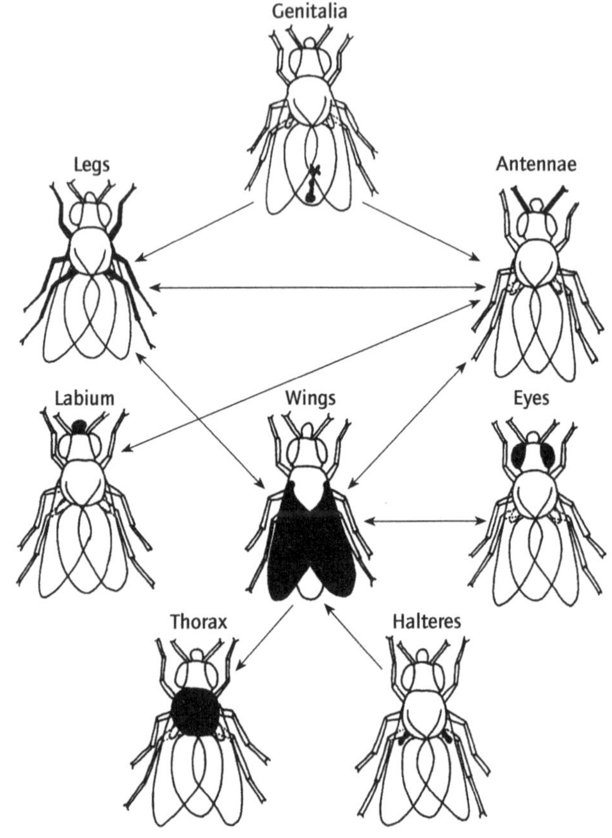

Arrows indicate the direction of spontaneous changes in the state of determination of structures derived from discs. For example, antennae discs to legs as shown in Box 15.10.

For instance, leg discs have been observed to change to give wings, antennae or mouthparts but not eyes, halteres or genitalia.

Note that all the different cells of the disc must undergo a spontaneous change and assume a new developmental fate, but they must retain local identity since they continue to give rise to structures which show regional organization. Such a spontaneous change is referred to as *transdetermination*.

DNA in the presence of another protein, extradenticle, which is the product of the *Drosophila* homeodomain gene *Exd*. Further details of the homeobox and the structure of the homeodomain are given in Fig. 15.21.

15.4.2.2 *The genetic hierarchy involved in regional organization in* Drosophila

Each region of the adult insect is developed from a specific part of the larva, and although the imaginal discs are not differentiated their positional information is clearly fixed. The segmental relationships are shown in Fig. 15.22. Note that each thoracic segment has a specific structure. Segment T2 normally has wings and T3 has modified wings – the sensory structures, halteres. There are also regional subdivisions within the segments.

A clear model of the determination of the regional identity of the larva is now becoming clear. The many genes that are involved can be arranged in hierarchical groups.

- *Maternal effect genes*. These genes control the formation of gradients of morphogenic proteins. Some code for morphogenic substances, e.g. *bicoid* gene which codes for an anterior organizer.

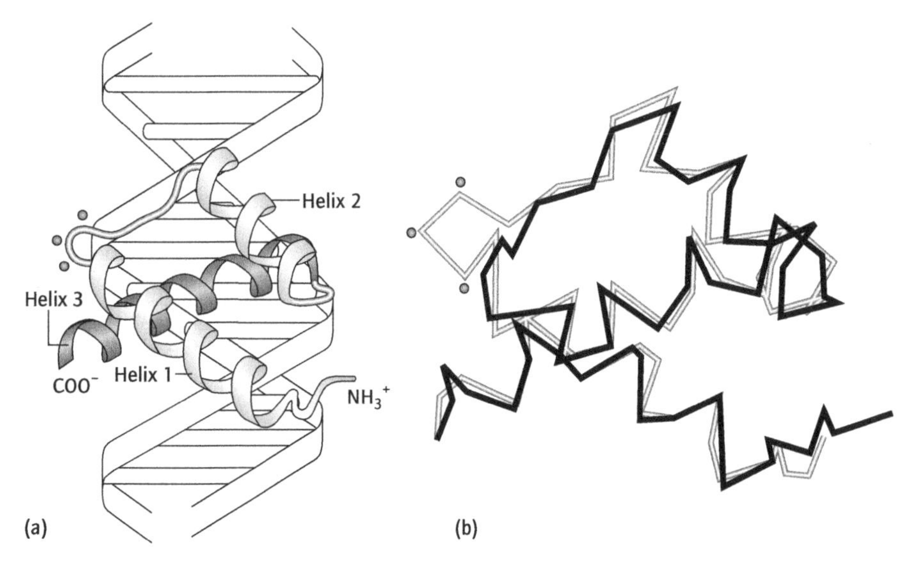

Fig. 15.21 The highly conserved genetic structure of an Hox gene is a consequence of its function as a DNA-transcription regulatory gene. (a) The homeodomain region of the protein has a three-dimensional structure that allows it to bind with the DNA double helix. Because of this the homeodomain regions of Hox genes are highly conserved. (b) A superimposition of the polypeptide backbone of the homeodomain of *Drosophila* engrailed and a yeast Hox gene product MAT α2. Note how very similar the two gene products are despite billions of years of invertebrate evolution. (From Gerhart and Kirschener, 1997.)

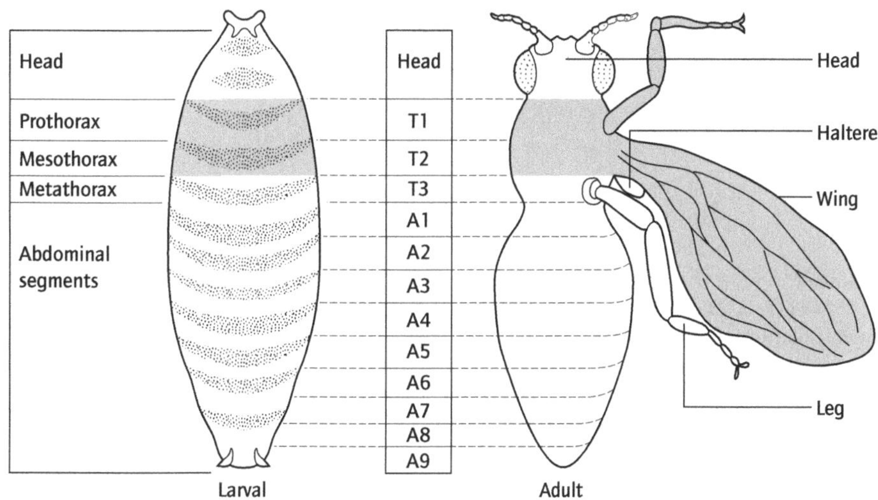

Fig. 15.22 The segmental structure of the larvae and adults of *Drosophila*. The three thoracic segments can be distinguished by the different appendages. Note the band of bristles on each segment of the larva. The modifications of this banding pattern permit the effects of some mutations to be interpreted in terms of changes in segmental architecture. (After Gilbert, 1990.)

Others code for substances involved in the fixing of bicoid protein in anterior regions of the oocyte cytoplasm.

- *Gap genes*. Gap genes are so named because, in the absence of the gap gene products, there are major omissions in the regional structure of the embryo. The gap genes are activated or suppressed by the specific concentration of morphogenic substances to which the nuclei are exposed (i.e. a genetic response to the gradient system). They are thus regulated by maternal effect gene products.
- *Pair rule genes*. The expression of the pair rule genes is regulated by the gene products of the gap genes. The pair rule genes divide the embryo into bands two segment primordia wide.
- The *segment polarity* and *homeotic genes*. The segment polarity genes further subdivide the embryo into segmental regions and, together with the pair rules genes, determine which of the *homeotic* genes are active and hence control the structural identity and morphology of each segmental region.

The invertebrate models used by experimental embryologists have frequently been explained by the supposition that gradients were created in the early embryo of unknown morphogenic substances. The developmental fate of individual cells was frequently shown to be consistent with a model based on such a gradient system. Experiments on echinoderm embryos have, for instance, been explained in this way (see Box 15.7).

The genetic basis for such a system is now being worked out in great detail using *Drosophila*. The results of this analysis are having a profound influence on our understanding of the determination of regional identity throughout the animal kingdom.

In insect embryos, such as *Drosophila*, each segment, primordium, has a unique identity long before the onset of regionally specific differentiation. The regional identity is influenced by maternal genes that establish a prepattern of four localized components in the unfertilized egg (Fig. 15.23a). The maternally provided prepattern in turn creates a zygotic pattern consisting of at least seven unique band-like domains along the anteroposterior axis (Fig. 15.23b) and four along the dorsoventral axis (Fig. 15.23c). The application of modern molecular techniques into the expression and regulation of genes involved in this

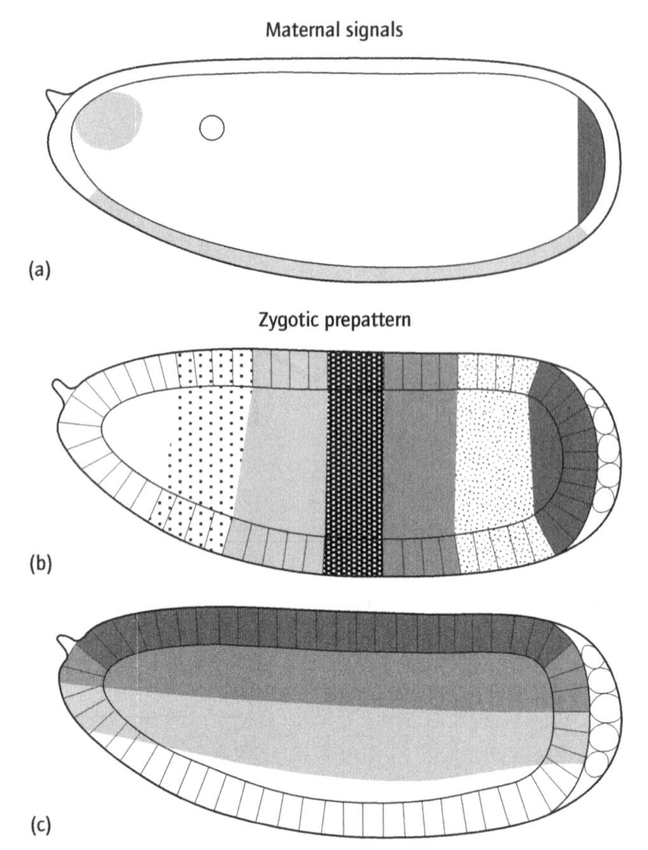

Fig. 15.23 A model for the development of regional pattern in *Drosophila*. (a) The prepattern created by maternal genes in the unfertilized eggs has four localized components. (b) The anteroposterior organization of the zygote comprises at least seven domains. (c) Regional organization along the ventrodorsal comprises at least four domains. (After Nusslein-Volhard, 1991.)

system is leading to a very rapid increase in knowledge, and some findings are summarized in more detail in Box 15.9 and Box 15.10.

15.4.2.3 Colinearity of genes and regional organization in the animal kingdom

The Hox genes of *Drosophila* are a set of genes in the homeodomain/homeotic gene, family with a precise role in the definition of anterior/posterior regional identity. Each of the eight *Drosophila* Hox genes is essential for the regionally specific organization of gene expression. Mutations of Hox genes therefore cause the regional identity to be lost. Consequently, normally dissimilar segments come to be more alike. The phenotypic expression of the loss of regional identity is of course very striking as in the case of the *Ubx* mutations which cause wings to develop where halteres would normally occur – the famous four-winged dipteran condition (see Box 15.10). These phenotypic expressions of gene malfunction are not simply random monstrosities but they reflect the basic organization of the embryo. Each Hox gene controls the expression of other genes in specific regions of the embryo. The eight *Drosophila* Hox genes are: *lab*, Labial; *pb*, Proboscipedia; *Dfd*, Deformed; *Scr*, Sex combs reduced; *Antp*, Antennapedia; *Ubx*, Ultrabithorax; *AbdA*, AbdomenA; *AbdB*, AbdomenB. The genes do not occur at random in the *Drosophila* genome but are clustered together in a relatively short region on one chromosome. What is even more striking is the finding that they occur in the same linear order as the regions they control. The discovery of the colinearity of the Hox genes and the segmental regions they specify was a great step forwards. The significance of this finding became even more striking with the discovery that the Hox genes of *Drosophila* have homologues (i.e. equivalent genes sharing very similar base sequences outside the conserved homeobox region) in other organisms including chordates, vertebrates and man. Moreover, the homologous genes of chordates also control regional identity and are also found on the chomosome in a linear sequence that matches that of the regions they control (for details see Box 15.10).

It may be supposed that this pattern reflects a common evolutionary origin from some regionally organized ancestor with a linear sequence of regulatory genes controlling the expression of that regional identity. Over evolutionary time there has been some divergence and duplication of Hox genes but the pattern has remained clear. The invertebrate chordate *Amphioxus* has a sequence of 12 Hox genes and the homology between these and the eight Hox genes of *Drosophila* has been established revealing duplications in the linear sequence. The subsequent evolution of the vertebrates from their chordate ancestors has been accompanied by multiple duplication of the entire sequence so that now four series of homologous Hox genes are found in the mouse.

Phylogenies of animals based on 18S ribosomal RNA phylogenies suggest a three-branched tree for bilateral animals comprising three clades, the deuterostomes (including within it the vertebrates) and two great protostome clades, the lophotrochozoans and ecdysozoans. An analysis of the structure of the Hox genes from a range of invertebrates reflects these supposed relationships since the posterior Hox genes of a brachiopod and a polychaete annelid are found to be similar whereas a distinct posterio-Hox gene is shared by a priapulid, a nematode and an arthropod. The authors of this study suggest that 'the ancestors of the two major protostome lineages had a minimum of 8 to 10 Hox genes' and the period of Hox-gene duplication following this diversification 'occurred before the radiation of each of the three great bilaterian clades'.

Although the *Drosophila* Hox genes were first named by reference to the regions they control, it has proved useful to adopt a common terminology and increasingly the *Drosophila* homologues are referred to as *Drosophila* Hox1, Hox2, etc.

The dorsal–ventral axes of invertebrates and vertebrates are also being found to share a common genetic regulatory mechanism which has lead some authors to revise the idea that the dorsal (antineural) and ventral (neural) surfaces of chordates and protostome invertebrates are homologous. It is perhaps too soon to embed this idea in an invertebrate textbook as a 'fact' but readers of this volume should be aware that the traditional distinction that is made between vertebrates and invertebrates

Box 15.9 Genetic control of regional organization in *Drosophila*. I: Regional identity

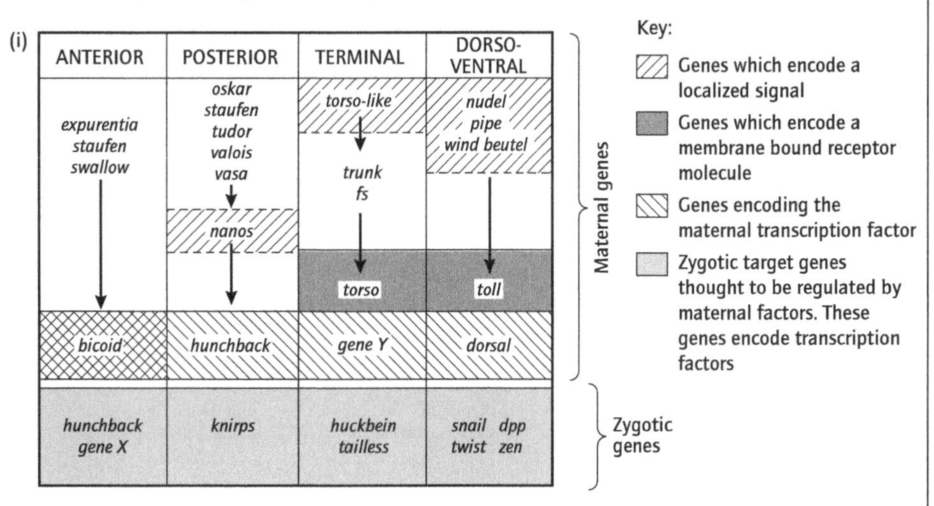

Cascade hierarchies (after Nusslein-Volhard, 1991).

1 Maternal effect genes

Maternal effect genes are mutations that affect the regional differentiation of embryos developing from the eggs of the affected flies.

In *Drosophila* maternal effect genes controlling four systems of axis determination are known. The four axes are:

I ANTERIOR AXIS SYSTEM: controls the segmental organization of the head and thorax.

II POSTERIOR AXIS SYSTEM: controls the segmental organization of the segmented abdomen.

III TERMINAL AXIS SYSTEM: controls the non-segmented telson and acron regions at the posterior.

IV DORSOVENTRAL AXIS SYSTEM: controls the pattern along the dorsoventral axis.

Each system of axis determination is controlled by a cascade-like hierarchy of interacting genes. At a high level in the cascade sequence are genes which encode a localized signal; at a lower level are genes which encode a maternal transcription factor that is asymmetrically and regionally distributed, and at the lowest level are target genes of the zygote that respond to the maternal gene effects and encode transcription factors. The transcription factors may interact with the pair rule and segment polarity genes as explained below, giving further regional specificity.

The four cascade hierarchies are summarized in Fig. (i); note that in the anterior system the *bicoid* gene encodes both the localized signal and the maternal transcription factor. (For completeness some of the genes are listed but are not referred to in the text.)

I The anterior polarity system

The offspring of homozygous *bicoid* (bcd/bcd) females completely lack head structures. They develop as telson | abdomen | telson as shown in Fig. (ii a). The gene product of the wild-type bicoid gene is an anterior morphogen; flies that lack this gene produce abnormal eggs (no head structures). The gene influences the transcription of a zygotic gene, one of the so-called gap genes *hunchback (hb)*. Mutations at this locus are also recognized by specific regional abnormalities.

II The posterior polarity system

Several maternal effect genes have been found which if absent in the mother, lead to the formation of larvae with deficient abdominal regions. These posterior maternal effect genes have been found in turn to activate the transcription of the gene *nanos*. The wild-type gene *nanos* transcribes a species of mRNA that codes for a protein repressing *hunchback* translation. The appearance of *nanos*-deficient embryos is seen in Fig. (ii) (b). This scheme is set out in Fig. (iii).

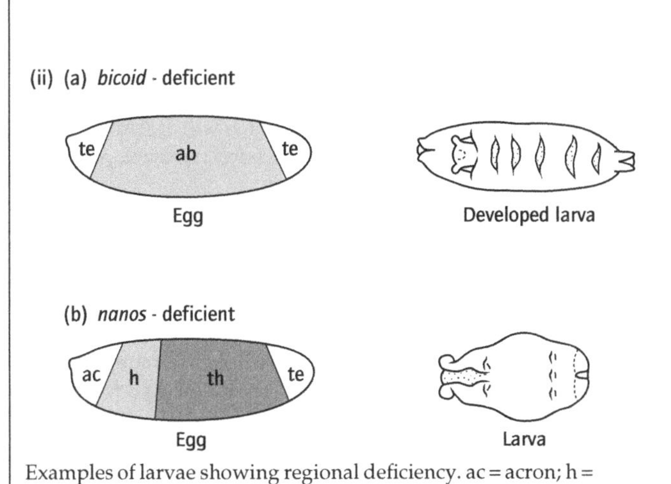

Examples of larvae showing regional deficiency. ac = acron; h = head; th = thorax; te = telson; ab = abdomen. (After Gilbert, 1990.)

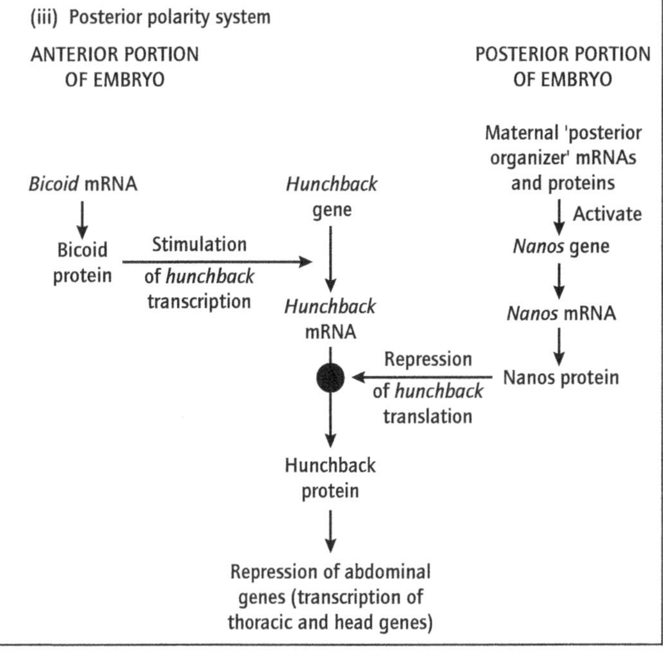

Continued p. 396

Box 15.9 *(cont'd)*

(iv)

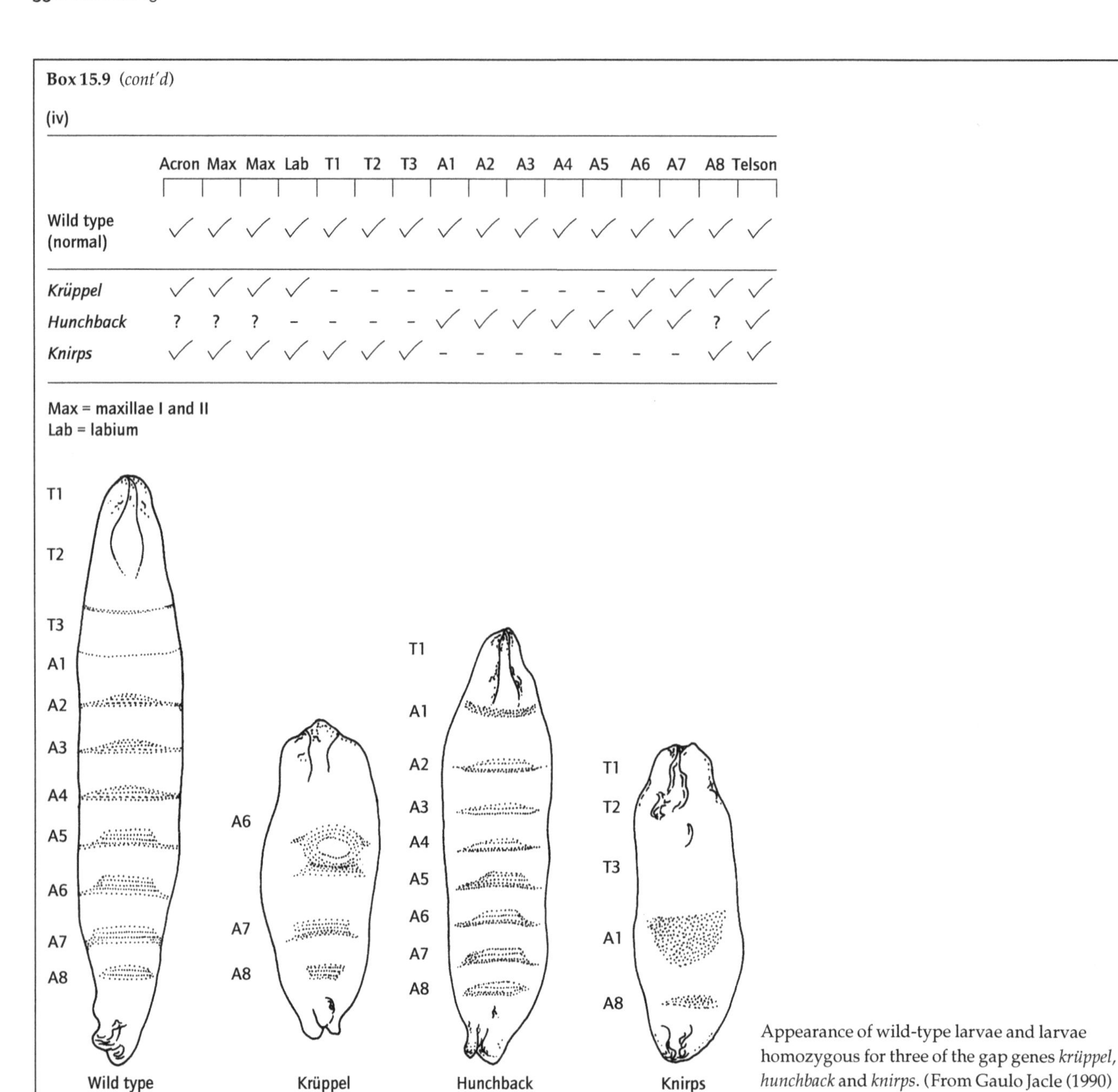

	Acron	Max	Max	Lab	T1	T2	T3	A1	A2	A3	A4	A5	A6	A7	A8	Telson
Wild type (normal)	✓	✓	✓	✓	✓	✓	✓	✓	✓	✓	✓	✓	✓	✓	✓	✓
Krüppel	✓	✓	✓	✓	-	-	-	-	-	-	-	-	✓	✓	✓	✓
Hunchback	?	?	?	-	-	-	-	✓	✓	✓	✓	✓	✓	✓	?	✓
Knirps	✓	✓	✓	✓	✓	✓	✓	-	-	-	-	-	-	-	✓	✓

Max = maxillae I and II
Lab = labium

Appearance of wild-type larvae and larvae homozygous for three of the gap genes *krüppel*, *hunchback* and *knirps*. (From Gaulo Jacle (1990) and Weigel *et al.*, 1990, after Gilbert, 1990.)

2 The gap genes

The gap genes are defined as a series of mutations causing specific regions of the resulting embryos to be missing. Figure (iv) shows some of the gap genes and the segmental regions missing in the homozygous mutant larvae.

3 Pair-rule genes and segment polarity genes

Pair-rule genes have the effect of dividing the embryo into a series of bands or stripes that correspond to the 15 segmental boundaries of the animal.

There are at least eight of these genes acting early in development whose activity is controlled by the gap genes, and others acting later in development. An important feature of the genes is their sensitivity to promoter and repressor substances that leads to a stabilized pattern of transcription.

Segment polarity genes are transcribed from only one band of nuclei in each segment region thus further refining the regional specificity in the embryo.

Three different types of segment pattern mutation are illustrated in Fig. (v); in each case the shaded regions represent the areas where the specific gene products coded by the wild-type gene are transcribed. In homozygous mutant forms these regions are deleted, i.e. no gene product transcription has occurred and the specified regions are missing.

Continued

Box 15.9 (*cont'd*)

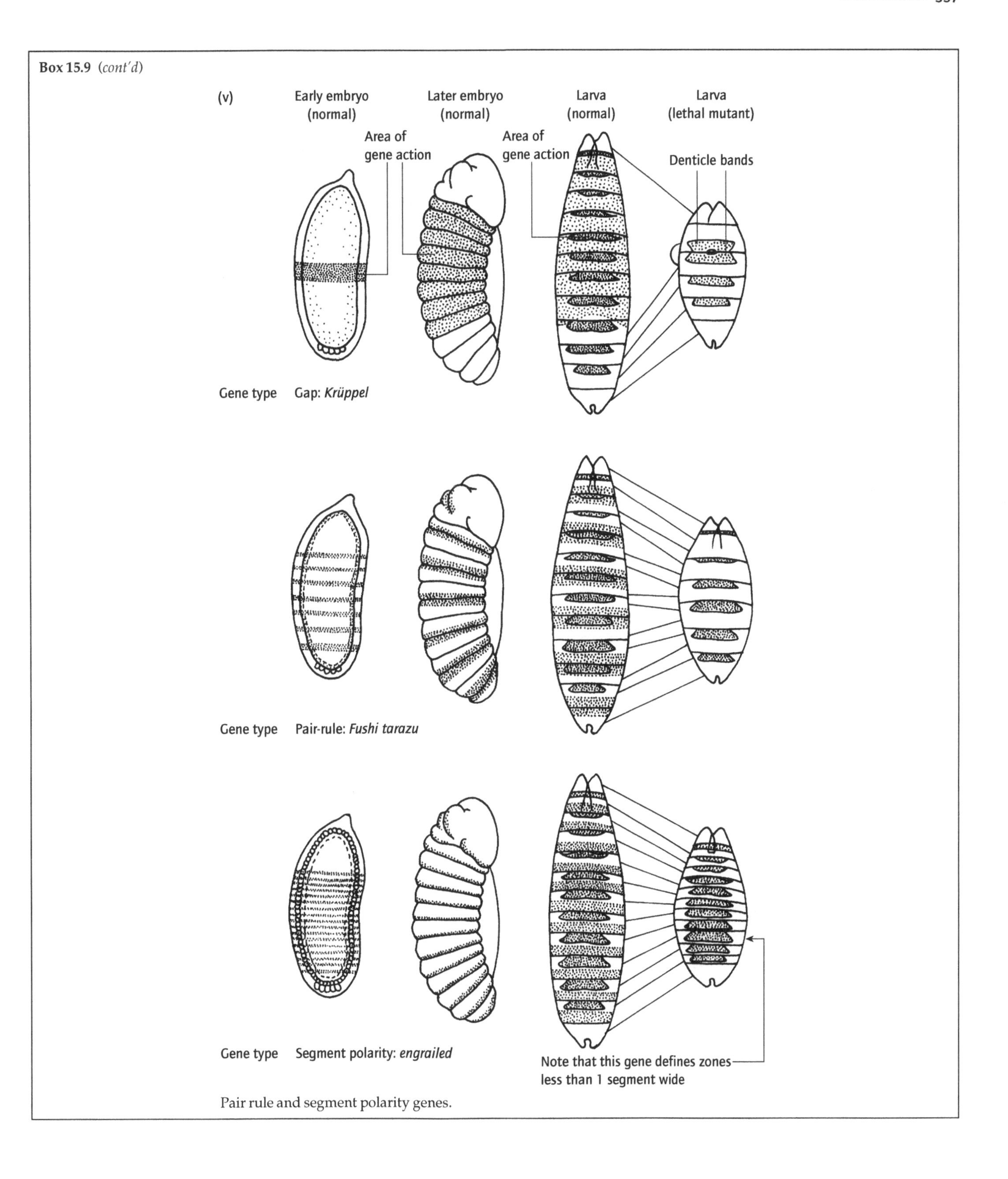

Pair rule and segment polarity genes.

Box 15.10 Homeobox (HOX) genes and the control of regional organization

1 Segmentation and regional organization of *Drosophila*

The dipteran *Drosophila* is constructed of a fixed number of segments. Each segment having a defined structure in the adult sometimes manifested by the development of the segment specific limbs and other ectodermal structures. The limbs and other ectodermal structures are developed from *imaginal discs* at the imaginal moult/metamorphosis. The regional identity of the larval segments is manifested primarily in the patterns of bristles at the segmental boundaries.

The body regions are:

Head including mandibular, maxillary and labial head segments

Thorax including:

T1 Leg1 – no wing

T2 Leg2 – Wing

T3 Leg3 – haltere

Abdomen: A1 to A8

Thus the body consists of a complex 'Head' and three thoracic segments and eight abdominal segments (i).

Comparison of larval and adults segments in *Drosophila*.

The three thoracic segments can be distinguished by their appendages:

T1 (prothoracic) has legs only;

T2 (mesothoracic) has wings and legs;

T3 (metathoracic) has halteres and legs.

2 Homeotic mutants of *Drosophila*

Mutants of the fruit fly *Drosophila* were discovered by Bateson as long ago as 1894 that had the effect of causing adjacent segments of the flies to develop a similar morphology – hence the term coined by Bateson *homeotic*. One such homeotic mutant for instance is *Ultrabithorax* which causes the third abdominal segment to develop a wing and not a haltere. Another homeotic mutant is *Antennapedia*, mutation of which is associated with the formation of a thoracic leg, not an antenna as expected on the head (ii).

In normal development the proteins encoded by these genes maintain the differences between segments and specify the regional identity of the segments. Mutations, i.e. incorrect DNA sequences and hence defective gene transcript proteins, lead to failures of regional specification.

Two major groups of homeotic genes occur on the 3rd chromosome of *Drosophila*:

The *Antennapedia* complex

The *Bithorax* complex

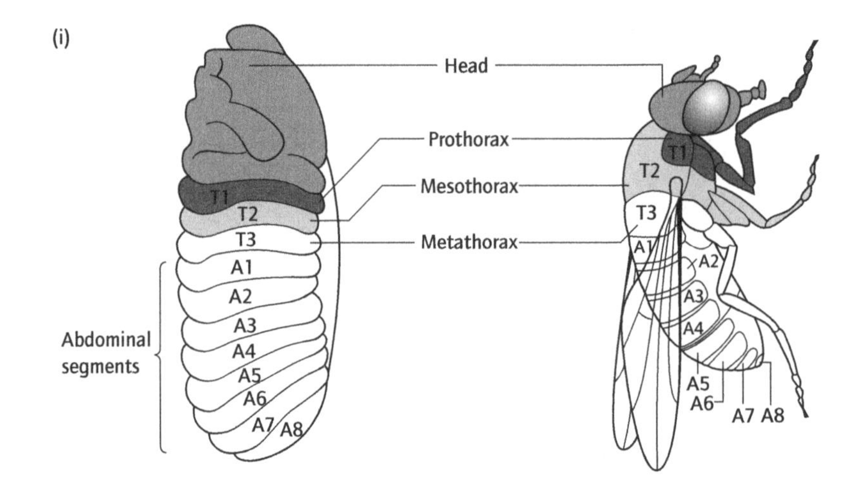

(ii)

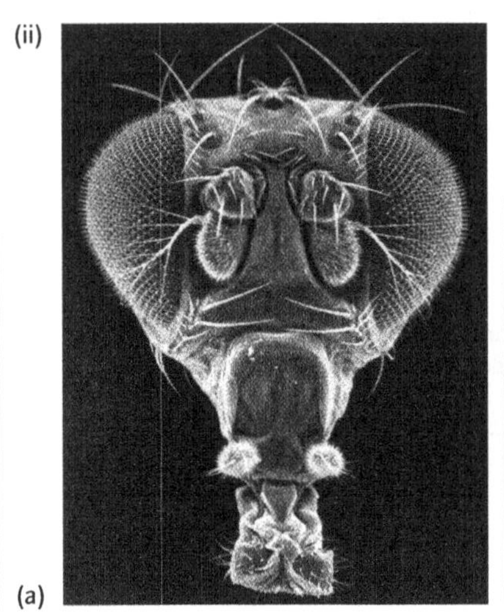

(a)

(b)

(a) Head of a wild-type fly.

(b) Head of a fly containing the *Antennapedia* mutation.

Continued

Box 15.10 *(cont'd)*

Regulatory mutations in the *Ubx* gene in the wild-type T2 produces a wing and T3 a balancing organ or haltere. In *bithorax* mutations the anterior compartment of T3 is transformed into an anterior compartment of a wing. The *postbithorax* mutation transforms posterior T3 to wing while *Haltere mimic* transforms T2 into an extra haltere. When *bithorax* and *postbithorax* are combined the result is a four-winged fly.

The role of the homeotic genes is a subject of intense research. *Ultrabithorax Ubx* mutations cause the entire third thoracic segment to develop as if it were a second thoracic segment, i.e. two wings. Mutations in the bithorax complex can cause a partial change, for instance *anterobithorax abx* and *bithorax bx* mutations cause the anterior half of the haltere to develop as a wing while the posterior half remains haltere-like and in contrast *postbithorax pbx* mutations cause the posterior half to become wing-like. It was formerly thought that

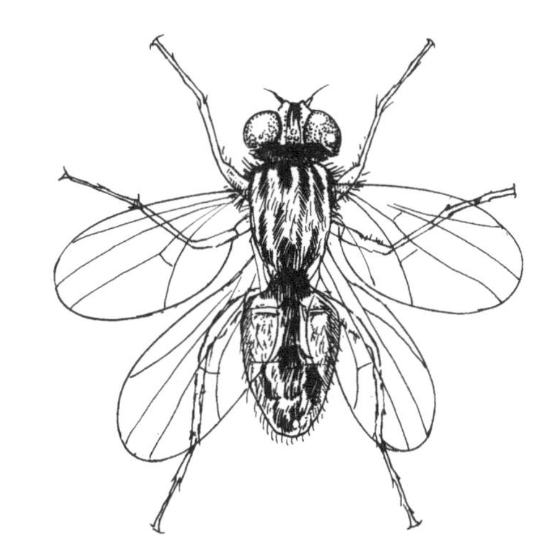

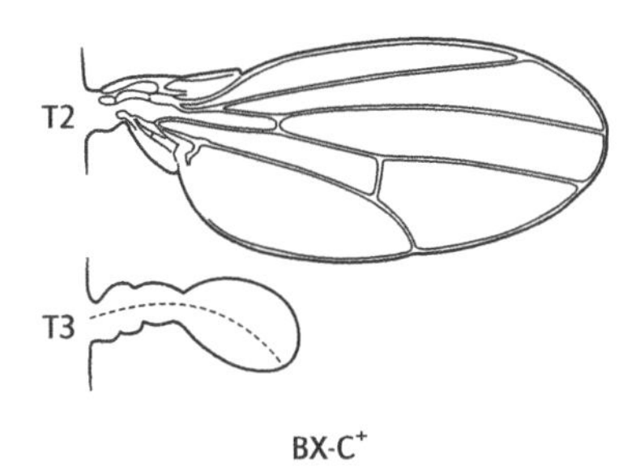

BX-C⁺

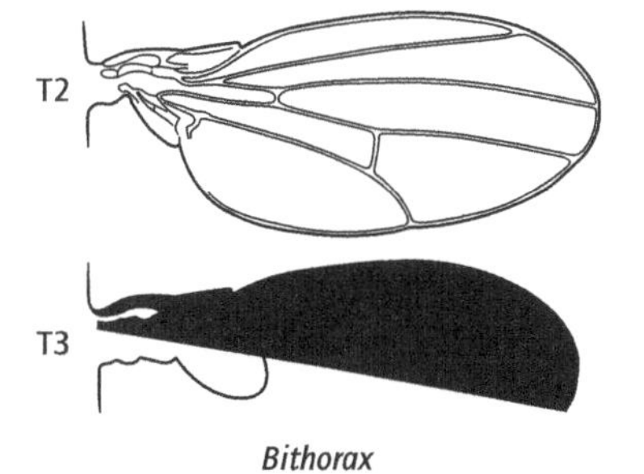

Bithorax

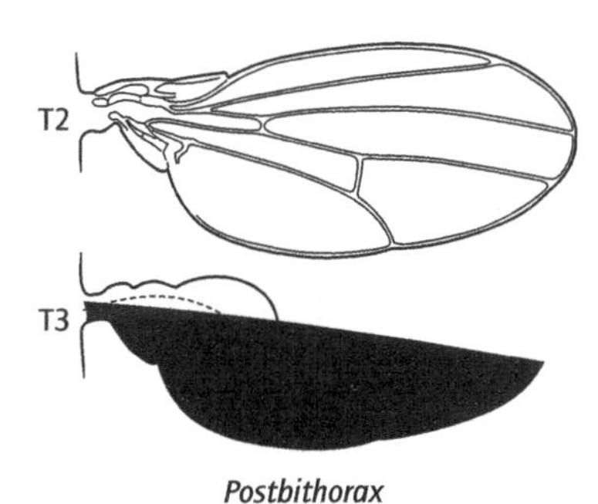

Postbithorax

Haltere mimic

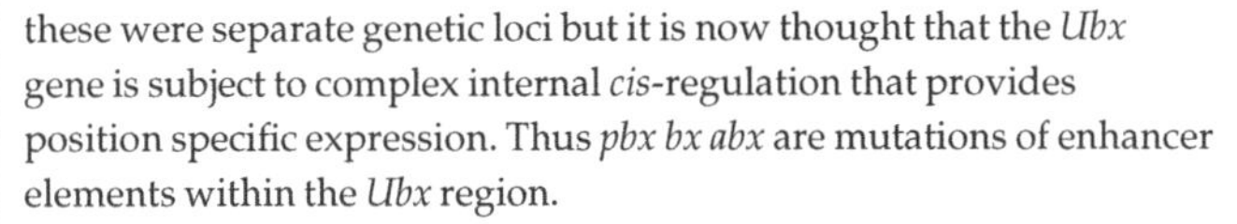

these were separate genetic loci but it is now thought that the *Ubx* gene is subject to complex internal *cis*-regulation that provides position specific expression. Thus *pbx bx abx* are mutations of enhancer elements within the *Ubx* region.

3 Highly conserved nature of homeotic genes

The homeotic genes are *regulator genes* i.e. they *regulate the expression of other target genes*. They are characterized by a highly conserved 60-amino acid sequence, the *homeodomain*. This homeodomain allows the homeotic genes to be recognized as a class of genes encoding a specific type of protein involved in regulation of other gene expression.

Homologous genes are found in virtually all organisms and are now known as Hox genes. The *Drosophila* homeotic genes are recognized as a subset (Hox genes) of the larger class of homeobox genes.

Continued p. 400

Box 15.10 *(cont'd)*

4 Regional organization, gene contiguity and pattern in diverse organisms

The linear sequence of structures of the adult insect body (see **1** above) is reflected in the contiguous sequence of eight homeotic (Hox) genes in *Drosophila*. As they are usually written the genes that are expressed in the most anterior regions are at the left and the genes expressed in the most posterior regions are to the right.

The *Drosophila* Hox genes are:

(i) Antennapedia complex, i.e. *Labial – lab; Proboscipedia – pb; Deformed – Dfd; Sex combs reduced – Scr; Antennapedia – Antp.*

(ii) Bithorax complex, i.e. *Ultrabithorax – Ubx; Abdominal-A Abd-A; Abdominal B – Abd-B; Caudal cad.*

These genes occur in a linear sequence on chromosome three and their homologous genes in other organisms are organized in the same pattern (with some duplication – see figure below).

Note: the vertebrate scheme of notation has largely taken over naming the genes in sequence in the series HoxA, HoxB, HoxC and HoxD, e.g. mouse Hox a1 ≡ *Drosophila* lab, mouse Hox a2 ≡ Drosophila pb, etc.

These findings are really astounding and open up a new era in the study of evolutionary processes and a new science 'the evolution of developmental processes' is emerging. Duplications seem to have occurred when the vertebrates diverged from the chordate ancestors such as *Amphioxus* and again when the jawed fish arose from the agnathan ancestors (see Holland, P.W.H., Garcia-Fernanez, J., Williams, N.A. & Sidow, A. 1994. Gene duplications and the origins of vertebrates. *Development* (supplement), 125–133).

5 Anterior–posterior gradients and regional organization of the egg – maternal effect genes *bicoid* and *nanos*

bicoid mRNA at the future anterior end of the oocyte
nanos mRNA at the future posterior end of the oocyte

This pattern of distribution is influenced by the effects of the oocyte on the overlying follicle cells. This is because the secretion of a protein in the oocyte encoded by the gene *gurken* affects the overlying follicle cells and causes them to become posterior polarized cells (PPC). The PPC in turn make secretions that affect the distribution of the *bicoid* and *nanos* mRNAs. Genes of this kind are called 'maternal effect

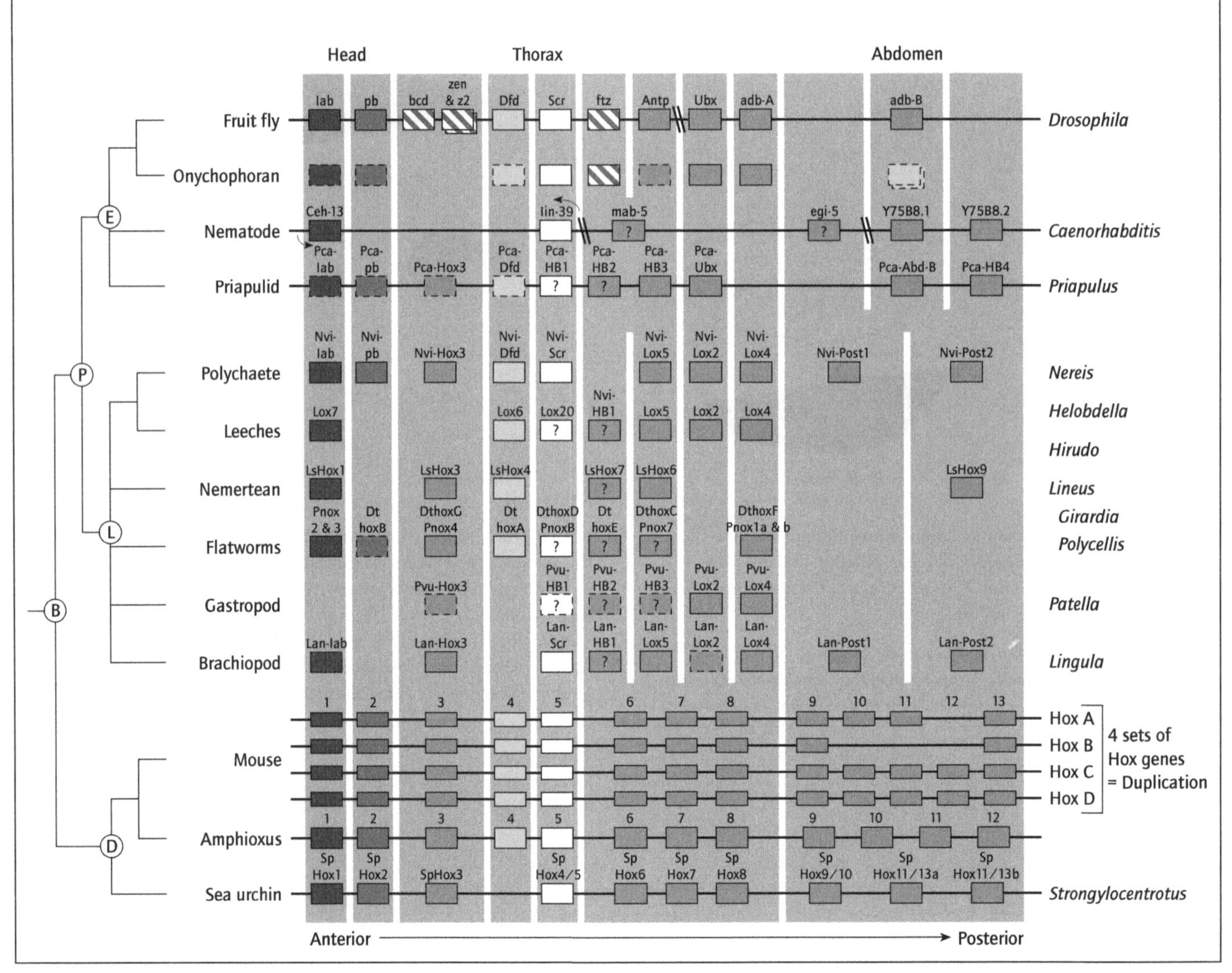

Continued

Box 15.10 *(cont'd)*

genes' because the deficiencies of eggs with the mutant phenotype cannot be rescued by fertilization with normal wild-type sperm.

- Bicoid transcription factor acts by activating and suppressing transcription of other genes in the anterior of the embryo.
- Nanos transcription factor acts by binding to certain mRNA species and blocking their translation.
- Nanos transcription factor blocks translation of *hunchback*. *Hunchback* is another maternal effect gene the mRNA of which is accumulated in the oocyte cytoplasm and distributed rather evenly. In addition bicoid actively stimulates transcription of *hunchback*. The gradients set up in the egg cytoplasm in turn control the activity of a number of other genes in a hierarchical cascade.

6 Gap genes, pair rule genes and segment polarity genes
The understanding of regional control in *Drosophila* was stimulated by the work of J. Nusslein-Volhard who was the first female to be awarded a Nobel Prize for her work. She scanned lethal mutants of *Drosophila* for the effects on the larvae, cataloguing all the mutant mutations according to the effects of the genes on the pattern of bristles on the larvae. This was painstaking work involving cataloguing thousands of mutants. Now the work is seen as a breakthrough, more and more of the genes have been sequenced and the pattern of expression of the genes in the embryo has been shown to be crucial in the progressive definition of the regions of the embryo and future adult. Ultimately these genes control the expression of the homeotic genes.

The genes can be categorized into a sequence of activity defining a progressively more complex segmental form. The genes also show that the regional organization of the larvae includes definition not just of segments but of anterior and posterior compartments within segments.

The gap genes
Larvae with the mutant genotype have a phenotype in which certain regions of the body seem to be missing. Having identified and sequenced the genes it is possible to show the pattern of expression. The eight gap genes are *huckbein – hkb; tailless – tll; giant – gt; empty spiracles – ems; orthodenticle – otd; krüppel – kr; knirps; kni*. Each gene is expressed in a relatively narrow band at a specific location in the embryo in relation to the gradients of the maternal effect genes and dependent transcription factors.

(Recall the gradient hypotheses developed to explain echinoderm development. See Box 15.7.)

Pair rule genes
These genes are so named because the mutant phenotypes are missing either odd-numbered or even-numbered segments but not both.

The transient non-repeating pattern of eight bands of gap gene expression is sufficient to set up a new pattern of eight pair rule genes. These genes are each expressed in a repeating pattern. The initial pattern involves three primary pair rule genes *runt; hairy; even-skipped – eve*. Although each gene is expressed in a sevenfold repeating pattern the bands of expression are slightly displaced one to another. Thus for example *runt* is in T1 and T2, *hairy* is expressed at the border between T2 and T3. Other genes are expressed in relation to this pattern, e.g. *fushi-tarazu – ftz*, Japanese for few bristles!

Segment polarity genes
A further spatial hierarchy is set up involving stripes of activity of a large number of genes in a pattern at a frequency greater than that of the 14 segments. Among the best known are *engrailed – en; wingless – wg; and hedgehog – hh*. The *en* gene is expressed in 14 stripes at the posterior boundary of each segment.

Activation and control of homeotic (Hox) genes
The activity of the Hox genes is controlled by the spatial expression of the gap genes combined with the effects of the pair rule and segment polarity genes which organize the structure of each defined segment.

The initial polarity of gene products in the egg – the anterior-posterior gradient – has defined a sequence of highly structured segments and, each segment now has a unique identity. This leads to the expression of different homeotic (Hox) genes in cells set aside in the imaginal discs in these regions and hence each segment comes to have its own characteristic structure.

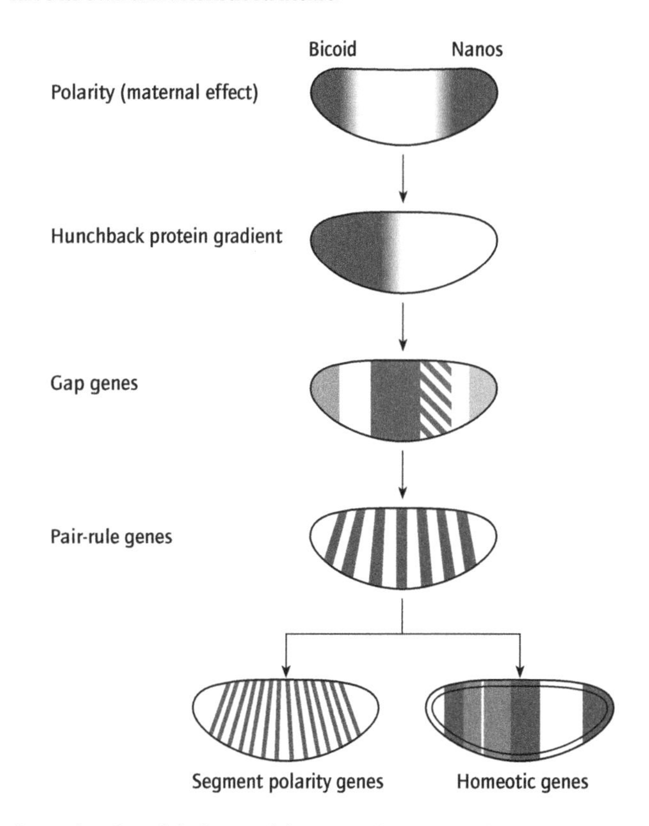

Generalized model of *Drosophila* pattern formation. The pattern is established by maternal effect genes that form gradients and regions of morphogenetic proteins. These morphogenetic determinants create a gradient of hunchback protein that differentially activates the gap genes that define broad territories of the embryo. The gap genes enable the expression of pair-rule genes, each of which divides the embryo into regions about two segment primordia wide. The segment polarity genes then divide the embryo into segment-sized units along the anterior–posterior axis. The combination of these genes defines the spatial domains of the homeotic genes that define the identities of each of the segments. In this way, periodicity is generated from non-periodicity, and each segment is given a unique identity.

Continued p. 402

Box 15.10 *(cont'd)*

7 Specification of dorsal and ventral polarity

We have discussed the progressive organization of regional identity along the anterior to posterior axis. In a similar way the dorsoventral polarity can be traced to polarity in the egg cytoplasm and the progressive development of regional organization in the embryo leading to specification of dorsoventral organization and the processes of gastrulation.

This pattern of organization was also stimulated by Nusslein-Volhard's work on embryonic abnormalities in lethal mutations.

The polarity can be traced to the expression of the gene *dorsal*. The dorsal gene transcript is everywhere but is incorporated into the nuclei of blastoderm cells only in the future ventral region of the embryo. The molecular biology/genetics of this is complex involving a number of other genes but can be traced ultimately to the asymmetrical position of the oocyte nucleus.

The nuclei that receive *dorsal* gene transcript are ventralized and will be involved in gastrulation movements. The targets of the dorsal gene transcripts are *rhomboid; twist* and *snail* and dorsal gene transcript also blocks the transcription of other genes such as *tolloid* and *decapentaplegic*.

A key feature of homeotic (Hox) genes is the ability of the gene transcript proteins to bind to DNA at specific recognition sites and hence control the transcription of other gene products.

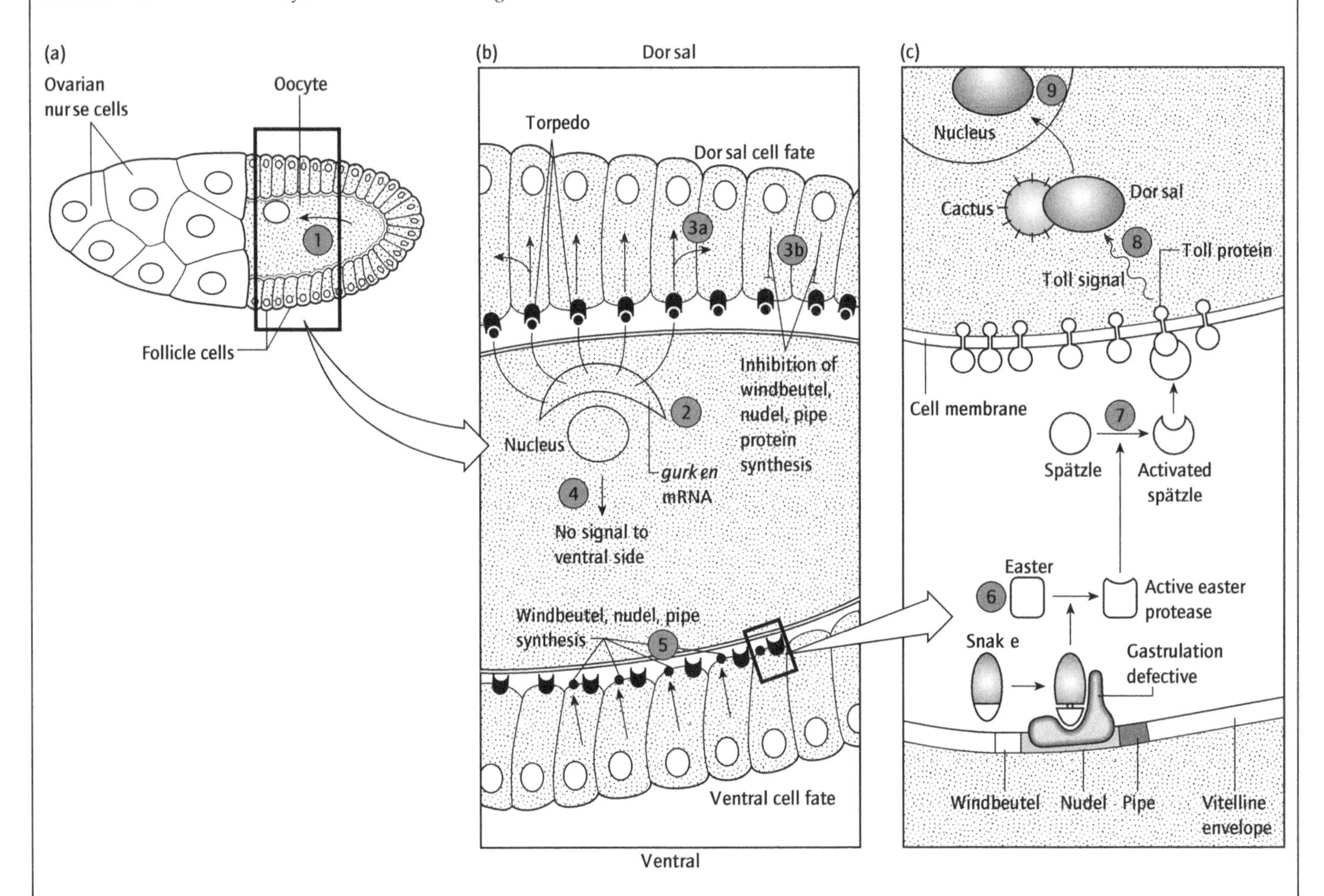

Key:

1 Oocyte nucleus travels to anterior dorsal side of oocyte. It collects *cornichon* and *gurken* mRNA
2 *cornichon* and *gurken* messages translated. The gurken protein is received by torpedo proteins during mid-oogenesis
3a Torpedo signal causes follicle cells to differentiate to a dorsal morphology
3b Synthesis of windbeutel, nudel, and pipe proteins inhibited in dorsal follicle cells
4 Cornichon and gurken proteins do not diffuse to ventral side
5 Ventral follicle cells synthesize windbeutel, nudel, and pipe proteins
6 Ventral follicle proteins absorb snake and gastrulation-defective proteins to effect splitting of easter zymogen, making active easter protease only on ventral side
7 Easter splits spätzle, which binds to toll-receptor protein
8 Toll signal causes phosphorylation and degradation of cactus protein, releasing it from dorsal
9 Dorsal protein enters the nucleus and ventralizes the cell

is a matter of convenience only and may expect to see increasing evidence of shared developmental pathways in the next few years and the invertebrate–vertebrate dichotomy that appears to be supported by the title of this book does not have any basis in phylogeny.

15.5 Larval development and metamorphosis

The ciliated larvae of marine invertebrates are adapted for a pelagic life and the ciliated bands which provide their locomotory power are not adequate for the larger adults (see Chapter 10). Metamorphosis in these animals therefore frequently involves the loss of these ciliated bands and a transition to a mode of life in which muscle cells provide the locomotory forces.

In gastropods, metamorphosis of the veliger larva is accomplished progressively with a gradual reduction of the velum, which eventually becomes unable to support the developing snail and is replaced by the foot as the chief locomotory organ. Metamorphosis of the bivalve molluscs is often more rapid, with a sudden loss of the velum.

The conflicting demands of larval locomotion and the requirement for development towards the adult stage are also illustrated by the polychaete annelids, in which segments are added progressively during embryonic or larval life (see Section 15.2.2). The segments are derived from paired segment blastemas: ventral bands in the posterior region of the larva. The mesoderm cells in the blastema are derived from paired mesentoblasts formed from the 4d cell during typical spiral cleavage (see Section 15.2.1, Fig. 15.13 and Box 15.2).

The segment blastema has mesodermal and ectodermal elements. The two bands of mesoderm-producing cells proliferate a series of blocks of tissue in which the coelom is formed; in these animals, unlike the echinoderms described below, the mesoderm is not derived from the cavity of the archenteron and is termed a 'schizocoel'. The segments are produced by co-ordinated organogenesis of the mesodermal and ectodermal tissues in which the ventral nerve cord plays a major inductive and organizational role. As each newly proliferated segment develops in front of the pygidium, it increases the mass of the larva. In most cases each segment is provided with locomotory or flotation devices during the pelagic phase.

The development of the Echinodermata involves one of the most dramatic forms of metamorphosis in the animal kingdom. A fully developed echinoderm larva is bilaterally symmetrical; the dominant symmetry of the adult echinoderm, however, is pentaradial, although there is sometimes a secondary bilateral symmetry imposed on this (see Section 7.3.2). The coelomic pouches of echinoderms are derived from lateral outpushings of the tip of the archenteron some time after the completion of gastrulation (see Fig. 15.12). Their formation is therefore 'enterocoelic'. A study of living echonoderms suggests that there were primitively three paired coelomic pouches: the axocoel, hydrocoel and somatocoel. The development of these coelomic spaces and the subsequent metamorphosis is illustrated in Fig. 15.24. In most living echinoderms, the right-hand axocoel and hydrocoel are reduced or totally suppressed. The left hydrocoel becomes subdivided into the hydrocoel and in outgrowth forming the stone canal and the hydropore (Fig. 15.24, i–iv).

These primitive coelom anlagen are represented in a 9-day pluteus larva in Fig. 15.24(v). The left- and right-hand somatocoels spread over the stomach and form the body cavities of the adult. The left hydrocoel creates a pentaradial fluid-filled anlagen from which the water vascular system of the adult develops. The mouth of the future adult forms in the centre of the hydrocoel and this establishes the oral surface of the urchin. The oral–aboral axis of the adult is therefore approximately along the left–right axis of the pluteus larva (Fig. 15.24vi). The developing sea-urchin or starfish appears as an imaginal disc which is unfurled at metamorphosis when the ectodermal epithelium of the larva shrinks and is discarded (Fig. 15.24vii, viii).

Animals with an exoskeleton cannot grow and develop in a progressive manner, but must proceed through a series of moults. At each moult the old skeleton is discarded and the body expands prior to hardening of the new external covering. This is well illustrated by the Crustacea. Their development frequently involves a series of morphologically distinct larval stages which finally moult to give the adult form (see Box 14.5 and Chapter 8). In the Crustacea growth will often continue through a sequence of moults after the adult morphology has been reached. In the winged insects, however, there is no further growth once the definitive adult stage has been reached.

In marine invertebrates the larvae are responsible for establishing the young adult in a suitable environment for its subsequent development. This is an important stage in the life history of the animal and for organisms which, as adults, are sedentary or sessile, the choice of a suitable substratum may be critical.

Marine biologists sometimes observe clouds of newly metamorphosed larvae settling apparently at random into substrata in which they will not survive; but this is not usually the case. There are many studies which have revealed the precision with which pelagic larvae are able to 'choose' a suitable substratum. The processes involved include:

- Behavioural sequences which bring the larva into contact with a suitable substratum.
- Delayed metamorphosis in the absence of suitable substrata.
- Discrimination and selection of a preferred substratum.
- Gregarious behaviour and the chemosensory detection of surfaces previously inhabited by adults or larvae of their own kind.

The larvae of the marine mollusc *Mytilus edulis*, for instance, exhibit a complex sequence of behavioural changes during their development.

The larvae of many species have been shown to be highly selective of the substrata on to which they will settle. Box 15.11 shows how multiple-choice experiments reveal differences in the choice of settlement surface by larvae of closely related species of the tube-living *Spirorbis* worms, the adults of which are characteristically found on different substrata.

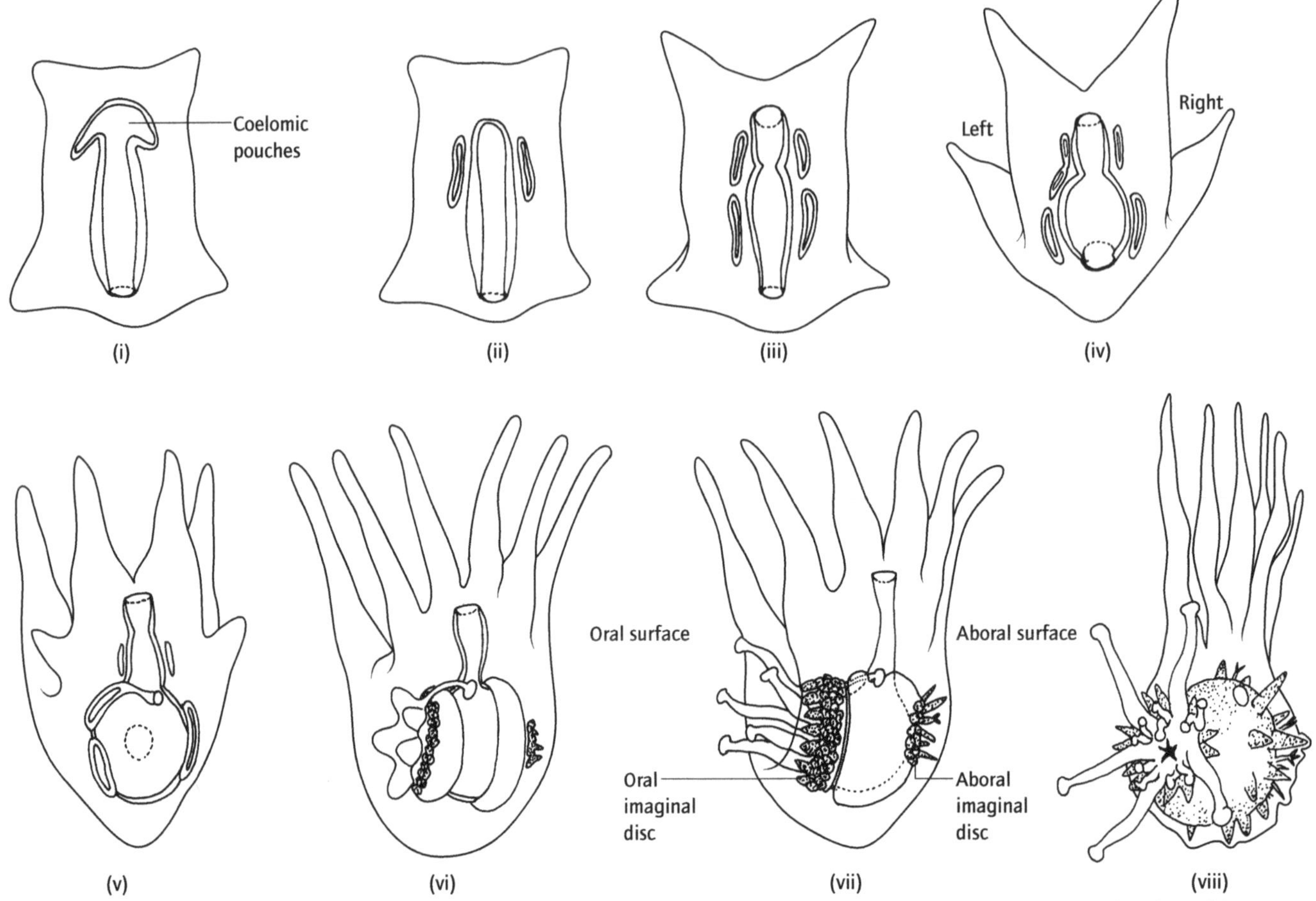

Fig. 15.24 Development of the coelomic system of a sea-urchin and metamorphosis to the adult form. (i) Formation of the coelomic pouches from outpushings at the tip of the archenteron after the completion of gastrulation. (ii) The paired coelomic pouches. (iii), (iv) Subdivision of the pouches. (v) Differentiation of the second left coelom as hydrocoel and stone canal–hydropore complex, and expansion of the third left and right coelomic pouches. This stage is reached in *Psammechinus* after about 9 days of planktonic development. (vi) Organization of the pentaradial water vascular ring defining the oral surface of the developing echinoderm and establishing the oral–aboral axis. (vii) Progressive development of oral and aboral imaginal discs supported by expansion of the ciliated locomotory bands of the pluteus. (viii) *Psammechinus* embryo shortly before the completion of metamorphosis. Larval tissues will be discarded and the adult anlagen united to form the definitive sea-urchin.

The cypris larvae of barnacles also show a remarkable and complex behaviour prior to settlement, which enables them to choose good sites for metamorphosis. They respond to the texture of the surface (they prefer rough or pitted surfaces), but above all they respond to the presence of other barnacles, barnacle larvae, or the remnants of older barnacles of their own species. The chemical nature of these substances is the subject of intensive research as they may provide the basis for the construction of a new generation of biologically based 'non-toxic' substances that could be incorporated into antifouling paints.

15.5.1 The development and metamorphosis of insects

The phylum Uniramia includes the large group of animals commonly referred to as insects (the subphylum Hexapoda). There are, within this assemblage, animals with three different developmental patterns. The myriapod-like classes do not have wings, e.g. collembolans and thysanurans, and these develop gradually through a series of moults. Their morphology never undergoes any striking change and they cannot be said to undergo metamorphosis (Fig. 15.25a): they are ametabolous. A primitive feature of their development is the total cleavage of the eggs of many species, although some show an interesting characteristic in which cleavage eventually resembles that of the winged insects, but is at first total.

The other insects show a more or less dramatic metamorphosis during their development and have a fixed number of larval instars prior to the adult form. The development in more advanced insects can be categorized in a number of ways. In some, the so-called 'long germ band insects', the sequence of adult segments is established at an early stage in

Box 15.11 Metamorphosis and substratum choice by marine larvae

1 Many benthic organisms are found on a specific substratum which is characteristic for that species. Sometimes several closely related species occur sympatrically but on different substrata. The observed distribution of the adults is due to discrimination by the larvae. This has been demonstrated experimentally for several small, tubiculous polychaetes of the genus *Spirorbis*:

(i) *Spirorbis borealis* on *Fucus serratus;*
(ii) *Spirorbis tridentatus* on bare rock;
(iii) *Spirorbis corallinae* on *Corallina officinalis.*

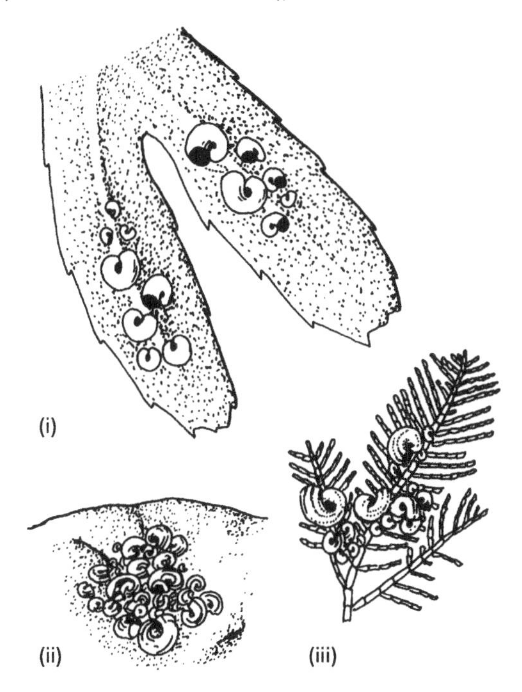

When presented with different substrata in a two-way choice experiment' the larvae of these different species show clear differences in their preferences (see Table B.2).

Table B.2 Experiments on the choice of settlement substratum by larvae of different species of the polychaete *Spirorbis* (from data of De Silva, 1962).

Species	Substratum	Total number of larvae
Spirorbis borealis	*Fucus serratus*	1297
	Corallina officinalis	18
	Fucus serratus	457
	Filmed stone	295
Spirorbis tridentatus	*Fucus serratus*	0
	Filmed stone	52
	Corallina officinalis	0
	Filmed stone	55
Spirorbis corallinae	*Corallina officinalis*	63
	Fucus serratus	2

2 The cypris larva of an acorn barnacle (see below) is very choosy about a site for settlement.

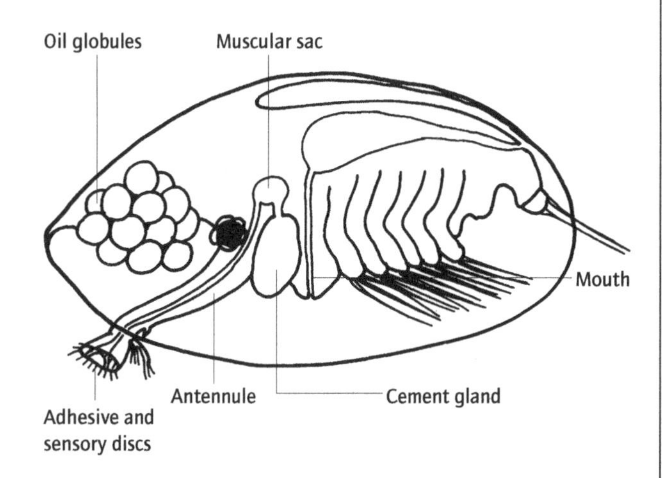

Before finally choosing to settle down the larva moves around a potential surface, showing exploratory behaviour and it may swim away and choose another site.

3 There are many factors which may increase the probability that a barnacle larva will settle on a particular surface.

Some factors stimulating settlement and metamorphosis:

(a) A rough surface;
(b) A pitted or grooved surface;
(c) Remains of old barnacle tests;
(d) The presence of newly settled cyprids.

Of these the most important are those due to the presence of other barnacles.

An unattractive surface can be rendered attractive by soaking the surface in extract of barnacle tissue. The attractive substance is a protein which can be detected by the cypris larva as a single molecular layer.

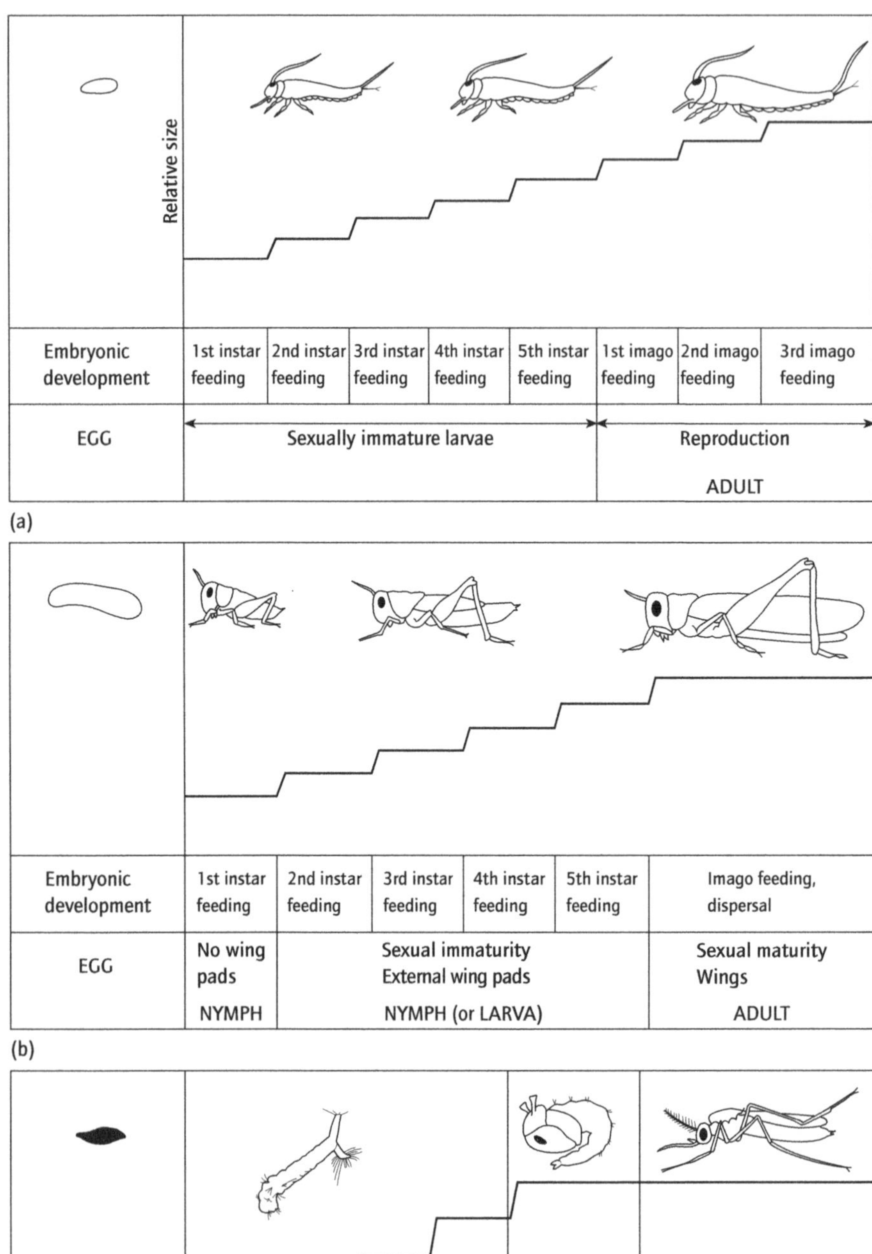

Fig. 15.25 Diagrammatic representation of growth and development in insects. (a) Ametabola: no metamorphosis. In this example growth continues after breeding condition has been reached.
(b) Hemimetabola: incomplete or partial metamorphosis. In this example the wings appear as external wing pads from the second instar. They are fully formed after the fifth and final moult. An increase in size occurs shortly after each moult.
(c) Holometabola: complete metamorphosis. In this example, there are four larval instars, a modified larval or pupal instar and an adult phase. Note growth is restricted to the period after each larval moult.

the development of the egg. Not all insects exhibit such a precocious pattern of segment formation, and segments are added progressively in a manner more like that of polychaetes and chordates. These insects are said to show 'short germ band development'. Already the pattern of expression of homeotic genes in these insects is being established to show how the same genes are involved in segment formation in the two models.

In several orders of insect, the pre-adults, called 'nymphs' or, if aquatic, 'naiads', have external wing buds and metamorphosis is not extreme. These insects, which include the dragonflies and grasshoppers, are sometimes referred to as the 'exopterygotes' and their development is said to be 'hemimetabolous'. In the case of the grasshoppers and locust (illustrated in Fig. 15.25b), the nymphs occupy a similar niche to the adults and there is no major reorganization at the metamorphosis from the last larval instar to the winged adult. When the juveniles are aquatic there may be a rather marked change in the morphology associated with the niche divergence between adults and juveniles. This metamorphosis, however, is not as dramatic as it is in those insects which have internal wing buds (the endopterygotes) where development is said to be holometabolous. In these a transitional pupal stage occurs between the final larval instar and the adult phase, as shown in Figs 15.25c and 8.33. The pupa can be best interpreted as a much modified terminal larval instar. The larvae of the holometabolous insects belong to a number of different types, as illustrated in Fig. 15.26; they are often referred to by the general names caterpillar, grub, maggot, etc. As explained in Chapter 14 and Fig. 15.25c, the larvae are specialized non-dispersive feeding stages, whereas the adults are specialized for dispersal and reproduction. The pupa is a stage during which locomotion and feeding are suspended while major reorganization of the body structure occurs.

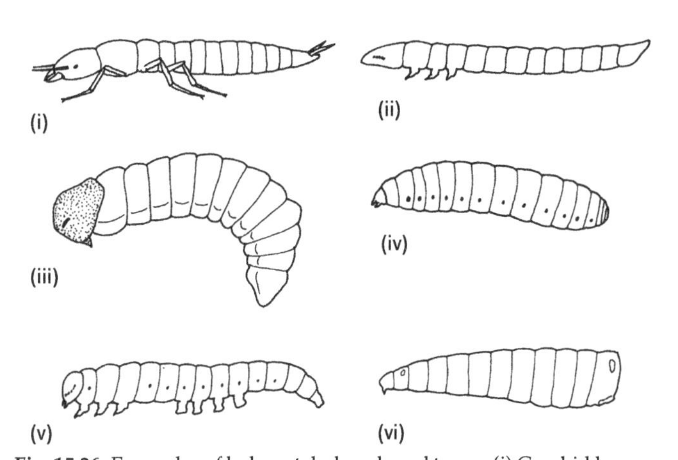

Fig. 15.26 Examples of holometabolous larval types. (i) Carabid larva – Coleoptera (beetles); (ii) elaterid larva or wireworm – Coleoptera; (iii) curculonid or weevil – Coleoptera; (iv) bee-grub – Hymenoptera (honey bee); (v) caterpillar larva – Lepidoptera (butterfly); (vi) maggot larva – Diptera (blow-fly). In each case the morphology of the larva is very different from that of the adult.

15.6 Regeneration

15.6.1 Introduction

Regeneration can be defined as the capacity to replace, by compensatory growth and differentiation, parts of the body which are accidentally lost or which are autotomized. The ability to regenerate missing parts of the body in this way is a prominent feature of the biology of many of the soft-bodied invertebrates such as, for example, sponges, cnidarians, flatworms, nemertines, annelids and some echinoderms. Such animals also exhibit asexual reproduction by fission (see Chapter 14) and the two processes are obviously related. Invertebrates with hardened external coverings such as the arthropod groups, the aschelminth phlya and the molluscs have poor powers of regeneration and do not usually reproduce asexually by fission. Regeneration in the arthropods is usually restricted to limb regeneration, which takes place when the animals moult.

Regeneration involves a number of processes which are similar to those taking place during normal development. These include:

- Proliferation of undifferentiated cells, as in a blastula, and the construction of a blastema.
- Pattern formation and the organization of cells in a spatial hierarchy.
- Differentiation and the expression of pattern.

Regeneration therefore provides a convenient model for the investigation of developmental events. In some animals regeneration also involves de-differentiation, which is not a feature of normal development.

In order for regeneration to occur, it is essential that the organism responds to the loss of components of the body, and the response must involve both the proliferation of a segment blastema and the development of an appropriate pattern in the cells proliferated by it.

15.6.2 The origin of the regeneration blastema

When regenerative growth occurs the new cells must be derived either from a reserve population of previously undifferentiated, totipotent cells or by de-differentiation from previously differentiation cells. Considerable controversy has arisen over which of these two alternatives is involved. The Cnidaria have particularly good powers of regeneration and they have a pool of interstitial cells, from which the different cells, such as the cnidoblasts, are normally derived and constantly replaced (Fig. 15.27).

In *Hydra*, the interstitial cells (or I cells) form a pool of mobile reserve cells which normally congregate in the ectoderm prior to asexual reproduction by budding. A wound provokes a similar response and constitutes a competing locus of attraction

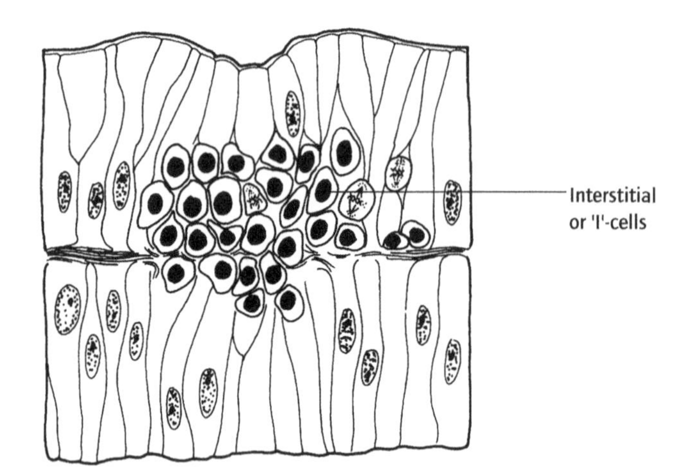

Fig. 15.27 Cross-section of a cnidarian showing interstitial cells and the proliferation prior to regeneration.

for the I cells, which form the regeneration blastema. In *Hydra*, though not in all cnidarians, I cells continue to proliferate throughout life and there is therefore a constant supply of cells for regeneration. The idea that I cells are a self-maintaining population of reserve cells essential for regeneration is, however, an oversimplification. The I cells can be destroyed by chemical means without the loss of regenerative ability, and fragments of *Hydra* which do not normally contain I cells are capable of regeneration. Moreover, in suitable media; differentiated cells in explants of *Hydra* tissues can differentiate into interstitial cells, multiply, and they are then capable of redifferentiation to cells of a different type. The normal route to differentiation may be via the population of reserve cells, but this is not the only route.

Undifferentiated reserve cells, called 'neoblasts', are also implicated in the phenomenal regenerative prowess of the free-living flatworms. These animals have been utilized as favourable material for the study of regeneration for more than a hundred years. Transverse section of the planarian leads to reconstruction of two complete flatworms (Fig. 15.28a) and, similarly, small fragments, including sagitally sectioned ones, will reconstitute a complete worm (Fig. 15.28b). The first stage of regeneration is the formation of a wound blastema and its subsequent invasion by neoblasts. The role of these reserve cells has been demonstrated by irradiation and transplantation experiments. Irradiation with X-rays at 3000 rad can prevent neoblast proliferation, but this does not kill the organism. Such an irradiated animal will fail to regenerate; if, however, a fragment of a flatworm which has not been irradiated is implanted host regeneration can occur. Moreover, if the cells of the implanted tissue can be identified, for instance if they have a different colour, the regenerated fragment has the characteristic colour of the implanted tissue (Fig. 15.29).

Totipotent reserve cells are not universally involved in regeneration, and indeed a prominent role for such cells, such as occurs in the planarians, may be rather unusual.

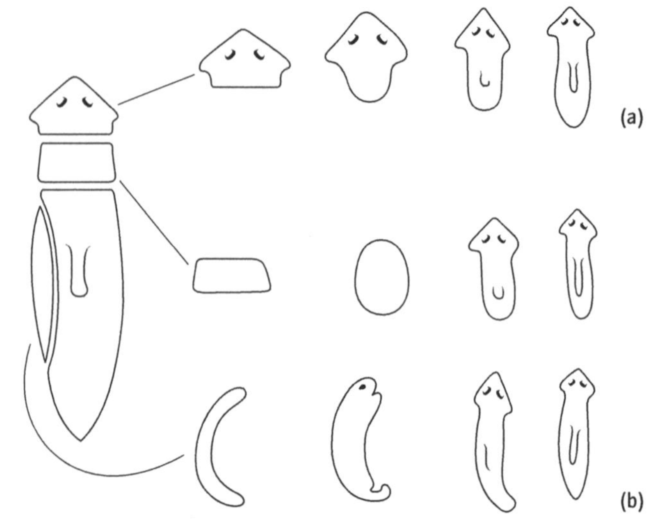

Fig. 15.28 Regeneration in flatworms. (a) Simple transection and subsequent regeneration of the missing posterior or anterior parts by the anterior and posterior fragments. (b) Even small fragments of fragments can reconstitute an entire flatworm, with the original tissue retaining its regional identity.

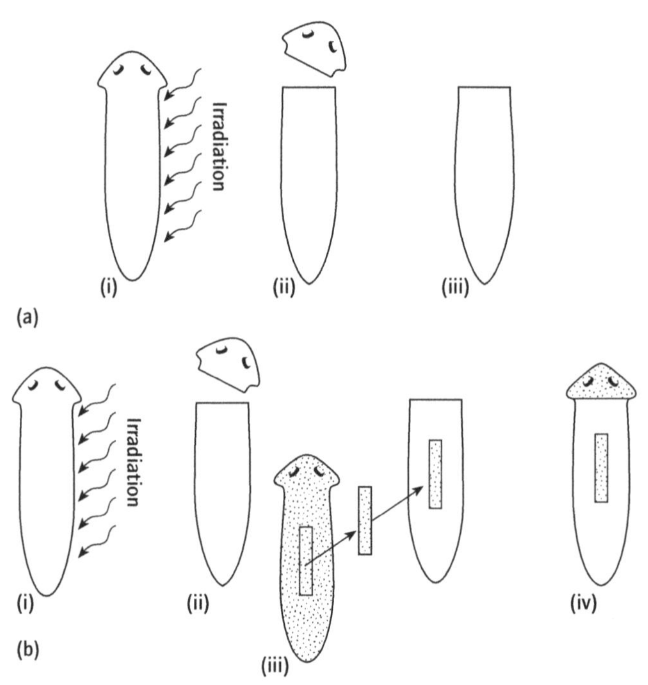

Fig. 15.29 Evidence for the special role of 'neoblast' cells in planarian regeneration. (a) (i) A specimen is irradiated with high-energy X-rays; (ii) the anterior is removed; (iii) the posterior fragment fails to regenerate. (b) (i), (ii), as in (a); (iii) a small fragment from a genetically distinct specimen is implanted into the mid-region; (iv) the head can now regenerate; it has the genetic character of the implanted fragment.

In annelids the formation of the regeneration blastema does not involve a population of distinct totipotent reserve cells, but rather de-differentiation and redeployment of differentiated cells derived from ectodermal and mesodermal layers. Loss of caudal segments, for example, is followed by wound healing which involves the migration of coelomocytes to the damaged surface and reconstitution of the growth zones which appear as bands of characteristic cells with spherical nuclei and a prominent nucleolus. In nereid polychaetes differentiation of the segment anlagen requires the presence of a cerebral growth hormone, and formation of the segmental ectodermal components (chaetal sacs and parapodia) occurs close to the ventral nerve. There is likely to be an inductive influence on segment formation and a mechanism for gene regulation in the vicinity of the ventral nerve cord.

15.6.3 Regeneration and regional organization

Many soft-bodied invertebrates are able to reconstitute a complete regionally organized structure from a very small fragment of the body (see Fig. 15.28 and Fig. 14.3a,b) which shows the asexual reproduction of a polychaete from spontaneously autotomized fragments.

What all these examples have in common is the reconstitution of a complete pattern from part of a pattern. In each case the fragment retains its original polarity and compensatory growth re-establishes the fragment in its original position in the body plan. In Fig. 15.26, a head fragment of a planarian replaces the missing caudal region and a caudal fragment will replace the missing head. Each area of the body has a regional identity within the whole, and this regional identity is retained during regeneration. In segmented animals the regional identity is more precisely defined and each segment has an identity in a linear hierarchy. In insect embryos there is a fixed number of segments and on each segment specific structures will develop. The genetic control of regional identity is now beginning to be unravelled in great detail in the insect *Drosophila melanogaster*, and is discussed in detail in Section 15.4 above. There are genes which act in a regional way and specify regional identities with a resolution below that of the segment, so that distinct subsegmental boundaries are delineated in the developing insect larva.

The arthropods are not able to replace lost segments but many of the annelids, which are also segmented animals, are able to do so. In many (perhaps all) annelids each segment is unique and also forms part of a single, integrated whole. One way in which this is expressed is the number of segments which, in most annelids, is fixed. In polychaetes of the family Nereidae, caudal regeneration requires the presence of a cerebral hormone but this seems to be a secondary adaptation related to the strictly selemparous reproduction of these animals (see Sections 14.3 and 14.4). Nevertheless the rate of segment proliferation is still subject to positional control (Box 15.12). The ancient and deep seated role of Hox genes in the organization of regional identity in bilateria opens the way to a re-interpretation of this older information though a study of Hox-gene expression in the annelids and other groups. We can expect great steps forward in our understanding of the evolution of body forms through study of developmental processes.

15.6.4 Regeneration, growth and reproduction

Regeneration presents a demand on the resources of an organism and there is likely to be associated with it an enhanced resource allocation to somatic functions and a consequent reduction for processes of sexual reproduction. There is then a potential conflict between regeneration and sexual reproduction, and regulatory mechanisms which effectively control this antagonism between the two processes are to be expected. In particular, semelparous organisms engaged in the build up of reproductive tissues should not divert resources to regenerative growth unless there is some compensatory increase in survivorship, fecundity or offspring survivorship.

A mechanism of this sort is to be found in the semelparous nereid worms which are, as shown above, capable of compensatory regenerative growth following the loss of posterior segments. As these worms approach maturity, resources are committed irrevocably to reproduction and they will not survive the single breeding episode. In these circumstances, segments regenerated during the final stages of reproduction would have little value. An endocrine mechanism ensures that regeneration does not occur during this stage of the life cycle. During sexual maturation secretion of the cerebral hormone is gradually reduced, and reduction in the level of circulating hormone allows the final stages of gametocyte maturation (see Section 14.4) to be completed. At the same time, the reduction in hormone level initiates somatic changes associated with reproduction. The same hormone is essential for caudal regeneration, and sexually mature nereid worms are not able to replace lost caudal segments. Polychaete worms which breed several years in succession do not have this endocrine mechanism. Regenerative growth proceeds even in mature specimens, and indeed the reproductive value of regenerated segments remains high as they can contribute to reproductive output in subsequent years.

15.7 Conclusion: invertebrate development and the genetic programme

The experimental study of invertebrate development can be traced back to the late nineteenth century. It has therefore developed in parallel with the study of genetics, the discovery of the molecular basis of cell heredity and the rapidly developing field of molecular biology. An outstanding challenge for the future is the unification of these disciplines, and the invertebrates provide a wealth of convenient model systems. Considerable progress has been made, and much of the material presented in this chapter can now be reappraised from a molecular biological point of view.

The messenger molecules of the animal cell are mRNA sequences decoded from DNA sequences in the nucleus. In the early sections of this chapter, we learned that DNA sequences of the zygote nucleus are, in almost all animals, passed entire to the daughter cells during early cleavage. These mRNA molecules can be investigated by the techniques of modern molecular biology. During oogenesis (see Chapter 4), maternal mRNA

Box 15.12 Positional information and caudal regeneration in annelids

1 *Segment number.* Annelids are composed of the following body regions (see also Fig. 4.51):

(a) The prostomium.
(b) A specific number of segments.
(c) A postsegmental pygidium.

In some polychaetes, and in leeches, the number of segments is quite small, e.g.
Clymenella torquata – 22 segments.
Ophryotrocha puerilis – 25 segments.
In others, the number is much higher but nevertheless the segment number may be a species-specific character.

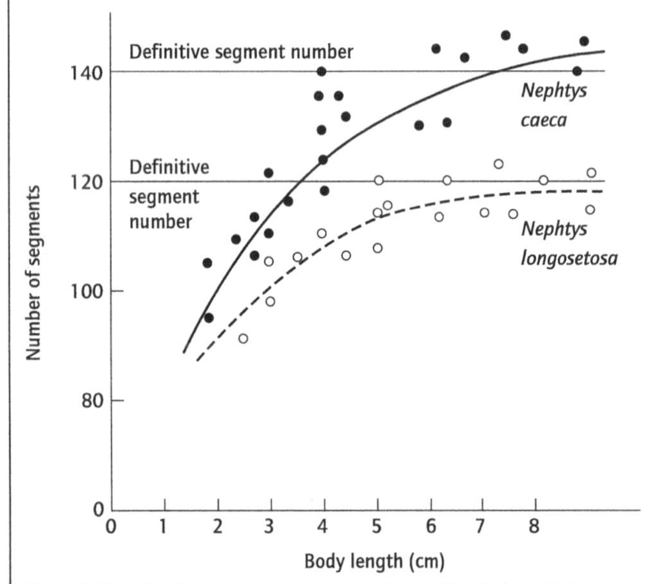

The relationship between segment number and body length in two closely related species of *Nephtys*.

In young animals the rate of segment proliferation is high, but in older animals growth is increasingly due to segment enlargement. The high rate of segment production characteristic of young worms is restored by amputation of tail segments. After segment amputation and wound healing a new segment proliferation zone is established. In *Nereis* the rate of segment proliferation is then directly proportional to the number of segments removed. (After Golding, 1967.)

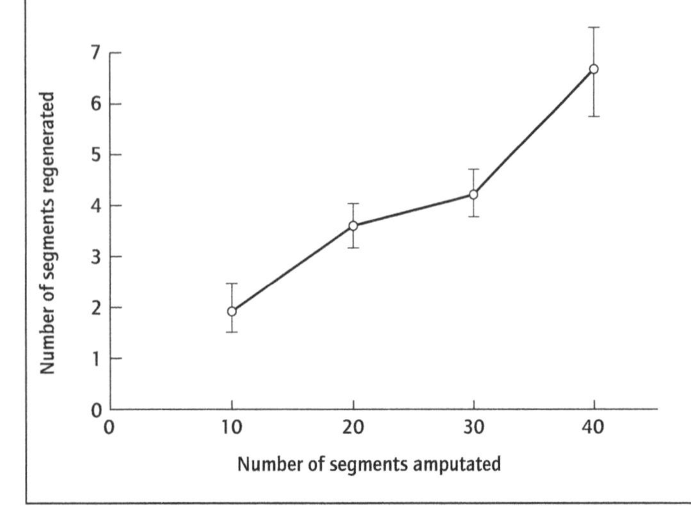

2 *Segment identity.* Each segment of the annelid body behaves as if it were part of the integrated whole. In all annelids each segment has its own particular structure and identity. This is particularly obvious in the tubiculous worm *Chaetopterus variopedatus* illustrated in Fig. 9.4a. A single segment of this apparently complex worm is able to regenerate anterior and posterior segments to re-form an entire worm!

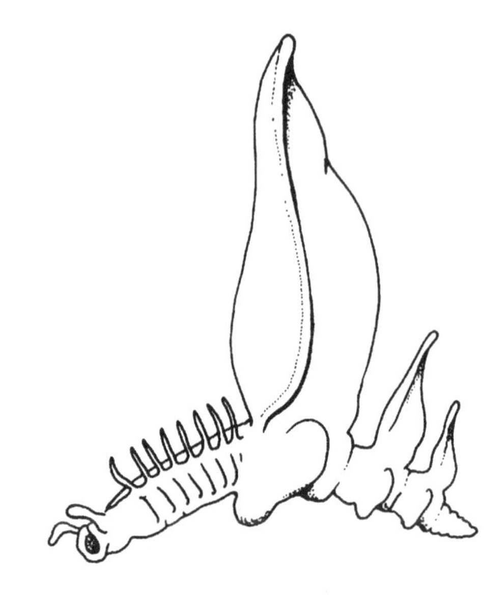

A stage in the regeneration of a complete worm from an isolated fan segment of *Chaetopterus variopedatus*.

The specific regional identity of each segment is also evident during regeneration of fragments of the worm *Clymenella torquata* which has, as an adult, exactly 22 segments.

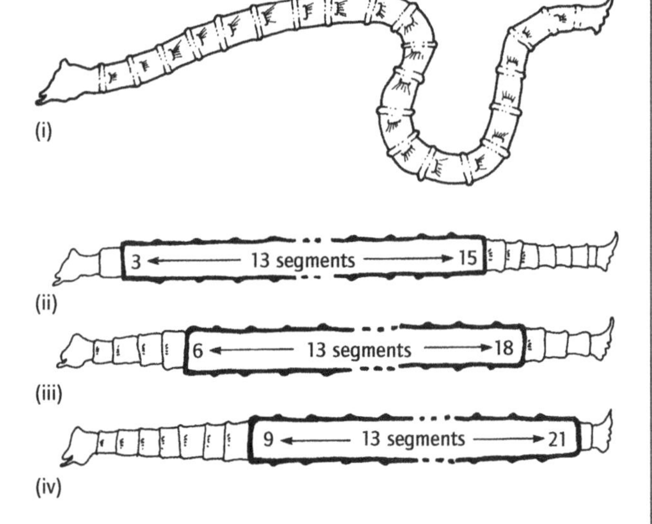

(i) The intact worm. Compensatory regeneration of fragments of *Clymenella* each 13 segments long but taken from different regions of the body.
(ii) Segments 3–15.
(iii) Segments 6–18.
(iv) Segments 9–21.

Note that in each case the segments are re-established in their original position in the hierarchy.

Continued

Box 15.12 *(cont'd)*

3 *Morphallaxis*. In some polychaetes loss of cephalic segments causes a morphological rearrangement of the remaining segments – a process sometimes referred to as *morphallaxis*. It is as if the segments have redefined their position in an organized hierarchy.

This phenomenon, observed in the fan-worms, e.g. *Sabella* is illustrated below. Fan-worms are likely to lose the crown of feeding tentacles because of predation by fish, and these tentacles can be replaced.

Sabella has a prostomium with a complex crown of tentacles, a peristomium, a fixed number of thoracic segments with a distinctive arrangement of the parapodia, and a large number of similar abdominal segments (i). No more than three anterior segments are regenerated. If more than three are lost, posterior segments lose their chaetae and are converted into thoracic segments (ii–iv).

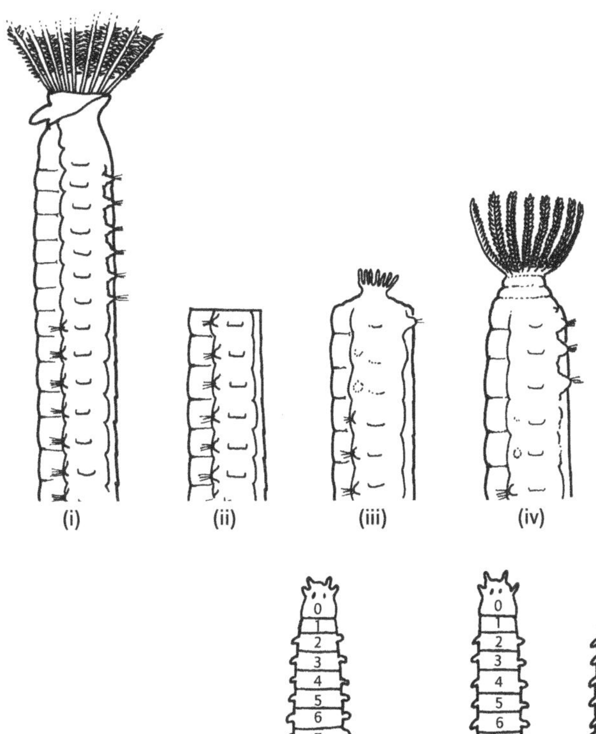

(i) The intact *Sabella*.

(ii) Loss of prostomiun, peristomium and all thoracic segments.

(iii) Early stages of regeneration and morphallaxis.

(iv) Regeneration showing crown of tentacles, peristomium and one thoracic segment being formed. The appropriate number of abdominal segments are converted into thoracic segments. In this way the feeding tentacles are replaced most quickly but it is also clear that each segment has its characteristic structure because of its position in an anterior/posterior gradient.

4 *A model of regeneration*. Many observations are compatible with the following simple model:

(A) The prostomium has positional value 0.

(B) The pygidium has a positional value equal to 1+ the species-specific segment number.

(C) The segment blastema exists at the anterior face of the pygidium.

(D) The rate of segment proliferation is a function of the difference between the value of the last segment and that of the pygidium and is zero when that value is unity.

(E) Segment proliferation continues until the difference between the positional value of the oldest segment and the pygidium is unity.

(F) Loss of caudal segments results in the reformation of the pygidium with its specifically high positional value. This model is illustrated for the polychaete *Ophryotrocha*. It is applicable to normal embryonic growth (a) and to regenerative growth (b).

The experiment illustrated in (c) suggests that the nerve cord carries the positional information. Deflection of the nerve cord (in, say, segment 9) causes the formation of an additional pygidium. This supernumerary tail will now proliferate segments according to the above rules.

Similar two-tailed worms are occasionally encountered in the wild and can also be formed by grafting fragments of two worms together. In every case the rate of proliferation by each pygidium follows the rules for normal growth and is proportional to the difference between the positional value of the last segment and that of the blastema.

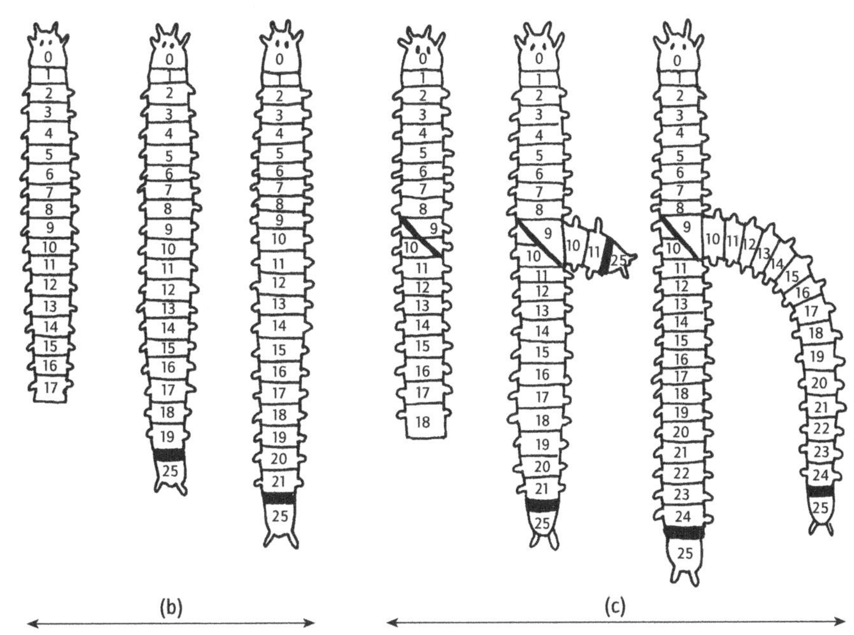

(a) Normal growth of *Ophryotrocha*. The prostomium has value 0. The pygidium has value 25. Growth establishes a sequence of segments by intercalation. (b) Regenerative growth in *Ophryotrocha*. After caudal ablation a number discontinuity exists between the last segment and the reconstructed pygidium. (c) Induction of a supernumerary pygidium by surgical interference with the ventral nerve cord. (After Pfannenstiel, 1984.)

molecules are stored in the oocyte cytoplasm. These masked mRNA molecules are freed from inhibition at fertilization and they provide the material for early protein synthesis. Even more importantly this leads to the construction of regulatory proteins which in turn control further gene transcription processes.

Ultimately, the genome of the zygotic nucleus becomes the source of genetic information. However, the crucial role of the oocyte cytoplasm remains at the centre of differentiation and the creation of order. We have seen in a wide range of invertebrates that the function of a cell nucleus and the messages that are ultimately transcribed or translated are determined by its history in the embryo.

Sometimes specific cytoplasmic substances seem to call into play specific patterns of enzyme production (see the development of the tunicate *Styela*, for instance). In other examples, contacts between cells seem to call forth functional responses. Evidence for these interactions was presented, for example, in experiments with mollusc and sea-urchin embryos. Some of the most exciting developments are taking place through studies of insect embryos, especially that of the fruit fly *Drosophila melanogaster*, where a wealth of genetic information can be combined with a convenient experimental model. It is certain that invertebrate embryos will continue to supply some of the best material with which to unravel the complexities of animal development.

15.8 Further reading

Berril, N.J. 1971. *Developmental Biology*, McGraw-Hill, New York.

Brookbank, J.W. 1978. *Developmental Biology: Embryos, Plants and Regeneration*. Harper & Row, New York.

Browder, L.W. 1984. *Developmental Biology*, 2nd edn. Saunders, New York.

Carroll, S., Grenier, J. & Weatherbee, S. 2001. *From DNA to Diversity*. Blackwell Science, Oxford.

Davidson, E.H. 1968. *Gene Action in Early Development*. Academic Press, New York.

Epel, D. 1977. The program of fertilisation. *Sci. Am.*, **237**, 129–138.

Gerhart, J. & Kirschner, M. 1997. *Cells, Embryos, and Evolution*. Blackwell Science, Oxford.

Gehring, W.J. 1985. The molecular basis of development. *Sci. Am.*, **253**, 137–146.

Gilbert F.S. 1990. *Developmental Biology*, 3rd edn. Sinauer Associates, Massachusetts.

Gilbert F.S. 1997. *Developmental Biology*, 5th edn. Sinauer Associates, Massachusetts.

Ingham, P.W. 1988. The molecular genetics of embryonic pattern formation in *Drosophila*. *Nature*, **335**, 25–34.

Nishida, H. & Sawada, K. 2001. *Macho-1* encodes a localised mRNA in ascidian eggs that specifies muscle cell fate during embryogenesis. *Nature* **409**, 724–729.

Nusslein-Volhard, C. 1991. Determination of the embryonic axes of *Drosophila*. *Development, Supplement*, **1**, 1–10.

Oppenheimer, S.B. 1980. *Introduction to Embryonic Development*. Allyn & Bacon, New York.

Reverberi, G. 1971. *Experimental Embryology of Marine and Freshwater Invertebrates*. Amsterdam.

Rosa de, R., Grenier, J.K., Andreva, T., Cook, C.E., Adoutte, A., Akami, M., Carrol, S.B. & Balavoine, G. 1999. Hox genes in brachiopods and priapulids and protostome evolution. *Nature*, **399**, 772–776.

Slack, J.M.W. 1983. *From Egg to Embryo*. Cambridge University Press, Cambridge.

Stearns, L.W. 1974. *Sea Urchin Development: Cellular and Molecular Aspects*. Dowden, Hutchinson & Ross, Pennsylvania.

Whittaker, J.R. 1987. Cell lineages and determination of cell fate in development. *Am. Zool.*, **27**, 607–622.

CHAPTER 16

Control Systems

Most of the earlier chapters in this section of the book have concentrated each on a single functional system – feeding, locomotion, respiration and the like – yet it is fundamental to our central thesis that selection acts not on individual attributes in isolation but on whole organisms. All the genes carried by individual animals succeed or fail together. In this chapter, we will consider systems that are important in the control of these diverse functions and that contribute greatly to the integrity of the life of the organism.

Sensory systems are adapted for information gathering *by means of which animals monitor changes in their internal and external environments. The nervous system provides the means of* communication *within the body and is responsible for the* integration *of sensory data and the recognition of features of significance. It also exercises the higher function of* control *– spontaneously initiating activity, generating patterns of behaviour that are appropriate for the particular species and modifying responses to stimulation in the light of previous experience. Finally, neural and endocrine signals* regulate *the functions of the musculature and other effectors.*

Ethology – the scientific study of animal behaviour – provides a valuable unifying focus for neurobiological research, even when the latter is immediately concerned with cellular or molecular phenomena. Such a perspective guards against the impoverishment which results when 'the whole animal is treated essentially as if it were a neuromuscular preparation' (Pantin). On the other hand, observations which restrict themselves to the behaviour of the whole animal can be compared with the study of a 'black box', such as a calculator or computer. Much can, indeed, be learnt about the properties of the apparatus by observing it in operation and by comparing input and output, etc. However, neurobiology extends this study by 'opening the box' to study its internal structure and the properties of its various components, and the science of neuroethology investigates the mechanisms by which these components generate and control its functions.

Because many invertebrate systems and their components are particularly amenable to investigation, they provide valuable 'models' for fundamental research and their use has led to many of the most dramatic advances in neurobiology. Outstanding in this regard is the giant axon *or nerve fibre of the squid Loligo, but large neurones with cell bodies up to 1 mm in diameter are common in molluscs and it is almost routine for a neurobiologist to insert as many as four electrodes into a single cell for purposes of recording, current injection, etc.*

The range of complexity shown by nervous systems (e.g. those of Hydra and Octopus) is greater than that of any other category of systems with which we are acquainted. The exhaustive description of the structure of some of the simpler model systems, and of the functional relationships of their components, is rapidly becoming a reality, whereas such knowledge remains only a distant dream with respect to vertebrates. Examples include the entire nervous system of the nematode, Caenorhabditis elegans, and the peripheral cardiac and stomatogastric ganglia of crustaceans.

Invertebrate nervous systems are, typically, highly stereotyped in organization. Many individual neurones *can be identified from specimen to specimen, and their structure, physiology and roles investigated. Indeed, their homologues can, in some cases, be recognized across quite major taxonomic boundaries, e.g. in locusts, moths and flies. The availability of identified neurones plays a large part in the appeal of invertebrate preparations for studies on the neural basis of behaviour. Indeed, the success of this work is now causing a real problem on account of the multiplicity of cells described and the different ways in which they are classified.*

The fruit fly, Drosophila, is uniquely placed to contribute to the study of the genetics of neural function and behaviour. *Typically, the many mutants which have been produced have first been identified by their behavioural defects, but then provide material with which the morphological, biochemical and genetical bases of behaviour can be investigated. For example, the mutant shaker has been known for more than 40 years and the genetic codes for its defective K^+ channels have recently been deciphered. Caenorhabditis has also been extensively used for this purpose. It has six chromosomes (flies have four) and only 3000–5000 genes. Mutants can be maintained in nutritive media even when they are severely defective (e.g. paralysed) and reproduction is still possible because one of the sexes is a self-fertilizing hermaphrodite. According to a count as early as 1984, 228 mutations were known, affecting 14 genes, just among those rendering the animal insensitive to touch. In the future, invertebrate systems may even be of value as models for the study of* neurological diseases *– for example, when Drosophila is genetically engineered to produce the aberrant protein* *associated with Huntingdon's disease, rapid degeneration of photoreceptor cells is observed.*

Although model systems have greatly facilitated progress in numerous areas of research, they provide only a fragmentary understanding of the animal kingdom and may be misleading if generalizations based on their study are applied too widely. Studies on a greater diversity of organisms guard against these dangers and provide a truly comparative account with the possibility of providing insights into the evolutionary origins of neural mechanisms.

16.1 Potentials

16.1.1 Membranes

According to the fluid mosaic model, the plasma membrane of the cell consists largely of a bilayer of lipid molecules whose hydrophilic poles extend outwards. It is usually separated from its neighbour bounding an adjacent cell by an electron-lucent, intercellular space about 20-nm across. The bilayer forms a major barrier to the diffusion of ions, etc. Embedded within it are protein and glycoprotein molecules, many of which span it completely. They are thus well placed to act as 'channels' and 'pumps' by means of which ions can cross the membrane. Some components are anchored in position in specialized regions of the cell surface which are differentiated for reception of stimuli, nervous transmission, etc. but many float freely within the membrane.

The concentrations of ions within the cytoplasm and intercellular fluid, respectively, are affected by passive movements like diffusion, and the process of active transport. The membranes of nerve fibres have a 'sodium pump' responsible for a net outward transfer (efflux) of Na^+ ions and the uptake of K^+. This movement occurs against the electrochemical gradient (see below), and is dependent on metabolic energy. The pump consists of an enzyme known as a $Na^+ - K^+$-activated ATPase, so called because of its ability to catalyse the breakdown of ATP and thus tap its energy.

16.1.2 Potentials

Foremost among the range of techniques now involved in the study of nervous systems are those involving intracellular electrodes and oscilloscopes, by means of which electrical potential differences can be recorded. A recent development is the 'patch-clamp' technique, by means of which a minute fragment of cell membrane can be attached to the orifice of a microelectrode and the function of individual membrane channels studied.

The presence of potential differences across the plasma membrane, a general characteristic of living cells, has its basis in the differential distribution of ions, and the differential permeability of the membrane. Usually, Na^+ and K^+ ions are most important. In most inactive cells, the membrane is permeable mainly to K^+ ions, which are concentrated inside the cell. Their diffusion across the membrane will lead to a build up of positive charge outside the cell and of negative charge inside. This resting potential across the membrane will impede the further efflux of K^+ and the system will come into electrochemical equilibrium. This mechanism can be readily demonstrated by modifying the level of K^+ ions in the bathing medium and observing the effect. However, it has also proved possible to squeeze out the cytoplasm from the squid axon (Box 16.1) and replace it with experimental saline. The natural imbalance of K^+ ion concentrations can thus be reversed and the preparation dutifully performs, in accordance with the hypothesis, by reversing the polarity of the resting potential!

During nervous activity, it is usually Na^+, mainly present outside the cell, which is the permeable ion – the inside of the membrane now becomes positively charged at the point of equilibrium. In addition to the effects on membrane potential of influences impinging on the cell from without (see below), changes can occur spontaneously, due to the operation of special channels through which Ca^{2+} ions leak into the cell. Such simple changes in permeability are the basis of much of the signalling that takes place within the nervous system.

16.1.3 Transduction

Sensory cells and neurones receive a diversity of signals, involving different forms of energy, which must be converted into a 'common currency' so that they can interact. Electrical energy in the form of changes in potential differences across cell membranes constitutes this currency and the process of conversion is called 'transduction'.

Two main mechanisms of transduction are known at present and both involve receptor molecules embedded in the plasma membrane (Fig. 16.1). Mechanical stimuli impinging on receptor cells alter the permeability of the ion channels formed by such molecules and lead to changes in membrane potential. Some aspects of the sense of taste and the actions of 'fast' chemical transmitters (Section 16.2.3) have a similar basis.

In contrast, olfactory and some gustatory stimuli, light, 'slow' transmitters and many hormones activate receptor molecules whose effects on membrane potentials are mediated by G-proteins and intracellular second messengers (Fig. 16.1). The second stage in this process involves amplification of the signal, although the 'gain' may be modest (in the eye of the horseshoe crab, *Limulus*, about eight molecules of G-protein are activated per molecule of the receptor, rhodopsin, compared with hundreds in vertebrate rods). The identity of the second messenger in most invertebrate photoreceptor cells investigated (*Limulus*, squid, flies) is generally thought to be $InsP_3$ (inositol triphosphate), although details of its action remain controversial. The second messengers cAMP (cyclic adenosine monophosphate) and arachidonic acid are involved in the cellular mechanisms underlying learning in *Aplysia* (Box 16.2). Transduction processes mediated by G-proteins are remarkably homogeneous throughout the animal kingdom, as shown by the ability of activated rhodopsin from molluscs to set in motion the biochemical cascade involved in vertebrate vision.

Box 16.1 Giant fibre system of the squid, *Loligo*

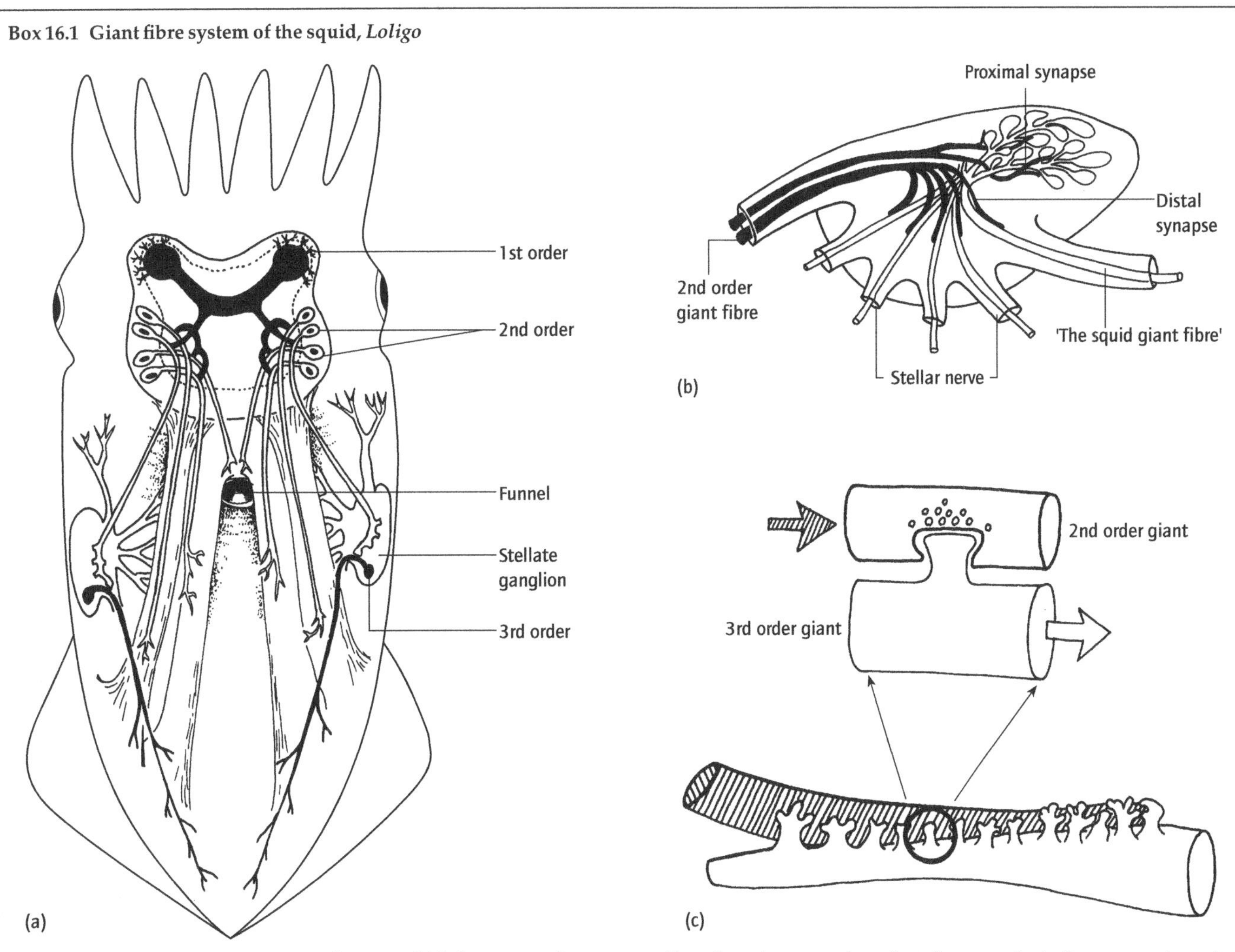

(a) Two first-order giant neurones send axons which fuse across the midline and innervate several second order cells situated in the posterior part of the brain. Most of the latter are motoneurones and control muscles of the head and funnel directly, but two have axons which extend to the stellate ganglion, where they synapse with the third-order giant fibres. (After Young, 1939.) (b) Stellate ganglion. Each second-order giant forms 'distal' synapses with the third-order fibres and another giant fibre forms 'proximal' junctions on these also. Each third-order fibre develops by fusion of the fibres of a number of cells and one is present in each stellar nerve (only five nerves shown). The most posterior, and largest fibre is the giant fibre exploited with such good effect by neuroscientists.
(c) Synaptic contacts formed by second-order and third-order giant fibres. Physiological coupling between the two elements is mediated by a multiplicity of junctions, at each of which the post-synaptic process is indented into the pre-synaptic. Arrows show direction of impulse flow.

16.1.4 Conduction of graded potentials

A major role of nervous elements is the ability to carry information from one point in the body to another – a function achieved by the conduction of changes in membrane potential. Since a nerve fibre has the properties of an electric cable, local changes in potential will spread along it. Such conduction is called 'passive' or 'electrotonic spread', and the potential changes involved are both graded and decremental. Their magnitude depends on the level of the initial potential and the effect will gradually fade as current leaks through the membrane – it is about halved during conduction down the axon of the photo-receptor cell in the fly eye. An analogy would be the effect of throwing a stone into a lake. The size of the ripples depends on the size of the stone and on the distance away from the source. Many nerve cells – 'non-spiking neurones' (see below) – seem to function in this way entirely and all neurones have regions that work like this.

16.1.5 Action potentials

In most cases, partial depolarization of an axon triggers a rapid change involving loss and reversal of potential difference (Fig. 16.2). This action potential or, popularly, 'spike', shows a threshold effect since it is initiated only when the stimulus reaches a

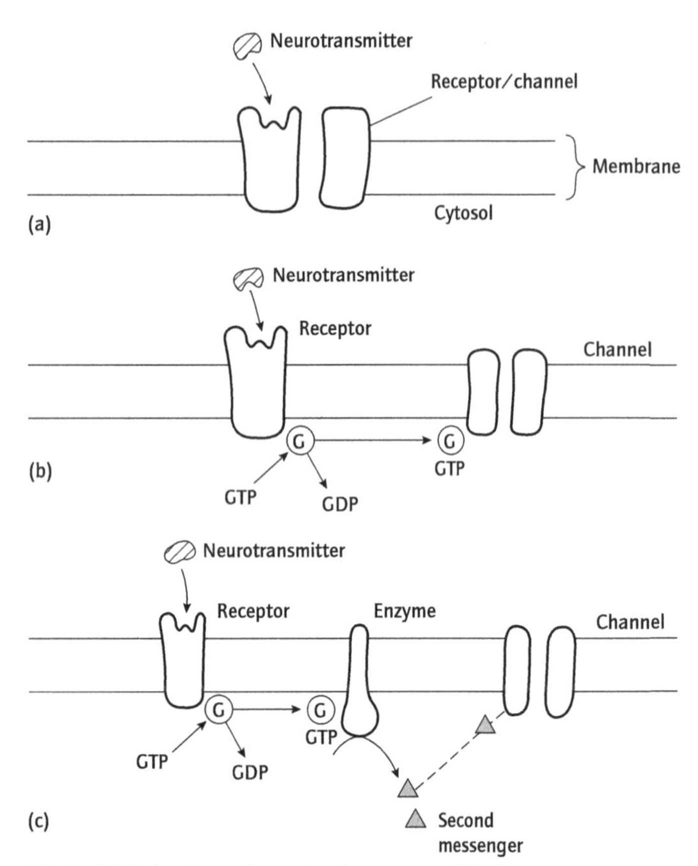

Fig. 16.1 Mechanisms of transduction as exemplified by neurotransmitter actions: (a) stimulus impinging on receptor has a direct effect on ion channels; (b) stimulus activates receptor which influences ion channels through the mediation of a G-protein; (c) action of the G-protein is, in turn, mediated by an intracellular 'second messenger'. (From Aidley, 1998.)

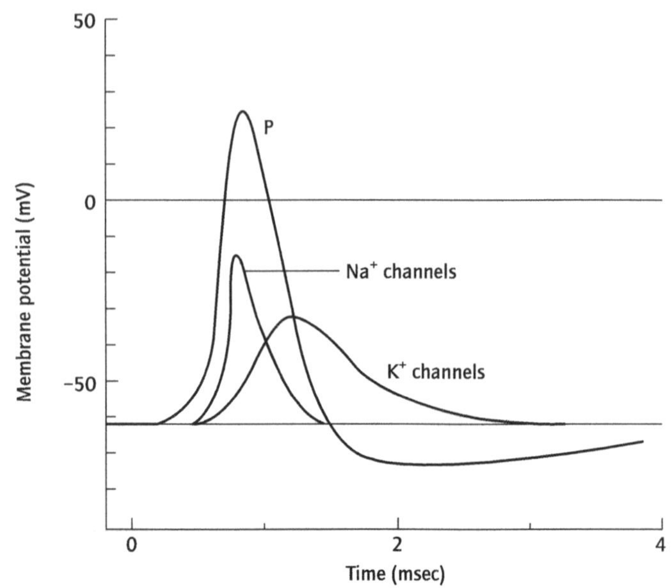

Fig. 16.2 Action potential in the squid axon. Graph shows changes in membrane potential occurring at any given point with time, and changes in the number of open Na^+ and K^+ channels. The channels are ion-selective and voltage gated. The Na^+ channels are presumed to change their form when depolarized, opening 'gates' through which Na^+ ions enter the axon. Na^+-channel opening is a transient phenomenon and inactivation rapidly follows. Voltage-gated K^+ channels open after the Na^+ and the diffusion of K^+ ions restores the resting potential (only one out of every 10^7 K^+ ions present is required per impulse). The discovery of the basis of the action potential led to the award of the Nobel Prize in 1963. (After Hodgkin and Huxley, 1952.)

certain minimum level. It is an all-or-nothing response, since further increases in intensity of stimulation fail to enhance it. In life, action potentials are triggered by spontaneous potentials, or those generated by transduction (see above), which spread passively from their sites of origin into an area of the membrane capable of responding actively. Depolarization then spreads to adjacent areas of membrane, stimulating them into action, and so on. Consequently, an action potential is actively propagated along the membrane. It is not decremental but self-perpetuating. An appropriate analogy is a trail of gunpowder ignited at one end with a match.

The velocity of conduction is dependent largely on the diameter of the fibre and consequently, many invertebrates have *giant fibres* to mediate escape reactions, etc. The marine annelid *Myxicola* holds the record for size with axons more than 1 mm in diameter, conducting at 20 m s^{-1}. In some cases, the speed of conduction is greatly increased by insulating lengths of the fibre with glial wrappings (Section 16.2.2). Such conduction is saltatory – it jumps from one region of uninsulated membrane to the next, being regenerated at each of these. Earthworm giant fibres have two such 'hot spots' on the dorsal surface in each segment; shrimp axons – the 'gold medallists' of the animal world for conduction velocity (200 m s^{-1} at 20°C) – have them at the branch points of the axons.

16.1.6 Spiking and non-spiking neurones

Non-spiking neurones are commonly associated with receptors, such as the photoreceptors of insect eyes, and interneurones controlling motoneurones, as in the hydromedusa *Polyorchis*. Non-spiking interneurones in insects typically lack the two main sets of branches, mainly associated with the receipt and the transmission of information, respectively, possessed by spiking cells. The only known non-spiking motoneurones are those innervating the muscles of the body wall in the nematode *Ascaris*.

We do not know 'why' some neurones use graded potentials, whereas others employ action potentials. Photoreceptor cells of both types are present in the eyes of the giant clam, *Tridacna*; all interneurons controlling flight in the locust are spiking, whereas many which influence limb movements are non-spiking. It cannot be explained simply on the basis of the length of the nerve fibres involved. Propagated potentials are doubtless indispensable for fibres extending to the extremities of limbs in large

Box 16.2 Cell biology of learning

The same basic types of learning can be observed throughout the animal kingdom, but only recently have their mechanisms been investigated, by Kandel and others, in terms of cell biology.

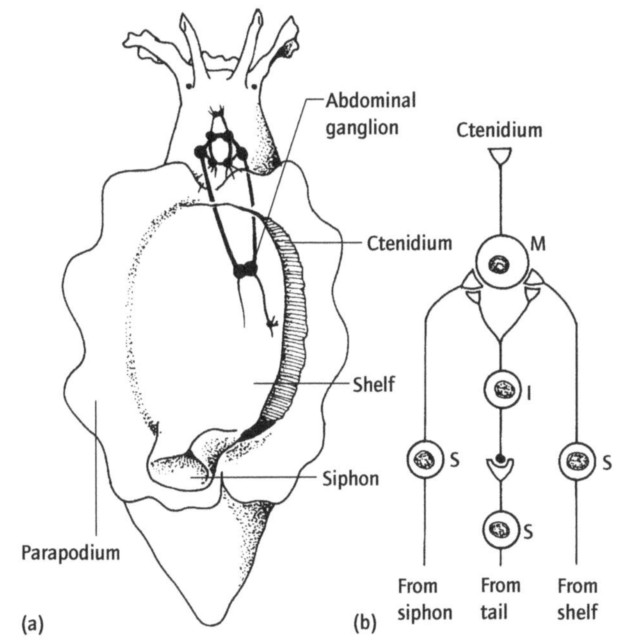

(a) View of nervous system and respiratory chamber of *Aplysia* from above. (b) Synaptic connections made by neurones within the abdominal ganglion. S, sensory neurones; I, interneurone; M, motoneurone. (After Kandel & Schwartz, 1982.)

The mantle cavity in the mollusc *Aplysia*, has a protective shelf and a siphon. Gentle touch activates a sensory pathway from each, and the neurones involved synapse directly on the motoneurones controlling a defensive contraction of the ctenidium (a monosynaptic reflex arc – see Section 16.8.3). Sensory neurones innervating the tail stimulate interneurones which synapse on the sensory terminals.

The pre-synaptic terminals of the sensory neurones are the site of changes responsible for behavioural modification – i.e. for learning. Frequent stimulation of the respiratory structures, causing the repeated invasion of the terminals by action potentials, leads to a gradual inactivation of the Ca^{2+} channels (Section 16.2.3), a reduction in the quantity of transmitter released per stimulus, and a diminished behavioural response – the simplest form of learning, called '*habituation*'.

If an intense, noxious stimulus is applied to the tail, the response to a wide variety of other stimuli, including gentle touching of the respiratory structures, is enhanced. This form of non-associative learning, known as *sensitization*, is mediated by the release of the transmitter 5-HT (5-hydroxytryptamine) and peptides by the interneurones, which initiates a cascade of metabolic effects within the sensory terminals.

5-HT → Synthesis of cAMP → Activates protein kinase (an enzyme) → Phosphorylates K^+ channels ↘

Enhanced transmitter release ← Greater Ca^{2+} influx ← Action potentials longer lasting ← K^+ channels inactivated

Conditioning is a form of *associative* learning (remember Pavlov's dogs?). Thus, if a series of mild stimuli applied to the gill shelf are, in each case, immediately followed by the noxious stimulation of the tail, the animal will start to respond far more vigorously to shelf stimulation alone, than to siphon stimulation for example. Such pairing of stimuli in the way described leads to the release of 5-HT from the interneurones, and to cAMP synthesis within *all* the sensory terminals, *at the same time* as the level of Ca^{2+} is elevated within the terminals of the stimulated, shelf sensory neurones *only*, by the arrival of action potentials. Ca^{2+} is thought to activate cAMP which thus has a potency in enhancing transmitter release beyond that seen during sensitization. It is noteworthy that the *dunce* mutant of *Drosophila* (which has a learning deficit) has an abnormal cAMP metabolism.

The cAMP-mediated mechanism is only one of those thought to underlie conditioning in *Aplysia*. The forms of learning described above are *short*-term – their memory endures only for minutes or at most hours. However, training programmes can be undertaken which result in the development of *long-term memory* (more then three weeks in *Aplysia*). This depends not on transient metabolic effects but on *structural changes*. The number and size of synaptic sites, and the number of synaptic vesicles, are all significantly increased by long-term sensitization and reduced by habituation.

invertebrates. However, graded signals are employed by the barnacle photoreceptor with fibres up to 1-cm long, whereas many neurones with far shorter fibres, including some amacrine cells (Section 16.2.1) produce action potentials, as do tiny oocytes, gland cells, etc.

Many cells using graded potentials engage in almost continuous discharge of transmitter, and small changes in membrane potential (as little as 0.3 mV with respect to the photoreceptor cells of the fly) can modify this in either direction. Even when present in the small numbers typical of invertebrate systems, they can provide smoothly graded control, appropriate, for example, for controlling the variable postural activities of limbs, whereas larger numbers of spiking cells with summed outputs are thought to be required to produce the same effect. With action potentials, the message is apparently 'scarcely more complex than a succession of dots in the morse code' (Adrian, 1932). However, the combination of frequency, pattern and duration of spike activity probably provides a highly sophisticated system of coding. Intermediates between the two extremes are common. Cells may spike under acute stimulation, but function as non-spiking neurones at other times. They may have a small region of the membrane which generates a spike, but which is then conducted to the further terminals in a decremental fashion; voltage-gated Ca^{2+} channels along the axon may sustain a potential, which nevertheless remains graded; similar channels in the terminal membrane may amplify a graded signal on arrival.

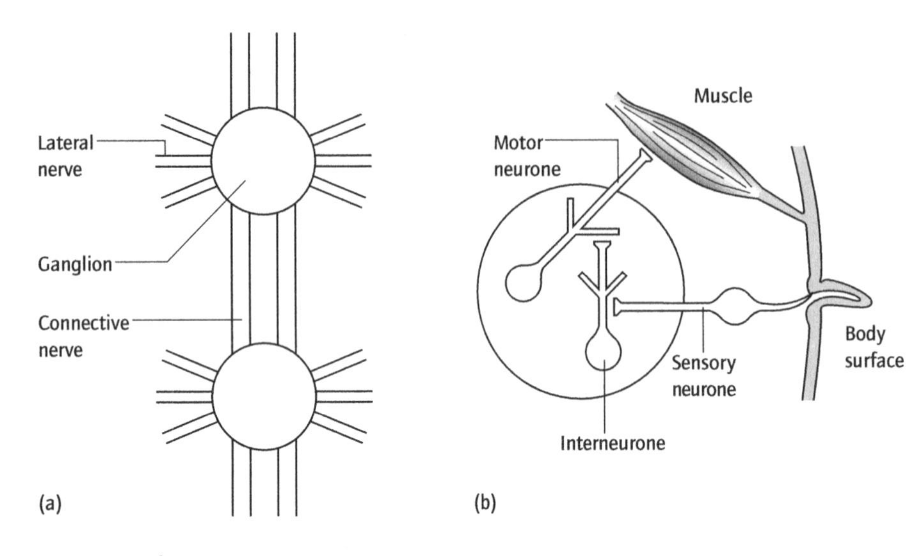

Fig. 16.3 Basic organization of the nervous system, as in arthropods: (a) two segmental ganglia, with lateral and connective nerves; (b) the three different types of neurones. (From Simmons & Young, 1999.)

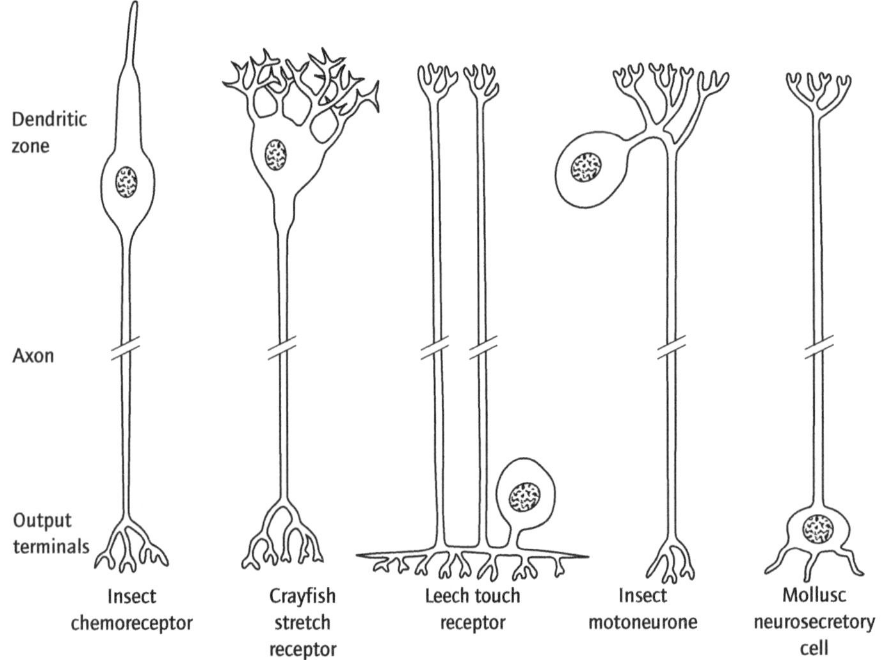

Fig. 16.4 Dynamic polarity is shown by neurones of diverse morphologies and functions; the position of the cell body is irrelevant to the pattern of functional organization. Some neurones are monopolar, with a single process arising from the cell body; others are bi- or multipolar.

16.2 Neurones and their connections

16.2.1 Neurones

The structure of neurones (i.e. nerve cells) is studied by light and electron microscopy and these are frequently combined with techniques that enable coloured dyes (e.g. Procion yellow) or dense materials (e.g. cobalt ions – see Box 16.6) to be injected into the cytoplasm, revealing the distribution of neuronal processes and the contacts made with other cells.

Neurones are characterized by the possession of elongate processes and the ability to conduct electrical potentials. They are traditionally classified as sensory (afferent) neurones which convey information into the central nervous system; motor neurones (efferent) which carry messages from the centre to the effectors (muscles, glands, etc.); and interneurones which link the previous two types (Fig. 16.3) (local interneurones are often distinguished from relay cells with longer processes). We can add neurosecretory cells that release hormones into the blood stream. However, cells with the characteristics of both sensory and motoneurones, and other combinations, have also been described.

Nerve cells are often described in terms of a common pattern of functional morphology (Fig. 16.4). The input (or dendritic) zone is the sensitive receptor region of sensory neurones, and the synaptic region of other cells. Graded receptor or synaptic potentials arise here and spread across the membrane. Spiking neurones have an impulse generation zone, where graded potentials which reach the threshold, trigger the generation of action potentials. The axon conducts potentials to the nerve

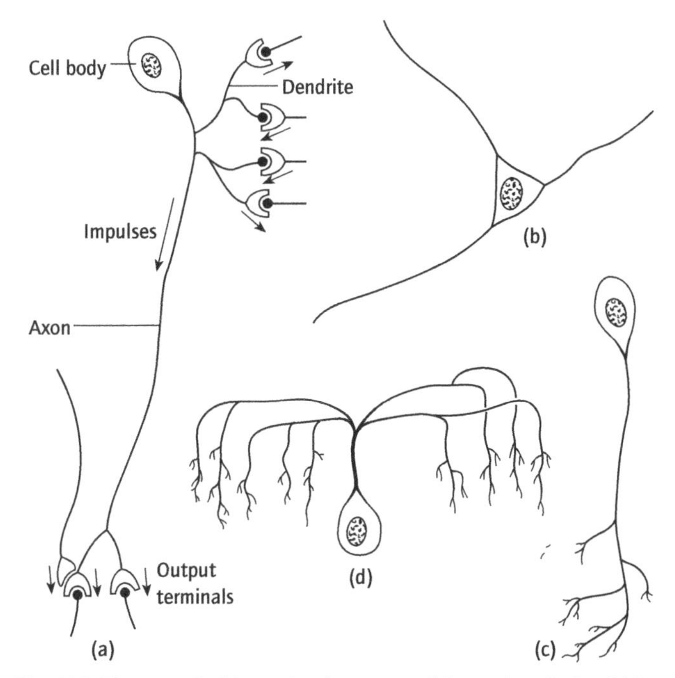

Fig. 16.5 Neurones lacking a simple pattern of dynamic polarity. (a) In motor neurones of the crustacean stomatogastric ganglion, and many others, the dendrites both receive and transmit information. Similarly, the output terminals may also be post-synaptic to other fibres. However, impulse traffic along the axon is, in nature, one way only. (b) Cnidarian isopolar neurone. (c) Neurone of polychaete corpora pedunculata. In neurones like these, dynamic polarity is apparently absent. (d) Amacrine cells, e.g. from the insect optic lobe, lack a major process identifiable as an axon.

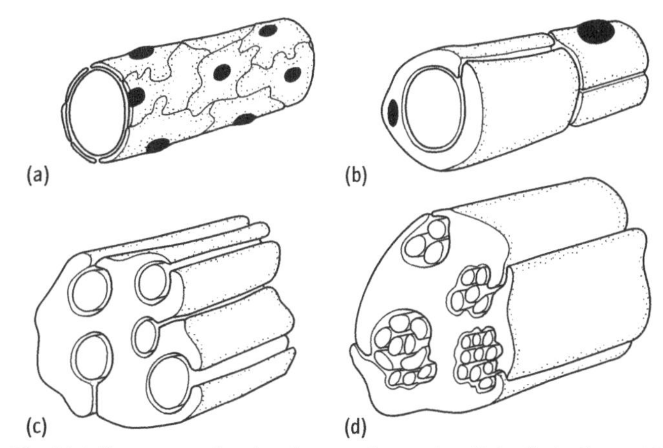

Fig. 16.6 Four types of ensheathment of axons by glial cells in the squid stellar nerve. (From Abbott *et al.*, 1995; after Villegas & Villegas, 1968.)

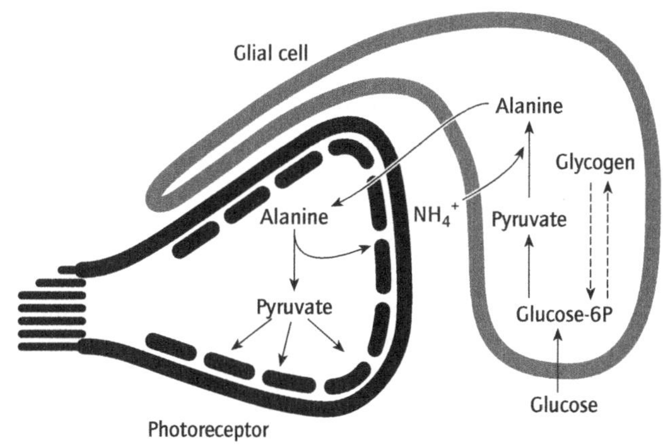

Fig. 16.7 Metabolic interactions between a photoreceptor and a neighbouring glial cell in the bee retina. (From Tsacopoulos *et al.*, 1994.)

terminals from which synaptic transmission occurs (i.e. communication with another cell).

This concept of the dynamic polarity of neurones requires modification in that many dendrites are sites of information transmission as well as reception, and pre-synaptic terminals may receive information as well as transmit it (Fig. 16.5). Furthermore, some nerve cells have axons which probably conduct impulses in both directions and others (amacrine cells) lack a process recognizable as an axon and have branches that are probably capable of independent activity.

16.2.2 Glial cells

The second category of cells present in the nervous system are called 'neuroglia' (Fig. 16.6). The cells have a mechanical role – supporting, separating and ensheathing nervous elements; an electrical role – insulating nerve fibres and enhancing the rate of nervous conduction; and a metabolic role – controlling the ionic environment within the nervous system and breaking down neurotransmitters following their release from neurones.

A remarkable example of neurone–glia cooperation is provided by the compound eye of the bee (Fig. 16.7). To sustain their activity when exposed to light, the photoreceptors (i.e. sensory neurones) metabolize pyruvate, which they produce from the amino acid alanine by a reaction which generates ammonium ions. These ions are discharged from the receptors, stimulate the uptake of glucose by adjacent glial cells and its conversion to alanine – the latter is then transferred to the receptors to support their metabolism.

Although glial cells cannot produce action potentials, they are, in exceptional cases, far more active participants in distinctively neural functions than previously realized. Those associated with the giant axons of the squid have receptors for pyruvate, the neurotransmitter (see below) discharged by the axons during electrical activity. Activation of these receptors causes the discharge of acetylcholine by the glial cells and the cells respond to their own transmitter by increasing their permeability to K^+, elevating their resting potential and thereby perhaps facilitating the uptake of the K^+ ions released into the intercellular space during neural activity.

As noted above (Section 16.1.5), the velocity of conduction of action potentials is increased by glial wrappings. Myelin wrappings, formed by glial cells, of the type found in vertebrates are only rarely present in invertebrates – as in oceanic copepods.

Their presence accelerates the initiation of escape responses from 6 to 2 ms and this may be a significant factor in their domination of open-ocean communities.

16.2.3 Chemical synapses

Most neurones transmit information by means of the secretion of chemical substances called 'neurotransmitters', but whereas conventional gland cells disseminate their secretory products far and wide, neurones usually administer them in a highly localized and selective manner at specialized junctions with other cells, called 'synapses'.

Chemical synapses are most frequently formed by the terminal regions of nerve fibres. The contacts may involve bulbous endings or swellings ('varicosities') formed along the length of the fibre. Many synapses have such a highly characteristic ultrastructure (Fig. 16.8) that the existence of a functional contact is usually presumed merely on the basis of observation with the electron microscope, although this correlation remains a subject for debate. It is also noteworthy that a functional 'synaptic contact' between two cells may in fact be mediated by thousands of junctions (25 000 in the lobster stomatogastric ganglion) visible with the electron microscope.

The fundamental mechanisms of synaptic function have been investigated using the synapses of the giant fibre system of the squid (Box 16.1c), since both pre- and post-synaptic terminals are large enough to be impaled with microelectrodes. Llinas and others have shown that the arrival of an action potential leads to the opening of voltage-gated Ca^{2+} channels in the pre-synaptic membrane, and it is the elevated level of Ca^{2+} within the terminals which is the main cause of transmitter release, and not depolarization as such.

Following release, transmitters diffuse across the synaptic cleft and bind to protein receptor molecules embedded in the post-synaptic membrane or may reach more distant targets. 'Fast transmitters' activate receptors (e.g. many acetylcholine receptors) that function as selective ion channels leading to a change in membrane potential. Such effects are typically extremely rapid, involving a synaptic delay as brief as 0.4 ms, and last for only a few ms. Alternatively, a receptor for 'slow' transmitters (e.g. octopamine, neuropeptides) may introduce a far longer delay and have longer lasting effects on ion channels through the mediation of a second messenger (Section 16.1.3).

A single cell will usually receive a multiplicity of synaptic inputs, involving different transmitters, inducing the generation of excitatory (depolarizing) and inhibitory (usually hyperpolarizing) post-synaptic potentials, respectively. Compounds which are used as neurotransmitters may also act as neuromodulators – so called because they have less clear-cut effects. For example, a modulator may, by itself, have no apparent effect on a cell, but may sensitize the cell to another stimulus (see Section 16.9.4). The actions of transmitters and modulators are rapidly terminated by enzymic degradation (acetylcholine, peptides) or re-uptake into the pre-synaptic terminals (amines, amino acids).

One advantage of chemical synapses is that they can amplify a signal. For example, transmitter (probably histamine) released from the photoreceptor neurones of the fly in response to depolarization induces hyperpolarization in the post-synaptic neurones: the change in post-synaptic potential is 7–14 times greater than that of the pre-synaptic membranes. Another attribute of chemical synapses that is crucial to the function of the nervous system is their great flexibility – indeed there is evidence that this is the basis for various types of learning (Box 16.2).

16.2.4 Neurotransmitters

In recent years, an explosion of information has resulted from the application of the techniques of biochemistry and molecular biology. Neurotransmitters, hormones, receptors, etc. can be chemically identified and their genes sequenced, and antibodies to the molecules involved can be produced.

Neurotransmitters fall into three major classes based on their pattern of synthesis, storage and discharge. Within the first group is acetylcholine – which is also well known in vertebrates – and octopamine which (as a major transmitter) seems to be an invertebrate speciality. These compounds are produced largely in nerve fibre terminals, by fairly simple synthetic pathways, with the aid of the appropriate enzymes. They are stored mainly within synaptic vesicles which are usually 30–50 nm in diameter (Fig. 16.8). Transmitter release is generally thought to occur by fusion of the vesicle membrane at specialized sites marked by pre-synaptic thickenings or bars, with the subsequent discharge of the vesicle contents into the synaptic cleft – i.e. by exocytosis. Release apparently occurs at low levels even in resting terminals, but the arrival of an action potential provokes exocytosis of large numbers of vesicles simultaneously.

The second class of transmitters is composed of neuropeptides. The first neuropeptide to be identified in invertebrates was proctolin, a compound of five amino acid residues (Arg-Tyr-Leu-Pro-Thr), that was isolated from the cockroach by Staratt and Brown in 1975. Peptides are often present in individual species as peptide families (Table 16.1) whose members are closely related in chemical composition and biological activity, and extensions to the families come to light when other animals are investigated. Yet other members (analogues) can be synthesized artificially. Recent years have witnessed an explosion of knowledge of neuropeptides – between 1984 and 1989, the number identified in insects alone rose from four to over 40. The latter date saw the isolation of the first peptide in nematodes and, over the following ten years, nearly 50 were firmly identified, all members of the FMRFamide family (Table 16.1). Work in this area was greatly accelerated by the progress and completion of the *Caenorhabditis* genome sequencing project.

Peptides are synthesized in the neuronal cell bodies by the rough endoplasmic reticulum and packaged into secretory granules, usually 70–200 nm in diameter, by the Golgi apparatus. The initial product is a large precursor, or propeptide, which may contain many copies (up to 28 in the case of FMRFamide)

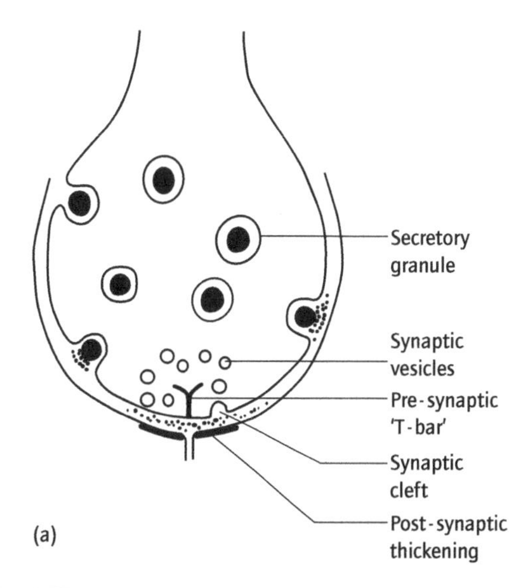

Fig. 16.8 Ultrastructure of chemical synapses. (a) Figure of arthropod synapse, showing pattern of secretory release. (After Golding, 1988.) Many synapses lack one or more of the features shown. (b) Electron micrograph of a synapse in the cerebral ganglion of the earthworm, *Lumbricus*. A diffuse pre-synaptic thickening is present (arrow). g, secretory granules; v, synaptic vesicles; bar, 100 nm. (From Golding & May, 1982.) (c) Diversity of secretory granule types in nerve fibre terminals in the neuropile of the earthworm, *Lumbricus*. Arrow, granule exocytosis; bar, 200 nm. (From Golding & Pow, 1988).

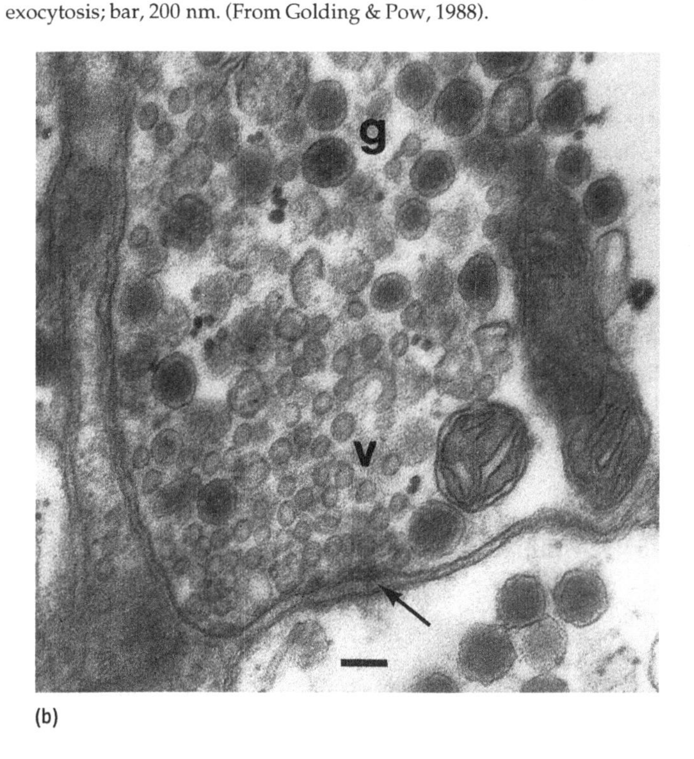

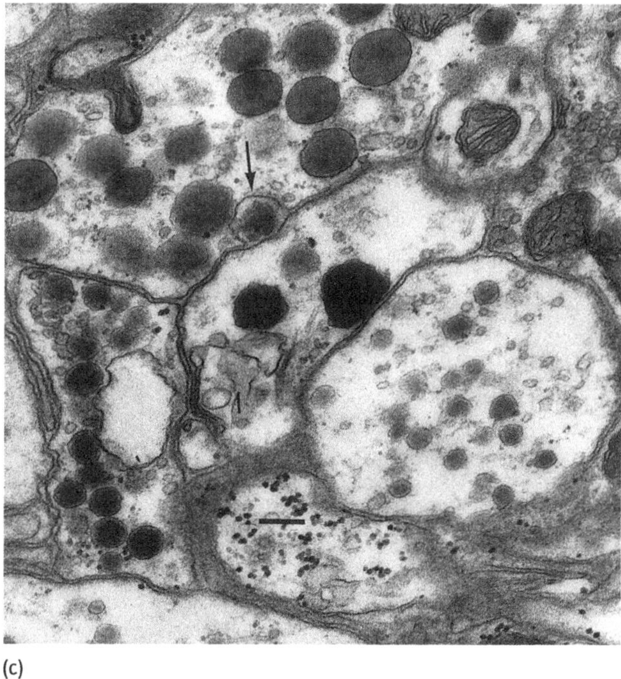

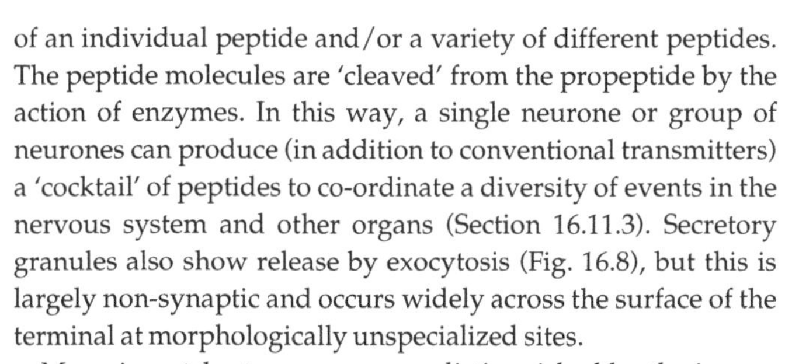

of an individual peptide and/or a variety of different peptides. The peptide molecules are 'cleaved' from the propeptide by the action of enzymes. In this way, a single neurone or group of neurones can produce (in addition to conventional transmitters) a 'cocktail' of peptides to co-ordinate a diversity of events in the nervous system and other organs (Section 16.11.3). Secretory granules also show release by exocytosis (Fig. 16.8), but this is largely non-synaptic and occurs widely across the surface of the terminal at morphologically unspecialized sites.

Many invertebrate neurones are distinguished by the immunological affinity of their peptide secretory products with vertebrate peptides (such compounds and their receptors are even present in protists and sponges). One of the most remarkable examples of phyletic conservation is provided by a peptide – the head activator – first identified in the nervous system of cnidarians. The same peptide is present in mammalian brains where, as in *Hydra*, it regulates mitosis in neurone precursor cells.

The third class of neurotransmitters is made up of gases, most notably nitric oxide (NO). They cannot be stored in vesicles or granules and are produced by enzymic action when the cells of origin are activated. Diffusing freely into adjacent cells, NO acts by stimulating the synthesis of cyclic GMP. The photoreceptors and (second-order) monopolar cells of the insect retina provide an interesting example of bidirectional synaptic transmission (Fig. 16.9). The photoreceptors signal to the monopolar cells by the release of the transmitter histamine; when stimulated, the monopolar cells produce NO which apparently acts as a retrograde transmitter, signalling back to the receptors.

Table 16.1 FMRFamide*, first identified in molluscs, and some of its family of peptides identified in the nematode, *Caenorhabditis elegans*. (After Brownlee & Fairweather, 1999.) The amino acid sequences are shown, together with the phonetic equivalents.

F M R F amide Phe-Met-Arg-Phe-NH_2
P N F L R F amide Pro-Asn-Phe-Leu-Arg-Phe-NH_2
K P S F V R F amide Lys-Pro-Ser-Phe-Val-Arg-Phe-NH_2
S P R E P I R F amide Ser-Pro-Arg-Glu-Pro-Ile-Arg-Phe-NH_2
S D P N F L R F amide Ser-Asp-Pro-Asn-Phe-Leu-Arg-Phe-NH_2
E A E E P L G T M R F amide Glu-Ala-Glu-Glu-Pro-Leu-Gly-Thr-Met-Arg-Phe-NH_2

* Commonly referred to as 'Fer-merf-er-mide'.

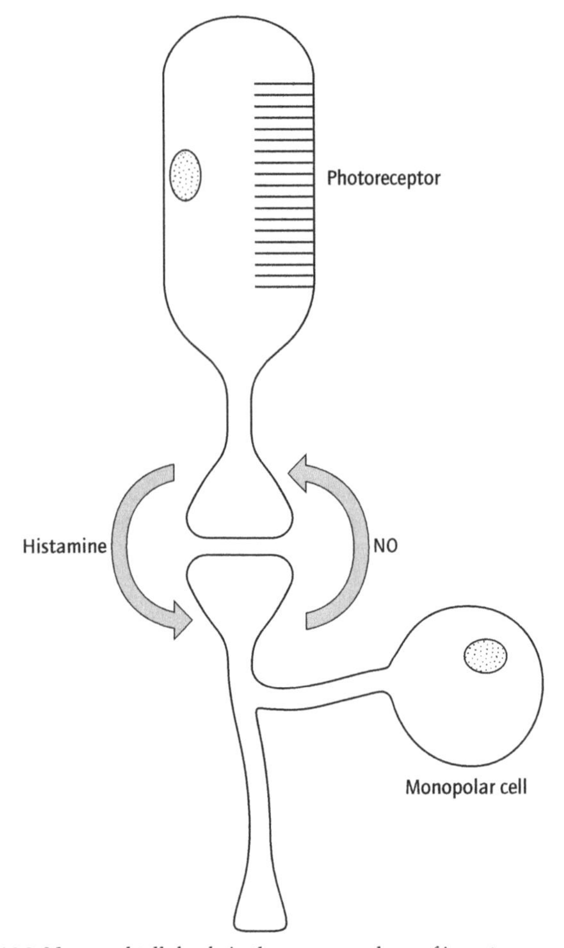

Fig. 16.9 Neuronal talk-back: in the compound eye of insects, photoreceptors provide synaptic input to monopolar cells by releasing the classical transmitter histamine. It is postulated that monopolar cells provide feedback to the receptors by releasing NO. (After Bicker, 1998.)

16.2.5 Electrical synapses

One of the great controversies in the history of biology concerned the mechanism of synaptic transmission. It is ironical that when the matter had apparently been finally resolved in favour of the chemical hypothesis, mainly as a result of work on peripheral junctions in vertebrates, electrical synapses were discovered in the crayfish (see Box 16.8) by Furshpan and Potter, in 1959. Electrical synapses enable one cell to stimulate another directly, without intervention of transmitters and their receptors. Transmission depends on the presence of gap junctions* at which protein molecules (called 'connexons'), forming hollow cylinders through which current can flow, span the intercellular space allowing depolarization in one partner to spread directly to the other. Transmission is rapid, involving a fraction of the synaptic delay encountered at chemical synapses. In many cases, transmission can proceed in either direction. In others, the synapses are rectifying – they are sensitive to the direction of the trans-synaptic potential, providing a pathway with low resistance in one direction, but shutting down when the potential is reversed.

In some cases, electrical synapses are the means by which a number of cells are integrated into a single functional unit. The effect may be to synchronize the output of a group of cells – examples include receptor cells within the units ('ommatidia') in the locust eye and neurosecretory cells which secrete spawning hormone in molluscs (also many gland and muscle cells). Neurones with giant fibres are usually linked with electrical synapses, giving the system a rapidity of function approaching that of a single, huge cell.

Embryonic neurones are extensively coupled by gap junctions, many of which are lost as ion channels and chemical synapses develop.

16.3 Organization of nervous systems

16.3.1 'Neuroid' systems

No true nervous system is formed in sponges. However, these organisms have contractile ability, both of the entire body and just the exhalent oscula. Some of them 'back-flush' to clear their canals and some eject their larvae by vigorous, apparently co-ordinated contractions. Many responses to stimulation are probably due only to the spread of mechanical effects – contraction of myocytes stretching adjacent cells, and so on. However, in the hexactinellid *Rhabdocalyptus*, Mackie and his colleagues have shown that electrical impulses are conducted throughout the sponge, in an all-or-nothing manner at 0.26 cm s^{-1}, by the syncytial trabecular tissue. The impulses are propagated directly to the choanocytes where they bring about shut-down of pumping.

* Gap junctions are not to be confused with 'tight junctions' where fusion of the outer leaflets of the membranes occurs. Tight junctions prevent diffusion through the intercellular space and are the structural basis of blood–brain barriers, by means of which, in insects and cephalopods (as in vertebrates), the chemical environment of the nervous system is regulated.

Epithelial conduction with a velocity of 3–35 cm s^{-1} is important in hydromedusae and in urochordates. In some cases, it enables the epithelium to be used like a single receptor organ with an enormous surface area, as in the hydromedusan *Sarsia*, where mechanical stimulation anywhere on the outer surface of the bell initiates an action potential which spreads, via gap junctions between the cells, across the whole epithelium. Typically, epithelial conduction provides input for the nervous system (e.g. via electrical synapses with epidermal sensory neurones in the urochordate *Oikopleura*). It is also employed in motor control to spread excitation across a sheet of muscle fibres or throughout a ciliated epithelium (e.g. of the urochordate pharyngeal basket).

16.3.2 Nerve nets

A nerve net is a diffuse, two-dimensional plexus of bi- or multipolar neurones (Fig. 16.10) and such systems constitute the characteristic grade of organization of cnidarian nervous systems. Their elements are located between the bases of the cells of the epithelium, above the mesoglea. They may be present in either or both of the two cell layers and, in the latter case, they may communicate across the mesoglea. Nerve nets show a conduction velocity of 10–100 cm s^{-1}.

A distinctive feature is the way activity spreads out in every direction, from any point of stimulation. Early investigators cut tortuous patterns from sheets of cnidarian body wall and demonstrated that activity could spread through them. Some nets are *through-conducting*, whereas in others, conduction is *decremental* – it fades out with increasing distance from the point of initiation (Section 16.8.3).

A nerve net can show regional differentiation. For example, the mesenteries of sea anemones have numbers of large, bipolar neurones, with a vertical orientation, which provide a rapid, through-conducting pathway. In *Hydra* subsets of neurones within each net are distinguished by their production of different peptides (Section 16.2.4) (a cell can express first one peptide and then another). Furthermore, two distinct nerve nets may be present in the same epithelium. Scyphozoan medusae (e.g. *Aurelia*) have a number of ganglia distributed around the margin of the bell. The ganglia receive sensory input from the 'diffuse nerve net', but provide motor output to the swimming muscles, via the more rapidly conducting 'giant fibre nerve net' – the first net to be described, by Schafer in 1879. The two nets cover much

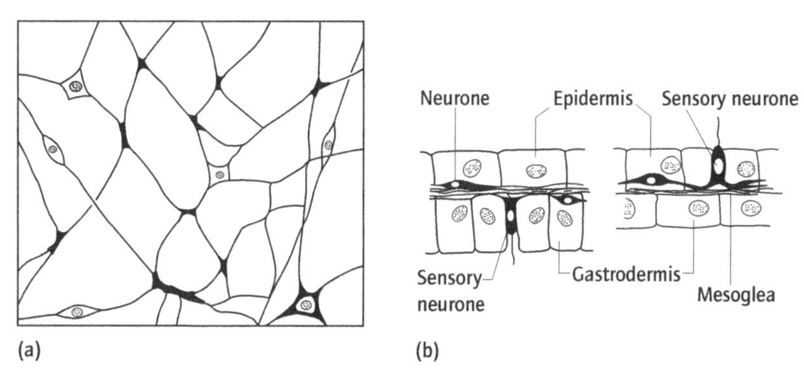

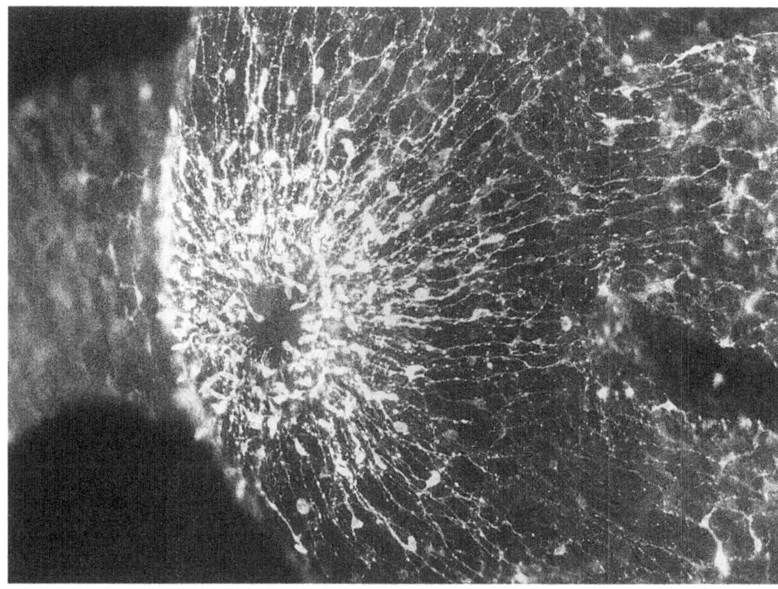

Fig. 16.10 Nerve nets in cnidarians. (a) Surface view of a cnidarian nerve net composed mainly of isopolar and bipolar neurones (details of synapses not shown). (b) Body wall of *Hydra* showing nerve nets in both epidermis and gastrodermis. Sensory neurones are restricted to the gastrodermis in the column (left), and to the epidermis in the tentacles (right). (After Bode *et al.*, 1989.) (c) Immunofluorescence staining for a neuropeptide fragment (arg-phe-amide – see Table 16.1), showing part of the nerve net of *Hydra*. Note the concentration round the mouth. (From Grimmelikhuijzen, 1985.)

the same area of the lower surface of the bell, but do not communicate directly.

Nerve nets in Anthozoa and Scyphozoa differ markedly from those in Hydrozoa. In the former, neurones within a particular net communicate by chemical, mainly symmetrical (i.e. two-way) synapses; electrical synapses and epithelial conduction are not present. In Hydrozoa, electrical coupling between the constituent cells of a net (and epithelial conduction) is widespread. Transmission between nets is by chemical, mainly polarized synapses in all three groups.

The fast chemical transmission encountered (e.g. across symmetrical synapses in Scyphozoa, with a synaptic delay of 1 ms) suggests that 'classical' transmitters are involved (Sections 16.2.3 and 4); furthermore, Westfall and her colleagues have recently provided evidence for the presence of the amines dopamine and 5-hydroxytryptamine. In contrast, well over a dozen neuropeptides (Fig. 16.10c) have been isolated from the sea anemone, *Anthopleura elegantissima*, alone.

The patterns of behaviour shown by cnidarians such as anemones include feeding, swimming, climbing onto the shells of commensal hermit crabs, intraspecific aggression, 'walking' on the tentacles, moving the pedal disc, and burrowing. Their complexity and integration may seem surprising in view of the low level of organization of the nervous system, but evidently neural adaptations in these and other cnidarians are remarkably elegant and effective.

In platyhelminths, a number of nerve nets are present, associated with different tissue layers, and have the same functional characteristics as those described above. The nets have connections at a deeper level with nerve cords. Nerve nets are probably important in other groups – e.g. in echinoderms, the foot in molluscs, and the innervation of the gut (even in vertebrates).

16.3.3 Central and peripheral nervous systems

An important distinction is that between the central nervous system, consisting of nerve cords and ganglia, and the peripheral nervous system consisting of nerves and receptors. The central nervous system acts as the destination for sensory input, the centre of synaptic integration, and the source of motor control, whereas peripheral nerves act mainly as conducting pathways (nerve nets combine both roles). Diminutive, peripheral ganglia are intermediate in that, in life, they are always connected to and influenced by the central nervous system, but are capable of a great measure of independent activity. Examples include the ganglia of the small, pincer-like organs of defence (pedicellariae) of some echinoderms, the cardiac ganglia of crustaceans and the parapodial ganglia of polychaete annelids.

It used to be thought that virtually all sensory neurones had their cell bodies in the periphery in invertebrates, but it is now clear from many groups that many lie in the central nervous system. A curious feature in nematodes, and some platyhelminths and echinoderms, is that instead of nerve fibres running from the motoneurones to the muscle blocks, the muscle cells have long, axon-like processes which extend to form postsynaptic terminals at the surface of the nerve cord.

In hydromedusae, two circular nerve cords are situated close to the margin of the bell. The inner one is medullary in character, with nerve cell bodies (mainly bipolar) evenly scattered throughout, whose fibres extend parallel to each other. Medullary cords in platyhelminths may be arranged according to the orthogonal pattern (Fig. 16.11a), regarded as being the basic plan for the protostome invertebrates. The anterior regions of the nerve cords may be merely thickened, but in higher platyhelminths and nemertines a well-differentiated cerebral ganglion (or 'brain') is formed.

In most annelids, pairs of ectodermal ganglionic rudiments in each segment become linked by connectives and typically form a single internal ventral nerve cord running the length of the body (Fig. 16.11b). Pairs of nerves extend to the periphery in each segment. Anteriorly, a supraoesophageal ganglion develops and many polychaetes have well developed sensory tentacles, eyes and olfactory organs providing input to different parts of the brain. Arthropods also have segmented nerve cords, but in many groups the tendency for segmental rudiments to become incorporated into compound ganglia is taken far further (Fig. 16.11c). A well-developed brain is commonly formed, with regions comprising the proto-, deutero- and tritocerebrum, respectively. Nervous systems in molluscs consist of a number of ganglia, connected by commissures and connectives, from which nerves extend to the periphery. Features of interest include the figure-of-eight arrangement of the connectives (Fig. 16.11d) – a consequence of the process of torsion (see Section 5.1.3.5) – in primitive gastropods. In many groups, ganglia tend to be concentrated to form a circumoesophageal ring.

Nematodes have a number of longitudinal, medullary nerve cords (two main and four smaller in *Ascaris*) linked by a nerve ring round the oesophagus. In echinoderms, as exemplified by starfishes, the ectoneural system retains a primitive epidermal position (Fig. 16.11e) and combines sensory and motor roles. A radial nerve extends down the lower surface of each arm and there is a circumoral ring. A deeper, exclusively motor hyponeural system is present, and an aboral, apical system is of major importance in crinoids. Ascidian tunicates have a single, cerebral ganglion situated midway between the two siphons, from which nerves run to the periphery. In the 'tadpole larva', a hollow neural tube thought to be homologous with that of vertebrates extends back from the ganglion along the tail.

16.3.4 Making the right connections

However great their final complexity, nervous systems develop from simple epithelia. Fibres extending out from the embryonic cells have tips forming amoeboid growth cones complete with pseudopodia, etc. The formation of the 'right' connections is crucial to the future function of the nervous system (Section 16.9), and the mechanisms involved have been investigated in insects such as grasshoppers, since their embryonic ganglia are

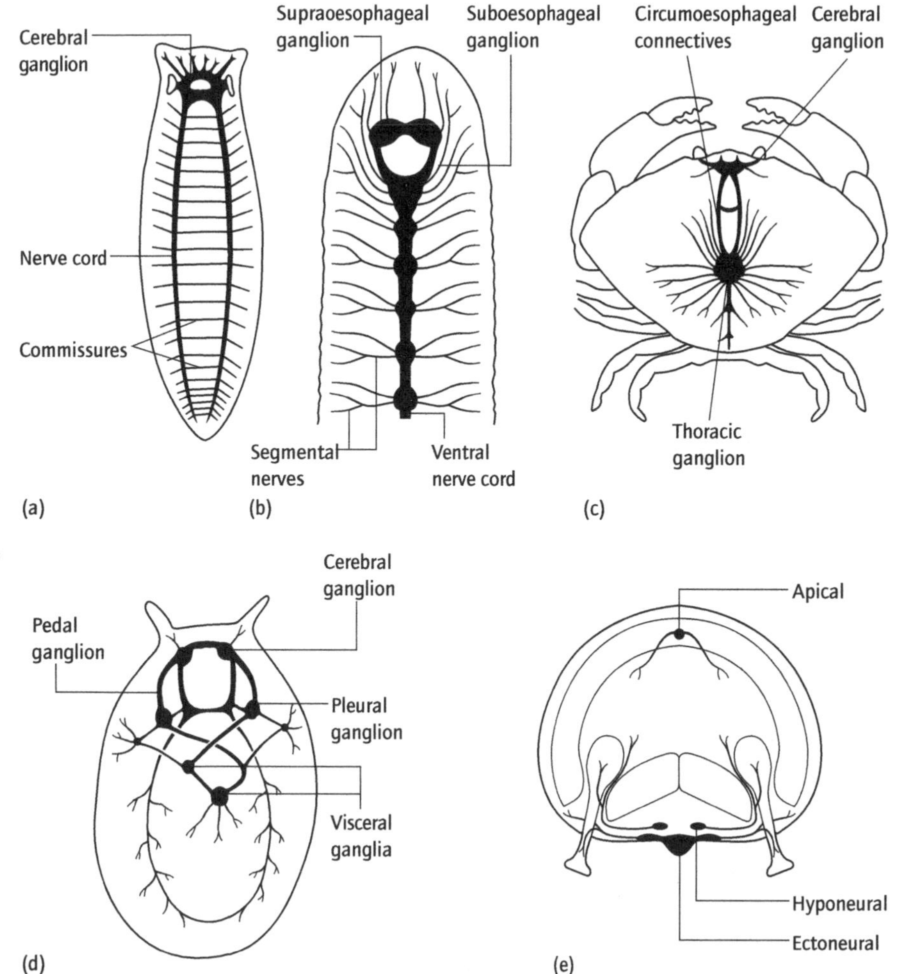

Fig. 16.11 Diversity of invertebrate nervous systems. (a) Orthogonal pattern of paired nerve cords and serially arranged commissures in a platyhelminth. (b) The segmented ventral nerve cord in leeches is made up of well-differentiated ganglia, containing the cell bodies, and connectives composed of nerve fibres. The suboesophageal ganglion is derived from a number of segmental ganglia. (c) Highly condensed central nervous system in a decapod crustacean. (d) The mollusc, *Patella*, showing the principal ganglia and torsion of the visceral connectives. The most primitive molluscs have nervous systems little advanced beyond those of platyhelminths, whereas some cephalopods have brains and behaviour as complex as those of fishes. (e) Starfish arm, as viewed in cross-section. Nerve fibres from the ectoneural system terminate on the surface of the hyponeural system but there is little or no direct contact between the two systems.

fairly transparent and contain limited numbers of cells. The lineage of individual cells can be traced from the 61 neuroblasts (embryonic neural cells) present in each of the ventral ganglia and their growing axons can be reliably recognized.

The outgrowth of fibres from different cells in ganglia of grasshopper embryos occurs in a distinct sequence, and the first to complete this process are called 'pioneer fibres'. Whether growing from one ganglion to the next, or extending from epidermal sensory cells into the central nervous system, pioneer fibres follow stereotyped pathways. In the periphery, it is important that connections are established before the development of obstacles, such as the differentiation of limb-segment boundaries in arthropods.

The growth cones appear to be guided by a variey of 'cues' provided by their surroundings, but seem to follow basement membranes and glial cells, particularly. They are influenced by at least four types of chemical signals: attractants and repellants that diffuse from distant sites of origin and set up a concentration gradient; and attractants and repellants that act on contact (Fig. 16.12). A commonly encountered pattern is the growth of axons towards the mid-line of the ganglion, marked by a special group of glial cells. The fibres either cross the mid-line (to form commissures) and then run longitudinally, never to cross it again,

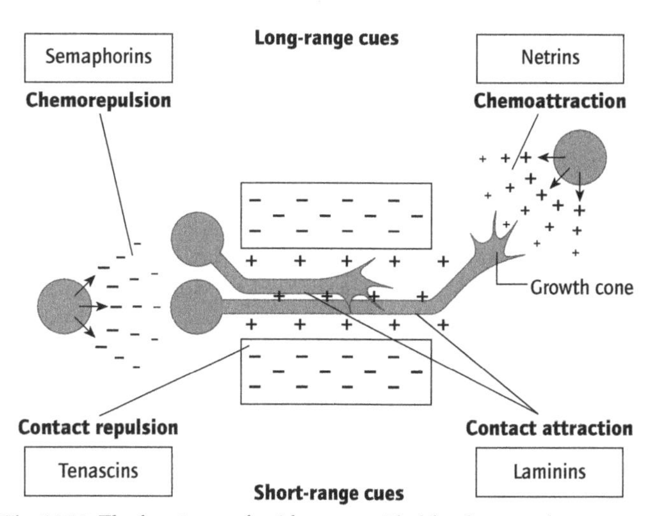

Fig. 16.12 The four types of guidance provided for the growth cones of axons, withe examples of some of the compounds involved. (From Tessier-Lavigne & Goodman, 1996.)

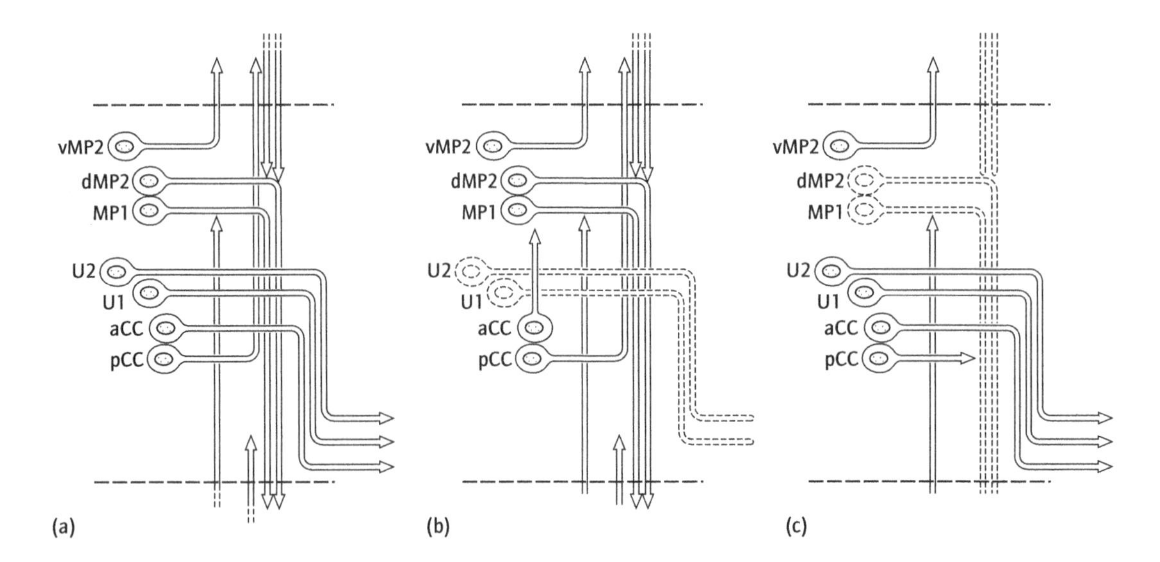

Fig. 16.13 (a) Development of neuronal pathways in the locust segmental ganglion on the seventh of 20 days of development. (b) Following removal of the pioneer neurones U1 and U2; aCC goes astray. (c) Removal of the pioneer neurones MP1 and dMP2 from this and adjacent ganglia – pCC goes astray. (After Bastiani *et al.*, 1985.)

or turn longitudinally without crossing it. They are guided by diffusable attractants (e.g. netrins) and by repulsive compounds (e.g. the roundabout (*robo*) protein, produced by the mid-line). Some axons are therefore deflected from the mid-line, whereas commissural fibres, thought to have low levels of receptors for *robo*, cross it. In doing so, they are stimulated to produce receptors so that they cross it only once. In *robo* mutants of *Drosophila*, axons wander freely across the mid-line and back again.

Fibres growing out later often follow the pioneers, since in the absence of the latter, in experiments in which individual embryonic cells are removed by laser-beam microsurgery, they go badly astray (Fig. 16.13). If the development of a cell is experimentally delayed, its axon will still follow the right route even though many additional features have appeared in the mean time. Although a given identified neurone may show significant differences in the completed patterns of its dendrites, etc. from one individual to the next, it forms the same synaptic connections in virtually every case. Having served their purpose, pioneer fibres either die or take on a more conventional role.

The influence of peripheral organs on neural development varies considerably. Although the various ganglia in the nerve cord in the insect embryo possess the same number of cells, some cells in the abdominal ganglia are lost, whereas their counterparts in the thorax are developed to control the limbs. This development is independent of peripheral structures and occurs even if the limbs are removed. In contrast, in the leech *Hirudo*, certain neurones proliferate in the ganglia of sexual segments in response to the influence of the genital apparatus (although only these segments have the ability to respond in this way). Some cells, whose homologues in other segments supply the body wall, innervate the genital organs in sexual segments and have special synaptic input appropriate to this role. If these organs are removed during embryonic development, the cells develop 'conventional' targets and sources of control – the peripheral organs in this case not only attract motor axons but must also 'specify' the connections to be made by the dendrites of the motor units in the central nervous system.

Similarly, olfactory cells on the antenna of the male moth, that are sensitive to the sex pheromone produced by the female (Section 16.10.5), send fibres into the antennal lobe of the brain where they induce the formation of a prominent, sex-specific, synaptic complex. Schneiderman and his colleagues have shown that replacement of a female antennal rudiment by that of a male results in both the development of such a complex in the brain of the female and in male-specific behaviour – an upwind flight – in response to the pheromone.

Similar recognition processes are probably involved during repair and regeneration of nervous systems. Cut axons can sometimes seek out and re-fuse with their severed parts; regenerating motor fibres may initially form synapses on inappropriate muscles, but such connections are eliminated and the 'correct' ones restored; fibres growing out from the brain of the platyhelminth *Notoplana* can find their correct targets (and exercise control over behaviour – see Section 16.3.8) even if the brain is transplanted into the tail!

Each stage of life of endopterygote insects has its own, largely distinctive behavioural repertoire and metamorphosis involves profound changes in the relevant neural circuits. The hormone ecdysone (Section 16.11.5) binds selectively to the nuclei of redundant neurones and induces their death; the dendritic trees of many other cells, and their synaptic connections, are extensively remodelled (see Box 16.3).

16.3.5 Differentiation of neuropile

At least three patterns of histological organization of nerve cords

Box 16.3 Remodelling the nervous system

During embryology, some cells develop, whereas others are *redundant* and are eliminated. In the nematode *Caenorhabditis*, hermaphrodite worms have 302 neurones, whereas males retain an extra 79 in connection with reproductive functions. This process of *programmed cell death* is genetically determined. The cells which die have the same developmental 'programme' within their genetic make-up as their surviving siblings, but in their case, an inbuilt 'suicide programme' is switched on. One of the most interesting mutants in *Caenorhabditis* (of gene *ced-3*) is defective in this regard. Cells which normally die survive and form functional connections – apparently the animal doesn't even notice!

Some motor neurones in *Caenorhabditis* show a complete reversal in the direction of flow of information during larval development, whereas their morphology is virtually unchanged. In the first-stage larva, they receive synaptic input dorsally and their output is to ventral muscles; in later stages and the adult, the situation is reversed.

In endopterygote insects (e.g. Lepidoptera), almost all cells making up the larval body are destroyed during metamorphosis and adult tissues develop *de novo* from imaginal discs. In contrast, most neurones present in the adult central nervous system are derived from cells present in the larval nervous system. However, during metamorphosis, the nervous system is extensively remodelled, in both gross morphology (Fig. a) and cell structure, for a life which is radically different with respect to the animal's environment, body structure, sensory capacity, locomotion and behaviour. Much of this remodelling is triggered by changes in the endocrine regime which governs metamorphosis (Section 16.11.5).

Proliferation and differentiation of most motor neurones in Lepidoptera is completed during embryogenesis, and these cells are functional in the larva. They vary widely in their fate (Fig. b). Some with a larval-specific function are eliminated in the pupa. Others with a role in the control of behaviour associated with adult emergence are programmed to die shortly afterwards. Yet others are extensively remodelled in their branching patterns and connections to serve adult function. Typically, the classical transmitter produced during larval life is retained in the adult, whereas the neuropeptide(s) may change.

In contrast, many neuroblasts due to give rise to sensory neurones and interneurones enter a state of arrested development and only complete proliferation in the larva. Non-functional at this stage (Fig. c) and showing only rudimentary differentiation, they do not develop fully until metamorphosis.

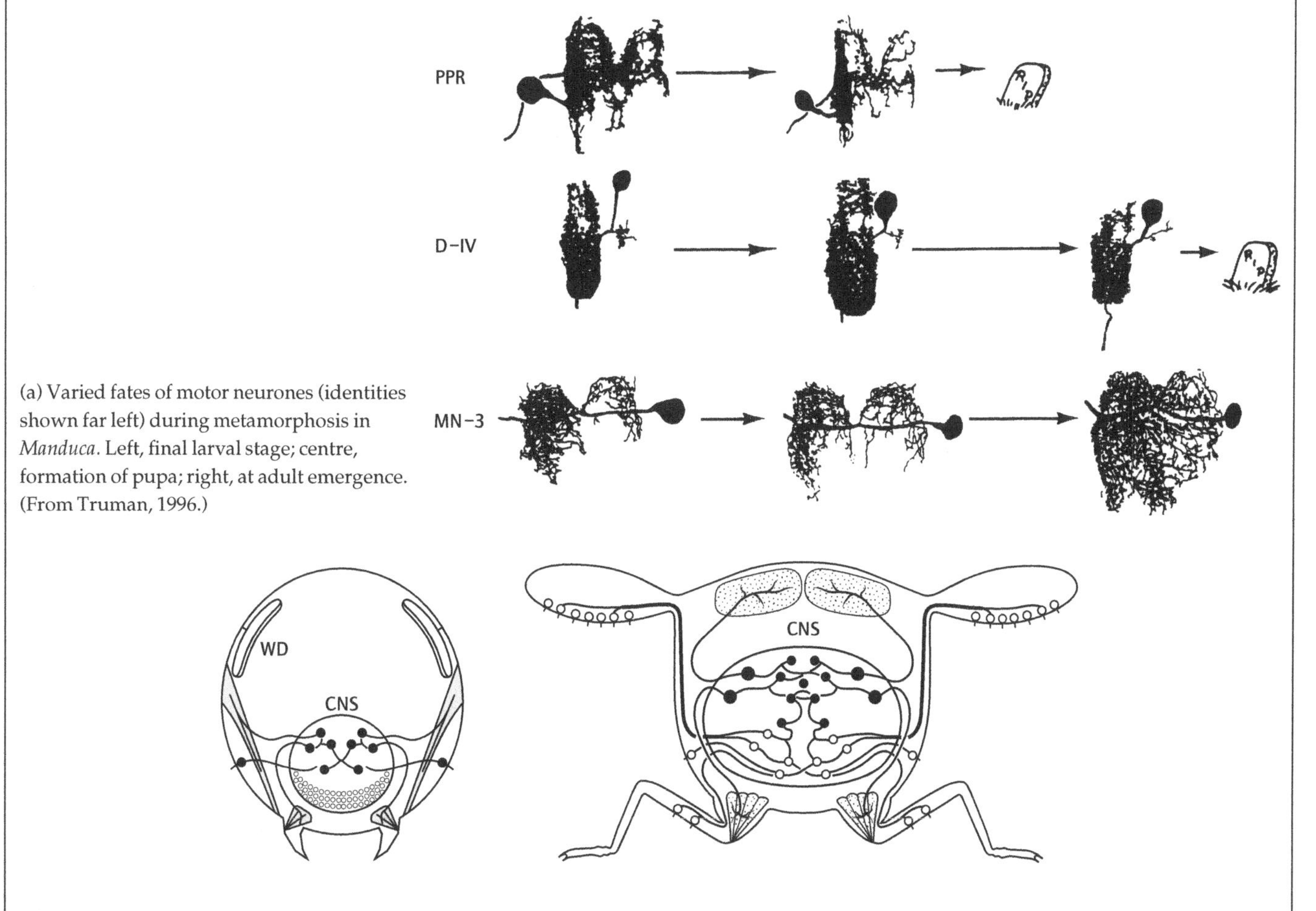

(a) Varied fates of motor neurones (identities shown far left) during metamorphosis in *Manduca*. Left, final larval stage; centre, formation of pupa; right, at adult emergence. (From Truman, 1996.)

(b) Contrasting patterns of development of typical motor neurones (filled circles) and sensory neurones (open circles), respectively, in endopterygote insects, showing larva (left) and adult (right). (From Truman, 1996.)

and ganglia may be distinguished (Fig. 16.14). Most commonly, regions of *differentiated neuropile* are present. Neurone cell bodies, often enveloped by glial cells, are segregated within the peripheral region or rind. Their branching processes contribute to the complex weft of nerve fibres forming the core or neuropile where the great majority of synaptic contacts are made, although some are formed on the cells (e.g. in molluscs).

A parallel array of neurones as seen in the nerve cords of nematodes and the inner nerve ring of hydromedusae. Cell bodies are not segregated histologically and synaptic contacts are made between adjacent fibres. In nematodes, neurones typically each have a single process (unbranched in *Caenorhabditis*; with two branch points at most in *Ascaris*) and the possible synaptic partners of a given fibre are restricted to those adjacent to it (these constitute the 'neighbourhood' of the fibre).

Lastly, an advanced islet pattern of organization is present in the brain of *Octopus*. Cell bodies are not only present as a rind, but are also scattered within the neuropile – a pattern comparable to that of vertebrate grey matter.

16.3.6 Segmental arrays

Many elongate animals (e.g. annelids) whose locomotory patterns involve the production of waves of contraction travelling up or down the body, organize their muscles as a series of blocks or somites which can be activated in sequence. Within the nervous system, each segment is equipped with a 'standard set' of neural components, which thus show serial repetition along the length of the body (of course, different segments may also have their own distinctive features – see Section 16.3.4). The relevant units and their interconnections have been identified in some invertebrates and a 'wiring diagram' of the system controlling swimming in the leech is given here (Fig. 16.15).

The set will usually include motoneurones with axons to the muscles, local interneurones to generate the appropriate pattern of activity (Section 16.9.2) and relay interneurones co-ordinating activity in adjacent segments – in the leech by inhibitory fibres. Not shown in the figure are sensory cells to generate proprioceptive feedback from the muscles and others whose activation initiates swimming.

Serial repetition is not resticted to segmented animals. The radial nerves of echinoderms and the axial cords in the arms of cephalopods show this effect, which involves groups of cells, areas of neuropile, and bilateral nerves extending to the periphery. In the nematode *Ascaris*, there are five sets of motoneurones, each of eleven cells, serially arranged along the body.

16.3.7 Hot lines

When an earthworm with its front end extended from the burrow encounters the proverbial 'early bird', intersegmental relays as described above, conducting at about 3 cm s^{-1}, are poorly

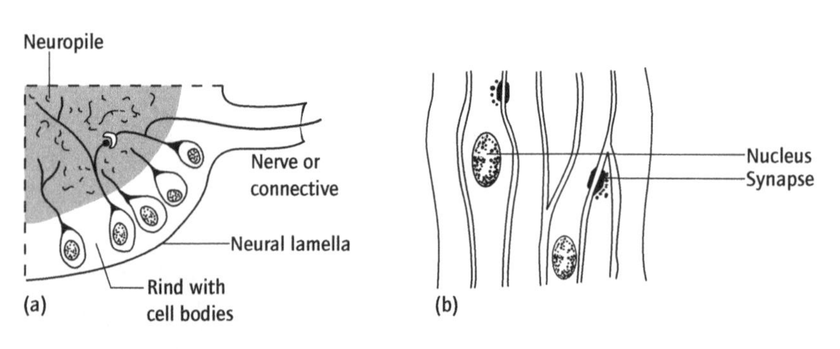

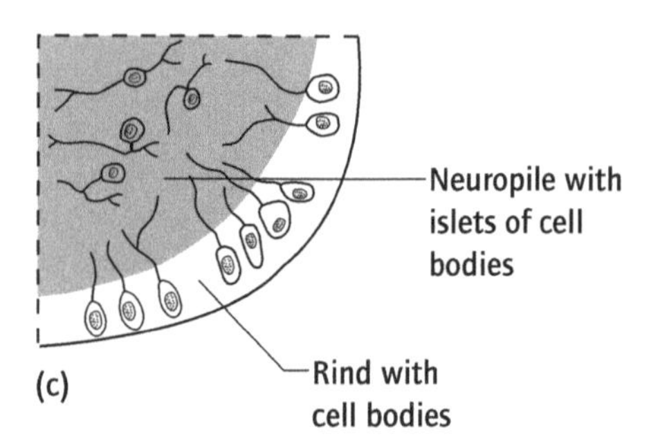

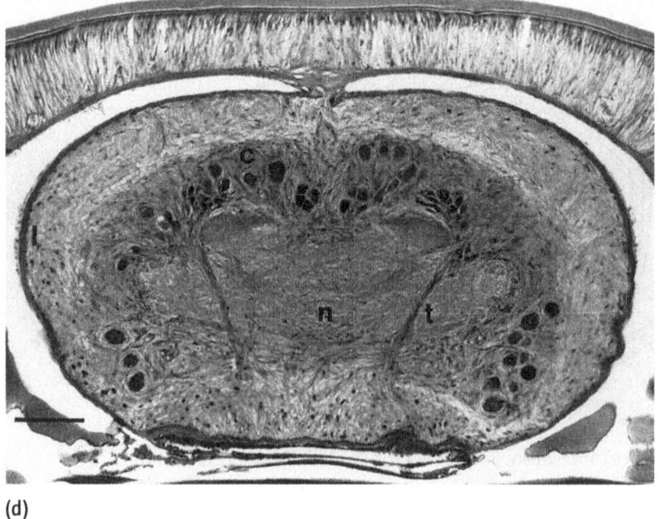

Fig. 16.14 Patterns of differentiation of neuropile. (a) Differentiated neuropile. (b) Parallel array. (c) Advanced islet. (d) Transverse section of the cerebral ganglion of the polychaete annelid *Nereis*. c, neuron cell bodies; l, neural lamella or brain capsule; n, tract of neurosecretory axons extending to neurohaemal area on the brain floor (cf. Fig. 16. 43); bar, 100 μm. (From Golding & Whittle, 1974.)

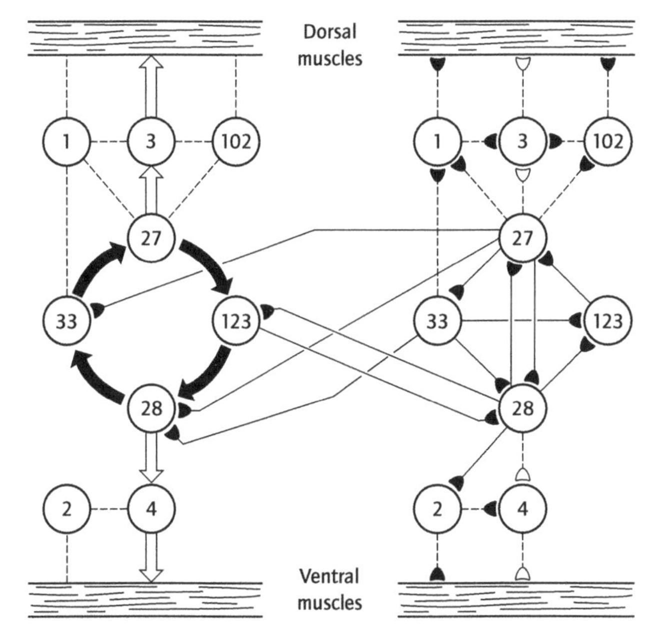

Fig. 16.15 Segmentally replicated set of neurones (identified by numbers) controlling swimming in the leech. The anterior ganglion (left) shows the oscillating pattern of nervous activity in the quartet of interneurones, which ensures that the excitatory motoneurones 3 and 4, inducing contraction of the dorsal and ventral muscles, respectively, are activated alternately. The posterior ganglion shows the synaptic connections within and between the ganglia. Clear bulbs, excitatory; filled bulbs, inhibitory chemical synapses; a rectifying electrical synapse is also formed between cells 33 and 28. (After Friesen *et al.*, 1976.)

adapted to bring about a rapid escape response – a through-conducting pathway is required. Such pathways frequently consist of 'giant' axons (Section 16.1.5) originating, for example, in earthworms from a series of cells (one per ganglion), some of which have fused longitudinally to form multicellular syncytia, others being connected by electrical synapses. Maximal contraction of all segments of the tail of the freshwater oligochaete *Branchiura* can be achieved in 7 ms, said to be the fastest escape response known in invertebrates.

In hydromedusae, a bundle of giant fibres, coupled by electrical junctions, permits the very rapid spread of excitation to all parts of the inner nerve ring and, in consequence, the symmetrical contraction of the bell required for swimming. A ring giant axon in *Amphigona* synapses with eight motor fibres extending over the muscles on the lower surface of the bell. Low levels of synaptic stimulation of the motor fibres evoke slowly conducting action potentials (Section 16.1.5) of low amplitude that are dependent on Ca^{2+} channels and that generate the slow pulsations of the bell used in feeding. When the escape response is evoked, acute synaptic stimulation of the motor fibres generates rapidly conducting, Na^+-dependent potentials, of high amplitude, that cause vigorous contractions of the bell, each of which propels the animal five times as far as a feeding pulsation. The properties of the neurones ensure that the Ca^{2+} and Na^+ spikes do not interact. This work by Mackie and his colleagues provided the first demonstration of the capacity of single cells to generate two types of action potential and provides a striking example of cnidarian parsimony.

In other animals, through-conducting pathways exert general stimulatory or inhibitory influences on segmental systems whose patterns of activity are controlled in detail by intersegmental relays (see Fig. 16.33). Of course, many fibres within nerve cords are intermediate between the two extremes described above and extend across a few segments.

16.3.8 Brain power

In higher invertebrates particularly, anteriorly situated ganglia are especially well developed and complex. This tendency is part of the more general process of encephalization, or head development.

First, there can be no doubt that this phenomenon is largely related to the inevitable concentration of sense organs at the front end of mobile, bilaterally symmetrical animals. The role of the brain in the processing and integration of sensory input, and of responding to the latter, is well illustrated in insects. The protocerebrum with its large optic lobes receives information from the eyes, the deuterocerebrum from the antennae with their well developed chemoreceptors, and the tritocerebrum from the anterior region of the alimentary canal. A prominent feature of the protocerebrum in many arthropods, as of the cerebral ganglia of higher polychaetes and even platyhelminths, is the presence of corpora pedunculata or 'mushroom bodies'. They consist of large numbers of tiny neurones (less than 3 μm in diameter in insects) whose bundled axons form the 'stalks'. They are thought to be involved in the integration of sensory information provided by the different cephalic sense organs, together with that from other parts of the body, and in associative learning.

A second example of 'front-end business' concerns the presence of neural machinery controlling feeding behaviour which, for example, is abolished by removal of the supra- and/or suboesophageal ganglia in annelids. The physiology of feeding in the blowfly has been described in great detail by Dethier (1976) in his book *The Hungry Fly*. Detection of food substances by chemoreceptors on the tarsal (distal) segments of the legs, and then by those on the tip of the proboscis, results in the passage of nerve impulses to the suboesophageal ganglion, where interneurones and then motoneurones are activated inducing eversion of the proboscis and sucking movements by the pharyngeal pump.

Third, anterior ganglia act as higher centres in exercising overriding control of the reflex and spontaneous activities of 'lower' levels. During locomotion in the platyhelminth *Notoplana*, food is passed directly to the mouth, whereas in the stationary animal it is passed first for inspection to the front end. A decerebrate specimen does not show the latter behavioural modification, nor is feeding inhibited when the gut is completely full. In crayfish, nerve fibres from the brain can block startle responses (see Box 16.8) and influence the development of habituation by actions on segmental synaptic junctions. These findings show that the brain

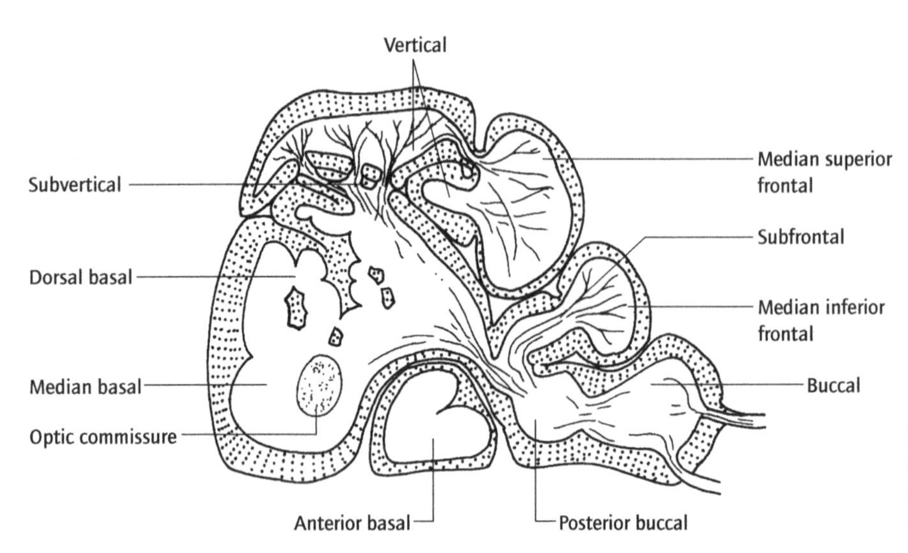

Fig. 16.16 Supraoesophageal region of the brain of *Octopus*, as viewed in a longitudinal section, showing some of the 25 functionally unique lobes. (After Young, 1963.)

receives information from a range of senses and takes 'strategic decisions' controlling the reflex actions of lower centres, although it is the latter which possess the machinery for organizing the activity in detail. In a wide variety of invertebrates, decerebrate animals are generally characterized by hyperactivity.

Fourth, higher centres have a role in controlling the animal's state of 'arousal' or 'mood', which has widespread effects within the nervous system. For example, neurones in the locust eye can have their sensitivity to stimuli greatly enhanced by arousal mediated by nervous input from the brain; motor centres are stimulated simultaneously.

Finally, the brain plays a key part in higher nervous functions and advanced forms of behaviour. Honey bees can be trained to visit as many as nine different sources of food, and to remember the appropriate time of the day for each. Complex 'memory traces' of this kind are stored in the corpora pedunculata – i.e. not in the main sensorimotor pathway as may be the case with simple forms of learning (see Box 16.2) but in a higher centre. A bee can learn to associate an odour with a sugar reward after just a single experience and will retain the memory for several days. Injections of minute amounts (as little as 5 nl) of neurotransmitters into the corpora pedunculata indicates that memory formation is a distinct process from memory retrieval. For example, dopamine administered at different times during training and testing is found to have no effect on the first process but to impair the second.

In *Octopus* (Fig. 16.16), work by Young has shown that different parts of the brain are involved in the process of learning in relation to different types of sensory information. Visual pathways lead from the optic to the superior frontal, and thence to the vertical lobes, whereas tactile learning involves the inferior frontal, subfrontal and vertical lobes. Within each of these pathways, the information is relayed to a series of centres – for example, four centres for visual input have been identified within the vertical lobes. There are also fibres running back from the vertical to the optic lobes. Memory traces are present at more than one level in any given system. Thus, removal of the vertical lobes substantially impairs memory of a visual process, but some impressions of the latter still remain in the other lobes. Furthermore, memory within one side of the brain, of a lesson learnt by unilateral presentation of stimuli, is shared over a period of hours with the other side.

Contrary to what may have been inferred from previous comments on learning (Box 16.2), higher nervous functions apparently require enormous numbers of neural units. For example, the 'highest' centres in *Octopus*, the vertical lobes, contain 25 million 'microneurones', many without axons. The lobes facilitate learning processes and this may be accomplished by the formation of a coded representation of a situation, and be the basis of the animals' notable abilities to generalize from one subject to a similar one.

16.4 Receptors

16.4.1 A fundamental characteristic

Sensitivity to environmental influences is a general characteristic of living cells and is shown even in the absence of obvious structural differentiation. One of the best known extraocular photoreceptors consists of a pair of neurones originating in the terminal abdominal ganglion of the crayfish, discovered by Prosser in 1934. The cells are responsible for photonegative behaviour, but also receive information from mechanoreceptors in the tail.

In addition to this generalized sensitivity, most animals develop a range of specialized receptor cells, which often form parts of multicellular sense organs (see Box 16.4). In the great majority of receptors, the sensitive part of the cell is structurally differentiated by the presence of cilia or microvilli (see Fig. 16.17) and in some cases, both. In many cases, sensory cilia are modified in structure, although this seems to bear little relation to the sensory modality involved. The significance of these organelles is presumably related to the expanded surface area they provide. In some cases (e.g. the crayfish eye), intramem-

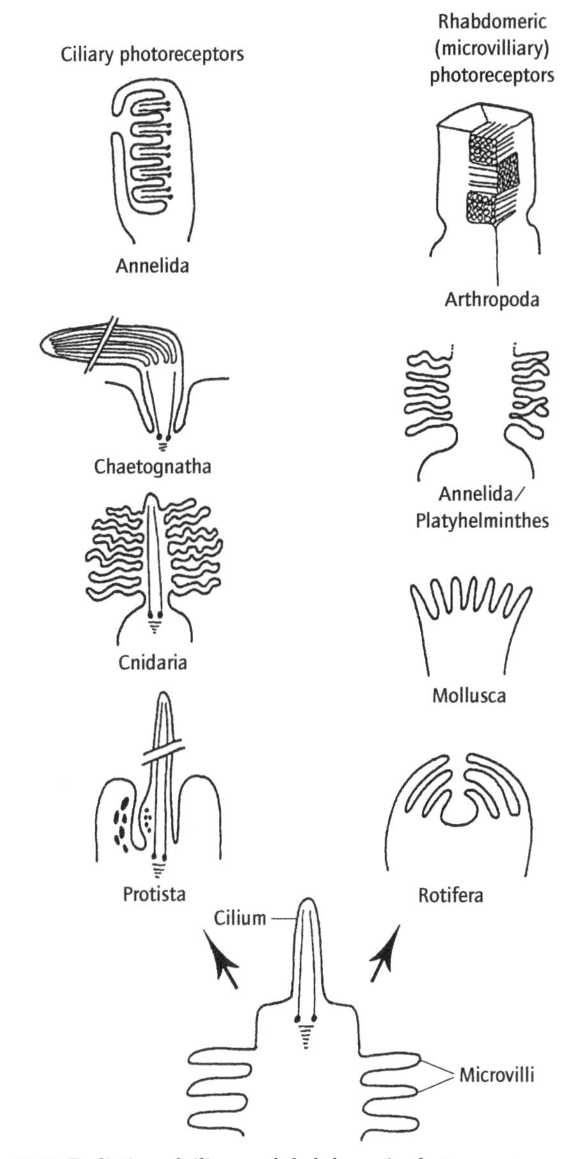

Fig. 16.17 Radiation of ciliary and rhabdomeric photoreceptors (largely after Eakin, 1968).

branous particles, visible with the electron microscope and presumed to consist of visual pigment, are concentrated within the microvillar membranes, implying that the latter are the locus of sensitivity. In others (e.g. *Drosophila*), the complement of particles within the microvilli resembles that of surrounding areas of the plasma membrane, suggesting a more diffuse sensitivity.

16.4.2 Classification of receptors

Aristotle recognized what we call 'the five senses' and these have been joined by others, some quite exotic (e.g. honey bees display a finely tuned sensitivity to magnetic fields). Types of senses are called 'sensory modalities'. Contemporary classifications are based on the physical character of the stimulus concerned

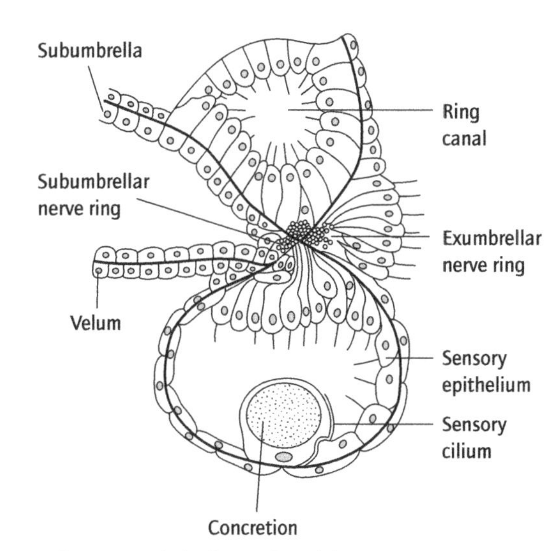

Fig. 16.18 Statocyst of a hydromedusa (after Barnes, 1980).

(e.g. light, mechanical, chemical, etc.). However, we may also distinguish between exteroceptors, sensitive to external influences, interoceptors which respond to internal factors, and proprioceptors signalling movements or positions of muscles, joints, etc.; or between phasic receptors responding to changes in the environment, and tonic ones whose activity is related to the absolute level of stimulation – many receptors are a combination of the two.

Sensitivity to one modality may be exploited to provide information about another. For example, receptors sensitive to gravity called statocysts are specialized mechanoreceptors (Fig. 16.18). They are widespread in invertebrates, for example, being associated with the margin of the bells of jellyfish (see Fig. 3.17) and present in *Octopus*. The organ consists of a vesicle containing one or more dense bodies, the statoliths. As the statolith moves under the influence of gravity, it touches or distorts sensory cilia and activates the receptor cells. In the urochordate *Oikopleura*, the statolith is replaced by a melanin droplet which, being lighter than water, floats upwards and in this way indicates the direction of gravity.

The role of statocysts was shown in shrimp by Kreidl, in 1893. These animals shed and replace the sand grains which form their statoliths when they moult. By providing iron filings instead of sand, Kreidl was able, with the use of a magnet, to make the shrimp swim upside-down, on their sides, etc. as the magnetic field simulated that of gravity. Statocysts can detect acceleration, but tubular systems in which movement of fluid activates receptors, comparable to the semicircular canals of vertebrates, are also present in some invertebrates (e.g. *Octopus*).

16.4.3 Specialists and generalists

Receptors for a given sensory modality have in common a heightened sensitivity to the form of energy involved, as shown by a lower threshold, and more intense response to stimulation.

Box 16.4 Insect sensilla

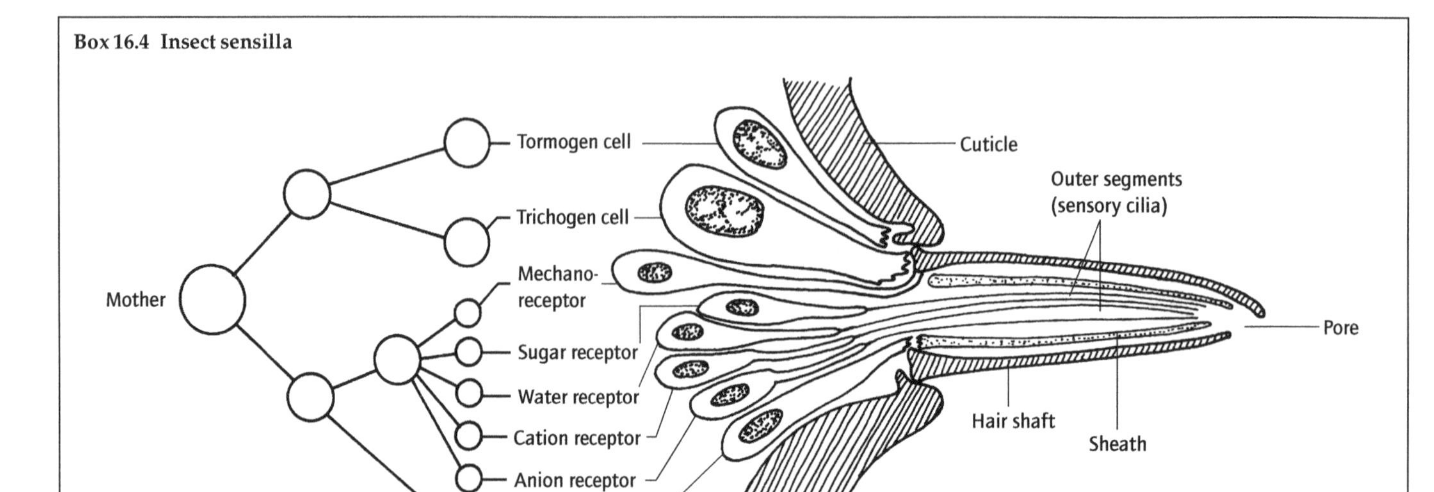

(a) Development and structure of an insect sensillum. The tormogen cell secretes the socket, the trichogen cell the hair shaft (both contribute to the lymph); the sheath cell forms the sheath around the cilia. Note that an individual sensillum may incorporate receptor cells for a diversity of senses. Not shown are nerve fibres extending from the receptors towards the central nervous system. (After Dethier, 1976; and Hansen, 1978.)

Insects have combined receptor neurones with cuticular hairs, pegs, plates, pits etc. to produce an astonishing array of sensory organs called *sensilla* – the body just bristles with such 'antennae'. All the cells composing a sensillum are produced by division from one mother cell.

Chemoreceptors, whether taste or olfactory, show close similarities with those of other animals. The dendrite (or 'inner segment') of each bipolar receptor neurone extends to the base of the organ, where it gives rise to one or more cilia (the 'outer segment'). One or many pores (up to 15 000 per hair) provide access by which chemical substances can reach the cilia via the special fluid, or 'lymph', in which the organelles are bathed. Odours used for conspecific signalling in moths (i.e. pheromones – Section 16.10.5), whose receptor neurones are located on the antennae, are rapidly inactivated by binding to a special protein in the lymph and then degraded by enzymes.

Mechanoreceptors are well developed and are ciliated, as in the great majority of other animals. Movement of the hair distorts the cilium and stimulates the receptor cell. *Campaniform sensilla*, each consisting of a single neurone associated with a ridged, cuticular cap, detect stress within the cuticle. *Chordotonal sensilla* provide information regarding the positions or movements of joints. They also provide the sensory components for *tympanic organs* – auditory organs which may be associated with widely separated parts of the insect body. Such organs consist of modified parts of the tracheal system, and have air sacs with tympanic membranes whose vibrations in response to sound stimulate the receptors associated with them. A single ('cyclopean') ear is present in the mid-ventral thoracic wall of the praying mantis. It is sensitive to ultrasonic frequencies and may provide a measure of protection against insectivorous bats.

Insects apparently lack specialized organs sensitive to gravity and acceleration, and depend on the many sense organs associated with joints, etc. to provide the relevant information. However, flies have *halteres* (Fig. b), which are homologues of the second pair of wings (flies only have one pair) and function like gyroscopes, signalling rotation in space during flight. Dumb-bell-shaped, they oscillate up and down, aided by a hinge joint at the base. They are powered by homologues of the power muscles of the wings and adjusted by 11 tiny muscles (which have input from the eyes) corresponding to those which adjust the wing beat for steering. 335 strain receptors are present at the base and other sensillae are located on the head.

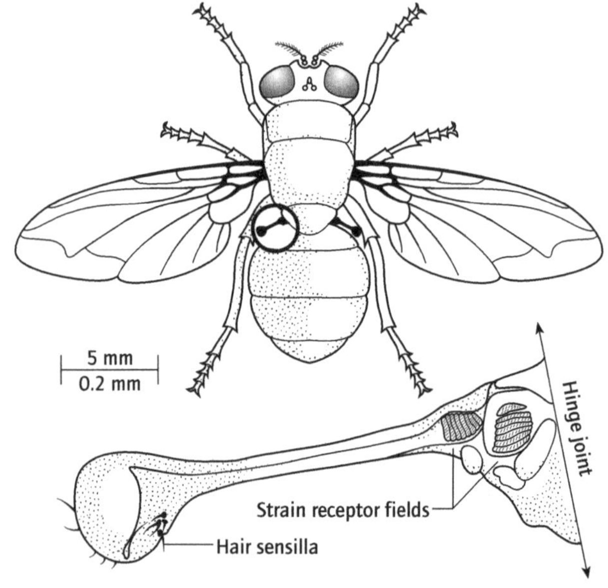

(b) Haltere of blow fly, *Calliphora vicinia*. (After Hengstenberg, 1998.)

However, whereas some are specialists, others are generalists. Olfactory specialists each have a highly restricted spectrum of response to odours, with, for example, an acute sensitivity only to a single compound, a pheromone (Section 16.10.5), secreted as a chemical signal by another member of the species, or only to odours associated with food. For example, about 50 cell types are present in *Drosophila*, in each of which only one or a small number of olfactory genes are expressed. Olfactory generalists respond to a wider variety of stimuli within the modality, as in *Caenorhabditis*, where only 15 cell types are present, each with many different receptor molecules. However, each generalist may have its own pattern of sensitivity, enabling the animal to recognize any given substance by the unique combination of receptors activated.

16.4.4 Intensity coding

The information provided by receptors is not usually just 'on' or 'off', but also 'how much', and in spiking neurones this is encoded in the frequency of the impulses generated in the nerve fibres, which is proportional to the log of the stimulus intensity, or to some comparable function. The range of stimulation intensity to which the organism is sensitive is extended by the control of receptor sensitivity exercised by centrifugal fibres (impulse traffic is directed towards the periphery) such as innervate the cephalopod statocyst (Fig. 16.19). This feature is particularly well developed in invertebrates. Lastly, different cells can operate across different parts of a wider range. In a classic study, Nicholls and others have shown that mechanoreceptors in the body wall of the leech, with identified cell bodies in each ganglion of the ventral cord, are of several different types. 'Touch' (T) cells are sensitive even to water currents in the medium. 'Pressure' (P) receptors start to respond when a pressure of about 7 g is applied by means of a probe. Finally, 'nociceptor' (N) units are activated only in response to severe damaging stimuli.

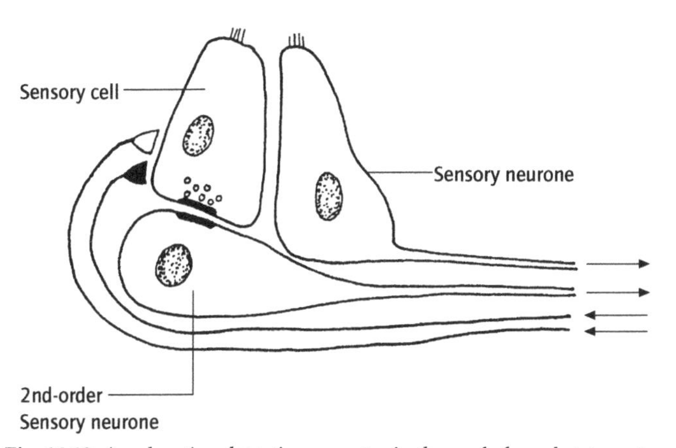

Fig. 16.19 Acceleration detection receptor in the cephalopod statocyst. The organ contains both sensory neurones and sensory cells, the latter communicating with the central nervous system by forming chemical synapses with second-order sensory neurones. Each of the different types of components receive both excitatory (clear bulb) and inhibitory (filled bulb) efferent synapses (see Section 16.6.6), as shown for the sensory cells. (After Williamson, 1989.)

16.4.5 Sensory cells and sensory neurones

Many sense organs in vertebrates are made up of sensory cells lacking nerve fibres, which transmit information across chemical synapses to the terminals of sensory neurones whose axons extend to the central nervous system (despite the direction of information flow, the neurones are often (confusingly) called 'primary', and the peripheral cells, 'secondary'). In other cases, the receptor is itself a neurone and this is characteristic of the great majority of sensory systems in invertebrates. Exceptions include the Langerhans receptor in the urochordate *Oikopleura*, and photoreceptors in the cnidarian *Tamoya*. The acceleration detection system in the statocyst of cephalopods incorporates both sensory cells and sensory neurones (Fig. 16.19).

16.5 Vision

16.5.1 Visual pigments

Vision is the sensory modality which provides us with most information about the world and the same applies to many of the most advanced invertebrates, as shown by the refinement of their eyes.

Sensitivity to light, a small band within the electromagnetic spectrum, is conferred by the possession of photoreceptor molecules – i.e. visual pigments – which absorb radiant energy, with the release of free energy. Rhodopsin is a compound of the carotenoid retinal, a vitamin A derivative, and the protein opsin. It is nearly universal, being found in organisms as diverse as the alga *Volvox* and man. Exposure to light converts the retinal from its 11-*cis* isomer to the 11-*trans* configuration, and then leads to bleaching of the pigment due to the dissociation of retinal and opsin. This change affects ion channels in the photoreceptor cell membrane, through the mediation of a second messenger (see Section 16.1.3). Rhodopsin is regenerated by enzymic synthesis.

Rhodopsin exists in a variety of forms, depending on both the type of retinal and that of opsin. Fireflies active at dusk emit yellow light as signals to other members of the same species, whereas those active after dark produce green light – and the maximal sensitivities of the retinal pigments in the different species are precisely tuned accordingly.

Colour vision is important in many crustaceans, insects and arachnids (surprisingly, *Octopus* is colour blind) and is based on the possession of several types of receptors with pigments sensitive to different parts of the spectrum. For example, bees have receptor neurones sensitive to yellow–green (540 nm), blue (440 nm) and ultraviolet light (340 nm), respectively. Butterflies are also sensitive to red light. In flies, the majority of receptors exhibit a curious dual sensitivity (blue–green *and* ultraviolet) due to the presence of both rhodopsin and another pigment. The other receptors show a range of spectral sensitivities, partly on account of the presence of screening pigments which cut out some wavelengths so that the receptors only respond to the others. The retina of the stomatopod crustacean *Pseudosquilla* has at least ten spectral types of receptor cells (we get by on three!).

The orientation of polarized light can also be detected by some eyes. In flies, units at the margins of the compound eyes are specialized for this function. The pattern of polarization of the sky enables bees to navigate with reference to the position of the sun even when this is obscured by cloud. *Octopus* may use this ability to 'see through' the silvery camouflage of fish.

16.5.2 Ciliary and rhabdomeric eyes

The great majority of photoreceptors belong to one or the other of two types, as shown by Eakin (Fig. 16.17). In one class, the probable sensory organelles are cilia, or derivatives of ciliary membranes (as in vertebrates). In the second type, the organelles are microvilli, whose ordered arrays where formed are called rhabdomes. It is noteworthy that dendrites bearing microvilli often possess a cilium or ciliary rudiment at the apex, suggesting that the ciliary complex may have a role in rhabdome development. Furthermore, receptors with definitive ciliary and rhabdomeric organelles at different levels in the same cells have been reported in serpulid polychaete annelids; in the cnidarian *Polyorchis*, microvilli intermingle with elements derived from ciliary membrane; and a few others too seem to be mixed.

Many photoreceptor membranes of either type show a massive, daily breakdown and renewal. In the nocturnal spider *Dinopus*, the rhabdome occupies only 15% of the receptor cell volume during the day, but within one hour after sunset this has increased to 90%, increasing the catch of photons from 6 to 74%. The surplus membrane is destroyed within two hours after sunrise.

Most ciliary units are 'off' receptors. Darkness leads to Na^+ channel opening (usually), depolarization, and transmitter release from the terminals. In contrast, rhabdomeric receptors typically show 'on' responses – the above effects result from exposure to illumination. The distinction is illustrated in the clam *Pecten*, whose eyes have an upper layer of ciliary ('off') receptors and a lower layer of rhabdomeric ('on') cells. Exceptions to the rule include the 'off' response of microvilliary receptors in the gastropod mollusc *Onchidium* and those of both microvilliary and ciliary types in salps (urochordates). It will be interesting to learn how the mixed ones work. Ciliary receptors such as those of sabellid fan worms and those of *Pecten* seem well adapted to mediate 'shadow reflexes' – defensive reactions to sudden decreases in illumination. However, rhabdomeric 'on' receptors are involved in such a reflex in barnacles. Decrease in illumination hyperpolarizes the receptors, blocking the release of their inhibitory transmitter and thus allowing the activation of the innervated neurones in the brain.

The two types of photoreceptors were long regarded as having a phylogenetic significance – ciliary receptors being representative of cnidarians and deuterostomes; microvilliary receptors of the protostomes. However, many exceptions are now known: starfish have microvilliary receptors; the eyes of *Pecten* include both; the right larval ocellus in the platyhelminth *Pseudoceros* consists of several microvilliary receptors, whereas the left has three microvilliary, plus one ciliary cell.

Many flagellates have part of the flagellum (a large cilium) specialized for receptor function and a stigma (pigment spot) confers directional sensitivity. Most extraordinary of all, some dinoflagellates have ocelloids – i.e. parts of the single cell resemble a simple eye with a cornea, lens, and crystelline retina-like layer, backed by a pigment cup, and claimed to be able to form images!

16.5.3 Ocelli and eyes

Photoreceptor organs in invertebrates show a wide spectrum of grades and patterns of organization (Figs 16.20 and 16.21). The simpler organs are called ocelli and only provide information regarding the intensity and direction of light.

Eyes which form well focused images are called 'camera eyes'. They depend, first, on the presence of an expanded sheet of receptors constituting a retina (which may either be direct (everted) or inverted [Fig. 16.21]), and second, on the presence of a device for focusing the light, although when the lens lies adjacent to the surface of the retina and the retina is thick, image formation must be poor. High-performance vision is of critical importance to predators such as jumping spiders, which need to be able to judge the speed of an object (all spider eyes are of the pigmented eye-cup plus lens type).

The camera eye reaches its climax in *Octopus* and its relatives (Fig. 5.25b). It comes complete with eyelids, an adjustable pupil, a moveable lens and extrinsic muscles by which, with the contributions of input from the statocyst, vision can be fixed on an object of interest, irrespective of body movements. Cephalopods hold the record for the largest eyes. Those of *Architeuthis*, the giant squid, can measure 40 cm in diameter and the retina could conceivably contain up to 10^{10} receptors (cf. 10^8 in man). The eye of *Nautilus* is a puzzle, lacking both lens and cornea. It functions liked a 'pin-hole' camera, but resolution and sensitivity are poor.

16.5.4 Lenses and mirrors

The convex cornea is responsible for much of the refraction of light in terrestrial species (e.g. spiders). In aquatic animals, lenses must do all the work and, being spherical, would be expected to have the defect of spherical aberration. Matthiessen (1886) suggested that the lens has a high refractive index at its core, falling to lower values at the periphery, and indeed such lenses are found in cephalopods, some gastropod molluscs, some copepod crustaceans, and alciopid polychaetes – as well as fishes. Obviously, 'there really is only one way to make a decent lens, using biological material'! (Land, 1984).

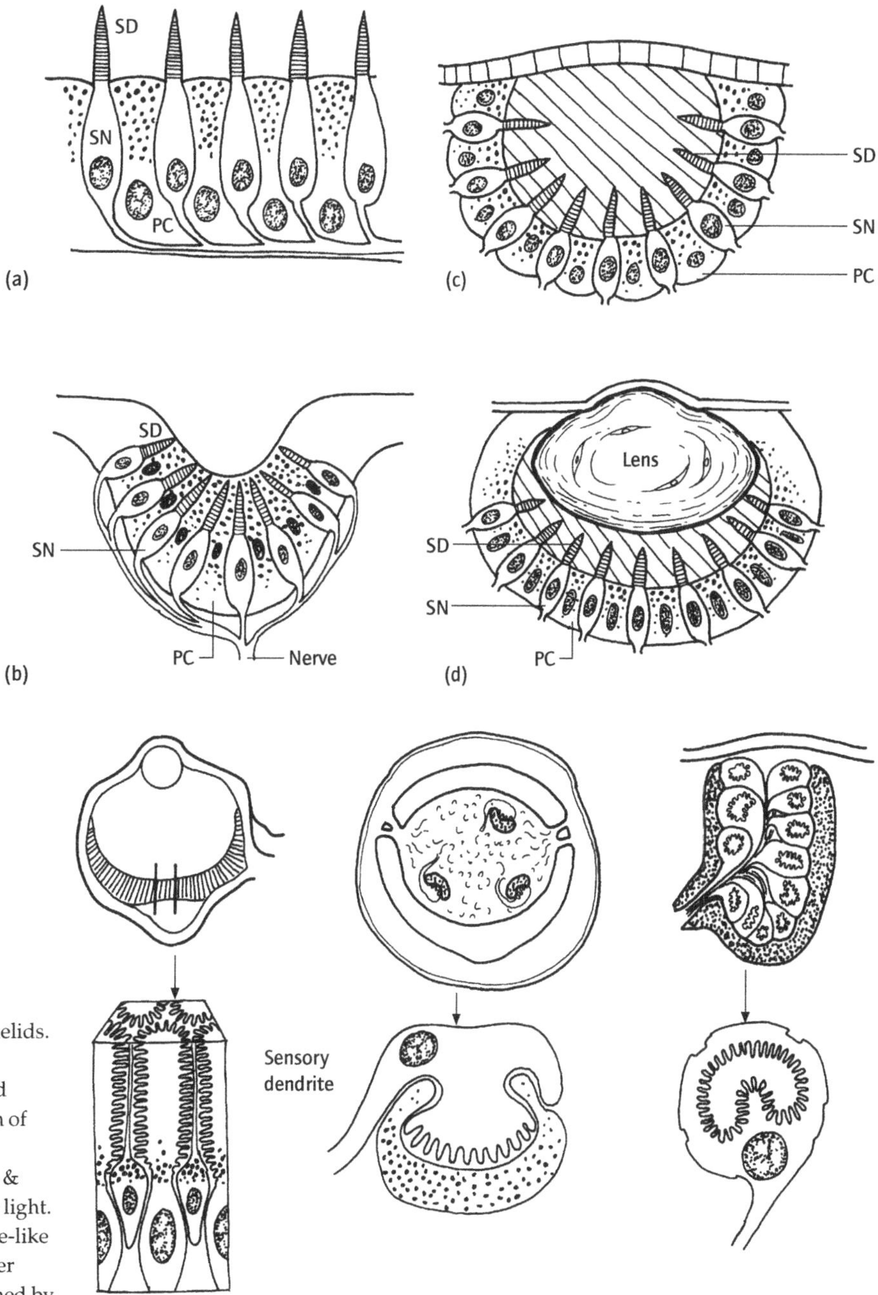

Fig. 16.20 Grades of organization of increasing complexity shown by pigment-cup ocelli and eyes. Such sequences can be based on examples from each of many phyla – as many as '40–65 phyletic lines of complexity' (Salvini-Plawen & Mayr, 1977) have been identified. Examples in cnidarians range from areas of pigmented epithelium in *Leukartiara* (a), through rudimentary ocelli in *Bougainvillia* (c), to eyes with lenses in *Tamoya* (d). In molluscs, eyes range from the simple, open eye-cups of limpets (b, *Patella*), or the five-celled ocelli of some nudibranchs, through intermediates in *Nerita* (c) and *Valvata* (d), to the eyes of cephalopods, with 20 million receptors, which bear comparison with our own. PC, pigment cell; SD, sensory dendrite; SN, sensory neurone. Dendrites bear cilia or microvilli.

Fig. 16.21 Direct and inverted photoreceptors in annelids. (a) Direct – eye-cup of the polychaete *Vanadis* (after Hermans & Eakin, 1974), with receptor poles directed towards the light. (b) Inverted – eye-cups in the brain of the polychaete *Armandia* each with a single sensory neurone and one-celled pigment cup (after Hermans & Cloney, 1966); receptor poles directed away from the light. (c) Eye-cup with sensory neurones, each with vacuole-like 'phaosome', in the leech *Hirudo* corresponds to neither category. (After Hess, 1897.) Phaosomes may be formed by both ciliary and microvilliary photoreceptors.

Mirrors are important in focusing the light within many eyes, either by themselves or in combination with lenses. The mirror may redirect the light back through the receptors, virtually doubling the effective illumination (some spiders); or one layer of receptors receives light focused by the lens, and a second layer after it has been refocused by the mirror (some crustaceans).

16.5.5 Compound eyes

Compound eyes are organs consisting of a number of virtually identical units, or ommatidia, arranged in geometrical array. They are sometimes called mosaic eyes since the image is formed by the contributions of all the ommatidia, each of which provides only a small part. They are thought to be well adapted for detecting movement, but their resolving power (except in flies) is only about 1° at best, as compared with 1′ in vertebrates and *Octopus*. Whereas the retinas of pigmented eye-cups are concave and form inverted images, compound eyes and their retinas are always convex and the images erect (Fig. 16.22). Definitive compound eyes are found in many arthropods, but are not present in spiders, millipedes, etc. Eyes of this type are also present on the tentacles of sabellid fan-worms, in the

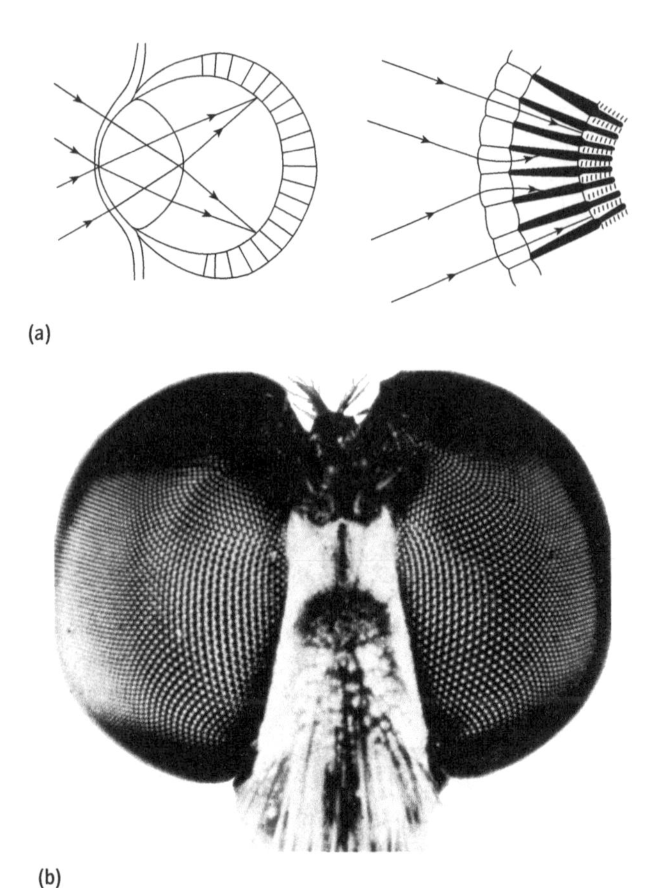

Fig. 16.22 (a) Contrasting patterns of organization of an eye-cup (left) and a compound eye (right). (b) Head of a robber fly. Anterior large facets of the compound eyes are approximately 60 µm in diameter. (Courtesy, Professor M. Land, University of Sussex.)

mollusc *Arca*, and form the 'optic cushions' situated at the tips of the arms in starfish.

The acuity of compound eyes depends on the angular separation of the ommatidia. Whereas the eye of *Daphnia* has 22 ommatidia, each with an angular separation from its neighbours of 38°, the crab *Leptograpsus* has several thousand, with a separation of 1.5°, and dragonflies have nearly 30 000. The resolving power of the latter would be far greater, just as a mosaic with small components is superior to that with large ones, but this is correlated with a loss of sensitivity. Just as vertebrate eyes have a fovea, some crab eyes have fine-grained bands, made up of ommatidia with a small angular separation. A compromise between resolution and sensitivity must always be struck in the organization of eyes.

In arthropods, each ommatidium is an elongate unit consisting, typically, of a corneal lens, a crystalline cone, a number of receptor (retinular) cells (four in *Daphnia*, eight in crabs and flies, and ten to fifteen in *Limulus*), and associated pigment cells (Fig. 16.23a). The retinular cells are arranged like the segments of an orange. Regularly arranged microvilli project from along the inner surfaces of each cell and constitute a rhabdomere, and the rhabdomeres within an individual ommatidium together make up its rhabdome. Many compond eyes show some overlap between the fields of vision of adjacent ommatidia, but in the crustacean *Phronima*, each ommatidium of the medial eyes can accept light from a field 4° across, whereas the angle of separation of the ommaditia is only 0.5°. However, perhaps the most extraordinary compound eyes are those of stomatopod crustaceans. Whereas we have two images of an object, these animals have six, since each eye has three bands of ommatidia looking out in the same direction! These last two examples make it clear that no analysis of the capabilities of compound eyes is adequate that fails to take account of the remarkable 'computing' capacity of the brain in analysing and interpreting the information provided by the receptors (see Section 16.6).

Types of compound eyes in arthropods differ in the degree and nature of the functional isolation of the ommatidia. The apposition eyes (Fig. 16.23a) of bees, for example, are adapted for high light intensities. Pigment cells screen the rhabdome from all light apart from that entering via the lens of that particular ommatidium.

Superposition eyes (Fig. 16.23b) such as those of shrimps and moths have a wide 'clear zone' interposed between the crystalline cone and the rhabdome, and the latter is shorter. They function like apposition eyes in conditions of high illumination, maximizing the resolution of the eye. However, in dim light, extension of pigment cells or migration of pigment within cells which remain fixed in position exposes the rhabdomes to light entering via surrounding ommatidia. The rays of light from a given source impinging on different ommatidia are redirected to amplify the input to a single rhabdome – an arrangement which maximizes the sensitivity of the eye in dim light, albeit at the expense of resolution. The movements of pigment are controlled by antagonistic transmitters and/or hormones.

In the eyes described above, there is no possibility of keeping separate the information contributed by different retinular cells within a single ommatidium, since the rhabdome is 'closed' – the microvilli interdigitate. Furthermore, each cell synapses on each of a small cluster of neurones (called a 'cartridge') situated immediately beneath the the ommatidium. A different situation is encountered in the eyes of flies whose mechanism of neural superposition represents the summit of development of arthropod optics (Box 16.5).

16.6 Sensory processing

16.6.1 Making sense of the world

Information regarding the external and internal environments must not only be gathered and encoded as changes in membrane potential in receptors, but processed – modified and transformed in ways that are adaptive to the animal concerned. For example, in the locust, the simple ocelli on the top of the head produce a poorly focused image and show massive convergence – information from about 1000 receptors in each is funnelled through a comparatively small number (25) of second-order neurones. During flight the ocelli provide a rapid, overall assess-

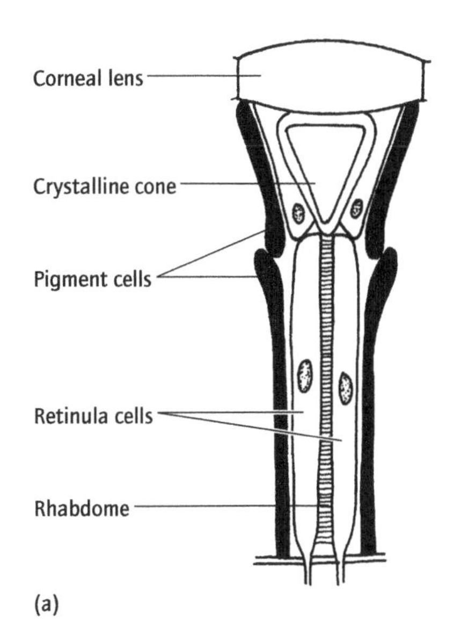

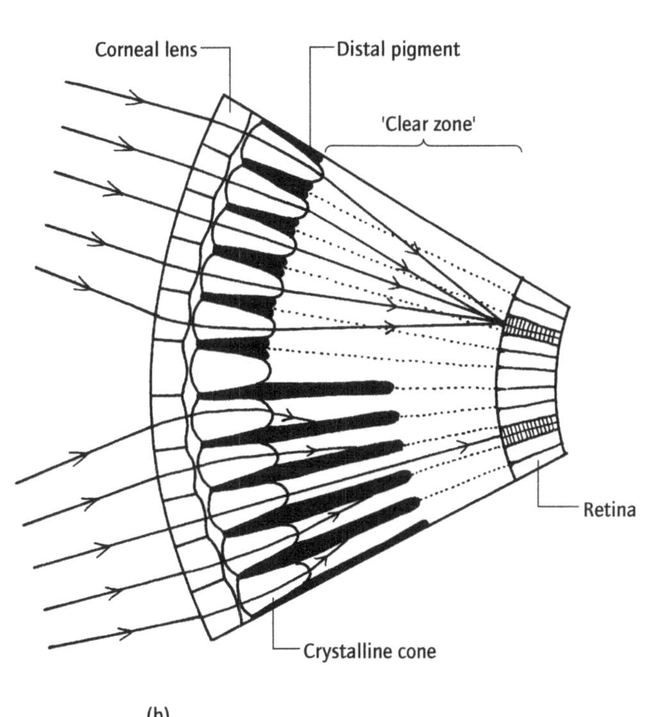

Fig. 16.23 (a) Longitudinal section of an ommatidium of an apposition, compound eye. (After Wigglesworth, 1970.) (b) Paths of light rays within a superposition eye when light adapted (lower), and dark adapted (upper).

Box 16.5 Neural superposition eyes

Eyes with neural superposition optics provide resolving powers up to 100 times greater than that of other compound eyes. Within a given ommatidium, the rhabdome is 'open', the rhabdomeres remaining separate. Each rhabdomere receives light from a distinct angle and the axon of its cell extends to a cartridge shared not with others within the same ommatidium, but with those of surrounding ommatidia which share the same optical axis. Consequently, rays of light from a single source impinge on six, differently placed retinular cells in adjacent ommatidia. Nerve fibres from these cells follow a spiral path and all synapse on the same cartridge in the optic ganglion.

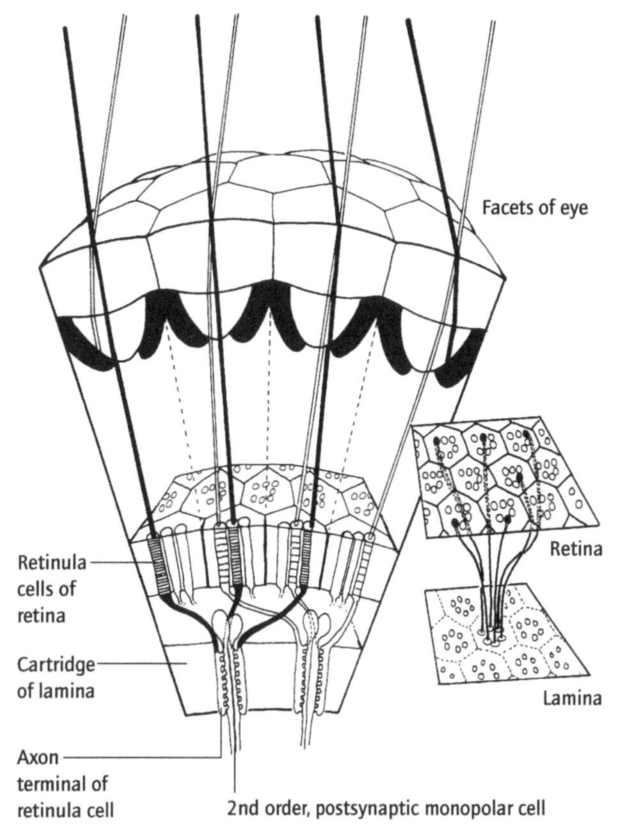

Paths of light rays from two points in the visual field (open and dense lines, respectively), their projection to retinular cells in adjacent ommatidia, and the convergence of the fibres of the receptors they activate upon optic cartridges in the laminar. Right, light rays from a single point impinge on different retinular cells in neighbouring ommatidia, but the axons of the latter converge upon a single cartridge. Only retinular cells 1–6 of the eight present in each ommatidium usually contribute to neural superposition. (After Strausfeld & Nassel, 1981.)

ment of the position of the horizon, whereas, as we shall see, the lateral, compound eyes can distinguish finer points of detail.

The sophistication of the sensory and neural machinery of arthropods is apparent in the recognition of particular features or combinations of features, out of a multitude of similar alternatives. 'An ant leaving a newly discovered food source at the base of a landmark . . . periodically turns back and faces the landmark . . . takes several "snapshots" from different vantage points' (Judd & Collett, 1998) and uses them to find its way back subsequently. Bees and wasps show similar behaviour. Some South American orchid bees visit daily a succession of the same flowers, each of whose positions they must remember, following a stereotyped route over 20 km long.

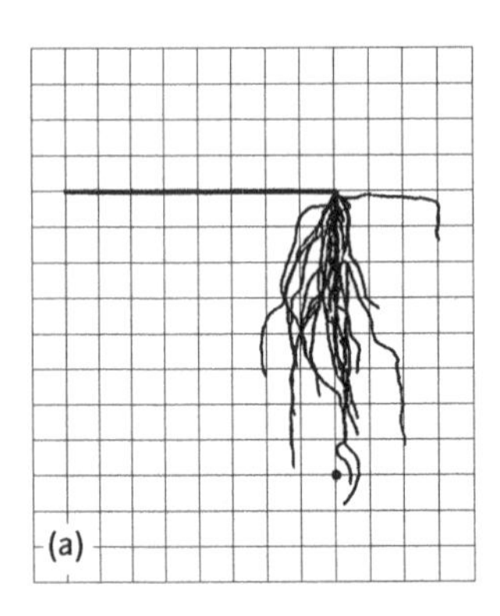

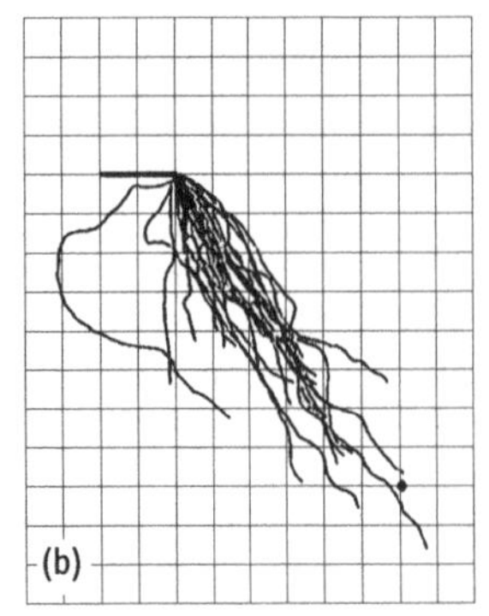

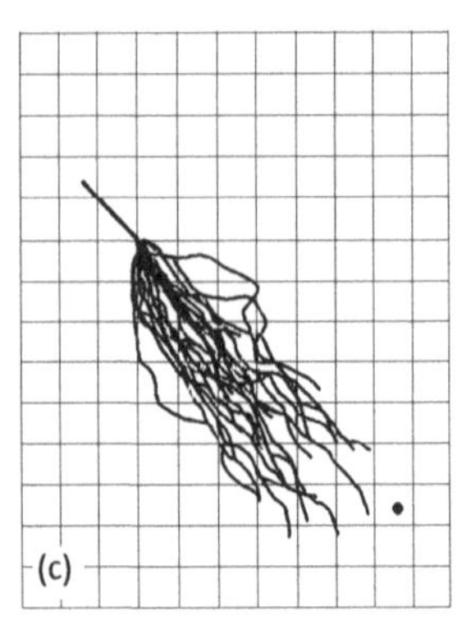

Fig. 16.24 Path integration: ants were trained to leave the nest (solid spot) and visit a source of food (far left, top) via open ground and a channel. (a) They are able to retrace their steps; (b) if the channel is shortened, they will (at least in part) adjust the angle at which they strike out from the channel; (c) similarly, if the channel is rotated. (From Collett *et al.*, 1998.)

In the absence of visual landmarks, ants can find their way back home by 'path integration' (Fig. 16.24). They can apparently set a course on the basis of an inner record of the directions in which they have travelled and the distance for each. The appropriate vector must be continually updated with movement over the ground.

One of the most remarkable examples of sensory processing is that by honey bees which deduce the direction and distance of a food source on the basis not only of the 'waggle (figures of eight) dance' executed by other workers at the entrance to the hive (as shown by Von Frisch in 1950), but also by their 'song' – the sounds emitted by their vibrating wings. Bees with the mutation called 'diminutive wings' are unable to recruit other workers in this way.

16.6.2 Sensory systems

Information flows into the central nervous system through sensory pathways, being modified en route at a series of centres of synaptic interaction (Fig. 16.25). A series of modules – virtually identical synaptic centres – is present at a given level. They are often connected by local interneurones and are thus able to influence each other. The best known of such interactions results in lateral inhibition, first described in the classic work on *Limulus* by Hartline. The response of a single ommatidium when it alone is exposed to a given level of illumination is greater than when the whole area is illuminated, since the cell is then subject to the inhibitory influence of its neighbours. Sensory systems also show a hierarchy of organization – information passes to progressively higher and higher centres.

The optic pathway in insects involves three distinct regions of the brain – the laminar, medulla and lobular regions of the optic lobe (Fig. 16.26) – in addition to the protocerebrum. Each ommatidium within the eye has a module or *cartridge* of second-order neurones in the laminar, and then a *column* of third-order cells in the medulla. This principle of retinotopy means that a pattern of illumination falling on the eyes is reflected repeatedly in morphologically ordered patterns of activity at four different levels of the visual system. 'The complexity of the insect retina [i.e. optic lobe] is something stupendous, disconcerting, and without precedents in other animals. When one considers the inextricable thicket of compound or faceted eyes; when one penetrates the labyrinth of neurones and integrating fibres of the three great retinal segments . . . one is completely overwhelmed.' (Ramon y Cajal, 1937.)

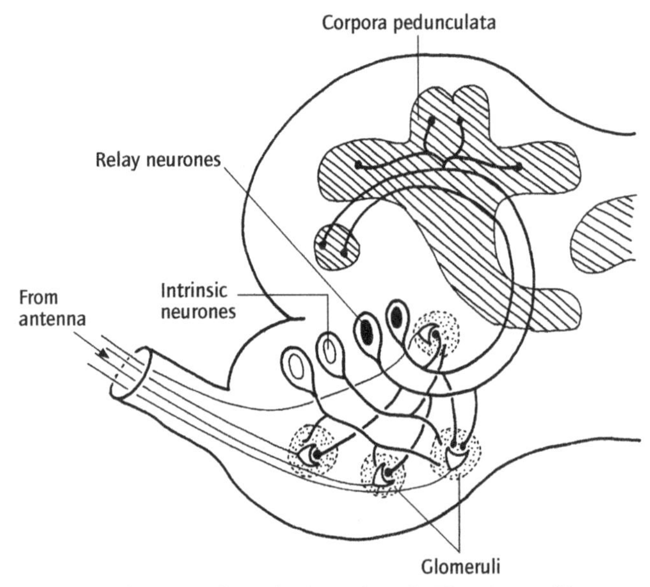

Fig. 16.25 Olfactory pathway in the cockroach. The 'glomeruli' are synaptic centres consisting of triads, involving input terminals, intrinsic neurones, and relay neurones conveying the processed information to higher centres. (After Boeckh *et al.*, 1975.)

16.6.3 Convergence and divergence

The convergence which characterizes insect ocelli is typical of many sensory systems. A related idea is that of the sensory field – the array of receptors which provide sensory input to a cell or centre in a nervous pathway. For example, the medial giant fibre of an earthworm has an anterior sensory field – its main input is from receptors situated anteriorly – whereas the pair of lateral giant fibres have mainly posterior fields. The two fields overlap in the mid-region of the body.

Sensory pathways are also characterized by divergence, since information from a single receptor, or group of receptors, is conveyed into the central nervous system via multiple, or parallel

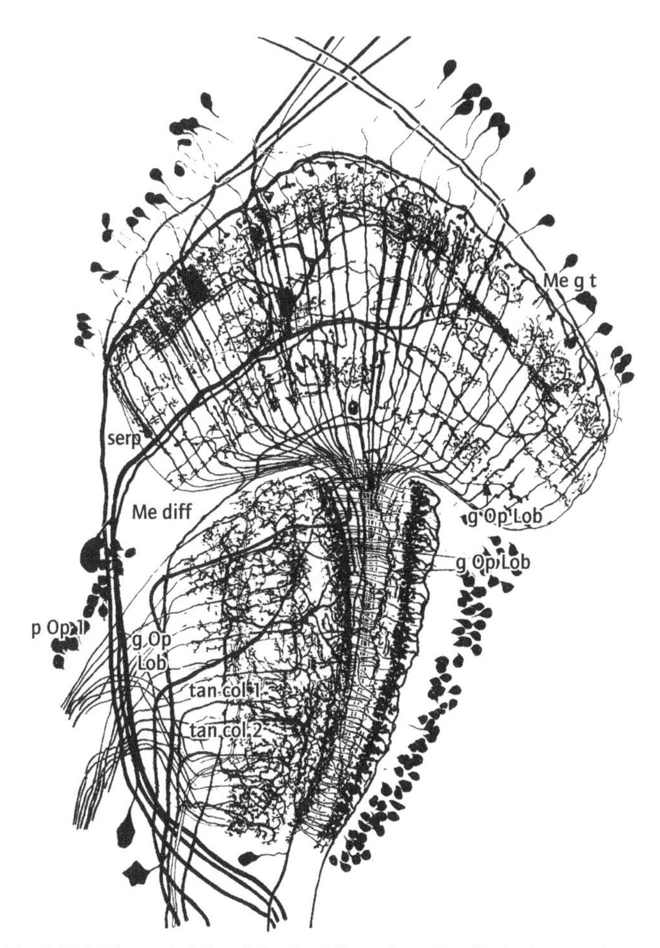

Fig. 16.26 The optic lobe of the fly, *Musca domestica*. In the lamina (not shown), all the cell types, and most of the connections they make, have now been identified and the majority studied electrophysiologically. Great progress has also been made in exploring the medulla (upper part of figure) and lobula (lower part). (From Strausfeld & Nassel, 1981.)

pathways. Such pathways can be used to extract and segregate different types of information. Divergence is also characteristic of sensory systems in that, although a system is usually responsive to only a single sensory modality, it provides output to a number of motor centres and thus influences several different types of behaviour.

16.6.4 Labelled lines

In contrast to signals which are differentiated by the chemical diversity of the transmitters, etc. involved, many different items of information transmitted by the nervous system involve apparently identical patterns of electrical changes and the same chemical mediators. The significance of the message depend on the neural elements involved, which are therefore said to be 'labelled'.

This principle can be seen in operation in the escape response of the direction sensitivity in relation to auditory input in the cricket, and photoreception in the leech, which depend on the stimulation of an ipsilateral interneurone and the inhibition of a contralateral one.

The lunging attack of a toad, a natural predator of cockroaches and crickets, creates a current of air which is detected by sensory hairs on the anal cerci of the insect (removal of the cerci reduces the chance of escape). The hairs are arranged in a number of columns, which are sensitive to wind from different directions. The different columns form distinct combinations of connections with giant interneurones extending up to the thorax, and the various giants indirectly stimulate different motoneurones. For example, giant number 5 is activated only by puffs of air from the rear quadrant on the same side of the body, and excites the slow depressor motoneurones (and muscles) on that side to induce a turning motion away from the source of the air. When the cerci are rotated experimentally, the animal is 'fooled' and reacts as if the puffs were coming from a different direction.

Direction sensitivity to touch in the leech depends on a combination of labelled lines and impulse frequency. The four neurones in each segment are maximally sensitive to touch in positions at right angles to each other. Touch at any point between two cells causes a response by each which bears a precise mathematical relationship to the angle between the cell and the point of contact. The responses of the two combine to specify the point of contact – creating a vector effect which results in a bending movement of the body away from the touch.

16.6.5 Feature extraction

A single photon is adequate to produce a measurable 'bump' in the membrane potential of a photoreceptor cell, and a single molecule can have a similar effect on a chemoreceptor. Seeing that an individual may possess millions of receptors, it is clear that the animal must filter out from its sensory input those features of significance, or salience, and regard the rest as redundant. Directionally sensitive movement detection is an example of feature extraction whose neural basis has been subject to intensive study (see Box 16.6).

Arthropods show characteristic patterns of behaviour, such as the claw-waving, sexual displays of fiddler crabs, which are often very striking and stereotyped. They act as 'sign stimuli' or 'releasers', which elicit particular behavioural responses but mean nothing to members of other species. Apparently, the sensory and nervous systems are organized to detect, conduct and amplify such signals, whereas others are ignored.

16.6.6 Centrifugal control

Information flow in sensory pathways is not one-way – nerve fibres may be directed back from any level to more peripheral centres (i.e., centrifugally – see Fig. 16.19) and mediate, for example, negative feedback effects. The movement detector neurones in the locust are influenced by both inhibitory and stimulatory centrifugal fibres from the brain. When the locust

Box 16.6 Directionally sensitive movement detection

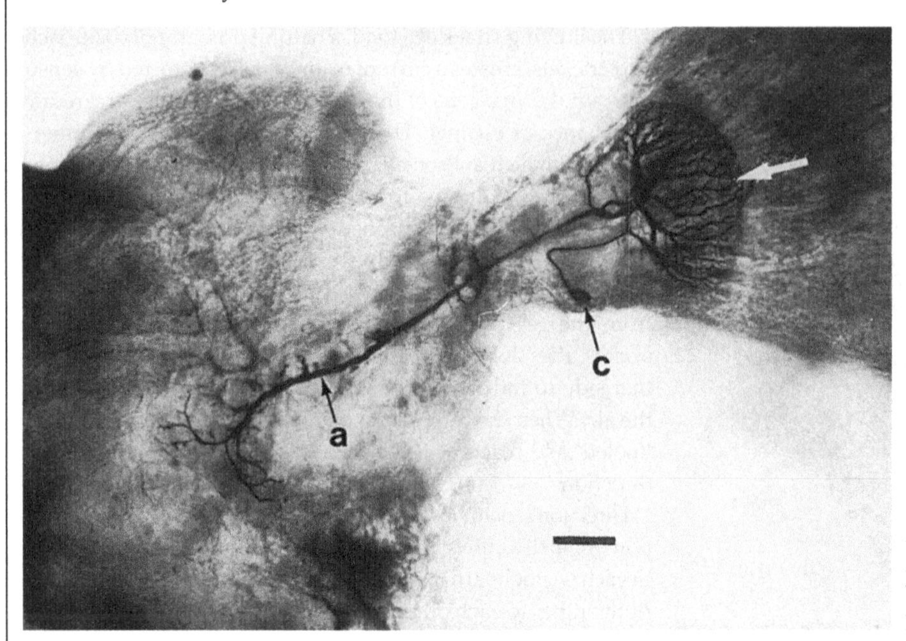

(a) LGMD neurone in the optic lobe of the locust injected with cobalt chloride solution. Arrow, fan-like array of dendrites in the optic lobe; c, cell body; a, axon extending into the brain (left); bar, 100 μm. (Courtesy, Dr Claire Rind, University of Newcastle upon Tyne.)

As described above (Section 16.6.2), a whole series of retinotopic projections extend into the optic lobe of insects. In flies, the H1 neurone is a 'directionally sensitive movement detector' – it is only activated by objects moving forward over the eyes. This depends on the sequential stimulation of receptors (retinular cells) 1 and 6 of individual ommatidia (see Box 16.5) – in that order, but not the reverse.

In the locust, neural pathways from the retina feed into the 'lobular giant movement detector' (LGMD) neurone (Fig. a), which has a large, fan-like array of dendrites so that it can receive stimuli from units covering any part of the visual field. Work by E. C. Rind has included making recordings of neural activity while showing the locust a video of a well-known space movie. The LGMD neurone is stimulated only if an object in the visual field is *moving*. Units pre-synaptic to the LGMD neurone are phasic – each fires once and then becomes quiescent. Consequently, only a moving object, which excites a whole series of units, will provide a succession of stimuli onto the LGMD neurone and induce its activation. Furthermore, the neurone shows vigorous and increasing levels of activity only in response to *approaching* objects (Fig. b) when the image of the object on the retina expands with increasing speed. Clearly, the system is finely tuned to the detection of impending collision, since even minor deviations (2–3°) from such a course greatly reduces the response.

The effect is thought to depend on multiple inputs from the medulla which stimulate the LGMD, but inhibit each other (Fig. c). A computer network has been modelled in accordance with this pattern and performs obligingly!

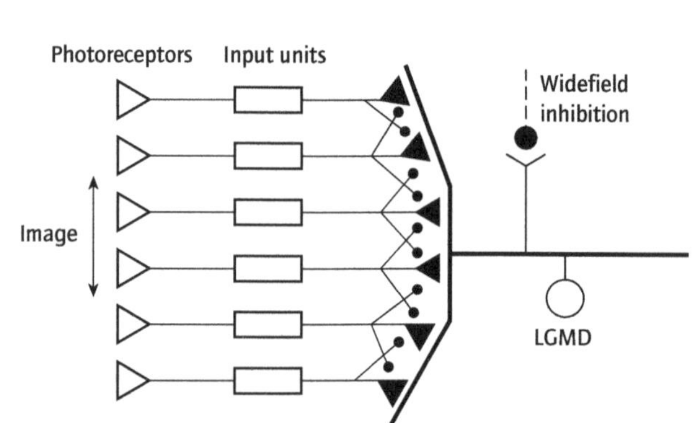

(c) Diagram of the neural circuits thought to underlie movement detection by the LGMD neurone. Input units stimulate (filled cones) the LGMD, but inhibit (filled circles) each other. (From Simmons & Young, 1999.)

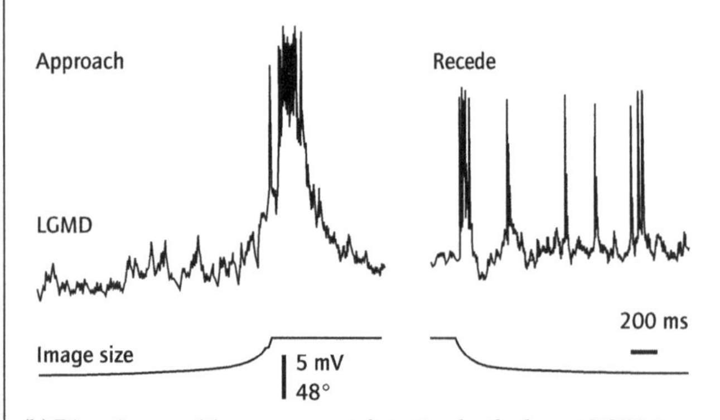

(b) Direction sensitive movement detection by the locust LGDM neurone: intracellular recording (upper) in response to object approaching, and then receding, at 3.5 m s^{-1} (From Rind, 1996.)

moves its head rapidly, the 'descending contralateral movement detector (DCMD) neurones', which are controlled by the LGMD cells (see Box 16.6), are inhibited, since otherwise any small stationary object would move across the visual field and activate the system. Conversely, the system can be rendered more sensitive to visual stimulation when the organism enters a heightened state of arousal.

16.7 Spontaneity

16.7.1 Neural initiation

A regrettable side-effect of the major contribution to neurobiology made by studies of reflex arcs has been the tendency to regard nervous systems as being as dependent on external stimulation for their activity as a switched-off computer. Nothing could be further from the truth, as even the most superficial observation of the behaviour of animals – from *Hydra* to *Octopus* – will reveal. The capacity of the nervous system for spontaneous activity – its role in initiating events – is as crucial as its ability to respond to changes in the environment.

Endogenous initiation of activity can result as development of the nervous system is completed. However, much spontaneous activity is rhythmical and, when it occurs at short intervals of time, may be due to spontaneous potential generation (Section 16.1.2). Synchronized, rhythmical bursts of activity ('brain waves') involving large numbers of neurones are well known in vertebrates and man. They are widespread in invertebrates also (Fig. 16.27) and generally show a high frequency (above 50 Hz) except in *Octopus*, in which they are, intriguingly, like vertebrates (below 25 Hz). Other rhythms have far longer periods and a different basis in cell biology (Box 16.7).

16.7.2 Movement

In *Hydra*, series of spontaneous contractions of the epidermis occur, each series resulting in the withdrawal of the body to form a small ball. They are caused by series of 'contraction pulses' – electrical impulses which can be recorded from the surface of the body. Pulses are initiated in the epidermal nerve net in the subhypostomal region, since an animal treated twice with colchicine, which is selectively toxic to nerve cells, becomes entirely quiescent. Once initiated, the pulses are propagated in a through-conducting manner, probably in both the nerve net and the epidermal epithelium.

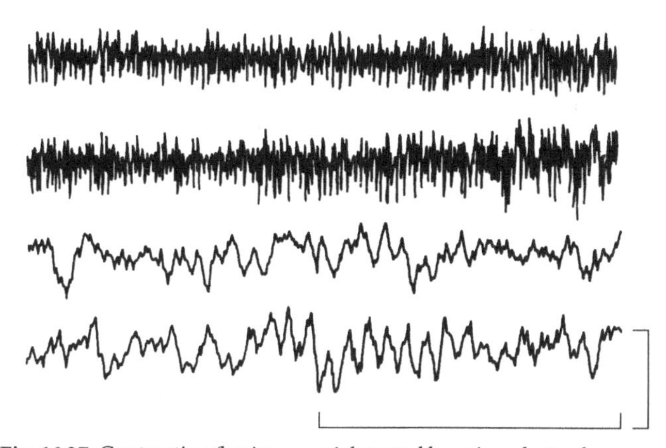

Fig. 16.27 Contrasting 'brain waves' detected by microelectrodes positioned on the surface of the brain of a crayfish (upper two traces) and a frog (lower traces). Bars, 1 s and 50 μV. (From Bullock & Basar, 1988.)

Spontaneous activity on the part of the marginal ganglia of scyphozoan medusae generates the rhythmical contractions of the bell involved in swimming. One of the ganglia takes the lead for a time and then another will take over. The leader generates action potentials at intervals of, say, 2 s for an animal 2-cm across or 20 s for a 20-cm specimen. These spread widely across the under surface of the bell in the 'giant fibre nerve net' (Section 16.3.2). They stimulate the swimming muscles and also reset the other ganglia so that they do not themselves fire and interfere with the rhythm.

After activity there is need for rest, even for 'busy bees'. Honey bees enter a state of profound rest at night, with remarkable similarities to the phenomenon of sleep. Movement, muscle tone, body temperature, sensitivity to stimuli – all are reduced. In contrast to the pattern in mammals, a condition comparable to 'deep sleep' is more pronounced towards the end of the sleep phase.

16.7.3 Autonomic functions

The annelid *Arenicola* lives in a burrow in the sand (Fig. 9.30), which it ventilates to aid respiration (Section 11.4.5). It shows periodic pumping movements, at intervals of about 40 min, which cause a current of water or bubbles of air to flow over its gills. A complex sequence of movements is involved – first locomotion to the tail-end of the burrow; then anteriorly directed, peristaltic waves of contraction of the body wall, associated with some anterior movement; lastly, the direction of ventilation is briefly reversed. This activity is not a reflex response to lack of O_2 (although it is modulated by the latter), but is generated by multiple pacemakers, co-ordinated in life, present in the ventral nerve cord. Similarly, spontaneous feeding movements with a shorter time interval (about 7 min) arise from nervous elements in the oesophagus.

Together with respiratory movements, the heart beat is one of the most obvious rhythmical functions of animals. In ascidian urochordates, the activity is myogenic and has its primary source in the muscle cells – the heart is unusual in lacking innervation. In insects, the heart beat is myogenic, but is modulated both by local nervous influences and by hormones released from the ventral nerve cord (e.g. during flight); in leeches, a peptide neuromodulator (Section 16.2.3) helps to maintain the capacity of the multiple hearts for spontaneous activity, and excitatory and inhibitory transmitters regulate the timing of the beat. In contrast, the heart beat is neurogenic in crustaceans and has its source in the activity of the cardiac ganglion. This small

Box 16.7 Molecular biology of clock function

Early investigations of the basis of clock function in *Aplysia* involved the administration of anisomysin, which blocks protein synthesis within the cell by binding to a subunit of the ribosomes. Introducing a brief pulse of the substance (but only during the dark phase of the cycle) stops the clock temporarily, so that it 'runs slow'. Jacklet (1981) concluded that 'the daily synthesis of protein is a general requirement for circadian clocks'.

Investigations of circadian rhythms in *Drosophila* led to the discovery of the period gene by R. J. Konopka & S. Benzer, in 1971, and several other 'clock genes' have since been identified. We now know that the molecular basis of clock function is essentially the same in organisms as diverse as *Neurospora* and mammals and involves *intracellular* oscillation – it is apparently never dependent on cell-to-cell communication (unlike more rapid oscillations in neural activity – for example, see Fig. 16.34). A *positive* element which stimulates gene activity (and prevents it fading out altogether) is coupled to an element which provides *negative feedback*.

A model for circadian clock function in *Drosophila* includes the following features (Fig. a). Activity of the *clock* and *cycle* genes leads to synthesis of positive elements, the proteins CLK and CYC. These form *heterodimers* (made up of one CLK and one CLC molecule) which activate genes whose products are actually responsible for rhythmical behaviour and metabolism.

CLK and CYC heterodimers also activate the *period* and *timeless* genes. Once their products, PER (a protein made up of approximately 1200 amino acid residues) and TIM, are present in sufficient quantity, they too heterodimerize and, in so doing, become capable of entering the nucleus where they block the actions of CLK/CYC (and thus inhibit off their own production) (Fig. b).

Why is it that the negative feedback effect of PER/TIM does not just stabilize the system? The reason, probably, relates to the delay involved in the accumulation and dimerization of PER and TIM. This will cause an overshoot effect, choking off the actions of CLK/CYC. Once PER/TIM is broken down – and this process also introduces a delay – a new cycle can begin. Adjustment of the rhythm results from the action of light in accelerating the breakdown of TIM.

Various mutants of the *period* gene have been identified – for example, with rhythms of 19 hours (*per*^{s}), 29 hours (*per*^{l}), or which are arhythmic (*per*^{o}) (Fig. c) – and, curiously, all of them also show modifications of the rhythm, 55 seconds long, of the courtship song. The gene has been sequenced and mutants have been restored to normal rhythmicity by genetic engineering!

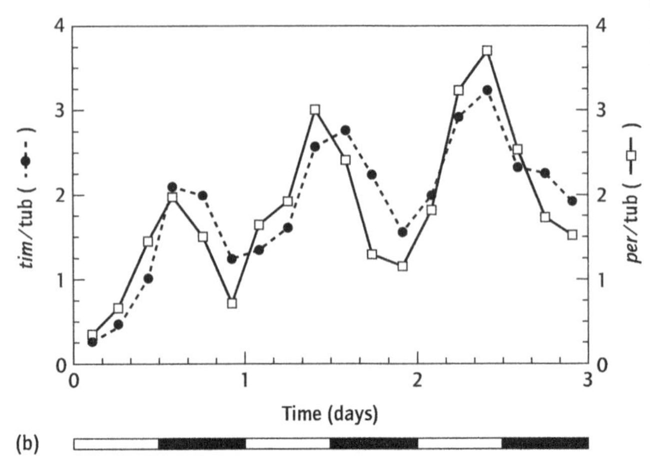

(b) Rhythmical production of *tim* RNA (filled circles) and *per* RNA (open squares) under free-running conditions (constant darkness); normal pattern of illumination shown below. Note co-ordinated expression of *tim* and *per*. (From Sehgal *et al.*, 1995.)

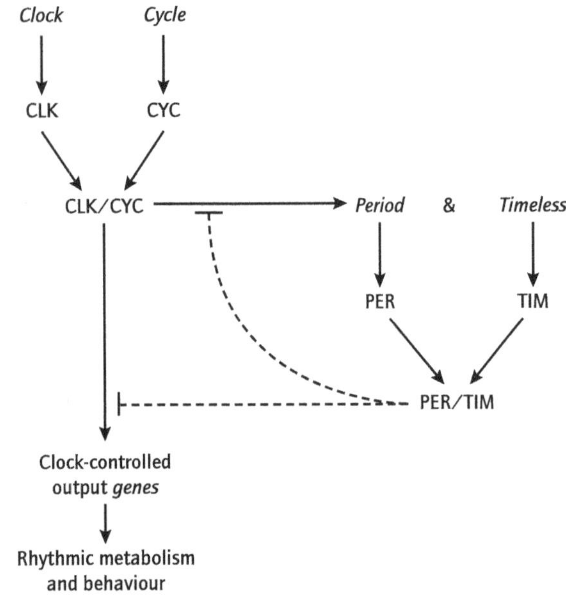

(a) Model of circadian clock function in *Drosophila*. Broken lines and bars indicate inhibitory feedback. (After Dunlap, 1999.)

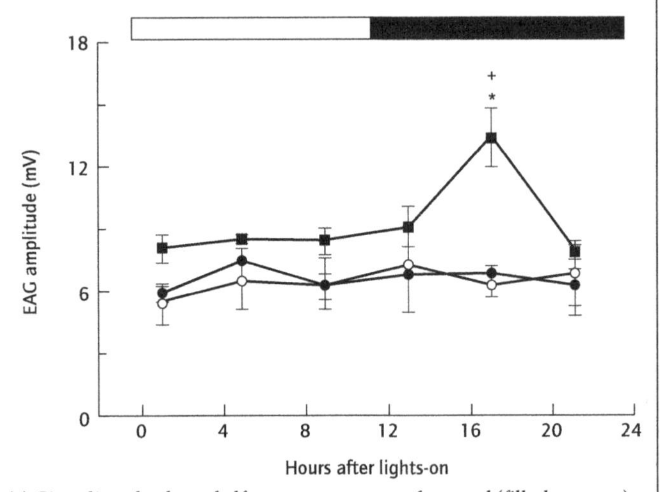

(c) Circadian rhythm of olfactory responses of normal (filled squares) *Drosophila* during the first-day free-running in constant light; open and filled circles show absence of clock mechanism in flies with mutant *per* and *tim* genes, respectively. The bar indicates the normal pattern of illumination. (From Krishnan *et al.*, 1999.)

ensemble of neurones represents the nervous system in microcosm. All the cells show spontaneous activity, although in life they are co-ordinated by one which acts as a pacemaker. The ganglion is subject to the influence of higher centres (i.e. by synaptic input from the central nervous system), to the feedback effects of sensory neurones associated with the heart, and to endocrine influences.

16.7.4 Biological clocks

Some periodic phenomena imply that a 'clock' mechanism is in operation, keeping track of the time of day or the passage of time. A classic study was the work of Pittendrigh on the circadian (*circa dies* – about a day) rhythm of pupal eclosion in *Drosophila pseudoobscura*.

Clocks are indispensable to animals such as bees and ants which navigate by the sun. Such animals can maintain an appropriate course over periods of time, making adjustments to compensate for the apparent movement of the sun. Many organisms have the ability to respond to changing day length, for example, by modifying their pattern of development or reproduction with the approach of autumn (e.g. see Section 14.4.4). This involves the capacity to 'measure' the length of day and/or night and a notable feature of such mechanisms is that they are temperature-compensated (i.e. independent), remarkable in poikilothermic invertebrates, since major errors in the measurement of time would otherwise result.

Circadian rhythms are entrained by the daily cycle of illumination and thus synchronized with the latter. However, if the animal is maintained under constant conditions, these rhythms persist for several days with only a small change in the length of each cycle. The length of this 'free-running' rhythm in the cockroach *Leucophaea* is affected by the length of the 'daily' cycles to which the animals are exposed during development and cannot be adjusted later. Clock mechanisms are aften 'masked' until favourable conditions allow the expression of the on-going oscillation.

In the mollusc *Aplysia*, rhythmical activity is widespread in the central nervous system, and it continues for several weeks even when ganglia are removed from the body and kept in organ culture. Different neurones have their own rhythms, but in life these are synchronized by a 'master clock' located within the 'D cells' in the eye (not the photoreceptors, but cells which receive synaptic input directly from them). A circadian rhythm of activity can be detected within the isolated eye and the cells exert their influence not by a hormone as was once thought, but via axons which extend throughout almost the entire nervous system. In contrast, autonomous circadian clocks are present throughout *Drosophila*. The oscillations of a set of neurones in the optic lobe govern the well-known locomotory rhythm, whereas clock function by the chemosensory cells of the antennae is independent and controls a rhythm in olfactory function (see Box 16.7) (similarly, a clock in the brain of *Limulus* modulates the sensitivity of the compound eyes).

16.8 Neural bases of behaviour

16.8.1 Independent effectors

Some actions are undertaken by cells as adaptive responses to stimulation which they themselves receive – they are independent effectors. Some gland and muscle cells come into this category; chromatophores in many organisms respond directly to light by virtue of the presence of rhodopsin-like molecules; and co-ordination of the beat of cilia is brought about by mechanical interactions between adjacent organelles.

Protists are of necessity independent effectors, yet show mechanisms of control identical to some of those operating in nervous function. Encounter with an obstacle anteriorly by *Paramecium* induces membrane depolarization, an unusual example of a graded action potential (Section 16.1.5) mediated by Ca^{2+} channels located in the ciliary membranes, and causes a brief pause in swimming. With more intense stimulation, Ca^{2+} influx activates Na^{+} channels and reversal of ciliary beat. Stimulation posteriorly increases K^{+} permeability, elevates the resting potential and accelerates the beat. The graphically named *dancer* mutant has hypersensitive Ca^{2+} channels.

Nerve fibres are known to form synaptic junctions with cnidocytes (stinging cells in cnidarians – see Fig. 3.14) and doubtless mediate the influence of the nutritional state of the organism on cnidocyte function. They are probably also responsible for the propagation of discharge across a field of the cells as observed in anemones. However, cnidocytes can also act independently: a cell's chemo- and mechanoreceptors combine, when stimulated, to generate a receptor potential; Ca^{2+} enters via voltage-gated channels in the plasma membrane; and 'exocytosis' of the cyst results in the latter's discharge.

16.8.2 Units of behaviour

Most effectors are not independent. Behaviour typically owes its origin to activity generated spontaneously within the nervous system or represents a response to stimulation. Some units of behaviour called reflexes are stereotyped, relatively simple motor actions each evoked by a specific stimulus; the action varies in strength and/or extent depending on the intensity of the stimulus.

Other behavioural sequences are called 'fixed action patterns' or, now more commonly, 'motor patterns'. Such patterns are species-characteristic, stereotyped actions, often of considerable complexity, and do not vary with the strength of stimulation. They may be spontaneous, but if evoked by stimulation, this acts only as a trigger. In consequence, quite different stimuli can evoke the same pattern – for example, crayfish adopt a characteristic defensive posture in response to a wide variety of threatening stimuli. Motor patterns are genetically determined and the expression of a precise 'programme' built into the 'wiring' of the nervous system. Feedback influences have no role in the control of the pattern, as such.

We may conclude that the two concepts differ mainly in the role ascribed to stimuli. Stimuli have a role in reflexes comparable to that of the finger movements of a person playing the piano, whereas in motor patterns, they correspond at most to the action of switching on a recording.

16.8.3 Reflex actions

Many cnidarians respond to light touches by a slow, local contraction, the extent of which depends on the intensity, number, and frequency of stimuli. The classic explanation, developed by Pantin, depends on the process of synaptic facilitation within the nerve net, by means of which a second or later impulse arriving at a synapse triggers a response which the first was unable to evoke. However, different mechanisms underlie the same phenomenon in the hydrozoan *Cordylophora*, in which it is mediated by epithelial conduction (Section 16.3.1), and in the coral *Porites* (Fig. 16.28).

Like the knee-jerk response in man, the reflex involving the crayfish stretch receptor is monosynaptic, the connection between the receptor and the motor neurone being a single chemical synapse (Fig. 16.29). It contributes to the maintenance of posture, since if its associated muscle is stretched, activity in the receptor and then in the motor neurone which it drives will induce muscle contraction counteracting the stretching effect. The circuit also has a role in implementing instructions from the central nervous system to the musculature.

We have seen that many escape or startle responses involve giant axons (Section 16.1.5). Such elements are not present in leeches, but a fast, through-conducting pathway which mediates rapid contraction of the whole body is present in the nerve cord. It consists of a series of 'S-cells', one per segment, whose longitudinally directed axons are coupled to those in adjacent segments by electrical synapses (Fig. 16.30). The system illustrates the importance of electrical (and therefore fast) synapses (Section 16.2.5) in circuits mediating escape responses. The various types of mechanoreceptor within the body wall also form monosynaptic reflex arcs with motoneurones in the same segment.

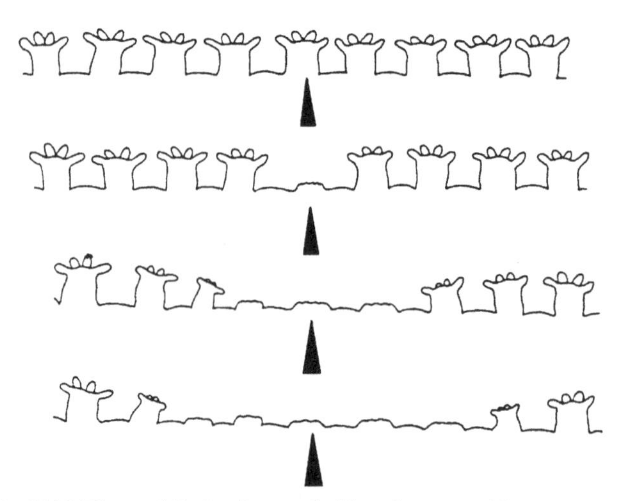

Fig. 16.28 The coral *Porites* shows a facilitated response (the terms 'decremental' and 'incremental' are also used) to successive, electrical stimuli, despite the propagation of potentials throughout the colonial nerve net with each and every stimulus. (After Shelton, 1975.)

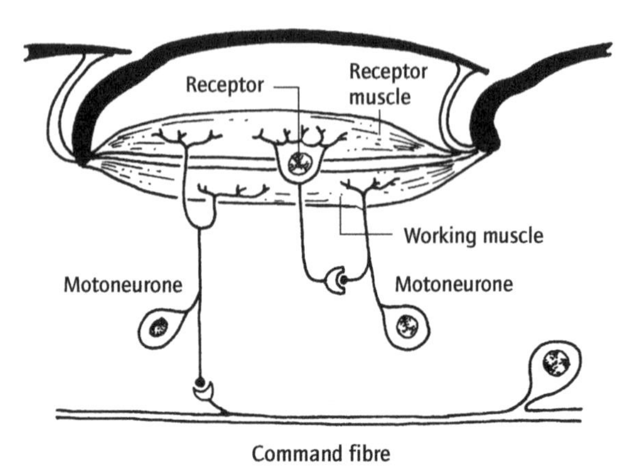

Fig. 16.29 Neuronal circuits involving the stretch receptor of the crayfish. If contraction of the 'working' and receptor muscles due to activation of the motoneurone is impeded by external factors, the receptor is activated, leading to further stimulation of the muscle. The receptor also mediates the monosynaptic stretch reflex. (Inhibitory innervation of the receptor cell by centrifugal fibres is not shown.) (After Kennedy, 1976.)

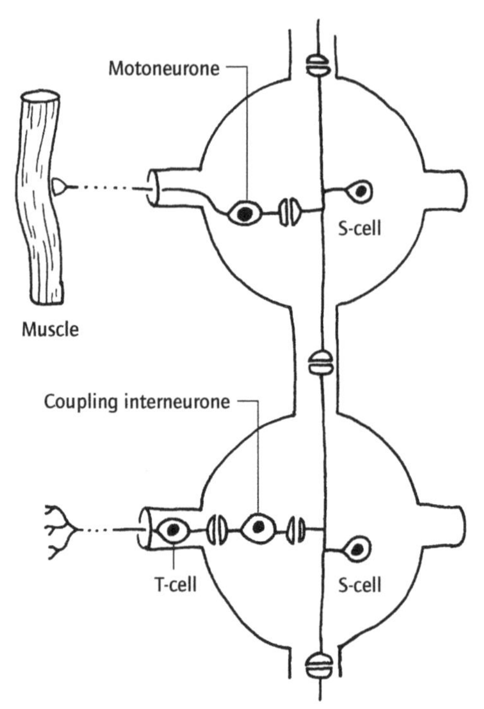

Fig. 16.30 Neuronal circuits involved in the startle response in the leech. Transmission from the sensory T-cells through the coupling interneurones, a series of S-cells, and into the motoneurones, is achieved exclusively by electrical synapses – only the neuromuscular junctions are chemically transmitting. Only one side of two adjacent ganglia is shown. Note the classical pattern for a reflex of the involvement of sensory, inter- and motor neurones.

Box 16.8 Crayfish tail-flips

Crayfish tail-flips have been regarded as typical motor (fixed action) patterns. Sudden, threatening stimuli intense enough to evoke a single impulse in the giant fibres initiate a rapid, highly stereotyped, escape response – a 'tail-flip' – caused by the simultaneous contraction of the flexor muscles of the abdominal segments. The action is completed too rapidly to permit its modification by feedback.

Activity in the giant fibres inhibits the motoneurones to the extensor muscles (antagonists of the flexors), the extensor muscles themselves, and the muscle stretch receptors of the extensors, thus reliably preventing contraction of the extensors, and their interference with the tail-flip, until maximum force is exerted by the flexors. The giant fibres are thus acting as *command neurones* (Section 16.9.1) whose output brings into play the various co-ordinated components of the action.

However, after the abdomen is flexed, there is a rapid re-extension of the body by contraction of the extensor muscles. This 'recovery stroke', an invariable part of the natural sequence, is *not centrally generated*, but is a *chain reflex effect*. The muscle receptors, once freed from their inhibition, are stimulated by being stretched and activate the extensors by a reflex arc (Section 16.8.3). Furthermore, the flips that follow the first are not controlled by the giant fibres but by a parallel, non-giant pathway. The latter receives the same sensory input but has more slowly conducting components and incorporates directional information propelling the animal away from the source of the threat.

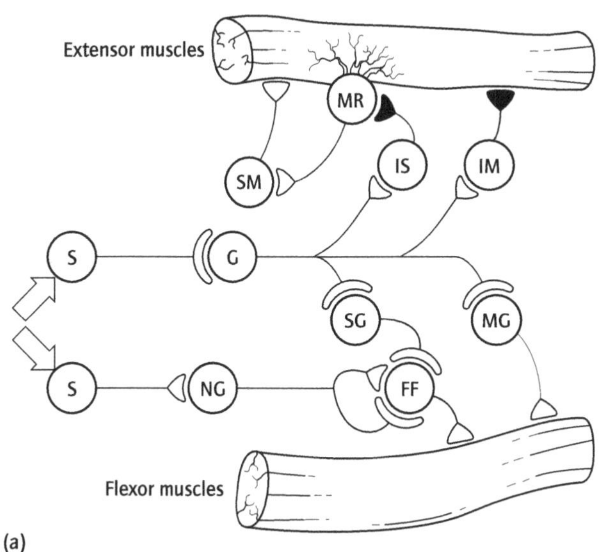

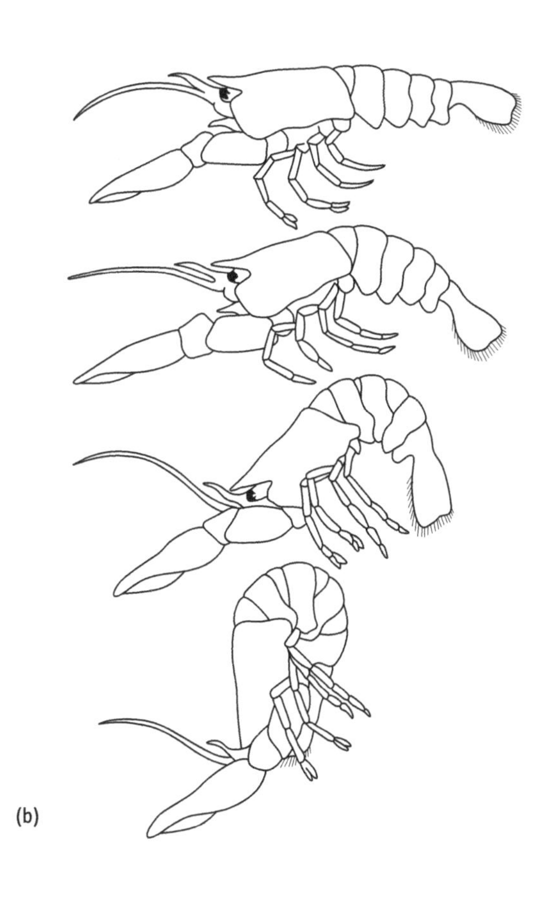

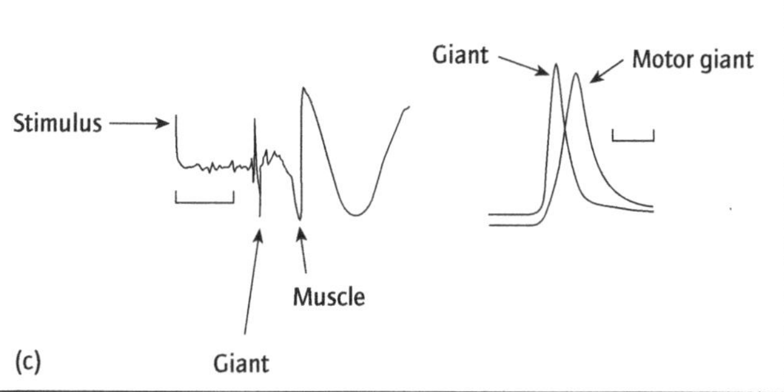

(a) Simplified neuronal circuits involving both giant neurones and non-giant pathways, present within each abdominal segment, controlling crayfish tail-flips. Chemical synapses shown by bulbous terminals, electrical synapses by flat terminals. The systems are activated by environmental stimuli (arrows). FF, fast flexor motoneurones; G, giant command neurone; IM, inhibitory motoneurones; IS, inhibitory input to stretch receptor; MG, motor giant neurones; MR, muscle stretch receptor; NG, non-giant neural pathways; S, sensory neurones; SG, segmental giant neurones; SM, stimulatory motor neurone. The synapses between the giant fibre and the motor giants were the first electrical synapses to be discovered (see Section 16.2.5). (b) Crayfish tail-flip. (After Wine & Krasne, 1982.) (c) Left: Electrical activity in the giant command neurone, about 7 msec after a stimulus is applied to the animal, is followed by activity in the flexor muscles after about 10 msec. The short latency is typical of escape responses. Bar, 5 msec. Right: Activity in giant neurone is followed by activity in motor giant with a negligible delay (about 0.1 msec), on account of coupling with an electrical synapse. Bar 1 msec. (From Simmons & Young, 1999.)

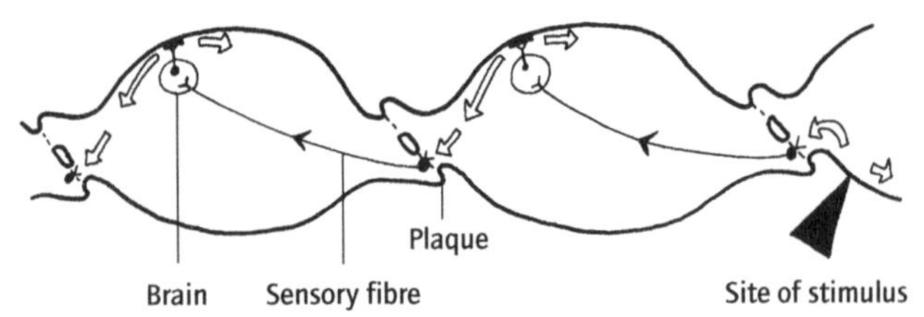

Fig. 16.31 Reflex arcs within each of a series of *Salpa* zooids. Adjacent zooids are connected by epidermal plaques. The cells on the output side of a plaque have an ultrastructure resembling that of pre-synaptic terminals. Facing them are 6–12 ciliated receptor neurones. Conduction of pulses throughout the chain may induce its dissociation. (After Anderson & Bone, 1980.)

Perhaps the most curious reflexes ever described is shown by some salps (urochordates), which form elongate chains of up to 20 'zooids' interconnected by epidermal plaques (Fig. 16.31). Excitation is mediated by a sensory pathway from plaque to the brain, motor innervation of the epidermis, epithelial conduction to the next plaque, and so on.

16.8.4 Motor patterns

The inherent capacity of the central nervous system to generate and co-ordinate integrated patterns of motor actions has been demonstrated in many groups of animals. 'The withdrawal of the mantle and closure of the valves in a clam, *Mya*, . . . copulatory movements . . . the flight rhythm, walking . . . respiratory movements . . . heart beat . . . swimming beat of jellyfish have all been shown with varying degrees of rigor [*sic*] to be centrally controlled' (Bullock, 1977).

The sea slug *Tritonia* reacts to touch by a predatory starfish with a bout of swimming lasting about 30 s. This involves alternate contractions of the dorsal and ventral muscles, and could conceivably involve either a centrally generated pattern, or a chain of reflexes – the initial stimulus evoking contraction of the dorsal muscles, which by sensory feedback through reflex arcs stimulates the ventral muscles and so on. Investigations have shown that the normal, characteristic sequence of nervous activity in the motoneurones can be provoked in brains kept *in vitro* by an initial stimulus applied to the severed sensory nerves. Clearly, in this case, even if sensory feedback is provided by the muscles, it is not essential for pattern generation (see Box 16.9).

16.8.5 Variations upon two themes

Many units of behaviour share at least some of the characteristics of both reflex actions and motor patterns (Box 16.8). In the locust, a single stretch receptor cell associated with the hinge of each wing is activated at the end of each upstroke. Here at least is a system which we might expect to work by means of a series of reflex arcs, but this is not the case, as shown by the classic work of D.M. Wilson. Even isolated thoracic ganglia, deprived of sensory input, can show the pattern of activity associated with flight (as can larvae before the development of wings). Instead, the receptors are restricted to modulating the timing of motoneurone activity in the current cycle and can also reset the rhythm.

The LGMD neurones of the locust (see Box 16.6) provide reliable synaptic output to the pair of DCMD cells which descend to the third thoracic ganglion, where each innervates two further identified neurones ('C' and 'M') (Fig. 16.32). Each C neurone synapses onto the motoneurones of two antagonistic muscles found in the 'thighs' of the large legs – these are the extensor and flexor muscles for the tibia (the lower part of the leg).

First, the C cell induces both 'cocking' of the leg, locking the tibia under the insect, and 'co-contraction' of the two antagonistic muscles which distorts the elastic cuticle of the femur. Second, a positive feedback influence from cuticular stress receptors acts on the motoneurones to reinforce extensor contraction. Last, feedback activity in another set of receptors, evoked by co-contraction, stimulates the M cell. If this coincides with continuing input from the DCMD, etc. the M cell is activated, inhibiting the flexor motoneurones and triggering the jump as the energy stored in the distorted cuticle is released. Clearly, sensory feedback is an integral part of the mechanism by which the jump is controlled and ensures that external stimuli cannot evoke the jump unless the mechanism is fully primed (see Section 10.6.2.4).

In conclusion, we must note the irreverence of Nature with respect to our categories. The nervous system clearly does have the inherent capacity to generate integrated motor programmes. However, some fixed action patterns are 'more fixed than others' – for example, some rapid withdrawal responses in annelids mediated by the giant fibres are graded, whereas others are all or nothing – and feedback effects often constitute one of the main ways in which patterns of behaviour achieve the goals to which they appear to be directed.

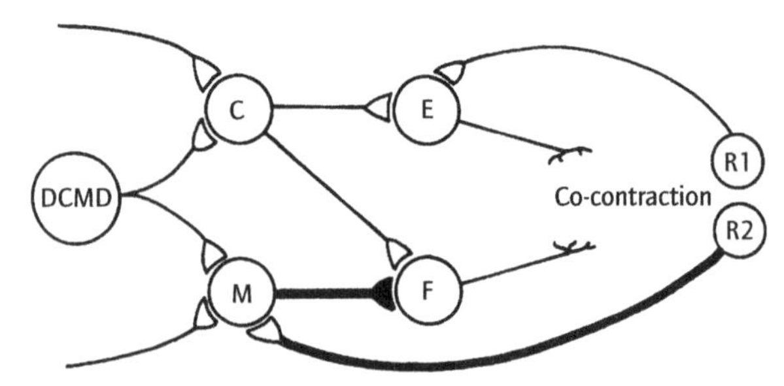

Fig. 16.32 Neuronal circuits controlling the locust jump. DCMD, movement detector neurone; C, C neurone; M, M neurone; E, F, motoneurones to the extensor and flexor muscles, respectively; R1, R2, sensory neurones mediating the feedback effects of co-contraction. Heavy lines, nervous pathways leading to the inhibition of the flexors. (After Pearson, 1983.)

16.9 Organization of motor output

16.9.1 Command neurones

Systems of neurones controlling motor output, like sensory systems, often show a hierarchy of organization and this is well illustrated by the circuits controlling the abdominal appendages (pleopods) in crayfish (Fig. 16.33). Command neurones were first discovered in crustaceans by Wiersma, who observed that activation of particular, single cells will evoke co-ordinated patterns of behaviour. A command neurone is a decision-maker. Just as sensory pathways show divergence and provide output to a multiplicity of motor centres (Section 16.6.3) so command cells within motor systems are the focus of convergence and integration of input from a diversity of sense organs. The resulting activity of the cell, or lack of it, determines whether or not the motor action is initiated. Such activity is therefore both necessary and sufficient to generate the action; it does not usually generate the motor rhythm, but provides tonic stimulation for the pattern-generating neurones situated at the next level in the hierarchy.

In a virtually complete description of the neural pathway, from sensory reception to patterned motor output, responsible for controlling swimming in the leech, neurones representing two levels of command function have been identified. A pair of 'trigger cells' in the brain (suboesophageal ganglion) receives direct input from over 150 epidermal mechanoreceptor cells situated throughout the body and controls a group of 'gating cells' present in each segmental ganglion. Once stimulated, the gating cells can show sustained activity without the support of the trigger cells. In contrast, gating cells evoke activity in the pattern generating cells (which make up the next level in the hierarchy) only so long as they, themselves, remain active.

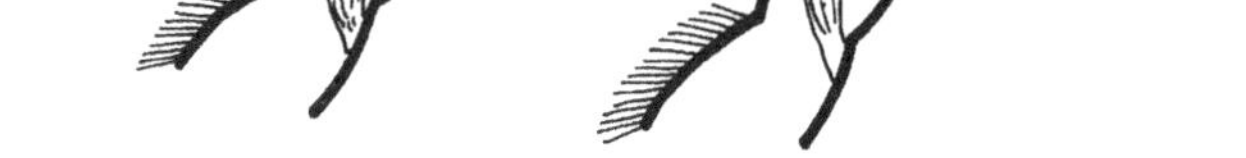

Fig. 16.33 Hierarchy in neuronal circuits controlling motor output of crayfish pleopods (cells in two adjacent segmental ganglia are shown). Three levels can be distinguished, namely, command neurones, pattern-generating interneurones, and motoneurones. More than one neurone of each type is usually present, including both stimulatory and inhibitory command cells. (After Stein, 1971.)

In many motor systems, no single cell, or group of cells, seems to correspond to a command unit (e.g. Fig. 16.32). Furthermore, although Mauthner's cells in teleosts and aquatic amphibia are command neurones, the concept has little general relevance to vertebrates.

16.9.2 Central pattern generators

Command cells where present determine whether or not a system is active, but the generation of patterns of motor output is neither their role, nor typically that of motoneurones, but of interneurones interposed between the two (Fig. 16.33). In the crayfish, one non-spiking interneurone in each segment shows regular, spontaneous oscillations in membrane potential. Depolarization and transmitter discharge stimulate the motor neurones of the muscles responsible for the power-stroke of the pleopod, but inhibit the motoneurones responsible for the recovery stroke. The role of the interneurone in pattern generation can be shown by accelerating or retarding the oscillation experimentally, when the rhythm is found to be reset; it does not just resume a pattern in time with the previous one. Other interneurones mediate the intersegmental co-ordination responsible for the metachronal rhythm of pleopod beating.

This hierarchy is not always so well defined. For example, CV1 cells can reasonably be regarded as command neurones for the central pattern generator (CPG) of feeding activity in the mollusc *Lymnaea*. However, feedback activity from the CPG onto CV1 changes the activity of the latter, modulating the character of its effects on the CPG. In this case, the command neurones are also part of the pattern generator.

Pattern generation by a neural network requires a source of activity, which may be provided by command input or may arise spontaneously (Section 16.1.2) from one or more of the network's constituent units. Excluding external influences, the character of the output of a CPG depends on both the properties of the individual neurones (e.g. their rates of intrinsic activity) and the nature of the synaptic interactions between them (see Box 16.9).

These principles are shown by the system controlling the heart beat in the leech. The CPG, one of the simplest and most common types encountered in invertebrates, is an oscillator consisting essentially of two neurones that show alternate activity (Fig. 16.34). The two units are spontaneously active and reciprocally inhibitory. Any tendency for one of them to establish a stable dominance over the other is prevented in two main ways. First, the synaptic output of an active cell declines with time, diminishing its influence on its partner; and, second, an automatic rebound effect within the inhibited cell – whereby, for example, hyperpolarization results in the delayed activation of certain ion channels which then depolarize the cell – enables the cell to resume activity.

Modulation of a CPG by the actions of neurotransmitters, on the properties of the neurones and of their synaptic interactions,

Box 16.9 Versatility of central pattern generation

The pattern generated by a neural network, and the activity of the component parts of the latter, can be *modulated* by synaptic and endocrine influences.

Intrinsic modulation is a feature of the central pattern generator (CPG) for swimming in *Tritonia* (Fig. a). Sensory stimulation activates a single command neurone, the dorsal ramp interneurone (DRI), on each side of the brain. The DRI does not generate the swim pattern, but stimulates six other interneurones to do so. None of them can generate rhythmical activity individually; the pattern is generated by their mutual interactions. 5-HT, released during activity by the dorsal swim interneurones, has multiple effects, reconfiguring the network to enable the interneurones to generate the swim pattern instead of others for which they are responsible. The phenomenon is called *intrinsic modulation* because the agents are part of the neural system involved.

In the lobster, neural networks within the stomatogastric ganglion form three distinct CPGs controlling muscles in the different parts of the stomach. For example, the pattern of muscle contractions in the pyloric region is generated by one interneurone and 13 motoneurones (the predominance of cells of the latter type in a CPG is unusual). At least nine different transmitters, released by nerve fibres originating in other ganglia, influence the output of the ganglion (Fig. b), enabling the stomach to deal with different foods at the various stages of digestion. Other networks generate the oesophageal and gastric mill patterns, each controlling muscle contractions in a different part of the gut. When a pair of pyloric suppressor neurones, which provide synaptic input to the ganglion, are activated, the pattern generators are reconfigured to produce a single integrated rhythm which drives swallowing movements (Fig. c). Such *extrinsic modulation* involves the influence of components situated outside the neural system involved.

In a similar way, a CPG in the abdominal ganglia of the moth *Manduca* drives a characteristic pattern of behaviour in the larva and pupa, by means of which the old cuticle is shed. The adult displays different pattern which it uses to dig itself out of the underground pupal chamber, but reverts to the larval/pupal pattern if the connectives between thoracic and abdominal ganglia are severed. Clearly, the underlying neural circuitry of the larva is retained by the adult, but is modulated to produce a novel, and more appropriate, sequence of behaviour*. These observations show that neural network forming CPG's are not necessarily 'hard-wired' – the output may depend on the 'programme'.

* In many other such cases, the circuitry is physically remodelled (see Box 16.3).

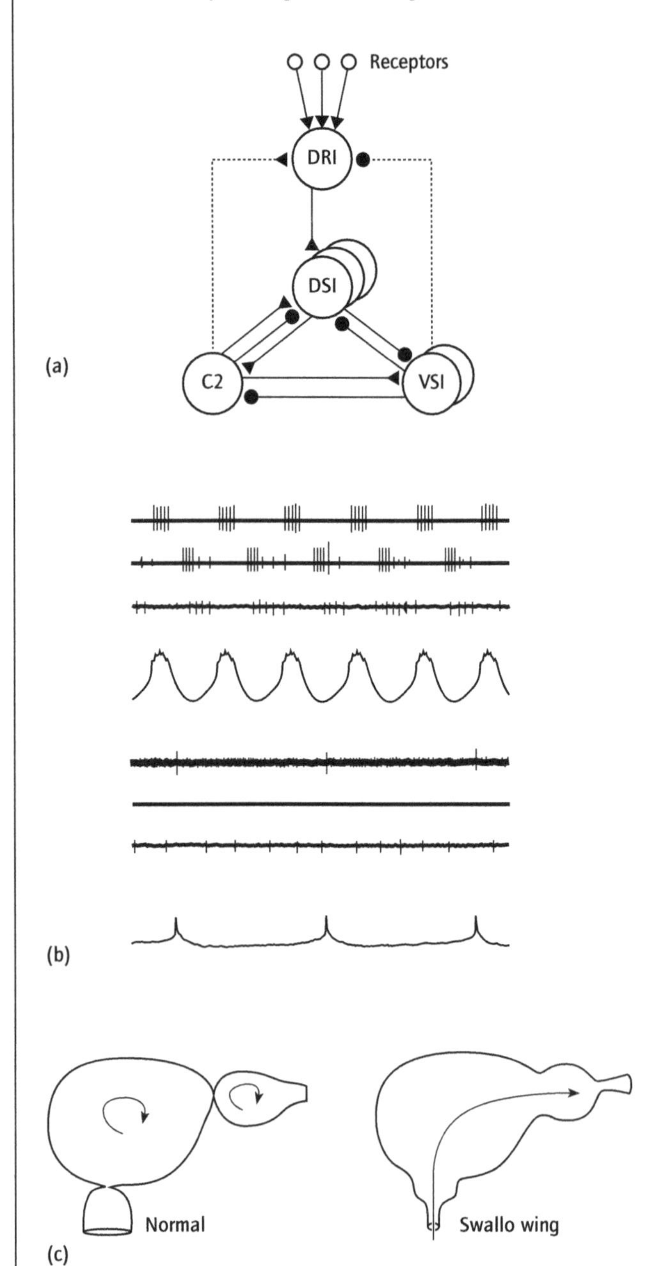

(a) Motor pattern generator for escape swimming in the sea slug, *Tritonia*. The dorsal ramp interneurone (DRI) provides tonic stimulation; the swim pattern is generated by three dorsal swim interneurones (DSI), two ventral swim interneurones (VSI) and one C2 interneurone. 5-HT produced by the DSIs modulates the functional characteristics of the system. (From Simmons & Young, 1999; after Frost & Katz, 1996.) (b) Effect of synaptic input from the commissural and oesophageal ganglia on the pattern of electrical activity generated by the pyloric CPG of the lobster stomatogastric ganglion: upper traces, normal pattern (three extracellular recordings from different motor neurones; one intracellular recording from the interneurone); lower traces, activity when deprived of normal synaptic input. (From Harris-Warrick & Flamm, 1986.) (c) Reconfiguration of the stomatogastric ganglion by extrinsic modulation: networks for the oesophageal, pyloric and gastric mill rhythms can operate independently, or can be reconfigured to generate an integrated rhythm which mediates the action of swallowing. (From Simmons & Young, 1999; after Meyrand *et al.*, 1994.)

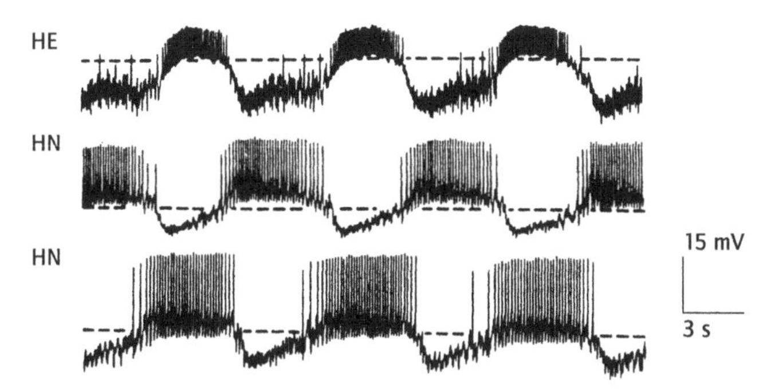

Fig. 16.34 Generation of the pattern controlling heart beat in the leech. Simultaneous intracellular recordings of rhythmical activity by two mutually inhibitory interneurones (HN) and of a motoneurone (HE) subject to the inhibitory control of the upper interneurone. (From Arbas & Calabrese, 1987.)

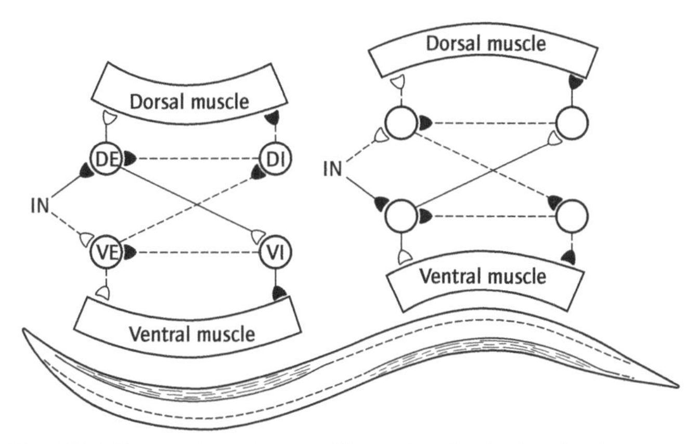

Fig. 16.35 Neuronal circuit controlling swimming in *Ascaris* which involves the alternate contraction of dorsal (left) and ventral muscles (right). IN, interneurones; DE, VE, DI and VI, dorsal and ventral excitatory and inhibitory motoneurones, respectively. Clear bulbs, excitatory; filled bulbs, inhibitory synapses. (After Stretton *et al.*, 1985.)

enables a diversity of patterns to be generated (Box 16.9). Similarly, patterns can be modified by varying the identity of the cells involved in their generation. Thus Burrows has found that a single motoneurone supplying a leg muscle in the locust is subject to the influence of at least 12 interneurones, which can be involved in different combinations. The presence of systems of interneurones capable of generating a variety of patterns of activity enables individual motoneurones to be used in different ways and avoids unnecessary duplication of fibres providing neuromuscular innervation.

16.9.3 Motoneurones

Although generation of motor pattern frequently comes within the province of interneurones, many motoneurones show mutual interactions and are thus also involved in this process. In the nematode, *Ascaris*, five large interneurones extending the length of the body provide tonic stimulation for swimming. The excitatory motoneurones supplying the dorsal muscle in each 'segment' of the body both stimulate the inhibitory neurones for the ventral muscle, and through them, inhibit the ventral excitatory neurones (Fig. 16.35). Activation of ventral excitatory cells has similar effects on dorsal neurones. Activity oscillates between the dorsal and ventral neurones with a frequency and intensity dependent on the degree of stimulation by the interneurones, evoking contractions alternately in their muscles. Other cells form relays between adjacent regions of the body and co-ordinate their activity.

The prediction that pattern generation by motor units would necessitate duplication of the latter seems to be borne out in the case of *Ascaris*, since different types of motoneurones may control forward and backward swimming. It probably matters little to the animal because of its limited repertoire of actions, and this may apply to many motor systems in invertebrates.

The rule that the same set of motoneurones is used for different purposes, depending on the input they receive, is not generally followed by units involved in escape reactions. In squid, the giant fibres mediate the 'jet-propelled' startle reaction, whereas smaller fibres control muscle contractions involved in respiratory movements. In the medusa *Aglantha*, the escape reaction consists of one to three violent contractions of the bell, mediated by eight giant motor fibres running up beneath the subumbrellar surface and by smaller diameter, lateral motoneurones. The normal swimming contractions involve just the latter. Similarly, the lateral giant fibres of the crayfish excite not only (via a pair of central giants) the five to nine 'fast flexor' motoneurones used in flexible escape responses, but also a pair of giant motor neurones in each segment which are reserved exclusively for the tail-flips (Box 16.8).

16.9.4 Neuromuscular innervation

The control of muscular activity by motor neurones is mediated by the discharge of chemical neurotransmitters at synapses (neuromuscular junctions) (Section 16.2.3) formed between nerve terminals and muscle fibres. Patterns of innervation of muscles show wide variations. The presence or absence of propagated action potentials within individual muscle fibres, and/or transmission via gap junctions between fibres, is also of great importance.

In contrast to vertebrate somatic muscle, invertebrate muscles typically have a complex polyneuronal innervation and 'decision-making' is often delegated to the periphery where conflicting influences collide within the innervated structure. In crabs, inhibitory fibres discharge γ-aminobutyric acid (GABA) both at synapses with the excitatory terminals (*pre*-synaptic inhibition), blocking release of the excitatory transmitter glutamate, and at synapses on the muscle fibres (*post*-synaptic inhibition) to damp down any excitation that gets through (Fig. 16.36).

The extensor tibiae muscle in the locust is controlled by four nerve fibres. Excitatory fibres and an inhibitory fibre re-

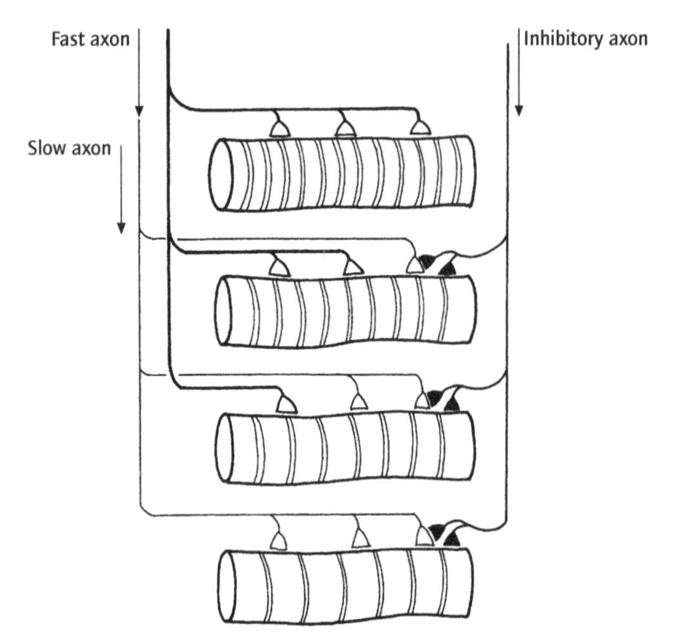

Fig. 16.36 Control of muscle contraction in Crustacea. Individual muscles in crabs are composed of fibres with different structural and functional characteristics. Some fibres readily generate action potentials and show a rapid development of tension, whereas others use mainly graded potentials and contract more slowly. A single, fast-conducting axon sends branches mainly to the various fast muscle fibres; its synapses rapidly become fatigued upon repetitive stimulation. A fine-diameter fibre directs its terminals more towards slower muscle fibres and upon repetitive stimulation shows facilitation. Thus the degree of contraction is controlled, not by recruitment of extra motor units as in vertebrates, but by a graded control of the muscle fibres.

lease the transmitters glutamate and GABA, respectively. In addition, octopamine, released by terminals of a DUM (dorsal unpaired median) neurone, has a modulatory action. On its own it seems to have little effect, and its primary role is to confer on the muscle, which otherwise generates the more prolonged tension required for the maintenance of posture, the ability to respond to motor excitation by more intense contraction and rapid relaxation appropriate for locomotion. Glutamate and the peptide proctolin (Section 16.2.4) act as co-transmitters for some insect muscles. Glutamate alone is released at low levels of neural activity, but both are discharged during repetitive stimulation. Whereas glutamate causes depolarization and twitch-like contractions by the muscle, proctolin induces the build-up of a sustained contraction without affecting the membrane potential.

The adductor muscles of bivalve molluscs show great tenacity and can hold the valves closed for hours or even days – a valuable defence against predation. Contraction is induced by activation of excitatory motoneurones which release acetylcholine, but once tension has developed, the muscle enters a state of 'catch' (during which O_2 consumption is low) whose maintenance is independent of further stimulation. Relaxation is brought about by the activation of inhibitory fibres whose transmitter is apparently 5-hydroxytryptamine, although the catch-relaxing peptide (CARP) may also be involved in the control of such muscles.

In insects with direct flight muscles, impulses in the nerve fibres 'beat time' for the contractions. In contrast, the specialized, indirect flight muscles of, for example, flies have a frequency as high as 1000 beats s^{-1}, whereas the number of action potentials in nerve and muscle fibres is only five to ten. The neural influence has just a tonic effect and puts the muscle into an 'active state'. Contraction and relaxation are internal 'reflex actions' of the muscle fibres, triggered by stretch (see Section 10.6.2.2). Synaptic neuromuscular systems in insects can operate at over 500 cycles s^{-1}, but asynchronous systems have the advantage of economy in structure and function.

In brittlestars, 'juxtaligamental nerve cells' innervate connective tissues whose rigidity they can apparently control. Transmitter-like substances change the properties of the ground substance and allow the collagen fibres within it either to slip or be held rigid. The mechanism is presumably an adaptation to deal with the heavy, calcareous ossicles present in echinoderms. A remarkable motor system!

16.10 Chemical communication

16.10.1 Spectrum of chemical communication

Neurotransmitters are the secretory products of neurones which are typically discharged in close proximity to other cells, whose activity they influence. In consequence, neural action is highly specific in both space and time. In contrast, hormones are chemical messengers which influence other, distant cells to which they are carried in the blood stream or other body fluids. Many hormones are produced by neurones – neurosecretory cells (Box 16.10) – whereas others are secreted by non-neural cells which are often aggregated to form glands. Glands secreting hormones are called endocrine because they secrete their products internally, into the circulation, whereas exocrine glands such as mucus- or enzyme-secreting glands discharge their products externally, often into ducts.

Whereas the function of a typical neurone may be compared with the use of the telephone, endocrine function resembles the use of a megaphone. Hormones often have effects that are widespread within the body, but this does not mean that they cannot show great specificity of action. They are dependent for their effects on the possession by the target cells of specific receptor molecules. For example, different categories of chromatophores, even those having the same coloured pigment, may be sensitive to different hormones.

Other substances, called 'pheromones'*, are secreted into the environment and have effects of biological significance on other individuals of the same species. The phenomenon is

* Terminology: semiochemicals mediate interactions between organisms; semiochemicals are known as pheromones when the interactions are between members of the same species, and allelochemicals between different species.

Box 16.10 Neurosecretion

Ernst Scharrer postulated that some nerve cells – *neurosecretory cells* – combine the properties of neurones and endocrine gland cells. This idea was pursued in invertebrates by Berta Scharrer, Hanstrom and others, who showed that such cells are widespread within these animals. Studies on insects, particularly, led the way in providing conclusive evidence for the theory.

Neurosecretory cell bodies are typically clustered within the central nervous system to form a ganglionic nucleus. Axons extend from the cell bodies and form swollen terminals in close association with blood spaces. The terminals may be aggregated to form a discrete body, a *neurohaemal organ*, or spread across a *neurohaemal area* at the surface of a ganglion or a nerve. Neurosecretory material is produced in the cell bodies, transported down the axons and stored in the terminals. Viewed with the electron microscope, the material consists of the minute secretory granules present in other neurones (Section 16.2.4).

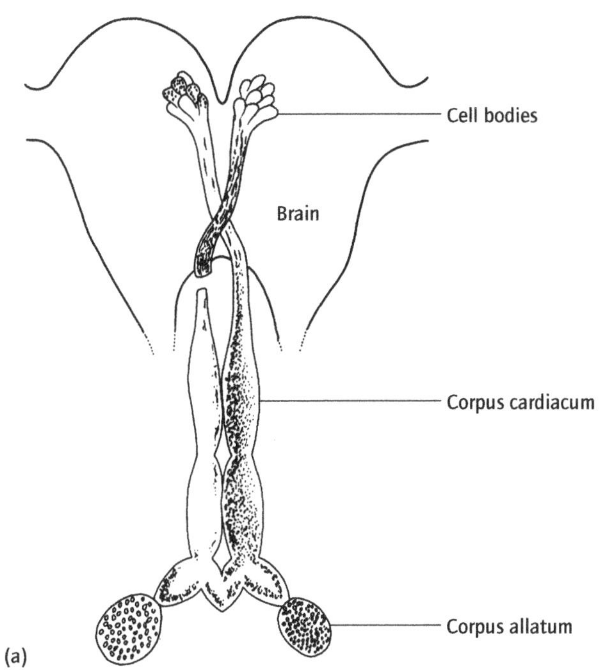

(a) A classic study of invertebrate neurosecretion. Section of the axons carrying secretory material from the cell bodies in the brain of the cockroach *Leucophaea* causes an accumulation of material above the level of the cut and depletion beneath it. The corpus cardiacum and corpus allatum are neurohaemal organs for cerebral neurosecretory cells and also contain their own ('intrinsic') gland cells. (From Scharrer, 1952.)

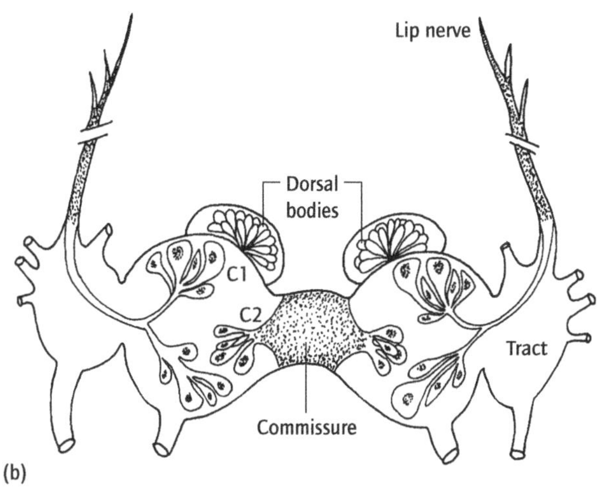

(b) Cerebral ganglia of the mollusc *Lymnaea*, showing fibre tracts leading from two types of neurosecretory cells to neurohaemal areas at the surface of the commissure and lip nerves, respectively. C1, 'light green cells' controlling growth; C2, 'caudodorsal cells' controlling spawning; the 'dorsal bodies' are non-nervous endocrine glands which secrete a gonadotrophic hormone. (After Joosse and Geraerts, 1983.)

Activity within neurosecretory cells can arise either spontaneously or via synaptic excitation. It leads to the passage of action potentials down the axons, influx of Ca^{2+} ions into the terminals and the discharge of hormone into the blood stream. Neurosecretory release occurs by the process of exocytosis and this was first established by invertebrate studies, particularly that on the blow-fly, *Calliphora*, by Normann in 1965.

More recent findings have blurred the distinction between neurosecretory cells and 'ordinary' neurones. Neurones resembling neurosecretory cells have been described in cnidarians and platyhelminths which lack vascular systems. In higher invertebrates, some neurones have now been described which share the *function* of classical neurosecretory cells (i.e. they secrete hormones), but lack the distinctive cytology of the latter. Conversely, other neurones have the *cytology*, but not the *function* of typical neurosecretory cells – they do not release their products into the circulation, but extend to make direct contact with the target cells. Individual neurones may release a transmitter from synaptic terminals in the central nervous system and release the same substance from neurohaemal terminals into the blood where it acts as a hormone. An example is the role of octopamine in the control of dominant/submissive posture in the lobster.

The essential similarity of neurosecretory and other neurones is shown by the fact that the secretion of peptides, originally thought to be restricted within the nervous system to neurosecretory cells, is now known to be a characteristic of many and perhaps all neurones, including those producing conventional transmitters.

Continued p. 452

Box 16.10 *(cont'd)*

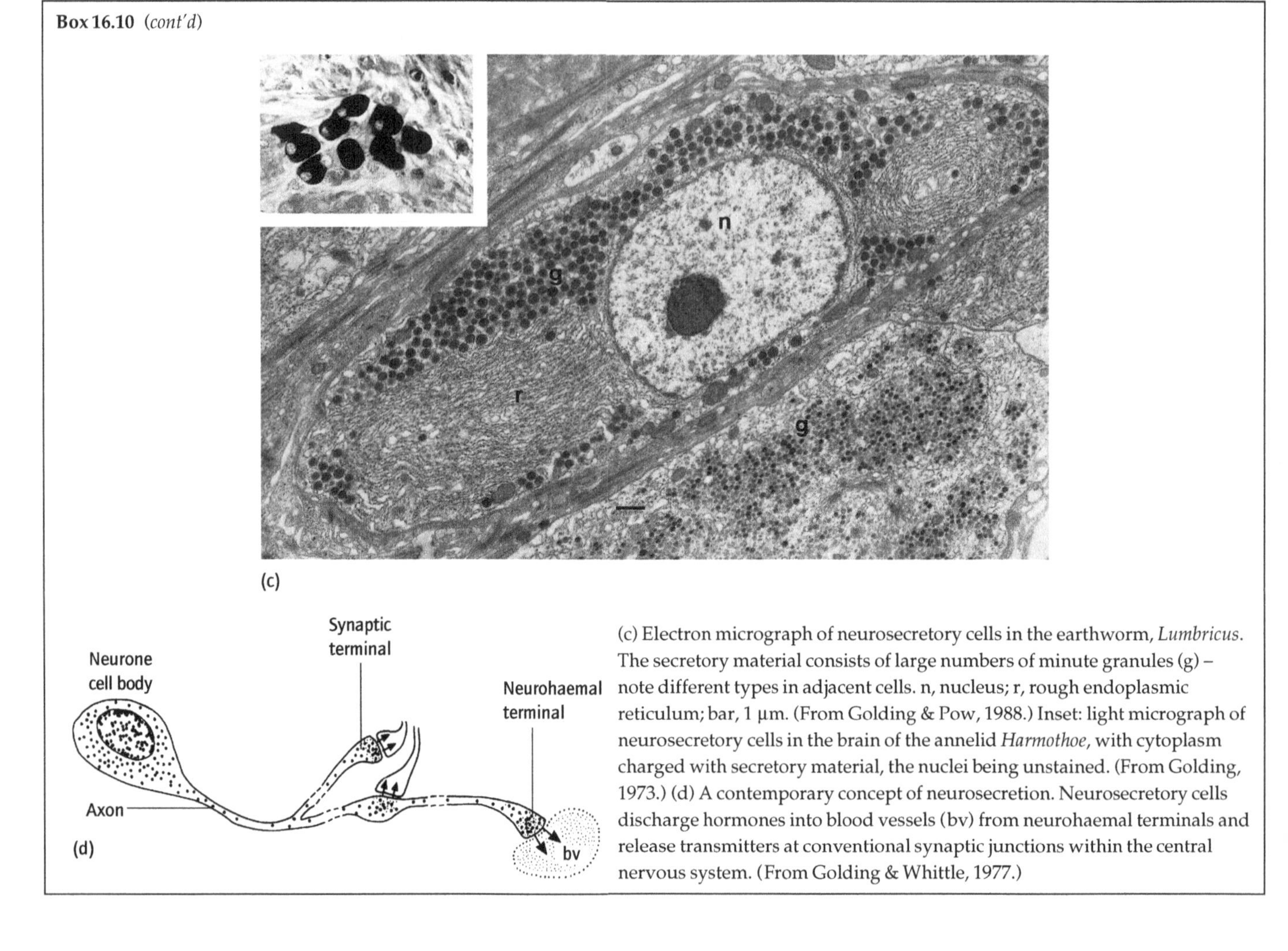

(c) Electron micrograph of neurosecretory cells in the earthworm, *Lumbricus*. The secretory material consists of large numbers of minute granules (g) – note different types in adjacent cells. n, nucleus; r, rough endoplasmic reticulum; bar, 1 μm. (From Golding & Pow, 1988.) Inset: light micrograph of neurosecretory cells in the brain of the annelid *Harmothoe*, with cytoplasm charged with secretory material, the nuclei being unstained. (From Golding, 1973.) (d) A contemporary concept of neurosecretion. Neurosecretory cells discharge hormones into blood vessels (bv) from neurohaemal terminals and release transmitters at conventional synaptic junctions within the central nervous system. (From Golding & Whittle, 1977.)

encountered from the protist level, mediating aggregation of individuals of amoeboid slime moulds, to the social insects – 'walking batteries of exocrine glands' (Wilson, 1975).

16.10.2 Quality and quantity

A number of criteria must be met to establish the endocrine status of an organ. In summary, surgical removal of the organ, but not of other organs (the latter is the control for the experiment) must abolish the effect its hormone is thought to evoke. Replacement therapy must be successful – i.e. reimplantation of the organ (but not of other organs), or injection of extracts, should restore the effect. This requirement can also be met by the parabiosis or grafting technique (Fig. 16.37). Alternatively, cells or organs can be maintained *in vitro*. Introduction of the gland or its extracts into the culture can then be used to assess its action. Last, the substance involved should be extracted and purified from homogenates and its chemical structure determined.

In order to determine the quantity of hormone present in a gland, tissue extracts or the blood, sensitive assays must be

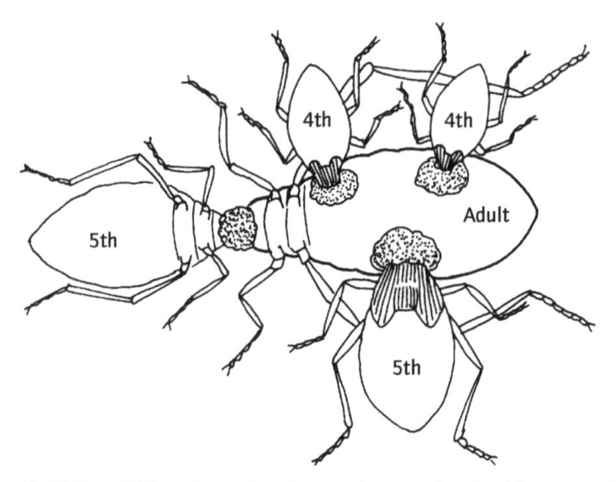

Fig. 16.37 Two fifth and two fourth stage larvae joined, with wax seals, in parabiosis and sharing the circulation with an adult *Rhodnius*. The larvae provide moulting and juvenile hormones and induce the adult to moult again, when it shows regression to a more juvenile form. (After Wigglesworth, 1940.)

developed. In the case of ecdysone (Section 16.11.5), the *Calliphora* test used by Karlson depends on the ability of the hormone to induce pupation in ligated fly larvae. Radioimmune assay has been extensively used more recently. Antibodies to an ecdysone–protein complex will bind radioactive ecdysone. The extent to which they do so depends on the amount of unlabelled ecdysone present (since this will compete for the binding sites) and thus provides a measure of the latter.

16.10.3 Neuroendocrine systems

Nervous and endocrine systems do not exist in the body in functional isolation from each other, but form integrated, complex neuroendocrine systems. Neurosecretory cells are frequently important in this regard (Box 16.10). In a reflex pathway consisting of sensory reception, processing and integration by interneurones, and control of effectors, the last step may be mediated by hormones. In higher invertebrates, endocrine systems may show a hierarchy of organization, with the products of 'first-order' endocrine cells controlling the secretory activity of 'second-order' elements, and so on. Feedback loops can be formed at various levels to regulate the activity of the system (Fig. 16.38). The nervous system may also control non-neural endocrine glands via synaptic contacts (Section 16.2.3).

16.10.4 Mechanisms of hormone action

The actions of peptide hormones (e.g. PTTH – Section 16.11.5) are mediated by receptor molecules within the plasma membranes of the target organs. Receptor activation leads to changes in the concentration of a second messenger (Section 16.1.3) and, typically, activation of metabolic pathways within the cells. In contrast, steroids penetrate inside the cell, partly by diffusion and partly by a special uptake mechanism. In the case of ecdysone, the moulting hormone of arthropods, the functional receptor is located in the nucleus and consists of a heterodimer of what is nominally the receptor molecule (EcR) and the ultraspiracle protein (USP) (by themselves, the two proteins bind ecdysone poorly (EcR) or not at all (USP)). The activated receptors then stimulate the genetic machinery.

Flies provide valuable material with which to investigate the mechanism of steroid action because their salivary glands contain 'giant' chromosomes ten times longer, and one hundred times thicker, than normal. During development, various bands within the chromosomes become greatly thickened and such 'puffs' are sites of gene activity involving intense RNA synthesis (Fig. 16.39). It has been shown experimentally that ecdysone induces the appearance of a specific sequence of early and late puffs identical to that normally observed prior to moulting. As analysed by Ashburner and his colleagues, the earliest puffs appear within 5 min of exposure to ecdysone and are caused by the hormone alone. Later puffs are due to the actions of protein products of the genes associated with the first puffs. Single or small groups of bands in the chromosomes can be dissected out and the nucleotide sequences of ecdysone-responsive genes identified.

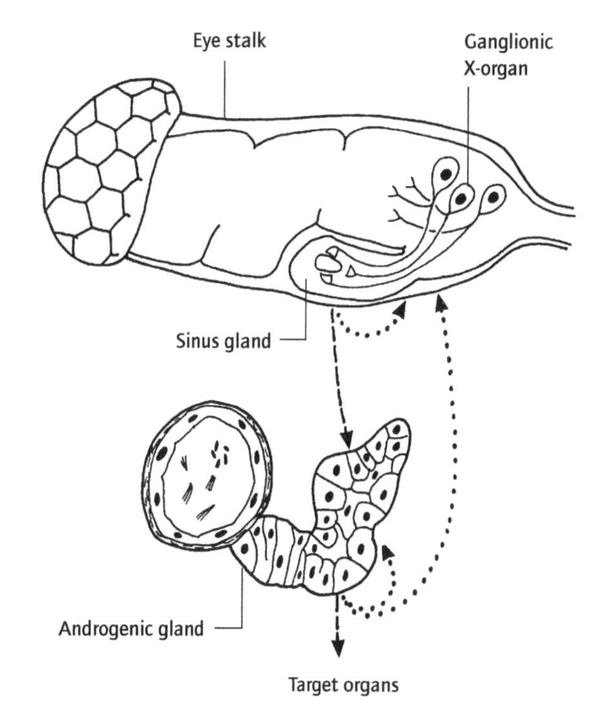

Fig. 16.38 Neuroendocrine system in crustaceans. Under the influence of environmental and internal stimuli, a number of hormones are produced in neurosecretory cells (forming the ganglionic X-organ) in the eyestalk. They are transported down axons to a neurohaemal organ (the sinus gland) and released into the blood. One hormone exerts an inhibitory influence on the androgenic glands which are attached to the vas deferens. The androgenic gland hormone stimulates spermatogenesis and the development of secondary sexual characteristics (e.g. large chelae). Potential short-loop and long-loop feedback effects of the hormones are also shown.

16.10.5 Pheromones

The first pheromone to be identified was bombykol (Fig. 16.40), a substance secreted by females of the silk moth, *Bombyx mori*, the culmination of work by Butenandt and his colleagues involving the extraction of half a million abdominal glands. Amazingly, a single molecule is thought to have a detectable effect on an olfactory receptor on the antenna of the male, and 200 molecules are sufficient to evoke a behavioural response. The male moth sets off upwind with a characteristic zigzag flight pattern (Fig. 16.41). Many hundreds of pheromones have now been identified in insects alone and the pace has accelerated with the introduction of the coupled GC (gas chromatography)–electroantennogram system (Fig. 16.42). In order to identify which of the enormous range of volatile compounds present in, for example, the body of female moths, have a role as pheromones, a sample is passed through the GC column to separate the compounds; then split and presented

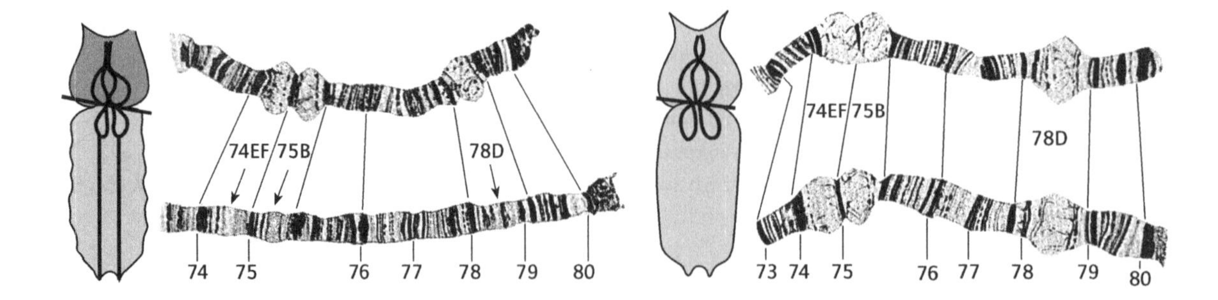

Fig. 16.39 Ligation early in the last larval stage (before ecdysone has been discharged into the circulation) results in a normal puffing pattern (upper left chromosome) anterior to the ligature, but not posterior to it (lower left). Ligation later in the stage fails to prevent puffing (chromosomes on right). (From Becker, 1962.)

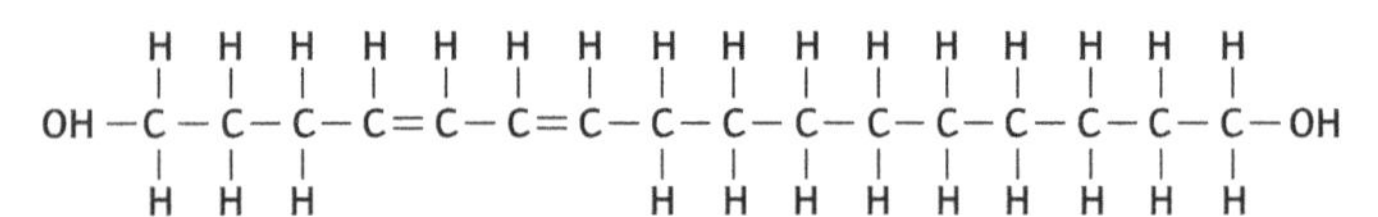

Fig. 16.40 Molecular structure of the pheromone bombykol.

simultaneously to the GC detector and an antenna (or an olfactory receptor cell) connected to recording electrodes. Comparison of the output enables the pheromone to be pinpointed and it is subsequently identified by mass spectroscopy. Portable electroantennogram sensors are now available to monitor levels of pheromones – used for pest control (see Section 16.12) – in glasshouses!

Recent research has established the importance of blends of compounds. A combination of chemicals may be required to elicit the initial effect – for example, as a sex attractant. Alternatively, one compound may have a primary, long-range action (e.g. as an attractant) and another, a secondary, short-range effect (e.g. as a copulatory releaser). For example, bombykol is secreted in combination with another, related compound – bombykal – which has inhibitory effects on flight. At close range, males of some moths are themselves stimulated to secrete sex pheromones evoking responses by the female.

Pheromones are of great importance in the regulation of caste structure in social insects. The best known example is the secretion of the queen substance by the mandibular glands of honey bee queens. Its two components – 9-oxydecanoic acid and 9-hydroxydecanoic acid – have 'primer' activities, inhibiting both gonadial development within the workers, and their construction of special cells in which larvae would develop as queens. When the queen is removed from the hive, these inhibitions are lifted and sexually active females are produced to replace her. The same substance also has 'releaser' activities, being important as a sex attractant, and during swarming and settling.

The relationship of many symbiotic organisms is established by the use of allelochemicals. Furthermore, in insects a substance acting as a pheromone to members of the same species may act as an allelochemical to those of another, for example, by attracting parasites. Parasites may similarly exploit the endocrine signals of their hosts. 'Conformers' tune in to these signals in order to synchronize their development with that of the host. 'Regulators' manipulate the endocrine regime of the host to their own advantage. For example, parasites may secrete juvenile hormone (Section 16.11.5) into the host's circulation, prolonging the larval stage of the host and perpetuating optimal nutritional conditions for themselves. Other relationships between chemical mediators include the neurosecretory peptide which activates synthesis of sex pheromones in moths, and the use of various compounds both as neurotransmitters (or hormones) and pheromones.

16.11 Roles of endocrine systems

Hormones are involved in the control of a wide variety of processes in invertebrates – a small and diverse selection is described below.

16.11.1 Control of reproduction and senescence in *Nereis*

The pattern of endocrine control in nereids (ragworms) is comparatively simple and similar patterns are shown by nemertines

Fig. 16.41 Upwind flight of a male moth (continuous line), following a plume of pheromone emitted by a female. (From Baker, 1990; after Doving, 1990.)

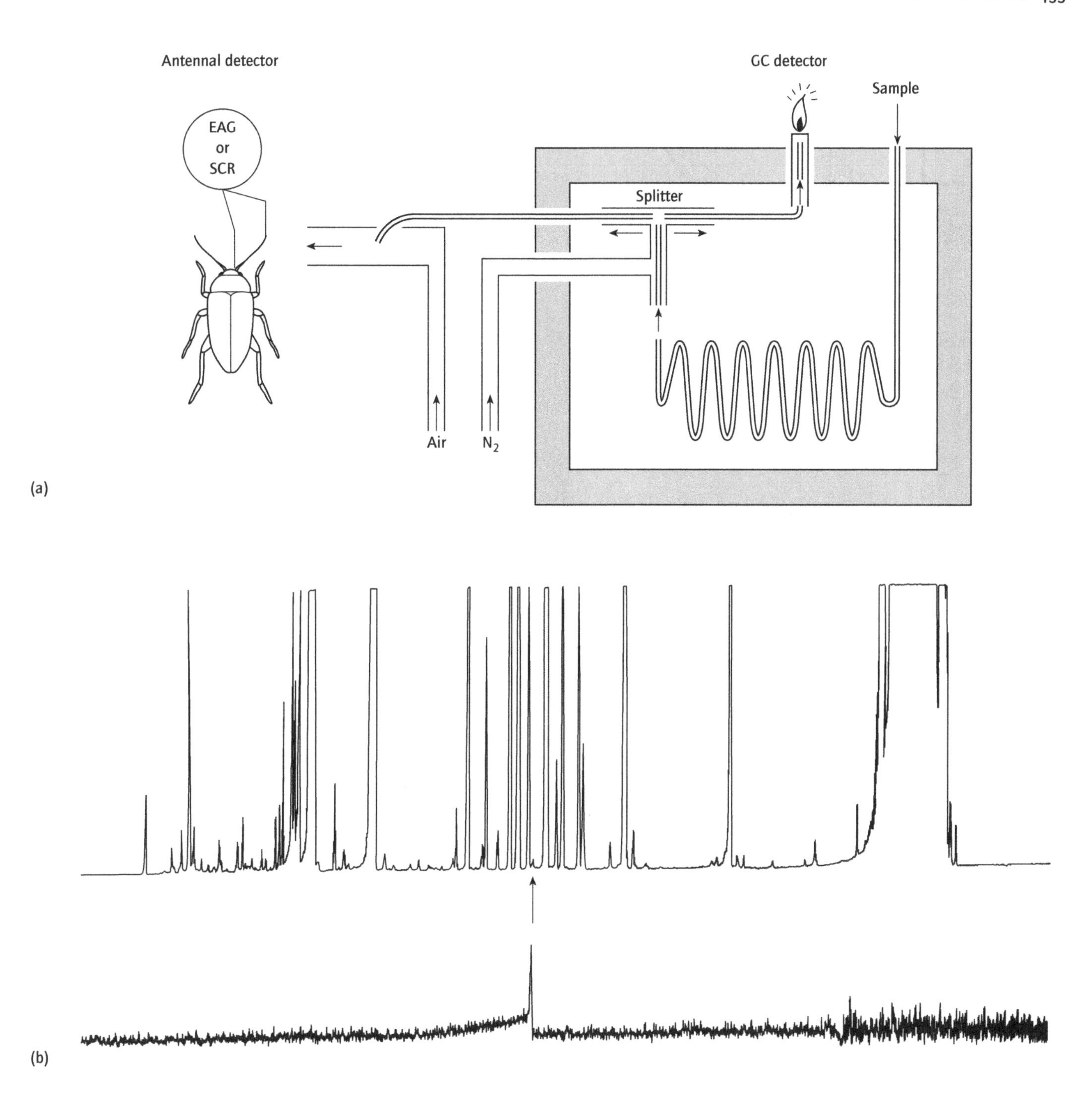

Fig. 16.42 (a) Coupled GC–electroantennograph system. (From Angelopoulos *et al.*, 1999.) (b) Coupled recordings of (upper) GC of volatiles from fennel, and (lower) responses of olfactory cell of raspberry beetle. (Courtesy, Dr Christine M. Woodcock, IACR-Rothamsted.)

and some other invertebrates (but not by most other annelids). As first shown by Durchon, removal of the brain (Fig. 16.43) results in precocious sexual maturation. The final stage of the life cycle, in most species, sees the metamorphosis of the body into the heteronereis form (Fig. 4.57) adapted for swarming and spawning in the surface waters of the sea. This process too is inhibited by the brain hormone. In contrast, the hormone is indispensable for the proliferation of body segments (see Sections 15.6.3 and 15.6.4).

Decerebration at earlier stages of life leads to oocyte degeneration and incomplete metamorphosis. This suggests that the hormone has a dual action, incorporating a trophic effect sustaining somatic growth and the early stages of gametogenesis, and an inhibitory effect on processes appropriate to the final stages of development. A decline in the rate of secretion allows the co-ordinated onset of metamorphosis and gamete

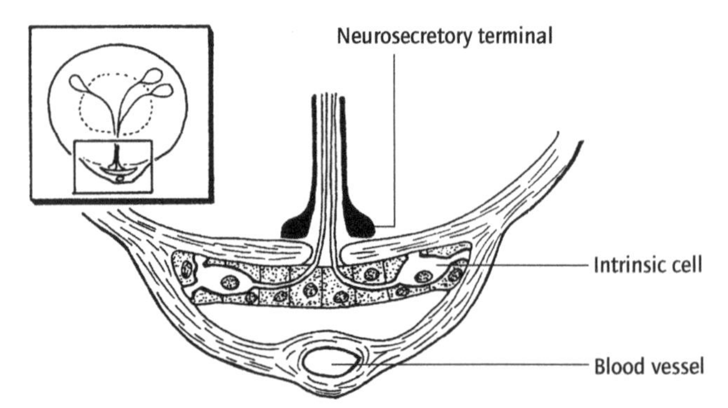

Fig. 16.43 Brain (inset) and infracerebral gland in the annelid *Nereis*. Neurosecretory cells in the brain have axons which terminate on the brain floor; others (intrinsic cells) are located in the gland. (After Golding, 1992.)

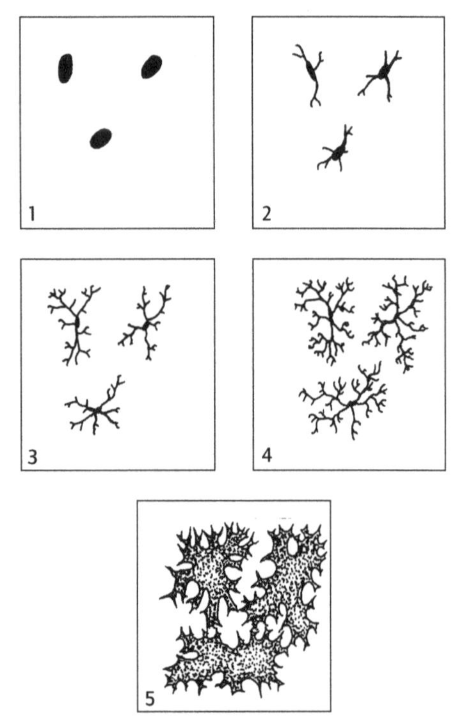

Fig. 16.44 Chromatophores showing the arbitrary scale by which the extent of pigment dispersion is assessed.

maturation. A substance produced by the maturing gametes reduces the endocrine activity of the brain, which will accelerate gamete development and so on, providing a positive feedback effect.

Sexual maturity is accompanied by somatic senescence, as shown by the abandonment of feeding and the loss of the ability to regenerate posterior segments. Spawning is rapidly followed by death. Experimental elevation of the level of brain hormone, by regular implantation of brains taken from young animals, blocks spawning, restores feeding and regeneration, and perpetuates life indefinitely.

16.11.2 Regulation of colour change in crustaceans

Some functions are controlled not by the secretion of a single hormone, but by a balance of antagonistic factors. Sugar metabolism, and salt and water balance, are often regulated in this way, which provides for the rapid reversal of an endocrine effect.

In crustaceans, colour change is physiological in character, since it involves no change in the quantity of pigment, but is mediated by the movement (or 'migration') of pigment granules within cells called 'chromatophores' (Fig. 16.44). The cells remain fixed in position and shape. Hormones controlling chromatophores are typically released (some of them from the sinus gland in the eye-stalk – Fig. 16.38) in response to environmental stimuli (background illumination, etc.), but some show circadian or tidal rhythms. Two antagonistic hormones are usually involved, one inducing pigment dispersion and the other pigment concentration. However, in some cases a multiplicity of factors affect chromatophores of a single colour, and different patterns of coloration of the body can be produced. Hormones also influence migration of retinal pigments which mediate changes in the physiology of vision in the compound eye (Fig. 16.22b).

16.11.3 Integration of behaviour with spawning in molluscs

Peptide secretion by neurosecretory cells usually involves the discharge, not of a single substance, but of a 'cocktail' of compounds which is well adapted to exert a co-ordinated influence on neural networks generating behaviour and the function of other organ systems. In *Aplysia*, the egg-laying hormone is discharged into the blood stream by the bag cells and induces gamete discharge by the gonads. It also stimulates the cerebral ganglia to initiate egg laying behaviour. Another bag cell product, α-bag cell peptide, stimulates the nerve terminals within the neurohaemal complex, ensuring their maximum participation. Several stimulate particular neurones (Fig. 16.45) in the abdominal ganglion, whose activity influences the function of the heart, gills, and reproductive tract.

16.11.4 Initiation of an endocrine cascade in starfish

Spawning can be induced in mature starfish by injecting extracts of radial nerves into the body cavity. Strangely, the peptide hormone involved (gonad stimulating substance or GSS) is present throughout the radial nerves and nerve ring. The sites of synthesis – probably the supporting cells of the nerves, just beneath their surface – and release, and whether the coelomic fluid is used to transport it to the tissues, remain matters of debate.

GSS induces gamete maturation in ovaries or ovary fragments *in vitro* and this seemed to indicate that its effect must be direct. In fact, GSS stimulates the production and discharge of a second substance by the follicle cells of the gonad (Fig. 16.46). This meiosis-inducing substance (MIS) was shown by Kanatani to be the fairly simple compound 1-methyladenine.

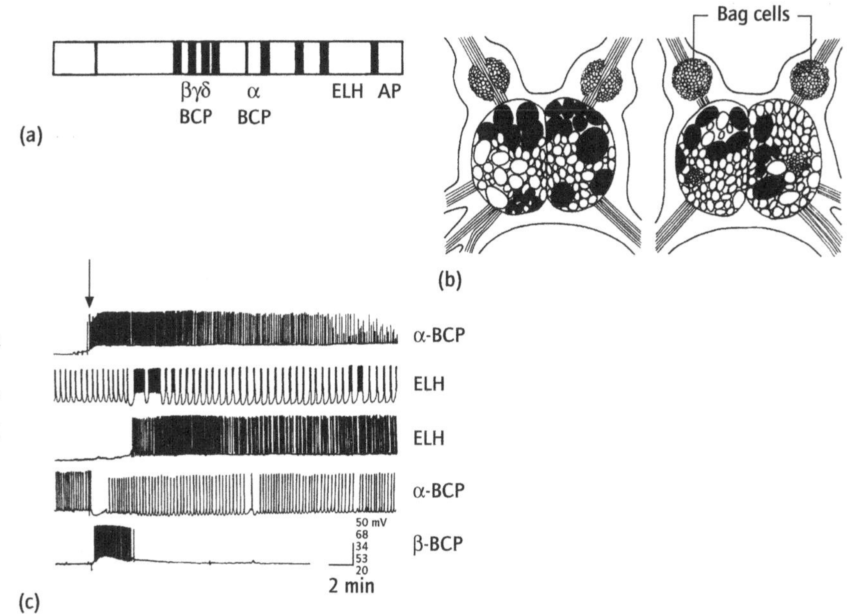

Fig. 16.45 Bag cells, their peptides and their effects in the abdominal ganglion of *Aplysia californica*. (a) Propeptide and the known secretory products within it (see Section 16.2.4): α-, β-, γ- and δ-bag cell peptides; egg-laying hormone; and acidic peptide. Note that other sequences may also be produced. (b) Location of the bag cell clusters and of identified neurones (filled cells) that respond to their secretory products. Left, dorsal; right, ventral. (c) Effects on the electrical activity of particular neurones of various bag cell peptides administered at the time arrowed. (From Geraerts *et al.*, 1988; after Mayeri & Rothman, 1985.)

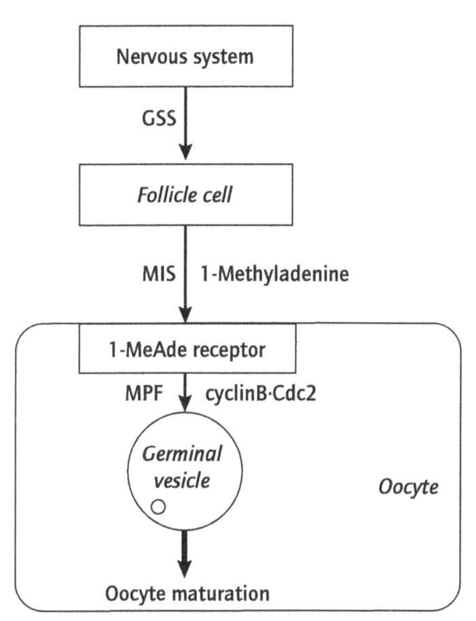

Fig. 16.46 Endocrine cascade controlling oocyte maturation in starfish, showing primary (GSS), secondary (MIS, 1-methyladenine) and tertiary (MPF, cyclin B–cdc2 protein complex) mediators. (From Kishimoto, 1999.)

MIS has a curious combination of endocrine and para-endocrine roles. It diffuses across the intercellular space, binds to receptor molecules in the oocyte membrane and stimulates the production of a third compound, maturation promoting factor (MPF). MPF induces oocyte maturation and breakdown of the follicular envelope, allowing the gametes to be expelled. MIS also diffuses into the coelomic fluid, where it stimulates muscle contractions which aid spawning, and in some species evokes brooding behaviour. MIS induces sperm activation and spawning in males.

What started as endocrinology of starfish reproduction went on to make a significant contribution to the fundamentals of knowledge of cell division. MFP now stands for 'M-phase promoting factor'. It consists of a complex of cyclin B and the protein cdc2 and controls cell division in all eukaryotic cells. In the immature oocyte, the inactive form of MPF predominates. Activation by 1-methyladenine of a G-protein-linked receptor activates MPF via two distinct pathways: it stimulates an activator of MPF and suppresses an inactivator of this compound. A secondary positive feedback action of activated MPF on its activators and inactivators enhances the effect, and there is a similar, tertiary action of germinal vesicle contents.

16.11.5 The endocrine orchestra of insects

The importance of the brain was first demonstrated by S. Kopec, from 1917 to 1923. The larva of the gypsy moth *Lymantria dispar* will only pupate if exposed to the influence of the brain for a certain minimum period of time, the head critical period. The hormone involved, prothoracicotrophic hormone (PTTH, Box 16.11) is indispensable for moulting at each stage in the life cycle. In the bug *Rhodnius* (Fig. 16.37), stretching of the abdomen by ingestion of a blood meal provides the trigger for PTTH release (the animal can be 'fooled' by blowing it up with air). In the moth *Manduca sexta*, a critical weight has to be attained and discharge is 'circadian gated' – a clock mechanism (Section 16.7.4) in the brain permits release only during the night.

In *Rhodnius*, PTTH continues to be secreted after the head critical period during larval development, and again in the adult, indicating that it has roles that are as yet unknown.

The role of the prothoracic glands was demonstrated in experiments using pupae of the cecropia silk moth. Isolated abdomens fail to moult when active brains are implanted within them,

Box 16.11 The Holy Grail of invertebrate endocrinology – the insect brain hormone

Approximately 70 years were to elapse – and millions of insect brains were to be used – between the classic work of Kopec (Section 16.11.5), which established the dependence of moulting and development in insects on the endocrine activity of the brain, and the elucidation of the chemical structure of the hormone (now known as PTTH) involved.

The structure of PTTH (Fig. a), its propetide and its gene were determined in *Bombyx* by H. Ishizaki and his colleagues in 1990. It consists of a homodimer of two identical peptide chains, each of 109 residues, linked by disulphide bonds (molecular weight approximately 30 kDa). As in *Manduca* (Fig. b), it is produced by two cells situated laterally on each side of the brain and released from the corpora allata. It is noteworthy that whereas it took several decades of dedicated research to identify the hormone, a few months sufficed to develop a culture of genetically engineered *E. coli*, 400 ml of which produces the same amount of peptide as that present in 10 000 pupal brains. Both a 'big PTTH' and a 'little PTTH' (22–28 kDa and 4–7 kDa, respectively) may be present in *Manduca*.

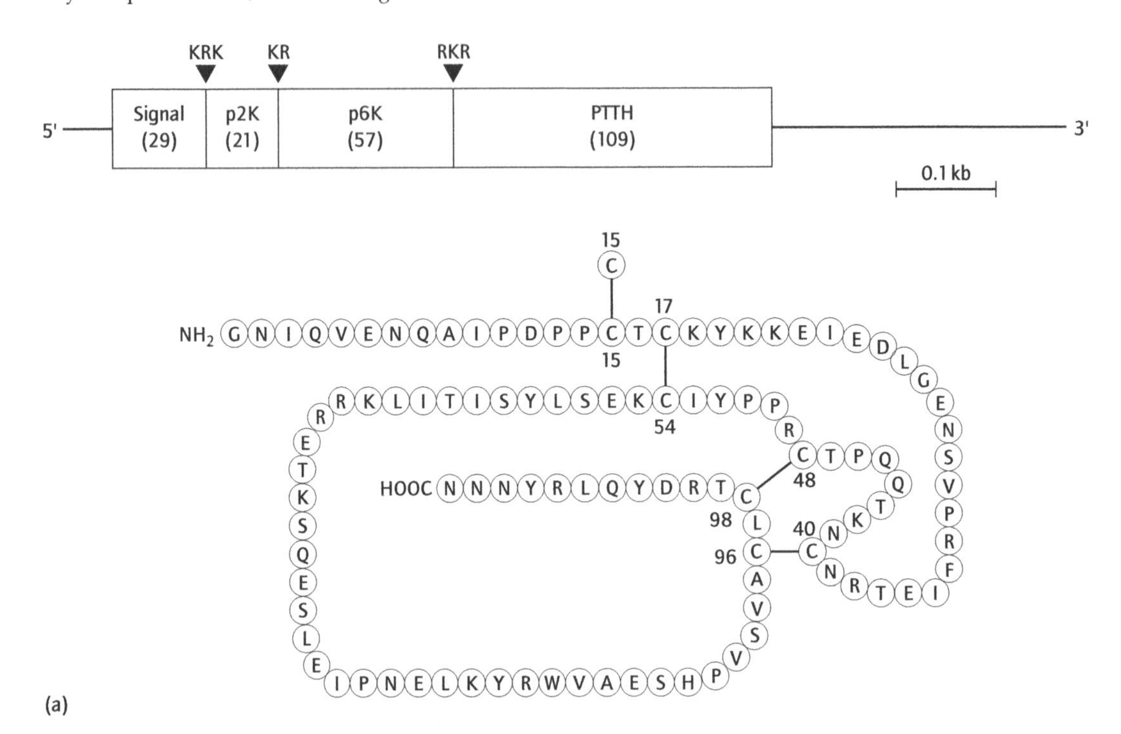

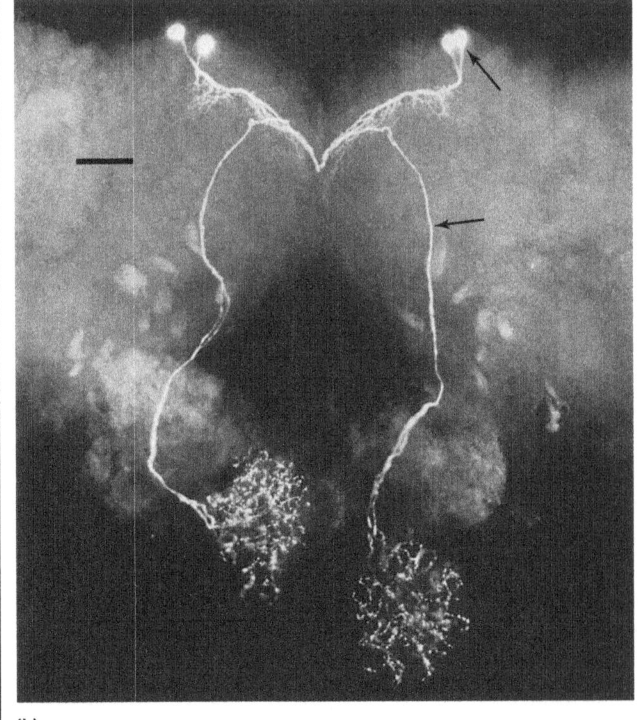

(a) Upper, prepropeptide (the large molecular weight precursor) of PTTH in *Bombyx mori*, showing the cleavage sites. (From Kawakami *et al.*, 1990.) Lower, amino acid sequence of the PTTH subunit. The functional hormone consists of a homodimer, i.e. of two subunits linked (at position 15) by a disulphide bond. (From Ishibashi *et al.*, 1994.) (b) Immunofluorescence staining of cells secreting 'big PTTH' in the pupa of *Manduca*. Neurone cell bodies in the brain and their axons are both arrowed; the latter form branching terminals in the paired corpora allata (lower). Bar, 100 μm. (From Watson *et al.*, 1989.)

Fig. 16.47 Molecular structure of (a) ecdysone and (b) juvenile hormone I.

whereas if both active brains and inactive prothoracic glands are used, development proceeds. This indicates that PTTH acts by stimulating the prothoracic glands or their homologues to secrete a second hormone – the steroid ecdysone, or one of the other members of the family of compounds known as the ecdysteroids (Fig. 16.47). Karlson and his colleagues used 500 kg of silkworm pupae to obtain 200 mg of the pure substance, which in 1965 became the first invertebrate hormone to be chemically identified.

PTTH acts by inducing Ca^{2+} influx into the prothoracic gland cells, a rise in cAMP and thus to synthesis of ecdysone from cholesterol. Levels of ecdysone in the blood rise sharply as a result. The ecdysone peak inhibits synthesis, by the epidermis, of proteins for the inner (endocuticular) layer of the old cuticle and stimulates apolysis (separation of epidermis and cuticle), epidermal mitosis and formation of a new epicuticle. The subsequent decline in ecdysone level is necessary to allow later stages in moulting, such as the partial digestion of the old cuticle and deposition of the inner layers of the new one. Release of EH (see below) is also now unblocked and ecdysis rapidly follows.

A third hormone, juvenile hormone (JH) (Fig. 16.47), of which there are three main forms, is secreted by the gland cells of the corpus allatum. Its role is most clearly indicated by experiments in which the corpus allatum is removed from, or transplanted between, individuals at different stages of development (Fig. 16.48) and by parabiosis (Fig. 16.37). JH is the *status quo* hormone – its presence ensures that, when moulting is triggered by ecdysone, an instar of the same type as the preceeding one will result.

In exopterygotes such as the locust, when ecdysone is secreted in the absence of JH, adult development is initiated. The situation is inevitably more complex in endoptygotes and the larva-to-pupa moult in *Manduca* (Fig. 16.49) is commonly thought to require two peaks of ecdysone. The small commitment peak, on days 3–4 of the 5th larval instar, is initiated by three surges of PTTH over a period of about 20 h. In the absence of JH, ecdysone reprogrammes the tissues so that many larval genes are permanently repressed. During days 7–9, a larger, 'prepupal peak' of ecdysone (again stimulated by PTTH), but now in the presence of JH, results in moulting to form a pupa. During the pupal stage, high titres of ecdysone in the virtual absence of JH stimulates moulting to form the adult – if JH is administered experimentally, a second pupal stage results.

Both ecdysone and JH have a variety of functions in addition to those mentioned here, for example, in embryonic development, gametogenesis and the regulation of caste in social insects.

Eclosion hormone (EH) has been most closely studied by Truman and his colleagues. In *Manduca*, it is secreted initially by the ventral ganglia, but by the brain/corpus cardiacum/corpus allatum complex during the pupa-to-adult moult. Together with ecdysis-triggering hormone secreted by Inka cells situated near the spiracles, EH triggers the complex sequence of actions involved in ecdysis – the shedding of the old cuticle. For example, at the close of the pupal stage, the moth sheds its old cuticle,

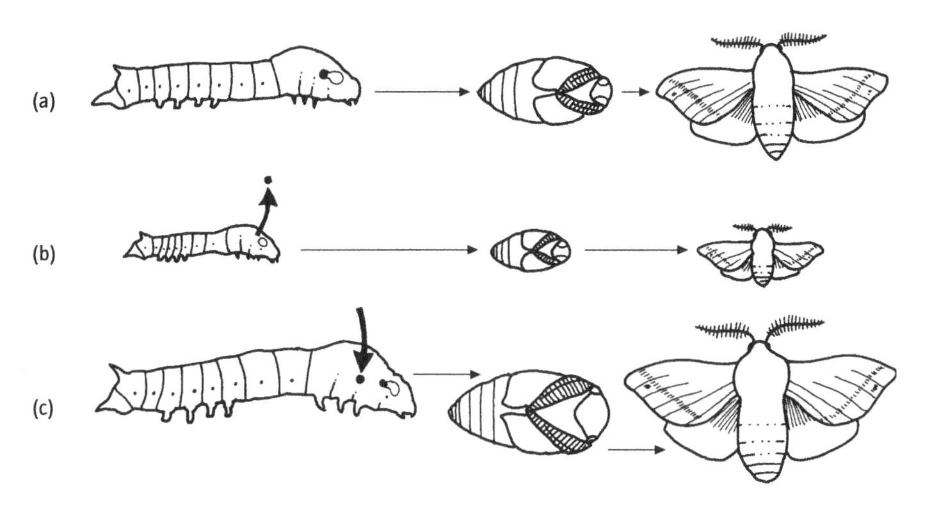

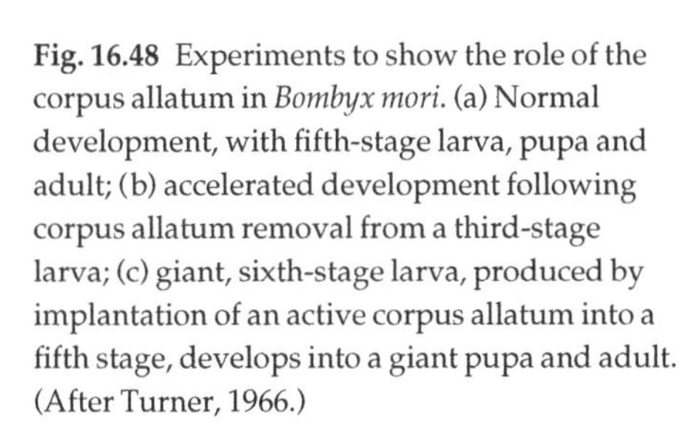

Fig. 16.48 Experiments to show the role of the corpus allatum in *Bombyx mori*. (a) Normal development, with fifth-stage larva, pupa and adult; (b) accelerated development following corpus allatum removal from a third-stage larva; (c) giant, sixth-stage larva, produced by implantation of an active corpus allatum into a fifth stage, develops into a giant pupa and adult. (After Turner, 1966.)

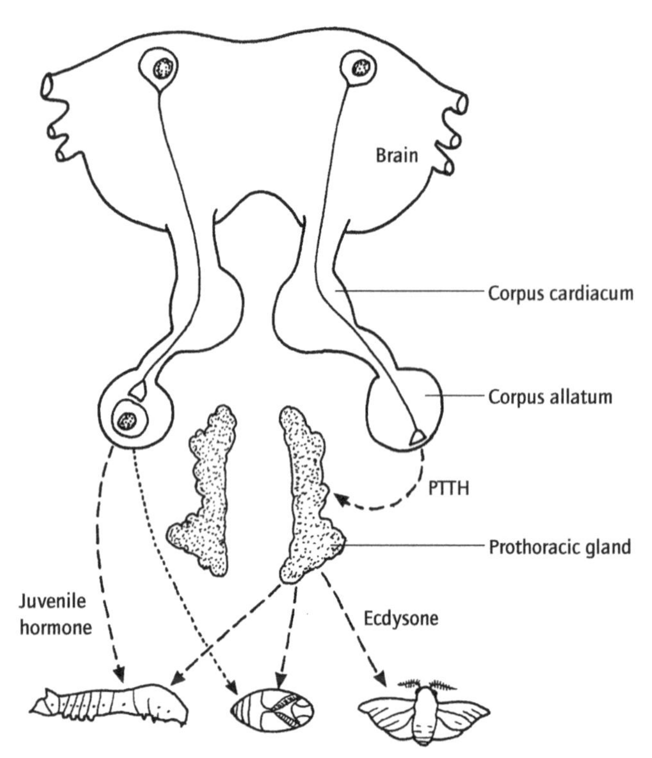

Fig. 16.49 Neuroendocrine control of development in the endopterygote insect *Manduca*. Neurosecretory cells in the brain (one shown, right), with terminals in the corpus allatum, secrete the hormone PTTH which controls the prothoracic glands. Other neurones (one shown, left) control the gland cells of the corpus allatum by, at least in part, the local action of their secretory products. The progression from larva–pupa–adult is controlled by ecdysone and juvenile hormone.

crawls upwards out of the soil and, reaching the surface, finds a perch and starts wing inflation behaviour. Freedom from confinement in the soil stimulates the ventral ganglia (already primed by exposure to EH) to secrete cardioacceleratory peptides to aid wing inflation and bursicon to influence inflation behaviour and then to promote tanning of the cuticle.

Crustaceans have the same moulting hormone, 20-hydroxyecdysone, as insects, secreted by non-nervous endocrine glands, the Y-organs. Hormone production is controlled by neurosecretory cells in the eye-stalk (Fig. 16.38). Whereas PTTH in insects has a stimulatory action, the eye-stalk peptide inhibits the Y-organs so that eye-stalk removal accelerates moulting or even results in a series of rapid moults. However, cAMP is used as the second messenger in each case (Section 16.10.4). Lastly, new research by Laufer and his colleagues has shown that, in crustaceans, a JH-like compound is secreted by the antennary glands and has roles similar to those of JH in insects.

16.12 Applications

In view of the importance of control mechanisms in the lives of animals, it is not surprising that the nervous and endocrine systems are the main targets of agents designed to disrupt the physiology of invertebrate pests, as they are of venoms produced by natural predators (e.g. spiders). There can be little doubt that future years will see a dramatic expansion in the applications of knowledge of control systems to problems of economic importance.

16.12.1 Golden age – dark age

Compounds isolated from plants were the first to make a major impact as insecticides. Nicotine acts as a neurotransmitter mimic, activating acetylcholine receptors and disrupting nervous activity. It has been superseded by synthetic analogues (e.g. imidacloprid). Pyrethrum continues to be widely used – pyrethroids target voltage-gated Na^+ channels in nerve fibres (Fig. 16.2). Azadiractin, a product of the neem tree, attracted much attention during the 1990s.

Müller discovered the insecticidal activity of the synthetic compound DDT, a chlorinated hydrocarbon, in 1939, for which he was awarded the Nobel Prize in Physiology or Medicine in 1948. This substance also stimulates Na^+ channels. Its efficacy was astonishing, even leading to predictions that, for example, 'mosquito-transmitted diseases will disappear' (Fig. 16.50), and during the following forty years nearly four billion kilograms of chlorinated hydrocarbons were used worldwide!

Concerns about safety and environmental damage (DDT is highly persistent) gathered pace in the 1950s and 1960s. Most notable was the publication of Rachel Carson's book, *Silent Spring*, in 1962, in which it was stated that 'it is our alarming misfortune that so primitive a science has armed itself with the most modern and terrible weapons and . . . turned them against the earth'. Far from ushering in the 'golden age' predicted when artificial pesticides were first introduced, the 1940s to 1960s came to be regarded as 'the dark ages of pest control' (Newsom, 1980). Severe restrictions or outright bans started to be introduced in the early 1970s.

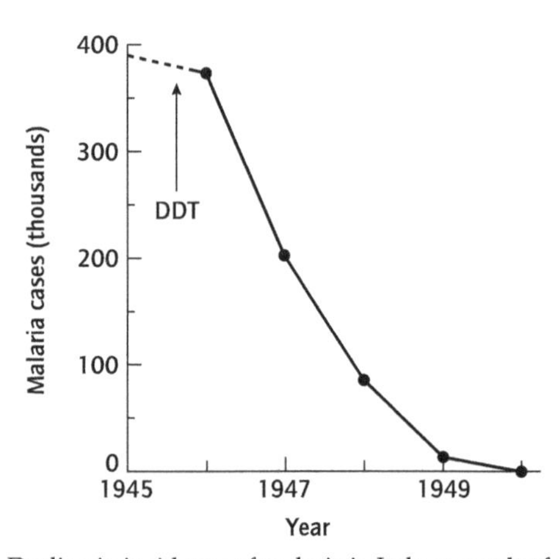

Fig. 16.50 Decline in incidence of malaria in Italy as result of application of DDT. (From Casida & Quistad, 1998; after Müller, 1959.)

Box 16.12 Chemical weapons, ancient and modern: mechanisms of action of principal insecticides

- *DDT and pyrethroids*: activation of voltage-gated Na^+ channels.
- *Other chlorinated hydrocarbons (e.g. endosulphan, lindane)*: block action of inhibitory transmitter GABA on Cl^- channels, inducing hyperexcitation.
- *Nicotine (and synthetic analogues)*: activate acetylcholine receptors.
- *Organophosphorus compounds and methylcarbamates*: inhibit acetylcholinesterase and induce hyperexcitation.
- *Avermectins (isolated from the mould* Streptomyces*)*: complex molecules (structure not shown) which activate Cl^- channels (GABA-, glutamate- and voltage-gated), causing paralysis; ivermectin is a major anthelmintic and blocks pharyngeal pumping in nemertodes by this action.
- *Bisacylhydrazines*: non-steroidal agonists (i.e. mimics) of ecdysone, developed particularly by Rohm and Haas Company scientists; they induce a condition equivalent to ecdysone excess ('hyperecdysonism') causing premature – and lethal – initiation of moulting. RH-5992 (MIMIC^R etc.) is highly toxic to lepidopteran larvae, but not to 'friendly' insects.
- *Juvenile hormone analogues*: one class of insect growth/development regulators used to control domestic fleas, preventing development to the adult; insecticidal activity in carpets, etc. persists for months or years.

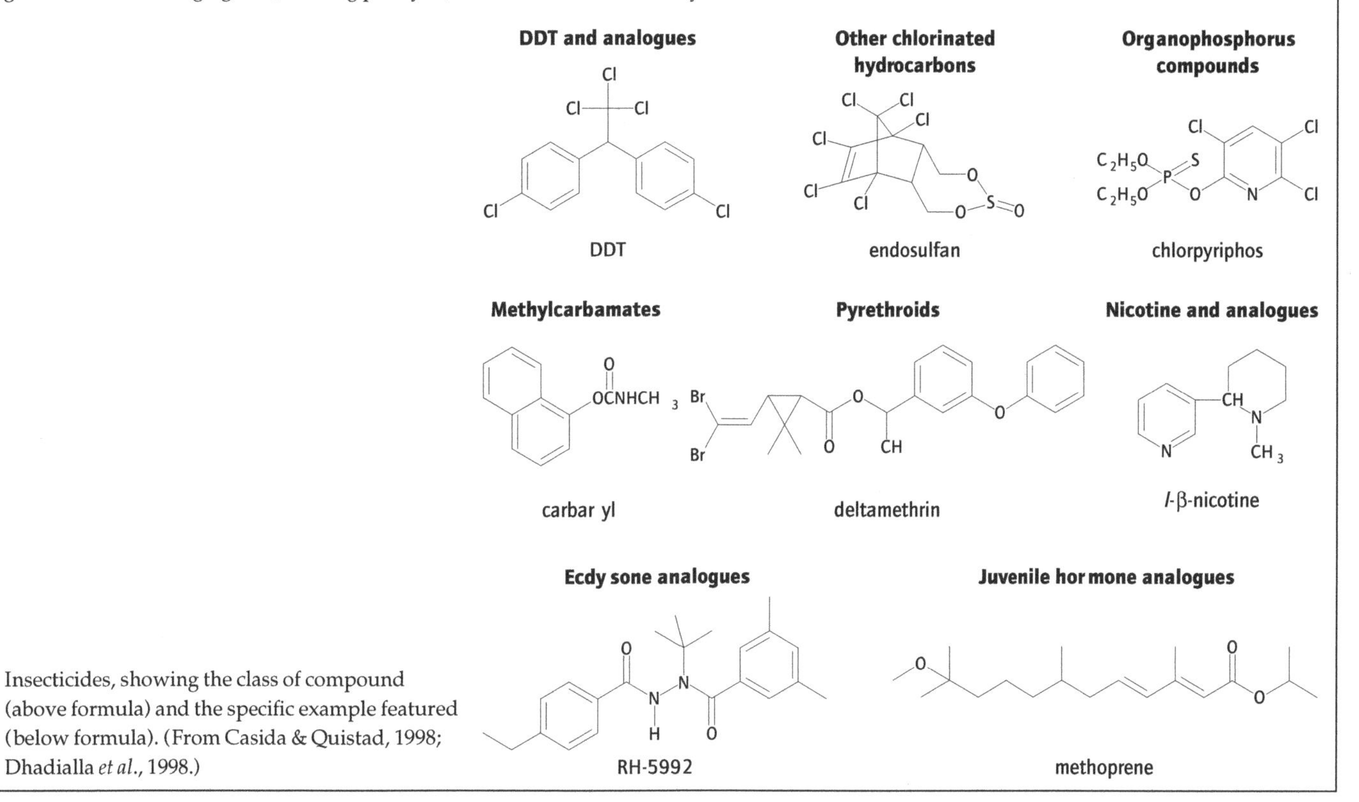

Insecticides, showing the class of compound (above formula) and the specific example featured (below formula). (From Casida & Quistad, 1998; Dhadialla *et al.*, 1998.)

The search for alternative chemical insecticides was a major focus of scientific endeavour in the latter part of the twentieth century (Box 16.12). This was driven not only by concerns about safety, but also by the facility with which insects (with their short life cycles) develop resistance. For example, replacement of just a single amino acid – alanine – with serine or glycine at position 302 in the GABA receptor molecule confers resistance to dieldrin.

Typically, the most useful compounds are analogues of naturally occurring compounds – i.e. their actions mimic those of the latter. Analogues may or may not be chemically related to the natural compound; they may be agonists and activate the relevant receptor molecules, or antagonists and block them. Several analogues of JH (Section 16.11.5) (e.g. methoprene) are already used to control insect pests and, recently, a range of ecdysone analogues has been discovered (ecdysone and JH have also long been used to enhance silk production by *Bombyx mori*).

16.12.2 Biorational pest control

Control of pests by exploiting the properties of their own molecules is potentially the most environmentally acceptable method of control. Such compounds are often specific to, for example, insects, non-toxic to vertebrates and biodegradable. Ecdysone (Section 16.11.5) is produced by many plants in far greater quantities than those found within arthropods – there is little doubt that this represents a natural 'pest management strategy', although some insects have responded by developing detoxification procedures. The introduction of the relevant genes into crop plants is of potential value (but highly controversial!). The application of neuropeptides in pest control faces a major problem of delivery, since these compounds do not penetrate the insect cuticle and they are degraded in the gut. Analogues that are more stable than the native compounds might withstand digestion. Another possibility involves the introduction of peptide genes into baculoviruses of lepidopterans by genetic engineering (see below).

One company alone markets pheromones (Section 16.10.5) for more than 70 species of insects of economic importance. An obvious, dramatic use involves spraying the field with a sex attractant so that the males cannot find mates – for example, the application of glossyplure for control of the pink bollworm in cotton. However, pheromones are most effectively used as just part of a programme of integrated pest management (IPM). For, example, they can be used in traps to monitor the development and scale of an infestation, so that pesticides are employed only when really needed. However, if a pesticide or disease fungus is combined with a pheromone in traps, insects can be killed without the application of chemicals to the crops.

A combination of semiochemicals may be deployed in what are called 'push–pull' or 'stimulo-deterrent diversionary strategies'. For example, a pheromone is used to attract predators or parasitoids of the pest species to the crop and, in addition, a 'trap crop' which produces allelochemicals attractive to the pest can be planted nearby.

Undoubtedly the most acceptable way of using hormones for pest control uses the good offices of their natural enemies. Some parasitoid wasps lay eggs on or in the body of their insect hosts but, before doing so, inject a non-lethal venom which suppresses the rise in ecdysone levels and thus blocks moulting (Fig. 16.51).

16.12.3 Genetically modified viruses

Baculoviruses are attracting increasing attention in relation to insect pest control, since they are non-polluting and non-toxic to vertebrates – indeed, they are typically genus- or species-specific. Their major drawback is their slow killing speed. This is being addressed by genetic engineering, so that the virus provokes the sustained secretion of, say, PTTH or EH (Section 16.11.5), by the insect. Initial results, with respect to such hormonal peptides, are disappointing, but viruses engineered to induce the production of scorpion toxins – which activate neuronal Na^+ channels – by the host are more effective, particularly when used in combination with low doses of pyrethrum.

In some cases, viruses prolong the life of the (feeding) larval stages of their hosts and, curiously, genetic engineering to remove the relevant gene improves the kill time. Although the release of recombinant (i.e. GM) organisms into the environment is highly controversial, gene deletion is generally regarded with less caution than gene insertion and will probably provide the first GM baculoviruses to be approved by the regulatory authorities for commercial use.

16.12.4 Compound eye technology

Strikingly different examples of applied biology are based on knowledge of arthropod compound eyes. Light entering the rhabdome (the light sensitive rod composed of microvilli – see Section 16.5.5) at a slight angle is reflected repeatedly off its walls because of its higher refractive index, i.e. the rhabdome acts as a light guide (Fig. 16.52). This principle underlies the development of fibre optic transmission systems which span the globe. The organization of compound eyes and the associated neural

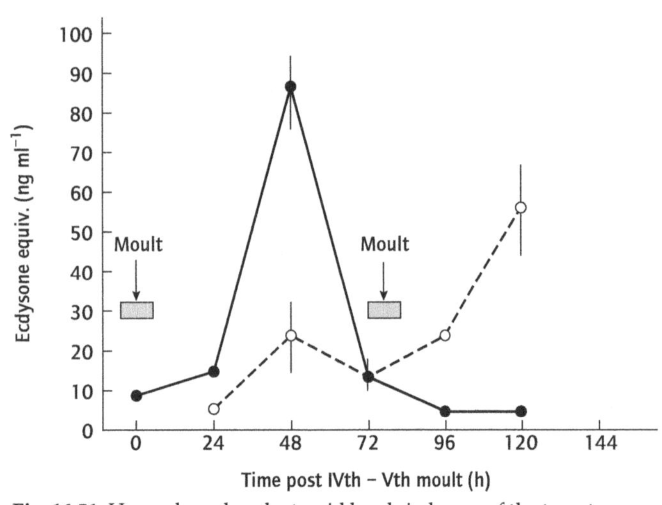

Fig. 16.51 Haemolymph ecdysteroid levels in larvae of the tomato moth, *Lacanobia oleracea*; filled circles, unparasitized; empty circles, infested with the ectoparasite wasp, *Eulophus pennicornis*. Moulting was blocked in parasitized specimens. (From Weaver *et al.*, 1997.)

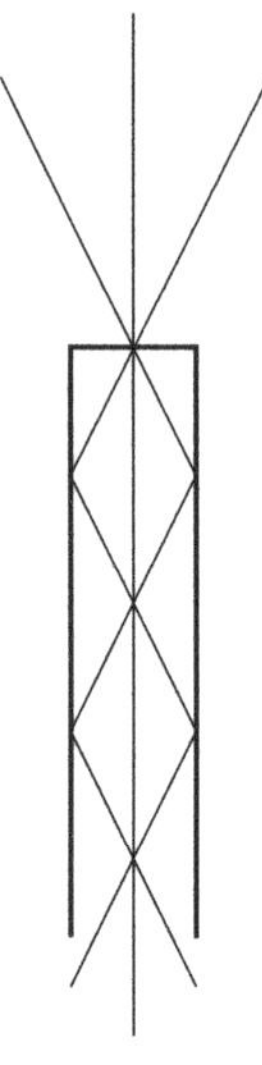

Fig. 16.52 Rhabdomes of ommatidia in compound eyes act as waveguides, internally reflecting rays of light closely aligned to the axis of the rhabdome. (After Van Hateren, 1989.)

circuitry has also inspired the design of eyes and guidance system for small robots which are able to navigate around obstacles to reach a goal.

16.13 Conclusion

We will conclude our survey of invertebrate control systems with two examples which exemplify the progress made in recent years towards explaining behaviour and physiology in terms of neural mechanisms.

The soil nematode, *Caenorhabditis elegans*, has been studied in great detail by Brenner and his colleagues at Cambridge. The worm is almost transparent, only 1-mm long, and when cultured in the laboratory has a life cycle of 3.5 days at 20°C. The structure of the nervous system has been reconstructed in its entirety from photographs of serial sections, taken with the electron microscope.

The hermaphrodite has exactly 302 neurones, each with a predictable position and shape. Furthermore, the 'wiring diagram' of the system is now complete – all of the synaptic contacts made by each of the cells are known. About 600 gap junctions, 5000 chemical synapses and 2000 neuromuscular junctions are formed, although their number and position may vary.

The embryological cell lineage of every neurone (and all the other cells) has been followed, and the stages at which particular neurones differentiate during development are known, as are the changes made in synaptic connections.

A study of the marine mollusc, *Pleurobranchia californica*, by Davis and others in Santa Cruz, California, has been very successful in uncovering the neural mechanisms underlying principles of behaviour which may be observed in almost any animal.

Different aspects of behaviour are organized as a hierarchy. Thus the escape response takes precedence over feeding – and stimuli for the former activate inhibitory synapses on the single 'command' neurone which controls feeding. Motivation for feeding shows the usual dependence on recent consumption and sensory pathways activated by the latter inhibit the feeding command neurone. Competing stimuli evoke variable responses depending on their relative strengths and the animal must have a capacity for decision-making, since it usually does only one thing at a time. This has its basis in the integration of sensory input within the nervous system. For example, if a stimulus for feeding is accompanied by one for withdrawal of the oral veil, the former prevails, since two neurones become active which block the motor output responsible for withdrawal. In contrast, if the food stimulus is weak, or if the animal has fed, different circuits are activated and withdrawal takes precedence. The endocrine system is also involved, since the spawning hormone exerts an influence on neurones in the buccal gangion, inhibiting feeding. Behaviour is both orderly and adaptive and in this it contrasts with the variety of competing influences that impinge on animals.

The results of invertebrate neuroscience have uncovered many of the patterns of neural organization which underlie the behavioural and physiological characteristics of animals. Nevertheless, much remains to be done, as indicated (to take just one example) by evidence provided by the completion of the *Caenorhabditis* genome project that as many as 50 members of the nuclear receptor family (of which ecdysone is one – see Section 16.10.4) are functional in this animal.

16.14 Further reading

Aidley, D.J. 1998. *The Physiology of Excitable Cells*, 4th edn. Cambridge University Press, Cambridge.

Bullock, T.H. & Horridge, G.A. 1965. *Structure and Function in the Nervous Systems of Invertebrates*. Freeman, San Francisco.

Breidbach, O. & Kutsch W. (Eds) 1995. *The Nervous Systems of Invertebrates: An Evolutionary and Comparative Approach*. Birkhauser, Basel.

Eaton, R.C. (Ed.) 1984. *Neural Mechanisms of Startle Behaviour*. Plenum Press, New York.

International Society for Neuroethology web site: www.neurobio.arizona.edu/isn/

Laufer, H. & Downer, R.G.H. 1988. *Endocrinology of Selected Invertebrate Types*. Alan Liss, New York.

Manning, A. & Dawkins, M.S. 1998. *An Introduction to Animal Behaviour*, 5th edn. Cambridge University Press, Cambridge.

Simmons, P.J. & Young, D. 1999. *Nerve Cells and Animal Behaviour*, 2nd edn. Cambridge University Press, Cambridge.

CHAPTER 17

Basic Principles Revisited

17.1 Basic physiological features of phenotypes

Animals require resources as building blocks to make new tissue (somatic and reproductive) and replace spent somatic tissues, and also as fuel to power these processes. Unlike autotrophic organisms, that can elaborate organic requirements from inorganic constituents plus an energy source such as sunlight, animals require to *feed* from other organisms (considered in Chapter 9). With the advent of this *heterotrophy*, there evolved a need to be able to move to find and capture resources and to avoid being eaten by other animals. Once locomotion evolved (see Chapter 10) there would have been co-evolutionary pressure on predators and their prey to move more effectively. Locomotion also requires power (Chapter 11). As well, there is a need to invest resources in various processes and structures that provide a defence against exploitation by other organisms (addressed in Chapter 13).

Macromolecules acquired in the food are broken down to their subunits by an enzymically-mediated process known as *digestion*, prior to their absorption into the tissues for either (a) utilization in the *resynthesis* of macromolecules (*anabolic processes*) or (b) breakdown to release energy to power these processes (*catabolic processes*). Resources in excess of requirements, particularly amino acids and 'spent' tissue proteins, are also catabolized and *excreted* (see Chapter 12).

The major fuel for the metabolism (= anabolism + catabolism) of organisms is carbohydrate, usually monosaccharide, but it can be stored prior to use as polysaccharide (often glycogen) and/or as fat. Energy is released from this fuel by catabolic oxidation. However, after the origin of life but prior to the origin of photosynthesis, when the earth's atmosphere was devoid of O_2 (Chapters 1 and 2), this must have occurred *without* O_2 by the transfer of electrons to organic components, which in reduced form accumulated as end products. With the advent of an oxygenated atmosphere, more complete oxidation became possible and more efficient processes evolved; this required the supply of O_2 from the outside world and the removal of CO_2 which forms as an end product in the process. *Aerobic respiration* is now most widespread in the animal kingdom, but *anaerobic* respiration is still found as the major process of catabolism in some species. A universal compound, important in the short-term storage and transfer of the energy released in these catabolic processes, is the phosphorylated form of adenine (a nucleotide). Adenine triphosphate (ATP) is generated from the diphosphate form, ADP, by the reactions associated with respiration. It returns to this state after releasing its energy and is recycled. Respiratory processes were considered in Chapter 11.

These fundamental features of physiology are summarized in Fig. 17.1. Note that the resource input to the organism is limited by the resource-acquiring processes and structures of feeding. Hence, even when the availability of food in the environment is unlimited, the amount that can be made available to metabolism is finite and limiting. The more that is invested in one metabolic demand, the less that is available for others. The rates and routes of use are controlled by enzymes and are therefore ultimately specified by genes. The way the limited resources are allocated crucially influences the biology of organisms at a number of levels: allocations between anabolism and catabolism influence the *physiology* of the organism; allocations between different activity

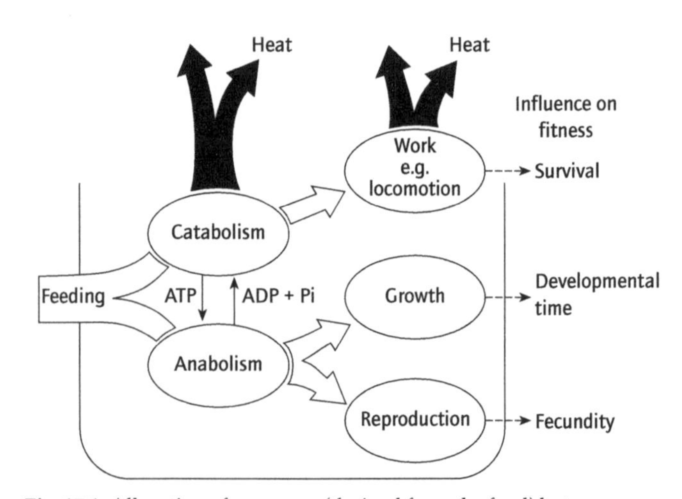

Fig. 17.1 Allocation of resources (derived from the food) between anabolic and catabolic processes is the basis of animal physiology (after Calow, 1986).

demands influence the *behaviour* of the organism; allocations of the limited products of anabolism between different structures influence the *form* of the animal; and between somatic and reproductive structures the *reproductive* and *life-cycle* biology of the organism. Moreover, the extent to which these different demands are supported will importantly influence survivorship and fecundity and hence fitness. Therefore, according to Darwinian principles, those gene-determined patterns of resource allocation and utilization (often called *strategies*) that maximize the transmission of the genes that code for them will be selectively favoured. Hence, though the physiologies of all organisms are based upon a common organization, they will have been 'tuned' (to at least some extent) by natural selection according to the ecological circumstances in which they operate – and this is what is meant by *adaptation*.

It will be clear from this description of organisms that, functionally, they operate as integrated wholes. There are trade-offs in metabolic investments that might lead to trade-offs between the components of fitness; an increased investment in locomotion might enhance survival but mean that less resource is available for reproduction. The principle of organismic integration is also important for morphological structures since the development of one structure must be compatible with others. There are two consequences of this. First, there have to be proximate (immediate) *controls* over the development of form, the expression of behaviour and of physiological function. The control systems (see Chapter 16) involve chemical and electrical signals. Second, there is ultimate selection for integration; a gene-determined trait has to be compatible with the organismic environment in which it is expressed as well as bringing survival and reproductive gains by interaction with the external environment. Hence, genes and the traits they specify are selected *within the context of the organism*. This *organismic* or *holistic* orientation, which differs to some extent from the 'gene pool philosophy' that views organisms as collections of dissociable, 'selfish' genes, is the one we used in the preceding chapters.

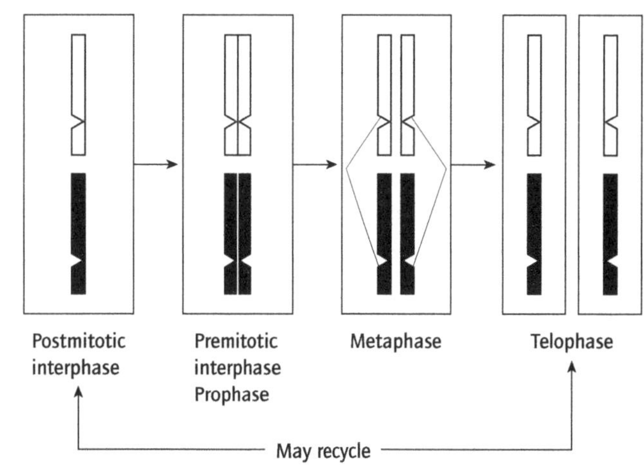

Fig. 17.2 The behaviour of a pair of chromosomes in mitosis (after Paul, 1967).

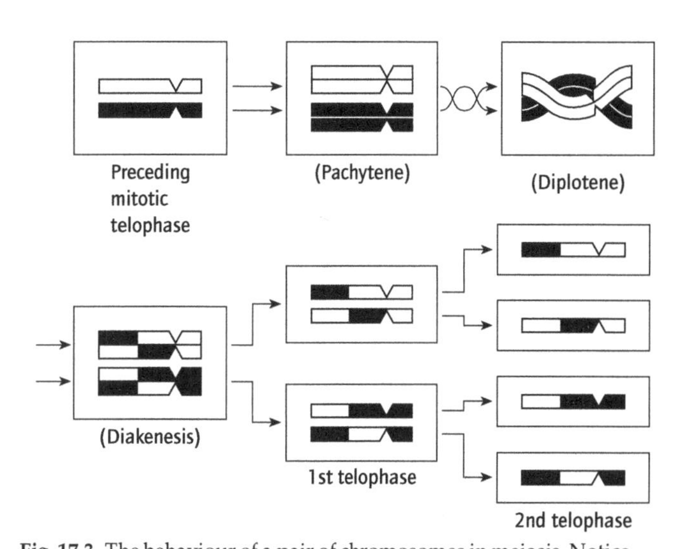

Fig. 17.3 The behaviour of a pair of chromosomes in meiosis. Notice that the crossing-over processes mean that the chromosomes of the progeny do not contain genes in the same arrangement as the parent (after Paul, 1967).

17.2 The primacy of replication and reproduction

A very important, perhaps *the* most important, investment of resources by organisms is into reproduction. These are used to form the propagules that are the vehicles of genetic transmission.

Sometimes, as noted earlier (Section 14.2), a cell or group of cells separates from the parent with a perfect replica of its genome. At the heart of this process, asexual reproduction, is cell division by mitosis (Fig. 17.2). The cell or cells in the propagule is/are formed by mitosis and these, in turn, reproduce a replica organism by mitosis.

Alternatively, single germ cells (gametes) are produced, usually in specialized organs (the gonads), within the organisms. These must usually fuse with other germ cells before a new organism is reproduced. The germ cells are formed from a process of cell division which results in cells that have only half the DNA and number of chromosomes of the ordinary somatic cells. This is meiosis (Fig. 17.3). The full chromosome complement is reinstated by fusion with other germ cells (syngamy, fertilization) and since these are often derived from different parents, the progeny contain a mix of two genetic programmes. This is known as *sexual reproduction*. Processes of reproduction were considered in detail in Chapter 14.

17.3 Ontogeny

By definition the products of reproduction are smaller and usually simpler than the parent. The products of sexual reproduction are unicellular whereas the parents have many, functionally differentiated cells. Hence *development*, or *ontogeny*, must involve cell division to achieve a size increase (*growth*) and cell specialization (*differentiation*). Since specific cells occur in specific places in an organism there has to be *patterning*, and since adults have complex shapes, whereas the products of reproduction are usually more or less spherical, there has to be some shaping (*morphogenesis*).

Cell division (*cleavage* of the fertilized egg) occurs by the process of mitosis. This faithfully replicates the original genome and has been described above. Some of the early embryologists (particularly August Weismann, 1834–1914) thought that *differentiation* involved the progressive jettisoning of unwanted parts of the hereditary material in specific cell lines. But an understanding of mitosis, and the realization that some invertebrates can be fully regenerated from small pieces of somatic tissue, suggest that every somatic cell contains a more-or-less complete copy of the original genome. Hence, differentiation must involve switching, whether on or off, of specific parts of the programme in different cells. In this way, though containing the same genetic programme, different cells come to produce different proteins and hence function in different ways.

The reasons for, and mechanisms underpinning, the control of gene expression in invertebrates (eukaryotic organisms) are quite different from those we see operating in bacteria (prokaryotes). Bacterial gene control acts to adjust the cell's activities to the environment prevailing at any one time. To do this they either switch gene expression on, via activators, or off, via repressors. All the gene switches operate by either blocking or enhancing the binding of RNA polymerase to specific promotors (a promotor is a site to which RNA polymerase attaches to initiate transcription). In invertebrates (and all other eukaryotes) the control of gene expression in a cell is less to do with responding to that cell's immediate environment and more to do to with acting to regulate the body as a whole. Therefore changes in eukaryote gene expression (1) act to maintain homeostasis (maintenance of a constant internal environment) in the body, and (2) mediate the decisions that produce the body. In the case of the latter this entails making sure that the appropriate genes are expressed, in a tightly prescribed order, in the appropriate cells at the correct time during development. It turns out that many genes are activated only once. Their actions not uncommonly produce irreversible effects. This 'once-and-for-all' expression of the genes that determine the developmental programme in eukaryotes is radically different from the environmentally-mediated, reversible responses found in bacteria. Furthermore there is a limit to how complex the bacterial regulatory scheme can become. Only a limited number of switches can fit onto or near one promotor site. Consequently in eukaryotes this physical limitation is overcome by 'control-at-a-distance.' Where many regulatory sequences scattered around the chromosomes are able to influence the transcription of a particular gene. This mechanism incorporates two novel features not found in bacteria: (1) a group of proteins whose function is to aid the binding of RNA polymerase to the promotor; (2) two 'groups' of modular regulatory proteins that bind to distant sites. While most gene regulation in eukaryotes takes place at the initiation of transcription, there are also a number of posttranscriptional control processes.

Our understanding of pattern formation in invertebrates has advanced dramatically over the past few years. This has been due to concentrated efforts in unravelling the regulation of development in two animals that contain roughly the same amount of DNA, a representative complex animal, the fruit fly *Drosophila* and a representative simple animal, the nematode *Caenorhabditis*. What is remarkable is how much commonality exists, even if we include studies on the representative mammal, the mouse.

The principal feature of pattern formation is induction. This is the ability of cells to influence and alter the developmental trajectories of adjacent cells via the production of chemicals called morphogens. Exactly where any particular cell will end up involves the addition to that cell of a 'label'. For example in *Drosophila*, attaching the positional labels that determine segmentation is by means of chemical gradients set up within the egg, based on the instructions of maternal genes. The relative positions of structures within segments is determined by an organized assembly of genes, referred to as homeotic genes. It now turns out that homeotic genes, containing a homeobox domain, are not just restricted to *Drosophila* but are now thought to occur in all animals.

Morphogenesis involves a combination of cell movement and differential growth. The first process has been well described for the early stages of development in sea-urchins (Fig. 17.4a and see Chapter 15) – principally because the embryo is transparent, enabling cells inside to be observed and even recorded with time-lapse cinematography. At an early stage the embryo consists of a hollow ball of cells – the blastula. After this stage, some cells have to move inside to form internal organs, like the gut. This process is known as gastrulation and occurs in two stages. First there is inward movement of individual cells and then wholesale invagination. Migration begins when cells lose contact with each other, move into the blastocoel (inner cavity of blastula) and then migrate along the inner surface by extensions, filopodia, that pull them along. The invagination occurs in two stages: a slow phase of inward bending followed, after a lag, by rapid invagination. The first phase probably occurs by changes in adhesiveness of cells; they adhere less to each other but remain attached to a basement membrane. This causes them to become pear-shaped and could cause inward bending as shown in Fig. 17.4b. The rapid phase probably occurs by the cells on this intucked region forming filopodia that contract after making contact with the roof of the blastocoel. These processes, involving cell–cell membrane interactions and active migration by filopodia, are probably fairly typical of morphogenetic changes in general. A major question, which remains unresolved, however, is how such processes are controlled to result in a completely organized individual.

Finally, differential growth of organs, external and internal, can cause shaping. Julian Huxley (in his *Problems of Relative Growth*, 1932) discovered that if you plot the logarithm of the size of one part of an organism against the logarithm of the size of another (or of the size of the whole organism) the points often

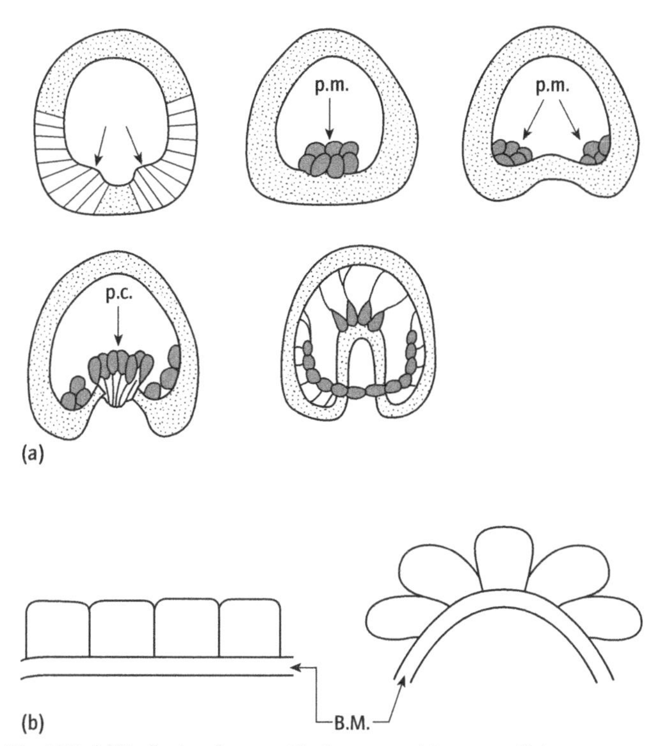

Fig. 17.4 (a) Early development in the sea-urchin. p.m. = Primary mesenchyme cells – these migrate to the 'corners' of the embryo using filopodia; p.c. = pear-shaped cells. (After Gustafson & Wolpert, 1967.) (b) Change in shape of cells at point of invagination. B.M. = Basement membrane.

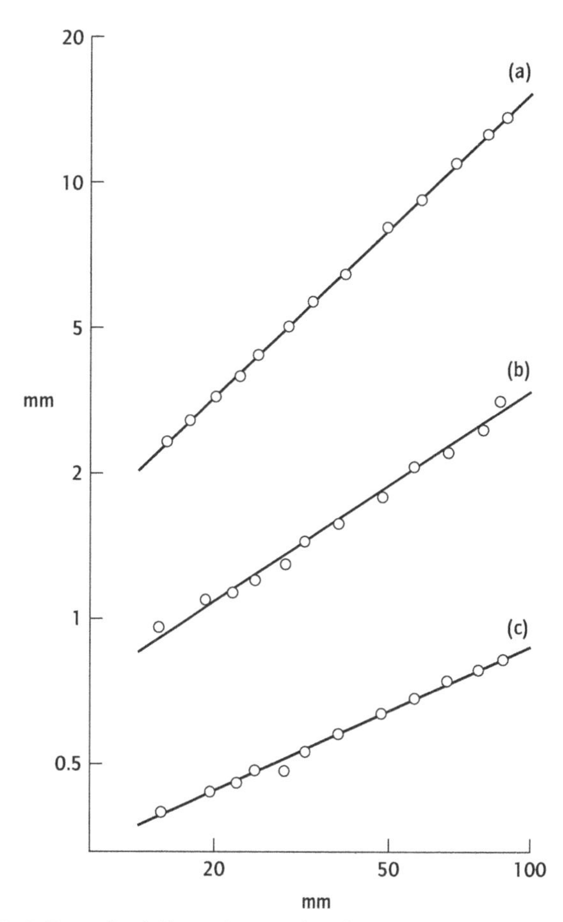

Fig. 17.5 Example of allometric growth in the insect *Carausius morosus*. Length of posterior part of prothorax (a), width of head (b), and diameter of eye (c) plotted against total length of body. Co-ordinates are logarithmic. (After Wigglesworth, 1972.)

fall on straight lines (Fig. 17.5). Equal intervals on a logarithmic scale represent equal multiplication factors so that these straight lines suggest that one organ multiplies itself (grows) at some specific rate relative to another. Clearly allometric relationships will have a profound effect on the shaping of an organism.

Here, then we have summarized the basic processes of development – differentiation, pattern and morphogenesis – and made some comments about how they might be controlled. In Chapter 15 we discussed the details of these processes as they apply to invertebrates.

17.4 Ontogeny and phylogeny

Since both ontogeny and phylogeny are *apparently* progressive it is tempting to assume that the latter occurs by additions to the former. Then ontogeny *recapitulates* phylogeny – a view expressed by Ernst Haeckel (late nineteenth century). Ontogeny of the 'higher' form is, according to this view, supposed to repeat the adult forms lower down the scale of organization. Von Baer, a contemporary of Haeckel, argued instead that no higher animal repeats any adult stage of lower animals but, because development always proceeds from undifferentiated to differentiated state, the initial phases of development must be conserved in different phyla. Hence, it is these embryonic forms, rather than adults, that are repeated in different phyletic lines, and this is certainly consistent with the principles of development enunciated in the last section.

Terminal addition fits neatly into the Lamarckian theory of acquired characters, since the latter are usually acquired later in life and added on to existing structures. Mendelian genetics, on the other hand, undermines this view, for a mutation can bring a change at any stage of development. Indeed it is now known that genes control rates of development and changes in these can cause profound alterations by either arresting development (removing the 'old' adult) or causing it to go on beyond the normal end point. Walter Garstang (1868–1949) was one of the first zoologists to recognize the importance of genetics for development and the possibilities for evolutionary change allowed by these kinds of alteration to the rate or timing of development.

Figure 17.6 classifies the complete spectrum of evolutionary shifts that might arise in this way. Each square contains a *developmental trajectory*, i.e. index of shape against size or age. The solid lines show the ancestral trajectory and the broken lines the descendant. *Start* = initiation of development; *stop* = cessation of development. In the top line development is decelerated or truncated, and the changes are referred to as paedomorphosis.

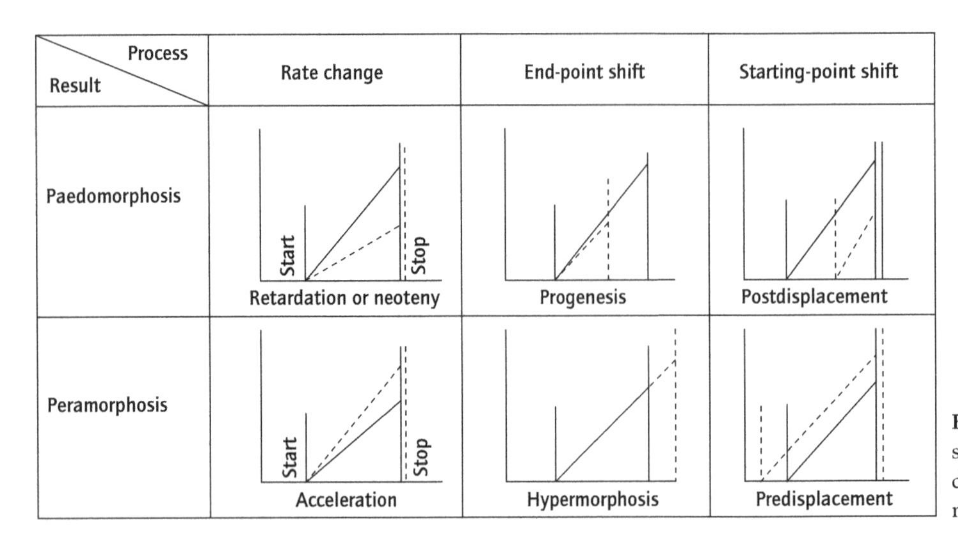

Fig. 17.6 Classification of possible evolutionary shifts in development. Abscissae = developmental time or size; ordinates = morphological changes (after Calow, 1983).

Either by reduced rates of development (neoteny) or truncation (progenesis) embryonic, larval or juvenile features turn up in adults. The bottom line shows the reverse and is referred to as peramorphosis.

Walter Garstang was particularly interested in larval forms and thought that paedomorphosis had had a dominant effect in evolution. And indeed paedomorphosis could have been important in a number of instances (Chapter 2). For example, among living species sexual maturity is attained at a small size in some male echiurans, crustaceans (and fish) which attach 'parasitically' to the much larger (sometimes by several orders of magnitude) females. The evolution of six-legged insects from many-legged ancestors might well have occurred by paedomorphosis (Fig. 17.7). Finally, a widely accepted theory of vertebrate evolution suggests that they are derived from some larval stock resembling the free-swimming 'tadpole' larva of present-day tunicates (see Fig. 7.30).

Yet accelerative shifts are also a possibility. Consider the evolutionary trends in the patterns of sutures found on the shells of ammonites that are depicted in Fig. 17.8. Here the descendants' slopes are steeper than those of the ancestors so there has been *acceleration*. Also, sutural growth of descendants goes beyond that of ancestors, suggesting *hypermorphosis* (Fig. 17.6). Thirdly, the descendants' trajectories are 'higher' than those of the ancestor so the descendants begin with something of a start and this suggests some *predisplacement* (Fig. 17.6).

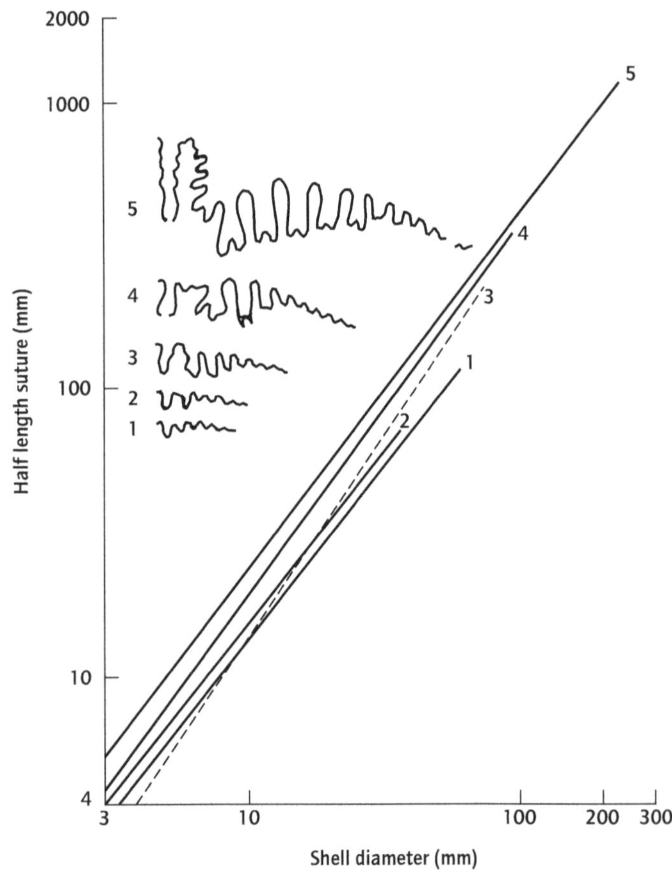

Fig. 17.8 Mixture of developmental changes in ammonites. 1 = Ancestor; 2–5 = descendants. Older parts of the stage are to the right. (After Newell, 1949; see also Calow, 1983.)

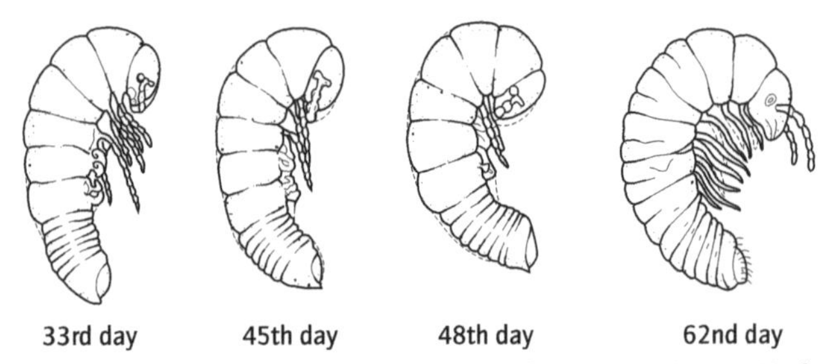

Fig. 17.7 Stages in myriapod development. The six-legged stage, 33rd day, is sometimes used as evidence for a paedomorphic origin of insects (see also Gould, 1977).

To summarize, then: ontogeny does not recapitulate phylogeny. Development has to do certain common things irrespective of taxa. It has to do them on the basis of processes common to all animal life – mitosis, differential expression of a genome, cell movement, differential growth. All of this points to the fact there has to be, as von Baer suspected, some commonality in early development. Nevertheless, small modifications in the *rates* and timings of these processes, which in principle can be effected by small genetic changes, are likely to have had profound effects on phylogenesis. Such changes might indirectly have been instrumental in causing some of the quantal jumps between levels of organization alluded to above (Chapters 1 and 2), a view that was certainly advocated by Walter Garstang.

17.5 Size and shape – scaling

Size and body shape are linked by geometry. These relationships are referred to as scaling relationships. One approach to the study of scaling is to define the geometrical rules behind the relationships and to understand both ontogeny and phylogeny in these terms. This was initiated earlier this century by a famous scholar and zoologist called D'Arcy Thompson in his book *On Growth and Form* (1917). Another approach is to consider the implications of the scaling relationships for the way organisms function. This illustrates how form might constrain function and how these constraints might be escaped by changes in shape of the kind described in the previous section. We illustrate each approach in turn with an example in the sections that follow.

17.5.1 The equiangular spire

A well-known example of the D'Arcy Thompson approach involves snail shells and geometrical properties of the so-called equiangular spire.

If you take a snail with a planospiral shell, such as *Planorbis* (similar in shape to the Stylommatophora in Fig. 5.16), draw a radius from its centre to its edge and construct tangents with the edge of each whorl along the radius, then these tangents remain parallel – which also means that the angle between them and the radius remains constant. Indeed the angle between any tangent around the edge of the shell and a radius will be constant. This is what is meant by an equiangular spire. Equiangularity means that the shape of the shell is remaining constant as the snail grows.

This constancy in shape as size increases is possible only for a limited range of shapes. The geometry of this was explored by D'Arcy Thompson. The outer line of the planorbid shell describes a spiral. Moving an angle θ round the spiral we find that the radius r is given by:

$$r = r_0 W^{\theta}$$

where W is a constant. What this means is that r increases by a constant factor for each full rotation; so if we plot the value of r for each whorl along a single radius (i.e. full rotation) on a logarithmic scale, against whorl number we should get a straight-line relationship. This is why the equiangular spire is often also called a logarithmic spire.

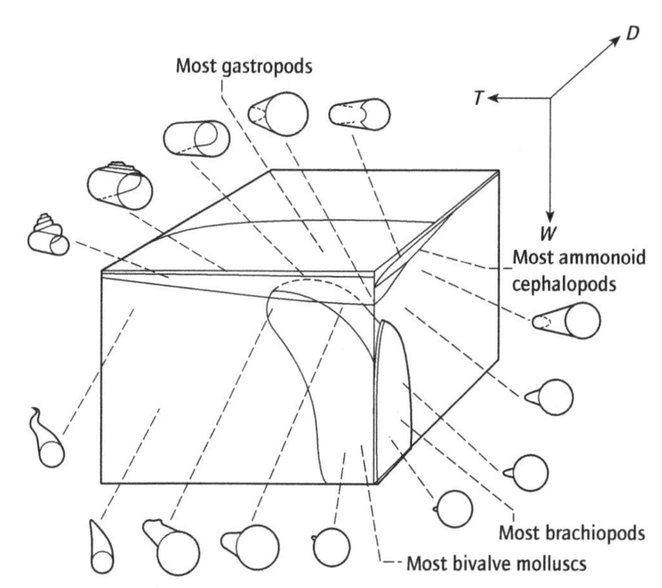

Fig. 17.9 How the range of possible shell forms is related to values of T, D and W – see text for further explanation. (After Raup, 1966.)

Now if the spiral does not lie in a plane, but rises helically, like a *Helix*, as we move along the spiral through an angle of θ we find that the spiral rises a distance of y along its axis. If growth is not to change the shape of the shell then:

$$y = T(r - r_0)$$

where T is another constant – *indicating the extent to which the shell rises in a spire.*

Now consider the radius within each whorl; where the radius of the shell is r and the radius within each whorl is r'. Again to keep the shape of the shell constant, r' must grow in proportion to r, such that

$$r' = r(1 - D)/(1 + D)$$

where D is yet another constant – *in fact showing how far the inner edge of the shell is from the axis of the spiral.*

To an important extent, the relative values of W, T, D define the range of shapes that might derive from the equiangular spire. Fig. 17.9 from D.M. Raup shows the ranges actually found in various animals. Most gastropods have low values of W, but T and D vary a lot. Bivalves have high Ws but fairly low Ts and Ds. Nautiloids and most of the extinct ammonoids have plane spirals in which $T = 0$ and Ws are low.

17.5.2 Form and function

The scaling relationship that we have encountered most frequently in the preceding pages is the one relating to surface and volume. This is fundamental to the functioning of invertebrates because they interact with their environment at surfaces – in taking up nutrients and O_2 and, to some extent, in moving – yet metabolism takes place in the volume of biomass. The relationship between surface and volume and the constraints that this puts on, for example, the relationship between the rate of uptake of food and O_2 have been treated earlier (Section 11.6.1). The argument is that since area grows in two dimensions, whereas biomass grows in three, the activities of the latter will be constrained by the former in a way that is predictable from the geometry.

This tension between volume and surface relationships, together with difficulties of the diffusion of nutrients and gases through solid biomass, put limits on the size that can be achieved by globularity (being one evolutionary pressure for multicellularity; Section 3.1) and solidity (being another evolutionary pressure for the evolution of complex body surfaces [Section 11.4.4], body cavities [Section 2.3.1] and blood systems [Section 11.4.2]).

There are other scaling relationships of this type; for example that relate to the strengths of exoskeletons, the support that limbs need to give to bodies, the relationship between the size of the female reproductive system and the size of her propagules. These are treated in more detail in several good books on scaling listed in Section 17.6.

17.6 Further reading

Bennett, A.F. 1997. Adaptation and the evolution of physiological characters. In: Dantzler, W.H. (Ed.) *Handbook of Physiology. Section 13. Comparative Physiology*, Vol. I, Chapter 1, pp. 3–16. Oxford University Press, Oxford.

Calder, W.A. III 1995. *Size, Function and Life History*. Dover, New York.

Gould, S.J. 1977. *Ontogeny and Phylogeny*. Harvard University Press, Cambridge, Massachusetts.

Hoffman, A.A. & Parsons, P.A. 1997. *Extreme Environmental Change and Evolution*. Cambridge University Press, Cambridge.

Huxley, J.S. 1932. *Problems of Relative Growth*. Methuen, London.

Kozlowski, J. & Weiner, J. 1997. Interspecific allometries are by-products of body size optimisation. *Amer. Nat.*, **149**, 352–380.

Lewin, B. 1998. *Genes VI*. Oxford University Press, Oxford.

McGowan, C. 1994. *Diatoms to Dinosaurs. The Size and Scale of Living Things*. Island Press, New York.

McKinney, M.L. & McNamara, K.J. 1991. *Heterochrony: The Evolution of Ontogeny*. Plenum Press, New York.

McNamara, K.J. 1988. Patterns of heterochrony in the fossil record. *Trends Ecol. Evol.*, **3**, 176–180.

McNamara, K.J. (Ed.) 1995. *Evolutionary Change and Heterochrony*. Wiley, Chichester.

McNeill Alexander, R. 1999. *Energy for Animal Life*. Oxford University Press, Oxford.

McMahon, T.A. & Bonner, J.T. 1983. *On Size and Life*: W.H. Freeman & Co., New York.

Maynard Smith, J. 1986. *The Problems of Biology*. Oxford University Press, Oxford.

Peters, R.H. 1983. *The Ecological Implications of Body Size*. Cambridge University Press, Cambridge.

Raff, R.A. 1996. *The Shape of Life*. University of Chicago Press, Chicago.

Raup, D.M. 1966. Geometrical analysis of shell coiling: general problems. *J. Palaeontol.*, **40**, 1178–1190.

Schmidt-Nielsen, K. 1984. *Scaling: Why is Animal Size so Important?* Cambridge University Press, Cambridge.

Thomson, K.S. 1988. *Morphogenesis and Evolution*. Oxford University Press, New York.

Weibel, E.R., Taylor, C.R. & Bolis, L. (Eds) 1998. *Principles of Animal Design. The Optimisation and Symmorphosis Debate*. Cambridge University Press, Cambridge.

West, G.B., Brown, J.H. & Enquist, B.J. 1997. A general model for the origin of allometric scaling laws in biology. *Science*, **276**, 122–126.

Glossary

abdomen term applied to the posterior zone of any body divided into three distinct regions of which the anterior region is the head (cf. **thorax**)

aboral surface of body opposite to that bearing the mouth, i.e. in animals with the mouth in the centre of the upper or lower surface

acclimation change in physiology as a result of the exposure of an organism to a changed environment

acoelomate without a body cavity, other than those of the gut lumen and within organ systems

acron anterior, non-segmental region of an arthropod

acrosome filamentous tube(s) at anterior end of spermatozoon (q.v.) that contacts and fuses with the cell membrane of the ovum (q.v.) during fertilization (q.v.)

active transport movement of solutes against a concentration gradient by a process that uses energy

adductor muscle muscle closing, or holding closed, the valves of a shell

adenosine triphosphate (ATP) the main energy-carrying molecule in living organisms

ageing irreversible, deteriorative changes in individuals with time causing increased vulnerability and reduced vitality [= senescence]

amoebocyte cell capable of amoeboid movement

anabiotic see **cryptobiotic**

anaerobic without air; often used to describe those energy-yielding metabolic (q.v.) processes that proceed without oxygen

annulate descriptive of cylindrical organism, organ, etc. of which the external surface is divided into a chain of rings or 'annuli' by furrows, giving the appearance of segments

anoxic without free oxygen

antennae filiform and often elongate chemosensory appendages on the head of some arthropods and polychaetes, of onychophorans, etc.

apodeme internally directed process of the exoskeleton of an arthropod

apomixis see **asexual reproduction**

apterygote wingless

arrhenotoky form of parthenogenesis (q.v.) in which unfertilized eggs develop into (haploid) males whilst fertilized eggs develop into (diploid) females [= haplodiploidy]

article one of the units comprising a jointed appendage; i.e. the rigid section between two adjacent joints

asexual reproduction form of multiplication not involving meiotic reduction division and fusion of gametes (q.v.) (cf. **sexual reproduction**) [= apomixis]

autotomy self-induced amputation of an appendage or of a body region, as, for example, a means of escape from a predator or during budding (q.v.) or fission (q.v.)

basement membrane amorphous sheet underlying an epithelium (q.v.), composed of a type of collagen (q.v.) and carbohydrate

benthic appertaining to the bottom or substratum of aquatic systems (cf. **pelagic**)

bilateral symmetry symmetry in which the body can be divided into two, and only two, mirror-image halves

bioturbation disturbance of benthic (q.v.) sediments by animal activity

bipectinate the state of a mollusc ctenidium (q.v.) that retains the primitive condition of having gill filaments on either side of the central axis (cf. **monopectinate**)

biradial state in which each of the four quadrants of a spherical organism or embryological stage is the same as that of the opposite quadrant but differs from the adjacent ones

biramous having two branches (cf. **uniramous**); used of an arthropod limb

blastema group of undifferentiated dividing cells from which differentiated cells may be derived (see **differentiation**)

blastocoel cavity within the blastula (q.v.)

blastopore aperture through which the embryonic gut in a gastrula (q.v.) communicates with the external environment

blastula hollow ball of cells formed from a zygote (q.v.) by cleavage (q.v.) during embryology

budding form of asexual multiplication (q.v.) in which a new individual begins life as an outgrowth from the body of the parent; it may then separate to lead an independent existence, or remain connected or otherwise associated to form a colonial (q.v.) organism

bursa copulatrix sac receiving spermatozoa (q.v.) during the act of copulation (q.v.)

carapace protective exoskeletal shield covering all or part of the dorsal and lateral surfaces of an arthropod

catabolism see **metabolism**

cephalic appertaining to the head

cephalization development of a head during phylogeny (q.v.) or ontogeny (q.v.)

cephalothorax region of the body of some crustaceans formed by the fusion of head and thorax (q.v.)

cerci pair of variously shaped appendages on the last abdominal segment of many insects

chaeta small, stiff, projecting, chitinous (q.v.) bristle of annelid, pogonophoran and echiuran worms (cf. **seta**)

chelate state of an arthropod limb that terminates in a pair of pincers or forceps

chemoautotrophy mode of bacterial nutrition in which the organism can synthesize all its nutritional requirements by chemosynthesis (q.v.)

chemosynthesis synthesis of organic compounds according to the general equation

$$CO_2 + 2H_2X \rightarrow (CH_2O) + H_2O + 2X$$

using the energy (and often the reducing power) released by the oxidation of inorganic materials such as Fe^{2+}, CH_4, NH_3, NO_2^-, H^+, S, etc. [= chemolithotrophy] or of pre-existing organic compounds (acetate, formate, etc.); carried out only in bacteria under anaerobic conditions (cf. **photosynthesis**)

chitin a nitrogen-containing polysaccharide

chloroplast eukaryote organelle, occurring in several protists and most plants, in which photosynthesis takes place

choanocytes characteristic flagellated cells of poriferans

chromatophore a pigment-containing cell

chromosome thread-like structure in the cell nucleus that contains genetic information

cilia projecting propulsive organelles that beat in a single plane by means of a stiff propulsive and a lax recovery stroke; usually relatively short and occurring as several to many per cell; non-motile forms are specialized for sensory reception; found only in eukaryotes (q.v.) (cf. **flagella**)

cirrus eversible (q.v.) copulatory organ (cf. **penis** and **gonopod**)

clade group of organisms all of which share the same ancestral form (cf. **grade**)

cleavage mitotic cell divisions subdividing a zygote (q.v.) into a multicellular but undifferentiated embryo (q.v.); cell growth does not occur during this process

cleidoic descriptive of an egg (q.v.) enclosed within a protective coat

clitellum secretory epithelium forming a cocoon, especially in clitellate annelids

cloaca dilated terminal region of gut receiving the discharge of some other organ system(s)

cnidocyte cnidarian cell containing nematocysts (q.v.)

coelom fluid-filled cavity within tissues of mesodermal (q.v.) origin and enclosed by a mesodermal membrane (cf. **pseudocoel** and **haemocoel**); range from large hydrostatic body cavities, bounded by a peritoneum (q.v.) (see **schizocoel** and **enterocoel**), to, by definition, the epithelium (q.v.) lined spaces within mesodermal organs

coelomocyte cell suspended in coelomic (q.v.) fluid

coelomoduct (a) a blind-ending osmoregulatory/excretory gland and duct of mesodermal (q.v.) origin (cf. **nephridium**), or (b) open mesodermal ducts from coelomic (q.v.) body cavities to the exterior which serve for the discharge of gametes or of coelomic fluid

cohort group of individuals in a population all born at approximately the same time

collagen a fibrous protein usually associated with connective tissue or cuticular lattices

colonial descriptive of organisms produced asexually which remain associated with each other; in many animals, retaining tissue contact with other polyps (q.v.) or zooids (q.v.) as a result of incomplete budding (see also **modular**); also used to describe those sexually-produced individuals that form semi-permanent aggregations in space

commensal descriptive of an organism that lives in close proximity to another of a different type, e.g. in its burrow, without so far as is known affecting it

commissure cross-connection between nerve cords or ganglia (q.v.) [= connective]

compound eye single eye formed from few to many individual optical units, ommatidia, each with its own lens, field of view, receptor cells, etc. (cf. **ocellus**)

contractile vacuole intracellular, membrane-bound vacuole, concerned with osmoregulation, that fills with liquid and suddenly contracts, expelling the liquid to the exterior

copulation act of sperm transfer by an organ of one individual into the body or a duct/sac of another; a widespread, but not the only, precursor to internal fertilization

corona locomotor apparatus of rotiferans consisting of rings of cilia (q.v.)

coxal sacs eversible (q.v.), thin-walled vesicles associated with the base of the legs of some uniramians and serving to take up environmental water; similar vesicles also occur in onychophorans

cryptic 'hidden', e.g. by virtue of resemblance to the surrounding area

cryptobiotic capable of entering a resistant state of suspended animation during periods of environmental adversity (usually lack of water) and therefore able to inhabit temporary aquatic environments; also termed anabiotic

ctenidia gills of a type confined to molluscs comprising (primitively) banks of filaments issuing from each side of a central axis

cuticle non-cellular and resistant body covering secreted by the epidermis (q.v.) and moulted periodically; often ornamented and/or thickened locally into plates

cyst encapsulated, desiccation-resistant stage in life history

cytochromes respiratory enzymes, located in mitochondria, of similar structure to haemoglobin

definitive host host in which a parasite (q.v.) reproduces sexually (q.v.)

degrowth shrinkage of a starved animal

depolarize to decrease the electrical potential difference, (usually) across the cell membrane (cf. **hyperpolarize**)

deposit feeding consumption of detrital (q.v.) materials and/or associated organisms on and/or in the substratum

dermis non-muscular layer of body wall of mesodermal (q.v.) origin located beneath the epidermis (q.v.)

determinate development see **mosaic development**

detritus particulate decomposed or decomposing organic matter associated with water or the substratum, together with those microbial organisms colonizing it

deuterostome state in which the blastopore (q.v.) does not form the mouth, although it may form the anus (cf. **protostome**); also used to describe the animals displaying this state

diapause resting phase in life history in which metabolic activity is low and adverse conditions can be tolerated

differentiation process whereby totipotent embryonic cells become specialized to carry out different functions

dimorphic occurring in two distinct (usually morphological) forms (cf. **monomorphic** and **polymorphic**)

diploid having two of each chromosome in each somatic (q.v.) cell (cf. **haploid** and **polyploid**)

direct development development without a larval (q.v.) stage (cf. **indirect development**)

eclosion emergence of an insect, particularly that of the adult from a pupa (q.v.)

ectoderm outer germ layer, i.e. that covering the gastrula (q.v.) (cf. **endoderm** and **mesoderm**)

ectotherm an organism that derives its heat from the environment rather than from its own metabolism

egg general term for the initial stage of animal development – ovum (q.v.), zygote (q.v.), or a complex of cells or developing embryo, food reserves, etc. contained within a common shell or capsule
embryo early stage of development hatching from, or retained within, the egg (q.v.) and not capable of independent life
endoderm inner germ layer, i.e. that forming the embryonic gut of the gastrula (q.v.) (cf. **ectoderm** and **mesoderm**)
endopod the inner branch of a biramous (q.v.) arthropod limb (cf. **epipod** and **exopod**)
enterocoel coelom (q.v.) formed by the evagination of pouches from the embryonic gut (cf. **schizocoel**)
epiboly spreading of mobile yolk-free cells over yolky ones in gastrulation (q.v.)
epidermis outermost cellular covering of body, formed from ectoderm (q.v.)
epifaunal descriptive of benthic (q.v.) animals associated with the substratum surface (cf. **infaunal**)
epipod process arising from the basal article(s) (q.v.) of an arthropod limb (styli (q.v.) are probably epipods) (cf. **endopod** and **exopod**)
epithelium sheet or tube of tissue (q.v.) covering a free surface, e.g. forming the lining of a cavity
eukaryote having internal membrane-bound nucleus and organelles within the cell (cf. **prokaryote**); all organisms except bacteria are eukaryote
eutely state in which cells do not divide in the adult; the latter therefore has a fixed (and usually small) number of cells and growth occurs only by cell enlargement
eversible capable of being protruded by being turned inside out; extension is usually achieved hydraulically (q.v.) and retraction by muscular action
evolution origin and subsequent change over time
exhalent descriptive of the outward respiratory or feeding current (cf. **inhalent**)
exopod outer branch of a biramous (q.v.) arthropod limb (cf. **endopod** and **epipod**)
fate map formal description of the spatial distribution of cells or regions in a zygote (q.v.), embryo (q.v.) or imaginal disc (q.v.) that would normally give rise to the different parts of an organism
fermentation enzymatically mediated breakdown of organic compounds, and the generation of ATP, under anoxic (q.v.) conditions
fertilization process of fusion of gametes (q.v.) to create a zygote (q.v.)
filopodia filament-like projections of cell cytoplasm
fission form of asexual multiplication (q.v.) involving division of the body into two or more parts, each or all of which can grow into new individuals
flagella projecting propulsive organelles that beat in a rotary or corkscrew manner; usually relatively long and occurring as one or two per cell; found only in eukaryotes (q.v.) [bacterial flagella are of a fundamentally different form] (cf. **cilia**)
flame cells ciliated cells at proximal end of a protonephridium (q.v.)
flosculae minute projecting sense organs, terminating in a number of papillae, found on the trunk of some priapulans, loriciferans, kinorhynchs and rotiferans
fore-gut anterior section of gut of ectodermal (q.v.) origin and lining (cf. **mid-gut** and **hind-gut**)
fumarole region of earth's crust out of which issues heated water and reduced substances
furca paired processes, of variable form, issuing from the telson (q.v.) of crustaceans
gamete haploid germ cell capable of fusing with a germ cell of the opposite sexuality to form a zygote (q.v.) (see **spermatozoon** and **ovum**)
ganglion discrete body of nervous tissue containing neurones (q.v.)
gastrula embryological stage succeeding the blastula (q.v.) in which the single layer of cells is converted into a two-layered state by cell migration, ingrowth, etc., and in which mesoderm (q.v.) cells are proliferated
gastrulation process whereby a gastrula (q.v.) is derived from a blastula (q.v.)
gemmule multicellular asexual propagule (q.v.) of some poriferans contained within a protective coat
gizzard muscular region of gut in which food may be ground
glia accessory cells associated with neurones (q.v.) [= neuroglia]
gonochoristic having separate sexes [= dioecious] (cf. **hermaphroditic**)
gonopod modified leg serving as a copulatory organ (cf. **cirrus** and **penis**)
gonopore orifice through which gametes are discharged
grade group of animals sharing the same type of bodily organization but without having inherited it from a common ancestral form (cf. **clade**)
haemocoel body cavity formed by blood sinuses, often deriving from the blastocoel (q.v.) (cf. **coelom** and **pseudocoel**)
haplodiploidy see **arrhenotoky**
haploid having only one of each chromosome in each somatic (q.v.) or sex cell (cf. **diploid** and **polyploid**)
hermaphroditic capable of producing both ova (q.v.) and spermatozoa (q.v.), either at the same time or sequentially (cf. **gonochoristic**)
heterotrophic mode of nutrition that requires the intake of preformed organic compounds
hind-gut posterior section of gut of ectodermal (q.v.) origin and lining (cf. **fore-gut** and **mid-gut**)
histolysis breakdown of tissues
homiotherm an organism whose body temperature is maintained at a more or less constant level (cf. **poikilotherm**) [= homoiotherm and homeotherm]
hydraulic operated by water pressure
hydrostatic descriptive of skeletal systems in which muscular forces are transmitted by water within body cavities or tissues
hyperpolarize to increase the electrical potential difference, (usually) across the cell membrane (cf. **depolarize**)
hypostome mound of tissue bearing the mouth in cnidarians
hypoxia conditions of low oxygen availability
imaginal disc group of undifferentiated cells in a larva (q.v.) from which a particular organ system will develop
indeterminate development see **regulative development**
indirect development development via a larval (q.v.) stage (cf. **direct development**)
infaunal descriptive of benthic (q.v.) animals that live buried or in burrows within the substratum (cf. **epifaunal**)
inhalent descriptive of the inward respiratory or feeding current (cf. **exhalent**)
instar one of several larval (q.v.) stages separated from the other such stages by a moult
integument non-muscular layers of body wall
intermediate host see **secondary host**
interstitial (a) appertaining to the spaces (interstices) within sediments, or, when used of cnidarian cells, (b) totipotent epidermal cells
introvert eversible (q.v.) and retractable anterior region of body
iteroparous breeding several times per lifetime (cf. semelparous)
jaws hard structures, often protractable, located in the anterior fore-gut or, in arthropods, alongside the preoral cavity (q.v.) that obtain

and/or macerate food materials; often occur as a left/right pair, but may be dorsal/ventral or in 3s, 4s, 5s or more; sometimes used defensively (see **mandibles**)

larva a juvenile phase differing markedly in morphology and ecology from the adult

lecithotrophic development at the expense of internal resources (i.e. yolk) provided by the female parent; used especially of marine larvae (cf. **planktotrophic**)

lemnisci tubular sacs associated with the proboscis of acanthocephalans and at least some bdelloid rotiferans; they are essentially infoldings of epidermis that probably function as hydraulic reservoirs

littoral intertidal

lophophore a system of hydraulically operated, ciliated, feeding tentacles, formed as outgrowths of the body wall, that surround the mouth but not anus

lorica vase-shaped protective case formed by thickened cuticle (q.v.)

macroevolution genesis of taxonomic variety; i.e. evolutionary changes at or above the species level (cf. **microevolution**)

macromeres large, yolk-filled cells in the early embryo (q.v.) (cf. **micromeres**)

macrophagous feeding on relatively large food particles (cf. **microphagous**)

Malpighian tubule blind-ending, tubular, excretory diverticulum of gut

mandibles most anterior pair of mouthpart appendages of many arthropods; usually short, stout, unjointed structures forming biting or chewing jaws (q.v.); also used for the jaw elements of some polychaetes

mantle cavity region of environment enclosed within the confines of the shell of molluscs and brachiopods, within which are located the respiratory and feeding organs respectively

maxilla primary mouthpart of an arthropod additional to and located posterior to the mandible (q.v.); also used for some jaw elements in eunicid polychaetes

medusa one of the two body forms of cnidarians: pulsatile, usually pelagic (q.v.), disc-, bell- or umbrella-shaped and often gelatinous (cf. **polyp**)

mesenchyme diffuse connective-tissue cells set in a jelly-like matrix

mesoderm germ layer elaborated between the ectoderm (q.v.) and endoderm (q.v.)

mesoglea (or **mesogloea**) thick or thin, cellular or acellular layer of jelly-like material between outer and inner cell layers of coelenterates

mesosome second body region of tripartite oligomeric (q.v.) animals; its body cavity, the mesocoel, may support a lophophore (q.v.) (cf. **prosome** and **metasome**)

metabolism chemical processes occurring in organisms to break down structures and substances (catabolism) and to build them up (anabolism)

metachronal rhythm pattern of synchronized movement of cilia (q.v.) or of multiple limbs in which the movement of each element has a fixed phase relationship to the others

metameric with a body largely comprising a linear series of from several to many segments (q.v.) (cf. **monomeric** and **oligomeric**)

metamorphosis drastic change in body form required to convert a larva (q.v.) into the adult

metanephridium open nephridium (q.v.) with an extracellular duct (cf. **protonephridium**)

metasome third body region of tripartite oligomeric (q.v.) animals; its coelom (q.v.), the metacoel, forms the main body cavity (cf. **prosome** and **mesosome**)

microevolution changes in gene frequencies observed within a single population over time (cf. **macroevolution**)

micromeres small cells, without yolk, in the early embryo (q.v.) (cf. **macromeres**)

microphagous feeding on small or minute food particles (cf. **macrophagous**)

microtriches see **microvilli**

microvilli numerous small finger-like projections of the free surface of cells responsible for absorption and, in specialized form, for sensory reception; in cestodes, they are termed microtriches

mid-gut region of gut of endodermal (q.v.) origin and lining (cf. **fore-gut** and **hind-gut**)

mimicry resemblance to an object or to another organism potentially resulting in concealment by virtue of 'mistaken identity'

mixonephridium see **nephromixium**

modular descriptive of a colonial animal that consists of repeated and connected, asexually produced units (or 'individuals') (see **polyp**, **zooid**, and **colonial**)

monomeric with a body not partitioned or divided internally into segments (q.v.) (cf. **oligomeric** and **metameric**)

monomorphic with only a single body form (cf. **dimorphic** and **polymorphic**)

monopectinate descriptive of an advanced mollusc ctenidium (q.v.) with gill filaments on only one side of the central axis (cf. **bipectinate**)

mosaic (= **determinate**) **development** development in which the cells of the embryo (q.v.) have their developmental fate fixed at an early embryological stage (by inheritance of maternal cytoplasm) so that the early embryo comprises a fixed pattern in which there is little capacity for the replacement of missing elements (cf. **regulative development**)

mucus mixture of mucoprotein (mucopolysaccharide bound to protein) secreted by mucous cells

naiad aquatic nymph (q.v.) of certain insects differing rather more from the adult form as a result of specific adaptations for aquatic life; e.g. for the capture of aquatic prey and/or for the uptake of dissolved respiratory gases

nanoplankton plankton (q.v.) of 2–20 μm size in largest dimension

natural selection evolutionary mechanism proposed by C.R. Darwin based on differential survival and reproductive success in resource-limited environments

nekton pelagic (q.v.) animals capable of making progress against natural water flow (cf. **plankton**)

nematocyst intracellular organelle of cnidarians with an eversible (q.v.) coiled thread, used for prey capture, defence, etc.; contained within cnidocytes (q.v.)

neoteny see **paedomorphosis**

nephridium osmoregulatory/excretory organ of ectodermal (q.v.) origin (cf. **coelomoduct**)

nephromixium metanephridium (q.v.) like organ with regions of both ectodermal (q.v.) and mesodermal (q.v.) origin [= mixonephridium]

neurone cell specialized for the conduction of electrical signals and the transmission of information [= nerve cell]

neuropile region of nervous system in which nerve fibres and their terminals form synapses (q.v.)

neurosecretory cell neurone (q.v.) with a glandular function, usually producing hormones

notochord elastic dorsal skeletal rod, derived from highly vacuolated cells bound within a common sheath, characterizing the chordates

nymph juvenile insect differing little from the adult, except in size and in the development of organ systems found only in the adult (e.g.

wings and gonads); characteristically their wing buds develop externally (as distinct from insect **larvae**)

ocellus a simple light-sensitive organ (cf. **compound eye**)

oligomeric with a body comprising a few (two or three) segments (q.v.) (cf. **monomeric** and **metameric**)

ontogeny the course of development of an individual organism from zygote (q.v.) to adult

opisthosoma posterior region of the body of those chelicerates in which the body is visibly divided into two distinct sections (cf. **prosoma**); sometimes used in a comparable fashion in other types of animals with two body regions

oral appertaining to the mouth

organ one or more tissues (q.v.) comprising a structural and functional unit

osphradium chemoreceptory tissue or organ in the mantle cavity (q.v.) of molluscs

ostia pores; for example through which water enters the body (in poriferans) or through which blood enters the heart (in animals with an open blood system)

oviparous egg laying

ovipositor tubular organ of some insects used to place eggs in specific microhabitats

ovum female gamete (q.v.)

paedomorphosis juvenilization process whereby either the adult retains juvenile features ('neoteny') or the organism becomes reproductively mature whilst still a juvenile in form and age ('progenesis')

parasite an organism that lives within or attached (permanently or temporarily) to another and causes it harm

parenchyma diffuse tissue (q.v.) of vacuolated cells that often fills the space between epidermis and gut in acoelomate (q.v.) animals

parthenogenesis form of asexual multiplication (q.v.) in which the ovum (q.v.) develops into a new individual without fertilization (q.v.)

pectines comb-like opisthosomal (q.v.) sense organs of scorpions

pedicellariae compound, articulating spines that function as forceps or pincers in certain echinoderms

pelagic appertaining to the water mass of an aquatic system (cf. **benthic**)

penis erectile (not eversible – q.v.) copulatory organ (cf. **cirrus** and **gonopod**)

periostracum proteinaceous covering of a mollusc or brachiopod shell

pericardial cavity cavity within which a heart is situated

peristalsis waves of contraction of circular and longitudinal muscles passing along a tubular organ or organism and having a propulsive effect

peritoneum the mesodermal bounding membrane of a coelomic (q.v.) body cavity

phagocytosis process whereby pseudopodia (q.v.) of an amoeboid cell flow around a particle to engulf it within a vacuole

pharyngeal cleft hole in the wall of the pharynx (q.v.) that extends right through the body to open at the surface; serves to permit the discharge of water taken in through the mouth

pharynx region of fore-gut (q.v.), located posterior to the buccal cavity and anterior to the oesophagus

photoperiodism ability to exhibit physiological responses consequent on changes in relative daylength

photoreceptor a cell sensitive to light

photosynthesis synthesis of organic compounds using the energy in sunlight, through the chlorophyll molecule, according to the general equation

$$CO_2 + 2H_2X \rightarrow (CH_2O) + H_2O + 2X$$

In the oxyphotosynthesis of some bacteria and all photosynthetic eukaryotes, X = oxygen; in the anoxyphotosynthesis of some other bacteria, X is never oxygen – it is often, but not always, sulphur, H_2X then equalling H_2S

phylogeny the course of evolutionary descent and relationship

pinnule a small side branch of a tentaculate organ

pinocytosis ingestion of small liquid droplets by a cell

plankton pelagic (q.v.) organisms that effectively are suspended in the water and cannot make progress against its movement (cf. **nekton**)

planktotrophic feeding, at least in part, on materials captured from the plankton (q.v.); used especially of marine larvae (cf. **lecithotrophic**)

plasmodium multinucleate amoeboid mass bounded by a single cell membrane

pleopods the abdominal (q.v.) appendages of many crustaceans, often used in swimming

plexus network

poikilotherm an organism whose body temperature varies with that of its environment (cf. **homiotherm**)

polymorphic occurring in more than two distinct body forms (cf. **monomorphic** and **dimorphic**)

polyp one of the two body forms of cnidarians: a sedentary (q.v) or sessile (q.v.) cylinder attached aborally (q.v.) and with a ring of tentacles around the mouth; often form colonial (q.v.) systems; sometimes used interchangeably with **zooid** (q.v.)

polyphyletic a group of organisms derived from more than one ancestral form

polyploid having more than two copies of each chromosome in each somatic (q.v.) cell (cf. **haploid** and **diploid**)

preoral cavity space in front of the mouth in which the mouthparts function or external predigestion takes place

primary host see **definitive host**

proboscis general term for any trunk-like process on the head or anterior body associated with feeding

proglottid serially repeated body unit of a cestode

prokaryote lacking internal membrane-bound organelles and a nuclear membrane within the cell (cf. **eukaryote**); bacteria are prokaryote

propagule reproductive body that separates from the parent; it may be multicellular (vegetative) or cellular (gametic); if cellular, it may be produced by meiosis (sexual) or by mitosis or aberrant forms of meiosis not leading to genetic reduction (asexual)

prosoma anterior region of the body (which includes the head) of those chelicerates in which the body is visibly divided into two distinct sections (cf. **opisthosoma**); sometimes used in a comparable fashion in other types of animals with two body regions

prosome first body region of tripartite oligomeric (q.v.) animals – its body cavity is the protocoel (cf. **mesosome** and **metasome**); and also the term used for the prosoma (q.v.) of copepods

prostomium anterior, non-segmental region of an annelid

protoeukaryote the hypothetical, probably phagocytic (q.v.) host cell which, together with various endosymbiotic prokaryotes (q.v.), formed the first eukaryote (q.v.) cell

protonephridium blind-ending nephridium (q.v.) with an intracellular duct (cf. **metanephridium**)

protostome the state in which the blastopore (q.v.) forms the mouth (cf. **deuterostome**); also used to describe animals that display this state

pseudocoel any body cavity that is not a coelom (q.v.)

pseudocopulation close association during gamete (q.v.) discharge of mating pairs of animals with external fertilization; the ova (q.v.) are

thus fertilized (q.v.) immediately on leaving the female gonopore(s) (q.v.)
pseudofaeces faecal-like pellets of material taken out of suspension in the water by filter feeders but subsequently rejected (i.e. particles collected but not ingested)
pseudopodium temporary lobular protrusion of protoplasm formed during the movement, phagocytosis (q.v.), etc. of amoeboid cells
pupa non-motile transitional stage in the development of some insects occurring between the larval stages and the adult
pygidium posterior, non-segmental region of an annelid
radial cleavage type of cell division in which the cleavage plane is parallel or perpendicular to the polar axis of the blastula (q.v.) (cf. **spiral cleavage**)
radial symmetry symmetry about any plane passed perpendicular to the oral/aboral axis
ramus a branch (e.g. of a limb)
regeneration replacement by compensatory growth and differentiation (q.v.) of lost parts of an organism
regulative (= **indeterminate**) **development** development in which the embryo (q.v.) is able to compensate for missing cells and still produce a normal larva or adult, because the developmental fate of its cells is fixed only at a late stage (cf. **mosaic development**)
rejuvenate make young again
respiratory pigment molecule that combines reversibly with oxygen and so functions as a carrier or store
rhabdites see **rhabdoids**
rhabdocoel general term for groups of turbellarian flatworms that have a simple gut without lateral branches or diverticula
rhabdoids rod-like structures, of uncertain function, in epidermis (q.v.) of flatworms and flatworm-like animals; some arise from gland cells and are termed 'rhabdites'
rhabdome ordered array of photoreceptive microvilli (q.v.); e.g. within a compound eye (q.v.)
rostrum loose term for any median anterior projection of the body
scalids cuticular and epidermal projections, of a variety of forms (including spiniform, club-shaped, feathered and scale-like), disposed in whorls around the introvert (q.v.) of kinorhynchs, loriciferans, priapulans and larval nematomorphs; with sensory, locomotory, food capture or penetrant function
schizocoel coelom (q.v.) formed within blocks of mesodermal (q.v.) tissue by cavitation (cf. **enterocoel**)
sclerite a plate comprising part of an exoskeleton
sclerotized chemical hardening (and darkening) of areas of cuticle (q.v.); results from a tanning process
secondary host host in which a parasite (q.v.) reproduces not at all or only asexually (q.v.)
sedentary tending not to move far
segment a semi-independent, serially repeated unit of the body; segmentation may affect only the body wall and associated structures or almost the whole body
semelparous breeding only once and then dying
septum membrane separating one region of the body from another
sessile permanently attached to a substratum; not capable of locomotion
seta bristle-like projection of cuticle (q.v.), with or without cellular material (cf. **chaeta**)
sexual reproduction form of multiplication in which there is exchange of chromosome material during meiosis, and in which gametes (q.v.) combine in the process of fertilization (q.v.) (cf. **asexual reproduction**)
skeleton a system for the transmission of muscular forces and/or for providing support for the body
somatic appertaining to the body as distinct from the sex cells
spermatheca sac in which a recipient animal stores spermatozoa (q.v.) prior to discharge of its ova and their subsequent fertilization (q.v.)
spermatophore a packet of spermatozoa (q.v.) enclosed within some protective covering
spermatozoon male gamete (q.v.), usually capable of active locomotion [= sperm]
spinneret external nozzle through which silk-producing glands discharge
spiracle opening at the body surface of part of a tracheal (q.v.) system
spiral cleavage type of cell division in which the cleavage plane is oblique to the polar axis of the blastula, there being alternate clockwise and anti-clockwise rotation about the polar axis during the sequence of transverse divisions following the 4-cell stage (cf. **radial cleavage**)
spontaneous generation notion that living organisms could arise directly and spontaneously from non-living materials (e.g. mud)
squamous epithelium epithelium (q.v.) composed of flattened cells
statoblast multicellular asexual propagule (q.v.) of some bryozoans contained within a protective coat
statocyst organ sensitive to gravity and/or acceleration
sternite ventral element of the segmental exoskeleton of arthropods
stolons stalk- or root-like structures by which animals may be connected to each other or to the substratum or from which asexual buds may be liberated
stylet hard, pointed, dart-like structure used for penetration of cells or tissues
styli minute, paired, unjointed appendage-like processes associated with the bases of the legs in some myriapods, and present in an equivalent position on some of the abdominal segments, and rarely also on those of the thorax, of most apterygote insects
subchelate descriptive of an arthropod limb in which the terminal article is reflexed back over the penultimate article to form a distally hinged grasping organ
subumbrella the lower, usually concave surface of a medusa (q.v.)
superficial cleavage pattern of cleavage (q.v.) in which the zygote (q.v.) gives rise to a syncytium (q.v.); the nuclei of the syncytium move towards the surface; and cell boundaries are then organized around the nuclei
suspension feeding capture and consumption of materials suspended in water; capture is usually effected by some form of filter
symbiosis an intimate association between two dissimilar organisms which interact with each other; one is usually dependent on the other
synapse junction, across which information is transferred, between two cells at least one of which is a neurone (q.v.); the transmitting cell is 'presynaptic', the receiving one 'postsynaptic'
syncytium multicellular structure in which cell boundaries are partially or completely absent, ranging from cytoplasmic masses containing many nuclei without any apparent separation into the component cells through to networks of almost complete cells which are in cytoplasmic continuity through intercellular bridges
tegument syncytial (q.v.) external epithelium of parasitic platyhelminths
teleological purposeful or goal-directed
telson posterior, non-segmental region of an arthropod
tentacle any slender, flexible, projecting structure; often sensory, sometimes used for food capture

tergite dorsal element of the segmental exoskeleton of arthropods
test external, or almost external, protective body covering, usually composed of a number of elements
thorax term applied to the middle zone of any body divided into three distinct regions of which the anterior region is the head (cf. **abdomen**)
tissue associated cells of the same (or of a few) type(s) performing the same function; usually bound together by intercellular material (cf. **organ**)
tonic sustained
totipotent cells of a multicellular organism that are capable of differentiating (q.v.) into any specialist cell
trachea tube conveying air from the external environment directly to the tissues
tracheoles terminal capillary-like distributaries of a trachea (q.v.)
triploblastic embryonic condition in which three tissue layers – ectoderm (q.v.), mesoderm (q.v.) and endoderm (q.v.) – can be recognized
trochophore early larval stage of many marine animals characterized by a complete, double preoral band of cilia
tubicolous tube-dwelling
ultrafiltration passage of fluid under pressure through a semipermeable membrane
umbrella the upper, usually convex surface of a medusa (q.v.)
uniramous having a single branch (cf. **biramous**); used of an arthropod limb
uropods the last pair of abdominal (q.v.) appendages of decapod crustaceans, which together with the telson (q.v.) form a tail fan
urosome term applied to the opisthosoma (q.v.) of copepods
vital force mysterious, non-physical force once thought to give 'life' to organisms and to direct development and evolution
viviparity development of an embryo (q.v.) within the body of the parent using, in part, resources passing directly from parent to embryo
warning coloration distinctive, bright, contrasting scheme (e.g. black and yellow; black and red) often associated with noxiousness or toxicity in potential prey species
zoochlorellae general name given to symbiotic chlorophyte algae found within the tissues of various, mainly freshwater, invertebrates
zooid a modular individual in a colonial (q.v.) system produced by repeated incomplete budding (q.v.); applied to all such animals other than cnidarians
zooxanthellae general name given to symbiotic dinoflagellate algae found within the tissues of various, mainly marine, invertebrates
zygote single cell produced by union of a sperm (q.v.) and an ovum (q.v.) at fertilization

Illustration Sources

Abbott, N.J., Williamson, R. & Maddock, L. 1995. *Cephalopod Neurobiology*. Oxford University Press.

Agelopoulos, N., Birkett, M.A., Hick, A.J., Hooper, A.M., Pickett, J.A., Pow, E.M., Smart, L.E., Smiley, D.W.M., Wadhams, L.J. & Woodcock, C.M. 1999. *Pesticide Science*, **55**, 225–235.

Aidley, D.J. 1998. *The Physiology of Excitable Cells*. (4th ed.), Cambridge University Press.

Alexander, R. McN. 1979. *The Invertebrates*. Cambridge University Press, Cambridge.

Alldredge, A. 1976. *Sci. Am.*, **235** (1), 94–102.

Anderson, D.T. 1964. *Embryology and Phylogeny in Annelids and Arthropods*. Pergamon Press, New York.

Anderson, P.A.V. & Bone, Q. 1980. *Proc. R. Soc. Lond(B)*, **210**, 559–574.

Arbas, E.A. & Calabrese, R.L. 1987. *J. Neurosci.*, **7**, 3945–3952.

Atkins, D. 1933. *J. Mar. Biol. Assoc., UK*, **19**, 233–252.

Atwood, H.L. 1973. *Am. J. Zool.*, **13**, 357–378.

Austin, C.R. 1965. *Fertilisation*. Prentice Hall Inc., New Jersey.

Baehr, J.C., Porcheron, P. & Dray, F. 1978. *C.R. Acad. Sci. (Paris)*, **287D**, 523–525.

Baer, J. & Joyeux, C. 1961. Classe des Trématodes. In: Grassé, P.-P. (Ed.) *Traité de Zoologie*, **4**, *Platyhelminthes, Mésozoaires, Acanthocéphales, Némertiens*, pp. 561–692. Masson, Paris.

Baker, A.N., Rowe, F.W.E. & Clark, H.E.S. 1986. *Nature, Lond.*, **321**, 862–864.

Baker, T.C. 1990. In Døving, K.B. (Ed). *Proceedings of the* 10th *International Symposium on Olfaction and Taste*, pp. 18–25.

Barnes, R.D. 1980. *Invertebrate Zoology*, 4th edn. Saunders, Philadelphia.

Barnes, R.S.K. & Hughes, R.N. 1982. *An Introduction to Marine Ecology*. Blackwell Scientific Publications, Oxford.

Bastiani, M.J., Doe, C.Q., Helfand, S.L. & Goodman, C.S. 1985. *Trends Neurosci.*, **8**, 257–266.

Bayne, B.L., Thompson, R.J. & Widdows, J. 1976. In: B.L. Bayne (Ed.) *Marine Mussels: Their Physiology and Ecology*. Cambridge University Press, Cambridge.

Becker, G. 1937. *Z. Morph. Ökol. Tiere*, **33**, 72–127.

Becker, H.J. 1962. *Chromosoma*, **13**, 341–384.

Belk, D. 1982. In: Parker, S.P. (Ed.) *Synopsis and Classification of Living Organisms*, **2**, 174–180. McGraw-Hill, New York.

Bergquist, P.R. 1978. *Sponges*. Hutchinson, London.

Berrill, N.J. 1950. *The Tunicata*. Ray Society, London.

Bicker, G. 1998. *Trends in Neuroscience*, **21**, 349–355.

Biscardi, H.M. & Webster, G.C. 1977. *Exp. Gerontol.*, **12**, 201–205.

Blower, J.G. 1985. *Millipedes*. Brill, Leiden.

Bode, H.R., Heimfeld, S., Koizumi, O., Littlefield, C.L. & Yaross, M.S. 1989. *Am. Zool.*, **28**, 1053–1063.

Boeckh, J., Ernst, K-D., Sass, H. & Waldow, U. 1975. In: Denton, D. (Ed.) *Olfaction and Taste*, **V**, 239–245. Academic Press, New York.

Boss, K.J. 1982: In: Parker, S.P. (Ed.) *Synopsis and Classification of Living Organisms*, **1**, 945–1166. McGraw-Hill, New York.

Boxshall, G.A. & Lincoln, R.J. 1987. *Phil. Trans. Roy. Soc. Lond. (B)*, **315**, 267–303.

Brill, B. 1973. *Z. Zellforsch.*, **144**, 231–245.

Brownlee, D.J.A. & Fairweather, I. 1999. *Trends in Neuroscience*, **22**, 16–24.

Buchsbaum, R. 1951. *Animals Without Backbones*, Vol. 1. Pelican, Harmondsworth.

Bullock, T.H. & Basar, E. 1988. *Brain Res. Rev.*, **13**, 57–76.

Bullough, W.S. 1958. *Practical Invertebrate Anatomy*, 2nd edn. Macmillan, London.

Cain, A.J. & Sheppard, P.M. 1954. *Genetics*, **39**, 89–116.

Calkins, G.N. 1926. *The Biology of the Protozoa*. Baillière Tindall & Cox, London.

Calow, P. 1985. Causes de la mort i costos d'autoproteccio. In: *Biologia Avui*. Fundacio Caixa de Pensions, Barcelona.

Calow, P. 1986. In: Peberdy, R. & Gardner, P. (Eds) *The Collins Encyclopedia of Animal Evolution*, pp. 90–91. Equinox, Oxford.

Calow, P. & Read, D.A. 1986. In: Tyler, S. (Ed.) *Advances in the Biology of Turbellarians and Related Platyhelminthes*, pp. 263–272. D.W. Junk, Dordrecht.

Campbell, R.D. 1967. Tissue dynamics of steady-state growth in *Hydra Littoralis*. II. Patterns of tissue movement. *J. Morphol.*, **121**, 19–28.

Carpenter, W.B. 1866. *Phil. Trans. Roy. Soc. Lond.*, **156**, 671–756.

Casida, J.E. & Quistad, G.B. 1998. *Annual Review of Entomology*, **43**, 1–16.

Caullery, M. & Mesnil, F. 1901. *Arch. Anat. Microsc.*, **4**, 381–470.

Clark, A.H. 1915. *US Natn. Mus. Bull*, **82**, Vol. 1(1), 1–406.

Clark, R.B. 1964. *Dynamics in Metazoan Evolution*. Clarendon Press, Oxford.

Clarke, K.U. 1973. *The Biology of the Arthropoda*. Arnold, London.

Clarkson, E.N.K. 1986. *Invertebrate Palaeontology and Evolution*, 2nd edn. Allen & Unwin, London.

Clement, A.C. 1962. *J. Exp. Zool.*, **149**, 193–215.

Cloudsley-Thompson, J. 1958. *Spiders, Scorpions, Centipedes and Mites*. Pergamon Press, London.

Cohen, A.C. 1982. In: Parker, S.P. (Ed.) *Synopsis and Classification of Living Organisms*, **2**, 181–202. McGraw-Hill, New York.

Collett, M., Collett, T.S., Bisch, S. & Wehner, R. 1998. *Nature*, **394**, 269–272.

Colwin, L.H. & Colwin, A.L. 1961. *J. Biophys. Biochem. Cytol.*, **10**: 231–254.

Conway Morris, S. 1979. *Ann. Rev. Ecol. Syst.*, **10**, 327–349.

Conway Morris, S. 1985. *Phil. Trans. Roy. Soc. Lond. (B)*, **307**, 507–582.

Conway Morris, S. 1995. A new phylum from the lobster's lips. *Nature, Lond.*, **378**, 661–662.

Corliss, J.O. 1979. *The Ciliated Protozoa*, 2nd edn. Pergamon Press, Oxford.

Cottrell, G.A. 1989. *Comp. Biochem. Physiol. (A)*, **93**, 41–45.

Cuénot, L. 1949. In: Grassé, P-P. (Ed.) *Traité de Zoologie*, **VI**, 3–75. Masson, Paris.

Danielsson, D. 1892. *Norw. N-Atlantic Exped. (1876–1878) Rep. Zool.*, **21**, 1–28.

Davies, I. 1983. *Ageing*. Edward Arnold, London.

Dehorne, A. 1933. *Bull. Biol. Fr. Belgique*, **67**, 298–326.

Dethier, V.E. 1976. *The Hungry Fly*. Harvard University Press, Cambridge, Mass.

Dhadialla, T.S., Carlson, G.R. & Le, D.P. 1998. *Annual Review of Entomology*, **43**, 545–569.

Dixon, A.F.G. 1973. *Biology of Aphids*. Studies in Biology No. 44. Edward Arnold, London.

Dunlap, J.C. 1999. *Cell*, **96**, 271–290.

Durchon, M. 1967. *L'endocrinologie chez le Vers et les Molluscs*. Masson, Paris.

Eakin, R.M. 1968. *Evol. Biol.*, **2**, 194–242.

Elner, R.W. & Hughes, R.N. 1978. *J. Anim. Ecol.*, **47**, 103–116.

Epel, D. 1977. *Sci. Am.*, **237** (5), 129–138.

Fewkes, J. 1883. *Bull. Mus. Comp. Zool., Harvard*, **11**, 167–208.

Fingerman, M. 1976. *Animal Diversity*, 2nd edn. Holt, Rinehart & Winston, New York.

Fox, H.M., Wingfield, C.A. & Simmonds, B.G. 1937. *J. Exp. Biol.*, **14**, 210–218.

Fraser, J.H. 1982. *British Pelagic Tunicates*. Cambridge University Press, Cambridge.

Fretter, V. & Graham, A. 1976. *A Functional Anatomy of Invertebrates*. Academic Press, London and New York.

Friesen, W.Q., Poon, W. & Stent, G.S. 1976. *Proc. Nat. Acad. Sci. (USA)*, **73**, 3734–3738.

Frost, W.N. & Katz, P.S. 1996. *Proceedings of the National Academy of Sciences USA*, **93**, 422–426.

Funch, P. & Kristensen, R.M. 1995. Cycliophora is a new phylum with affinities to Entoprocta and Ectoprocta. *Nature, London*, **378**, 711–714.

Gage, J.D. & Tyler, P.A. 1991. *Deep-sea Biology*. Cambridge University Press, Cambridge.

Geraerts, W.P.M., Ter Maat, A. & Vreugdenhil, E. 1988. In: Laufer, H. & Downer, R.G.H. (Eds) *Endocrinology of Selected Invertebrate Types*, pp. 141–231. Liss, New York.

George, J.D. & Southward, E.C. 1973. *J. Mar. Biol. Assoc. UK*, **53**, 403–424.

Gibson, P.H. & Clark, R.B. 1976. *J. Mar. Biol. Assoc. UK*, **56**, 649–674.

Gibson, R. 1982. In: Parker, S.P. (Ed.) *Synopsis and Classification of Living Organisms*, pp. 823–846. McGraw-Hill, New York.

Gilbert, S.C. 1990. *Developmental Biology*, 3rd edn. Sinauer Associates, Massachusetts.

Gilbert, L.E. 1982. *Sci. Am.*, **247** (2), 102–107B.

Gilbert, L.I. 1989. In: Koolman, J. (Ed.) *Ecdysone: From Chemistry to Mode of Action*, pp. 448–471. Thieme, Stuttgart.

Glaessner, M.F. & Wade, M. 1966. *Palaeontology*, **9**, 599–628.

Gnaiger, E. 1983. *J. Exp. Zool.*, **228**, 471–490.

Golding, D.W. 1967. *J. Embryol. Exp. Morph.*, **18**, 79–80.

Golding, D.W. 1973. *Acta Zool. (Stockh.)*, **54**, 101–120.

Golding, D.W. 1988. *New Scientist*, **119**, 52–55.

Golding, D.W. 1992. In: Harrison, F.W. & Gardiner, S. (Eds) *Microscopic Anatomy of Invertebrates*, **7**, 153–179. Liss, New York.

Golding, D.W. & May, B.A. 1982. *Acta Zool. (Stockh.)*, **63**, 229–238.

Golding, D.W. & Pow, D.V. 1988. In Thorndyke M.C. & Goldsworthy G.J. (Eds) *Neurohormones in Invertebrates*. Cambridge University Press, pp. 7–18.

Golding, D.W. & Whittle, A.C. 1974. *Tissue & Cell*, **6**, 599–611.

Golding, D.W. & Whittle, A.C. 1977. *Int. Rev. Cytol. Suppl.*, **5**, 189–302.

Goodrich, E.S. 1945. *Q.J. Microsc. Sci.*, **86**, 113–393.

Gordon, D.P. 1975. *Cah. Biol. Mar.*, **16**, 367–382.

Grassé, P.-P. (Ed.). 1948. *Traité de Zoologie*, **XI**. Masson, Paris.

Grassé, P.-P. 1961. Classe des Dicyémides. In: Grassé, P.-P. (Ed.) *Traité de Zoologie*, **4**, *Platyhelminthes, Mésozoaires, Acanthocéphales, Némertiens*, pp. 707–729. Masson, Paris.

Grassé, P.-P. (Ed.). 1965. *Traité de Zoologie*, **IV**. Masson, Paris.

Green, J. 1961. *A Biology of Crustacea*. Witherby, London.

Grimmelikhuijzen, C.J.P. 1985. *Cell & Tissue Research*, **241**, 171–182.

Gupta, B.L. & Berridge, M.J. 1966. *J. Morphol.*, **120**, 23–82.

Gustafson, T. & Wolpert, L. 1967. *Biol. Rev.*, **42**, 442–498.

Hackman, R.H. 1971. In: Florkin, M. & Scheer, B.T. (Eds) *Chemical Zoology*, **6**, 1–62. Academic Press, New York.

Hansen, K. 1978. In: Hazelbauer, G.I. (Ed.) *Taxis and Behaviour Receptors and Recognition*, **5B**, 231–292. Chapman & Hall, London.

Hardy, A.C. 1956. *The Open Sea. The World of Plankton*. Collins, London.

Harris-Warwick, R.M. & Flamm, R.E. 1986. *Trends in Neurosciences*, **9**, 432–437.

Hedgpeth, J.W. 1982. In: Parker, S.P. (Ed.) *Synopsis and Classification of Living Organisms*, **2**, 169–173. McGraw-Hill, New York.

Hengstenberg, R. 1998. *Nature*, **392**, 757–758.

Hermans, C.O. & Cloney, R.A. 1966. *Z. Zellforsch.*, **72**, 583–596.

Hermans, C.O. & Eakin, R.M. 1974. *Z. Morph. Tiere*, **79**, 245–267.

Hescheler, K. 1900. In: Lang, A. (Ed.) *Lehrbuch der Vergleichenden Anatomie der Wirbellosen Thiere*, 3rd edn. Fischer, Jena.

Hess, R. 1887. *Z. Wiss. Zool.*, **62**, 247–283.

Higgins, R.P. 1983. *Smithsonian Contrib. Mar. Sci.*, **18**, 1–131.

Hines, A.H. 1979. In: Stancyk, S.E. (Ed.) *Reproductive Ecology of Marine Invertebrates*, pp. 213–234. University of South Carolina Press, Columbia SC.

Hodgkin, A.L. & Huxley, A.F. 1952. *J. Physiol.*, **117**, 500–544.

Holland, N.D., Grimmer, T.C. & Kubota, H. 1975. *Biol. Bull.*, **148**, 219–242.

Holt, C.S. & Waters, T.F. 1967. *Ecology*, **48**, 225–234.

Hughes, T.E. 1959. *Mites or the Acari*. Athlone, London.

Hummon, W.D. 1982. In: Parker, S.P. (Ed.) *Synopsis and Classification of Living Organisms*, **1**, 857–863. McGraw-Hill, New York.

Hyman, L.H. 1940. *The Invertebrates*, Vol. I: *Protozoa through Ctenophora*. McGraw-Hill, New York.

Hyman, L.H. 1951. *The Invertebrates*, Vol. II: *Platyhelminthes & Rhynchocoela*. McGraw-Hill, New York.

Imms, A.D. 1964. *A General Textbook of Entomology*, 9th edn, revised reprint. Methuen, London.

Ishibashi, J., Kataoka, H., Isogai, A., Kawakami, A., Saegusa, H., Yagi, Y., Mizoguchi, A., Ishizaki, H. & Suzuki, A. 1994. *Biochemistry*, **33**, 5912–5919.

Ito, Y. 1980. *Comparative Ecology*. Cambridge University Press, Cambridge.

Jägersten, G. 1973. *The Evolution of the Metazoan Life Cycle*. Academic Press, New York.

Jeannel, R. 1960. *Introduction to Entomology*. Hutchinson, London.

Joose, J. & Geraerts, W.P.M. 1983. In Saleudin, A.S.M. & Wilbur, K.M. (Eds) *The Mollusca*, Vol. 5. Academic Press, New York.

Jouin, C. 1971. *Smithsonian Contributions in Zoology*, **76**, 47–56.

Jones, A.M. & Baxter, J.M. 1987. *Molluscs: Caudofoveata, Solenogastres, Polylacophora and Scaphopoda*. Brill, Leiden.

Jones, J.D. 1955. *J. Exp. Biol.*, **32**, 110–125.

Jones, M.L. 1985. In: Conway Morris, *S. et al.* (Eds) *The Origin and Relationships of Lower Invertebrates*, pp. 327–342. Clarendon Press, Oxford.

Joosse, J. & Geraerts, W.P.M. 1983. In: Saleudin, A.S.M. & Wilbur, K.M. (Eds) *The Mollusca*, Vol. 5. Academic Press, New York.

Joyeux, C. & Baer, J-G. 1961. Classe des Cestodes. In: Grassé, P.-P. (Ed.) *Traité de Zoologie*, **4**, *Platyhelminthes, Mésozoaires, Acanthocéphales, Némertiens*, pp. 347–560. Masson, Paris.

Kandel, E.R. & Schwartz, J.H. 1982. *Science*, **218**, 433–443.

Kawakami, A., Katoaka, H., Oka, T., Mizoguchi, A., Kimura-Kawakami, M., Adachi, T., Iwami, M., Nagasawa, H., Suzuki, A. & Ishizaki, H. 1990. *Science*, **247**, 1333–1335.

Kennedy, D. 1976. In: Fentress, J.C. (Ed.) *Simpler Networks and Behaviour*. Sinauer, Sunderland, Massachusetts.

Kershaw, D.R. 1983. *Animal Diversity*. University Tutorial Press, Slough.

Kishimoto, T. 1999. *Encyclopedia of Reproduction*, Vol. 3, pp. 481–488.

Koolman, J. 1990. *Zool. Sci.*, **7**, 563–580.

Kozloff, E.N. 1990. *Invertebrates*. Saunders, Philadelphia.

Krebs, J.R., Erichsen, J.T., Webber, M.I. & Charnov, E.L. 1977. *Anim. Behav.*, **25**, 30–38.

Krishnan, B., Dryer, S.E. & Hardin, P.E. 1999. *Nature*, **400**, 375–378.

Kudo, R.R. 1946. *Protozoology*, 3rd edn. Thomas, Springfield, Illinois.

Lacaze-Duthiers, F.J.H. de. 1861. *Ann. Sci. Nat. (Zool.)*, **15**, 259–330.

Lamb, M.J. 1977. *Biology of Ageing*. Blackie, Glasgow.

Lemche, H. & Wingfield, K.G. 1959. *Galathea Rep.*, **3**, 9–71.

Lester, S.M. 1985. *Mar. Biol.*, **85**, 263–268.

Lewis, J.G.E. 1981. *The Biology of Centipedes*. Cambridge University Press, Cambridge.

Lewis, J.G.E. 1987. In Stearns, S.C. (Ed.) *The Evolution of Sex and its Consequences*. Birkhauser Verlag, Basel.

McArthur, V.E. 1996. The Ecology of East Anglian Coastal Lagoons. PhD Thesis, University of Cambridge.

McFarland, W.N., Pough, F.N., Cade, T.J. & Heiser, J.B. 1979. *Vertebrate Life*. Macmillan, New York.

MacKinnon, D.L. & Haws, R.S.J. 1961. *An Introduction to the Study of Protozoa*. Clarendon Press, Oxford.

McLaughlin, P.A. 1980. *Comparative Morphology of Recent Crustacea*. Freeman, San Francisco.

Manton, S.M. 1952. *J. Linn, Soc. (Zool.)*, **42**, 93–117.

Manton, S.M. 1965. *J. Linn. Soc. (Zool.)*, **45**, 251–483.

Marcus, E. 1929. *Klassen und Ordnungen des Tierreichs*, **5**, 1–608.

Margulis, L. & Schwartz, K.V. 1982. *Five Kingdoms*. Freeman, San Francisco.

Marion, M.A.-F. 1886. *Arch. Zool. Exp. Gén. (2)*, **4**, 304–326.

Marshall, A.J. & Williams, W.D. (Eds) 1972. *Textbook of Zoology. Invertebrates*. Macmillan, London.

Mayeri, E. & Rothman, B.S. 1985. In: Selverston, A.I. (Ed.) *Model Networks and Behavior*, pp. 285–301. Plenum, New York.

Meglitsch, P.A. 1972. *Invertebrate Zoology*, 2nd edn. Oxford University Press, Oxford.

Meyrand, P., Simmers, A.J. & Moulins, M. 1994. *Nature*, **351**, 60–63.

Millar, R.H. 1970. *British Ascidians*. Academic Press, London.

Miller, R.L. 1966. *J. Exp. Zool.*, **162**, 23–44.

Miyan, J.A. & Ewing, A.W. 1986. *J. Exp. Biol.*, **116**, 313–322.

Moore, R.C. (Ed.). 1957. *Treatise on Invertebrate Paleontology, Part L. Mollusca*, **4**. University of Kansas Press, Lawrence.

Moore, R.C. (Ed.). 1965. *Treatise on Invertebrate Paleontology, Part H. Brachiopoda*. University of Kansas Press, Lawrence.

Morgan, C.I. 1982. In: Parker, S.P. (Ed.) *Synopsis and Classification of Living Organisms*, **2**, 731–739. McGraw-Hill, New York.

Morgan, C.I. & King, P.E. 1976. *British Tardigrades*. Academic Press, London.

Mortensen, T. 1928–51. *A Monograph of the Echinoidea*. 5 vols. Reitsel, Copenhagen.

Müller, P. 1959. *The Insecticide Dichlorodiphenyltrichoroethane and its Significance*. Vol. 2. Berkhaüser, Basel. 570pp.

Muscatine, L. *et al.*, 1975. *Symp. Soc. Exp. Biol.*, **29**, 175–203.

Newell, N.D. 1949. *Evolution*, **3**, 103–240.

Nichols, D. 1962. *Echinoderms*, 3rd edn. Hutchinson, London.

Nichols, D. 1969. *Echinoderms*, 4th edn. Hutchinson, London.

Noble, E.R. & Noble, G.A. 1976. *Parasitology*. Lea & Febiger, Philadelphia.

Nusslein-Volhard, C. 1991. *Development*. Suppl. **1**, 1–10.

Ohnishi, T. & Sugiyama, M. 1963. *Embryologia*, **8**, 79–88.

Olive, P.J.W. 1980. In: Rhoads, D.C. & Lutz, R.A. (Eds), *Skeletal Growth in Aquatic Organisms*. Plenum Press, New York.

Olive, P.J.W. 1985a. *Symp. Soc. Exp. Biol.*, **39**, 267–300.

Olive, P.J.W. 1985b. In: *Syst. Association*, series **28**, 42–59. Oxford University Press, Oxford.

Oschman, J.L. & Berridge, M.J. 1971. *Federation Proceedings, Federation of American Societies for Experimental Biology*, **30**, 49–56.

Pashley, H.E. 1985. The foraging behaviour of Nereis diversicolor (*Polychaeta*). PhD thesis, University of Cambridge.

Pearson, K.G. 1983. *J. Physiol. (Paris)*, **78**, 765–771.

Pennak, R.W. 1978. *Fresh-water Invertebrates of the United States*, 2nd edn. Wiley, New York.

Pfannestiel, H.D. 1984. *Wilhem Roux's Arch. Dev. Biol.*, **194**, 32–36.

Phillipson, J. 1981. In: Townsend, C.R. & Calow, P. (Eds) *Physiological Ecology*, pp. 20–45. Blackwell Scientific Publications, Oxford.

Pierrot-Bults, A.C. & Chidgey, K.C. 1987. *Chaetognatha*. Brill, Leiden.

Pringle, J.W.S. 1975. *Insect Flight*. Oxford University Press, Oxford.

Rice, M. 1985. In: Conway Morris, S., George, J.D., Gibson, R. & Platt, H.M. (Eds) *Origins and Relationships of Lower Invertebrates*. Clarendon Press, Oxford.

Rind, E.C. 1996. *Journal of Neurophysiology*, **75**, 986–995.

Ritter-Zahony, R. von. 1911. *Das Tierreich*, **29**, 1–35.

Robbins, T.E. & Shick, J.M. 1980. In: *Nutrition in the Lower Metazoa*, Pergamon Press, Oxford.

Ruppert, E.E. & Barnes, R.D. 1994. *Invertebrate Zoology*, 6th edn. Saunders, Fort Worth.

Russell-Hunter, W.D. 1979. *A Life of Invertebrates*. Macmillan, New York.

Sanders, D.S. 1982. *Insect Clocks*, 2nd edn. Pergamon Press, Oxford.

Sanders, H.L. 1957. *Syst. Zool.*, **6**, 112–128.

Satterlie, R.A. & Spencer, A.N. 1987. In: Ali, M.A. (Ed.) *Nervous Systems in Invertebrates*, pp. 213–264. Plenum, New York.

Savory, T.H. 1935. *The Arachnida*. Edward Arnold, London.

Scharrer, B. 1952. *Biol. Bull. (Woods Hole)*, **102**, 261–272.

Schepotieff, A. 1909. *Zool. Jb. Syst.*, **28**, 429–448.
Schmidt-Nielsen K. 1984. *Scaling: Why is animal size so important?* Cambridge University Press, Cambridge.
Sebens, K.P. & De Riemer, K. 1977. *Mar. Biol.*, **43**, 247–256.
Sedgwick, A. 1888. *Q.J. Microsc, Sci.*, **28**, 431–493.
Sehgal, A., Rothenfluh-Hilfiker H., Hunter-Ensor, M., Chen, Y., Myers, M.P. & Young, M.W. 1995. *Science*, **270**, 808–810.
Shelton, G.A.B. 1975. *Proc. R. Soc. Lond. B.*, **190**, 239–256.
Sheppard, P.M. 1958. *Natural Selection and Heredity*. Hutchinson, London.
Shick, P.M. & Dykens, J.A. 1985. *Oecologia*, **66**, 33–41.
Sibly, R.M. & Calow, P. 1986. Physiological Ecology of Animals: an Evolutionary Approach. Blackwell Scientific Publications, Oxford.
Silva, P.H.D.H. de. 1962. *J. Exp. Biol.*, **39**, 483–490.
Simmons, P.J. & Young, D. 1999. *Nerve Cells and Animal Behaviour*, 2nd edn. Cambridge University Press.
Sleigh, M.A., Dodge, J.D. & Patterson, D.J. 1984. In: Barnes, R.S.K. (Ed.) *A Synoptic Classification of Living Organisms*, pp. 25–88. Blackwell Scientific Publications, Oxford.
Smart, P. 1976. *The Illustrated Encyclopedia of the Butterfly World*. Hamlyn, London.
Smyth, J.D. 1962. *Introduction to Animal Parasitology*. English Universities Press, London.
Smyth, J.D. & Halton, D.W. 1983. *The Physiology of Trematodes*. Cambridge University Press, Cambridge.
Snodgrass, R.E. 1935. *Principles of Insect Morphology*. McGraw-Hill, New York.
Snow, K.R. 1970. *The Arachnids: An Introduction*. Routledge & Kegan Paul, London.
Southward, E.C. 1980. *Zool. Jb. Anat. Ontog.*, **103**, 264–275.
Southward, E.C. 1982. *J. Mar. Biol. Assoc., UK*, **62**, 889–906.
Spengel, J.W. 1932. *Sci. Res. Michael Sars N. Atlantic Deep Sea Exped.*, **5** (5), 1–27.
Stein, P.S.G. 1971. *J. Neurophysiol.*, **34**, 310–318.
Sterrer, W.E. 1982. In: Parker, S.P. (Ed.) *Synopsis and Classification of Living Organisms*, **1**, 847–851. McGraw-Hill, New York.
Stiasny, G. 1914. *Z. Wiss. Zool.*, **110**, 36–75.
Strausfeld, N.J. & Nassel, D.R. 1981. In: Autrum, H. (Ed.) *Handbook of Sensory Physiology*, Vol. VII/6B. Springer-Verlag, Berlin.
Stretton, A.O.W., Davis, R.E., Angstadt, J.D., Donmoyer, J.E. & Johnson, C.D. 1985. *Trends Neurosci.*, **8**, 294–299.
Strumwasser, F. 1974. *Neurosciences Third Study Program*, 459–478.
Tessier-Lavigne, M. & Goodman, C.S. 1996. *Science*, **274**, 1123–1233.
Treherne, J.E. & Foster, W.A. 1980. *Anim. Behav.*, **28**, 1119–1122.
Trench, R.K. 1975. *Symp. Soc. Exp. Biol.*, **29**, 229–265.
Trinkaus, J.P. 1969. *Cells into Organs*. Prentice-Hall, New Jersey.
Trueman, E.R. & Foster-Smith, R. 1976. *J. Zool., Lond.*, **179**, 373–386.
Truman, J.W. 1988. *Adv. Ins. Physiol.*, **21**, 1–34.
Truman, J.W. 1996. In Gilbert, L.I., Tata, J.R. & Atkinson, B.G. (Eds) *Metamorphosis: Postembryonic Reprogramming of Gene Expression in Amphibian and Insect Cells*, pp. 283–320. Academic Press, San Diego.
Tsacopoulos, M. 1994. *Journal of Neuroscience*, **14**, 1339–1351.
Turner, C.D. 1966. *General Endocrinology*. Saunders, Philadelphia.
Valentine, J.W. & Moores, E.M. 1974. *Sci. Am.*, **230** (4), 80–89.
Van Hateren, J.H. 1989. In Stavenga, D.G. & Hardie, R.C. (Eds) *Facets of Vision*, pp. 74–89. Springer, Berlin.
Villegas, G.M. & Villegas, R. 1968. *Journal of General Physiology*, **51**, 44–60.
Wallace, M.M.H. & Mackerras, I.M. 1970. In: C.S.I.R.O., *The Insects of Australia*, pp. 205–216. Melbourne University Press, Melbourne.
Warner, G.F. 1977. *The Biology of Crabs*. Elek, London.
Waterman, T.H. 1960. *The Physiology of Crustacea*. Academic Press, New York.
Watson, R.D., Spaziani, E. & Bollenbacher, W.E. 1989. In: Koolman, J. (Ed.) *Ecdysone: From Chemistry to Mode of Action*, pp. 188–203. Thieme, Stuttgart.
Weaver, R.J., Marris, G.C., Olieff, S., Mosson, J.H. & Edwards, J.P. 1997. *Archives of Insect Biochemistry and Physiology*, **35**, 169–178.
Weeks, J.C., Jacobs, G.A. & Miles, C.I. 1989. *Am. Zool.*, **29**, 1331–1344.
Welsch, U. & Storch, V. 1976. *Comparative Animal Cytology*. Sidgwick & Jackson, London.
Wenyon, C.M. 1926. *Protozoology*. Baillière, Tindall & Cox, London.
Whittington, H.B. 1979. In: House, M.R. (Ed.) *The Origin of Major Invertebrate Groups*, pp. 253–268. Academic Press, London.
Widdows, J. & Bayne, B.L. 1971. *J. Mar. Biol. Assoc., UK*, **51**, 827–843.
Wigglesworth, V.B. 1940. *J. Exp. Biol.*, **17**, 201–222.
Wigglesworth, V.B. 1972. *Principles of Insect Physiology*, 7th edn. Chapman & Hall, London.
Williamson, R. 1989. *J. Comp. Physiol. (A)*, **165**, 847–860.
Wine, J.J. & Krasne, F.B. 1982. In: Sandeman, D.C. & Atwood, H.L. (Eds) *The Biology of Crustacea*, Vol, 4, 241–292. Academic Press, New York.
Wright, A.D. 1979. In: House, M.R. (Ed.) *The Origin of Major Invertebrate Groups*, pp. 235–252. Academic Press, London.
Wrona, F.J. & Davies, R.W. 1984. *Can. J. Fish. Aquatic Sci.*, **41**, 380–385.
Yager, J. & Schram, F.R. 1986. *Proc. Biol. Soc. Wash.*, **99** (1), 65–70.
Young, J.Z. 1939. *Phil. Trans. Roy. Soc. Lond. B.*, **229**, 465–503.
Young, J.Z. 1962. *The Life of Vertebrates*, 2nd edn. Clarendon Press, Oxford.
Young, J.Z. 1963. *Nature (Lond)*, **198**, 636–640.
Zullo, V.Z. 1982. In: Parker, S.P. (Ed.) *Synopsis and Classification of Living Organisms*, **2**, 220–228. McGraw-Hill, New York.

Index

Note: Page numbers in *italics* refer to figures; those in **bold** refer to tables.

Lightning Source UK Ltd.
Milton Keynes UK
UKOW06f0205210115

244788UK00001B/1/P